Precision Agriculture

Related Society Publications

Proceedings of the First Workshop Soil-Specific Crop Management. A Workshop on Research and Development Issues. 1993. Robert, P.C., R.H. Rust, and W.E. Larson (ed.) Minneapolis, MN. April 14-16, 1992. ASA, CSSA, SSSA, Madison, WI.

Proceedings of the Second International Conference on Site-Specific Management for Agricultural Systems. 1995. Robert, P.C., R.H. Rust, and W.E. Larson (ed.) Bloomington/Minneapolis, MN. March 27-30, 1994. ASA, CSSA, SSSA, Madison, WI.

For information on these titles, please contact the ASA, CSSA, SSSA Headquarters Office, ATTN: Book Orders, 677 South Segoe Road, Madison, WI 53711-1086. Phone: (608) 273-8080. Fax: (608) 273-2021

PROCEEDINGS

OF THE

THIRD INTERNATIONAL CONFERENCE ON

PRECISION AGRICULTURE

June 23-26, 1996

Minneapolis, Minnesota

Editors

P.C. Robert, R.H. Rust, and W.E. Larson

Conducted by the
PRECISION AGRICULTURE CENTER
Department of Soil, Water, and Climate
University of Minnesota
St. Paul, Minnesota

Published by
American Society of Agronomy
Crop Science Society of America
Soil Science Society of America

Cover art provided by authors

American Society of Agronomy, Inc.
Crop Science Society of America, Inc.
Soil Science Society of America, Inc.
677 South Segoe Road, Madison, Wisconsin 53711 USA

Library of Congress Catalog Card Number: 96-80281

Printed in the United States of America

TABLE OF CONTENTS

WORKSHOP - PRECISION NITROGEN MANAGEMENT - LATEST RESEARCH RESULTS

SESSION I - NATURAL RESOURCES VARIABILITY

SESSION III - ENGINEERING TECHNOLOGY

PREFACE

Since the 1994 site-specific management for agricultural systems conference, field applications of precision agriculture - precision farming, site-specific crop management, etc., - by producers and agribusiness have increased rapidly. At the same time, research conducted by a variety of scientists in the US and abroad grew significantly as attested by papers presented at this conference. Both, however, are indicating that optimum applications of the concept of what was initially call 'Farming by Soil Type' , about twelve years ago, is much more complex that initially thought and requires much more research and development in management prescriptions, machinery, sensors, and software.

It has been said that new technologies go through three phases: excitement, chaos, and rebirth. Precision agriculture is still in a first development phase or infancy but will definitively change agricultural management by bringing technology and information age to the farm. It is the foundation for the next agricultural revolution.

This book contains the proceedings of the Third International Conference on Precision Agriculture held in Minneapolis (Bloomington) June 23-26, 1996. Previous conferences were held in Minneapolis, April 14-16, 1992 and March 27-30, 1994 . The proceedings provide an overview of recent and current research and applications related to various aspects of precision agriculture, namely, soil resources, managing variability, technology, profitability, environment, and technology transfer. Keynote address papers, located at the beginning of each section, provide an introduction or a state of the art review of each aspect. These are followed by volunteer oral and poster papers. Papers presented during the pre-conference workshop on 'Precision Nitrogen Management-Latest Research Results' are located at the beginning of the book. Abstracts of oral and poster presentations without manuscripts are at the end of each section.

Fifteen workgroups met to list and rank research and development needs in precision agriculture. The summary of the workgroup recommendations are presented in Appendix I. The list of participants is in Appendix II and the list of exhibitors in Appendix III.

On behalf of all participants, we wish to express our gratitude to sponsoring organizations for their support and to ASA-CSSA-SSSA for publishing this document. We also wish to express our appreciation to all speakers and poster participants for their presentations and manuscripts, and to all participants who made the conference a success. We look forward for a fourth conference July 19-23, 1998 in Saint Paul, Minnesota.

P.C. Robert,
R.H. Rust,
W.E. Larson, co-editors

ACKNOWLEDGMENTS

Organizing Committee

P.C. Robert, Chair *
E.L. Anderson, MES, Univ. MN
P.E. Fixen, Potash and Phosphate Institute
W.E. Larson *
R.H. Rust *

Editorial Committee

P.C. Robert, Chair *
R.R. Allmaras , USDA-ARS, St. Paul
W. E. Larson *
D.J. Mulla *
R. H. Rust *

Special thanks are expressed to Karen Mellem (*) for editing and organizing the manuscripts in a camera-ready copy.

* Department of Soil, Water, and Climate, Univ. MN

Sponsors

    The University of Minnesota:
        Precision Agriculture Center
        Department of Soil, Water, and Climate
        Minnesota Extension Service

    American Society of Agronomy, Crop Science Society of America,
        Soil Science Society of America

    ASAE- The Society for Engineering in Agricultural, Food, and
    Biological Systems

    American Society for Photogrammetry and Remote Sensing

CONTRIBUTORS

Ag-Chem Co./SOIL TEQ, Inc. Minnetonka, MN

Agri-Growth, Inc. Hollandale, MN

Cargill, Inc. Minneapolis, MN

CASE-IH Corporation, Racine, WI

Cenex-Land O Lakes, St. Paul, MN

Deere & Co. Moline, IL

Monsanto, Inc. St. Louis, MO

New Holland North America, Inc. New Holland, PA

Northrup King Co. Minneapolis, MN

Pioneer Hi-Bred, International. West Des Moines, IA

Tyler Industries, Inc. Benson, MN

U.S. Department of Agriculture:

    Agriculture Research Service

    Cooperative State Research, Education,
    and Extension Service

    Natural Resources Conservation Service

WORKSHOP

PRECISION NITROGEN MANAGEMENT -

LATEST RESEARCH RESULTS

Relationship of Nitrogen and Topography

K.R. Hollands

CENTROL Crop Consulting
Twin Valley, Minnesota

In the majority of cases, crop production fields in the Red River Valley have nitrogen levels that vary with topography. The high areas normally have a higher amount of residual nitrogen and the low areas have usually been very low in residual nitrogen. In the past, there has **not** been an efficient way of spreading these variable fields accurately.

Now with the introduction of GPS into agriculture and new mapping software, these variable applications can be made accurately and economically. Presently many growers are using grid soil sampling, however with grid points being 350-500 feet apart in rolling fields, some variability will undoubtedly be overlooked.

That was the case in a field which was grid soil sampled in the fall of 1995. There was enough rise and fall in this field that in some cases 3 consecutive grids were located in low areas with high areas between them. Obviously, these grid locations were missing much of the variability due to their location and distance between points. After review, it was decided to sample additional grid points alternating between high and low areas based solely on visual judgment when traveling across the field.

Since nitrogen levels seemed to follow the topography, it was logical to produce a topography map to compare with the nitrogen levels of the field. The finished 3-D topography map displayed the surface variation of this field. When comparing the topography map and the nitrogen map the nitrogen levels matched the topography.

Next year, many fields will be separated into various elevation ranges. These elevation ranges will be soil tested separately and the results from each will then be assigned to similar elevations within a certain elevation grouping. Since all elevations are geo-referenced, mapping software will then be used to develop a spreader map for a variable rate spreader. More uses for these topography maps will arise as the technology expands.

OBJECTIVES

In the past if producers questioned a soil test, one of their options was to split their fields and take separate soil samples from the high elevations and the low elevations. In all cases when fields were split, the low areas had less nitrogen than the higher areas. See example from 1994 for Knutson Brothers of Fisher, MN.

KNUTSON BROTHERS
1994 HIGH/LOW NITROGEN

FIELD	0-6" N	6-24" N	24-42" N	TOTAL
FIELD 1 HIGH	19	42	21	82
FIELD 1 LOW	17	15	9	41
FIELD 2 HIGH	18	15	24	57
FIELD 2 LOW	7	9	6	22
FIELD 3 HIGH	13	48	92	153
FIELD 3 LOW	20	24	40	84
FIELD 4 HIGH	22	60	66	148
FIELD 4 LOW	18	12	9	39
FIELD 5 HIGH	27	33	30	90
FIELD 5 LOW	13	15	6	34
FIELD 6 HIGH	28	18	27	73
FIELD 6 LOW	18	12	6	36

The nitrogen averaged 2.3 times higher in the higher elevations than the lower areas. Therefore, the research objectives of this project were to 1) identify how well nitrogen values corresponded to the topography of a crop production field, and 2) develop a spreader map for a variable rate spreader based on the results.

MATERIALS AND METHODS

The study was organized into three phases:
1. Field Operations - Soil sampling using a hydraulic probe mounted in a pickup.
   - Elevation data collection using laser equipment.
2. Mapping - Making nutrient maps, spreader maps and topography maps
3. Interpretation - Nitrogen levels evaluated with varying elevations

METHODOLOGY - FIELD OPERATIONS

The study was carried out in the fall of 1995. The field was the south 1/2 of the Northeast Quarter, Section 18 of Roome township. This field is located south of Fisher, MN in Polk County. The field was planted to wheat in 1995 and will be sugarbeets in 1996. This field had some rise and fall which was quite obvious to see with the naked eye, however was still fairly flat ground.

Twenty pre-set grid points for this 83.2 acre field were mapped and copied on a disk to bring to the field. Using GPS for guidance across the field, flags were placed on the centers of all grid points. A pickup followed, equipped with a 42" hydraulic soil sampling probe and was used to take 6-7 probes from each grid location. There was enough rise and fall in this field that in some cases 3 consecutive grids were located in low areas with high areas between them. Obviously, these grid locations were missing much of the variability due to their location and distance between points. Total costs of soil sampling this field was approximately $12.00 per acre with grids averaging 4.16 acres. (Figure 1).

A nitrogen soil test level map was developed to better understand the layout of the field. The white shaded areas represent where nitrogen soil test levels were the highest in the field. These areas required little or no applied nitrogen. The darker shaded areas represent medium nitrogen levels and required a medium amount of applied nitrogen. The gray shaded areas represent low nitrogen soil levels which required large amounts of nitrogen. (Figure 2)

After reviewing the results, it was decided to go back to this field and set additional grid points down the same GPS trail alternating between high and low areas. Soil test results of previously sampled grid points which were positioned on either high or low areas were used again and not resampled. Sixteen previously sampled points were used and 17 new points were added for a total of 33 grid points to produce this new high-low pattern. These points were positioned according to visual judgment of when the next highest or lowest point of elevation had been reached. This new method more accurately reflected the true nitrogen levels in this field. Total costs of soil sampling would have been $19.00 per acre with grids averaging 2.52 acres. (Figure 3)

A nitrogen soil test map was developed to better understand the layout of the field. The lighter shaded areas represent where the nitrogen soil test levels were the highest in the field. These areas required little or no applied nitrogen. The darker shaded areas represent medium nitrogen levels and required a medium amount of applied nitrogen. The gray shaded areas represent very low nitrogen levels and required a considerable amount of applied nitrogen. Notice that this map depicts some higher nitrogen areas which the 20 grid field did not. These were the higher elevation areas which did not receive sampling the first time. Even though the 20 grid pattern had done an acceptable job of determining most of the variability, this new method was more correct. The difference between the total nitrogen requirements of the 2 patterns was only 277 #/A of Urea (46-0-0). The 33 grid pattern matched the nitrogen requirements more precisely than the 20 grid pattern and required less nitrogen. (Figure 4)

Since nitrogen levels seemed to follow the topography, it was logical to produce a topography map to compare with the nitrogen levels of the field. A laser transmitter and a receiver were obtained to take elevation readings across the entire field. The receiver was connected to a mast which was mounted in the back of a pickup. The transmitter was placed in the lowest area of the field. Using GPS with a 13 by 26 grid pattern and the laser equipment simultaneously, elevation readings were taken every 100 feet and logged on a computer spreadsheet. The GPS trail shows the way the pickup traveled through the field. (Figure 5)

There were 338 elevation readings which varied 2.47 feet from the highest to the lowest point in the field. Elevations were logged to the hundredth of a foot. The spreadsheet was then displayed as a 3-D surface map. The finished topography map displays the surface variation of this field. (Figure 6) When comparing the topography map (Figure 6) and the nitrogen map (Figure 4) the nitrogen levels of this field matched the topography very closely. All of the higher elevations are where the nitrogen was the highest in the field. The lower elevations are where nitrogen was the lowest in the field.

Grid Soil Tested

FIGURE 1

Nitrate-N (0–42" lb/a)

1	11	48	85	5
47	27	1	111	-4
2	-6	-15	113	-13
9	-13	-9	239	55

4.1 acre grids - Red River Valley Field - Fall 1995

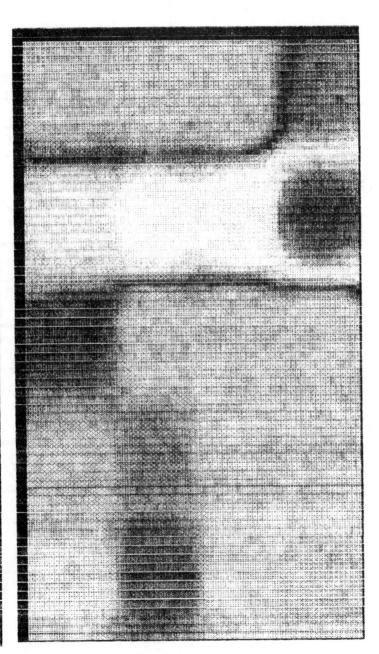

FIGURE 2

1995 Field - 20 Grids

8

Topography/Grid Sample

Available Nitrogen (0–42" lb/a)

FIGURE 3

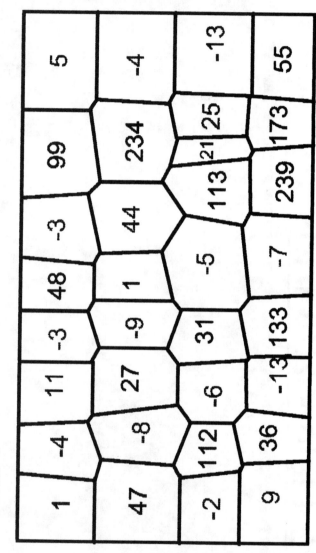

FIGURE 4

1995 Field - 33 Grids

GPS Trail of 338 Elevation Points

FIGURE 5

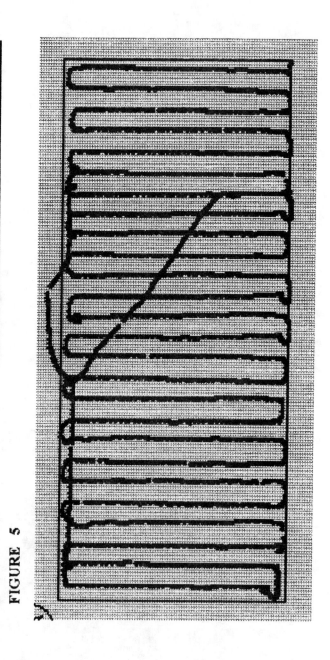

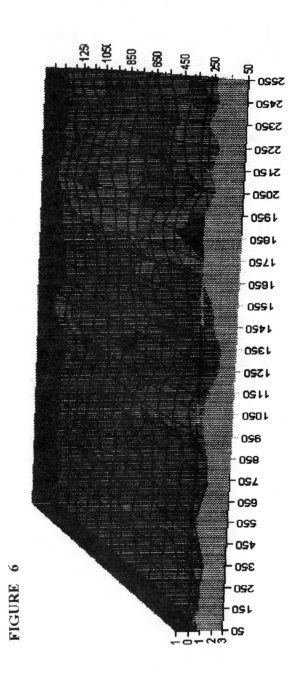

1995 Field - Topography Map

FIGURE 6

CONCLUSION

Even though the nitrogen variability was pin-pointed more accurately with the 33 grid (high/low) pattern, the extra cost may not be acceptable. Nitrogen levels were found to fluctuate with the topography of the field. After elevation readings are collected, nitrogen levels within a range of elevations may be somewhat predictable.

In the future, many fields will be separated into various elevation ranges. Each determined elevation range will then be soil tested. Each of the soil test results will then be assigned to similar elevations within a certain elevation grouping. These different nitrogen values in each different elevation grouping could be determined each year by soil testing. Since all of these points are geo-referenced, mapping software will then be used to develop a spreader map for a variable rate spreader. The total monetary investment involved could be less than the cost of a grid soil sampled field.

Once a topography map of a field is developed, the same map file could be used year after year for elevation sampling. Instead of normal grid soil samples being taken every 3rd or 4th year (usually before sugarbeets), this elevation sampling could be done more often with the potential for a greater impact on the profitability of all crops in the rotation. Better nitrogen management, brought forth from variable rate nitrogen applications between sugarbeet crops, should improve sugarbeet quality. More uses for these topography maps will arise as the technology expands.

Examples of future uses of topography maps:

1. Elevation Sampling   5. Variable Yield Goals
2. Variable Rate Spreading   6. More Efficient Drainage Evaluations
3. Variable Seeding Rate   7. Comparisons To Yield Maps
4. Variable Herbicide Rates 8. Comparisons to Satellite Imagery

Laser generated topography maps and elevation soil sampling may better reflect nitrogen requirements of rolling fields for all crops grown in the Red River Valley and other areas. Topography sampling should not be considered a replacement for grid soil sampling, but an alternative to it for special field situations.

Soil Sampling for Site-Specific Nitrogen Management

Richard B. Ferguson
University of Nebraska
Department of Agronomy - South Central Research & Extension Center
Clay Center, NE

Carol A. Gotway
University of Nebraska
Department of Biometry
Lincoln, NE

Gary W. Hergert
University of Nebraska
Department of Agronomy - West Central Research & Extension Center
North Platte, NE

Todd A. Peterson
Pioneer Hi-Bred International
West Des Moines, IA

ABSTRACT

Variable rate application of nitrogen (N) has the potential to increase N use efficiency of crops and reduce N loss to the environment. However, accurate application maps are required for variable rate application to be effective. In a three year study in Nebraska, grid soil sampling was evaluated as a means of generating the N rate map. Even on an apparently uniform site, improvements in the accuracy of the N rate were obtained with variable rate application, even with a relatively coarse grid density. However, the grid density required for accurate N rate maps will vary from soil to soil. Also, grid sampling annually for residual soil nitrate is not likely to be economically feasible. Ultimately, it is probable that accurate N rate maps will be produced with information from a variety of sources of spatial data which can be obtained at greater densities and at less cost than grid soil samples, such as yield maps and remotely sensed images, as well as from selective soil sampling.

INTRODUCTION

Generation of the application map is the most important component of variable rate nitrogen (N) application. As with a fixed rate recommendation for the entire field, the recommendation must be based on valid data, with correct interpretation of the influence of soil test levels or other information upon which the N recommendation is based. An accurate application map is necessary before valid judgment can be made of the potential benefits of VRT. Some knowledge of the minimum information required to produce an accurate application map is needed before any economic evaluation of VRT can be made. It is therefore critical that research be conducted to evaluate what is required for accurate N application maps - what information, and at what spatial and temporal densities, is critical to generate a valid map.

In Nebraska, the current approach used by the University of Nebraska for making

N recommendations for corn relies on three variables - expected yield, residual soil nitrate-N, and organic matter content, used in the following equation:

$$N\ REC = 35+1.2*EY-8*NO3PPM-0.14*EY*OM$$

where N REC is pounds of N per acre, EY is expected yield in bushels per acre, NO3PPM is the average root zone residual nitrate-N in ppm, and OM is percent organic matter (Hergert, et al, 1995). This equation is used as the starting point for site-specific N recommendations. Because of the large influence of climate on yield from year to year, most research studies of variable rate N application in Nebraska have used a uniform expected yield, and site-specific information about soil organic matter and residual nitrate-N.

Because of variability in yield and precipitation from year to year, accurate N recommendations for corn in Nebraska rely on annual soil sampling for residual nitrate-N. Non-mobile nutrients and soil attributes which vary little from year to year can be assessed less frequently, perhaps every 3 - 4 years. Many producers in Nebraska soil sample annually for residual nitrate-N because of the increased accuracy of the N recommendation and resulting increased profit and reduced loss of N to groundwater. However, it is not likely that annual grid soil sampling for residual nitrate-N will be acceptable economically. Consequently, an approach is needed to develop N application maps with the same accuracy as those developed from grid soil sampled residual nitrate-N data, but which can be obtained with less time and expense.

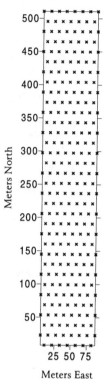

Figure 1. Soil sampling pattern and density.

This paper describes results from variable rate N research conducted in Nebraska, and suggests potential approaches for developing accurate N application maps without annual grid soil sampling.

MATERIALS AND METHODS

A field study was initiated at three locations in Nebraska in 1993, evaluating approaches to variable rate application of N for corn on furrow-irrigated fields. Data from one of these locations will be discussed in this paper, located in Clay county, on a Crete silt loam soil (fine, montmorillonitic, mesic Pachic Argiustolls). The study area is approximately 90 m by 500 m within a 16.2 ha field. The previous crop in all years has been corn.

Figure 1 illustrates the soil sampling pattern and density for the study area. Single cores were collected at the midpoint of cells 6.1 m (8 rows) wide and 15.2 m long, in an offset grid design. The resulting sample density was 54 cores/ha. Cores were 3.8 cm in diameter and 0.9 m deep. The surface 0.2 m of each core collected in the fall of 1993 was analyzed for organic matter (OM), pH, zinc (DTPA-Zn), phosphorus (Bray-1 P), and NO_3-N. The lower 0.7 m of each core was analyzed for NO_3-N only. Cores were collected in an offset grid, resulting in a triangular pattern which provided information similar to a regular

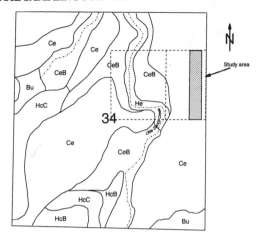

Figure 2. Soil survey map of study site, Clay county, Nebraska.

grid at less cost (Parkhurst, 1984; Yvantis, et al, 1987). In subsequent years, soil cores were collected in the fall at the same locations, but analyzed for NO_3-N only.

Nitrogen rate maps were developed each year using the University of Nebraska N recommendation algorithm for corn, using spatial information about OM and NO_3-N, and a uniform expected yield of 11.3 Mg/ha (180 bu/acre). SURFER and SAS software was used for mapping and statistical analysis. The crop was harvested each year with a yield mapping combine; in 1993 using a research combine in cooperation with Dr. Mark Schrock, Kansas State University, and in 1994 and 1995 using a John Deere 9600 equipped with an Ag-Leader Yield Monitor 2000. In all years yield position was determined by real time DGPS.

RESULTS AND DISCUSSION

N Recommendation Map

Figure 2 shows the soil series map for the study area and surrounding field. The entire study area is mapped as Crete silt loam, even though there is a substantial range of soil organic matter (OM) in the study area (Figure 3). The study area is generally quite uniform and level, and in some respects a situation not likely to benefit from variable rate N application due to the apparent uniformity of the site. Figure 4 illustrates the pattern of residual soil nitrate-N each year of the study to a depth of 0.9 m. The average NO_3-N varied little from year to year, with relatively small ranges in each year. Nitrate-N concentrations for individual cores ranged from 1.5 to 7.5 mg/kg with a mean of 4 in 1993, from 2.5 to 9.3 with a mean of 5 mg/kg in 1994, and from 1.1 to 10 with a mean of 4.4 mg/kg in 1995. In two of the years (1993 and 1994), there is a significant correlation between OM and soil NO_3-N (Table 1) - generally NO_3-N is higher in the central portion of the study area where OM is also highest. Figure 5 illustrates the resulting N rate maps for each year. Since the N rate maps are based on a fixed expected yield and spatial OM and NO_3-N, the patterns exhibited in the rate maps reflect those of the OM and NO_3-N maps - the N rate is lowest where OM and NO_3-N are highest. Since NO_3-N has a relatively small range within years,

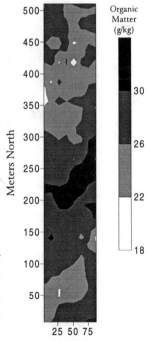

Figure 3. Organic matter, g/kg.

Table 1. Pearson correlation coefficients for selected relationships.

Factor A	Factor B	Pearson Correlation Coefficient (R)	Prob > R
1995 Yield Index	Organic Matter	0.5	0.0001
Phosphorus	Zinc	0.48	0.0001
1994 Yield Index	1995 Yield Index	0.429	0.0001
1993 Root Zone NO$_3$-N	Organic Matter	0.415	0.0001
1994 Root Zone NO$_3$-N	Organic Matter	0.369	0.0001
1994 Chlorophyll	1994 Yield Index	0.311	0.0001
1994 Broken Stalks	1994 Root Zone NO$_3$	0.294	0.0001
1994 Broken Stalks	Organic Matter	0.271	0.0001
1995 Root Zone NO$_3$-N	Organic Matter	0.157	0.0121

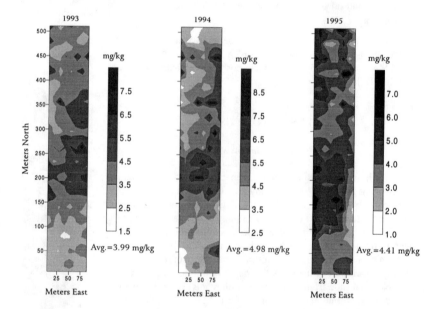

Figure 4. Root zone nitrate-N (mg/kg), 1993 - 1995.

varies little from year to year, and is correlated with OM, the N application maps generally follow OM trends and vary little from year to year.

The effect of sampling density on the resulting N rate map is shown in Figure 6 and Table 2. Figure 6 illustrates actual data compared to an interpolated map using all sample points and an interpolated map from 1/8 the sample points (simulating a grid density of 6.75 cores/ha, or 2.7 cores/acre). Using SURFER, residuals were created for each sample point from the two interpolated maps. Residuals were mapped in 15 kg/ha

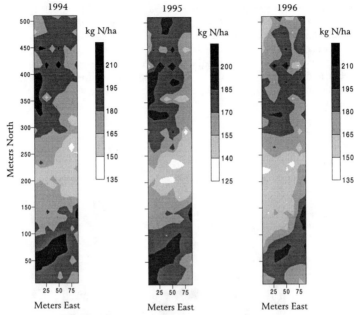

Figure 5. Nitrogen rate (kg/ha), 1994 - 1996.

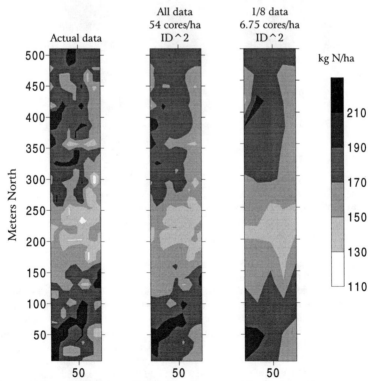

Figure 6. Recommended N rate (kg/ha), comparing actual data, inverse distance squared interpolated, and inverse distance squared with 1/8 the original sample density.

increments (not shown), and the percent of area in each residual range calculated using the area analysis option in SURFER. The patterns for all three maps are similar. Even with increased interpolation and smoothing with fewer sample points, the variable rate map at the lower density is still an improvement over fixed rate application. Table 2 shows that 47% of the area was correctly fertilized with VRT at the low sample density, while 29% was correctly fertilized at the fixed rate (170 kg N/ha). Approximately 94 % of the area was fertilized within 15 kg N/ha of the correct rate with VRT at the low density, while 70% of the area was within 15 kg N/ha of the correct rate when applied uniformly.

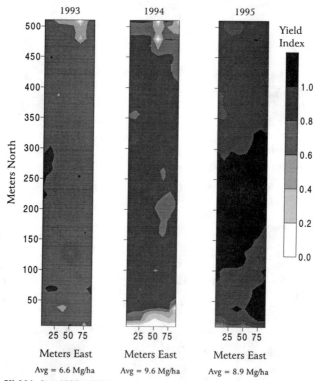

Figure 7. Yield index, 1993 - 1995.

Relationship of N Application Map to Other Parameters

Yield maps from 1993 - 1995 are shown in Figure 7, expressed as an index of maximum yield for that year. Grain yield in 1995 is significantly correlated to OM (Table 1), with highest yields in the central portion of the study area. Yield patterns in 1994, and to a lesser extent in 1993, were influenced by broken stalks due to high winds at critical growth stages. Stalk breakage was evaluated spatially in 1994 (Figure 8) and showed some negative correlation to OM. This pattern is consistent with observations from other studies, which have found stalk breakage due to high wind to increase with factors which enhance rapid crop growth, such as N rate and crop rotation. The yield map observed in 1995 is more likely representative of yield potential of the study area in a typical year.

Table 2. Deviation from 1995 recommended N rate at sample points with two application methods - uniform and variable rate, using ID^2 interpolation with reduced density dataset.

Deviation From Actual Rate at Sample Point (kg N/ha)	Fixed N Rate (170 kg N/ha) (% of area)	Inverse Distance Squared, 1/8 data set (% of area)
+30	1.8	0.7
+15	7.9	5.8
0	28.7	47.1
-15	33.0	40.6
-30	25.1	5.1
-45	3.4	0.7

Patterns of phosphorus and zinc in soil were quite dissimilar to other patterns observed (Figure 9). These nutrient levels were significantly correlated (Table 1) to each other, but not to other soil or crop parameters. These patterns are the result of livestock confinement at a farmstead formerly located at the northern end of the study area, 40 years or more earlier. Concentrations of P and Zn above the critical level have no positive impact on yield, as evidenced by the lack of influence of elevated P or Zn on the yield maps in any year (Figure 7).

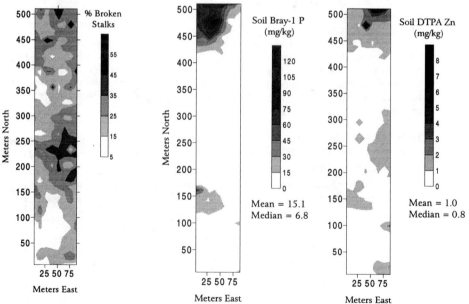

Figure 9. Phosphorus and zinc distribution.

Figure 8. Broken stalks (%), July 1994.

Potential Sampling Approaches

Although grid soil sampling can provide an accurate basis for variable rate N application, the cost and labor associated with annual sample collection make it likely that other approaches will ultimately be used to create N application maps. Grid soil sampling as a starting point, or selective sampling based on some strategy, may be components of these approaches.

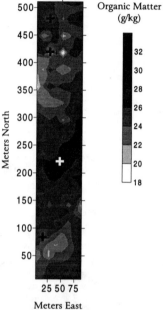

Initial grid at "economically acceptable" density followed by re-sampling every 3-4 years

This approach, although currently practiced commercially for non-mobile nutrients such as P and K, is not likely to be a valid practice for N by itself. However, it may serve as a starting point for use with other maps, such as nutrient removal maps developed from yield maps, and modeled N loss maps, in the generation of N application maps. Such an approach might be adequate for this study site, since NO_3-N varied relatively little within or among years.

Selective sampling according to soil survey

This approach may be useful for some soils and not others. As shown in Figure 2, the study area is mapped as one series but exhibits a fair range in OM and N recommendation levels. Sampling according to mapped soils would provide no direction for fields such as this one. However, future soil surveys may contain more detail which could increase their usefulness for site-specific management. Soil surveys will also not reflect variability due to management or land modification. Some of the variation in OM at this site is likely due to leveling for furrow irrigation. Consequently, it is likely that, for some soils, soil surveys may provide useful information for generating N rate maps, while not for other soils.

Figure 10. Selective sampling strategy for residual NO_3-N: two locations in areas of avg. OM, one in high OM, one in low OM.

Table 3. Deviation from recommended N rate at sample points - uniform application, and variable application, based on spatial OM and 1993 residual N map adjusted by selective sampling of field for 1994 recommendations.

Deviation From Actual Rate at Sample Point (kg N/ha)	Fixed N Rate (% of area)	Variable N Rate - Residual N adjusted by selective sampling (% of area)
+30	1.3	0.7
+15	15.4	18.1
0	26.8	55.7
-15	41.2	25.5
-30	14.9	0
-45	0.2	0

Initial grid followed by selective sampling

With this approach, a detailed grid would be collected once to characterize the field. After that, the N recommendation map could be adjusted according to samples collected at strategic points. Figure 10 illustrates one possible approach of sampling the study area according to OM level. Table 3 illustrates that when the residual N map from 1993 is adjusted according to the 1994 NO_3-N concentration at selected points in the field, chosen to represent a range in potential NO_3-N concentration, the result is still a more accurate fertilization method than applying a uniform rate, with considerable less investment than grid sampling. In this example, 56% of the field received the correct rate, while only 27% of the field received the correct rate with uniform application.

Remote sensing/yield map/soil sample combinations

A remotely sensed image of the study field, Figure 11, was obtained in July 1995. Patterns observed in this image of the study area are very similar to those seen in the OM and N rate maps, as well as the 1995 yield map. Early variable rate N research has suggested that yield maps or topographic information can improve the accuracy of the N application map (Kitchen, et al, 1995;. Vetsch, et al, 1995). Yield maps, integrated over several years to account for

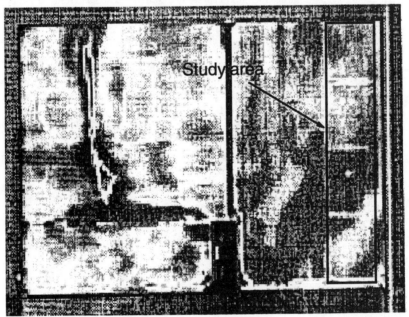

Figure 11. Remotely sensed image of study area, July 1995.

climatic variation, as well as remotely sensed data may serve as a less costly means to provide information from which N rate maps can be developed. Remotely sensed images have the advantage of evaluating the crop during the growing season, allowing correction of developing N deficiency with high clearance applicators or by fertigation with center pivots. Combinations of spatial information obtained from yield mapping, remote sensing, and selective or grid soil sampling are the most likely approaches which will be used to generate N rate maps.

SUMMARY

Grid soil sampling can provide an accurate means of developing variable rate N application maps. The Clay county study site, even though it is a relatively uniform site, can be more accurately fertilized by using variable rate rather than uniform application. The sampling density required for acceptable accuracy will vary from field to field, with more variable fields requiring greater sampling densities to create accurate maps. Current recommendation procedures based on annual residual soil nitrate-N testing are not likely to be widely adopted to grid sampling for variable N application because of the time and expense required. Although considerable additional research is needed, combinations of yield maps, remotely sensed images, and soil sampling appear to be the most likely means by which accurate N rate maps will be generated in the future.

REFERENCES

Hergert, G.W., R.B. Ferguson, and C.A. Shapiro. 1995. Fertilizer Suggestions for Corn. Univ. of Nebraska Cooperative Extension NebGuide G74-174A.

Kitchen, N.R., D.F. Hughes, K.A. Sudduth, and S.J. Birrell. 1995. Comparison of variable rate to single rate nitrogen fertilizer application: Corn production and residual soil NO_3-N. *In* P.C. Robert et al.(eds.) Proceedings of the 2nd Intl. Conference, Site-Specific Management for Agricultural Systems.

Parkhurst, D.F. 1984. Optimal sampling geometry for hazardous waste sites. Environ. Sci. Technol. 18:521-523.

Vetsch, J.A., G.L. Malzer, P.C. Robert, and D.R. Huggins. 1995. Nitrogen specific management by soil condition; Managing fertilizer nitrogen in corn. *In* P.C. Robert et al (eds.) Proceedings of the 2nd Intl. Conference, Site-Specific Management for Agricultural Systems.

Yvantis, E.A., G.T. Flatman, and J.V. Behar. 1987. Efficiency of kriging estimation for square, triangular and hexagonal grids. Mathematical Geology 19:183-205.

Assessment of Plant Nitrogen in Irrigated Corn

W.C. Bausch
H.R. Duke

USDA-Agricultural Research Service
Water Management Research
AERC-CSU Foothills Campus
Ft. Collins, CO

C. J. Iremonger

Soil and Crop Sciences Dept.
Colorado State University
Ft. Collins, CO

ABSTRACT

Applying small amounts of nitrogen (N) fertilizer to irrigated corn during the growing season "as needed" by the crop has potential for reducing the amount of N applied to a particular field. For this N management scheme to be practical, procedures must be developed that can rapidly assess the plant N status in large fields. This paper shows that plant N in irrigated corn can be reliably estimated using a previously developed N Reflectance Index (NRI) calculated from measured green (G) and near-infrared (NIR) canopy reflectance. The field study site consisted of 12 contiguous plots with four imposed N treatments, each replicated three times. SPAD chlorophyll meter measurements, whole plant sampling for N analysis, and canopy reflectance measurements were made throughout the 1995 corn growing season. Measured plant total N (all leaves) and estimated plant total N based on the NRI compared favorably as long as the soil background was obscured. Consequently, the NRI is assumed to be a better indicator of N sufficiency than the N Sufficiency Index calculated from SPAD measurements because a plant community consisting of many plants is monitored instead of a single point on a single corn leaf. Furthermore, assessment of plant N status is rapid and can be easily mapped with GIS tools to show areas that are N deficient.

INTRODUCTION

Blackmer and Schepers (1995) demonstrated use of a Minolta[1] Soil Plant Analysis Development (SPAD) chlorophyll meter as an effective tool to schedule N fertigation in irrigated corn using procedures outlined by Peterson et al. (1993). Not only the "as needed" but also the "where needed" must be assessed on a routine basis to effectively utilize this N management scheme. Sequentially clamping the SPAD chlorophyll meter on the same relative leaf of 30 representative plants and

---

[1]Brand names given do not imply endorsement by the authors, ARS, or CSU.

recording the average within a field plot is not time consuming; however, attempting to characterize the plant N distribution within large fields would be a laborious and time consuming process.

Remote sensing of canopy reflectance has the advantage of sampling a plant community rather than a single plant and of rapidly assessing its spatial variability in a field. Blackmer et al. (1996) presented results of spectroradiometer-measured canopy reflectance acquired in late August [dent growth stage, R5 (Ritchie et al., 1989)] which showed reflectance peaks in the green and near-infrared portions of the electromagnetic spectrum with reflectance differences in these two regions due to plant N status. Their results were similar to those presented by McMurtrey et al. (1994) for spectrophotometer measured leaf reflectance of excised corn leaves at the V12 (12 mature leaves) growth stage. Blackmer et al. (1996) compared relative reflectances of individual wavebands and combinations of wavebands at R5 with relative grain yields and reported that the ratio of the data for the 550- to 600-nm interval to the 800- to 900-nm interval provided equal or superior predictability compared with any individual wavelength or with the NIR/red ratio (760- to 900-nm interval divided by 630- to 690-nm interval) used by Walburg et al. (1982).

Based on results presented by McMurtrey et al. (1994) and others that have reported nutrient deficiency effects on leaf reflectance, Bausch and Duke (1996) developed an N Reflectance Index to monitor plant N status of irrigated corn from measured green (G, 520 to 600 nm) and near-infrared (NIR, 760 to 900 nm) canopy reflectance. The N Reflectance Index was defined as a ratio of the NIR/G for an area of interest to the NIR/G for a well N-fertilized reference, i.e., an area that is never N deficient. Comparison of the N Reflectance Index (NRI) to the N Sufficiency Index (NSI) which is calculated as the ratio of SPAD meter measurements for an area of interest to SPAD meter measurements for the reference area (Peterson et al., 1993) produced a near 1:1 relationship. Consequently, indication of the "need" for nitrogen fertilizer using the NRI is nearly identical to that from the NSI, i.e., NRI ≤ 0.95 indicates an N deficiency. The purpose of this paper is to show that the NRI can be used routinely to assess plant N status and its variability in an irrigated corn field.

MATERIALS AND METHODS

The experiment was conducted at the Agricultural Research, Development, and Education Center, Colorado State University (CSU), northeast of Ft. Collins, CO. Results reported herein are part of a long-term water/nitrogen interaction study. Data utilized were collected during the 1995 growing season on 12 plots with four nitrogen treatments (preplant, sidedress, prescription, and reference), each replicated three times. Field dimensions for the area of interest were 85 m by 140 m. Each plot was 28 rows (0.76 m spacing) wide by 46 m long with a north/south row orientation. Field plot numbering was from west to east, north to south, i.e., plots 25 to 28 were in the first rep, plots 33 to 36 (second rep), and plots 41 to 44 (third rep).

Each field plot, except for the reference N plots, was sampled at nine locations before planting to determine residual N in the soil profile. Nitrogen recommendations for each plot were based on procedures described by Mortvedt et al. (1995) using composited soil samples from the top 0.6 m depth and a 11 Mg/ha yield goal; these are presented in Table 1 for the three N treatments. Urea ammonium nitrate (UAN, 32% N) was applied at a depth of 0.15 m on 10 May, day of year (DOY) 130, at the recommended rate to the preplant N treatments, approximately one-half the recommended rate (86 kg/ha) to the sidedress N treatments, 36 kg/ha (approximately 30%) to the prescription N treatments, and 168 kg/ha to the reference N treatments. An additional 86 kg/ha N (UAN) was applied on DOY 188 (V6 growth stage) to the sidedress treatments with one fertilizer shank midway between rows. Additional N (UAN) was applied to the prescription N treatments on DOY 173, 207, and 222 at 28, 19, and 39 kg/ha, respectively, when the mean NSI for the treatments was < 0.95. The reference N treatments had additional N (UAN) applied on DOY 173 and 207 (39 and 19 kg/ha, respectively) to insure adequate N availability. Nitrogen applied to the prescription N and reference N treatments on DOY 207 and 222 was applied with a custom built applicator mounted on a high-clearance tractor by dribbling the fertilizer on the soil surface between rows; it was incorporated via irrigation the following day.

Table 1. Nitrogen recommendations for the three replications of the three N treatments.

Replication	N Treatment					
	Preplant		Sidedress		Prescription	
	Plot No.	Amount (kg/ha)	Plot No.	Amount (kg/ha)	Plot No.	Amount (kg/ha)
1	26	173	25	173	27	122
2	35	182	34	175	33	128
3	42	161	41	167	43	117

Pioneer brand hybrid 3790 was planted on DOY 132 (12 May) at ≈ 84 000 seeds/ha. Emergence did not occur until DOY 153 (2 June) due to a cool, wet spring. Corn tasseled on DOY 220 (8 Aug.); a hard freeze on 21 Sept. (DOY 264) killed the corn in its R5 (dent) growth stage.

The field was furrow irrigated using surge technology. Irrigations occurred on DOY 194, 208, 215, 223, 229, 236, 243, 250, and 257. Depth of application was based on estimated crop evapotranspiration (Et) using procedures defined in SCHED (Buchleiter et al., 1988), the USDA-ARS irrigation scheduling computer program. This model utilizes the modified Penman equation to calculate alfalfa reference Et; crop coefficients are time-based. Reference Et was calculated from climatic data measured by an automated weather station located on the research farm.

Minolta SPAD 502 chlorophyll meter measurements were taken at least once per week during the growing season starting at the V1 growth stage. Sampling procedures outlined by Peterson et al. (1993) were used. Thirty measurements were made on representative plants in each plot; the average SPAD value, as calculated by the meter, was recorded. The most recently mature leaf (leaf collar exposed) was used at each measurement date until tasseling (VT) at which time the ear leaf was measured. The NSI was calculated from SPAD data, i.e., SPAD measurements for a particular field plot were divided by the mean SPAD measurements for the reference plots. Ten representative plants also were removed at random from each plot at the V3, V6, V12, R1, and R3 growth stages for biomass and N analysis. Plants were separated into leaves, stems, and ears (R growth stages) with each component composited into a single, respective sample.

Plant samples were oven dried at 65°C for 24 h and ground. Subsamples ($\approx$ 0.1 g) were weighed and placed in a LECO, model CHN-1000, analyzer. The analyzer used the Dumas method (Bremner and Mulvaney, 1982) which is a dry oxidation (combustion) technique. Analyzer output was expressed as total N (%).

Corn canopy radiance and incoming irradiance were measured simultaneously with a mobile boom-mounted data acquistion system similar to that described by Bausch et al. (1990), except the boom was mounted on a high-clearance tractor for field access. The data system consisted of an instrument platform with two Exotech 100BX four-band radiometers and a HarvestMaster 286LX portable field computer to sample and store radiometer sensor voltages. The down-looking radiometer measured target radiance and was fitted with 15° field of view (FOV) optics; it was locked in its nadir view angle, i.e., pointed perpendicular to the crop surface. The other radiometer looked upward to measure irradiance; its FOV was 180°. Height of the down-looking radiometer above ground was $\approx$ 10 m; the viewed spot was circular with a diameter of 2.6 m. A third Exotech four-band radiometer (15° FOV) was positioned on a vertical support near the side of the tractor; its view angle was 75°. This radiometer's view was perpendicular to the crop rows and was located 1 m above the crop canopy. Its viewed spot was an ellipse with a major axis of 5.2 m and a minor axis of 1.2 m. The center of the ellipse was within 1 m of the center of the circular spot viewed from the nadir position. Height of the 75° view radiometer was adjusted before each measurement session. Radiant energy was measured in the green, red, and near-infrared spectral wavebands (0.52-0.60, 0.63-0.69, and 0.76-0.90 μm, respectively). Measurements were acquired twice each week around solar noon between the V1 and R5 growth stages. The high-clearance tractor traveled the border rows between field plots, thus, the center of the down-looking radiometer was over the eighth row from both sides of the plot. A data point was acquired when the operator pushed the trigger button; thus, point spacing varied between three and five meters.

Position (latitude and longitude) of each data point was determined with Ashtech Ranger GPS equipment; the rover antenna was positioned directly above the down-looking radiometer. The GPS base station was located on the research farm; its antenna was located over a surveyed point which had an accuracy of at least 1 cm. GPS satellite data were recorded every 2 s by the base and rover receivers; the data were post processed to achieve 1 m accuracies on GPS positions.

Canopy reflectance was calculated for each waveband using the procedure described by Neale (1987). When the computer program calculated canopy reflectances from the measured radiometer voltages, the post-processed GPS positions were combined with the canopy reflectance data to produce a single data file. Reflectance data acquired on days when clouds affected the measurements were excluded. The NRI for each data point for the three N treatments was calculated by ratioing the NIR/G for that particular point to the mean NIR/G from the reference treatments.

Measurement dates were compared to obtain matched times between the NRI and the NSI, as well as the NRI and whole plant leaf total N. Corresponding measurement dates for the NRI and the NSI occurred three times on the same day, ± one day four times, ± two days one time, and ± three days one time. Match dates between the NRI and whole plant leaf total N occurred on the same day two times and ± one day three times.

RESULTS AND DISCUSSION

Temporal plots of the NSI and the NRI for the second replication of the prescription, sidedress, and preplant N treatments (field plots 33, 34, and 35, respectively) are shown in Fig. 1 for nadir and 75° view angle acquired canopy reflectance data. Mean NRI values on any given day for each N treatment field plot were calculated from 20 to 25 individual data points; error bars were calculated as the mean ± one standard deviation. Individual NRI data points represent a composite reflectance consisting of sunlit and shaded leaves from 25 to 35 plants depending on crop height as well as sunlit and shaded soil depending on growth stage. The NSI value was based on point measurements from 30 leaves averaged by the SPAD meter. Depending on the radiometer view angle and the N treatment, the correlation between the NSI and the NRI was different at times and showed similar trends at times. The NRI calculated from nadir acquired canopy reflectance has a significant soil reflectance component until the vegetation effectively covers the soil surface, which is between the V11 and V13 growth stages for corn. Consequently, changing soil surface moisture content changes soil color and its reflectance which introduces variability not associated with actual plant conditions (Bausch, 1993). At the V9 growth stage, the 75° view angle radiometer "saw" only vegetation; this was supported by 35 mm slides taken with a camera located at the radiometer position. A 135-mm focal length lens was used on the camera to make its FOV similar to the radiometer's FOV.

Comparisons between measured plant total N (all leaves) and estimated plant total N (all leaves) are shown in Fig. 2. Plant total N was estimated from mean NRI values calculated from both nadir and 75° view canopy reflectance data . These data represent the same field plots shown in Fig. 1. Relationships between plant total N (all leaves) and the NRI were developed using data acquired in 1994 and 1995 from small plots that had six N rates applied to them. Equations were developed to estimate plant total N (TN, % DW) before VT and after VT. For the nadir view, these relationships were $TN(\%) = 3.98 \times NRI-1.09$, $r^2 = 0.71$ (before VT) and $TN(\%) = 6.32 \times NRI-3.55$, $r^2 = 0.89$ (after VT), using data that represented the V9-V15 growth stages and the R1-R2 growth stages, respectively. Corresponding estimating equations for the 75° view angle were $TN(\%) = 4.57 \times NRI-1.60$, $r^2 = 0.82$ and $TN(\%) = 4.80 \times NRI-2.20$, $r^2 = 0.92$. Estimated plant total N appeared to be somewhat better for the 75° view than for the nadir view angle. Since measured

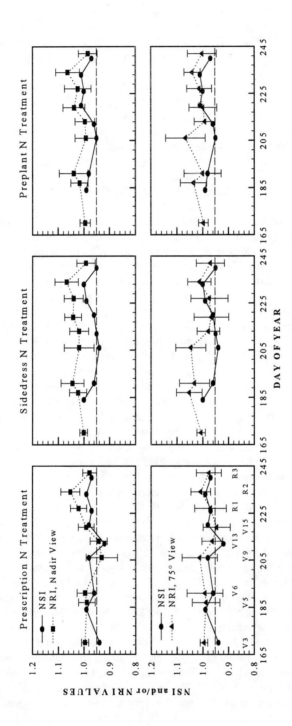

Fig. 1. Temporal plots of NSI and NRI from the nadir view (top row) and the 75° view (bottom row) for field plots 33, 34, and 35.

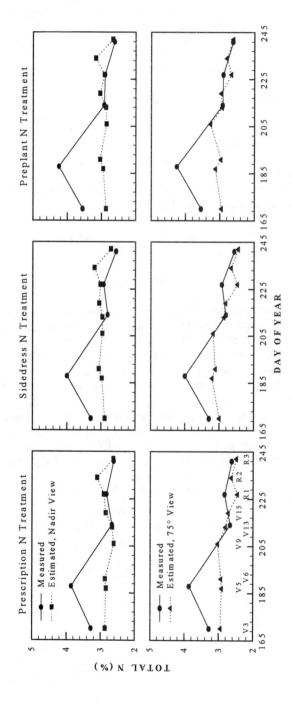

Fig. 2. Measured and estimated total N from the nadir (top row) and 75° view (bottom row) for field plots 33, 34, and 35.

data at the V9 growth stage were not available for these comparisons, one can only presume that estimated TN at V9 from 75° viewed canopy reflectance is somewhat representative of the measured data. Estimated TN around V5 or V6 (DOY 188 and 191) and before are not representative of measured TN for either view because of the predominate soil reflectance.

Spatial variability of plant N status within the nine contiguous field plots at the V9 (DOY 206) and V13 (DOY 213) corn growth stages is portrayed by Fig. 3 for the two view angles. Plot numbers and their nitrogen treatment designation were listed in Table 1. Thiessen polygons were constructed around each data point and assigned the value of the data point. Polygon shading was based on NRI ≤ 0.95 and NRI > 0.95. The light shaded areas (NRI ≤ 0.95) represent an N deficiency. The nadir view on DOY 206 indicates a larger N deficient area than the 75° view; this is primarily caused by soil representing a predominate portion of the viewed scene. Differences between the two views on DOY 213 are less pronounced. Soil background influence on nadir acquired canopy reflectance at the V13 growth stage was minimal. Some of the difference between the two views may be related to the 75° view radiometer "seeing" more sunlit vegetation on one side of the field plot as opposed to the other side of the plot while data were being collected.

CONCLUSIONS

The comparisons between measured and estimated plant total N (all leaves) indicate that the NRI represents the plant N status in corn when the soil background is obscured by the crop canopy. Based on the data presented, a nadir viewing radiometer is useful when the corn canopy effectively obscures the soil surface which normally occurs at the V12 to V13 growth stage, whereas, an oblique viewing radiometer (75°) can detect differences by the V9 growth stage. The NRI calculated from canopy reflectance measurements presumably provides a better assessment of N sufficiency than the NSI calculated from SPAD measurements because it monitors a plant community rather than a small area composed of single points on 30 single leaves. Spatial assessment of N deficient areas in a field can be accomplished very rapidly with remote sensing and mapped with GIS tools. Thus, nitrogen can be applied "as needed" and "where needed" by the crop to reduce broad across-the-field N applications.

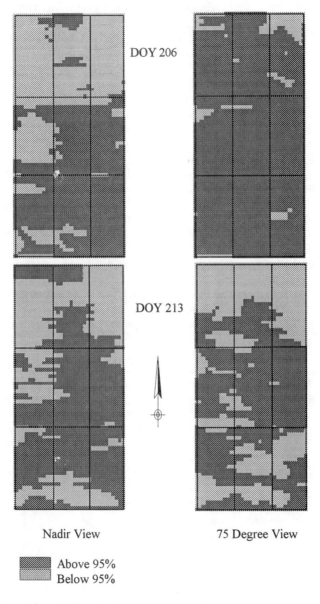

DOY 206

DOY 213

Nadir View   75 Degree View

Above 95%
Below 95%

Fig 3. Spatial variability of plant N status in irrigated corn at V9 and V13 growth stages (DOY 206 and 213, respectively) as determined by the NRI from nadir and 75° view acquired canopy reflectance data. Light shaded areas are N deficient.

REFERENCES

Bausch, W.C. 1993. Soil background effects on reflectance-based crop coefficients for corn. Remote Sens. Environ. 46:213-222.

Bausch, W.C., D.M. Lund, and M.C. Blue. 1990. Robotic data acquisition of directional reflectance factors. Remote Sens. Environ. 30:159-168.

Bausch, W.C. and H.R. Duke. 1996. Remote sensing of plant nitrogen status in corn. Accepted by Trans. of the ASAE.

Blackmer, T.M. and J.S. Schepers. 1995. Use of a chlorophyll meter to monitor nitrogen status and schedule fertigation for corn. J. Prod. Agric. 8:56-60.

Blackmer, T.M., J.S. Schepers, G.E. Varvel, and E.A. Walter-Shea. 1996. Nitrogen deficiency detection using reflected shortwave radiation from irrigated corn canopies. Agron. J. 88:1-5.

Bremner, J.M. and C.S. Mulvaney. 1982. Nitrogen - Total. p. 595-624. *In* A.L. Page, R.H. Miller, and D.R. Keeney (ed.) Methods of Soil Analysis, Part 2-Chemical and Microbiological Properties, 2nd Ed. Agron. No. 9.

Buchleiter, G.W., H.R. Duke, and D.F. Heermann. 1988. User's guide for USDA-ARS irrigation scheduling program SCHED. USDA Agricultural Research Service (unpublished), 30 pp.

McMurtrey III, J.E., E.W. Chappelle, M.S. Kim, J.J. Meisinger, and L.A. Corp. 1994. Distinguishing nitrogen fertilization levels in field corn (Zea mays L.) with actively induced fluorescence and passive reflectance measurements. Remote Sens. Environ. 47:36-44.

Mortvedt, J.J., D.G. Westfall, and R.L. Croissant. 1995. Fertilizer suggestions for corn. Service in Action no. 0.538. Cooperative Extension, Colorado State Univ., Ft. Collins.

Neale, C.M.U. 1987. Development of reflectance-based crop coefficients for corn. Ph.D. Dissertation, Colorado State Univ., Ft. Collins. 170 pp.

Peterson, T.A., T.M. Blackmer, D.D. Francis, and J.S. Schepers. 1993. Using a chlorophyll meter to improve N management. NebGuide G93-1171-A. Cooperative Extension, Institute of Agriculture and Natural Resources, Univ. of Nebraska, Lincoln.

Ritchie, S.W., J.J. Hanaway, and G.O. Benson. 1989. How a corn plant grows. Iowa State Ext. Serv. Spec. Rep. 48, Iowa State Univ., Ames. 21 pp.

Walburg, G., M.E. Bauer, C.S.T. Daughtry, and T.L. Housley. 1982. Effects of nitrogen nutrition on the growth, yield, and reflectance characteristics of corn canopies. Agron. J. 74:677-683.

Remote Sensing to Identify Spatial Patterns in Optimal Rates of Nitrogen Fertilization

A. M. Blackmer
S. E. White

Department of Agronomy
Iowa State University
Ames, Iowa

ABSTRACT

Aerial photography was used with yield mapping to identify spatial patterns in nitrogen (N) sufficiency within cornfields having various N rates applied in strips. These spatial patterns were used to map crop response to the N. This technique should be helpful in efforts to develop a new generation of N fertilizer recommendations needed for variable-rate application technologies.

INTRODUCTION

Maps indicating spatial patterns in optimal rates of N fertilization are required for effective use of variable-rate application technologies during corn production. Although it commonly is assumed that such maps can be generated directly from maps of yields from previous seasons, mounting evidence suggests that yield data alone do not provide reliable assessments of N fertilizer needs. Extensive studies in Iowa, for example, showed no correlation between current recommendations based on yields and observed optimal rates of N fertilization (Blackmer et al., 1992).

One problem with recommendations based solely on yields is that they fail to consider variability in amounts of plant-available N supplied by soils. Another problem is that such recommendations fail to consider variability in percentages of fertilizer N lost from the rooting zone before it is needed by the crop. Recommended rates of N fertilization should be based on differences between amounts of plant-available N supplied by specific soils and amounts of N needed by crops on these soils.

Maps of observed yield *responses* to applied N should provide a more reliable basis for developing maps showing the rates at which fertilizers should be applied. Maps of yield response, however, are difficult to develop. One reason is that a single measurement of yield response must involve yield measurements on a series of plots having different rates of N added. Another reason is that soil-N availability and many other yield-affecting factors (water availability, weeds, insects, etc.) tend to vary in different spatial patterns, so meaningful assessments of yield response require that all the treatments be replicated. This poses a dilemma on spatially variable soils because increasing the size of the experimental area adds more yield variability not associated with treatments. Spatial variability within an experimental area, therefore, adds uncertainty to each individual assessment of yield response.

Combines equipped with on-the-go yield monitors and global positioning systems may offer a practical way to make enough observations to construct maps of yield response. It has not been demonstrated, however, that current systems have the required precision when used on plots small enough to assess patterns of yield response in spatially variable soils. A system that requires a 100 m swath to attain the necessary level of precision, for example, could not be used to map yield response in a field where important differences in fertilizer requirement were caused by undulating topography in which crests were separated by an average distance of 100 m. Inability to resolve differences in fertilizer requirements in highly variable soils could limit the value of variable-rate fertilization in situations where this new technology is most needed.

Recent studies have shown that N deficiency in corn can be detected by analyzing color photographs taken from airplanes (Blackmer & Schepers, 1996; Blackmer et al., 1995, 1994, 1996). Deficiencies of N result in leaves having less chlorophyll, less photosynthesis, and increased reflectance of radiation normally absorbed during photosynthesis. Persons with normal color vision see the deficiencies as a lighter-green color. The perceived color of corn varies with many factors other than N deficiencies, so assessments of N deficiency must be based on comparisons with plants known to have adequate N under otherwise identical conditions. The assessments of deficiency, therefore, are based on a measurement of crop-color response to N and require experimental designs similar to those needed for measuring yield response.

The objective of this report is to describe an ongoing research project utilizing aerial photography of canopy color, yield monitoring, and other new technologies to develop a new generation of N fertilizer recommendations for use with variable-rate application technologies. An underlying assumption is that better methods of characterizing spatial patterns in response of corn to N fertilization is a critical step in developing such recommendations. Results from a single field are presented to illustrate the nature of these studies and offer some preliminary observations.

METHODS

Strip plots were established for 1995 in a 5.4-ha field within the Lester-Fluvaquents-Wadena soil association in Greene County, Iowa. Soil survey maps indicate 6 different soil series (Fig. 1a) and expected corn yields that ranged from

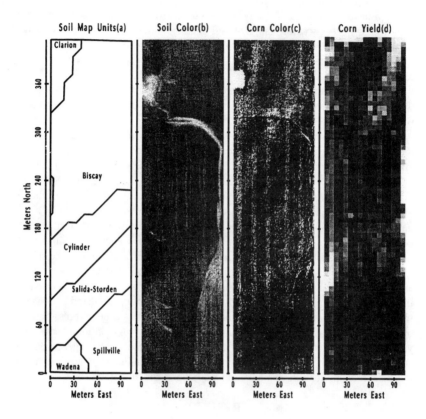

Fig. 1. Soil survey map (a), aerial photograph of soil immediately after planting
(b), aerial photograph of the crop canopy in mid August (c), and grain yield
map (d) for a 5.4-ha field planted to corn. Plots from which soil and tissue
samples were collected are marked on the photograph of the soil (b). The
eight-row strips receiving the lowest rate of N fertilization can be identified
by their lighter color in the canopy photograph (c). Yield levels are
proportional to blackness in yield map (d).

3.1 Mg/ha for the Salida-Storden complex to 7.6 Mg/ha for the Spillville loam.
Except for applications of N, the field was managed by the producer using his
normal practices. These include harvesting with an eight-row combine (76 cm
between rows) equipped with a yield monitor and differential global positioning
system. The planter sprayed 34 kg N/ha as urea-ammonium-nitrate solution in a 30-
cm band over the row. The preceding crop was soybean, and no fertilizer N was
applied in the fall or before planting.

The field contained 17 eight-row strips to be harvested as individual swaths of the combine. Fifteen of these were divided into five blocks of three strips. Fifteen small (12-m length, width of the block) plots were positioned (see Fig. 1b) so as to have maximal uniformity of soil characteristics within each plot and to capture the widest possible range of soil characteristics among the plots. When corn plants were 30 cm tall, soil samples (composite of 48 cores 3.2 cm in diameter) were collected to a depth of 30 cm for determination of nitrate and other relevant soil characteristics.

Soon after the soil samples were collected urea-ammonium-nitrate solution was injected midway between rows to a depth of 15 cm to give N treatments (including N applied at planting) of 56, 112, and 168 kg N/ha. The N treatments were randomly assigned to the eight-row strips within each block. Each knife of the applicator had independent flow control, and rate of fertilizer flow through the knives was adjusted for rate of applicator travel by using ground-sensing radar.

The field was photographed using color film and a LMK-1000 aerial mapping camera with a 15-cm distortion-free lens from a height of 730 m shortly after the corn was planted and in mid August. The pictures were digitized on a grid having about 94,800 cells/ha. Samples for the end-of-season cornstalk test (Binford et al., 1992b) were collected to spot check N status of the corn within each N treatment of each small plot where soil samples had been collected in late spring.

Grain yields as measured by the combine were corrected to 15.5% moisture and mapped on a grid size of 5,400 cells/ha. A Unix ARCINFO geographic information system was used to study relationships among digitized soil survey maps, digitized photographs of the soil soon after planting, digitized photographs of the crop canopy in mid August, and maps of yields. Maps of color response and yield response were generated by creating a geographically correct dataset consisting of the difference between measured values resulting from an incremental increase in N application. This dataset was kriged and smoothed.

RESULTS

The spring of 1995 had unusually large amounts of rainfall that delayed planting by about three weeks. An intense rainfall event shortly after planting resulted in a surface flow of water across the strips (320 - 340 m north in Fig. 1b) and along the rows on the east side of the field. This flow fanned out and covered much of the southeastern corner of the field. The soil photograph showed an area of light-colored soil marking a portion of the cornfield previously occupied by a barn (360 m north, 5 m east in Fig. 1b). A band of relatively dark soil includes the area mapped as Cylinder loam. Diffuse streaks in soil color going lengthwise the field were caused by uneven distribution of residue following tillage.

The mean concentration of nitrate in the surface 30-cm layer of soil across the 15 plots was 10 mg N/kg soil. The nitrate concentrations were within the range of 7 to 13 mg N/kg in all plots except one, which was 18 mg N/kg. The highest nitrate concentration occurred on a plot near the site of the old barn and probably should be attributed to the residual effects of an old barnyard. The observed concentrations of nitrate were substantially below the 25 mg N/kg normally considered optimal for corn production (Magdoff et al., 1990; Binford et al.,

1992a). The observed concentrations, however, are consistent with those expected following above-average spring rainfall.

The aerial photograph of the crop canopy (Fig. 1c) revealed light-colored strips that correctly identify the 8-row strips receiving only 56 kg N/ha. Differences in color between strips receiving 112 and 168 kg N/ha were not readily detectable; an observation suggesting that the 112 kg N treatment avoided N deficiencies at essentially all locations within the field. At some sites lack of color contrast between the 56-kg rate and higher rates indicated that this rate corrected all deficiencies. This rate appeared to be adequate north of the 320 m mark in Fig. 1c; the sharpness of the boundary suggests that the northern end of the field probably had a different cropping history than did the remainder of the field, even though cropping history had been similar for several years.

The aerial photograph of the crop canopy (Fig. 1c) showed spatial patterns in color clearly unrelated to N treatments. An area of relatively light color seems to approximately coincide with the Salida-Storden complex, which is rated as the poorest soil for corn production. An area of relatively dark color within the area mapped as Biscay seems to indicate that the plants detected a difference in soil type that was not indicated by the soil survey map. At this site, as at many sites studied by this technique, crop color often varied in spatial pattern somewhat differently than would be expected from soil survey maps.

Yield maps (Fig. 1d) generated by using data from the combine showed patterns generally similar to those observed in crop color; the areas of darkest green color tended to have the highest yields. Strips receiving only 56 kg N/ha tended to have lower yields than did strips receiving 112 kg N/ha, but these differences usually disappeared in portions of the field where the higher N rate did not induce a response in canopy color. Although strips having 112 and 168 kg N/ha seemed to be different in many small parts of the field, the yield differences between these treatments were not consistent enough to correctly identify these treatments from the yield map.

Data from the combine indicated mean yields of 6.4 Mg/ha for the 56 kg/ha treatment, 7.9 Mg/ha for the 112 kg/ha treatment, and 7.4 Mg/ha for the 168 kg/ha treatment. Analysis of mean yields for individual strips showed a significant (alpha = 0.05) yield difference between the first and second levels of N application and no significant difference in yields between the first and third or second and third levels of N. The yield decrease between the second and third N levels was significant at alpha = 0.20. Yield decreases from excesses of N could be expected because a period of low-moisture stress occurred near the time of silking and because high rates of N application frequently depress yields under such situations.

Mean stalk nitrate concentrations at the end of the season were 0.08 g N/kg for the 56 kg N/ha treatment, 1.04 g N/kg for the 112 kg N/ha treatment, and 2.60 g N/kg for the 168 kg N/ha treatment if the one plot near the site of the old barn is excluded. These findings indicate that the lowest rate was not adequate to maximize yields, the middle rate was near optimal, and the highest rate was excessive (Binford et al., 1992b). Stalk nitrate concentration for the plot near the site of the old barn was 3.12 g N/kg with only 56 kg N/ha applied and clearly was different from the remainder of the field. Point estimates of N sufficiency that were provided by the cornstalk test showed considerable variability and were poorly

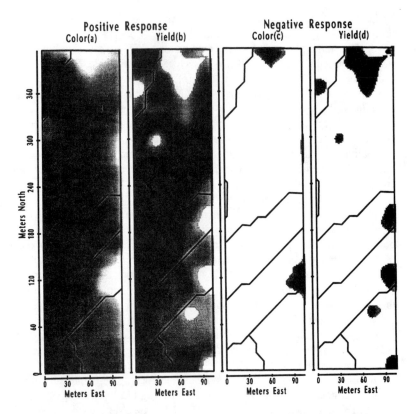

Fig. 2. Spatial patterns in intensity of crop responses (color and yield) to increasing rate of N fertilization from 56 to 112 kg/ha. Magnitude of the response is proportional to the intensity of shading in this figure. A positive color response indicates that the canopy had a darker green color with a higher N rate.

related to soil mapping units. Such variability should be expected after examining the canopy photograph, which revealed complex spatial patterns in canopy color that were unrelated to fertilizer treatments applied in strips.

Spatial patterns were more clearly defined in the color photograph (Fig. 1c) than in the yield map (Fig. 1d). This should be expected because the canopy color is illustrated by using many more cells/ha. Moreover, the yield maps have errors caused by nonuniform or "bunchy" flow of corn through various parts of the combine. This nonuniform flow may be caused in part by small areas of uneven soil that jar the combine or alter its rate of travel. This problem results in measured yield being assigned to a slightly incorrect location. Although the photographs can reveal spatial patterns within plots of the size normally needed to measure yields, they do not indicate yields. The aerial photographs, therefore, aid in interpretation of yield maps but they do not provide a substitute for yield maps. Information concerning actual yield response (i.e., Mg/ha) is essential for selecting economically optimal rates of fertilization.

Maps showing spatial patterns of crop response to increasing rate of N fertilization from 56 kg N/ha to 112 kg N/ha are presented in Fig. 2. Positive and negative responses are mapped independently to illustrate spatial patterns and intensity of both positive and negative effects. The patterns of crop color and crop yields tend to agree that the first increase in rate of fertilization gave a positive response in a large area in the middle of the west half of the field (Fig. 2a and b) and a negative response in areas at the north end and along the east edge of the field (Fig. 2c and d). Some areas showed lack of agreement. The color response, for example, was much greater in the area of the old barn (360 m north 5 m east in Fig. 2).

Disagreements between color response and yield response should be expected under some conditions. Under conditions of extreme deficiencies of N, for example, an increment of added N can increase yields in situations where the higher yielding corn is still deficient and shows strong symptoms of N deficiency. Color response and yield response would show opposite trends if additions of N intensify drought-stress at silking time, reduce the number of kernels formed, and decrease the amount of N translocated from leaves as the grain develops. Under these conditions, the net result of added N is more green leaves and less grain at the end of the season. Such observations underscore the need for ground-truthing aerial photographs with yield data or tissue-test data. They also illustrate that studies of contradictions between color photographs and yield maps may provide significant information that cannot be obtained from only the photographs or yield maps.

Canopy color showed positive responses to increasing N rate from 112 to 168 kg N/ha (Fig. 3c) in areas of the field having negative yield responses (Fig. 3d). This probably is explained by the extra N intensifying drought stress during the silking period. The maps showed that yield responses to this increment of N were negative over much of the field. Data in Fig. 3 illustrate that when comparing high rates of N spatial patterns in color response provide little useful information unless they are compared with spatial patterns in yield. Comparisons of spatial patterns in color response and yield response, however, help to identify probable reasons for differences in yields. These comparisons also give information as to how land area in the field should be divided for effective studies of the most important factors affecting N fertilizer requirement.

Data from this field in 1995 clearly support previous observations that yields are not reliable indicators of N fertilizer needs. Some of the best yield responses to the first incremental increase in N rate occurred in an area of the field having the lowest yield potential. Negative yield responses were produced by this increment in areas having some of the highest yields. The finding that yields were decreased by increasing N rates from 112 to 168 kg N/ha underscores the need to base N fertilizer recommendations on observed responses to N fertilization.

Spatial patterns of N response usually showed no obvious relationship to soil mapping units in the field studied. Additional studies on this soil association are necessary to determine if patterns of yield response can be predicted from measurable soil characteristics. Studies in other fields have shown patterns of yield response that vary with measurable differences in soil characteristics, and additional studies are needed to determine whether these patterns are reasonably stable across a normal range of weather conditions. Although it will never be possible to adjust

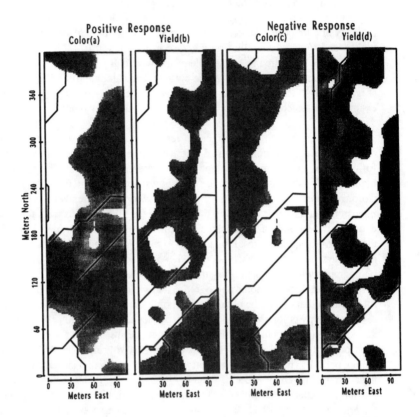

Fig. 3. Spatial patterns in intensity of crop response (color and yield) to increasing rate
of N fertilization from 112 to 168 kg/ha. Magnitude of the response is
proportional to the intensity of shading.

N rates for weather conditions after fertilization, it should be possible to identify the
rates most likely to maximize profits for the producer within the range of weather
conditions normally encountered.

CONCLUSIONS

Aerial photographs of cornfields taken in August show spatial patterns in color
that reveal crop-sensed differences in soil properties. These patterns can be compared
with soil survey maps and with aerial photographs of soils to determine if the spatial
patterns correspond to obvious differences in soil characteristics or soil mapping units.

Some of these patterns may be caused by earlier management practices and explained by considering management history.

Application of N fertilizer treatments in strips enables characterization of spatial patterns in response of the crop to N. Information provided by photographs of canopy color can aid in interpretation of yield maps generated from data collected by combines because the photographs have much greater power to resolve spatial patterns of crop response that occur within small areas. The aerial photographs cannot replace yield maps, however, because measurements of actual yield increases are necessary to determine economically optimal rates of fertilization. Use of remote sensing with yield monitoring, soil testing, and tissue analysis should aid efforts to map spatial patterns in optimal rates of N fertilization within fields. Use of this method on selected fields within a region should help develop a new generation of N fertilizer recommendations that can be used to guide variable-rate fertilizer applicators in all fields within the region.

ACKNOWLEDGMENTS

Funding for this study was provided in part by the Iowa Corn Promotion Board and Agri Industries.

REFERENCES

Binford, G.D., A.M. Blackmer, and M.E. Cerrato. 1992a. Relationships between corn yields and soil nitrate in late spring. Agron. J. 84:53-59.

Binford, G.D., A.M. Blackmer, and B.G. Meese. 1992b. Optimal concentrations of nitrate in cornstalks at maturity. Agron. J. 84:881-887.

Blackmer, A.M., T.F. Morris, and G.D. Binford. 1992. Predicting N fertilizer needs for corn in humid regions: Advances in Iowa. p. 57-72. In B.R. Bock and K.R. Kelley (ed.) Predicting N fertilizer needs for corn in humid regions. Bull. Y226. National Fertilizer and Environmental Research Center, Tennessee Valley Authority, Muscle Shoals, Alabama 35660.

Blackmer, T.M., and J.S. Schepers. 1996. Aerial photography to detect nitrogen stress in corn. J. Plant Physiol. 148:440-444.

Blackmer, T.M., J.S. Schepers, and G.E. Meyer. 1995. Remote sensing to detect nitrogen deficiency in corn. p. 505-512. In P.C. Robert et. al., (ed.) Proceedings of site-specific management for agricultural systems. Second international conference. ASA, CSSA, SSSA, Madison, WI.

Blackmer, T.M., J.S. Schepers, and G.E. Varvel. 1994. Light reflectance compared to other N stress measurements in corn leaves. Agron. J. 86:934-938.

Blackmer, T.M., J.S. Schepers, G.E. Varvel, and E.A. Walter-Shea. 1996. Nitrogen deficiency detection using reflected shortwave radiation from irrigated crop canopies. Agron. J. 88:1-5.

Magdoff, F.R., W.E. Jokela, R.H. Fox, and G.F. Griffin. 1990. A soil test for nitrogen availability in the Northeastern United States. Commun. Soil Sci. Plant Anal. 21:1103-1115.

Variability of Corn Yield and Soil Profile Nitrates in Relation to Site-Specific N Management

M. W. Everett
F. J. Pierce

Crop and Soil Sciences Department
Michigan State University
E. Lansing, MI

ABSTRACT

This study examined the spatial variability of corn yield (*Zea mays* L.) and soil NO_3^--N over a three year period within a field to assess the potential for site-specific N management. The structure and extent of the spatial variability of corn yield and soil profile NO_3-N and the lack of sufficient correlation among and between them in space and time do not support the use of site-specific N management in this field.

INTRODUCTION

Matching fertilizer N applications to crop need by location within fields should improve use-efficiency of N fertilizers and reduce the potential for NO_3^-N leaching. The realization of these benefits by farmers depends to a large extent on the adequacy of site-specific N recommendations. Since the basis for fertilizer N recommendations includes some combination of yield goal, N mineralization potential, and residual profile NO_3^-N in the spring (Jansson & Persson, 1982; Olsen & Kurtz, 1982; Stanford & Legg, 1984; Black, 1993), the spatial variability of these parameters must be known, non-random, of sufficient magnitude, and manageable. Additionally, the spatial structure should not have a strong temporal variability since N recommendations are prescriptive and based on long-term averages.

The nitrogen cycle in soil is dynamic and both spatially (Cahn et al., 1994; Cambardella et al., 1994; Robertson et al., 1994) and temporally variable (Jokela and Randall, 1989; Legg & Meisinger, 1982; Meisinger, 1984; Francis et al., 1993). A wide range of spatial dependence (1 to 201 m) has been reported for soil profile NO_3^-N (White et al., 1987; Van Meirvenne & Hofman, 1989; Cahn et al., 1994; Cambardella et al., 1994), however overall relationships of soil profile NO_3^-N indicate that only a moderate degree of spatial dependence exists. Robertson et al. (1994) indicated that net N mineralization was highly spatially correlated at a range of 103.8 m in a cultivated site, whereas in an uncultivated experiment, spatial dependence was low (r^2=0.43 and a range of 10.7 m).

While soil profile NO_3^-N is a contributing factor in corn yield variability and NO_3^-N leaching (Bundy & Malone, 1988; Jokela & Randall, 1989; Angle et al., 1993; Liang et al., 1991; Binford et al., 1992), Hoeft et al. (1992) concluded that the release of mineralizable N and/or the availability of residual NO_3^-N was often difficult to predict and difficult to relate to increased corn yields in the Midwest.

In general, little is known about the spatial variability of these N recommendation parameters.

While many aspects of corn production and N dynamics are well documented, comparatively little is known regarding the co-variability of these parameters on a field scale and the extent to which that variability can be managed. This knowledge is essential if site-specific N fertilizer management is to be feasible. This study examined whether the spatial and temporal variability of corn yield and soil profile NO_3-N within a field supports the use of site-specific management of N fertilizers.

MATERIALS AND METHODS

Field studies were conducted from 1992 to 1994 in a 22.6 ha farm field located in the floodplain of the Kalamazoo River in southwest Michigan (42° 40'54" N, 85° 30'54" W). The field contains five soil mapping units primarily with loamy sand and sandy loam textures (Fig. 1). A water table is present within 1.2 m during the growing season and overhead irrigation was applied when needed to supplement rainfall. The field has been cropped to continuous corn using a tillage system consisting of fall chisel plowing followed by spring secondary tillage with a disk and harrow. Plant populations, fertilizer, varieties and pesticide applications varied annually.

Four transects oriented N-S were located randomly within each irrigation zone (Fig. 1). Each transect was divided into grid points spaced 30.5 m apart. A total of 60 grid points locate the experimental units.

Corn yields, plant populations, and grain moisture were measured at each grid point by sampling corn plants from two 9.1 m rows spaced 76 cm apart centered on each grid point. Composite soil samples were collected at each grid point in June and after harvest in the fall in 1993 and 1994 for determination of NO_3-N and NH_4-N. Two 6.4 cm diameter cores were collected from the center portion of each grid point to a depth of 1.2 m and sectioned into 0.3 m increments and composited by depth. Soil was collected on 8 June and 18 December, 1993 for all transects. In 1994, the farmer had applied fertilizer N to all but transect 2. Therefore, on 7 June, 1994 only transect 2 plus an additional transect (5) (Fig. 1) were sampled. On November 11, 1994, soil samples were collected on all five transects. Air-dried soil samples were extracted in duplicate with a 10:1 solution:soil ratio of 1.0 M KCl (Keeney & Nelson, 1982) and the filtrate analyzed colorimetrically for NO_3-N and NH_4-N using a injection flow analysis.

Descriptive statistics and correlations (SAS, 1990) were calculated for all parameters collected over the three year period for the whole field, each individual transect, and the three main soil map units. General spatial analysis was performed using methods described by Robertson et al. (1994) using GS[+] (Gamma Design Software, 1993).

RESULTS AND DISCUSSION

There are three geographic scales evident in Fig. 1 - soil map units, transects within irrigation zone, and whole field. The results of this study are summarized

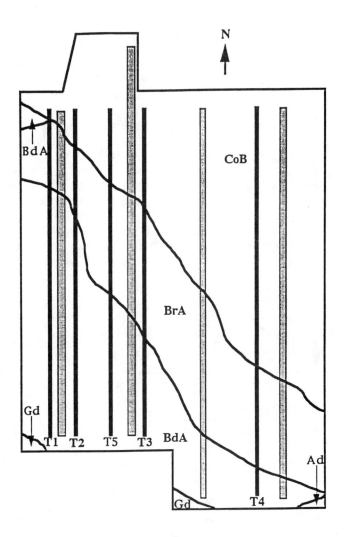

MAP UNIT	DESCRIPTION	% COMP.
CoB	Coloma loamy sand	43
BrA	Brady sandy loam	31
BdA	Bronson sandy loam	19
Gd	Gilford sandy loam	6
Ad	Adrian muck	1

Fig. 1. Layout of experimental field showing irrigation access lanes, transects and soil mapping units.

by each scale for corn performance and soil profile NO_3-N. The main points to be made here are as follows:

Corn Performance

On a whole field basis, corn yields were highest and least variable in 1994 (12.9±0.5 Mg ha^{-1}) and were similar in 1992 (9.9±1.3 Mg ha^{-1}) and 1993 (9.4±0.8 Mg ha^{-1}) (Table 1). Spatial dependence of corn yield, as measured by the fit (r^2) of the spherical model to the semivariogram, was not detected for the whole field (n=60) for any year or for the three year average yield (Table 1). Correlations of grain yield between years were weak (R< 0.5). Corn grain moisture was lowest in 1994; 21.3 ± 3.2 % compared to 34.7 ± 4.3 %, and 25.0 ± 1.5 % in 1992 and 1993, respectively, even though harvest dates were later in 1992 and 1993, and showed considerably more spatial dependence than corn grain yield. Correlations between grain yield and moisture were strong and negative (-.49 to near -1.0). Delayed maturity, whatever the cause, generally reduced corn grain yield in a given year. Correlations of corn grain moisture between years were high and positively correlated (R ranging from 0.77 to 0.84). Corn plant populations at harvest were lowest in 1994 (6.1 ± 0.48 plants ha^{-1} 10^{-4}), in part due to lower seeding rates, and

Table 1. Descriptive statistics and geostatistics for corn grain yield, grain moisture, and plant population at harvest for the whole field (n=60) or the period 1992-1994.

YEAR	AVG	STD	MIN	MAX	C_o[§]	C_o+C	C/(C_o+C)[¶]	Range[#]	RSS[††]	r^2
							Spherical Model Parameters			
		Corn Yield						m		
		— Mg ha^{-1} —								
1992	9.9	1.3	5.9	11.7	0.54	1.83	0.70	85.1	11.7	0.13
1993	9.4	0.8	5.4	11.5	1.32	1.32	0.00	7.7	9.1	0.02
1994	12.9	0.5	11.5	13.7	0.33	1.48	0.78	19.7	12.7	0.02
92'-94'	10.8	0.8	8.2	12.1	0.09	0.60	0.85	17.4	2.3	0.02
		Grain Moisture								
		— % —								
1992	34.7	4.3	30.3	42.9	2.86	10.83	0.74	335.5	294.3	0.34
1993	25.0	1.5	22.3	29.3	0.49	3.53	0.86	447.6	18.2	0.50
1994	21.3	3.2	16.4	29.6	0.01	11.30	1.00	156.6	556.6	0.31
92'-94'	27.0	2.4	23.5	33.8	1.30	7.51	0.83	348.7	178.6	0.34
		Plant population								
		— plant ha^{-1} 10^{-4} —								
1992	6.7	0.49	5.5	7.8	0.19	0.25	0.25	195.7	0.	0.07
1993	6.6	0.34	5.4	7.8	0.80	0.14	0.43	505.0	0.0	0.23
1994	6.1	0.48	4.3	7.1	0.17	0.24	0.32	7.7	0.2	0.00
92'-94'	6.5	0.27	5.5	7.1	0.06	0.08	0.22	7.7	0.0	0.00

[§]C_o = nugget variance
[¶]C/(C_o+C) = structural variance C as a proportion of model sample variance (C_o+C)
[#]range = distance (m) over which structural variance is expressed
[††]RSS = residual sum of squares

higher in 1992 (6.7 ± 0.49 plants ha^{-1} 10^{-4}) and 1993 (6.6 ± 0.34 plants ha^{-1} 10^{-4}). Plant populations varied within a few thousand plants ha^{-1}, showed no spatial dependence, and were not highly correlated with corn grain yield in any of the three years sampled (R < 0.43) or with corn grain moisture from 1992-1994. Plant populations were essentially a non-factor at any scale of this analysis.

Within transects, corn yields varied both in magnitude and variability within a transect (Table 1, Fig. 2). Transects 1 and 2 were located closest to the trees at the west and south ends of the field and tended to show a decline in corn yield moving from north to south (Fig. 2). Spatial dependence of corn yield varied by year and transect within year (Table 1). Spatial dependence was strongest in 1993 for all transects, with r^2 ranging from 0.62 to 0.82. For 1992 and 1994, spatial dependence was not strong with the exception of T4 in 1992 and T1 in 1994. When corn yields were averaged for the three years, 1992 to 1994, spatial dependence was strong for T1 and T2, moderate for T4, and not detected for T3. Where spatial dependence was strong, the structural variance (C) was a significant portion of the total variance (C/C$_0$+C), ranging from 0.59 to 1.0. Corn grain yields were negatively correlated with grain moisture but varied with transect and with year. Correlations of corn yields between years were not strong, generally < 0.50, with highest correlations within transect 2 (0.50 to 0.83). Spatial dependence for all transects was detected only when determined for the three year average, with highest values for Transects 3 and 4 (r^2 = 0. 58 and 0.66, respectively).

On the soil map unit basis, corn yields were consistently higher and grain moisture lower in all 3 years for map unit CoB than map units BrA and BdA, with average grain yields of 11.2, 10.8 and 10.1, respectively, and average grain moisture contents of 25.5, 26.6, and 29.7 %, respectively, for the 3 year period. However, map unit BdA is bordered by woods in the southwest portion of the field and yields appear to be directly influenced by the trees (Fig. 1). There was little spatial dependence of crop performance within a map unit and correlations of corn yield between years was low in map units CoB and BrA (-0.28 to 0.32) but higher for map unit BdA (0.49 to 0.60).

In general, there was little spatial dependence detected for corn yield, grain moisture, or plant population on a field or soil map unit basis, as none of the models fit the semivariogram. However, there was strong spatial correlation for grain yield and grain moisture within a given transect within a given year or for the three year average, although there is no consistency for any transect among years. At this scale of measurement, there appears to be considerable heterogeneity in corn performance, with the variability having a significant random component. Lack of spatial dependence makes it difficult to develop site specific yield goals for this field. Since yield goal is a basis for fertilizer recommendations, site specific management would be difficult. In general, there were few correlations between plant components with the exception of corn grain moisture.

Soil Profile NO$_3$N

On a whole field basis, soil NO$_3$N co-varied by soil depth, location in the field, and sampling date (Table 2). With a few exceptions, soil NO$_3$N concentrations were low and decreased with depth in the spring (8 June, 1993) and moderate to high and increased with depth after the growing season (18 December, 1993 and 11 November, 1994). For most of the field, low NO$_3$N concentrations indicate little residual NO$_3$N remaining from the previous year. Higher NO$_3$N in the surface 30 cm represents N mineralized by 8 June.

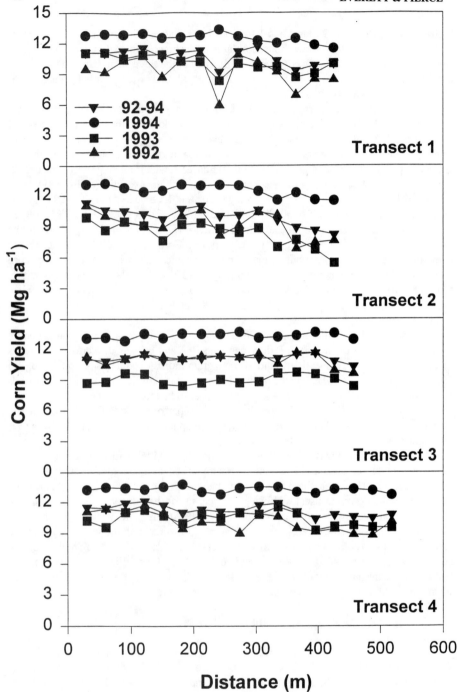

Fig. 2. Spatial distribution of corn along four transects within the experimental field for the period 1992-1994.

Table 2. Descriptive statistics and geostatistics of soil profile NO_3-N -sampled 8 June and 18 December, 1993 and 11 November 1994 for the whole field (n-60).

Date	Depth	Soil Profile NO_3-N				C_o[1]	Spherical Model Parameters				
		Avg	Std	Min	Max		C_o+C	$C/(C_o+C)$[#]	Range[††]	RSS[†††]	r^2
	cm	------- mg kg^{-1} -------							m		
7 June	0-30	5.2	2.4	3.0	16.5	1.2	1.5	0.24	166.5	0.3	0.17
1993	30-60	4.8	1.5	2.8	10.8	0.9	4.7	0.80	303.5	1.3	0.86
	60-90	3.4	1.3	1.8	10.4	2.0	5.4	0.62	102.3	9.3	0.26
	90-120	3.2	1.0	1.7	5.3	0.8	5.8	0.87	65.4	17.6	0.14
Dec 18	0-30	7.5	2.3	3.6	15.2	0.4	2.1	0.81	79.4	2.2	0.18
1993	30-60	7.4	2.9	3.6	22.1	0.4	1.1	0.64	114.7	0.2	0.44
	60-90	8.0	3.7	4.0	18.8	0.0	0.5	1.00	99.4	0.1	0.32
	90-120	8.1	3.6	3.9	21.7	0.2	0.5	0.60	74.9	0.1	0.14
Nov 11	0-30	3.9	1.3	2.0	9.7	0.2	0.5	0.68	117.2	0.0	0.55
1994	30-60	5.8	5.2	2.0	27.3	2.4	5.3	0.55	128.9	2.5	0.58
	60-90	8.4	8.5	2.5	49.0	4.6	13.0	0.64	67.6	21.2	0.29
	90-120	7.1	5.6	2.5	42.5	1.4	6.0	0.76	92.8	10.7	0.32

C_o = nugget variance
[#]$C/(C_o+C)$ = structural variance C as a proportion of model sample variance (C_o+C)
[††]range = distance (m) over which structural variance is expressed
[†††]RSS = residual sum of squares

Spatial dependence of soil NO_3N varied with sampling date and soil depth, with r^2 for the spherical model fit ranging from 0.14-0.86. The highest and most consistent spatial correlation was detected for the 30-60 cm soil depth. Corn yields were not highly correlated with soil profile NO_3N, ranging from -0.01 to -0.27. Correlations of soil NO_3N between sampling dates were also low.

Soil NO_3N concentrations on the transect scale are illustrated in Fig. 3 for Transect 2 and for the 18 December sampling date in Table 3. Both years show that baseline levels of soil NO_3-N are reached by spring and that residual soil NO_3N remain after harvest. Therefore, residual soil nitrates are leached overwinter and early spring and not available to the succeeding corn crop. Estimated NO_3N leaching losses over the winter of 1993-94 were estimated to range from 0 to 89 kg N ha^{-1} in transect 2. Given the high NO_3N concentrations in Transect 2 on 11 November, 1994 and leaching to 1993 and 1994 baseline levels, the estimated leaching losses exceed 400 kg N ha^{-1} at certain transect points and average 33.8±84.5 kg N ha^{-1}.

The spatial dependence of soil NO_3N varied by soil depth and date of sampling, ranging from very weak on 7 June, 1994 to strong on 18 December, 1993 (Table 3). Spatial dependence of overwinter leaching loss estimates for Transect 2 for the 1993-1994 winter was low. Correlations between corn yield and soil profile NO_3N for transect 2 were -0.57, -0.53 and 0.11 for 8 June, 1993, 18 December, 1993 and 11 November, 1994

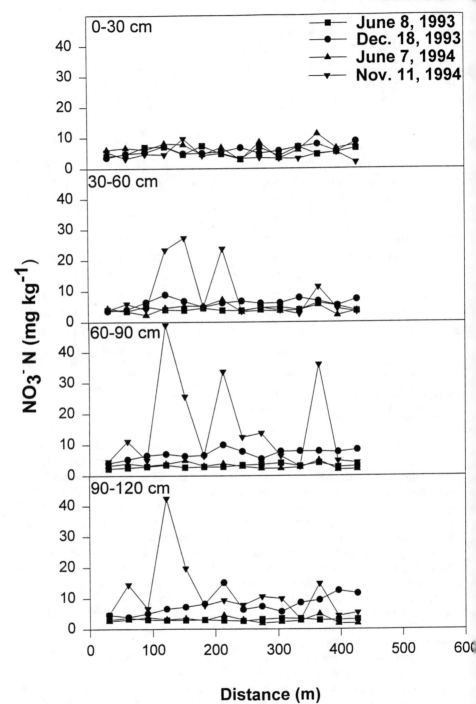

Fig. 3. Soil profile NO$_3$-N for Transect 2 for four dates in 1993 through 1994.

Table 3. Descriptive statistics and geostatistics of soil profile NO₃N sampled 8 June and 18 December, 1993 and 11 November 1994 for Transect 2 (n=14).

DATE	Depth	Soil Profile NO₃N				$C_o^{¶}$	Spherical Model Parameters				
		Avg	Std	Min	Max		C_o+C	$C/(C_o+C)^{\#}$	Range[††]	RSS[†††]	r^2
	cm	—— mg kg⁻¹ ——							m		
7 June 1993	0-30	7.7	3.6	3.2	16.5	1.2	1.8	0.35	345.7	0.8	0.12
	30-60	4.8	1.3	2.8	7.2	0.0	1.8	1.00	61.1	1.3	0.18
	60-90	3.4	1.0	1.8	5.6	0.0	2.4	1.00	80.9	2.5	0.28
	90-120	3.5	1.0	2.3	5.7	1.9	11.7	0.83	127.0	6.3	0.84
18 Dec 1993	0-30	5.9	1.4	3.6	8.7	5.4	16.3	0.67	152.4	28.1	0.63
	30-60	6.3	1.4	3.6	8.9	0.4	2.0	0.80	102.3	0.3	0.69
	60-90	7.0	1.5	4.0	10.0	0.0	1.2	1.00	88.8	0.3	0.63
	90-120	8.0	3.1	3.9	15.1	0.0	6.3	1.00	88.0	0.1	0.82
7 June 1994	0-30	6.6	2.1	2.8	11.2	3.0	5.5	0.46	71.3	11.8	0.03
	30-60	4.2	1.3	2.2	7.3	1.5	2.1	0.32	125.6	0.9	0.16
	60-90	3.2	0.9	2.0	5.0	0.8	0.9	0.11	510.9	0.1	0.01
	90-120	3.0	0.9	1.8	5.0	0.8	0.9	0.11	30.4	0.6	0.05
11 Nov 1994	0-30	4.4	1.7	2.0	9.7	2.6	3.4	0.23	98.8	1.3	0.10
	30-60	9.0	5.7	2.6	27.3	47.9	99.9	0.52	142.5	1267.6	0.45
	60-90	15.4	8.9	3.0	49.0	120.5	236.3	0.49	71.3	2.4x10⁴	0.02
	90-120	11.5	8.7	3.2	42.5	85.6	95.3	0.10	61.0	1789.0	0.00

$^{¶}C_o$ = nugget variance
$^{\#}C/(C_o+C)$ = structural variance C as a proportion of model sample variance (C_o+C)
[††]range = distance (m) over which structural variance is expressed
[†††]RSS = residual sum of squares

sampling dates and 1993 and 1994 yields, respectively. However, the correlations for the other transects were much lower.

In general, spatial dependence of soil profile NO₃N was low for the whole field or soil map unit basis. In certain cases, soil profile NO₃N within transects were spatially correlated but this was not consistent among transects. Therefore, there is a large random component to the spatial variability of soil profile NO₃N.

SUMMARY AND CONCLUSIONS

To what extent do the data in this study support the use of site-specific management of N in this field? First, it is clear that N leaching is a problem in this field, although not everywhere, and would favor the use of site-specific management as a potential solution. Second, while significant variability existed in corn yield, the variability was not spatially structured and varied temporally, with little correlation between years. Therefore, delineating site specific yield goals would be difficult. Third, soil profile NO₃N is highly variable, not sufficiently spatially correlated, and little correlation to yield or between sampling dates. Additionally, residual soil NO₃N is leached rather completely overwinter and is not a factor in soil fertilizer recommendations in the spring. Therefore, soil profile NO_3-N would be difficult to manage site-specifically. Additionally, soil map units were rather heterogeneous and may not be a basis for delineation of soil management zones. While we admit that our sampling design and intensity may not have revealed the true spatial structure of the field, sampling costs to achieve this result may preclude this as an option for farmers.

REFERENCES

Angle, J. S., C. M. Gross, R. L. Hill, and M. S. McIntosh. 1993. Soil nitrate concentrations under corn as affected by tillage, manure, and fertilizer applications. J. Environ. Qual. 22:141-147.

Binford, G. D., A. M. Blackmer, and M. E. Cerrato. 1992. Relationship between determining yield of field-grown maize. Crop Sci. 32:1220-1225.

Black, C.A. 1993. Soil fertility: Evaluation and control. Lewis Publ. Boca Raton, FL.

Bundy, L.G., and E.S. Malone. 1988. Effect of residual profile nitrate on corn response to applied nitrogen. Soil Sci. Soc. Am. J. 52:1377-1383.

Cahn, M.D., J.W. Hummel, and B.H. Brouer. 1994. Spatial analysis of soil fertility for site-specific crop management. Soil Sci. Soc. Am. J. 58:1240-1248.

Cambardella, C.A., T.B. Moorman, J.M. Novak, T.B. Parkin, D.L. Karlen, R.F. Turco, and A.E. Konopka. 1994. Field-scale variability of soil properties in central Iowa soils. Soil Sci. Soc. Am. J. 58:1501-1511.

Francis, D. D., J. W. Doran, and R. D. Lohry. 1993. Immobilization and uptake of nitrogen applied to corn as starter fertilizer. Soil Sci. Soc. Am. J. 57:1023-1026

Gamma Design Software. 1993. Geostatistical analysis. Gamma Design Software, Plainwell, MI.

Hoeft, R. G., H. M. Brown, D. Mengel, D. J. Eckert, and M. L. Vitosh. 1992. Predicting N fertilizer needs for corn in the humid regions: Advances in the Midwest. In B.R. Bock and K.R. Kelley (eds.). Predicting N fertilizer needs for corn in humid regions. Bull. 1226. National Fertilizer and Environmental Research Center, Tennessee Valley Authority, Muscle Shoals, Alabama.

Jansson, S.L., and J. Persson. 1982. Mineralization and immobilization of soil nitrogen. pp. 229-248. In F.J. Stevenson, J.M. Bremner, R.D. Hauck, and D.R. Keeney (eds.) Nitrogen In agricultural soils. Agronomy 22. ASA-CSSA-SSSA, Madison, WI.

Jokela, W.E., and G.W. Randall. 1989. Corn yield and residual soil nitrate as affected by time and rate of nitrogen application. Agron. J. 81:720-726.

Keeney, D R., and D.W. Nelson. 1982. Nitrogen-Inorganic forms. pp. 643-693. In A.L. Page, D.E. Baker, R. Ellis, Jr., D.R. Keeney, R.H. Miller, and J.D. Rhoades (eds.). Methods of soil analysis. Part 2. Chemical and microbiological properties. Second Edition. Agronomy 9. ASA-CSSA-SSSA, Madison, WI.

Liang, B.C., M. Remillard, and A.F. MacKenzie. 1991. Influence of fertilizer, irrigation, and non-growing season precipitation on soil nitrate-nitrogen under corn. J. Environ. Qual. 20:123-128.

Legg, J.O. and J.J. Meisinger. 1982. Soil nitrogen budgets. pp. 503-566. In F.J. Stevenson, J.M. Bremner, R.D. Hauck, and D.R. Keeney (eds.). Nitrogen In agricultural soils. Agronomy 22. ASA-CSSA-SSSA, Madison, WI.

Meisinger, J.J. 1984. Evaluating plant-available nitrogen in soil- rop systems. pp. 391-413. *In* J.D. Beaton, C.A.I. Goring, R.D. Hauck, R.G. Hoeft, G.W. Randall, and D.A. Russel (eds). Nitrogen In crop production. ASA-CSSA-SSSA, Madison, WI.

Olsen, R.A. and L.T. Kurtz. 1982. Crop nitrogen requirements, utilization, and fertilization. pp. 567-599. *In* F.J. Stevenson, J.M. Bremner, R.D. Hauck, and D.R. Keeney. Nitrogen In agricultural soils. Agronomy 22. ASA-CSSA-SSSA, Madison, WI.

Robertson, G.P., J.R. Crum, and B.G. Ellis. 1994. The spatial variability of soil resources following long-term disturbance. Oecologia. 96:451-456.

SAS Institute. 1990. SAS/STAT user's guide. Version 6. SAS Inst., Cary, NC.

Stanford, G. and J.O. Legg. 1984. Nitrogen and yield potential. pp. 263-272. *In* J.D. Beaton, C.A.I. Goring, R.D. Hauck, R.G. Hoeft, G.W. Randall, and D.A. Russel (eds.). Nitrogen In crop production. ASA-CSSA-SSSA. Madison, WI.

Van Meirvenne, M. and G. Hofman. 1989. Spatial variability of soil nitrate nitrogen after potatoes and its change during winter. Plant Soil. 120:103-110.

White, R.E., R.A. Haigh, and J.H. Macduff. 1987. Freqency distribution and spatially dependent variability of ammonium nitrate concentrations in soil under grazed and ungrazed grassland. Fert. Res. 11:193-208.

A Landscape-scale Assessment of the Nitrogen and Non-Nitrogen Benefits of Pea in a Crop Rotation

C. Stevenson
C. van Kessel

Dept. of Soil Science
University of Saskatchewan
51 Campus Drive
Saskatoon, SK, Canada

ABSTRACT

A landscape-scale study was established to examine the N and non-N rotation benefits of pea to wheat across a hummocky terrain. Factors related to soil N availability, an N effect, were responsible for lower wheat seed yield in depressions of the pea-wheat rotation. In the wheat-wheat rotation, grassy weed infestation, a non-N effect, was the most important landscape-scale control of seed yield.

INTRODUCTION

The most recognized benefit of a pulse crop to a succeeding cereal is the improvement in yield. For example, seed yield of barley (*Hordeum vulgare* L.) and wheat (*Triticum aestivum* L.) following pea (*Pisum sativum* L.) increased by > 20% compared to the seed yield of a cereal crop following a cereal (Wright, 1990; Evans et al., 1991; Smiley et al., 1994). Factors responsible for the rotation benefit are those affecting soil N availability (N benefit; e.g., N mineralized from legume residue) and those that are non-N related (non-N benefit; e.g., pest problems).

Rotation studies often are conducted on uniform, level fields. However, most fields are characterized by hummocky or undulating features. In landscapes with this form of topography, water redistribution is a fundamental control of nutrient cycling, soil productivity, and crop yield (Pennock et al., 1987). Lower slope positions (e.g., footslopes) typically have the greatest soil water and nutrient content, and generally produce the highest crop yield (Halvorson and Doll, 1991; Fiez et al., 1994; Stevenson et al., 1995). The benefit of pea in a rotation also is likely variable and responsive to topography. Our objective was to assess the landscape-scale variability of N and non-N rotation benefits of pea in a crop rotation.

MATERIAL AND METHODS

A pea-wheat and wheat-wheat rotations were established on two adjacent 100 m by 100 m areas in 1993 at Birch Hills, Saskatchewan, Canada. A 100 sampling point systematic grid with 10 m spacing between each point was superimposed in each rotation. Sampling points were characterized as shoulders, low-catchment footslopes, or high-catchment footslopes (depressions) (Fig. 1). Pea

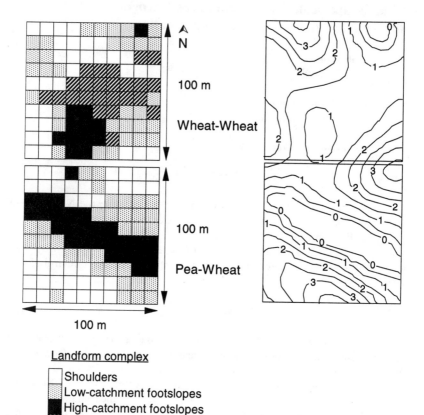

Landform complex

☐ Shoulders
▦ Low-catchment footslopes
■ High-catchment footslopes
▨ Levels

Fig. 1. A. Spatial pattern of landform complexes (shoulders are not shaded; low-catchment footslopes are dotted; high-catchment footslopes are filled) at the Birch Hills site. B. Contour of the landscape within each crop rotation. Elevations were assigned relative to the lowest point in the sampling grid which was assigned an elevation of 0.0 m.

and wheat in the first rotation phase (1993), and wheat in the second rotation phase (1994), were managed with practices typical of those for the region. In particular, urea, at a rate of 45 kg N ha^{-1}, was banded prior to sowing wheat in the second phase of the rotations.

In 1993, the A-value method was used to determine N_2 fixation by pea, with canola (*Brassica napus* L.) as the reference crop (Rennie and Rennie, 1983). The N content and yields of pea residue also were determined in 1993. In the second phase of the rotations, soil water and inorganic N content were determined just prior to seeding. Leaf disease severity (tan spot: *Pyrenophora tritici-repentis* [Died.]Drechs., and septoria leaf blotch: *Septoria avenae* Frank f. sp. *triticea*, Johns., *Septoria tritici* Rob. in Desm., and *Septoria nodorum* [Berk.]Berk.) was assessed at anthesis, in the second phase of both rotations. Lesions on the flag and upper leaves were rated using a 0 to 11 scale. At harvest in 1994, quackgrass (*Agropyron repens* L.) and wild oat (*Avena fatua* L.) infestations were assessed in both rotations. A visual rating method (0 = none, 1 = moderate, and 2 = severe yield losses) was used to assess weed infestations. Nitrogen-15 labelled pea residue was used to determine the recovery of N from pea residue by harvest in the second phase of the pea-wheat rotation. Also, ^{15}N-labelled fertilizer was used to calculate the A-value (an indicator of soil N supplying power) of wheat residue in the second phase of both rotations (Senaratne and Hardarson, 1988). The N content and yields of wheat residue and seed were determined at harvest in the second phase of both rotations. Measurements made in each rotation were conducted at all 100 sampling points.

RESULTS

The percentage N that pea derived from N_2 fixation was 19% higher in the shoulders and footslopes than in the depressions (Table 1). An opposite landform effect was observed for the residue yield of pea. As a result, 13 less kg N ha^{-1} were fixed by pea growing in the shoulders as compared to footslopes and depressions. If the N contribution of legume residue to the subsequent crop is related to N_2 fixation, we would expect it to be greatest in the lower slope positions.

Six percent of the N (7 of the 130 kg ha^{-1}) accumulated by wheat in the pea-wheat rotation was derived from pea residue. This proportion did not vary among landform complexes (Table 2). Also, the recovery of pea residue N in the subsequent wheat crop was small relative to the amount that was returned to the soil, and insensitive to the effect of landform complex (Table 2). As a result, the 38% increase for the N accumulated by the succeeding wheat crop in the depressions was not related to the actual N contribution from pea residue total N repeated in (Table 3). Our results showed that 6 of the 7 kg N ha^{-1} derived from pea residue was directly attributable to the N_2-fixing activity of pea. However, the additional N contribution of N_2 fixation in the footslopes and depressions did not correspond to the actual N contribution from pea residue in following year. Others

Table 1. Dinitrogen fixation and yield of pea residue at harvest among three landform complexes at Birch Hills, Saskatchewan in 1993.

Landform complexes	n	N_2 fixation		Yield
		%	kg ha^{-1}	kg ha^{-1}
Shoulders	40	84	64	3929
Footslopes	33	80	75	4427
Depressions	27	63	79	5459
SD		16	33	1621

Table 2. The N accumulated in pea residue that was returned to soil in 1993, and the recovery of N from pea residue and N accumulation in the subsequent wheat crop at harvest in 1994 among three landform complexes at Birch Hills, Saskatchewan.

Landform complexes	n	Pea residue N	Wheat N		
			Recovery‡	Ndfr†	Total accumulation
		kg ha^{-1}	%	kg ha^{-1}	
Shoulders	40	47	14	6	107
Footslopes	33	68	12	8	116
Depressions	27	85	11	8	154
SD		33	3	4	41

†N derived from pea residue.
‡ The recovery of N from pea residue.

have found that most of the N from pulse crop residues is incorporated into the soil organic matter in a more recalcitrant form, thus being relatively unavailable to succeeding crop (Bremer and Van Kessel, 1992; Jensen, 1994).

In the pea-wheat rotation, wheat seed yield in the depressions was 18% lower than in footslopes and shoulders (Table 3). Similarly, a 19% reduction in seed yield occurred in the depressions of the wheat-wheat rotation. The controls responsible for these landform effects differed between rotations.

In both rotations, soil water content, the A-value, residue yield and total N accumulation were the highest in high-catchment footslopes, whereas leaf disease severity and harvest index (seed yield / total yield) was the lowest in these areas (Table 3). In the pea-wheat rotation, soil inorganic N was significantly higher in the depressional areas as compared to the low-catchment areas. This difference for soil inorganic N was not significant in the wheat-wheat rotation (Table 4). Analysis of covariance was used to explain the lower seed yield in the high-catchment footslopes as compared to other landform complexes. Soil water content, soil inorganic N, and the A-value residue yield covaried with the landform effect on residue yield (Table 4). Of these covariates, soil water content explained a major portion of the covariance with residue yield. Residue yield was the most important covariate related to the landform effect on seed yield (Table 4). Leaf disease severity covaried with neither the landform effect on residue yield nor seed yield. In the wheat-wheat rotation, the overall yield was 58% lower than in the pea-wheat rotation (Table 3). This yield reduction was due to 35% greater leaf disease severity in the wheat-wheat as compared to the pea-wheat rotation, and the competitive effect of grassy weeds in the wheat-wheat rotation (Table 3). Because

Table 3. Soil water and inorganic N content just before seeding, the A-value at harvest, wheat leaf disease severity at anthesis, and grassy weed infestation, and total (residue plus seed) N content, and yields of wheat at harvest in three landform complexes and two crop rotations at Birch Hills, Saskatchewan in 1994.

Landform complex	n	Soil water content	Soil inorganic N	A-value	Leaf diseases	Grassy weeds	Total N content	Residue yield	Seed yield	Harvest index
		% (g g⁻¹)	kg ha⁻¹	kg N ha⁻¹	0–11 scale	0–2 scale		kg ha⁻¹		
Pea-wheat										
Shoulders	40	5.6	39	379	6.5	0	107	5206	2874	0.36
Footslopes	33	9.8	65	473	6.1	0	116	5818	2657	0.32
Depressions	27	19.5	103	580	5.3	0	154	7135	2256	0.24
SD		4.6	28	182	1.1	0	37	1406	616	0.04
Wheat-wheat										
Shoulders	32	10.6	32	230	9.4	0.6	73	4653	1684	0.27
Footslopes	32	13.5	60	385	9.2	0.6	85	5393	1758	0.24
Depressions	12	17.9	54	504	8.6	1.4	69	5788	1387	0.19
SD		6.2	38	178	1.1	0.6	27	1430	540	0.06

Table 4.   Mean square and associated significance (P) for the covariance of soil water content and inorganic N just before seeding, the A-value of wheat residue at harvest, and wheat leaf disease severity at anthesis with the landform effect for wheat residue yield at harvest, and the covariance residue yield and leaf disease severity with wheat seed yield at harvest in the pea-wheat rotation at Birch Hills, Saskatchewan in 1994.

Source	df	Mean square	P	%†
		Covariance with residue yield		
Soil water content	1	82213740	< 0.001	82
Soil inorganic N	1	9607992	0.017	9
A-value	1	6671477	0.046	7
Leaf diseases	1	2232614	0.245	2
Landform	2	100228	0.940	
Low vs. high‡	1	55357	0.854	
Error	93	1627476		
		Covariance with seed yield		
Residue yield	1	2067263	0.009	88
Leaf diseases	1	287001	0.320	12
Landform	2	6671899	< 0.001	
Low vs. high	1	6493397	< 0.001	
Error	95	287804		

† The percentage of the sum of squares for a particular covariate divided by the total sum of squares for all covariates.
‡ Contrast comparison of the low vs. the high-catchment footslopes.

Table 5. Mean square and associated significance (*P*) for the covariance of soil water content just before seeding, the A-value of wheat residue at harvest, and grassy weed infestation at harvest with the landform effect on wheat seed yield at harvest in the wheat-wheat rotation at Birch Hills, Saskatchewan in 1994.

Source	df	Mean square	*P*	%†
Soil water content	1	256258	0.291	4
A-value	1	597893	0.109	10
Grassy weeds	1	5229829	< 0.000	86
Landform	2	290658	0.283	
Low vs. high‡	1	464358	0.156	
Error	70	226236		

† The percentage of the sum of squares for a particular covariate divided by the total sum of squares for all covariates.
‡ Contrast comparison of the low vs. the high-catchment footslopes.

leaf diseases did not vary among landform complexes, the 0.8 unit increase in grassy weed infestation in the depressions covaried with the landform effect on seed yield in the wheat-wheat rotation (Table 5). Therefore, wheat following wheat did not respond as much to the greater soil N supplying power (the A-value) in the footslopes and depressions as in the pea-wheat rotation.

The expected N contribution associated with the greater N_2 fixation by pea and return of N from pea residue in the footslopes and depressions was not realized in the N derived from pea residue of a subsequent wheat crop growing in those same slope positions (Tables 1 and 2). In the depressions, however, wheat accumulated 42 kg ha^{-1} extra N in its biomass as compared to the other landform complexes. This meant that the mineralization of N sources other than pea residue (e.g., native soil organic matter and unaccounted N from pea roots) allowed wheat to accumulate more N in the depressions. This meant that the N contribution from pea residue was not responsible for the landscape-scale control of yield in the succeeding wheat crop. Apparently, pest problems that were eliminated by including pea prior to wheat in a crop rotation made wheat more responsive to differences in soil water and N availability among landform complexes. The non-N related factor responsible for the overall yield reduction in the wheat-wheat rotation also was related to the lower seed yields in the depressions of the same rotation. Future investigations should confirm whether the effect of crop rotation on the landscape-scale controls of crop yield in producers fields varies with different cropping histories and growing conditions (Fiez et al., 1994).

Researcher and producers have become interested in the variable rate applications of N fertilizer across different management units (e.g., landform complexes) in a field; site-specific or precision farming. Our results suggest that a wheat crop following pea would require lower N fertilizer rates in the depressions to reduce the negative effect of excessive soil N supply. In the wheat-wheat rotation, an N fertilizer strategy to alleviate the competitive effect of grassy weeds may have been considered. However, it is not certain what fertilizer rates in the depressions would have provided a competitive advantage to wheat growing in the depressions. Future variable-rate studies will have to address the complexity of decisions when N and non-N related factors affect the demand for N in different management units in a field.

Our results also demonstrate that crop rotation may affect site-specific weed management strategies. In the wheat-wheat rotation, for example, the application of a grassy weed herbicide, or higher rates of it, would have been necessary in the depressions. Whereas, grassy weed herbicides may not have been necessary across the entire area of the pea-wheat rotation. Therefore, crop rotation may influence a number of the possible strategies associated with site-specific farming.

ACKNOWLEDGEMENTS

Funding was provided by the Agriculture Development Fund of Saskatchewan, the Saskatchewan Pulse Crop Development Board, the Natural Sciences and Engineering Research Council of Canada, and the Canadian Wheat Board.

REFERENCES

Bremer, E. and C. van Kessel. 1992. Plant-available nitrogen from lentil and wheat residues during a subsequent growing season. Soil Sci. Soc. Am. J. 56:1155–1160.

Evans, J., N.A. Fettell, D.R. Coventry, G.E. O'Connor, D.N. Walsgott, J. Mahoney, and E.L. Armstrong. 1991. Wheat response after temperate legumes in south-eastern Australia. Aust. J. Agric. Res. 42:31–43.

Fiez, T.E., B.C. Miller, and W.L. Pan. 1994. Winter wheat yield and grain protein across varied landscape positions. Agron. J. 86:1026–1032.

Halvorson, G.A. and E.C. Doll. 1991. Topographic effects on spring wheat yields and water use. Soil Sci. Soc. Am. J. 55:1680–1685.

Jensen, E.S. 1994. Availability of nitrogen in ^{15}N-labelled mature pea residues to subsequent crops in the field. Soil Biol. Biochem. 26:465–472.

Pennock, D.J., B.J. Zebarth, and E. de Jong. 1987. Landform classification and soil distribution in hummocky terrain, Saskatchewan, Canada. Geoderma 40:297–315.

Rennie, R.J. and Rennie, D.A. 1983. Techniques for quantifying N_2 fixation in association with nonlegumes under field and greenhouse conditions. Can. J. Microbiol. 29:1022–1035.

Senaratne, R. and Hardarson, G. 1988. Estimation of residual N effect of faba bean and pea on two succeeding cereals using ^{15}N methodology. Plant Soil 110:81–89.

Smiley, R.W., R.E. Ingham, W. Uddin, and G.H. Cook. 1994. Crop sequences for managing cereal cyst nematode and fungal pathogens of winter wheat. Plant Dis. 78:1142–1149.

Stevenson, F.C., J.D. Knight, and C. van Kessel. 1995. Dinitrogen fixation in pea: controls at the landscape- and micro-scale. Soil Sci. Soc. Am. J. 59:1603–1611.

Wright, A.T. 1990. Yield effect of pulses on subsequent cereal crops in the northern prairies. Can. J. Plant Sci. 70:1023–1032.

The Feasibility of Variable Rate N Fertilization in Saskatchewan

M.P. Solohub
C. van Kessel
D.J. Pennock

Dept. of Soil Science
University of Saskatchewan
51 Campus Drive
Saskatoon, SK, Canada

ABSTRACT

The feasibility of variable rate N fertilization was assessed during a single growing season at Birch Hills, Saskatchewan, Canada (105° 2' W, 53° 3'10 N). The experiment was carried out across a 120 m x 620 m wheat field, which was divided into twelve 10 m x 620 m strips. Each strip was further broken down into 10 m x 10 m grid cells. Each of these grid cells was assigned to one of three possible management units based on their topographical characteristics. Prior to fertilization, initial soil testing was carried out; ten surface soil samples from each management unit were collected and analyzed for available N and P. The spring nutrient data, along with two separate growing season precipitation scenarios, were used to develop the fertilizer recommendations. At seeding each of the strips was fertilized in accordance to one of the two growing season precipitation scenarios or at a constant, or conventional, rate. Soil mineral N and available moisture were measured at ninety benchmark sites prior to seeding, weekly through the crop vegetative state, and following harvest. These benchmark sites were evenly distributed among management units and fertilizer rates. Crop yield was also measured at each of the ninety benchmark sites. Preliminary results indicate that fertilizer responses varied between management units. These observations suggest that fertilizer application can be manipulated on the basis of landscape - scale management units to minimize inputs, while maintaining yield.

INTRODUCTION

In order to implement Variable Rate Fertilization (that is, to vary inputs in response to changes in soil properties) the spatial variability within a given field must be assessed. Assessment of soil variability across a landscape is possible since changes in soil properties are predictable and have a large non - random component (Hall and Olson, 1991). This concept of systematic spatial variability was first widely proposed by Milne in 1936. Milne suggested that soils along a landscape are related much the same as links in a chain. Soils at one location influence the surrounding soils by influencing drainage conditions, erosion and deposition, leaching, and translocation and redeposition of chemical constituents. This early work is the background of the catena concept (Milne 1936). Since the conception of Milne's catena concept several authors have improved upon it by attempting to model water movement both along and across the catena (Aandahl 1948, Hugget 1975, Pennock et al.1987). By understanding the effect landscape position has on soil physical processes it is possible for researchers to make accurate predictions about soil physical, chemical and hydrologic properties, which in turn can be used to make predications about variability in soil productivity across a landscape.

Implemenation of Variable Rate Fertilization

There are three different approaches to Variable Rate Fertilization; the *low - tech approach,* the *medium - tech approach,* and the *high - tech approach.* The low - tech approach has been implemented by farmers for many years. With this approach the farmer simply makes a double pass across certain portions of the field during fertilization, or manure is selectively spread across certain portions of the field. With the medium - tech approach the rate at which fertilizer is administered can be adjusted from the cab of the tractor. Given that the operator is aware of differences in nutrient requirements across a field he or she can adjust the fertilizer rates on the go. The medium - tech approach was utilized in this research. The high - tech approach incorporates geographical informational systems, global positioning systems, and computer activated fertilizer spreaders. Soil data and historical yield data are incorporated to produce a fertilizer rate map. This fertilizer rate map, which is tagged to G.P.S. coordinates, is subsequently downloaded into a tractor mounted computer. From here the fertilizer rate is automatically adjusted in accordance to the digital rate map.

THE VARIABLE RATE EXPERIENCE AT
BIRCH HILLS, SASKATCHEWAN

The focus of this paper is an experience with Variable Rate Fertilization at Birch Hills, Saskatchewan (105° 2' W, 53° 3'10 N). This site is within Blaine Lake - Hamlin Association; the soils are Black Chernozems (Udic Haploboroll) developed on a rolling lacustrine landscape. The research was conducted at the landscape - scale, nitrogen fertilizer rates were varied across the landscape and nutrient and moisture status were monitored throughout the growing season. Finally yields were measured across the landscape and the economic advantages of variable rate fertilization were assessed.

Landscape - Scale Research

The research conducted at Birch Hills is a departure from the traditional small plot work which has characterized agricultural research. The experimental site occupied a 620 m x 120 m area within the farmer's field. After a detailed topographic survey was conducted a management unit map of the area was developed based on its topographic characteristics. This map is an assemblage of 10 m x 10 m grid cells, where each grid cell is defined by its topographic

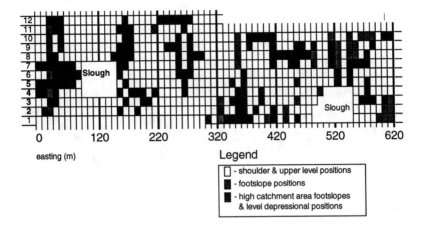

Fig. 1. Management Unit Map at Birch Hills, Saskatchewan; the squares are the 10 m x 10 m grid cells.

characteristics. This map was superimposed onto the research area prior to fertilization (Fig. 1).

Variable Nitrogen Rates

The two premises that Variable Rate Fertilization was based on at this site are: 1) productivity varies across landscapes, and 2) these differences in productivity are controlled by differences in soil nitrogen (N) and soil moisture across a landscape. It was also reasoned that nutrient requirements would depend on growing season precipitation. This uncertainty in growing season precipitation resulted in the development of two fertilizer scenarios, referred to as the "wet " year scenario and the "dry" year scenario.

In the "wet" year scenario it was reasoned that growing conditions would be optimal at the shoulder positions, and that excess moisture in the lower depressional positions would limit crop growth. In this scenario 90 kg ha^{-1} of actual N was applied to the shoulder positions and 30 kg ha^{-1} of actual N was applied to the lower depressional areas in the field. In the "dry" year scenario the converse was assumed to be true, that is crop growth would be limited by moisture at the shoulder positions and crop growth would be optimal in the depressional positions. In this scenario 30 kg ha^{-1} of actual N was applied to the shoulder positions and 90 kg ha^{-1} of actual N was applied to the depressional positions. Both of these scenarios were compared to a constant, or conventional, fertilizer rate. In the conventional treatment the field was fertilized at a constant rate of 60 kg ha^{-1}, regardless of landscape position.

The field was seeded and fertilized with a 10 m wide air seeder. The implement traveled in an east - west direction, for this reason the fertilizer treatments were imposed onto the management unit map in an east - west direction. As a result, the research plot was comprised of twelve 10 m x 620 m strips. Each

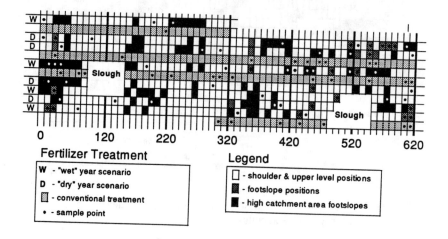

Fig. 2. Management Unit / Treatment Map

of these strips was seeded in accordance to either a "wet" year scenario, a "dry" year scenario, or at a "conventional" rate. (Fig. 2).

NUTRIENT AND MOISTURE STATUS THROUGHOUT THE GROWING SEASON

Soil nutrient status and soil moisture were monitored at 90 sampling points, which were laid out prior to seeding and fertilization (figure 2). These 90 sampling points were distributed equally among the different management unit / fertilizer treatment combinations. Surface samples (0 - 15 cm) were obtained at ten randomly selected replicates from each of the management units prior to fertilization. These samples were analyzed for sodium bicarbonate extractable phosphorus, and soil moisture. Mineral N and soil moisture were determined at each of the ninety sampling points prior to fertilization, weekly through the growing season, and in the fall. Each sampling point was sampled to 90 cm (0 - 15 cm, 15 - 30 cm, 30 - 60 cm, and 60 - 90 cm) in the spring and fall, and to 30 cm (0 - 15 cm and 15 - 30 cm) throughout the growing season.

Pre - seeding sodium bicarbonate extractable phosphorus showed a significant landscape effect (Kruskal - Wallis H test, = 0.05). The phosphorus levels were highest in the lower level and depressional positions followed by the footslopes; the shoulder positions contained the least sodium bicarbonate extractable phosphorus (Fig. 3). Based on this data, especially the low phosphorus levels in the shoulder positions, a rate of 25 kg ha^{-1} of seed - placed phosphorus was applied across the field.

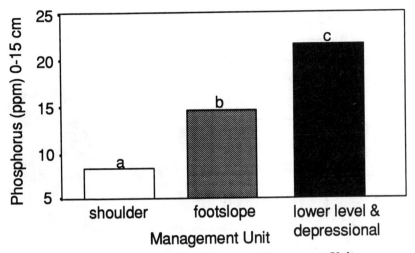

Figure 3: NaHCO$_3$ Extractable Phosphorus vs. Management Unit.

Soil organic carbon levels did not appear to be related to landscape position. No significant differences between soil organic carbon levels at the management unit level were detected (Kruskal - Wallis H test, = 0.05). The relatively high levels of organic carbon indicates that these are good quality soils (Fig. 4). Also, the lack of significant differences across the landscape indicates that this field is homogeneous with respect to soil organic carbon.

Soil moisture varied both spatially and temporally throughout the growing season. The shoulder positions contained significantly more soil moisture than did the lower level and depressional positions, regardless of sampling date (Kruskal - Wallis H test, = 0.05). Precipitation was considered to be slightly below normal and soil moisture stores were being depleted to a greater extent in the shoulder positions through the month of July. It is possible that crops in the shoulder positions may have experienced more moisture stress, as compared to those in the lower level and depressional positions, through this time period (Fig. 5).

A significant difference in soil mineral N across the landscape was not detected at the 95% confidence level (Kruskal - Wallis H test). There did, however, appear to be a temporal trend in soil mineral N levels. Mineral N levels increased following the June 1st seeding and fertilization operation to a maximum on June 25th, following June 25th the levels of mineral N decreased slowly over time (Fig. 6). This increase in soil mineral N over the 25 days following fertilization was likely due to an increase in soil temperatures, and thus, a resultant increase in the net mineralization rate. The decrease in mineral N throughout the remainder of the growing season was likely due to crop use and non - crop N losses.

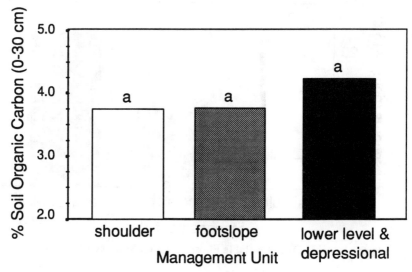

Figure 4: Soil Organic Carbon vs. Management Unit.

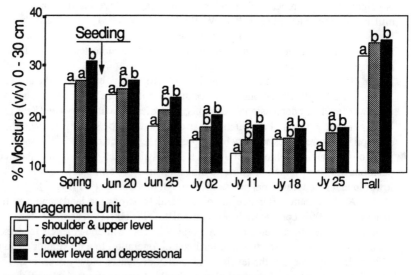

Figure 5: Spatial and Temporal Variations in Soil Moisture.

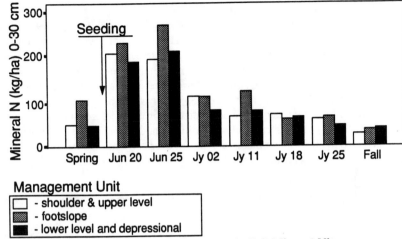

Figure 6: Spatial and Temporal Variations in Soil Mineral Nitrogen.

CROP YIELDS ACROSS MANAGEMENT UNITS

Yield samples were obtained at the same 90 sampling points that were established for nutrient and moisture sampling (Fig. 2). One meter square yield samples were obtained from each of the sampling points, each sample was taken to be representative of the grid cell from which it was obtained. Preliminary results indicate that grain yields did not vary with fertilizer treatment; this is particularly evident in the lower level and depressional positions. This lack of fertilizer response in the lower level and depressional positions may be due to a large N supply power in these areas; this possibility should be investigated in an attempt to gain a better understanding of the N dynamics across management units. It should be noted, however, that grain yields appeared to be approximately 10 bushels/acre greater in the lower level depressional positions.

ECONOMIC ASSESSMENT OF VARIABLE RATE FERTILIZATION

The ultimate success of Variable Rate Fertilization is dependent upon the economic advantage of implementing this technology into a farm based management plan. The four main factors that control the profitability of this approach are; 1) the value of the commodity, 2) the savings in fertilizer costs, 3) the change in crop yield, and 4) the cost associated with implementing Variable Rate Fertilization. Obviously if commodity prices are extremely low and increases in yield, or reductions in input costs are marginal, it is unlikely that Variable Rate Fertilization will be implemented.

At Birch Hills, based on grain yields and input costs for the different fertilizer rate / management unit combinations, the value of the crop minus the cost of fertilizer was calculated. The net profits obtained for the different fertilizer rate/management unit combinations were compared to the net profits obtained when the field was fertilized at a conventional rate of 60 kg N/ha. In 1995, at Birch Hills, net returns over constant rate fertilization were not maximized for either the "wet" year or "dry" year scenarios. Maximum net returns were realized, however, for the fertilizer rate / management unit combination where shoulders were fertilized at a rate of 60 kg N/ha and lower level and depressional positions were fertilized at 30 kg N/ha. Based on $5.00/bu wheat and $0.30/lb. N the most profitable fertilizer rate/management unit combination yielded an increase in profit of $3.26/acre over conventional rate fertilization. If, for example, fertilizer N costs increased to $0.50/lb. an increase in profit, over conventional rate fertilization, of $4.77/acre would be attainable.

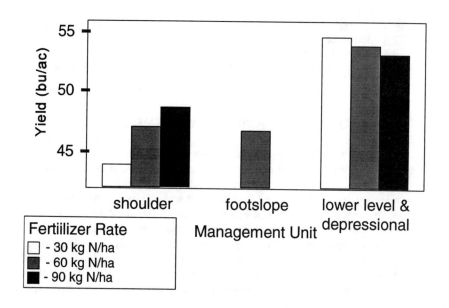

Fig. 7: Grain Yield vs. Nitrogen Rate at the Management Unit Level.

SUMMARY

Variability in the soil productivity factors measured at Birch Hills was limited. Soil organic carbon and mineral N did not appear to vary significantly across management units. A strong landscape control was noted, however, for sodium bicarbonate extractable phosphorus and soil moisture. Although the Birch Hills site appeared to be relatively homogeneous, Variable Rate Fertilization yielded an optimum economic return of $3.00 - $5.00/acre.

ACKNOWLEDGEMENTS

The authors would like to gratefully acknowledge the Canada - Saskatchewan Agricultural Greenplan Agreement and the Saskatchewan Wheat Pool for their generous financial contributions. We would also like to thank Flexicoil Limited for providing an airseeder and tractor during the 1995 seeding operation.

REFERENCES

Aandahl, A.R. 1948. The characterization of slope positions and their influence on the total nitrogen content on a few virgin soils in western Iowa. Soil Science Society of America Proceedings 13: 446 - 454

Hall, G.F. and C.G. Olson. 1991. Predicting variability of soils from landscape models. p. 9 - 24. *In* Spatial Variabilities of soils and Landforms. SSSA Special Publication No. 28.

Hugget, R.J. 1975. Soil landscape systems: A model of soil genesis. Geoderma 13: 1-22.

Milne, G. 1936. Normal erosion as a factor in normal soil development. Nature 138: 548.

Pennock, D.J., B.J. Zebarth, and E. deJong. 1987. Landform classification and soil fertility studies. Canadian Journal of Soil Science 40:146-156.

Precision Center Pivot Irrigation For Efficient Use of Water and Nitrogen

R.G. Evans
S. Han
M.W. Kroeger

Biological Systems Engineering Department
Washington State University
Prosser, WA

Sally M. Schneider

USDA-ARS
Prosser, WA

ABSTRACT

Precision applications of irrigation, fertilizer and pesticides, accounting for field variability, for self-propelled (SP) irrigation systems is discussed. Site-specific soils and climatic data are linked through detailed GIS to SP irrigation, plant growth and irrigation scheduling models to produce field prescription maps. Prototype precision control systems have been developed, installed and field tested. The hardware, software and communications systems for precision applications with SP irrigation work well, however, the major limitation is the inability to rationally develop coherent prescriptions.

INTRODUCTION

Approximately one fourth of all irrigation, or about 60% of all sprinkler irrigated lands, in the US utilize self-propelled (SP) irrigation systems. About 25% of all irrigated lands (about 190,000 out of 810,000 ha) in Washington are irrigated with SP systems, of which about 50 to 60 thousand hectares annually produce potatoes. In the Pacific Northwest (PNW), SP irrigation systems have allowed agricultural development of sandy soils (very low organic matter) with large differences in topography and other crop growing parameters within the same field. While producing high quality crops, many of these fields are subject to water and wind erosion of soils and are highly susceptible to leaching of soluble materials, such as nitrate-nitrogen (NO_3-N) and other agrichemicals (e.g., herbicides).

A committee of the National Research Council (1994) recently completed a report, **Soil and Water Quality: An Agenda for Agriculture**, that specifically stated: "Much greater progress [in improving water quality] could be made if producers had better tools and information to refine the management of their farming systems." In our opinion, one of the greatest constraints to managing for improving water quality is the inability of agricultural producers to control inputs in ways that account for the positional and temporal variability in growing conditions across a field.

Field spatial variability is beginning to be addressed through what is often called site-specific crop management (SSCM) which includes a multitude of technologies. These technologies include precision planting, precision land application of fertilizers and yield monitoring. Much of the emphasis has been placed on precision placement of fertilizers to match varying needs of different areas in a field. Most of these technologies are also primarily directed at rainfed agriculture where the results may be less than optimum because of the inability to manage soil water. SSCM of water and nutrients involves a number of technologies to assess variability, management, control, location determination, and geographic information systems (Schueller, 1991). Irrigation in arid areas provides unique opportunities to manage all production inputs much more precisely than under rainfed conditions and effectively reduce the risk of leaching agrichemicals to the ground water. SP systems are particularly amenable to SSCM because of their automation and large area coverage with a single pipe lateral.

The concept of varying agrichemicals, fertilizer and water applications by SP systems to meet the specific needs of a crop in each unique zone within a field has been discussed for many years, but its implementation has been limited by a lack of suitable technology. However, the necessary "components" for SSCM on irrigated lands have become a reality with the advent of: 1) global positioning systems; 2) geographic information systems; 3) real-time yield monitoring; 4) improved techniques for remote sensing of soil and crop status (Schmugge, 1981; Gallo et al., 1985); and, 5) improved computers, communications, smart sensors and monitoring systems to provide adequate feedback and control capabilities. These technologies have been and are becoming available at a rapid rate and declining cost.

The next step is the integration of irrigation, fertilizer and pest management strategies into systems that optimize total management practices for temporal and spatial variability. It is now possible to quickly and precisely implement decisions (prescriptions) regarding the location and timing of nutrient, water, and pesticide inputs to meet the needs of the crop in discrete areas in a field throughout the season. However, **the major research question** is the determination of what a prescription should be and how it should be formulated. The identification and quantification of contributing factors and their interactions that influence a real-time prescription are difficult. A prescription must specify when, where, and how much of a specific input or cultural operation should be applied to optimize yield quality and quantity while protecting the environment, especially in the area of water quality. Prescription building requires collection and integration of detailed knowledge about magnitude and extent of variability of climatic parameters, soil characteristics, pests, diseases and other production inputs followed by the systematic processing of large amounts of data into coherent and rational decisions that optimize management on that site. When variability of field soil and plant sampling is needed, geostatistical methods can be used to optimize the sampling size, and to interpolate data at unsampled locations (McBratney & Webster, 1983; Han et al., 1993). There must also be the ability to implement decisions quickly and easily since climatic variations and pest outbreaks require timely prescriptions to apply water and chemical inputs on a daily and seasonal basis.

Microprocessor controlled self-propelled CP and LM systems linked to a central integrating computer provide a unique platform as well as control capabilities for precision crop management and are an effective and economical means to deliver SSCM under irrigated conditions. With appropriate controls and decision making tools, these systems can be managed to account for spatial variations in water, fertilizer and pesticides requirements. Most SPs in the PNW are used in potato-wheat-corn-alfalfa cropping sequences, and are commonly used to apply chemicals (e.g., nitrogen).

PREVIOUS STUDIES

In 1994, we completed a 2 year study funded by the Washington Department of Ecology on three commercial center pivot irrigated fields, a 51 ha system and two at 41 ha each, on sands and fine sandy loam soils (Evans et al, 1995a, 1995b; Han & Evans, 1994). Topographic surveys were conducted and the fields were sampled in 1993 on a 61- x 61- meter grid for soil texture, NO_3-N, phosphorus, and pH. Additional follow-up soil sampling was conducted in specific areas. The grower provided information on sprinkler "packages", irrigation and fertilization timing. Sprinkle application uniformity tests were also conducted. A GIS database was created for the three fields (Evans & Han, 1994a). These data were used as the baseline for further investigation of NO_3-N leaching problems in the fields. We also have transformed a DOS based center pivot irrigation model/program (CPIM: Evans et al, 1993) which has been tested and linked to the GIS. A mathematical model to characterize wind distortion of individual sprinkler water distribution patterns has also been developed (Han et al, 1994).

In cooperation with the USDA-ARS, a PC ARC/INFO (GIS) and SIMPOTATO (a potato growth simulator: Hodges et al., 1992) interface was developed to study the potato yield and nitrogen leaching distributions (Han et al, 1995b). The most important distributed input parameters were identified as irrigation water/nitrogen distribution, soil types and initial soil nitrogen for each soil layer (Evans and Han, 1994b). Utilizing field data and irrigation and nitrogen management from the grower, the PC ARC/INFO-SIMPOTATO interface provided a useful tool for simulating the best specific crop and nutrient management practices that would be possible without implementation of SSCM for water and nitrogen (field treated on the average). For example, if application uniformities were increased from 90 to 100% over the entire field (assuming all other inputs remain the same), simulated potato yields increased by 3.9% and NO_3-N leaching decreased by 2.8%. Reducing water and nitrogen inputs by 5% with a 90% uniformity of application would result in a 1.1% simulated yield decrease while leaching would decrease by about 27%. This study clearly showed that greater increases in yields and reductions in leaching could be achieved if irrigation and nutrient applications could be optimized in specific areas of the field (e.g., SSCM) (Evans & Han, 1995a).

CURRENT RESEARCH

An ongoing project, initiated in 1995, is providing a focus to bring together three major current research efforts in the Pacific northwest (PNW): 1) on-going

GIS-center pivot model-crop simulation model linkage with a strong emphasis in water and nutrient management by Washington State University; 2) on-going USDA-ARS (Prosser) programs on yield monitoring, crop modeling, and global positioning technology; and, 3) current research and development by Nelson Irrigation Corp., a major manufacturer of sprinklers, on a new generation of equipment suitable for implementing irrigation SSCM.

A GIS database has been created for two adjacent commercially-farmed center pivot irrigated fields northwest of Richland, WA on sandy soils. Topographic surveys were conducted and the two fields were soil sampled in 1995 on a 61- x 61-m grid for texture, NO_3-N, phosphorus, potassium, sulphur and pH in 30 cm increments to 90 cm to characterize the site (P, K, and S were sampled for top 30 cm only). Additional follow-up soil sampling was conducted in specific areas as warranted. In 1995, both fields were planted to winter wheat followed by buckwheat after wheat harvest. Yields of the wheat and the buckwheat were monitored and yield maps developed and entered as separate layers into the respective field GIS. Figure 1 presents example maps from the GIS data base for one field.

Ground penetrating radar (GPR) in cooperation with researchers at Oregon State University was used with the GPS to help locate 6 passive fiberglass wick soil solute samplers ("PCAP": Boll et al,. 1992) in each field. The PCAPs were installed (115 cm deep) to collect and quantify leachate for analyses. Directly above each PCAP, two frequency domain reflectometry (FDR) probes (Campbell Scientific CS615-L) were installed horizontally at 30 and 90 cm. Tipping bucket rain gauges (also used to collect samples of the applied water), neutron scattering access tubes, and vacuum extractor cups were located adjacent to each PCAP. The FDR probes and rain gauges were continuously monitored by in-field dataloggers to provide information on application amounts, timing and progress of wetting fronts in the soil. Neutron probe sites and hand-held FDR probes (top 30 cm) were used to site-specifically monitor soil water levels on at least a weekly basis. The grower has provided information on historical and current irrigation and fertilization timing and amounts. An automated agricultural weather station has also been installed on the site and data are collected hourly by a computer in the farm office.

Initial procedures and software to generate management maps have been developed. The GIS database is used to spatially correlate several layers and to geostatistically extrapolate between measured values. These data are linked through the GIS to existing computer simulation models for center pivot irrigation, plant growth and nutrient uptake, and subsurface conditions, allowing for input, manipulation, query, and output of large amounts of spatial data to generate optimum water and nutrient application rate maps.

The amounts of water and nitrogen applied to each area are based on site-specific soil water monitoring, irrigation scheduling programs based on climatic data, and periodic plant sampling. These prescriptions are compared to the output of real-time crop simulators that utilize dynamic data on soil and environmental data contained in the field GIS. Each day, the application prescriptions for water and nitrogen, also stored as a data layer within the GIS, is delivered by computer-controlled spray nozzles according to the position of the SP lateral within the field.

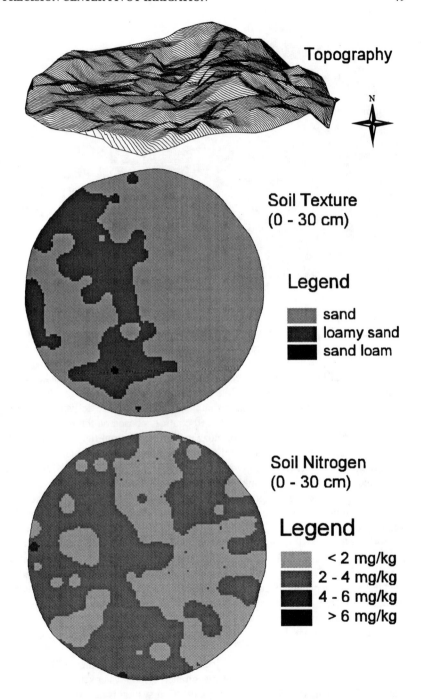

Figure 1. Example GIS maps of topography (3-dimensional) and 2-dimensional maps of soil texture and the residual soil nitrates for the top 30 cm of the soil.

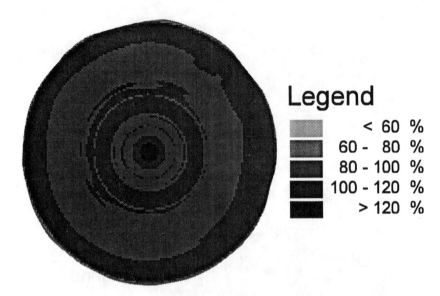

Fig. 2. Example of simulated water distribution in the GIS database generated from the CPIM model for one of the center pivots in the study shown as percent of he average depth applied.

The CPIM model has been calibrated for the sprinkler packages and hydraulic characteristics of each of the two CP systems. The implementation of an optimum irrigation water application rate map requires precise delivery of water to specific field locations. The computerized control is based on the results (e.g., the required application depth based on nozzle flow rates) from the hydraulic model which calculates the actual water applications to each area. Fig. 2 shows simulated water distribution over a center pivot irrigated field where the CPIM model has been calibrated for the existing sprinkler package.

The first step to integrate these pieces into a decision support system that generates the management maps which are interfaced with the hardware to produce a SSCM system for center pivot irrigation has been completed. In cooperation with our private/industrial partners, we have produced a working prototype of an integrated SSCM system for SP irrigation. The communications and control systems to implement SSCM on one of the center pivots have been developed, installed and field tested.   The necessary communications protocols and computer interfaces with the control system, valving and sprinklers to implement site-specific management maps for water and fertilizer (N) applications were developed and field tested. Considerable commercial resources have been committed to this aspect of the project.

An important aspect of this project has been the development of a prototype control and communications system using commercial soil water sensors that provide real-time data to the field GIS. This data collection system is necessary to drive the development of real-time simulator based prescriptions for site specific SP system operation. Initial work on the development of real-time sensor (e.g., soil water status) feedback and communications protocols was completed in late 1995 and is being implemented in 1996.

Nelson Irrigation Corporation of Walla Walla has installed a totally new sprayer "package" with pressure regulators and a solenoid on each spray head on the test CP. The test pivot was divided into zones of 2 to 4 heads each. Each of the 30 zones is turned on or off by independent cycling controllers that communicate over a 2-wire RS485 bus along the pivot lateral. Prescribed irrigation amounts are achieved by cycling the sprinklers. A master controller on the RS485 bus commands each controller to implement a cycle ratio as dictated by the CP and growth models. The bus master is able to store information for an entire pivot revolution. The RS485 bus master is interfaced to a remote PC via spread-spectrum radio modems. Application uniformity tests were also conducted. Real time pivot information collected by the RS485 master (pivot lateral direction/location [magnometers], pressure transducers), and in field information (temperature, precipitation and soil water status) are also linked to the remote PC via spread spectrum radio modems. This information is used by the computer as part of the decision process. The office computer runs computer models that utilize the GIS data bases, current soil water status data, climatic data, and develops real-time "management maps" or prescriptions. The prescriptions are then communicated to the CP for implementation. The grower has the capability to over-ride any prescription or change the decision criteria at any time.

Since it is necessary to assess current practices in order to ascertain if an improvement has been made by the SSCM, an adjacent SP is used as the control. Evaluation of the effectiveness on the various systems and how they can be improved is currently underway. The process is directed towards assessing the ability of the controlled SP to deliver a prescription as programmed. Catch can tests have been conducted with and without automated control. Remote sensing (fixed wing aircraft) was initiated on the two wheat circles in 1995 using specific band multi-spectral video, regular photographic and IR cameras in conjunction with Soil Test, Inc. of Moses Lake, WA .

We have also developed methodologies, procedures and the communications systems for real-time processing of yield monitoring data by the farm office computer as a harvester moves through the field (Han et al, 1995a, 1995c). These were developed for combine harvesters on wheat and buckwheat and potato harvesters in cooperation with the yield monitor manufacturers. We believe that yield maps are an important part in characterizing spatial variability across a field, but they must be correlated with locations and magnitude of damage from pests and diseases as well as soil fertility, topography, soil texture and other factors to develop an integrated and useful assessment.

In 1996, we are cooperating with the USDA-ARS systems group at Prosser and are installing a separate pipe lateral and spray booms for herbicide and fungicide applications on two small T-L linear move sprinkler systems near Paterson, WA. The same communications and control technology developed for site-specific water applications can be directly ported over to pesticide applications. We will be evaluating: 1) lateral hydraulics and uniformity; 2) nozzle sizing, pressures and other operational parameters; 3) appropriate flow rates, mixing ratios, adjuvents and stickers for the herbicides; 4) zone size per controller; and, 5) application boom placement.

Work was initiated in 1995 by a MS graduate student (Biosystems Engineering) to begin the process of developing an expert system for potatoes in the PNW. A EX-SYS expert system shell was purchased for this purpose. A carefully and scientifically composed questionnaire (in collaboration with the Washington Potato Commission) has been sent to 409 Washington potato growers and is intended to assess grower access to computers, their concepts of how they would use SSCM technology, and to gain additional farmer input and help identify and quantify relevant management practices and costs. The information from the questionnaire will be integrated into the expert shell using systems engineering methodologies along with results from current research into the start of a decision support system that generates the management maps which must be interfaced with the hardware to produce a working prototype SSCM system for SP irrigation. This aspect of the project is relying heavily on the previous work of Goodell et al (1993) in California on cotton and Stevenson et al (1994) in Wisconsin based on the use of their WISDOM model for potatoes. Much of this effort is being done on a workstation and will be ported over to a PC environment.

Economic analyses of SSCM of center pivot applications of water and nitrogen is being conducted in the Department of Agricultural Economics at Washington State University. A Masters student is performing a full cost economic budgeting assessment of the probable effects of SSCM of center pivots on the profitability of PNW potato and maize production. This budgeting analysis focuses on the change in costs and returns which occur with the new technology by comparing them to a traditionally managed control circle which is located adjacent to the circle having the SSCM equipment installed. Further insight into the economics of adoption of SSCM of center pivot irrigation systems is being gained through a survey of Washington potato growers by soliciting information on their willingness to adopt SSCM technologies.

Original plans called for potatoes to be planted in 1996 in both of the fields. However, the grower was unable to obtain a suitable contract and has decided to plant spring wheat in one and winter wheat in the other. Both will be followed by a second buckwheat crop to be harvested in October. This is not a major problem except that potato information for the expert system and economic studies will come from other fields and be extrapolated to SSCM situations. Testing and comparing the SSCM on the wheat will verify and demonstrate the concept at a much lower economic risk to the grower. We will be monitoring some nearby potato circles to provide needed data. Yields will be monitored on both circles for both the wheat and buckwheat and the resulting yield maps registered into the GIS in 1996.

SUMMARY

The conclusion of this research, thus far, is that the hardware, software and communications systems to deliver a prescription work well, and that the major limitation lies in the ability to interpret spatially variable data and develop rational and coherent SSCM prescriptions. Wide variations in soil types, soil chemical properties, subsurface conditions, topography, drainage, insect/weed/disease problems, soil compaction, weather patterns, irrigation system operation and maintenance, wind distortion of sprinkler patterns, as well as external factors such as herbicide drift, make the determination of a single set of management practices for SPs almost impossible. These may also be impacted by crop variety, tillage and crop rotations. This variability greatly increases the probability that NO_3-N will leach beyond the crop root zone, especially in areas where water accumulates.

Higher yields and improved quality may result from more optimal water, nutrient, and pesticide applications in specific areas of each field. If farmers adopt SSCM technology based primarily on economic self-interest, improved ground water quality will be an added benefit. Alternatively, farmers may adopt such technology to avoid the possibility of new regulations and penalties associated with potential negative impacts on ground water. However, due to the random variability in water application distributions due to wind, start-stop operations of the SP machines, and sprinkler pattern variations combined with the low cost of water and nitrogen fertilizers, it is probably not economically feasible to site-specifically manage only for water and/or nitrogen. Consequently, we believe that it is necessary and desirable to expand the same technology into other cultural areas that will attract grower interest.

REFERENCES

Boll, J., T.S. Steenhuis, and J.S. Selker, 1992. Fiberglass wicks for sampling of water and solutes in the vadose zone. J. Soil Sci. Soc. of Am. 56:701-707.

Evans, R.G., S. Han, L.G. James, and M.W. Kroeger. 1993. CPIM - a computer simulation program for center pivot irrigation systems. ASAE Paper No. 93-3065, ASAE, St. Joseph, MI.

Evans, R.G., and S. Han, 1994a. Field-scale GIS soil database creation. ASAE Tech. Paper No. 94-3078. Presented at 1994 Annual International Summer Meeting of ASAE. Kansas City, MO. June 20-23. 23 p.

Evans, R.G., and S. Han, 1994b. Mapping the nitrogen leaching potential under center pivot irrigation. ASAE Tech. Paper No. 94-2555. Presented at 1994 Annual International winter Meeting of ASAE. Atlanta, GA. December 12-15, 1994. 13 p.

Evans, R.G., S. Han, and S.L. Rawlins, 1995a. GIS capabilities and limitations for precision farming. ASAE Tech. Paper 95-3236, Presented at Annual Meeting of ASAE. Chicago, IL, June 18-22. 14 p.

Evans, R.G., S. Han, M.W. Kroeger, 1995b. Distribution uniformity evaluation for chemigation with center pivots. Transactions of the ASAE. 38(1):85-92.

Gallo, K. P, C. S. T. Daughtry, and M. E. Bauer. 1985. Spectral estimation of absorbed photosynthetically active radiation in corn canopies. Remote Sens. Environ. 17:221-232.

Goodell, P.B., J.F. Strand, and M. Ostergard, 1993. Delivering expert systems to agriculture: Experiences with CALEX/Cotton. AI Applications. 7 (2&3): 14-20.

Han, S., C.E. Goering, M.D. Cahn, and J.W. Hummel. 1993. A robust method for estimating soil properties in unsampled cells. Transactions of the ASAE. 36(5):1363-1368.

Han, S., and R.G. Evans, 1994. An integrated system for identifying BMPs under center pivot irrigation. ASAE Tech. Paper 94-3519, Presented at ASAE International Winter Meeting, Atlanta, GA. Dec. 12-15, 1994. 17 p.

Han, S., R.G. Evans, and M.W. Kroeger, 1994. Sprinkler distribution patterns in windy conditions. Transactions of the ASAE. 37(6):1481-1489.

Han, S., S.L. Rawlins, and R.G. Evans, 1995a. A Bitmap method for determining harvest width in yield mapping. ASAE Tech. Paper 95-1333, Presented at Annual Meeting of ASAE. Chicago, IL, June 18-22. 15 p.

Han, S., R.G. Evans, T. Hodges, and S.L. Rawlins, 1995b. Linking a GIS with a potato simulation model for Site-Specific Crop Management. J. of Environmental Quality 24(4):772-777. July-August.

Han, S., S.L. Rawlins, S.M. Schneider, and R.G. Evans, 1995c. Yield mapping with differential GPS. ASAE Technical Paper PNW95-302. Presented at 50th Annual Regional Conference, Pacific Northwest Section, ASAE. Harrison Hot Springs, British Columbia, Canada. October 1-3. 10 p.

Hodges, T., S.L. Johnson, and B.S. Johnson. 1992. A modular structure for crop simulation models: implemented in the SIMPOTATO model. Agronomy Journal 84:911-915.

McBratney, A. B., and R. Webster. 1983. How many observations are needed for regional estimation of soil properties? Soil Sci. 135:177-183.

National Research Council, 1994. Soil and water quality: An agenda for agriculture. Academy Press. Washington D.C.

Schmugge, T. 1981. Remote sensing of soil moisture with microwave radiometers. ASAE Paper No. 81-4503, St. Joseph, MI.

Schueller, J. K. 1991. In-field site-specific crop production. In Automated Agriculture for the 21st Century: Proceedings of the 1991 Symposium, ASAE, St. Joseph, MI, p. 291-292.

Stevenson, W.R., J.A. Wyman, K.A. Kelling, L.K. Binning, D. Curwen., T.R. Connell, and D.J. Heider, 1994. Development of prescriptive crop and pest management software for farming systems involving potatoes. Proc. 2nd Natl. IPM Symp. April 19-22. p 42-52.

Spatially Varied Nitrogen Application Through a Center Pivot Irrigation System

B. A. King

Biological and Agricultural Engineering Department
University of Idaho
Aberdeen, Idaho

J. C. Stark

Department of Plant, Soil and Entomological Sciences
University of Idaho
Aberdeen, Idaho

I. R. McCann

Department of Agricultural Mechanisation
Sultan Qaboos University
Sultanate of Oman

D. T. Westermann

USDA Agricultural Research Service
Northwest Irrigation and Soils Research Laboratory
Kimberly, Idaho

ABSTRACT

Nitrogen fertilizer was applied through a center pivot irrigation system in variable amounts according to a chemigation map consisting of four management zones. Actual nitrogen application was monitored in each zone using catch cans and water sample analysis. The results show that the spatially varied nitrogen application was achieved with the same accuracy as that of conventional uniform application.

BACKGROUND

In-season application of nitrogen fertilizer through irrigation systems is often advantageous compared to other application methods. This practice generically known as chemigation can increase nitrogen use efficiency and yield, particularly in high nitrogen use crops with shallow root zones such as potatoes. With potatoes, high soil nitrogen levels prior to tuber initiation can delay tuber initiation and promote vegetative growth, decreasing tuber size and yield. The potential leaching of nitrogen is also reduced by in-season nitrogen application, particularly on coarse textured soils.

Crop nutrient requirements typically vary across a field due to variability in residual soil nutrient levels and crop production potential resulting from varying topography, soil depth and soil texture. Conventional chemigation with center pivot irrigation systems treats the field uniformly with regard to irrigation and chemigation management. King et al., (1995) modified a 3-span linear move irrigation system to provide step-wise variable water application along the system lateral, thus effectively providing two-dimensional control of water and chemical application. Variable water application is achieved by replacing the single sprinkler package with a dual sprinkler package sized to provide 1/3 and 2/3 of the original application rate along the system length. Each sprinkler package has on-off control in multiple sprinkler zones along the system length. This arrangement provides for step-wise variable application depths of 0, 1/3, 2/3 and full along the system lateral. Position determination of a center-pivot irrigation system is accomplished by measuring the angle of the first tower. Spatially variable chemical application is achieved in the same step-wise proportions as water application by adjusting the chemical injection rate proportional to total system flow rate to maintain constant chemical concentration in the applied water.

The objective of this study was to install the variable rate irrigation system on a full-scale commercial center pivot to test and document system operation. Nitrogen fertilizer was applied through the center pivot irrigation system in spatially variable amounts according to a chemigation map consisting of four management zones. Nitrogen application was monitored at several locations in each management zone using catch cans and water sample analysis to evaluate system performance.

Methods and Materials

A 9-tower, 354-m long commercial center pivot irrigation system was equipped with the site-specific irrigation control system described by King et. al. (1995) in the spring of 1995. The center-pivot system has a design capacity of 8.2 mm/d and is equipped with 138 kPa pressure-regulated Nelson R30 sprinklers with D6 rotator plates. The field top soil is predominately silt loam with soil textures ranging from silty clay to loamy sand. Field topography is characterized by short length moderate slopes under the outer third of the system. This outer area contains rock outcrops and varying soil depths having variable yield. The sloping areas are susceptible to excessive runoff as a result of soil surface seal formation which further reduces yield due to reduced infiltration.

The field was soil sampled in the fall of 1994 to develop a soil resource data base upon which to begin development of variable rate irrigation and chemigation application decisions. Soil samples were collected on a diagonal grid arrangement with 61 m spacing. Four soil samples were collected to a depth of 0.6 m on a 5 m radius about the grid point in each principal direction. The four samples were composited in 0.3 m depth increments and analyzed for soil texture, organic matter, pH, and residual nitrogen, phosphorus and potassium. At two locations in the field, ten additional grid points on a 15.2 m spacing were soil sampled to obtain information on the short range variability in the measured parameters.

Residual soil nitrogen in the top 0.6 m of the soil profile was mapped using block kriging (Deutsch & Journel, 1992) to estimate residual nitrogen at unknown locations. Residual soil nitrogen in the 0.6 m soil profile was computed following the University of Idaho fertilizer recommendations for spring wheat (Tindall et al., 1991). The spatial structure was modeled using a spherical semivariogram model with a range of 67 m, sill of 336.2 $(kg/ha)^2$ and nugget of 87.8 $(kg/ha)^2$.

The relative yield potential of the field was estimated based on normalized vegetative index and qualitative ground truth observations of spring wheat growth during the 1994 growing season. Digitized images of visible red (625 - 675 nm) and near infrared (725 - 775 nm) reflectance and crop canopy temperature (8 - 12 μm) were obtained from a commercial company (Intermountain Technologies, Pocatello, Idaho). The digitized images were collected using aerial videography from low altitude fixed-wing aircraft. Normalized vegetative index (NDVI) was computed as:

$$NDVI = (IR - R)/(IR + R) \tag{1}$$

where IR is the digitized near infrared reflectance and R is the digitized visible red reflectance. The computed NDVI ranges from -1 to 1 with higher values indicating greater standing biomass and yield (Guoxiang & Dawei, 1990). Image analysis and mapping were performed using a geographic information system (IDRISI, Clark University). The computed NDVI image and residual soil nitrogen map were resampled to obtained equivalent raster scales.

A histogram of the computed NDVI values for the 1994 spring wheat crop is shown in Fig. 1. Qualitative field observations of crop growth revealed that locations with NDVI values below 0.25 were rock outcrops and shallow soils with very limited crop growth and essentially zero yield. Locations with NDVI values greater than 0.63 were areas having wild oat infestations resulting from inadequate herbicide control. In the absence of quantitative yield information between these two extremes, relative yield potential was taken as a straight-line approximation between these two NDVI values.

Total nitrogen requirement for the 1995 spring wheat crop was computed based on estimated yield potential across the field and University of Idaho fertilizer guidelines (Tindall et al., 1991). The nitrogen chemigation requirement for the field was computed based on the producer's yield goal of 8064 kg/ha, relative yield potential for the field based on the estimated relationship to NDVI, total nitrogen requirements as a function of yield goal from the University of Idaho fertilizer guide and a preplant uniform nitrogen fertilizer application of 67 kg/ha. A histogram of the computed nitrogen chemigation requirements for the field are shown in Fig. 2. Approximately 21 percent of the field did not require additional nitrogen application. This was a result of elevated residual soil nitrogen in low yielding areas of the field due to years of uniform nitrogen fertilization and the 67 kg/ha preplant nitrogen fertilizer application. The histogram of Fig. 2 shows that there was basically only two nitrogen chemigation management zones required, zero and 90 kg/ha. Nitrogen chemigation requirements from 1 to 80 kg/ha constituted a relative small portion of the field.

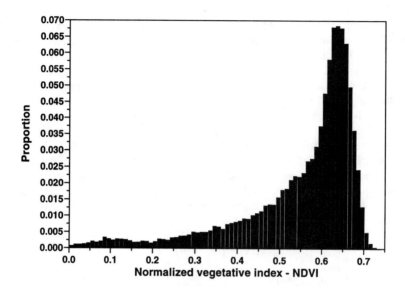

Fig. 1. Histogram of computed normalized vegetative index for the 1994 spring wheat crop.

The variable rate center pivot irrigation system is capable of applying step-wise relative chemical applications of 0, 1/3, 2/3 and full through the application of water in these same relative amounts and maintaining a constant concentration of the chemical in the applied water. The nitrogen chemigation requirement map was constructed by grouping the nitrogen chemigation requirements shown in Fig. 2 into four management zones having relative proportional applications of 0, 1/3, 2/3 and 1. The resulting nitrogen chemigation application map is shown in Fig. 3. Management zones 1, 2, 3 and 4 required 0, 27, 54, and 81 kg N/ha. The nitrogen chemigation map could have been constructed with a finer degree of control by making the total nitrogen application in multiple passes.

The variable rate chemigation application map that was actually implemented is shown in Fig. 4. The field was divided into eight sectors which alternately received variable and uniform nitrogen chemigation. This arrangement was imposed to evaluate and document the performance of the variable rate chemigation system by comparing nitrogen application in the two treatments. The polar coordinate grid system depicted in Figs. 3 and 4 represents how the map is implemented by the variable rate control system. The cell size under the last span of the center pivot is 38.1 m (125 ft) radially by 6 degrees in angular dimension. This cell size corresponds to an area of 0.13 ha (0.32 acres) and gets smaller closer to the pivot point.

The nitrogen fertilizer source used was urea-ammonium nitrate (URAN) which is 32% nitrogen by weight of which 7.8% is nitrate, 7.8% ammonia and 16.4% urea. The volume of URAN required to implement the nitrogen chemigation map based on the chemigation management zone values shown in Fig. 3 was 8742

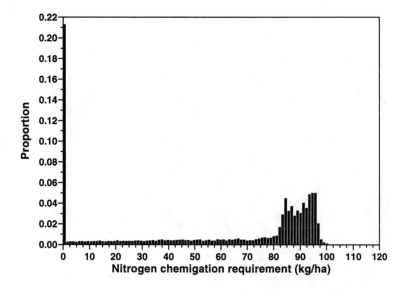

Fig. 2. Histogram of computed nitrogen chemigation requirements for the 1995 spring wheat crop.

L (2309 gal). However, the delivery truck could only hold 6168 L (1629 gal) which the producer deemed sufficient since our computed nitrogen requirement was greater than he traditionally applied. Thus, the nitrogen chemigation application was reduced by 29.5 percent to accommodate this change as reflected in Fig. 4.

Catch cans were placed at select locations in the field to measure irrigation application depth and capture water for nitrate and ammonium concentration analysis. Twenty-four sets of four catch cans each were placed in the field at the locations designated by a "+" in Fig. 4. Each set of four cans was radially aligned with the pivot point with a spacing of 2 m between cans. The sample bottles were placed in the catch cans a few hours ahead of the center pivot system and collected within a few hours after the center pivot passed. The samples were immediately frozen and remained so up to 1 to 2 days before analysis.

RESULTS AND DISCUSSION

The nitrate nitrogen and ammonium nitrogen concentrations in the applied irrigation water along with respective sample location relative to the pivot point are shown in Figs. 5 and 6, respectively. The concentration of ammonium nitrogen averaged 18% less that of the nitrate nitrogen despite the fact that they are in equal proportions in the nitrogen source. This 18% difference is largely attributed to volatilization of the ammonia from the irrigation spray due to the pH (7.98) of the irrigation water. The background concentration of nitrate and ammonium in the irrigation water source was less than 1 mg/L and considered insignificant.

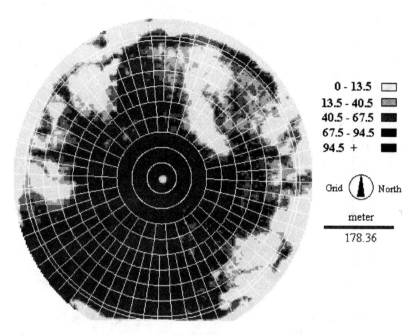

Fig. 3. Computed nitrogen chemigation requirements, kg N/ha, grouped by ratios of 0, 1/3, 2/3 and full.

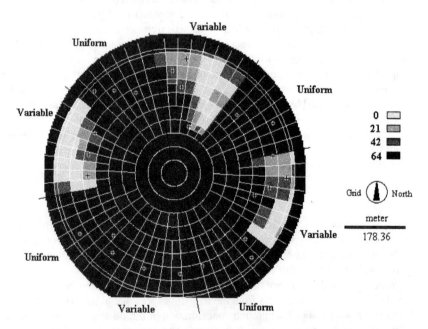

Fig. 4. Executed variable rate nitrogen chemigation map in kg N/ha.

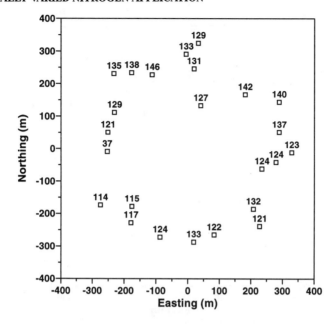

Fig. 5. Average measured nitrate nitrogen concentration in the applied irrigation water at each sample location relative to the pivot point, mg/L.

The chemigation test started when the pivot end was located at approximately -348,-61 easting and northing relative to the pivot point and progressed in the counter clockwise direction. The measured concentration of nitrate and ammonium in the applied water at location -253, -9 was greatly reduced because the URAN supply tank was not equipped with a suction feed which allowed air into the suction side of the chemigation pump reducing the injection flow rate. For this reason, this nitrogen sample location was omitted from further analysis. The measured concentration of nitrate and ammonium at location 18,-287 was greater than for nearby locations because system total flow rate was over-estimated. The center pivot system endgun is off at this location and its flow rate was estimated to be 14.6% of total system flow. Since this location is in a variable rate treatment sector, the URAN injection flow rate was reduced 14.6% to maintain constant concentration in the irrigation water. Since the endgun is not pressure regulated, accurate calculation of flow rate requires that the pressure at the endgun be known on a real time basis. If this feature were available, then more uniform chemical applications could be achieved. The measured nitrate and ammonium concentration at location -113,227 is elevated because the endgun is normally turned off for a couple of degrees of rotation in this general location. Since this location is in a uniform rate treatment sector, the URAN injection flow rate was not reduced according to flow but remained constant at the predetermined injection rate. Thus, nitrogen concentrations increased when total system flow decreased.

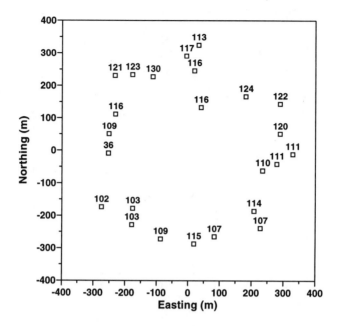

Fig. 6. Average measured ammonium nitrogen concentration in the applied irrigation water at each sample location relative to the pivot point, mg/L.

The measured nitrogen concentrations at locations -177,234 and -233, 231 appear slightly elevated which is attributed to reduced total system flow because of increased elevation at the system end. This general location has the highest elevation compared to the rest of the field. The measured nitrogen concentrations at the beginning of the chemigation test were the lowest measured. The exact cause of this is unknown since the URAN injection rate was set at the beginning of the test and never adjusted. System speed was decreased 16% when the end of the system was approximately at -150,-320 but system speed does not affect injection flow rate, only application depth.

The statistical hypothesis of equal mean concentration of nitrate and ammonium for the variable rate and uniform rate treatment sectors were tested using the SAS TTEST procedure. Summary statistics are shown in Table 1. The mean concentration of nitrate and ammonium between the two treatments are not significantly different (p=0.537 and p=0.785, respectively). The coefficient of variation is less for the variable rate application treatment. This is largely due to changes in total system flow rate being accounted for by proportional changes in chemical injection flow rate. The higher coefficient of variation for the uniform treatment is due to the endgun turning on and off and the unexplained lower nitrogen concentrations measured at the beginning of the test. The results demonstrate that the variable rate control system can apply spatially variable amounts of chemicals with accuracy equivalent to that of conventional uniform application.

Table 1. Summary statistics of measured nitrogen concentration in applied irrigation water.

Parameter	Uniform Application			Variable Application		
	n	mean mg/L	cv	n	mean mg/L	cv
Nitrate	44	129.4	0.097	48	128.1	0.054
Ammonium	44	114.1	0.089	48	113.7	0.043

The average measured irrigation application depth at each sample location is shown in Fig. 7. Comparison of the measured application depths with the target application map of Fig. 4 shows excellent agreement with two exceptions. The measured application depth at location -84,-265 and location 235,-62 are below the target application depth and less than that for nearby locations for the same management zone and system speed. The reason for this is unknown. There were no observed system failures or malfunctions that would account for these reduced applications. Excluding these two identified anomalies and sample locations on management zone boundary transitions (Fig. 4), the average application depths in the variable rate treatment were 4.3 mm, 8.9 mm and 13.8 mm in management zones 1, 2, and 3, respectively. These average measured application depths are very near the target proportions of 1/3, 2/3 and full application. These results show that the variable rate irrigation system is capable of applying step-wise variable target amounts in these proportions. Based on the average application depths in each management zone and an average nitrate nitrogen concentration of 128.1 mg/L which represents 24.38% of the total nitrogen applied, the nitrogen applied in each management zone (neglecting the unaccounted loss in measured ammonium) was 22.6, 46.7 and 64.7 kg N/ha for zones 1, 2 and 3, respectively. These application amounts are nearly equivalent to the target amounts shown in Fig. 4.

SUMMARY AND CONCLUSIONS

A commercial center pivot irrigation system was retrofitted for variable rate irrigation and chemigation. Nitrogen fertilizer was applied through the irrigation system in variable amounts according to a chemigation map consisting of four management zones. Actual nitrogen application was monitored at several locations in each management zone using catch can and water sample nitrate and ammonium analysis. Statistical comparison of the mean nitrate and ammonium concentration in water applied to uniform and variable rate treatments show no significant difference. The average depth of irrigation water applied to each management zone in the variable rate treatment was nearly equivalent to the target amount. These results show that the variable rate irrigation and chemigation system can apply step-wise spatially varied amounts of nitrogen with equivalent or greater accuracy than that of conventional uniform application.

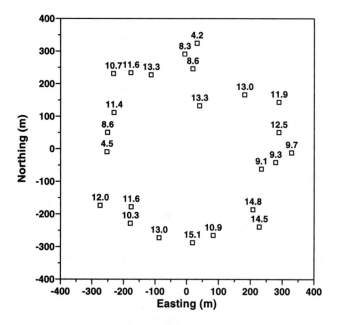

Fig. 7. Average measured irrigation application depth at each sample location relative to the pivot point, mm.

REFERENCES

Deutsch C.V., and A.G. Journel. 1992. GSLIB geostatistical software library and user's guide. Oxford University Press, Inc., New York, NY.

Guoxiang, L., and Z. Dawei. 1990. Estimating the production of winter wheat by remote sensing and unified ground network. I. system verification. In Applications of remote sensing in agriculture, Clark, J.A. and M.D. Steven (eds.). Butterworths, London, UK.

King, B.A., R.A. Brady, I.R. McCann, and J.C. Stark. 1995. Variable rate water application through sprinkler irrigation. Site-specific management for agricultural systems. ASA. Madison, WI. pp. 485-493.

Tindall. T.A., J.C. Stark, B.D. Brown, and D.T. Westermann. 1991. Idaho fertilizer guide: irrigated spring wheat in southern Idaho. CIS No. 828, University of Idaho, College of Agriculture. Moscow, Idaho.

SESSION I

NATURAL RESOURCES VARIABILITY

Spatial Variability of Soil Properties on Two Center-Pivot Irrigated Fields

Shufeng Han
Robert G. Evans
Biological Systems Engineering Department
Washington State University, Prosser, WA

Sally M. Schneider
Stephen L. Rawlins
USDA-Agricultural Research Service, Prosser, WA

ABSTRACT

The knowledge of soil variability within a field is fundamental to the development of site-specific management strategies to increase crop production and reduce the potential pollutant transport to the groundwater. In March 1995, 738 soil samples were collected from two center-pivot irrigated fields in South Central Washington State on a 61- by 61-m grid at 30-cm depth intervals to a maximum depth of 90-cm. The soil samples were analyzed for texture, nitrate-N, pH, P, K, and sulfate-S. Large spatial variability was observed for most of the soil properties from classical statistical analysis and geostatistical analysis. Most of the empirical probability distribution functions for the measured soil properties were neither normally nor lognormally distributed. Both kriging and nonparametric distance-weighting techniques were used to generate soil texture and nutrient distribution maps from the soil sampling data. These distribution maps, when combined with yield and topographical maps, can be used to define different management zones in the fields for site-specific management of water and nutrients.

INTRODUCTION

Spatial variability of soil properties results in spatially varying crop yields at the field level. The knowledge of soil variability within a field is fundamental to the development of site-specific management strategies to increase crop production and reduce the potential pollutant transport to the groundwater. Numerous studies have examined spatial variation of soil physical and chemical characteristics (Campbell, 1978; Meirvenne & Hofman, 1989; Agbu & Olson, 1990; Bahri et al., 1993; Berndtsson & Bahri, 1995), and soil-water properties (Nielsen et al., 1973; Greminger et al., 1985). Traditionally, soil variability has been characterized by random sampling and classical techniques of statistical analysis, such as regression, to predict properties at unsampled points. Further study showed this method to be inadequate for observations that are auto-correlated in either time or space. Recent developments in geostatistical theory enable spatial relationships of sample values to be considered in interpolation (Matheron, 1963; Burgess & Webster, 1980a,b). McBratney and Webster (1983) found that the actual precision achieved by applying geostatistical theory to soil sampling in their studies was from 3 to 9 times greater than would have been judged by using classical methods.

Central to the geostatistical theory are the semi-variogram and kriging. Kriging, as a Best Linear Unbiased Estimates (BLUE), is now extensively used to estimate soil properties at unsampled locations. However, the fundamental assumptions made on the statistical properties of the data must be carefully checked before using the technique. These assumptions, known as the Intrinsic Hypothesis of Matheron's regionalized variable theory (Matheron, 1971), cannot always be satisfied with soil data (Davidoff et al., 1986). Besides its computational complexity, kriging does not always produce the best estimates (Cooke et al., 1993).

The objectives of this study were (i) to examine the spatial variability and the size of the spatial dependence of soil texture and nutrients on center-pivot irrigated fields; and (ii) to generate soil texture and nutrient distribution maps by using kriging and nonparametric distance-weighting techniques.

MATERIALS AND METHODS

Study Fields, Sampling, and Measurements

The study fields are two center-pivot irrigated circles located in south central Washington State near the Columbia River. The two circles are next to each other, with a total area of 90-ha. The USDA classification of the soil is Quincy loamy sand (Xeric Torripsamments, mixed, mesic), 2 to 15 percent slopes. Crop production in the area is dependent on irrigation, and the typical crops in the rotation include potato, wheat, buckwheat, corn, and alfalfa. The two study fields were treated equally before soil sampling in regard to agriculture management and mineral fertilizers application. At the time of soil sampling, no crop was planted on the two fields.

In March 1995, 738 soil samples in the two study fields were collected roughly on a 61- by 61-m grid at 30-cm depth intervals to a maximum depth of 90-cm. Locations of the sampling sites were determined by GPS equipment.

Some important soil properties that influence the potato production were measured in a commercial laboratory. These include: texture, nitrate-N, pH, P, K, and sulfate-S. Soil texture, nitrate-N, and pH were analyzed for each of the three soil layers (0- to 30- cm, 30- to 60- cm, 60- to 90- cm). Soil P, K, and sulphate-S were analyzed only for the top soil layer (0- to 30- cm). This resulted in a total of 15 data sets.

Data Analysis

Classical statistical analysis was performed to determine the sample mean, median, standard deviation (SD), minimum, maximum, and coefficient of variation (CV) for each of the 15 data sets. The Shapiro-Wilk statistic, W, was calculated for each data set to test whether it is normally distributed. A large deviation from normality does not warrant the application of the kriging technique.

The Intrinsic Hypothesis of Matheron's regionalized variable theory (Matheron, 1971) assumes data stationarity; in other words, no significant trend should exist in the data before using kriging. It is highly unrealistic to assume the stationarity for soil data. A robust technique is to remove trend first, and then work with the residual, a small scale high-frequency variation. A median-polishing detrending algorithm was used in this study. The algorithm has been presented by several authors (Emerson and Hoaglin, 1981; Cressie et al., 1990), and modified by Han et al. (1993).

After data detrending, the experimental semi-variances, $\hat{\gamma}(h)$, for a soil property Z were calculated by:

$$\hat{\gamma}(h) = \frac{1}{2m(h)} \sum_{i=1}^{m(h)} [z(x_i) - z(x_i+h)]^2 \tag{1}$$

where m(h) is the number of pairs of observations separated by distance h.

The experimental semi-variances were used to fit a specific semi-variogram model. A number of semi-variogram models are useful in soil science (McBratney and Webster, 1986). Of these models, the spherical model is most commonly used, and is expressed as:

$$\begin{cases} \gamma(h) = c_0+c[\frac{3}{2}(\frac{h}{a})-\frac{1}{2}(\frac{h}{a})^3] & for \ \ 0 < h \le a \\ \gamma(h) = c_0+c & for \ \ h > a \end{cases} \tag{2}$$

where c_0 is the nugget, $c_0 + c$ is the sill, and a is the range.

From c_0, c, and a, a mean correlation distance (MCD) is calculated by (Han et al., 1994):

$$MCD = \frac{3}{8} * \frac{c}{c_0+c} a \tag{3}$$

The MCD represents the "relatedness" of the soil property in question, or the size of the spatial dependence of the soil property.

Finally, both punctual kriging and nonparametric distance-weighting techniques were applied to each data set to generate soil texture and nutrient distribution maps.

RESULTS AND DISCUSSION

Basic Statistics

Descriptive statistics for the measured soil properties without data detrending were given in Table 1. Most of the empirical probability distribution functions for the different soil properties displayed several *probable outliers* as identified

Table 1. Descriptive statistics† of the measured soil properties without data detrending.

Soil Property	N	n	Mean	Median	SD	Min.	Max.	CV	W	D	Norm. distr.‡	P
Sand Content (%)												
0- to 30- cm	243	243	86.1	87.6	6.35	64.4	97.8	7.38	.910	.141	Rejected	
30- to 60- cm	241	238	91.0	91.8	4.28	75.8	99.6	4.71	.936	.100	Rejected	
60- to 90- cm	240	238	93.5	94.2	3.46	80.4	99.6	3.70	.914	.132	Rejected	
Clay Content (%)												
0- to 30- cm	243	243	1.6	1.6	0.93	0.0	4.8	58.01	.960	.104	Rejected	
30- to 60- cm	241	241	0.8	0.8	0.73	0.0	3.0	91.00	.882	.146	Rejected	
60- to 90- cm	240	239	0.5	0.4	0.57	0.0	2.2	105.70	.836	.172	Rejected	
Nitrate-N (mg/kg)												
0- to 30- cm	246	242	2.6	2.3	1.17	0.8	6.4	44.93	.906	.151	Rejected	
30- to 60- cm	242	238	2.3	2.0	1.28	0.5	7.9	56.21	.848	.226	Rejected	
60- to 90- cm	244	236	2.6	2.0	1.70	0.4	8.5	64.59	.854	.214	Rejected	
pH (pH unit)												
0- to 30- cm	246	245	7.1	7.1	0.38	6.1	8.3	5.37	.971	.100	Not rejected	0.01
30- to 60- cm	242	241	7.9	8.0	0.43	6.4	8.6	5.44	.936	.117	Rejected	
60- to 90- cm	244	241	8.3	8.4	0.32	7.3	8.7	3.87	.859	.182	Rejected	
P (mg/kg)												
0- to 30- cm	246	245	24.0	24.0	5.81	7.0	42.0	24.23	.978	.066	Not rejected	0.01
K (mg/kg)												
0- to 30- cm	246	244	182.8	180.0	37.83	100.0	300.0	20.70	.951	.105	Rejected	
Sulphate-S (mg/kg)												
0- to 30- cm	246	245	4.2	4.0	2.65	1.0	14.0	63.01	.899	.133	Rejected	

† N = original number of samples; n = number of samples after removing *probable outliers*; SD = standard deviation; CV = coefficient of variation, %; W = Shapiro-Wilk statistic; D = Kolmogorov statistic; P = significance level.

‡ Shapiro-Wilk statistic (W) was used for the normal distribution test.

by box plots. The number of probable outliers varied from 0 to 8 (Table 1). These outliers were excluded in the calculation of descriptive statistics. Sand content and pH displayed a small CV (less than 10%). On the other hand, clay content, nitrate-N, and sulphate-S displayed a large CV (greater than 50%), which indicated a large spatial variability for these soil properties. Both P and K exhibited a moderate spatial variability. Normal distribution test showed that only P and pH (0- to 30- cm) were normally distributed, and all other soil properties were not normally distributed. Although lognormal distributions have been found for many geochemical elements (Bahri et al., 1993), a log-transformation of the studied data sets did not improve their normality.

To remove the possible trends and improve the normality of the original data sets, the median-polishing detrending algorithm was applied to each of the 15 data sets. The detrending process was performed separately for each field, since samples were taken with two grid patterns, one for each field. Only row and column effects were removed from the original data sets to avoid negative values of residuals. A normal distribution test of the measured soil properties after data detrending was performed (Table 2). Although the distributions of many soil properties were still not normal, data detrending, in general, can slightly improve the normality. This was indicated by reduced CV/D and increased W in Table 2, as compared with CV, D, and W values in Table 1.

Table 2. Normal distribution test[†] of the measured soil properties after data detrending.

Soil Property	CV	W	D	Norm. distr.[‡]	P
Sand Content (%)					
0- to 30- cm	5.80	.942	.140	Rejected	
30- to 60- cm	4.05	.928	.131	Rejected	
60- to 90- cm	3.21	.919	.149	Rejected	
Clay Content (%)					
0- to 30- cm	59.42	.983	.095	Not rejected	0.01
30- to 60- cm	84.66	.950	.125	Rejected	
60- to 90- cm	135.93	.921	.186	Rejected	
Nitrate-N (mg/kg)					
0- to 30- cm	40.76	.943	.122	Rejected	
30- to 60- cm	55.87	.875	.170	Rejected	
60- to 90- cm	61.13	.894	.124	Rejected	
pH (pH unit)					
0- to 30- cm	5.06	.956	.131	Rejected	
30- to 60- cm	5.00	.951	.167	Rejected	
60- to 90- cm	3.78	.902	.182	Rejected	
P (mg/kg)					
0- to 30- cm	21.73	.978	.084	Not rejected	0.01
K (mg/kg)					
0- to 30- cm	18.29	.951	.124	Rejected	
Sulphate-S (mg/kg)					
0- to 30- cm	66.03	.968	.096	Rejected	

† CV = coefficient of variation, %; W = Shapiro-Wilk statistic; D = Kolmogorov statistic; P = significance level.

‡ Shapiro-Wilk statistic (W) was used for the normal distribution test.

Semi-variograms

Only the omni-directional semi-variances were considered in this study, since the data detrending process would eliminate some anisotropy of the data. Soil texture and nutrients semi-variograms were shown in Fig. 1 and 2, respectively. Table 3 listed the c_0, c, a, and MCD values (Eq. 2 and 3) for each of the soil properties.

Significant nugget variances existed in all data sets. Nugget arises from a contribution of measurement errors and variation over distances much smaller than the closest sampling interval. Clay content (60- to 90- cm), pH (0- to 30- cm, 30- to 60- cm), and Sulphate-S displayed pure nugget effects, indicating

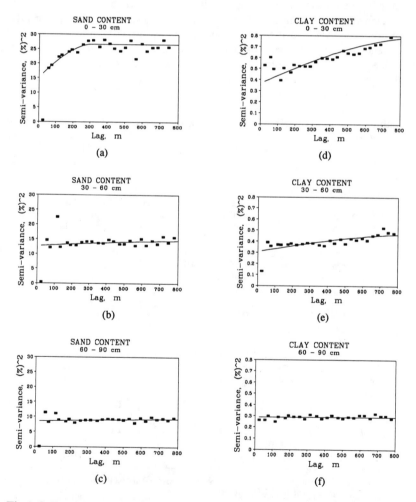

Fig.1 Soil texture (sand, clay) semi-variograms after data detrending.

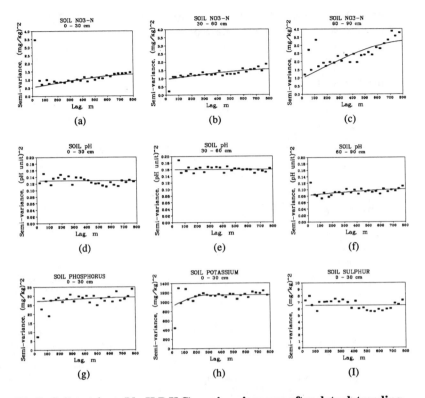

Fig.2. Soil nutrients(N,pH,P,K,S) semi-variograms after data detrending.

there were no spatial structures evident over the range of h for which these semi-variograms were defined. More samples at a smaller sampling interval are required to obtain reliable semi-variograms for these soil properties.

The MCD values in Table 3 indicated that the size of the spatial dependence for sand content, P, K, and possibly pH is between 20- to 60- m. Clay content and nitrate-N had much larger MCD values; they were correlated over a long distance (greater than 100 m).

Distribution Maps

To generate soil texture and nutrients distribution maps, each field was divided into 6.1- by 6.1- m cells. A cell value is calculated by the average of four point estimates from the centers of each quarter of the cell.

Both punctual kriging and nonparametric distance-weighting techniques were applied to estimate point values. The correlation between the kriged map and the distance-weighted map for each property was compared on a cell by cell basis (Table 4). Both methods generated similar distribution maps, as indicated by the large correlation coefficients.

Table 3. Semi-variogram parameters[†] and the MCD[‡] values of the measured soil properties after data detrending.

Soil Property	c_0	c	a (m)	MCD (m)
Sand Content (%)				
0- to 30- cm	15.23	11.41	309.4	47.2
30- to 60- cm	12.69	1.75	900.7	39.6
60- to 90- cm	8.52	0.75	900.7	26.3
Clay Content (%)				
0- to 30- cm	0.37	0.42	900.7	178.6
30- to 60- cm	0.31	0.16	900.7	114.1
60- to 90- cm	0.29	----	-----	-----
Nitrate-N (mg/kg)				
0- to 30- cm	0.52	0.83	900.7	205.2
30- to 60- cm	0.94	0.62	900.7	132.4
60- to 90- cm	0.95	2.36	901.0	239.3
pH (pH unit)				
0- to 30- cm	0.13	----	-----	-----
30- to 60- cm	0.16	----	-----	-----
60- to 90- cm	0.08	0.02	900.7	63.2
P (mg/kg)				
0- to 30- cm	26.89	2.50	900.7	27.5
K (mg/kg)				
0- to 30- cm	905.18	243.67	283.5	21.1
Sulphate-S (mg/kg)				
0- to 30- cm	6.53	----	-----	-----

† c_0 = nugget; c = sill - nugget; a = range.
‡ MCD = mean correlation distance.

CONCLUSIONS

Large spatial variability of soil texture and nutrients exist within the two center-pivot irrigated fields. The magnitude of variability, in ascending order, is: pH, sand content, K, P, nitrate-N, sulphate-S, and clay content.

Most of the empirical probability distribution functions for the measured soil properties are neither normally nor lognormally distributed.

Both kriging and nonparametric distance-weighting techniques can be used to generate soil texture and nutrient distribution maps from the soil sampling data.

Table 4. Correlation between the kriged map and distance-weighted map for each property, compared on a cell by cell basis.

Soil Property	Correlation Coefficient
Sand Content (%)	
0- to 30- cm	0.9513
30- to 60- cm	0.8580
60- to 90- cm	0.8148
Clay Content (%)	
0- to 30- cm	0.9174
30- to 60- cm	0.8737
60- to 90- cm	0.7804
Nitrate-N (mg/kg)	
0- to 30- cm	0.9219
30- to 60- cm	0.8651
60- to 90- cm	0.9398
pH (pH unit)	
0- to 30- cm	0.6605
30- to 60- cm	0.7497
60- to 90- cm	0.7350
P (mg/kg)	
0- to 30- cm	0.7229
K (mg/kg)	
0- to 30- cm	0.8410
Sulphate-S (mg/kg)	
0- to 30- cm	0.7171

REFERENCES

Agbu, P.A. and K.R. Olson. 1990. Spatial variability of soil properties in selected Illinois mollisols. Soil Sci. 150:777-786.

Bahri, A., R. Berndtsson, and K. Jinno. 1993. Spatial dependence of geochemical elements in a semiarid agricultural field: I. Scale properties. Soil Sci. Soc. Am. J. 57:1316-1322.

Berndtsson, R. and A. Bahri. 1995. Field variability of element concentrations in wheat and soil. Soil Sci. 159:311-320.

Burgess, T.M. and R. Webster. 1980a. Optimal interpolation and isarithmic mapping of soil properties. I. The semi-variogram and punctual kriging. Jour. of Soil Sci. 31:315-331.

Burgess, T.M. and R. Webster. 1980b. Optimal interpolation and isarithmic mapping of soil properties. II. Block kriging. Jour. of Soil Sci. 31:333-341.

Campbell, J.B. 1978. Spatial variability of sand content and pH within single contiguous delineations of two soil mapping units. Soil Sci. Soc. Am. J. 42:460-464.

Cooke, R.A., S. Mostaghimi, and J.B. Campbell. 1993. Assessment of methods for interpolating steady-state infiltrability. Trans. ASAE 36(5):1333-1341.

Cressie, N., C.A. Gotway, and M.O. Grondona. 1990. Spatial prediction from networks. Chemometrics and Intelligent Laboratory Systems 7:251-271.

Davidoff, B., J.W. Lewis, and H.M. Selim. 1986. A method to verify the presence of a trend in studying spatial variability of soil temperature. Soil Sci. Soc. Am. J. 50:1122-1127.

Emerson, J.D. and D.C. Hoaglin. 1983. Analysis of two-way tables by medians, In: D.C. Hoaglin, F. Mosteller and J.W. Tukey (Eds.), Understanding Robust and Exploratory Data Analysis, Wiley, New York. pp 166-210.

Greminger, P.J, Y.K. Sud, and D.R. Nielsen. 1985. Spatial variability of field-measured soil-water characteristics. Soil Sci. Soc. Am. J. 49:1075-1089.

Han, S., C.E. Goering, M.D. Cahn, and J.W. Hummel. 1993. A robust method for estimating soil properties in unsampled cells. Trans. ASAE 36:1363-1368.

Han, S., J.W. Hummel, C.E. Goering, and M.D. Cahn. 1994. Cell size selection for site-specific crop management. Trans. ASAE 37(1):19-26.

Matheron, G. 1963. Principles of geostatistics. Econ. Geol. 58:1246-1266.

Matheron, G. 1971. The theory of regionalized variables and its applications. Les Cahiers du Centre de Morphologie Mathematique, No. 5., Fontainebleau: Centre de Geostatistique.

McBratney, A.B. and R. Webster. 1983. How many observations are needed for regional estimation of soil properties? Soil Sci. 135:177-183.

McBratney, A.B. and R. Webster. 1986. Choosing functions for semi-variograms of soil properties and fitting them to sampling estimates. Jour. of Soil Sci. 37:617-639.

Meirvenne, M.V. and G. Hofman. 1989. Spatial variability of soil nitrate nitrogen after potatoes and its change during winter. Plant and Soil 120:103-110.

Nielsen, D.R., J.W. Biggar, and K.T. Erh. 1973. Spatial variability of field-measured soil-water properties. Hilgardia 42:215-259.

Effects of Long-term Cultivation on a Morainal Landscape in Alberta, Canada

B.D. Walker

Land Resource Unit
Agriculture and Agri-Food Canada
Edmonton, Canada

K. Haugen-Kozyra

KHK Consulting Services
Edmonton, Canada

C. Wang

Eastern Cereals and Oilseeds Research Centre
Agriculture and Agri-Food Canada
Ottawa, Canada

Agriculture and Agri-Food Canada established a network of benchmark sites across Canada to monitor agricultural soil quality. One of these sites, located in Alberta at 52°25′35″ N and 110°07′35″ W, was paired with uncultivated (native) land nearby. The proximity and similarity of terrain enabled an assessment of changes in soil attributes resulting from 80 years of cultivation.

The landscape at both sites features hummocky to undulating moraine. The soils are mainly Typic Cryoborolls on slopes and Argic Cryaquolls in depressions. Hilltops have thin (native) or "eroded" (cultivated) profiles. Both sites were characterized and sampled in detail, using a transect sampling design suited to the distinct landforms. The compared data sets include both sample analysis and field measurements.

Dramatic differences in pH of cultivated Ap versus native Ah horizons emphasize the impact of cultivation across the landscape. Cultivated hilltops have pH values almost 2.5 units higher than occur on lower slopes and in depressions. In the native landscape there is only about 1.0 unit difference. Organic C, C:N ratios, and available K also show landscape differences caused by cultivation. These data show that site specific management could be beneficial. Moisture variability, however, would be an important factor to consider in evaluating a precision farming system.

INTRODUCTION

Agriculture and Agri-Food Canada established 23 benchmark sites across Canada, from 1989-92, to monitor agricultural soil quality. The sites were selected to represent common production systems on typical landscapes in major

agro-ecological regions.   One of these sites was established in 1990 on
cultivated land at 52°25′35″ N and 110°07′35″ W near Provost, Alberta (Walker
and Wang 1994). A parcel of uncultivated land 1.6 km away was characterized
and sampled in 1991, using the same methodology.

The similarity of the two sites provided an opportunity to examine soil
attributes that approximated conditions prior to cultivation, and, by comparison,
to assess changes brought about by 80 years of cultivation. Attributes featured
in this paper include pH, organic C, C:N ratio, and available K.

MATERIALS AND METHODS

Sites

The Provost sites are located in east-central Alberta, at 52°25′35″ N and
110°07′35″ W, about 300 km southeast of Edmonton. The cultivated site
encompasses an area of 8.8 ha. The uncultivated (native) site, located 1.6 km
northwest of the cultivated site, covers roughly 11 ha of pasture land.

Both sites occur on hummocky to undulating moraine. Parent materials
are moderately calcareous, CL-L textured, glacial till overlain by a thin capping
of L textured, waterlaid material on slopes and in depressions. The upland soils
are mainly Typic Cryoborolls (Orthic Dark Brown Chernozemic). Hilltops have
thin topsoil (native site) or are "eroded" (cultivated site). Wet mineral Argic
Cryaquolls (Humic Luvic Gleysols) typify the depressions.

The sites occur in the southern fringe of the Aspen Parkland Ecoregion
of the Prairies Ecozone (ESWG 1995). The regional climate features short,
warm summers and long, cold winters with continuous snow cover. The nearest
long-term climate station, at Macklin, SK (52°20′N 109°57′W), has a mean
annual temperature of 1.7°C (AES n.d.). Mean summer temperature is 15.0°C
and mean winter temperature is -12.5°C. The moisture regime is semiarid; the
mean annual precipitation of 394 mm is characterized by a distinct summer
maximum, but with a growing season deficit of about 300 mm.

Ecologically, the area is transitional between the drier treeless grassland
to the south and aspen parkland to the north. Grassland plant communities are
dominant and associated with drier segments of the landscape. Groves of aspen
(*Populus tremuloides*) occur in moister sites such as shallow depressions and
north-facing slopes, and account for about 15% of the land cover. Upland shrub
communities account for another 10-15% cover. Wetland depressions, usually
ringed with willows and dominated by vegetation such as sedges, account for
about 15% of the terrain.

Native landscapes like the one studied here are rare in the region. The
native site was grazed, "fairly heavily" in the words of the owner, from 1906
until the early 1970's. Grazing was described as "light" from the mid 1970's to
the mid 80's. No domestic grazing has occurred since the late 1980's.

Most land in the region has been cultivated for several decades; native
vegetation has been replaced with cereal and oilseed crops. The field containing
the cultivated site was broken for cropping in 1912. The crop rotation was

wheat-fallow prior to 1991, canola-wheat-fallow since then. The plow was the principal tillage tool, drawn by horses until 1940, tractor thereafter. Deep-tillage cultivators replaced the plow as the main tillage implement in about 1950. Harvesting changed from threshing machine to combine in 1947. Since then crop residues have been left on the field and tilled into the soil. Use of chemical fertilizers (mainly 11-48-0) and herbicides began *circa* 1950, high nitrogen fertilizers (e.g. 34-0-0 and anhydrous ammonia) in 1977.

Sampling Design

The Provost sites, characterized with detailed (1:2,000 scale) topographic and soil surveys, have distinct relief (Fig. 1). Soil patterns repeat in such landscapes. With the repetition comes a degree of predictability about many soil attributes. A stratified random sampling method using transects (Wang 1982) was designed to sample the repeating landscape patterns. Transects were laid out from hilltops to depressions at several locations within each landscape (Fig. 1). Sampling points were staked out at 10 m intervals along each transect. Nine transects with 64 points represented the cultivated landscape, 7 transects with 61 sampling points depicted the native site.

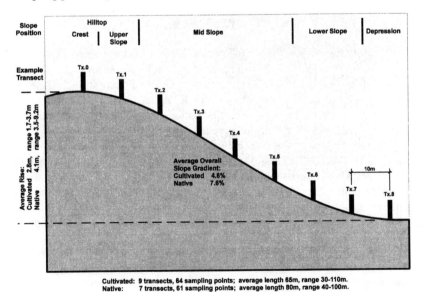

Fig. 1.   Schematic of transect layout and the average topographic features of the cultivated and native sites.

Soil and landscape features including slope position, soil series, and A horizon depth were recorded for each sampling point. Slope position was reported as one of five classes: 1) crest, 2) upper slope, 3) mid slope, 4) lower slope, and 5) depression. Samples of the topsoil were collected at each sampling point. Subsurface horizons from about the 50-60 cm depth were collected at one-quarter of the sampling points. At the native site, B horizon samples were taken at most points where the A horizon was less than 10 cm thick. Bulk density was determined gravimetrically, by the Kubiena box sampling method, on the topsoil and B horizon samples. In addition, two pedons representing the dominant soils at each site were described and sampled by horizon.

Analytical Methods

The soil samples were air-dried and roller-ground to separate the fine earth fraction (<2mm) from coarse fragments. Analyses important to this discussion included: soil reaction (CaCl$_2$ pH) by pH meter in a 1:2 soil to 0.01 M CaCl$_2$ solution (Sheldrick 1984); total C by LECO induction furnace (Sheldrick 1984); inorganic C (CaCO$_3$ equivalent) by the inorganic carbon manometric (calcimeter) method of Bascombe (1961); total N using a semi-micro version of the Kjeldahl-Wilforth-Gunning method (AOAC 1955); organic C by difference between total and inorganic C; available P by the sodium bicarbonate extraction method (Sheldrick 1984); and available K by the pH7, 1M NH$_4$OAc extraction method (Sheldrick 1984).

RESULTS

Data for selected attributes are summarized for three slope positions at each site in Table 1. Upper slope data (n = 7 and 9 for cultivated and native sites, respectively) are similar to those for the crest. For many attributes, the two slope positions can be aggregated into a single "hilltop" position (Fig. 1). Lower slope data for the cultivated (n = 18) and native (n = 11) sites resemble those of cultivated mid slopes and native depressions, respectively.

An obvious difference between the two sites is thickness of the uppermost A horizon (Table 1). Cultivated Ap's average 11 cm across the landscape; Ah's at the native site average 4 cm on crests, 7 cm on upper slopes, 9 cm on mid slopes, and 11 cm on lower slopes and in depressions. The difference across hilltops and mid slopes impacted on some of the comparisons.

In order to compare C and N contents in layers of equal depth, a "simulated plow layer" (simulated Ap) was calculated based on native site data for crest, upper, and mid slope positions. First the mass of soil per unit area (kg m^{-2}) was calculated using thickness and bulk density of each thin Ah horizon and its underlying Bm to a total depth of 11 cm. Then organic C and total N contents of the two layers were factored in, totaled, and converted back into relative contents (%) for the simulated Ap. This procedure also provided an estimate of the dilution effect of subsoil materials on the primordial plow layer after initial cultivation.

Dramatic differences in pH of cultivated Ap versus native Ah horizons emphasize the impact of cultivation across the landscape (Fig. 2). pH of cultivated hilltops averages almost 2.5 units more than in lower slopes and depressions. In the native landscape there is only about 1.0 unit difference.

Table 1. Descriptive statistics for selected attributes at three of the five slope positions on the cultivated and native sites.

Slope Position: Variable	n	Mean	Std. Dev.	Range	Median	n	Mean	Std. Dev.	Range	Median
Crest:		*Cultivated Site*					*Native Site*			
A horizon thickness (cm)	9	11	2	8-16	11	8	4	1	3-6	5
pH (CaCl$_2$)	9	7.5	0.4	6.5-7.8	7.6	8	5.9	0.3	5.6-6.3	5.9
Organic C (%)	9	1.80	0.30	1.38-2.33	1.84	8	6.39	1.98	4.73-10.17	5.65
Total N (%)	9	0.18	0.03	0.12-0.23	0.18	8	0.53	0.11	0.40-0.70	0.49
C:N Ratio	9	10.1	0.7	9.5-11.5	10.1	8	11.9	1.6	10.3-14.5	11.5
Available K (ug g soil^{-1})	9	283	66	216-374	253	8	625	113	515-885	605
Mid Slope:		*Cultivated Site*					*Native Site*			
A horizon thickness (cm)	25	11	1	9-13	11	28	9	4	3-23	9
pH (CaCl$_2$)	25	6.2	1.1	4.8-7.7	6.0	28	5.8	0.6	4.8-7.3	5.6
Organic C (%)	25	2.62	0.56	1.76-3.59	2.62	28	6.66	2.02	3.10-11.82	6.62
Total N (%)	25	0.23	0.04	0.17-0.31	0.23	28	0.54	0.16	0.28-1.05	0.54
C:N Ratio	25	11.1	0.8	9.7-12.8	11.2	28	12.4	1.1	9.5-14.4	12.5
Available K (ug g soil^{-1})	25	450	185	193-974	393	28	509	138	211-774	474
Depression:		*Cultivated Site*					*Native Site*			
A horizon thickness (cm)	7	11	2	8-13	12	7	11	2	8-13	10
pH (CaCl$_2$)	7	5.1	0.2	4.9-5.4	5.0	7	5.0	0.2	4.8-5.5	5.0
Organic C (%)	7	3.56	0.19	3.32-3.83	3.57	7	7.69	1.22	6.02-9.38	7.96
Total N (%)	7	0.33	0.01	0.31-0.34	0.34	7	0.65	0.07	0.56-0.75	0.66
C:N Ratio	7	10.8	0.4	10.3-11.3	10.7	7	11.8	1.2	10.6-14.0	11.5
Available K (ug g soil^{-1})	7	753	124	518-860	794	7	695	149	561-997	628

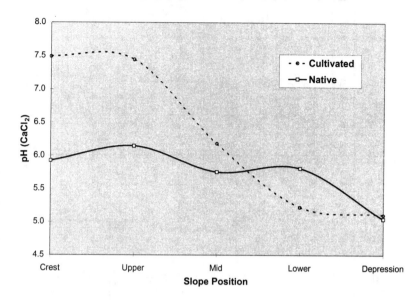

Fig. 2.   pH variability in A horizons across cultivated and native landscapes.

The difference in organic C content between native and cultivated sites is much greater than differences among slope positions within each site (Fig. 3). Cultivated Ap horizons have roughly one-third the organic C content of native Ah horizons. A fairer comparison should include the "simulated plow layer" calculated from native site data. Results for the simulated Ap (Sim-Native, Fig. 3) suggest that the primordial Ap horizon on hilltops and mid slopes had about half the organic C content of the native Ah due to dilution by the underlying subsoil. Total N content (%) mimicks the carbon results.

There is a modest difference in C:N ratios between cultivated and native sites with the greatest difference across the hilltops (about 2-3 units, Fig. 4). Differences in C:N ratios among slope positions at either site are not significant. C:N ratios in the simulated Ap closely follow those of the native Ah.

Analysis of available P data are incomplete. Preliminary results indicate low values across hilltops and on mid slopes at both sites, with much higher values on lower slopes and in depressions.

Available K results for the topsoil show that the two sites differ substantially across hilltops, but are reasonably similar through mid slope, lower slope, and depressional positions (Fig. 5).

DISCUSSION

The first plowing undoubtedly had a significant impact on the hilltop and many mid slope soils in this landscape. Even shallow plowing depths of 3-4 inches (personal communication from farm operator Bill Carter) probably mixed the native Ah horizon with some of the underlying B at many locations. Carbon, nitrogen, and nutrients would have been diluted in this primordial Ap horizon. This trend likely continued over the ensuing decades, especially with the advent of tractor power and the deep-tillage cultivator, and was probably intensified by erosion and mineralization processes. Within several years the new Ap horizon likely began to include calcareous C horizon material in places.

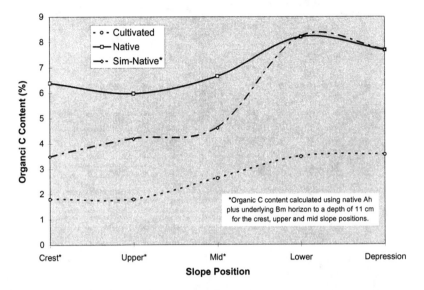

Fig. 3.   Comparison of organic C content of cultivated Ap, native Ah, and simulated Ap (Sim-Native) horizons by slope position.

A statisitical perusal of the data sets (Table 1) creates a first impression about the effects of 80 years of cultivation. For several attributes, like pH and available K, cultivation has amplified the differences between slope positions. For others, differences between slope positions have been somewhat reduced (e.g. organic C) or are negligible (e.g. A horizon thickness) after cultivation. Perhaps most interestingly, the variability or "noise" in the data for several attributes at any slope position is less for the cultivated condition.

Most hilltops in the cultivated landscape, when in fallow, are clearly visible by their grayish colors in contrast to darker soils on adjacent slopes. Such knolls (hilltops) are dominated by calcareous soils with a rudimentary Apk-Ck profile. Calcareous topsoil has also been moved downward to some mid slope locations. With carbonate contents of 1 to 11%, pH values averaging 7.5 are not surprising in such soils. While the alkaline soil reaction and the presence of carbonates are not intrinsically detrimental to crop growth, interactions with nutrients like P may be problematic. In contrast, the acidic pH's that dominate the lower slope and depressional locations are borderline for optimum crop growth, and may become an issue if acidification progresses.

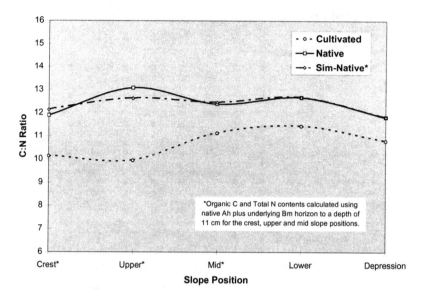

Fig. 4.   Comparison of C:N ratio in cultivated Ap, native Ah, and simulated
Ap (Sim-Native) horizons by slope position.

The wide range in pH at the cultivated site affects herbicide management
for weed control.   Herbicides belonging to the chemical family Sulfonylurea
(e.g. GLEAN®) degrade slowly at pH 7.5 and greater, and can leave residues
that are harmful to some crops for several years after application in high pH soils
(Ahrens 1994).   Further, the Sulfonylurea herbicides tend to be moderately
mobile at high pH.   In contrast, herbicides of the Imidazolinone family (e.g.
PURSUIT®) bind more readily to organic matter and clay below pH 6.5,
decreasing their degradation.   In addition, anaerobic conditions, which may
occasionally occur in depressional areas, severely hamper degradation of
Imidazolinone herbicides (Ahrens 1994).

The grayish colored hilltops of the cultivated site also identify the soils
with the lowest organic C content in the landscape. This is a visual indication of
the lower carbon content throughout the cultivated landscape (Fig. 3).   Three
processes that have been intensified by cultivation are likely responsible for the
reduction in soil carbon: dilution, erosion and mineralization. In a related study
at these sites, loss of organic C at each slope position was evaluated according to
net soil erosion rates calculated from ^{137}Cs data[1]. The results showed that tillage
dilution plus erosion accounted for the loss of soil C on hilltops. Mineralization
alone explained losses on lower slopes and in depressions.   Mid slope microsites
had varying effects of all three processes. Dilution caused by the redistribution
of soil materials within the landscape could not be accounted for in the study.

[1]   Walker, B.D., K. Haugen-Kozyra, N.M. Finlayson, and C. Wang. In prep. Comparison of cultivated
and native soils in a morainal landscape in east central Alberta. Research Br., Agriculture and Agri-
Food Canada, Edmonton, AB.

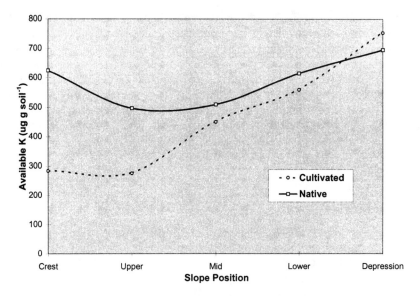

Fig. 5.   Comparison of available K in cultivated Ap and native Ah horizons
     by slope position.

     C:N ratios augment the carbon picture. The overall difference between
native (including Sim-Native) and cultivated sites (Fig. 4) affirms faster organic
matter mineralization under cultivation. Most affected by cultivation and the
amount and characteristics of C inputs is the light fraction (LF) organic matter,
which normally has a wide C:N ratio (Gregorich and Ellert 1993). Plant residue
returned in the cropping system is likely lower than in the native system,
resulting in little LF organic matter in the cultivated soils.
     The difference in C:N ratios on the hilltops illustrates the role of plant
residue. Under cultivation, return of crop residue to the soil has likely been the
least on hilltops due to lower yields over the long term. Also, fresh plant residue
and LF organic matter were likely more susceptible to erosion from hilltops. In
the native landscape, well established grass communities and minimal erosion
on hilltops promoted entrapment and accumulation of plant residue for decay.
     The image of erosion and dilution of nutrients on cultivated hilltops is
supported by the available K results (Fig. 5). Plant available K levels are usually
considered to be high in Alberta soils as illustrated by the native Ah horizons.
While K levels are still moderate to high at the cultivated site, the significantly
lower values across hilltops reflect dilution of underlying subsoil, and erosion of
both topsoil and plant residues. Data from the two native site pedons indicate
that available K in the B and C horizons is about 50% and 20-30%, respectively,
of levels in the Ah. Dilution by cultivation was therefore likely a factor, as was
shown for organic C and total N. Further, the highest levels of available K occur
in the cultivated depressions which implies deposition of K-enriched material,
both plant residues and soil, eroded from surrounding slopes.

Based on the few attributes presented here, the cultivated landscape would undoubtedly benefit from site specific agronomic management. This would be especially critical for herbicide application. With its distinctive hilltops, some activities could be undertaken based only on visual assessments. For example, manuring of hilltops could be easily undertaken, and would increase carbon, improve soil structure, and help reduce erosion. It would be more convenient, however, to use the latest electronic technology for managing some variables, in particular for identifying and dealing with areas of low pH. But for any precision management system developed for this landscape, moisture is an overshadowing variable. Questions about the best placement of inputs to improve economic returns remain to be answered.

ACKNOWLEDGMENTS

Thanks to Agriculture and Agri-Food Canada, National Soil Conservation Program and the Environmental Sustainability Initiative for funding; Dennis, Bill, and John Carter and families for the opportunity to study their land; and WW. Pettapiece and D.C. Penney for reviewing the manuscript.

REFERENCES

AES (Atmospheric Environment Service). No date. Canadian climate normals. Temperature and precipitation. 1951-1980. Prairie provinces. Env. Can., Can. Climate Prog., Downsview, ON. 429 pp.

Ahrens, William H. (ed.). 1994. Herbicide handbook. Seventh edition. Herbicide Handbook Committee, Weed Sc. Soc. Amer., Champaign, IL.

AOAC (Association of Official Agricultural Chemists). 1955. Official methods of analysis. Eighth ed. Washington, D.C.

Bascombe, C.L. 1961. Calcimeter for routine use on soil samples. Chem. Ind., Part II: 1826-1827.

ESWG (Ecological Stratification Working Group). 1995. A national ecological framework for Canada. Agriculture and Agri-Food Canada, Research Branch, Centre for Land and Biological Resources Research; and Environment Canada, Environment Conservation Serv., State of the Environment Directorate, Ottawa/Hull. 126 pp + map.

Gregorich, E.G. and B.H. Ellert. 1993. Light fraction and macroorganic matter in mineral soils. P. 397-407 *In* M.R. Carter (ed.) Soil sampling and methods of analysis. Lewis Publishers, Boca Raton, FL.

Sheldrick, B.H. 1984. Analytical methods manual. LRRI Contrib. No. 84-30, Agric. Can., Research Br., Ottawa, ON.

Walker, B.D. and C. Wang. 1994. Benchmark site documentation: 05-AB (Provost, Alberta). Centre for Land and Biological Resources Research, Agriculture and Agri-Food Canada, CLBRR Contribution No. 95-03. 22 pp.

Wang, C. 1982. Application of transect method to soil survey problems. Tech. Bull. 1984-4E. Agric. Can., Research Br., Ottawa, ON. 34 pp.

Relating Corn/Soybean Yield to Variability in Soil and Landscape Characteristics

B. R. Khakural
P. C. Robert
D. J. Mulla

Department of Soil, Water and Climate
University of Minnesota
St. Paul, Minnesota

ABSTRACT

Relationships between landscape characteristics, soil properties, and corn/soybean yields were studied across a glacial till landscape in west central Minnesota. Decreases in corn/soybean yields were observed at eroded crest/backslope relative to foot or upper toeslope positions. Depth to free $CaCO_3$, surface available P, relative elevation, and slope gradient were found to be the most important variables affecting corn and soybean yields.

INTRODUCTION

Soil properties vary across landscapes due to differences in topographic variables, parent material, and soil development. Systematic variation of soil properties across landscapes are controlled by geomorphic and hydrologic processes acting on the landscape. Erosion and sedimentation processes exert a strong influence on soil properties (Kleiss, 1970; Malo et al., 1974). Soil properties such as solum thickness, thickness of A horizon, organic matter content, soil pH, cation exchange capacity, bulk density, soil profile wetness and plant nutrients vary with landscape position (Brubaker et al., 1993; Buol et al., 1989; Kleiss, 1970; Malo et al., 1974; Mulla et al., 1992). Slope aspect, slope curvature and slope position affect soil water storage (Hanna et al., 1982; Sinai et al., 1981).

Variation in soil fertility and hydrologic properties across landscapes also affects crop yield. Greater crop yields were obtained in footslope positions compared to the backslope and sideslope positions in western Iowa (Spomer & Piest, 1982). Increased yields on footslopes were attributed to deposition of soil, organic matter, and nutrients from upslope positions and additional plant available water. In Nebraska, increased corn (*Zea mays* L.) yields were reported on the upper and lower interfluve and footslope positions while decreased yields were reported on upper and lower linear slopes (Jones et al., 1989). Soils on lower landscape positions have been found to contain more available water than on upper landscape positions (Hanna et al., 1982).

The pioneering work of Ruhe and Walker (1968) has led to a wealth of information relating variability in crop yields to landscape position. Our understanding of the effects of landscape position on soil nutrient availability is relatively poor in comparison. Variations in landscape characteristics and soil properties across landscapes and their effects on soil productivity and crop yield need to be assessed for site-specific management of plant nutrients. The objectives of this research were: 1) to study variations in soil properties, landscape characteristics, and corn/soybean yields across a glacial till landscape in west central Minnesota, 2) to study the relationships between landscape characteristics, soil properties, and corn/soybean yields, and 3) develop soil management maps for crop yield based upon pattern or variation for the most significant soil or landscape parameter.

MATERIALS AND METHODS

Site Selection

This study was conducted nearby Cyrus, Minnesota. Cyrus receives an average rainfall of 62.2 cm annually and has a mean annual temperature of 6.1 °C. A 40 m by 426 m hillslope area including summit, backslope, sideslopes, and depressional positions, was selected. The experimental site includes soils with different erosion phases (slightly, moderately, and severely eroded), and drainage characteristics. There is a 3.6 m difference in elevation between the lowest and highest points in the landscape. Barnes (fine-loamy, mixed Udic Haploboroll) and Langhei (fine-loamy, mixed calcareous, frigid Typic Udorthent) loams (2-6% eroded) are the dominant upland soils. They are formed in calcareous loamy glacial till (USDA-SCS, 1972). The Langhei soil has a thin, grayish, calcareous surface layer and occurs on the crests of the hills or on the upper part of the slopes. The Barnes soil has a relatively thicker darker colored surface layer and occurs on more uniform slopes. The dominant soils at the foot and toeslope positions are Flom (fine-loamy, mixed, frigid Typic Haplaquoll) and Parnell (fine, montmorillonitic, frigid Typic Argiaquoll) loams, respectively. The Flom soils are formed in slight depressions while the Parnell soils are formed in enclosed potholes. The experimental field is surface ditch drained.

Crop Management

A cropping system with a corn-soybean (*Glycine max* L.) rotation was practiced. The field was chisel plowed in the fall (1993) and tandem disked in the spring (1994). In the fall of 1993, a fertilizer mix of N, P, and K was broadcast at the rate of 11, 56, and 56 kg ha^{-1}, respectively. Corn variety Ciba 4123 was planted on 10 May 1994 at the rate of 68,000 plants ha^{-1}. A starter fertilizer with N, P, K, Zn was banded at the rate of 11, 54, 39, and 1.1 kg ha^{-1} respectively, at planting. Anhydrous ammonia was applied at the rate of 129 kg N ha^{-1}. After harvesting corn, the experimental field was moldboard plowed in the fall of 1994 and worked with a field cultivator in the spring of 1995. A fertilizer mix of N, P, and K was broadcast on 5 May 1995 at the rate of 12, 58, and 67 kg ha^{-1} respectively. Soybean variety Ciba 3103 was planted on 21 May 1995 at the rate of 66 kg ha^{-1}. Two center rows of corn/soybean were harvested from each transect using a plot combine. Yields were recorded for every 15 m segment.

Elevation Survey and Field Data Collection

Elevation survey of the experimental site was conducted using a Survey Theodolite (Geodimeter 136 Geotronics AB, Danderyd, Sweden). Four south-north oriented parallel transects 426 m long and 10 m wide were selected for soil description and sampling. At each transect, depth of A horizon, and depth to free $CaCO_3$ were recorded at 15 m intervals. Soil profile (0 to 1 m) water content was monitored with a neutron probe (Model 503, Campbell Pacific Nuclear, Martinez, CA) throughout the growing season at 15 day intervals. A total of 14 neutron probe access tubes were installed in two transects. They were distributed across the soil-landscape based on landscape position and soil differences.

Soil sampling and Laboratory Procedure

Soil samples were collected from each transect at 30 m intervals. The first year (1994), soil samples were collected on 6 June from four different depths (0-15, 15-30, 30-60, 60-100 cm) to determine particle size, pH, organic matter, and available phosphorus and potassium. Only surface (0-15 cm) samples were collected on 25 May during the second year (1995) for determining soil pH, and available P and K. Samples were crushed and passed through a 2 mm sieve. The weight of coarse fragments (> 2 mm) was recorded. Soil core samples were also collected from neutron probe installation sites for determining soil bulk density and water content to calibrate neutron probe readings.

Particle size analysis was performed using a hydrometer method (Gee & Bauder, 1986). Soil organic matter content was determined using a modified Walkley Black method (Nelson & Sommers, 1986). Soil pH was measured in a 1:1 soil:water suspension, Available P was extracted by the Bray and Kurtz (1945) method and measured with a spectrophotometer. Exchangeable K was extracted with 1 $\underline{N}$ ammonium acetate (at pH 7.0) and determined with a flame photometer.

Calculation of Topographic Variables

Topographic variables such as slope gradient, slope aspect, profile curvature, and plan curvature were calculated using ANUDEM and TAPES-G (Terrain Analysis Program for the Environmental Sciences-Grid) programs (Center for Resource and Environmental Studies, Australian National University).

Statistical Analysis

Correlation and multiple regression analyses were performed to relate corn or soybean yields, soil properties and topographic variables (SAS, 1985). Both backward, and stepwise multiple regression procedures were used to select models for predicting corn/soybean yields (Freund & Littell, 1991). The criteria for variable addition or deletion was p = 0.05.

RESULTS AND DISCUSSION

Soil Properties and Crop Yield Variation Across the Landscape

Corn and soybean yields averaged 10.7 and 3.8 Mg ha^{-1} in 1994 and 1995, respectively (Table 1). Average yields were relatively high due to the high level of management, above average precipitation and warmer growing season temperatures. Cyrus received 64 and 85 cm of precipitation annually in 1994 and 1995, respectively. Coefficients of variation for corn and soybean yields were quite small (7.1 and 7.4). This may be due to spatial averaging caused by harvesting in 15 m blocks. Soil properties were more variable than crop yields partly because they represent point samples. The variability in soil properties was greatest for surface coarse fragments and least for soil pH. Depth to CaCO$_3$ and surface available Bray P also had relatively high coefficients of variability.

The depth of dark colored A horizon varied from 15 to 20 cm at the eroded crest, backslope and sideslope positions to as high as 75 to 80 cm at footslope positions (Table 1, Fig. 1a). The depth to free CaCO$_3$ varied from 0 (exposed at the surface) at the severely eroded crest/sideslope to greater than 150 cm at the footslope/toeslope positions (Fig 1b). Surface (0-15 cm) organic matter contents were highest at footslopes and lowest at the eroded crest and at toeslope

Table 1. Summary of simple statistics for selected soil properties, topographic variables and crop yields.

Variable	Mean	Std Dev	Range	CV (%)
Corn yield (Mg ha^{-1})	10.7	0.76	9.1-12.1	7.1
Soybean yield (Mg ha^{-1})	3.8	0.28	3.2-4.4	7.4
Depth to free CaCO$_3$ (cm)	59.2	44.37	0-150	71.4
A horizon thickness (cm)	32.5	13.80	15-80	42.5
Available P$_A$ (mg kg^{-1}) 1994	24.6	17.60	1-81	71.6
1995	30.1	16.42	10-95	54.5
Available K$_A$ (mg kg^{-1}) 1994	196.6	51.24	122-370	26.1
1995	168.6	35.70	101-263	21.2
Water storage $_{AV}$ (cm)	30.2	4.44	25-38	14.7
Organic matter$_A$ (%)	5.5	1.50	2.9-9.9	27.1
PH$_A$	7.5	0.59	6.3-8.3	7.9
Relative elevation (m)	1.9	1.21	0.0-3.7	64.6
Slope gradient (%)	2.2	1.12	0.4-4.8	50.4
Coarse fragments$_A$ (%)	1.9	2.16	0.0-9.8	112.2
Coarse fragments$_{AV}$ (%)	4.2	2.31	0.0-9.8	55.8
Sand$_A$ (%)	29.3	5.02	16-39	17.1
Sand$_{AV}$ (%)	32.7	8.84	9-47	27.0
Clay$_A$ (%)	26.7	6.00	18-38	22.5
Clay$_{AV}$ (%)	27.4	3.50	18-34	12.8
Silt$_A$ (%)	44.1	7.37	28-59	16.7
Silt$_{AV}$ (%)	39.9	8.33	26-62	20.9

†$_A$ = Surface horizon (0-15 cm) $_{Av}$ = Average for 0-100 cm

positions (Fig. 1c). However, average organic matter contents in the 0-1 m soil profile were highest at the toeslope position in the landscape (data not presented). Soils at the lowest position of the landscape are calcareous at the surface and have buried A (Ab) horizons. This may be due to deposition of eroded sediments from upslope or related to past management.

Generally, soil profile (0-1 m) silt and clay contents were greater and sand content was less at the foot or toeslope positions (or concave depressions) than at summit, backslope, and sideslope positions (data not presented). Soil surface pH varied from slightly acid (6.1-6.5) to moderately alkaline (7.9-8.3) (Table 1, Fig. 1d). Moderately alkaline surface pH was observed on eroded crest/backslope positions and at the toeslope positions. Soils at the sideslope and footslope had neutral (6.5 to 7.3) to slightly alkaline (7.3 to 7.8) surface pH.

Available soil test P and K levels varied greatly across the landscape (Fig. 2). Low (5-10 mg kg^{-1}) and medium (10-15 mg kg^{-1}) levels of Bray available P were observed at the eroded crest and at the toeslope positions with moderately alkaline surface pH (Fig. 2a, 2b), respectively. Bray soil test P levels were high (15-20 mg kg^{-1}) to very high (> 20 mg kg^{-1}) at the footslope and sideslope positions with slightly acid

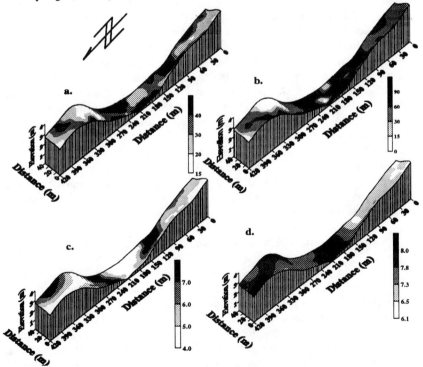

Fig. 1. Spatial variability in depth of A horizon (a), depth to free CaCO$_3$ (b), surface organic matter content (c), and surface pH (d) across the studied landscape.

to mildly alkaline (6.1 to 7.3) surface pH. In general, available K was greater at the footslope and toeslope positions than in upland summit, backslope and sideslope positions (Fig. 2c, 2d). Available K levels were high (120 - 160 mg kg^{-1}) to very high (> 160 mg kg^{-1}) throughout the entire soil-landscape studied.

Corn and soybean yields were least at the lower toeslope position and at the eroded crest and sideslope positions (Fig. 3). Soils at these locations were calcareous at the surface with thinner A horizons, moderately alkaline surface pH, and low surface organic matter contents (Fig. 1). Greatest corn yields were observed at footslope and upper toeslope positions (Fig 3a). Decreased corn yield at the eroded crest/sideslope may be due to poor soil water availability, nutrient imbalance, and limitations to root growth. Decreased corn and soybean yield at the lower toeslope position may be due to excess water and nutrient imbalance. Excessive amount of $CaCO_3$ and mildly alkaline soil pH may have affected the fertility of the soil by decreasing P and micro nutrient availability. Water storage was greatest at the lowest point of the landscape and least at the eroded crest and backslopes (data not presented)

Soil Properties in Relation to Landscape Characteristics

Tables 2 and 3 show the correlation between soil properties, topographic variables, and corn/soybean yields. Relative elevation had a positive correlation with soil profile (0 to 1.0 m) sand content and a negative correlation with soil profile silt content, clay content and water storage (Table 2). It had a negative correlation with surface available P, available K, A horizon thickness, and depth to free $CaCO_3$. Soil profile sand content was positively correlated, but clay content was negatively correlated, with slope gradient. There was a significant negative correlation ($p \geq 0.05$) between slope gradient and surface pH, available P and K in 1995, but no significant correlation was observed with these variables in 1994 (Tables 2 and 3). Surface available K, A horizon thickness, depth to $CaCO_3$, soil water storage, and soil profile clay content were positively correlated, while soil profile silt content was negatively correlated with profile curvature. Surface pH, surface clay content, and surface coarse fragments were positively correlated, but surface silt content was negatively correlated, with slope aspect. (Table 3).

Corn/soybean Yield in Relation to Soil and Landscape Characteristics

Corn yield was positively correlated with depth to free $CaCO_3$, depth of A horizon, surface (0-15 cm) available P, slope gradient, and water storage in the 0-1 m soil profile (Table 2). It was negatively correlated with relative elevation, surface coarse fragments, and slope aspect. Soybean yield was positively correlated with depth to free $CaCO_3$, surface silt content, surface organic matter content, slope gradient, and profile curvature (Table 3). It was negatively correlated with surface soil pH, surface coarse fragments, surface clay content, and slope aspect. Larson (1986) also reported an inverse correlation for small grain yield with depth of $CaCO_3$ and pH, and a positive correlation with organic matter.

Multiple regression models for predicting corn/soybean yields are presented in Table 4. Depth to free $CaCO_3$ (including a quadratic term), surface available P, available K, relative elevation, and slope gradient explained 65% of the variability in

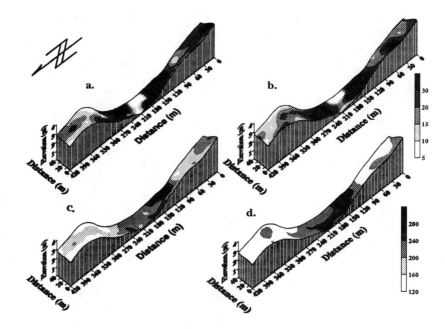

Fig. 2. Spatial variation in surface available P (a-b) and K (c-d) across the
studied landscape for 1994 and 1995.

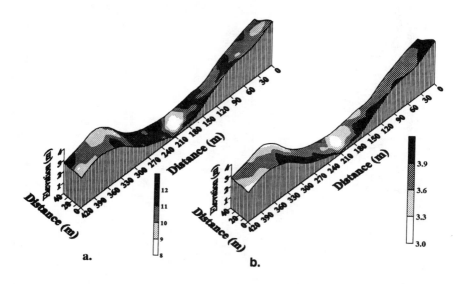

Fig. 3. Spatial variation in corn (a) and soybean (b) yields across the studied
landscape for 1994 and 1995, respectively.

Table 2. Correlation coefficients (r values) between soil properties, topographic variables, and corn yield in 1994.†

	CY	DC	DA	PA	KA	WS	OMA	PHA	RE	SG	SASP	PRC	CFA	SAV	CAV	SiAV
CY	1.00	0.57**	0.36**	0.41**	0.14	0.31*	0.23	0.04	-0.42**	0.28	0.28	0.25	-0.28	0.04	0.07	-0.07
DC		1.00	0.64**	0.41**	0.14	0.49**	0.14**	-0.04	-0.47**	-0.10	-0.18	0.27*	-0.29*	-0.34*	0.31*	0.24**
DA			1.00	0.33*	0.40**	0.43**	0.42**	-0.01	-0.42**	-0.24	-0.12	0.26	-0.26*	-0.65**	0.21	0.60**
PA				1.00	0.70**	0.28*	0.37**	-0.27*	-0.28*	-0.02	-0.38*	0.27*	0.41**	0.03	0.15	-0.07
KA					1.00	0.55**	0.35**	-0.31*	-0.49**	-0.23	-0.27	0.35*	0.10	-0.38*	0.23*	0.33*
WS						1.00	0.16	0.19	-0.91**	-0.26	-0.04	0.50**	-0.45**	-0.44**	0.39**	0.30*
OMA							1.00	-0.19	-0.21	0.08	-0.18*	0.22	0.41**	-0.31*	0.05	0.31*
PHA								1.00	-0.20	-0.25	0.54**	-0.10*	-0.13	-0.11	-0.17	0.17
RE									1.00	0.18	0.06	-0.53**	-0.10	0.40**	-0.30*	-0.30*
SG										1.00	-0.20	0.08	0.30*	-0.14	0.06	0.24
SASP											1.00	-0.30	0.05	0.37	-0.23*	-0.04
PRC												1.00	-0.30	-0.14	0.38**	0.04**
CFA													1.00	-0.10	-0.06	-0.92**
SAV														1.00	-0.32	-0.06
CAV															1.00	-0.32
SiAV																1.00

† CY = Corn yield, DC = Depth of dark colored A horizon, PA = Bray's available P (0-15 cm), KA = Available K (0 - 15 cm), DA = Depth to free calcium carbonate, WS = Water storage (0 - 1 m), OMA = Organic matter content (0 - 15 cm), pHA = Soil pH (0-15 cm), RE = Relative elevation, SG = Slope gradient, SASP = Slope aspect, PRC = Profile curvature, CFA = Coarse fragments (0-15 cm), SAV = Average sand content (0 - 1 m), CAV = Average clay content (0 - 1 m), SiAV = Average silt content (0 -1 m)
* = Significant at the 0.05 level ** = Significant at the 0.01 level

Table 3. Correlation coefficients (r values) between soil properties, topographic variables, and soybean yield in 1995.†

	SY	DC	DA	PA	KA	WS	OMA	pHA	RE	SG	SASP	PRC	CFA	SA	CA	SiA
SY	1.00	0.39**	0.23*	-0.07	0.03	0.05	0.29**	-0.45**	0.02	0.48**	-0.29*	0.33*	-0.48**	0.10	-0.29	0.30
DC		1.00	0.64**	0.08	0.31*	0.52**	0.14**	-0.10	-0.47**	-0.10	-0.18	0.27*	-0.29*	-0.20*	0.01	0.13
DA			1.00	0.18	0.34**	0.37**	0.42**	-0.18	-0.42**	-0.24*	-0.12	0.26	-0.26*	-0.39**	0.10	0.19
PA				1.00	0.65**	0.61**	0.02	0.15	-0.62**	-0.35*	0.13	0.24	0.20	-0.06	0.10	-0.02
KA					1.00	0.61**	0.23	-0.04	-0.58**	-0.29	-0.12	0.35*	-0.05	-0.18	-0.01	0.14
WS						1.00	0.15	0.09	-0.91**	-0.26	-0.10	0.49**	0.14	-0.12*	0.20	0.07
OMA							1.00	-0.25	-0.21	0.08	-0.10	0.22	0.41**	-0.47**	0.02	0.31*
pHA								1.00	-0.10	-0.30	0.50**	-0.05	-0.17	0.37*	0.17	0.38*
RE									1.00	0.18	0.06	-0.53**	0.41**	0.13	-0.22	0.08
SG										1.00	-0.20	0.08	-0.14	0.15	-0.16	0.03
SASP											1.00	-0.30	0.30*	-0.08	0.31	0.29*
PRC												1.00	-0.05	-0.13	-0.06	0.14*
CFA													1.00	0.42*	0.28	-0.50**
SA														1.00	-0.09	-0.61**
CA															1.00	-0.74**
SiA																1.00

† SY = Soybean yield, DC = Depth of dark colored A horizon, PA = Bray's available P (0-15 cm), KA = Available K (0 - 15 cm), WS = Water storage (0.1 m), OMA = Organic matter content (0-15 cm), pHA = pH (0-15 cm), RE = Relative elevation, SG = Slope gradient, SASP = Slope aspect, DA = Depth to free calcium carbonate, PRC = Profile curvature, CFA = Coarse fragments (0-15 cm), SA = Sand content (0 - 15 cm), CA = Clay content (0 - 15 cm), SiA = Silt content (0-15 cm)
* = Significant at the 0.05 level ** = Significant at the 0.01 level

Table 4. Regression models for predicting corn and soybean yield.†

Crop	Regression model	Model R^2
Corn	$CY = 11.272 + 0.0199* D_C - 0.000081 (D_C)^2 + 0.0185* P_A$ $- 0.0077 * K_A - 0.292 * RE + 0.127 * S_G$	0.65^{**}
Soybean	$SY = 2.906 + 0.0083* D_C - 0.0000376 * (D_C)^2 + 0.00395* P_A$ $+ 0.093 * RE + 0.104 * S_G + 1.550 * PRC$	0.78^{**}

† CY = Corn yield (Mg ha⁻¹), SY = Soybean yield (Mg ha⁻¹), D_C = Depth to free $CaCO_3$ (cm),
OM_A = Surface (0 - 15 cm) organic matter (%), pH_A = surface pH, P_A = Surface available P (mg kg⁻¹),
K_A = Surface available K (mg kg⁻¹), RE = relative elevation (m), S_G = Slope gradient (%), S_{ASP} = Slope
aspect (NE = 1, NW = 2, SW = 3 SE = 4), PRC = Profile curvature
** = Significant at the 0.01 level

corn yield. Depth to free $CaCO_3$ (including a quadratic term), surface available P, relative elevation, slope gradient, and profile curvature explained 78% of the soybean yield variability. Depth to free $CaCO_3$ was observed to be the most significant variable affecting to crop yields, explaining 38 and 46% of the variability in corn and soybean yields, respectively. Earlier studies from South Dakota (Gollany et al., 1992) and Montana (Burke, 1984) also identified depth to $CaCO_3$ as the most important variables in predicting corn and small-grain yields, respectively. Depth to $CaCO_3$ was positively correlated with A horizon thickness and soil water storage in the 1.0 m soil profile (Tables 1 and 2).

Management Groupings

The experimental field was divided into three management zones based on corn yield, soybean yield or depth to free $CaCO_3$ (Fig. 4a-c). Depth to free $CaCO_3$ was used in management groupings because it was identified as the most significant predictor of corn/soybean yields in the multiple regression analysis. The frequency distributions for corn yield, soybean yield or depth to $CaCO_3$ were examined, and cutoff values were used representing the mean plus or minus one half standard deviation. Three yield classes used for dividing the experimental field into management zones were: less than 10 Mg ha⁻¹ (low), 10 - 11 Mg ha⁻¹ (medium), and greater than 11 Mg ha⁻¹ (high) for corn, and less than 3.6 Mg ha⁻¹ (low), 3.6 - 4.0 Mg ha⁻¹ (medium), and greater than 4.0 Mg ha⁻¹ (high) for soybean, respectively. Three $CaCO_3$ depth classes used in management grouping were: shallow (< 40 cm), moderate (40 - 80 cm), and deep (> 80 cm).

Management zones based on corn or soybean yields differed significantly in mean corn or soybean yields, respectively (Table 5), which is not surprising since these groupings were based upon maximizing differences in yield. However, the yield class grouping does not maximize differences in soil properties. Management zones based on soybean yield showed small differences in these soil properties. For both management grouping systems based on yield classes, depth to free $CaCO_3$ was significantly different between management zones. Surface organic matter content was significantly less, but surface pH was significantly greater for zone 1 (low yield potential) than for zones 2 (medium yield potential) and 3 (high yield potential). There were no significant differences in available soil test P and K levels between management zones developed from soybean yields. When management zones were

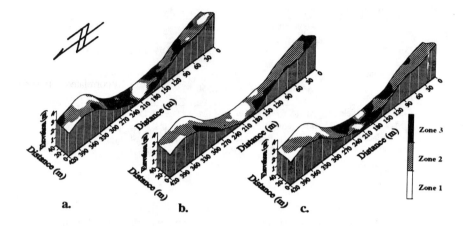

Fig. 4. Spatial locations for soil management groups based on corn yield (a), soybean
yield (b), or depth to free $CaCO_3$ (c).

Table 5. Comparison of mean soil properties and corn/soybean yields in management zones
divided on the basis of corn yield, soybean yield, or depth to $CaCO_3$ †

Zone	Yield class	Measured yield Corn	Soybean	OM	pH	P	K	D_A	D_C	RE.	S_G
		---- Mg ha^{-1} ----		%		---mg kg^{-1}---		---- cm ----		m	%
				Based on corn yield							
1	Low	9.6a		4.7a	7.5ab	20a	197a	24a	17a	2.1a	1.6a
2	Medium	10.6b		5.6b	7.3a	21a	193a	35b	55b	2.2a	2.2b
3	High	11.5c		5.9b	7.6b	31b	201a	35b	87c	1.3b	2.5b
				Based on soybean yield							
1	Low		3.45a	4.7a	7.9a	34a	172a	26a	30a	1.6a	1.6a
2	Medium		3.82b	6.0b	7.3b	28a	165a	38b	74b	1.8a	2.2a
3	High		4.18c	5.6b	7.2b	29a	172a	29a	70c	2.2a	3.1b
				Based on depth to $CaCO_3$							
1	Low	10.1a	3.51a	5.0a	7.9a	16a	175a	24a	12a	2.0a	2.1a
2	Medium	10.8b	3.87b	5.6ab	7.2b	24b	189a	30b	55b	2.3a	2.4a
3	High	11.3c	3.84b	6.0b	7.5b	36c	244b	49c	133c	0.7b	1.9a

†Means within a column and management grouping system followed by the same letter are not
significantly different at the p = 0.05 level.
OM = Organic matter, D_A = A horizon thickness, D_C = Depth to free $CaCO_3$.

based on corn yields, there were no significant differences in available K levels between
zones, but zone 3 had significantly greater available P than zones 1 and 2.

Management zones based on depth to free $CaCO_3$ had greater differences for soil
properties such as A horizon thickness, available P and K than management zones
based on corn or soybean yield classes. They were significantly different in depth to
free $CaCO_3$, A horizon thickness, available P and K. Surface organic matter content was
significantly less in zone 1 than in zone 3. Surface pH was significantly greater for
zone 1 than for zones 2 and 3. Management zones based on depth to $CaCO_3$ differed

significantly in corn yields, but there was no significant difference in soybean yields between zones 2 and 3.

SUMMARY AND CONCLUSION

Landscape position and slope characteristics were strongly correlated with spatial patterns in soil, horizon development, soil fertility, and crop yield in west central Minnesota. More of the variability in yield was controlled by permanent soil features such as depth to $CaCO_3$ than by variability in soil fertility. Soil sampling for variations in soil fertility would have explained only small portion of the variability in crop yield for this study. The experimental field was divided into three management zones based on corn yield, soybean yield and depth to free $CaCO_3$. The management grouping system based on corn/soybean yields maximized differences in yield between management zones. However, it did not maximize differences in soil properties. The management grouping system based on depth to $CaCO_3$ showed greater differences in soil properties between management zones than grouping systems based on crop yields and may be more useful in fertility management. Dividing a field into management zones based upon differences due to yield monitoring may obscure differences in soil fertility. Sampling and management strategies for precision farming should consider the relationships between landscape characteristics, soil development, soil fertility, and soil productivity.

ACKNOWLEDGMENT

The authors wish to thank Mr. Jay Dudding for his help with field work and Dr. Jay Bell for his assistance with calculation of topographic variables. The cooperation of collaborating farmer, Mr. Gary Paynes is gratefully appreciated. This project was funded in part by the USDA-NRCS grant no. 68-6322-3-25 and the Minnesota Corn Research and Promotion Council.

REFERENCES

Bray, R. H., and L. T. Kurtz. 1945. Determination of total organic and available forms of phosphorus in soils. Soil Sci. 59:39-45.

Brubaker, S. C., A. J. Johnes, D. T. Lewis, and K. Frank. 1993. Soil properties associated with landscape positions. Soil Sci. Soc. Am. J. 57:235-239.

Buol, S. W., H. D. Hole, and R. J. McCracken. 1989. Soil genesis and classification. Third edition. Iowa State University Press, Ames, Iowa.

Burke, T. H. 1984. Evaluating selected soil morphological, classification, climatic, and site variables that influence dryland small grain yield on Montana soils. M. S. thesis, Montana State Univ., Bozeman.

Freund, R. J. and R. C. Littell, 1991. SAS Systems for Regression. SAS Series in Statistical Applications. Second edition. SAS Inst. Inc., Cary, NC.

Gee, G. W. , and J. W. Bauder. 1986. Particle-size analysis. p. 383-441. *In* A. Klute (ed.) Methods of soil analysis. Part 1. 2nd ed. Agron. Monogr. 9. ASA and SSSA, Madison, WI.

Gollany, H. T., T. E. Schumacher, M. J. Lindstrom, P. D. Evenson, and G. D. Lemme. 1992. Topsoil depth and desurfacing effects on properties and productivity of a Typic Argiustoll. Soil Sci. Soc. Am. J. 56:220-225.

Hanna, A. Y., P. W. Harlan, and D. T. Lewis. 1982. Soil available water as influenced by landscape position and aspect. Agron. J. 74:999-1004.

Jones, A. J., L. N. Mielke, C. A. Bartles, and C. A. Miller. 1989. Relationship of landscape position and properties to crop production. J. Soil Water Conserv. 44:328-332.

Kleiss, H. J. 1970. Hillslope sedimentation and soil formation in northeastern Iowa. Soil Sci. Soc. Am. Proc. 34:287-290.

Larson, M. H. 1986. The influence of soil series on cereal grain yield. M. S. Thesis. Montana State Univ., Bozeman.

Malo, D. D., B. K. Worcester, D. K. Cassel, and K. D. Matzdorf. 1974. Soil landscape relationships in a closed drainage system. Soil Sci. Soc. Am. Proc. 38:813-818.

Mulla, D. J., A. U. Bhatti, M. W. Hammond, and J. A. Benson. 1992. A comparison of winter wheat yield and quality under uniform versus spatially variable fertilizer management. Agric.Ecosys. Environ. 38:301-311.

Nelson, D. W., and L. E. Sommers. 1986. Total carbon, organic carbon, and organic matter. p. 539-577. In A. L. Page (ed.) Methods of soil analysis. Part 2, 2nd ed., Agron. Monogr. 9. ASA and SSSA, Madison, WI.

Ruhe, R. V. , and P. H. Walker. 1968. Hillslope models and soil formation: I. Open systems. p. 551-560. In J. W. Holmes (ed.) Trans. Int. Congr. Soil Sci. 9th, Adelaide. 1968. Vol. 4. Elsevier, New York.

SAS, 1985. Statistical analysis system. User's Guide : Statistics, Version 5 edition. Statistical Analysis System Institute Inc. Cary, NC.

Sinai, G., D. Zaslavsky, and P. Golany. 1981. The effect of soil surface curvature on moisture and yield - Beer Sheba observations. Soil. Sci.132:367-375.

Spomer, R, G., and R. F. Piest. 1982. Soil productivity and erosion of Iowa loess soils. Trans. ASAE 25:1295-1299.

USDA-SCS. 1972. Soil Survey. Pope County, Minnesota. U. S. Government Printing Office. Washington, D. C.

Analysis of Spatial Factors Influencing Crop Yield

K. A. Sudduth
USDA-Agricultural Research Service
Cropping Systems and Water Quality Research Unit
Columbia, Missouri

S. T. Drummond
S. J. Birrell
N.R. Kitchen
University of Missouri
Columbia, Missouri

ABSTRACT

The spatial relationship between crop yields and soil and site parameters was modeled using several methods. Yield maps estimated by projection pursuit regression and neural network analysis agreed well with measured yields. These methods also allowed generation of response curves for estimated yield as a function of each of the input parameters. These response curves were useful for investigating the relationship between yields and individual soil and site parameters.

BACKGROUND

Understanding the functional relationship of crop yield to other spatial factors is a basic need for successful site-specific crop management (SSCM). A first approximation to this relationship can be obtained with conventional nutrient recommendation procedures (for example, Buchholz, 1983). However, these recommendation procedures are generally based upon response data averaged over a large geographic area, thus diluting the precision of the response relationship. To apply inputs with the precision needed for SSCM, it could be more appropriate to develop individual response functions for particular soils or soil associations, or perhaps even for a particular field, or for similar areas within a field.

Another shortcoming of the current nutrient recommendation procedures is they necessarily assume that all factors limiting yield are included in the recommendation process. When the procedures are applied on a point-by-point basis within a field, there may be areas in which crop growth and yield are limited by other factors, such as water availability. In these portions of the field, the current recommendation procedures will not accurately relate crop yield to spatial soil and site parameters.

The use of analysis techniques to predict yield from input parameters is of importance in developing SSCM methods and recommendation procedures, but even more important is the ability to use the techniques and models to understand yield response (or sensitivity) to changes in critical factors. One approach to developing such an understanding of these relationships is through the application of crop growth models (Hoogenboom et al., 1993). Another approach is based on empirical analysis of multivariate spatial data. We previously used the empirical approach to study the spatial relationship between soil properties and yields for a research field in central Missouri (Drummond et al., 1995). For this paper, we have expanded on the previous study and applied this approach to additional datasets.

OBJECTIVES

The overall objective was to study the relationship between yields and soil properties (i.e. nutrient availability, organic matter) and site properties (i.e. elevation) on a spatial basis. Specific objectives were to: (1) develop and/or evaluate methods for predicting spatial crop yields, and (2) use the model results to investigate the sensitivity of crop yield to variations in soil and site parameters.

MATERIALS AND METHODS

Data were collected on two fields, 36 ha and 28 ha in size, located near Centralia, in central Missouri. The soils of the area are characterized as claypan soils, primarily of the Mexico-Putnam association (fine, montmorillonitic, mesic Udollic Ochraqualfs). These soils are poorly drained and have a restrictive, high-clay layer (the claypan) occurring below the topsoil. The two fields were managed in a high yield goal, high input, minimum till corn-soybean rotation. Fertilizer and chemical inputs were applied at a single rate.

Data were obtained for one field (Field 1) in 1993 (corn), 1994 (soybean), and 1995 (grain sorghum). Grain sorghum was planted in this field in 1995, rather than corn, because an excessively wet spring delayed planting until mid-June. Yield data for the other field (Field 2) were obtained only in 1995 (soybean). Conditions for crop production were quite different between the three years. The 1993 growing season was characterized by heavy and frequent rains, with an annual precipitation of 157 cm. Yield reductions were observed in lower portions of the landscape, due to excess water. The 1994 precipitation of 82 cm was only slightly below average, but less than 5 cm of rainfall was received in July and August and crops experienced drought stress during much of the growing season. In 1995, precipitation was 115 cm, with an excessively wet planting season which again caused stand problems and some yield reductions in the lower portion of the landscape.

Data Acquisition

Data obtained on the study fields included grain yield, elevation, and a number of soil properties. Grain yield measurements were obtained using a full-size combine equipped with a commercial yield sensing system and global positioning system (GPS) receiver, using data collection and processing techniques described by Birrell et al. (1995). Detailed topographic data were obtained on each field, using a total station surveying instrument and standard mapping procedures.

Based on our previous work (Sudduth et al., 1995), topsoil depth above the claypan was estimated from soil conductivity. A mobile measurement system described by Kitchen et al. (1996) was used to obtain root-zone soil conductivity data with a commercial electromagnetic induction (EM) sensor. The actual depth of topsoil was measured at a set of randomly selected calibration points and a regression between topsoil depth and the inverse of soil conductivity was developed (Field 1: $r^2 = 0.90$, std. err. = 6.7 cm; Field 2: $r^2 = 0.89$, std. err. = 9.4 cm). These regressions were then applied to convert the EM data to topsoil depth.

Field 1 was soil sampled on a 30 m grid in the spring of 1995. A hand soil probe was used to collect soil cores to a 20 cm depth. Three soil cores obtained within a 1 m radius of each sample position were combined, oven dried and analyzed by the University of Missouri Soil and Plant Testing Services Laboratory. Soil properties measured were phosphorus, potassium, pH, organic matter, calcium and magnesium. Cation exchange capacity (CEC) and magnesium saturation were calculated according to standard procedures (Buchholz, 1983). Field 2 was soil sampled on a 25 m grid in the spring of 1996. Procedures were identical to those for Field 1, except that 8 cores were combined at each sample position.

Data Analysis

Yield and topsoil depth data were analyzed using geostatistics, and appropriate semivariogram models and parameters were used to krige the data to a grid with a 10 m cell size. Data from the grid cell centered closest to each soil sampling point was extracted and combined with the soil sample data for analysis. If any data was missing for a grid cell, that cell was eliminated from the analysis. The whole-field datasets ranged in size from 301 to 436 observations (Table 1).

Additional datasets were created for analysis by dividing each field into 5 sub-field areas on the basis of elevation and topsoil depth. It was thought that the relationship of yield to soil and site parameters might be more predictable within these areas than across the entire field. The two relatively static parameters of elevation and topsoil depth were chosen because previous analysis indicated that these had the most consistent impact on yields of all the measured parameters in the dataset. To create the sub-field areas, each field was first divided into areas of low (<25 cm), medium, and high (>50 cm) topsoil depth. The medium and high topsoil depth areas were then subdivided into the lower 1/3 of the landscape and the higher 2/3 of the landscape (Fig. 1). The sub-field datasets ranged in size from 14 to 232 observations (Table 1).

Standard correlation and stepwise multiple linear regression (SMLR) analyses were completed on each whole-field and sub-field dataset. Another analysis method used was projection pursuit regression (PPR). This nonparametric regression method requires only a few general assumptions about the shape of the regression surface, in contrast to the linearity constraints of SMLR (Friedman and Stuetzle, 1981). In PPR, the regression response (yield in this case) is modeled as the sum of a set of general (nonlinear) smooth functions of linear combinations of the independent (soil and site property) variables. In the SMLR and PPR analyses, yield data for each field-year were regressed against seven soil and site variables – phosphorus, potassium, pH, organic matter, topsoil depth, CEC, and elevation. The

Table 1. Correlations between yields and soil and site properties for whole-field and sub-field areas.

Field/Year	Area	# Obs	P	K	pH	Organic Matter	CEC	Ca	Mg	Mg Sat.	Topsoil Depth	Elevation	Slope	Curvature
Field1 1993	all	318	-0.08	-0.01	0.07	0.06	-0.15*	-0.04	-0.17*	-0.16*	-0.13*	-0.01	-0.05	-0.08
	lt	16	-0.46*	-0.74*	0.64*	-0.27	-0.60*	-0.20	-0.45*	-0.21	-0.28	0.48*	-0.06	-0.25
	mtle	67	0.14	0.11	0.44*	0.31*	-0.25*	0.07	-0.18	-0.05	0.06	0.14	-0.22*	-0.08
	htle	43	-0.03	-0.10	-0.26*	0.08	0.07	-0.18	-0.12	-0.21	0.15	0.04	0.08	0.08
	mthe	170	-0.09	-0.03	-0.01	-0.02	-0.16*	-0.12	-0.21*	-0.25*	-0.10	-0.14*	-0.04	-0.09
	hthe	22	-0.61*	-0.10	-0.31	-0.22	-0.31	-0.33	-0.67*	-0.77*	-0.41*	-0.47*	-0.33	0.17
Field1 1994	all	344	0.09	-0.05	0.13*	-0.14*	-0.16*	-0.08	-0.12*	-0.10*	0.39*	-0.22*	-0.10*	0.11*
	lt	17	0.45*	0.41*	-0.57*	0.03	0.13	-0.25	-0.08	-0.27	0.13	-0.24	-0.10	-0.09
	mtle	71	0.57*	0.31*	0.27*	0.14	-0.06	0.00	0.07	0.23*	0.08	0.03	0.01	0.14
	htle	47	0.40*	-0.08	0.10	-0.10	0.12	0.27*	0.33*	0.43*	0.52*	-0.65*	-0.07	0.01
	mthe	187	0.11	0.03	0.33*	-0.17*	-0.25*	-0.03	-0.21*	-0.21*	0.07	-0.02	-0.22*	0.14*
	hthe	22	0.17	-0.20	0.31	0.36*	-0.34	0.02	-0.09	0.08	0.09	0.52*	-0.40*	-0.40*
Field1 1995	all	301	0.06	-0.02	0.05	0.16*	-0.06	-0.07	-0.05	0.01	-0.03	0.26*	-0.09	-0.06
	lt	14	-0.28	-0.31	0.17	0.16	-0.20	-0.18	-0.09	0.08	-0.03	0.08	-0.01	-0.02
	mtle	72	-0.47*	-0.55*	-0.04	-0.32*	-0.44*	-0.61*	-0.36*	-0.16	0.14	-0.03	-0.22*	-0.19
	htle	40	-0.44*	-0.38*	-0.16	-0.11	-0.35*	-0.62*	-0.39*	-0.27*	-0.36*	0.34*	0.48*	-0.17
	mthe	161	-0.00	-0.05	0.03	0.18*	0.07	0.06	0.01	0.04	0.05	0.24*	-0.10	-0.01
	hthe	21	-0.01	-0.44*	-0.01	0.32	-0.11	-0.12	-0.06	0.01	0.12	0.11	-0.10	-0.54*
Field2 1995	all	436	0.05	-0.37*	-0.10*	-0.43*	-0.48*	-0.52*	-0.44*	-0.34*	-0.09*	0.49*	-0.08*	-0.16*
	lt	68	0.19	-0.28*	-0.12	-0.25*	-0.42*	-0.49*	-0.35*	-0.24*	0.27*	0.57*	-0.25*	-0.18
	mtle	81	-0.52*	-0.28*	0.06	-0.21*	-0.11	-0.02	-0.10	-0.01	-0.33*	0.48*	0.52*	-0.29*
	htle	26	-0.38*	-0.42*	0.04	-0.21	-0.32	-0.27	-0.34*	-0.28	-0.44*	-0.21	-0.41*	-0.22
	mthe	232	-0.05	-0.05	-0.24*	-0.09	-0.09	-0.24*	-0.01	0.11	0.37*	-0.02	0.05	-0.15*
	hthe	29	-0.01	-0.76*	-0.22	-0.56*	-0.66*	-0.73*	-0.61*	-0.52*	-0.62*	-0.43*	0.27	0.18

Area designations: lt = low topsoil depth; mtle = medium topsoil, low elevation; htle = high topsoil, low elevation; mthe = medium topsoil, high elevation; hthe=high topsoil, high elevation. Starred correlations are significant at the 0.10 level.

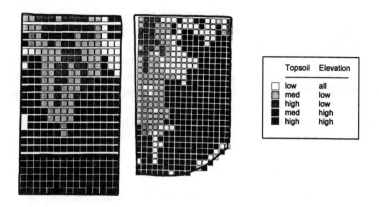

Topsoil	Elevation
low	all
med	low
high	low
med	high
high	high

Fig. 1. Sub-field areas classified by topsoil depth and elevation.

other original variables were not used in this analysis to reduce problems associated with colinearity. On these fields, we found that calcium, magnesium, and magnesium saturation were highly correlated with CEC, and slope was correlated with topsoil depth.

A neural network analysis was also used to model the data. Neural networks are computing systems composed of simple, highly interconnected processing units which respond to the sum of inputs from all connections in accordance with an activation function (Hopfield and Tank, 1986). For this study, a backpropagation network (BPN) with a sigmoid activation function was chosen. With this nonlinear activation function, the learning process for the network was nonlinear and nonparametric. The BPN was designed with three layers; an input layer consisting of seven input neurons, a hidden layer consisting of ten hidden neurons, and an output layer consisting of a single output neuron. Each layer was fully connected with the following layer in a feed-forward arrangement. Relatively few hidden neurons were used since this allowed training to be completed in a reasonable time frame, and also because a smaller number of neurons tends to limit the ability of the network to overfit the data. The dataset was randomly divided into training and testing sets for cross-validation as a further means to guard against overfitting. The BPN was trained with each field-year dataset in a separate session using the same seven input variables as were used in the SMLR and PPR analyses.

RESULTS AND DISCUSSION

Yield patterns for Field 1 varied considerably from year to year. Visual comparison of yield maps and soils maps from this field allowed us to associate some, but not all, yield patterns with soil variations. For example, the spatial pattern of 1994 yields showed some similarity to the pattern of topsoil depth variation across the field (Drummond et al., 1995).

Whole-field statistical correlations of yields to soil and site parameters (Table 1) were difficult to interpret in a way that yielded meaningful information.

We concluded that these problems were likely due to a complex and nonlinear functional relationship between yield and soil properties, which was not well represented by the correlation statistic. Also, the form of the yield response function was likely different from region to region within the field, due to different factors controlling the expression of yield. In some cases, correlations calculated on a sub-field basis were more significant than those calculated on a whole-field basis (Table 1). However, this was not always the case, and it was apparent that linear correlation analysis was not the best approach to discerning the functional relationships between yield and soil and site properties.

Regression and Neural Network Analysis

SMLR analysis was applied to the yield and soil datasets, both for the entire field and for the five sub-field areas defined above. The best significant models selected for each field included from 4 to 6 variables, with r^2 values ranging from 0.13 to 0.43 (Table 2). Elevation was the only model variable common to all four datasets. Topsoil depth, organic matter, and phosphorus were also significant variables in a majority of cases. Although the SMLR analysis gave some insight into the relationships between yield and soil properties, the linearity constraints imposed by this model meant that it could not accurately describe the data.

Nonparametric regression analysis by PPR provided significantly better estimates of yield than did SMLR analysis using the same input datasets (Table 2). Estimated yield maps based on PPR compared well with actual yield maps (Fig. 2). The best PPR estimations were obtained for field-years with well-defined, relatively large-scale yield patterns (Table 2, Fig. 2). For example, the areas of highest yield for Field 1 in 1994 were reproduced well, as were the areas of lowest yield for Field 2 in 1995. The correspondence of PPR estimates and actual yields was weaker when the spatial structure of actual yield was less well-defined, as was the case for Field 1 in 1993 and 1995.

The BPN approach also showed promise for estimating yields (Table 2). Predicted yield maps were similar to actual yield maps, and to the maps predicted by PPR for Field 1 in 1994 and Field 2 in 1995 (Fig. 2). The BPN maps reproduced the same major features as did the PPR maps, although at a lower level of accuracy. In general, the BPN-derived maps showed less localized variation in yield than those created with PPR. Yields estimated by BPN for Field 1 in 1993 and 1995 were less accurate. Some improvement in BPN yield estimates might be possible with additional refinement of network training parameters and procedures.

Table 2. Model statistics for estimation of yield data as a function of seven soil and site parameters, using different analysis methods.

	Field 1 1993	Field 1 1994	Field 1 1995	Field 2 1995
Stepwise multiple linear regression (SMLR)				
r^2	0.13	0.25	0.15	0.43
std. error, Mg/ha	0.65	0.25	0.60	0.41
Projection pursuit regression (PPR)				
r^2	0.51	0.68	0.56	0.77
std. error, Mg/ha	0.48	0.16	0.43	0.26
Backpropagation neural network (BPN) analysis				
r^2	0.12	0.55	0.20	0.52
std. error, Mg/ha	0.64	0.19	0.57	0.38
Calibration dataset				
number of data points	318	344	301	436
mean yield, Mg/ha	7.30	1.63	5.25	2.20

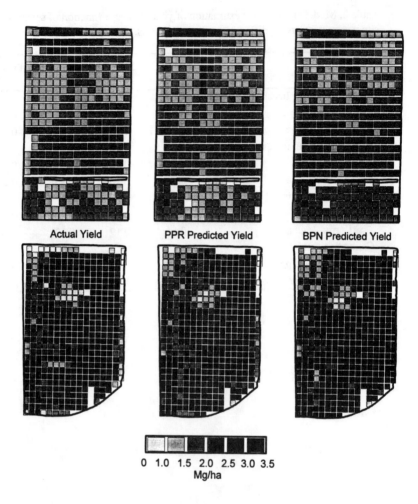

Fig. 2. Measured, PPR predicted, and BPN predicted soybean yields for Field 1 in
   1994 (top) and Field 2 in 1995 (bottom).

Yield Response Curves

The response of the PPR and BPN yield estimates to variations in the input
parameters was investigated on a point-by-point basis. Sensitivity analyses were
conducted by holding all but one of the model input parameters constant and
varying the other parameter from a minimum to a maximum value. All response
curves for each sub-field area were then combined into a mean response curve for
that area. For generation of the response curves, all variables were normalized to
a field-year mean of zero and unity standard deviation. This facilitated comparison
of yield responses to the different soil and site parameters.

Response curves generated by the PPR and BPN analyses were similar overall. For example, the general response to elevation for the 1995 Field 2 yield data was positive for both methods (Fig. 3). For this and most other parameters, the PPR curves exhibited more local variation than did the BPN curves. Since r^2 values for the PPR analyses were higher than those for the BPN (Table 2), this may reflect better modeling of actual trends in the data. On the other hand, further investigation of the results is needed to be sure that the PPR analysis is not overfitting the data.

The PPR response curves appeared to successfully model major yield-limiting factors, as shown by 1995 data from Field 2. For example, higher soybean yields were related to increases in elevation within the field (Fig. 3), with the strongest response found in the sub-field areas of lower elevation. This trend was caused by the excess rainfall in the spring of 1995, which caused significant problems with crop stands in the low-lying areas of the field. Yield decreases indicated at the highest elevations were likely caused by the presence of a tree line which reduced yield in that area.

Yield response to topsoil depth was large and positive in the low and medium topsoil areas of the field (Fig. 4). The response was negative in the high topsoil areas because the locations of greater topsoil accumulation were generally the same locations where standing water reduced crop stand early in the season. Yield response to higher levels of phosphorus (Fig. 4), potassium, and lime were generally negative for this field-year. Most areas of the field were sufficient in these nutrients, and the negative relationship may have been due to mining of nutrients in the more productive areas of the field.

Differences could be observed when comparing the 1995 soybean response curves for Field 2 (Fig. 3 and 4) to the 1994 soybean response curves for Field 1 (Fig. 5). For example, in low-elevation areas the response of soybean yield to changes in elevation was negative for 1994, but positive for 1995. Since crop growth was water-limited in 1994, the run-on areas at low elevation were at an advantage; while in 1995 similar run-on areas had crop growth limited by excess water. The phosphorus response was also different between the two field-years, with a positive response to higher P observed in the lower landscape areas during 1994. Soil test P levels were lower on Field 1, and significant areas of the field were within the responsive range. The low- elevation portions of the field showed greater response, since crop growth was not limited by low water availability in those areas.

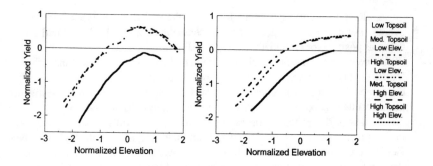

Figure 3. Comparison of PPR-estimated (left) and BPN-estimated (right) soybean yield responses to elevation for Field 2 in 1995.

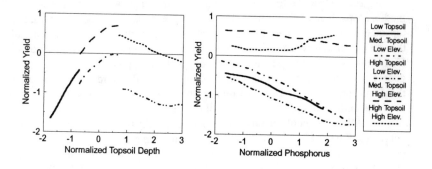

Figure 4. PPR-estimated soybean yield responses to topsoil depth (left) and soil test phosphorus (right) for Field 2 in 1995.

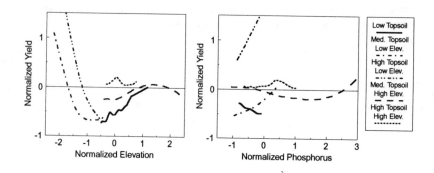

Figure 5. PPR-estimated soybean yield responses to elevation (left) and soil test phosphorus (right) for Field 1 in 1994.

CONCLUSIONS

The process of understanding yield variability was made difficult by the number of inter-related factors that affect yield. Correlation analysis was not particularly useful in understanding yield variability, due to complex nonlinear relationships between yield-limiting factors. Dividing the field into smaller sub-field areas based upon topsoil depth and elevation did not measurably improve the ability of this method to explain yield variability, but other methods of sub-field classification could be investigated in hopes of finding regions that respond homogeneously to input factors. Stepwise multiple linear regression was not able to accurately model yields. As with standard correlation, this was likely due to linearity constraints.

Prediction capabilities were highest for the nonlinear, nonparametric methods. Yield response curves developed with these methods agreed in general with our observations of yield-limiting behavior on these fields. These response curves will be useful for studying the interactions between multiple critical factors and crop yields. The fact that similar predictions and response curves were obtained from two dissimilar methods (PPR and BPN) provides some degree of confidence that the methods are modeling the association between yield patterns and soil properties in a reasonable manner. Further study and refinement of both methods is needed, to optimize their use for yield response investigation.

REFERENCES

Birrell, S.J., K.A. Sudduth, and S.C. Borgelt. 1995. Crop yield mapping: comparison of yield monitors and mapping techniques. Comp. Elect. Agric. 14:215-233.

Buchholz, D.D. 1983. Soil Test Interpretations and Recommendations Handbook. Univ. of Missouri, Coll. of Agric., Dept. of Agronomy, Columbia, MO.

Drummond, S.T., K.A. Sudduth, and S.J. Birrell. 1995. Analysis and correlation methods for spatial data. Paper No. 951335. ASAE, St. Joseph, MI.

Friedman, J.H., and W. Stuetzle. 1981.Projection pursuit regression. J. Am. Stat. Assoc. 76:817-823.

Hoogenboom, G., H. Lai, and D.D. Gresham. 1993. Spatial yield prediction. Paper No. 933350. ASAE, St. Joseph, MI.

Hopfield, J.J., and D.W. Tank. 1986. Computing with neural circuits: a model. Science 233:625-632.

Kitchen N.R., K.A. Sudduth, and S.T. Drummond. 1996. Mapping sand deposition from 1993 midwest floods with electromagnetic induction measurements. J. Soil Water Cons. 51(4):336-340.

Sudduth, K.A., N.R. Kitchen, D.F. Hughes, and S.T. Drummond. 1995. Electromagnetic induction sensing as an indicator of productivity on claypan soils. *In* Proc. Second Intl. Conf. on Site-Specific Management for Agricultural Systems, pp. 671-681. Am. Soc. Agron., Madison, WI.

Consistency And Change In Spatial Variability Of Crop Yield Over Successive Seasons: Methods Of Data Analysis

R.M. Lark
J. V. Stafford

Silsoe Research Institute
Wrest Park
Silsoe
Bedford   MK45 4HS
United Kingdom

ABSTRACT

Automated pattern recognition by cluster analysis is proposed as a tool for interpreting the complex information on temporal and spatial variability which is contained in a sequence of yield maps of one field. The method was evaluated on yield maps for three successive seasons on one experimental field. It was shown that a few general patterns of season to season variation could be identified (including consistently above average yields). A tentative interpretation of some of these patterns was possible.

INTRODUCTION

Since Mercer and Hall (1911) and other early studies, it has been known that yields of crops exhibit spatial variability within fields. There is increasing interest in whether this variability implies that there could be advantage in managing arable fields in a spatially variable way rather than treating them according to their average conditions (see various papers in Robert et al, 1994). In particular, inputs such as fertilisers might be used more efficiently and with reduced environmental impact if their dose rates were varied spatially to match local requirements.

Determining local requirements for fertiliser or other management activities is not simple. One possibility is that a map of yield variability might be a useful source of information. To this end, several yield mapping systems have been devised for combinable crops (eg. Vanischen and De Baerdemaker, 1991; Stafford et al, 1996). As the crop is harvested, positions are obtained from the GPS satellite system while the rate of mass flow of grain within the combine is measured by a sensor. This information is then processed to produce a map of yield.

A yield map, however, represents retrospective information on the accumulated effects of many spatially variable factors. Some interpretation is necessary, for only when the principal sources of variation have been identified can a management response be defined. An *a priori* categorisation of possible sources of variation may be useful.

Some sources of variability will be *ephemeral*, a unique pattern in one season (eg. patches of wind-borne disease or deer damage). This variation in a yield map may not be relevant in the following season.

Other sources of variation may be *medium term*, persistent over a few seasons only. For example, the quantity and quality of organic residues will differ between parts of a field previously used to grow different crops. If such effects are predictable they may be relevant to management in the subsequent season.

Other sources of variation may be *consistent factors* with *season-dependent expression*. For example, the significance of available water capacity of the soil will depend on the weather. Such sources of variation may be unpredictable, and so difficult to manage.

Still other sources of variation may be *consistent factors of the environment expressed in most seasons*. For example, headland compaction may reduce yields in most years and certain weed species may occur in persistent patches (Miller et al, 1995). Such consistent sources of variation are most likely to be amenable to spatially variable management.

It would seem likely that yield maps will be most useful when several seasons' data can be compared. A set of yield maps contains a lot of information; a first step in its interpretation would be to identify a few general patterns of temporal variation in yield such as "regions of consistently high yield" or "regions with high yield in all but dry years". If this can be achieved by automated pattern recognition then subsequent interpretation may be assisted, for example, by comparing soil conditions in the different regions.

In this paper, the use of cluster analysis is described for identifying generalised patterns of temporal variation in yield. The techniques are applied to sequences of yield maps for three fields. In one case, the classes identified are compared to information on soil variation in the field.

MATERIALS AND METHODS

Data sets

Yield maps were obtained during the harvests of 1993-1995 for Cashmore Field at Silsoe Research Institute using the system described by Stafford et al (1996). Additional data were available. These were local weather records (precipitation and Class A pan evaporation), soil map of the field to soil series (described in Table 1) and measurements of soil properties made at 100 sample sites in March 1995 (percent moisture, organic matter, pH and mineral nitrogen for depths 0 to 20 cm and 20 to 80 cm).

Cluster analysis

Conventional clustering (non-hierarchical) establishes a set of k classes among a set of N objects with measurements of n variables. The classes are groups of objects which resemble each other with respect to the variables. The analysis is achieved by minimising:

$$\sum_{i=1}^{N} \sum_{j=1}^{k} d_{i,j}^2 \, \mu_{i,j} \qquad (1)$$

where $d_{i,j}$ is a measure of resemblance between object i and class j (often the Euclidean or Mahalanobis distance in n-dimensional space) and $\mu_{i,j}$ is a value for the *membership* of object i in class j. In conventional analysis:

$$\mu_{i,j} \in \{0, 1\} \qquad (2)$$

and

$$\sum_{j=1}^{k} \mu_{i,j} = 1.0 \qquad (3)$$

That is to say an object belongs to one and only one cluster (Webster & Oliver, 1990; Chapter 11 discuss the methods).

Increasingly, environmental scientists have come to regard such a "crisp" approach to classification as artificial, given the essentially continuous variability of natural phenomena (Burrough, 1989). A limited number of "patterns" may be recognised in such data and individual objects resemble these to differing degrees. This view of classification can be accommodated by a "fuzzy generalisation" of cluster analysis (see Bezdek et al, 1984). This replaces condition (2) above with (4) below and retains condition (3).

$$0.0 \leq \mu_{i,j} \leq 1.0 \qquad (4)$$

Membership is now a continuous variable representing a grade of resemblance between an object i and a "fuzzy" class j such that $\mu_{i,j} = 1.0$ denotes complete resemblance. A more comprehensive account is given by McBratney and Moore (1985).

Table 1 : Soil Map Units on Cashmore Field*

Soil Series	Parent material	Comments
Lowlands	Colluvium over Lower Greensand	well drained
Hallsworth	Drift over Lower Greensand	
Nercwys	Drift over Lower Greensand	
Evesham	Gault clay	calcareous clay
Bardsey	Gault clay	fine loam, non-calcareous
Enborne	Alluvium	gley
Fladbury	Alluvium	gley

* See Stafford et al (1996)

Bezdek et al (1984) give an algorithm for fuzzy clustering (FCM) which, for specified k and a "fuzziness" exponent m (which denotes how "broad" or "narrow" the classes can be), obtains an optimal set of membership values in each of k classes for N objects. Each class, j, can be described by its *centre* - a vector of values of the n variables representing the "typical" class member (and mathematically equivalent to the vector of weighted means of the n variables for all N objects where the weight for object i is its membership in class j, $\mu_{i,j}$).

Data analysis

For each sequence of yield maps, n is the number of maps. The N objects are points in the field for which n separate yield values are recorded. Yield values for each harvest were standardised to zero mean and unit variance. Cluster analysis was then carried out using the FCM algorithm of Bezdek et al (1984) for $k = 2, 3$, ..., 10. Both Euclidean and Mahalanobis similarity measures were used. The Normalised Classification Entropy (NCE) (see McBratney & Moore, 1985) was calculated for each analysis and that one selected for which NCE is a minimum. If NCE did not show a single clear minimum value, a parsimony rule was applied and the clustering into k groups selected for which $NCE_{k-1} > NCE_k \approx NCE_{k+1}$. If two or more minima occurred that for which k is smallest was selected.

The output of the selected analysis includes a set of k membership values for each object. Following the procedure of McBratney et al (1992), sets of membership values were interpolated at nodes on a 5 m square grid. The procedure ensures that conditions (3) and (4) above are preserved. Delaunay triangulation with linear interpolation was used to create the grid. A map showing the class of maximum membership was generated from these data. Maps of membership in any one class could also be produced.

The map of classes of maximum membership was overlaid on the map of soil series. The class of maximum membership and the soil map unit were recorded for 500 randomly located sites in a contingency table. The association between these categories and between clusters and soil parent material was tested by a chi-squared approximation.

The cluster of maximum membership was identified for each of the 100 soil sampling sites. Differences among the mean values of the classes (maximum membership) were tested by a random effects ANOVA and the intra-class correlation of the properties (proportion of variance attributable to class differences).

RESULTS

The cluster centres are shown in Fig. 1. The standardised yield values corresponding to the center of each cluster are shown for each year. Figure 2 shows a map of the cluster of maximum membership at points across the field.

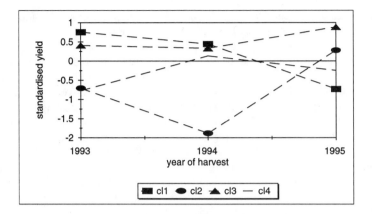

Fig. 1. Centers of fuzzy clusters from yield maps for three seasons.

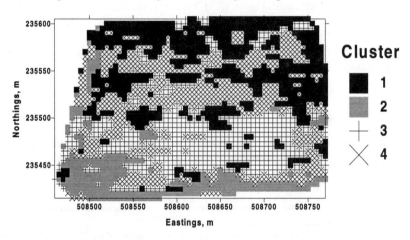

Fig. 2. Spatial distribution of the cluster of maximum membership.

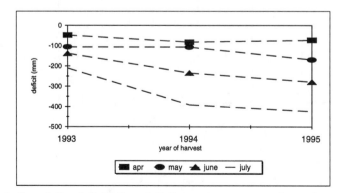

Fig. 3. Cumulative deficit of precipitation over pan evaporation at the end of successive months for years of harvests.

Figure 3 shows the cumulative deficit of pan evaporation over precipitation at the ends of April, May, June and July in each of the three years.

One cluster (3) has center elements which correspond to yields above average in all three years (and the highest of all clusters in 1995). Another cluster (1) has centre values corresponding to the highest yields in 1993 and 1994 but the lowest in 1995. The behaviour of this cluster might be tentatively related to the seasonal differences in evaporation/precipitation balances, suggesting that cluster 1 represents the limiting effect of dry conditions in Spring and Summer 1995.

Table 2 shows the contingency table for soil series (as mapped) and the cluster of maximum membership at 500 locations. The top figure in each cell is the number of locations corresponding to it and the lower figure is the standardised residual from the expectation under a null hypothesis of no association (Minitab, 1988). A positive residual indicates an association. Table 3 shows the X^2 statistic for the full table (sum of squared standardised residuals) and the p value for the null hypothesis assuming an approximation to χ^2. The same statistics are shown for the cluster/soil parent material table. X^2 for this statistic is a component of that for Table 2.

There is strong evidence that the cluster of maximum membership is significantly associated with soil series and that this is primarily explained by the association with soil parent material (X^2 for the combined table is 90% that of the full table).

Table 2 : Cashmore field. Contingency Table of Soil series and Cluster of maximum membership for 500 random locations

(see text for explanation)

		Soil series						
		Lowlands	Hallsworth	Nercwys	Evesham	Bardsey	Enborne	Fladbury
Cluster of maximum membership	1	97 4.1	9 0.6	33 2.7	4 -2.7	10 -3.2	3 -3.9	0 -1.5
	2	3 -4.0	1 -1.0	0 -2.6	9 2.0	8 -0.3	26 7.2	5 5.0
	3	30 -3.7	1 -2.2	10 -2.1	17 1.2	54 6.1	28 2.1	2 0
	4	75 1.7	13 2.2	24 0.9	15 0.4	13 -2.5	10 -2.3	0 -1.5

Table 3 : Results of χ^2 analysis on Table 2 and combined and sub-tables

Null hypothesis	X^2	degrees of freedom	p
no association of cluster of maximum membership with:			
soil map unit	251.4	18	<0.001
map units grouped by parent material	226.4	6	<0.001
soil map units over:			
Lower Greensand	6.5	6	0.17
Gault Clay	10.8	3	0.01
Alluvium	3.2	3	0.16

**Table 4 : Results of random effects ANOVA on soil properties
by cluster of maximum membership**

Soil property	p value for null hypothesis	r_i (intra-class correlation)
percent water (0-20 cm)	<0.001	0.35
percent water (20-80 cm)	<0.001	0.36
pH (0-20 cm)	0.12	0.04
pH (20-80 cm)	0.13	0.04
organic matter (0-20 cm)	0.04	0.08
organic matter (20-80 cm)	0.32	<0.01
mineral nitrogen (0-20 cm)	0.22	0.02
mineral nitrogen (20-80 cm)	0.21	0.02

Having rejected the null hypothesis for clusters/parent material, association of clusters and series *within* each parent material was tested (see Table 3). A null hypothesis of no association is accepted for series over alluvium and over Lower Greensand but rejected for series over Gault Clay.

The tables for parent material and series within parent material are not shown. However, the former indicates an association of clusters 1 and 4 with series primarily over the Lower Greensand and 2 and 3 with the series over the other materials. Within Gault Clay, clusters 2 and 4 are associated with the Evesham map unit and cluster 3 with the Bardsey unit.

Table 4 shows that soil moisture (both depths, in March) is significantly related to cluster of maximum membership with over a third of the variance accounted for. For 0 to 20 cm, the cluster means (percent water) and 95% confidence intervals are 20.6 ± 0.76, 27.5 ± 6.9, 24.6 ± 1.3 and 22.0 ± 1.3 respectively. Of the remaining properties only soil organic matter (0 to 20 cm) is significantly related to the cluster of maximum membership but only eight percent of variance is explained.

DISCUSSION AND CONCLUSIONS

On Cashmore field, the areas of high yield in 1993 and 1994 are represented by clusters 1 and 3, which are associated respectively with some of the soils over Lower Greensand and the lighter of the map units occurring over Gault Clay (Bardsey series).

Highest relative yield in 1995 is represented by cluster 3 (mainly associated with Gault Clay and the Bardsey map unit) and lowest yields were represented by cluster 1 (mainly associated with the Lower Greensand). It is also noted that the cluster of maximum membership at a site explains a significant proportion of soil moisture at both depths sampled and that the mean percent moisture is lowest for sites where cluster 1 has maximum membership. These results are consistent with the tentative observation that the between-season variation represented by cluster 1 reflects the evaporation/precipitation deficits in the seasons. It is possible, then, that a region has been identified in which soil moisture is the major limiting factor on yield. Cluster 4 may also represent sites limited by available moisture in 1995 when its central relative yield was lower than in previous seasons.

It has been shown that automated classification enables some interpretation of the complexity of yield variation by identifying areas within which the crop performs similarly in successive seasons. These can then be related to soil and environmental variables and some advances may be made in identifying limiting factors.

ACKNOWLEDGEMENTS

The contribution of B. Ambler and H. C. Bolam to data collection at Silsoe is acknowledged. The soil survey of Cashmore field was conducted by Professor J. Catt of Rothamsted Experimental Station. Funding from MAFF and the EC is acknowledged.

REFERENCES

Bezdek, J.C., R. Ehrlich, W. Full. 1984. FCM: The fuzzy c-Means clustering algorithm. Computers and Geosciences, 10:191-203.

Burrough. P.A. 1989. Fuzzy mathematical methods for soil survey and land evaluation. J. Soil Science, 40: 477-492.

Mercer, W.B., A.D. Hall. 1911. The experimental error of field trials. J. Agric. Sci. 4:107-132.

McBratney, A.B., A.W. Moore. 1985. Application of fuzzy sets to climatic classification. Agric. and Forest Meteorology, 35: 165-185.

McBratney, A.B., J.J. De Gruitjer, D.J. Brus. 1992. Spacial prediction and mapping of continuous soil classes. Geoderma, 54: 39-64.

Miller, P.C.H., J.V. Stafford, M.E.R. Paice, L.J. Rew. 1995. The patch spraying of herbicides in arable crops. Proc. Brighton Crop Protection Conference, Nov. 1995, British Crop Protection Council, 3:1077-1086.

MINITAB.   1988.   Minitab reference manual, Release 6.   Minitab Inc.,
    Philadelphia.
Robert, P. C., R.H. Rust, W.E. Larson (eds). 1994. Proceedings of site-specific
    management for agricultural systems. 2nd International Conference, 27-30
    March 1994, Minneapolis.
Stafford, J.V., B. Ambler, R.M. Lark, J. CattJ. 1996. Mapping and interpreting the
    yield variation in cereal crops. Computers and Electronics in Agriculture,
    14 (2/3): 101-119.
Vanischen, R., J. DeBaerdemaeker. 1991. Continuous wheat yield measurement
    on a combine. pp. 346-355.   *In* Proceedings ASAE Symposium on
    Automated Agriculture for the 21st Century, 16-17 December 1991,
    Chicago, Ill., ASAE, St. Joseph, MI
Webster, R., M.A. Oliver. 1990. Statistical Methods in Soil and Land Resource
    Survey. Oxford University Press, Oxford.

Multivariate Analysis as a Tool for Interpreting Relationships Between Site Variables and Crop Yields

A.P. Mallarino
Department of Agronomy
Iowa State University
Ames, Iowa

P.N. Hinz
Department of Statistics
Iowa State University
Ames, Iowa

E.S. Oyarzabal
Ceregen, a unit of Monsanto Co.
Urbandale, Iowa

ABSTRACT

Precision farming technologies allow for collection of large amounts of data from producers' fields. This research used factor analysis to study relationships among many site variables and corn yields. Correlated variables were grouped into soil fertility, weed control, and conditions for early growth factors. Their importance in explaining the yield variability differed greatly among fields.

BACKGROUND

Newly introduced precision farming technologies and related practices allow for collection of large amounts of data from producers' fields. Soil chemical and physical properties, climatic data, incidences of diseases, pests or weeds, and crop yields are the most common variables recorded using these technologies. The data usually are georeferenced and can be organized into several layers of information. The collected data often are processed into maps by means of a variety of gridding and interpolation techniques to depict the spatial distribution of the variables. Further analysis of these data is useful for understanding relationships among site variables and between these variables and crop yields at a scale that could not be achieved before. These relationships could be studied by various statistical procedures.

Simple correlation and regression statistical analyses are usually applied to the data. The relationships between georeferenced measurements can be further studied by a combination of graphical and numerical analytical tools referred to as geographical information systems (GIS) or by spatial statistical methods. Classic and spatial statistics are useful to study relationships between variables although the basic assumptions and the focus differ. The correlation analysis emphasizes relationships between variables independently of the spatial distribution. The spatial statistical methods emphasize the spatial correlation and spatial distribution of variables.

Simple correlation analyses show that many site variables often, but not always, are correlated with crop yields and that often some variables are correlated among themselves. If correlated site variables are used in a multiple regression analysis to explain crop yields, the correlations make it difficult to interpret the regression equation (Bowerman and O'Connell, 1990). The problem is that the value of the regression coefficient for one variable changes depending on what other variables are used in the equation. Moreover, tests of significance of the coefficients become unreliable when variables are highly correlated. Multivariate analysis techniques could partly circumvent the problems created by correlated variables and could facilitate the interpretation of potentially complex relationships.

The problem caused by correlated variables can be minimized by grouping variables so that the correlation of two variables from the different groups is small and that for two variables from the same group is large. Each group can be represented by a new variable which is created from the variables in the group. These new variables can be used as independent variables in a multiple regression equation. The correlations among the new variables will be small if the groupings are successful, and the interpretation problems of multiple regression will be minimized. Groups of correlated variables can be defined by using principal component analysis and factor analysis (Johnson and Wichern, 1982). Factor analysis often is more successful at identifying groups of correlated variables because it is an analysis of covariances whereas principal component analysis is an analysis of variances.

The objectives of this study were to illustrate the use of multivariate analysis to determine how corn yields from various producers' fields were related to several site variables. This was accomplished by using factor analysis to form groups of correlated site variables, defining new variables that represent the groups, and using these variables in multiple regression analyses to explain yields.

METHODS

Data were collected from three cornfields. Soil types, rainfall, and management practices varied considerably among fields. Tillage practices and the corn hybrid were uniform within each field, although soil types varied within fields. Moreover, parts of some fields had different management histories (i.e., different crops and soil or crop management practices). This situation is common in the Corn Belt because of increases of farm and field sizes.

Soil samples and various soil and plant measurements were collected from each field following a grid-point sampling scheme. Composite (8 cores) soil samples (0 to 15-cm depth) were collected from 9-m^2 sampling points at the intersection of grid lines spaced 15 m. The total area sampled at each field ranged from 3 to 6 ha. There were 176, 117, and 258 sampling points for Fields 1, 2, and 3, respectively. Soil tests included organic matter, pH, and available P (Bray-1), K (ammonium acetate), and N-NO_3. Other measurements on each sampling area were visual indexes of microtopography of the sampling area (level, concave, or convex), of the area covered by crop residues from previous years, of broadleaf weeds, and of grass weeds; height of corn plants 4 and 8 weeks after emergence; and corn stands. To avoid the usual uncertainty about the accuracy of yield measurements collected with yield monitors from small areas, grain yields from the sampling areas were collected by hand.

Simple correlation analyses and factor analyses were performed for measurements collected from each field. Factors were extracted with the factor procedure of the SAS package (SAS Inc., 1988) using the principal factor analysis method. The initial factor loadings were rotated using the Promax oblique (nonorthogonal) rotation method. The factors shown correspond to those with eigenvalues greater than one. New variables, called latent variables, were created by standardizing and averaging the variables with largest loadings within each factor. To study the relationships between the underlying factors and yield, multiple regression models were fit for each field. Grain yield was the dependent variable and the latent variables were the independent variables. The models were of the form $Y = b_0 + b_1 F_1 + b_2 F_2 + b_3 F_3 + \epsilon$, where Y represents estimated corn yields, b_0 to b_3 are coefficients, F_1 to F_3 are the latent variables created based on the factors, and ϵ represents residual error.

RESULTS AND DISCUSSION

The three fields sampled represent commonly found cornfields in the Corn Belt. The mean values and the variability for most measurements were different across fields (Table 1).

Table 1. Descriptive statistics for measurements made in three cornfields.

Field	OM	P	K	pH	NI	RE	H1	H2	GR	BL	CS	Y
						Mean						
1	5.4	88	243	6.5	45	0	23	118	12	4	62948	11.2
2	3.7	20	107	6.6	26	37	46	111	5	1	62266	10.1
3	5.6	45	213	6.2	51	54	32	109	6	6	64852	12.5
						Maximum						
1	6.6	124	385	6.8	188	5	37	286	60	20	87150	15.1
2	4.5	40	160	7.3	45	46	96	168	20	10	72087	13.1
3	8.2	100	476	6.8	152	95	48	147	75	90	75315	17.7
						Minimum						
1	3.0	56	147	6.1	8	0	12	53	0	0	43037	6.0
2	2.9	9	76	6.1	12	30	7	54	0	0	39809	5.3
3	3.3	12	114	5.3	11	5	19	65	0	0	46265	7.3
						Standard deviation						
1	0.5	13	32	0.2	28	1	7	27	9	6	7332	1.7
2	0.4	5	18	0.3	9	3	21	27	8	2	5373	1.6
3	0.8	17	59	0.3	28	18	7	23	8	11	4620	1.4

† OM=organic matter (%); P=soil P, K=soil K, and NI=soil N-NO$_3$ (mg/kg); RE=residue, and GR and BL=grass or broadleaf weeds (area covered, %); H1 and H2=corn height (cm); CS=corn stand; and Y=grain yield (Mg/ha).

Many site variables were significantly correlated among themselves and with corn yield. The variables involved in high correlations, however, varied markedly among fields and variables correlated in one field were not correlated in another. For example, soil P and K were correlated in Field 1 but were not correlated in Field 2. Soil K was negatively correlated with yield in Field 1 but was positively correlated with yield in Field 2. Similar observations apply to correlations between other measurements. There were various degrees of correlation among soil, plant, and yield measurements within each field. As an example, Table 2 shows the correlation coefficients for Field 2.

Table 2. Simple correlation coefficients for the variables measured in Field 2.

VAR[†]	\multicolumn{12}{c}{Variable[†]}

VAR[†]	OM	P	K	pH	NI	TO	RE	H1	H2	GR	BL	CS
P	0.05											
K	0.28	-0.01										
pH	-0.36	0.43	-0.55									
NI	-0.43	0.08	-0.02	-0.06								
TO	0.27	0.11	0.11	-0.12	-0.07							
RE	-0.02	0.03	-0.01	0.10	0.11	-0.03						
H1	0.15	0.19	0.47	-0.18	0.03	0.02	0.05					
H2	0.29	0.08	0.78	-0.62	0.09	0.14	0.00	0.62				
GR	0.11	-0.09	-0.05	-0.13	-0.36	-0.25	-0.18	0.03	0.04			
BL	0.04	-0.04	-0.04	-0.13	-0.27	-0.15	-0.15	-0.01	0.03	0.82		
CS	0.07	-0.28	0.35	-0.47	0.10	0.07	-0.01	0.19	0.44	-0.14	-0.13	
Y	0.34	-0.09	0.77	-0.62	0.01	0.04	0.02	0.44	0.82	-0.08	-0.05	0.59

† VAR=variable, OM=organic matter, P=soil P, K=soil K, NI=soil $N-NO_3$, TO= microtopography, RE=residue, H1 and H2=plant height, GR and BL=grass or broadleaf weeds, CS=corn stand, and Y=grain yield.
‡ Correlations greater than 0.18 differ significantly ($P<0.05$) from zero.

The lack of a consistent correlation between any two variables across different fields should not be surprising, and this result has been observed before (Pierce et al., 1994; Mallarino, 1994 and 1996). The values of the variables are usually the result of intrinsic variation in soil properties and management practices. For example, different histories of fertilization and different soil types could explain the lack of an overall correlation between two soil tests across fields. Also, it is possible that a particular variable is not related to yields in a field because the range of variation within that field is above or below the range in which it influences yields. This does not preclude that the correlation between any two variables could be similar across fields. For example, in areas with little history of fertilization, significant correlations could be expected between soil P, K, texture, and/or organic matter and between these and crop yields. The apparently unreasonable correlation found for some variables should not be surprising either. For example, the negative correlation found in Field 2 between soil pH and yield could represent a true effect of high pH in reducing corn yield, could (most likely given the range in pH values) be that soil pH is correlated to a nonmeasured variable that really influenced yields negatively, or could represent random error.

Groups of correlated variables were defined for each field using factor analysis. The basis for selecting the measured variables included within each latent variable is in partial correlation coefficients that often are called factor loadings. As an example for this presentation, Table 3 shows the rotated factor reference structure for Field 2. The variables with large (absolute value) factor loadings are more likely to represent an unobservable common factor. The fact that two or more variables are grouped in a latent variable suggests a possible common factor that makes them vary together within a field. The signs of the factor loadings provide information of how these variables relate when representing the common factor. There are no established or clear rules that help decide what is a "large" factor loading. In some instances, the latent variable is well defined and the factor loadings for some variables are clearly larger than those for other variables. In other instances, however, the latent variable is not as well defined.

Table 3. Rotated factor loadings for Field 2.

Variable	Factor 1	Factor 2	Factor 3
	------------ Coefficients ------------		
Organic matter	0.42	0.06	0.22
Soil P	0.34	0.71	-0.05
Soil K	0.70	-0.12	-0.04
Soil pH	-0.34	0.63	-0.10
Soil NO_3	-0.07	-0.11	-0.48
Microtopograhy	0.22	0.08	-0.15
Residue cover	0.02	0.07	-0.19
Corn height 1	0.64	0.20	0.01
Corn height 2	0.83	-0.08	-0.01
Grass weeds	0.00	-0.02	0.89
Broadleaf weeds	-0.01	-0.02	0.82
Corn stand	0.23	-0.42	-0.19

The latent variables identified for these fields are shown in Table 4. The interpretation of each latent variable is an important aspect of factor analysis and no general rules can be provided. In this study, the latent variables could be easily interpreted in some cases but not in others. The underlying common factor (an intrinsic soil property, a management practice, a climatic variable, or a combination of these factors) may not be readily obvious but the results provide a basis for speculation. For example, we interpreted the latent variable 3 in Fields 2 and 3 as "weed control" because it was mostly composed of the variables broadleaf and grass weeds. We interpreted the latent variable 1 in Field 1 as "conditions for early growth" because components with high factor loadings included plant height, corn stand, grass weeds, and soil pH. In Field 2, however, a similar interpretation was given to the latent variable that was composed of plant height, soil K, and soil organic matter. We believe that in these fields there was an underlaying factor that was influencing early growth of corn, although it was represented by different measurements. A more easily understood common factor related to soil fertility was interpreted for the latent variable composed of soil P, K, and organic matter in Field 3. The reason for the association of corn stand with soil P and soil pH in Field 2, however, is not obvious.

Table 4. Variables with largest factor loadings grouped into the latent variables.

	Latent variable		
Field	1	2	3
1	Height 1 and 2 Corn stand Grass weeds Soil pH	NA[†]	NA
2	Height 1 and 2 Soil K Organic matter	Soil P Soil pH Corn stand	Grass weeds Broadleaf weeds
3	Height 1 and 2 Soil pH Residue cover	Soil P Soil K Organic matter	Grass weeds Broadleaf weeds Soil nitrate

† NA = Not applicable (eigenvalues smaller than one).

The fact that several groups of correlated site variables could be identified for each field does not necessarily means that they explain yield variability. The R^2 values of multiple regression models of grain yield on the identified latent variables ranged from 0.28 to 0.71 across fields (Table 5). This result demonstrates that high variation in measured site variables does not necessarily explain highly variable crop yields. The results also show that the factors that are related to crop yields often vary among fields. Interpretation of the signs of the coefficients requires study of signs and relative weights of the factor loadings of the variables that compose each latent variable. For example, the negative signs for the "weed control" latent variables in Fields 2 and 3 seem reasonable because a high index for broadleaf or grass weed suggested a weed problem. This effect could have been overlooked by simple observation of a table of simple correlations because of the poor negative correlation between these variables and corn yield. The negative sign for the "soil fertility" latent variable in Field 2 seems unreasonable but could be explained by the positive correlation between soil P and pH and the negative simple correlations between these two soil variables and yields (Table 2). Further specific interpretations for each field are beyond the objectives of this report.

The lack of consistent correlations between variables across fields and in the proportion of yield variability accounted for by the latent variables can be explained by several reasons. One reason could be that corn yields were affected by one or more nonmeasured variables. Another reason could be the different mean values and ranges for the variables in different fields. Of course, cause and effect relationships should not be directly drawn from these relationships. The models and the latent variables that are significantly related to yield can be used to understand the reasons for yield variability, and this understanding can, in turn, be used to manage the fields better.

Table 5. Coefficients and statistics of multiple regression models relating grain yield with the latent variables for three cornfields.

Field	Intercept	Latent variable			R^2	$P > F$
		1	2	3		
1	11.23 (0.09)[‡]	1.71 (0.12)	NA[†]	NA	0.58	0.001
2	10.08 (0.08)	1.46 (0.12)	-0.69 (0.11)	-0.16 (0.09)	0.71	0.001
3	12.48 (0.07)	0.54 (0.09)	0.23 (0.09)	-0.77 (0.11)	0.28	0.001

† NA = not applicable (eigenvalues smaller than one).
‡ Numbers in parentheses are standard errors of the estimates.

CONCLUSIONS

This research illustrated the use of multivariate analysis for analyzing several layers of information from producers' fields. Factor analysis provided a rational criterion for including and arranging correlated variables in multiple regression models relating crop yields with site variables. The results showed, however, that only part of the yield variability could be explained by the contribution of the measured site variables, that the site variables involved in significant relationships varied among fields, and that high variation in some measured site variables does not necessarily explain highly variable crop yields. The results also show that the choice of site variables to be measured is very important because the variables that may explain yield variability probably are different across fields.

REFERENCES

Bowerman, B.L., and R.T. O'Connell. 1990. Linear statistical models. An applied approach. 2nd. edition. PWS-Kent Publishing Co. Boston, MS.

Johnson, R.A., and D.W. Wichern. 1982. Applied multivariate analysis. Prentice-Hall Inc. Englewood Cliffs, NJ.

Mallarino, A.P. 1994. Spatial variability of phosphorus and potassium in no-till corn and soybean fields. p. 115-121. *In* Proceedings, The Integrated Crop Management Conference. 30 Nov.-1 Dec. Iowa State Univ. Extension. Ames, IA.

Mallarino, A.P. 1996. Evaluation of optimum and above-optimum phosphorus supply for corn by analysis of young plants, leaves, stalks, and grain. Agron. J. 88:377-381.

Pierce, F.J., D.D. Warncke, and M.W. Everett. 1994. Yield and nutrient variability in glacial soils of Michigan. p. 133-150. *In* P.C. Robert et al. (ed.). Proceedings, 2nd. Int. Conf. on Site-Specific Management for Agricultural Systems. March 27-30, Minneapolis, MN. ASA, SSSA, ASA. Madison, WI.

SAS Institute. 1988. SAS/STAT User's Guide, Release 6.03 Edition. 1028 pp. SAS Institute Inc., Cary, NC. SAS Institute Inc., 1988.

Directed Soil Sampling

S. Pocknee
B.C. Boydell
H.M. Green
D.J. Waters
C.K. Kvien

NESPAL/University of Georgia
Coastal Plains Experiment Station
Tifton, GA

ABSTRACT

A predisposition to use systematic grid sampling techniques as a basis for fertilizer management currently exists in precision agriculture. Grid sampling techniques have a number of undesirable characteristics which make their use for commercial within-field soil mapping questionable. Directed sampling refers to a simple technique of incorporating prior knowledge about soil variability in to the sampling design to match sampling distribution and intensity with known soil patterns.

BACKGROUND

Grid Sampling

Systematic grid sampling refers to a process whereby a field is divided into many smaller uniform cells. These cells are individually sampled and the results are combined with information about the position of each sample to form field maps for the attributes measured. There are many variations of the basic cell pattern including square, rectangular, offset (diamond, hexagonal, triangular), and stratified systematic unaligned (Berry and Baker, 1968; Congalton, 1988; Petersen and Calvin, 1986; Webster and Oliver, 1990; Wollenhaupt *et al.*, 1994). There are two basic methods of sampling the cells - point sampling and cell sampling (Wollenhaupt *et al.*, 1994). Systematic sampling procedures are well documented and the designs seem well accepted in research circles.

Problems with Grid Sampling

There are a number of factors which limit the usefulness of grid sampling designs for commercial within-field soil mapping. These are summarized below.

1. There is little pedological rationale for the grid size used. Economics and precedent generally dictate the number of samples taken and hence the grid cell size. The risk of missing soil boundaries with large inter-sample spacings is discussed in Burgess and Webster (1984). Wösten *et al.* (1987) made the point that

it is not possible to recommend any one particular scale when mapping soils, but that an evaluation is required to consider the desired accuracy of data and the associated costs. Several researchers have indicated that smaller grid spacings are required for adequate within-field soil parameter characterization (Wollenhaupt et al., 1994) and yet large 1 and 2 ha grid spacings are still the norm in commercial circles.

2. Grid point sampling can be unduly biased by localized soil irregularities. Grid point sampling has been more widely adopted relative to grid cell sampling because of its ease of practical implementation. It also has the advantage of producing smooth, contoured maps with intuitive appeal to users. Unfortunately, due to the small radius from which cores are taken for each sample, it is possible for a sample to be totally unrepresentative of the bulk of the soil within the area it will ultimately represent. A grid intersection which randomly lands on the site of an old tree stump or the place where last years gypsum was piled will cause a much larger area to be incorrectly mapped based on this irregularity. If this were to happen enough times in a mapped area, the resultant map could theoretically be worse than the result from simple random sampling.

3. Grid sampling is a slave to its own uniformity. Soil attributes do not vary uniformly at any scale. To properly characterize a field a uniform grid sampling design would need to be at least at the scale of the smallest soil unit of a which characterization is desired. If it is desirable to site-specifically manage soil units of down to half hectare in size then the grid size needs to be at least this small just to avoid missing such soil units completely. In practice the grid size would need to be even smaller so that several points are taken from the smallest unit. This ensures that the units influence on the resulting map is correctly placed. In other parts of the field, larger soil units will have been over sampled resulting in waste of sampling resources. The possibility of stratifying the gridding scheme so each part of a field receives a grid of appropriate size has been mentioned (Wollenhaupt et al., 1994), although the ancillary information this requires about spatial patterns may be better used as discussed later in the paper.

4. Grid sampling can be unduly biased by systematic features. This issue has been raised before and the hazards of using a pattern synchronous with repeating, thin, elongated features have been well demonstrated (Berry and Baker, 1968; Congalton, 1988; and Wollenhaupt et al., 1994).

The primary argument for retaining grid sampling patterns (besides ease of commercial implementation) is to discern patterns about which nothing is known and for which no guide as to spatial extent exists. A systematic sampling design uniformly spread over a field may result in the "chance" finding of some feature not recognizable in any other way except though soil testing. In the commercial agricultural context this argument seems shaky. If a spatial feature is recognizable by no other manner than soil testing, then its importance in overall operations would seem questionable. The resources involved in finding it may be better allocated elsewhere.

This paper is concerned with re-visiting an alternative to grid sampling and evaluating it in the light of the recent technological advances that have so aided the precision agriculture movement.

DIRECTED SAMPLING

Concept

The concept behind directed sampling is far from novel. Essentially, the idea is to divide a field into smaller units as required by, and based on, the patterns present within the field and then to sample each of these units individually. This technique is recommended in almost every soil testing handbook. The following excerpts from State Extension soil testing publications are indicative of instructions given throughout the USA (and indeed, the world).

Variations in soil types, slope, drainage, or past management may require that smaller areas be sampled, resulting in three or mc·e composite samples per field.
- Georgia

Keep in mind that each sample should represent only one general soil type or condition. If the field you are sampling contains areas that are obviously different in slope, color, drainage, and texture and if those areas can be fertilized separately, submit a separate sample .... for each area.
- North Carolina

Non-uniform areas should be subdivided on the basis of obvious differences such as slope position or soil type.
- Ohio

Sample different soil types separately. Thus hill slopes, well-drained valley floors, and poorly drained areas should all be sampled separately. Such areas as forage and wheat, with different management histories, should be sampled separately. In this example, a separate soil sample should be taken from each of the four following sampling areas: forage on poorly drained soil, forage on well-drained soil, wheat on well-drained valley floor soil, and wheat on hill slope soil.
- Oregon

Implementation of this widely recommended procedure has been impeded in the past due to the practical limitations of both gathering and implementing the information. However, with the advent of GPS, GIS, and VRT, many of the limitations are gone. With appropriate base maps and information, a GIS can output an optimized set of sampling locations. GPS guidance removes the prior problems of uncertainty in exact within-field position. The previously limiting factor of practical inability to treat sub-field units individually is being steadily removed by advances in VRT.

The current challenge appears to be developing appropriate methodologies that combine local expert knowledge with other information sources to produce appropriate and meaningful sampling maps.

Despite its widespread recommendation, the authors were unable to find a generic name for this sampling pattern/methodology. Although such a pattern is termed "directed sampling" here, terms such as "guided sampling", "advised sampling" or "mission sampling" would probably suffice equally (if a name is required at all). "Sampling by soils" is appropriate but currently carries connotations of grid sampling. "Farming by soils" (Carr *et al.*, 1991; Wollenhaupt and Bucholz, 1993) is also appropriate but implies considerations outside the realm of soil sampling.

Theoretical considerations for directed sampling

The theoretical basis for directed sampling as the authors see it is based on three main points.

1. If a soil unit is totally homogeneous, only one soil sample from anywhere within its boundaries is required to totally characterize the entire unit.

2. If a soil unit is not homogeneous but the relationship of all parts within it are exactly known to all other parts, only one soil sample from anywhere within its boundaries is required to totally characterize the entire unit, *provided the location of that sample is known* (Fig 1).

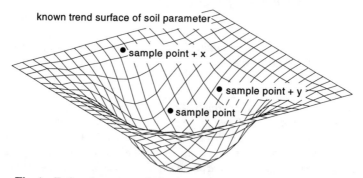

Fig. 1. Estimating unsampled points using known intra-soil unit parameter relations as a guide.

3. If the interface relationship between two homogeneous soil units is known, a single sample in each of the two soil units should be sufficient to characterize both soil units and their interface (Fig 2).

The aim of an optimal soil sampling pattern is to divide a field into homogeneous sub-units or sub-units with quantified internal structure and then to sample once for each unit. In reality, no soil unit is totally homogeneous. The aim then, must be to divide fields into units which are homogeneous for practical

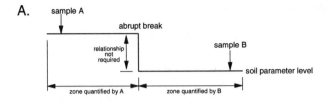

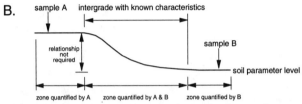

Figure 2 Estimating unsampled points using known inter-
soil unit parameter relations as a guide. A. abrupt boundary.
B. gradual boundary.

agronomic purposes and then bulk an appropriate number of sub-samples to find
the average of each "homogenous" unit. Measurement on a bulked sample of soil
will only give an unbiased estimate of the mean of replicate observations if the soil
property is strictly additive (Webster and Burgess, 1984). It is also implicit that
measurements are those that can be taken on a disturbed sample. The number of
sub-samples should be based on the variability judged to be present within each
unit. This variability may be quantified using appropriate base maps and an
appropriate tool such as regional variance in relevant spectral data from an aerial
photograph. This sub-sampling is vital and is one of the strengths of this procedure
relative to grid point sampling.

Directed sampling has the potential to produce vastly more accurate soil
maps relative to conventional random sampling. Because each analysed sample is
representative of many dispersed sub-samples and any interpolation is guided by
ancillary information, the strategy can, at worst, produce results no less accurate
than conventional random methods. This is in contrast to grid point sampling which
can theoretically lead to less accurate maps than simple random sampling.

Geostatistical methods such as co-kriging are potentially very useful for this
sampling design and analysis but the complexity of the procedure will be
significantly increased. Patterns will often be different for different parameters
hence there is a need to create and combine many different map layers if the
sampling pattern is to be optimized. Because of the need to handle many related
polygonal surfaces within and between maps, GIS is a sensible tool for combining
and manipulating this data.

Base maps for directed sampling

Directed sampling is based on prior knowledge of spatial soil patterns. Below are a suggested list of maps which may reveal patterns in certain soil parameters (either directly or after manipulation into their various derivatives).

- NRCS soil maps
- management history maps
- past yield maps
- digital elevation maps (DEM's)
- ground penetrating radar maps

- aerial crop images
- past grid sampling maps
- aerial soil color images
- electromagnetic induction maps
- expert information maps

The accuracy of the sampling pattern hinges completely on the strength of the prior knowledge. A base map for directed sampling should consist of information which is reliably expected to relate to the attribute being assessed. The optimal base map for any given purpose is unlikely to exist but will need to be customized from several sources. One of the main points the authors wish to convey in this paper is the importance of site-specific choice of base maps. Flexibility in deciding which information is relevant to the final sampling plan is paramount to the success of this methodology. It is probable that the most relevant information resides not in any remotely sensed image but in the knowledge-base of local experts (farmers/consultants/extension personnel). A field image may be best used simply to provide a geographically referenced template onto which expert knowledge can be scribed.

Directed sampling methodology

Three broad steps to the directed sampling method are listed below with brief comments for each.

1. Gathering and manipulation of relevant base maps. Base maps may need to be re-sampled, rectified, geo-referenced, co-ordinate transformed, subset, classified, and/or enhanced. Manipulation of maps into relevant derivatives such as drainage class maps from DEMs (Bell *et al.*, 1994) may be valuable.

2. Combination of maps and manipulation of resultant map. GIS overlay procedures are implemented to combine relevant maps. Non-sensical or non-manageable regions are eliminated.

3. Determination and output of sample locations. Stratification of sub-samples based on region size and estimated relative variability. Positioning of intra-region sub-samples, either randomly or based on some other rational plan.

A DIRECTED SAMPLING EXAMPLE

A simplified example is given here to demonstrate some of the methodology and some of the strengths and weakness' of the technique. The example comes from a 6.5 ha production agriculture field (31°32'30'' N, 83°38'44'' W) in Tift County, Georgia. This field is the subject of varied and ongoing spatial variability research by NESPAL and collaborating researchers. The topsoil of this field was sampled on a 20m x 20m square grid in March 1995. All individual samples were bulked from six sub-samples placed strategically around the sample point to account for row variability. Ten regions were randomly selected within the field for more intense sampling resulting in a total of 213 samples. The soil samples were analysed for standard chemical parameters. This data is used as a reference for evaluation of alternative sampling strategies.

Method

For brevity, the discussion of method here necessarily entails some use of semi-technical GIS terminology. Some terminology is specific to the software utilized (ERDAS IMAGINE, Earth Resource Data Analysis System, 1995)[1]. However, the processes are common to most geographic information systems.

A bare ground aerial photograph of the field was taken in January 1996 with standard 35mm Kodachrome slide film. The photograph was digitized with a slide scanner and the image was rectified and geo-referenced using ground control points surveyed with precision GPS equipment. The field boundary was determined by manually tracing an 'area of interest' around the bare soil region. The image was then 'subset' (cropped) to the field boundary to remove all extraneous spectral data. A contrast stretch was applied to the image to enhance contrast between intra-field regions (Fig 3).

Figure 3 Comparison of original aerial image (left) and contrast enhanced image (right) (gray scale images do not fully convey patterns evident in color images).

The next step was to classify the image into "homogeneous" regions. This was accomplished by assembling class signatures (training sets) from representative regions of the image in areas of known texture. In this case it was thought necessary only to form two classes, termed "clay" and "sand". After classification the image was 'clumped' (a process to identify regions of contiguous, identical pixels) and individual regions smaller than the arbitrary figure of 0.05 ha were 'eliminated' (a process to assimilate small regions into larger surrounding regions). A smoothing routine was then used to reduce the jagged nature of the sand-clay boundaries. At the end of this process the field had been divided into several regions, each region classed as either sand or clay.

Although the number of sub-samples pulled from a region should be strictly based on the variability present, concerns of practical implementation and economics frequently override this. It is normal to pull approximately 20 soil cores from a field this size in South Georgia. Having divided the image into internally "homogeneous" regions the bare soil image was then used to determine an estimate of within-region variability. This was done by calculating the number of 'clumps' per unit area present in each region of the initially classified map before 'elimination'. This estimate, along with the area of each region, was used to calculate the number of samples to be pulled from each region using the formula below:

$$no.\ samples for\ region_i = \frac{area_i \times var_i}{\sum\limits_{i=1}^{n} area_i \times var_i} \times sample\ number$$

where sample number = 20, var_i denotes the estimate of variability in region I, and n is the total number of regions in all classes. Once the number of samples per region was determined the 'classification accuracy assessment' module of IMAGINE was used to randomly distribute sample sites within the individual regions. Soil test values at each sample site were estimated by using the result from the nearest 1995 soil test site. This procedure resulted in ten sites being averaged (to simulate bulking) for both the sand and clay classes and the results were re-entered into the GIS. The resultant soil P and pH maps are displayed in Fig 4 along with corresponding maps constructed from all of the 1995 sample points using kriging.

Discussion

It is obvious that the directed sampling strategy worked quite well for P. Excellent agreement is evident given that only 2 samples were required for analysis versus 213 for the kriged map. Given that a textural influence on P can be expected in the coastal plain region this agreement is to be expected. The strategy performed less admirably in the case of pH. In this case both classes were found to have a pH of 5.8 - the overall field average. It appears that a base map founded on soil color is of little use for predicting spatial distribution of pH in this field at current levels. This example reinforces the principle that the base maps used must be predictive of the spatial patterns of the parameters being measured. As different maps will be better or worse indicators for different parameters, the final map must be a combination of as many base maps as required.

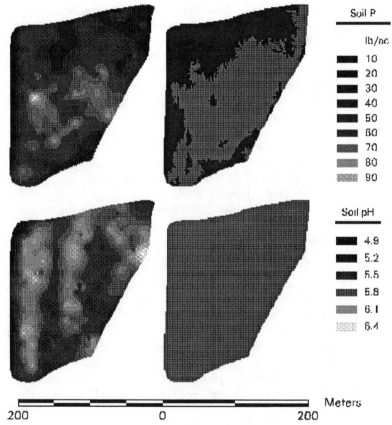

Figure 4 Comparison of kriged maps on left (213 samples) and directed sampling maps on right (2 samples) for P (top) and pH (bottom).

1. Products mentioned are intended as additional reader information and are not a product endorsement over or to the exclusion of similar products.

REFERENCES

Bell, J.C., R.L. Cunningham, and M.W. Havens. 1994. Soil drainage class probability mapping using a soil-landscape model. Soil Sci. Soc. Am. J. 58:464-470.

Berry, B.J.L. and A.M. Baker, 1968. Geographic Sampling. p. 91-100. *In* Berry, B.J.L. and D.F. Marble, (eds.), 1968. Spatial analysis: A reader in statistical geography, Prentice-Hall, Englewood Cliffs, N.J.

Burgess, T.M. and R. Webster. 1984. Optimal sampling strategies for mapping soil types. II. Risk functions and sampling intervals. J. Soil Sci. 35:655-665.

Carr, P.M., G.R. Carlson, J.S. Jacobsen, G.A. Nielson, and E.O. Skogley. 1991. Farming soils, not fields: A strategy for increasing fertilizer profitability. J. Prod. Agric. 4:57-61.

Congalton, R.G., 1988. A comparison of sampling schemes used in generating error matrices for assessing the accuracy of maps generated from remotely-sensed data. Phot. Eng. and Remote Sens. 54:593-600.

Earth Resource Data Analysis System. 1995. ERDAS Field Guide. 3rd Ed. Earth Resour. Data Anal. Syst., Atlanta, GA.

Petersen, R.G. and L.D. Calvin. 1986. Sampling. p33-51. *In* A. Klute (ed.). Methods of soil analysis. I. Physical and mineralogical methods. 2nd Ed. American Society of Agronomy. Madison, WI.

Webster, R. and T.M. Burgess. 1984. Sampling and bulking strategies for estimating soil properties in small regions. J. Soil Sci. 35:127-140.

Webster, R. and M.A. Oliver. 1990. Statistical methods in soil and land resource survey. Oxford University Press, New York. p 42-46.

Wollenhaupt, N.C., and D.D. Buchholz. 1993. Profitability of farming by soils. p. 199-211. *In* Robert, P.C.et al. (Eds). Proceedings of soil specific crop management. A workshop on research and development issues. ASA-CSA-SSSA. Madison, WI.

Wollenhaupt, N.C., R.P. Wolkowski, and M.K. Clayton, 1994. Mapping soil test phosphorus and potassium for variable rate fertilizer application J. Prod. Agric. 7:441-447.

Wösten, J.H.M., M.H. Bannink, and J. Bouma. 1987. Land evaluation at different scales: You pay for what you get! Soil Surv. and Land Eval. 7:13-24.

Improved Soil Mapping using Electromagnetic Induction Surveys

Dan B. Jaynes

National Soil Tilth Laboratory
USDA-ARS
2150 Pammel Dr., Ames, IA

Precision farming technologies rely on accurate field maps of the soil characteristics that affect yield. These maps are best produced at a spatial resolution that is comparable to the scale of application of chemicals, seed, or other inputs (boom width or even distance between individual applicators). Unfortunately, generation of these maps is expensive and laborious since intensive soil sampling and laboratory analyses are required. What is needed is an accurate, fast, inexpensive method of producing soil maps at a level of resolution that is comparable with current and future application technology.

Soil electrical conductivity as measured by electromagnetic induction (EMI) has been used successfully to characterize soils. The electrical conductivity of a soil is determined by a combination of soil water content, dissolved salt content, clay content and mineralogy, and soil temperature (McNeill, 1980b). In many fields, a single property (e.g. salinity) is the primary factor directly controlling soil electrical conductivity. Thus, once the correlation between electrical conductivity and this property is established, an EMI survey can be used to map this soil attribute quickly and cheaply. For example, EMI meters have been successfully used to measure soil salinity (Rhoades & Corwin, 1981; Cameron et al., 1981; Lesch et al., 1992), soil water content (Kachanoski et al., 1988), to map groundwater contaminant plumes associated with elevated chloride, sulphate, and nitrate levels (Greenhouse & Slaine, 1983; Drommerhausen et al., 1995), and measure clay content (Williams & Hoey, 1987).

EMI has also been used to determine soil properties it cannot measure directly. EMI has been used to determine soil cation exchange capacity and exchangeable Ca and Mg (McBride et al., 1990), depth to claypans (Doolittle, et al., 1994), field scale leaching rates of solutes (Slavich & Yang, 1990), spatial pattern of groundwater recharge (Cook et al., 1989; Cook et al., 1992), herbicide partition coefficients (Jaynes et al., 1994) and yield (Jaynes et al., 1995). These studies were successful because the parameter of interest either influenced a soil property (e.g. water content) that affects the EMI reading directly or because the parameter is associated with pedogenic processes that create properties EMI responds to. For example, Jaynes et al. (1994) were able to estimate the herbicide partition coefficient (K_d) for atrazine using EMI measurements nearly as well as they could using the standard approach of estimating K_d from measured organic carbon contents. This was possible not because EMI responds to K_d directly but because in the soils studied, high K_d values are due to high organic carbon levels and organic carbon accumulates in the poorly drained soils that have higher clay and water contents - properties EMI does respond to.

There are three ways in which EMI surveys can be used in field mapping. First, an EMI survey can be used as a reconnaissance map to aid in subsequent soil sampling. Areas of contrasting electrical conductivity can be targeted for soil sampling and these areas are easily delineated by EMI. This may reduce the number of samples required to characterize a field or improve the characterization by sampling both the typical as well as the extreme areas of the field. Second, EMI data can be used as co-regional data to refine maps of sparsely-sampled soil properties using cokriging. This technique uses the concept of co-regionality to improve the spatial mapping of a difficult and sparsely measured property such as soil organic carbon using the easily and intensively measured electrical conductivity. Third, EMI data can be used as direct surrogate measures of a second soil property. For example, Jaynes et al. (1994) used this approach to map atrazine sorption affinities using EMI surveys.

EXAMPLE STUDIES

Reconnaissance mapping

Data from a recent study by Cambardella et al. (1994) will be used to illustrate reconnaissance mapping by EMI to guide subsequent soil sampling. The objective was to map the spatial distribution of soil organic carbon (SOC) across a 6.25 ha field. SOC is an important indicator of soil fertility as well as an important factor controlling herbicide efficacy and fate (Bailey & White, 1970). Its content in soils can easily vary by more than four fold across a field.

The study area was located in central Iowa on soils developed from glacial till. The primary characteristic distinguishing the soils is drainage class, which ranges from well drained on uplands to very poorly drained in local depressions. Jaynes et al. (1994) found a positive correlation between EMI conductivity and soil organic carbon measurements suggesting the EMI measurements may be useful in mapping organic carbon spatial distributions.

Prior to planting in 1992, a grid was laid out across a 250 by 250 m area in the southern half of the field. The main grid spacing was 25 m in both the east-west and north-south directions. Within 1 m of each grid point, three 6-cm diameter cores were taken to a depth of 15 cm and composited for analysis. The mass of organic carbon was measured using dry combustion methods. Details of grid establishment and SOC analysis are described in Cambardella et al. (1994).

Several days after soil sampling, EMI measurements were made with an EM38 electromagnetic induction meter (Geonics Limited, Ontario, Canada[1]). A single reading was made at each grid point with the meter in the vertical dipole orientation and suspended 20 cm above the ground. Readings were made in this configuration so that the results could be used directly in a subsequent transect study where the meter was towed across the field on a boom. The EM38 integrates

[1]Trade and company names are used for the benefit of readers and do not imply endorsement by the USDA.

over an area approximately equal to its length of 1 m and over a depth of approximately 3 m, although the measurement is mostly influenced by properties in the 0 to 1.5-m depth increment (McNeill, 1980a). A vertical rather than horizontal dipole orientation for the meter was used so that slight variations in the height of the meter above the ground surface would have minimal effect on the measurements (McNeill, 1980a).

To test the advantages of EMI reconnaissance mapping, SOC data from only 9 of the grid locations were used to estimate SOC distribution. This limited number of samples was used to represent the typical practice of sparse sampling to map the spatial distribution of soil properties. Two methods were used to select the 9 "samples". First, the perimeter grid locations were not used to avoid any edge effects. Then the field was divided into 9 equal 0.56 ha cells and the center grid value within each cell was considered an SOC "sample" (Fig. 1). SOC at the remaining grid locations was interpolated by linear kriging (Golden Software, Golden, CO). These estimates were then compared with the SOC values at the remaining 112 grid locations.

In the second approach, all 121 EMI measurements were used to draw a map of electrical conductivity. The distribution of electrical conductivity within the grid is shown in Fig. 1. Low values of conductivity were located in the northwest, east-central, and south central portions of the site. High values were measured in the southeast and southwest corners and the central through north-central portion of the site. Based on this conductivity map, 9 locations were chosen from both high and

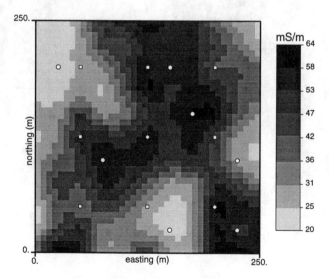

Fig. 1. Map of electrical conductivity measured by EMI and grid locations used to estimate soil organic carbon. Circles are cell-centered grid points, squares are reconnaissance determined grid points, and diamonds are locations used in both.

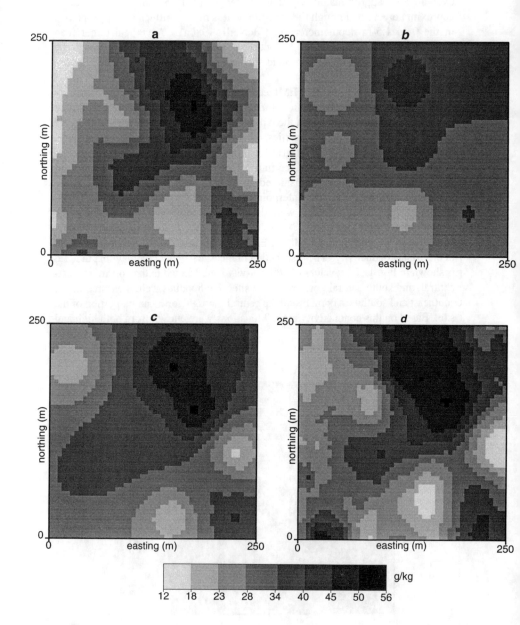

Fig. 2. SOC distribution estimated from: a) measured SOC at all 121 locations, b) SOC data at 9 cell-centered grid points, c) SOC data at 9 grid points identified from EMI mapping, and d) the same points as in c and cokriging using all EMI data.

low conductivity areas for SOC "sampling" (Fig. 1). SOC data from these 9 locations were used to interpolate SOC values at the remaining 112 grid locations as before. Again, the estimated SOC values were then compared to the unused 112 measured SOC values.

The interpolated distributions of SOC based on either cell or EMI reconnaissance directed sample locations is shown in Fig. 2b and c. These estimates can be compared directly to the measured distribution of SOC determined using all 121 grid locations (Fig 2a). The map developed from the reconnaissance EMI data more closely resembles the map developed using all of the measured SOC data than the map developed from cell sampling. In particular, the reconnaissance derived map captures the high SOC content in the northeast quarter of the site and the low content in the east-central, south-central and northwest corner. Neither map completely reproduces the complexity of the SOC spatial pattern.

Overall the SOC map derived from cell sampling over predicted the average SOC content within the grid by 12% and underestimated the total variance by 83%. Correlation between the estimated and measured SOC distribution was 0.487 with a standard error of the estimate of 9.5 g/kg (about 1 and 1/2 gray-level divisions in Fig. 2). Similarly the SOC map estimated from the EMI reconnaissance over predicted the field average SOC by 22% but underestimated the variance by 52%. The correlation between estimated and measured SOC improved to 0.704 with an unchanged standard error of the estimate of 9.5 g/kg. Thus, in this case SOC estimated by selecting sample locations based on the EMI reconnaissance map gave a more faithful representation of the true SOC distribution than the estimate based on samples drawn from the center of regularly spaced cells.

Co-regional Variable

The same SOC data was used to illustrate using EMI conductivity as a co-regional variable for improving SOC estimation by cokriging. Cokriging is an interpolation method that takes advantage of the correlation between two co-regional variables. The more intensively sampled variable can be used to better predict the distribution of the lesser sampled variable (Journel & Huijbregts, 1978).

Cokriging was used to estimate SOC values using the same 9 "measured" SOC values from the EMI reconnaissance map and the EMI conductivity data at the remaining 112 grid locations. Variogram and cross-variogram development was conducted as described in Journel and Huijbregts (1978). Experimental and fitted variograms for EMI conductivity are shown in Fig. 3a. A combined nugget and gaussian model was found to fit the experimental variogram well. (Fig. 3b and 3c). The model fit the experimental variograms well in each case with a range of 60 m for each and a nominal nugget value.

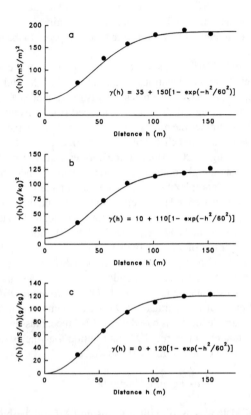

Fig. 3. Experimental and fitted variograms: a) electromagnetic induction (EMI) conductivity data, b) soil organic carbon (SOC) content and c) cross variogram of EMI conductivity and SOC content.

Data from all 121 grid points was used in calculating these variograms. While in a true application of this method the SOC data would not be available for variogram construction, our studies of the spatial behavior of both SOC and conductivity show that their relationships are consistent across the soils found within the study site. Thus, variograms developed at other locations should be transferable. Developing variograms with the full data set and then using only a subsample to estimate SOC distribution allows for a best case test of using cokriging with EMI data for SOC estimation.

The cokriged estimate of SOC distribution is shown in Fig. 2d. The cokriged distribution captures both the spatial distribution of high and low SOC and the complexity of the distribution. Overall the cokriged estimate still over predicts the average SOC content measured at the 121 grid points by 16% and under predicts the variance in the measured SOC by 22%. The correlation between this estimate and measured SOC content is 0.762 with a standard error of the estimate of 5.9 g/kg, a 38% reduction compared to using just the 9 SOC values. Thus, cokriging with the EMI conductivity data and the 9 "measured" SOC values has improved the standard error of the estimate from about 1 and 1/2 gray level divisions in Fig. 2 to less than 1.

Surrogate measure

EMI conductivity can also be used as a surrogate variable for a more difficult to measure soil property. For example Doolittle et al. (1994) developed a relationship between EMI measured conductivity and the depth to a clay pan in a Missouri soil. Specifically they found

$$D = 1662 * 10^{-0.033\,C} \qquad\qquad [1]$$

where D is the depth to clay (cm) and C is electrical conductivity (mS/m) measured with an EM38 in the vertical dipole orientation. Thus, mapping depth to clay layer should be possible using Eq. [1] in conjunction with an EMI survey.

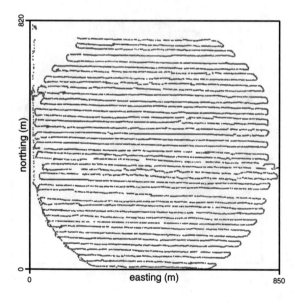

Fig. 4. Locations where electrical conductivity was measured by electromagnetic induction using an EM38 pulled on a boom behind an ATV equipped with a global positioning system.

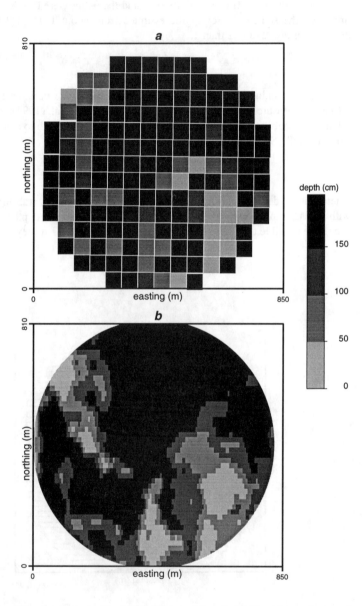

Fig. 5. Depth to clay layer in Kansas field a) measured at 168 grid locations and b) estimated from Eq. [1] and measurements of soil electrical conductivity made by electromagnetic induction.

We tested this approach on a 53 ha irrigated wheat field in Stafford Co., Kansas. This field is the focus of an intensive variable rate N project at Kansas State University (Redulla et al., 1996). The soils within the field were developed in wind blown sand overlying a clay layer. The depth to the clay varies across the field and may affect the water holding capacity and thus the yield. The field was grid sampled at a 55 m interval and depth to the clay layer within the top 152 cm was measured.

In the spring of 1995, electrical conductivity was mapped by EMI using an EM38 towed on a wooden boom behind a 5-wheel all terrain vehicle equipped with a global positioning system (Ambuel et al., 1991). Electrical conductivity was recorded along with position at 2 s intervals while driving transects across the field (Fig. 4). The conductivity measurements were then transformed into depth to clay estimates using [Eq. 1] after correcting the conductivity measurement to account for the height of the meter above the ground (McNeill, 1980a).

The resulting estimates for depth to clay are compared to the measured values in Fig. 5. Both maps show shallow depth to clay in the northwest, southeast and southern regions of the field. They also show depth to clay exceeding 152 cm in the northern half of the field. The EMI derived map gives much more detail than the grid sampled map since the former was developed from 9700 point measurements while the grid map is based on only 164 observations. Perhaps more importantly, the EMI derived map took only about 1/2 day to produce while the grid sampled map took considerably longer for the soil sampling and laboratory analysis.

CONCLUSIONS

EMI techniques have the potential to gather geo-referenced data quickly and cheaply. This data may be used in several ways to improve our knowledge of the spatial patterns of soil properties across fields. These approaches include using EMI measurements for pre-sampling reconnaissance, as secondary variables for improving estimates of primary variables through geostatistical methods, and as surrogate measures of parameters more costly to measure.

ACKNOWLEDGEMENTS

The author thanks C. Cambardella for providing the SOC data, G. Kluitenberg and J. Havlin for providing access to the Kansas field and the depth to clay layer data, and J. Cook for assisting with the GPS.

REFERENCES

Ambuel, J., T.S. Colvin, and K. Jeyapalan. 1991. Satellite based positioning system for farm equipment. SAE Trans., J. Commercial Vehicles, 100-2:323-329.

Bailey, G.W., and J.L. White. 1970. Factors influencing adsorption, desorption and movement of pesticides in soil. Res. Rev. 32:29-92.

Cambardella, C.A., T.B. Moorman, J.M. Novak, T.B. Parkin, D.L. Karlen, R.F. Turco, and A. Konopka. 1994. Field-scale variability of biological, chemical, and physical soil properties in central Iowa soils. Soil Sci. Soc.

Am. J. 58:1501-1511.

Cameron, D.R., E. DeJong, D.W.L. Read, and M. Oosterveld. 1981. Mapping salinity using resistivity and electromagnetic inductive techniques. Can. J. Soil Sci. 61:67-78.

Cook, P.G., M.W. Hughs, G.R. Walker, and G.B. Allison. 1989. The calibration of frequency-domain electromagnetic induction meters and their possible use in recharge studies. J. Hydrol. 107:251-265.

Cook, P.G., G.R. Walker, G. Buselli, I. Potts, and A.R. Dodds. 1992. The application of electromagnetic techniques to groundwater recharge investigations. J.. Hydrol. 130:201-229.

Doolittle, J.A., K.A. Sudduth, N.R. Kitchen, and S.J. Indorante. 1994. Estimating depths to claypans using electromagnetic induction methods. J. Soil and Water Cons. 49:572-575.

Drommerhausen, D.J., D.E. Radcliffe, D.E. Brune, and H.D. Gunter. 1995. Electromagnetic conductivity surveys of dairies for groundwater nitrate. J. Environ. Qual. 24:1083-1091.

Greenhouse, J.P., and D.D. Slaine. 1983. The use of reconnaissance electromagnetic methods to map contaminant migration. Ground Water Monitoring Rev. 3(2):47-59.

Jaynes, D.B., T.S. Colvin, and J. Ambuel. 1995. Yield mapping by electromagnetic induction. p.383-394. In P.C. Robert et al. (ed.) Site-specific management for agricultural systems. Proc. 2nd Inter. Conf. Minneapolis, MN. 27-30 Mar. 1994. Am. Soc. Agron., Inc., Madison, WI.

Jaynes, D. B., J.M. Novak, T.B. Moorman, and C.A. Cambardella. 1994. Estimating herbicide partition coefficients from electromagnetic induction measurements. J. Environ. Qual. 24:36-41.

Journel, A.G. and Ch.J. Huijbregts. 1978. Mining geostatistics. Academic Press, New York, NY.

Kachanoski, R.G., E.G. Gregorich, and I.J. Van Wesenbeeck. 1988. Estimating spatial variations of soil water content using noncontacting electromagnetic inductive methods. Can. J. Soil Sci. 68:715-722.

Lesch, S.M., J.D. Rhoades, L.J. Lund, and D.L. Corwin. 1992. Mapping soil salinity using calibrated electromagnetic measurements. Soil Sci. Soc. Am. J. 56:540-548.

McBride, R.A., A.M. Gordon, and S.C. Shrive. 1990. Estimating forest soil quality from terrain measurements of apparent electrical conductivity. Soil Sci. Soc. Am. J. 54:290-293.

McNeill, J.D. 1980a. Electromagnetic terrain conductivity measurement at low induction numbers. Tech. Note TN-6. Geonics Ltd. Mississauga, Ont., Canada.

McNeill, J.D. 1980b. Electrical conductivity of soils and rocks. Tech. Note TN-5. Geonics Ltd. Mississauga, Ont., Canada.

Redulla, C.A., J.L. Havlin, C.J. Kluitenberg, N. Zhang, and M.D. Schrock. 1996. Variable nitrogen management for improving groundwater quality. (this issue). In P.C. Robert et al. (ed.) Site-specific management for agricultural systems. Proc. 3rd Inter. Conf. Minneapolis, MN. 23-26 June 1996. Am. Soc. Agron., Inc., Madison, WI.

Rhoades, J.D., and D.L. Corwin. 1981. Determining soil electrical conductivity - depth relations using an inductive electromagnetic soil conductivity meter. Soil Sci. Soc. Am. J. 45:255-260.

Slavich, P.G., and J. Yang. 1990. Estimation of field scale leaching rates from chloride mass balance and electromagnetic induction measurements. Irrig. Sci. 11:7-14.

Williams, B.G., and D. Hoey. 1987. The use of electromagnetic induction to detect the spatial variability of the salt and clay contents of soils. Aust. J. Soil Res. 25:21-27.

The Factor of Time : A New Consideration in Precision Farming

Mingchu Zhang
M. Nyborg

Department of Renewable Resources
University of Alberta
Edmonton, Alberta

E. D. Solberg

Agronomy Branch
Alberta Agriculture and Rural Development
Edmonton, Alberta

ABSTRACT

One factor that needs to be emphasized in Precision Farming is time, which implies that the nutrients applied in soil are not necessarily synchronized with crop requirement. Consequently, nutrients are exposed to any possible path of loss. The result of this not only reduces nutrients efficiency to the crop but also increases environmental contamination. However, this problem can be overcome by use of controlled release fertilizers. For example, we have conducted experiments which showed that there was less nitrogen loss through denitrification in controlled release urea as compared with conventional urea. Also, we found that barley (*Hordeum vulgare* L.) roots proliferated around granules when controlled release urea was applied into soil. With this proliferation, nitrogen uptake by crops is expected to be maximized. Over all, this paper stresses the importance of the factor of time in Precision Farming, and that controlled release fertilizers are effective sources to match fertilizer availability in soil with crop demand.

INTRODUCTION

One of the research focuses in the Precision Farming is site-specific fertilization. By using state of art computer technologies, variable rate fertilizers, mainly nitrogen, are applied based on the variation of soil fertility, as such nitrogen fertilizer efficiency is improved and NO_3^- leaching is minimized.

With the current site-specific fertilization practice, only space variability is considered. As a matter of fact, the pattern of nutrient uptake by plants are sigmoidal, meaning that plants absorb more nutrients at certain a period of time than they do in any other growth stages. One time fertilizer application before seeding or at seeding only provides quick available nutrients to plants, but does not guarantee that plants will absorb all applied nutrients. Often there is a 'time lag' period between the time of fertilizer application and the time that plant roots approach and absorb the applied nutrients. In that period, the applied nutrients are vulnerable to loss, e.g., leaching. Under such circumstances, considering only space variation in fertilization is not enough, the factor of time also needs to be emphasized.

Controlled release is a technology aiming to deliver chemicals (medicine, herbicide, or pesticide) at a predetermined rate to a target over a period of time. To deliver nutrients to plant roots, fertilizer granules can be coated with polymers so that the nutrients are released in a desired rate over a time period. With polymer coating, fertilizers, firstly, are not exposed to the soil environment after application, therefore the chance for loss is minimized. Secondly, the coated fertilizer meters nutrients in a rate which could match plant uptake pattern (researchers are still working on this subject), hence increasing the nutrient use efficiency and minimizing negative environmental impact.

The objective of this paper is to report some of our research with controlled release fertilizers, and with hope, the paper can illustrate the future potential importance of controlled release fertilizers in Precision Farming.

METHODS

Two experiments were conducted with the controlled release fertilizers.

Experiment 1

A laboratory incubation investigation was conducted to determine the release of nitrogen from controlled release fertilizers and subsequent denitrification in soil water at 33 kPa water potential, and 33 kPa saturation (7.5 days in 33 kPa, and 7.5 days in saturation). The treatments were: nil; conventional urea fertilizer; thin-coated urea; thick-coated urea; and guanidine sulfate - ammonium mixture (G.S.). There were 3 replicates. Each of the 0.5 L containers held 400 g of soil to which was added 80 mg of fertilizer N (13 to 14 granules). Each granule was inserted 25 mm deep into the soil. Lids include provision for aeration. One third of the contents was removed at 15, 30, and 60 d intervals. The intact granules were removed from the soil samples and the remaining soil was spread thinly and quickly dried at room temperature. Mineral nitrogen (NH_4- N and NO_3- N) were extracted (using 2 M KCl) and determined (Technicon AutoAnalyzer II, 1973a, 1973b). Results were statistically analyzed by ANOVA.

Experiment 2

This greenhouse experiment was conducted to examine the interaction of plant roots with coated fertilizers. One granule of fertilizer and one barley seed were put in each container (8-cm dia., 5.2-cm depth). Fertilizer granules and barley seeds were placed exactly in the same place in all containers for ease of sampling later. Two hundred grams of soil was placed in the pot. There were 6 treatments: nil, urea, coated urea I, coated urea II, mono-ammonium phosphorus (MAP), and coated MAP. Treatments of nil, urea, and MAP were replicated 4 times. Considering the variation in the nutrient release rate of individual granules, treatments of coated urea I, coated urea II, and coated MAP were replicated 8, 8, and 10 times, respectively. At sampling (28 days after seeding), a 2-cm dia. × 2-cm seeding), a 2-cm dia. × 2-cm depth soil core, centered on the specific area of fertilizer granule application, was taken from each of the treatments. The roots were washed from the core. The remaining roots in the container were separated from shoots and washed. Samples of shoots and roots were dried at 65°C for 48 h and weighed.

RESULTS

Experiment 1

After 15 days of incubation in soil at 33 kPa water potential, nitrogen release from fertilizer granules was the greatest from non-coated urea, followed by thin-coated urea, then thick-coated urea and finally G.S. (Table 1). The release of nitrogen from the thin-coated and thick-coated urea, and G.S. was showed by the increase of mineral nitrogen concentration in soil with time.

Incubation with alternating 33 kPa and saturation, denitrification occurred in almost all treatments except at the first 15 days with controlled release fertilizers (Table 2). In the first 15 days in which soil was in 7.5-day 33 kPa water potential and then saturation, denitrification occurred in non-coated urea but not in the coated urea and G.S. . The nitrogen release from G.S. was dependent on hydrolysis (Zhang et al. 1992). The mineral nitrogen content in G.S. treatment was higher with saturation than in 33 k P at 15 days. As soil mineral nitrogen increases with time so did the apparent denitrification lessen from all treatments.

Experiment 2

Root proliferation around coated granules was observed in the treatments of coated urea I, coated urea II and coated MAP, but not in non-coated urea and non-coated MAP (Table 3). The total root mass was not different among the treatments, but the mass of roots from the sampling core was greater in coated urea than non-coated urea. Root mass from the core tended to be greater with coated MAP than with MAP, but the difference was not significant. The proportion of root mass in the sampling core was greater in all of the coated fertilizers.

Table 1. Mineral nitrogen released from fertilizer granules in 33 kPa water potential

Treatments	Mineral nitrogen in soil		
	15 days	30 days	60 days
	ug N /g soil		
Nil	59a[+]	69a	88a
Urea	278d	218c	267c
Thin-coated urea	208c	231c	264c
Thick-coated urea	146b	216c	217b
Guanidine sulphate	68a	98b	253c

[+]Letters in common indicating difference not significant at P=0.05 using LSD.

Table 2. Comparison of mineral nitrogen in incubated soil with fertilizer nitrogen
    under 33 kPa water potential and saturation.

Treatments	Soil water content	Mineral nitrogen in soil			Avg of
		15 d††	30 d	60 d	all times
		ug N / g soil			
Nil	FC	59	69	88	72
Nil	FC-Sat‡	63	26	23	37
Urea†	FC	219	149	179	182
Urea	FC-Sat	175	§	149	162
Thin-coated urea	FC	148	163	175	162
Thin-coated urea	FC-Sat	169	127	91	129
Thick-coated urea	FC	86	148	182	139
Thick-coated urea	FC-Sat	84	62	148	98
GS	FC	9	29	165	68
GS	FC-Sat	51	30	106	62
LSD (0.05)					21

† net values used for fertilizer treatments(i.e. fertilizer treatments minus Nil treatment)
†† d = days
‡ Sat = saturation
§ missing data

Table 3. Root mass from sampling core and its proportion to the total root mass
    in the pot

Treatments	Root proliferation around granule	Total root mass core	Root mass from sampling in the pot	Root mass from sampling core to total root mass in the pot
		mg	mg	%
Nil	no	104.2 a[+]	1.3 a	1.2 a
Urea	no	112.0 a	2.2 a	1.9 ab
Coated urea I	yes	119.9 a	5.3 c	4.5 d
Coated urea II	yes	127.7 a	4.2 bc	3.3 c
MAP	no	133.6 a	4.3 bc	3.1 bc
Coated MAP	yes	118.4 a	4.7 c	4.1 cd

[+]Letters in common indicating difference not significant at P=0.05 using LSD.

DISCUSSION

The mineral N released from fertilizers was markedly slowed when polymer-coated urea and guanidine sulfate were used. For example, thick-coated urea released only 43% of applied N in 15 d, or 57% less as compared to the conventional urea. Considering the amount of substrate for nitrification and subsequent denitrification or leaching, coating slowed urea release into soil hence reducing N losses. Using the same coated urea, Nyborg et al. (1993) found that when compared with non-coated urea, fall applied thick-coated urea had 24% less mineral N and 42% less NO_3-N 6 or 6.5 months after October application.

The short growing season (80 to 120 day) of the Prairie Provinces of Canada forces most farmers to produce spring-sown cereal or oil seeds. By estimation, some 30-50% of the fertilizer N is applied the previous fall instead of at spring sowing. The poor performance of fall application is closely connected with nitrification and subsequent denitrification in saturated top soil perched on frozen subsoil in late winter or early spring (Heaney et al. 1992, and Nyborg et al. 1990). However, as showed by Experiment 1, denitrification was reduced by using controlled fertilizers. Giving the natural condition of western Canada, fall application of N fertilizer will continue to be a routine farming practice. The use of controlled release fertilizer has the potential to dramatically reduce the losses from fall applied N fertilizer.

Root proliferation around coated granules was found in the Experiment 2. Miller and Ohlrogge (1958) found general root proliferation in fertilizer band volumes. Drew and Saker (1975) reported that growth of one barley root branch grew faster than other branches by supplying 0.01 mM NO_3^-. Our work illustrated similar phenomenon but with the coated fertilizers.

One time application of N fertilizer before or at seeding leaves soluble N in the soil. The peak N uptake period of plants, however, often occurs some 20 or more days later than the time of application. Under such circumstance, N applied in soil is ready for all possible losses. In the case of controlled release fertilizers, the nutrients are slowly released, thereby avoiding the exposure of soluble nutrients in soil before crop uptake. Also, the low concentrations of nutrients from fertilizers in soil are hospitable to the tender roots of young plants. This was shown by the root proliferation around granules. By developing such relationship with roots, coated fertilizers are expected to provide little chance for loss.

In conclusion, to maximize the fertilizer use efficiency, the factor of time has to be dealt with in Precision Farming. Coated fertilizers provide a tool to cope with such a issue. Given consideration of both space and time variables, we will more precisely manage fertilizers.

ACKNOWLEDGMENTS

Research fund was provide by Alberta Agricultural Research Institute and Sherritt, Inc.

DISCUSSION

The mineral nitrogen released from fertilizers was markedly slowed when polymer-coated urea and guanidine sulfate were used. For example, thick-coated urea released only 43% of applied nitrogen in 15 d, or 57% less as compared to the conventional urea. Considering the amount of substrate for nitrification and subsequent denitrification or leaching, coating slowed urea release into soil hence reducing nitrogen losses. Using the same coated urea, Nyborg et al. (1993) found that when compared with non-coated urea, fall applied thick-coated urea had 24% less mineral nitrogen and 42% less NO_3-N 6 or 6.5 months after October application.

The short growing season (80 to 120 day) of the Prairie Provinces of Canada forces most farmers to produce spring-sown cereal or oil seeds. By estimation, some 30-50% of the fertilizer nitrogen is applied the previous fall instead of at spring sowing. The poor performance of fall application is closely connected with nitrification and subsequent denitrification in saturated top soil perched on frozen subsoil in late winter or early spring (Heaney et al. 1992, and Nyborg et al. 1990). However, as showed by Experiment 1, denitrification was reduced by using controlled fertilizers. Giving the natural condition of western Canada, fall application of nitrogen fertilizer will continue to be a routine farming practice. The use of controlled release fertilizer has the potential to dramatically reduce the losses from fall applied nitrogen fertilizer.

Root proliferation around coated granules was found in the Experiment 2. Miller and Ohlrogge (1958) found general root proliferation in fertilizer band volumes. Drew and Saker (1975) reported that growth of one barley root branch grew faster than other branches by supplying 0.01 mM NO_3^-. Our work illustrated similar phenomenon but with the coated fertilizers.

One time application of nitrogen fertilizer before or at seeding leaves soluble nitrogen in the soil. The peak nitrogen uptake period of plants, however, often occurs some 20 or more days later than the time of application. Under such circumstance, nitrogen applied in soil is ready for all possible losses. In the case of controlled release fertilizers, the nutrients are slowly released, thereby avoiding the exposure of soluble nutrients in soil before crop uptake. Also, the low concentrations of nutrients from fertilizers in soil are hospitable to the tender roots of young plants. This was shown by the root proliferation around granules. By developing such relationship with roots, coated fertilizers are expected to provide little chance for loss.

In conclusion, to maximize the fertilizer use efficiency, the factor of time has to be dealt with in Precision Farming. Coated fertilizers provide a tool to cope with such a issue. Given consideration of both space and time variables, we will more precisely manage fertilizers.

ACKNOWLEDGMENTS

Research fund was provide by Alberta Agricultural Research Institute and Sherritt, Inc.

REFERENCES

Drew, M.C., and L.R. Saker. 1975. Nutrient supply and the growth of the seminal root system in barley. Localized compensatory increase in lateral root growth and rates of nitrate uptake when nitrate supply is restricted to only part of the root system. J. Exp. Bot. 26:79-90.

Heaney, D.J., M. Nyborg, E. D. Solberg, S.S. Malhi, and J. Ashworth 1992. Overwinter nitrate loss and denitrification potential of cultivated soils in Alberta. Soil Biol. Biochem. 24:877-884.

Miller, M.H., and A.J. Ohlrogge. 1958. Principles of nutrient uptake from fertilizer bands. 1. Effect of placement of nitrogen fertilizer on the uptake of band-placed phosphorus at different soil phosphorus levels. Agron. J. 50:95-97.

Nyborg, M., S.S. Malhi, and E. D. Solberg. 1990. Effect of date of application on recovery of N15-labeled urea and potassium nitrate. Can. J. Soil Sci. 70:21-31.

Nyborg, M., E. D. Solberg, and M. Zhang. 1993. Polymer-coated urea in the field: mineralization, nitrification, and barley yield and nitrogen uptake. Dahlia Greidinger Memorial International Workshop on Controlled/Slow Release Fertilizers. Haifa, Israel March 7-12 1993.

Technicon AutoAnalyzer II. 1973a. Ammonia in water and waster water. Industrial Method No. 98-70W. Technicon Industrial Systems, Tarrytown, New York.

Technicon AutoAnalyzer II. 1973b. Nitrate and nitrite in water and waster water. Industrial Method No. 100-70W. Revised January, 1978. Technicon Industrial Systems, Tarrytown, New York.

Zhang, Z. M. Nyborg, K.M. Worsley, and D.A. Gower. 1992. Guanidine sulphate: slow release of mineral nitrogen during incubation in soil. Commun Soil Sci. Plant Anal 23:431-439.

Soil Property Contributions to Yield Variation Patterns

C. A. Cambardella
T. S. Colvin
D. L. Karlen
S. D. Logsdon
E. C. Berry
J. K. Radke
T. C. Kaspar
T. B. Parkin
D. B. Jaynes

USDA-ARS
National Soil Tilth Laboratory (NSTL)
Ames, IA

ABSTRACT

Multiple linear regression analysis was used to relate crop yield patterns observed over a seven year period (1989-1995) with patterns in soil properties measured in 1994 for one field in central Iowa. Our results emphasize the importance of soil structure in defining water relations in agricultural systems.

BACKGROUND

Technological advances in recent years have led to the development of accurate, inexpensive yield monitors and equipment for variable-rate application of fertilizer and agricultural chemicals. Detailed maps delineating yield spatial patterns are confirming that crop yields aren't uniform across agricultural fields. Colvin et al. (1996) describe the yield patterns for corn and soybeans in rotation after six consecutive years within a single field. They found that certain locations within the field had either consistently high, consistently low, or erratic yields when compared to whole-field averages. These patterns in yield are probably controlled by soil properties and moderated by the interaction of climate and weed, insect, disease, and management pressures (Jaynes and Colvin, 1996).

Understanding the spatial distribution of soil attributes that affect yield pattern is important for the development of soil-specific management systems for crop production that minimize environmental contamination while maintaining economic viability. Many of the soil properties that influence productivity exhibit high spatial and temporal variability (Cambardella et al., 1994). Quantification of this variability is complicated by the practice of arbitrarily delineating the boundaries of agricultural fields with little regard for variation in soil type, landscape characteristics, or drainage class.

The objective of this study was to functionally relate crop yield patterns observed over a 7 year period (1989-1995) with patterns of soil properties measured in 1994 for one field in central Iowa. We will use information from this analysis in the development of soil quality estimates relative to crop production for this field site.

METHODS AND MATERIALS

Field Histories

Data was collected from a 16 ha field in Boone County, Iowa that has been cropped to a 2 year corn and soybean rotation since 1958 (Karlen et al., 1995). Tillage has been chisel plow and field cultivator since 1981. Total N-P-K application for 1981 through 1995 was 168-38-93 kg ha^{-1}. The field is located in a Clarion-Nicollet-Canisteo soil association that developed in calcareous glacial till deposited during the Cary substage of glaciation about 14,000 years ago.

Sampling Design

Crop yield was measured for seven consecutive years starting with corn in 1989. Yield data was collected in 20 m X 2 m increments from eight transects along the E-W axis of the field with a small commercial combine (Colvin, 1990). The yield data was used to identify thirteen sampling locations having consistently high, low, or erratic yields (Colvin et al., 1996). Soil parameters were measured to a depth of 15 cm at these locations in 1994. Bulk density (BD), organic C (OC), aggregate size distribution as percent macroaggregation (MACRO) and aggregate mean weight diameter (MWD) following wet sieving, soil texture (TEX), and gravimetric water content (SM1 for 0-7.5 cm depth increment; SM2 for 7.5-15 cm depth increment) were quantified for soil core samples brought back to the laboratory. Volumetric water content (TDR), cone index (CONE), uniformity coefficient (UC), infiltration (INFILT), runoff (RUNOFF), field respiration (RESP), electromagnetic conductance (EM), earthworm numbers (WORMS), and weed density (WEEDS) were estimated in situ for each of the 13 sampling locations.

Regression Analyses

We used two multiple linear regression procedures, stepwise linear regression (STEP) and maximum r^2-improvement regression (MAXR) (SAS,1985) to analyze the effect of the soil properties (independent variables) on crop yield (dependent variable) variability. We determined the suite of soil parameters that contributed the most to yield variation for any given year. The MAXR analysis was constrained by setting the alpha level at 0.05 and requiring the r^2 to be equal to or greater than 0.95 before accepting the model. This means the model was accepted as valid when all of the parameters in the model had a probability level of 0.05 or less and the r^2 was at least 0.95. The STEP analysis was not constrained by r^2 and we set the probability level at 0.05 for parameter entry into the model and required the probability level to be 0.05 in order for the parameter to remain in the model.

We used the regression coefficients for each model to predict crop yield for the corresponding year. The difference between observed and predicted yield was evaluated in order to determine which model was the most accurate predictor of yield for that year.

Table 1. Predicted and observed yields for 1989-1995

	1989	Yield	1990	Yield	1991	Yield
	Corn	Bu/Ac	Beans	Bu/Ac	Corn	Bu/Ac
Plot	PRED	OBS	PRED	OBS	PRED	OBS
LL*	129	102	56	18	158	71
LL	109	103	15	16	46	58
LL	129	133	33	31	101	90
LH	119	117	6	0	98	107
LH	150	152	0	0	10	11
LH	59	60	29	32	52	37
HL	163	160	53	59	189	186
HL	171	167	60	54	182	182
HL	167	167	61	57	184	174
HH	178	180	58	60	218	218
HH	166	170	63	61	207	220
HH	182	183	52	58	191	195
	1992	Yield	1993	Yield	1994	Yield
	Beans	Bu/Ac	Corn	Bu/Ac	Beans	Bu/Ac
Plot	PRED	OBS	PRED	OBS	PRED	OBS
LL*	24	30	104	78	23	19
LL	30	30	88	87	19	17
LL	30	33	92	92	36	36
LH	51	51	0	0	53	52
LH	44	44	0	0	50	52
LH	28	25	53	54	20	21
HL	48	50	87	89	53	53
HL	49	51	87	87	49	48
HL	49	49	88	87	53	52
HH	44	45	88	88	47	48
HH	48	46	91	91	46	46
HH	48	46	66	66	52	50

* LL = consistent low yield; LH = erratic low yield;
HL = consistent high yield; HH = erratic high yield

Table 1. Continued - -
Predicted and observed yield 1989-1995

Plot	1995 Corn PRED	Yield Bu/Ac OBS
LL*	42	85
LL	90	84
LL	125	127
LH	165	163
LH	160	160
LH	67	69
HL	146	147
HL	144	147
HL	170	165
HH	158	159
HH	159	162
HH	153	155

* LL = consistent low yield; LH = erratic low yield;
HL = consistent high yield; HH = erratic high yield

Table 2. Regression parameters yield vs. soil variables for 1989-1995

	1989	MAXR		1993	STEP
	Corn	r^2=.9946		Corn	r^2=.9979
Variable	Coeff	Pr > F	Variable	Coeff	Pr > F
MWD	- 3.75	0.0077	SM1	- 1.34	0.0482
MACRO	3.35	0.0228	MWD	- 2.41	0.0010
WEEDS	- 4.16	0.0452	WEEDS	- 6.28	0.0016
INTER	154.96	0.0015	TDR	5.83	0.0028
			SLOPE	9.18	0.0090
			SAND	0.88	0.0131
			BD	55.86	0.0164
			INTER	- 93.42	0.0454
	1990	MAXR		1994	MAXR
	Beans	r^2=.9682		Beans	r^2=.9893
Variable	Coeff	Pr > F	Variable	Coeff	Pr > F
BD	55.12	0.0021	BD	16.33	0.0490
UC	48.17	0.0043	MACRO	1.93	0.0001
MACRO	- 3.31	0.0161	MWD	- 1.21	0.0001
WORMS	0.84	0.0296	SM2	1.28	0.0010
INFILT	132.91	0.0227	SAND	O.56	0.0032
INTER	-128.35	0.0051	INTER	- 70.36	0.0016
	1991	MAXR		1995	MAXR
	Corn	r^2=.9900		Corn	r^2=.9921
Variable	Coeff	Pr > F	Variable	Coeff	Pr > F
CONE	76.67	0.0016	MACRO	2.24	0.0001
UC	203.52	0.0001	SM1	- 10.01	0.0001
MACRO	- 13.80	0.0004	SM2	11.07	0.0001
WORMS	2.48	0.0053	SLOPE	- 17.76	0.0001
INFILT	414.02	0.0023	INTER	43.68	0.0034
INTER	-313.57	0.0007			
	1992	MAXR			
	Beans	r^2=.9518			
Variable	Coeff	Pr > F			
BD	39.90	0.0004			
EM	0.93	0.0005			
TDR	0.87	0.0117			
INTER	- 59.43	0.0018			

RESULTS AND DISCUSSION

The MAXR regression analysis most accurately predicted yield six out of the seven years compared with the STEP analysis. Stepwise regression analysis predicted yield most accurately for 1993, a year in which precipitation levels were very high in Iowa. In all cases, the regression models did a better job of predicting soybean yield than corn yield. This may be due to the fact that the soil parameters were quantified in a year when the field was cropped to soybeans. Yield predictions were poorest for the sites where yield was consistently low for the seven year period (Table 1). For the consistently high and erratic sites, the models did a fairly good job of predicting yield.

Aggregate size distribution contributed significantly to yield variability in seven out of the seven years, and bulk density, soil moisture, and soil texture in four out of seven years (Table 2). This result emphasizes the importance of soil structure in defining water relations in the soil-plant system. The hierarchical construct proposed by Tisdale and Oades (1982) to describe the agradation and degradation of aggregate structure in soil encompasses both the aggregates themselves and the pore volume that exists between and within the aggregates. Therefore, aggregate size distributions integrate the effect of relatively static soil characteristics, such as texture, mineralogy and organic matter content, with dynamic soil attributes, such as percent water-filled pore space, soil matric potential, and surface-seal formation. This translates into direct and indirect affects on soil water relations and plant-available water, which in turn, can affect yield.

SUMMARY AND CONCLUSIONS

Detailed maps delineating the spatial pattern of crop yield are now possible as a result of recent advances in computer technology coupled with state-of-the-art machinery for planting and harvesting. A more quantitative understanding of soil attribute spatial pattern is needed in order to make informed decisions about how to change agricultural management practices in response to yield patterns. The goal of this study was to functionally relate crop yield patterns observed over a seven year period (1989-1995) with patterns in soil properties measured in 1994 for one field in Boone County Iowa. Information from this analysis will be used in the development of soil quality estimates relative to crop production for this field site.

Multiple linear regression analysis was used to analyze the effect of soil properties on yield variability on a year-by-year basis. The regression coefficients for each best-fit model were used to predict crop yield and the results were compared to observed yields for the corresponding year.

In all cases, the models did a better job of predicting soybean yield than corn yield. Yield predictions differed most from observed values for the sites where yield was consistently low for the seven year period. The models predicted yield fairly accurately for all other sites.

We are interpreting the results of this study with great caution since there are severe limitations to applying multiple linear regression analysis when the number of dependent variables exceeds the number of observations. Predicted and observed values for yield are given for rough comparisons only. The most significant result of this study was the fact that aggregate size distributions contributed significantly to yield variability in seven out of seven years. This result, even if interpreted strictly qualitatively, emphasizes the importance of soil aggregate structure in defining water relations in agricultural systems.

These studies are continuing at this field site and at other field sites throughout Iowa. The spatial pattern of crop yield differs from year to year because of the interaction of spatially and temporally variable soil parameters and climate. Systematic grid sampling for soil attributes and yield has been ongoing for a field site in central Iowa since 1992. The field is sampled once a year after harvest. This study will capture longer-term changes in spatial patterns of soil attributes and how they correlate with changes in yield pattern. In addition, we are grid sampling another field site intensively across the growing season in order to capture the seasonal dynamics of spatial pattern for soil parameters. In the future, we will be integrating climate variables into these models and exploring the use of more sophisticated mathematical tools to better define the interaction of soil and climate variables and their combined effect on crop yield.

REFERENCES

Cambardella, C.A., T.B. Moorman, J.M. Novak, T.B. Parkin, D.L. Karlen, R.F. Turco, and A.E. Konopka. 1994. Field-scale variability of soil properties in central Iowa soils. Soil Sci. Soc. Am. J. 58:1501-1511.

Colvin, T.S., D. B. Jaynes, and D. L. Karlen. 1996. Six-year field variability in central Iowa. (in review, TSAE)

Colvin, T. S. 1990. Automated weighing and moisture sampling for a field plot combine. Appl. Eng. Agric. 6:713-714.

Jaynes, D. B. and T. S. Colvin. 1996. Spatiotemporal variability of corn and soybean yield. (accepted, Agronomy Journal)

Karlen, D.L., M. D. Duffy, and T. S. Colvin. 1995. Nutrient, labor, energy, and economic evaluations of two farming systems in Iowa. J. Prod. Agric. 8:540-546.

SAS Institute. 1985. SAS user's guide. Statistics. Version 5 ed. SAS Inst., Cary, NC

Tisdall, J. M. and J. M Oades. 1982. Organic matter and water-stable aggregates in soils. J. Soil Sci. 33:141-163.

Predicting Corn Yield Across a Soil Landscape in West Central Minnesota Using a Soil Productivity Model

B. R. Khakural
P. C. Robert

Department of Soil Water and Climate
University of Minnesota, St. Paul

A. M. Starfield
Department of Ecology
University of Minnesota, St. Paul

ABSTRACT

A productivity index model was developed for predicting corn yield using soil and landscape characteristics, climatic and management data. Performance of the model was evaluated by comparing model predictions with measured corn yields. Preliminary testing of the model showed promising results. Spatial distribution of soil productivity index and predicted corn yield across a soil landscape in West Central Minnesota followed similar trends with the measured corn yield.

INTRODUCTION

Crop productivity depends on soil, crop management, and climatic factors. The knowledge of soil productivity and crop yield variability is essential for determining site-specific rates of plant nutrients. There is a need for new and more precise approach for predicting soil productivity based on soil, landscape, climatic and management information rather than on yields. The objectives of this research were : 1) to develop a model for predicting soil productivity and corn yield using soil properties, climatic and management information, 2). to test and use the productivity model to predict variation in soil productivity and corn yield across a soil landscape in West Central Minnesota.

MATERIALS AND METHODS

Model Development

A productivity index model was developed to corn yield using soil and landscape characteristics, climatic and management data. Flow chart of the model is presented in Fig. 1. A set of soil and landscape characteristics such as organic matter, available water capacity, permeability, bulk density, soil pH, depth to root restricting layer, drainage characteristics, and slope gradient were used to calculate soil productivity indices (SPI). These soil properties and landscape characteristics were selected on the basis of: multiple regression analysis of selected soil properties with average corn yield or Minnesota crop equivalency ratings (Anderson et al.,

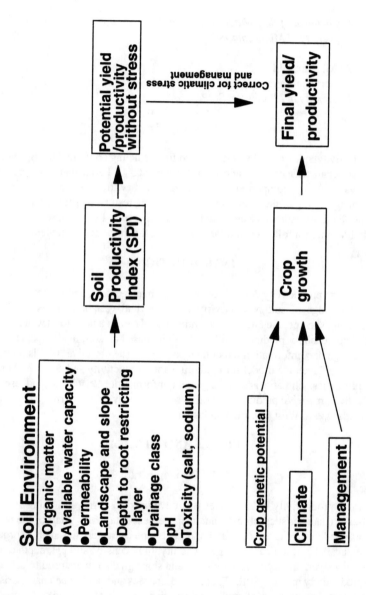

Fig. 1. Flow chart of soil productivity model.

1992) and earlier research work on productivity index (Kiniry et al., 1983; Sinclair, 1992; Walker, 1976). The soil properties, landscape characteristics and rating system listed are tentative and can be changed or modified depending on results of field studies conducted at the Cyrus and Lamberton sites.

Unstressed corn yield was calculated by regressing (relating) SPI with maximum corn yield (Fig. 2) . Soil profile water regime and water stress were simulated (Hanks, 1974; Shaw, 1983; Timlin et al., 1986). Actual corn yield for a particular year was calculated from unstressed yield by correcting for water stress (drought stress) (Fig. 3), early frost and bad management decisions such as late planting.

A frame modeling approach was used to simulate soil profile water regime and water stress (Fig 4). Frame-based modeling is an approach in which sets of simple models are constructed to operate within well defined constraints or frames. Frame model can use qualitative rules along with simple quantitative modules.

Model Testing

Soil productivity indices were calculated using soil information obtained from the state soil survey database (3 SD) of the Natural Resource Conservation Service. SPI values were compared with average corn yield (as reported in county soil surveys), and Minnesota crop equivalency ratings (Anderson et al., 1992). Measured and SPI model predicted corn yields were compared for Ves (*fine-loamy, mixed, mesic Typic Haplustoll*) and Normania (*fine-loamy, mixed, mesic Aquic Haplustoll*) soils at the University of Minnesota-Southwest Experiment Station, Lamberton, MN.

Model Application

The soil productivity model was applied to a soil-landscape (420 X 40 m) at Cyrus, Pope County, MN. Soil productivity indices and corn yields were estimated at 30 m intervals along four transects which are 10 m apart. Soil surface pH and organic matter contents were measured from soil samples collected at each grid point. Bulk density measurements were determined from soil samples collected from 14 selected sites representing the experimental area. Available water capacity (AWC) was estimated from sand, clay, organic matter, and bulk density (Rawls and Brakensiek, 1985). Permeability measurements were obtained from State Soil Survey Database (3 SD). Slope gradient (%) was calculated from topographic survey data (10 X 15 m grid). Weather data such as daily precipitation, air temperatures and pan evaporation were collected at the experimental site or nearest weather station.

RESULTS AND DISCUSSIONS

Relationships between SPI and average corn yield and Minnesota crop equivalency rating are presented in Fig. 5a and 5b, respectively. There was significant correlation between SPI and average corn yield from County Soil Survey ($R^2 = 0.74$) and Minnesota crop equivalency rating ($R^2 = 0.80$).

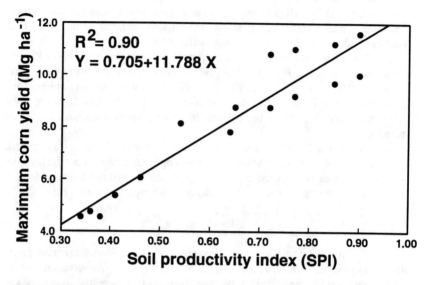

Fig. 2. Relationship between soil productivity index and maximum corn yield.

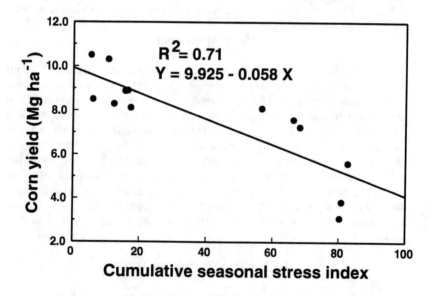

Fig. 3. Relationship between cumulative stress index (drought) and corn yield.

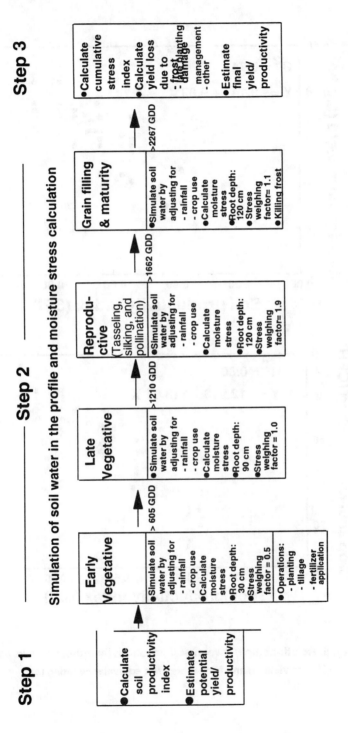

Fig. 4. Different steps and frames used in soil productivity model.

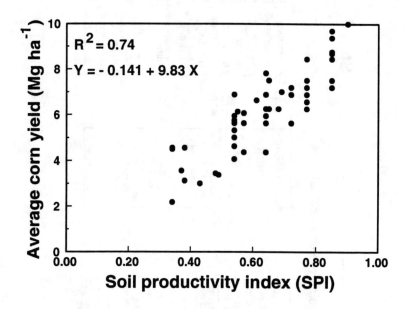

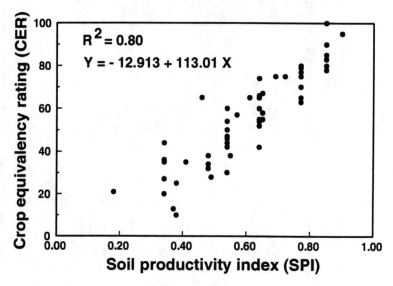

Fig. 5. Relationship between soil productivity index and average corn yield (a) and Minnesota crop equivalency rating (b).

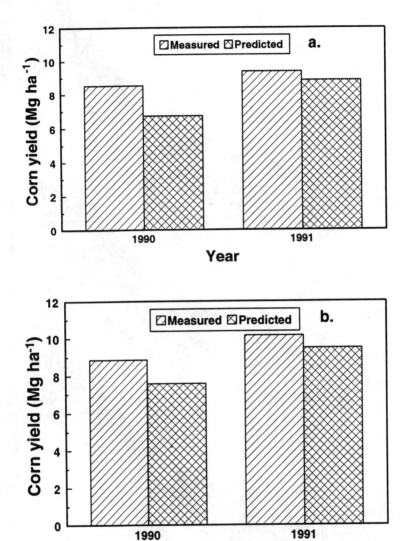

Fig. 6. Measured and SPI model predicted corn yields for Ves (a) and Normania (b) soils at Lamberton, MN.

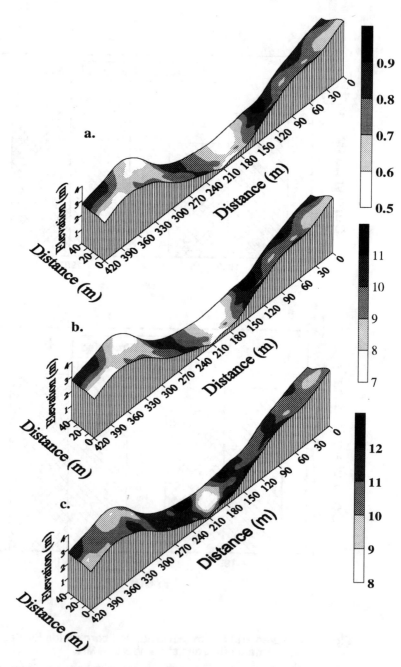

Fig. 7. Spatial variation in SPI (a), SPI model predicted corn yield (b),
and measured corn yield (c) across a glacial till landscape in
west central Minnesota. SPI ranges from 0 to 1. Measured and
predicted corn yields are in Mg/ha.

Measured and SPI model predicted corn yields for Ves and Normania soils at the University of Minnesota-South West Experiment Station, Lamberton, MN are shown in Fig. 6a and 6b, respectively . There was good agreement between measured and model predicted corn yield during 1990 and 1991 growing seasons in Ves and Normania soils.

Spatial distribution of SPI, model predicted corn yield, and measured corn yield across a glacial till soil-landscape are presented in Fig. 7a, 7b, and 7c, respectively. Distribution of SPI and predicted corn yield across the soil- landscape followed similar trends with the measured corn yield. The SPI model identified low yielding areas which are located at the eroded crest/side slopes and at lower toeslope positions in the landscape. However, corn yields predicted at these locations are relatively lower compared to the measured yields. The model needs to be further tested on other landscapes and calibrated under different soil and climatic conditions using large numbers of data sets. A thorough testing and validation of the model with existing data sets from Minnesota and neighboring states is required.

SUMMARY AND CONCLUSION

Preliminary testing and application of the model on a soil-landscape showed promising results. Distribution of SPI and predicted corn yield across the soil landscape followed similar trends with the measured corn yield . The SPI model can be a valuable tool in predicting crop yields across soil-landscapes and site-specific management of plant nutrients. It needs further testing and validation using existing data sets.

ACKNOWLEDGMENT

This project was funded in part by the USDA-NRCS grant no. 68-6322-3-25 and the Minnesota Corn Research and Promotion Council.

REFERENCES

Anderson, J.L., P.C. Robert, and R.H.. Rust. 1992. Productivity factors and crop equivalent ratings for soils of Minnesota. University of Minnesota Agric. Exp. Stn. Tech. Bull. AG-BU-2199-F.

Hanks, R.J. 1974. Model for predicting plant yield as influenced by water use. Agron. J. 66:660-665.

Kiniry, L.N., C.L. Scrivner, and M. E.Keener. 1983. A soil productivity index based upon predicted water depletion and root growth. Res. Bull. 1051. Mo. Agr. Exp. Sta., Columbia.

Rawls, W.J., and D.L. Brakensiek. 1985. Agricultural management effect on soil water retention. p. 115-117. In D. G. DeCoursey (ed.) Proc. of the National Resources Modeling Symp., Pingree Park, Colorado. 16-21 Oct. 1983. ARS 30. USDA-ARS, Beltsville, MD.

Sinclair, H.R. Jr. 1992. Soil rating for plant growth. Unpublished report. National Soil Survey Center. Lincoln, NE.

Shaw, R.H. 1983. Soil moisture and moisture stress prediction for corn in a western corn belt state. Korean J. Crop Sci. 28:1-11.

Timlin, D.J., R.B. Bryant, V.A. Snyder, R.J. Wagenet. 1986. Modeling corn grain yield in relation to soil erosion using a water budget approach. Soil Sci. Soc. Am. J. 50:718-723.

Walker, C.F. 1976. A model to estimate corn yield for Indiana soils. M. S. thesis, Purdue University. Lafayette, Indiana.

Nutrient Mapping Implications of Short-Range Variability

S.J. Birrell
Dept. of Agricultural and Biological Engineering
University of Missouri
Columbia, Missouri

K.A. Sudduth
USDA-Agricultural Research Service
Cropping Systems and Water Quality Research Unit
Columbia, Missouri

N.R. Kitchen
Dept. of Soil and Atmospheric Sciences
University of Missouri
Columbia, Missouri

ABSTRACT

Successful site specific nutrient application depends on accurate soil nutrient maps, which are generally developed from grid samples. The implication of short range variability for soil nutrient mapping is investigated. Interpretation of soil nutrient maps must consider the level of confidence associated with estimated values.

INTRODUCTION

The economic and environmental benefits of site specific nutrient application will only be realized if spatial variability across the field is accurately determined. Generally the spatial variability of soil nutrients is determined by grid sampling. While increasing the sampling intensity is the best way to improve the accuracy of soil nutrient maps, the maximum sampling intensity is limited by cost. Studies have addressed the effect of different grid sampling sizes, determining the degree of misapplication by comparing maps developed from coarser grids containing a subset of the original sampling grid to the reference map developed using all the samples in the original grid (Motz and Searcy, 1993). However, this method assumes that the recorded values for the finest grid are an accurate representation of the actual value at each grid intersection, and does not account for micro- and meso-variability of soil nutrient levels.

Although the interpolation method (Kriging, inverse distance, etc.) can have a significant effect on the maps developed, varying interpolation parameters such as number of neighbors and search radius can have a greater effect than using different interpolation procedures. Nutrient recommendation decisions based on soil nutrient maps should not only consider the magnitude of the value for each cell, but also the reliability of the estimate, and whether any differences are significant or merely a result of uncertainty in the estimates.

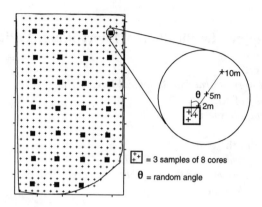

Figure 1 Soil sampling pattern.

METHODS

A 28 ha field was sampled on a 25 m grid, with additional samples taken at the intersection of every fourth row and column (100 m grid) of the original sample grid (Figure 1). Two additional samples were taken co-incident with the original 100 m samples (for a total of 3 samples). Three other samples were obtained 2 m, 5 m and 10 m from the grid point in a straight line on a random bearing (Figure 1).

Each sample consisted of 8 cores to a depth of 6-7 inches obtained within a radius of 50 cm from the sample point. Twenty randomly selected samples were split into sub-samples (duplicates) before submission the to the University of Missouri Soil Testing Laboratory for analysis. Although the samples were collected in a sequential manner, the sample identification numbers were randomized to remove the possibility of systematic bias in laboratory analysis. Unfortunately, results from two additional fields sampled in a similar manner had to be eliminated because statistical analysis proved that the small scale samples (additional co-incident samples, 2 m, 5 m and 10 m samples) were biased. This was traced to systematic error in the laboratory analysis on the particular day the small scale samples were tested.

The data was used to create 6 different data sets for analysis. The first data set consisted of all the data (25 m grid), including all small scale samples. A second set consisted of the 100 m grid samples with all small scale samples. The third data set consisted of all 3 co-incident samples of the 100 m grid. The final three data sets were obtained by assigning one of the 3 co-incident samples on the 100 m grid to each set.

STATISTICAL ANALYSIS

Classical Statistics

General ANOVA methods are based on the assumption that the data samples are independent, but this assumption is violated when there is a spatial relationship between samples. However, the small scale samples were taken on a 100 m grid which is almost at the limit of spatial dependence. Therefore, classical statistical methods were used during preliminary data analysis on the different data sets. During classical analysis, all small scale samples taken at a particular grid point were considered to have the same "location". The 2, 5, and 10 m points were considered to have different "distances", the co-incident samples were different "samples", and the duplicate sub-samples were termed "sub-samples". The different data sets were analyzed using unbalanced ANOVA models (location, distance, sample, sub-sample) to investigate the relative contribution of the different sources of variation.

Geostatistical Analysis

The different data sets were analyzed using geostatistical methods. The semi-variograms of the different data sets displayed no drift provided the semi-variogram was restricted to 2/3 of the maximum lag to prevent edge effects. Therefore, the data was assumed to satisfy the intrinsic hypothesis. The distribution of the complete data set was slightly skewed.

The classical semi-variogram (Matheron, 1963) is affected by skewed data.

$$\gamma(h) \equiv \frac{1}{|2N(h)|} \sum (Z(s_i) + Z(s_j))^2$$

Therefore, the robust semi-variogram estimator developed by Cressie and Hawkins (1980) was used.

$$\gamma(h) \equiv \frac{(\frac{1}{|2N(h)|} \sum |(Z(s_i) - Z(s_j))|^{1/2})^4}{0.457 + \frac{0.0484}{|N(h)|}}$$

The semi-variograms were developed using various lag interval widths to compare the effect on semi-variogram estimation. Theoretical semi-variogram models were fitted using linear (linear model) and non-linear (spherical and exponential models) least squares estimation procedures in SAS. The model parameters were estimated using both a non-weighted and a weighted procedure where weighting was based upon the number of paired observations at each lag distance. The model with the smallest error sum of squares was selected as the best fit model.

A significant advantage of Kriging is that the estimation variance is provided as well as the estimated value. Assuming a normal population distribution, the confidence interval (95%) around the estimate can be determined using

$$(\bar{Z} + (1.96)\sigma_o / \sqrt{n}, \quad \bar{Z} - (1.96)\sigma_o / \sqrt{n},)$$

Simple Kriging was used to develop maps of the Kriged estimate, upper confidence interval and lower confidence interval on a 10 m grid, for each of the different data sets. The correlation between the different maps was determined on a cell by cell basis.

RESULTS

All the pH values for co-incident soil samples were within 10% of their mean value at each location. Approximately 80 percent of the potassium samples were within 10% of their mean value and the rest were within 25% of the mean. However, for phosphorous only 50% of the samples were within 10% of the mean value, 42% of the samples were within 20% and the rest were within 30% of their respective means. Therefore, short range variability can significantly affect individual cell estimates and the maps developed.

The frequency distributions of the different nutrients for the complete data set were slightly skewed. However, the distribution curves for the reduced data sets varied from slightly skewed distributions to highly irregular distribution patterns, although they were samples from the same population. Therefore, even the frequency distributions were affected by the number of samples and short range variability.

Classical Statistics

The assumptions upon which classical ANOVA methods are based were violated. However, the results provided an indication of the relative contribution of the different sources of variation. As expected location within the field was highly significant (<1% level). Distance was a significant parameter for potassium at the 5% level, and not significant for most other soil measurements (pH, neutralizable acidity, calcium, magnesium, potassium). The sample was significant for phosphorous at the 1% level and distance was only significant at the 5% level. Therefore, for phosphorous the location of the individual coring positions for each sample (meso-variability) was more significant than the relative position (macro-variability) of the small scale sampling site.

Geostatistical Analysis

The semi-variograms of pH, P and K for the full data set (Figure 2), were calculated using the classical and robust semi-variogram estimation equations. The range of spatial influence was approximately 100 m for this data set. However, the range of the spatial structure was not maintained when the reduced data sets were used. Many of the reduced data sets exhibited pure nugget variance, and semi-variograms produced from the co-incident data sets did in some cases display significantly different characteristics. Therefore, the co-incident data sets did not have enough data points for the reliable characterization of semi-variograms.

The classical semi-variogram consistently exhibited a higher sill value than the robust semi-variograms. However, the latter tended to have a greater spatial range. The width of the lag interval did not have a significant effect on the determination of the semi-variogram parameters when the weighted least squares procedure was used. However, lag interval width did have a significant effect on the error sum of squares and correlation coefficient between the experimental and fitted semi-variogram model. The correlation between the experimental and theoretical semi-variogram increased as the lag interval width increased. The lag

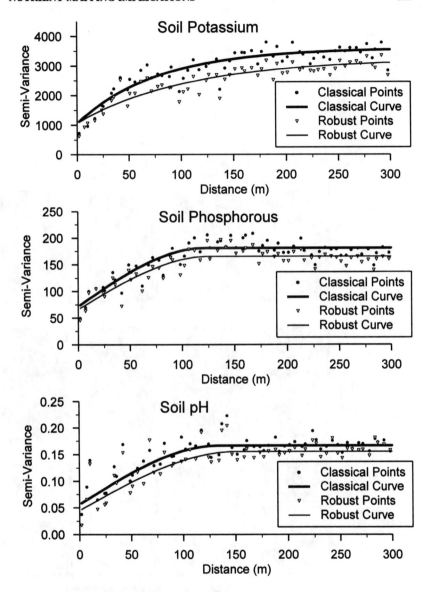

Figure 2 Estimation of semi-variograms using classical and robust procedures (complete data set).

interval did affect the estimated semi-variogram parameters with the non-weighted least squares procedure, particularly with short lag widths, when the number of paired observations could be very low. The relative effect of lag width on the correlation coefficient was more pronounced for the non-weighted least squares procedure than the weighted methods, due to the increased amount of averaging at each lag interval which occurred as the interval width increased.

212

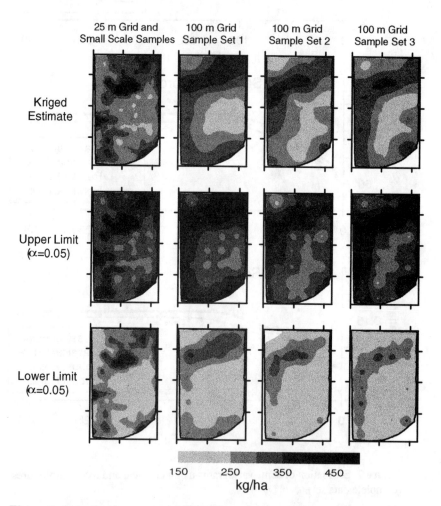

Figure 3 Kriged estimates (top) and 95% confidence intervals (middle and bottom) for soil potassium, developed from the complete data set (left) and the three co-incident 100 m grid sample sets (right).

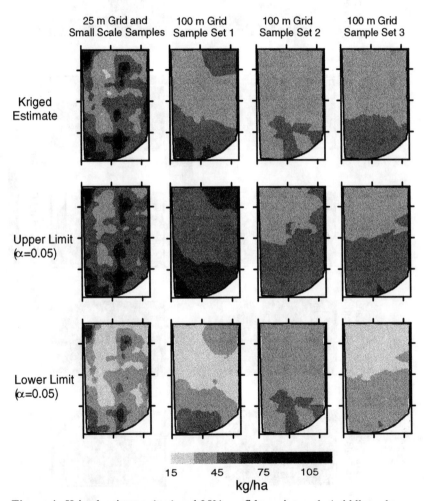

Figure 4 Kriged estimates (top) and 95% confidence intervals (middle and bottom) for soil phosphorous, developed from the complete data set (left) and the three co-incident 100 m grid sample sets (right).

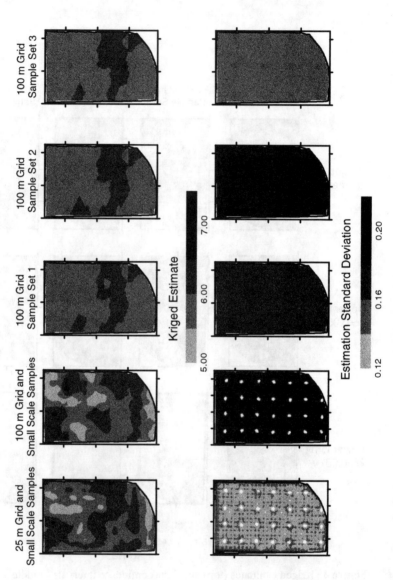

Figure 5 Kriged estimate (top) and estimation standard deviation (bottom) for soil pH, developed from the complete data set (left), 100 m grid samples (2nd left) and the three co-incident 100 m grid

Table I Correlations between Kriged maps developed from the 6 data sets.

	25m Grid Samples[1]	100m Grid Samples[1]	100m Grid Samples[2]	100m Grid Set 1	100m Grid Set 2	100m Grid Set 3
Soil pH						
25m Grid Samples[1]	1.00	0.56	0.70	0.28	0.27	0.33
100m Grid Samples[1]	0.56	1.00	0.63	0.05	0.05	0.11
100m Grid Samples[2]	0.70	0.63	1.00	0.30	0.28	0.37
100m Grid Sample Set 1	0.28	0.05	0.30	1.00	0.98	0.96
100m Grid Sample Set 2	0.27	0.05	0.28	0.98	1.00	0.96
100m Grid Sample Set 3	0.33	0.11	0.37	0.96	0.96	1.00
Phosphorous						
25m Grid Samples[1]	1.00	0.67	0.48	0.42	0.31	0.35
100m Grid Samples[1]	0.67	1.00	0.83	0.75	0.56	0.63
100m Grid Samples[2]	0.48	0.83	1.00	0.89	0.68	0.73
100m Grid Sample Set 1	0.42	0.75	0.89	1.00	0.72	0.76
100m Grid Sample Set 2	0.31	0.56	0.68	0.72	1.00	0.97
100m Grid Sample Set 3	0.35	0.63	0.73	0.76	0.97	1.00
Potassium						
25m Grid Samples[1]	1.00	0.63	0.61	0.60	0.54	0.58
100m Grid Samples[1]	0.63	1.00	0.90	0.81	0.84	0.84
100m Grid Samples[2]	0.61	0.90	1.00	0.92	0.93	0.92
100m Grid Sample Set 1	0.60	0.81	0.92	1.00	0.75	0.75
100m Grid Sample Set 2	0.54	0.84	0.93	0.75	1.00	0.89
100m Grid Sample Set 3	0.58	0.84	0.92	0.75	0.89	1.00

[1] Includes all small scale samples (3 samples at 100m grid and 2, 5, 10m from original point)
[2] Includes small scale samples at grid point (3 samples at 100m grid)

The potassium Kriged maps using the different data sets showed the greatest similarity. The maps from the co-incident 100 m data sets displayed similar trends (Figure 3), and were consistent with the patterns obtained using the complete data set. The range between the upper and lower confidence limits were also similar using the different data sets.

The phosphorous maps (Figure 4) were the least similar. The Kriged maps from the co-incident data sets were significantly different from the map developed using the complete data set and from each other. Similarly, the confidence limits were not consistent among the maps. The lower confidence limit map shows that a significant portion of the field would require a high fertilizer application rate, while the upper map shows that a relatively small portion of the field would respond to fertilizer. Therefore, there is a high probability that fertilizer application based on the Kriged estimates would result in misapplication in some areas.

Soil pH was relatively less variable than phosphorous. The maps based on 100 m grids were similar to the original and to each other (Figure 5). The 100 m grid map which included the small scale samples identified most of the trends seen in the original grid. The maps based on the co-incident samples showed the same basic soil pH levels as the original, but considerable local variability was missed. Although the estimation maps were similar, the different sampling patterns resulted in very different estimation variances. The estimation standard deviation for the complete data set generally was less than 0.14, which is of the same approximate magnitude as the standard deviation of the laboratory analysis. For the 100 m data, the estimation standard deviation depended on the data set used, which means that the semi-variogram developed was not adequately modeled from this data.

The correlation between the different maps was determined on a cell by cell basis (Table 1). The correlation between the original map and those from 100 m grid maps was highest for potassium, whereas the correlations for pH and phosphorous were more variable and generally lower than those for potassium.

CONCLUSIONS

The interpretation of soil nutrient maps must consider the level of confidence associated with estimated values, and the maps must not be automatically accepted as a true representation of the actual conditions in the field. The acceptance of interpolated maps without questioning their quality could compromise the nutrient recommendation procedure. Also, it will not be possible to accurately evaluate the spatial relationship of soil nutrients and yields if the nutrient patterns are not accurately determined.

REFERENCES

Cressie, N., and D.M. Hawkins. 1980. Robust estimation of the variogram. J. of the Int. Assoc. For Mathematical Geology, 12: 115-125.

Matheron, G. 1963. Principles of geostatistics. Economic Geology, 58:1246-1266.

Motz, D.S., and S.W. Searcy. 1993. Interpolation methods for spatially variable data. ASAE paper 93-3561, ASAE, St Joseph, MI

Mapping Techniques and Intensity of Soil Sampling for Precision Farming

S.B.Mohamed
E.J.Evans
R.S.Shiel

Departments of Agriculture & AES
University of Newcastle
Newcastle upon Tyne
NE1 7RU
United Kingdom

ABSTRACT

Three methods were used to delineate areas of a field requiring different rates of P fertiliser. There was little difference between methods, though the Fault Procedure, particularly at low sample density (4 ha^{-1}), gave the smallest error rate. As the sample density increased, the loss of accuracy increased relatively slowly. The impact of precision farming on fertiliser use and cost:benefit analysis is considered.

BACKGROUND

A uniform application of fertiliser across a whole field is still the predominant practice. Where there is spatial variation in soil characters this results in over- or under-fertilisation of part of the area. Over-fertilisation may have a negative impact on drainage water quality and reduces profit margins, while under-fertilisation may restrict crop yields and quality. Variable rate application avoids these problems but requires knowledge of the variability of soil parameters within each field and whether this is random or patterned.

Classical statistics are not appropriate to provide the information on spatial features of a data set such as the location of the high or low values, the trend or degree of continuity hence geostatistical analysis becomes the most common approach taken to study spatial dependence (Vieira et al., 1981; Vauclin et al., 1983; Webster & Oliver,1990). Geographical information systems (GIS), also are a management tool for evaluation and presentation of spatial variation (Smith et al., 1991).

Oliver (1987) described the semi-variogram as the central tool of geostatistics which can quantify the scale and intensity of spatial variation and provide information for local estimation by kriging as well as for optimisation of sampling intensity. Directional trends can also be determined by calculating semivariance along different orientations.

The most convenient way to represent variation is by means of maps of soil properties, but the reliability of current mapping procedures need to be examined.

Soil sampling to determine availability of nutrients and other properties must adequately identify within - field variation. Intensive sampling based on a grid is the most common method. This approach is highly effective, but expensive due to labour and analysis costs.

The greatest accuracy of maps based on the smallest number of samples is required to correctly fertilise the soil and to avoid unnecessary sampling and analytical costs. Cost will be directly proportion to sample number and can be compared with efficiency of attribution in a procedure similar to a fertiliser cost:benefit analysis (Hall, 1921).

Objectives

1. To examine the geographical variability in field soil.
2. To discover if an accurate view of the spatial distribution of soil and crop properties could have been obtained with fewer samples, and hence to estimate the optimum sampling density.
3. To determine the most appropriate mapping method for the available data.

Methods and Techniques

To make an accurate contour map, regularly spaced data is needed, such a dataset can be produced from irregularly spaced field samples by interpolation. There are many different methods of structuring data. All involve interpolation of measured data to grid cells according to their position in the field (co-ordinates of each sample point are a prerequisite for any interpolation).

Methods of gridding and drawing contour maps vary and the "accuracy" of any map will depend on values attached to the options within each interpolation method, the intensity of the irregular point data (number of points/unit area), the total number of measured values and presence of 'faults' or discontinuities, across which the contouring is not constrained to match. All these methods are to some extent arbitrary, some are crude and others can be inappropriate for the type of data (Webster & McBratney, 1987).

Unimap, (UNIRAS A/S., 1989), is a powerful GIS package which supports several interpolation methods: Bilinear, Polynomial, Fault, Minimum curvature, Bicubic and Kriging.

Bilinear approximates existing points onto a grid and interpolates by bilinear interpolation. In this a quadratic function is used to achieve smoothing of variation between points. Values are refined by distance weighing methods. "Faults" or discontinuities are recognised. It is the fastest method and preferred if there is a very large number of data points (more than 1000).

Polynomial fits a fifth order bivariate polynomial to a triangulated set of data after development of a network of triangles covering the surface, starting with pair of points closest together, and working outwards from their midpoint. This method does not recognise "faults". It is preferred when the surface is highly irregular.

Fault is a variation of bilinear interpolation which uses the exact position of the irregular data points rather than estimating a value for the cell mid point, and also recognises "faults". It should be used when discontinuities are known to be present.

Minimum curvature estimates the value of a bivariate function at the grid nodes on a 2D regular grid. A maximum of 200 irregular data points is allowed when this method is used and it may not be very accurate for sparse data.

Bicubic is a grid resembling method which interpolates a new and more (or less) dense gridnet from a regular grid of data points (either read in or interpolated by one of the methods above).

Kriging is a two stage process : In the first stage, the spatial structure of the data is determined from the data semivariogram which is computed from the average difference of all measured values per distance group in various directions. It produces a graph from which the **sill**, **nugget** and **range** values can be interpreted. In the second stage, Kriging is supplied with a model to the data semivariogram for interpolation which is less arbitrary than the other methods (Webster & McBratney, 1987; Isaaks & Srivastava, 1989; Webster & Oliver,1990). Kriging incidentally can test for randomness in the distribution; it has been evaluated by many workers and is becoming a common technique for estimating soil properties and making better fertiliser recommendations (James & Wells, 1990; Mulla, 1991). For this reason we are using it as a standard method.

We neglected using the following: Polynomial because it does not recognise "faults" and requires a highly irregular surface; Minimum curvature deals with small numbers of data points and has poor accuracy for sparse data; Bicubic method requires interpolated data.

Two adjoining fields (Fig. 1) in northern England (55°17'N 1°39'30"W) with a long common boundary were sampled intensively in 1991. Data is presented here primarily for North field (17 ha) from which 244 samples were taken on a grid ~20x40 m. The soil P was extracted by the Olsen method (MAFF 1986) and results placed into one of three index categories - 0 containing 0 to 9, 1 containing 10 to 15 and 2 containing >15 mg P/dm^3 (MAFF 1988). These indices form the bases for P fertiliser recommendations in England and Wales. Soils with a different index will usually receive a different rate of fertiliser application. To examine the effect of sample spacing, maps were produced using each method at 20x40, 40x40 and 60x40 m. The second and third groups were obtained by deleted every second and every second and third sample respectively. This is similar to the method of Voltz & Webster's, (1990), eliminating each measured data point in turn and estimating the value as cross-validation for the mapping technique used. An intuitive map, base on visual inspection of all 244 data points is also presented. Lines were drawn at equal distances between points with smooth curves joining segments. There were 14 isolated points which have been placed in the same class as the predominant surrounding points.

To choose the "best map" realisation method, the actual measured data of available P was superimposed on the contour maps and the proportion of posted data points attributed to the "correct class" calculated. Taking the kriged 20x40 grid map as standard, maps at other grid spacings and produced by other mapping techniques were overlaid to determine the area of agreement. Soil P index areas in common, or disputed between the pairs of maps were recorded and reported as percentage of the area of the whole field. No attempts were made to go beyond 40x60 m spacing because deriving a variogram for fewer than 50 data points is generally not recommended (Wollenhaupt et al, 1994).

RESULTS

The semivariogram (Fig. 2) demonstrates non random - i.e. structured - soil variation and long ranges of influence. The sampling densities all utilised grid spacings which were less than the range. The resulting Kriged map at 20x40 m shows large contiguous areas in each index category (Fig. 1).

The three grid spacings for all mapping methods resulted in visually similar patterns with agreement over 80% or more of the area with the Kriged 20 x 40 m map, and disagreement by no more than one index category. (Table 1).

Reducing the number of samples to one third decreased the accuracy of the maps by only 15% or less using any of the three methods (Fig. 3). It is clear from these results that reducing sample number did not affect map accuracy to the same extent, and therefore the efficiency of the method, based on a cost:benefit approach, was improved.

The proportion of wrongly attributed data points was smallest with hand contouring and increased with grid size in all map realisation methods (Table 2).

Overall, "Fault" is the most effective technique under the conditions of this study probably because it uses the exact position of irregular data while the other methods do not. Isaaks and Srivastava (1989), concluded that different methods may be better for different estimation criteria.

Table(1) Percent areas in North field with soil of P index 0, 1 or 2 based on different sampling intensities

Technique and Intensity	Index	0	1	2
Kriging 20x40 m	0	28.62	8.13	....
V	1	4.48	43.30	2.47
Kriging 40x40 m	2	....	4.51	8/49

Table (2) Data points (% in brackets) falling outside the correct class ranges of available P with different mapping densities and methods based on 244 points.

Spacing (m)	Kriging	Bilinear	Fault
20X40	39 (16)	49 (20)	23 (9)
40X40	61 (25)	57 (23)	39(16)
40X60	70 (29)	70 (29)	59(24)

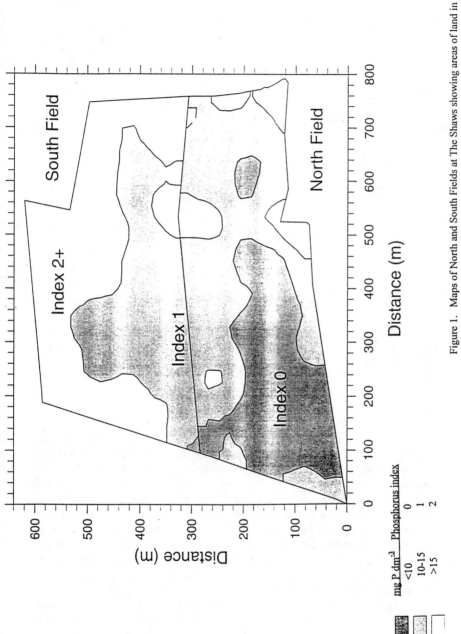

Figure 1. Maps of North and South Fields at The Shaws showing areas of land in each P index category.

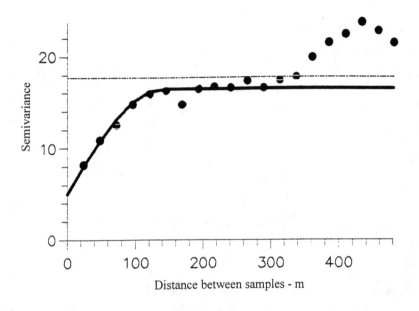

Fig. 2. Semivariogram of available phosphorus from soil samples in North Field,
The Shaws.

DISCUSSION

The large variogram ranges (60 m and above) suggest that sampling intervals of less than 50 m are not necessary. Variogram range values above 100 m in agronomically important soil properties was also reported by Mulla, (1993) and Webster and McBratney, (1987). They suggested an optimum sampling interval of 50-100 m as adequate for the mapping of most of these properties. The lowest sampling density examined in this study (40 x60 m spacing) was adequate to characterise the within-field variation, a result similar to Mulla & Hammond (1988) as reported by Hammond (1993). They concluded that a 60 m grid was adequate for developing soil maps, but that significant detail was lost at 120 m spacing. In practice 120 m would be inappropriate for most mapping techniques unless fields are very large (>70 ha) or several fields were considered at once.

There is little to choose between methods of map realisation at a given sampling density in terms of the % of correctly attributed points. They all correctly identify more than 80% of within-field variation, but Fault seems to be the most accurate method at low sample density. Wollenhaupt et al. (1993) also concluded that the number of soil samples collected is more important than the technique used to map the data as reported by Sawyer (1994). None of the methods improved on hand contouring in terms of percentage of points correctly attributed, but automatic

Figure 3. Agreement (%) between a map based on Kriging samples at 20x40 m and other spacings and methods.

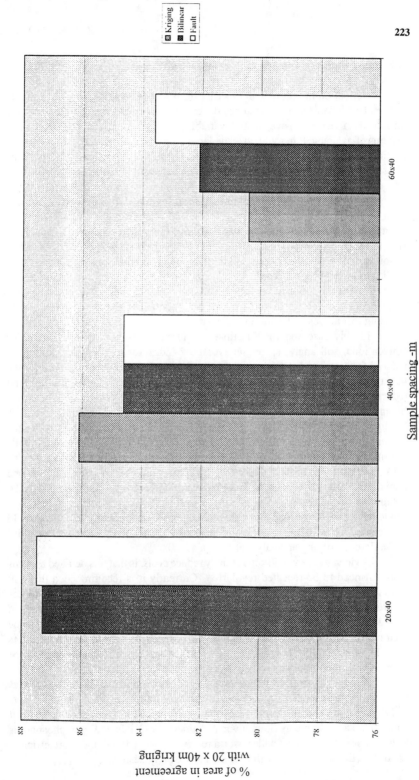

contouring with a computer is more attractive as it can handle large datasets quickly.

Reduction in accuracy in this case decreases less rapidly than the number of samples. The very small accuracy difference with decreased sample density is attributed to the extensive spatial variability which exhibited well-developed patterns of available P in the field. The large range of influence indicates uniform portions of the field having a specified level of soil test values which can be classified into similar categories, and hence a smaller number of samples can adequately represent each area. The large areas of each soil index would make machine application of differential rates of fertiliser application practical. Results from the South field (data not presented), showed an increase in correctly attributed points of soil available P due to fewer categories (only two) compared to the North field (three categories). Accuracy of fertiliser application is also affected by pattern of soil variation. Wollenhaupt et al., (1994), concluded that variable rate fertiliser application would not be expected, on average, to fertilise more than 60% of a field correctly when soils occur in more than one soil test category. However, we obtained 70% or more with two or three categories through selecting appropriate mapping techniques and sampling intensity.

In this case the total fertiliser requirement is not less than that for conventional soil analysis, as the average P index for the field is one but the reduced fertiliser needed on the small area of index 2 does not compensate for the larger amount needed on the substantial area of index zero. There is therefore no reduction in fertiliser use to offset the additional cost of analysis, though there may be an increase in yield as a result of the application of the appropriate rate of fertiliser to the areas with a low soil index. This in fact was the justification for the original introduction of soil analysis (Cooke, 1967). One third of the field falls into index 0, which would justify the application of an additional 50 kg P_2O_5/ha (MAFF, 1988). The basis of expected growth response to P for cereals across one third of the field is equivalent to 75 kg/ha over the whole field (Cooke, 1967). To obtain this it is necessary to carry out four soil analyses per hectare, so that each would have to cost less than the value of 20 kg of wheat (currently $3.30). Arnold & Shepherd's (1990) review suggests a response of only 50 kg/ha across one third of the field, justifying an analytical cost of $2.10.

A view, which would substantially reduce costs, is that this detailed analysis does not need to be repeated regularly. Currently it is recommended that soil analysis be carried out every fifth year (MAFF, 1988). Furthermore the differences in fertility between areas of the field should progressively be reduced or removed by the differential fertilising, this is not occuring with current practices as can be seen by the phosphate isolines which do not respect the current field boundaries (Fig. 1). If variability is reduced in future then less intensive sampling will be necessary, further reducing costs.

Perhaps of greater importance is the environmental benefits such as reduction of the chance of movement of some P into water systems which may cause potential problems of surface water eutrophication and hence, growth of undesirable aquatic plants (Lemunyon & Gilbert, 1993). In addition, phosphorus is the major nutrient with smallest world reserves (Jacob, 1953) and more efficient use of this limited resource will become increasingly desirable.

CONCLUSION

Optimisation of fertiliser inputs can lead to both economic and environmental benefits through careful implementation of mapping techniques. The full benefits of these techniques can only be obtained through a better understanding of the relationship between sampling intensity, cost and agronomic benefits. It is impossible to generalise, as variability in soil properties determines minimum sample density, while the proportion of high and low fertility areas affects yield benefit and fertiliser costs. In this example, map realisation was most efficient by the Fault procedure and map accuracy decreased much more slowly than did sample number to 4 samples/hectare. Assessing the value of environmental benefits such as reduced P run off and improved water quality are more difficult to quantify.

ACKNOWLEDGEMENTS

We wish to thank John and Peter Moore of The Shaws, Felton for generously allowing us access to their land over an extended period.

REFERENCES

Arnold, P.W., and M.A. Shepherd. 1990 Phosphous and potassium requirements of cereals. HGCA Research Review 16. HGCA: London

Cooke, G.W. 1967 The control of soil fertility Crosby Lockwood: London.

Hall, A.D. 1921. Fertilisers and Manures. John Murray: London.

Hammond, M.W. 1993. Cost analysis of variable fertility management of phosphorus and potassium for potato production in central Washington. P. 213-228. *In* Soil Specific Crop Management. P.C. Robert et al. (Eds.) Proceedings of First Workshop of the ASA-CSSA-SSSA, 1992. Madison, WI.

Isaaks, E.H., and R.M. Srivastava. 1989. Introduction to Applied Geostatistics. Oxford University Press, New York. pp. 561.

Jacob, K.D. 1953. Phosphate resources and processing facilities. p.117-166. *In* Fertiliser Technology and Resources. K.O. Jacob (Ed.) Agronomy Monograph III. Academic Press, New York.

James, D.W., and K.L. Wells. 1990. Soil sample collection and handling: Technique based on source and degree of field variability. p. 25-43. *In* Soil testing and plant analysis, 3rd edn. (Westerman, R.L. Ed.). Soil Science Society of America, Book Series No.3. Madison, WI.

Lemunyon, J.L., and R.G. Gilbert. 1993. The concept and need for a phosphorus assessment tool. J. Production Agric. 6: MAFF (1986) The Analysis of Agricultural Materials. Reference Book 427. HMSO London.

MAFF 1988 Fertiliser recommendations. Reference Book 209. HMSO: London.

Mula, D.J. 1991. Using geostatics and GIS to manage spatial patterns in soil fertility. p. 336-345. *In* Proceedings of the 1991 Symposium on Automated Agriculture for the 21st Century. ASAE Publ. 11-91. Am. Soc. Agric. Eng., St. Joseph, MI.

Mulla, D.J. 1993. Mapping and managing spatial patterns in soil fertility and crop yield. p. 15-26. *In* Soil Specific Crop Management. P.C. Robert et al. (Eds.). Proc. of First Workshop of the ASA-CSSA-SSSA, 1992. Madison, WI.

Oliver, M.A. 1987. Geostatistics and its application to soil science. Mulla and Hammond 1988. Soil Use and Management. 3:1, 8-20.

Sawyer, J.E. 1994. Concepts of variable rate technology with considerations for fertilizer application. Journal of Production Agriculture 7:2, 195-201.

Smith, S.M., H.E. Schreier, and S. Brown. 1991. Spatial analysis of forage parameters using geographic information system and image-analysis techniques. Grass and Forage Science 46, 183-189.

UNIRAS A/S 1989. UNimap 2000 Users Manual. Soborg, Denmark.

Vauclin, M., S.R. Vieira, G. Vachaud, and D.R. Nielsen. 1983. The use of co-Kriging with limiting field soil observations. Soil Science Society of America Journal 47: 175-184.

Vieira, S.R., D.R. Nielsen, and J.W. Biggar. 1981. Spatial variability of field measured infiltration rate. Soil Science Society of America Journal. 45: 1040-1048.

Voltz, M., and R. Webster. 1990. A comparison of kriging, cubic, splines and classification for predicting soil properties from sample information. Journal of Soil Science 41, 473-490.

Webster, R. and A.B. McBratney. 1987. Mapping soil fertility at Broom's Barn by simple kriging. Journal of the science of Food and Agriculture 38: 97-115.

Webster, R., and M.A. Oliver. 1990. Statistical Methods in Soil and Land Resource Survey. Oxford University Press.

Wibawa, W.D., D.L. Dludlu, L.J. Swenson, D.G. Hopkins, and W.C. Dahnke. 1993. Variable fertilizer application based on yield goal and soil map unit. Journal of Production Agriculture 6: 255-261.

Wollenhaupt, N.C., R.P. Wolkowski, and M.K. Clayton. 1994. Mapping soil test phosphorus and potassium for variable-rate fertilizer application. Journal of Production Agriculture 7: 441-448.

The Use of Grid Soil Sampling to Measure Soil Nutrient Variation Within Intensively Managed Grass Fields in the UK

M. A. Froment
A. G. Chalmers
S. Peel

ADAS Bridgets Research Centre
Martyr Worthy Winchester
Hampshire
United Kingdom

C. J. Dawson

Chris Dawson Associates
Ox Carr Lane Strensall
York
United Kingdom

ABSTRACT

In the UK, intensive grassland receives large nutrient inputs from applications of inorganic fertilisers and organic manures. As part of a project investigating the impact of dairy systems on environmental pollution, topsoil nutrient status was measured within a sample of permanent grass fields, providing an opportunity to quantify within-field variation in soil total nitrogen (N), extractable phosphorus (P), potassium (K) and magnesium (Mg) and soil pH. In the UK soil nutrient status is known to vary significantly between fields, but variability within fields is rarely considered when making fertiliser decisions. Data is presented for five contiguous fields in a single block of grassland (27.2ha) sampled to a depth of 0-30cm, on a 50m grid pattern. Data from grid points were statistically manipulated to provide nutrient contour maps, by kriging, using 'Surfer' computer software. There were large differences in all soil nutrients within fields. Some of the within-field differences may be explained by cow behaviour, as the highest levels of N, P and K were on the leeward side of a small wood. Soil Mg results also appeared to be influenced by previous management, particularly cow grazing. The results from this study clearly demonstrate that large within-field variation in soil nutrient status can occur in grassland fields. This is likely to be due to non-uniform distribution of dung and urine and/or differential application of fertilisers and manures, variation in nutrient offtake in grass and variation in soil type. Grid soil sampling techniques may be useful in identifying this variation, and varying fertiliser application within fields offers the prospect of reduced environmental pollution and more efficient use of nutrients. The methodologies required to provide reliable nutrient maps in grassland have not been fully investigated and the economic and environmental implications of such an approach under UK conditions have not yet been determined. However, based on the results obtained here, they could be large.

BACKGROUND

Grassland is the most important farm crop in the UK, covering in excess of half of the total farmed area of 11 million hectares. Grassland is grown as short term 'leys' in rotation with arable crops, but mainly as permanent pasture. Grassland systems dominate in the higher rainfall areas in the north and west of the UK and the most productive pastures are often associated with intensive lowland dairy farms which rely on high quality grazed grass in summer and grass silage for winter feeding.

Intensive grassland for silage production receives high inputs of both inorganic and organic nitrogen (Burnhill *et al.*, 1995). Current recommendation systems for nitrogen fertiliser use do not rely upon soil analysis, but an assessment of the requirement of the crop to be grown, making standard allowances for soil nitrogen residues from previous cropping, grass management and manuring (Anon., 1994a). Soil analysis is used for providing recommendations on the amounts of P and K required under grazing or cutting systems. Mg and soil pH are routinely measured, but as grass growth is not susceptible to Mg deficiency, this nutrient is usually only applied when soil Mg is at very low levels or when associated with specific livestock disorders. Soil acidity is corrected by liming, particularly where grass is grown in rotation with arable crops.

National surveys carried out for the UK Ministry of Agriculture have indicated the level of variability in the pH, organic matter and soil nutrient status between fields (Table 1). Guidelines on the methodologies for whole field sampling are available, but variability within grass fields is rarely considered when making fertiliser decisions. As average field size increases in the UK and the agricultural industry responds to economic and environmental pressures to use inorganic and organic fertilisers more effectively and responsibly, there is an increasing need to also consider within-field variation in soil nutrient status. As part of the 'MIDaS' project (Peel, 1995) investigating the impact of dairy systems on environmental pollution, soil nutrient status was measured within a sample of permanent grass fields, providing an opportunity to quantify within-field variation in soil total N, extractable P, K and Mg and soil pH.

METHODS

Data is presented for soil total N, extractable P, K and Mg and soil pH measured in five contiguous fields in a single block of permanent grassland (Table 2) at a depth of 0-30cm, using a regular 50m square grid pattern. Sampling took place in December 1993 and will be repeated using the same sampling plan in 1997. A 0-30cm sampling depth was used to ensure that most of the soil N was included. Total soil N rather than soil mineral nitrogen was measured in this study, as it provides a better estimate of the overall level of N fertility as a consequence of previous management. In the UK, recommendations on P and K fertiliser inputs to grassland are normally based on soil sampling to 7.5cm or 15cm. Soil type within fields was predominantly shallow calcareous silty clay loam overlying chalk, described as Andover and Panholes soil series (Jarvis *et al.*, 1984), and pasture had been managed as grazed grass with occasional cutting. The soil sample at each 'grid point' comprised four cores forming a square pattern 15-20cm around the sample point. Samples were taken using an Eijelkamp gouge auger. Samples were analysed using standard analytical procedures (Anon., 1986). Data from grid points were statistically manipulated to provide a nutrient contour map for the 27.2ha block, by kriging, using 'Surfer' computer software (Anon., 1994b).

Table 1. Soil pH, organic matter (OM) and nutrient status of ley arable (LA - at 0-15cm depth) and grass (G - at 0-7.5cm depth) soils, in England and Wales (% of samples in each category).

	pH (in water)					
	<5.0	5.0-5.4	5.5-5.9	6.0-6.5	6.5-6.9	7.0-7.4
ley arable	0	4	12	27	12	23
grass	2	17	33	24	6	4

	Soil OM (%)					
	<2.0	2.0-4.9	5.0-7.9	8.0-9.9	10.0-12.9	>13.0
ley arable	4	56	31	6	2	1
grass	0	18	40	23	13	6

	Soil P mg/l					
	0-9	10-15	16-25	26-45	46-70	71-100
ley arable	9	21	35	26	7	3
grass	16	24	29	22	6	3

	Soil K mg/l					
	0-60	61-120	121-240	241-400	401-600	601-900
ley arable	4	30	51	11	4	0
grass	4	36	48	10	2	0

	Soil Mg mg/l					
	0-25	26-50	51-100	101-175	176-250	251-350
ley arable	2	16	38	24	7	12
grass	0	3	35	35	15	11

Source: Church, B.M. and Skinner, R.J. (1986).

Table 2. Field areas and histories of fields sampled in December 1993.

Field name	Indiana	Virginia North	Virginia South	Tennessee North	Tennessee South
Area (ha)	6.9	4.7	5.4	4.2	6.0
No. grid points	28	21	23	15	24
Year sown	1967	1979	1979	1973	1973

RESULTS AND DISCUSSION

Soil analysis and mapping

There were large differences in soil total N, P, K and Mg from the individual grid sampled points within the grazing block. The ranges and distribution in soil analysis results for the different nutrients are shown in Figure 1 and kriged maps demonstrating the within field variation in total N, P, K, Mg and soil pH are shown in Figures 2-5.

Soil nutrient levels for most elements were highest in Indiana field, the oldest sward. Indiana and Virginia North are closer to the dairy cow buildings and, historically, are likely to have been more intensively grazed than more distant fields. These results clearly demonstrate the impact of previous field management on soil nutrient status, particularly in grazed grass systems where the effects of grazing or cutting practices can be compounded from season to season.

Soil maps created from grid sampled P and K data suggested that there were similarities in the distribution between the two nutrients within the block of grassland fields and that like total N, these were related to previous field management. P and K levels declined towards the southern boundary and areas of high nutrient concentration, such as in the lee of the wood in Indiana, were evident on both maps. Data for soil P, K and Mg was skewed, with several high values, which in the case of P and K, related to 'hot spots' in Indiana field.

Soil Magnesium also declined towards the southern boundary. Whilst the topsoil in the entire block is a silty clay loam, the depth of soil overlying the chalk depends on the location of a shallow dissecting valley where depth increases. At the northern end soil depth is only 15-30cm overlying chalk (Andover soil series), whilst at the southern boundary this increases between 30cm to 80cm (Panholes soil series). At the sampling depth used in this study of 0-30cm, it would have been expected that soil Mg would have increased rather than decreased towards the southern boundary, as low soil Mg is associated with shallow chalk soils (Jarvis et al., 1984). The results suggest dung and urine from intensive cattle grazing by dairy cows, which also receive supplementary concentrate feeds high in Mg, may explain the high levels of Mg reported on the shallow chalk areas. Soil pH ranged from 6.8 to 8.3, with 95% of values within the range of 7.8 to 8.3. Mean soil pH was 8.0. Further studies examining soil depth to chalk, or measurement of pH may improve

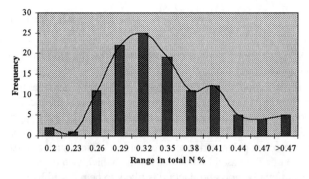

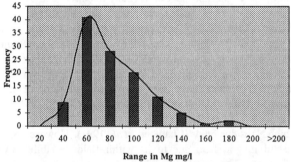

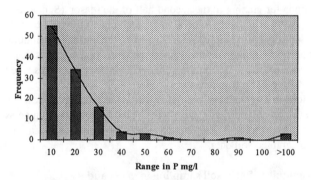

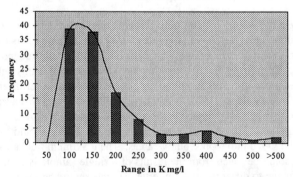

Fig. 1. Frequency distribution of data from individual grid points for total nitrogen (%), magnesium (mg/l), phosphorus (mg/l), and potassium (mg/l).

understanding of the key factors involved in influencing results from nutrient mapping and cropping potential.

Comparing the variation within the 27.2ha block it can be seen that there was greater variation in soil P than total N, although soil P levels in some localised areas were high. Some of these within-field differences may be explained by cow behaviour, as the highest levels of N and P were on the leeward side of a small wood, likely to have been a popular resting area for cattle. In areas such as these, nutrient addition would be increased by dung and urine, which also discourages grazing by cattle. Additionally, grass production is likely to be reduced by shade limiting growth and therefore the potential for nutrient offtake in herbage.

Soil N ranged from 0.15% to 0.54%, with results from individual grid points appearing to follow a normal distribution about the mean. Previous studies investigating N losses through leaching in grassland (Cuttle, 1992), have shown that there can be large differences within fields and that these can occur over short distances. These nutrient 'hot spots' have been associated with dung or urine patches. Results from soil mineral nitrogen measurement (SMN), although expensive, are sometimes used in the UK as a basis for advice on N rates in arable crops, particularly in high N residue fields (Chambers *et al.*, 1991). However, SMN has been less successful in grassland due to its within-field variability over comparatively short distances. The technique has however been used successfully to adjust N applications for grazing in the second half of the season (Scholefield and Titchen 1995). SMN is unlikely to be a viable technique on which to base variable within-field nitrogen use in grassland. However, the normal distribution in total N, which is cheaper to measure than soil mineral nitrogen, suggests that it might be a more stable measurement and have some value in advisory terms when used in conjunction with other information.

Whilst previous management of old fields within the grazing block had a large effect on the levels of soil nutrients, there were also differences within areas previously managed as whole fields, emphasising the complexity in the interpretation of grid sampled soil data.

Economics of grid soil sampling in grassland

In the UK there has been a general trend towards increasing field size to ease management and facilitate the use of larger machinery and large numbers of cattle, and single grassland blocks of the size mapped in this study are not uncommon. Many arable farmers are investigating the possibilities of using soil nutrient maps derived from grid soil sampling as a basis for variable fertiliser application (Anon., 1994c, Budden, 1994). Many large arable producers will also be considering the role of such an approach in the grassland areas they manage.

Under current advisory guidelines for soil sampling in the UK, fields should be initially assessed and divided into representative areas. The maximum field area upon which advice should be given from the results of a single sample is 10ha. Additionally, areas known to differ in some important aspects should be sampled separately. Under whole field sampling a minimum of 25 subsamples are collected using a systematic sampling pattern.

Soil analysis costs in the UK are an important component in the cost of

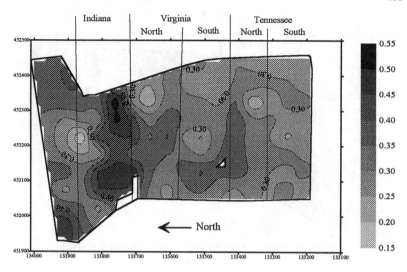

Fig. 2. Field map for soil total N (%) derived by Kriging

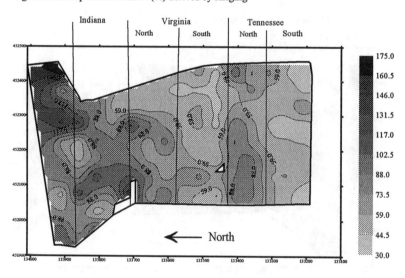

Fig. 3. Field map for extractable Magnesium (mg/l) derived by Kriging

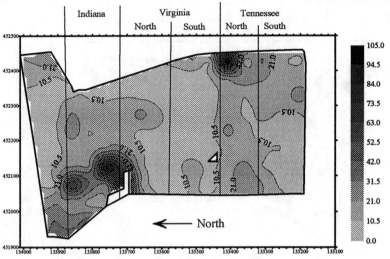

Fig. 4. Field map of extractable phosphorus (mg/l) derived by kriging

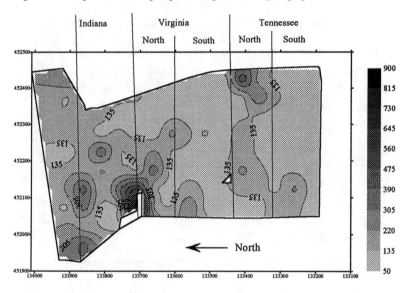

Fig. 5. Field map of extractable potassium (mg/l) derived by Kriging

producing field maps of soil nutrient status. Soil analysis costs for P, K, Mg and pH are approximately US$10 per sample and therefore, at the 50m grid spacing used in this study, analytical costs would have been US$40/ha. A 50m spacing might not be commercially acceptable to farmers and in arable crop systems a 100m spacing has been adopted. Appropriate sampling techniques for grid soil sampling approaches in grassland have not been determined under UK conditions. It is known that there can be significant variation in nutrient status in grassland and that this can occur over relatively short distances and further research will be required to determine whether variation in soil nutrients can be quantified by a more refined sampling approach.

Despite the high cost of grid soil sampling, the results from this study have highlighted the potential benefits from such an approach. There is potential for significant savings in fertiliser costs in P and K by limiting applications in the Indiana field. P and K fertiliser applications under grazed grass systems are low, but if grass in areas of high nutrient status were cut for forage which has significantly higher fertiliser requirements, there could be savings of up to US$60 ha/year. The benefits from limiting fertiliser additions in areas of high nutrient status also have environmental benefits.

Whilst accepting the technical difficulties in accurately mapping nutrients in grassland fields and its cost, benefits are likely to accrue over several seasons. Detailed grid mapping in one season also provides a rational basis for subsequent field division for targeted soil sampling in following years.

CONCLUSIONS

The results from this study clearly demonstrate that large within-field variation in soil N, P, K and Mg can occur in grassland fields in the UK. This is likely to be due to several factors including variation in soil type or depth, non-uniform distribution of dung and urine from grazing animals, and/or differential application of fertilisers and manures, and variation in grassland productivity resulting in differential nutrient offtakes. Grid soil sampling techniques may be useful in identifying this variation, and varying fertiliser application within fields offers the prospect of reduced environmental pollution and more efficient use of nutrients. As average field size increases in the UK there is a need to evaluate whether alternative techniques such as grid soil sampling provide an improved system on which to offer fertiliser advice. In the UK, developments of mapping systems in other arable crops are progressing rapidly and variable within-field fertiliser application, based on soil nutrient maps derived from grid soil sampling, are being used for the application of P, K, Mg and lime. However, the sampling methodologies that should be applied to obtain reliable nutrient maps in grassland are unproved. Division of large grazing blocks into smaller areas based on previous experience and local knowledge, coupled to established systematic sampling procedures, remain a reliable option for farmers. The economic and environmental implications of a mapping approach have not yet been determined, but it offers the potential for a more rational use of fertiliser which could, based on the results obtained here, provide large benefits.

ACKNOWLEDGEMENTS

The 'MIDaS' project is funded through joint UK government and industry funding by ADAS, BOCM Pauls, Borregaard Ligno Tech, Hydro Agri, Linbury Trust, Livestock Systems, the Maize Growers Association, Ministry of Agriculture, Fisheries and Food, Trident Feeds and Wessex Water. Thanks are also due to Rachel Barton for assistance in the preparation of the manuscript.

REFERENCES

Anon. 1986. The analysis of agricultural materials. MAFF ADAS Reference Book 427. London: HMSO.

Anon. 1994a. Fertiliser recommendations for Agricultural and Horticultural Crops. Sixth edition. MAFF ADAS Reference Book 209. London: HMSO.

Anon. 1994b. Surfer. Contouring and 3D surface mapping, user's guide. Golden Software Inc., Colorado, USA.

Anon. 1994c. Field mapping and spreading by GPS. Arable Farming. May 1994. p59.

Budden, A.L. 1994. A soil analysis sampling method to determine pH, P, K and Mg fertiliser application using variable rate technology. Sampling to make Decisions. Aspects of Applied Biology 37 pp. 281-282.

Burnhill, P.M., A.G. Chalmers, and J. Fairgrieve. 1995. The British Survey of Fertiliser Practice. Fertiliser use on farm crops 1994. pp33 MAFF, Edinburgh. HMSO. (ISBN 0 11 495304 X).

Chambers, B.J., M.A. Shepherd, and L.V. Vaidyanathan. 1991. Research and advisory applications of soil mineral nitrogen measurements. Soil Use and Management, 7:66.

Church, B.M., and R.J. Skinner. 1986. The pH and nutrient status of agricultural soils in England and Wales. Journal of Agricultural Science (Cambridge) 107: 21-28.

Cuttle, S.P. 1992. Spatial variability and the use of ceramic cup samplers to measure nitrate leaching from pastures. pp. 71-74. In Nitrate and Farming systems. Aspects of Applied Biology 30.

Jarvis, M.G., R.H. Allen, S.J. Fordham, J. Hazelden, A.J. Moffat, and R.G. Sturdy. 1984. pp. 84. In Soils and their use in South East England. Bulletin No. 15 Soil Survey of England and Wales, Harpenden.

Peel, S., A.G. Chalmers, and P.J.A. Withers. 1995. The MIDaS project: Developing environmentally acceptable dairy systems for the UK. pp. 118. In Proceedings of the Symposium on Applied Research for Sustainable Dairy Farming. Research Station for Cattle, Sheep and Horse Husbandry, Lelystad, The Netherlands May 31-June 2 1995.

Scholefield, D. and N.M. Titchen. 1995. Development of a rapid field test for soil mineral nitrogen and its application to grazed grassland. Soil Use and Management 11:33-44.

Soil-Test Variability in Adjacent Iowa Fields

D. L. Karlen
C. A. Cambardella
T. S. Colvin

USDA-ARS
National Soil Tilth Laboratory (NSTL)
Ames, IA

ABSTRACT

Small-scale spatial variability of selected soil-test parameters for two adjacent fields in central Iowa is discussed. We used semivariance analysis to determine the distance to which parameters were correlated and to estimate the strength of that correlation. Distinct differences in spatial dependence patterns were observed for the two farming systems.

BACKGROUND

Precision agriculture requires three primary components: (I) global positioning to know where equipment is located, (II) real-time mechanisms for controlling nutrient, pesticide, seed, water, or other crop production inputs, and (III) databases that provide information needed to develop an appropriate input response to various site-specific conditions. Technologies associated with requirements I and II are well advanced compared to development and interpretation of databases that are needed to obtain the full economic and environmental benefits of precision agriculture. Assessing soil fertility is one of the first operations needed to develop databases for precision agriculture.

Franzen and Peck (1995) stated that addressing soil heterogeneity is one of the oldest challenges facing farmers and agricultural researchers. They also acknowledged that long ago, extensive soil sampling was recognized as a basis for site-specific fertilizer application. However as fertilizers became less expensive, it was simpler to treat entire fields than to complicate the process with spot application. As a result, soil sampling protocol was generally based on finding a central tendency for a field or field segment. The development of computer-controlled fertilizer application equipment and global positioning systems have renewed challenges to develop efficient sampling and mapping procedures that accurately define spatial variability (Wollenhaupt et al., 1994; Franzen & Peck, 1995; Pierce et al., 1995).

Several anthropogenic (controllable) and nonanthropogenic factors affect soil fertility (Karlen & Sharpley, 1994) and contribute to the spatial variability encountered within a sampling area (Cambardella et al., 1994). Landscape position, especially through its impact on water relations, is a major nonanthropogenic factor that affects not only soil fertility parameters but also crop growth and yield (Pierce et al., 1995). Tillage, crop rotation, and application of manure are controllable factors that can have a major impact on physical, chemical, and biological parameters within a soil.

Recent studies conducted to determine appropriate soil-test sampling protocol for precision agriculture have suggested that grid points should not be spaced more than 66 to 100 m apart (Wollenhaupt et al., 1994; Franzen & Peck, 1995; Pierce et al., 1995). However, those studies were not conducted in fields with grid points located closer than 25 m, and there was no indication that animal manure or municipal sludge was being applied to any of the sites. The objective for this study was to examine spatial variability in soil test parameters within 1.5- and 2.3-ha segments of adjacent 16 ha fields in central Iowa that had distinctly different long-term management histories. A 15-m grid spacing was chosen to correspond to plot dimensions for which crop yield was being measured.

METHODS AND MATERIALS

Field Histories

The 1.5 ha sampling area was located in the center portion of a 16 ha field that had been managed for the past 25 y in a 5-year corn (*Zea mays* L.), soybean [*Glycine max* (L.) Merr.] corn, oat (*Avena sativa* L.), and mixed hay rotation with ridge-tillage for row crops and manure/municipal sludge as the primary nutrient source (Alternative System). Total N-P-K application for 1984 through 1993 was 1859-511-1780 kg ha^{-1}. The 2.3 ha area was located within an adjacent 16 ha field that has been in a 2-year, conventionally tilled and fertilized corn and soybean rotation since 1958 (Karlen et al., 1995) (Conventional System). Total N-P-K application for 1984 through 1993 was 810-187-478 kg ha^{-1}.

Sampling Design

Three, 5-cm diameter soil cores were collected within a 1-m radius at each grid point with a hydraulic soil sampler. Cores were fractionated into 0 to 5-, 5 to 10-, 10 to 20-, and 20 to 30-cm depth increments and composited. Samples were air dried, crushed, passed through a 2 mm sieve, and analyzed for pH, Bray extractable P, exchangeable K, Ca, and Mg, and total organic C and N. Corn grain yield was measured in 15 m increments, centered on each grid point with a modified commercial combine (Colvin, 1990). The Conventional System grid was sampled in 1992 and the Alternative System grid in 1993. To compensate for temporal differences in yield as a result of this, relative grain yield for each grid point was computed by dividing grid point yields by the average for each sampling grid.

Statistical Analyses

Data analysis for each of the grids was done in three stages: (i) normality tests were conducted; (ii) distributions were described using traditional summary statistics (mean, standard deviation, and CV); and (iii) semivariograms were defined and differences in nugget and total semivariance and range examined for the variables . Normality tests were conducted using PROC UNIVARIATE and non-normal data were log-transformed to stabilize the variance (SAS Institute, 1985). Geostatistical software (GS+, Gamma Design Software, St. Plainwell, MI) was used to analyze the spatial structure of the data and to define the semivariograms. Semivariance was calculated for log-transformed data. The active lag distances used for calculation of semivariance were 122 m and 182 m for the Alternative and Conventional System respectively. Selection of models for semivariograms was made based on visual fit and the r^2 values for the regression. Isotropic models were fitted in all cases.

RESULTS AND DISCUSSION

Almost every soil-test parameter showed significant farming system differences at every sampling depth (Table 1). This was not unexpected since nutrient management, crop rotation, and tillage practices were distinctly different for the two fields. In addition, the results may reflect temporal variability since the two systems were sampled in different years. Time of sampling probably affected crop yield more than the soil parameters. When the yields were adjusted for seasonal differences, relative yields between the two systems weren't significantly different (Table 1).

The soil properties displayed differences in spatial dependence, as determined by semivariance analysis (Table 2). Spherical isotropic models were defined for most of the soil variables measured at all depths for both sites. Notable exceptions were soil nitrate-N for the Alternative site and extractable calcium and magnesium for the Conventional site, which were defined by linear models. For spherical models, semivariance increases with distance between samples (lag distance) to a constant value (sill or total semivariance) at a given separation distance (range of spatial dependence). For linear models, as the space between the two samples increases, the difference between them also increases. If the linear model has a slope that is close to zero, then the total variance equals the nugget variance, and the variables are described as randomly distributed. The semivariogram for soil nitrate-N at the Alternative site exhibited a slope close to zero, suggesting nitrate-N is randomly distributed at this site. Calcium and magnesium at the Conventional site both had positively increasing linear slopes.

The nugget variance expressed as a percentage of the total semivariance was used to define distinct classes of spatial dependence for the soil variables. If the ratio was less than or equal to 25%, the variable is considered strongly spatially dependent; if the ratio was between 25 and 75%, the variable was considered moderately spatially dependent; and if the ratio was greater than 75%, the variable was considered weakly spatially dependent (Cambardella et al. , 1994).

The semivariance parameters for soil variables for the two farming systems were distinctly different although there was no consistent spatial pattern at either site with depth of sampling (Table 2). Soil pH, total organic C and total N were strongly spatially correlated at all depths for the Conventional system (Table 2). Previous research conducted on conventionally managed corn/soybean systems showed similar trends (Cambardella et al., 1994). For the Alternative system, Table

1. Descriptive statistics for Conventional and Alternative System Grids

	******** Mean	Conv. SD	***** CV	******** Mean	Alt. SD	***** CV
pH						
0-5 cm	5.80	1.0	18	7.62	0.3	3
5-10 cm	5.73	1.1	19	7.52	1.0	13
10-20 cm	5.90	1.1	18	7.65	0.5	6
20-30 cm	6.18	1.0	16	7.58	1.1	14
P						
0-5 cm	30.21	16.9	56	57.51	37.4	65
5-10 cm	18.05	10.0	56	27.85	25.6	92
10-20 cm	10.30	6.8	66	9.66	9.2	95
20-30 cm	7.47	5.2	70	6.65	5.5	83
K						
0-5 cm	160	81	51	240	69	29
5-10 cm	102	54	53	122	40	33
10-20 cm	91	49	54	116	27	24
20-30 cm	95	33	35	110	27	24
TC						
0-5 cm	2.72	1.4	52	5.08	0.7	14
5-10 cm	2.46	1.4	57	4.12	1.0	25
10-20 cm	2.44	1.4	58	3.86	0.9	24
20-30 cm	2.42	1.6	65	3.23	0.9	29
TN						
0-5 cm	0.22	0.1	44	0.40	0.1	13
5-10 cm	0.20	0.1	47	0.31	0.1	23
10-20 cm	0.19	0.1	48	0.28	0.1	21
20-30 cm	0.18	0.1	49	0.22	0.1	29
Ca						
0-5 cm	3338	2473	74	4689	762	16
5-10 cm	3152	2400	76	4647	1014	22
10-20 cm	3282	2237	68	4794	922	19
20-30 cm	3883	2542	66	4533	1044	23
Mg						
0-5 cm	290	102	35	409	99	24
5-10 cm	271	83	31	392	121	31
10-20 cm	277	74	27	399	115	29
20-30 cm	347	123	35	413	126	31
NO3						
0-5 cm	69.86	43.0	62	28.08	6.3	22
5-10 cm	48.34	36.2	75	22.98	7.0	31
10-20 cm	20.53	13.1	64	21.11	4.4	21
20-30 cm	10.50	5.3	51	19.23	5.0	26
NH4						
0-5 cm	18.10	9.9	55	19.51	13.4	69
5-10 cm	23.16	15.6	68	11.57	5.4	47
10-20 cm	15.85	9.1	57	8.65	4.4	51
20-30 cm	10.12	8.2	81	6.94	4.2	61
Rel Yield	100	20.2	20	100	19.1	19

Table 2. Semivariance parameters for Conventional and Alternative System Grids

	Ratio	Conv. Class	Range	Ratio	Alt. Class	Range
pH						
0-5 cm	0.000001	S*	278	0.10	S	74
5-10 cm	0.000001	S	256	59.0	M	36
10-20 cm	0.000001	S	260	0.042	S	59
20-30 cm	0.000001	S	316	–	R	>122
P						
0-5 cm	54.8	M	78	7.4	S	97
5-10 cm	31.4	M	47	16.2	S	88
10-20 cm	48.4	M	59	17.9	S	70
20-30 cm	0.035	S	57	13.7	S	67
K						
0-5 cm	24.1	S	37	27.4	M	40
5-10 cm	62.8	M	40	51.8	M	91
10-20 cm	42.7	M	55	8.8	S	78
20-30 cm	–	R	>182	24.2	S	102
TC						
0-5 cm	4.7	S	78	23.8	S	140
5-10 cm	4.3	S	320	27.5	M	40
10-20 cm	16.7	S	262	0.20	S	39
20-30 cm	15.5	S	324	–	R	>122
TN						
0-5 cm	6.1	S	84	43.5	M	129
5-10 cm	9.4	S	275	42.3	M	38
10-20 cm	14.6	S	234	5.6	M	66
20-30 cm	20.8	S	252	22.6	M	96
Ca						
0-5 cm	24.1	S	102	0.036	S	67
5-10 cm	–	R	>182	59.8	M	36
10-20 cm	–	R	>182	0.20	S	62
20-30 cm	31.7	M	203	36.3	M	131
Mg						
0-5 cm	–	R	>182	0.10	S	66
5-10 cm	64.0	M	120	5.3	S	64
10-20 cm	–	R	>182	7.7	S	70
20-30 cm	–	R	>182	11.1	S	85
NO3						
0-5 cm	63.5	M	210	–	R	>122
5-10 cm	70.2	M	168	–	R	>122
10-20 cm	59.1	M	73	0.30	S	42
20-30 cm	–	R	>182	–	R	>122
NH4						
0-5 cm	72.7	M	42	19.1	S	31
5-10 cm	–	R	>182	52.7	M	27
10-20 cm	56.4	M	260	1.2	S	36
20-30 cm	17.9	S	69	24.0	S	34
Rel Yield	0.1	S	72	0.10	S	46

* S = Strong, M = Moderate, R = Random

Bray's P and extractable magnesium were strongly spatially correlated at all depths and soil ammonium-N was strongly spatially correlated for three out of four of the depths. The strong spatial dependence of P and ammonium-N is likely a result of long-term manure application at this site. It has been hypothesized that strongly spatially dependent soil variables may be controlled by intrinsic variations in soil characteristics such as texture and mineralogy, whereas extrinsic variations, such as fertilizer application and tillage, may control the variability of less strongly correlated soil properties (Rao and Wagenet, 1985; Cambardella et al., 1994). The results presented here suggest that this may be true only for conventional systems with one or two crops, intensive tillage, and high chemical inputs. Extrinsic factors, such as crop rotation and manure application, appear to play a larger role in controlling strong spatial dependence for systems with greater than two crops, less-intensive tillage, and low chemical inputs.

The range values were consistently higher for the Conventional system compared with the Alternative system (Table 2). For the strongly spatially dependent properties at the Conventional site, range distances encompassed several map units and were generally greater than 200 meters (Table 2). These results suggest the spatial pattern may be related to landscape position rather than soil type at this site. For the strongly spatially dependent properties at the Alternative site, the range values were always less than 100 meters. Tillage and nutrient-management protocols likely contributed substantially to the observed small-scale spatial variability at this site.

SUMMARY AND CONCLUSIONS

This study was undertaken to examine the small-scale spatial variability of selected soil-test parameters for two adjacent field sites with known long-term soil and crop management histories. The study demonstrated significant farming system effects, although they were expected because of distinct differences in crop rotation and nutrient management practices. Semivariance analysis was used to determine the distance to which soil-test parameter values were spatially correlated and to estimate the strength of that correlation. Distinct differences in spatial dependence patterns were observed for the two farming systems. For the Conventional system, soil properties exhibiting strong spatial dependence were spatially related across distances that spanned several map units and were correlated with intrinsic soil characteristics. Strong spatial dependence occurred at a smaller scale for the Alternative site and appeared to be related predominantly to tillage and nutrient management strategies.

These results have important implications for how and when site-specific information should be collected and interpreted. Most of the spatial variability at the Conventional site could have been captured by using relatively coarse grid-spacing, certainly larger than the 60-100 m recommended in the literature. The Alternative system would require a finer grid-spacing to describe the smaller-scale spatial variability occuring at this site. The strong small-scale spatial dependence exhibited at the Alternative site may suggest temporal variability as well, since the soil biochemical processes responsible for nutrient transformations are known to vary across the growing season. On the other hand, strong spatial dependence at the

Conventional site appears to be related to intrinsic soil characteristics that don't change substantially across the growing season.

REFERENCES

Cambardella, C.A., T.B. Moorman, J.M. Novak, T.B. Parkin, D.L. Karlen, R.F. Turco, and A.E. Konopka. 1994. Field-scale variability of soil properties in central Iowa soils. Soil Sci. Soc. Am. J. 58:1501-1511.

Colvin, T.S. 1990. Automated weighing and moisture sampling for a field plot combine. Appl. Eng. Agric. 6:713-714.

Franzen, D.W., and T.R. Peck. 1995. Sampling for site-specific application. pp. 535-551. *In* P.C. Robert, R.H. Rust, and W.E. Larson (ed.) Site-Specific Management for Agricultural Systems. ASA-CSSA-SSSA, Inc., Madison, WI.

Karlen, D.L., and A.N. Sharpley. 1994. Management strategies for sustainable soil fertility. pp. 47-108. *In* J.L. Hatfield and D.L. Karlen (ed.) Sustainable Agriculture Systems. CRC Press, Inc., Boca Raton, FL.

Pierce, F.J., D.D. Warncke, and M.W. Everett. 1995. Yield and nutrient variability in glacial soils of Michigan. *In* P.C. Robert, R.H. Rust, and W.E. Larson (ed.) Site-Specific Management for Agricultural Systems. ASA-CSSA-SSSA, Inc., Madison, WI.

Rao, P.S.C, and R. J. Wagenet. 1985. Spatial variability of pesticides in field soils: Methods for data analysis and consequences. Weed Sci. 33 (Suppl. 2):18-24.

SAS Institute. 1985. SAS user's guide. Statistics. Version 5 ed. SAS Inst., Cary, NC

Wollenhaupt, N.C., R.P. Wolkowski, and M.K. Clayton. 1994. Mapping soil test phosphorus and potassium for variable-rate fertilizer application. J. Prod. Agric. 7:441-448.

Characterization and Immobilization of Cesium -137 in Soil at Los Alamos National Laboratory

N. Lu
C. F.V. Mason
J.R. Turney

CST-7 Environmental Science and Waste Technology
Chemical Science & Technology Division
Mail Stop J 514
Los Alamos National Laboratory, NM

ABSTRACT

At Los Alamos National Laboratory (LANL), cesium-137 (^{137}Cs) is a major contaminant in soils of Technical Area 21(TA-21) and is mainly associated with soil particles ≤ 2.00 mm. The ^{137}Cs was not leached by synthetic groundwater or acid rainwater. Soil erosion is a primary mechanism of ^{137}Cs transport in TA-21. The methodology that controls soil erosion can prevent the transport of ^{137}Cs.

INTRODUCTION

The TA-21 at Los Alamos National Laboratory is situated on DP Mesa and was used primarily for research and production of plutonium metal from 1946 to 1978. Solid Waste Management Unit (SWMU) 21-011(k) is located on the south wall of DP Canyon. The soils in SWMU 21-011(k) contain multiple contaminants which originated from discharges of acid-waste treatment facilities and spills from holding tanks. The initial radiological survey in 1992 and 1993 showed that americium-241, cesium-137, plutonium-239+240, and strontium-90 exceeded their respective soil Screening Action Levels (SAL). The radioactive levels of ^{137}Cs in surface soil range from 1,578 to 119,362 Bq kg^{-1} soil, which is 8 to 632 times as much as its SAL (188.7 Bq kg^{-1})(LA-UR-94-228, 1994).

Cesium-137 interacts strongly with micaceous clay minerals (Maule & Dudas, 1989) and organic matter in soils (Comans & Hockley, 1992). It was reported (Robbins et al., 1992) that about 90% of Chernobyl ^{137}Cs fallout in the sediments of lakes was strongly bound with clay minerals and was significantly transferred on non exchangeable sites of clay minerals. Irreversible sorption of ^{137}Cs on soil clay becomes significant over time scales of weeks to months. In this case, Cs migrates to interlayer sites of clay minerals, from which it is not easily released (Comans & Hockley, 1992). Cesium-137 was unlikely to be leached from contaminated soils, even under prolonged high rainfall (Kirk & Staunton, 1989). An average vertical migration rate of ^{137}Cs in an unsaturated soil layer was 1.0 mm yr^{-1} when an average rate of movement of soil water was 2500 mm yr^{-1} during the forty years after the explosion of the atomic bomb at Nagasaki, Japan in 1945 (Mahara, 1993). More than 95% of ^{137}Cs deposited as fallout was above a depth of

10 cm in the soil, and no ^{137}Cs was detected below 40 cm.

Understanding the characterization and reaction of ^{137}Cs in soil helps us to search for proper methods of controlling transport of ^{137}Cs and stabilizing it. The objectives of the research were to investigate the characterization of ^{137}Cs in the soils; to examine the leachability of ^{137}Cs in soil; and to test a potential approach for stabilization of ^{137}Cs. Five samples, including 1) relatively high level radioactive contaminated soil (Hot Spot), 2) surface soil (Surface), 3) wet sediment (Sediment), 4) drainage channel water, and 5) pine bark, were studied.

MATERIALS AND METHODS

Sampling, Soil Properties and Radiological Analysis

Five samples were taken from SWMU 21-011(k) area. Relatively high level radioactive contaminated soil was collected from a discharge point (Hot Spot) close to the top of DP Canyon wall. The surface sample was collected from soil (0 to 15 cm deep) close to the bottom of DP Canyon. The wet sediment (0 to 15 cm deep) was collected at the outfall of a drainage channel which passes through the hot spot area and drained into the canyon. The water sample was separated from the wet sediment sample by centrifuging. The bark sample was collected from a Ponderosa pine [*Pinus ponderosa* (Dougl. ex Laws)] tree at approximately 1-m height from the ground, which is growing close to the bottom of DP Canyon. The soil samples were air dried in a fume hood, passed through a 4-mm U.S. standard sieve and stored for use. Soil samples at particle-size fraction of 4.0 - 2.0 mm, 2.0 - 0.053 mm and < 0.053 mm were prepared by passing 100 g of soil ($\leq$ 4.0 mm) consecutively through a 2.0-mm and a 0.053-mm sieves. These samples were used for determination of ^{137}Cs distribution in soil particle size fractions. The pine bark sample was dried in an oven at 60°C for 72 h, ground and passed through a 0.5-mm standard sieve.

The pH of soil samples ranged from 7.7 to 8.2, which was measured on soil: water of 1:1 ratio. The Hot Spot and Sediment soils consist of 10% fine gravel, 50% or more of sand, 25% silt, and 10% clay, and low organic carbon content, ranged from 6 to 23 g kg^{-1} soil. Organic carbon contents of the soils were determined by dichromate oxidation method (Nelson and Sommers, 1982).

The gamma activity of soils and pine bark were measured as follows: five grams of air-dried soil were placed in a counting container and analyzed by gamma spectroscopy. Measured activities of ^{137}Cs were decay corrected and, hence, activities reported in this paper correspond to initial activities on March 7, 1995 (sampling time, t = 0).

Column Leaching

Hot Spot and Sediment soils were leached using solutions of synthetic groundwater (pH 8.0) (Elless et al., 1994), synthetic acid rainwater (pH 4.1), 0.5 M ammonium acetate (NH$_4$OAc, pH 7.0), and a mixture of 0.5 M sodium citrate-0.1 M sodium bicarbonate-0.1 M sodium dithionite (CBD, pH 7.2), respectively.

The glass funnels contained porous stone as base, with an 80-ml capacity, and were utilized as columns to perform the leaching process. Seventy grams of soil with particles of $\leq$ 4.0 mm were loaded in each column. The soil column was slowly saturated by introducing solution from a single tube at the center-top of the column with an application rate of 0.005 cm min^{-1} during the first 2 hours. Upon appearance of effluent at the bottom of the column, the application rate of leach solution was adjusted to 0.01 cm min^{-1} and remained constant during the 10-day period. A duplication was made from each treatment and a total 16 of the columns were conducted at room temperature. After leaching the columns for 10 days, the soils in the columns were removed, air dried and crushed to $\leq$ 2.0 mm. The gamma activities of the air-dried soils were analyzed using gamma spectroscopy.

A Sand Barrier

A *Plexiglas* column, with a length of 15 cm and an inside diameter of 9 cm, was used to construct a "sand barrier" which consisted of 5 layers: the top layer of 2-cm height fine gravel (>2 mm), the second layer of 1.5-cm height coarse sand (0.5 to 2.0 mm), the third layer of 7-cm height fine sand (<0.5 mm), the fourth layer of 1.5-cm height coarse sand, and, finally, a fifth layer of 2-cm height fine gravel layer at the bottom of the column. A screen (<4.0 mm mesh) was placed on the top and the bottom of the sand barrier (Fig. 1). Before the sand barrier was constructed, the gravels and the Ottawa sands were washed with deionized water several times and air dried. After the barrier was built up, the sand barrier was moistened by passing through about one liter of deionized water.

Thirty-eight grams of Hot Spot soil (particles of < 2.0 mm) were added to 500 ml of fresh synthetic groundwater to give a mixture of 7.6% turbidity. The mixture was stirred vigorously for 30 min. and was completely suspended. A subsample was taken for initial radioactivity determination when the suspension was still stirring. Following this, 400 ml of the suspension was poured on top of the sand barrier. All filtered water was collected at the bottom of the column and mixed thoroughly. The liquid was taken for radioactivity determination. The sand barrier was allowed to stand for 6 to 8 hours. After that, each layer was removed from the column and air dried. The radioactivity of each layer was determined by the same method used for radiological analysis of the soils.

Sorption and Desorption of Cesium-137

Laboratory sorption and desorption experiments were conducted with illite and zeolite resins of A-51®, IE-95®, and EP-9174® to evaluate the sorption and desorption of ^{137}Cs on these clay minerals. The particle size of the clays was less than 0.5 mm. Carrier-free $^{137}CsCl$ (non-radioactive Cs was not used as a carrier) was completely dissolved in fresh synthetic groundwater (pH 8.0). The initial activity of ^{137}Cs in the solution was 0.05 mCi L^{-1}. The clays and ^{137}Cs labeled groundwater at a ratio of 1:30 were transferred into 50-ml centrifuge tubes and continuously shaken at a constant temperature of 20 ± 1°C. Sorption was allowed to proceed from 6 to 480 hours. The samples were taken after 6, 24, 120, 240, and 480 hours, and subsequently centrifuged at 17,000 g for 1 hour to separate the liquids from solids.

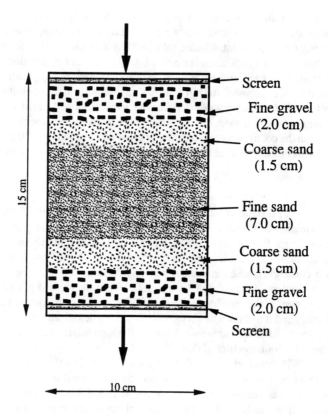

Fig. 1. A sand barrier structure.

Desorption studies were performed with the solids obtained from the sorption experiment. The solids were sequentially extracted with cesium-free fresh synthetic groundwater for the same periods of time used in the sorption experiment. The extracts were separated from solids by centrifuging.

The gamma activities of the liquids, collected from both sorption and desorption experiments, were analyzed using gamma spectroscopy. Measured activities of ^{137}Cs were decay corrected and, hence, activities reported here correspond to initial activities at the time that the sorption experiment was started.

RESULTS AND DISCUSSION

Cesium-137 Distribution

The highest level of ^{137}Cs occurred in the Hot Spot soil and the quantity of ^{137}Cs in the air-dried soil was 13,722 Bq kg^{-1}, which was 3.5 times as much as in the Sediment from the outfall of the drainage channel (4,155 Bq ^{137}Cs kg^{-1}), and was 25 fold as much as in Surface sample (152 Bq kg⁻¹) collected from the location close to the bottom of DP Canyon (Fig. 2). The results show that ¹³⁷Cs is a major contaminant in SWMU 21-011(k) area and ¹³⁷Cs in Hot Spot soil is 73 times as much as its SAL and 22 times of its SAL in the Sediment.

After water was separated from the wet sediment by centrifuging, ¹³⁷Cs in the water could not be detected by the gamma spectroscopy. Furthermore, ¹³⁷Cs in pine bark could not be detected by the gamma spectroscopy (Fig. 2). The results suggest that ¹³⁷Cs was neither released from the soil particles into water nor accumulated in the pine bark. The level of ¹³⁷Cs in the soil close to the bottom of the canyon was low, perhaps ¹³⁷Cs did not migrate to the roots of pine tree.

Wind can cause soil erosion and is one mechanism of ¹³⁷Cs transport. However, ¹³⁷Cs concentration in the sediments from outfall of the drainage channel were much greater than in the surface soil from the bottom of the canyon. Cesium-137 in the sediments may come from water assisted runoff (erosion) of soil particles from Hot Spot area, moving along the drainage channel to DP Canyon. Soils on the top of DP Canyon are often eroded with summer storm runoff down to the canyon. In Los Alamos, about 36% of the annual precipitation falls from storms during July and August. It was reported (Purtymun, 1974) that during the summer of 1967, rainfall carried about 88,000 kg of suspended sediments and transported down to the canyon in about 36,800 m³ of water. With this large amount of sediment runoff, about 2.59 x 10⁶ Bq of gross alpha emitter and about 4.1 x 10⁵ to 1.1 x 10⁷ Bq of

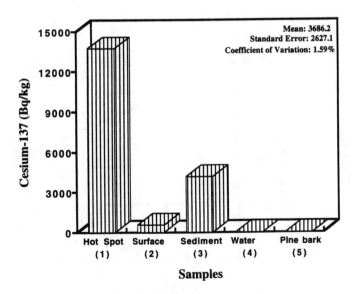

Fig. 2. Cesium-137 in air-dry soils, water and pine bark from SWMU 21-011(k).

gross beta were transported with suspended sediments to the bottom of the Canyon (Purtymun, 1974). Cesium-137 is still found further down the Canyon.

As shown in Fig. 3, radioactivity of ^{137}Cs decreased linearly as the soil particle size increased. The relationship between ^{137}Cs activity and particle size follows an equation of Y=19,602.9 - 2,158.4X with correlation coefficient (r) of - 0.9998, where Y is the activity of ^{137}Cs in Bq kg^{-1}, X is the soil particle size in mm. According to the percentage of each particle size fraction, the ^{137}Cs concentration in each fraction was calculated using the data shown in the Fig. 3. In 1 kg of air-dried Hot Spot soil, the fine gravel (11.5 wt. %) contained 1,255 Bq of ^{137}Cs, the sand fraction (48.8 wt. %) contained 7,503 Bq of ^{137}Cs, and silt-clay fraction (37.4 wt. %) contained 7,266 Bq of ^{137}Cs. fractions. These results suggest that ^{137}Cs in the soil from SWMU 21-011(k) is mainly associated with the sand and silt-clay fractions.

The silt and clay fractions of Los Alamos soil are largely montmorillonite and illite which are weathering products of the tuff (Staritzky, 1949). Montmorillonites and illites have strong ability to hold monovalent cations such as Cs^+, K^+, NH_4^+ and Rb^+ against the effects of leaching by rainfall (Borchardt, 1977; Fanning & Keramidas, 1977). These cations are weakly hydrated and lose their shell of hydration more easily than other cations, thus they may enter interlayer of illite preferentially. Once within the interlayer these ions may become "fixed" over time scales of weeks (Comans & Hockly, 1992). When this happens, the Cs ions are not exchangeable. Therefore, ^{137}Cs bound to these clay minerals is not leached by rainfall. Our data also showed that ^{137}Cs was not leached by synthetic groundwater and synthetic acid rainwater.

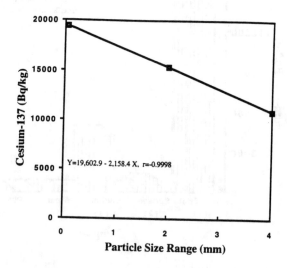

Fig. 3. Cesium-137 distribution in the particle size fractions of the Hot Spot soil from SWMU 21-011(k).

Sand-size (0.05 to 2 mm) fraction of Los Alamos soil consists of largely quartz and alkali feldspar crystals and crystal fragments and small amounts of rock fragments of tuff, pumice, rhyolite, and some mafic minerals (Purtymun, 1974). The crystals and rock fragments of tuff, pumice have large surface areas to adsorb ^{137}Cs.

Column Leaching of Cesium-137

In general, the leachability of ^{137}Cs from the soil was low. Cesium-137 was not leached from the soil using either synthetic groundwater or synthetic acid rainwater after 10 days of leaching. However, using NH_4OAc, about 9% of ^{137}Cs in Hot Spot soil and 30% of ^{137}Cs in Sediment was removed, while about 11% to 15% of ^{137}Cs was removed from the soil using CBD (Fig. 4).

Ammonium acetate solution was used to examine how much ^{137}Cs ions in the soil exchanges with NH_4^+ ions and so removed from the soils (Simard, 1993). Ammonium ions have the same hydrate size as Cs ions, while other cations (e.g. Ca^{2+}, Mg^{2+}) have larger hydrate size than Cs^+ or NH_4^+ ions (Bohn et al., 1985). Thus, NH_4^+ ions may move in to the exchange sites to replace Cs^+ ions. However, when Cs^+ ions are fixed in the interlayer of illite, even NH_4^+ ions could not displace Cs^+ ions from the interlayer. Cesium ions may be removed from the external surface of the minerals by NH_4^+ ions.

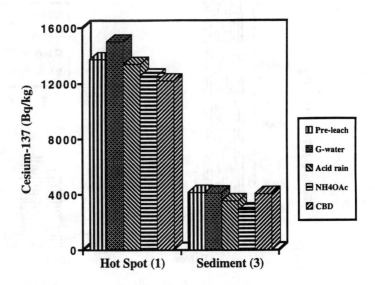

Fig. 4. Cesium-137 in the Hot Spot and Sediment soils from SWMU 21-011(k), pre- & post-leach.

The CBD solution was used to remove ^{137}Cs-bonded to Fe-oxide without disrupting clay structure (Mehra & Jackson, 1960). Free iron oxides, occur as discrete particles, nodules or as coatings on the soil particles. Free hematite (Fe$_2$O$_3$), geothite (FeOOH) and amorphous iron oxides are selectively dissolved by CBD (Mehra & Jackson, 1960). Therefore, the complex-reducing reagent of CBD removes mainly coated oxides bonded ^{137}Cs on the minerals during the leaching process. However, the percentage of ^{137}Cs removed was not directly related to the total Fe content in the soil. It is assumed that Fe(OH)$_3$ is the representative empirical formula for the Fe oxide (Tessier et al., 1979), and the ^{137}Cs bond to Fe oxide originates only from Fe oxide. Solutions of CBD may remove large amounts of Fe with small amounts of ^{137}Cs, depending on the soil properties and the fractions with which ^{137}Cs associated.

Sand Barrier and Clay Sorbents

The initial concentration of ^{137}Cs in the suspension was 15,475 Bq kg^{-1} per liter synthetic groundwater. After 400 ml of the suspension was passed through the sand barrier, the ^{137}Cs in the filtered water could not be detected (Fig. 5, Fig. 1). However, the ^{137}Cs level was 528 Bq kg^{-1} in the fine gravel at the top layer of the sand barrier, and 5,177 Bq kg^{-1} in the coarse sand at the second layer of the barrier,

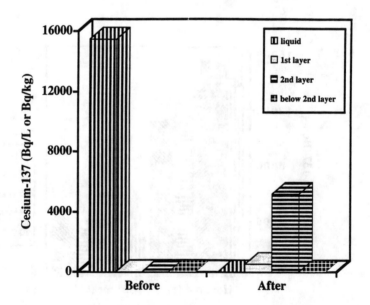

Fig. 5. Cesium-137 in the liquid and in the sand layers of the barrier before and after processes.

and no ^{137}Cs was found below the second layer of the barrier (Fig. 5, Fig. 1). The radioactivity balance from the sand barrier was 92.2% of recovery for all ^{137}Cs labeled soil particles. The error of 7.8% was due to experimental error. The result indicates that the sand barrier blocks the fine particles with which ^{137}Cs was mainly associated in the SWMU 21-011(k) soils.

The principle of the sand barrier is based on the filtration step for drinking water treatment. A typical sand filter is commonly used to remove suspended particles and colloidal matter in stream water. The sand filter is constructed with a smaller size of sand in the top surface, the larger size of graded gravel in the bottom (Cleasby, 1990). When the suspension was poured into the sand barrier, the soil particles in suspension are transported to the top surface of the barrier by sedimentation and moved to sandy layer by mass flow. The particle removal efficiency depends on the particle concentration in the water, particle density, filtration rate and filter pore diameter (Cleasby, 1990).

For clay sorbents the sorption data is plotted in Fig. 6 as the percentage of ^{137}Cs sorbed on clays vs. time. The amounts of ^{137}Cs adsorbed by the zeolite resin, A-51®, IE-95®, and EP-9174®. do not differ significantly from each other, average values of ^{137}Cs adsorbed by zeolite resins are shown in Fig. 6. The results show initially rapid adsorption, followed by slower uptake. For zeolite resins, about 95% to 99% of ^{137}Cs in the synthetic groundwater was sorbed after 6 hours with vigorous shaking. The percent ^{137}Cs sorbed did not change significantly during the remainder 474 hours (Fig. 6). The sorption of ^{137}Cs to illite was slower than to zeolite resins. During the first 6 hours with vigorous shaking, about 76% of ^{137}Cs in the synthetic groundwater was sorbed. A total of 93% ^{137}Cs was adsorbed to illite after 480 hours (Fig. 6). The desorption results show less than 1% of ^{137}Cs was desorbed from zeolites and about 3% of ^{137}Cs was desorbed from the illite after an additional 20 days with vigorous shaking (data not shown). This suggests that zeolites and illite are effective sorbent materials. Zeolites sorbed more efficiently than illite.

The unique feature of zeolite group is the presence of a structural channel which provides high CEC (100 to 300 meq/100g) (Zelazny & Calhoun, 1987). The zeolite resins used in this study are basic zeolites. If Cs ions are fixed in the basic zeolites, the structural channel is too narrow to allow other cations to exchange with Cs ions (Zelazny & Calhoun, 1987). It was reported that ^{137}Cs was sorbed strongly on zeolite tuffs from Yucca Mountain, Nevada and ^{137}Cs did not release from these tuffs (Thomas, 1987).

Soil erosion is a primary mechanism of ^{137}Cs transport at TA-21. Preventing soil erosion is a possible approach to control the migration of ^{137}Cs. At the top of DP Canyon, ^{137}Cs in Hot Spot area may be fixed using some cover materials because the area of Hot Spot is not large (<100 m^2) or deep (< 30 cm). On the slopes of the canyon, ^{137}Cs may be controlled using a erosion control web, such as "Miramat ™" which is a three-dimensional web of bonded polypropylene or PVC monofilaments and it can shield the soil particles from rain, wind and surface runoff while it allows natural revegetation to establish (GEOCIV, 1995). At the bottom of the slope, spreading of ^{137}Cs contaminated soil particles may be controlled by a barrier. Laboratory study shows that the sand barrier blocked ^{137}Cs labeled soil particles successfully. Clay minerals of zeolite or illite can be used in the barrier for stabilization of ^{137}Cs.

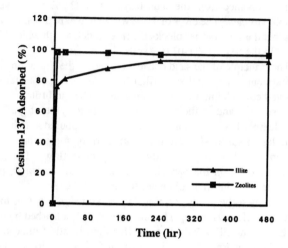

Fig. 6. Percentage of cesium-137 adsorbed on zeolite and illite.

CONCLUSIONS

At LANL, DP Canyon was subject to radioactive contaminant releases, resulting in the soils containing multiple radioactive contaminants. The results show that ^{137}Cs is a major contaminant in the soil and is mainly associated with silt, clay and sand fractions. Because ^{137}Cs has been "fixed" in the soil, it can not be leached by synthetic groundwater or acid rainwater. Soil erosion is a primary mechanism of ^{137}Cs transport in DP Canyon. It is possible to use a sand barrier to prevent that transport. Zeolites are effective materials for stabilizing ^{137}Cs in the barrier. The results from our studies are in agreement with those reported by previous investigators in Europe, Asia, and North America.

ACKNOWLEDGMENTS

This work was funded by Garry Allen, Field Unit-1 project leader, Environmental Restoration Program, Los Alamos National Laboratory. The authors wish to thank Inés Triay, leader of the Environmental Science and Waste Technology Group at the Los Alamos National Laboratory, for providing background studies and laboratory space for the execution of this research. Thanks to Janey Headstream and Malcolm Fowler for conducting many of the ^{137}Cs radiological analyses. Thanks to Charles R. Cotter and Pam Gordon for technical assistance. Thanks to Jennifer Murphy for providing information about TA-21. The authors especially appreciate Dr. Hajek, professor of soil mineralogy, Department of Agronomy and Soils, Auburn University for providing illite.

REFERENCES

Bohn, H.L., B.L. McNeal, and G.A. O' Connor. 1985. p 104-132. *In* H.L. Bohn (eds.) Soil chemistry. John Wiley & Sons, New York.

Borchardt, G.A. 1977. p 293 -330. *In* Dixon, J.B., and S.B. Weed (eds.) Minerals in soil environments. Soil Sci. Soc. Am. Madison, WI.

Cleasby, J.C. 1990. p 455-554. *In* F.W. Ponties (ed.) Water quality and treatment- a handbook of community water supplies. 4th ed. McGraw-Hill.

Comans, R.N.J., and D.E. Hockley. 1992. Kinetics of cesium sorption on illite. Geochim. Cosmochim. Acta. 56:1157-1164.

Elless, M.P., S.Y. Lee, and M.E. Timpson.   1994.   Physicochemical and mineralogical characterization of transuranic contaminated soils for uranium soil ID. *In* Uranium in Soils Integrated Demonstration (USID) Midyear. June 9, 1994. Washington, D.C.

Fanning, D.S., and V.Z. Keramidas. 1977. p 195-258. *In* Dixon, J.B., and S.B. Weed (eds.) Minerals in soil environments. Soil Sci. Soc. Am. Madison, WI.

GEOCIV. 1995. Soil stabilization products. Houston, Texas.

Kirk, G.J.D., and S. Staunton. 1989. On predicting the fate of radioactive cesium in soil beneath grassland. J. Soil Sci. 40:71-84.

Los Alamos National Laboratory report. Phase Report IC(TA-21), 1994, LA-UR-94-228.

Mahara, Y. 1993. Storage and migration of fallout strontium-90 and cesium-137 for over 40 years in the surface soil of Nagasaki. J. Environ. Qual. 22:722-730.

Maule, C.P., and M.J. Dudas. 1989. Preliminary identification of soil separates associated with fallout [137]Cs. Can. J. Soil Sci. 69:171-175.

Mehra, O.P., and M.L. Jackson. 1960. Iron oxide removal from soil and clays by a dithionite-citrate system buffered with sodium bicarbonate. *In* Clays clay minerals, Proc. 7th Conf., p. 317-327 Natl. Acad. Sci.- Natl. Res. Council Publ.

Nelson, D.W., and L.E. Sommers. 1982. Total carbon, organic carbon, and organic matter. p. 539-577. *In* Page A.L. et al. (eds.) Methods of soil analysis. Part 2, 2nd ed. Agron. Monogr. 9. ASA and SSSA, Madison, WI.

Purtymun, W.D. 1974. Los Alamos National Laboratory report. LA-5744, UC-11.

Robbins, J.A., G. Lindner, W. Pfeiffer, J. Kleiner, H.H. Stabel, and P. Frenzel. 1992. Epilimnetic scavenging of Chernobyl raduonuclides in Lake Constance. Geochim. Cosmochim. Acta. 56:2339-2361.

Simard, R.R. 1993. p 39-42. *In* M.R. Carter (ed.) Soil sampling and methods of analysis. Canadian Soc. of Soil Sci. Lewis Publishers Inc.

Staritzky, E. 1949. Los Alamos Scientific Laboratory report. LA-741.

Tessier, A., P.G.C. Campbell, and M. Bisson. 1979. Sequential extraction procedure for the speciation of particulate trace metals. Analytical Chemistry. 51:844-851.

Thomas, K.W. 1987. Los Alamos National Laboratory report, LA-10960-MS,

Zelazny, L.W., and F.G. Calhoun. 1987. p 453-470. *In* Dixon, J.B. and S.B. Weed (eds.) Minerals in soil environments. Soil Sci. Soc. Am. Madison, WI.

Within-Field Spatial Variability of Soil Nutrients and Corn Yield in a Montreal Lowlands Clay Soil

M.C. Nolin

Agriculture and Agri-Food Canada
Ste-Foy, Quebec

S.P. Guertin

MAPAQ
St-Hyacinthe, Quebec

C. Wang

Agriculture and Agri-Food Canada
Ottawa, Ontario

ABSTRACT

A study of the within-field variability of soil characteristics of a 10-ha mostly homogeneous field of the Montreal Lowlands Area has shown moderate variability (CV=15-35%), a clear pattern of spatial continuity for most variables studied by variography/kriging and a good relationship with soil types (1:3000 scale map), digital elevation model and on-the-go corn yield mapping making site-specific crop management *a priori* useful in this area.

INTRODUCTION

Conventional farm practices manage fields as a homogenous, uniform unit. Although a field may have considerable variation in its characters, such as different soil types, landscape features and previous management histories, the inputs supplied for crop production are applied at constant rates (Borgelt *et al.* 1989). Such management can create inefficiencies by over-treating and under-treating portions of a field increasing field management cost, decreasing net economic returns, contributing to surface and ground water pollution and globally wasting energy (Robert 1993).

Varying input rates within field according to the spatial variability of soil or site conditions is the basic concept behind site-specific crop management (SSCM). The science underlying this alternative approach to ensure sustainable crop production is not new (Goering 1993), but is has not been widely practised until recently because standard application equipment was designed to apply one rate in the field and because precise, accurate and efficient positioning techniques were also lacking. With the development of new technologies such as geographic information systems (GIS), guidance systems, sensors and computer-controlled application equipment, inputs rates or blends can now be varied accurately and easily over relatively short distances during application (Wollenhaupt *et al.* 1994).

Investments required to implement SSCM will be limited to areas or products where there is a good potential for profit: large sized farms, small set of cash grains and specialty crops needing high input level, agriculture area where high supporting services are available, *etc.* (Nowak 1993). In Quebec, Montreal Lowlands Area (MLA) offers, at this point of view, a certain potential for the application of SSCM. However, it is also hypothesized that the applicability of SSCM is restricted in very homogeneous field (minor variation in soil or landscape characteristics) which is often the case in the MLA. The question is how important should the soil variation be to ensure profitability of SSCM. At this time, few studies have been conducted to document soil and fertility test variability at the field scale in this agricultural area.

OBJECTIVES

The objectives of this study were: 1) to evaluate the spatial variability of selected soil physico-chemical properties of the topsoil layer (0-0.15 m) within a homogeneous field of the MLA, 2) to evaluate the ability of an intensive soil survey (1:3 000 scale map) to subdivide this field into smaller management units showing more uniformity than the whole field and 3) to study the relationships between soil characteristics, soil types, topography (digital elevation model) and 1995 corn yield (on-the-go yield mapping).

MATERIALS AND METHODS

Site Description

The research work was conducted on a 10-ha field located on an almost flat land (0-2% slopes) at St. Antoine-sur-Richelieu, in southern part of Quebec, close to Montreal. It belongs to the St. Lawrence Lowlands terrestrial ecoregion of Canada (Ecological Stratification Working Group 1995). Soils were mainly formed over clayey marine deposit of the Champlain Sea which was slightly reworked afterwards by the Proto-St. Lawrence River. Mean annual precipitation of this region is around 1000 mm. Climatic moisture index (Precipitation - Potential Evapotranspiration for the May to August period) is between -150 to -200 mm (Agronomic Interpretations Working Group 1995). About 2700 Corn Heat Units are available during growing season at this site. A 5-ha soil Quality Monitoring benchmark site has been established in this field since May 1989 (Nolin *et al.* 1995).

Management Information - 1995

Since 1980, the field has been under a rotation system (corn-soybean-small grain). On 6 May 1995, 80-0-0 kg ha^{-1} N-P-K fertilizer was broadcast and incorporated. In the afternoon, Eradicane™ herbicide (8 L ha^{-1}) was applied and incorporated by a field cultivator. Pioneer Brand 3921 hybrid corn (*Zea mays* L.) was planted at a seeding rate of 74 150 seeds ha^{-1} on 7 May. Fertilizer supplying 31-35-42-0.4 kg ha^{-1} N-P-K-B was banded at seeding. On 8 June, Partner™ was sprayed at a rate of 1 L ha^{-1} for additional weed control. N liquid fertilizer was applied at a rate of 125 kg ha^{-1} on 18 June. Corn was harvested on 18 October and corn stover was incorporated by plowing to a depth of approximately 0.25 m on 1 November.

Soil Survey

An intensive soil survey (scale 1:3000) was conducted at the beginning of the study. The experimental field was stratified into seven soil map units (11 delineations) mainly on the basis of topsoil texture (Fig. 1). Three soil series were identified in this field. Description, classification and distribution of these soil series and mapping units are given in Table 1.

Table 1. Description, classification and distribution of soil map units within the 10-*ha* experimental field

Soil Series	U.S. Soil Taxonomy	Map symbol	Surface texture	Area (%)
Du Contour	*Coarse-loamy over clayey, mixed, mesic Humic Endoaquept*	CT-scl	sandy clay loam	1.2
Du Jour	*Fine-clayey, mixed, mesic Aeric Humaquept*	DJ-cl	clay loam	3.9
		DJ-sic	silty clay	5.9
		DJ-c	clay	19.2
Providence	*Very-fine clayey, mixed, mesic Aeric Humaquept*	PV-sicl	silty clay loam	3.7
		PV-sic	silty clay	41.5
		PV-c	clay	24.6

Grid Sampling and Soil Analysis

Soil samples (*n*=130) were collected on a 30-m x 30-m grid in early November 1994, a few days after the harvest. At each grid point, five cores (one in each quadrant and at the centre of the 2-m radius sampling circle), 0.15-m deep, were composited to obtain a representative soil sample of the surface layer.

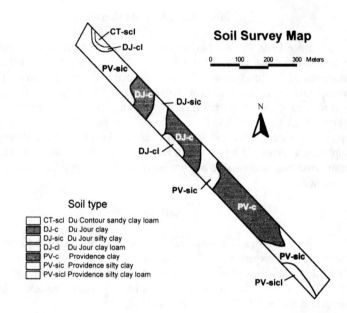

Fig. 1 Soil map of the experimental field

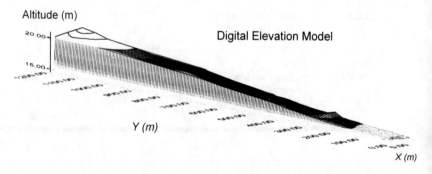

Fig. 2 Three-dimensional surface contour of the experimental field generated from DEM.

Soil samples were air-dried, ground, sieved and analyzed according to the standardized laboratory procedures proposed for soil testing by CPVQ (Conseil des Productions Végétales du Québec 1993); reference codes used to designate these methods are indicated in parentheses. Sand, silt and clay contents were determined by the hydrometer method (GR-1), organic C by wet oxidation (MA-1), pH in 1:1 water-soil ratio (PH-1), buffer pH by the SMP (Schoemaker-MacLean-Pratt) method (PH-2), P, K, Ca, Mg, Cu, Zn, Mn, Al and Fe extractable with Mehlich III (ME-1), total N (AZ-1), NH_4-N (AZ-2) and NO_3-N (AZ-3) with Technicon auto-analyzer. A Minolta SPAD-502 chlorophyll meter was used to measure the chlorophyll content (leaf greenness) of earleaf collected 15 days after silking (130 sampling sites x 3 plants x 3 replicates) in order to identify the field sites with different N responses.

Digital Elevation Model (DEM)

A ground survey was conducted on a 30-m grid across the 10-ha site using a level and a metric rod. The measured elevations were interpolated into a regular 5-m grid using the optimal Delaunay triangulation method available in the Surfer (V6.0) software (Golden Software Inc. 1995). This interpolation method was chosen because it is very effective at preserving break lines (microtopography) when accurate data are collected. Previous statistical analysis of monitoring data showed that minor changes in relief can result in large spatial variations of soil quality indicators (Wang *et al.* 1995). Altitude ranged from 15 to 25 m. Landscape is nearly level (slopes <1%) with a medium ridge (2 to 5% slopes) at the NE part of the field and several low ridges with side slopes of 1% running east-west (Fig. 2).

Yield Mapping

Yield information was obtained using a combine instrumented with an Ag Leader™ Yield monitor 2000™ and a DMC moisture sensor. These data were coupled with real-time differential global positioning system (DGPS) data gathered at the 2-5 meter level of accuracy. Yield data were subjected to quality control to reject anomalies caused by operators recording errors, combine grain flow transportation delays and cutterbar utilization variations (Eliason *et al.* 1995).

Statistical Analysis

Mean, median, coefficient of variation, skewness, kurtosis and simple correlation coefficient were determined using the SYSTAT for Windows (V5.0) statistical program (Systat Inc. 1992). Geostatistical analysis (semivariance and block kriging) were performed using GS$^+$ (V2.3) software (Gamma Design Software 1995).

RESULTS AND DISCUSSION

Within-Field Spatial Variability

A statistical summary of soil characteristics and corn yield for the whole field is presented in Table 2. The experimental field showed a low level of variability (CV's <15%) for thickness of Ap horizon, clay, pH, buffer pH, Al, corn yield, grain moisture and SPAD-502 meter values. These three last yield indicators exhibited CV values less than 5%. Corn yield ranged from 7.21 to 8.53 Mg ha^{-1}. The other soil characteristics showed a moderate level of variability (CV's 15-35%). Phosphorus soil tests varied from 21 (poor) to 121 (excessively rich) mg kg^{-1} while soil test K ranged from 37 (poor) to 590 (excessively rich) mg kg^{-1} (CPVQ 1996). Statistical distribution of studied variables were relatively normal as indicated by skewness and kurtosis indices near zero.

Spatial continuity means that data close to each other are more likely to have similar values and a strong spatial dependence than data that are far apart (Isaaks and Srivastava 1989). Within-field spatial continuity of soil characteristics, particularly for soil fertility indicators, are needed to efficiently and accurately implement SSCM. If values appear to be randomly located (pure noise), the best estimators of the value taken by a variable at any point within the field remains the field sample mean or the median and the best way to manage this field is applying a uniform field-average fertilizer treatment.

Geostatisticians normally define the spatial continuity through the semivariogram and consider kriging as the optimal method of interpolation because it gives the best linear unbiased estimate of the value of a variable at a given point in minimizing the error variance (Isaaks and Srivastava 1989). Semivariograms were computed for each soil and yield property. Parameters for the best fitting theoretical models (linear, linear with sill, spherical, exponential or gaussian) are presented on Table 3. Almost all studied variables showed spatial continuity in the field as indicated by high R^2 values (>0.90). Exponential and spherical models were the most often selected. The range (A_0), which measures the maximum distance over which properties remain spatially correlated, are relatively short (≤150 m) for clay, silt, NH$_4$-N, Ca, P, corn yield and SPAD-502 meter values. Corn yield range was particularly short (51 m). Very high range value may signal local trend (drift) in the property. That could be the case for thickness of Ap horizon, Mg, Cu and Mn. Another useful parameter to analyze for evaluating geostatistical analysis efficiency is the nugget variance (C_0) which represents the random variation (sampling and analytical errors, short-range spatial variation within the spacing of observations, *etc.*) while the structural variance (C) represents systematic variation. The greater the magnitude of nugget variance, the less precise interpolation estimates can be made from the data set. The ratio C_0/C_0+C is a consistent indicator of the importance of nugget variance in percentage of the total variation (sill). Several soil characteristics showed a very low ratio (<20%). Yield indicators (corn yield, grain moisture and SPAD-502 meter values) showed the highest ratio (44.9-61.9%).

Table 2. Statistical summary of soil characteristics, corn yield, grain moisture and SPAD-502 meter values for the whole field

Variables	Statistical parameters [x]						
	n	m	md	CV (%)	r	m_3	m_4
Thickness of Ap *(cm)*	130	29.4	29.0	12.9	20.0-39.0	0.20	0.09
Clay *(%)*	130	44.3	45.6	13.6	20.0-55.4	-1.41	2.34
Silt *(%)*	130	39.1	40.2	17.9	14.6-58.1	-0.35	1.29
Sand *(%)*	130	16.6	13.9	59.3	5.7-65.4	2.46	6.89
pH (1:1 water-soil)	130	6.00	5.97	5.6	5.39-7.18	0.56	0.27
Buffer pH (SMP)	130	6.20	6.22	4.4	5.65-7.00	0.07	-0.29
Organic C *(%)*	130	2.76	2.60	26.1	1.19-4.90	0.39	-0.23
Total N *(%)*	129	0.25	0.24	29.6	0.08-0.48	0.15	0.12
NO_3-N *(mg kg^{-1})*	129	10.0	9.6	35.7	3.5-18.7	0.40	-0.68
NH_4-N *(mg kg^{-1})*	129	6.4	6.1	28.8	3.0-13.6	1.07	1.70
C.E.C. *(cmol kg^{-1})*	130	32.0	32.1	15.9	15.5-48.0	-0.40	2.14
Ca *(mg kg^{-1})*	130	3058	3000	32.2	700-6900	0.70	2.03
Mg *(mg kg^{-1})*	130	486	465	34.4	115-960	0.25	-0.26
K *(mg kg^{-1})*	130	347	356	31.0	37-590	-0.17	-0.32
Cu *(mg kg^{-1})*	130	2.18	2.00	34.1	1.00-4.10	0.82	-0.15
Zn *(mg kg^{-1})*	130	2.92	2.80	31.5	1.00-6.70	1.06	2.55
Mn *(mg kg^{-1})*	130	25.3	21.0	64.9	4.0-74.0	1.00	0.35
Fe *(mg kg^{-1})*	130	244	230	24.9	130-380	0.38	-0.56
Al *(mg kg^{-1})*	130	1295	1260	10.3	1030-1680	0.80	0.08
P *(mg kg^{-1})*	130	52	48	35.6	21-121	0.92	1.09
Corn yield *(Mg ha^{-1})*	2428	7.85	7.86	4.0	7.21-8.53	0.01	-0.73
Grain moisture *(%)*	3482	22.4	22.4	4.4	19.8-27.1	0.28	0.17
SPAD-502 meter	130	56	56	4.1	50-62	0.11	-0.07

[x] n = sample size; m = mean; md = median; CV = coefficient of variation; r = range; m_3 = skewness; m_4 = kurtosis.

Table 3. Statistical parameters describing isotropic thoretical models fitted to experimental semivariograms

Variables	Lag (m)	Model	Nugget C_0	Sill C_0+C	Ratio C_0/C_0+C	Range A_0	R^2
Thickness of Ap	30	Linear with sill	9.26	20.83	0.445	931	0.942
Clay	50	Spherical	10.8	31.7	0.341	76	0.945
Silt	30	Spherical	12.2	41.2	0.296	73	0.929
pH (1:1 water-soil)	30	Spherical	0.0185	0.1025	0.180	165	0.990
Buffer pH (SMP)	30	Spherical	0.0136	0.0868	0.157	287	0.996
Organic C	30	Spherical	0.0810	0.6890	0.118	343	0.988
Total N	30	Spherical	0.0007	0.0066	0.106	308	0.986
NO_3-N	30	Exponential	5.36	16.69	0.321	198[x]	0.922
NH_4-N	29	Exponential	0.230	3.455	0.067	29[x]	0.915
Ca	30	Spherical	64000	846200	0.076	150	0.983
Mg	40	Spherical	13800	36740	0.376	736	0.928
K	30	Linear with sill	2550	14370	0.177	574	0.988
Cu	30	Spherical	0.059	0.872	0.068	786	0.983
Zn	30	Exponential	0.120	0.728	0.165	137[x]	0.981
Mn	30	Spherical	16	516	0.031	1032	0.982
Fe	30	Spherical	10	4058	0.002	326	0.988
Al	30	Spherical	2150	22070	0.097	262	0.994
P	30	Exponential	35.1	290.6	0.121	39[x]	0.933
Corn yield	5	Exponential	11.15	24.82	0.449	17[x]	0.715
Grain moisture	5	Exponential	0.647	1.045	0.619	110[x]	0.936
SPAD-502 meter	30	Exponential	3.010	5.162	0.583	42[x]	0.766

[x]   For exponential model, it is often proposed that range = 3 A_0.

Map Units Variability

Mean and coefficient of variation of soil characteristics and corn yield indicators for each map unit delineated by an intensive soil survey are presented in Table 4. It clearly indicates that stratification of the field by soil type reduces variability (CV) and gives significantly different management area. For example, CV's for K is reduced from 31 to 24% (weighed average), P from about 36 to 28%, organic C from 26 to 16%, Ca from 32 to 24%. PV-sicl map unit showed the

Table 4. Statistical summary of soil characteristics, corn yield, grain moisture and SPAD-502 meter values by soil map units

Variables		Soil map units [x]						
		PV-sicl (n=5)	PV-sic (n=53)	PV-c (n=29)	DJ-cl (n=9)	DJ-sic (n=7)	DJ-c (n=23)	CT-scl (n=4)
Thickness of	m^y	34.6	29.3	30.7	29.8	27.9	27.4	26.8
Ap (cm)	CV^y	7.3	12.3	12.7	11.0	14.6	11.4	4.7
Clay	m	36.0	44.7	49.7	33.9	47.3	44.6	27.0
(%)	CV	7.9	7.0	5.1	14.8	4.1	6.6	20.9
Silt	m	55.6	43.4	34.6	29.7	41.1	36.8	23.6
(%)	CV	4.1	7.1	6.2	20.5	0.7	10.6	35.2
Sand	m	8.4	11.9	15.7	36.4	11.6	18.6	49.4
(%)	CV	10.1	30.1	17.3	28.4	17.2	27.6	28.2
pH in water	m	6.73	6.12	5.98	5.75	5.73	5.87	5.50
	CV	1.6	5.1	4.4	4.8	3.0	3.6	2.0
Buffer pH	m	6.73	6.34	6.10	6.30	5.91	5.90	6.20
(SMP)	CV	0.4	3.2	3.3	1.8	2.9	2.9	0.8
Organic C	m	1.97	2.41	3.02	2.02	3.25	3.70	1.84
(%)	CV	17.1	14.1	18.7	33.9	7.6	13.4	17.2
Total N	m	0.19	0.22	0.28	0.16	0.30	0.35	0.12
(%)	CV	4.8	16.4	17.8	47.7	10.4	14.7	29.3
NO_3-N	m	7.17	9.60	9.17	7.01	11.24	13.80	6.94
(mg kg^{-1})	CV	43.9	33.6	32.0	37.3	27.9	19.5	22.7
NH_4-N	m	7.35	6.36	7.43	4.28	5.71	6.01	4.42
(mg kg^{-1})	CV	20.5	21.7	26.1	33.2	29.2	33.1	13.0
C.E.C.	m	32.9	33.6	33.1	23.0	32.0	32.3	18.5
(cmol kg^{-1})	CV	0.9	13.8	10.2	15.5	5.5	6.8	14.9
Ca	m	3804	3462	3138	1772	2707	2830	1013
(mg kg^{-1})	CV	2.4	29.4	24.0	31.1	10.1	20.4	30.8
Mg	m	719	611	442	332	399	336	205
(mg kg^{-1})	CV	4.7	19.9	15.8	51.4	11.3	25.2	43.1
K	m	385	314	448	233	346	362	185
(mg kg^{-1})	CV	6.0	30.8	16.9	40.8	6.0	25.2	17.9
Cu	m	3.98	2.61	2.00	1.83	1.64	1.46	1.53
(mg kg^{-1})	CV	2.1	24.4	17.7	26.9	11.0	16.0	18.8
Zn	m	4.04	3.33	2.76	1.69	2.74	2.73	1.30
(mg kg^{-1})	CV	21.8	30.7	15.1	20.0	7.8	15.5	26.6
Mn	m	62.8	31.8	27.9	12.7	11.3	9.3	18.3
(mg kg^{-1})	CV	11.8	49.3	33.2	34.9	31.0	52.2	50.5
Fe	m	324	274	189	284	223	204	323
(mg kg^{-1})	CV	3.5	19.1	18.8	20.4	12.3	13.3	11.1
Al	m	1190	1216	1386	1240	1313	1391	1343
(mg kg^{-1})	CV	5.1	5.2	9.3	11.2	7.1	11.0	3.5
P	m	91	56	44	55	46	39	71
(mg kg^{-1})	CV	10.8	33.0	20.2	25.9	37.1	30.9	16.4
Corn yield	m	8.23	7.88	7.80	7.68	8.03	7.82	7.39
(Mg ha^{-1})	CV	1.1	3.3	3.2	3.6	2.9	3.3	0.9
Grain moisture	m	21.8	22.1	22.8	21.9	22.4	22.6	22.1
(%)	CV	1.1	2.8	3.4	2.5	1.4	1.8	2.1
SPAD-502	m	57.2	56.1	55.8	55.0	57.7	56.5	52.2
meter	CV	1.7	3.8	4.1	5.5	4.1	3.1	2.3

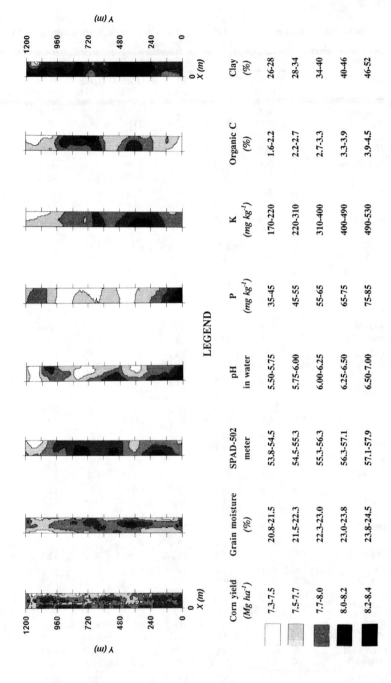

Fig 3. Spatial maps of corn yield indicators and selected soil characteristics derived with block kriging.

LEGEND

Corn yield (Mg ha^{-1})	Grain moisture (%)	SPAD-502 meter	pH in water	P (mg kg^{-1})	K (mg kg^{-1})	Organic C (%)	Clay (%)
7.3-7.5	20.8-21.5	53.8-54.5	5.50-5.75	35-45	170-220	1.6-2.2	26-28
7.5-7.7	21.5-22.3	54.5-55.3	5.75-6.00	45-55	220-310	2.2-2.7	28-34
7.7-8.0	22.3-23.0	55.3-56.3	6.00-6.25	55-65	310-400	2.7-3.3	34-40
8.0-8.2	23.0-23.8	56.3-57.1	6.25-6.50	65-75	400-490	3.3-3.9	40-46
8.2-8.4	23.8-24.5	57.1-57.9	6.50-7.00	75-85	490-530	3.9-4.5	46-52

Table 4. Statistical summary of soil characteristics, corn yield, grain moisture and SPAD-502 meter values by soil map units

Variables	Soil map units [x]						
	PV-sicl (n=5)	PV-sic (n=53)	PV-c (n=29)	DJ-cl (n=9)	DJ-sic (n=7)	DJ-c (n=23)	CT-scl (n=4)

[x]   PV-sicl = Providence silty clay loam; PV-sic = Providence silty clay; PV-c = Providence clay;
      DJ-cl = Du Jour clay loam; DJ-sic = Du Jour silty clay loam; DJ-c = Du Jour clay;
      CT-scl = Du Contour sandy clay loam.
[y] m = arithmetic mean; CV = coefficient of variation (%).

highest corn yield average (8.23 Mg ha^{-1}) and CT-scl, located on the NE ridge, the lowest one (7.39 Mg ha^{-1}). It is worth noting here that the least productive soil type has a yield potential representing 90% (based on 1-yr data) of the maximum yield recorded in this field (*i.e.* 8.23 Mg ha^{-1}). The other map units got about 95% of the maximum yield. The occurrence in the field of soil areas showing significantly different potential productivity cannot justify the implementation of SSCM in this case. However, variability of soil tests and other soil characteristics seems sufficiently high to make SSCM (*a priori*) useful in this relatively uniform field.

Site Characteristics and Corn Yield Relationships

Simple correlation coefficients (r) between corn yield, grain moisture, SPAD-502 meter values, altitude and soil characteristics are presented in Table 5. Soil solution acidity, measured by pH (water) as well as by other closely related variables (*e.g.* Ca) affected corn yield more than any other soil variables (r=0.43). Liming is required to achieve and maintain optimum yield potential. But application of excessive amounts of limestone will not be economical and can induce decrease of nutrient availability and crop growth (Borgelt *et al.* 1989). That could be the case for PV-sicl map unit. Liming needs as established by buffer pH (CPVQ 1996) differed significantly for each map unit: 10 Mg ha^{-1} for DJ-c, 7.8 Mg ha^{-1} for PV-c and 5.8 Mg ha^{-1} for DJ-cl, making SSCM useful. Relationships between corn yield and SPAD-502 meter values were significant (r=0.34) while no significant relationship was found between SPAD-502 meter values or corn yield and fall sampled NO$_3$-N, indicating that fall sampling was not a good estimator of residual NO$_3$-N in the MLA. Grain moisture was mainly related to organic matter content as indicated by the positive correlation with organic C (r=0.48) and total N (r=0.46). A high content of K seems also to explain grain moisture at harvest (r=0.47) while this relationship was negative for Mg (r=-0.35). Many soil characteristics showed a high negative correlation with altitude (r≤-0.50). Kriged maps (Fig. 3) of selected soil characteristics (pH, P, K, organic C and clay) and corn yield indicators (corn yield, grain moisture, SPAD-502 meter values) indicated that spatial patterns were strongly correlated each other as well as with soil map units (Fig. 1) and topography (Fig. 2).

Table 5. Correlation between soil characteristics, corn yield, grain moisture, SPAD-502 meter and altitude values

Variables	Corn yield	Grain moisture	SPAD-502 meter	Altitude
Thickness of Ap	0.24 **	-0.05	0.11	-0.41 ***
Clay	0.23 **	0.22 *	0.17	-0.30 ***
Silt	0.33 ***	-0.30 ***	0.18 *	-0.26 **
Sand	-0.37 ***	0.08	-0.23 **	0.37 ***
pH in water	0.43 ***	-0.10	0.24 **	-0.53 ***
Buffer pH	0.24 **	-0.34 ***	0.06	-0.22 *
Organic C	0.04	0.48 ***	0.17	-0.13
Total N	0.11	0.46 ***	0.19 *	-0.20 *
NO_3-N	0.06	0.04	0.14	0.09
NH_4-N	0.04	0.06	-0.02	-0.24 **
C.E.C.	0.41 ***	-0.05	0.33 ***	-0.55 ***
Ca	0.45 ***	-0.14	0.34 ***	-0.58 ***
Mg	0.25 **	-0.35 ***	0.07	-0.20 *
K	0.23 **	0.47 ***	0.24 **	-0.68 ***
Cu	0.22 *	-0.30 ***	0.04	-0.47 ***
Zn	0.35 ***	-0.07	0.20 *	-0.60 ***
Mn	0.24 **	-0.19 *	0.10	-0.64 ***
Fe	0.00	-0.38 ***	-0.07	0.13
Al	-0.16	0.44 ***	-0.06	-0.17
P	0.22 **	-0.11	0.16	-0.19 *
Corn yield	1.00	-0.07	0.34 ***	-0.25 **
Grain moisture	-0.07	1.00	0.06	-0.18 *
SPAD-502 meter	0.34 ***	0.06	1.00	-0.18 *
Altitude	-0.25 **	-0.18 *	-0.18 *	1.00

*, **, *** Significant at $P \leq 0.05$, $P \leq 0.01$ and $P \leq 0.001$, respectively $(n=130)$.

CONCLUSIONS

The Montreal Lowlands Area meets many proposed criteria justifying the applicability of SSCM. However, it is often hypothesized that SSCM is restricted in very homogeneous soil landscapes like those currently encountered in this agricultural area. The study of the within-field variability of selected soil physico-chemical properties of the topsoil layer of a 10-ha homogeneous field (clayey Aeric Humaquept, 0-2% slopes) showed moderate variability (CV=15-35%), particularly for fertility indicators.

An intensive soil survey (scale of 1:3000) allowed delineation of the field into seven map units mainly on the basis of topsoil texture which ranged from sandy clay loam to clay. This map reduced significantly spatial variability of soil characteristics. Variography and kriging showed a clear pattern of spatial continuity for all variables studied. These patterns were in good agreement with soil map units and fitted well with the contour map produced by interpolation with linear triangulation method. On-the-go corn yield mapping using a combine instrumented with yield and grain moisture sensors and with a DGPS providing 2 to 5 m position accuracy in real time showed a good relation between corn yield and soil solution acidity. Applying lime on a soil specific basis seems to be the first step in implementing efficient SSCM to improve overall productivity of the field. Grain moisture at harvest was related to organic C, K and Mg.

Although soil types did not show significant difference in corn yield, spatial variability of nutrients and the relationship between soil characteristics and corn yield observed in this relatively uniform field seem *a priori* sufficiently high to make SSCM useful in MLA. Summarizing the costs and returns associated with the application of SSCM techniques in this field will verify this assumption. An intensive soil survey map (scale $\geq 1:10\ 000$) combined with grid sampling remain indispensable to establish a permanent management unit framework to accurately and efficiently support SSCM.

ACKNOWLEDGEMENTS

The authors would like to acknowledge the farmer cooperators Camille and Paul-André Girouard for their time and access to their fields. We also acknowledge the technical assistance of Pascale Jean, Mario Deschênes and Claude Lévesque. We would also like to acknowledge Jean-Marc Cossette for GIS application as well as Réal Larose and Pierre-Yves Gasser (Ag-Knowledge) for their contribution in yield mapping. This work was funded by the Program on Energy Research and Development (PERD).

REFERENCES

Agronomic Interpretations Working Group. 1995. Land suitability rating system for agricultural crops. 1. Spring-seeded small grains. Edited by W.W. Pettapiece. Tech Bull. 1995-6E. Agriculture and Agri-Food Canada,

Ottawa, 90 p., 2 maps.

Borgelt, S.C., S.W. Searcy, B.A. Stout, and D.J. Mulla. 1989. A method for determining spatially variable liming rates. Paper 89-1034 A.S.A.E. 22 p.

Conseil des Productions Végétales du Québec (CPVQ). 1993. Méthodes d'analyse des sols, des fumiers et des tissus végétaux. Publication 93-0158. AGDEX 533. Québec.

Conseil des Productions Végétales du Québec (CPVQ). 1996. Grilles de référence en fertilisation. 2ième édition. AGDEX 540. Québec. 128 p.

Ecological Stratification Working Group. 1995. A national ecological framework for Canada. Agriculture and Agri-Food Canada and Environment Canada. Ottawa, Hull. 125 p. + 1 map. (1:7 500 000 scale).

Eliason, M., D. Heaney, T. Goddard, M. Green, C. McKenzie, D. Penney, H. Gehue, G. Lachapelle, and M.E. Cannon. 1995. Yield measurement and field mapping with an integrated GPS system. Pages 49-58. In P.C. Robert, et al.(Eds.) Site-specific management for agricultural systems. ASA/CSSA/SSSA Minneapolis, MN.

Gamma Design Software.1995. Geostatistics for the environmental sciences. Version 2.3. Plainwell, Michigan. 165 p.

Goering, C.E. 1993. Recycling a concept. Described as early as 1929, site-specific crop management surfaces again because of advances, environmental awareness. Agric. Eng. 74(6): 25.

Golden Software Inc. 1995. SURFER for Windows. Version 6. Contouring and 3D surface mapping.

Isaaks, E.H. and R.M. Srivastava. 1989. An introduction to applied geostatistics. Oxford University Press. New York. 561 p.

Nolin, M.C., C. Wang, M.J. Deschênes and C. Lévesque. 1995. Description des sites repères 17-QU & 18-QU. Rapport #1. Données de base. Programme d'évaluation de la qualité des sols. Agriculture and Agri-Food Canada. CRTRB Contrib. No 95-105. 32 p.

Nowak, P.J. 1993. Social issues related to soil specific crop management. P. 269-285. In P.C. Robert et al. (Eds.) Proceedings of soil-specific crop management. ASA/CSSA/SSSA, Madison, WI.

Robert, P.C. 1993. Soil-Specific management: soil survey opportunities and challenges. Pages 219-229 in J.M. Kimble, ed. Proceedings of the eight international soil management workshop: Utilization of soil survey information for sustainable land use. SCS/NSSC/USDA.

Systat Inc. 1992. Systat for Windows, Version 5. Statistics. Evanston Il. 750 p.

Wang, C., M.C. Nolin and J. Wu. 1995. Microrelief and spatial variability of some selected soil properties on an agricultural benchmark site in Quebec, Canada. Pages 339-350 in P.C. Robert, R.H. Rust and W.E. Larson, eds. Site-specific management for agricultural systems. ASA/CSSA/SSSA Minneapolis, MN.

Wollenhaupt, N.C., R.P. Wolkowski and M.K. Clayton. 1994. Mapping soil test phosphorus and potassium for variable-rate fertilizer application. J. Prod. Agric. 7(4): 441-448.

Landscape Position and Surface Curvature Effects on Soils Developed in the Palouse Landscape

B. N. Girgin
B. E. Frazier

Washington State University
Pullman, WA

ABSTRACT

Palouse region of eastern Washington is characterized by complex rolling hills with high erosion susceptibility. Various aspect and slope classes along with different soil types also create complex patterns in soil fertility and crop productivity. Division of fields into different units and addressing each unit as a separate management zone has been gaining importance over the recent years. Landscape modeling is one of the tools that helps define management zones based on the spatial variability of the soil and topographic characteristics. In addition to comprehensive models, there is an increasing demand for simpler techniques to assist planners with field scale, day-to-day land management. The objective of this study was to develop a simple landscape model within a Geographical Information Systems (GIS) framework to evaluate the effects of spatial variability of topographic factors on soil genesis. For this purpose, a commercial wheat farm was chosen as research site and a Digital Elevation Model (DEM) of the site was prepared. A FORTRAN algorithm was employed to calculate landscape parameters such as slope, aspect, profile and contour curvature. GIS overlay of these values were geo-referenced and combined with other data layers such as soil maps and air photos. Soil samples collected on three different transects and representative pits were opened for further evaluation of soil characteristics. A horizon thickness, depth to Bt and E horizons were measured for all sampling locations. Results indicate that spatial distribution of E horizon can be estimated by surface curvature, slope and aspect. Study also shows that contrasting soils that are in close proximity to each other too close to be separated on conventional soil maps can be detected with the help of landscape parameters. Bigger map units that extend over several hillslope positions can be further divided to smaller units to receive separate agricultural management based on the soil moisture characteristics defined by these landscape parameters.

Topographic Effect and Its Relation to Crop Production in an Individual Field

J. Krummel
H. Su

Argonne National Laboratory
Argonne, IL

ABSTRACT

Topographic variation within individual fields is one of the key controlling factors in determining the spatial and temporal dynamics of available soil moisture for agricultural crops. Understanding the relationship between topography, soil moisture and crop production could assist in the application of management strategies that employ precision agriculture technology. A study was conducted during the growing season in an individual corn field in northern Illinois that measured the spatial variation in soil moisture with a time domain reflectometer (TDR). An empirical approach was used to determine the topographic gradient of soil moisture during part of the growing season. We used the measured soil moisture data and other field information to simulate corn growth and yield from upland and lowland sites in the field. In addition, we evaluated the potential corn yield variation within the field and its relation to soil moisture dynamics and topographic gradients. Results suggested that (1) soil moisture was correlated with topographic gradient and (2) soil moisture variation across the topographic gradient was associated with simulated corn yield from upland and lowland sites. Results may imply that variation in crop yield within a field can be quantified in response to spatial and temporal variation in topography and soil moisture to improve management strategies that employ precision farming technologies.

Relationship of Electromagnetic Induction Measurements to Detailed Soil Observations in a 32-Hectare Field, Clarion-Webster Area, North-Central Iowa

T. E. Fenton
D. B. Jaynes

Iowa State University
Ames, IA

ABSTRACT

Stationary electromagnetic induction measurements (Em) made at grid points spaced at 25m intervals on a 250m by 250m grid. Soils were identified at each of the points and selected properties recorded. Eight traverses spaced at 50 m intervals were selected in the 32 ha field. Soil observations were made at 25m intervals within each traverse. Geo-referenced on-the-go Em measurements were obtained for the entire field. Spatial distribution of the Em values were compared with the soil observations and the soil map constructed from those observations to better define the inclusions within the larger scale published soil map. The relationships of the Em measurements to soil properties are discussed in detail.

Apparent Spatial Variability of Crop Model Parameters as Estimated from Yield Mapping Data

Y. Pachepsky
A. Trent
B. Acock

USDA-Agricultural Research Service
Beltsville, MD

ABSTRACT

Crop models can be used to develop site-specific management prescriptions if the! spatial variability of the essential parameters of the models is quantified. We have developed a technique to estimate semivariograms of the essential model parameters from yield mapping. The technique consists of (a) calibrating the! model at several reference points: (b) finding the essential parameters from the! sensitivity analysis (c) assuming initial semivariograms for the essential parameters, (d) kriging the model's parameters with these sernivariograms, (e) calculating yields for the kriged values, (f) stopping calculations if the correspondence between the calculated and measured yield maps is acceptable, otherwise (g) changing semivariograms to improve the correspondence between measured and calculated yields and returning to step (d). The Marquardt-Levenberg algorithm and ;genetic algorithms have been V used to perform step (9). Test cases have been developed for wheat using WHNSIM model and for the soybean crop using GLYCIM. The available water content and the initial soil nitrogen content have been chosen as the essential parameters. Yield maps have been generated using these models and the reasonable semivariograms. The technique has allowed us to reconstruct these semivariograms both before and after a random noise was added to the generated model parameters and/or to the yield. The robustness of the technique is currently under study.

A Study on Pattern Development of Nitrate Leaching and Potato Yield for a Farm Field

J. Verhagen

Agricultural University
Waginingen, Netherlands

ABSTRACT

Decisions for operational farm management should preferably be based on quantifiable knowledge on processes of the soil-crop system. Site Specific farm management should be based on the expected spatial development of the soil-crop system during the growing season. Processes in the soil-crop system can be quantified using dynamic simulation models. For a farm field in the Netherlands spatial patterns of crop growth were created via interpolation of 65 point simulations with a validated dynamic simulation model using 30 years of weather data. The patterns resulting from the multi-annual simulations were compared and two prototype patterns were extracted. The occurrence of these patterns is related to farm management practices and weather conditions. Risk assessment of leaching potential versus crop production was made on basis of the derived patterns.

SESSION II

MANAGING VARIABILITY

Moving From Precision to Prescription Farming:
The Next Plateau

S. L. Rawlins

APPROPRIATE SYSTEMS
2368 Eastwood Avenue
Richland, WA

ABSTRACT

The emergence of a number of technologies at affordable prices, has made it possible to precisely apply spatially-variable inputs to farmers' fields. At present the prescriptions for these inputs are typically empirical, based primarily upon grid sampling and soil tests. Such prescriptions draw upon only a small fraction of the existing scientific knowledge of processes controlling crop growth. These empirical prescriptions work well for P, K, lime and other inputs that don't leach or volatilize, but the primary variables controlling crop yield are more often water, nitrogen, pests or diseases or other factors that require within-season management. Developing prescriptions for real-time management of these inputs will require the use of far more knowledge about the processes limiting crop yield at any time at any area of the field than we're now using. Crop simulation models provide a means to package a huge reservoir of scientific knowledge about these processes in a form that can be used for prescribing inputs. But to be used, these simulators will need to be validated in the field, and will require much denser site-specific environmental data sets to feed them than we now have. Fortunately, the technologies to bridge these gaps appear to be here or on the near horizon. But a coordinated effort of a team composed of government, academia, and industry members will be required to move to this next plateau.

INTRODUCTION

This paper presents a strategy to convert a much larger proportion of the scientific information being produced by basic research scientists into knowledge products that can be used directly to prescribe spatially-variable inputs. It consists of packaging this information in the form of computer code that can be used directly by smart machines.

Agricultural managers at all levels need increased information on the consequences, in terms of profit, risk and environmental hazard, of alternative management decisions. Science should be in the business of providing this information. But existing agricultural research is largely constrained by disciplinary boundaries. Some disciplinary research has led to important breakthroughs such as double-cross hybrid corn, new short-strawed wheat varieties or bovine growth hormone. These kinds of breakthroughs, in the form of seeds, chemicals, or other capsulized products, are immediately transferable to users. They are analogous to the *golden nuggets* the first prospectors pick up in a virgin mine field. Sooner or later the nuggets become more difficult to find, and real progress requires *hard rock mining* -- that is incremental progress on a number of fronts, the sum of which adds up to substantial increases in productivity.

Golden nuggets can result from research confined to single scientific disciplines. Hard rock mining, on the other hand, requires an interdisciplinary approach that considers the system as a whole, balancing the research program to fill those gaps which result in the greatest improvement in the performance of the entire system. The tools of Systems Science (Heller and Rawlins, 1986; Rawlins, 1988) are important in guiding the research and interfacing the members of interdisciplinary teams into cohesive programs to develop integrated solutions to important problems.

Only a small fraction of disciplinary research results in products that can be immediately adopted in agriculture. Most of it contributes to an ever growing storehouse of information on the physical, chemical and biological processes involved in agricultural systems. In his book MEGATRENDS, John Naisbitt states that this storehouse of scientific information was doubling approximately every two years. Our libraries were beginning to burst at the seams. He comments that the shear mass of this information reserve often makes it more practical for a research scientist to repeat an experiment than to attempt to find the results of previous research in the literature. He states that "we are drowning in information". But at the same time we are "starving for knowledge". Knowledge is information in a form that can be brought to bear on the solution of important problems.

BACKGROUND

In 1981 the U. S. agricultural research system was receiving serious criticism from the Congress and from other sources, accusing it of being in disarray, that there was no leadership within the system, and that different parts of the system were working independently for their own good, not solving problems for the tax-paying public. Phillips (1981) of the Congressional Office of Technology Assessment proposed a solution:

> "This is a time when the food and agricultural problems need to be identified, the research participant identified, and the organization of each participant evaluated to determine if it can effectively carry out its role. Only then will the research community be able to merit and obtain the resources needed to meet the food and agricultural challenges of the future."

In other words, Phillips thought that agricultural research should focus on identifying and solving problems, functioning as a coordinated system, not a collection of independent parts. It was fortuitous that the newly appointed ARS Associate Deputy Administrator responsible for the ARS research program, Dr. T. J. Army, felt the same way. In 1966, he and a colleague from the International Minerals and Chemical Corporation (Army and Smith, 1966), stated:

"Some time ago we gave up thinking in terms of plant genetics
alone or the production of a particular crop, and have extended our
thoughts to the systems of cropping, feeding, processing, and
distribution. The total interaction of seed, plant population, soil
moisture, pesticides, and growth regulators all add up to what we
term the "systems approach."

A number of us on the ARS National Program Staff were embarrassed by
the criticism we were receiving, and with Dr. Army's support, developed a plan
(ARS, 1983)[1] to treat ARS research as a holistic system. A major goal of the
planning process was to stem the tide of criticism by transforming ARS into a
problem-solving agency -- from one characterized by *technology push*, where
scientists alone set priorities, to one concerned more with *market pull*; where
scientists were responsible for *excellence*, but customers were involved in
determining the *pertinence* of research objectives.

It was clear to all of us that the problems facing agriculture were complex,
and that they would require not only more, but a different kind of research --
research characterized more by *synthesis* and linked interdisciplinary teamwork
rather than *analysis* and disciplinary scientists working alone. To accomplish this
we established *systems research* as one of the six objectives in the Program. The
objective statement was:

"Develop the means for integrating scientific knowledge of
agricultural production, processing, and marketing into systems that
optimize resource management and facilitate transfer of technology
to users."

Later, ARS created the Agricultural Systems Research Institute in Beltsville,
MD to provide support for the systems objective (Heller and Rawlins, op. cit.), and
I moved across campus to become its chairman.

We saw simulators and expert systems, along with the required databases
and knowledge bases to support them, to be key means for both packaging
knowledge and transferring a much larger fraction of it to the farm. From the
beginning, Dr. Basil Acock, an ARS scientist involved in developing crop
simulators, then located in Mississippi, assumed a leading role in developing the
structure to do this.

Figure 1 illustrates the role he proposed that models and expert systems
could play in this process.

Looking only at the unshaded boxes, which effectively represent the present
system, we see that the major product of basic research, *articles read only by other
researchers* (many of which deal with the fundamental processes controlling crop
yield) has no direct pathway to the farmer. To transfer scientific knowledge that
could help the farmer make better management decisions we still rely heavily upon

---

[1]Current information about this plan is now available on the Internet World Wide
Web at URL: http://www.ars.usda.gov/programs.html

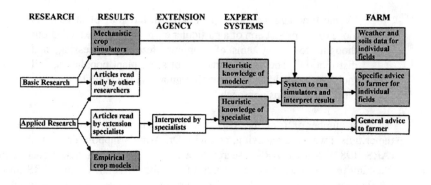

Figure 1.   The contribution of crop simulators and expert systems in transferring research
results to the farm. The unshaded boxes represent the components that exist in the
present system; the shaded boxes are new components proposed to be added.
(Graphic adapted from Acock, 1990).

applied field-plot research, largely carried out at research stations, that can only lead
to general recommendations. Specific management information for each field on
a farm will more likely be gained by the farmer himself, by his own field trials, or
by long-term experience.

In Dillman's (1986) terminology, we're very much in the "Community
Control", or "Mass Society" mode, where one solution is expected to fit all. He
points out that beginning in the early 1900s management was under "Community
Control". The only way to get new technology into the system was to place
Extension agents into the communities where they could influence the community
leaders. Essentially everyone in the community adopted a more or less uniform set
of practices that were roughly tailored to the local conditions. Beginning in the
1950s control shifted from the local community to the "Mass Society". This was
the era of standard recommendations for massive inputs of chemicals and seeds all
across the nation. Dillman points out that we are now approaching the diminishing
returns part of that input curve as we enter the "Information Age". In the
information age farmers are more interested in "what works on my farm" than
generalized recommendations. In this new era information has the potential of
substituting for high cost material inputs such as fuel, fertilizer and pesticides.

The shaded boxes in Figure 1 represent a new strategy, whereby basic
research scientists can join in the process of providing the specific advice required
in the information age. It begins with these scientists producing an additional
research product -- computer code. This product includes computerized crop
simulators, expert systems and databases. Scientists' work with other team
members through enhanced communication links that knit interdisciplinary teams
together. Instead of simply describing the new knowledge about how some basic
process affects the growth of a crop in printed papers, the team produces an

integrated system, that for any set of environmental inputs, computes the crop response. We call these *computerized decision support systems*. They have the potential not only to package existing scientific information into a form that it can be brought to bear on the solution of today's pressing problems, but they can also capture the judgement of real experts, including farmers, to assist others in making better decisions.

Theoretically, this all sounds possible, but implementing it has been extremely difficult. Old patterns of behavior are difficult to change. It was Max Planck (1950) who stated (in what has become to be known as Planck's Law):

> "New scientific truth does not triumph by convincing its opponents and making them see the light, but rather because its opponents eventually die, and a new generation grows up that is familiar with it."

In addition to *Planck's law* there were two other major barriers to overcome. First, few basic researchers have the resources or the broad span of knowledge required to build complete crop simulators. To obtain their participation and take advantage of their deep, but narrow, expertise, a plan was outlined to encourage basic research scientists to encode their unique knowledge in the form of simulator modules -- parts of the simulators that were within their knowledge domain -- and not try to build complete simulators. This required a model structure that would permit independent modules to be removed, modified or rewritten by different authors, and then be placed back into the overall model. This would permit simulators to be written by a team of experts, each contributing state-of-the-art knowledge, rather than by a single model builder, who would have to interpret the knowledge of each expert to include it in the model. To accomplish this end, the modules had to separate along scientific disciplinary lines that matched the knowledge span of individual scientists. Such a structure has been tentatively adopted.

The second barrier was that the scientist peer evaluation system depends upon publication in refereed journals. There were no refereed journals that accepted computer software as a publication. Basil Acock championed the idea of an electronic journal that would accept computer software in the form of models, model modules and expert systems as peer-reviewed journal articles. After many years of work, The American Society of Agronomy has agreed to sponsor such a journal on an interim basis, entitled the **Electronic Journal for Terrestrial Ecosystem Software**. This journal is now being developed on the Internet[2].

---

[2]The Internet home page is URL: http:\www.arsusda.gov/ejtes/. The Editor-in-Chief, Dr. Basil Acock, is now at the ARS Systems Research Lab in Beltsville, MD.

CUSTOM PRESCRIBED FARMING INITIATIVE

To provide a more direct means for farmers to use information from crop simulators to manage farm inputs, in March of 1983 I developed a concept paper[3] that spelled out a future scenario which I called *custom prescribed farming.* (Many of these same concepts were also published in Rawlins, 1984). The basic concept, which was neither new nor complicated, was analogous to the system dairymen use for tailoring feed rations to match the production of individual cows.

Even in 1983, some dairies were automated with sensors to identify individual cows and to measure milk production. Computers kept track of production and calculated feed rations for each cow, based partly upon computer models from dairy nutrition scientists. The models also handle other inputs, such as the cow's temperature or environmental factors that may have a bearing on her feed requirements. Smart machines delivered the prescribed ration to each cow while she was being milked, without personal intervention by the herd manager.

Here was a case where the fundamental knowledge of leading scientists, packaged in the form of computer models, was being used directly for herd management without having to be interpreted by specialists along the way. Forging this direct link to make it possible to bring the fundamental knowledge of key scientists to bear on management through smart machines was exactly what I wanted to do with crops.

Basic Concepts

The basic concepts of custom prescribed farming developed in the 1983 paper were as follows:

- Soil and crop information sensed on-the-go would be stored in a computer on a field vehicle, and would be used to help develop prescriptions for spatially-variable cultural practices including tillage, planting depth and density, as well as fertilizer, pesticide and herbicide application rates.

- The primary missing link was position-sensing technology to indicate where the field vehicle was located. Once this technology becomes available, site-specific data can be sensed and logged into the two-dimensional computer map.

- With position-sensing technology available, existing computer software would permit automatic guidance of the tractor. This would permit the tractor to follow the same tracks exactly without operator control, leaving

---

[3]Rawlins, Stephen L. Unpublished memo dated March 18, 1983 entitled The Use of Computers and Sensor Technology to Build Systems that Close the Gap Between Science Information and Farm Application. The memo, which was widely distributed to ARS and university scientists as well as to industry cooperators, is available now through @g INNOVATOR's @griculture Online Internet World Wide Web site at URL http://www.agriculture.com/contents/aginn/ai029608.html.

the operator free to observe and log field variables, such as weeds, insects and diseases, into the computer database simply by holding down a key programmed for that variable. Additional information such as soil water content with depth, canopy temperature, plant height, soil cover and other variables could be sensed on-the-go with automated sensors.

- Manually obtained data such as soil clay content, soil slope and aspect, soil organic matter and other such variables could be logged into the computer map for the field as relatively permanent data, and used to help develop prescriptions.

- Sensors were then being developed to measure the rate of grain flow from the cylinder of a combine. The addition of new sensors to measure the yield of other crops on the go was envisioned.

- The combination of all this information would make it possible to optimize the treatment of each square meter of the field rather than broadcast-treating the entire field. The possibility of treating small patches of weeds or insects before they grew to the point that it is economically feasible to broadcast-treat the entire field would make it possible to contain and eradicate weeds and other pests that now get out of control.

- On-the-go soil moisture sensing, the technology for which was then under development, would make it possible to automatically control the depth of the planter to make sure that seeds are placed in moist soil. This soil moisture content information, in addition to base data such as soil organic matter and clay content, would make it possible to automatically control the amount of soil-incorporated herbicide required to gain weed control, without causing damage to the growing crop.

- Sensor technology to remotely sense residue cover would make it possible to control the tillage operation to make certain that sufficient cover remained on the soil to control erosion. It would also contribute toward determining appropriate herbicide rates.

- Such a system would also make it possible to conduct automated field experiments. Fertilizer treatments, for example, could simply be programmed into the computer so that a series of different rates are randomly applied at various places within the field. When the crop is harvested, yields could be compared with fertilizer rates to generate response functions. Once such systems became practical for actual on-farm operation, yield trials could be automatically programmed into the computer each season, and the response functions could be used in conjunction with the projected cost of fertilizer and the sale price for the crop to determine the most profitable fertilizer rate to apply to a specific field.

•   This system could provide an alternative to printed papers as an outlet for scientific research. If the results of experimental work could be used as input into a crop model for a specific area, they could be used for decision support without all of the detail having to pass through the minds of extension agents and farmers along the way.

Progress on Key Concepts

This conference will document in detail the progress that has been made in precision farming technology. Variable-rate application of chemicals via truck-mounted applicators has been marketed since about 1986. Planters that allow variable-rate seeding are also available.

Yield sensors for most crops, cotton being a significant exception, as well as mapping software are now commercially available from a number of companies. Yield maps constructed by on-the-go monitoring of harvester output as a function of position are routinely made. But for crops such as cereal grains that are not planted in rows, maps from yield monitoring data must correct for variable swath width, as well as eliminate areas already harvested or outside the cropped area from being counted in the acreage accumulation. Han et al. (1995a) have recently developed a system that uses the GPS position to keep track of cut and uncut areas, and records yield data only once over any given area. Existing commercial mapping software does not automatically account for these factors.

Because many of the input data for crop simulation models are spatially referenced, geographic information systems (GIS) are effective tools to provide input, manipulation, query, and output of the large amount of spatial data required by them. In an initial study Han et al. (1995b) used a commercial GIS system linked to a potato crop simulation model. Data layers were established for each of the spatially-distributed input variables, and the simulator was then run for each area of the field that was significantly different. A new data layer was created containing the predicted spatial yield distribution, as well as nitrogen and water leaching distribution for the field.

On-the-go sensing of information for real-time feedback control of tillage, herbicide and insecticide application rate, or planting depth and density has yet to be adopted. Automatic guidance of chemical application machines is in the development stage with about 10-cm accuracy[4]. Until automatic guidance is routinely available, observation of conditions such as the location of specific weed, insect, disease, or other infestations within the field, and logging them into the computer database will usually require an additional person on the field machine.

On-the-go sensing and logging of information such as soil water content with depth, canopy temperature, plant height, soil cover and a number of other variables with automatic sensors, although possible in some cases, has not yet been implemented in a practical site-specific farming application. Manually-obtained data such as soil clay content, soil organic matter, and other such variables are used to help develop application maps. Field slope and aspect need not be measured manually, as anticipated in 1983, but can be generated from longitude, latitude and

[4]Personal communication, Mr. Don McGrath, Tyler Industries, Benson, MN.

elevation data logged by the GPS.

We're poised on the threshold of scientists publishing their knowledge of processes controlling crop growth in the form of computer code in a peer-reviewed electronic journal. Once this code is available, the next step will be to embed it into computer programs that prescribe spatially-variable inputs that meet crop requirements, and control the machines that apply these inputs. These programs and machines will not replace the grower's wisdom and judgement, but they will assist him in taking into account the complex factors controlling crop growth, much as computers assist pilots in operating modern aircraft. Until computers were available, the complexity of the mission a pilot could fly was constrained by his capability to sense conditions, make calculations in his head, and adjust the controls in time to avert a disaster. The same is now true for crop growers. Computers and smart machines can help growers handle infinitely more information and make decisions in time to avert negative consequences. Hungry customers for scientists computer code will help provide the incentive needed for them to produce it.

Near-Term Enhancements

Several possibilities exist for application of emerging technologies in the short term to achieve significant advances in management objectives.

Precision Guidance

Precision guidance of field machines could not only relieve the operator of some of the tedium of steering, it could permit machines to follow the same tracks exactly each time they enter the field. This controlled traffic pattern could confine soil compaction to narrow bands, leaving the remainder of the field untracked. These compact bands, which are never tilled, would act as all-weather *roadbeds*, providing access to fields so that they could be planted earlier, sprayed more often and harvested later.

Also, by confining compaction to these strips, other strips of the field could be tilled to provide an ideal seed bed that tended to shed water, so that it infiltrated primarily through still other strips on either side. These strips would be tilled to enhance water infiltration. By confining soluble fertilizer to the seedbed strips, water passing through the infiltration zones would not leach it.

In effect, this zonal tillage would permit the separation of water and chemical fluxes. This would allow the farmer, in effect, to have his cake and eat it too. He could have high fertility within the rootzone without subjecting it to leaching, even though more water was applied than the crop transpired. It would allow him to create a situation similar to that achieved by horticulturists by covering beds with plastic sheeting.

Variable-rate Water and Chemical Application Through Irrigation Machines

Although providing all-weather access to fields through permanent roadbeds could enhance the possibility of applying inputs where they are needed throughout the growing season, nothing compares with the capability that could be provided by self propelled irrigation systems equipped for variable-rate application technology for water and chemicals. In December, 1995, Precision Irrigation Systems, Inc. and the J. R. Simplot Company jointly announced a partnership to commercialize a variable-rate irrigation system[5]. In this conference, Evans et. al (1996) will report on an extensive research project to accomplish both variable rate application of water and chemicals with self-propelled irrigation systems. The capability to apply water or chemicals to any spot in the field throughout the growing season greatly enhances the manager's capability to meet the crop's needs in all areas of the field without excess.

Having the capability to apply inputs precisely where and when they are needed is a necessary, but insufficient condition for meeting but not exceeding the crop's requirements in all areas of a field. It's *precision farming*, but not *prescription farming*. To develop real-time field prescriptions we must know what process is limiting yield at any time in all areas of the field. This will require the application of a much greater proportion of the scientific knowledge of the processes controlling yield than is now available in the field.

Crop Simulator-based Prescriptions

A serious problem faced by today's farm managers is that the productive capacity of each area of a field varies from year to year, depending upon rainfall and other factors. An area that has low yield in a dry year, because the soil doesn't hold much water may have high yield during a wet year, because it has good drainage. We can prescribe inputs for variables like P, K, and lime (which are neither volatile nor leach), that can be managed on a whole-season or multi-season basis with empirical prescriptions based upon soil tests. But water, nitrogen, pests and diseases and other variables that dominate the control of crop yield, require within-season management.

To prescribe inputs that meet crop requirements in real time, we need to know what variable is limiting yield in each area of the field at all times, and have the capability to deliver inputs to reduce this limitation. The grower is not capable of eliminating the growth-limiting effects of some factors such as temperature, sunlight and, where irrigation is not available, rainfall. In these cases, he needs to know the maximum attainable yield under these constraints, and limit the application of inputs to grow a crop of this size.

Crop simulation models, at least theoretically, have the capability of predicting the yield at every area within the field using actual environmental data up to the present day, and some assumed scenario for environmental data from that time forward (Acock et al.,1982; Baker et al., 1983; Lemmon , 1986). But a serious barrier stands in the way of achieving this. To predict spatial yield distribution throughout the field, we must have far more information on the spatial distribution of soil properties and environment to drive the models. These data are required, first, to validate models in the field, to make certain that the processes coded into the simulator closely match those occurring in the field. But after that, they will be

---

[5]Press release dated December 27, 1995 issued by the J. R. Simplot Company, P. O. Box 912, Pocatello, ID 83204.

required each year to make real-time spatially-variable yield predictions.

Although this may seem to be an impossible barrier to overcome at this time, I'm encouraged by the remote sensing capabilities that will be available in the near future. The best of today's satellite-based remote sensing platforms provide information with a resolution (pixel size) of about 10 m. Even at this size, the quantity of data accumulated as they circle the earth is huge. The new generation of sensors will have a pixel size of about 1 meter. If they were to operate all of the time, the quantity of data accumulated would be 100 times greater than that for the 10 m pixel size. But the key is they won't operate all of the time. Instead of storing large volumes of data which the user must sort through to find what he needs, they will record scenes only upon demand. The future grower will simply place an electronic order for scenes at the latitude and longitude of his field. The scenes will be recorded and the data transmitted to him within about a day. The customer receives only the data requested, and the remote sensing provider does not have to maintain a huge library of data.

Putting this integrated system into place will not be easy, but given the technology advances we've seen during the past fifteen years, it is not unreasonable to expect that crop simulators and the data to drive them will be sufficiently advanced to at least begin using them for prescribing spatially and temporally variable inputs within the next fifteen years. By then, we'll also have a new generation of scientists who grew up with computers. I can't wait to see this.

CONCLUSIONS

Whether we will be able to bridge the gap between the scientific knowledge of the fundamental processes controlling the growth of crops and the application of this knowledge in the field is an open question. Simulation models offer one possibility for doing this. If written as a collection of modules, the primary knowledge of the scientist on the cutting edge of research can be encoded in a way that it can be brought to bear directly on the decision-making process in the field through smart machines. When we reach this point, crop growers will have the same advantages dairy managers have had for two decades, where fundamental scientific knowledge is brought to bear directly to support management decisions.

In the past the public sector has developed, validated and acquired data to parameterize models. Public sector research funding is shrinking. Support from the private sector will be required to maintain this effort. Perhaps the best support from the private sector will come from companies who become hungry customers for the new computer-coded knowledge produced by research scientists.

It's not beyond reason to expect that new companies will find opportunities to create businesses that focus on the acquisition, packaging and marketing of knowledge products. Given the new technologies on the horizon, and those that we cannot yet see, I think we can be optimistic that the knowledge transfer gap will be bridged. When this occurs, we'll make a quantum leap to the next plateau, where knowledge flows through simulators to smart machines on farms.

REFERENCES

Acock, B. 1990. Structuring Agricultural Research to Deliver Decision-support Products. *In* Proceedings of the Workshop on Computer-integrated Agriculture. pp. 43-48. Agricultural Research Institute, Bethesda, MD.

Acock, B., V. R. Reddy, F. D. Whisler, D. N. Baker, J. M. McKinion, H. F. Hodgesand K. J. Boote. 1982. The Soybean Crop Simulator GLYCIM Model Documentation. PB 85171163-AS. U. S. Department of Agriculture. Available from NTIS, Springfield, IL.

ARS. 1983. Agricultural Research Service: Program Plan. Miscellaneous Publications number 1429, ARS, U. S. Department of Agriculture.

Army, T. J., and M. E. Smith. 1966. What the Systems Approach Means to Your Job. Farm Technology, Vol 22. No.2.

Baker, D. N., J. R. Lambert and J. M. McKinion. 1983. GOSSYM: A Simulator of Cotton Growth and Yield. S. C. Agricultural Experiment Station Tech. Bull. 1089.

Dillman, Don A. 1986. Cooperative Extension at the Beginning of the 21st Century. The Rural Sociologist 6:102-119.

Evans, Robert G., Shufeng Han, Sally M. Schneider and Marty W. Kroeger. 1996. Precision Center Pivot Irrigation for Efficient use of Water and Nitrogen. Proceedings, 3rd International Conference on Precision Agriculture, American Society of Agronomy. (In Press)

Han, S., S. L. Rawlins, R. H. Campbell, and R. G. Evans. 1995a. A Bitmap Method for Determining Harvest Width in Yield Mapping, Presented at the June 18-23, 1995 ASAE Meeting, Paper No. 95-1333, ASAE, 2950 Niles Rd., St. Joseph, MI 49085-9659, USA.

Han, S., R. G. Evans, T. Hodges, and S. L. Rawlins. 1995b. Linking a GIS with a Potato Simulation Model for Site Specific Crop Management. J. of Environmental Quality 24:772-777.

Heller, Stephen R., and Stephen L. Rawlins. 1986. Agriculture Systems Research -- A New Initiative. *In* Human Systems Management. Elsevier Science Publishers B.V. (North-Holland) 6:289-296.

Lemmon, H. 1986. Cotton Crop Management Expert System -- COMAX. Science, 233:29-33.

Mayer, Andre; and Jean Mayer. 1974. Agriculture, the Island Empire. Daedalus, Vol 103, pp. 83-96.

Phillips, Michael J. 1981. OTA Assessment of the Research System. Agricultural Outlook Conference, USDA. Nov. 2-5, 1981.

Planck, M. 1950. Scientific Autobiography and Other Papers. Williams and Norgate, London, pp 33-34.

Rawlins, S. L. 1984. Practices for Conserving Natural Resources and Minimizing Adverse Environmental Impacts in Changing Agricultural Production Systems: Current State of the Art. *In* George W. Irving, Jr. (ed.), Changing Agricultural Production Systems and the Fate of Agricultural Chemicals. pp. 92-103. Agricultural Research Institute, Bethesda, MD. 1984.

Rawlins, Stephen L. 1988. Systems Science and Agricultural Sustainability. National Forum, Phi Kappa Phi (Summer Issue) pp. 19-22.

The Development of Management Units for Site-Specific Farming

B. L. McCann
D.J. Pennock
C. van Kessel
F. L. Walley

Department of Soil Science
University of Saskatchewan
Saskatoon, Saskatchewan, Canada.

ABSTRACT

Image analysis of black and white aerial photographs can be used as a cost effective method to delineate soil management units. In this study, extensive field sampling and laboratory analysis were used to characterize a soil landscape. The site was stratified into four management units by grouping the digital numbers on the scanned black and white aerial photograph into categories that reflect the changes in different soil properties across the landscape. The close relationship between the management units and the soil properties suggests that this technique is an effective method for stratifying landscapes into management units.

INTRODUCTION

Producers are looking to site-specific farming to increase the efficiency of their operations. It has long been recognized that the productivity on hilly terrain varies greatly from the knolls to the depressions. This variation in productivity reflects the different moisture and soil conditions that exist at the two locations. The goal of site-specific farming is to delineate management units - areas with similar productive potential - and apply the right amount of inputs to obtain the best returns.

The stratification of soil landscapes into management units is one of the major challenges facing both researchers and producers who are interested in site-specific farming. Over the years, a number of methods have been developed to stratify research sites into units with similar productive potential: (1) detailed soil surveys, (2) extensive sampling programs, and (3) collection of topographic data. These techniques have been effectively used for research applications but it is not economically feasible to use them for site-specific farming applications on cereal farms in Saskatchewan. These farms are large scale, low input, dryland operations. For example, field sizes typically range from 32 to 64 hectares (80 to 160 acres). Fertilizers inputs are relatively low by cornbelt standards: 35 to 60 kilogram per hectare (40 to 70 lbs/acre). Because of the large fields and relatively low inputs, the cost associated with detailed soil surveys, sample grids, and topographic surveys is too high in relation to the economic benefits.

The objective of this study, is to determine if image analysis of black and white aerial photographs can be used to stratify soil landscapes into management units. If successful, this technique would provide farm managers and agricultural consultants with an inexpensive method for mapping fields into management units for site-specific farming applications.

MATERIALS AND METHODS

Study Site Description

The study site is located 40 km north of Saskatoon, Saskatchewan, Canada near the community of Hepburn (SW 7-40-5-W3; or 52° 25' N, 106° 41' W). The site is situated on a glacial till landscape that is part of the Oxbow soil association (Acton and Ellis, 1978). Soils at the site are dominately Chernozemic (Udic Boroll) with significant Gleysolic soils (Typic Aquoll) in the depressional areas. Slope gradients at the site range from 5 to 10%. Surface drainage at the site is local with the depressional areas separated by linear ridges. These ridges dissect the site at different angles resulting in a hummocky surface pattern. The glacial till parent material has a texture that ranges from loam to clay loam.

Site Design

The research site covers an area that is 250 meters by 300 meters. The site encompasses several cycles within this knoll and depression landscape - one cycle extends from one knoll down into a depression and then to the top of the next knoll. A 9 by 11 point grid with a 25 meter sampling interval was used as a guide for the sample collection.

Surveying and Sample Collection

Once the grid positions were established within the landscape, the grids were surveyed and sampled. A Total Station was used to gather topographic and positional information for both the sampling points and the topographic inflection points within and around the grid periphery. At each grid point, soil cores were extracted and extruded using a truck mounted coring apparatus. Once extruded, the soil cores were described using the Canadian System of Soil Classification (Agriculture Canada, 1978). Samples were collected at 15 cm intervals to a depth of 60 cm. Finally, in an effort to assess *in situ* soil salinity, EM 38 reading were taken at each site (Geonics Limited, Ontario, Canada). (Mention of commercial products is solely to provide specific information and does not constitute endorsement by the University of Saskatchewan).

Before the soil data can be used for image analysis, the sampling points must be georeferenced. In this study, Global Positioning System Units (GPS) were used to establish the UTM coordinates for the corners of the sampling grids. In order to obtain an accuracy within 1 meter, Differential GPS measurements were taken using 2 Garmin II GPS units (Garmin International Inc. Lenexa, Kansas, USA). One unit acted as a base station and was set up on an obstruction-free site near the corner of the section. The second instrument was used as a field unit and was deployed to gather positional information for the grid corners and road intersections in the surrounding area. The two instruments gathered positional data simultaneously and stored it within the units. The data were subsequently downloaded into a DOS-based computer where the UTM coordinates were established for each point through post-processing.

Laboratory Methods

The soil samples collected in the field were subjected to a series of analytical procedures that provided information needed to characterize the soils. A number of chemical properties were examined: inorganic carbon, organic carbon, nitrate-nitrogen, pH, and electrical conductivity. Physical components include particle size distribution, moisture content, and moisture retention. Standard analytical techniques were used to gather all the information. The data generated from these procedures facilitated the classification and delineation of management units within the study areas.

Data Processing

The topographical data was used as the basis for the creation of the digital elevation model (DEM). The first step in this process utilized a Fortran program developed by Pennock et al. 1987 which is designed to convert the survey data to X, Y and Z values. The topographic mapping program Rockware was then used to create the DEM for the site (Rockware Incorporated, Wheat Ridge, Colorado). In order to create three dimensional depiction of the site, the graphics program 3D was employed (Wingle Hydrogeology, Golden, Colorado). This program was also used for draping the classified image over the DEM (Fig. 4).

Image Analysis

The image analysis techniques that were used in this study consists of 5 basic components: (1) image capture, (2) georeferencing, (3) enhancement, (4) analysis, and (5) classification. The first step in this process is image capture or the conversion of an aerial photograph to a digital image. The photograph for this site was obtained from Central Survey and Mapping Agency in Regina, Saskatchewan, Canada and was photographed at a scale of approximately 1:30,000. The photo was then scanned at a resolution of approximately 600 dots per inch. The area of interest on each photo was delineated and saved as a digital image. Digital images are composed of a two dimensional array of pixels or grid cells. For black and white digital images, the grey tone of each pixel is expressed as a digital number ranging from 0 to 255 with 0 representing black and 255 white.

Before a digital image can be utilized in the image analysis process, the inherent distortion must be corrected through image rectification (ERDAS, 1991). This process involves the projecting of the data onto a plane, and making it conform to a map projection system. The technique utilizes ground control points that act as reference points from which the image points are corrected and assigned UTM coordinates.

Image enhancement is the process in which spatial and spectral enhancement techniques are used to make an image more interpretable for a particular application (ERDAS, 1991). In this study, two enhancement procedures were used on the image to improve its quality: (1) low pass filtering, and (2) histogram stretching. A low pass filter was used to remove much of the random noise on the image caused by cultivation patterns and uneven straw distribution. Histogram stretching was used to increase the contrast of the image in an effort to facilitate classification.

Once the enhancement process was completed, the digital number for each sample point in the grid had to be established, analyzed, and grouped into categories. The sample grid was digitized and superimposed onto the enhanced image. The digital numbers for the sample sites were then recorded. Using the soil profile descriptions as a guide, a supervised classification was conducted on the digital image to stratify the landscape into four units based on its spectral characteristics. The spectral characteristics of these units were then compared to various soil properties using box and whisker diagrams. The examination of these relationships facilitated the adjustment of the categories to better reflect the pedological and hydrological conditions at the site. Once the categories had been finalized, colors were assigned to the different categories resulting in a GIS map of the management units for the study area.

RESULTS AND DISCUSSION

The grey tone pattern on black and white aerial photographs is often a reflection of soil properties that may be linked to productivity. The spectral properties of bare soil surfaces are largely governed by soil organic carbon, inorganic carbon, and moisture (Schreier et al., 1988). These properties can be either directly or indirectly linked to productivity. Because of these relationships, this study examined the possibility of stratifying soil landscapes into management units according to their spectral properties. Using the soil profile descriptions as a guide, the digital image for the research site was classified into four categories: (1) knolls, (2) mid slopes, (3) lower slopes, and (4) depressions (Fig. 1). These categories formed the basis for the management units at this site.

The four management units were compared with various soil properties to evaluate their differences in terms of productive potential. Because organic carbon is an indicator of soil fertility, the organic carbon levels were grouped according to the different management units (Fig. 2). In this box and whisker diagram, there is a clear trend in the organic carbon levels. For example, the knolls have the lowest levels outside of the depressions. These low levels are a reflection of the relatively dry growing conditions and the removal of soil as a result of erosion and cultivation. In the two management units immediately down slope from the knolls, the organic carbon levels are much higher than those in the knoll unit. Why the increase in organic carbon? These increases in organic carbon reflect the relatively moist conditions that have favored abundant plant growth and the subsequent accumulation of carbon (Pennock et al. 1994). Still further down slope, in the depressional areas, periodic flooding has increased moisture levels to the point where carbon accumulation has been retarded and hence, organic carbon levels are similar to those on the knolls (Pennock et al. 1994).

Soil nitrate levels have a similar relationship to the management units (Fig. 3). Like the relative values for organic carbon, the knolls have the lowest nitrate levels for any of the management units. These low values are likely a reflection of the low nitrogen mineralizing potential at these sites (Pennock et al. 1992). In the two management units just down slope from the knolls, the nitrate levels are higher (Fig. 3). These higher levels are largely a reflection of the relatively high nitrogen mineralizing potential at these sites.

This higher potential is primarily a function of the relatively high organic carbon levels. Moving still further down slope, the nitrate levels drop off in the depressional areas. Periodic flooding has created the anaerobic conditions that enhance loses of nitrogen through leaching and denitrification (Linn & Doran, 1984). These loses, combined with the lower nitrogen mineralizing potential at these sites, result in relatively low nitrate levels.

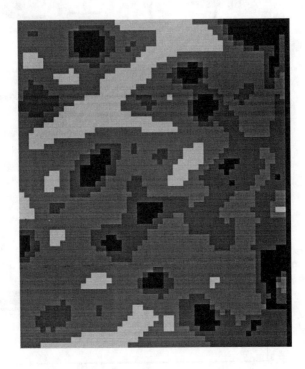

(Figure 1) Soil Management Units Derived from Digital Image.

In order to assess the relationship between the management units and the topography at the site, the GIS map was draped over the DEM (Fig. 4). It is evident that there is a close relationship between slope position and management units. For example, the knoll unit is associated with the upper most sections of the major ridges. These units are generally surrounded by mid-slope units which closely conform to the landscape curvature on the DEM. The lowest points in the landscape are classified as depressions. These units are always surrounded by the lower-slope units.

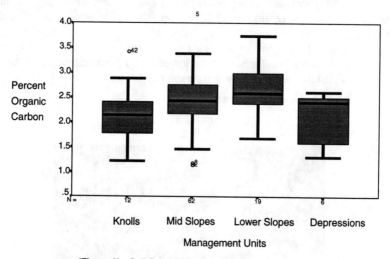

(Figure 2)   Soil Organic Carbon and Management Units

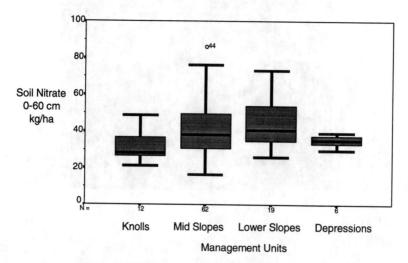

(Figure 3) Soil Nitrate and Management Units

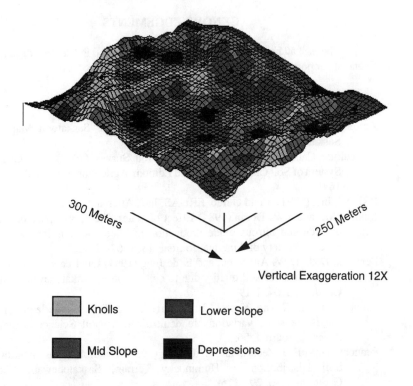

Vertical Exaggeration 12X

Knolls   Lower Slope

Mid Slope   Depressions

(Figure 4) Management Units Draped on the Digital Elevation Model

CONCLUSION

Image analysis of scanned panchromatic aerial photographs is an effective and inexpensive tool for delineating soil management units for site-specific farming. The level of accuracy and detail obtained with this method could otherwise be only achieved through a very detailed soil survey at considerable expense. Image analysis is not, however, a panacea. In order for the technique to be effective, the study area must be characterized with the use of existing soil survey information and ground truthing data. The 1:100,000 scale soil survey information that is available in Saskatchewan acts as a general guide for the broad characterization of soil landscapes. Ground-truth data, when collected in conjunction with a GPS, provides the information needed to locate and establish the spectral properties of specific soil management units within the soil landscape. Ground-truth data would not, however, need to be collected for all study sites with the same map unit. Once the spectral properties of a representative site were characterized, the information could easily be extrapolated to other sites with the same map unit. Utilizing image analysis in this manner should allow researchers and producers involved in site-specific farming to stratify their soil landscapes into management units that reflect the differences in the productive potential within the field.

ACKNOWLEDGEMENTS

The Saskatchewan Agricultural Development Fund is acknowledged for the financial support of this project.

REFERENCE LIST

Acton, D.F. and J.G. Ellis (1978) The Soils of the Saskatoon Map Area. Saskatchewan, Saskatoon, Saskatchewan.

Agriculture Canada Expert Committee on Soil Survey (1978). The Canadian System of Soil Classification. 2nd Edition. Agriculture Canada Publication 1646.

ERDAS, Inc. (1991). Field Guide. ERDAS, Inc., Atlanta, Georgia..

Linn, D.M., and J.W. Doran.(1984) Effect of water-filled pore space on carbon dioxide and nitrous oxide production in tilled and nontilled soils. Soil Science Society of America, Journal. 48: 1267-1272.

Pennock, D. J., D.W. Anderson, and E. de Jong (1994) Landscape-scale changes in indicators of soil quality due to cultivation in Saskatchewan, Canada. Geoderma. 64: 1-19.

Pennock, D. J.,C. van Kessel, R.E. Farrell, and R.A. Sutherland. (1992). Landscape-scale variations in denitrification. Soil Science Society of America, Journal. 56:770-776.

Pennock, D. J., B. J. Zebarth and E. de Jong (1987). Landform Classification and Soil Distribution in Hummocky Terrain, Saskatchewan, Canada. Geoderma. 40: 297-315.

Schreier, H., R. Wiart, and S. Smith (1988). Quantifying organic matter degradation in agricultural fields using PC-based image analysis. Journal of Soil and Water Conservation 43: 421-424.

Determination of Field and Cereal Crop Characteristics for Spatially Selective Applications of Nitrogen Fertilizers

M. Robert
A. Le Quintrec
D. Boisgontier
ITCF, Station expérimentale de Boigneville
France

G. Grenier
ENITA, Bordeaux
France

ABSTRACT

The aim of the experiment described in this paper was twofold. First, work was done to quantify the effects of factors potentially utilizable to design a reliable method for variable-site applications of nitrogen fertilizers. Secondly, the authors sought to carry out an initial trial and to compare it with a "classical" method of nitrogen fertilizer application.

OBJECTIVES

In France, consideration of environmental problems has lead to focusing on variable-site applications of nitrogen fertilizers rather than on variable-site applications of phosphate or potash. Over the last three years, **ITCF** (Technical Institute for **C**ereal crops and **F**orage) has carried out an experiment with a view to defining a reliable method of determining variable-site applications of nitrogen fertilizers on cereal crops. This long term experiment was built on previous knowledge.

Recently, some yield mapping systems and data recorders have become available. Thus, it should be possible to determine the spatial variability of crops or soils and to manage fertilization according to this variability (Schueller, 1992). However, the means of determining variable-site nitrogen fertilization rates according to such spatial variability have been, up to now, undefined.

Some methods have long existed allowing for the determination of the rate of application of nitrogen fertilizers for entire fields. Could such methods be made available for variable-site management, and, if so, what improvements on these methods would be required? This experiment was carried out to find some answers to these questions.

The first step in this work, started in 1992, was to characterize major factors which are responsible for yield variability (soil, plant, implement work, etc..). This step was also done in order to compare the classical application of nitrogen fertilizers (same quantity throughout the entire field) to a variable-site application.

Up until now, the rules for determining such variable-site application rates have been undefined, hence we used a pragmatic - and as yet unperfected -method based on experts' points of view (Carlotti, 1992).

MATERIALS AND METHOD

Soil and crop variability

The experiment was done near Malesherbes, at the edge of the well-known French wheat belt called "Beauce". Two different fields were chosen, because, for the owner of these fields, one of these appeared to be a "heterogeneous" field (Coudray 1), and the other a "homogeneous" field (Coudray 2).

On each field, a grid of 12 m in width and 12 m in length was applied (this grid corresponded to 111 cells for the first, and 126 cells for the other).

During the first year of the experiment, some measurements were taken to characterize soil properties and soil fertility. These measurements were taken on each cell (soil depth, stone-covered area, pH, organic matter content, etc.).

During the growing season and at harvest time, some measurements were also taken on plants and crops. These measurements were also taken on each cell (number of plants and number of tillers per square meter, gross yield, weight of one thousand grains, harvest humidity, etc..).

Crops were harvested with a small combine, designed for such an experiment. For yield measurements there was no use of a yield sensor. Such measurements were taken by weighing the grains harvested in each cell; and by measuring the real area harvested by the combine (four passes, 1.25 m in width and 9 m in length were harvested in each entire cell for the first year of the experiment). Other measurements were more expensive (such as N soil test after harvest and at mid-winter, biomass quantity at one leaf stage, etc..); hence only 20 measurements were taken on 20 random sites on each field.

Comparison between "classical" and "variable-site" applications of nitrogen fertilizers.

Starting from the second year of this experiment, each cell was divided into two sub-parts, the first one was called "classical" and the second, "variable-site". Each sub-part was 6 m in width and 12 m in length (only two passes of 1.25 m in width and 9 m in length were harvested on each sub-part).

On all "classical" cells, a nitrogen fertilizer rate was applied according to the theoretical yield level of the entire field (this theoretical level was calculated according to the most efficient method for such a determination currently used in France).

For each "variable-site" cell, a specific nitrogen fertilizer rate was applied according to the zone to which this cell belonged. Three different zones were defined, based on soil depth classification.

Results

In this paper, only major results are described and discussed.

Soil variability

The two fields had some very similar soil properties (tables 1a and 1b). Potash and phosphorus levels were sufficient, and it appeared unnecessary to take into account the variability of these elements.

Table 1a : Chemical measurements on Coudray 1

Coudray 1	Mean	Standard deviation	Minimum	Maximum	Coefficient of Variation
Organic matter (%)	2.13	0.12	1.91	2.45	5.4
Exchangeable Potassium ($^{0}/_{00}$)	0.33	0.04	0.27	0.49	12.7
Phosphorus ($^{0}/_{00}$)	0.29	0.06	0.20	0.54	20.5

Table 1b : Chemical measurements on Coudray 2

Coudray 2	Mean	Standard deviation	Minimum	Maximum	Coefficient of Variation
	2.21	0.11	1.97	2.54	4.9
Exchangeable Potassium ($^{0}/_{00}$)	0.38	0.04	0.30	0.51	10.8
Phosphorus ($^{0}/_{00}$)	0.24	0.04	0.16	0.40	17.6

Other measurements show some differences between these two fields (tables 2a and 2b). In Coudray 2, the soil depth was found to be greater than in Coudray 1. On the other hand, Coudray 1 had some parts with more than 30 % of the surface covered by stones, whereas there were no surface stones in Coudray 2.

In Coudray 1, the pH was very regular, unlike Coudray 2. The percentage of calcium carbonate was found to be greater in Coudray 1 than in Coudray 2.

The differences noted between Coudray 1 and Coudray 2 seemed relevant in explaining the "homogenous" and "heterogeneous" characterization of the fields by the owner: In case of drying conditions, yield of crops in Coudray 1 would be limited at some points by an insufficient water supply.

Table 2a : Specific soil properties of Coudray 1

Coudray 1	Mean	Standard deviation	Minimum	Maximum	Coefficient of Variation
Soil depth (cm)	49.1	13.7	30.8	113.8	27.9
Relative altitude (m)	0.82	0.50	0.00	2.02	60.2
Stone-covered area (%)	13.2	7.1	5.0	35.0	53.8
pH	8.1	0.0	8.0	8.2	0.6
Calcium carbonate (%)	5.3	2.6	2.1	14.5	49.4

Table 2b : Specific soil properties of Coudray 2

Coudray 2	Mean	Standard deviation	Minimum	Maximum	Coefficient of Variation
Soil depth (cm)	76.4	13.5	35.8	104.8	17.7
Relative altitude (m)	0.33	0.19	0.00	0.82	57.7
Stone-covered area (%)	-	-	-	-	-
pH	7.6	0.2	6.9	8.0	3.2
Calcium carbonates (%)	0.6	0.5	0.0	3.4	80.7

Crop variability

During the first three years of this experiment, climatic conditions were considered average (no drying conditions during spring, no excess moisture). So, yields were very similar between the two fields, there were no significant differences in terms of the coefficient of variation.

Table 3 : Yield characteristics for the three years and the two experimental fields

Yield (T/ha)		Mean	Standard deviation	1st decile	9th decile	Coefficient of Variation (%)
Coudray 1	1993 (barley)	64.5	3.2	60.5	68.7	5.0
(n=111)	1994 (wheat)	80.8	6.6	72.2	88.8	8.1
	1995 (wheat)	87.5	4.6	80.1	93.0	5.2
Coudray 2	1993 (barley)	69.0	2.6	65.7	72.3	3.7
(n=126)	1994 (wheat)	88.0	7.3	78.1	96.1	8.3
	1995 (wheat)	77.8	15.8	54.1	96.4	20.3
(n=60)	1995 (wheat)	90.9	6.9	79.1	97.9	7.6

(during crop year 1995, there was a stembreak attack, which affected 66 cells in Coudray 2. Having discounted these cells, results were found to be similar in Coudray 2 to Coudray 1 for the same crop year, and similar for crop years 94 and 95 in Coudray 2)

Statistical tools were used, such as variograms, to determine if measurements taken were spatially structured or not. In the case of this experiment, yield and yield components (such as the number of grains per square metre) showed such a spatial structure (table 4).

Results are similar to other results obtained in Western Europe (Delcourt H. et al, 1992).

Table 4 : Variogram of yield, calculated for each field and each crop year

	Crop	Type	nugget effect	sill	scale factor (m)	r^2
Coudray 1	1993 (Barley)	Spherical	0.60	1.05	40	0.99
	1994 (Wheat)	Spherical	0.25	1.10	50	0.98
	1995 (Wheat)	Spherical	0.40	1.10	50	0.99
Coudray 2	1993 (Barley)	Spherical	0.80	1.05	60	0.99
	1994 (Wheat)	Spherical	0.40	1.00	60	0.96

(for crop year 95, Variogram for Coudray 2 was not calculated, due to stembreak attack on 66 cells)

Yield variability showed a spatial structure, but we needed to know if this structure would be stable year after year, and what might be the factor(s) responsible for such a spatial structure.

To study spatial structure stability, we used two different approaches : the first was to compare, cell by cell, results from one year to results from another. This approach showed that such stability is very minor; for all comparisons the coefficients of correlation were less than 0.35 (Fig. 1).

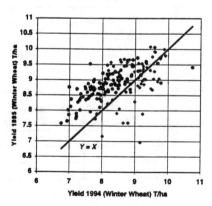

Fig. 1. Comparison between yield 94 and yield 95 (Coudray 1 + Coudray 2).

The second approach took into account the stability of yield maps. Yield maps drawn up in different years were compared, one to the other. The stability coefficient was calculated by noting the percentage of cells which belonged to the same zone on both maps. The other maps, built up with yield components such as grains/m², showed better stability (table 5).

Table 5 : The stability coefficient for yield and grains/m²

Comparison between :			Yield	Number of grains/m²
1993	and	1994	46.8 %	56.8 %
1993	and	1995	60.4 %	56.8 %
1994	and	1995	56.8 %	63.1 %

Relationship between crop variability and soil properties

Crop variability should be directly linked to soil properties (Chéry, 1995). To determine these relationships, the linear coefficient of correlation was calculated between yield and soil properties, and between yield components and soil properties. (tables 6a and 6b).

The soil depth was the most important factor and could be used to explain yield variability. The presence of stones was the second factor. For the crop year 1995, some results were excluded, due to disease on Coudray 2. But, after correction, the values of the coefficient of correlation were, nevertheless, small.

Table 6a : linear coefficient of correlation between yield and soil properties.

Coudray 1 + Coudray 2	Yield (crop year 93)	Yield (crop year 94)	Yield (crop year 95)
Soil depth	0.62	0.58	0.24
Relative altitude	- 0.36	- 0.22	- 0.08
Stone-covered area (%)	- 0.50	- 0.39	- 0.14

Table 6b : linear coefficient of correlation between grain/m² and soil properties

Coudray 1 + Coudray 2	Grains/m2 (93)	Grains/m2 (94)	Grains/m2 (95)
Soil depth	0.74	0.74	0.33
Relative altitude	- 0.52	- 0.27	0.02
Stone-covered area (%)	- 0.55	- 0.63	- 0.31

A stepwise linear regression was also computed. The threshold for introducing or eliminating a variable was fixed at 5%. In this case, soil depth proved to be the most important variable correlated with yield or yield components. The other variables were already eliminated from this regression (pH for example), or eliminated depending on the crop year (stone-covered area and relative altitude).

Results were somewhat better with yield components, such as grains/m², than with yield (table 7). But for crop year 1995, in spite of elimination of results from 66 cells on Coudray 2, poor results were obtained, probably due to disease problems on Coudray 2.

For the barley crop (crop year 1993), 3 variables were used to explain spatial variability : soil depth , stone-covered area and altitude. But, for the winter wheat crop, only soil depth was used to explain such variability. It was not the same crop, and the size of cells was different : 12 m in width and 12 m in length for crop year 1993, half-size for the following crop years.

Table 7 : Results of stepwise linear regression

Coudray 1 + Coudray 2	Formula	r^2	n
Yield	$Y93 = 0.07 * X1 - 0.16 * X2 - 1.59 * Z + 64.76$ $Y94 = 0.23 * X1 + 69.89$ $Y95 = 0.10 * X1 + 81.90$	0.44 0.34 0.06	237 237 170
Grain/m²	$G93 = 0.07 * X1 - 0.16 * X2 - 1.59 * Z + 64.76$ $G94 = 0.23 * X1 + 69.89$ $G95 = 0.10 * X1 + 81.90$	0.44 0.34 0.06	237 237 170

(Y93 = Yield for crop year 1993,   G 93 = Grain/m² for crop year 1993,
X1 = soil depth,   X2 = stone-covered area,   Z = relative altitude)

Definition of rules for variable-site fertilizer applications

The means to determine variable-site nitrogen fertilization rates according to such spatial variability have been, up to now, undefined. In accordance with previous results, we used soil depth to determine such variable-site applications.

At first, three classes in each field were defined, using a statistical classification method (table 8). This method minimizes the within-group variance and maximizes the between-group variance (this method does not take into account the spatial structure of soil variability).

Table 8 : Classes of soil depth used to delineate homogenous zones

	Coudray 1			Coudray 2		
	Class 1	Class 2	Class 3	Class 1	Class 2	Class 3
Soil depth (cm) extreme values mean value	27 to 44 38	45 to 64 50	65 to 113 75	28 to 57 47	58 to 79 70	80 to 105 88
surface (%)	46	36	18	9	44	47

For each "homogeneous" class, a specific nitrogen fertilizer rate was calculated according to the balance-sheet method, which is currently used in France. This method is based on a predicted balance of the mineral nitrogen in the soil, in which outputs would be equal to inputs. This balance is indicated below :

$$b*Y + N_f + L = N_i + (M_h + M_r) + X$$

where :

Outputs :	Inputs :
Y = grain yield target	N_i = initial mineral N at mid winter
b = N required per unit of grain	M_h = net mineralization from humus
N_f = residual mineral N at harvest	M_r = net mineralization from crop residues
L = possible leaching	X = nitrogen fertilizer

Beyond a certain depth, the yield target must be set at a fixed value. In the case of this experiment and with this balance-sheet method used, the yield target was limited to 9.5 T/ha, whereas soil depth was greater than 0.7 m (table 9). In the case of Coudray 2, the difference between class 2 and class 3 came from a difference in the initial mineral nitrogen quantity in the soil at mid-winter.

Yield targets, calculated with this method, greatly depended upon soil depth. Without target limitation, the three classes were separated by one metric ton per hectare.

Table 9 : Nitrogen fertilizer determination with the balance-sheet method, for crop year 94

1994	Coudray 1			Coudray 2		
	Class 1	Class 2	Class 3	Class 1	Class 2	Class 3
Grain yield target (T/ha)	7.5	8.5	9.5	8.5	9.5	9.5
Initial mineral N at mid-winter	20	25	30	30	40	50
Residual mineral N at harvest	20	20	30	20	30	30
Nitrogen fertilizer (kg N/ha)	185	210	245	205	235	225

During the growing season, sub-parts of cells identified as "variable-site" received nitrogen fertilizer, according to the class that these cells were part of. For the other sub-parts, identified as "classical", the nitrogen fertilizer rate was calculated using the same method, based on average values in each field. The nitrogen fertilizer rates were, respectively, 200 kg/ha for Coudray 1 and 200 kg/ha for Coudray 2.

Variable-site fertilization results

Surprisingly, for two years of this experiment, results for classical and variable-site applications appeared not to be significantly different. Considering each zone, yields and yield components were very close, wherever there had been differences in nitrogen fertilization rates applied (table 10).

Additionally, it was expected that yield differences should be 1 T/ha between zone 1 and zone 2 for both fields, and a 1 T/ha difference was also anticipated between zone 2 and zone 3 for Coudray 1. However, the differences found were less than 0.4 T/ha, and usually around 0.2 T/ha between zone 1 and zone 2 (they were around 0.7 T/ha between zone 2 and zone 3 for Coudray 1).

Lastly, it should be noted that differences observed between classes were similar in the case of classical applications and in the case of variable-site applications. Furthermore, in both cases, recorded yields were quite different from target yields.

Table 10 : Relationship between classical and specific-site application of Nitrogen fertilizer applications

Yield (T/ha)	Class 1			Class 2			Class 3		
Coudray 1	Cla		Varia	Cla		Varia	Cla		Varia
1994	7.96	NS	7.86	7.98	NS	8.01	8.69	NS	8.70
1995	8.65	0.09%	8.34	8.74	NS	8.78	9.06	2.76%	9.37
Coudray 2									
1994	8.44	NS	8.28	8.65	NS	8.47	9.02	2.84%	8.68
1995	9.15	NS	9.04	8.92	NS	9.24	9.25	NS	9.39

(Cla = Classical application / Varia = Variable-site application / N.S. = Non-significant test)

DISCUSSION

This experiment produced some results which corroborate other experiments done in the USA and Western Europe, such as low stability of yield maps, or as variability of soil fertility (Chéry, 1995; Chaney, 1990; Fiez et al., 1994; Mulla et al., 1992; Van Meirvenne, et al., 1990).

This experiment was a long-term experiment which measured numerous factors relating to soils, crops, implements and nitrogen fertilization. Results allowed for the characterization of in-field-variability, and produced some keys to the improvement of variable-site crop management.

Even though methods are available to determine the nitrogen fertilization rate for the entire field, these methods appear insufficient and cannot be directly applied to determine nitrogen fertilization rates in the case of variable-site management. This is due to the great imprecision in determining some variables used in these methods. Some variables are only estimated for the entire field, and some of them are not recorded.

Yield mapping appears to be insufficient to be used alone for specific-site management of nitrogen fertilization. Mapping should be complemented by the calculation of nitrogen fertilizer rates during the growing season, and the recalculation of nitrogen fertilizer rates according to crop development. Such calculations require some new sensors such as a biomass sensor, a humus mineralization sensor, a plant nitrogen content sensor, etc..(Juste, 1992).

Models designed for entire fields are not precise enough. In the Coudray experiment this point appeared clearly with the soil depth effect. From the experts' points of view, the increase soil in depth from 0.35 m to 0.5 m should correspond to an increase yield from 7.5 T/ha to 8.5 T/ha. However, it will be recalled that in the Coudray experiment only 0.2 T/ha increase was noted.

This experiment needs to be continued, because first results should be improved (or perhaps, contradicted) and, in any event, the hypothesis should be tested under other climatic conditions.

In light of these results, it should become possible to adjust the rules for variable-site nitrogen fertilization rates. In particular, these rules should be defined in accordance with some measurements concerning crop development, such as plant nitrogen content.

Nevertheless, the economic gains appeared to be lower than expected. Based on the Coudray experiment a farmer would probably gain no more than 40 $/ha for such variable-site management.

ACKNOWLEDGMENTS

This research was supported by grants from the French Ministry of Agriculture.

REFERENCES

Carlotti, B. 1992. Recueil des bases de préconisations de la fertilisation azotée des cultures. Ministère de l'Agriculture - Ministère de l'Environnement - Mission Eau Nitrates - CORPEN

Chaney, K. 1990. Effect of nitrogen fertilizer rate on soil nitrogen content after harvesting winter wheat. Jour. of Agric. Science, Cambridge, vol 114 : 171-176

Chéry, P. 1995. Variabilité de l'épaisseur de la couverture de sol. Conséquences pour le bilan hydrique hivernal d'un secteur de Petite Beauce.Thèse de Doctorat de l'Université de Nancy I.

Delcourt, H., J. De Baerdemaeker , J. Deckers. 1992. The spatial variability of the growth and the yield of winterwheat related to soil and soil nutrient maps. Presented at the colloquium "Ortung und navigation landwirtschaftlicher fahrzeuge". Weihensteplan, Germany, March 5-6.

Fiez ,T.E., B.C. Miller, W.L. Pan. 1994. Assesment of spatially variable nitrogen fertilizer management in winter wheat. J. Prod. Agri., vol 7, (1):86-93.

Juste, E. 1992. Diagnosis of nitrogen nutrition status of a winter wheat crop by "nitrate test" : relationship with the nitrogen dilution curve during the growth. Proc. 2nd ESA congress, Warwick Univ.

Mulla, D.J. A.U. Bhatti, M.W. Hammond, J.A. Benson. 1992. A comparison of winter wheat yield and quality under uniform versus spatially variable fertilizer management. Agriculture, ecosystems and environment, vol 38, (4):301-311.

Schueller, J.K. 1992. A review and integrating analysis of spatially variable control of crop production. Fertilizer Research. 33:1-34.

Van Meirvenne, M., G. Hofman, P. Demyttenaere. 1990. Spatial variability of N fertilizer application and wheat yield. Fertilizer Research. 23:15-23.

Remote Sensing Tools for Site-Specific Management

J.S. Schepers
T.M. Blackmer

USDA, ARS
Keim Hall
Lincoln, NE

T. Shah

Li-Cor Inc.
P.O. Box 4425
Lincoln, NE

N. Christensen

1001 S. 70th, Suite 210
Farmland Industries
Lincoln, NE

ABSTRACT

Grid sampling to generate a map from which to make variable rate fertilizer applications assumes that nutrient availability is a major reason for yield variability. The post-mortem aspects of a yield map leave many questions unanswered about the causes of yield variability. Remote sensing techniques represent a variety of tools to help make site-specific management decisions.

INTRODUCTION

The potential for making site-specific management decisions has attracted the attention of nearly all aspects of agriculture. Policy makers are even intrigued with the concept because it facilitates environmental stewardship and offers the potential to minimize environmental risks from agricultural activities.

The concept of making site-specific management decisions in agriculture follows after military applications to identify and characterize land-based activities. Agricultural applications go beyond describing the precise location of certain objects in that the "target" is usually moving and may be measuring something or making some kind of application. These agricultural applications generate enormous amount of information that must be merged in terms of time, location, and a variety of site-specific parameters. Making use of all this data is the challenge of the decade because the technology to make site-specific measurements in agriculture is clearly ahead of the science needed to use the information to make management decisions.

As producers consider site-specific management practices they typically begin with whatever is perceived to provide the greatest opportunity for increased profitability. Many times producers do not make a full-scale commitment to the concept unless consultants or someone in the agribusiness community provides special equipment, services, or technical assistance. The most common approach to site-specific nutrient management involves grid soil sampling and variable rate fertilizer application. This strategy attempts to compensate for variability in crop growth by adjusting the rate of nutrient application according to some predetermined criteria. The approach implies that nutrient availability is the primary manageable factor responsible for yield variability.

Yield mapping takes almost the opposite approach in that a yield map alone is not able to imply much of anything about the causes of yield variability. It simply provides a spatial record of yield and leaves the determination of variability up to the producer or others. In contrast, the grid sampling approach to site-specific management normally does not measure success in terms of reduced yield variability. It mostly relies on producer intuition to suggest that yield uniformity should be improved by using variable rate fertilizer applications. Average yield for the field is expected to improve and make the venture economical. Both the variable rate fertilizer application approach and yield mapping approach to site-specific management could benefit from techniques that offer mid-season information about spatial variability.

Producers who combine yield mapping and remote sensing technologies have available a series of field maps from which to identify management units that could be candidates for site-specific practices. This type of management strategy could result in a "smart sampling" map to use for collecting soil samples rather than using a computer generated design for grid sampling.

The objective of this study was to evaluate remote sensing techniques to provide information about the locations and possible causes of spatial variability in crop growth during the growing season of irrigated corn.

MATERIALS AND METHODS

A quarter-section center-pivot irrigation system (59 ha) under corn production near Shelton, Nebraska was used for this study. At the beginning of the project in 1994, soil samples (0-20 and 20-90 cm) were collected on a 12.2 by 24.4 m alternate grid. Soil samples were analyzed for total C, residual N, Bray P, Zn, and pH. Depth to sand and elevation were also determined for each sampling point. In 1995, remote sensing was initiated using the RESOURCE21™ protocol which involved taking a series of digital color images throughout the growing season. Field scouting after each flight was provided by Dr. Neal Christensen, Farmland Industries, Inc.

Canopy reflectance was measured in mid-August with experimental sensors provided by Li-Cor, Inc. The field of view for the sensors was 45°. Sensors were mounted in the nadir position (downward) at a height of 4.3 m (~1.5 m above the tassels) on a John Deere model 6500 high-clearance sprayer. A canopy reflectance map was generated by driving through the field at 16-row intervals (76-cm spacing) with data collection at 1-sec intervals. A yield map was generated using an AgLeader 2000 yield monitor on a John Deere 9600 combine provided in part by Deere and Company, Inc. Spatial position in the field was determined with an Ashtech AgNavigator using OmniStar satellite based differential correction.

RESULTS

An aerial photograph of bare soil color taken after planting in the spring of 1994 displayed patterns similar to those illustrated with the map of organic matter content generated from soils data. This was expected since soil color is correlated with organic matter content. Based on this relationship, it should be possible to calibrate an aerial photograph of bare soil by determining the organic matter content of soils from sites representing a range of selected colors. This technique makes it possible to generate an organic matter map from soil color that can be used to estimate the spatial variability in mineralization. In a general way, relative soil color can also provide information about residue cover, past erosion, soil slope, and areas with topsoil deposition. Any or all of these features may be candidates for site-specific management practices.

Estimates of relative soil fertility status for nutrients related to organic matter content can also be made from aerial photographs. Field calibration is necessary as noted above in the case of organic matter content. This approach should help identify areas where a nutrient deficiency may be expected but will not identify areas receiving high application rates of fertilizers, sewage sludge, and animal wastes.

RESOURCE21™ images of the crop generated during the early part of the growing season revealed many of the same landscape features and areas of probable infertility as noted in the bare soil image. Especially noteworthy in early-season crop canopy images were depression areas where excess water reduced crop growth and in one area appeared to induce a K deficiency. As the season progressed, the crop appeared to grow out of whatever had been limiting growth in many areas, but other areas began to show some type of stress. Field scouting revealed one of these stressed areas had a higher than average insect infestation.

Aerial images provide a unique view of crop damage caused by mechanical implements. Weeds typically proliferate in areas where canopy cover never developed or somchow was destroyed. These areas have reflectance patterns different than corn and are thus easy to identify in an aerial photograph or digital image. In total, the field used for this study had 1.0 ha destroyed in 1995 when the 12-row wide implement occasionally veered off track during the first cultivation and destroyed 9.1-m wide strips of young corn plants.

Images generated as part of the RESOURCE21™ program were collected from an altitude of ~2400 m. Cameras were designed to capture images in distinct reflectance bands and digital data combined into a "calibrated vegetation map." Procedures used to combine the digital data are proprietary.

Scientists traditionally calculate a normalized difference vegetative index (NDVI) using data from the red and near infrared reflectance bands. Chlorophyll meters also use specific wavelengths within these wavebands, but these meters measure light transmittance through individual leaves rather than reflectance from the crop canopy.

Light in the red waveband is either absorbed by chlorophyll, transmitted through the leaf blade, or reflected back into the atmosphere. As proportionately more red light is absorbed by chlorophyll in leaves, less is available for transmittance or reflection. This phenomena results in an inverse relationship between crop N status and reflectance in the red portion of the spectra. The same type of relationship exists between chlorophyll meter readings because of the strong positive relationship between leaf N concentration and chlorophyll content. Green light is absorbed to a lesser extent by chlorophyll. Therefore, its reflectance is negatively correlated with leaf N concentration because of the positive relationship with chlorophyll content. This is why it is feasible to measure reflectance of green light to evaluate crop N status. A caution is necessary because in contrast to leaf reflectance, measuring canopy reflectance in the green portion of the spectrum may be positively or negatively correlated with leaf chlorophyll content depending on percent crop cover and soil brightness. Research is still needed to develop and evaluate more effective mathematical relationships that transform reflectance data into meaningful indications of nutrient status and crop health.

Measuring crop N status with a chlorophyll meter is no match for using techniques that can evaluate an entire field at one time. The trade-offs involve speed of data collection by monitoring an entire field at one time using remote sensing techniques compared to the accuracy that may be sacrificed by monitoring a diverse area compared to a select group of plants. Short-range reflectance sensors (i.e., Li-Cor Incorporated) mounted on a high-clearance vehicle offer a compromise by measuring reflectance from a known group of plants. Reflectance data generated while driving through a field can be manipulated to generate a map of crop N status using NDVI or other relationships. The most appropriate relationships to describe crop stresses may change throughout the growing season depending on the type of stress anticipated. Considerable research is needed to determine what types of remote sensing data are needed to identify specific types of crop stress.

Yield maps can be an exciting management tool, but knowing how to use the technology remains somewhat of a mystery. This is because yield maps nicely identify where production problems existed, but provide little information about the causes of yield variability. Yield maps undoubtedly have a role to play in designing a "smart sampling" strategy. Because water plays a dominant role in crop growth and grain yield, climatic conditions therefore play a major role in yield patterns. Some causes of yield variability (i.e., low fertility, shallow rooting depth, low water holding capacity) tend to be perpetual from year to year but others vary depending on factors related to leaching potential, seedbed establishment, and denitrification.

SUMMARY

The strength of remote sensing lies in the opportunity to learn more about crop growth variability while the crop is still growing. Benefits can be realized by combining this information with grid sampled soil analysis and yield maps in developing an integrated crop production program. Detecting a potential problem early in the growing season can be viewed as a type of preventive medicine if the problem can be "fixed" by some type of remedial treatment. In other cases, problem areas may require treatment or modifications in management practices that can only be made when no crop is growing. Some causes of crop growth variability may not have an obvious remedy.

The Effects of Mapping and Scale on Variable-Rate Fertilizer Recommendations for Corn

C.A. Gotway

Department of Biometry
University of Nebraska
Lincoln, Nebraska

R.B. Ferguson

Department of Agronomy
South Central Research and Extension Center
University of Nebraska
Clay Center, Nebraska

G.W. Hergert

Department of Agronomy
West Central Research and Extension Center
North Platte, Nebraska

ABSTRACT

Soil samples from two research sites in Nebraska were used to evaluate the effects of scale and interpolation method on variable-rate fertilizer application recommendation maps. Soil nitrate, phosphorus, and zinc determinations were made from grid samples ranging in density from 104 to 0.86 cores per hectare. Map accuracy varied greatly depending on the nature of the spatial variability in soil parameters, but was much greater for site-specific recommendations than for those based on whole-field averages.

BACKGROUND

Correct determination of the amount of fertilizer to be applied to a particular location in the field is central to the environmental and economic benefits of variable-rate fertilizer application. Until remotely-sensed information can be reliably used to infer field nutrient requirements, grid soil sampling is likely to continue to be the most effective means to generate maps of nutrient levels and fertilizer application maps for variable-rate application. However, because of the cost and labor intensity associated with grid soil sampling, there is a tradeoff between the information attained and the cost involved. Research studies using grid soil sampling for variable-rate nutrient application have focused on the need to keep the number of samples to a minimum, while still attaining an acceptable level of accuracy. Hammond (1992) found that a 60 m grid was adequate for developing nutrient level maps, compared to distances of 30 and 120 m. Franzen and Peck (1994) found that a distance of 64.8 m between samples in a square grid was acceptably accurate compared to a distance of 24.6 m, and was preferable to a distance of 98.4 m. Wollenhaupt et al. (1994) recommended a two-stage grid sampling approach using an initial 90 m grid, with additional sampling on a finer grid if areas of optimum or lower soil test values are identified.

In conjunction with the sampling density, the interpolation method used to produce soil nutrient or fertilizer recommendation maps also affects map accuracy. Wollenhaupt et al. (1994) compared several interpolation methods, including inverse-distance-squared and kriging, for mapping soil phosphorus and potassium levels in two Wisconsin fields. The results of their study suggested that for data on a 31.8 m grid, maps produced using inverse-distance-squared interpolation were more accurate than those obtained by kriging when compared to control maps produced by Delaunay triangulation. For larger grid spacings, all of the mapping approaches considered in their study seemed to produce maps of equal accuracy when compared to the control. In a similar study of soil nitrate and organic matter in Nebraska, Gotway et al. (1996) used a prediction-validation approach to evaluate the accuracy of kriging and inverse-distance interpolation methods. Their results suggested that, regardless of the sampling density used, the accuracy of interpolation methods depended on the coefficient of variation of the data.

In both of these studies, the accuracy of the soil nutrient maps, and not the variable-rate fertilizer application rates derived from them, was emphasized. In this paper, we present the results of a study that used selected data from two grid-sampled research sites in Nebraska to evaluate the effects of grid density and interpolation methods on variable-rate nutrient and fertilizer recommendations.

MATERIALS AND METHODS

Two variable-rate nitrogen research sites were soil sampled in the fall of 1993 and the spring of 1994 at the initiation of two research studies. Soil samples were collected in a systematic grid with sample arrangement and density in the field depending on the location. Sites were located in Buffalo County (Hord silt loam [fine-silt, mixed, mesic Cumulic Haplustolls] and Blendon loam [coarse-loamy, mixed, mesic Pachic Haplustolls]); and Lincoln County (Cozad silt loam-fine-silty, mixed [calcareous], mesic Typic Haplustolls). Single cores, 3.8 cm in diameter, were collected at the center of grid cells. The cell sized varied with location, but was typically 12-15 m long by 6 m (8 rows) wide. Cores were collected to a depth of 0.9 m in two increments: 0-0.2 m and 0.2-0.9 m. Surface soil samples were analyzed for Bray-1 P, organic matter, DTPA-Zn, pH, and NO_3-N. Subsoil samples were analyzed for NO_3-N. Core densities ranged form 104 samples per hectare at the Lincoln County site to 34 samples per hectare at the Buffalo County site.

Selected data from these sites were used to evaluate the effects of sample grid density and interpolation method on the resulting maps of soil parameters and site-specific fertilizer recommendations. Data from the finest grid density were used to evaluate the accuracy of maps created with data from less-dense grids. The geostatistical analyses performed in this study (using GEOEAS) were relatively basic, since the goal was to produce maps and fertilizer recommendations quickly as this is typically the need in practice. Large scale trends in soil parameters (notably present in the Lincoln County soil phosphorus semivariogram) were ignored. Semivariogram estimation and modeling was especially difficult for soil parameters measured from the Lincoln county coarse grid (11.5 samples/ha) since there were only 25 data points available for semivariogram estimation. In such cases, it was difficult to clearly select the best semivariogram model and the model actually used in mapping was chosen rather arbitrarily.

Maps were created and area analyses were performed using SURFER software. Maps created with data from less-dense grids were produced by interpolation onto a regular grid, with spacings of 12 m by 12 m for the Buffalo County site, and 3 m by 5 m for the Lincoln County site. All search neighborhoods used in mapping were restricted to include only the 16 nearest neighbors within the range of the semivariogram determined from geostatistical analyses. The recommended N rates were based on spatial organic matter, spatial residual nitrate-N, and uniform expected yield, and recommendations for other nutrients were based solely on their spatial distributions (Hergert et al. (1995).

RESULTS AND DISCUSSION

The effects of grid-sampling density on soil nutrient maps and fertilizer recommendations were evaluated at both locations. Figure 1 illustrates the two sampling densities compared for each of the two sites. Each of the denser grids is the actual density at which cores were collected at that site. Each of the coarser grids is a subset of the full grid, and was selected to form a regular grid consisting of 10 to 25% of the full-grid cores. The coarser grids were selected in this way in order to more closely reflect those commonly used in commercial applications.

Table 1 shows the recommended fertilizer rates for nitrogen, phosphorus, and zinc for the Buffalo County site. Site-specific rates were calculated from (non-interpolated) data obtained with the fine grid (34 samples/ha) and also from interpolated maps (made by kriging and inverse-distance-squared (ID-2) interpolation) based on the coarser grid (0.86 samples/ha). Rates obtained using uniform application in a whole field management (WFM) strategy were also calculated for comparison. Using kriging, 34.5% (N), 37.8% (P), and 18.6% (Zn) of the field received a different fertilizer recommendation with the coarser grid, while 40% (N), 36.6% (P), and 14.0% (Zn) of the field was fertilized incorrectly using ID-2. Figure 2 illustrates the resulting Bray-1 P and DTPA-Zn maps and data postings for the Buffalo County site. This site shows evidence of an earlier animal confinement area in the southwest corner, and possibly a second confinement area in the northwest corner. Both the kriging and ID-2 maps based on the coarse grid were able to detect these areas of elevated concentration, as well as a third area that was more centrally located. The smoothing tendencies of kriging and the bulls-eye pattern characteristic of ID-2 are very pronounced in both the phosphorus and the zinc maps.

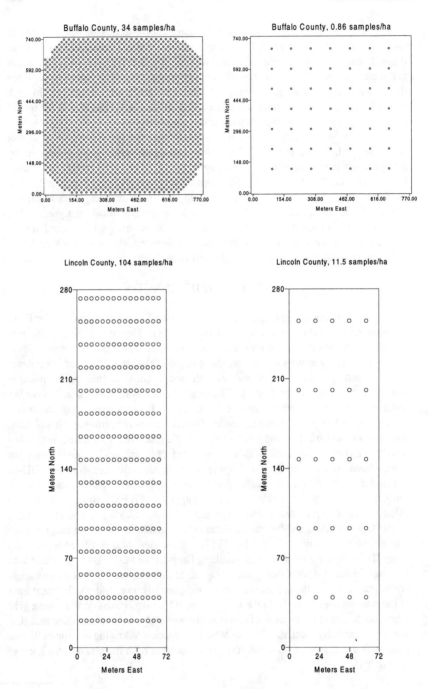

Figure 1. Grid Density Comparison, Buffalo and Lincoln Counties.

Table 1. Buffalo County Recommended Fertilizer Rates.

	Samples per hectare						
	34	0.86					
N Rate		Kriging		ID-2		WFM (208 kg/ha)	
(kg/ha)	% area	% area	%Diff.	% area	% Diff.	% area	% Diff.
0-80	0.4	0.0	0.4	0.0	0.4	0.0	0.4
80-100	0.2	0.0	0.2	0.1	0.1	0.0	0.2
100-120	0.2	0.7	-0.5	0.8	-0.6	0.0	0.2
120-140	0.7	2.2	-1.5	2.0	-1.3	0.0	0.7
140-160	1.7	4.2	-2.5	3.4	-1.7	0.0	1.7
160-180	5.5	4.7	0.8	3.9	1.6	0.0	5.5
180-200	18.9	16.0	2.9	16.6	2.3	0.0	18.9
200-220	40.1	52.7	-12.6	56.4	-16.3	100.0	-59.9
220-240	28.3	17.8	10.5	15.7	12.6	0.0	28.3
240-260	4.2	1.6	2.6	1.1	3.1	0.0	4.2
ITotall			34.5		40.0		
Avg/ha	209	205		205		208	

	Samples per hectare						
	34	0.86					
P Rate		Kriging		ID-2		WFM (0 kg/ha)	
(kg/ha)	% area	% area	%Diff.	% area	% Diff.	% area	% Diff.
0.0	27.2	46.1	-18.9	45.5	-18.3	100.0	-72.8
44.9	54.6	51.0	3.6	53.8	0.8	0.0	54.6
89.8	18.2	2.9	15.3	0.7	17.5	0.0	18.2
ITotall			37.8		36.6		
Avg/ha	40.8	25.5		24.7		0.0	

	Samples per hectare						
	34	0.86					
Zn Rate		Kriging		ID-2		WFM (208 kg/ha)	
(kg/ha)	% area	% area	%Diff.	% area	% Diff.	% area	% Diff.
0.0	87.7	97.0	-9.3	94.7	-7.0	100.0	-12.3
3.4	12.1	3.0	9.1	5.3	6.8	0.0	12.1
5.6	0.2	0.0	0.2	0.0	0.2	0.0	0.2
ITotall			18.6		14.0		
Avg/ha	0.42	0.10		0.18		0.0	

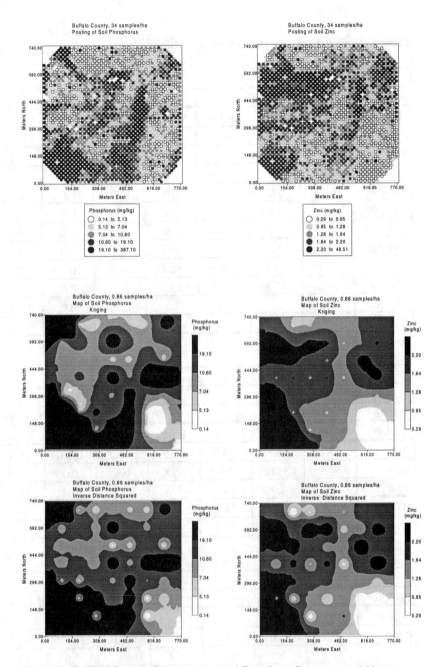

Figure 2. Buffalo County Phosphorus and Zinc Data Postings and Maps.

Table 2 shows the results for the same comparisons obtained from the Lincoln County data. For this site, with kriging, 58.0% (N), 11.4% (P), and 3.6% (Zn) of the field received a different fertilizer recommendation with the coarser grid, while 40% (N), 26.2% (P), and 3.6% (Zn) of the field was fertilized incorrectly using ID-2. Although, many factors influence mapping accuracy, one reason for the larger error using kriging for N rates at this site may be the large relative nugget effect in the semivariogram of the N data from the coarser grid. With larger sample spacings, there is not much information about nature of the semivariogram at small distances, and for some spatial distributions, this can lead to larger relative nugget effects which result in much smoother maps. Such smoothing is easily seen in Figure 3, which displays the NO_3-N and N rate postings and maps for the Lincoln County site. For visual ease, the data postings on the fine grid were also contoured using a cubic spline function. Past management at the Lincoln County site resulted in significant variation in nitrate-N. The pattern may be related to fencing confining cattle on this site in the past. The coarser sampling grid was unable to detect this pattern of elevated NO_3-N, and consequently N application maps resulting from the two grid-sampling densities differ considerably. It is interesting that, in this case, the bulls-eye pattern around data points produced by ID-2 interpolation actually helped to create a more realistic N rate map than that produced using kriging.

SUMMARY AND CONCLUSIONS

The accuracy of nutrient and/or fertilizer application maps really depends on several factors, but two of the most important factors appear to be the spatial distribution of the soil parameter concentrations and the actual locations of the grid samples. For relatively uncomplicated spatial distributions, a few samples may be all that is necessary to adequately characterize a field. But much depends on the actual locations of the grid samples, since for a given spatial distribution, it is possible to obtain samples that can infer the variation reasonably well, and other samples with which important features are not delineated. In many cases where the spatial distribution is rather complex, much finer grid densities than those currently used commercially are required to produce accurate maps of nutrient levels or fertilizer application maps.

The nature of the spatial distributions also affects the accuracy of interpolation methods. Kriging doesn't do as well in cases with high relative nugget effects. Inverse-distance might not do as well in cases where there are smooth large scale trends in the data because of its tendency to put rings around data points (e.g., Lincoln County phosphorus). Further research is needed to identify what parameters might be good predictors of required sampling density and appropriate interpolation technique.

Although these conclusions may appear to lead to circular sampling recommendations such as "if you know the nature of the spatial distribution, then you know where to sample in order to accurately infer the distribution," there are some general guidelines that can make any sampling strategy more effective.

Table 2. Lincoln County Recommended Fertilizer Rates.

N Rate	Samples per hectare						
	104	11.5					
		Kriging		ID-2		WFM (215 kg/ha)	
(kg/ha)	% area	% area	%Diff.	% area	% Diff.	% area	% Diff.
0-100	3.1	0.0	3.1	0.0	3.1	0.0	3.1
100-120	2.2	0.0	2.2	0.0	2.2	0.0	2.2
120-140	2.2	0.0	2.2	0.0	2.2	0.0	2.2
140-160	2.2	0.0	2.2	0.3	1.9	0.0	2.2
160-180	5.3	0.1	5.2	1.0	4.3	0.0	5.3
180-200	5.8	0.0	5.8	4.2	1.6	0.0	5.8
200-220	21.8	47.6	-25.8	36.8	-15.0	100.0	-78.2
220-240	28.0	23.8	4.2	31.2	-3.2	0.0	28.0
240-260	25.3	28.6	-3.3	25.9	-0.6	0.0	25.3
260-280	4.0	0.0	4.0	0.6	37.5	0.0	4.0
lTotall			58.0		40.0		
Avg/ha	215	226		226		215	

P Rate	Samples per hectare						
	104	11.5					
		Kriging		ID-2		WFM (0 kg/ha)	
(kg/ha)	% area	% area	%Diff.	% area	% Diff.	% area	% Diff.
0.0	52.9	58.6	-5.7	66.0	-13.1	100.0	-47.1
44.9	46.7	41.4	5.3	34.0	12.7	0.0	46.7
89.8	0.4	0.0	0.4	0.0	0.4	0.0	0.4
lTotall			11.4		26.2		
Avg/ha	21.3	18.6		15.3		0.0	

Zn Rate	Samples per hectare						
	104	11.5					
		Kriging		ID-2		WFM (208 kg/ha)	
(kg/ha)	% area	% area	%Diff.	% area	% Diff.	% area	% Diff.
0.0	98.2	100.0	-1.8	100.0	-1.8	100.0	-1.8
3.4	1.8	0.0	1.8	0.0	1.8	0.0	1.8
5.6	0.0	0.0	0.0	0.0	0.0	0.0	0.0
lTotall			3.6		3.6		
Avg/ha	0.06	0.0		0.0		0.0	

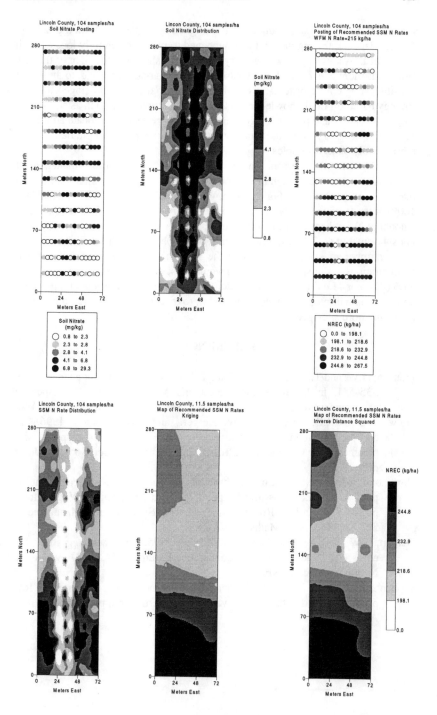

Figure 3. Lincoln County Nitrate and N Fertilizer Recommendations.

First, start with an offset, triangular, or unaligned systematic grid. In the Lincoln County data set, the regular grid just happened to miss the centrally located high nitrate area. An offset grid where every other row is shifted to the right of the row below it is more likely to pick up at least part of this variation. Second, to the extent possible, supplement systematic sampling with a smaller, second-stage sampling strategy based on field knowledge. Use additional knowledge such as soil series, cropping history, and prior management history to preferentially locate additional samples. If kriging is to be used, randomly select certain places in the field to take more closely-spaced samples.

Despite the errors associated with site-specific management recommendations, they are, in most cases, still vastly superior to their whole-field management counterparts. For example, WFM would recommend 208 kg/ha N for 100% of the Buffalo County field, where in fact only 40% of the field required this amount of N fertilizer. For Lincoln County, WFM would not recommend applying phosphorus, but 46.7% of the field required phosphorus fertilizer. Even with the small number of samples taken and the smoothing done on the soil nutrient concentrations, a site-specific approach would prescribe phosphorus application for 34% (ID-2) or 41.4% (kriging) of the field. The accuracy of nutrient and fertilizer applications will most likely continue to improve as recommendation algorithms are made more site-specific.

REFERENCES

Franzen, D.W. and T.R. Peck. 1994. Sampling for site-specific management. pp. 535-551. In P.C. Robert, R.H. Rust, and W.E. Larson, eds. Site Specific Management for Agricultural Systems, Soil Sci. Soc. Am., Madison, WI.

Gotway, C.A., R.B. Ferguson, G.W. Hergert, and T.A. Peterson. 1996. A comparison of kriging and inverse-distance methods for mapping soil parameters for variable rate nitrogen application. Soil Sci. Soc. Am. J., in press.

Hammond, M.W. 1992. Cost analysis of variable fertility management of phosphorus and potassium for potato production in central Washington. p. 213-228. In P.C. Robert et al., eds. Soil Specific Crop Management, Soil Sci. Soc. Am., Madison, WI.

Hergert, G.W., R.B. Ferguson, and C.A. Shapiro. 1995. Fertilizer suggestions for corn. Nebguide G74-174-A, Cooperative Extension, IANR, University of Nebraska, Lincoln, NE.

Wollenhaupt, N.C., R.P Wolkowski, and M.K. Clayton. 1994. Mapping soil test phosphorus and potassium for variable-rate fertilizer application. J. Prod. Agric., 7: 395-448.

Spatial Prediction for Precision Agriculture

B.M.Whelan
A.B. McBratney
R.A. Viscarra Rossel

Australian Centre for Precision Agriculture
The University of Sydney
Sydney, NSW 2006, Australia.

Spatial prediction methods are widely used to estimate data values at unsampled locations in spatial data sets. A classification of the more common techniques is presented. The applicability and performance of these methods in the context of Precision Agriculture is outlined with reference to sample size and intensity. Local kriging with a local variogram is introduced as a means of retaining the spatial detail in intensively gathered yield data.

INTRODUCTION

A primary requirement of a Site-specific Management System is the combination of an appropriate sampling strategy with a spatial prediction method that together provide a sufficiently detailed representation of the true spatial variability of relevant crop and soil properties. At present, depending on the variable of concern, sampling intensity is dictated by convenience and the trade-off between resolution and cost. As a consequence, a variety of field information which is currently obtained by conventional sampling techniques is generally sparse in comparison with yield data generated by real-time sensors. Spatial prediction methods can be used to extend the information available from data at a sparse or uneven set of locations by estimating the values of variables at the scale of interest. Such prediction methods usually assume that the variable in question varies more-or-less continuously in space. This paper will discuss spatial prediction methods in the context of Precision Agriculture with a view to making recommendations on appropriate techniques taking into account the number and intensity of observations within a field.

CLASSIFICATION OF SPATIAL PREDICTION METHODS

Laslett et al., (1987) presented a simple taxonomy of spatial prediction methods. They grouped the techniques under three main headings; namely, global or local, interpolating or non-interpolating, and smooth or non-smooth, predictors.

Global and Local Predictors

Global methods use all the data for prediction at a point and therefore, for large data sets, tend to be computationally expensive (Laslett *et al.*, 1987), although approximate versions are often used in practice. Local predictors use only 'neighbouring' points and are often based on partitioning the region containing the data sites into small elements and fitting simple functions to each element. Hence local methods may be preferred, especially on large data sets, where a single model may be inappropriate.

Interpolators and Non-Interpolators

Spatial prediction methods whose principle requires the prediction to exactly reproduce the data values at sites where data is available are said to act as interpolators. However, if measurement errors are known to be large, Laslett *et al.* (1987) suggest that this constraint may be relaxed a little. This principle may break down and not apply if there are replicate values which do not agree at a point so that only one value, such as the average, is honoured.

Kriging is usually thought to be an interpolator, although if a nugget effect exists, that is if the nugget effect is other than zero, the kriged surface will be discontinuous at the data sites. In such non-interpolating methods $f(x,y)$ can be made into an interpolator $f^*(x,y)$ by setting

$$f^*(x,y) \quad = \quad z \text{ if } (x,y) \text{ is a data site,}$$
$$= \quad f(x,y) \text{ otherwise}$$

Laslett et al. (1987) indicate that definitions of interpolators should exclude such cases, otherwise the terms 'discontinuous / continuous interpolator' might be applied.

Smoothers and Non-smoothers

A smoother is a spatial predictor whose predicted surface and the first partial derivatives thereof are continuous. A non-smooth predictor is one for which the discontinuity of the predictor or its partial derivatives is readily detected by the eye, whereas discontinuity of second and higher derivatives is not usually detected. Despite these definitions, Laslett et al. (1987) indicate that the concept of smoothness of a spatial predictor is somewhat subjective.

Potentially a whole variety of prediction techniques may be used: *inter alia,* global means and medians; local moving averages; inverse-square distance interpolation; Akima's interpolation; natural neighbour interpolation; quadratic trend; Laplacian smoothing splines; and various forms of kriging. For soil data, Laslett *et al.,* (1987) claim that all methods showed some deficiencies. Interpolation methods generally being very poor predictors, while of the non-interpolating methods, Laplacian smoothing splines and kriging performed best.

SAMPLE SIZE AND INTENSITY

There are two principal considerations in the use of spatial predictors for Precision Agriculture once the minimum area of interest (MAI) is decided. An appropriate predictor will depend on both the sample size and the sampling intensity.

Sample sizes of less than 10 will not allow any kind of localised spatial prediction and estimates of the mean for the field will be rather poor. Ten to 100 observations are probably inadequate for formal spatial statistical techniques and this is why methods such as inverse-square distance are in use. The quality of the estimation has to be questioned however. With more than 100 observations, relatively sophisticated techniques such as kriging and Laplacian smoothing splines (Laslett et al, 1987) will work well.

For between 100 and 500 observations, kriging with a local neighbourhood of points and a global variogram is often employed. At this sample size the assumption of a global variogram is born of necessity but once sample sizes get above 500 it seems wasteful to assume a single variogram within the field and local variograms can be estimated for moving neighbourhoods (Haas, 1990b).

A way of expressing the sampling intensity is *via* the number of observations per MAI. Sparse, moderate and intense categories are defined as 0.0001–0.01, 0.01–1 and > 1 observations per MAI, respectively. The purpose of spatial prediction is to convert x observations per MAI to a least 1 prediction per MAI. For sparse sampling intensities this represents a 100 to 10 000 fold increase in the number of predicted data points which would seem to stretch belief in the validity of the predictions. Here the prediction task seems too large and spatial prediction is probably inappropriate.

For moderate sampling intensities up to a 100-fold increase in the number of predicted data points is required which seems appropriate for valid prediction. In fact, at the most intense end of this range, and for intense sampling, one might ask why prediction is required. In many circumstances it may be useful to have more than one prediction per MAI but more importantly it may be necessary to move the location of predictions. This is particularly the case of yield maps where the observation intensity may be 10 per MAI but these observations are linearly clustered within the area and it would be beneficial to obtain an even intensity of observations.

CASE STUDY 1. SOIL pH – Moderate Sampling Intensity
(100 observations per hectare)

Laslett et al.(1987) examined soil pH data (0-10 cm) sampled on a 10m grid within a 1 ha field in Samford, Queensland, Australia to test and compare spatial prediction techniques. The performance of global means and medians; moving averages (1NN & 3NN); inverse square distance interpolation; Akima's interpolation; natural neighbour interpolation; quadratic trend; Laplacian smoothing splines; and ordinary kriging was tested. Comparisons were made by assessing the predictions against 64 observations made within the grid and withheld from model construction. A summary of their results is shown in Table 1.

Table 1. Comparison of spatial prediction techniques for soil pH (0.01M CaCl₂)

Method		Type		PSS
Laplacian smoothing splines	global	non-interpolator	smoother	3.02
Global kriging	global	non-interpolator	smoother	3.07
(with global variogram)				
Inverse square	global	interpolator	smoother	3.08
Local kriging	local ?	non-interpolator	smoother	3.09
(with global variogram)				
Akima interpolator	local	interpolator	smoother	3.40
Quadratic trend	global	non-interpolator	smoother	3.68
Global means	global	non-interpolator	smoother	4.66

PSS prediction sum of squares

Laslett et al. (1987) claimed that all methods showed some deficiencies. Interpolation methods were generally very poor predictors, while of the non-interpolating methods, Laplacian smoothing splines and kriging performed best. From their results isarithms of soil pH were contoured to compare the spatial prediction of inverse-square distance interpolation (Fig. 1a) and ordinary global kriging (Fig. 1b).

Global kriging shows a much smoother map but an examination of the two methods' predictions for some assessment points along a transect (Fig. 2) shows that neither inverse-square distance interpolation or ordinary global kriging appear to predict the pH of the field very well in this case.

In a subsequent study, Laslett & McBratney (1990) compared interpolators, Laplacian smoothing splines, intrinsic random functions, and universal kriging fitted by restricted maximum likelihood (REML). The kriging technique fitted with the REML consistently proved to be the best method. Several authors (including Laslett et al, 1987; Weisz et al., 1995; Brus et al., 1996) have found inverse-square distance interpolation to be perform reasonably when compared with kriging for modest sample sizes (100 or less), but most criticise the method for the lack of an error estimate.

CASE STUDY 2. Sorghum Yield – Intense sampling intensity (1400 observations per hectare)

Real-time sensors that intensively sample variables such as crop yield produce large data sets containing a wealth of information on small-scale spatial variability. By definition, Precision Agricultural techniques should aim to preserve and utilise this detail. Kriging using a global variogram may prove too restrictive in its representation of local spatial correlation in such instances whereas local variogram estimation and kriging may be used to preserve the true local spatial variability in the predictions.

Kriging using local variograms requires the calculation of a variogram model for the observations in a defined neighbourhood (Fig. 3). Defining the neighbourhood, which will influence the model, should be conditioned on the inclusion of ≥100 observation points to ensure a reasonably robust estimation (Webster & Oliver, 1992). Variogram (and drift) estimation (Figure 4) for the neighbourhood around the point to be estimated is followed by ordinary (or universal) kriging using the points in the neighbourhood (Haas, 1990a).

(a)

(b)

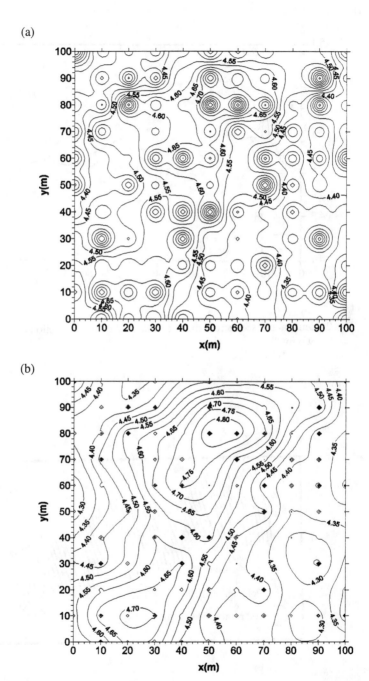

Fig. 1. Soil pH (0.01M CaCl$_2$) isarithm maps constructed by (a) inverse-square distance interpolation (b) global kriging.

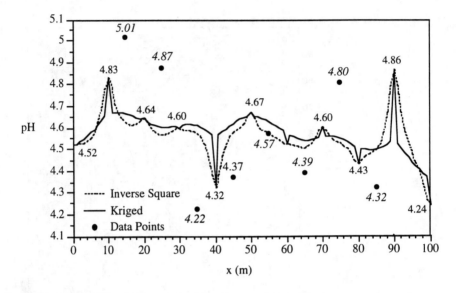

Fig. 2. Identical transect across maps 1a and 1b showing prediction performance of the two methods.

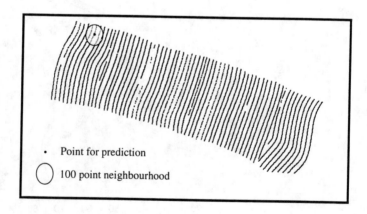

Fig. 3. Raw sorghum yield data map and example of one neighbourhood definition for local variogram estimation.

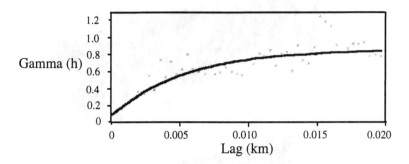

Fig. 4. Local variogram model for the neighbourhood in Figure 3.

In this study, the grain yield of a portion of a sorghum crop in Biniguy, NSW, Australia is examined. The crop, which traverses 2 contour banks, is predicted on a 2-metre grid using local mean, inverse-square distance, local kriging with a global variogram and local kriging with a local variogram (Figure 5). A neighbourhood defined as the nearest 100 observations to each prediction point is used in all methods.

The local mean map (Figure 5a) shows the gross features of the yield in 3 categories. It distinguishes only general areas of high and low yield from the field mean and recognises the influence of only 1 contour bank. Inverse-square distance (Figure 5b) has the prediction surface passing through the known data points (not necessarily smoothly) resulting in 'spots' on the map where changes in variability are poorly represented. Figure 5c, local kriging with a global variogram, shows a similar map because of areas in the field where the spatial variability differs markedly from that described by the global variogram model. Local kriging with a local variogram (Figure 5d) best preserves the observed variability across the field and this is represented in solid areas of colour where the local variance is low. This can be clearly seen when comparing Figure 5c with a map of the co-efficient of variation across the field (Figure 6).

As local kriging using local variograms results in the spatial variability about each prediction point being defined by an individual variogram, an error prediction can be calculated and the variance for an area around each kriged point determined. This is unique among the prediction methods used here and allows the 95% confidence limits of the yield to be mapped (Figure 7). These maps in their entirety are an impossible depiction as the total yields do not match that harvested from the field, but they do represent the spatial distribution of limits for individual points in the field.

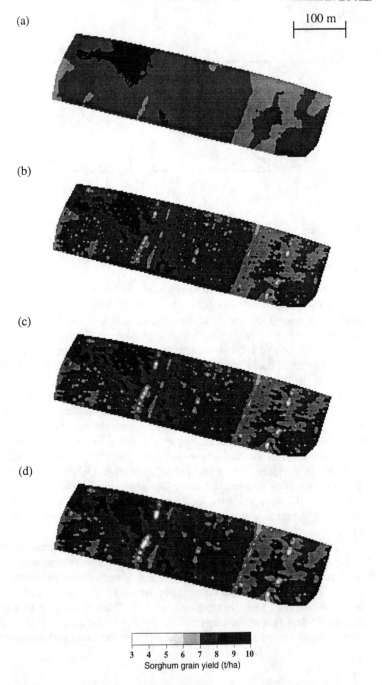

Fig. 5. Sorghum yield maps constructed by (a) local mean (b) inverse-square distance interpolation (c) global kriging with a global variogram (d) local kriging with a local variogram.

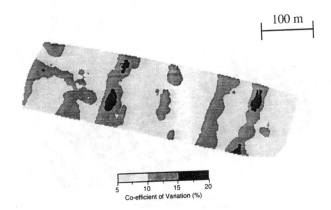

Fig. 6. Co-efficient of variation for sorghum yield constructed from the local kriging with a local variogram predictions.

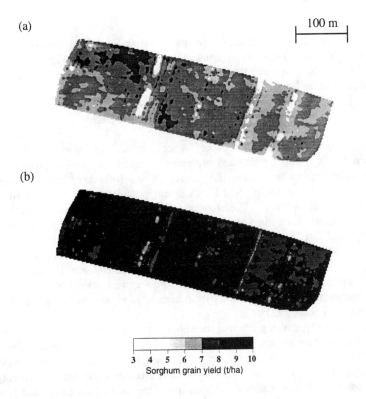

Fig. 7. 95% confidence limits for sorghum yield constructed from the local kriging with a local variogram predictions (a) lower 95% (b) upper 95%.

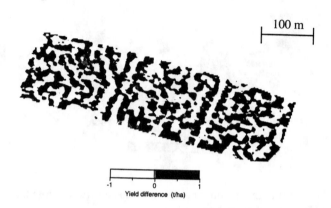

Fig. 8. Yield differences between predictions obtained by local kriging with a local variogram and the local mean.

Further conformation of the spatial variability information retained by local kriging using a local variogram can be ascertained from Figure 8 where yield differences of up to 1 t/ha are evident between prediction obtained by this method and the local mean prediction.

SOME FURTHER ISSUES

By necessity this short paper has not addressed several issues. Firstly, the data used for any spatial prediction procedure should be of known precision and that precision can then be built into the spatial predictor. Because of imprecision in variable measurement and within-field location, interpolators (exact spatial predictors) are generally not optimal. Therefore a considerable amount of pre-processing and deconvolution is required to obtain yields of known accuracy from real-time sensed crop yield data.

Secondly, recent work has shown that prediction of moderately intensively observed attributes can be improved by using ancillary information such as soil maps and topographic attributes derived from digital elevation models. The geostatistical technique of co-kriging and the hybrid method of regression-kriging are appropriate here (Odeh et al., 1995).

Thirdly, spatial prediction need not be at points but prediction can be made onto 'blocks' which represent management units, e.g., 5 metres by 5 metres. These blocks can overlap. A map on a 1 metre raster can then, for example, represent the average of a relevant quantity over the 5 metre by 5 metre block centred on that point (Burgess and Webster 1980). Geostatistical methods appear the most advanced for such predictions, particularly if an estimate of prediction accuracy is required.

PROVISIONAL RECOMMENDATION

The preceding results and discussion suggest that the appropriate prediction method will require conditions on both sample size and sampling intensity. Preliminary recommendations are given in Table 2.

Software for spatial prediction for precision agricultural applications should support the decisions and methods contained in Table 2.

Table 2.   Provisional recommendations for spatial prediction methods for Precision Agriculture in relation to sample size and intensity.

	Sampling Intensity (No. per minimum area of interest)		
Sample size	Sparse 0.0001-0.01	Moderate 0.01-1	Intense >1
<10	NA	NA	NA
10-100	NA	?IS	?IS/NR*
101-500	NA	GS	GS/NR*
>501	NA/GS	LGS	LGS/NR*

NA - not applicable - don't do it

?IS - inverse square or some informal prediction method but there may be problems with the accuracy of the estimates

GS - a geostatistical method such as ordinary kriging or universal kriging with a global variogram or Laplacian smoothing splines

LGS - a local neighbourhood method kriging or Laplacian smoothing splines

NR* - spatial prediction will only be necessity if the sampling is uneven.

CONCLUSIONS

Prediction methods should be employed in Precision Agriculture to accurately represent the spatial variability of moderate to intensely sampled field attributes and maintain the principle of minimum information loss. To this end, the sampling intensity should guide the prediction method utilised.

Local kriging using a local variogram appears well suited for use as a spatial prediction method for real-time sensed crop yield data. Mapping attribute values using this procedure and depicting the changing variability within a whole field should be of benefit in the process of determining whether a field warrants differential treatment and to what degree.

REFERENCES

Brus, D.J., J.J. de Gruijter, B.A. Marsman, R. Visschers, A.K. Bregt, A. Breeuwsma, and J. Bouma. 1996. The performance of spatial interpolation methods and choropleth maps to estimate soil properties at points: a soil survey case study. Environmetrics 7:1–16.

Burgess, T.M., and R. Webster, 1980. Optimal interpolation and isarithmic mapping of soil properties. II. Block kriging. J. Soil Sci. 31:333–341.

Haas, T.C. 1990a. Kriging and automated semivariogram modelling within a moving window. Atmospheric Environment 24A:1759–1769.

Haas, T.C. 1990b. Lognormal and moving window methods of estimating acid deposition. J. Am. Stat. Assoc. 85:950–963.

Laslett, G.M., and A.B. McBratney. 1990. Further comparison of spatial methods for predicting soil pH. Soil Sci. Soc. Am. J. 54:1553–1558.

Laslett, G.M., A.B. McBratney, P.J. Pahl, and M.F. Hutchinson. 1987. Comparison of several spatial prediction methods for soil pH. J. Soil Sci. 38:325–341.

Odeh, I.O.A., A.B. McBratney, and D.J. Chittleborough. 1995. Further results on spatial prediction of soil properties from landform attributes derived from digital elevation models: hetertopic cokriging and regression-kriging models. Geoderma 67:215–226.

Webster, R., and M.A. Oliver. 1992. Sample adequately to estimate variograms of soil properties. J. Soil Sci. 43:177–192.

Weisz, R., S. Fleischer, and Z. Smilowitz. 1995. Map generation in high-value integrated pest management: appropriate interpolation methods for site-specific pest management of Colorado potato beetle (Coleoptera: Chrysomeldiae). Journal of Economic Entomology 88:1650–1657.

Application of Simulation Models and Weather Generators to Optimize Farm Management Strategies

H.W.G. Booltink
J. Verhagen
J. Bouma

Wageningen Agricultural University
Department of Soil Science and Geology
The Netherlands

P.K. Thornton

International Fertilizer Development Center
Muscle Shoals, AL
USA

ABSTRACT

Well calibrated and validated simulation models can be used in real time to assess farm management practices in quantitative terms. An approach to predictive modeling is presented that can be used in combination with decision support systems and weather generators, which evaluates management scenarios on their economic merits as well as on their environmental impact.

INTRODUCTION

Many simulation studies have been carried out over the last few years that have investigated the relations between soil and crop conditions in past years. Studies on precision farming to date also appear to focus on past growing seasons (e.g. Bouma & Finke, 1993, and Sadler & Russell, 1996). However, precision farming is to be used by farmers who face a new growing season with unknown weather conditions.

Well calibrated and validated simulation models can be used to simulate past growing seasons by defining, in retrospect, what would have been the ideal form of management once weather conditions are known. We will refer to this approach as the *"retrospective mode"*. The results can be compared with actual management, and differences can be analyzed by considering weather, crop and soil conditions and management opportunities at any particular time during the growing season (e.g. Booltink & Verhagen, 1996).

Weather forecasts in combination with simulation models, can be used to predict soil and crop conditions for periods of several days, or several weeks, and even entire growing seasons, providing information on which management can be based, as illustrated by Thornton (1991). So far no studies within precision farming research are available where simulations models are used in this *"predictive mode"*.

To comprehend all aspects of precision farming and to understand the output of simulation models, a systems approach is necessary. Decision support systems are interactive computer-based systems that help decision makers utilize data and models to solve unstructured problems (Sprague & Carlson 1982). One such system is DSSAT version 3, the Decision Support System for Agrotechnology Transfer (Jones 1993; Uehara & Tsuji 1993). DSSAT3 was designed to allow users to (i) input, organize, and store data on crops, soils and weather, (ii) retrieve, analyze and display data, (iii) validate and calibrate crop growth models, and (iv) evaluate different management practices (Jones 1993). DSSAT3 is also equipped with utilities for managing and generating daily weather data as described by Pickering et al. (1994). The weather generator of Richardson and Wright (1984), WGEN, allows stochastic time series of weather to be generated when long-term weather records are not available.

This study proposes an approach to *"predictive"* modeling that combines crop models with decision support systems and weather generators, and which can be used for precision farming. Management practices will be evaluated on their economic merits as well as on the environmental impact.

Management can be based on the results of such simulation approaches, but it is crucial to have real field and weather data to continuously check and adjust the model as the growing season proceeds.

Materials and Methods

Soil

At the Van Bemmelenhoeve experimental farm in the Wieringermeer in the Netherlands, a field of approximately 6 ha was selected. A soil survey was conducted and the soil was classified as a Typic Udifluvent (Soil Survey Staff 1975). Geo-referenced data of the 65 soil profile descriptions within the experimental field were stored in a database and generalized using functional horizons. Functional horizons consist of combinations of genetic horizons that show identical behavior (Wosten et al. 1986; Finke 1993). Four functional layers could be distinguished within the experimental field. Basic characteristics of the four functional layers are presented in Table 1. Use of functional horizons is attractive because the vertical sequence of each point observation can be represented by only a few functional horizons, rather than a relatively high number of pedological horizons (Verhagen et al. 1995).

Van Uffelen et al. (1996) showed, through long-term spatial simulations, that two main soil types could be distinguished within the experimental field: a sandy part and a clayey part. In this study a representative profile from each of the two parts was selected for further analysis. A description in terms of functional layers and basic simulation characteristics, of the two soil profiles is presented in Table 2.

Table 1. Description of the functional layers and corresponding texture and hydraulic characteristics. BD refers to bulk density, SD to the standard deviation, Ksat to the saturated hydraulic conductivity, and θ_{sat} to the saturated water content

Layer	BD	SD	Organic Matter	Clay	Ksat	θsat	SD
	$(kg\ m^{-3})$		(%)	(%)	m day-1	$m^3\ m^{-3}$	
F1	1480	39	0 - 2	0 - 4	1.83	0.40	0.02
F2	1210	97	0 - 2	4 - 11	1.28	0.53	0.03
F3	1150	231	0 - 2	11 - 23	0.07	0.57	0.05
F4	1300	45	0 - 2	4 - 23	3.35	0.52	0.02

Table 2. Main soil physical properties of the two representative soil profiles, the clayey profiles (BEM 101) and the sandy profile (BEM 107).

Soil type	depth	functional layer	wilting point	field capacity	saturated water content	bulk density
	(m)		m^3m^{-3}	m^3m^{-3}	m^3m^{-3}	(kgm^3)
clayey	0 - 0.30	F4	0.050	0.394	0.440	1 300
(BEM 101)	0.30 - 0.40	F4	0.050	0.394	0.440	1 300
	0.40 - 0.60	F3	0.213	0.427	0.590	1 150
	0.60 - 0.90	F3	0.213	0.427	0.590	1 150
	0.90 - 1.20	F3	0.213	0.427	0.590	1 150
sandy	0- 0.30	F1	0.010	0.187	0.400	1 480
(BEM 107)	0.30 - 0.40	F1	0.010	0.187	0.400	1 480
	0.40 - 0.60	F1	0.010	0.187	0.400	1 480
	0.60 - 0.90	F1	0.010	0.187	0.400	1 480
	0.90 - 1.20	F1	0.010	0.187	0.400	1 480

Farm management

Fertilizer amounts in the Netherlands are based on the Dutch fertilizer recommendation which uses the total mineral nitrogen level (N_{min}) present in the rooting zone in early spring before fertilizing, to determine the application amount. The parameters for this fertilizer recommendation were derived from national trials. At the experimental site, the field average N_{min} level, measured in early spring, varied between 15 kg N ha^{-1} and 50 kg N ha^{l} over the field, which results in a fertilizer recommendation of: 100 ka ha^{-1} to be applied in early spring, 70 kg ha^{-1} as a second application, and a third application of 50 kg ha^{-1} during the vegetative growth phase in which the tassel is formed. According to the actual farm management practices in 1994 and 1995, the first application of 100 kg ha^{-1} was applied on April 20 (day of year 110), the second application of 70 kg ha^{-1} on May

2 (day of year 122) and the third application of 50 kg ha^{-1} on May 30 (day of year 150).

Winter wheat was sown on October 25 (day of year 298 of 1994) and harvest on August 11 (day 223). No irrigation was applied.

Economic data

A full economical analysis of the two farming systems is beyond the scope of this study. We therefore made a comparison based on the simulated differences between the conventional and the site-specific farming systems. Basic types of economic data for this comparison were supplied by the Research Station for Arable Farming and Field Production of Vegetables in the Netherlands and are summarized in Table 3.

Economic return *(ECRT)* in Dutch Guilder (Gld) is calculated according to:

$$ECRT = HWAH * GRPR + BWAH * BYPR - Nl \# M * APPR - NICM * NIPR \quad [1]$$

where *HWAH* is the grain yield at harvest (kg), *GRPR is* the grain price (Gld kg^{-1}), *BWAH* is the straw yield (kg), *BYPR* is the straw price (Gld kg^{-1}), *Nl#M* is the number of fertilizer applications, *APPR* are the base costs per fertilizer application (Gld app^{-1}), *NICM* is the amount of N-fertilizer applied (kg), and *NIPR is* the fertilizer price (Gld kg^{-1}).

Weather data

In predictive modeling weather is an important but unknown factor. Reliable weather forecasts generally do not predict more than five days ahead. However, to simulate the effect of different management practices, a daily weather forecast for a complete growing season is required. In this study weather generators, which were fitted to long-term historical weather records, were used to generate the required "missing" future weather data.

Rainfall is modeled using a two-stage third-order Markow model described by Jones and Thornton (1993) that was fitted to a thirty-year data set from a nearby climatological station. This rainfall generator, which is able to reproduce the

Table 3. Basic economic data for wheat, price level 1995. Gld refers to Dutch guilder, 1 Gld is equivalent with 0.6 US dollar, SD refers to the standard deviation. If no information on price fluctuations was known a fixed price was assumed.

	Units	Distribution	Average	SD
Grain yield	gld Mg^{-1}	normal	350.00	30.00
Harvest byproduct	gld Mg^{-1}	normal	85.00	8.00
N-fertilizer costs	gld kg^{-1}	fixed	1.04	---
N-costs per application	gld app^{-1}	fixed	24.00	---
Seed costs	gld kg^{-1}	fixed	1.22	---

monthly and annual rainfall variances observed in data sets from locations in both temperature and tropical zones, first determines whether any particular day is wet. This depends in part on whether there was rainfall on any of the three previous days. If "today" is determined to be wet, then the amount of rain that falls is determined by sampling from a gamma distribution, monthly parameters of which are calculated and interpolated to daily values using the 1 2-point Fourier transform described in Jones (1987).

Minimum and maximum temperatures and solar radiation are generated in response to whether today is wet or dry using WGEN, the weather generator developed by Richardson and Wright (1984).

Modeling

Simulation model

Within the DSSAT3, wheat growth and production were simulated with CERES-Wheat A detailed description of the model can be found in Ritchie (1986), Jones and Ritchie (1990) and Godwin and Jones (1991). The model allows the quantitative determination of wheat growth and yield.

The simulation of the soil water balance in the CERES models includes processes such as: infiltration and surface runoff, evaporation and transpiration, and drainage at the bottom of the soil profile Unsaturated flow of water from one soil layer to another is simulated as a capacity flow system, requiring only the soil water content at field capacity and permanent wilting point as hydraulic input parameters. The nitrogen sub-model simulates organic matter turnover rates with associated mineralization, immobilization of nitrogen, nitrification and denitrification, and urea hydrolysis and ammonia volatilization for different types of nitrogen such as organic fertilizers, green manure and synthetic fertilizers. Transport of nitrogen species occurs at the same rate as flow of water. Convective and/or dispersive transport of chemicals is not included. Crop growth is simulated in terms of phasic development with respect to the physiological age of the plant. Potential growth is dependent on photosynthetically active radiation (PAR). Interception of PAR is influenced by Leaf Area Index (LAI), row spacing, planting density, and photosynthetic efficiency of the crop. Actual biomass production is regulated by sub-optimal temperatures, soil-water deficits and nitrogen deficiencies. Partitioning of the photosynthetic products is governed by crop parameters and development stage of the crop. The model simulates all these processes with a daily time step.

Simulation procedure

For each soil type (sand and clay) within the experimental field, the same procedure was followed. Consisting of three steps:
   1. Model calibration
   2. Predictive modeling
   3. Scenario analyses

Model calibration

The basic soil physical characteristics, required by the CERES-model are as follows: saturated water content, water content at field capacity, and water content at permanent wilting point. These data, together with soil texture and nitrogen data, were obtained in previous studies by Verhagen et al. (1995) and Van Uffelen et al. (1996). Farm management details, rainfall, and temperature data were collected at the farm. Harvest yield data were collected by van Bergeijk and Goense (1996) who used site-specific harvesting techniques to measure grain and straw yield. Because no genetic coefficients for the variety used were available, a winter wheat cultivar was selected from the standard DSSAT data base, which best fitted the measured yields. By fine-tuning the soil physical characteristics the model was calibrated by comparing the following simulated and measured data:

1. grain yield
2. time series of soil mineral nitrogen at various soil depths
3. time final series soil water content at various soil depths

Predictive modeling

To explore the effect of different amounts of generated weather in combination with measured weather, the calibrated simulation model, was run for both main soil types. The first simulation used measured weather up until 5 days prior to the harvest date and generated weather for the last five days of the season. To allow a quantification of simulated differences in probabilistic terms, 20 replications of this first simulation using 20 different sets (of 5 days) of generated weather were carried out. The second step consisted of simulating 20 replications of measured weather up to 10 days prior to harvest, combined with generated weather to harvest. In total, 30 steps of 5 days (up to a maximum 150 days of predicted weather prior to harvest) were taken, resulting in 600 simulations for each soil type. Simulation results for both soil types were analyzed with respect to grain yield, N leaching and the residual amount of nitrogen in the soil after harvest. Economic returns were calculated according to Equation 1.

Scenario analyses

The predictive capability of simulation models has considerable potential for use in decision support systems. At the beginning of the growing season a farmer raises questions such as: when and how much should I fertilize? Because weather conditions for the season to come are unknown, answers should be presented in a way that takes into account the uncertainty of weather forecasts. To explore this *"predictive simulation mode"*, simulation results from the previous two steps were used to derive farm management practices for both main soil types. Using the automatic fertilizer option within DSSAT3, each time the nitrogen stress index, a number varying between 0 and 1, for no stress and maximum stress, respectively, exceeded 0.1, a fertilizer application of 25 kg N was applied. This procedure defines fertilizer applications in the optimum time window and optimizes the total amount of fertilizer necessary. Based on these results various management scenarios were defined for both soil types and simulated according to the predictive modeling procedure described above. Again results were analyzed in probabilistic terms, and economic returns were calculated.

RESULTS

Model calibration

Potential production of grain yield (production limited by the availability of solar radiation, temperature and genotype) was calculated as 10,416 kg ha[-1] for the 1994-95 growing season. The water limited grain yield (limited by the availability of solar radiation, temperature, genotype and water) for the clayey and sandy profile were both equal to the potential production level, indicating that no water stress occurred for winter wheat during the 19941995 growing season. All simulated deviations from the potential production are, therefore, due to nitrogen stress.

Calibration was performed by increasing the field capacity of the soil physical properties within CERES, resulting in a higher water storage capacity of the profile. Fig. 1 and Fig. 2 depict the simulated water contents for the clayey and sandy profile, respectively. While there is some scatter for both profiles, especially for the 0.15 -0.30 m layer, the simulations seem satisfactory. For the other layers

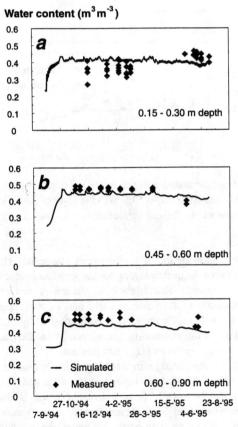

Fig. 1. Measured and simulated water contents at 3 depths within the clayey profile. **a** refers to a profile depth of 0.15-0.30m, **b** to the layer of 0.45-0.60m, and **c** to the layer of 0.60-0.90m minus the soil surface.

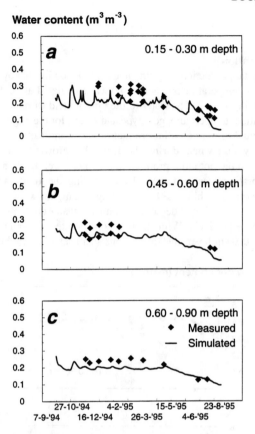

Fig. 2. Measured and simulated water contents at 3 depths within the sandy profile.
a refers to a profile depth of 0.15-0.30m, **b** to the layer of 0.45- 0.60m, and **c** to the
layer of 0.60-0.90m minus the soil surface.

the simulations are following the measured values reasonably. However, there is a
tendency, especially for the deeper layers, for the simulated values to underestimate
measured moisture contents. This might be explained by the fact that the capacity
water flow model used in CERES does not take into account capillary rise from the
ground water. Part of the differences is already masked by the increase in field
capacity brought about the calibration, but especially for the deeper soil layers in
the clayey soil capillary rise is an important process.

Measured and simulated grain yields (after calibration) and soil mineral N
status are presented in Table 4. Actual crop yields (water- and nitrogen- limited) are
very well predicted by the model. The soil mineral N status is also simulated well,
except for the February and May measurements in the sandy profile where a
differences of up to 40 kg ha[-1] can occur. However, there is considerable spatial
variability in these measurements. The simulation becomes more accurate again
at harvest.

Table 4. Measured and simulated variables for the clayey and sandy profile within the experimental field at the Van Bemmelenhoeve farm.

	Measured	Simulated
Grain yield (kg ha^{-1}):		
• potential	-	10 416
•water limited (clay)	-	10 416
•water limited (sand)	-	10 416
•actual (clay)	9 360	9 249
•actual (sand)	7 022	6 880
Soil mineral N (kg ha^{-1}) at day:		
• October 13 clay	110	116
• October 13 sand	144	144
• February 13 clay	48	66
• February 13 sand	14	29
• May 1 clay	46	63
• May 1 sand	19	59
• At harvest clay	86	80
• At harvest sand	122	125

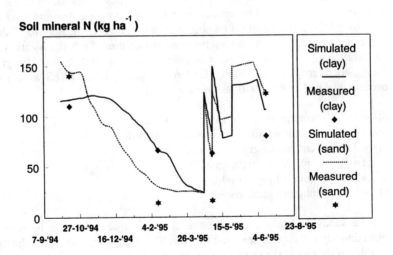

Fig. 3. Measured and simulated soil mineral N contents during the simulation period 1994.

Simulated and measured soil mineral N status for both soil types are presented in Fig. 3. After harvest in 1994, there is still a high amount of mineral N present in both soil types; which leaches over the winter season, finishing at a level of approximately 25 kg ha^{-1} for both soils. After the three applications of N the levels increase and at harvest in 1995 the amount is again considerable. In view of an average precipitation surplus of approximately 300 mm during the winter season in the Netherlands, the amount of mineral N present directly after harvest (leaching potential) should not exceed 35 kg ha^{-1}. Higher amounts than this will lead to high probabilities of exceeding the 50 g $_{NO}$ m^{-3} World Health Organization (WHO). Leaching for the 1994-95 winter season was calculated to be 130 kg ha^{-1} for the sandy profile and 50 kg ha^{-1} for the clayey profile. Leaching of nitrate in both profiles was thus exceeding the limits by a factor of between two and four. Because of the large N surplus directly after harvesting in 1995 high leaching probabilities could again be expected. Optimization with respect to fertilizer efficiency is thus necessary.

Predictive modeling

Figure 4a presents the simulated grain yields for the clayey and sandy profile using different time periods of predicted weather. When using 5 days of predicted weather prior to the harvest date the simulated mean is still close to the simulated value from the calibration study and the standard deviation is small, meaning that changes in weather at this stage do not have a big effect on final yield. When using longer periods of generated weather the standard deviation increases, while the mean stays rather constant for the sandy soil and decreases somewhat for the clayey soil. An increase in the standard deviation indicates that management risks are increasing (either in a positive or a negative sense). At approximately 40-50 days of predicted weather prior to harvest the standard deviations for both clay and sand are stabilizing This period corresponds approximately to the duration of the grain filling period (growth stage 6 in Fig. 4a). Growth stages within CERES-wheat have been defined by Ritchie (1986) as follows:

Stage 1   emergence to the terminal spikelet
Stage 2   terminal spikelet to end of leaf growth
Stage 3   end of leaf growth to anthesis
Stage 4   anthesis to begin grain filling
Stage 5   grain filling period
Stage 6   physiological maturity

For the sandy profile a slight increase in standard deviation can be seen at approximately 100 days of predicted weather. This corresponds with growth stage 2, which is the end of the vegetative growth phase.

When calculating the economic return for both soil profiles according to Equation 1 the same patterns can be seen (Fig. 5). It should be noted that the standard deviation of returns on the clayey soil, when using a full growing season of predicted weather (150 days), is much lower than that for the sandy soil indicating that economic risks on the clayey soil are lower.

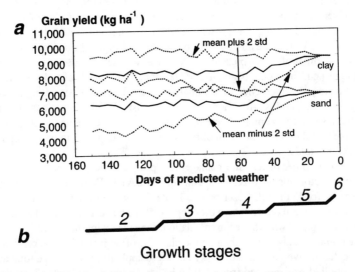

Fig. 4. Simulated grain yields for the clayey and sandy profile using an increasing amount of days of predicted weather (**a**). The solid lines represent the mean of 20 replications per point, the dotted lines depict the mean + 2 standard deviations (5-95 % probability levels). **b** represents the different wheat growth stages.

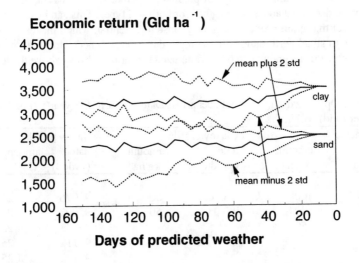

Fig. 5. Simulated economic returns for the clayey and sandy profile using an increasing amount of days of predicted weather, calculated according to Equation 1

Scenario analyses

After the calibration and predictive modeling work carried out above, it was clear that in order to prevent excessive nitrogen losses through leaching, the amount of mineral nitrogen in the soil at harvest should be less than approximately 35 kg ha^{-1}, given an average precipitation surplus of 300 mm. As already noted, no water stress occurred during the growing season, so that simulated deviations from potential production are all due to nitrogen stress. In Fig. 6 the nitrogen stress factor for both soil profiles is presented. Two peaks of severe nitrogen stress can be distinguished: first at the beginning of December 1994, and secondly at the end of growth stage 1 (emergence to terminal spikelet). The latter period in particular is very important and N stress here is a major cause of yield reductions. The period of stress is longer and more severe for the sandy soil, resulting in lower grain yields at harvest. Relieving some or all of this stress can be expected to have a significant impact on yields.

A number of realistic management practices were defined and simulated using the automatic fertilizer options within DSSAT3. We present only the optimal scenarios for both soils described in Table 5. For the clayey soil the number of applications was reduced to two dressings of 50 kg ha^{-1} each, a net reduction of 120 kg ha^{-1} compared with the recommendation. Since leaching probabilities are much higher on the sandy profile than on the clayey profile, instead of the original three applications, five dressings of 25 kg ha^{-1} were applied; however, the total amount of N applied was reduced from 220 kg ha^{-1} to 125 kg ha^{-1}. Figure 7 presents the simulated grain yield for the optimized scenarios for both soil profiles. Because of the better timing of fertilizer application, grain yields are approximately 2000 kg ha^{-1} higher on the sandy profile and slightly higher on the clayey profile (see also Fig. 4) compared with the recommendations.

When the number of days of predicted weather was increased, the mean grain yield decreased about 1000 kg ha^{-1} for both soil types. This can be explained by the fact that grain production is very close to potential, which makes it more sensitive to small perturbations in the input data (i.e. weather). However, on the sandy soil the mean grain yield of the optimized scenario is always above the level

Table 5. Optimized and original scenarios. Date refers to the year(yy) and day number (ddd) within the year.

Original		Clayey profile		Sandy profile	
date (yyddd)	amount (kg ha^{-1})	date (yyddd)	amount (kg ha^{-1})	date (yyddd)	amount (kg ha^{-1})
95110	100	95076	50	95076	25
95122	70	95103	50	95092	25
95150	50			95103	25
				95112	25
				95122	25
Total	220		100		125

Scenario analyses

After the calibration and predictive modeling work carried out above, it was clear that in order to prevent excessive nitrogen losses through leaching, the amount of mineral nitrogen in the soil at harvest should be less than approximately 35 kg ha^{-1}, given an average precipitation surplus of 300 mm. As already noted, no water stress occurred during the growing season, so that simulated deviations from potential production are all due to nitrogen stress. In Fig. 6 the nitrogen stress factor for both soil profiles is presented. Two peaks of severe nitrogen stress can be distinguished: first at the beginning of December 1994, and secondly at the end of growth stage 1 (emergence to terminal spikelet). The latter period in particular is very important and N stress here is a major cause of yield reductions. The period of stress is longer and more severe for the sandy soil, resulting in lower grain yields at harvest. Relieving some or all of this stress can be expected to have a significant impact on yields.

Table 5. Optimized and original scenarios. Date refers to the year(yy) and day number (ddd) within the year.

Original		Clayey profile		Sandy profile	
date (yyddd)	amount (kg ha^{-1})	date (yyddd)	amount (kg ha^{-1})	date (yyddd)	amount (kg ha^{-1})
95110	100	95076	50	95076	25
95122	70	95103	50	95092	25
95150	50			95103	25
				95112	25
				95122	25
Total	220		100		125

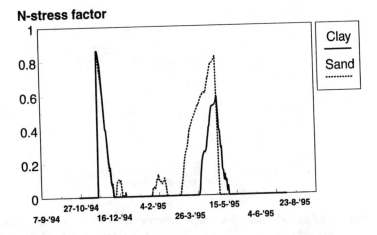

Fig. 6. Nitrogen stress factor for the clayey and sandy soil profile. A factor of 0 indicates no stress where as a factor of 1 indicates complete stress.

A number of realistic management practices were defined and simulated using the automatic fertilizer options within DSSAT3. We present only the optimal scenarios for both soils described in Table 5. For the clayey soil the number of applications was reduced to two dressings of 50 kg ha^{-1} each, a net reduction of 120 kg ha^{-1} compared with the recommendation. Since leaching probabilities are much higher on the sandy profile than on the clayey profile, instead of the original three applications, five dressings of 25 kg ha^{-1} were applied; however, the total amount of N applied was reduced from 220 kg ha^{-1} to 125 kg ha^{-1}. Figure 7 presents the simulated grain yield for the optimized scenarios for both soil profiles. Because of the better timing of fertilizer application, grain yields are approximately 2000 kg ha^{-1} higher on the sandy profile and slightly higher on the clayey profile (see also Fig. 4) compared with the recommendations.

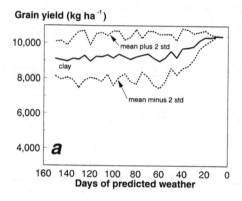

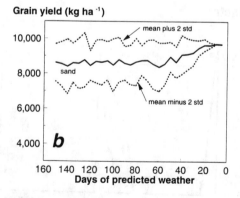

Fig. 7. Simulated grain yield for the alternative management scenarios for the clayey (**a**) and sandy (**b**) soil. The solid lines represent the mean of 20 replications per point, the dotted lines depict the mean $\pm$ 2 standard deviations (5-95 % probability levels).

When the number of days of predicted weather was increased, the mean grain yield decreased about 1000 kg ha^{-1} for both soil types. This can be explained by the fact that grain production is very close to potential, which makes it more sensitive to small perturbations in the input data (i.e. weather). However, on the sandy soil the mean grain yield of the optimized scenario is always above the level of the mean plus 2 standard deviations of the original scenario. The standard deviation remains more or less constant after 60 days for both soil types, indicating that the nitrogen stress at the end of the vegetative phase no longer has much effect on final simulated grain yield.

The amount of soil mineral N at harvest is below the critical value of 35 kg ha^{-1} for both soil types (Fig. 8). Because of the higher leaching sensitivity, the standard deviation for the sandy soil is somewhat higher than for the clayey soil. However, in terms of the mean amount of soil mineral N plus 2 standard deviations, the critical level N level is not exceeded with a 95 % probability for both soil types.

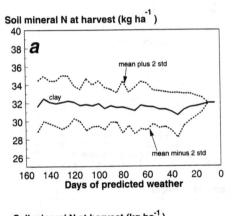

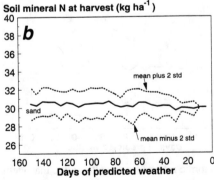

Fig. 8. Simulated soil mineral nitrogen for the alternative management scenarios for the clayey (a) and sandy (b) soil. The solid lines represent the mean of 20 replications per point, the dotted lines depict the mean + 2 standard deviations (5-95 % probability levels).

Although grain yields are only slightly higher on the clayey soil, the economic return increases considerably (Fig. 9), which is largely due to the reduced amount of fertilizer (125 kg N ha^{-1}) and one less application. On the sandy soil, on the other hand, an increased number of small applications results in significant higher yields and higher economic returns, assuming that the labor by the farmer is not a limiting factor and that the extra applications will not cause crop damage and thus yield reduction. Scenarios involving larger amounts of N in fewer applications resulted in more leaching and lower grain yields and reduced economic yields. Financial risks for the optimized scenarios are not significantly different from the original scenarios. Only the mean decreases when the number of days of predicted weather increases, indicating that using the 1994-95 growing season was highly conducive to wheat production at this site . Results of these analyses would no doubt be different for other growing seasons with other weather patterns.

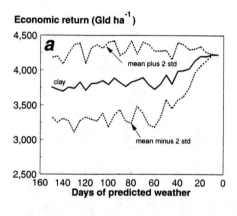

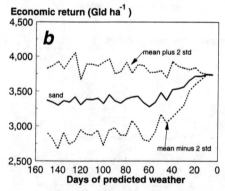

Fig. 9. Simulated economic returns for the alternative management scenario's for the clayey (a) and sandy *(b)* soil. The solid lines represent the mean of 20 replications per point, the dotted lines depict the mean + 2 standard deviations (5-95 % probability levels).

DISCUSSION AND CONCLUSIONS

This study demonstrates how well-calibrated, simulation models combined with weather generators fitted to long-term historical weather data, can be used as a predictive tool to support farm management decisions. The use of simulation models is especially suited to fine-tune fertilizer amounts and application times. The optimal scenarios achieved a reduction of 100 and 125 kg N-fertilizer ha^{-1} on the sandy and clayey soil profile, respectively, and a considerable reduction in the amount of soil mineral nitrogen at harvest, which is potentially leachable to the groundwater during the winter season. The sandy profile in the optimized scenarios also gave much better results in terms of yield and financial returns.

The study described above is merely a theoretical exercise in which the simulated scenarios have not yet been validated with field experiments. This implies that the outcomes are as good or as bad as the decision support system used, and the results should be interpreted as such. However, we feel that the results are promising, and that predictive modeling should be implemented in further field-studies using decision support systems and precision farming in the future.

ACKNOWLEDGMENT

The authors would like to thank Wageningen Agricultural University, the CT de Wit Graduate school for Production Ecology, and the International Fertilizer Development Center for providing support for P.K. Thornton's stay at in Wageningen. This study was also partly funded by the European Community projects EV5V-CT94-0480 and AIR3-CT94-1204. The authors would also like to thank H Oosterhuis, manager of the Van Bemmelenhoeve, and F Wijnands and A Krikke of the Research Station for Arable Farming and Field Production of Vegetables (PAGV) for providing the economic data.

REFERENCES

Bergeijk, J. van, and D. Goense. 1996. Soil tillage resistance as tool to map soil type differences.: Robert. et al. (eds). 3rd International conference on Precision Agriculture. June 23-26 Minneapolis, MN, USA.

Booltink, H.W.G., and J. Verhagen.1996. Using decision support systems to optimize barley management on spatial variable soil. *In* Teng, Kropff and van Laar. Systems Approaches for Agricultural Development (SAAD II). Kluwer, in press.

Bouma, J., and P.A. Finke, 1993. Origin and nature of soil resource variability. In: P.C. Robert, R.H. Rust, and W. Larson (eds.). Proceedings of the soil specific crop management workshop, April 14-16, 1992, Minneapolis, USA. pp. 3-13.

Finke, P.A. 1993. Field scale variability of soil structure and its impact on crop growth and nitrate leaching in the analysis of fertilizing scenario's. Geoderma, 60:89-107.

Godwin, D.C., and C.A. Jones. 1991. Nitrogen dynamics in soil-plant systems. *In* R.J. Hanks and J.T. Ritchie (eds.). Modelling plant and soil systems,

Monograph no. 31, Madison, WI. ASA, CSSA, SSSA publication.

Jones, J.W. 1993. Decision support systems for agricultural development. p 459-471. *In* F.W.T. Penning de Vries et al. (eds). Systems approaches for agricultural development, Kluwer Acad. Publ.

Jones, J.W., and J.T. Ritchie. 1990. Crop growth models. *In* G.J. Hoffman, T.A. Howell, and K.H. Solomon (eds). Management of farming irrigation systems. American Society of Engineers Monograph, St Joseph, MI .ASAE.

Jones, P.G.1987. Current availability and deficiencies in data relevant to agro-ecological studies in the geographic area covered by the IARCS. pp. 69-83. *In* A.H. Bunting (ed) Agricultural Environments. CAB International, Wallingford, UK.

Jones, P.G., and P.K. Thornton.1993. A rainfall generator for agricultural applications in the tropics. Agric. For. Meteorol., 63: 1-9.

Pickering, N.B., J.W. Hansen, J.W. Jones, C.M. Wells, V.K. Chan, and D.C. Godwin. 1994. Weatherman: A utility for managing and generating daily weather data. Agron. J., 86:332337.

Richardson, C.W., and D.A. Wright. 1984. WGEN: A model for generating daily weather variables. ARS-8. USDA-ARS, Washington DC, USA.

Ritchie, J.T. 1986. A user-oriented model of the soil water balance in wheat. pp. 203-305. *In* W. Day and R.K. Atkins (eds). Wheat growth and modelling. Plenum New York, USA

Thornton, P. 1991. Application of crop simulation models in agricultural research and development in the tropics and subtropics. International Fertilizer Development Centre. Alabama, USA pp 23.

Sadler, E.J., and G. Russell, 1996. Predicting crop yield variability. *In* Pierce, Robert, Sadler, and Searcy (Eds.) The state of site specific management for agriculture. Soil Sci. Soc. Am. Spec. Publ. Kluwer, in press.

Sprague, R.H. Jr, and E.H. Carlson. 1982. Building effective decision support systems. Prentice Hall, Inc. Englewood Cliffs, New Jersey USA.

Soil Survey Staff. 1975. Soil Taxonomy: A basic system of soil classification for making and interpreting soil surveys. USDA-SCS Agricultural Handbook 436, US Gov. Print Office, Washington DC.

Uehara, G., and Y. Tsuji. 1993. The IBSNAT project. pp 505-513. *In* F.W.T. Penning de Vries et al. (eds) Systems approaches for agricultural development, Kluwer Acad. Publ.

Uffelen, C. van, J. Verhagen, and J. Bouma. 1996. Comparison of simulated crop reaction patterns as a basis for site specific management. Agric. Sys. (in press)

Verhagen, A., H.W.G. Booltink, and J. Bouma. 1995. Site-specific management: balancing production and environmental requirements at farm level. Agricultural systems, 49:369384.

Wosten, J.H.M., M.H. Bannink, J.J. de Gruijter, and J. Bouma. 1986. A procedure to identify different groups of hydraulic conductivity and moisture retention curves for soil horizons. Journal of Hydrology, 86:133-145.

Strategies for Fertilizer Recommendations Based on Digital Agro Resource Maps

S. Haneklaus
D. Schroeder
E. Schnug

Institute of Plant Nutrition and Soil Science
Federal Agricultural Research Center
 Braunschweig-Völkenrode
Bundesallee 50, D-38116
Braunschweig, Germany

ABSTRACT

Fertilization today requires the responsible use of agroresources with special attention to its ecological effects. The new farm management concept *Local Resource Management* (LRM) which employs the techniques of precision agriculture, enables sustainable agricultural production in the sense of balanced fertilization. In LRM the distribution of fertilizers within the field is regulated according to the spatial variability of parameters of soil fertility and geomorphology. The variable distribution of fertilizers in the field avoids an uneconomic and ecological critical surplus as well as a minus of agroresources. The improved distribution of fertilizers is of special importance for nitrogen because a simple reduction of the amount applied will not necessarily reduce undesired nutrient losses into the environment because of the heterogeneity of soil fertility features in space.

The acceptance of LRM in agriculture will highly depend upon its practicability and therefore a sophisticated strategy for data collection is one key point in LRM. The geostatistical treatment and processing of soil data is performed by LORIS (Local Resource Information System), a GIS especially designed for the handling of geocoded agricultural data. LORIS enables the compilation, managing and computation of Digital Agro Resource Maps (DARMs).

In this paper ideas for efficient data collection under practical conditions are presented and examples for recommendations for a sustainable nitrogen fertilization which considers the spatial variability of soil characteristics in the field are demonstrated.

Constant key variables

Nitrogen is a major plant nutrient with the strongest influence on yield. In order to develop a concept for a spatially variable nitrogen fertilization it is necessary to allocate the determinant variables which influence the supply and remain almost constant within time. The most important key variables are the geomorphology and the soil type characterized by texture, soil depth and organic matter content. These three parameters show the strongest influence on the nitrogen dynamics of soils. The slope and aspect in the terrain are important parameters for lateral and vertical water and thus nitrogen fluxes so that digital elevation models which cover this aspect may become an integrative part for the design of spatially variable nitrogen fertilization.

More than 95% of the nitrogen is bound in the organic matter which is therefore an important nitrogen pool. The clay fraction is the most important parameter of soil texture and closely related with the water balance and especially potential vertical nitrogen movements.

Mapping of the spatial variability of key variables

The number of technical aids and agricultural implements for addressing the spatial variability of soil parameters is continuously increasing. But the existing procedure of data collection and methods for the calculation of the fertilizer recommendations are characterized by high costs, expenditure of labor and lack of security with regard to ecological as well as economical benefits. Under consideration of the prerequisites for geostatistical treatment of geocoded data it becomes evident that at least a sampling density of 50 * 50 m is necessary in order to calculate the spatial variability of soil parameters within the field. The required grid size may be reduced to 30 * 30 m depending on pedogenetical factors. Even automatic sampling procedures together with the rapid laboratory methods (Haneklaus, Vogel & Schnug 1994, Schnug & Haneklaus 1996) or chemical field tests, especially when done by service providers, cannot reduce the costs to an acceptable level.

Another approach actually used to address spatial variability better is the segmentation of fields into regular shaped subunits. But also this procedure is unsatisfying, because as shown many times the variability of only a 5 ha large subunit can easily comprise that of the whole surrounding landscape. The required size of one unit thus would have to be theoretically that of a single pedon which again is unrealistic just from the viewpoint of costs.

Sometimes databases with geological or soil information are existing from former times when costs for sampling information were not as important as today. A fine example for this is the so called 'Reichsbodenschaetzung' in Germany which was established during the 'Third Reich' when thousands of 'surveyors' were sent out to evaluate and map the German resources for food production. It is indeed an enticing idea to use these data as a first step into LRM but there are four main points making it very difficult to obtain real progress or benefit from it: These are the quality of the surveyors and thus their results, the accuracy of positions and boundaries, the farmers' fear of taxation and last but not least the maps containing only descriptive thematic information.

Remote sensing or sensors which operate on the go on agricultural implements are often believed to be the solution for overcoming the problem of efficient and economic data collection. When it comes to mapping the above key variables, sensor techniques and remote sensing may be an aid to capture geomorphology. Under good circumstances also the mapping of the organic matter content may work out that way, but at least the soil texture is fairly resistant to this method.

The most promising way at the moment is to employ the facilities of human skin, eyes and brain as biosensors in a process what should be called 'self surveying'. The idea behind 'self surveying' is to use labour already available on the farm to cut down costs and to employ reliable methods, which can be performed by nearly every member of staff after a short training period. In order to keep sources of variation small, the number of persons involved in the 'self surveying' process should be as small as possible - ideally one!

The methods employed are on-line or post processing DGPS for geocoding, the so called 'finger test' for texture and visual assessments for organic matter content and geomorphology. In landscapes with no expressed geomorphology the depth of the top soil layer is measured by means of a probe. It is not the task of the 'self surveyor' to estimate clay and organic matter content in absolute figures. Colour for organic matter and roughness or smoothness for texture are recorded by scores, for example from one to ten. After finishing 'self surveying' of a field, soil samples are taken representing each core value and analysed by conventional methods for clay and organic matter. Regressions of score values and laboratory data are used afterwards to calibrate the whole set of score values. The accuracy of this 'local calibration' procedure can be enhanced by increasing the number of replicates taken from different plots in the field which were rated the same core values.

Field 206a (Kassow)

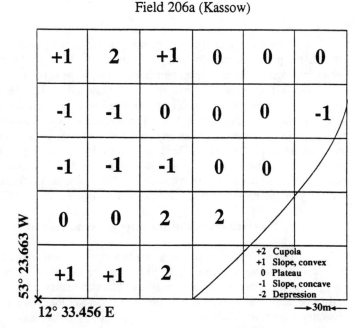

Fig. 1: Labels for geomorphology according to the VIP procedure in ARC-INFO

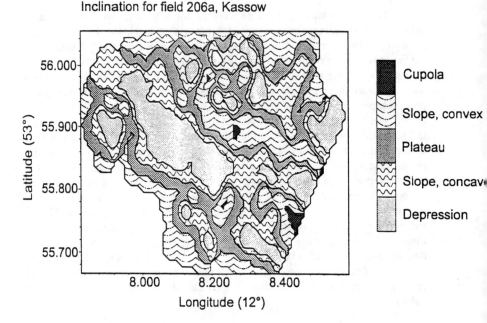

Fig. 2: Spatial variability of geomorphological features in the field

Geomorphology is assessed according to the scheme used by the VIP or HIGHLOW procedure in ARC-INFO (Anon. 1991) (Fig. 1). An example for geomorphological values in an eastern German landscape is given in Fig. 2.

By 'self surveying' with a minimum of efforts reliable DARMs for three of the most important soil features can be gathered in a very easy and economical acceptable way. Working on a 50 * 50 m grid the 'self surveyor' has to walk 20 km to cover a 100 ha of land recording on 400 individual plots, which should not take longer than two working days. The data collected by 'self surveying' can also be used to evaluate areas of pedological similarity in the field and to allocate 'monitor plots' which is a promising way to reduce the efforts for sampling for conventional soil analysis (Murphy, Haneklaus & Schnug 1994).

Determination of the nitrogen requirement

The determination of the nitrogen requirement is based on the prognosis of the plant growth and nitrogen mineralization in the soil. The calculation of the nitrogen requirement needs to integrate the soil characteristics, time of application and amount. In the example presented below the nitrogen fertilizer recommendation was designed on the basis of the key variables geomorphology, organic matter (OMC) and clay content in combination with the the longterm experiences of the farmer. This knowledge was transferred to spatially variable fertilizer maps in LORIS (Local Resource Information System) which is a GIS (Geographical Information System) specially designed for the needs of LRM.

All approaches modelling nitrogen in agroecosystems require reliable basic data about yield potential, soil characteristics and climatic conditions. The nitrogen mineralization rate can be estimated by the organic matter and clay content of the soil with minor errors (Richter, 1996). The spatially variable yield potential can be estimated from perennial yield maps. There are several systems available on the market which allow the geocoded on-line measurement of yields at a reasonable price level (Murphy, Schnug & Haneklaus 1995). Yield maps are also an effective instrument to gather information about the annual removal of plant nutrients which is essential, e.g., for the fertilizer recommendations of potassium and phosphorous (cp.. Schnug & Haneklaus 1995). Moreover yield mapping is a simple tool to monitor the efficiency of variable rate fertilization.

An important weak point of nitrogen simulation models is that they prognose only vertical nitrogen fluxes, while lateral fluxes are neglected (Engel 1996; Richter 1996). However, in areas with distinct differences in elevation lateral nitrogen fluxes show strong influence on the plant-available nitrogen and therefore need to be considered in the recommendations for nitrogen fertilization.

Summing it up it can be said that the major handicap for the use of nitrogen simulation models in agriculture, namely the localization of representative monitor plots, is solved by the self-surveying of fields. A simulation of lateral fluxes would further improve the accuracy of the fertilizer recommendation.

Changes in the nitrogen requirement proportionate to the variability of key variables

For the spatial variable nitrogen fertilization it is necessary to determine the full scale variability of the key variables (see above) in the field in a sufficient density. This information was so far gathered by tradtional cartography together with soil analysis and therefore restricted to scientific research projects. However, with the new approach of 'self surveying' it will be possible to adapt the fertilizer rate according to the spatial variability in the field with an acceptable expenditure. On the fields used for the spatial variable nitrogen fertilization on the Mariensee experimental farm of the FAL the total organic carbon content varied between 1.56 and 4.51% and the clay content ranged from 11.8 to 15.5%. The fertilizer recommendation was adjusted to these two key variables, because they are supposed to have the strongest impact on the nitrogen balance. Linked with the spatial variability of the available N concentrations in the monitor plots and coordinated with the local experience of the farm manager the nitrogen fertilization was stepwise reduced with increasing clay and organic matter contents (cp. Table 1). On one field the total nitrogen amount fertilized was kept constant however with a distribution in the field according to the spatial variability of the key variables while on the other field the nitrogen amount spread at variable rate was reduced by 20% (cp. Fig. 3).

Table 1: Strategy for nitrogen fertilization based on the spatial variability of the key variables clay and organic matter content (100% = 180 kg/ha N)

Clay content (%)	Organic carbon content (%)					
	1.56	2.17	2.75	3.34	3.63	4.51
11.8	120 (100)	110 (90)	100 (80)	90 (70)	80 (60)	70 (50)
12.5	110 (90)	100 (80)	90 (70)	80 (60)	70 (50)	65 (45)
13.2	100 (80)	90 (70)	80 (60)	70 (50)	65 (45)	60 (40)
13.9	90 (70)	80 (60)	70 (50)	65 (45)	60 (40)	55 (35)
14.6	80 (60)	70 (50)	65 (45)	60 (40)	55 (35)	50 (30)
15.4	70 (50)	65 (45)	60 (40)	55 (35)	50 (30)	40 (25)

For the fertilization the 'Kemistar' system consisting of a DGPS, spreader and on-board computer was used (Schröder & Haneklaus 1996).

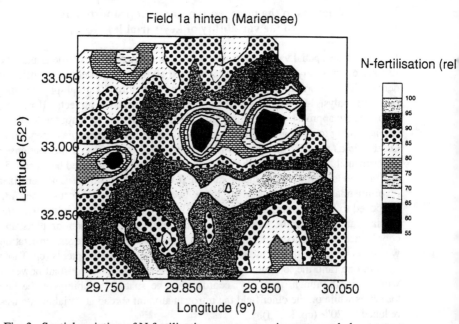

Fig. 3. Spatial variation of N fertilization acc. to organic matter and clay content in soil.

The two DARMs for organic matter and clay content were logically connected to a fertilzation map according to the scheme presented in Table 1 using the LORIS software. In addition the two strategies for N fertilization are tested in combination with and without growth regulators. The total nitrogen fertilization of 160kg N/ha that was uniformly applied is based upon the average available N content before fertilization and a yield potential of 9 t/ha for winter barley whereas for the spatially variable nitrogen fertilization this feature was integrated in the fertilizer strategy according to the variability of the monitor plots.

On another farm in Kassow (Mecklenburg-Vorpommern) where no sufficient analytical data were available the nitrogen fertilization strategy was designed according to the variability of geomorphological parameters which the farmer determined by self-surveying. The data were transformed to a DARM (see Fig. 1, 2) showing the geomorphological differences in the terrain. The nitrogen fertilization to winter oilseed rape was varied according to the scheme presented in Table 2. The average nitrogen fertilization of 180 kg/ha is the amount the farmer applied uniformly on one half of the field's area and that is based upon the analysis of mineral nitrogen in the soil and a yield potential of 3.5 t/ha. The nitrogen amount saved by spatially variable fertilization was 20%.

These two strategies for the available nitrogen fertilization based on the variability of the parameters geomorphology, organic matter and clay content together with the local experience of the farm manager are practical approaches to optimize the distribution of nitrogen in the sense of sustainable agriculture. The benefit of a nitrogen fertilization according to the spatial variability of soil fertility parameters will improve utilization of this macronutrient in agro-ecosystems reducing losses to the environment. It is also expected that the benefits of spatially maintained nitrogen fertilization will strongly increase if growth regulators have to be abandoned: In a cropping system with even distribution of fertilizers without growth regulators the maximum amount of nitrogen applied is guided by those parts of the field with highest nitrogen supply by the soil and which are therefore most vulnerable to lodging. In contrast areas with lower supply from the soil will not receive enough nitrogen to attain maximum productivity.

Table 2: Strategy for nitrogen fertilization based on the spatial variability of geomorphological parameters (100% = 180 kg/ha N)

Nitrogen fertilization (%)	Geomorphological feature
40	Depression
120	Cupola
70	Slope
100	Plateau

REFERENCES

Anonymous. 1991. Surface Modeling with TIN, Arc Info User's Guide. p. 6/14.

Engel. 1996. Standortgerechte N-Düngung mit Hilfe von GPS, Elektronik und Stickstoffsimulation. (Site specific N fertilization employing GPS, and nitrogen simulation.) Wissenschaftliche Mitteilungen der Bundesforschungsanstalt für Landwirtschaft Braunschweig Völkenrode. 2 (in press)

Haneklaus, S, W. Vogel, E. Schnug. 1994. Rubidium in soil and its determination by X-ray fluorescence spectroscopy. In: Defizite und Überschüsse an Mengen- und Spurenelementen. Proc. 14. Arbeitstagung "Mengen- und Spurenelemente", Jena 1994. pp. 29-36. Schubert Verlag. Leipzig. 1994.

Murphy, D.P., S. Haneklaus, and E. Schnug. 1994. Innovative soil sampling and analysis procedures for the local resource management of agricultural soils. Transactions of the 15th World Congress of Soil Science Vol. 6a, 613-630.

Murphy, D.P., E. Schnug, and S. Haneklaus. 1995. Yield mapping - a guide to improved techniques and strategies. Proc. 2nd Int. Conf. On Site-Specific management for Agricultural Systems, Minneapolis, March 27-30, 1994. ASA-CSSA-SSSA. Madison 1995.

Richter. 1996. Schätzgenauigkeit von N-Mineralisation, Ertrag und N-Bedarf in variablen Ökosystemen - Methodische Probleme bei der Übertragung der Ist-Analyse. (Variability and acccuracy of estimation for N-mineralization, yield, and N-requirement.) Wissenschaftliche Mitteilungen der Bundesforschungsanstalt für Landwirtschaft Braunschweig Völkenrode. 2 (in press)

Schnug, E., and S. Haneklaus. 1995. Challenges of 'Local Resource Management' for sustainable oilseed rape production. Proc. 9th Int. Rapeseed. Cambridge 1995.

Schnug, E., and S. Haneklaus. 1996. A rapid method for the indirect determination of the organic matter content of soils. Proc. 9th Int. Rapeseed. Congress Cambridge 1995.

Schröder,D, S. Haneklaus. 1996. Local Resource Management. Landwirtschaftliches Wochenblatt. 3:11-13.

Metering Characteristics Accompanying Rate Changes Necessary for Precision Farming

A. Bahri
Institut National de la Recherche Agronomique
(INRA), CRRA, BP: 589
Settat, Morocco

K. Von Bargen
M. F. Kocher
L. L. Bashford

University of Nebraska-Lincoln
Department of Biological Systems Engineering
Lincoln, NE

ABSTRACT

Agricultural machines used in precision farming must adjust application rates according to the needs of each cell within a field. Changing from an initial application rate to a new rate while the machine travels from one cell to another in the field is accompanied with some misapplication. The severity of this misapplication depends on the down-the-row delivery characteristics of the metering system and the magnitude of the rate change from cell to cell.

On-the-go rate change tests evaluated the down-the-row performance of an operator controlled metering system when increasing and decreasing wheat seeding rates by 10 and 20 kg/ha steps. The transition time from one cell to another ranged from 3 to 9 s depending upon the magnitude of the application rate change.

The difference between the initial and final seeding rate was based on a simple index. This separation index was based upon the initial and final down-the-row seeding rate distributions. When the separation index was greater than or equal to zero, the difference between the initial and final application rate was considered to be suitable for precision farming. The separation criterion was always satisfied with 20 kg/ha rate changes. For 10 kg/ha rate changes, the separation index was negative in most cases. This indicated that rate changes of 10 kg/ha or less were unlikely to provide detectable rate differences as the metering rate variability exceeded the magnitude of the 10 kg/ha rate change.

INTRODUCTION

Precision farming requires that a machine deliver the correct application rate at each field site. Some machines may require mechanical design changes to incorporate sensors and controllers to provide adequate rate modulation (Robert et al., 1992).

When evaluating grain drills for the ability to apply a uniform rate of seed or fertilizer, the Prairie Agricultural Machinery Institute of Canada (PAMI) considers a coefficient of variation (CV) of 15% or less as acceptable variation among row units for grain and fertilizer (PAMI, 1987). Most grain drills have less variability than the criterion. However, seed type, lateral slope, and slope in the direction of travel alters the variability in seeding and fertilizing application rates for many grain drills. The PAMI tests do not provide a down-the-row measure of variability. Bashford (1993) found the down-the-row CV when metering soybeans ranged from 12 to 40%, and from 12 to 20% for wheat using drills with external fluted metering mechanisms. He concluded that these CV values are probably too high to be acceptable for site specific crop management (SSCM).

Metering systems designed to apply uniform application rates must be controlled for spatially variable application. Incremental changes in application rates may not be possible with such applicators (Gaultney, 1989). Metering systems for granular material applicators are designed to apply a constant application rate per area with variable field speed. With SSCM, these metering mechanisms must be revaluated for their ability to deliver varying rates from site to site within a field. Application rate changes must be made on-the-go. Set points defined by a control map must be achieved quickly (Schueller, 1992). A transition phase accompanies the change from one application rate to another.

There is a transition time when changing the application rate while a machine is traveling across a field. For this time duration, a varying application rate occurs during the rate transition. Ollilia et al. (1990) used a Field Grid Sense system to spatially control field inputs for specific site requirements. With this system, 49.4 seconds were required to adjust to a new application rate. At a field speed of 8 km/h, the applicator traveled about 110 meters before attaining the new rate. Schueller (1991) reported that one spatially-variable dry fertilizer applicator had a transition time of 4 seconds. At 8 km/h, the machine traveled 8.9 meters before the new rate was achieved.

The effect of this transition on the application rate was addressed in this study for a conventional grain drill modified for on-the-go variable rate application. Specifically, the down-the-row and among metering unit variabilities were determined for a grain drill. Following the variability investigation, seeding rate changes were made, and the transition from an initial to a final rate was determined.

METHODS AND MATERIALS

A grain drill equipped with internal fluted cup metering units was used in the investigation. It was equipped with a roller chain drive from a ground drive to the metering system. An electric motor, rotary actuator, with the speed controlled by a rheostat replaced the ground drive in order to vary the application rate on-the-go. The open loop rotary control system was actuated by the tractor operator. This rotary control system provided seeding rate change increments of 10 or 20 kg/ha (Bahri, 1995). Rheostat positions were matched to rates of 70, 80, 90, 100, 110, and 120 kg/ha.

Two types of tests were conducted. First, the metering variability down-the-row was determined. Winter wheat seed was collected on micro cell mats (Astroturf door mats) 0.48 m long (Bahri, 1995). The drill was operated on a paved parking lot over a length of 60 micro cells at a speed of about 6 km/h.

Second, the ability of the drill metering and variable rate control system to deliver variable seeding rates was evaluated over the same mats for a distance of about 30 m. Wheat seeds delivered from three metering units were collected on the line of micro cells. Before passing over the micro cells, the drill was operated to reach steady state operation. This rate continued to be delivered over the micro cells for a distance of 10 m. At this point the operator changed the application rate, and the transition from the initial to the final rate was determined (Bahri, 1995).

Because variability occurs in both the initial and final delivery rates, the concept of Separation (S) was used to describe the overlap or coincidence of the two distributions of the rates. Separation serves as a simple criterion to identify an acceptable difference in two rates for SSCM.

Separation is the difference between the final rate at minus one standard deviation and plus one standard deviation for the initial rate for an increasing rate change, Fig. 1. For example, when S = 0, as illustrated in Figure 1, 84% of the final rate distribution (lightly shaded area) overlaps only 16% of the initial rate distribution (dark shaded area).

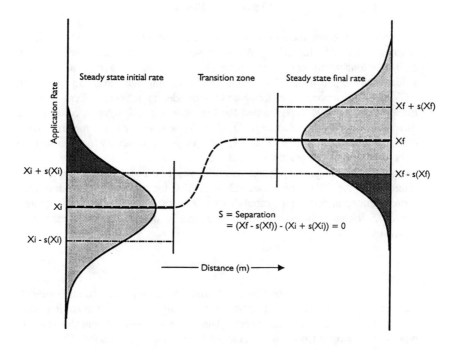

Figure 1.   A graphical representation of the separation of an initial and a final seeding rate. Depicted is the situation where S = 0.

Separation for a rate increase is:

$$S = [X_f - s(X_f)] - [X_i + s(X_i)]$$

For a rate decrease,

$$S = [X_i - s(X_i)] - [X_f + s(X_f)]$$

where:
X_i = The mean of the initial steady state rate
X_f = the mean of the final steady state rate
$s(X_i)$ = the standard deviation of X_i
$s(X_f)$ = the standard deviation of X_f

RESULTS AND DISCUSSION

Metering Uniformity

Typical down-the-row and among metering units seeding variability is illustrated in Table 1 for an 80 kg/ha wheat seeding rate. The data were collected on 0.48 m long micro cells. When combining all data, the 6 metering units over the 8 micro cells for a distance of 3.84 m, the mean seeding rate was 81.53 kg/ha with a CV of 2.56 % among the metering units. Considering individual micro cells, the highest rate was 104.52 kg/ha and the lowest was 58.53 kg/ha, a considerable difference for precision farming. Down-the-row CVs for the 6 metering units considered ranged from 10.4 to 18.5 %. Down-the-row variability for other seeding rates were similar and are given in Table 2.

When among metering unit variability was based upon data from individual micro cells, the CV ranged from 7.3 to 21.3 % for the 80 kg/ha seeding rate. Similar among metering unit variability was observed for the other seeding rates, Table 3. It can be noted that the CVs appear to decrease as the seeding rate increases.

Rate Changes

The seeding rate profile, Figure 2, was typical for a 10 kg/ha rate increase. Each data point represents the combined seeding rate from 3 metering units collected on a 0.48 m long micro cell. The distance traveled before the new rate was attained ranged from 8 to 15 m at a 6 km/h travel speed (Bahri, 1995). The corresponding response time was 4.8 to 9.0 s.

At the 20 kg/ha rate change, the response time was faster. The average travel distance before reaching the new rate was 5 m (Bahri, 1995). Further, it was more difficult with the control system used to make the 10 kg/ha increment changes than the 20 kg/ha changes.

Table 1.   Seeding rate for a nominal 80 kg/ha setting

Cell	Meter						Among metering units		
	1	2	4	4	5	6	Mean	SD	CV
	kg/ha								%
1	79.4	92.0	96.2	87.8	96.2	96.2	91.3	6.7	7.3
2	100.3	75.3	92.0	66.9	75.3	79.4	81.5	12.3	15.1
3	83.6	83.6	87.8	83.6	67.0	71.1	79.4	8.4	10.5
4	71.1	75.3	83.6	71.1	92.0	96.2	81.5	10.8	13.3
5	96.2	75.3	71.1	66.9	104.5	62.7	79.4	16.9	21.3
6	62.7	96.2	71.1	96.2	62.7	92.0	80.1	16.4	20.5
7	83.6	75.3	92.0	58.5	79.4	71.1	76.7	11.4	14.9
8	75.3	87.8	79.4	92.0	71.1	87.8	82.2	8.2	10.0
Down the row									
Mean kg/ha	81.5	82.6	84.2	77.9	81.0	82.1	81.5	2.1	2.6
SD kg/ha	12.4	8.7	9.6	13.8	15.0	12.8			
CV %	15.3	10.4	11.4	17.7	18.5	15.6			

Table 2.   Down-the-row variability

Seeding Rate, kg/ha	Coefficient of Variation, %	
	lowest	highest
70	12.3	18.5
80	10.4	18.5
90	11.1	18.9
100	10.7	17.5
110	8.2	12.8
120	10.5	12.2

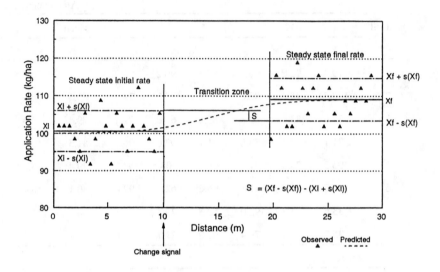

Figure 2. Rate overlap with a rate increase from 100 to 110 kg/ha, S = -3.06

Table 3. Variability Among Metering Units

| Seeding Rate | Coefficient of Variation, % | | |
| | 0.48 m Micro cell length | | 3.84 m distance |
kg/ha	lowest	highest	composite
70	9.2	19.4	3.51
80	7.3	21.3	2.56
90	7.4	21.0	4.57
100	9.3	21.7	3.98
110	4.5	15.1	2.48
120	4.8	13.3	1.32

Table 4.   Separation of initial and final rate
               distributions

Separation, S	Final and Initial Rate Overlap
kg/ha	% final $\cap$ % initial
10	84% $\cap$ 0
5	84% $\cap$ 2
0	84% $\cap$ 16
-5	84% $\cap$ 50
-10	84% $\cap$ 84

Separation

The single value separation index is a simple measure of the difference between an initial and a final rate. Both the rate difference and rate variability are considered. Separation measures the non-coincidence (lack of overlap) in the initial and final seeding rates for an on-the-go rate change. Consider an example situation where the standard deviation of both distributions is 5 kg/ha. When S = 0, the rate difference was considered a true rate difference for SSCM, and is illustrated in Figure 1. Table 4 illustrates the coincidence for several values of separation.

A typical 10 kg/ha rate change is illustrated in Figure 2. The separation index was -3.06, indicating considerable overlap in the distribution of the initial and final rate. In general, separation values for the 20 kg/ha rate change were greater than those for the 10 kg/ha changes, Table 5. For good variable rate adjustment for SSCM, an incremental change greater than 10 kg/ha is indicated for the rotary actuated metering system. The separation width (S $\geq$ 0) was always satisfied with 20 kg/ha rate changes. For the 10 kg/ha rate change, S was negative in most cases, thus, indicating that 10 kg/ha rate changes did not provide a real rate change.

CONCLUSIONS

In general when considering seeding rates on a micro cell basis (0.48 in length), seeding rates were quite variable with the internal fluted metering mechanism investigated. The down-the-row CVs ranged from 10 to 19 %. This variability makes it difficult to discern a small rate difference when adjusting rates on-the-go in SSCM.

The open loop rotary control system for variable rate adjustment would provide initial to final seed application rate differences for incremental changes greater than 10 kg/ha. A separation index based upon the standard deviations of the initial and final seeding rates provided a quantifiable measure of the difference between the rates. When 10 kg/ha rate changes were made, the separation index was small and negative in four of seven tests.

Table 5.   Separation (S) for indicated seeding rate changes

Rate change	X_i	$s(X_i)$	X_f	$s(X_f)$	S
			kg/ha		
70 to 80	69.17	4.02	80.73	3.80	3.74
90 to 100	85.49	5.65	96.36	7.42	-2.20
100 to 110	97.89	5.69	111.15	8.26	-0.69
100 to 110	100.97	5.32	108.77	5.73	-3.06
80 to 70	81.24	6.88	68.49	4.71	1.16
100 to 90	101.46	6.64	90.24	5.67	-1.06
110 to 100	110.03	5.20	100.44	3.90	0.49
70 to 90	70.19	4.45	90.75	6.25	9.86
80 to 100	82.09	4.71	98.74	5.99	5.95
80 to 100	82.60	4.69	97.55	8.48	1.78
100 to 120	100.95	6.72	122.20	10.89	3.64
90 to 70	89.73	4.60	69.51	3.02	12.60
100 to 80	100.95	5.74	82.77	7.17	5.27
120 to 100	119.14	5.88	103.33	7.43	2.50

REFERENCES

Bahri, A. 1995. Modulating wheat seeding rate for site specific management. Unpublished Ph.D. dissertation. University of Nebraska-Lincoln, NE.

Bashford, L.L. 1993. External flute seed metering evaluation related to site specific farming. ASAE Paper No. 93-8517. ASAE, St. Joseph, MI.

Gaultney, L.D. 1989. Prescription farming based on soil property sensors. ASAE Paper No. 89-1036. ASAE, St. Joseph, MI.

Ollila, D.G., J.A. Schumacher, and D.P. Froehlich. 1990. Integrating field grid sense system with direct injection technology. ASAE Paper No. 90-1628. ASAE, St. Joseph, MI.

PAMI. 1987. Evaluation Report 519. Tye Series V1114-5360 No Till Drill. Prairie Agricultural Machinery Institute, Humboldt, Saskatchewan, Canada.

Robert, P.C., R.H. Rust, and W.E. Larson. 1992. Proceedings of Soil Specific Crop Management. April 14-16, 1992. pp 181-195.

Schueller, J.K. 1991. Evaluating spatially-variable applicators. SOLUTIONS September/October 1991. pp. 44-45.

Schueller, J.K. 1992. A review and integrating analysis of Spatially-Variable Control of crop production. Fertilizer Research 33: 1-34.

Using Precision Farming Technologies for Improving Applied On-Farm Research

E. S. Oyarzabal
Ceregen, a Unit of Monsanto Co.
Urbandale, Iowa

A. P. Mallarino
Department of Agronomy
Iowa State University
Ames, Iowa

P. N. Hinz
Department of Statistics
Iowa State University
Ames, Iowa

ABSTRACT

This research demonstrates the use of data generated with precision farming technologies for comparing management practices on farmers' fields. Corn yield variability, treatment effects, and relationships between yields and several site variables were analyzed at various scales. Yield responses to the treatments are reported for two cornfields and for different areas within each field.

BACKGROUND

Applied agricultural experimentation involves the comparison of current and innovative products, technologies, systems, or recommendations. Field tests usually are repeated in several locations (environments) over a period of years using experimental designs. New products, technologies, systems, or management practices are recommended after statistical analyses confirm the advantages over existing practices. Obviously, the recommendations are extrapolated to large geographical areas and not all the environmental conditions are explored during the testing period.

New technologies available to farmers allow for georeferencing of measurements such as soil tests, scouting counts, various agronomic observations, and crop yields. These measurements can be collected using conventional methods or by newly developed remote sensing devices. After several cropping seasons all these layers of information will generate extensive farm databases. The data produced with these technologies will eventually allow the farmers to fine-tune the general recommendations and to produce more precise local recommendations.

On farm research using strip plots is an accepted methodology for complementing traditional small-plot research, for generating local recommendations, and for demonstrating management practices. It is used by farmers, dealers, industry, and universities (Rzewnicki et al., 1988; Shapiro et al.,

1989). The collection of yields is achieved by using combines equipped with weighing devices or with common combines and weigh wagons. In this study precision farming techniques (including yield monitors) were used to compare several agronomic treatments in many Iowa fields.

In early spring, soils managed with no-tillage usually are colder than those managed with conventional tillage. Early crop growth often is slower with no-tillage because the usually lower soil temperatures (due to the residue cover) affect plant growth, nutrient availability, nutrient uptake, and other physiological processes. It has been suggested that starter fertilizer (applied with the planter close to the seeds) would improve nutrient availability, early plant growth, and grain yields of corn. This management practice was tested in three Iowa fields. Applying soil residual herbicides for weed control at planting time is the most current recommendation for corn production in the Midwest. Chemical companies are continually searching for new herbicide molecules that combine higher weed activity with lower use rates. A new broadleaf herbicide for corn, halosulfuron, was compared with atrazine, which is the conventional standard herbicide for corn. This comparison was tested in six Iowa fields. The objective of this study was to demonstrate the use of data generated with precision farming techniques for treatment comparisons in farmers' fields.

MATERIALS AND METHODS

Different agronomic treatments (starter vs. no starter, new herbicide vs. standard herbicide) were compared in six Iowa cornfields in 1995. Data processing has been completed for two approximately 20-ha fields that will be used as examples for this presentation. One field (Field A) was in Jasper County and the other (Field B) was in Benton County. Corn was planted with conventional tillage in Field A and with no-tillage in Field B. Field B was managed with no-tillage during the last 6 years. The previous crop was soybean for both fields. Nitrogen, P, and K fertilizers were fall applied in Field A. Field B received starter fertilizer (in selected strips) at planting and sidedressed N fertilizer (146 kg N/ha) when the corn was at the four-leaf stage. Corn was planted in both fields in rows spaced 0.75 m at rates of 66,600 and 71,600 seeds per ha for Fields A and B, respectively. The size of the experimental area was 11.5 ha in each field. These areas were located at least 30 m from fences and end rows.

Two treatments were tested in Field A: a new herbicide (a premix of halosulfuron and acetochlor) and a conventional herbicide (a premix of atrazine and acetochlor). The treatments were applied to 13.5-m wide and 247-m long strips, and were replicated 10 times across the experimental area following a completely randomized block design. To avoid border effects and odd number of combine trips only the center 9-m of each strip was used for the analyses. Three treatments were tested in Field B: the same two herbicides (new and standard) used in Field A plus starter fertilizer (a 7-21-7 N-P-K mixture applied at a rate of 46.7 L/ha) and the standard herbicide without starter fertilizer. The treatments were applied to 9-m wide and 274-m long strips, and were replicated eight times across the experimental area following a completely randomized block design.

Table 1. Descriptive statistics for soil tests and other variables measured on two cornfields by grid sampling.

Field	Variable†	Mean	Minimum	Maximum	STD
A	P (mg/kg)	43	16	86	16.8
	K (mg/kg)	169	111	302	34.5
	pH	6.4	5.6	7.0	0.3
	Org. matter (%)	3.7	2.6	4.8	0.5
	N-NO$_3$ (mg/kg)	6	2	28	4.4
	Ca (mg/kg)	2102	1360	3200	328.9
	Mg (mg/kg)	198	48	456	81.7
	Corn height (cm)	87	39	112	12.1
	BL control (%)	100	96	100	0.8
	GR control (%)	99	92	100	1.7
B	P (mg/kg)	30	7	62	12.3
	K (mg/kg)	136	84	239	31.5
	pH	6.5	5.8	7.2	0.3
	Org. matter (%)	4.4	3.0	7.8	0.6
	N-NO$_3$ (mg/kg)	11	5	20	2.8
	Ca (mg/kg)	2499	2080	4200	268.8
	Mg (mg/kg)	290	92	540	66.2
	Corn height (cm)	96	56	116	13.3
	BL control (%)	99	83	100	2.8
	GR control (%)	100	96	100	0.6

† BL = broadleaf weeds and GR = grass weeds (100 = total control).

Several samples and measurements were collected from the experimental areas at planting and during the growing season following a systematic grid-point sampling scheme. The lines of the grid were spaced 27 m. The grid lines and the strips of the experimental designs were arranged so that sampling areas were always located over the same treatment (the new herbicide in Field A and the standard herbicide without starter fertilizer in Field B). The sampling area at each point was 20-m^2 in size. Composite soil samples (ten cores from a 15-cm depth) were collected from each sampling area. The soil samples were analyzed at the Iowa State University Soil Testing Laboratory for N-NO$_3$, P by the Bray-1 method, K, Ca, and Mg by the ammonium acetate method, organic matter by the Walkley Black method, and pH. Estimates of weed control (visual rating using a scale from 0 to 100) and corn height (measured with a ruler) were recorded from each sampling area 60 days after planting. Table 1 shows some descriptive statistics for these measurements.

Corn yields were measured and recorded using yield monitors equipped with real-time differential global positioning systems (DGPS) receivers. The grain in Field A was harvested with a six-row John Deere 9500 combine equipped with an impact flow rate sensor (Ag Leader 2000, Ag Leader Technology, Ames, IA) and a Leika DGPS receiver (Leika, Torrance, CA). In this case, the differential correction was obtained through the U.S. Coast Guard AM signal. The grain in Field B was harvested with a six-row New Holland TR-97 combine equipped with a flow rate sensor similar to that used in Field A and a Rockwell DGPS receiver (Rockwell International, Cedar Rapids, IA). In this case, the differential correction was achieved with an FM signal (DCI, Cupertino, CA).

The yield data set was subjected to several quality control procedures. The spatial accuracy was checked by georeferencing several key positions in the field with a hand held DGPS receiver (ShortStop, Monsanto Company, St. Louis, MO) and cross checking those points with the yield data location produced with the DGPS systems mentioned before. Also, the data set was cross referenced with georeferenced agronomic observations performed through the growing season. Because the experimental areas were located in the center of the fields, the yield data were unaffected by border effects.

The yield data from each experimental unit (i.e., each strip) were averaged for small areas defined by the grid soil-sampling scheme and the combine trips. Although the sampling technique followed a point sampling scheme, the soil test results and other measurements were assigned to 729-m^2 cells. The width of each cell (across the direction of crop rows) was equivalent to six combine trips, and it had a length of 27 m. The area of each strip (the smallest unit receiving a treatment) included two combine trips and each trip covered a 4.5-m wide area (six corn rows). In turn, each combine trip included ten subdivisions whose length coincided with the length of the grid-sampling cells. Thus, the yield data analyzed consisted of 20 values for each strip, each value representing the average yield of 121.5-m^2 areas. The two trips of the combine were considered as samples within experimental units and the subdivisions along each trip were considered as subsamples. The individual data points recorded by the yield monitors were not directly considered for the statistical analyses because of the known lack of precision of yield monitors over very short distances.

The data were processed by using a combination of numerical, graphical, and Geographical Information Systems (GIS) tools (spreadsheets, AutoCAD, and ARC/INFO). Maps representing all layers of information (soil-test values, evaluations of weed control and corn height, and yield data) were generated using ARC/INFO and ArcView. Digitized and georeferenced soil type maps (Natural Resources Conservation Service soil mapping units) were added as another layer of information. Tables with data suitable for statistical analysis were exported for analysis with the Statistical Analysis System package (SAS Institute, Inc., 1988). The statistical analyses performed included analyses of variance for completely randomized block designs, analyses of components of variance, and regression analyses.

RESULTS AND DISCUSSION

The results showed that halosulfuron and atrazine herbicides had similar effects on weed control and corn yields at both fields. Weed control was excellent for both herbicides even though the fields represented different tillage systems (conventional and no-tillage) and had markedly different management histories. Halosulfuron, however, is a more active compound that requires less active ingredient per unit area and has highly positive toxicological and environmental profiles. In Field A, corn yields were 8.35 Mg/ha for halosulfuron and 8.30 Mg/ha for atrazine. In Field B yields were 7.27 Mg/ha for halosulfuron and 7.33 Mg/ha for atrazine.

The starter fertilizer applied to Field B increased corn yields significantly compared with the control. Mean yields were 7.03 Mg/ha for the no-starter treatment and 7.33 Mg/ha for the starter treatment. The results suggest that starter fertilizer was advantageous to corn under the wet and cool conditions of spring in 1995.

Because on-farm trials usually involve larger experimental units than conventional small-plot trials, it is of interest to estimate the contribution to the overall variability of different parts of the field. Georeferenced data collected using grid sampling and yield monitors are useful to achieve this objective. Components of variance were estimated for categories of yield data referred to as strips, samples and subsamples (see methods), and are shown in Table 2.

Table 2. Analyses of variance of corn yields (Mg/ha) for two strip trials.

Field	Source†	Degrees of freedom	Mean square	$P>F$	Components of variance
A	Blocks	9	0.2706		
	Treatments	1	0.2041	0.149	
	Strips	9	0.0819		-0.003
	Samples	20	0.1358		-0.001
	Subsamples	360	0.1493		0.149
B	Blocks	7	1.8778		
	Treatments	2	3.9666	0.005	
	herbicides	1	0.3244	0.002	
	starter	1	7.1484	0.437	
	Strips	14	0.5062		0.004
	Samples	24	0.4165		0.016
	Subsamples	432	0.2531		0.253

† The mean squares of strips (the blocks by treatments interaction) was the error term for F tests of treatment effects. See methods for the meaning of strips, samples, and subsamples. The contrasts for herbicides and starter effects in Field B are nonorthogonal.

The "strips" component estimates the average variance among strips across the field (20 in Field A and 24 in Field B) after removing block and treatment effects. The "sample" component estimates the average variance among samples (two combine trips) within strips. The "subsample" component estimates the average variance among subsamples (10 areas 121.5-m^2 each along the direction of the rows) within combine trips and strips. The mean variation across strips and combine trips was negligible compared with the variation along the strips (after removing block and treatment effects). This very low variability among strips and combine trips together with the relatively few degrees of freedom explain the small negative estimates of variance in Field A. It is remarkable that most of the yield variation occurring in these fields was due to yield differences along the strips.

Part of the yield variability along the strips could be the result of variation in soil characteristics. For this reason, we attempted to estimate how treatment effects would change for different areas of the fields having relatively uniform soil test values. Two procedures were used, both of which were based on data from the grid-sampling cells. Each of the grid sampling cells included a complete replication of the treatments. In one procedure, the yield data for each field was divided into three classes according to the soil-test values (three groups for each nutrient). The criteria for creating the soil-test groups included agronomic considerations and the inclusion of at least three complete replications per group. Treatment effects were assessed by analyses of variance (one for each nutrient) which considered these soil-test groups and the interaction of the treatments with soil-test group as additional sources of variation. The groups were considered as repeated measures within the experimental units of the main treatments. A significant "soil-test group" main effect suggests that crop yields differ for areas having different soil-test values. A significant interaction between the main treatments and the soil-test groups suggests that the effects of the applied treatments on corn yields differ for areas of the field having different soil-test values.

In the other procedure, treatment effects were assessed by analyses of variance (one for each nutrient) which included two additional sources of variation. One was a continuous variable representing soil-test values of a nutrient. The other was the interaction of the main treatments with the continuous variable. The model and F tests were based on sequential sums of squares and estimates for the continuous variables were obtained after removing block and treatment effects. A significant "continuous soil-test variable" suggests that crop yields differ for areas having different soil-test values, and that this relationship could be represented by a linear equation. A significant interaction between the main treatments and the continuous soil-test variable suggests that yield differences between the applied treatments differ for areas having different soil-test values. Both procedures essentially achieve the same objective but are based on slightly different assumptions, express the results in different ways, and the interpretation may differ.

The results of these two procedures showed that corn yields did not differ among different parts of Field A having different soil test values and that the lack of treatment effects shown in Table 1 would be consistent over the entire field. For Field 2, however, both procedures showed that corn yields differed among areas having different values of soil P, pH, Ca, and Mg. The statistical significance of treatment differences for these areas did not differ, which was indicated by

Table 3. Relationships between grain yields and soil-test values for two cornfields.

Soil test	Field A		Field B	
	Coefficient†	P>F	Coefficient	P>F
P (mg/kg)	-1.61	0.22	-13.21	0.001
K (mg/kg)	-0.44	0.54	1.16	0.214
pH	-5.14	0.96	-293.15	0.001
Org. matter (%)	-30.84	0.57	-45.10	0.286
N-NO$_3$ (mg/kg)	3.33	0.56	8.51	0.389
Ca (mg/kg)	-0.04	0.66	0.23	0.019
Mg (mg/kg)	-0.27	0.38	1.27	0.010

† Linear regression coefficients after removing block and treatment effects.

nonsignificant interactions of the treatments with the soil-test groups or the continuous soil-test variable. Table 3 shows the linear regression coefficients of relationships between grain yields and soil-test values after removing block and treatment effects.

As an example of the procedure based on soil-test groups as repeated measures, Table 4 shows the mean yields of each treatment for areas of Field B having different soil-test P, pH, Ca, and Mg (the variables that were significantly related with yields). The yields were lower or higher in different parts of the field but the relative differences between the treatments were similar. It is obvious that the significant linear relationships between yields and some soil tests shown in Table 3 are not very clear for groups of low, medium, or high soil-test values for some nutrients shown in Table 4 (for example, for soil pH). This apparent disagreement between the procedures could be expected, however, because of the arbitrary criteria used in creating the groups. These differences should be considered when interpreting the results.

The overall results of analyses to study how treatment differences changed for different areas of the fields show that, for these fields and this set of treatments and measured variables, large variation in some site variables did not necessarily affect yields or differences between the tested treatments. Moreover, some of the apparent significant relationships should not be interpreted as "cause and effect" relationships. For example, corn yields were significantly lower for parts of the field with high soil-test P (Tables 3 and 4) but this result seems absurd because it is very unlikely that even the highest values observed for this variable (Table 1) would inhibit corn yields. Similarly, it is unlikely that the differences observed in soil-test Ca would affect yields because they were higher than levels normally considered adequate for corn. It is likely that high levels of these variables are related with other nonmeasured variables that were actually responsible for the yield reductions. Although these conclusions may surprise many, they are consistent with results of multivariate analyses of corn yields and many site variables that we (Mallarino et al.) present in this publication for other fields, and with other research (Pierce et al., 1994).

Table 4. Mean corn yields for areas of Field B having different soil-test values for
nutrients that were significantly related with yields.

Soil test	Group†	Halosulfuron and starter	Atrazine and starter	Atrazine and no starter	Mean
		---------------------------- Mg/ha ----------------------------			
P	<21	7.40	7.46	7.21	7.36
	21-30	7.34	7.34	7.09	7.25
	>30	7.15	7.27	6.90	7.11
pH	<6.5	7.34	7.46	7.09	7.30
	6.5-7.0	7.15	7.15	6.96	7.09
	>7.0	7.27	7.40	7.15	7.27
Ca	<2400	7.27	7.40	6.90	7.19
	2400-	7.15	7.21	7.02	7.13
	>2700	7.53	7.46	7.21	7.40
Mg	<280	7.34	7.34	6.96	7.21
	280-350	7.21	7.27	7.02	7.17
	>350	7.46	7.46	7.27	7.40

† See methods for criteria used to create the groups.

Of course, these negative or positive relationships could be fortuitous (could be just random results for this field). They also could be real, however, and this shows the potential advantages of site-specific crop management for comparing . Other complex statistical analyses (including multivariate analysis and spatial statistical methods) could help understand the relationships between correlated variables and yields. It is apparent, however, the one very useful way of knowing if the levels of a particular nutrient are directly affecting yields in a field is by applying replicated fertilization treatments that include at least a control and a rate deemed sufficient to achieve maximum yields.

CONCLUSIONS

The results showed that a combination of traditional on-farm strip trials and intensive collection of data using precision farming technologies can be used to obtain more complete information about management practices. The large variability usually observed within fields could be used to achieve many combinations of environments and treatments which, in theory, would allow for better understanding of the effects of agronomic treatments on crop yields. As shown by the two examples discussed in this presentation, however, this possibility would not always mean that the measurement of many variables and detailed measurement of crop yields will be useful to study agronomic practices and does not preclude the need of replications in many fields and years.

ACKNOWLEDGMENTS

The authors want to thanks Daniel VanSteenhuyse and Dennis Gardner for their cooperation with the field work. Special thanks to Todd Norton from Rockwell International for providing data for this study.

REFERENCES

Pierce, F.J., D.D. Warncke, and M.W. Everett. 1994. Yield and nutrient variability in glacial soils of Michigan. p. 133-150. *In* P.C. Robert et al. (ed.). Proceedings, 2nd. Int. Conf. on Site-Specific Management for Agricultural Systems. March 27-30, Minneapolis, MN. ASA, SSSA, ASA. Madison, WI.

Rzewnicki, P.E., R. Thompson, G.W. Lesoing, R.W. Elmore, C.A. Francis, A.M. Parkhurst, and R.S. Moomaw. 1988. On-farm experiment designs and implications for locating research sites. Am J. Alt. Agric. 3:168-173.

Shapiro, C.A., W.L. Kranz, and A.M. Parkhurst. 1989. Comparison of harvest techniques for corn field demonstrations. Am. J. of Alt. Agric. 4:59-64.

SAS Institute. 1988. SAS/STAT User's Guide, Release 6.03 Edition. 1028 pp. SAS Institute Inc., Cary, NC. SAS Institute Inc., 1988.

The Impact of Variable Rate N Application on N Use Efficiency of Furrow Irrigated Corn

G.W. Hergert
University of Nebraska WCREC
North Platte, NE

R.B. Ferguson
University of Nebraska SCREC
Clay Center, NE

C.A. Gotway
University of Nebraska
Lincoln, NE

T.A. Peterson
Pioneer Hi-Bred International
West Des Moines, IA

ABSTRACT

Nitrogen management strategies that enhance nitrogen use efficiency (NUE) and reduce nitrate leaching are required for furrow irrigated corn. Variable rate N application could improve NUE and reduce leaching by minimizing high nitrate areas in the field. Research was initiated in 1994 to establish relationships among soil chemical and physical properties, irrigation parameters, grain yield, soil nitrate and NUE for furrow irrigated corn which received uniformly applied ammonia, variably applied ammonia or variably applied ammonia minus 15%. Soil sampling showed highly variable soil nitrate at each site. Significant grain yield differences were shown between uniform versus variable rate. Both variable rate treatments produced larger decreases in residual soil nitrate and variability than the uniform N rate. NUE was high for all treatments in 1994 and no differences were shown. In 1995 variable rate minus 15% had significantly higher NUE than the uniform N application. The research indicates that VRT can increase NUE and reduce nitrate variability by creating more uniformly low nitrate levels in the field.

INTRODUCTION

Because N is the major limiting nutrient is most soils for grain production it will continue to be used to sustain production (Mengel, 1990; Swanson, 1982). Current environmental and economic considerations, however, demand improved N use efficiency (NUE). This is especially true for furrow irrigated soils in river valleys in Nebraska which have shown increasing nitrate levels since the mid 1950's (Engberg and Spalding, 1978). Soils in these valleys are generally shallow with coarse textured subsoils and groundwater 2 to 3 m beneath the soil surface. Much of this land is cropped to continuous corn (*Zea mays* L.). Progress has been made in many of these areas in Nebraska to reduce groundwater nitrate (Ferguson, et. al., 1994) due to the adoption of Best Management Practices (BMP's) based on University research and the creation of management areas monitored by Nebraska's Natural Resource Districts.

Nitrogen management strategies are needed that enhance NUE and reduce nitrate leaching under furrow irrigated corn. Variable rate N application holds the promise of improving NUE and reducing leaching by minimizing high nitrate areas in the field that are subject to leaching when over-irrigation occurs or if rainfall is above normal. Although increased adoption of BMPs has resulted in more efficient N fertilizer use, NUE in irrigated corn is often 50% or less. Accounting for the variability of N requirements within a field should allow for more efficient fertilizer N use, improve NUE and reduce potential loss of nitrate to the environment. Research in Nebraska (Hergert, et. al., 1995b) investigated the spatial variability of nitrate-N in 43 cultivated fields across the state. Samples were collected on a 30 m grid to a depth of 1.2 m and showed an average coefficient of variation (CV) of 52% for cumulative nitrate-N. This degree of variability suggests that further increases in NUE could be attained if N fertilizer rate were varied across the field with variable rate application.

Inherent inefficiencies associated with furrow irrigation tend to accentuate nitrate leaching in many situations. A typical infiltration profile in a furrow irrigated field is illustrated in Fig. 1. Depending upon furrow length, the upper end of the field is over-irrigated and the lower end is under-irrigated, consequently more leaching occurs at the upper end of the field. Root zone and intermediate vadose zone samples collected from the mid-Nebraska water quality demonstration project show a similar influence of furrow irrigation on accumulated nitrate (Ferguson, et. al., 1992).

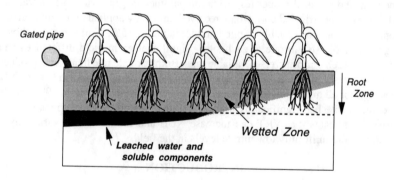

Fig. 1 A typical furrow irrigation infiltration profile.

In Nebraska about half of the 1.7 million hectares of irrigated corn are grown under furrow irrigation. Almost 75% of the N fertilizer applied to corn is from anhydrous ammonia (1994-1995 Nebraska Agricultural Statistics, 1995). A variable rate anhydrous ammonia applicator built by SoilTeq® has been used for research and demonstration plots in Nebraska since 1992.

A research project was initiated in 1994 with the objective of establishing relationships among soil chemical and physical properties, irrigation parameters, grain yield, soil nitrate and NUE for furrow irrigated corn which received uniformly applied ammonia versus variably applied ammonia. This report will present general results from three locations during 1994 and 1995 and specific detailed information from one of the sites on the influence of variable rate N application on a spatial N balance and apparent NUE.

EXPERIMENTAL METHODS

Three variable rate N experiments were initiated in 1994 on furrow irrigated sites. One was in west central Nebraska (Lincoln county) on a Cozad silt loam (coarse, silty, mixed mesic Fluventic Haplustoll), one in south central Nebraska (Clay county) on a Crete silt loam (fine, montmorillonitic, mesic Pachic Argiustoll) and one in central Nebraska (Buffalo county) on an Alda loam (coarse, loamy, mixed, mesic Fluvaquentic Haplustoll). A typical N treatment application map is shown in Fig. 2. Three N management regimes replicated five times were used: (1) fixed uniform rate N based on an average expected yield goal for the area, average soil organic matter content, an average soil nitrate of the plot area using the UNL N recommendations (Hergert, et. al., 1995a); (2) variable rate N based on an average expected goal, varying soil nitrate, and soil organic matter determined from grid sampling and (3) variable rate N application minus 15%.

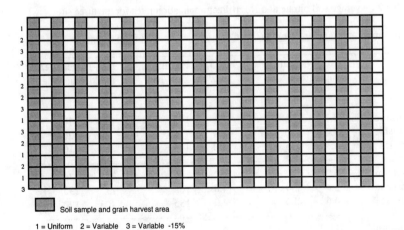

Soil sample and grain harvest area

1 = Uniform   2 = Variable   3 = Variable -15%

Fig. 2. Experiment map for N treatment, soil sampling and grain harvest used at the three locations.

Treatment strips were one planter width and ranged from 5 to 6 m wide. Field length was 280 m at the Lincoln county site, 503 m at the Clay county site and 312 m at the Buffalo county site. A transect of soil samples was taken down the middle of each of the treatment strips using a fixed spacing ranging from 20 to 30 m depending on the field. Grain was combine harvested from 3 rows 10 m long at site 1 and a subsample of grain was taken for N analysis (Fig. 2). The soil samples were from the middle of the area harvested for grain. The Clay and Buffalo county sites were harvested with an 8-row John Deere® 9600 combine equipped with an Ag-Leader® Yield Monitor 2000. Grain subsamples for N were not taken. Initial average soil test parameters based on samples for the 3 sites given in Table 1. The variability in residual nitrate was high (Fig. 3). Anhydrous ammonia was applied at the 6 to 8 leaf stage at all sites both years. Average N rates applied for the 3 treatments are shown in Table 2.

Table 1. Selected soil analyses for the three sites.

Location	Texture	Samples	pH	OM	Bray 1P	DTPA Zn
				g/kg	———— mg/kg ————	
Lincoln Co.	sil	225	8.0	15	20	2.0
Clay Co.	sil	255	6.1	27	15	1.0
Buffalo Co.	sil	158	8.1	19	24	0.9

Table 2. Average soil nitrate and N fertilizer application rates for the three sites.

N treatment	Lincoln County		Clay County		Buffalo County	
	NO_3-N mg/kg	Fertilizer kg N/ha	NO_3-N mg/kg	Fertilizer kg N/ha	NO_3-N mg/kg	Fertilzer kg N/ha
-- 1994 --						
Uniform	5.7	165	4.1	165	10.0	110
Variable	7.4	160	3.9	170	8.4	120
Var -15%	7.1	135	4.0	145	9.6	110
-- 1995 --						
Uniform	5.2	270	5.2	160	7.4	130
Variable	5.4	250	4.8	165	7.0	135
Var -15%	4.6	220	4.9	135	8.3	120
-- 1996 --						
Uniform	11.9	-	4.5	-	6.9	-
Variable	10.6	-	3.9	-	6.9	-
Var -15%	9.9	-	3.9	-	6.1	-

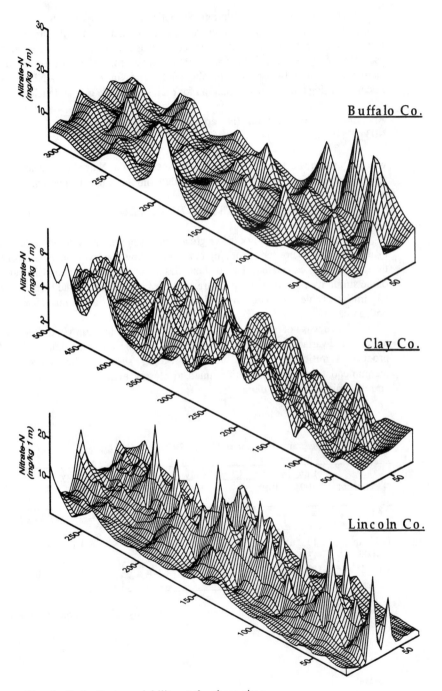

Fig. 3. Soil nitrate variability at the three sites.

The influence of variable rate N application on NUE was measured in several ways. The first was a comparison of the change in soil nitrate from spring to fall for the 3 treatments. The second was a comparison of apparent N use efficiency (ANUE) values. Data were available from all 3 sites for the first comparison but only for the Lincoln county site for ANUE. ANUE was calculated using the following equation for each soil sample grid and yield point:

$$ANUE = (N\ uptake(grain)/(N\ applied + residual\ NO3_{spring} - residual\ NO3_{fall}))\times100\%,$$

where N uptake was total N removed in harvested grain, N applied was applied fertilizer and $NO3_{spring}$ and $NO3_{fall}$ were residual nitrate-N in a 1 m depth.

RESULTS AND DISCUSSION

Nitrogen rates did not differ greatly between uniform and variable rate (Table 2). This is not too surprising considering how the 'average' N rate for a treatment strip is calculated. The mathematics of simple averaging explain much of the similarity especially when the treatment average initial nitrate levels were so similar (Table 2). The lower N from the variable rate minus 15% was noticeable.

There was not a consistent significant difference in grain yields between uniform and variable rate N application (Table 3). The variable minus 15% did produce significantly lower yields, however, showing that current N recommendations have been fine-tuned to closely match N requirements of current grain production demands.

Table 3. Average grain yields for 1994 and 1995.

N treatment	Lincoln County		Clay County		Buffalo County	
	1994	1995	1994	1995	1994	1995
						Mg/ha
Uniform	13.2 a[1]	11.1 b	10.3 a	9.0 a	6.9 a	4.4 b
Variable	13.1 a	11.3 a	10.4 a	9.0 a	6.9 a	4.9 a
Var - 15%	13.1 a	11.1 b	10.1 b	8.9 b	6.9 a	4.4 b

[1]Means followed by the same letter are not significantly different at the 5% level of probability.

Table 4. Influence of N treatment on change in soil nitrate (0-1 m) at the <u>three</u> sites.

N treatment	Lincoln County 93-94	94-95	Clay County 93-94	94-95	Buffalo County 93-94	94-95
				kg/ha	in	1 m
Uniform	-6	+78	+11	-8	-31	-7
Variable	-25	+63	+11	-4	-18	-1
Var -15%	-30	+64	+11	-7	-27	-24

The change in residual soil nitrate as influenced by treatments showed an encouraging pattern (Table 4). In 1994 at the Lincoln county site, both variable rate treatments produced significantly lower nitrate than the uniform N treatment. During 1995, N was over-applied due to a malfunction of the radar gun on the applicator. The over-application was 60 kg ha^{-1} for the uniform application, 44 kg ha^{-1} for the variable and 36 kg ha^{-1} for the variable minus 15% application. The higher rates and lower than expected yields were reflected in increased soil nitrate (Table 4).

At the Clay County site there was little change in soil nitrate either year, but since the initial level was low, the treatments reflect the influence that proper N management can have on keeping residual nitrate levels low (Table 2). At the Buffalo County site there was a significant decrease in residual nitrate in 1994 with the uniform treatment showing the largest decrease. In 1995 the variable rate minus 15% was significantly lower than the other two treatments and nitrate levels continued to decline (Table 2). With improved irrigation and N management a goal of decreasing residual nitrate levels to 3 to 4 mg kg^{-1} uniformly throughout the field should help reduce nitrate leaching significantly.

To determine whether variable rate N application can accomplish the goal more effectively than a uniform rate, the frequency distributions and simple statistics of soil nitrate can be analyzed. Data for the Lincoln county site are shown in Fig. 4. Nitrate distributions were less variable in the fall of 1994 compared to the fall of 1993 especially following the variable minus 15% rate application. Even with the over-application of N in 1995, the variation of both variable rate applications was significantly lower than that of uniform application again providing some evidence that variable rate is decreasing 'hot spots' of nitrate.

Apparent NUE values for the experiment were high (Table 5). Growing conditions during 1994 were excellent and yields and N removed in grain were high (Table 3). Yields in 1995 (and N removed in grain) were good, but the over-application of N decreased ANUE. The values still reflect high NUE values, however. The recovery of residual nitrate in the fall 1995 sampling helped the overall NUE, especially for the variable minus 15% application (Table 5). Since N applied for the variable minus 15% treatment was closer to an optimum N rate than other treatments, the effect on NUE was evident. The data

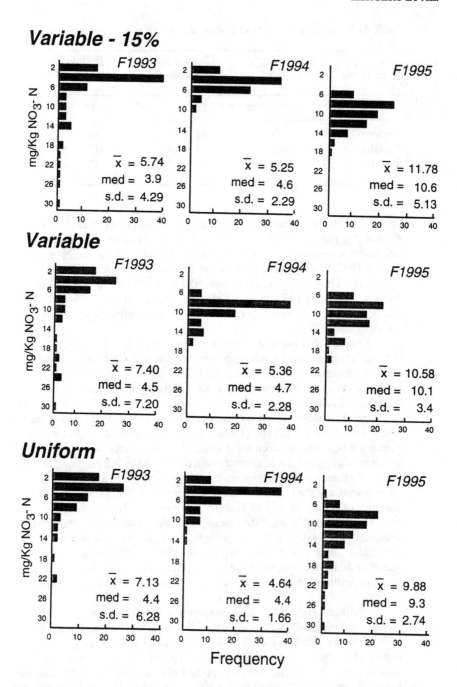

Fig. 4. Frequency distributions of soil nitrate at the Lincoln county site for the three N management treatments.

Table 5. Apparent N use efficiency for treatments at the Lincoln County site.

N treatment	1994	1995
	———— % ————	
Uniform	88.8 a[1]	64.4 b
Variable	84.2 a	69.7 b
Var -15%	88.4 a	80.0 a

do confirm the potential that variable rate N application has in improving NUE, decreasing long-term potentially leachable nitrate and decreasing nitrate-N levels in Nebraska ground water.

REFERENCES

Engberg, R.A., and R.F. Spalding. 1978. Groundwater quality atlas of Nebraska. Resource Atlas No. 3, Cons. and Survey Div., Univ. of Nebraska.

Ferguson, R.B., A.P Christiansen, D. Gosselin, M. Kuzila, E. Barnes, and T. Murphy. 1992. Root zone and intermediate vadose zone nitrate accumulation as influenced by nitrogen and irrigation management. Agron. Abstr., pp.277.

Ferguson, R.B. J.S. Schepers, G.W. Hergert, T.A. Peterson, J.E. Cahoon and C.A. Gotway. 1994. Variable Rate Nitrogen Application for Irrigated Agriculture: Opportunities for Groundwater Protection. pp. 207-216. National Symposium on Protecting Rural America's Water Resources: Partnerships for Pollution Solutions. Wash. D.C. Oct. 1994.

Hergert, G.W., R.B. Ferguson and C.A. Shapiro. 1995a. Fertilizer Suggestions for Corn. NebGuide G74-174 (Revised). Univ. NE Coop. Ext. Serv.

Hergert, G.W., R.B. Ferguson, C.A. Shapiro, E.J. Penas and F.B. Anderson. 1995b. Classical statistical and geostatistical analysis of soil nitrate-N. pp. 175-186. In P. Robert et al (ed.) Site-Specific Management for Agricultural Systems--Second Intl. Conf. Minneapolis, MN. Mar, 1994. ASA, CSSA, SSSA. Madison, WI.

Mengel, K. 1990. Impacts of intensive plant nutrient management on crop production and environment. In Trans. 14th International Congress of Soil Science, Plenary Papers, Contents, Author Index: 42-52. Intl. Soc. Soil Sci, Kyoto, Japan.

1994-1995 Nebraska Agricultural Statistics. 1995. NE Dept. of Agriculture. NE Agricultural Statistics Service. Lincoln, NE.

Swanson, E.R. 1982. Economic implications of controls on nitrogen fertilizer use. pp. 773-790. In F.J. Stephenson (ed.) Nitrogen in agricultural soils. Agronomy 22. ASA, SSSA.

Properties of Polyolefin Coated Urea (MEISTER) and Programmed Nitrogen Supply to Agricultural Plants

Sadao Shoji
5-13-27, Nishitaga
Taihaku-ku, Sendai, Japan

Nobumasa Kosuge
Chisso Corporation
7-3, Marunouchi, 2-Chome
Chiyoda-ku, Tokyo, Japan

Sayoko Miyoshi
Chisso Corporation,46-70
Sakinohama, Nakabaru
Tobata-ku, Kitakyushu, Japan

ABSTRACT

MEISTER is one of the excellent controlled availability fertilizers. It is granulated urea coated by a resin with controlled water permeability and contains urea-N of 40 %. Its dissolution is primarily temperature-dependent so that it is accurately predicted. Programmed nitrogen supply to agricultural plants is conducted using a computer according to the following information; 1) an amount of fertilizer nitrogen to supplement the shortage of natural N uptake by the plant to obtain the target yield, 2) recovery of MEISTER N by the plant, and 3) mean soil or air temperature data. The programmed MEISTER application can contribute to innovating fertilization methods and farming systems, and controlling agro-environmental degradation.

MEISTER developed by Fujita et al. (1977; 1983), is one of the excellent controlled availability fertilizers. It is divided into two groups; ordinary (linear) dissolution and delayed (sigmoid) dissolution. Its dissolution in the soil is primarily temperature-dependent so that it is accurately predicted by a computer using temperature data(Gandeza et al.,1991).

Programmed nitrogen supply meeting the plant nitrogen demand can be achieved by selecting a single or blended MEISTER products whose dissolution characteristics are synchronized to plant uptake (Shoji et al.,1991). In Japan the programmed MEISTER application has contributed to innovating fertilization methods and farming systems and increasing consumption of MEISTER products in agricultural plants (70 % in paddy rice). On the other hand, the vast majority of controlled availability fertilizers is used in non-agricultural sectors in USA (Trenkel,1996).

PROPERTIES OF MEISTER PRODUCTS

MEISTER is granulated urea (diameter; about 3.5 mm) coated by a resin with controlled water permeability and contains urea-N of 40 %. The resin coating is a blend of polyolefin-type resin (PO), ethylene vinyl acetate (EVA) and talc mineral. It is approximately 60 um in thickness and occupies 10 % of total weight of MEISTER(Fujita et al.,1977; 1983)

Dissolution rate of MEISTER products is controlled by water permeability of the resin coating by changing the ratio of water-impermeable PO to water-permeable EVA. For example, when the ratios are 50:50 and 100:0, the duration of dissolution (number of days to dissolve 80 % of urea-N from MEISTER particles in water at 25°C) are approximately 100 and 1300 days, respectively.

Temperature dependency, Q_{10}, is also important for programmed nitrogen supply to agricultural plants. For example, if this value is too large, the dissolution rate is too high resulting in excess nitrogen supply to the plants in hot summers. Therefore, the Q_{10} value of 2 was selected to meet the chemical reactions occurring in the plants and microbial activity in the soil.(Q_{10}=2 means that the rate of MEISTER dissolution doubles for every 10 C rise in temperature). This objective is achieved by incorporation of platy talc mineral to the resin. MEISTER with Q_{10} value of 2 can be produced by adding talc to the resin in the range of 50-60 % by weight. It is considered that water penetrates through voids formed around the talc plates and this process(detour water movement)can govern the temperature dependency.

Recent development of MEISTER with delayed or sigmoid dissolution property has stimulated innovation of fertilizer application methods and farming systems in Japan. The sigmoid dissolution is controlled by adding some organic compounds to the resin.

PROGRAMMED NITROGEN SUPPLY TO AGRICULTURAL PLANTS

MEISTER products show dissolution characteristics such as linear and sigmoid dissolution, temperature dependency of dissolution, and variations in dissolution rate or duration of dissolution. Thus in order to obtain a target yield of an agricultural plant programmed nitrogen supply by single basal fertilization is enabled by selection of most suitable MEISTER products. It is processed using a computer according to the following information:

1. An amount of fertilizer nitrogen to supplement the shortage of natural nitrogen absorbed by the plant at each growth stage,
2. Recovery of MEISTER nitrogen by the plant at each growth stage, and
3. Mean soil or air temperature data.

For example, the amounts of fertilizer nitrogen needed by dent corn to obtain the yield of 6 ton/ ha(seeds) at Tohoku University Farm are 10, 74 and 12 kgN/ha at the early, middle and late growth stages, respectively (total nitrogen absorbed by the plant is 160 kg N/ha). Recoveries of MEISTER nitrogen are 60 % at the early growth stage and 80 % at the middle and late growth stages. Mean soil or air temperature data for every 10 days is generally used. A blend of a linear dissolution product with short duration (for the early growth stage) and a sigmoid dissolution product with long duration (for the middle and late growth stages) is mostly suitable for the programmed nitrogen supply not only to corn but also to most agronomic crops.

Dissolution rate of MEISTER under field conditions is accurately predicted using soil or air temperature (Gandeza et al.,1991). Therefore, it is possible to simulate nitrogen uptake from MEISTER by the plant (Shoji et al.,1991).

Accurate controlled dissolution of MEISTER enables placing a large amount of MEISTER with seeds or plant roots. This placement was named "co-situ placement" by Shoji and Gandeza (1992) in order to differentiate from the early years of fertilizer use seed placement or contact placement using a small amount of conventional fertilizer to give the crop a rapid start (Randall & Hoeft ,1988). The central concept of co-situs placement is to apply nutrients in the intensive rooting zone with dissolution pattern of fertilizers synchronized to the plant demand over the whole growing season. This new placement has been effectively used for rice farming in Japan.

The co-situ placement using MEISTER remarkably increased recoveries of fertilizer nitrogen by rice. For example, the recovery of basal nitrogen can be increased from 20 -30 % with broadcast application of conventional nitrogen fertilizers to 70-80 % with co-situ application of MEISTER (Shoji & Kanno,1994).

INNOVATIVE FARMING SYSTEMS USING MEISTER

MEISTER products with excellent dissolution properties have been contributing to innovative farming systems as listed below (Shoji & Kanno, 1994); For lowland rice:
1.   No-till farming by direct seeding and single basal fertilization, and
2.   No-till farming by transplanting of rice seedlings with single basal fertilization to nursery boxes (no extra MEISTER application to paddy).

For upland and horticultural plants:
1.   No-till farming by single basal fertilization (e.g., corn)
2.   Multi-cropping by single basal fertilization (e.g.,lettuce-Chinese cabbage), and
3.   Single basal fertilization to nursery pots to supply nutrients for the whole growing season (e.g., strawberry for 300 days and cyclamen for 200 days).

The new farming systems can significantly contribute not only to decreasing the farming cost but also to controlling agro-environmental degradation. For example, the new no-till transplanting rice culture decreased the farming cost by 65 % as compared to that of the conventional rice culture and notably increased the recovery of fertilizer nitrogen as described before. Maximizing plant uptake of fertilizer nitrogen is considered to be most useful in controlling nitrogen fertilizer pollution.

REFERENCES

Fujita,T., C.Takahashi, M.Ohshima, T.Ushioda, and H.Shimizu.1977. Method of producing coated fertilizers. United States Patent 4,019,890.

Fujita,T., C.Takahashi, S.Yoshida, and H.Shimizu. 1983. Coated granular fertilizer capable of controlling effects of temperature upon dissolution-out rate. United States Patent 4,369,055.

Gandeza,A.T., S.Shoji, and I.Yamada. 1991. Simulation of crop response to polyolefin-coated urea:1. Field dissolution. Soil Sci. Soc.Am. J. 55:1462-1467.

Randall, G.W., and R.G.Hoeft. 1988. Placement methods of improved efficiency of P and K fertilizers: A review. J.Prod. Agric.1:70-79.

Shoji, S., and A.T.Gandeza. 1992. New concept of controlled release fertilization. p 1-7. *In* S. Shoji and A.T. Gandeza (ed) Controlled release fertilizers with polyolefin resin coating. Konno Printing Co., Sendai, Japan.

Shoji, S., and H. Kanno. 1994. Use of polyolefin-coated fertilizers for increasing fertilizer efficiency and reducing.nitrate leaching and nitrous oxide emissions. Fert.Res.39:147-152.

Shoji, S., A.T.Gandeza, and K.Kimura.1991 Simulation of crop response to polyolefin-coated urea: 2. Nitrogen uptake by corn. Soil Sci. Soc. Am. J. 55: 1468-1473.

Trenkel, M.E. 1996. Controlled release and stabilized fertilizers - present situation and outlook for agriculture. FAO. Text.

Yield Mapping; Errors and Algorithms

B. S. Blackmore
C. J. Marshall

Centre for Precision Farming
School of Agriculture Food and Environment
Cranfield University
Silsoe, Bedford. England

Since the advent of differential global positioning systems into the agricultural sector, positional information and yield data from a combine harvester have been used to produce yield maps. These maps are central to the new type of arable management called Precision Farming, which promotes the better use of information to improve the management of variability on the farm. As the yield maps highly influence the decision making process, it is incumbent on the vendors of yield mapping equipment to ensure that the maps do actually represent the variation in the yield and not other systemic errors. Two main errors have been identified in many yield maps; the lag time between detachment and sensing of the grain, which offsets the yield position along the route of the combine, and the unknown width of crop entering the header. This paper sets out to analyze these and other errors and suggests that a technique called Potential Mapping should be used to overcome the unknown crop width error.

YIELD MAP DATA

During harvest, the data required for a yield map is produced by recording the spot yield within a crop as it is harvested and the position of the combine harvester at that same time. The spot yield data is taken from a flow sensor, usually mounted in or around the clean grain elevator. Borgelt, (1992) described various techniques that are currently being used, ranging from volumetric measurement to mass flow, to change in capacitance, each with their own advantages and disadvantages.

The positional data can be retrieved from a number of different systems, but the *de facto* standard at present is the Differential Global Positioning System. Figure 1 shows the positions at which the combine recorded the yield data. (For an in-depth report on DGPS in agriculture, see Computers and Electronics in Agriculture, Vol. 11, Special issue on GPS.)

These two data sets are combined together into randomized triplets and stored as Comma Delineated ASCII or ADIS, with the syntax: Latitude, Longitude and Yield. Some systems may include further data such as date, time, quality of GPS signal etc.

YIELD MAP PRODUCTION

Point data

The first yield maps were produced by classifying and colouring each data point according to the value of the yield. Although accurate, this system treats every data point as independent, which we know is not the case, as yield is spatially related. The data set is also quite dense but it can still lead to areas that are either missed out or the symbols used start to overlap each other (see Fig. 2).

The data that comes directly from the combine is often found to be highly variable, and as such, of limited management use. Hence, some form of smoothing is used to be able to recognise trends across the field. The degree of smoothing required is dependent on the interpretation of the yield map.

Figure 3 shows the usual process involved in preparing a yield map. Firstly, the two data sets, current position and spot yield are brought together on the combine and stored on a secondary storage device, such as a smart card or floppy disk. The data set is then transferred to the PC where fundamental data processing operations occur, such as removal of outliers and coordinate conversion. The data must then be regularised into a grid that can be used for contouring. Finally, expert knowledge is used to interpret the data and represent it in such a form that is useful to the farm manager.

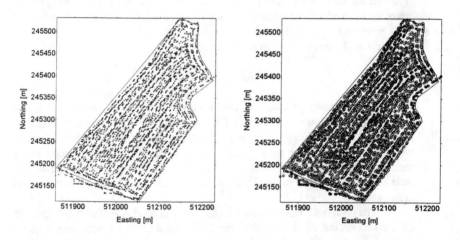

Fig. 1. GPS data.

Fig. 2. GPS data and shaded symbols.

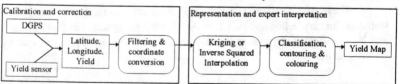

Fig. 3. Yield map data processing.

KRIGING

Figure 4 shows a yield map that has been interpolated by kriging. This is a method of estimation which is defined as a Best Linear Unbiased Estimator (BLUE). It is linear because its estimates are weighted linear combinations of the available data. (Isaaks & Srivastava, 1989)

$$\hat{y}_0 = \sum_{i=1}^{i=n} w_i y_i \tag{1}$$

where y_i, i=1..n are the available sample values and y_0 is the unknown true value. This estimator is unbiased if we set the side condition;

$$\sum_{i=1}^{n} w_i = 1 \tag{2}$$

Using this side condition and Lagrange multipliers μ, we can make the variance a minimum. This minimum error variance is based on $(n+1)$ variables. By partial differentiation with respect to all the variables and equating to zero, the following are obtained.

$$\sum_{j=1}^{n} w_j \tilde{C}_{ij} + \mu = \tilde{C}_{i0} \text{ for } i = 1 \cdots n \quad \sum_{i=1}^{n} w_i = 1 \tag{3}$$

These equations are easily solved once the covariances $\tilde{C}_{ij}$ have been chosen and can be written in the following matrix form.

$$\text{C.W} = \text{D} \tag{4}$$

Where **C** are covariances between the sample points, **W** are the weights and **D** is the covariance of the sample points with the unknown value being estimated. **C** is a measure of the statistical clustering in yield, between samples as opposed to the geometrical distance. **D** is similarly a measure of the statistical closeness between each one of the sample points and the point being estimated.

Hence it is seen that the kriging estimator is a best (i.e. minimum variance) linear unbiased estimator, and as such, will be influenced by already inherent errors in the yield data set. It should be noted that kriging is best suited to sparse data sets such as soil and mineral maps or yield data with missing samples that require interpolation.

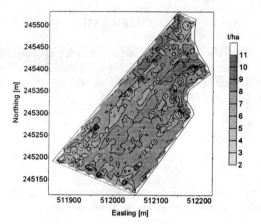

Fig. 4. Kriged contour map representing yield variation

Inverse squared interpolation

For comparison, an inverse distance method would give weights that are inversely proportional to the geometric distance from the unknown position. To generalise this even further the power p of the distance may be used. This gives a whole family of estimates of the following form:

$$\hat{y}_0 = \frac{\sum_{i=1}^{n} \frac{1}{d_i^P} y_i}{\sum_{i=1}^{n} \frac{1}{d_i^P}} \tag{5}$$

All of these estimates are linear and unbiased *but* do not have the added advantage of minimum variance of the kriging estimate. Nor do they account for redundant sample points, but they are still influenced by the inherent errors in the yield data set.

ERROR PROPAGATION

Main errors

There are six main groups of errors that have been identified and their ranking has been intuitively estimated with regard to their effect on the yield maps:
- Unknown crop width entering the header during harvest
- Time lag of grain through the threshing mechanism
- The inherent 'wandering' error from the GPS
- Surging grain through the combine transport system
- Grain losses from the combine
- Sensor accuracy and calibration

Pre and Post harvest accuracy

There are three approaches to improving the accuracy of yield maps.

Firstly, to improve the accuracy of data collection at source. This must be the most preferential action. For yield mapping some of the errors listed above can be accounted for but others are more difficult, due to the nature of the combine and the way in which it is used.

Secondly, use an 'expert filter' for remedial error filtration or correction. An expert filter is an expert system that includes knowledge about the field, combine, crop, GPS, and other characteristics, which assesses the data and filters out those points that an expert would not consider reasonable. e.g. yield points outside the field boundary due to GPS errors. The unpublished thesis by Rands (1995), describes such a system and identifies absolute, relative and trend outliers, in seven distinct classes, each of which have multiple causes and combinations. They are format errors, absolute outliers, distance from last reading, Dilution Of Precision (DOP), numbers of satellites, yield readings and internal errors. Some of these errors can be corrected, others must be deleted from the data set before interpolation occurs.

Thirdly, the use of Potential Mapping to compensate for the unknown crop width error.

Remedial correction

This paper deals with the first two groups of errors and explains the use of Potential Mapping as these errors are seen as the most significant in the yield mapping context. The GPS errors are also significant, but their characteristics have been assessed by Saunders, et. al. (1996).

The data points that reflect the yield values have errors in them because of the unknown width of the crop entering the combine header. This is due to the grain sensor's associated electronics that calculate the output from the combine in tons per hectare. The mass is calculated from the output of the grain sensor over time and the area is deduced from the width of the combine header multiplied by the distance travelled over the same time interval.

Next, the clean grain takes a finite amount of time to work its way through the combine's threshing mechanism until it reaches the grain sensor, usually mounted in the clean grain elevator. Experience has shown that this time lag can be highly variable. (Reitz & Kutzbach 1996)

These harvester errors should be kept to a minimum or even eliminated so as not to detract from the underlying variable trends. Some combines fitted with yield mapping systems have additional features fitted to them to try and over come these errors. Yield recording can start and stop automatically when the header is raised and lowered into work. This overcomes the problem that the combine covers some areas, especially the headlands during transport, more than once. The second feature is a set of buttons in the cab that allows the driver to key in the current width of crop entering the header. Even with this system the resolution is still 1 in 4, or a quarter of the header width. Research by Reitz and Kutzbach (1996) investigated the automatic assessment of crop width by ultrasonic methods, but it believed by the authors that by the use of Potential Mapping, this factor is not needed.

EXAMPLES OF ERRORS

Unknown Crop Width

Each point in Fig. 1 represents the position of the combine at a certain time. At that same time the spot yield is recorded in tons per hectare . The mass (y_T) is calculated by summating the instantaneous mass (y_t) of clean grain passing the sensor during the time interval since the last reading. $(t \in [t_l, t_m])$ over the time period T.

$$y_T = \sum_{t \in [t_l, t_m]} y_t$$

(6)

The distance travelled is calculated by summating the pulses P_t from a proximity switch on the wheel (or some other device)

$$d_T = \sum_{t \in [t_l, t_m]} P_t$$

(7)

The area measurement is calculated by knowing the header width (W_{header}) which is then multiplied by the distance traveled (d_T) since the last reading.

$$A_T = W_{header} \cdot d_T$$

(8)

The crop density is then recorded and multiplied by appropriate constants to give tons per hectare since the last reading.

$$y_i = \frac{y_T}{A_T}$$

(9)

The problem arises when it can be seen that the header does not always have a full width of standing crop entering it. The equation (8) assumes that the header width is always full. If we let W_{crop} be the width of the crop, then any lateral movement by the driver that narrows the crop width (W_{crop}) entering the header will give an inherent error e_{width} proportional to the gap.

$$e_{width} = \frac{W_{header} - W_{crop}}{W_{header}}$$

(10)

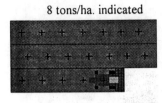
8 tons/ha. indicated

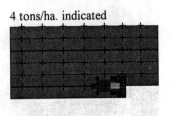
4 tons/ha. indicated

Fig. 5.   a)Header full of crop   b) Header half full of crop

Consider the following example. If a known plot of land of one hectare has on it eight tons of clean grain in a standing crop, then the combine should reflect this by indicating 8 t/ha. This is calculated by measuring the forward distance travelled every 1.2 seconds (giving 1.2m) and multiplying it by the width of the header (5m) to give area covered since last reading.(A = 6m², eqn.8). The mass since the last reading (y_T = 4.8 kg) is then divided by the area covered to give 4.8 kg in 6m². which gives a spot yield of y_i = 0.8 kg/m²=8 t/ha.

Now, this calculation assumes the header to be full at all times. In practice we know that this is difficult to achieve and rarely happens. This leads to an instantaneous error for the data point that does not show cumulatively. i.e. the tonnage from the field is correct.

This can be shown if we consider the combine working in a field with half of the header full of crop at all times. (See Fig. 5)   This gives *twice* as many readings with *half* the yield for each - one error negates the other. But in yield mapping using kriging it is imperative that the instantaneous yields are as accurate as possible.

Let Y_a be the yield from the plot when $\left(w_{header} = w_{crop} \right)$ then from eqn. 6

$$Y_a = \sum_1^n y_i = \sum_1^{10,000} (0.8) \approx 8 \text{ t / ha}$$

but when $\left(w_{header} = 2. w_{crop} \right)$ then we get twice as many points with only half the indicated yield, hence Y_b becomes;

$$Y_b = \sum_1^n y_i = \sum_1^{20,000} (0.4) \approx 8 \text{ t / ha}$$

Fig. 6. Finishing the field.

Thus over the whole field the indicated yield is correct, but the indicated instantaneous yield is only half of what it should be $\left(e_{width} = 0.5\right)$. Thus if this data were kriged then the resulting map would indicate a yield of 4 t/ha.

When the yield map is then interpolated using either kriging or inverse squared methods, it assumes that each data point is correct and it interpolates between them. Once again it can be seen that if the combine header was run at half capacity across the whole field then just by interpolating between these points is going to indicate half of the yield that was actually there. This situation will not occur as the driver is not likely to harvest the whole field at half rate. But what does often happen is when a land is nearly finished the driver must take a final pass down the field with only a small amount of standing crop entering the combine header.(See Fig. 6) This *is* likely to happen on each finish, especially in irregular fields. Therefor if kriging, inverse squared or any other type of straight forward interpolation methods are used a low yielding stripe appears on the yield map where the combine finished the land. (See Fig. 7) These and other misleading artifacts can be seen in the yield map that artificially reduce the apparent yield, even when the actual yield is good.

Many yield maps produced in this way exhibit these characteristic lines of low yield parallel to long field boundaries. In effect these yield maps tell us more about the way in which the field was harvested than about the variation of the yield itself.

Time Lag

Furthermore, many yield maps exhibit a low indicated yield where the combine enters a standing crop.(Fig. 7) These artifacts can be identified by the alternate low spots at each end of the field and are corroborated by isolating the direction of combine travel and harvesting pattern from Fig.1. This artificial feature is because of the incorrect assessment of the time lag within the combine. This is due to the fact that when the combine first starts up the sieves, elevators and other temporary storage spaces are empty and take time to fill up. Even more importantly is the time that it takes for the clean grain to pass through the combine mechanisms

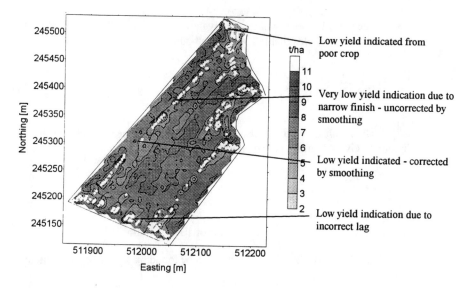

Fig. 7 Example of yield map with outliers superimposed

before reaching the clean grain sensor. As the combine is moving any time lag will appear as a positional shift. This in itself can be easily rectified, by moving the yield back by a set number of positions, but there are still three other problems.

Firstly if the time lag is not assessed correctly then all the yield will be offset by an incorrect amount. Secondly, if the time lag varies depending on the situation, then a constant offset will not be good enough. Thirdly, if the data is only recorded when the header is down, then yield data is lost at the end of each land when the header is lifted while grain is still flowing through the threshing mechanism.

POTENTIAL MAPPING

Potential Mapping is the process of summating data within a known area. For yield mapping, this area can be defined after the harvest from a standard georeferenced source and uses the positional information from the DGPS to locate the spot yield. The yield data itself must be recorded as mass since the last reading (kg), as opposed to mass per area (t/ha). All of the spot yields are then summated and divided by the potential area from the GIS. In this manner the area information can be found without the assumption that the header is full at all times.

Method

The randomized triplets Easting, Northing and Yield (E_i, N_i, y_{Ti}) where $i \in [1, n]$ and n is the number of elements in the data set, are read in from file, classified into a potential area and summated with others that also fall within that area. Let the potential area be bounded by k and l for the Eastings and r and s for the Northings. (See Fig. 8)

The yield $\left(Y_{potential}\right)$ within any given potential area (PA) can be calculated by summating the yields that fall within its bounds.

$$Y_{potential} = \sum_{E_i \in [E_k, E_l]} \sum_{N_i \in [Nr, N_s]} y_{Ti} \tag{11}$$

The result is the total mass within the potential area, which can then be divided to give kilograms per square metre.

$$y_{potential} = \frac{Y_{potential}}{PA} \tag{12}$$

This can be multiplied by a factor of 10 to give t/ha

The total yield $\left(Y_{Total}\right)$ for the field can be found by summating the potential yields

$$Y_{Total} = \sum Y_{potential} \tag{13}$$

It can be seen that Potential Mapping does have a smoothing effect, proportional to the size of the potential area, but more importantly, it has removed the harvesting artifact due to the unknown crop width (W_{crop}). At present, regular potential areas with an equal aspect ratio have been used to demonstrate the principle, but they also give a pronounced edge effect when the grid size becomes coarse. It is suggested that quadtree or tessellation algorithms could be used to

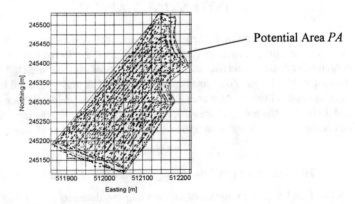

Fig. 8. Randomised triplet data with superimposed potential areas

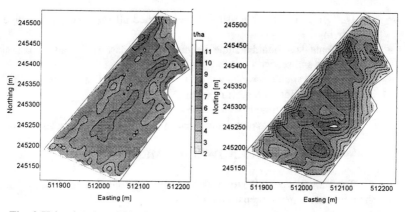

Fig. 9 Kriged and smoothed   Potential Mapped (30m x 30m)

overcome this effect. Although the smoothing effect is beneficial, Potential Mapping assumes that the whole field has been harvested and is sensitive to missing data and data that has been shifted, such as the time lag problem. Nevertheless, it is felt that Potential Mapping in this particular situation of having a very rich data set with the area calculation problem, is a better alternative to the normal interpolation techniques, such as kriging.

INTERPRETATION

The interpretation of a yield map is also very important, as poor choice of map representation can be very misleading. Larscheid and Blackmore (1996), identified the main factors that should be considered when producing a yield map. Inevitably as these maps are used for managerial purposes, spatial trends are important. Figure 4 shows a yield map that has too many perturbances to be able to take into account when considering the practical implications of management. This requires some type of smoothing of the data so that only the important highs and lows in the crop are brought out. Again this requires expert knowledge and an understanding of the use of the yield map, as different uses will require different interpretations. Figure 9 (left), shows the same data set as Fig. 4, but it has been modified by a smoothing algorithm. The spatial variability has been smoothed out but the harvesting errors still show through. Figure 9 (right) is the Potential Mapped data set with the same interpretation, but shows a distinctively different yield pattern, this time without the harvest anomalies. In this particular data set it would appear that the natural shape of the yield variability is actually in a NW-SE direction. Compare this with the kriged data set showing the SW-NE direction that it was harvested in.

CONCLUSIONS

- For the very rich yield data sets recorded from harvesters where the crop width cannot be ascertained - use Potential Mapping.
- For sparse data sets (with perhaps missing data) use kriging.

- Data from the combine should be recorded all the time the combine is moving.
- Combine data should include weight gain since last reading and date, hours, minutes and seconds
- Combines should have an automatic switch that indicates when harvesting (not switch off recording)
- The grain flow time lag should be corrected dynamically
- Potential Mapping can be improved by the use of a quadtree structure

ACKNOWLEDGEMENTS

The authors would like to thank Massey Ferguson and Shuttleworth Farms for their assistance in the preparation of this paper.

NOMENCLATURE

μ	Lagrange multiplier (-)
A_T	Area covered over time period T (m²)
C	Covariance matrix between the sample points (t/ha)²
$\tilde{C}_{ij}$	Covariance between two points (-)
$d_i^{\,p}$	Geometric distance to the power p (m)
d_T	Distance travelled over time period T (m)
D	Covariance matrix of the sample points with unknown yield (t/ha)²
E , N	Easting (m), Northing (m)
T	Time period between samples (s)
y_0	Unknown value (t/ha)
$\hat{y}_0$	Estimate of unknown (t/ha)
y_i	Sample point (t/ha)
$y_{potential}$	Yield in potential area (t/ha)
$Y_{potential}$	Summated mass in potential area (kg)
y_t	Instantaneous yield (kg)
y_T	Summated yield during period T (kg)
Y_T	Total yield in the field (t)
w_i	Weighting value (-)
W	The weighting matrix (-)

REFERENCES

Reitz, P., and H.D. Kutzbach 1996. Investigations on a particular yield mapping system for combine harvesters. Computers and Electronics in Agriculture vol. 14 pp. 137-150.

Borgelt, S.C. 1992. Sensing and measurement technologies for site specific management. p 141-157. *In* P.C. Robert et. al. (eds.). Proceedings of soil specific crop management: A workshop on research and development issues. ASA-CSSA-SSSA, Madison, WI.

Rands, M. 1995. The development of an expert filter to improve the quality of yield mapping data. Unpublished MS thesis, Silsoe College, Cranfield University.

Saunders, S.P., G. Larscheid, B.S. Blackmore and J.V. Stafford. 1996. A method for direct comparison of differential global positioning systems suitable for precision farming. (This conference).

Larscheid, G. and B.S. Blackmore. 1996. Interaction between farm managers and information systems with respect to yield mapping. (This conference).

Isaaks, E.H., and R.M. Srivastava. 1989. Applied Geostatistics. Oxford University Press.

Grain Yield Stability in Continuous Corn and Corn-Soybean Cropping Systems on a Sandy Landscape

J.A. Lamb
J.L. Anderson
G.W. Rehm

Dept. of Soil, Water, and Climate
University of Minnesota
St. Paul, MN

R.H. Dowdy

USDA-ARS
University of Minnesota
Dept. of Soil, Water, and Climate
St. Paul, MN

ABSTRACT

Year-to-year consistency of crop yields within a farm field is needed to use site-specific management. A five-year study was conducted from 1991 to 1995 to determine if patterns of grain yields are similar over a number of years and if grain yields from one or more years can be used to predict grain yields for subsequent years. The experimental site was located at the Northern Cornbelt Sand Plain Management Systems Evaluation Area near Princeton, Minnesota. The research area was 1.78 ha with soils mapped as a Zimmerman fine sand (mixed, frigid, argic, Udipsamment) or a Cantlin loamy fine sand (sandy, mixed, frigid, typic, Udipsamment). Either continuous corn or corn and soybean in a corn-soybean rotation was grown from 1990 through 1995 after a previous history of alfalfa from 1981 through 1989. Cultural practices were applied uniformly to each 1.78 ha site each year. Each 1.78 ha was divided into 60 grid cells (15 m X 18 m) and grain yields, corrected to 15.5 (corn) or 13.0 (soybean) percent moisture, were determined by hand harvesting an area (two rows 6 m. long) within each of the 60 grid cells. Grain yields were not spatially consistent from year to year in either the continuous corn or corn-soybean cropping systems. Areas with better grain yields were not consistent from year to year, and conversely, poor production areas were not found in similar locations each year. Only 0.4 to 42 percent of the grain yield variability for one year is accounted for by a knowledge of the grain yields from a previous year. The lack of grain yield stability on a sandy soil raises serious questions for the potential for utilization of this information. The data indicate that the use of grain yield maps for fertilizer recommendations on a site specific basis will require a much longer term database than the normally recommended five years, unless there is a construct of inputs that explains the grain yield patterns each year.

INTRODUCTION

Spatial and temporal grain yield variability has been accepted by agricultural producers for a long time. Most farmers recognize that yield variability exists when they harvest. Crop yield variability across a field can be influenced by weather, pests, and soil properties. Until recently, technology on the harvester has not been available to measure, reference, and record grain yields to document the location and magnitude of the variability.

Soil variability in the landscape has been studied and mapped by a soil survey program of the USDA-NRCS for a number of years (Soil Survey Staff, 1992). This mapping is based upon soil physical, chemical, and morphological properties as well as geomorphy in the landscape. These county-wide surveys have documented a substantial variability among soils within production fields (Larson & Robert, 1991) caused by variation in one or more of the major soil properties (nutrient availability, cation exchange capacity, organic matter, texture, hydraulic conductivity, and structure). Working with farmers in northeast Nebraska, Spitze et al. (1973) measured corn yield as affected by soil type. Although management practices were constant, yields were affected by soil type under both irrigated and dryland conditions.

Patterns of grain yield variability are important for implementation of variable rate fertilizer applications because yield goals or expected yields influence nitrogen (N), phosphorus (P), and potassium (K) fertilizer recommendations in much of the Midwest. One suggested approach to establishing a yield goal on a whole field basis has been to take the yield from the last five years, delete the least and greatest and use the average of the three remaining grain yields to set the yield goal for future fertilizer application (Rehm & Schmitt, 1989). To adopt this concept in a site specific manner, two requirements are: 1) the yield information must be obtained in a site specific manner, and 2) the grain yield pattern in a field should be stable over a number of years. Yield stability requires that areas with similar yield potential (low or high) should be in the same geographic locations over a series of years. In order to measure yield stability, this study was conducted to determine: 1) if yield spatial patterns are similar over a number of years, and 2) if grain yield from one or more years can be used to predict grain yields for subsequent years.

MATERIALS AND METHODS

This study was conducted from 1991 through the 1995 growing season at the Northern Cornbelt Sand Plains Management Systems Evaluation Area located near Princeton, Minnesota (45 °, 31 ', 34 " north latitude and 93°, 37 ', 8" west longitude). This research site was part of an outwash sand area identified as the Anoka Sand Plain. The research areas were 1.78 ha and the major soil series are Zimmerman fine sand and Cantlin loamy fine sand. The Zimmerman soil is a mixed, frigid, Argic Udipsamment while the Cantlin is a sandy, mixed, frigid, Typic Udipsamment. The difference in elevation within the experimental area is 3 m.

Alfalfa was grown at this site from 1981 through 1989. Two cropping systems were used, continuous corn and a corn-soybean rotation. Both years of the corn-soybean rotation were grown each year. Cultural practices (fertilizer, seeding rate, herbicide, and irrigation) used for corn and soybean production were applied uniformly to each 1.78 ha site.

Grain yields were determined by hand harvesting an area (two rows 6 m long) within each of 60 grids (15 m X 18 m) located in each research area. Grain yields were corrected to 15.5 percent moisture for corn and 13.0 percent for soybean.

Yield data were analyzed for stability over years by using ranked correlations similar to one used in plant breeding research to test stability of genotypic traits over different years and locations (Falconer, 1981). For simplicity, plant breeders will use this analysis to determine if plant performance for a selected trait measured from field studies conducted in one location is similar at other locations. If the genotypes are similar (stable) in each environment then the interaction of genotype and environment would be small. In the plant breeding example, there is replication of genotypes and thus an error term to test the genotype by environment interaction. To adapt this analysis to yield stability, a location in the field is analogous to a genotype and year is the environment. An error term for the interaction of location and year is not available in this adaptation. If the location (genotype) term is highly significant, then the yields would be considered stable since the interaction is used as error. Since there is no error term to test the interaction between location and environment, ranked correlation of grain yields by location over the five years was performed. The higher the correlation the more stable grain yields will be in the field over years. The analysis of variance and correlation analysis were computed using the GLM and CORR procedures in SAS (SAS Institute, 1988).

RESULTS AND DISCUSSION

Grain Yields: The mean continuous corn grain yields ranged from 4.2 Mg/ha in 1993 to 8.5 Mg/ha in 1991. The difference between the least and the greatest grain yields of continuous corn within the research area was 4.5 Mg/ha in 1991, 2.8 Mg/ha in 1992, 2.8 Mg/ha in 1993, 3.2 Mg/ha in 1994, and 3.6 Mg/ha in 1995. These differences were reasonably consistent over the years and are representative of yields for those years in larger production fields. The relative grain yields for each grid cell in each year based on the highest yield equaling 100 % are shown in Figure 1. Visual inspection indicates distribution of the yields across the cropping area changes from year to year.

Mean grain yields for the corn in the corn-soybean rotation from 1991, 1993, and 1995 ranged from 5.1 to 9.2 Mg/ha. The soybean yields were 1.6 Mg/ha in 1992 and 2.5 Mg/ha in 1994. The range of corn yield within a year was 3.3 Mg/ha in 1991, 2.8 Mg/ha in 1993 and 1995. The soybean yield ranges within a year were 1.0 and 1.1 Mg/ha in 1992 and 1994, respectively. The distributions are visually displayed in Figure 2.

Mean soybean grain yields in the soybean-corn rotation ranged from 1.6 to 2.2 Mg/ha. The yields within years ranges were 1.1, 1.0 and 1.4 Mg/ha in 1991, 1993, and 1995, respectively. Corn yields were 7.9 and 7.7 Mg/ha in 1992 and 1994 and ranged 2.8 and 3.0 Mg/ha in 1992 and 1994. Figure 3 shows the relative

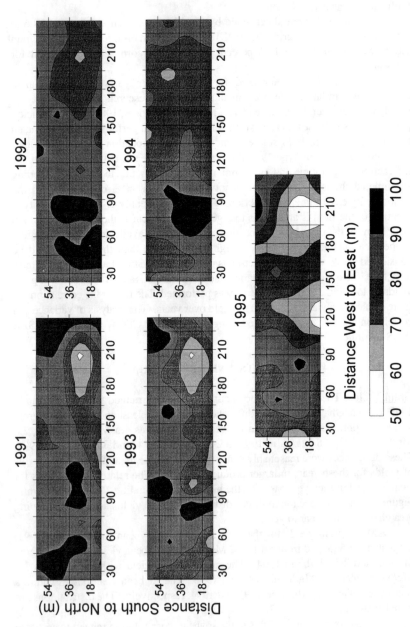

Figure 1. Continuous corn relative grain yield (%) from 1991 to 1995.

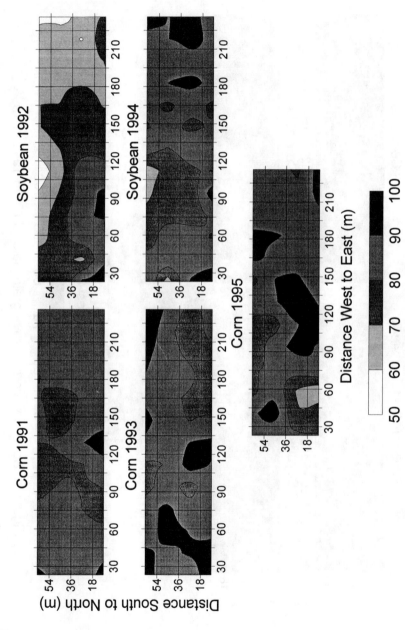

Figure 2. Corn - soybean relative grain yield (%) from 1991 to 1995.

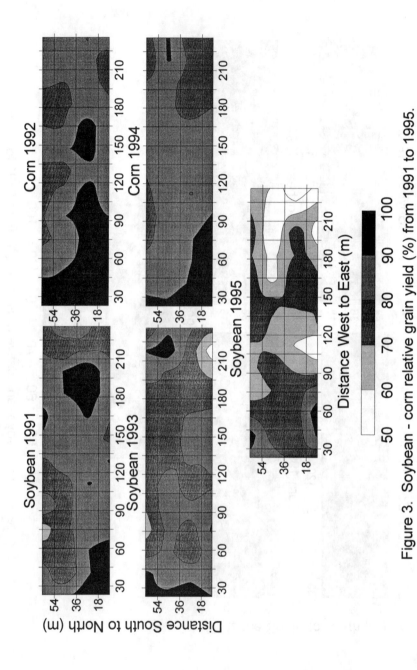

Figure 3. Soybean - corn relative grain yield (%) from 1991 to 1995.

yield distributions for 1991 to 1995.

Grain yield stability: The stability analysis showed that years and locations were highly significant for all three cropping systems. The effects of year on grain yield were approximately 100, 2000, and 1000 times greater than the location or spatial effect for the continuous corn, corn-soybean, and soybean-corn cropping systems, respectively. Significance of the location effect indicated that the relationship was significant for corn and soybean grain yields over years to the location of the grid cell in the field. The best statistic to determine stability would be the test of a location by year interaction. This, however, is not possible since the location in the field cannot be replicated. Rank correlations of location between each year pair were performed to evaluate yield stability over time. The ranked correlations ranged from 0.19 to 0.65, for continuous corn, 0.06 to 0.44 for corn-soybean, and 0.08 to 0.63 for soybean-corn cropping systems. Although most of these correlations are significant, they only account for 4 to 42 % of the year-to-year (years paired) variability in continuous corn grain yield, 0.4 to 19 % for corn-soybean, and 0.6 to 40 % for soybean-corn cropping systems. If the spatial patterns were very similar over years, the ranked correlations (r^2) should be much greater.

Use of yield map for recommendation: Ranked correlations were used to determine if yields measured across a landscape in either one year or a series of years could be used to predict yields across that same landscape in subsequent years for all three cropping systems. The r value was 0.68 for continuous corn, 0.36 for corn-soybean and 0.53 for soybean-corn cropping systems when four years (1991 through 1994) were averaged and then correlated with the fifth year (1995). This accounts for 46 %, 13 %, and 28 % of the variability in the 1995 grain yield from the continuous corn, corn-soybean, and soybean-corn cropping systems, respectively. When the worst grain yield year (1993) or the worst and the best (1993 and 1991) were dropped from the correlation analyses, the relationship decreased, $r = 0.60$ and $r = 0.57$, respectively, for continuous corn. No change was noted for the soybean-corn system, $r = 0.52$ and $r = .50$, respectively, while in the corn-soybean system dropping 1993 increased the correlation to $r = 0.44$. Dropping the best year 1991 in the corn-soybean system decreased the correlation to $r = 0.35$. The use of additional years would increase the correlations but not to the point that prediction was consistently useful. The use of several years worth of grain yield information in intensively managed corn production was not as precise as needed in determining yield goals for future management decisions.

The results of this study demonstrated the instability of corn and soybean grain yields from an intensively managed field. Yield data collected for a four year period to predict the following year were not satisfactory for establishing fixed yield goals or documenting changes in yield potential in production fields.

REFERENCES

Falconer, D.S. 1981. Introduction to Quantitative Genetics. 2nd Ed. Longman Inc., New York, NY.

Larson, W.E., and P.C. Robert. 1991. Farming by soil. p. 103-112. *In* R. Lal and F.J. Pierce (eds.) Soil Management for Sustainability. Soil and Water Conservation Society, Ankeny, IA.

Rehm, G.W., and M.A. Schmitt. 1989. Setting realistic crop yield goals. Minnesota Extension Service Clean Water Folder AG-FS-3873. 2 p.

SAS Institute. 1988. SAS/STAT user's guide. Version 6.03 ed. SAS Inst. Cary, NC.

Soil Survey Staff. 1992. Soil Survey Manual. USDA-SCS. U.S. Gov. Print. Office. Washington D.C.

Spitze, D.C., G.W. Rehm, and R.S. Moomaw. 1973. An effective educational approach for working with adult farmers. J. Agron. Ed. 2:36-39.

Yield Indices For Corn Response To Applied Fertilizer: Application In Site-Specific Crop Management

R.G. Kachanoski
G.L. Fairchild
E.G. Beauchamp

Land Resource Science
University of Guelph
Guelph, Ontario, Canada

ABSTRACT

On-the-goal yield monitors are now widely available. However, it is not clear how to use this Site Specific Crop Management (SSCM) tool in a effective manner for N fertilization. The objectives of this study were to examine different yield indices as predictors of maximum economic rate of N (MERN) application; to develop a predictive relationship for MERN based on the most useful of these indices; and to evaluate the potential usefulness of these indices in SSCM. Two large, independent historical data sets containing 202 (from 1962-1986) and 56 (from 1986-1990) field trials were used to develop the indices. Data sets were divided into three geographical groupings. Yield potential (Ymax) and most economic yield (Ye) were highly correlated to each other, and both were poor predictors ($r^2 < 0.15$) of MERN. Check yield (Yc, yield with no N applied) explained more of the variation in MERN (20-56%) than either Ymax or Ye. The maximum yield increase ΔYmax=Ymax - Yc, and the economic yield increase, ΔYe = Ye-Yc, were both strongly correlated (r=0.70 to 0.88) with MERN. However, the relationship is non-linear. An equation was derived to describe the relationship based on the correlation between coefficients describing the fertilizer response. Field experiments were carried out to demonstrate the use of ΔY measurements to obtain the site specific map for N fertilizer application. The procedure requires the establishment of strip check plots with no N fertilizer. The study indicates that on-the-go yield measurements from a single rate of applied N fertilizer cannot be used to estimate the map for fertilizer N application.

INTRODUCTION

A serious limitation to our understanding of how to apply Soil Specific Crop Management, SSCM, is the requirement for site management maps that relate strongly to crop response to applied fertilizer. A current research challenge in SSCM is to obtain these management maps. Yield mapping has been suggested for use in SSCM, particularly for nitrogen (N) fertilizer application (Robert et al. 1992; Blackmore, 1995; Reetz, 1994; Reichenberger, 1995). Recent years have seen the development and then proliferation of yield monitors. Yield monitors on combines are now commonly discussed in the farm and trade magazines (Reetz,

1994; Reichenberger, 1995; Blackmore, 1995) and are commercially available from at least one major combine manufacturer in 1995. The availability of inexpensive differential global positioning systems and "add-on" commercial yield monitors makes this technology a reality today.

Yield monitors collect the data necessary to describe the yield distribution in the field in detail. However, the application of this data to create management maps necessary for SSCM and the application of nitrogen fertilizer remains a subject of debate in research. Of particular concern is that yield is often found to be poorly correlated to fertilizer requirements. Yield distribution data collected on-the-go from combines will be of little utility in creating SSCM maps to predict fertilizer application rates without a yield index that relates strongly to recommended fertilizer rates.

The objectives of this research were (1) to examine different yield indices as predictors of maximum economic rate of nitrogen (MERN), utilizing large historical data sets of corn yield response to applied nitrogen fertilizer from Ontario, Canada, and (2.) to develop a predictive relationship for MERN based on the most useful of these indices.

Materials and Methods

Data sets from approximately 250 field N response trials conducted across southern Ontario from 1962 to 1992 were obtained from a comprehensive review of nitrogen requirements for corn done by Beauchamp et al. (1987) and from field trials conducted from 1986-1990. A quadratic polynomial equation was fit to the data from each of the field trials, of the form

$$Y = A + BN - CN^2 \qquad (1)$$

where Y= yield, N=rate of applied N fertilizer and A, B and C are regression coefficients .

The nitrogen rate for maximum yield (Nmax) and the maximum economic yield (MERN) were obtained by setting the first derivative of equation (1) equal to zero and the price ratio (R), respectively, and then solving for N. The price ratio was taken to be the estimated price of N per kg divided by the estimated price of grain corn per kg., a price ratio of 5 was used. The economic yield (Y_e), and maximum yield (Y_{MAX}) were obtained by setting N = MERN and Nmax, respectively, in equation (1). Economic yield increase over check yield (ΔY_e) and maximum yield increase over check yield (ΔY_{MAX}) were obtained by subtracting check yield from Y_e and Y_{MAX}, respectively. The correlation between MERN and the yield indices check yield (Y_c), Y_e, Y_{MAX}, ΔY_e, ΔY_{MAX} was examined to determine the utility of the indices as predictors of MERN.

The data sets were grouped into three classes or regions of origin; Southwestern, Central, and Eastern Ontario. The data set classes were significantly different. In addition, the Southwestern Ontario data class was further subdivided into preplant N applications and sidedress N applications. The four classes of data will be used in this analysis.

RESULTS AND DISCUSSION

In the analysis of this data the quadratic equation (1) was chosen because it adequately represented the data and tended to give the highest coefficient of determination (r^2) for the individual data sets (Beauchamp et al., 1987). Beauchamp et al. (1987) also fit a square root and a log model to the data. However, the MERN and Y_e values calculated by each model were highly correlated ($r>0.9$), thus the trends in the data sets can be interpreted using equation (1).

Y_{MAX} (yield potential) explained only 7.4%, 0.7%, 2.2% and 10.3% of the variability of the MERN, for the four 1962-1986 data sets, respectively, and only 1.3% of the variability for the 1986-1990 data set.

Y_e was, as expected, related to Y_{MAX} with correlation greater than $r=0.98$ in three of the 1962-1986 data sets and the 1986-1990 data set and $r=0.94$ in the fourth 1962-1986 data set. Y_e was approximately 90 to 95% of Y_{MAX} . However, the relationship between Y_{MAX} and Y_e did not increase the ability to predict the rate of fertilizer required to give the most economic yield. The correlations between Y_e and MERN are as poor as between Y_{MAX} and MERN . Y_e explained only 7.4%, 0.1%, 7.7% and 15.0% of the variability of the MERN, for the four 1962-1986 data sets, respectively, and 1.5% of the variability for the 1986-1990 data set.

The poor correlation between the yield indices Y_{MAX} , Y and MERN suggests that recommended rates of N fertilizer would not be very well predicted by yield data in either form.

The correlation analysis indicated a very strong relationship between Y_c and MERN. Y_c explained 24.8%, 56.4%, 47.3% and 19.6% of the variation in the MERN in the 1962-1986 data sets and 51.1% of the variability in the 1986-1990 data set. In all five data sets, Y_c explained more of the variation in the MERN than did Y_{MAX} or Y_e. This suggests that in Ontario, the nitrogen supplying capacity of the soil and thus check yield are so different from location to location or from one year to the next, that it significantly reduces the usefulness of the indices Y_{MAX} or Y_e as predictors of the MERN.

For other nutrients, crop yield is often not correlated well with soil test values because of variations among locations attributable to climate and soil. In these cases, yield increase over check yield (ΔY) has been used and has the advantage that direct economic interpretations can be made. ΔY_{MAX} explained 77.1%, 69.8%, 49.5% and 56.6% of the variation in the 1962-1986 data sets and 74.1% of the variability in the 1986-1990 data set. ΔY_e explained 74.7%, 66.8%, 49.5% and 60.6% of the variation in the 1962-1986 data sets and 73.0% of the variation in the 1986-1990 data set. The relationship, however, is non-linear. The high correlation between ΔY_e , ΔY_{MAX} and the MERN, and the lower correlation of Y_c with Y_{MAX} and Y_e illustrate the poor ability to predict the MERN from only Y_m and Y_e . However, ΔY_e , ΔY_{MAX} appear to be reasonable predictive indices of the MERN and would be more useful in interpreting yield response data.

An examination of the data indicated that the B and C coefficients of the quadratic response curves (eq. 1) of the 1962-1986 and 1986-1990 trials were also highly correlated (r= 0.95, Southwestern Ontario (sidedressed), and r=0.92, Southwestern Ontario (preplant), Central and Eastern Ontario for the 1962-1986 data sets; r= 0.96, 1986-1990 data set). This is quite a strong relationship since the data come from such a large number of trials and years and is consistent across two independent historical data sets.

The fact that B and C are correlated explains the presence of the relationship between ΔY and MERN. If

$$\Delta Y_N = B{\cdot}N - CN^2 \tag{2}$$

$$C = \alpha B \tag{3}$$

then

$$B = \Delta Y/(N - \alpha N^2) \tag{4}$$

$$C = \alpha \Delta Y/(N - \alpha N^2) \tag{5}$$

substitution of equations (4) and (5) into

$$MERN = \frac{B - R}{2 C} \tag{6}$$

where R = the price ratio, gives

$$MERN = \frac{1}{2\alpha}\left[1 - \frac{R(N - \alpha N^2)}{\Delta Y_N}\right] \tag{7}$$

Equation (7) indicates that the MERN can be estimated from a measurement of the yield increase over check yield, ΔY_N, for any given rate of fertilizer, N. Setting N=MERN gives a unique relationship between MERN and maximum economic yield gain ΔY_e, that depends only on α.

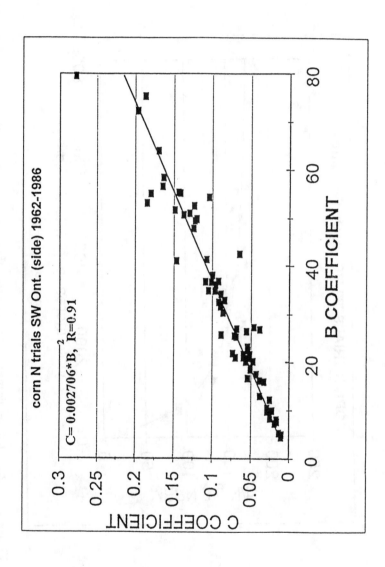

Figure 1. Relationship of quadratic regression coefficients B and C for a subset of the N fertilizer trials.

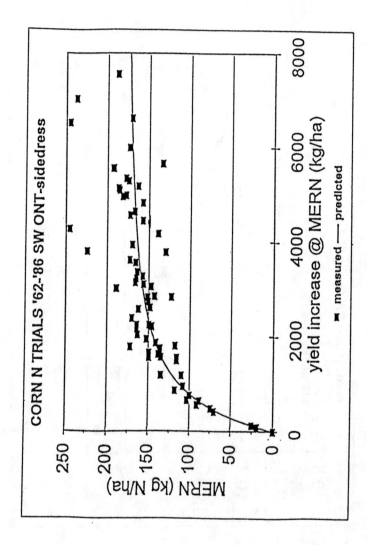

Figure 2.   Predicted versus measured relationship of maximum economic yield increase and maximum economic N rate, for corn in SW Ontario.

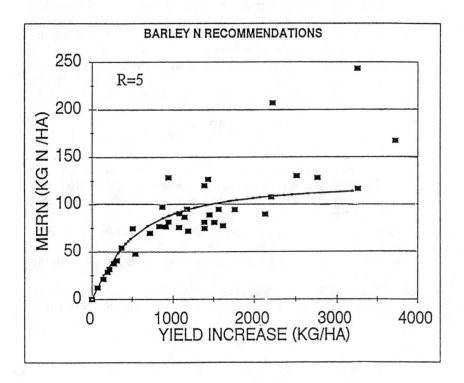

Figure 3. Predicted versus measured relationship of maximum economic yield increase and maximum economic N rate, for barley in Ontario.

Figure 1 shows the relationship (r^2=0.91) between C and B for all of the corn grain fertilizer response trails for sidedress application of N fertilizer for 1962-1986 (South-Western Ontario). The relationship suggests α =0.00276. A comparison of measured and predicted ΔY_e vs MERN using $\alpha = 0.00276$ and equation (7) is given in Fig. 2.

Similar relationships have been examined for N response trials of other crops. The relationships for ΔY_e vs MERN for barley in Ontario is shown in Fig. 3.

Research is currently underway in Ontario to examine the usefulness of using measurements of ΔY_N to predict the spatial patterns of MERN within fields. The concept is that farmers would establish strip check plots (zero N applied) across the field and yield monitors would be used to estimate ΔY_N changes along the strip by comparing the check yield to yield with fertilizer on either side of the check strip. Methods of extrapolating the information using digital topography analysis and remote sensing imagery are being examined, along with an estimation of the minimum number of strips needed for different soil/landforms in Ontario.

REFERENCES

Beauchamp, E.B. , P.G. Newdick, and R.W. Sheard. 1987. Nitrogen requirements for corn in southern Ontario. (120 p.). Department of Land Resource Science, University of Guelph, Guelph, Ontario. CANADA. N1G 2W1.

Blackmore, S. 1995. Precision farming: An introduction. Outlook on Agriculture. 23:275-280.

Reetz,H.F. Jr. 1994. Site-specific nutrient management systems for the 1990's. Better Crops. 78:(4):14-19.

Reichenberger, L. 1995. Precision farming pays it way. Farm Journal. April 1995. p.12-13.

Robert, P.C. , R.H. Rust, and W. E. Larsen (Eds). 1992. Proc. Soil specific crop management, April 14-16, 1992, Minneapolis, Minn., ASA, Madsion, WI. 395 p.

An Investigation into the Relationship Between Yield Maps, Soil Variation and Crop Development in the UK

J. Clarke
M.A. Froment

ADAS Boxworth Research Centre
Boxworth
Cambridge
United Kingdom

J. Stafford
M. Lark

Silsoe Research Institute
Wrest Park
Silsoe, Bedford
United Kingdom

ABSTRACT

This paper reports on a research project being carried out at three sites in England investigating the use of yield mapping technologies in the management of combinable arable crop rotations. The aims of the project are to investigate ways of interpreting yield maps in the light of information on spatially variable soil and crop factors within the field. The final goal is development of strategies to apply appropriate fertiliser inputs on a spatially variable basis. The study will involve investigation of season-to-season variation in yield maps and identification of key parameters which can be measured in order to improve cropping advice. Two fields have been established on each of 3 sites, two at agricultural research centres and one at a commercial farm. Yield maps, obtained by using Massey Ferguson combine harvesters fitted with yield monitors and GPS equipment, were available for 2 or 3 years prior to the start of the project. In 1995, in addition to mapping for yield, soil chemical and physical properties, crop performance and incidence of weeds, pests and diseases were monitored at 21 sampling points within each field. Initial analyses have indicated that around half of the variability in yield over the period of yield mapping was attributable to a component which was consistent between seasons. The consistent pattern is being analysed todetermine whether factors such as differences in soil type, aspect and proximity to boundaries are responsible. The corollary of the above is that half of the variability was due to ephemeral effects. From data collected in the first season, simple correlations do not exist between the variables measured and yield at any of the sites. Fields will be mapped and monitored for two more seasons. Following the analyses, variable application rate experiments will be set up on 2 commercial farms.

BACKGROUND

In the UK, combine harvesters fitted with yield mapping equipment have been introduced on around 100 farms in the last 3-5 years, although the economic benefits of yield monitors were acknowledged much earlier (O'Callaghan, 1988). The potential of yield mapping as a component in managing fertiliser inputs was investigated in northern England in the early 1990's (Schnug et al., 1993). Both yield maps and soil nutrient maps are now regularly featured in the farming press where they are advocated as the first step in the 'Precision Farming' systems of the future (Anon., 1994a). The yield mapping combine harvester market is dominated by Massey Ferguson in the UK, who have lead the development of this technology since the late 1980's. There are believed to be between 70 and 80 combines with yield mapping capability in the UK, more than in any other European Union (EU) member state and the number is expected to increase as the cost of the technology continues to fall.

Linked to yield mapping, soil nutrient mapping services are now becoming more widely available (Budden, 1994, Anon., 1995a) and machinery manufacturers are testing prototype equipment for variable application of seeds, fertilisers and pesticides (Anon., 1994b). Some researchers believe that the rapid rate of technology development has outpaced our understanding and our ability to effectively utilise it. In response to some of these concerns, a research project was initiated to answer some of the questions as to how fertiliser inputs might be used variably within fields, taking into account information from yield mapping and soil and crop monitoring.

Although yield maps give an overall picture of spatial variability within a field, interpretation of the variability is complex. The complexity is confounded when yield maps of one site in successive seasons are compared as shown by Stafford et al. (1996). A rational approach to understanding variability must involve analysis of a sequence of yield maps in order to identify permanent, long-term and transient patterns of variability.

The farmers perspective

The production of yield maps only has a value to farmers if it results in an economic benefit. The sales of yield mapping combines indicate that many farmers are confident of achieving a reasonable return. Discussions with farmers have indicated that there are many reasons for the adoption of this technology and that these reasons are not necessarily confined to the development of within field variable application of fertilisers.

UK farming is based upon high input : high output intensive systems and increasingly the achievement of improved and consistent crop quality. In the UK about one third of the growing costs of combinable arable crops are attributable to the variable costs (seeds, fertilisers and pesticides), so reductions in inputs, whilst important, are not necessarily the primary factor influencing the farm budget. Reasons given by farmers for adopting yield mapping technology include, an ability to estimate weight of grain harvested ex-field prior to despatch to the grain co-operative thus avoiding the costs of independent weighbridge charges, identification of low yielding areas within fields and farms that can be entered into the EU set-aside programme, identification of low yielding areas, such as those associated with shade or susceptible to damage by wildlife from woodlands adjacent to fields, or from construction projects e.g. pipeline developments across fields, to justify land rent reductions or compensation payments. Whilst yield is important, yield trends within a field in any one year and generally from season to season, are often seen as more important than absolute data on yield level.

Economic perspective

A typical budget for winter wheat production within the UK is shown in Table 1. It can be seen that fertiliser costs represent 41% of total variable costs and that nitrogen is the highest cost fertiliser input. Nitrogen is the most important fertiliser element influencing yield in the UK and the potential savings from variable application compared to whole field application have been estimated at up to US$45/ha (Froment *et al.*, 1995). Phosphate and potash fertiliser is routinely applied on an annual basis in the UK and rates are based upon estimated offtakes in grain and straw.

Currently most arable farmers set-aside 10% of their total arable land area each season, for which they receive government payments on both the land set-aside and a payment per ha for most other arable crops grown. These areas can be whole fields or field margins. Farmers prefer to set-aside lower yielding areas of their farms and in doing so, there are major economic benefits in identifying lower yielding areas.

Table 1. UK financial budget for winter wheat production in 1995 (US$/ha)

Factor	US$	% of total costs
OUTPUT	**1717**	
Sale value (8t/ha)	1317	
Area payment (ha)	400	
VARIABLE COSTS	**409**	33
Seed (160kg)	81	
Nitrogen (200kg)	*110*	
Phosphate (80kg)	*40*	
Potash (60kg)	*18*	
Agrochemical sprays	160	
FIXED COSTS	**848**	67
Labour	285	
Machinery and power	270	
Rent/overheads	293	
NET MARGIN	**460**	

Based on: Nix (1995) and Anon. (1995a)

Environmental factors

There is continuing pressure on farmers, through government legislation, to limit some fertiliser and agrochemical use to the crops actual requirements. Limiting the amount of nitrate leaching to groundwater from agricultural land is a high priority for the UK government (Anon., 1993) and attention is now turning to phosphate pollution of water courses. Modelling studies have indicated that targeting nitrogen application within fields could reduce potentially leachable nitrogen in residues post harvest by 10-30% (Froment *et al.*, 1995).

METHODS

Background details of field sites used in this study

Three sites on farms have been chosen for this study, two on ADAS research centres at ADAS Boxworth, Cambridgeshire and ADAS Bridgets, Hampshire and one on a commercial farm at Yokefleet, Humberside (see Figure 1). At each site two fields were chosen. The sites represent a contrast in terms of both soil and climatic conditions within the main arable farming areas of the UK. At Boxworth, the soils are derived from chalky boulder clay, a clayey drift containing chalk fragments. The soils are difficult to work and have impervious sub-soils. At Bridgets the soils are derived from clay with flints or loess material, overlying chalk. At Yokefleet, on the north bank of the river Humber, the soils are silty clays formed from alluvial deposition. Mean rainfall ranges from 525 mm at Yokefleet, 550 mm at Boxworth and up to 810 mm at Bridgets.

Figure 1. Location of Boxworth, Bridgets and Yokefleet field sites

Management of crops

Standard crop management was applied uniformly to the whole field at each site in terms of cultivation, seedrate, fertilisers and agrochemicals. Crops followed combinable crop rotations and are managed to a high standard aimed at minimising any effects from weeds, pests or disease. Field names and rotational details are shown in Table 2.

Table 2. Location, field name and rotational details of sites used in the study

Site/field name	Year			
	1992	1993	1994	1995
Boxworth				
Knapwell	beans	wheat	wheat	wheat
Top Pavements	oilseed rape	linseed	wheat	wheat
Bridgets				
New Hampshire	wheat	wheat	barley	oilseed rape
Mississippi	barley	barley	barley	oilseed rape
Yokefleet				
Duncroft	beans	wheat	set-aside	oilseed rape
Staddlethorpe	peas	wheat	set-aside	oilseed rape

Point location

Using visual interpretation maps from the 1992-94 harvests, 21 sampling points were selected within each field. These points were divided into three 'bands' comprising of 7 sampling points each. Each band was identified as high (about >120% of mean yield), intermediate (90-110% of mean yield) or low (<80% of mean yield). Points were located more than 24m away from field boundaries and other fixed objects. Final positioning of each point ensured that they fell at a point between tramlines (tractor wheelings, normally at 24 or 12m spacing). Points were marked by dead reckoning and the use of magnets buried below plough depth to allow relocation following farm operations such as ploughing. The position of the points was also mapped using a backpack system (Stafford and LeBars, 1996). Each point was taken to represent the centre of a square 10m x 10m from which samples were taken according to a pre-determined protocol.

Crop and soil assessments at sampling points

At each of the 21 sampling points detailed measurements of both the soil and crop are being undertaken. These assessments began in 1995. These include; soil pH, extractable P, K and Mg, % organic matter and % calcium carbonate at 0-15cm, and 0-90cm in 30cm depth increments and soil mineral nitrogen, total N and particle size distribution, in 30cm increments to 90cm depth. Crop assessments have included plant population, dry matter yield in spring and summer, crop lodging, yield of grain and straw at harvest and distribution of nitrogen in dry matter, grain specific weight and Hagberg Falling Number in cereals, oil% and glucosinolate in oilseed rape. Visual assessments of the crop have been carried out at regular intervals throughout the growing season, to verify that site management has been effective in controlling weeds, pests and disease.

At the Boxworth site, neutron probes have been installed and these are being monitored to measure changes in soil moisture status at critical growth stages. At both the Boxworth and Yokefleet sites, soil compaction is being measured during the winter period to a depth of 0.5 metres using a soil penetrometer.

Yield maps have been produced for all fields at each of the three sites using Massey Ferguson combine harvesters fitted with GPS and yield-monitoring equipment.

Analysis of yield variation

Comparisons have been made between yield maps of several seasons for each field. Data were extracted comprising yield values for each season at corresponding locations within the field. Yields within each season were then standardised to unit variance and zero mean. A principle components analysis (Webster and Oliver, 1990) was conducted. This identifies orthogonal linear combinations of the original yield values which account for the maximum variation in the original data. In these examples the first principle component appeared to represent yield variations which were consistent between seasons. Non-hierarchical clustering was then applied to the data sets. This will identify 'clusters', i.e. groups of locations within the field where the yield over all seasons are similar. Such clusters may correspond to locations where yield is consistently high or low, or where some particular pattern of variation between seasons is apparent. Fuzzy clustering was used, since the yield variation of interest is essentially continuous. See Lark and Stafford (1996) for a fuller account of this application of fuzzy clustering.

RESULTS AND DISCUSSION

Yield maps

Preliminary analysis of data from four fields (Knapwell, Top Pavements, New Hampshire and Mississippi) has been completed using principle component analysis and fuzzy clustering.

At the Boxworth site, about 40% of variation in yield over all seasons is accounted for by the first principal component in Knapwell field, for which the weights were very similar and about 55% of the variation in Top Pavements. In Knapwell, the normalised classification entropy suggests that a meaningful grouping of the data into four classes is possible and of these, one class represents consistently above-average yield. However, in Top Pavements, the normalised classification entropy suggested that a meaningful grouping of the data into classes was not possible. For this data set the variation was continuous rather than clustered and therefore principal components analysis will be the most informative technique.

At the Bridgets site, the first principal component of the data set in Mississippi accounted for about 40% of the variation over all seasons. There was good evidence for the occurrence of four classes in this data set, representing four distinct patterns of variation in standardised yield over the three years of data studied (see Figures 2 and 3). In New Hampshire field, the first principal component of this data set also accounted for about 40% of variation over four seasons. There was evidence for three or four classes among the data, one of which corresponded to consistently above-average yield.

Figure 2. Output of fuzzy cluster analysis on yield maps of Mississippi field[#]

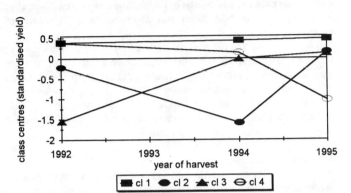

Output of fuzzy cluster analysis on yield maps of Mississippi field. Note that yields are standardised to zero mean and unit variance within each year. The graph shows the typical or 'central' pattern of between-season variation in yield for each cluster. Cluster 1 for example, represents consistently above-average yield in all three years.

Figure 3. Spatial distribution of the fuzzy cluster of maximum membership, Mississippi field

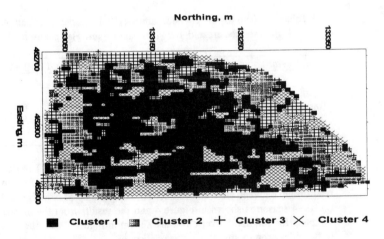

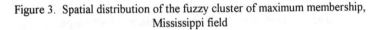

These results are preliminary only, but it is notable that in all but one case, despite the considerable complexity represented by the variation in the sequences of yield maps, variation could be generalised to the behaviour of a few classes.

Crop and soil monitoring

As only one years monitoring is so far available, only limited data analysis has been undertaken. However, this reveals few obvious factors causing the differences in yield. The preliminary findings at Bridgets indicate few differences in any of the crop and soil variables measured in Mississippi field. In New Hampshire, there were differences in % calcium carbonate (0-

15cm) and soil particle size distribution (0-30cm), with higher levels of calcium carbonate reflecting areas of the field with shallower soils and lower yields. The depth of soil overlying the chalk is less in New Hampshire than in Mississippi, but in both fields soil depth to chalk is considered to be important.

At Yokefleet, there were few meaningful relationships between the variables measured and final yield, however at Duncroft there was a significant positive relationship between yield and both soil phosphorous (0-30cm depth) and soil mineral nitrogen (0-90cm). Relationships between soil magnesium (0-30cm) and yield were confounded with the clay and fine silt content of the soil. It is possible these soil physical effects may influence soil structure, drainage or soil moisture characteristics.

Intensive grid soil sampling

At the ADAS Bridgets site, where soil type within fields is predominantly shallow calcareous silty clay loam overlying chalk, described as Andover and Panholes soil series (Jarvis *et al.*, 1984), it was evident from assessments at the 21 assessment points, that soil depth varied significantly within fields and influenced the level of soil nutrients (see Table 3). Both fields at this site were subsequently intensively mapped to 1m depth using a 20m x 20m regular grid pattern, to determine the depth of soil to underlying chalk and the depth of distinct soil horizons within the profile.

Table 3. Mean topsoil chemical analysis, 0-15cm depth, of samples for high, intermediate and low yield bands, New Hampshire field

Band mean	pH	P (mg/l)	Mg (mg/l)	CaCO$_3$%
High	8.36	22.3	41.4	7.7
Intermediate	8.34	15.4	36.0	23.2
Low	8.40	17.4	33.3	26.6

In New Hampshire, visual comparisons of yield in the 1994 season, when the field was sown with spring barley, suggested that there may be a relationship between yield and soil depth. A yield map of New Hampshire for the 1994 season is shown in Figure 4 and the results from the intensive grid sampling for soil depth over chalk are shown in Figure 5. Further analysis is to be undertaken to investigate this relationship.

Whilst soil depth over chalk appears to be important at Bridgets, at each of the sites studied so far, different factors appear to be having the largest effect on crop performance and the yield map. At Yokefleet, soils are also important, but here it is the way the soils were formed by deposition from alluvial flooding and the clay content of different soil horizons which has influenced the results. At Boxworth, where the soils are relatively uniform clay soils, shade from field margins and aspect seems to be more important.

Further investigations and analysis are being undertaken over the next two seasons and variable fertiliser application rate experiments will be set up on commercial farms. Whilst it is clear that soil, crop and yield factors cannot be simply correlated and that contrasting approaches will be required at different geographical sites, preliminary results have been encouraging.

Figure 4. Yield map from New Hampshire field from the 1995 season

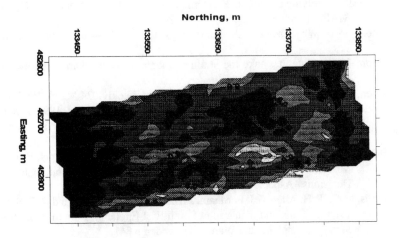

Figure 5. Depth of soil (cm) overlying chalk in New Hampshire Field

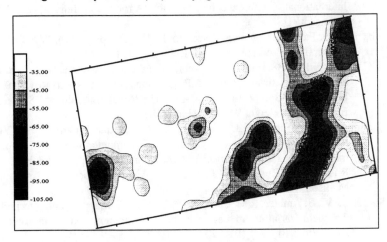

ACKNOWLEDGEMENTS

The 'Yield Mapping' project is funded through joint UK government and industry funding under the LINK Technologies for Sustainable Farming Systems programme. Co-funders are the UK Ministry of Agriculture, Fisheries and Food, Massey Ferguson (UK) Ltd., Yokefleet Farms Ltd. (John Fenton) and Crowmarsh Battle Farms Ltd. (Philip

Chamberlain). The assistance in particular of Mark Moore of Massey Ferguson is gratefully acknowledged.

REFERENCES

Anon. 1993. Solving the Nitrate Problem. Progress in Research and Development. MAFF, London.

Anon. 1994a. When the sky's the limit. Crops. 16 April 1994 Vol. 12 No. 7. Pp. 6-7.

Anon. 1994b. Satellites take the strain in mapping seeds. Farming News. 14 October 1994. P. 25.

Anon. 1995a. New nutrient map service. Farmers Weekly. 24 March 1995. P68.

Froment, M., P. Dampney, G. Goodlass, C. Dawson, and J. Clarke. 1995. A review of spatial variation of nutrients in soil. Internal report for MAFF, London. 68 pp.

Budden, A.L. 1994. A soil analysis sampling method to determine pH, P, K, and Mg fertiliser application using variable rate technology. Sampling to Make Decisions. Aspects of Applied Biology 37 pp 281-282.

Jarvis, M.G., R.H. Allen, S.J. Fordham, J. Hazelden, A.J. Moffat, and R.G. Sturdy. 1984. In Soils and their use in South East England. Bulletin No. 15 Soil Survey of England and Wales, Harpenden p 84.

Lark, RF.M. and J.V. Stafford. 1996. Consistency and changes in spatial variability of crop yield over successive seasons: Methods of data analysis. Third International Conference on Precision Agriculture, June 23-26, 1996, Minnesota USA.

Nix, J. 1995. Farm Management Pocketbook. 25th edition (1995). Wye College, University of London, Ashford, Kent, England.

O'Callaghan, J.R. 1988. Engineering Applications and Developments. In Towards an Agro-Industrial Future. Proceedings of the sixth Royal Show International Symposium pp 45-50. The Royal Agricultgural Society of England, Stoneleigh, Warwickshire, UK.

Schnug, E., D.P. Murphy, E.J. Evans, S. Hanekelaus, and J. Lamp. 1993. Yield mapping and application of yield maps to computer aided local resource management. In Soil Specific Crop Management pp 87-93 Eds. P.C. Robert, R.H. Rust, & W.E. Larson. American Society of Agronomy, Madison, WI USA.

Stafford, J.V., B. Ambler, R.M. Lark, and J. Catt. 1996. Mapping and interpreting the yield variation in cereal crops. Computers and Electgronics in Agriculture 14 (2/3) 101-120.

Stafford, J.V. and J.M. LeBars. 1996. A GPS backpack systgem for mapping soil and crop parameters in agricultural fields. Journal of Navigation 49 (1) 9-21.

Webster, R. and M.A. Oliver. 1990. Statistical methods in soil and land resource survey. Oxford University Press, Oxford England.

Precision Agriculture for Potatoes in the Pacific Northwest

S. M. Schneider
S. L. Rawlins

USDA-Agricultural Research Service
Prosser, WA

S. Han
R. G. Evans

Biological Systems Engineering Department
Washington State University
Prosser, WA

R. H. Campbell

HarvestMaster, Inc.
Logan, UT

ABSTRACT

Precision management of center pivot irrigated fields requires a knowledge of spatial variation within the field. Yield represents the integration of a multitude of processes taking place in the field, and is a reasonable place to begin to identify significant areas of variability. We mapped potato yields in five commercial center pivot fields (240 ha total size) in south central Washington using the HM-500 yield monitor developed by HarvestMaster, Inc. A pair of spread spectrum radio modems was used to transmit real-time yield data from the harvester to the mobile office. This allowed a real-time display of the raw yield data on the computer in the mobile office, permitting problems to be immediately detected without having to have an observer on the harvester. Substantial spatial variability of potato yields, both within and between fields was observed. The yield maps will be used to identify high and low yielding areas to focus further precision management research efforts.

INTRODUCTION

Precision management requires knowledge of the spatial and temporal variability of factors determining crop productivity within fields. The interaction of the many factors is complex, not simply additive. A reduction in the level of a particular plant process might be compensated for by an increase in another process. Yield is the result of the crop's integration of the variability over the entire growing season. It is not surprising, then, that yield mapping is often one of the first steps taken in precision management.

Most yield monitoring has been conducted on grain and seed crops (Robert et al., 1995). This paper reports on a system used to monitor yields of potatoes which is also applicable to other bulk crops such as sugarbeets, processing tomatoes, and grapes.

METHODS

A study was initiated in 1995 to develop and deploy water and nitrogen management prescriptions for center pivot irrigated fields (Evans et al., 1996; Han et al., 1996). The study was sited on a commercial potato farm located in south central Washington state. Five potato circles, 240 ha total size, were monitored for yield.

Yield Monitoring Equipment

The HM-500 (HarvestMaster, Inc., Logan, UT) potato yield monitoring system consisted of a signal conditioning and control unit (SCCU), belt speed and load cell sensors, differential global positioning system (DGPS), on-board field computer, and a spread spectrum radio frequency (RF) modem (Fig. 1).

The SCCU was a manufacturer's prototype specifically designed for flow measurement of bulk crops on field harvesters using conveyor systems. As a signal conditioner and control unit, this instrument made the measurements, multiplexed data from both the sensors and the DGPS receiver, and delivered all the information over an RS-232 serial channel to the on-board, hand-held computer. Three additional serial ports with 12 volt and 5 volt DC power are included on the SCCU to handle expansion of the monitoring system to accept data such as on-the-go product quality assessment as appropriate sensors become available.

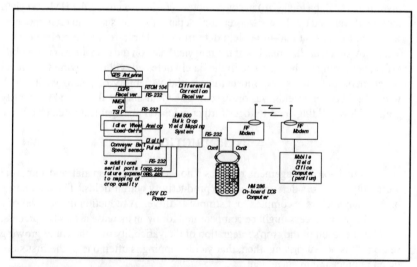

Figure 1. Potato Yield Monitor Components

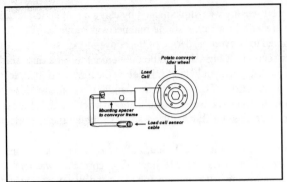

Figure 2. Load Cell Mechanism

The SCCU provides a precision excitation voltage to the load cells and converts the millivolt signal output to a weight reading. The load cell excitation and signal cabling utilized normal shielding and twisted pair design practices to minimize electromagnetic interference. The sampling rate was increased to 100 hz with appropriate low pass digital signal processing employed to minimize signal aliasing due to machine vibration, and to smooth weight fluctuations due to machine motion. Earlier work had been conducted at a continuous 10 hz sampling rate (Campbell et al., 1994), but laboratory tests showed a significant decrease in signal variability on vibrating platforms using the faster sampling rate.

The weighing sensors were a generic, bar type full bridge strain gauge load cell with 5" potato conveyor idler wheels side-mounted on the measurement end of the bar (Fig. 2). Each load cell idler wheel was spaced out from the side of the conveyor channel roughly 1" and cantilever mounted.

Normal loads carried by the idler wheel on the potato harvester were in the range of 9 to 14 kg (20 to 30 lb). Load cell overload protection was insured by using significantly over-rated load cells (227 kg or 500 lb full scale). The high gain, high resolution electronics of the SCCU compensated for the reduced system resolution of the higher capacity load cells. The high capacity of the load cells allowed the side (axial torque applied) mounting of the idler wheel without adversely affecting the performance of the load cell

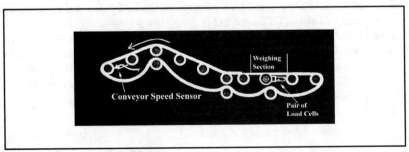

Figure 3. Conveyor Belt Diagram

Conveyor belt speed sensing was implemented by using a magnetic pickup sensor fixed adjacent to a shaft on which a single magnet was mounted (Fig. 3). The SCCU was programmed to convert the period of the conveyor belt speed sensor pulse to units of feet per second. It turned out to be convenient to set a threshold speed in the software, such that conveyor belt speeds below the threshold (0.25 feet per second) were taken as zero. This procedure handled the measurement adequately without going to the expense of a multi-tooth wheel and inductive pickup sensor such as is often used on flywheel and anti-skid brake measurement systems.

The on-board computer was a DOS-based, battery-operated, hand- held computer with a screen size of 24 characters x 16 lines. The computer was trickle charged from the SCCU, allowing it to operate for extended periods. During second to second operations, the program queried the SCCU for product flow rate (belt speed x weight per unit belt length). DGPS data were passed directly through the SCCU to the on-board computer, with the SCCU acting as the communications multiplexor. This design required only one of the serial ports on the on-board computer, leaving the other available for real time data transmission to a mobile office via spread spectrum RF modems. In this research system, the on-board computer handled the tasks of converting raw data to engineering units, with one set of screens specifically designed to accommodate the setting of calibration constants and system checkout.

The on-board computer processing program operated as follows. On receipt of a DGPS position packet at a user definable interval as short as one second, the on-board computer queried the SCCU for flow rate data. The SCCU's processor, keeping a running-average buffer of information being sampled at 100 hz, responded with the current flow rate. The on-board computer calculated yield by dividing the product flow rate by the area coverage rate computed from successive GPS fixes and swath width. Each piece of yield information was stored in a local file on a PCMCIA card in the on-board computer, as well as being re- transmitted via RF modem to a nearby mobile field office. The mobile office was equipped with a GPS base station, a pentium computer, and a radio modem to allow communication with the yield monitor. Real-time raw yield values were displayed on the mobile office computer. This allowed prompt detection of problems with the yield monitor without requiring an observer on the harvester

The GPS receiver built into the HM-500 system had an accuracy of about 5m, when operating in differential mode. The base station in the mobile office sent differential corrections to the GPS receiver in the HM-500 via the pair of RF modems. In this study, GPS time and position data were also recorded separately using a sub-meter accuracy PRO/XL receiver (Trimble, Inc., Sunnyvale, CA). The GPS position data in the yield file, generated by the HM-500, was used as a backup. GPS positions and time were recorded on the PRO/XL at 2s intervals without real-time correction. The base-station data were recorded on a notebook computer and used to post-process the PRO/XL data.

Data collection

Harvest began on September 21, 1995 and continued to October 17. Two, two-row harvesters, one equipped with a yield monitor, were intended to work in tandem across five circles (240-ha total). A two-row side digger was used which dug two rows and deposited the potatoes between two adjacent rows yet to be harvested. The yield monitor, computers, and radio modems were turned on and checked each morning before harvest began. The equipment ran undisturbed all day, unless a problem was detected in the data being received in the mobile office. Each evening, the system was shut down, the data files were downloaded, and memory cleared for the use the next day.

Data files and format. Two data files were created during the data collection process: a yield file generated by the HM-500 and a GPS position file generated by the PRO/XL. The yield file contained records collected approximately every 3s. Each record included the following items:
GPS hour (unsigned char)
GPS minute (unsigned char)
GPS second (unsigned char)
GPS latitude (double precision, real)
GPS longitude (double precision, real)
GPS altitude(double precision, real)
total potato weight during the time interval (floating point)
time interval between the current and the previous record (floating point)
potato flow rate (weight per unit time) (floating point)
load from load cell A (floating point)
load from load cell B (floating point)
belt rotational speed from belt speed sensor (floating point)
calibration factor for potato flow rate (floating point)
observations (unsigned char)

The observations code, with 8-bits storage capacity was, added to record any true/false events. Toggle keys on the on-board computer could be used to simultaneously register up to 8 events to record parameters that are difficult to measure, such as rocky, weedy, or rotten patches.

An ASCII file-format, one record per line, was easy to read, but required about 116 bytes to store a record. Instead, a binary file with IEEE notation was used, requiring only 60 bytes for one record. The new format saved about 50% of the disk space. A program was written to convert this binary file to an ASCII file after it was downloaded to the office computer. The on-board computer had a storage capacity of 1MB. At a 3s recording interval and using the new format, the computer can store up to 13h, roughly one day's data. If the ASCII format was used, files would have to be downloaded in the middle of the day, as well as in the evening.

The GPS position file recorded by the PRO/XL contained GPS positions and time. The PRO/XL has a storage capacity of 1MB. Using a 2s recording interval, it can store up to 15h data. It was downloaded every evening.

Data Processing

Several post-processing steps are required to generate potato yield maps from the yield file and the GPS position file.

Processing the yield file. Spurious yield records, such as negative potato weights, were removed from the yield file. The time stamps in the file were then adjusted to account for the time delay between digging and weighing. The original time stamps in the yield file were the times when each potato mass was recorded, not when it was dug. There was a significant time delay from when the potato was dug to when its weight was recorded. The time delay depended on the total travel distance of each potato mass and its travel speed. It was found that the total travel distance of each potato mass was nearly constant and the travel speed was about the same as the belt speed. Thus, the time delay could be calculated from the belt speed. This algorithm was more accurate than applying a fixed time delay which does not account for the start/stop operation of the machine. The time at which the potato was dug can then be calculated using the time it was weighed and the time delay.

Merging the yield file and the GPS file. The adjusted yield file and the GPS files were merged according to their common index, satellite time. The recording intervals in the two files were different, thus a linear interpolation of positions was used when the time stamps did not match exactly. Occasionally, the GPS may lose signals or erroneous positions may be recorded. These positions were evaluated and removed when appropriate.

Calculating the moving-average for the yield values. Each record in the merged file represents an instantaneous weight measurement. These 'point' measurements are usually not stable, as is the case with many grain monitors. The random errors can be canceled out by averaging several instantaneous measurements. A moving-average algorithm was used to average the weight measurements over a fixed distance of 15 m. This method is preferable to using a fixed number of points, considering the possible stop/start operation of the machine. For a normal travel speed, about 5 measurements were averaged to create one robust yield estimate.

Determining the actual harvest width. With two harvesters and a side-digger working the field, each two-row width dug by the harvester instrumented with the yield monitoring system, could include the two rows of potatoes that grew there, four rows of potatoes if the side-digger was used on one side or six rows, if the side-digger was used on both sides. Thus potato mass could be from a 2-row, 4-row, or 6-row swath. An average weight over the entire, straight travel path was calculated. This average weight could then be used to roughly determine the number of rows harvested. Visual inspection of the GPS track maps were also made. In general, the number of harvest rows could be accurately determined, although errors in interpretation were possible, especially on the short travel paths, such as the edges of the circles.

Generating the yield maps. Each field was divided into 6.1- by 6.1 m cells. The cell values were generated by a non-parametric distance-weighting algorithm as described by Han et. al (1993). Twenty nearest-neighbors were used for the estimation. This method, compared with other geostatistical methods (kriging), was easy to use, required less computing time, and produced about the same accuracy (Evans and Han, 1994). All yield weight values were normalized to the five-circle average. This was done to facilitate evaluation of the variability both within and between circles and to protect the privacy of the commercial grower's data.

RESULTS AND DISCUSSION

The GPS tracks of the instrumented harvester can be seen in Fig. 4. In circle B, the harvesters did follow in tandem, resulting in a fairly uniform coverage of the field by the instrumented harvester. In circles E, D, and C, the harvesters did not follow in tandem, leading to densely monitored areas and to areas not monitored at all.

The normalized yields are shown in Fig. 5. The impact of the density of monitoring on the yield estimation is obvious. In uniformly monitored areas, the nearest neighbors used to calculate cell yield values are relatively close to each other resulting in small patches of similar yield, with areas of variability readily evident. In the sparsely monitored areas, the nearest neighbors are far apart, resulting in a loss of resolution of variability. Large patches are estimated from the sparse data. This is most easily seen in the enlargement of circle D, in which the western side was monitored fairly uniformly, but monitoring on the eastern side was sparse in some strips. Another approach would be to not estimate yield at all if the distance between measurements exceeds a pre-selected threshold. The best solution to the problem would be to instrument all harvesters in the field.

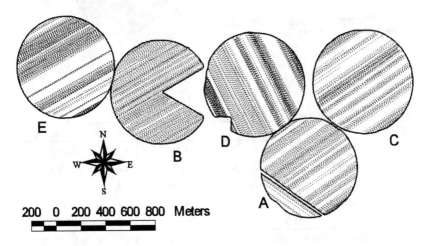

Figure 4. GPS tracks.

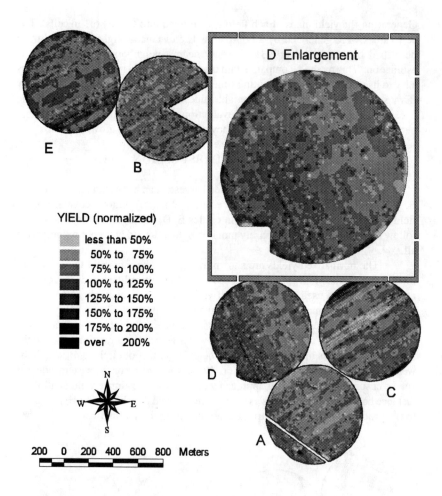

Figure 5. Potato Yield Maps

Table 1. Descriptive Statistics of Potato Yields

Circle	N[1]	Mean	SD	CV	Min.	Max.
A	2761	95.2	31.64	33.23	0	445.0
B	3971	88.8	26.87	30.26	0.4	324.8
C	3219	105.1	49.48	47.06	1.4	448.6
D	4987	111.8	38.47	34.41	0.7	685.0
E	4312	100.5	30.05	29.89	1.8	409.3

[1]N is the number of running averages calculated to populate the yield map,
Mean is the average yield for each circle in percent of the 5-circle average,
SD is the standard deviation, CV is the coefficient of variation in percent,
Min. and Max. are the minimum and maximum yield value for each circle.

An on-going problem in potato yield monitoring was the presence of rocks. Rocks were "harvested" along with the potatoes and had a large impact on yield estimations. Various approaches, such as an acoustic rock detector, have been discussed, but not yet tested. Yield in these circles was also impacted by the widespread occurrence of potato late blight. Presence of the disease was not mapped, therefore specific correlations between disease and yield cannot be made. Future investigations will include the mapping of significant pest problems. Soil clinging to the potatoes was not much of a problem in the sandy soils in these circles, but is reported to be a problem in heavier soils (Hess and Hoskinson, 1996). A weigh-table yield monitoring approach for use in heavier soils has been developed (Rawlins et al., 1995).

Descriptive statistics are shown in Table 1. The average yield/circle ranged from 88.8% to 111.8% of the 5-circle average. As can be seen from the minimum and maximum values for each circle, and the coefficient of variation (CV), the amount of variability in the yield data is great. The distribution of yield values was tested for uniformity in each of the five circles using the Kolmogorov statistic (SAS Institute, Inc., 1988). The assumption of uniformity could not be rejected in any of the circles.

CONCLUSIONS

Yield monitoring of potato, and any other crop harvested with a conveyor belt system, is feasible and contributes information important to the site-specific management of these crops. Large spatial variability in potato yields was observed. Maps of spatially and temporally variable soil characteristics, topography, microclimates, and insect, weed, and disease populations are now needed to identify and manage those factors and their interactions which have the greatest impact on crop productivity. Quantification of the time delay between when a potato is dug and when it crosses the load cells, one of the most challenging problems in post-processing of the data files, has been solved. This is important in order to tag the yield with the location that actually produced it. Without accurate positioning, correlations between yield and other spatial data would have little chance of success. Spatially variable data, collected on an appropriate grid and at appropriate time intervals, will be the foundation of site-specific management. While extensive ground sampling may not be economically feasible, remote sensing might be able to provide some of the inputs needed (Blackmer and Schepers, 1996). Many new technologies such as GPS, GIS, yield monitors, remote sensing, and precision application equipment are now available to agriculture and will significantly enhance growers' ability to remain profitable while continuing to be good stewards of the land and feeding a hungry world.

REFERENCES

Blackmer, T. M. and J. S. Schepers. 1996. Using DGPS to Improve Corn Production and Water Quality. GPS World, March 1996, p. 44-52.

Campbell, Ronald H., Stephen L. Rawlins, and Shufeng Han. 1994. "Monitoring Methods for Potato Yield Mapping", ASAE paper 941584, American Society of Agricultural Engineers, Atlanta, Georgia Meetings, December 13-16, 1994.

Evans, R.G. and S. Han. 1994. Field-scale GIS soil database creation. ASAE Paper No. 94-3078, ASAE, St. Joseph, MI.

Evans, R. G., S. Han, S. M. Schneider, and M. W. Kroeger. 1996. Precision Center Pivot Irrigation for Efficient Use of Water and Nitrogen. (this Proceedings). *In* Robert, P.C. (ed.) 1996. Proceedings of the 3rd Intern. Conf. on Precision Agriculture. Minneapolis, MN. June 23-26, 1996. ASA, CSSA, SSSA, Madison, WI.

Han, S., S. M. Schneider, R. G. Evans, and S. L. Rawlins. 1996. Spatial Variability of Soil Properties on Two Center Pivot Irrigated Fields. (this Proceedings) *In* Robert, P.C. (ed.) 1996. Proceedings of the 3rd Intern. Conf. on Precision Agriculture. Minneapolis, MN. June 23-26, 1996. ASA, CSSA, SSSA, Madison, WI.

Han, S., C.E. Goering, M.D. Cahn, and J.W. Hummel. 1993. A robust method for estimating soil properties in unsampled cells. TRANSACTIONS of the ASAE 36(5):1363-1368.

Hess, J. R. and R. L. Hoskinson. 1996. Methods for Characterization and Analysis of Spatial and Temporal Variability for Researching and Managing Integrated Farming Systems. (this Proceedings). *In* Proceedings of the 3rd Intern. Conf. on Precision Agriculture. Minneapolis, MN. June 23-26, 1996. ASA, CSSA, SSSA, Madison, WI.

Rawlins, S. L., G. S. Campbell, R. H. Campbell, and J. R. Hess. 1995. Yield Mapping of Potato. p 59-68. *In* Robert, P.C., R.H. Rust, and W.E. Larson (eds.) Proceedings of the 2nd Intern. Conf. on Site-Specific Management for Agricultural Systems. Minneapolis, MN. March 27-30, 1994. ASA, CSSA, SSSA, Madison, WI.

Robert, P.C., R.H. Rust, and W.E. Larson (eds.) 1995. Proceedings of the 2nd Intern. Conf. on Site-Specific Management for Agricultural Systems. Minneapolis, MN. March 27-30, 1994. ASA, CSSA, SSSA, Madison, WI.

SAS Procedures Guide, Release 6.03. 1988. SAS Institute, Inc. Cary, NC. 441pp.

Georeferencing Wild Oat Infestations in Small Grains: Accuracy and Efficiency of Three Weed Survey Techniques

C. T. Colliver
B. D. Maxwell

Montana State University
Bozeman, Montana

D. A. Tyler
D. W. Roberts

Ashtech Agriculture Division
Belgrade, Montana

D. S. Long

Northern Agricultural Research Center
Montana State University
Havre, Montana

The ability to manage weed infestations in a spatially precise manner will rely on efficient methods of mapping weed distributions and an effective means of predicting the change in distribution over time, as a result of weed seed dispersal and demographic processes. In 1995, we initiated a study in 2 small grain production fields in Montana to compare three different ground based methods for collecting georeferenced weed information for the creation of weed management maps to be used the following season. The first method included georeferencing wild oat seedling densities for creation of interpolated maps of seedling densities. The second method included georeferencing mature wild oat patch perimeters immediately prior to harvest, when the wild oat patches are most visible in the crop. The third method obtained weed patch location and infestation information during harvest from the combine by the operator or an accompanying individual. Three survey methods (Seedling density, perimeter observations, and observations from a combine at harvest) were evaluated on their potential to describe wild oat distributions for a specific area and on their efficiency to be conducted in the field. The area of infestation determined for the separate survey methods ranged from 68.6% infested for the interpolated seedling map, 66.5% infested for the perimeter map, and 90% infested for the combine map. Infestations described by the perimeter method provided the most detailed description of patch boundaries, while the combine method had the least detail. This relationship was consistent across both locations. The most efficient method of patch description was the combine survey, with the seedling survey method taking the most time per hectare causing it to be the least efficient. The perimeter survey resulted in the largest reduction in cost of weed management by significantly reducing the area requiring management below that of the combine survey method, and by not having the large expense associated with the seedling survey.

INTRODUCTION

The application of precision management techniques to weed control will require the development of methods to efficiently create accurate weed maps on the field scale. Classical and spatial statistics have been used to describe and map the patchy distributions of weeds in arable fields (Marshall, 1988; Wiles et al., 1991; Wilson and Brain, 1991; Mortensen et al., 1993; Dessaint and Caussanel, 1994; Johnson et al., 1995; Mortensen et al., 1995; Colliver and Maxwell, 1995; Rew and Cussans, 1995). These methods have allowed the estimation of possible savings to producers through reduced herbicide usage as was demonstrated by Mortensen et al. (1995) in no-till corn and soybeans (also see Audsley, 1993). Maxwell and Colliver (1995) calculated a 2 fold increase in annualized net return for a ten year period by simulating patch spraying where weed densities exceeded a threshold in cereals rather than whole field management. Rew et al., (1996) calculated a consistent savings over conventional practices from patch spraying quack grass that resulted from a field surveying technique used with tramlines. These studies demonstrate the economic advantage of precision weed management, but none provide methods to create georeferenced weed maps at the field scale for use with conventional farming practices.

Field scale weed survey techniques can be divided into three major categories; remote sensing (e.g. Hanson et al. 1995), ground-based digital image and reflectance analysis (e.g. Felton et al., 1991; Duff, 1993; Woebbecke et al., 1995) and GPS assisted scouting (e.g. Stafford et al., 1996). The first two categories circumvent or reduce the need for human assisted crop scouting. The automated techniques are directed at removing the need for human judgement through technology based weed recognition prior to, or during, the time of herbicide application (Felton et al., 1991; Hanson et al., 1995). Automated techniques are constrained by a restricted window of time when the weed can be recognized, or the technology is designed to work only when there are no crop plants present. An alternative approach is to create a georeferenced map and then apply the management using the historic map.

The third category, part of which we describe in this paper, uses ground-based GPS crop scouting or harvest surveying techniques to create georeferenced weed maps for each field. These maps can then be associated with weed dispersal models for prediction of following year weed distributions. The product is a weed map that can be used for site specific weed management at any phase of the weed life cycle. This approach also has several constraints. It requires spatially accurate GPS (<2m), advanced GIS manipulation and accurate weed dispersion and population dynamic models that must be interfaced with the GIS to produce the following year weed maps. This approach facilitates application of long-term economic weed density thresholds which are more likely to be accepted by producers (Maxwell, 1992).

The use of GPS assisted weed mapping will be fully implemented when both the survey methods and weed dispersion models are linked. Although little work has been conducted on weed dispersion, some annual weeds demonstrate predictable dispersal patterns (Ballaré et al., 1987b; McCanny and Cavers, 1988; Howard et al., 1991). Previous work with annual weeds suggests that natural seed dissemination, tillage and combine aided dispersal are the main factors that influence the location of future weed patches (Wilson, 1972; Sagar and Mortimer, 1976; Ballaré et al., 1987a; Howard et al., 1991), and all are mathematically predictable. The purpose of this paper is to describe three methods of weed patch mapping, present the type of information gained from each method and then compare the methods based on patch description accuracy and sampling efficiency.

MATERIALS AND METHODS

In June of 1995, a weed mapping project was initiated in an irrigated barley field near Amsterdam, Montana and in a recropped dryland spring wheat field near Chester, Montana. In each field three methods of georeferencing wild oat (*Avena fatua* L.) patch locations were used during the growing season; seedling scouting, patch perimeter observations at the panicle stage and harvest time observations. Ashtech[1] SuperCA remote receiver and base station GPS equipment were used with all three survey methods.

Seedling Sampling Method

At each location, an area of a field containing wild oat seedling patches was marked off. This area was semi-systematically sampled by walking approximately parallel transects and taking wild oat seedling density counts in a 0.0625 m^2 circular quadrat every 5 to 10 m. To maximize description of patch structure and location, extra quadrat counts were taken as wild oat seedling patches were intersected. Quadrat locations were georeferenced with a backpack GPS unit interfaced with a laptop computer and seedling densities were entered at each location. At the Amsterdam location, 230 quadrat counts were taken in a 50m by 100m area (0.5 hectares) and 231 counts were taken in a 100m by 230m (2.02 ha) area at the Chester site. Wild oat seedling density counts were taken at the four to five leaf stage of the crop.

Weed Patch Perimeter Method

Prior to harvest and after emergence of wild oat panicles above the crop (time of maximum wild oat visibility), wild oat patch locations were determined by walking around areas containing wild oat panicles. The perimeter of each patch was determined as the outermost occurrence of wild oat panicles. The patch perimeters were georeferenced using the backpack GPS and laptop computer. The area where patch perimeters were determined overlapped the area where seedling information was gathered in each field. An area approximately 14 hectares (34.6 acres) in size

[1]Ashtech Agricultural Division, 90 W. Central Ave., Belgrade, MT 59714, (406) 388-1993

was covered at the Chester site and about 1.36 hectares (3.2 acres) was covered at the Amsterdam location.

Harvest Mapping Method

The locations of mature wild oat patches were georeferenced from the combine during harvest in the same field areas where wild oat seedling and perimeter surveys were conducted. This was accomplished by tagging GPS location data as the combine header entered and exited wild oat patches. The patch identification tag was designed as a toggle to be turned on by the combine operator when wild oat plants were encountered, providing a continuous stream of tagged GPS locations until the toggle was switched off. The length of tagged GPS locations multiplied by the header width provide for an estimation of the area infested with wild oats at time of harvest. Combine patch observations were gathered by the combine operator at the Chester location and by the first author, accompanying the operator, at the Amsterdam site.

Data analysis

Interpolation

Geostatistical techniques were used to create interpolated maps of wild oat seedling density data. Seedling density data were transformed ($ln[z + 1]$) to remove heteroscedasticity prior to geostatistical analysis in GS+ version 1.21[2]. An anisotropic model was used during kriging, because of the presence of large scale trend in the mean.

Map construction and comparison

MapInfo Professional[3] was used to create all seedling density, patch perimeter and combine observation maps. The area containing wild oats was estimated for each survey method using MapInfo. The area of weed infestation for each survey method was determined for the Amsterdam site (Table 1). The common area of weed infestation was determined in the area where all three survey methods overlapped at the Amsterdam site. The common area, named the Amsterdam Overlay (Table 3), is the area where all three survey methods detected wild oats. Two combines were used during harvest at the Amsterdam site, but only one machine was equipped with GPS equipment resulting in missed weed presence information (i.e. skipped swaths). The calculation of area infested with wild oats for the combine method included those areas of skipped swaths directly adjacent to observed swaths which contained wild oats.

The three survey methods mapped three different sized areas at the Chester site. The area of infestation was calculated for each method separately, based on

---

[2]Gamma Design Software, P.O. Box 201, 457 East Bridge St., Plainwell, MI
[3]MapInfo Professional. One Global View, Troy, NY 12180-8399, (518) 285-6000

Table 1. The area of wild oat infestation for the three survey methods and the time to conduct each survey at the Amsterdam site.

Survey Method	Total Area Surveyed (ha)	Area Infested by Wild Oats (ha)	Survey Time ([hours:min] / ha)
Seedling	0.514	0.353 (69%)	9:14
Perimeter	0.514	0.342 (67%)	0:28
Combine	0.514	0.464 (90%)	--na--

Table 2. The relative area of wild oat infestation for the three survey methods and the time to conduct each survey at the Chester site.

Survey Method	Total Area Surveyed (ha)	Area Infested (ha)	Survey Time ([hours:min] / ha)
Seedling	2.02	0.620 (31%)	3:44
Perimeter	14.0	1.254 (9%)	0:10
Combine	3.10	1.431 (46%)	--na--

Table 3. The relative area of wild oat infestation for the two regions of survey area overlay and the estimated area of infestation by each survey method within each overlay.

Overlay Map	Survey Methods Included	Total Area of Common Overlap (ha)	Area of Common Infestation (ha)	Area Estimated to be Infested by Individual Survey Method in Common Survey Area (ha)		
				Seedling	Perimeter	Combine
Amsterdam Overlay	Seedling Perimeter Combine	0.514	0.289 (56%)	0.353 (69%)	0.342 (67%)	0.465 (90%)
Chester Overlay 1	Seedling Perimeter Combine	1.158	0.313 (27%)	0.486 (42%)	0.495 (43%)	0.705 (61%)
Chester Overlay 2	Perimeter Combine	3.096	0.542 (18%)	--na--	0.796 (26%)	1.431 (46%)

Table 4. The ability of the seedling and the combine survey methods to capture the same wild oat infestation area as estimated by the perimeter survey method for both the Amsterdam and Chester sites.

Farm Location (map)	Survey Method	Total Area (ha)	Patch Area Missed (ha)	Percent Accuracy
Amsterdam (Overlay)	Seedling	0.514	0.046	91%
	Combine	0.514	0.027	95%
Chester (Overlay 1)	Seedling	1.158	0.126	89%
	Combine	1.158	0.102	91%

Table 5. Calculation of scouting cost (scouting time multiplied by $25/hr), required treatment area, treatment cost and percent savings compared with a conventional broadcast herbicide application for a 64.8 ha (160 ac) field based on the wild oat infestation levels estimated for the Chester Overlay 1 site.

Survey Method	Scouting Time per 64.8 ha (hours:min)	Scouting Cost ($)	Treated Area (ha)	Treatment Cost ($)	Percent Savings (%)
Seedling	241:36	6040	27.2	1343	-131% ($4183 more)
Perimeter	10:24	260	27.8	1375	49% ($1565 less)
Combine	0	0	39.5	1952	39% ($1248 less)
None (broadcast application)	0	0	64.8	3200	0%

the total area covered for each survey (Table 2). The different areas covered by the survey methods allowed for construction of two overlay maps. Chester Overlay 1, consisted of seedling, perimeter and combine survey information. Chester Overlay 2, consisted of perimeter and combine survey information. The area of weed infestation by individual survey method, and the area of common infestation for all survey methods was calculated for Chester Overlay 1 and Chester Overlay 2 (Table 3).

Calculation of map accuracy and sampling time

The perimeter survey method provided the most detailed description of the actual wild oat infestation and was used as the "true" infestation for determination of map accuracy. Accuracy is defined as the ability of a survey method to describe or encompass the actual area of infestation. Map accuracy was calculated in MapInfo by subtracting the perimeter infestation area from the area of infestation for the seedling and combine surveys, respectively. The amount of perimeter infestation area remaining was measured, divided by the total perimeter infestation area and subtracted from 100% to produce a percent accuracy (Table 4). The accuracy calculations were based on the area where all three survey methods overlapped.

The scouting time for each survey method was estimated as hectares mapped per hour of weed surveying, based on the time spent on each survey method at the Chester site (Table 5), and was calculated for the time it would take to survey 64.752 ha (160 ac). Scouting time was considered to be 0 minutes when the survey method could be accomplished during another farm operation and not increase the amount of time spent on that particular operation (e.g. combine survey method). Scouting cost was calculated as $25 per hour multiplied by the scouting time for 64.752 ha. This information was combined with the area of infestation data from the Chester Overlay 1(Table 3), to calculate the costs and savings of using the individual survey methods in comparison with conventional broadcast management techniques (Table 5). Treatment cost was calculated by multiplying the cost per hectare of a typical post emergence wild oat herbicide, $49.40, by the percent area requiring treatment for each survey method for a 64.752 ha field (Table 5). The cost of applying this management practice to the entire 64.752 ha field was used as the base cost from which percent savings could be determined.

RESULTS

Sampling Methods Comparisons and Map Accuracy

The three surveying approaches produced maps with similar areas containing wild oats at the Amsterdam site. The kriged seedling map showed 68.6% of the Amsterdam study area infested with wild oats, the patch perimeter map showed 66.5% infested and the combine map showed 90% of the area infested (Table 1).

The 1.158 ha area at the Chester site where seedling, perimeter and combine observations overlapped (Chester Overlay 1, in Table 3), wild oats occupied 42% (0.4859 ha) using the seedling survey approach, 42.7% (0.4947 ha) using the perimeter survey approach and 60.8% (0.7047 ha) with the combine survey approach. The area of overlap for the perimeter and combine survey methods extended beyond the seedling survey area at the Chester site. The level of infestation was 25.7% (0.7957 ha) for the perimeter survey and 46.2% (1.431 ha) for the combine survey, over the extended area where both combine and perimeter observations overlapped (Chester Overlay 2, in Table 3). The overlay (Figure 1) of the perimeter and combine survey maps suggested that major wild oat patches were easily detected during combine operation while smaller patches were recorded with less frequency. The combine data contain long swaths of wild oat observations in areas where there were no weeds (Figure 1), this was due to operator error.

The accuracy of the seedling survey method and the combine method was compared with the perimeter survey data in the area where all three survey methods overlapped, for both the Amsterdam site and the Chester site (Table 4). The seedling and combine survey methods were able to account for 91% and 95%, respectively, of the weed patches present at the Amsterdam site. The seedling and combine survey data from the Chester site captured 89% and 91% of the weed infestations, respectively.

Scouting Time and Sampling Efficiency

The wild oat seedling survey method required 230 observations and over 4 hours 45 minutes survey time for 0.5154 hectares at the Amsterdam site (Table 1). Wild oat patch perimeters were walked with a backpack GPS in 36 minutes for 1.37 ha (Table 1) which overlapped the same area where seedlings were counted at the Amsterdam site (Figure 2). The collection of combine information required 10 minutes of setup time and 10 minutes of breakdown time on the day of harvest because the combine used was not fitted with GPS equipment. The combine survey approach has the highest efficiency with regard to increased time of sampling necessary to create a weed map.

The seedling survey method required 231 observations and 7 hours 32 minutes survey time for 2.02 ha at the Chester site (Table 2). Wild oat patches totaling 1.254 ha (8.96%) were outlined over 14 ha at the Chester site (Table 2), taking 2 hours and 15 minutes to complete (Figure 3). Collection of wild oat patch information during harvest required downloading of yield monitor-weed location data to a laptop computer once daily. Otherwise, no extra scouting was required.

The infestation data from Chester Overlay 1 (Table 3) was used to roughly estimate the cost of scouting necessary to create a map for each survey method and estimate the accompanying reduction in herbicide cost when the map is used for precision management over a 64.752 hectare area (Table 5). The seedling survey approach resulted in a loss of 131% ($4183) due to the amount of time necessary to survey the area. The perimeter survey method showed a decrease in the amount of herbicide cost (57%) and a reduced overall cost of weed management (49%). The combine method also resulted in a decrease in herbicide cost (39%) without the associated cost of patch scouting. However, the perimeter survey method resulted in the largest increase in reduced management costs and was 10% ($313 / 64.752 ha) more cost efficient than the combine survey method.

DISCUSSION

This work demonstrates how patch forming weeds can be mapped with GPS equipment to spatially describe weed infestations for precision management. Our work supports the findings of Thompson et al. (1991) that weed maps should be constructed from a variety of field scouting or automated sensing methods. We found seedling scouting to be the most time consuming technique and to lead to possible under-estimation of the actual level of infestation when compared with the other two survey methods. However, the potential of this method to identify the areas within infestd areas that exceed the density threshold for management and thereby further reduce the area requiring weed control.

Mapping weeds during other farm operations maybe the most efficient method, however, combine observations poorly defined patch boundaries leading to over-estimation of weed infestation levels. This over-estimation leads to an increase in the minimum area requiring weed control, thus an increase in weed control costs and in the case of herbicide application, more area receiving chemical where weeds are not present. The perimeter survey method provides the most detailed definition of patch shape for a weed like wild oat, and the opportunity for maximizing the precision and savings of weed control measures.

Weed scouting with backpack GPS equipment for both the seedling and perimeter survey methods can add considerable time to weed control efforts decreasing their cost efficiency. The efficiency is dependent on weed distribution with more aggregated populations being more efficient to map with backpack equipment. We calculated that for a single scout to cover 14 ha of a moderately infested field using the perimeter survey method would take 2 hours and 15 minutes. This comes to approximately 10 hours and 30 minutes.to cover 64.752 ha (160 ac). Suppose this area was to be precision "patch" treated with a post emergent wild oat herbicide costing $49.40 (US) per ha ($20 ac). The perimeter survey indicated 1.254 ha (8.9%) of this area contained wild oat patches. If the minimum required area to patch spray increased to 40% (5.6 ha), due to the size of the spray booms for example, the savings would amount to $415 for 14 ha, or $1920 for 64.752 ha (160 ac). These cost calculations do not include time and money saved by decreased water usage that would result from less area sprayed with a herbicide or the reduction of time spent in the field. These savings could be used to offset initial investment in GPS equipment and the time required to scout each field.

The ability to predict how weed patches move and expand over time will aid scouting techniques and possibly lead to a reduction in the number of times a field is scouted in future years. Combine dispersal of annual weed seed has previously been measured (Ballaré et al., 1987a; McCanny and Cavers, 1988; Howard et al., 1991) whereas actual quantification of weed patch movement has not. Because of the uncertainty associated with patch movement, the addition of a buffer region, or minimum required spray area, around georeferenced weed patches prior to patch spraying was suggested (Rew and Cussans, 1995). If the buffer region were based on the dynamics of patch movement and dispersion, the buffer size could be optimized for target weed species to minimize weed escapes and reduce the number of times patches must be surveyed.

Initially, intensive weed surveying may be necessary to produce an accurate weed patch map which can then be the basis for patch management and future weed scouting. Current methods of crop scouting do take note of the areas in the field with high pest infestations, but they fail to quantitatively and spatially assess the extent of the infestation (see Stafford et al., 1996 for exception). With the use of georeferenced historic maps, weed seedling scouting can be focused on areas where weed patches were present in past years and used to update the current year's map, allowing recommendations of post emergence weed management to be fine tuned for the current year's management practice.

ACKNOWLEDGMENTS

The authors wish to thank Carl and Janice Mattson, Ranger Dykema and Sam Hoffman for their cooperation, patience and input toward accomplishing the work presented here.

REFERENCES

Audsley, E. 1993. Operational research analysis of patch spraying. Crop Protection 12:111-119.

Colliver, C.T., and B.D. Maxwell. 1995. Detection and mapping of wild oat (*Avena fatua* L.) seedling patterns in a field. WSSA Abstracts 35:50.

Ballaré, C.L., A.L. Scopel, C.M. Ghersa and R.A. Sánchez. 1987a. The demography of *Datura ferox* (L.) in soybean crops. Weed Research 27:91-102.

Ballaré, C.L., A.L. Scopel, C.M. Ghersa and R.A. Sánchez. 1987b. The population ecology of *Datura ferox* in soybean crops. A simulation approach incorporating seed dispersal. Agriculture, Ecosystems and the Environment 19:177-188.

Dessaint, F., and J.-P. Caussanel. 1994. Trend surface analysis: a simple tool for modeling spatial patterns of weeds. Crop Protection 13(6):436-438.

Duff, P. 1993. Detectspray System. Proceedings of the Weed Science Society of America. 33:45.

Felton, W.L., A.F. Doss, P.G. Nash, and K.R. McCloy. 1991, A microprocessor controlled technology: To selectively spot spray weeds. *In* Automated Agriculture for the 21st Century: Proceedings of the 1991 Symposium, 16-17 December 1991, Chicago, Illinois. St. Joseph, MI, USA: ASAE publication 11-91.

Hanson, L.D., P.C. Robert, and M. Bauer. 1995. Mapping wild oat infestations using digital imagery for site-specific management. pp. 495-503. *In* Site specific management for agricultural systems, P.C. Robert, et al. (ed.), Agronomy Society of America.

Howard, C.L., A.M. Mortimer, P. Gould, P.D. Putwain, R. Cousens, and G.W. Cussans. 1991. The dispersal of weeds: Seed movement in arable agriculture. Brighton Crop Protection Conference - Weeds - 1991 6:821-828.

Johnson, G.A., D.A. Mortensen, L.J. Young and A.R. Martin. 1995. The stability of weed seedling population models and parameters in Eastern Nebraska corn (*Zea mays* L.) and soybean (*Glycine max* L.) fields. Weed Science 43:604-611.

Marshall, E.J.P. 1988. Field-scale estimates of grass weed populations in arable land. Weed Research 28:191-198.

Maxwell, B.D. 1992. Weed thresholds: the space component and considerations for herbicide resistance. Weed Tech. 6:205-212.

Maxwell, B.D. and C.T. Colliver. 1995. Expanding economic thresholds by including spatial and temporal weed dynamics. Proceedings of the British Crop Protection Conference -- Weeds. Brighton, U.K., 3:1069-1076.

McCanny, S.J. and P.B. Cavers. 1988. Spread of proso millet (*Panicum miliaceum* L.) in Ontario, Canada. II. Dispersal by combines. Weed Research 28:67-72.

Mortensen, D.A., G.A. Johnson, D.Y. Wyse, and A.R. Martin. 1995. Managing spatially variable weed populations. pp. 397-415. *In* Site specific management for agricultural systems, P.C. Robert, et al. (ed.) Agronomy Society of America, St. Joseph, MI.

Mortensen, D.A., G.A. Johnson, and L.J. Young. 1993. Weed distribution in agricultural fields. pp. 113-124. *In* Soil specific crop management. P.C. Robert, et al. (ed.) Agronomy Society of America, St. Joseph, MI.

Rew, L.J., G.W. Cussans, M.A. Mugglestone and P.C.H. Miller. 199_. A technique for mapping the spatial distribution of *Elymus repens* L., with estimate of the potential reduction in herbicide usage from patch spraying. Weed Research (accepted).

Rew, L.J. and G.W. Cussans. 1995. Patch ecology and dynamics - how much do we know? Proceedings of the British Crop Protection Conference -- Weeds. Brighton, U.K., 3:1059-1068.

Sagar, G.R. and A.M. Mortimer. 1976. An approach to the study of the population dynamics of plants with special reference to weeds. pp 1-47, *In* Applied Biology Volume I, T.H. Coaker, ed. Academic Press, London.

Stafford, J.V., J.M. Le Bars, and B Ambler. 1996. A hand-held data logger with integral GPS for producing weed maps by field walking. Computers and Electronics in Agriculture 14:235-247.

Thompson, J.F., J.V. Stafford, and P.C.H. Miller. 1991. Potential for automatic weed detection and selective herbicide application. Crop Protection 10:254-259.

Wiles, L.J., G.W. Oliver, A.C. York, H.J. Gold, and G.G. Wilkerson. 1992b. Spatial distribution of broadleaf weed in North Carolina soybean (*Glycine max*) fields. Weed Science 40:554-557.

Wilson, B.J. 1972. The dispersal of *Avena fatua* L. by wheat harvesting. Proceedings of the 10th British Weed Control Conference, 242-247.

Wilson, B.J., and P. Brain. 1991. Long-term stability of distribution of *Alopecurus myosuroides* Huds. within cereal fields. Weed Research 31:367-373.

Woebbecke, D.M., G.E. Meyer, K. Von Bargen and D.A. Mortensen. 1995. Shape features for identifying young weeds using image analysis. Transactions of the ASAE 38(1):271-281.

Spatially Variable Treatment of Weed Patches

J.V. Stafford
P.C.H. Miller

Silsoe Research Institute
Wrest Park
Silsoe
Bedford   MK45 4HS
United Kingdom

ABSTRACT

The amount of herbicide applied to cereal crops can be significantly reduced by targeting weed patches, providing both environmental and economic benefit. A patch spraying system is described which uses treatment maps generated from field weed maps to control a direct injection spraying system. System and agronomic evaluation has demonstrated the effectiveness of the system in reducing herbicide input by 40 to 60%.

INTRODUCTION

In an era of increasing concern over the environmental impact of farming operations and appreciation of the need to optimise use of inputs, the uniform application of agrochemicals is no longer acceptable. The technologies developed within precision agriculture provide scope for targeting herbicide more accurately and thus achieving a significant degree of optimisation of the use of herbicide. Weed control is essential for the successful production of arable crops with, typically, two or three applications being made to winter-grown small grain cereal crops. These weed control measures range in cost from $20 (£13) to $110 (£75) per hectare. Thus, if targeted application only reduces herbicide input by 50%, a significant saving in input costs could be made whilst reducing the environmental impact significantly in terms of residues on crops and leached agrochemical.

There is considerable spatial variability in the distribution of weeds across a field both in terms of plant density and weed species. This is due to a number of factors such as seed dispersal mechanisms, soil type and moisture content and surface aspect. It is in fact observed that many weed species grow in patches (eg. Wilson and Brain, 1990). The concept and implementation of patch spraying by Silsoe Research Institute is described in this paper.

THE CONCEPT OF PATCH SPRAYING

Targeted application of herbicide can be undertaken in two distinctly different ways. In the first, a real-time weed detection system mounted on the field vehicle detects "individual" weeds and passes that information to a control system that controls the spraying system mounted on the field vehicle. In the second approach, field maps of weed patches, generated by weed sensing systems (as yet undefined) at some time prior to the spraying operation, are used as the basis for sprayer control. The field weed map provides data on the patchiness of weed infestation which, together with information on the reliability of weed data and a weed treatment strategy, may be used to generate a treatment map. This treatment map provides essentially a two-dimensional control signal to the sprayer controller. A second essential sensing system is a means to provide dynamic position information such that the sprayer position may be continually related to the field treatment map. With a suitably divided sprayer boom, a control algorithm can then direct the spatial application of herbicide.

The field mapping approach has been implemented because it has a number of advantages over a real time detection and application system. These include:

(a)   decision making, including the selection of an appropriate herbicide formulation, on the basis of a "whole field" view of weed infestation compared with knowledge of weed infestation only within the "sensing volume" of the detection system;

(b)   the technically easier task of detecting weed patches with a number of different systems compared with detecting individual weeds with one system;

(c)   the flexibility to choose appropriate time windows for weed patch detection in terms of maximising weed to crop contrast and in terms of carrying out detection during normal field operations;

(d)   spraying of pre-emergence herbicides according to previously mapped weed patch distributions;

(e)   the feasibility of defining the goodness of the weed data in terms of error sources and magnitudes in order to modify the treatment strategy (Miller et al, 1995).

Implementation of the patch spraying concept requires the four sub-systems of weed patch detection, field mapping software, spatial location and precision application, together with a strategy for transforming field weed maps into treatment maps.

WEED PATCH MAPS

Weed Distribution

Weeds are distributed non-uniformly across agricultural fields. Moreover many species tend to occur in patches which remain relatively stable in size and location from year to year. For example Wilson and Brain (1990) reported that over a ten year period blackgrass grew in well defined patches on a commercially operated farm. The patches were stable and there was little evidence of new patches forming under conventional herbicide treatments. In a field of winter wheat, Marshall (1988) observed that three grass weed species grew in definite patches in spite of good weed control in some years. It would appear that the seedbank in the soil, plus the relatively localised re-distribution of weed seed by combine harvesting and cultivation lead to stability in weed patches (Rew et al, 1996).

Weed Patch Modelling

A simulation model has been developed by Day et al (1996) at Silsoe Research Institute to describe the dynamics of the weed population in a crop. It is a spatial model which cycles through the processes of weed germination, spray kill, seed production, weed/crop competition, natural seed dispersal, seed dispersal at harvest, seed dispersal during cultivation and seed survival. The model can be run over a number of years to indicate how the patchiness of weeds developed from an initial uniform distribution of weeds. The dispersal term in the model re-distributes to cells around the parent cell as a function of the distance between the centres of the source and sink cells. The function that has been used is a two-dimensional Gaussian whose standard deviation is the dispersal parameter and which approximates quite well to the natural seed dispersal of grass weeds. The model provides a valuable tool to assess the relationships between observed weed populations, the processes that may modify them in subsequent years and viability of the application of a patch spraying regime. An example of its use is shown in Fig. 1 where the effect of spraying threshold on predicted current cost for patch spraying for controlling blackgrass is shown.

Weed Patch Sensing

The field weed map approach is amenable to the use of various manual, semi-automatic and automatic methods of weed patch detection. Research at Silsoe Research Institute has concentrated on near-ground imaging (Thompson et al, 1990; Brivot and Marchant, 1996), aerial imaging and interpretation (Stafford & Miller, 1993), assisted manual surveying using high clearance survey vehicles (Rew et al, 1996), assisted manual surveying from field operations and assisted manual surveying by field walking (Stafford et al., 1996).

Weed patch detection is difficult because of the close similarity between the spectral characteristics of weed and crop species; the similarity causes problems even to experienced crop consultants and weed scientists. There are time windows within the growing season, however, when the contrast between weed and crop is heightened; these include at weed or crop florescence and at crop senescence. Weed detection is a two stage process; (a) green vegetative material has to be discriminated from soil, stone and other material, (b) weed species have to be discriminated from crop species. The first stage can be achieved relatively simply by generating a vegetation index image from images captured in the infra-red (> 700 nm) and in the chlorophyll absorption band (~ 680 nm). Thompson et al (1990) took this approach to produce binary images of weed infestation in small grain cereal crops early in the growing season. They then used row geometry and spacing to estimate the likelihood of vegetative patches being weed rather than

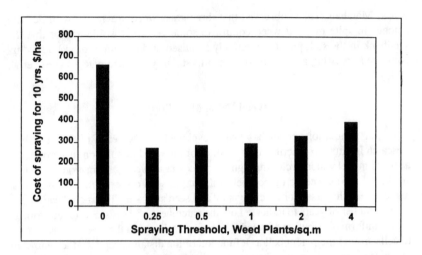

Fig. 1 : Modelling the effect of spray threshold on predicted current cost , $/ha

crop. Recent research at Silsoe Research Institute has studied methods to capture images over large areas by using radio-controlled model planes, identifying time and spectral windows of enhanced contrast and using image interpretation techniques to discriminate weed patches from cereal crops.

The ability to combine field location with manual weed recognition provides the basis for a number of approaches to weed map generation. Rew et al (1996) described the use of a high clearance mobile survey platform. Two operatives with push button units record presence/density of weed infestation as the platform is driven down field tramlines. The weed data was tagged automatically with location derived either by GPS or by a dead-reckoning system (Stafford and Ambler, 1994). An example of a field weed map obtained by this method is shown in Fig. 2.

A backpack GPS system and hand-held PC has been developed for logging weed patch positions during field walking (Stafford & Le Bars, 1996). The PC screen displays a field map showing boundaries and weed patches logged so far, together with a cursor indicating operator position derived from the GPS receiver. For recording weed patches, options are provided to input geometrically shaped patches or to walk irregular patches to define their position. Weed type and patch density are entered from a menu-based system.

A combine harvester fitted with a GPS system for yield mapping purposes also provided an opportunity to log weed patch positions particularly for those weeds which were easily seen at harvest time. An initial study where couch grass in a wheat crop was mapped, showed good agreement between a weed map generated by the combine harvester and one generated by the backpack system (Stafford & Le Bars, 1996).

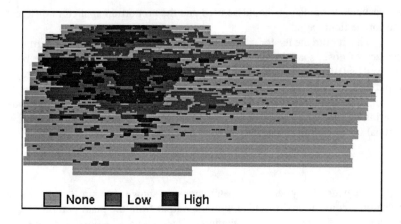

None   Low   High

Fig. 2: Field weed map generated by survey vehicle.
Infestation levels of blackgrass.

Treatment Maps

A treatment map is generated from a raw weed map by one of two approaches. If there is confidence in the weed data then a relatively simple editing program is used to coalesce small weed patches, dilate patches, and apply a herbicide application strategy that is species and infestation density dependent. The effect of dilation is to put a "guard ring" around each patch to allow for positioning and spray application inaccuracies associated with the field sprayer. In reality, there is considerable uncertainty in the field weed map in terms of positioning accuracy, definition of patch edge, weed infestation, species identification and operator mis-identification. An approach is therefore being developed at Silsoe Research Institute where probability maps are estimated for each of the factors and a joint probability weed patch distribution map is generated prior to incorporation of a herbicide application strategy.

THE PATCH SPRAYER

The experimental sprayer developed at Silsoe Research Institute is based on a 12 m boom with two parallel spray lines each with nozzles in groups of four at 0.5 m spacing. A direct injection system is used in which water from the spray tank is mixed with herbicide concentrate metered from a cylinder according to demand. The output from each boom section is separately controlled from a central control system via solenoid valves. With the two lines, four levels of control (including off) were available depending on the spray strategy implemented. For example, a base concentration could be supplied on one line with a different concentration on the second line. Thus, three levels of concentration could be applied. Alternatively, two pesticide formulations could be used and the proportion applied to different areas of the field varied.

The requirements for a spatially variable spraying system in terms of accuracy of dose, spatial resolution, dose range and dose resolution have been reviewed by Paice et al (1996). They considered the options for controlling sprayer output and concluded that direct injection was the most appropriate in terms of control, dose rate turn down ratio and environmental benefits. The latter include metering of concentrate only on demand, no dilute solution in the spray tank and minimal system washing and disposal of dilute concentrate.

Hydraulic System

The hydraulic system of the patch sprayer is shown in Figure 3. It is similar to that on a conventional machine with the output from the main pump fed to a pressure control valve which used compressed air acting on a flexible diaphragm to regulate a by-pass flow back to the tank. Herbicide was metered into the spray lines by the injection metering system (Frost, 1990) which mixed concentrate and water only on demand. The injection metering pumps on each of the two spray lines pumped water into the metering cylinders and displaced active chemical formulation through solenoid valves. The concentrate was mixed with the pressurised water supply in the manifolds from which each of the individual boom sections were fed. Injection metering pumps were controlled according to the number of boom solenoid valves open, the set dose rates and the forward speed of the sprayer. Feedback control of the flowrate was implemented by monitoring the rotational speed of the metering pumps together with the line pressures up-stream and down-stream of the pumps and relating these to a defined pump characteristic stored in the controller.

Control System

The sprayer control system was based on two single card micro-computers (one for each spray line) implementing a proportional plus integral plus derivative control algorithm.

Data was transferred around the sprayer by a data bus which was an implementation of the Controller Area Network (CAN) protocol (Bosch, 1991). A schematic of the data bus system is shown in Fig. 4. The data bus had six nodes. The PC carried out two functions: firstly, to receive control signals (GPS, wheel speed, driver input) to relate to the treatment map held in memory and issue control signals to the sprayer control system; and secondly to monitor performance of the system through logging data fed back from the sprayer controller.

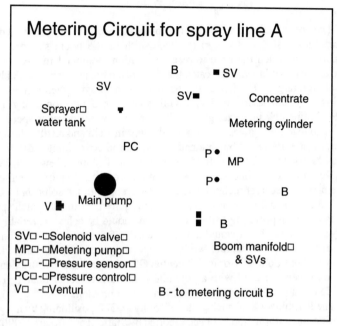

Fig. 3 : Patch sprayer hydraulic system

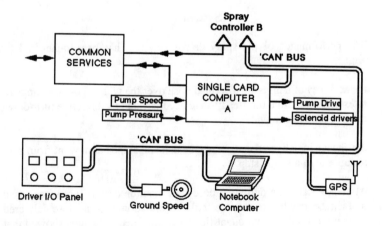

Fig. 4 : Patch sprayer data bus system

Position System

The specification for the experimental patch spraying system laid down a basic resolution of 2 m by 2 m (ie. the width of the boom sections). This specification indicated the need to resolve position to about 1 m. A review by Stafford and Ambler in 1994 indicated that GPS could not at that time reliably meet this resolution target. The initial approach to patch spraying therefore used dead reckoning location based on field tramline count and position along a given tramline obtained by integrating the output from a wheel or radar speed sensor system. Provision was made to calibrate the unit in-field and results showed that an accuracy of ± 1% could be obtained on level terrain over limited distances but that this degraded to ± 2.5% on fields with substantial slopes (Rew et al, 1996).

Due to the availability of a complete constellation of satellites and improvements in receiver technology and software, position resolution by GPS is now possible to 1 to 2 m. However, a further requirement for the patch spraying system is that the specified position resolution should be achieved reliably, both spatially and temporarily; it would not be acceptable for a spraying operation to be paused in the middle of a field whilst acceptable positioning is restored. The distribution of positioning error in differential GPS thus imposes a restriction on its use. Even kinematic GPS with a potential resolution of better than 1 m is not acceptable because of the cycle slip/re-initialisation problem.

The feasibility of enhancing the reliability of GPS positioning by integrating other positioning information has been reported by Stafford and Bolam (1996). The system proposed has the potential to provide reliable positioning to 1 to 2 m by the use of GPS, wheel speed sensors and a digital compass mounted on the sprayer.

EVALUATION

System

The performance of the spray delivery system has been quantified by monitoring:

(a)   the accuracy of delivering a defined dose over the widest possible range of rates so as to define the steady-state accuracy and the effective turn-down ratio of the system; and

(b)   the change in delivered dose when the demand at the boom went from zero to full rate in a single step.

At steady state, the delivered dose was monitored by using water based solutions of a tracer dye to simulate the active formulation with the dose delivered to nozzle positions on the boom quantified by spectrometry. Results showed that the delivered dose was within 5% of the target for forward speeds in the range 1.0 to 3.5 m s^{-1} and for rates of between 0.75 to 5.0 l ha^{-1} (Miller et al, 1995).

Dynamic response characteristics of the injection metering system have been quantified using a 4.0 M solution of sodium chloride to simulate the chemical formulation with the delivered dose to the mixing chambers quantified by electrical conductivity. The control algorithm on the sprayer is such that when there is no demand for pesticide to be delivered from a section of the boom, both water and pesticide flows are turned off. The largest step change in demand would be for the whole boom to be required to deliver pesticide at the full rate on entering a weed patch from an area requiring no pesticide. Simulating this demand situation with the salt solution showed that the largest deviation in delivered concentration was a reduction of 40% for a period of less than 0.3 seconds (Miller et al, 1995).

Agronomic

The experimental spraying system has been used to apply a range of herbicide formulations on commercial and experimental holdings mainly aimed at the control of grassweeds in cereal crops. Maps generated by a manual survey vehicle have been used as the basis for the sprayer control. Mapped weed areas have been edited to create a treatment map typically by adding a 4 m "guard ring" around each patch. Savings in the use of herbicide have been in the range 7 to 69% depending upon the weed distribution within the field.

CONCLUSIONS

A patch spraying system consisting of weed detection, mapping, location and precision application subsystems has been developed and evaluated at Silsoe Research Institute. It has been demonstrated that the targeting of herbicide to grass weed patches in cereal crops leads to an average reduction in herbicide use by 40 to 60% depending on the weed patch distribution within the field.

Whilst technology has developed to the point where a system can be built, it is recognised that development of weed maps is the critical path in the process. For research purposes, systematic manual surveying can generate weed maps but methods for automatic detection of weeds have yet to be developed to the point where they can be used in production agriculture.

ACKNOWLEDGEMENTS

The research reported in this paper has been partly funded by the UK Ministry of Agriculture, Fisheries and Food, the UK Home-Grown Cereals Authority and the European Commission. The work has involved collaboration between Rothamsted Experimental Station and Silsoe Research Institute, and the authors gratefully acknowledge the contribution of colleagues at both Institutes.

REFERENCES

Bosch. 1991. CAN Specification (V.2). Robert Bosch GmbH, Postfach 50, D-7000 Stuttgart 1, Germany.

Brivot, R, and J.A. Marchant. 1996. Segmentation of plants and weeds using infra-red images. Acta Horticulturae, 406 (accepted for publication).

Day, W, M.E.R. Paice, and E. Audsley. 1996. Modelling weed control under spatially selective spraying. Acta Horticulturae, 406 (accepted for publication).

Frost, A R. 1990. A pesticide injection metering system for use on agricultural spraying machines. Journal of Agricultural Engineering Research, 46, 55-70.

Marshall, E.J.P. 1988. Field-scale estimates for grass weed populations in arable land. Weed Research, 28, 191-198.

Miller, P.C.H., J.V. Stafford, M.E.R. Paice, and L.J. Rew. 1995. The patch spraying of herbicides in arable crops. Proceedings Brighton Crop Protection Conference - Weeds, pp. 1077-1085.

Paice, M.E.R., P.C.H. Miller, and W. Day. 1996. Control requirements for spatially selective herbicide sprayers. Computers and Electronics in Agriculture, 14, 163-177.

Rew, L.J., G.W. Cussans, M.A. Mugglestone, and P.C.H. Miller. 1996. A technique for surveying the spatial distribution of *Elymus Repens* and *Cirsium Arvense* in cereal fields and estimates of the potential reduction in herbicide use from patch spraying. Weed Research (in press).

Stafford, J.V. and B. Ambler. 1994. In-field location using GPS for spatially variable field operations. Computers and Electronics in Agriculture, 11, 23-36.

Stafford, J.V., and H.C. Bolam. 1996. Reliable positioning for precision agriculture operations. Proc. 3rd Int. Conference on Precision Agriculture, Minneapolis, USA, 23-26 June.

Stafford, J.V., J.M. Le Bars, and B. Ambler. 1996. A hand-held data logger with integral GPS for producing weed maps by field walking. Computers and Electronics in Agriculture, 14, (2/3), 235-247.

Stafford, J.V., and P.C.H. Miller. 1993. Spatially selective application of herbicides to cereal crops. Computers and Electronics in Agriculture, 9, 217-229.

Thompson, J.F., J.V. Stafford, and B. Ambler. 1990. Weed detection in cereal crops. ASAE Paper No. 90-1629.

Wilson, B.J. and P. Brain. 1990. Weed monitoring on a whole farm - patchiness and the stability of distribution of *Alopecurus Myosuroides* over a ten year period. Proceedings of European Weed Society Symposium - Integrated Weed Management in Cereals, 47-52.

Danish Research on Precision Agriculture; Methods for Estimating Yield Variations

S.E. Olesen

Department of Land Use
Danish Institute of Plant and Soil Science
Research Centre Foulum
P.O.Box 25,
DK-8830 Tjele, Denmark

S.E. Simmelsgaard
M.N. Andersen
E. Friis

Department of Soil Science
Danish Institute of Plant and Soil Science
Research Centre Foulum
P.O.Box 25
DK-8830 Tjele, Denmark

ABSTRACT

The present paper describes briefly a Danish research project on Precision Agriculture and presents a model-based method for estimating yield variations in the field. Model output based on measured soils characteristics showed acceptable correlation with measured yields. Model output showed a poor correlation with yield-meter data.

BACKGROUND

A Danish research project on precision agriculture was carried out in the period 1992-1995 in cooperation between the Danish Institute of Plant and Soil Science, the Danish Agricultural Advisory Service and Risø National Laboratory.
The main objectives of the project were to
- investigate the agricultural, environmental and economic prospects of using spatial and temporal variability of soil and crop parameters by site specific application of fertilizers and pesticides
- demonstrate site specific farming on ordinary farms

The more specific objectives were to investigate
- the spatial and temporal variability of physical and chemical soil parameters and their influence on crop yield
- the spatial and temporal variability of light reflectance and crop growth
- the spatial and temporal variability of weeds, and weed seeds in the soil

-      the agricultural potential of site specific fertilizer and pesticide application
-      the economical potential of site specific farming
-      the environmental potential of site specific farming

The present paper is focusing on activities related to development of methods for estimating yield variations. An overview of all the activities in the project is given in Olesen (1995) and Olesen and Simmelsgaard (1995).

MATERIALS AND METHODS

Experimental field

The results presented in this paper are based on two years (1993 and 1994) experimental work at the "Vindum" field. The field is about 10 ha (300 by 400 m) and was grown with winter wheat. The field was farmed in a traditional way and had even fertilization throughout the field. The nitrogen fertilization rate was 125 to 150 kg/ha, which was about 70 % of the normal recommended rate. No farmyard manure was applied. The textural classification of the mineral soil (USDA) is sandy loam.

Soil characteristics (static as well as dynamic) were measured in grid point in a 20*20 m grid. At all grid points samples were taken at the depth of 0-25 and 25-75 cm for laboratory analyses of soil texture, organic C, $CaCO_3$, pH and available plant nutrients. Pedological investigations by hand drills were carried out at a grid of 40 by 40 m. At 20 grid points samples were taken at 40, 60 and 80 cm depth and analyzed for moisture content at pF 2.0 and 4.2, bulk density, porosity and textural composition. About half of the samples were used for estimation of the complete retention curve.

Soil samples for determination of available nitrogen and nitrogen mineralization were taken at relevant times in the growing seasons. Soil moisture content was measured on a regular basis at five points using Time Domain Reflectometry (TDR) equipment. Meteorological data used in modelling were taken from a nearby station.

Crop growth and development were monitored at a regular basis by means of ordinary sampling and radiometric methods. Sub-plots around selected grid point were harvested by hand. The major part of the field was harvested by a Dronningborg Industries (Massey Ferguson) combine harvester, equipped with a receiver for the Differential Global Position System (DGPS) and a yieldmeter. Yield maps were based on position and yield measurements every 2-3 seconds. Furthermore, the weed density was examined at the 20 by 20 grid points.

Nitrogen response experiments were carried out at 6 locations covering the soil textural variability within the field. Grid point yields and yields from the nitrogen response experiments were used in combination with soils data for modelling crop water consumption and grain yield.

Yield modelling

Measured and kriged point data from the "Vindum" field in 1993 and 1994 (Kristensen et al., 1995, Debosz & Kristensen, 1995) and data from five nitrogen application experiments were analyzed to relate grain yield variability to nitrogen and water availability.

Yield decrease due to drought was modelled with the MVTOOL model (Olesen, 1992). MVTOOL runs on a daily basis and consists of empirical sub-models for crop development, soil water balance and crop yield. It comprises a range of agricultural crops including winter wheat. In MVTOOL's crop development module, growth stage and leaf area index are modelled from temperature sums while root penetration proceeds at a fixed daily rate from growth start until the maximum effective root depth is reached. Maximum effective root depth is defined for both crops and soil types and the lowest of the two alternatives is used in each case. The calculated growth stages for winter wheat may be altered according to user observations of growth start, end of flowering, yellow ripeness and harvest.

The processes in MVTOOL's soil water balance module include evaporation from bare soil and intercepted water, and transpiration. Modelling is of a conceptual nature, based on the field and wilting capacity concept with empirical models for the processes. Transpiration and evaporation are functions of potential evapotranspiration, leaf area index, and available water in the root zone and an upper root zone reservoir, which handles situations where only part of the root zone is rewetted. The field and wilting capacity are defined for two layers: The plough layer, which depth may vary, and the subsoil. Values are derived either from the soil type or from the textural composition of the two layers.

In the yield module, response to drought is modelled from an empirical relationship between the ratio of actual transpiration (E_{aT}) to potential (E_{pT}), and relative yield (Y_a/Y_{max}):

$$\frac{Y_a}{Y_{max}} = \prod_{i=1}^{n}\left[1 - \left(1 - \frac{E_{aT,i}}{E_{pT,i}}\right)\left(a + a_1 t_{s,i} + a_2 t_{s,i}^2\right)\right] \quad (1)$$

where i is day number and $t_{s,i}$ is the temperature sum at day i. The polynomial part of Eqn. (1) describes a growth stage dependent drought sensitivity which has been derived from irrigation experiments. For winter wheat: $a = -7.437E-05$, $a_1 = 1.204E-04$, and $a_2 = -7.263E-8$ valid in the interval $1 < t_s < 1656°C$ has been found by Plauborg et al. (1994). Input to MVTOOL thus comprises data describing the soil types and the crops. Daily climatic input to the system is potential evapotranspiration, which was obtained from interpolated values in a 40 by 40 km grid covering Denmark, and air temperature and precipitation, which was measured at a nearby climatic station.

Yield response to applied nitrogen may, in most cases, be described reasonably well by a second degree polynomial. Further, it may be assumed that other yield reducing factors decrease yield proportionally (same percentage) regardless of N-level. If so, the following simple hypothesis may be formulated:

$$Y = \left(a + b \cdot N_{app} + c \cdot N_{app}^2 \right) \cdot f_w \cdot f_p \cdots f_n \qquad (2)$$

where Y is the yield, N_{app} is the amount of applied nitrogen and f_i $(0 \le f \le 1)$ are yield reducing factors, water, pest, etc.

If only the water factor f_w $(f_w = Y_a / Y_{max}$, Eqn. (1)) is taken into account together with a factor to equalize year differences and if instead of applied nitrogen, the total plant available amount of nitrogen (N_{av}) is taken into account:

$$N_{av} = N_{app} + N_{min} + N_{mineralized} \qquad (3)$$

where N_{min} is the soil mineral N in spring and $N_{mineralized}$ is the mineralization during the growing season then Eqn. (2) may be reduced to the following model:

$$Y = \left(b \cdot N_{av} + c \cdot N_{av}^2 \right) \cdot f_w \cdot f_{year} \qquad (4)$$

assuming that the yield is zero if there is no available nitrogen. This model was utilized to describe the influence of nitrogen and water availability on crop yield variation.

To estimate the parameters in the model (Eqn.(4)), results from five nitrogen application experiments in 1993 and 1994 were used. Five N-levels were applied: 125, 150, 175, 200 and 225 kg N/ha in 1993 and 75, 150, 180, 210 and 240 kg N/ha in 1994. In the experiment with the highest yield level, f_w was calculated by MVTOOL as an average of 9 kriged areas, 10 by 10 meter. In the other four experiments, f_w was calculated from the relative differences in the yield level between the experiments. The differences in yield between the experiments may have been caused by other factors than water availability. In all except one experiment there was good agreement between f_w calculated with MVTOOL and the yield differences observed. N_{min} was also calculated for each experiments as an average of 9 kriged areas, 10 by 10 meter. Actual mineralized N was not measured directly, but was estimated from kriged values of potentially mineralizable N (N_{potmin}) as:

$$N_{mineralized} = d \cdot N_{potmin} \qquad (5)$$

where N_{potmin} was measured in the laboratory at 20°C over 28 days (Debosz & Kristensen, 1995), and d is a regression parameter.

RESULTS AND DISCUSSION

Climatic conditions

Both the 1993 and 94 seasons were characterized by unusually long drought periods with high evaporative demands and almost no rain. In 1993 the drought period started in April and lasted until beginning of July, a potential soil water deficit of more than 200 mm was reached. In 1994 a minor drought period occurred in April/May and in July there was almost no rainfall (Fig. 1). Precipitation in March 1994 was quite high which may have affected the early growth of the crop.

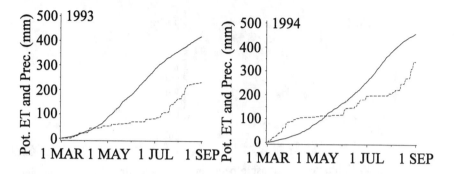

Fig. 1. Accumulated potential evapotranspiration (—) and precipitation (---) for the two seasons.

Calibration of MVTOOL

MVTOOL was calibrated using datasets from the 19 points at Vindum where crop spectral reflectance and final yield were measured in both 1993 and 1994 and soil water retention and texture were measured in 1993. Based on the observed growth stages and leaf area indices derived from spectral reflectance, dates for growth start, end of flowering, yellow ripeness and harvest were given as input for each year. Root growth rate was set to 40 mm/day and maximum effective root depth was set to the following values (Table 1) according to the soil type classification of the subsoil:

The soil type (JB No.) was determined according to the Danish soil classification system (Madsen & Holst, 1987), which is based on the textural composition with the JB No. increasing primarily with increasing clay content. Field and wilting capacity were estimated for both the plough layer and the subsoil from the measured water content at pF 2.0 and 4.2, respectively. The values for the root growth rate and maximum root depth were set to obtain a reasonable straightP line relationship between measured yield and simulated relative yield (Fig. 2). Furthermore, the line should pass through the origin (0,0), and the measured yield, corresponding to a simulated relative yield of 1.0, should be in the same range as the maximum yield normally observed for a good crop or the highest yield observed.

Table 1. Maximum effective root depth according to soil type.

Soiltype	Root depth (mm)
JB1	800
JB2-JB4	1500
JB5-JB7	1800
JB8-JB10	2000

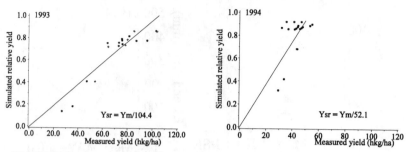

Fig. 2. Calibrated relationship between MVTOOL simulated relative yield and measured yield of grain d.m. for winter wheat in 1993 and 1994.

As MVTOOL has been developed for irrigation purposes it has not been adequately tested on clayey soil types (JB > 4). The setting of the root growth rate is crucial for the calculation of relative yield since it determines if drought stress will occur before the maximum effective root depth has been reached. The value of 40 mm/day did, in most cases, allow for sufficient water abstraction by root growth to prevent drought stress before maximum effective root depth was reached. The maximum effective root depths seem very large, although, effective root depths of 1.4 to 2.1 m for wheat have been reported in the literature (Andersen, 1986; Barraclough et al., 1989; Gajri & Prihar, 1985; Gregory et al., 1978; McGowan et al., 1984). However, these quite large root depths compensate for the fact that MVTOOL does not include calculation of a possible capillary water rise into the root zone. In two points in 1993 (Fig. 2) grain yields above 100 hkg d.m./ha were found, which indicate no or little growth limitation by drought. Since drought generally limits growth when about 50 % of the available water is used (Mogensen & Hansen, 1979) and since the potential soil water deficit reached 200 mm in 1993, it seems, that the crop had about 350-400 mm of available water at the two points. The capacity for available soil water to 1.0 m depth was 205 mm in the same two points. So either root depths were appreciable above 1.0 m or water was transferred by lateral flow. In 1993, MVTOOL could be calibrated to predict the measured yield reasonably well (Fig. 2).

At low yield levels, the predictions were not as good as at higher levels, which may be explained by MVTOOL's simulation of leaf area index. This does not decrease in response to drought stress and in turn, potential transpiration and drought stress level are overestimated during prolonged drought. For 1994 most simulations showed relative yields close to 1, although, measured yields were rather variable. Perhaps root growth was hampered by high precipitation and temporary (perched) ground water during spring or yield was affected by other factors.

Table 2. Values of potentially mineralizable N (N_{potmin}), mineralized N (N_{min}) and water factor (f_w) used in regression Eqn. (4).

Year	Experiment No.	N_{potmin} (kg/ha)	N_{min} (kg/ha)	f_w
1993	1	52.6	25.0	0.77
	2	39.7	26.1	0.63
	3	38.7	26.9	0.59
	4	44.3	30.2	0.60
	5	36.0	18.4	0.18
1994	1	51.3	26.3	0.68
	2	29.2	28.8	0.83
	3	29.7	27.6	0.84
	4	33.7	31.5	0.69
	5	44.7	38.1	0.40

Calibration of a nitrogen production function

The constants in Eqns. (4) and (5) were estimated by regression analysis. Firstly, approximate values of b and c in each year were estimated with the values for d and f_{year} set to one. Thereafter, the value for d was optimized by non-linear regression analysis. Values of 0.6 and 0.4 were found for 1993 and 1994, respectively. Values, used in the regression analysis, for N_{potmin}, N_{min} and f_w are given in Table 2.

The yield level in 1994 was considerably lower than in 1993. To apply the same model for both years, f_{year} was calculated as:

$$f_{year} = \frac{Y_{opt}(1994)}{Y_{opt}(1993)} = 0.59 \qquad (6)$$

where Y_{opt} is the yield at $\delta Y/\delta N = 0$.

The following general model (Eqn. (4)) could satisfactorily describe the experimental yield for both years:

$$Y\ (hkg/ha) = \left(0.958 \cdot N_{av} - 0.00174 \cdot N_{av}^2\right) \cdot f_w \cdot f_{year} \qquad (7)$$

$s = 4.7$ hkg/ha, CV = 7.9 %, where $f_{year} = 1$ for 1993 and $f_{year} = 0.59$ in 1994, $d = 0.6$ in 1993 and $d = 0.4$ in 1994 (Eqn. (5)). For $f_w = 1$ and $f_{year} = 1$ the model has it optimum at $N_{av} = 275$ kg N/ha and $Y = 132$ hkg/ha (Fig. 3).

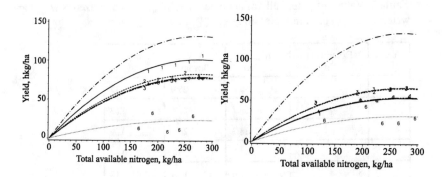

Fig. 3. Relationship between grain yield (85 % d.m.) and available nitrogen in the six nitrogen application experiments (1..6) in 1993 and 94. (—) simulated with the combined nitrogen and water model. Upper curve shows N-response with f_w and f_{year} set to unity.

Validation of the combined nitrogen and water model

To test the model, three types of data were used: **1.** yield and soil properties measured in the individual points, **2.** yield measured in the individual points but kriged soil properties, **3.** yieldmeter data and kriged soil properties from the whole area. The water factor (f_w) was obtained in a 20 by 20 m grid for each year by running MVTOOL with kriged values for soil texture, field capacity and wilting point capacity for the plough layer and the subsoil. **1.** Test of the model in 18 points, where available water capacity, texture, N_{min} and N-mineralization were measured together with yield measured by hand harvest of four 0.5 m² areas, showed a correlation between estimated yield and measured yield of 85 % in 1993

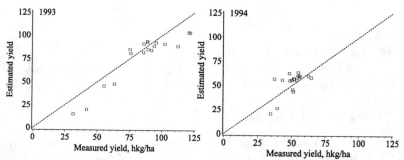

Figure 4. Relationship between grain yield (85 % d.m.) simulated with the combined nitrogen and water model on measured soil data and hand harvested yield in 1993 and 1994.

and 42 % in 1994 (Fig. 4). For both years, the correlation was 82 %. Although, not independent of the calibration, this approach may be compared to the study by Finke & Goense (1993), who were able to explain 68 % of the variation in a Dutch field. **2.** Test of the model in 35 points, where yield likewise was measured by hand harvest, but with kriged input data to the model, gave a correlation between estimated and measured yield of 52 % in 1993 and 31 % in 1994 (Fig. 5). **3.** Estimated yield from kriged input data from the whole area showed a poor correlation to kriged yieldmeter data on 29 % in 1993 and 14 % in 1994. In the analysis, points with an organic matter content of more than 1 % in the subsoil were left out. Also 5 outliers were left out (Fig. 6).

When both input data and yield were measured in the same point, yield estimated with the derived model showed a good correlation to measured yield, especially in 1993 In 1994, the crop development during the season was very different from 1993. Apparently, there have been other factors than drought and nitrogen, which have caused the yield variation.

When passing on from comparison of point measurements to kriged input data and then to yieldmeter data the correlation became smaller and smaller. Apparently, the smoothing, that took place in the kriging procedure, was too large. Also, it may be assumed, that yield differences are smoothened during the passage of the grain through the combiner. This, together with the uncertainty imposed by the GPS-positioning, indicate that the yieldmeter data were not sufficiently accurate to be used for testing the model.

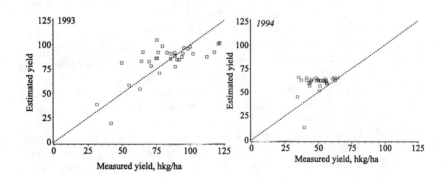

Fig. 5. Relationship between grain yield (85 % d.m.) simulated with the combined nitrogen and water model on kriged soil data and handharvested yield in 1993 and 1994.

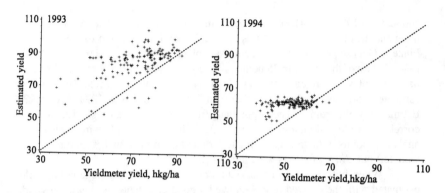

Fig. 6. Relationship between grain yield (85 % d.m.) simulated with the combined nitrogen and water model on kriged soil data and yieldmeter yield in 1993 and 1994.

REFERENCES

Andersen, A. 1986. Root growth in different soil types. Tidsskr.Planteavl, Beretning S-1827. Danish Inst. of Plant and Soil Science. København. 90 pp.

Barraclough, P.B., H. Kuhlmann, and A.H. Weir . 1989. The effects of prolonged drought and nitrogen fertilizer on root and shoot growth and water uptake by winter wheat. J. Agr. Crop Sci. 163, 352-360.

Debosz, K., and K. Kristensen. 1995. Spatial Covariability of N Mineralization and Textural Fractions in two Agriculutural Fields. The Danish Institute of Plant and Soil Science, SP-report No. 26, p. 174-180.

Gajri, P.R., and S.S. Prihar. 1985. Rooting, water use and yield relations in wheat on loamy sand and sandy loam soils. Field Crops Res. 12, 115-132.

Gregory, P.J., M. McGowan, P.V. Biscoe, and B. Hunter. 1978. Water relations of winter wheat. 1. growth of the root system. J. Agr. Sci. 91, 91-102.

Kristensen, K., S.E. Simmelsgaard, J. Djurhuus, and S.E. Olesen. 1995. Spatial Variability of Soil Physical and Chemical parameters. The Danish Institute of Plant and Soil Science, SP-report No. 26, p. 39-55.

Madsen, H.B., and K. Holst. 1987. Potentielle marginal jorder. Marginal jorder og miljoeinteresser. Miljoeministeriets projektundersoegelser 1986. Teknikerrapport nr. 1. (Potential marginal soils in Denmark. Ministry of Environment. Technical Report no. 1). (In Danish).

McGowan, M., P. Blanch, P.J. Gregory, and D. Haycock. 1984. Water relations of winter wheat. 5. The root system and osmotic adjustment in relation to crop evaporation. J. Agr. Sci. 102, 415-425.

Mogensen, V.O., and B.S. Hansen. 1979. Drought periods in Denmark 1956-1976. Calculations for field irrigation. Year Report, Royal Vet. And Agric. University, p. 25-42.

Olesen, J.E. 1992. MVTOOL version 1.00. A tool for developing MARKVAND. AJMET Research Note. Research Centre Foulum. 62 pp.

Olesen, S.E. (1995). Ed.: Proceedings of a Seminar on Site Specific Farming, at Koldkaergaard, Aarhus, Denmark. Danish Institue of Plant and Soil Science, SP-report 26, 204pp.

Olesen, S.E., and S.E. Simmelsgaard. 1995. Danish Research on Site Specific Farming. Proceedings of a Seminar on Site Specific Farming, at Koldkaergaard, Aarhus, Denmark. Danish Institue of Plant and Soil Science, SP-report 26, 28-38.

Plauborg, F., M.N. Andersen, and T. Heidmann (1994). MARKVAND. A PC-based software to support decisions on irrigation in Denmark. p. 1181-1186. *In* Jensen, H.E. et al.(Eds.). International Soil Tillage Research Organization. Proc. 13th Int. Conf. Vol. II.

Spatial Characterization of Corn Rootworm Populations in Continuous and Rotated Corn

M. M. Ellsbury
W. D. Woodson

USDA-Agricultural Research Service
Northern Grain Insects Research Laboratory
Brookings, South Dakota

S. A. Clay
C. G. Carlson

Plant Science Department
South Dakota State University
Brookings, South Dakota

ABSTRACT

Spatial variability in adult emergence for northern and western corn rootworms, *Diabrotica* spp., was characterized by grid sampling in rotated and continuous corn. Knowledge of the spatial distribution of rootworm populations should have application to precision farming for reduced pesticide inputs, provided that spatial variation of rootworm populations can be economically and reliably determined.

INTRODUCTION

Western corn rootworms (WCR), *Diabrotica virgifera virgifera* LeConte, and northern corn rootworms (NCR), *D. barberi* Smith and Lawrence, are among the most important pests of corn grown in the midwestern United States. Much research effort has been invested with limited success in defining temperature-driven predictive models for development of the soil-dwelling stages (Bergman and Turpin 1986) and emergence phenology of adult corn rootworms (Elliott et al. 1990). Spatially variable edaphic factors, such as soil type, soil moisture, snow cover, or crop residue, affect soil temperatures, and thus also affect the phenology of corn rootworm development. Because these edaphic conditions vary spatially, the distribution of corn rootworms also will likely vary spatially as a function of edaphic influences that mediate behavioral responses or act as mortality factors.

Spatial distributions of insect populations usually are considered nonrandom or heterogeneous (Liebhold et al., 1993) as the result of environmental influences and behavioral responses (Taylor 1984) that affect distributions, particularly of the relatively nonmobile stages of soil-dwelling insects. The result is aggregated or contagious distributions that make reliable, economical sampling or scouting strategies difficult to develop. Since the influence of environmentally mediated mortality factors may vary seasonally as well as spatially, the spatial distributions of agriculturally important insects also vary from one growing season to another. Although the spatial nature of insect population distribution can be inferred from statistical indices developed from frequency distributions, such as Moran's I statistic or Geary's c (Williams et al. 1992), these techniques do not measure spatial relatedness directly nor do they allow correlation of insect population distribution with spatially variable environmental factors.

Geostatistical analysis provides an alternative approach to the characterization of spatially variable ecological data (Rossi et al. 1992), particularly for insect pest populations (Rossi et al. 1993, Liebhold et al. 1993, Roberts et al. 1993). Geostatistical techniques have been used to characterize the spatial nature of populations of nymphs and adults of tarnished plant bugs, *Lygus hesperus* Knight (Schotzko & O'Keefe 1989) in lentils over the growing season. Geostatistical analyses of the soil-dwelling stages of the wireworm, *Limonius californicus*, have shown aggregated spatial distributions (Williams et al. 1992). Rossi et al. (1993) estimated NCR and evaluated economic risk for treatment decisions on a regional scale using geostatistical techniques applied to stochastic simulations of spatial variability in data taken from northwest Iowa.

Site-specific management decisions may be possible at the field scale on the basis of accurate knowledge of the spatial distributions of insect pest populations, particularly for the nonmobile stages of such pests. In the case of corn rootworms, an understanding of spatial variability in the distribution of the various life stages of this pest is needed on the field scale for definition of appropriate field sampling procedures for each stage. Application of geostatistical techniques should permit contour mapping of expected insect population density for unsampled field areas. Our objectives were to characterize spatial variability in adult corn rootworm emergence patterns and to determine whether geostatistical methods can provide estimates of adult population densities for northern and western corn rootworms in unsampled areas.

PROCEDURES

Study Sites

Spatial distribution of corn rootworm adult emergence was characterized in two quarter-section (~64.75 ha) corn fields during 1995. One field, located on South Dakota State University land in Brookings Co., SD, had a long history of cultivation in continuous corn, *Zea mays* L., for grain and silage production. This field received terbufos soil insecticide (10.1 kg a.i./ha) at planting time. The other field, located in northern Moody County, SD, was farmed in a 2-year rotation of corn and soybean, *Glycine max* L., without insecticde treatment.

Grid Sampling

Adult corn rootworm emergence was monitored by grid sampling each field using cages (0.5 m^2) similar to the design of Fisher (1984). The cages were installed during the week of 24 July 1995 in a hexagonal (offset) grid pattern.

Cages were centered over corn stalks in ~1 m of row cut to within 10 cm of the ground. Numbers of emerging northern and western corn rootworm adults were counted weekly in each cage from 28 July through 22 September 1995.

The hexagonal grid sampling pattern was chosen since this design has provided better estimates than other sampling designs for assessing the spatial structure of insect populations (Schotzko & O'Keefe 1990, Williams et al. 1992). Distances between sample points initially were laid out with the aid of a distance measuring wheel (Forestry Suppliers, Jackson, MS). Latitude and longitude of each sample point was recorded using a differentially-corrected global positioning system (Chervils Microcomputer Systems, Chervils, OR). Eleven sampling transacts ~75 m apart were established in each field. Sampling points within each transect also were ~75 m apart and were offset from points in adjacent transacts by~ 32.5 m to produce the hexagonal pattern. Sampling points that occurred in waterways, wet spots, or other unplanted areas of the field were relocated whenever possible to the nearest cultivated area of the field but were placed at least 10 m in from the edge of the uncultivated area. To provide some samples at distances closer than 75 m for geostatistical analysis, a cross-shaped pattern of sampling points was established in each field. Each arm of the cross contained 3 additional sample points aligned with the existing sample grid such that the minimum distance between points was ~25 m. The location of each sample point in latitude and longitude degrees was converted to coordinate values in meters using the latitude and longitude of the southwest corner of each field as the origin.

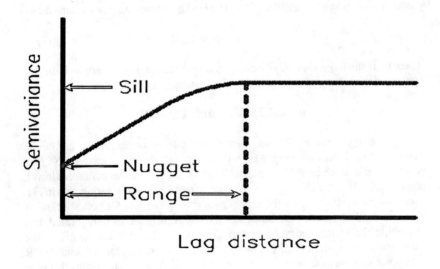

Figure 1. Generalized semivariogram showing the range of influence beyond which samples become independent, the sill value at which the graph levels off, and the nugget, indicative of variability due to experimental error and random effects.

Semivariograms for Spatial Characterization

Emergence data were transformed to log $(z+1)$ where necessary to stabilize variances. Spatial dependence among adult corn rootworm emergence data was characterized by semivariograms (Figure 1), calculated using a geostatistical software package, GS+ (Gamma Design Software, Plainwell, MI). Spherical or exponential models were fitted to semivariance values calculated by the program. Semivariograms plot lag distance between sample pairs against a semivariance statistic for samples at each lag distance. The semivariance is calculated as

$$\gamma(h) = [1/2N(\,h\,)]\sum\,[z_i - z_{i+h}\,]^2 \qquad (1)$$

where h is the lag distance between samples for variable z. For ecological data, such that for corn rootworm emergence, the semivariance is expected to increase as the lag interval increases out to a distance where spatial dependence ceases to be detectable. Certain features of semivariograms, that are important for ecological interpretation of spatial data from insect populations, are shown in Figure 1. The nugget (C_o), when present, is the distance along the y-axis from the origin to the y-intercept and is interpreted as variability due to experimental error and/or other random effects. Presence of a nugget suggests variability on a spatial scale smaller than that detectable by the original sampling protocol. The range is the lag distance beyond which samples are considered independent. The corresponding value of the semivariance at this point is termed the sill and is equivalent to the combination of the nugget effect and variability, (C), attributable to spatial dependence. The proportion of variability attributable to spatial dependence may be estimated as

$$\% \text{ variability} = C\,/\,(C_o + C) \qquad (2)$$

where C_o is the nugget and C is variability due to spatial dependence as indicated by the distance from the nugget to the sill along the y-axis of Figure 1.

RESULTS AND DISCUSSION

Both NCR and WCR adults were collected from the continuous corn field. Total emergence per cage ranged from 0 to over 100 beetles per cage for NCR in certain areas of each field. Numbers of WCR in the continuous corn were lower, ranging from 0 to a maximum of about 50 per sample unit. Semiovariograms in Fig. 2 and 3 show evidence of spatial dependence among samples for both species. A spherical model provided the best fit to the semivariogram of transformed data describing the spatial nature of western corn rootworm adult emergence in the continuous corn system (Fig. 2, $r^2 = 0.886$). The semivariogram of adult WCR emergence showed a nugget effect at about 0.504 (y-intercept), suggesting that only about 64% of the variability among sample pairs was explained by the spatial structure of this population. One possible interpretation of the nugget effect observed in the western corn rootworm data is that spatial structure existed in the population below the scale

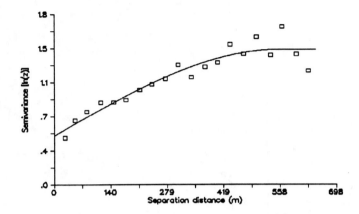

Figure 2. Semivariogram for western corn rootworm emergence based on data taken from continuous corn. Data were transformed to log $(z+1)$ and fitted to a spherical model ($r^2 = 0.894$).

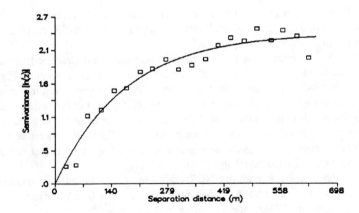

Figure 3. Semivariogram for northern corn rootworm adult emergence based on data taken from continuous corn. Data were transformed to log $(z+1)$ and fitted to an exponential model ($r^2 = 0.945$).

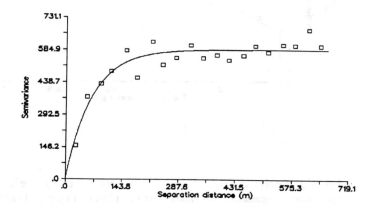

Figure 4. Semivariogram for northern corn rootworm adult emergence from rotated no-till corn. Data were fitted to an exponential model $(r^2 = 0.855)$.

of 75 m between grid sampling points and thus was not accounted for by the model. Total numbers of adult WCR trapped were spatially correlated to a separation distance of about 550 m.

An exponential model on transformed data best described the spatial structure of northern corn rootworm adult emergence in continuous corn (Fig. 3, of about $r^2=0.945$). No nugget effect was seen for the NCR population in continuous corn, indicating that spatial dependence accounted for most of the variation among samples in this population. Spatial dependence was evident out to about 515 m, although the semivariogram also appeared to level off initially at a range of about 210 m, suggesting that NCR emergence data were spatially correlated over a smaller range of influence than 515 m.

The semivariogram for NCR adult emergence sampled in rotated corn (Figure 4) was fitted to an exponential model $(r^2 = 0.855)$. Data were not transformed. Spatial dependence among samples was evident over a range of about 200 m for NCR in rotated corn. Lack of a nugget effect suggested that the sample grid spacing of 75 m provided adequate resolution for characterizing spatial variability in adult NCR emergence. Density of NCR adults ranged from 0 to over 115 insects per 0.5 m^2 cage. Significant WCR emergence was not recorded from the rotated corn, probably because this insect does not survive the two-year rotation.

Adult populations of both species of corn rootworms exhibited distinctive spatial structure in their emergence patterns. The reasons for this spatial variability are unknown at this time, but probably relate to differential influence of mortality factors acting at several possible points in the life cycle of corn rootworms. Behavioral responses during oviposition, related to variability in host plant quality or soil physical conditions in different areas of a field, could influence the spatial distribution of emerging adults. Differential mortality also may occur when newly hatched larvae are unable to locate host plant roots because of adverse soil physical conditions which may vary spatially depending on soil type, soil moisture conditions, and topography.

CONCLUSION

Geostatistical techniques, global positioning system technology, and geographic information systems offer new approaches to the characterization, sampling, and management of insect pest populations in relation to spatially variable mortality factors (Liebhold et al. 1993). Geostatistical techniques based on kriging (Isaaks and Srivastava 1989) can provide interpolated values for response variables, such as insect population densities, at points not actually sampled. Use of indicator kriging allows prediction of spatial probability that a response, expressed as an indicator variable, will exceed an established threshold. Applied to insect pest populations, this technique should allow prediction of areas in field where pest populations may be expected to exceed an action threshold. In the case of soil-dwelling pests, such as corn rootworms, geostatistical techniques should allow development of contour maps of expected insect population density or action thresholds. These thresholds may be expressed as indicator variables correlated to plant development, soil physical properties, or other environmental factors, more easily and economically sampled than are insect populations.

SUMMARY

Spatial distribution of adult emergence patterns were characterized as semivariograms produced from seasonal emergence data for adult northern corn rootworms, *Diabrotica barberi* Smith and Lawrence, and western corn rootworms, *D. virgifera virgifera* Leconte. Data were obtained by grid sampling from two fields of about 70 ha each, one in continuous, conventionally tilled corn and the other in rotated, no-till corn. Sampling was done using emergence enclosures subtending about 0.5 m² from furrow to furrow. Emergence enclosures were arranged in an offset grid pattern with about 75 m between enclosures within each grid row and 75 m between grid rows. One sampled field was planted to corn following corn on 76 cm rows and the other was corn following soybean on 54.6 cm rows. Emergence patterns of northern corn rootworms adults were spatially correlated in both the continuous corn field and the corn soybean rotation. Significant western corn rootworm emergence occurred only in the continuous corn field, even though this field had been treated with soil insecticide. Spatial variation in western corn rootworm emergence was evident in the continuous corn field. Knowledge of the spatial distribution of rootworm populations should have application to precision farming for reduced pesticide inputs, provided the spatial variation of rootworm populations can be economically and reliably monitored and predicted.

REFERENCES

Bergman, M. K., and F. T. Turpin. 1986. Phenology of field populations of corn rootworms (Coleoptera: Chrysomelidae) relative to calendar date and heat units. Environ. Entomol. 15: 109-112.

Elliott, N. C., J. J. Jackson, and R. D. Gustin. 1990. Predicting western corn rootworm beetle (Coleoptera: Chrysomelidae) emergence from the soil using soil or air temperature. Can. Entomol. 122: 1079-1091.

Fisher, J. R. 1984. Comparison of emergence of *Diabrotica virgifera virgifera* (Coleoptera: Chrysomelidae) from cut and uncut corn plants in artificial and natural infestations. J. Kansas Entomol. Soc. 57: 405-408.

Isaaks, E. H,. and R. M. Srivastava. 1989. An introduction to applied geostatistics. Oxford Univ. Press. New York.

Liebhold, A. M., R. E. Rossi, and W. P. Kemp. 1993. Geostatistics and geographic information systems in applied insect ecology. Annu. Rev. Entomol. 38: 303-327.

Roberts, E. A., F. W. Ravlin, and S. J. Fleischer. 1993. Spatial data representation for integrated pest management programs. Amer. Entomol. 39: 92-107.

Rossi, R. R., P. W. Borth, and J. J. Tollefson. 1993. Stochastic simulation for characterizing ecological spatial patterns and appraising risk. Ecol. Applic. 3: 719-735.

Rossi, R. E., D. J. Mulla, A. G. Journel, and E. H. Franz. 1992. Geostatistical tools for modeling and interpreting ecological spatial dependence. Ecol. Monogr. 62: 277-314.

Schotzko, D. G., and L. E. O'Keeffe. 1989. Geostatistical description of the spatial distribution of *Lygus hesperus* (Heteroptera: Miridae) in lentils. J. Econ. Entomol. 82: 1277-1288.

Taylor, L. R. 1984. Assessing and interpreting the spatial distributions of insect populations. Annu. Rev. Entomol. 29: 321-357.

Williams, L. III, D. J. Schotzko, and J. P. McCaffrey. 1992. Geostatistical description of the spatial distribution of *Limonius californicus* (Coleoptera: Elateridae) wireworms in the northwestern United States, with comments on sampling. Environ. Entomol. 21: 983-995.

Spatial Stability of Weed Patches in Agricultural Fields

R. Gerhards

Universität Bonn
Institut für Pflanzenbau
Katzenburgweg 5
BONN - GERMANY

D. Y. Wyse-Pester
D. A. Mortensen

Department of Agronomy
University of Nebraska
LINCOLN, NE

ABSTRACT

Intensive field surveys were conducted in eastern Nebraska to determine the extend to which common sunflower (*Helianthus annuus* L.), velvetleaf (*Abutilon theophrasti* Medik.), foxtail species (*Setaria spec.*) and hemp dogbane (*Apocynum cannabinum* L.) populations are spatially stable. Weed density was sampled prior to postemergence herbicide application at approximately 800 interrow and 800 intrarow locations (4.1 ha) on a regular 7 m grid. Spatial maps revealed that weed seedling distribution was significantly aggregated with large areas being weed free. Spatial pattern of weed patches was stable in location and size for all broadleaf species, but not for foxtail species in all four years. Postemergence herbicide applications were more effective at reducing weed infestation levels than preemergence applications. The weed management program used by the farmer has effectively prevented broadleaf populations from colonizing new field areas, while the foxtail distribution and density increased in each of the four years.

BACKGROUND

Environmental and social concerns about pesticide use have led to an interest in more sustainable farming systems with a reduced reliance on pesticides. Current research focusing on integrated pest mange practices has demonstrated that a reduction in herbicide use must be accompanied by other control and management technologies to decrease weed infestation level and fitness (Mortensen et al., 1995). To-date these approaches have been applied uniformly across fields. An alternate approach is to manage with rather than overcome spatial heterogeneity (Mortensen et al., 1995).

Weed seedling populations are often spatially heterogeneous (Marshall, 1988; Van Groenendael, 1988; Nordmeyer & Niemann, 1992; Wiles et al., 1992; Mortensen et al., 1993; Johnson et al., 1995) resulting in aggregated weed patches of varying density and areas with few or no weed seedlings. This heterogeneity is of considerable importance in weed population assessment and management.

Modern information systems technologies including powerful digital image analysis systems combined with global positioning systems (GPS) and geographic information systems (GIS) (Johnson et al., 1995; Gerhards et al., 1995) make it possible to collect, analyze and use spatially referenced weed density data. Gerhards et al., (1995) used digital maps of spatial weed distribution for site specific management of weeds in winter wheat. In the study, a GIS-based intermittent herbicide sprayer applied Isoproturon and Fluroxypyr only when weed density exceeded the economic threshold reducing herbicide use by 40-50%.

Few studies have attempted to quantify spatial stability of weed patches in agricultural fields. Wilson & Brain (1991) found that the pattern of blackgrass (*Alopecurus myosuroides* Huds.) patchiness persisted during a 10 year study at established infestations. Persistence of patchiness was attributed to the poor ability of blackgrass to colonize new locations when effective herbicides were applied. The pattern of patchiness was most stable in fields planted to cereals. Wyse-Pester et al. (1996) used Pearson, Spearman rank and chi-square correlation analysis to quantify the relationship between weed density at individual X,Y-coordinates in the sampling grid in four fields. While spatial stability was observed with the Spearman rank and chi-square methods, considerable variation in density at individual X,Y coordinates was observed. Such variation results from the sampling method, weed emergence, mortality, and propagule migration.

OBJECTIVES

The objective of this paper is to determine the extent of spatial stability and to describe the spatial attributes of weed patches in one farm field over a period of four years. In addition, the influence of pre- and postemergence herbicide use on the stability and spatial attributes of patches within and between years is considered.

SITE SPECIFIC DATA ACQUISITION

Common sunflower (*Helianthus annuus* L.), velvetleaf (*Abutilon theophrasti* Medik.), foxtail species (*Setaria viridis, S. glauca* L.) and hemp dogbane (*Apocynum cannabinum* L.) were sampled over four years (1992 - 1995) in a furrow-irrigated field managed by a practicing farmer (Johnson et al., 1995; Wyse-Pester et al., 1996). Soybean was planted in 1992 and corn in 1993, 1994, and 1995 in 76 cm spaced rows. Preemergence herbicides were applied in a 30 cm band. Weed density was sampled prior to postemergence herbicide application at approximately 800 interrow and 800 intrarow locations (4.1 ha) on a regular 7 m grid using a 0.38 m^2 frame. In 1994 and 1995 an additional assessment was made in mid-July.

A linear interpolation method was used to estimate weed seedling densities at unsampled positions and to draw maps of weed distribution (Gerhards et al., 1996). The distance between each sampling point (7 m) was divided into 20 equal units (35 cm) and density values of neighboring points within 10 m were used to estimate weed density at unsampled points. Density ranges were set at <0.1, 0.1-3, >3-15 and >15 weed seedlings m^{-2} and represented graphically with a grey scale.

Relative area (%) covered with each density class (grey level) was calculated using digital image analysis.

SPATIAL DISTRIBUTION AND STABILITY OF WEED PATCHES

Common sunflower

Mean seedling density ranged from 0.7 (1994 and 1995) to 2.4 seedlings m^{-2} (1993) in the maps of the undisturbed populations (Table 1). Preemergence herbicide application reduced average seedling density by about 60% in 1992, 1993 and 1995 but had no effect in 1994. After postemergence control, mean seedling density was significantly reduced in 1994 and 1995. Still, some plants survived at locations with high infestation levels. Assuming that those surviving plants reach maturity and produce seeds, they are of considerable importance to maintenance of the population (Table 2) since common sunflower is an annual with short-lived seeds (Hanf, 1990).

Hemp dogbane

Hemp dogbane is the only perennial in this study whose reproduction is largely vegetative. Size, location and density of hemp dogbane patches were highly stable over all four years (Figure 1). Those patches were located in one side of the field and covered between 8 and 23% of the field in all four years regardless of weed control practice. Infestation levels and location of patches were unaffected by herbicide treatments (Table 1 and 2, Figure 1)

Foxtails

Foxtail infestation level and distribution was least stable of the four species studied. Interrow seedling density increased from 0.1 seedlings m^{-2} in 1992 to 63.4 seedlings m^{-2} in 1995 (Table 1). 96% of the total field was free of foxtails in 1992 but only 1% was weed free in 1995 (Table 2). The significant increase in infestation level from 1992 to 1993 might have been caused by seeds that were imported from neighboring fields, survival of late emerging plants or a persistent weed seedbank. In 1994 and 1995, a large number of plants survived both herbicide treatments and will likely maintain high infestation level (Figure 2). A different control strategy including pre- and postemergence herbicide application and the implementation of a different crop rotation or other mortality events is necessary to manage this increasing infestation.

Velvetleaf

Velvetleaf was the dominant broadleaf species in this study (Table 1). At least 60% of the interrow field area were infested in each of the four years (Table 2). Still, spatial distribution of the two major patches was very stable in size and location (Figure 3). These results are consistent with those of Wilson & Brain (1991) who found that population of blackgrass (*Alopecurus myosuroides* Huds.)

Table 1. Mean weed density (seedlings m^{-2}) for common sunflower, velvetleaf, hemp dogbane and foxtail species before (date 1) and after (date 2) broadcast postemergence herbicide application; weed population was sampled in the interrow area (no preemergence herbicide and one cultivation in summer) and in the intrarow space (preemergence band application) using a 0.38 m^2 frame.

	Common sunflower				Hemp dogbane			
	1992	1993	1994	1995	1992	1993	1994	1995
				weeds m^{-2}				
Date 1, interrow	1.7	2.4	.7	.7	.4	.2	.4	.2
Date 1, intrarow (preemergence herbicide)	.5	.8	1.0	.3	.1	.4	.1	.1
Date 2, interrow (cultivation)	t	—	.003	.02	—	—	.07	.1
Date 2, intrarow	—	—	.07	.03	—	—	.3	.2

	Foxtail species				Velvetleaf			
	1992	1993	1994	1995	1992	1993	1994	1995
				weeds m^{-2}				
Date 1, interrow	.1	2.6	29.2	63.4	22.1	3.2	3.7	2.1
Date 1, intrarow (preemergence herbicide)	.01	1.5	18.8	8.5	6.8	1.7	2.3	1.7
Date 2, interrow (cultivation)	t	—	.1	4.2	—	—	.06	2.1
Date 2, intrarow	—	—	.1	1.8	—	—	.2	.8

t dashes indicate that no observations were made

Table 2. Weed free area (%) for common sunflower, velvetleaf, hemp dogbane and foxtail species before (date 1) and after (date 2) broadcast postemergence herbicide application; weed population was sampled in the interrow area (no preemergence herbicide and one cultivation) and in the intrarow space (preemergence band application) using a 0.38 m^2 frame.

	Common sunflower				Hemp dogbane			
	1992	1993	1994	1995	1992	1993	1994	1995
	weed free %				weed free %			
Date 1, interrow	61	59	75	69	77	87	78	86
Date 1, intrarow (preemergence herbicide)	76	70	76	81	92	83	91	90
Date 2, interrow (cultivation)	—[t]	—	99	99	—	—	92	89
Date 2, intrarow	—	—	96	97	—	—	77[t]	86

	Foxtail species				Velvetleaf			
	1992	1993	1994	1995	1992	1993	1994	1995
	weed free %				weed free %			
Date 1, interrow	96	35	10	1	21	40	40	42
Date 1, intrarow (preemergence herbicide)	99	58	5	21	34	47	46	47
Date 2, interrow (cultivation)	—[t]	—	98	62	—	—	92	50
Date 2, intrarow	—	—	97	69	—	—	79	68

[t] dashes indicate that no observations were made

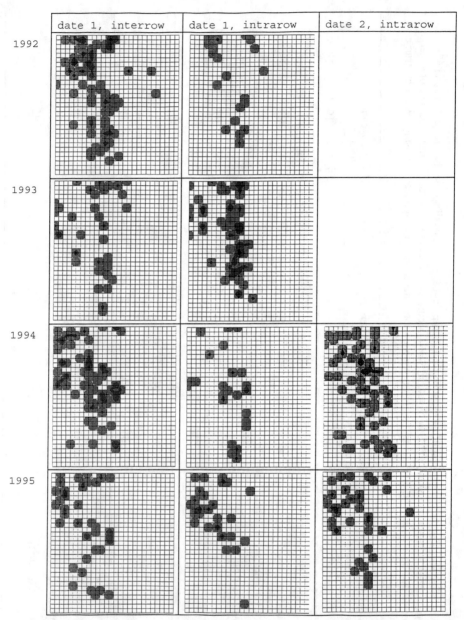

Fig.1. Distribution of hemp dogbane (*Apocynum cannabinum*) on a 4.1 ha field (217 by 189 m) plante to soybean in 1992 and corn in 1993, 1994 and 1995; weed seedling density was sampled every 7 r before (date 1) and after (date 2) postemergence application in the interrow area (no prememergenc herbicide) and in the intrarow area (preemergence band application); white = weed free, light grey = 0.1- seedlings m^{-2}, dark grey >3-15 seedlings m^{-2}, black = >15 seedlings m^{-2}.

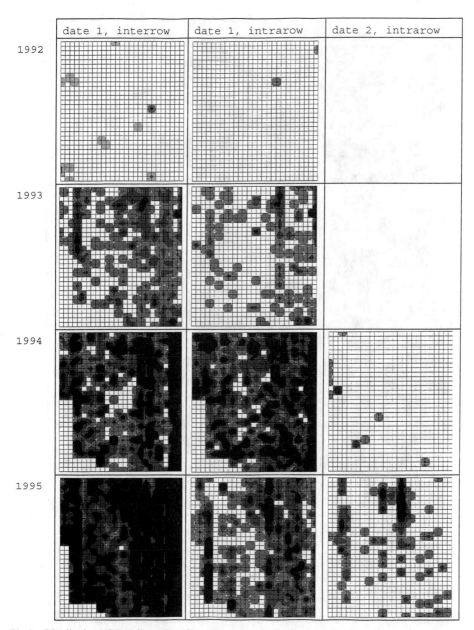

Fig.2. Distribution of foxtail species (*Setaria spec.*) on a 4.1 ha field (217 by 189 m) planted to soybean in 1992 and corn in 1993, 1994 and 1995; weed seedling density was sampled every 7 m before (date 1) and after (date 2) postemergence application in the interrow area (no prememergence herbicide) and in the intrarow area (preemergence band application); white = weed free, light grey = 0.1-3 seedlings m^{-2}, dark grey >3-15 seedlings m^{-2}, black = >15 seedlings m^{-2}.

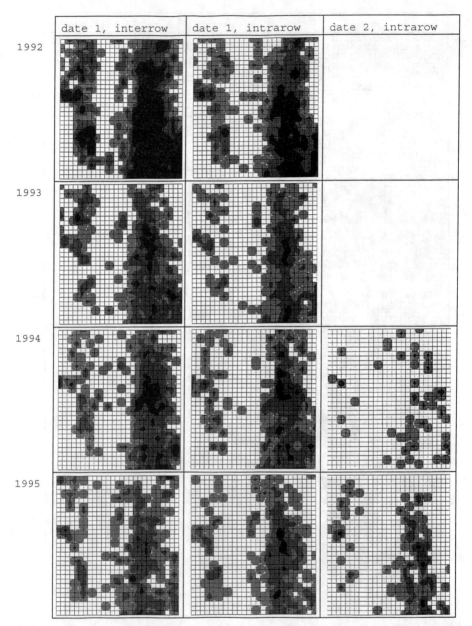

Fig.3. Distribution of velvetleaf (*Abutilon theophrasti*) on a 4.1 ha field (217 by 189 m) planted to soybean in 1992 and corn in 1993, 1994 and 1995; weed seedling density was sampled every 7 m before (date 1) and after (date 2) postemergence application in the interrow area (no prememergence herbicide) and in the intrarow area (preemergence band application); white = weed free, light grey = 0.1-3 seedlings m^{-2}, dark grey >3-15 seedlings m^{-2}, black = >15 seedlings m^{-2}.

persisted during a 10-year study in established infestations. The persistence in patchiness was attributed to the poor ability of blackgrass to colonize new locations when effective herbicides were applied each year.

After pre- and postemergence herbicides were applied in 1994 and 1995, an average of 0.2 and 0.8 seedlings m^{-2} survived and more than 20% of the field was still covered with this species. Velvetleaf is an annual with hard, large, long-lived seeds (Hanf, 1990). The amount of seeds produced by the surviving plants coupled with a persistent seed bank will probably maintain this population for many years.

CONCLUSION

The size and location of velvetleaf, common sunflower and hemp dogbane patches was stable over the four year study period. Hemp dogbane patch location, size and density was the most stable of the four species studied. Herbicide treatments had little effect on hemp dogbane infestation levels. A relatively large number of velvetleaf and some common sunflower seedlings survived in patches with high infestation levels even when effective pre- and postemergence herbicides were applied each year. Seed rain from those plants coupled with persistent seedbanks will probably maintain populations in these patch centers.

All broadleaf species showed a poor ability to colonize new areas when effective herbicides were applied each year. Further investigations in other fields, cropping systems and for other weed species are needed to gain a greater understanding of the extent of spatial stability in weed patches. More studies are also needed to decide if site specific weed management decisions could be based on instantenuous or historic weed distribution maps. These decisions need to be validated in fields using site specific control methods.

While a relatively small area of the field was free of any one weed species, GIS-based weed distribution maps still offer the potential for site specific and integrated weed management strategies. Preemergence spraying may be confined to those areas with high infestation levels and rates of postemergence applications could be reduced in locations with lower densities. For example, type and rate of herbicide could be selected according to a species distribution, with no or reduced rates of herbicides applied to patch edges and high efficacy treatments to patch centers.

The fact that weed patchiness persisted for all three broadleaf species over the four-year study indicates the present management strategy is containing but not reducing the broadleaf infestations. This is not the case for foxtail whose infestation level increased in each of the four years of the study. Alternative control strategies including different crop rotations and tillage practices as well as physical or biological control methods could be combined with these chemical controls to reduce infestation level in those patches. From this perspective, site specific weed control combined with integrated weed management strategies becomes ecologically important and economically feasible even in fields with high infestation levels.

504   GERHARDS ET AL.

REFERENCES

Gerhards, R., R. Hayer, M. Sökefeld, K. Schulze-Lohne, W. Kühbauch, W.Buchner, and M. Graff. 1995. Ein Verfahren zur teilschlaggerechten Unkrautkontrolle in Winterweizen. (Site specific weed control in winter wheat). Mitt. Ges. Pfl.bauwissenschaften 8:172-175.

Gerhards, R., M. Sökefeld, D. Knuf, and W. Kühbauch. 1996. Kartierung und geostatistische Analyse der Unkrautverteilung in Zuckerrübenschlägen als Grundlage für eine teilschlagspezifische Bekämpfung. (Mapping and geostatistical analysis of weed distribution in sugar beets). Journal of Agronomy and Crop Science, in press.

Hanf, M., 1990: Ackerunkräuter Europas. (Weeds in Europe).BASF, München.

Johnson, G. A., D. A. Mortensen, and A. R. Martin. 1995. A simulation of herbicide use based on weed spatial distribution. Weed Res. 35:197-205.

Marshall, E.J.P. 1988. Field-scale estimates of grass populations in arable land.Weed Res. 28:191-198.

Mortensen, D. A., G. A. Johnson, and L. J. Young. 1993. Weed distributionsin agricultural fields. Proc. ASA-CSSA-SSSA, Soil Specific Crop Management. 113-124.

Mortensen, D. A., G. A. Johnson, D. Y. Wyse, and A. R. Martin. 1995.Managing spatially variable weed populations. Proc. ASA-CSSA-SSSA, Soil Specific Management for Agricultural Systems. 397-415.

Nordmeyer, H., and P. Niemann. 1992. Möglichkeiten der gezieltenTeilflächenbehandlung mit Herbiziden auf der Grundlage von Unkrautverteilung und Bodenvariabilität. (Site specific weed control according to soil variability) Z. Pflkrankh. Pflschutz Sonderh. XIII:539-547.

Van Groenendael, J.M. 1988: Patchy distribution of weeds and some implications for modelling population dynamics: A short literature review. Weed Res. 28:437-441.

Wiles, L. J., G. W. Oliver, A. C. York, H. J. Gold, and G. G. Wilkerson. 1992.Spatial distribution of broadleaf weeds in North Carolina soybean (Glycine max) fields. Weed Sci. 40:554-557.

Wilson, B. J., and P. Brain. 1991. Long-term stability of distribution of Alopecurus myosuroides Huds. within cereal fields. Weed Res. 31:367-373.

Wyse-Pester, D. Y., D. A. Mortensen, and C. A. Gotway. 1996. Statistical methods to quantify spatial stability of weed population. North Cent. Weed Sci. Soc., in press.

Spatial Relationships of Soil Nitrogen with Corn Yield Response to Applied Nitrogen

T.W. Bruulsema

Potash & Phosphate Institute
Guelph, Ontario, Canada

G.L. Malzer
P.C. Robert
J.G. Davis
P.J. Copeland

Department of Soil, Water, and Climate
University of Minnesota
St. Paul, Minnesota, U.S.A.

ABSTRACT

Relationships between within-field spatial variability in soil nitrogen availability indices, landscape attributes and crop response to nitrogen fertilizer were examined in southwest Minnesota in 1994. Multiple regression models explained 30 to 61 percent of the variability in crop yield response.

INTRODUCTION

The objective of this study was to determine whether corn yield responses to nitrogen fertilizer could be predicted by a combination of measured soil nitrogen indices and landscape attributes. We evaluated a method of estimating nitrogen response curves across landscapes by interpolating corn yields to specific points where soil nitrogen indices had been measured.

METHODS

On-farm studies were conducted on three corn production fields in southwest Minnesota in 1994. Areas under study within the fields ranged from 5 to 6 hectares. Treatments were applied in late October 1993 in continuous strips the length of the plot, using four to six replicate strips per rate of nitrogen. Strip width was approximately 5 m, depending on field equipment used at each site. Nitrogen was applied as anhydrous ammonia using a radar controlled variable-rate applicator in order to ensure a constant rate of application both up and down slopes. Corn yields were measured on the center rows from each strip in 15 m segments, using a plot combine with a load cell.

Soils were sampled in the zero rate treatments, resulting in an approximate 30 m grid, both in October 1993 and in June 1994. October samples measured nitrate-N, total N (Kjeldahl), phosphate-borate N and hot KCl N (Gianello and Bremner, 1986) in a single core (5 cm diam.) to 30 cm depth. June samples were composites of nine cores (2 cm diam.) taken within a one m radius, to 30 cm depth. Dried soils were analyzed for the same indices as for the the fall samples, excepting total N. In addition, the fresh soils were analyzed for anaerobically mineralizable N (ANMN; Keeney, 1982) and biologically active soil organic N (ASN; Duxbury *et al.*, 1991).

Digital elevation models (DEM) of the field landscape were constructed using laser theodolite measurements taken on a spacing of ~10 m. Slope gradient and profile curvature (convex positive) were calculated on the basis of a 3 by 3 cell grid with node spacing of ~5 m (Moore et al., 1993; Pennock et al., 1987). An insolation index was computed as the sine of solar elevation angle relative to the south-facing slope angle, using 61° as median solar elevation relative to the horizontal for the spring period. Soil phototone was digitized from color film aerial photographs of bare soil taken in May 1994.

At each of the soil sample points within each field, corn yield for each level of N fertilizer was estimated by interpolation of the yields measured in segmented strips. Interpolation was by punctual kriging from the nearest three neighbours using GS+/386 software (Gamma Design Software, 1993). The choice of semivariance model used by the kriging interpolator was based on jack-knife cross-validation statistics (Table 1).

Weighted quadratic yield response models were fitted at each soil sample point using PROC GLM (SAS Institute, 1988). The inverse of the kriging variance for each yield estimate was used as the weight factor. Economic optimum nitrogen rate (EONR) was calculated from the first derivative of the quadratic model using a price ratio of 3.4 kg corn grain per kg of N fertilizer. Limits were imposed such that EONR was not allowed to be less than zero or greater than the highest rate of N applied. A second parameter calculated from the response curves was Δ Yield (Kachanoski *et al.*, 1996), here defined as the difference in yield at EONR versus that at zero N fertilizer.

Relationships between yield responses, soil nitrogen indices and landscape attributes were evaluated by the use of simple Pearson correlation statistics and by stepwise multiple regression. The stepwise procedure included independent variables only if significant at $P<0.20$.

Table 1. Mean, coefficient of variation (CV), and kriging SD for corn yield estimates at the soil sample points for each N rate treatment at each site.

Site	N Rate	Mean Yield	CV	Kriging SD[‡]	Jack-Knife[†] R^2
	kg ha^{-1}	t ha^{-1}		t ha^{-1}	
Hanska					
	0	10.9	7%	0.3	0.55
	67	12.1	4%	0.4	0.46
	101	12.2	5%	0.4	0.53
	134	12.5	4%	0.5	0.23
	168	12.5	4%	0.5	0.11
	202	12.2	3%	0.3	0.31
Hector					
	0	8.6	25%	1.2	0.73
	67	10.7	17%	1.6	0.55
	101	11.5	12%	1.3	0.57
	134	11.3	15%	1.0	0.61
	168	11.6	11%	1.2	0.41
	202	11.9	10%	1.1	0.39
Lake Crystal					
	0	8.8	19%	0.8	0.71
	67	10.8	10%	0.9	0.46
	134	11.1	6%	0.7	0.28
	202	11.2	7%	1.1	0.11

† Jack-knife R^2 refers to cross-validation at the locations of the harvested yield segments.

‡ Mean kriging SD over all sample points

RESULTS AND DISCUSSION

A typical site elevation map is depicted in Fig. 1, indicating the pattern of Δ Yield and the points at which soils were sampled. Other sites were similar in scale and sampling
pattern.

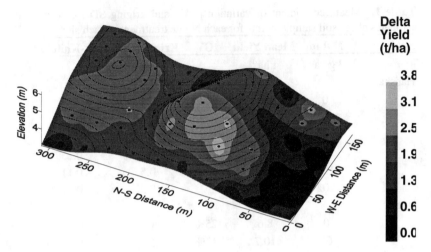

Fig. 1. Overlay of elevation, soil sample points and estimated Δ Yield (t ha⁻¹) at the Hanska site. Contour interval of elevation is 0.2 m.

Soil nitrogen indices measured in the spring were no better related to crop responses than were the fall samples (Tables 2 & 3). This is surprising considering that the spring samples were composites and, particularly for nitrate, that indices would be expected to relate more closely when measured closer to the time of crop nitrogen uptake. Considering both simple correlation coefficients and significance in multiple regression models (Tables 6 & 7), the most consistent soil nitrogen index across sites was total N. The indices measured on fresh soils, ANMN and ASN, were not consistent predictors of crop response, though they were significant at some of the sites (Tables 3 & 6).

Table 2. Pearson correlation coefficients for the relationship between economic optimum N rate (EONR) and measured soil N indices.

	Hanska	Hector	Lake Crystal
Fall Samples			
Nitrate-N	-0.31*	NS	-0.39**
PBN	NS	-0.27*	-0.29*
Hot KCl-N	NS	NS	-0.44**
Total N	-0.31*	-0.38**	NS
Spring Samples			
Nitrate-N	NS	NS	-0.31*
PBN	NS	NS	NS
Hot KCl-N	NS	NS	-0.34*
ANMN	NS	NS	NS
ASN	NS	NS	NS

Table 3. Pearson correlation coefficients for the relationship between yield response (Δ Yield) and measured soil N indices.

	Hanska	Hector	Lake Crystal
Fall Samples			
Nitrate-N	NS	-0.39**	-0.40**
PBN	NS	-0.36**	-0.62**
Hot KCl-N	NS	NS	-0.60**
Total N	-0.42**	-0.37**	-0.54**
Spring Samples			
Nitrate-N	NS	NS	-0.43**
PBN	NS	NS	-0.61**
Hot KCl-N	NS	NS	-0.64**
ANMN	NS	-0.32*	-0.40**
ASN	-0.32*	NS	NS

At the Hanska site, both EONR and Δ Yield were positively correlated to elevation and soil phototone (Tables 4 & 5). Responses to nitrogen tended to be larger at higher elevation and on soil of lighter color. At Lake Crystal, EONR was negatively related to slope gradient, indicating higher nitrogen requirements on level than on sloping land (Table 4). Profile curvature was positively associated with Δ Yield, indicating greater responsiveness to nitrogen on convex landforms than on concave (i.e. more response on knolls than in bowls).

Table 4. Pearson correlation coefficients for the relationship between economic optimum N rate (EONR) and landscape attributes calculated from the DEM.

	Hanska	Hector	Lake Crystal
Landscape Attribute			
Relative Elevation	0.37**	NS	NS
Slope Gradient	NS	NS	-0.26*
Profile Curvature	NS	NS	NS
Bare Soil Phototone	0.36**	NS	NS

Table 5. Pearson correlation coefficients for the relationship between yield response (Δ Yield) and landscape attributes calculated from the DEM.

	Hanska	Hector	Lake Crystal
Landscape Attribute			
Relative Elevation	0.44**	NS	NS
Slope Gradient	0.31*	NS	NS
Profile Curvature	0.32*	NS	0.43**
Bare Soil Phototone	0.51**	NS	NS

Sites varied in terms of the relationship of yield response and landscape attributes. For example, EONR was positively related to elevation at Hanska, negatively related at Hector and unrelated at Lake Crystal (Table 6). The Hector site had very high soil water content during much of the growing period, and nitrogen responses appeared to occur most strongly in the areas where soils were saturated.

In general, Δ Yield was more strongly related than EONR to soil N indices and topography (compare Table 2 with Table 3; Table 4 with Table 5; and Table 6 with Table 7). This implies that Δ Yield may be more predictable than EONR, which is encouraging because Δ Yield is more closely associated with the net return to fertilizer nitrogen use. The profitability of variable rate technology is more likely to be maximized by focusing on Δ Yield rather than EONR.

Table 6. Multiple linear regression coefficients for the dependent variable EONR derived from stepwise regression against soil and landscape attributes.

Attribute		Hanska	Hector	Lake Crystal
Intercept (kg N ha^{-1})		-590	430	48
Fall	Nitrate-N (mg kg^{-1})			-4
	Hot KCl-N (mg kg^{-1})			-5
	Total N (g kg^{-1})	-36	-63	
Spring	Soil Water (g kg^{-1})	170		53
	ANMN (mg kg^{-1})			1
	Nitrate-N (mg kg^{-1})			-4
Elevation (m)		41	-34	
Slope Gradient (°)				13
Profile Curvature (° 100 m^{-1})			43	
Soil Phototone		2		
R^2		**0.30**	**0.26**	**0.36**

Although the measured yields for the zero rate treatments were spatially closer to the points of estimation, the kriging SD for check (zero N rate) yield estimates were not substantially lower than those for other treatments (Table 1). This likely resulted from higher yield variability at the zero rate of nitrogen.

Table 7. Multiple linear regression coefficients for the dependent variable Δ Yield derived from stepwise regression against soil and landscape attributes.

Attribute		Hanska	Hector	Lake Crystal
Intercept (t ha^{-1})		-42	13	7
Fall	Nitrate-N (mg kg^{-1})		-0.4	
	PBN (mg kg^{-1})			-0.02
	Hot KCl-N	(mg kg^{-1})		0.2
	Total N (g kg^{-1})	-1	-2	-0.9
Spring	Soil Water (g kg^{-1})	3		1
	Hot KCl-N (mg kg^{-1})			-4
Elevation (m)			-1	-0.3
Profile Curvature (° 100 m^{-1})				0.5
Insolation Index		37		
Soil Phototone		0.08		
R^2		**0.46**	**0.39**	**0.61**

The strength of relationships between soil N indices, landscape attributes and yield responses was generally less than expected. Possible reasons include low yield variability and low yield response variability, particularly at the Hanska site (Tables 1 & 8). The 1994 growing season produced excellent yields but with less responsiveness to nitrogen than encountered in other years (Table 8). Another potential reason diminishing the strength of the relationships was differences in the scale of the parameters measured and estimated. The soil N indices represented a scale of ~2 m; the landscape attributes a scale of ~10 m, while the kriged yield estimates represent a scale of ~30 m.

Table 8. Mean and coefficient of variation (CV) for estimated economic optimum nitrogen rate (EONR) and yield response (Δ Yield) at the soil sample points within each site.

Site	EONR (kg N ha^{-1})		Δ Yield (t ha^{-1})	
	mean	CV	mean	CV
Hanska	100	60%	1.5	75%
Hector	113	61%	2.8	77%
Lake Crystal	129	40%	2.3	70%

Correlation and regression statistics likely over-estimate the actual strength of relationship, because of spatial autocorrelation (Dutilleul, 1993; Long, et al., 1993). In this study, all sample points were separated by at least 30 m, which minimized but did not completely avoid autocorrelation errors. However, while autocorrelation is of concern when inferring general principles, it may not impede the utility of the relationships for interpolating (mapping) within the bounds of the site, particularly for regularly spaced sample points.

ACKNOWLEDGMENTS

Funding provided by the Legislative Committee on Minnesota Resources is gratefully acknowledged. Technical assistance by T. Graff is greatly appreciated. Valuable comments and suggestions on approaches to the study were provided by W.H. Thompson, J.C. Bell, M.D. Tomer, and D.S. Long.

REFERENCES

Dutilleul, P. 1993. Modifying the t test for assessing the correlation between two spatial processes. Biometrics 49:305-314.
Duxbury, J.M., J.G. Lauren, and J.R. Fruci. 1991. Measurement of the biologically active soil nitrogen fraction by a ^{15}N technique. Agriculture, Ecosystems and Environment 34:121-129.
Gamma Design Software. 1993. GS+: Geostatistics for the Environmental Sciences. Plainwell, Michigan.
Gianello, C., and J.M. Bremner. 1986. Comparison of chemical methods of assessing potentially available organic nitrogen in soil. Comm. Soil Sci. Plant Anal. 17:215-236.
Kachanoski, R.G., I.P. O'Halloran, D. Aspinall, and P. von Bertoldi. 1996. Delta yield: Mapping fertilizer nitrogen requirement for crops. Better Crops With Plant Food 80:20-23.
Keeney, D.R. 1982. Nitrogen availability indices. In Page, A.L., R.H. Miller, and D.R. Keeney (eds.) Methods of Soil Analysis. Part 2: Chemical and Microbiological properties. ASA, Madison, WI.
Long, D.S., S.D. DeGloria, D.A. Griffith, G.R. Carlson and G.A. Nielsen. 1993. Spatial regression analysis of crop and soil variability within an experimental research field. In P.C. Robert, R.H. Rust, and W.E. Larson, (ed.) Proceedings of Soil Specific Crop Management: a Workshop on Research and Development Issues, April 14-16, 1992, Minneapolis, MN. American Society of Agronomy, Madison, WI.
Moore, I.D., P.E. Gessler, G.A. Nielsen, and G.A. Peterson. 1993. Terrain analysis for soil specific crop management. p. 27-56. In P.C. Robert, R.H. Rust, and W.E. Larson, (ed.) Proceedings of Soil Specific Crop Management: a Workshop on Research and Development Issues, April 14-16, 1992, Minneapolis, MN. American Society of Agronomy, Madison, WI.
Pennock, D.J., B.J. Zebarth, and E. De Jong. 1987. Landform classification and soil distribution in hummocky terrain, Saskatchewan, Canada. Geoderma 40:297-315.
SAS Institute. 1988. SAS/STAT User's Guide; Release 6.03 Edition. SAS Institute, Inc., Cary, NC, USA.

Using Yield Variability to Characterize Spatial Crop Response to Applied N

J.G. Davis
G.L. Malzer
P.J. Copeland
J.A. Lamb
P.C. Robert

Department of Soil, Water, and Climate
University of Minnesota
St. Paul, Minnesota

T.W. Bruulsema

Potash & Phosphate Institute
Guelph, Ontario

ABSTRACT

Yield monitors mounted on combines provide corn (*Zea mays* L.) producers with the capability of determining within-field yield variability. The usefulness of yield measurements in characterizing plant nutrient needs is based on the assumption that fertilizer requirement is proportional to yields; however, some recent studies in the upper Midwest have shown poor relationships between optimal N rates and grain yields. Therefore, a need exists for information demonstrating the efficacy of yield level as a criterion for delineating areas within fields having dissimilar N requirements. Four on-farm studies (4-6 ha) were conducted in southwest Minnesota in 1994 and 1995 in order to examine and characterize spatial variability in corn grain yield. Fertilizer N was applied as anhydrous ammonia in constant-rate strips (rates ranged from 0 to 200 kg N ha-1) across the fields. Yields were determined on 15-m segments within each treatment strip. Economic optimum N rates were determined from spatially-referenced yield response "sub-blocks", composed of a complete set of N rates within adjacent 15-m yield segments. Yield levels from check plots (zero N applied) were more accurate and precise predictors of spatial patterns of N requirement than were yield levels from well fertilized plots (168 kg ha-1 N). These results point out that producers are likely to have difficulty when using yield maps produced from well-fertilized soils to determine areas within fields having different N requirements. The results also emphasize the need for tools, in addition to yield maps, that can be used to determine these areas.

INTRODUCTION

Yield monitors have become widely available in recent years, giving producers unprecedented capabilities to visualize spatial yield variability within fields using yield maps. One use of these yield maps is to produce a "yield goal" map using yield maps from several years. The yield goal map may then be used with the appropriate recommendation algorithm to produce a fertilizer prescription map to variably apply nutrients such as nitrogen (N). This procedure involves an assumption that yield goal is an adequate representation of spatial nutrient needs, and that yield goal can reflect fertilizer use efficiency. Fertilizer N recommendations generally are made for whole fields, but these same recommendations are being applied to much smaller areas within fields. The objective of this study was to determine whether yield monitor maps should be used to make decisions concerning spatial fertilizer N recommendations.

METHODS

On-farm studies were conducted on four corn production fields in southwest Minnesota in 1994 and 1995. Areas under study within the fields ranged from 4.5 to 5 ha. Treatments were applied in constant-rate strips the length of the plot (243 to 304 m) as fall-applied anhydrous ammonia. A radar-controlled variable-rate anhydrous ammonia applicator was used compensate for variations in applicator speed and thus apply a constant rate of N. The strips were 4.6 m wide, the width of six corn rows. Six strips representing each of six N rate treatments (0, 67, 101, 134, 168, and 202 kg ha^{-1}) were arranged side-by-side to form a complete block (Fig. 1). Corn yields were measured on the center rows from each strip in 15 m segments. A N-rate "sub-block" was formed from an adjacent set of six N-rate treatments for each 15-m harvest segment. The N-rate sub-block formed the smallest spatially-referenced unit containing all the information used for later analyses.

A quadratic polynomial was fit to yield response data from each sub-block to determine spatial yield response characteristics. Other models were considered, but the quadratic model usually fit the data better (higher R^2) and discontinuous models such as the linear-plateau and quadratic-plateau failed to fit the data in some sub-blocks. The first parameter calculated from response functions was the economic optimum N rate (EONR), determined for each sub-block by setting the first derivative of the yield response function equal to a selected fertilizer-to-corn price ratio. The fertilizer-to-corn price ratio for 1994 was 3.25 ($0.26 kg^{-1} for N, $0.080 kg^{-1} for corn) and 3 for 1995 ($0.37 kg^{-1} for N, $0.12 kg^{-1} for corn). The EONR calculations were constrained to be greater than or equal to zero and no greater than the highest rate of N applied (202 kg ha^{-1}). The second parameter was profitability, here defined as the increase in crop value due to N fertilization. It was calculated by subtracting the crop value for the check plots and the cost for applied fertilizer from the crop value at the EONR. For these analyses, the costs of implementing variable-rate N management were not included, because these costs are rather unstable at present. A farm manager could

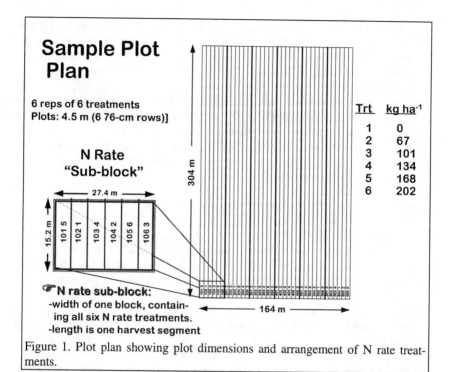

Figure 1. Plot plan showing plot dimensions and arrangement of N rate treatments.

use these results for comparisons, and insert prevailing local costs to evaluate the cost-effectiveness of variable-rate N application. The third parameter calculated from the response data was *delta yield* (Δ yield), defined as the difference in yield for a fertilized plot versus that at zero N fertilizer (Kachanoski et al., 1996). For this study the yield at 168 kg N ha[-1] was chosen as the yield for a well-fertilized plot, because University of Minnesota recommendations would suggest application of 146 kg N ha[-1] (Rehm et. al, 1993). To apply N in adequate but not excessive levels 168 kg N ha[-1] was judged to be the most appropriate of the N rates applied. The choice of the well-fertilized N rate for Δ yield calculations is important to the accuracy of these observations. Δ yield will be an adequate index of responsiveness if the selected well-fertilized rate is on a portion of the yield function that is neither rapidly increasing or decreasing. With no prior knowledge of the shape of the yield function, however, this choice is at best an educated guess.

Map comparisons were based on the product moment correlation (*r*), a covariance measure. It was calculated from the following equation as suggested by

$$r = \frac{\text{covariance(comparison map, reference map)}}{\text{st. dev. comparison map} \bullet \text{st. dev. reference map}}$$

Robinson and Bryson (1957). This calculation is equivalent to a Pearson product moment calculation from standard statistical texts, and is attractive because it is readily computed, and is relatively easily understood. However, any statistical inferences from these correlations rely on an assumption of independent random sampling, an assumption that is violated when comparing maps that are spatially autocorrelated (Unwin, 1981). Therefore, the correlation coefficients were used to indicate the similarity of mapped patterns, but no further statistical inferences were made.

Surface maps showing relative elevation of the field landscape were made using x, y, and z-coordinates taken on an approximate 10 m spacing using laser theodolite measurements. The relative spatial correspondence of patterns in measured yield parameters with relative elevation could be observed visually by overlaying the three dimensional elevation map with a two-dimensional contour map of the yield parameters.

RESULTS AND DISCUSSION

A map showing relative elevation for the Hanska, 1995 site is shown in Fig. 2. Yield parameters are overlayed on the elevation surface map in Figs. 3-7. Spatial patterns in yield associated with topography are observable in both the check plot yield map and in the map of 168 kg N ha^{-1} yield. Spatial patterns in yield level are more evident in the check plot yield map than in the 168 kg N ha^{-1} map. This result should be expected when areas within fields requiring additional N are fertilized. At higher rates of added N, factors other than N control yield levels.

Patterns in EONR levels also were related to topography (Fig. 5), although these patterns did not directly correspond to patterns in yield level. This is reasonable because the correspondence of yield level and EONR depends on the shape of the response function. Areas within the field requiring less N were associated with sideslopes, and areas requiring the highest N applications were associated with localized depressional areas. These depressional areas often had stunted corn growth and reduced grain yields due to flooded soils early in the growing season. However, local climatic conditions can reverse these trends. Depressional areas had the lowest EONR at Hanska 1994, but low areas had the highest EONR at the Hector 1994 site (maps not shown). These results were the result of somewhat more rainfall during the growing season at the Hector site (717 mm) compared to the Hanska site (684 mm) in 1994, and poorer drainage at Hector. Patterns of Δ yield and profitability were similar (Figs. 6-7) because both are related to the N responsiveness of the crop. However, the areas requiring higher levels of N (higher EONR) were not always the most responsive (high Δ yield) or most profitable areas. This outcome likely is due to dissimilar fertilizer use efficiencies caused by variable drainage patterns within a field.

Map comparisons between observed yield levels and EONR, Δ yield , and profitability by product-moment correlations are shown in Table 1. The best correlations were obtained between check plot yields and maps of profitability. This outcome is expected when yield levels are predictive of responsiveness to added N. Because of the visual similarity of patterns of Δ yield and profitability, it was logical to expect similar correlations between check plot yield and Δ yield as were obtained with check plot yield and profitability. However, correlations for

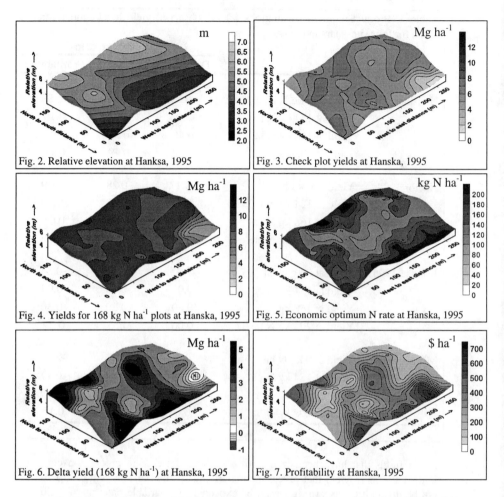

Fig. 2. Relative elevation at Hanksa, 1995

Fig. 3. Check plot yields at Hanska, 1995

Fig. 4. Yields for 168 kg N ha^{-1} plots at Hanska, 1995

Fig. 5. Economic optimum N rate at Hanska, 1995

Fig. 6. Delta yield (168 kg N ha^{-1}) at Hanska, 1995

Fig. 7. Profitability at Hanska, 1995

check plot yield and Δ yield comparisons were not as high as for check plot yield and profitability comparisons with Hector 1994 and Hanska 1995 data. These lower correlations were caused by greater variability in Δ yield within sub-blocks compared to profitability within sub-blocks. Variability in profitability within sub-blocks was restricted in the calculations, because a negative profitability was not allowed. It was reasoned that a producer would not be expected to apply N where it was not economically favorable, so sub-blocks having negative responsiveness were set to zero profitability.

Correlations between check plot yields and the three reference maps were superior to correlations between the more well-fertilized plots and reference maps. This observation agrees with visual observations of spatial maps where patterns observed in check plot maps tended to disappear in spatial maps of yields where higher rates of N were applied. The implication of these results is

Table 1. Product-moment correlations (*r*) for observed yield levels and reference maps for four site-years in the study.

Comparison map		Reference map		
		Delta yield	EONR	Profitability
Hanska 1994	Yield at 0 kg N ha^{-1}	-0.79	-0.34	-0.89
	Yield at 134 kg N ha^{-1}	-0.06	-0.11	0.15
	Yield at 168 kg N ha^{-1}	0.48	0.34	0.03
	Yield at 202 kg N ha^{-1}	0.32	0.49	0.11
Hector 1994	Yield at 0 kg N ha^{-1}	-0.47	-0.28	-0.79
	Yield at 134 kg N ha^{-1}	-0.15	-0.28	-0.19
	Yield at 168 kg N ha^{-1}	0.08	-0.05	0.01
	Yield at 202 kg N ha^{-1}	-0.35	0.20	-0.22
Hanksa 1995	Yield at 0 kg N ha^{-1}	-0.45	-0.36	-0.84
	Yield at 134 kg N ha^{-1}	0.07	-0.13	-0.04
	Yield at 168 kg N ha^{-1}	0.40	-0.20	-0.30
	Yield at 202 kg N ha^{-1}	0.00	0.33	0.27
Morgan 1995	Yield at 0 kg N ha^{-1}	-0.66	0.05	-0.77
	Yield at 134 kg N ha^{-1}	-0.14	-0.01	-0.17
	Yield at 168 kg N ha^{-1}	0.27	0.18	-0.09
	Yield at 202 kg N ha^{-1}	-0.07	0.30	0.16

that in fields that have been well-fertilized, the usual case in production corn fields, patterns in yield that reveal the underlying spatial N needs within a field will be difficult to discern. It should be expected that variable patterns of yield response will decrease as N application is increased to adequate but not excessive levels. However, if the purpose for yield monitoring is to create a prescription map to guide spatial N application, crop yield levels from zero-N test strips will be needed. These test strips will help reveal the inherent spatial N supply.

It should be emphasized that the whole-field correlations presented here give a single measure of association for an entire field. Similiarities between comparison maps and reference maps in one area of the field may well be masked by strong dissimilarities in other areas. A comparison method that tests within-field similarities in patterns will provide more information concerning these smaller-scale patterns.

It has been suggested that several years of yield maps be created before any fertilizer recommendations are made. Results from this study suggest that even with several years of yield maps, patterns revealing spatial N needs may not be clearly revealed. Yield levels vary with interactions of climate and factors that affect water status with the soil profile, but these are likely to complex interactions of water and nutrient dynamics that will have unpredictable effects on yield.

ACKNOWLEDGMENTS

Funding for this project was provided by the Legislative Committee on Minnesota Resources. Technical assistance by T.J. Graff and H. Dikici is greatly appreciated. Helpful suggestions and guidance on analysis approaches used in the study were provided by J.C. Bell and J.A. Thompson.

REFERENCES

Kachanoski, R.G., I.P. O'Halloran, D. Aspinall, and P. von Bertoldi. 1996. Delta yield: Mapping fertilizer nitrogen requirement for crops. Better Crops With Plant Food 80:20-23.

Rehm, G.W., M.A. Schmitt, and R. Munter. 1993. Fertilizing corn in Minnesota. Univ. of Minn. Ext. Serv. AG-FO-3790B

Robinson, A. H., and R. A. Bryson. 1957. A method for describing quantitatively the correspondence of geographical distributions. Annals Assoc. Amer. Geogr. 47:379-91.

Unwin, D. 1981. Map comparison . p.187-207. *In* Introductory spatial analysis. Methuen and Co. New York.

Variability of Soil Nitrate and Phosphate Under Different Landscapes

D.W. Franzen
L.J. Cihacek

Department of Soil Science
North Dakota State University

V.L. Hofman

Department of Agricultural Engineering
North Dakota State University

ABSTRACT

One factor affecting soil nitrogen (N) levels is landscape. Sampling by landscape was effective in reproducing nitrate-N variability from three fields originally sampled in a 33 m grid. Nitrate-N was more highly correlated to landscape than was soil phosphorus.

BACKGROUND

Literature Review

Soil nitrogen (N) is affected by topography, or landscape (Stevenson, 1982). Landscape affects soil N through its influence on soil water movement and microclimate. Soil water content may be related to landscape (Halvorson & Doll, 1991). Basing fertilizer N applications on yield goal and sampling based on the soil mapping unit has been shown to be effective in increasing crop productivity (Carr, et al., 1991), however, soil mapping units often contain inclusions of other soil series which may contribute to errors (Steinward, et al., 1996). Soil survey maps may not be accurate for an individual field, and even when boundaries are correct, these boundaries may not be strongly related to soil nutrient levels (Franzen & Peck, 1993). On the other hand, landscape position has been shown to be related to soil N levels (Fiez, et al., 1994; Nolan, et al. 1995) and may increase profitability of variable-rate N application more when combined with N recommendations based on yield goals for each landscape (Fiez, et al., 1994). Landscape sampling may reduce the number of samples necessary to represent nutrient levels in some fields compared with intensive grid sampling (Hollands, 1996).

Objectives

The objectives of this study were:

1.   To evaluate soil nitrate-N and phosphorus (P) levels using a dense grid at three locations and determine the relationship of landscape to soil nitrate-N and P variability.
2.   Compare nitrate-N and P information gathered from a 33 m grid soil sampling with landscape and less dense grid sampling results.

METHODS AND MATERIALS

Three fields were sampled in a grid pattern in the fall of 1995. The first site is a 16.2 ha field located near Colfax, ND and was cropped to corn during 1995. The second site is a 16.2 ha field located near Valley City, ND and cropped to sunflower in 1995. The third site is a 30.9 ha field near Mandan, ND on the North Dakota Soil Conservation District Area IV farm associated with the USDA-ARS Great Plains Research Laboratory. Both the Colfax and Valley City sites were sampled on a 33 m grid.

The Mandan site is divided into three fields. The east field is 10.9 ha, the center field is 11.7 ha and the west field is 8.3 ha. The west and east fields were cropped to spring wheat and the center field was cropped to sunflower. The east field was sampled in a 33 m grid and the center and west fields were sampled in a 46 m grid. Three to five sample cores were taken per sample in a diagonal pattern approximately 3 m apart to represent each grid sample location. Sample cores were taken at 0-15 cm and 15-60 cm. Olsen-P (Sodium bicarbonate extracted) and nitrate-N were analyzed on the surface sample and nitrate-N was analyzed on the 15-60 cm sample. Nitrate-N is reported in terms of kg/ha 0-60 cm.

The Mandan field was also mapped for relative elevation. A laser-leveler device was used in combination with a DGPS receiver to record location and relative elevation in a 33 m grid across all the Mandan field. Data was mapped using Surfer (Golden Software, Inc., Golden, CO). At Valley City and Colfax, fields were harvested too late in the fall to be mapped, but organic matter was used as a landscape related factor for comparison. Map parameters were inverse distance squared estimates using 8 nearest neighbors for the 33 and 66 m grids. Four nearest neighbors were used for the 100 m grid, and 2 nearest neighbors for the 3.3 ha grid. Correlation and conventional statistics were performed using SYSTAT for windows (SYSTAT, Inc., 1992, Evanston, IL).

RESULTS AND DISCUSSION

At Mandan, soil nitrate-N levels are shown in Fig. 1. The Mandan landscape map is shown in Fig. 2. Low nitrate-N in the center field was related to sunflower uptake. However, the high area of nitrate-N was due to residual N from the death of sunflower in the area because of standing water early in the growing season. Differences in nitrate-N in other areas of the field were related to other factors such as washing of manure into the field from the neighbor to the north.

Figure 1. Mandan nitrate-N 0-60 cm 1995.

Figure 2. Mandan relative elevation.

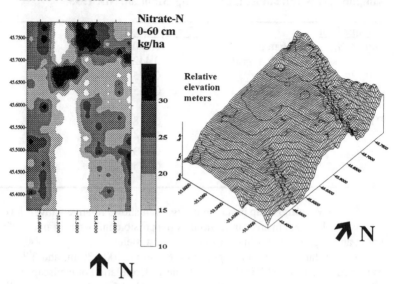

Figure 3. Mandan P levels, 1995.

Figure 4. Valley City 0-60 cm nitrate-N, 1995.

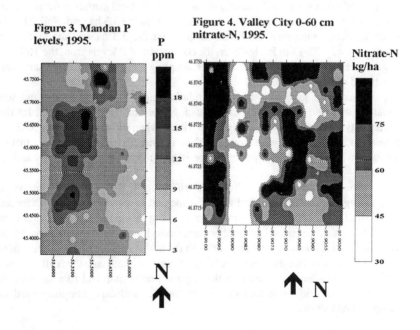

Table 1. Correlation of 33-43 m soil nitrate-N and P levels with different sampling grids and landscape sampling, Mandan, ND, 1995.

Comparison		Correlation (r)
Nitrate-N,	66-86 m grid	0.290
	100-130 m grid	0.442
	3.3 ha grid	0.225
	Landscape	0.755
P levels,	66-86 m grid	0.578
	100-130 m grid	0.224
	3.3 ha grid	0.583
	Landscape	0.575

The Mandan P map is shown in Fig. 3. It shows relatively low P levels in the east field, except in areas influenced by the possible influence of manure in the north, and higher P levels in the center and west fields.

Correlation of 33-43 m grid with 66-86 m, 100-130 m, and 3.3 ha grid estimated nitrate-N and P is shown in Table 1. Correlation of landscape sampled nitrate-N and P are also shown in Table 1. Nitrate-N from landscape sampling is more highly correlated with field nitrate-N levels than from sampling at the 66-86 m, 100-130 m and 3.3 ha grids. The 100-130 m grid is most highly correlated among grid sizes with the 33 m grid nitrate-N levels. Grid sampling for nitrate-N using a 66-86 m grid is similar to the estimates from the 3.3 ha grid. Soil P levels from landscape sampling, the 66-86 m grid and the 3.3 ha grid are similar in correlation with 33-43 m P. However, the correlation of P levels from the 100-130 m grid is relatively low compared to other sampling methods. Landscape sampling was based on 15 samples.

Taken off guard by the strong relationship between landscape and soil nitrate-N and also soil P levels at Mandan, elevations were not planned for the Valley City and Colfax sites until into the winter, when snow/mud prevented these measurements until next year. However, there is evidence based on organic matter patterns at these two sites that landscape is also related to soil nitrate-N and P levels.

Soil nitrate-N at Valley City is shown in Fig. 4. Soil P levels are shown in Fig. 5. Organic matter is related to landscape and is shown in Fig. 6. There are visual similarities between the nitrate-N map in Fig. 4 and the organic matter map in Fig. 6. Low soil nitrate-N generally occurs in the low organic matter areas, while high nitrate-N levels are found most often in the higher organic matter areas. There are fewer similarities between the organic matter map and the soil P map. Landscape sampling was based on the organic matter map and five samples. A comparison of grid maps and one based on the landscape sampling used in comparison is shown in Fig. 10.

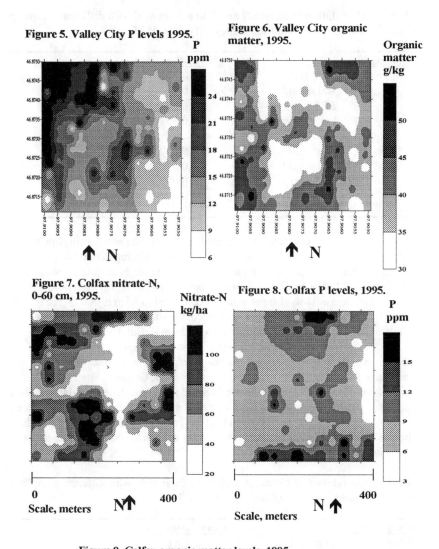

Figure 5. Valley City P levels 1995.

Figure 6. Valley City organic matter, 1995.

Figure 7. Colfax nitrate-N, 0-60 cm, 1995.

Figure 8. Colfax P levels, 1995.

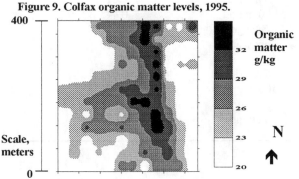

Figure 9. Colfax organic matter levels, 1995.

Table 2. Correlation of nitrate-N and P in a 33 m grid with different sampling grids and landscape sampling, Valley City, ND, 1995.

Comparisons		Correlation (r)
Nitrate-N,	66 m grid	0.501
	100 m grid	0.211
	3.3 ha grid	0.207
	Landscape	0.352
P levels,	66 m grid	0.750
	100 m grid	0.679
	3.3 ha grid	0.093
	Landscape	0.327

Estimated nitrate-N and P from landscape sampling based on organic matter levels and from a 66 m, 100 m and a 3.3 ha grid was compared to the 33 m grid values (Table 2). Landscape nitrate-N was more strongly correlated with 33 m nitrate-N than 100 m or 3.3 ha grid levels, but was not as highly correlated as a 66 m grid. Landscape P was also not as highly correlated as 66 m or 100 m grid P, with 66 m grid P being more highly correlated than the 100 m grid P. However, landscape P was more highly correlated with 33 m P than 3.3 ha grid P.

At Colfax, the nitrate-N levels are shown in Figure 7. Soil P levels are shown in Figure 8. The organic matter levels were also used as a basis for landscape sample mapping of nitrate-N and P levels. Organic matter from Colfax is shown in Figure 9. Estimated nitrate-N and P from landscape sampling based on organic matter levels was compared to the 33 m grid values (Table 3). Estimated nitrate-N and P from 66 m, 100 m and 3.3 ha grids were also compared to values from a 33 m grid. The landscape sampling was based on the organic matter map and five samples.

Table 3. Correlation of nitrate-N and P in a 33 m grid with different sampling grids and landscape sampling, Colfax, ND, 1995.

Comparison		Correlation (r)
Nitrate-N,	66 m grid	0.616
	100 m grid	0.448
	3.3 ha grid	0.061
	Landscape	0.320
P levels,	66 m grid	0.620
	100 m grid	0.370
	3.3 ha grid	0.173
	Landscape	0.157

Figure 10. Nitrate-N mapping using landscape and different grid sizes compared to original 33 m grid, Valley City, 1995.

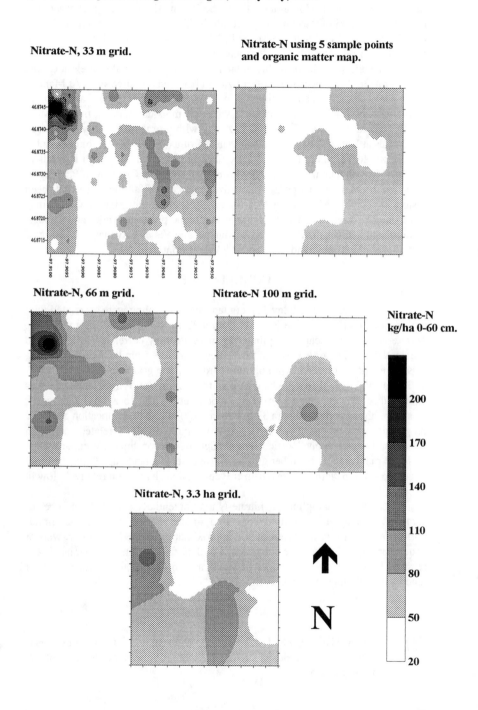

Landscape nitrate-N was more highly correlated with 33 m grid than a 3.3 ha grid, but not as correlated as the 66 m or 100 m grids. The 66 m grid nitrate-N was more highly correlated with 33 m nitrate-N than the 100 m grid. Landscape P was not as correlated with 33 m P as landscape nitrate-N. The 66 m P grid was more highly correlated with 33 m P than landscape P, 100 m P and 3.3 ha grid P. Landscape P was not as correlated with 33 m P as any grid pattern tested.

The difference in the ability of landscape to represent soil P levels at Colfax may be due to differences in past fertilization between the three sites. At Mandan and Valley City, annual applications of soil P may have been lower than at Colfax in the past. Normally, yield goals rise with soil moisture patterns from western North Dakota to east across the state. Along with a probable history of reduced yield goals at Valley City and Mandan than at Colfax, levels of P recommended at Colfax would be higher than at Valley City and Mandan. Although P levels at Colfax (Mean P, 9 ppm) are generally lower than at Valley City (Mean P, 16 ppm) or Mandan (Mean P, 10 ppm), P fertilizer rates at Colfax have probably been historically higher . More heavy P fertilizer applications in the past at Colfax may have masked original differences due to landscape position, while more modest starter P applications made regularly at Mandan and Valley City may have allowed a relationship between landscape and P levels to continue.

CONCLUSIONS

Soil nitrate-N and P levels from landscape sampling were correlated with soil nitrate-N and available P levels from a 33 m grid in three different fields. Nitrate-N from landscape sampling was more strongly correlated to 33 m grid nitrate-N than 66 m grids in one of three sites. Nitrate-N from landscape sampling was more highly correlated with nitrate-N from a 33 m grid than 100 m grids in 2 of 3 sites and higher than a 3.3 ha grid at all three sites. Soil P from landscape was similar in correlation with a 33 m grid as a 66 m grid at one location. Landscape P was more highly correlated with 33 m P than a 100 m grid at one location and a 3.3 ha grid at two locations. The 3.3 ha grid P was more highly correlated with 33 m P than landscape P at one location. The strength of relationship between landscape P and 33 m P compared to other grid size P suggests that landscape P may be more related to soil P patterns where fertilizer P applications have been relatively low in the past.

Landscape sampling for nitrate-N may potentially reduce the number of samples needed within a field compared to the number needed for similar correlation from a grid pattern, while approximating nirate-N levels as well or better than a grid. Further research needs to be done to determine whether landscape relationships hold over a number of years, and to determine the correlation of landscape with other nutrient levels.

REFERENCES

Carr, P.M., G.R. Carlson, J.S. Jacobsen, G.A. Nielsen, and E.O. Skogley. 1991. Farming soils, not fields: a strategy for increasing fertilizer profitability. J. Prod. Agric. 4:57-61.

Fiez, T.E., B.C. Miller, and W.L. Pan. 1994. Assessment of spatially variable nitrogen fertilizer management in winter wheat. J. Prod. Agric. 7:86-93.

Franzen, D.W., and T.R. Peck. 1993. Soil sampling for variable rate fertilization. pp. 81-90. *In* 1993 Illinois Fertilizer Conference Proceedings, Jan. 25-27, 1993. Springfield, IL. R.G. Hoeft, ed. Ill. Coop. Ext. Serv., Urbana, IL.

Halvorson, G.A., and E.C. Doll. 1991. Topographic effect of spring wheat yields and water use. Soil Sci. Soc. Am. J. 55:1680-1685.

Hollands, K. 1996. Relationship of nitrogen and topography. pp. 123-128. *In* 1995 Sugarbeet Research and Extension Reports, Vol. 26. N. Dak. St. Univ. Ext. Serv., Fargo, ND.

Nolan, S.C., T.W. Goddard, D.J. Heaney, D.C. Penney, and R.C. McKenzie. 1995. Effects of fertilizer on yield at different soil landscape positions. pp. 553-558. *In* Proceedings of Site-Specific Management for Agricultural Systems Second Annual Conference. March 27-30, 1994. Minneapolis, MN. P.C. Robert et al. (Eds.) ASA-CSSA-SSSA. Madison, WI.

Steinward, A.L., D.L. Karlen, and T.E. Fenton. 1996. An evaluation of soil survey crop yield interpretations for two central Iowa farms. J. Soil and Water Cons. 51(1):66-71.

Stevenson, F.J. 1982. Origin and distribution of nitrogen in soil. p. 1-42. *In* Nitrogen in Agricultural Soils. Agronomy Monograph No. 22. F.J. Stevenson, ed. ASA-CSSA-SSSA, Madison, WI.

Using Harvest Index To Locate Environmental Stress

P.J. Copeland
G.L. Malzer
J.G. Davis
J.A. Lamb
P.C. Robert

Department of Soil, Water, and Climate
University of Minnesota
St. Paul, Minnesota, U.S.A.

T.W. Bruulsema

Potash & Phosphate Institute
Guelph, Ontario, Canada

ABSTRACT

Harvest index (HI), the ratio of grain to total above-ground biomass, is under genetic and environmental control. Variability of HI within a field of a single hybrid or variety is due to the interaction of that environment and the crop. Our objective was to relate dry matter accumulation in corn (*Zea mays* L.) stover and grain to specific regions of stress in the field under different N application treatments. Deviations from the observed maximum HI indicated environmental stress was affecting the partitioning of dry matter to the grain in some parts of the field at each of six N rates. Vegetative growth was reduced at low N rates but this reduction was not consistent across sites or within a field. Landscape position may explain some of the variability in response of HI to N rate.

INTRODUCTION

Managing field variability is dependent on understanding the environmental conditions appearing in the field and whether they might be modified by the producer. Any adverse effects put stress on the plant system which responds by limiting resources to the organ growing most rapidly at the time of stress (see review by Donald and Hamblin, 1976). In an ideal, non-stressed, growing environment corn will partition up to 60% of its dry matter to grain yield (Prihar and Stewart, 1990) for a harvest index of 0.6. Harvest index (HI) is the ratio of economic product, or grain yield, to the above-ground biomass at harvest and is under genetic and environmental control (Snyder and Carlson, 1984). Snyder and Carlson (1984) suggest that the variability in HI observed across years reflects the effect of the environment on dry matter partitioning and accumulation.

Environment effects on HI include planting density (DeLoughery and Crookston, 1979), water and nutrient availability (Eck, 1984; Uhart and Andrade, 1995 ; Westgate, 1994), light availability (Schussler and Westgate, 1991), weed interference (Tollenaar et al., 1994), and insect damage (Spike and Tollefson, 1991). Under moderate levels of stress, HI has been found to be stable but under severe stresses or combinations of stresses HI has been reduced (Sinclair et al., 1990; Tollenaar et al., 1994; Spike and Tollefson, 1991; DeLoughery and Crookston, 1976).

Prihar and Stewart (1990) proposed the existence of a "genetic harvest index," the maximum HI attainable in a given environment. Deviations from the genetic HI which represent environmental stresses affecting dry matter accumulation in grain could be useful in evaluating management practices.

Our objective was to evaluate field spatial changes in HI as influenced by N fertilization and to determine if differential dry matter accumulation of grain and/or stover were related to specific areas of the field experiencing environmental stress.

METHODS

Field experiments were conducted for one year on each of four commercial dryland corn fields, approximately 5 ha in size, in south-central and southwestern Minnesota in 1994 and 1995. These regions of Minnesota are characterized by fine- and medium-to-fine-textured soils with poor to moderate internal drainage. Relief at each field was typical of landscapes formed in glacial till. Producer cooperators managed all field operations except N applications and corn harvest. Previous crop was soybean [Glycine max (L.) Merr.] at three sites and sugarbeet (Beta vulgaris L.) at the other.

The experimental design was a split-block in space with six replicates. Six N treatments, rates of anhydrous ammonia at 0, 67, 101, 134, 168, and 202 kg ha^{-1}, were randomized within each replicate and applied in the fall in strips 5.6 to 6.1 m wide and extending across the field (274 to 305 m). A uniform N rate in each strip was assured by applying anhydrous ammonia with a computerized flow control and variable metering system.

Harvest index was determined from biomass samples collected at physiological maturity at 30.4-m intervals (also called plots in this analysis) along each strip. Each sample, consisting of two representative plants at the collection point, was separated into ears and stover and dried at 60 C and dry weights recorded. Ears were shelled, and from the shelling percentage, weight of cobbs was determined and combined with stover weight. Harvest index of the biomass sample was calculated as:

$$HI = \frac{\text{grain dry weight}}{\text{total dry weight}}$$

Grain was harvested from the two middle rows of each strip by plot combine equipped with a ground distance monitor and computerized load cell.

Grain weight was recorded every 15.2 m but only the grain weights that corresponded with the biomass samples (every 30.4 m) were used in this analysis. Details on other procedures and analyses of these data are reported elsewhere in this publication (Malzer et al., 1996; Bruulsema et al., 1996; Davis et al., 1996).

Stover dry matter and total dry matter were determined from the harvest grain oven-dry weight and HI.

$$\text{Stover dry matter} = \text{harvest grain dry weight} \left(\frac{1}{\text{HI}} - 1 \right)$$

$$\text{Total dry matter} = \text{grain dry weight} + \text{stover dry weight}$$

Grain to stover ratio is related to HI by the following equation and is used to examine the relationship of reproductive or economic product to vegetative production.

$$\text{Grain to stover ratio} = \frac{(\text{total dry matter x HI})}{\text{stover dry matter}}$$

Harvest index values appeared to be normally distributed by visual inspection of frequency histograms so whole-field effects of N on HI were examined by analysis of variance and correlation procedures in software from the SAS Institute, Inc.(1988). Genetic HI and the Line of maximum grain to stover ratio were defined as the average of the top 5% of observations for that parameter for each site. Maximum yield was defined as the top 5% of yield values or a fitted line (by eye) of the highest values that are linear. The Line of minimum grain to stover ratio was defined as the lowest 5% of observations for the ratio or a fitted line of the lowest values that are linear.

Categories of stress were determined arbitrarily by the mid-line of the Lines of maximum and minimum grain to stover ratio and by the most efficient stover size, defined as the value of stover size that intersects the Lines of maximum yield and maximum grain to stover ratio at the same point (see Figures 2 and 3).

RESULTS AND DISCUSSION

Growing conditions were excellent during both years, although planting of corn occurred earlier in 1994 than in 1995. Precipitation was well-distributed in both years and slightly above expected historic levels. Severe moisture deficits would not be expected in these fine- to medium-textured soils.

Application of N had a significant effect on HI at three of four sites and on grain and stover production at all sites (Table 1). The interactions of blocking (REP) and segment, that is division of the field in a perpendicular direction to blocking, were generally not significant. At the whole-field level, this supports the assumption of random experimental error.

Table 1. Source of variation and significance of treatment effects on HI, grain dry weight, and stover dry weight by site.

Source of variation	Hanska 1994			Hector 1994			Hanska 1995			Morgan 1995		
	HI	Grain	Stover	HI	Grain	Stover	HI	Grain	Stover	HI	Grain	Stover
Rep (R)	**	NS	**	**	NS	**	NS	*	NS	NS	**	**
N rate (N)	NS	**	**	**	**	**	**	**	**	*	**	**
Segment (S)	*	*	*	NS	**	**	NS	**	**	**	**	*
R x S	NS	NS	NS	**	NS	NS	NS	NS	NS	NS	NS	NS
N x S	NS	NS	NS	**	**	**	NS	**	**	**	**	**

*,**Significant at the 0.05 and 0.01 probability levels respectively.

Table 2. Field means for grain dry matter, stover dry matter, and harvest index (HI) and field genetic HI[†].

Site	No. of obs. in mean	Grain	Stover	HI	Genetic HI
		-----Mg ha^{-1}-----			
Hanska 1994	360	10.2	7.4	0.58	0.62
Hector 1994	324	9.1	7.0	0.57	0.61
Hanska 1995	324	6.9	7.1	0.49	0.57
Morgan 1995	324	7.5	7.1	0.51	0.59

[†]Genetic HI represents the maximum HI achievable in a given envrionment.

Whole-field means for dry matter production and HI are shown in Table 2. Mean stover weight was consistent across sites but mean grain dry matter ranged from about 7 to 10 Mg ha^{-1}. Mean HI was higher in 1994 than in 1995, perhaps reflecting the longer grain-filling period associated with early planting in 1994. The genetic HI, or the maximum HI values observed, at all sites and the HI means for 1994 are in the range of harvest indices from studies of irrigated corn reported by Prihar and Stewart (1990). This suggests that environmental conditions were nearly ideal for dry matter partitioning to grain in some plots in each field and in a great number of plots in the 1994 experiments.

The variability of HI at each site is shown in Figure 1. The amount of total dry matter at the Hanska 1994 site was quite high (~17 Mg ha^{-1}) and clustered close to the genetic HI. Since these data consist of all N rates together (including zero N), there appears to have been very little stress, either during vegetative growth or during grain-filling. This, combined with the early planting in 1994 may account for the high yields observed at Hanska that year.

All other sites (Figure 1b-d) exhibited a much wider range in total dry matter, under 5 to above 20 Mg ha^{-1}, and HI, points that deviate below the genetic HI, particularly at the 1995 sites. It is noteworthy that HI alone reveals nothing about yield since high indices were recorded at very low weights for total dry matter. In such a case one might surmise that vegetative growth and kernel number were limited by environmental stress and, if it occurred in the check plots, that the stress was a lack of N availability.

At each site, HI was reduced at the highest levels of total dry matter accumulation (Figure 1). This may correspond to constant yield or a less-than-proportional increase in grain yield at greater than optimum stover size (Figure 2). This is demonstrated in Figure 3 with data from the Hanska 1995 site. As stover weight increases beyond 6.5 Mg ha^{-1}, maximum yield remains constant at ~8.9 Mg ha^{-1}. Above a stover size of ~10 Mg ha^{-1}, maximum yield appears to decrease. This appears to be typical of responses to N when water is deficient (Donald and Hamblin, 1976). This would be expected if high N rates allowed great amounts of stover to be produced. Such growth would require more water to sustain and result in stress during grain-filling. Eghball and Maranville (1993) found that root growth of corn was reduced under high N rates while Dale and Daniels (1995) showed that the probability of moisture stress occurring

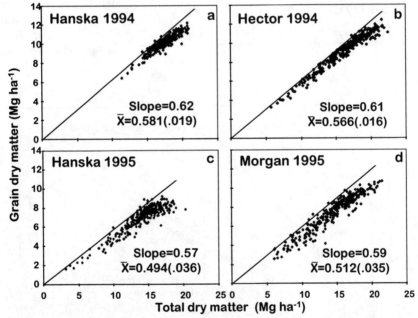

Figure 1a-d. Plots of grain and total dry matter, genetic HI (line through the origin) and field HI mean and standard error for each site. The line represents the upper bound or the maximum attainable HI in each environment. The slope is equivalent to HI. Figure 1c shows the data for the site used as the primary example in Figures 3 and 4.

differs greatly with individual soils. In a field with variable relief, N and moisture stresses may be expressed in different regions according to N level and soil texture.

Whole-field correlation of HI with N rate was ~0.3 for three sites and zero for Hanska in 1994 (Table 3). Of the four experiments, grain yields were highest at the Hanska 1994 site and the amount of N required to reach those levels was the lowest of the four sites. Stover dry matter was strongly negatively correlated with HI at Hanska in 1994, suggesting that high levels of stover production reduced HI while maintaining maximum grain yield. See Figure 2 for a diagram of these relationships. This reduction in HI at large stover size represents a decrease in efficiency of grain dry matter production and may be associated with high N application rates. Donald and Hamblin (1976) reported that negative correlations of HI with vegetative components in barley were much stronger at high N than at low N, where growth was limited and competition for light and water was weak.

Table 3. Correlation coefficients for harvest index with N rate and stover dry matter.

	N rate	**Stover dry matter**
Hanska 1994	-0.01	-0.72
Hector 1994	0.29	-0.35
Hanska 1995	0.31	-0.35
Morgan 1995	0.27	-0.16

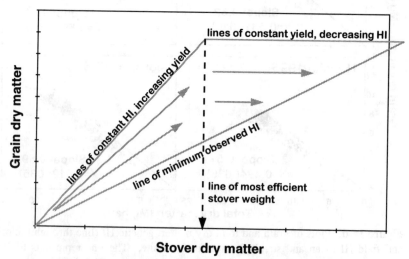

Figure 2. Diagram showing the relationships of yield, stover and HI. Maximum yield is dependent on stover size up to the most efficient stover size while HI remains constant. Above this stover size, maximum yield remains constant and HI decreases.

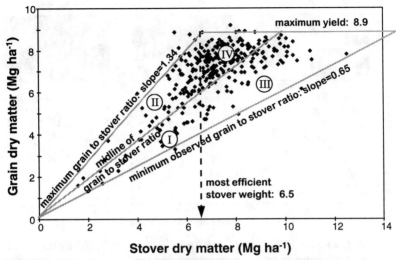

Figure 3. Grain and stover data for all N treatments at the Hanska 1995 site. Four cases of environmental effects defined by the grain to stover ratio midpoint and the most efficient stover weight correspond to: **I**) low HI *and* low stover, **II**) high HI *and* low stover, **III**) low HI *and* high stover, and **IV**) high HI *and* high stover.

Four different cases of effects are defined by high and low HI and high and low stover weight in Figure 3. In general, these correspond to early season stresses (low stover weight, *I* and *II*), late season stresses (low HI, *I* and *III*) and no stress (*IV*).

The spatial distribution of relative stress categories within different N treatments at the Hanska 1995 site is shown in Figure 4. Stress on vegetative growth and reproductive stages was evident in a large portion of the field in the check plots (Figure 4a). Addition of N (Figure 4b) alleviated much of the stress on vegetative growth and increased the range of yield up to the maximum of the field, but this does not appear to be consistent across the field. Further additions in N rate (Figure 4c-f) increased HI and stover size (more black area on map and less light-gray shading). However, a few plots in each of the higher levels of N gave very high stover size but with somewhat reduced yield. This may be due to late season moisture stress in large plants that reduces dry matter partitioning to the grain.

Except for the highest rate of N (Figure 4f), addition of N alone was not able to eliminate the combination of stress during vegetative growth and during reproductive stages from the Hanska 1995 site. The southeast corner of this field was a depression that flooded early in the season and showed both types of stress (white color in map) at all N rates except 202 kg ha^{-1}.

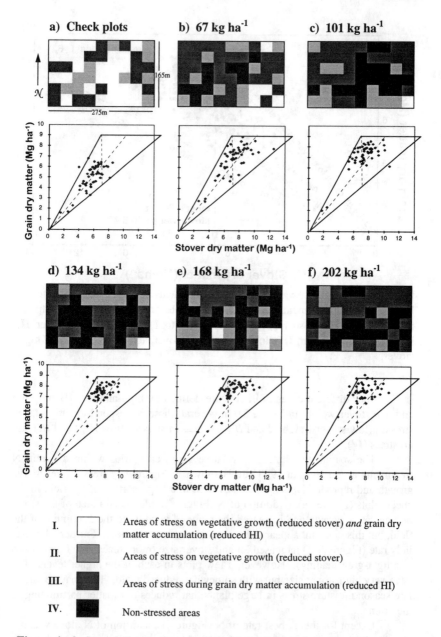

Figure 4a-f. Spatial distribution of relative stress categories with different N application rates at the Hanska 1995 site. Shading levels (none, light to dark gray, and black) correspond to graph sectors I to IV in Figure 3.

CONCLUDING REMARKS

Relating harest index data to plant N dynamics may aid our understanding of how to increase grain production efficiency with site-specific N management. A HI lower than the genetic HI indicates that external environmental stress factors were present in the field and restricted crop response potential. Knowing how and where these environmental factors manifest themselves could provide valuable information on if and how those areas might be managed differently.

Stress during vegetative growth was generally alleviated by addition of N. Some stress experienced during the grain filling was decreased by N but HI was reduced in some areas of the field at all N rates. This could occur at high N when high vegetative growth depletes soil water required for grain filling.

If such a N-soil water interaction was consistently associated with low organic matter, eroded shoulders for example, a producer might opt to reduce N rates in such areas even though the soils may have a low N supply and be very responsive to N. Landscape features were not consistently related to stress categories. This may be due to changing N dynamics and interaction with soil water at different N rates or perhaps other factors.

Further research might investigate plant N dynamics by sampling for total dry matter and plant N at silking (as per suggestion of Jackson et al., 1985) as well as at maturity to better understand harvest index and plant nitrogen use variability across a field.

ACKNOWLEDGMENTS

Funding for this project was provided by the Department of Energy via the Legislative Committee on Minnesota Resources. Technical assistance was provided by T. Graff and H. Dikici. These contributions are gratefully acknowledged.

REFERENCES

Bruulsema, T.W., G.L. Malzer, P.C. Robert, J.G. Davis, P.J. Copeland. 1996. Spatial relationships of soil nitrogen with corn yield response to applied nitrogen. (This publication.)

Dale, R.F. and J.A. Daniels. 1995. A weather-soil variable for estimating soil moisture stress and corn yield probabilities. Agron J. 87:1115-1121.

Davis, J.G., G.L. Malzer, P.J. Copeland, J.A. Lamb, P.C. Robert, and T.W. Bruulsema. 1996. Using yield variability to characterize spatial crop response to applied N. (This publication.)

DeLoughery, R.L. and R.K. Crookston. 1979. Harvest index of corn affected by population density, maturity rating, and environment. Agron. J. 71:577-580.

Donald, C.M. and J. Hamblin. 1976. The biological yield and harvest index of cereals as agronomic and plant breeding criteria. Adv. Agron. 28:361-405.

Eck. H.V. 1984. Irrigated corn yield response to nitrogen and water. Agron. J. 76:421-428.

Eghball, B. and J. Maranville. 1993. Root development and nitrogen influx of corn genotypes grown under combined drought and nitrogen stress. Agron. J. 85:147-152.

Jackson, W.A., W.L. Pan, R.H. Moll, and E.J. Kamprath. 1985. Uptake, translocation, and reduction of nitrate. In C.A. Neyra (ed.) Biochemical basis of plant breeding: Volume 2. Nitrogen metabolism. CRC Press, Boca Raton, Florida.

Malzer, G.L., P.J. Copeland, J.G. Davis, J.A. Lamb, P.C. Robert, and T.W. Bruulsema. 1996. Spatial variability of profitability in site-specific N management. (This publication.)

Prihar, S.S. and B.A. Stewart. 1990. Using upper-bound slope through origin to estimate genetic harvest index. Agron. J. 82:1160-1165.

SAS Institute, Inc. 1988. SAS/STAT® User's Guide, Release 6.03 Edition. Cary, NC.

Schussler, J.R. and M.E. Westgate. 1991. Maize kernel set at low water potential. I. Sensitivity to reduced assimilates during early kernel growth. Crop Sci. 31:1189-1195.

Sinclair, T.R., J.M. Bennett, and R.C. Muchow. 1990. Relative sensitivity of grain yield and biomass accumulation to drought in field-grown maize. Crop Sci. 30:690-693.

Snyder, F.W. and G.E. Carlson. 1984. Selecting partitioning of photosynthetic products in drops. Adv. Agron. 37:47-73.

Spike, B.P. and J.J. Tollefson. 1991. Response of western corn rootworm-infested corn to nitrogen fertilization and plant density. Crop Sci. 31:776-785.

Tollenaar, M., S.P. Nissanka, A. Aguilera, S.F. Wiese, and C.J. Swanton. 1994. Effect of weed interference and soil nitrogen on four maize hybrids. Agron. J. 86:596-601.

Uhart, S.A. and F.H. Andrade. 1995. Nitrogen deficiency in maize: I. Effects of crop growth, development, dry matter partitioning, and kernel set. Crop Sci. 35:1376-1383.

Westgate, M.E. 1994. Water status and development of the maize endosperm and embryo during drought. Crop Sci. 34:76-83.

Long-Term Yields From Individual Plots: Implications for Managing Spatial Variability

Paul M. Porter

Dept. of Agronomy and Plant Genetics
University of Minnesota
Lamberton, Minnesota

David R. Huggins
Catherine A. Perillo

Dept. Soils, Water, and Climate
University of Minnesota
Lamberton, Minnesota

Joseph G. Lauer
Edward S. Oplinger

Dept. of Agronomy
University of Wisconsin
Madison, Wisconsin

R. Kent Crookston

Dept. of Agronomy and Plant Genetics
University of Minnesota
St. Paul, Minnesota

ABSTRACT

Year-to-year (seasonal) variability in continuous soybean yield was three times greater and in continuous corn yield was four times greater than plot-to-plot (field) variability in three long-term trials conducted in MN and WI. With one exception, each of the four plots for both corn and soybean at each location produced the highest and lowest corn and soybean yields within a season at least one time during the 10-year period. In any one year, yield ranges (in any one field) of more than 25% of the average yield occurred in one-fourth of the growing seasons studied, but when averaged over 10-years there was no reason to suspect that differences of more than 10% existed.

OBJECTIVES

At each of three locations over a 10-year period (1986-1995) yield data from four continuous corn plots and four continuous soybean plots, as well as eight plots in a corn/soybean rotation, were evaluated to determine the relative amounts of spatial and temporal variability in fields that were considered to have uniform yield potentials at each location. The implications of managing the yield variability are discussed in relation to yield map interpretation.

MATERIALS AND METHODS

Several of the longest ongoing studies designed to evaluate corn-soybean cropping sequences in the northern Corn Belt were initiated in the early 1980s at two locations in Minnesota and one location in Wisconsin. The studies were established near Lamberton, MN in 1981 on a Webster clay loam, near Waseca, MN in 1982 on a Nicollet clay loam, and near Arlington, WI in 1983 on a Plano silt loam. Recommended practices for optimum production were followed. Details of the soil types, soil fertility, and fertilizers and pesticides are described by Crookston et al. (1991) and Meese et al. (1991).

The soils in each field at each location were considered to be uniform, and the topography of the fields was relatively flat with slopes less than 2%. Based on previous yields, there was no reason to suspect yield differences across each field at each location. Field size was approximately 1 ha. At all three locations there were 14 different cropping sequences. There were four replicates (plots) of each treatment at each location. This paper only deals with yields from 1986 through 1995 of the continuous corn, and continuous soybean, and corn/soybean plots.

Tillage at Lamberton and Waseca was conventional (moldboard plowed in the fall) and plots were arranged in a randomized complete block design. At Arlington, the original design included different tillage systems, N fertility levels, and the 14 cropping sequences. Only data from the conventional tillage system, averaged across all N fertility levels, are discussed in this paper.

Planting and harvest dates varied according to seasonal conditions at each location. In general, planting occurred between late-April and late-May, and harvest occurred between mid-September and late-October.

Corn was planted in 76-cm rows at all three locations. Soybean was planted in 20-cm rows at Arlington and 76-cm rows at Lamberton and Waseca. Corn and soybean were seeded at a rate of 60 000 to 75 000 and 400 000 to 450 000 viable seeds per hectare, respectively. At Lamberton, plots were 12 rows wide and 10 m long; harvest was from 8 m of four of the center rows. At Waseca, plots were 6 rows wide and 18 m long; harvest was from 15 m of the two center rows. At Arlington, plots were 12 rows wide and 9 m long; harvest was from 6 corn rows and 12 soybean rows. Plots were harvested with a plot combine. Grain yield moisture contents were adjusted to 155 g kg^{-1} and 130 g kg^{-1} for corn and soybean, respectively.

Data Analysis and Definitions

An analysis of variance for yield of each crop at each location was conducted using yield data from the four plots over the 10-year period (SAS, 1988). At each location, field average was the average yield of four plots each year. Field range was the maximum minus minimum yield of the four plots each year. Field standard deviation was the standard deviation of yield of the four plots each year. Ten-year average yield was the average yield from 1986 through 1995 of each plot. Ten-year range was the maximum minus minimum yield from 1986 through 1995 of each plot. Ten-year standard deviation was the standard deviation of yield from 1986 through 1995 of each plot. Field yields, ranges, and standard deviations were averaged over the 10-year period, and the 10-year yields, ranges, and standard deviations were averaged across the four plots. The average field standard deviation and the average 10-year standard deviation were measures of field and seasonal variability, respectively. Yield data were misplaced and data analysis was not possible for corn grown at Waseca in 1991 and for soybean grown at Waseca in 1991 and 1993.

RESULTS AND DISCUSSION

Analysis of variance for the plot and year effect at each of the three locations showed that year had a highly significant effect on both corn and soybean yields at each location (Table 1). This was not surprising, as yields across the Northern Corn Belt were below normal in 1988 because of generally hot, dry conditions and in 1993 because of cool, wet conditions (Changon, 1996; Minnesota Department of Natural Resources, 1989).

Table 1. Analysis of variance for the plot and year effect in a field at the three locations.

	Location	Plot effect	Year effect	CV	Avg. yield	--- LSD$_{(0.05)}$ --- Plot	Year
		------ Pr>F ------		%	-------- Mg/ha -------		
Corn							
	Lamberton	0.73	<0.01***	9.7	7.21	NS	1.014
	Waseca	0.05*	<0.01***	6.5	8.13	0.515	0.774
	Arlington	0.57	<0.01***	7.8	8.47	NS	0.956
Soybean							
	Lamberton	0.93	<0.01***	8.9	2.36	NS	0.303
	Waseca	0.55	<0.01***	9.5	2.47	NS	0.344
	Arlington	0.42	<0.01***	8.2	3.51	NS	0.417

*, *** Significant at the 0.05 and 0.001 probability levels, respectively.
NS = Not significant at 0.05.

Over the 10-year period the location in the field of the four soybean plots at Lamberton, at Waseca, and at Arlington did not influence the soybean yields, and the location in the field of the four corn plots at Lamberton and at Arlington did not influence corn yields (Table 1). The effect on yield of site in the field (plot) over the 10-year period was significant only for corn at Waseca, with the yield for plot #3 averaging 8.2% higher (8.56 vs. 7.91 Mg ha^{-1}) than the yields of plots #2 and #4 (Tables 1 and 2).

The fact that the location in the field did not influence yields over the 10-year period (except for corn at Waseca) was not surprising. The trials were conducted on uniform soils with little to no visible topographical differences, and from a practical perspective, the fields selected at each location were thought of as having a uniform yield potential.

Continuous Corn Yield Variability

Corn yields for each year over a 10-year period for each of four plots at the three locations are shown in Table 2, as are the average yield, range in yield, and standard deviation for the four plots at each location for each year (field average, field range, and field standard deviation, respectively) and the average yield, range in yield, and standard deviation over the 10-year period at each location for each plot (10-year average, 10-year range, and 10-year standard deviation, respectively). Also shown are the 10-year average yield, range, and standard deviation averaged across plots for each location and the field average yield, range, and standard deviation averaged across years for each location.

An example of how to interpret data in Table 2 is as follows. At Lamberton during 1986, the average corn yield of all four plots in the field was 8.8 Mg ha^{-1}. That year, the range between the highest and lowest yielding plots was 1.7 Mg ha^{-1} while the field standard deviation was 0.7 Mg ha^{-1}. Over the 10-year period, plot #1 produced an average yield of 7.08 Mg ha^{-1}, the range between the highest and lowest yielding season was 4.89 Mg ha^{-1}, and the standard deviation across years was 1.7 Mg ha^{-1}. Over the 10-year period, the field range between the maximum and minimum yielding plots within a year was 1.46 Mg ha^{-1}, and field variability as measured by standard deviation 0.62 Mg ha^{-1}. Across the four sites, the 10-year range between the maximum and minimum yielding seasons for each plot was 5.48 Mg ha^{-1}, and seasonal yield variability as measured by standard deviation was 1.75 Mg ha^{-1}.

At all three locations, each of the four plots produced the greatest corn yield at least one season (compared to the other three plots) during the 10-year period. Likewise, at both Lamberton and Arlington, each of the four plots produced the lowest corn yield within at least one season (compared to the other three plots) during the 10-year period (Table 2).

At Lamberton, seasonal variability in corn yield was 2.8 times that of field variability (1.75 vs. 0.62 Mg ha^{-1}). At Waseca, seasonal variability was 4.3 times that of field variability (2.46 vs. 0.57 Mg ha^{-1}). At Arlington, seasonal variability was 4.0 times that of field variability (2.34 vs. 0.59 Mg ha^{-1}).

Table 2. Continuous corn yields over time at four plots in fields near Lamberton MN, Waseca MN, and Arlington WI.

Plot	1986	1987	1988	1989	1990	1991	1992	1993	1994	1995	10-yr avg.	10-yr range	10-yr std. dev.	Avg. across years
							Mg ha^{-1}							
Lamberton														
1	8.7	7.5	3.9	8.8	7.9	7.4	8.0	3.9	6.9	7.8	7.08	4.89	1.7	
2	9.7	8.1	5.7	9.2	8.6	7.1	8.6	2.7	6.2	7.3	7.32	7.04	2.1	
3	8.7	8.0	6.5	9.0	8.8	7.0	6.4	4.3	6.8	8.0	7.35	4.71	1.5	
4	8.0	8.4	4.9	8.9	8.4	6.4	7.6	3.6	8.1	6.6	7.09	5.31	1.7	
Field avg.	8.8	8.0	5.3	9.0	8.4	7.0	7.6	3.6	7.0	7.4				7.21
Field range	1.7	0.9	2.6	0.4	0.9	1.0	2.3	1.6	1.9	1.4				1.46
Field std. dev.	0.7	0.4	1.1	0.2	0.4	0.4	1.0	0.7	0.8	0.6				0.62
								Avg. across plots:			7.21	5.48	1.75	
Waseca														
1	7.7	11.3	4.7	10.6	9.2	--	9.9	3.4	9.1	7.4	8.14	7.93	2.7	
2	7.1	10.0	3.7	10.5	8.5	--	9.2	3.9	9.8	8.3	7.90	6.78	2.5	
3	7.6	11.3	5.3	11.6	9.4	--	9.3	4.3	10.0	8.2	8.56	7.29	2.5	
4	7.8	11.1	4.6	9.6	9.2	--	8.3	4.6	8.4	7.6	7.92	6.52	2.2	
Field avg.	7.5	10.9	4.6	10.6	9.1	--	9.2	4.1	9.3	7.9				8.13
Field range	0.7	1.3	1.5	1.9	0.9	--	1.6	1.2	1.6	0.9				1.30
Field std. dev.	0.3	0.6	0.6	0.8	0.4	--	0.7	0.5	0.7	0.4				0.57
								Avg. across plots:			8.13	7.14	2.46	
Arlington														
1	10.4	11.9	5.6	9.3	9.5	9.0	9.0	5.6	8.9	7.8	8.7	6.3	1.9	
2	11.3	10.1	4.0	8.9	9.9	8.5	9.8	5.1	10.4	7.1	8.5	7.3	2.4	
3	10.6	9.9	3.5	9.3	9.0	8.3	8.7	5.4	10.5	7.9	8.3	7.2	2.3	
4	11.0	10.0	2.8	9.7	9.5	8.3	9.5	4.5	10.9	7.5	8.4	8.3	2.8	
Field avg.	10.8	10.5	3.9	9.3	9.5	8.5	9.2	5.1	10.2	7.6				8.47
Field range	0.9	2.1	2.8	0.8	0.9	0.8	1.1	1.2	2.0	0.8				1.35
Field std. dev.	0.4	1.0	1.2	0.3	0.4	0.3	0.5	0.5	0.9	0.4				0.59
								Avg. across plots:			8.47	7.28	2.34	

Continuous Soybean Yield Variability

Soybean yields for each year over a 10-year period for each of the four plots at the three locations are shown in Table 3. At each of the three locations, each plot produced the greatest soybean yield within at least one season (compared to the other three plots) during the 10-year period. Likewise, at Lamberton and Arlington, each plot produced the lowest soybean yield within at least one season (compared to the other three plots) during the 10-year period.

At Lamberton, seasonal variability in soybean yield was 2.8 times that of field variability (0.52 vs. 0.19 Mg ha^{-1}). At Waseca, seasonal variability was 2.2 times that of field variability (0.48 vs. 0.22 Mg ha^{-1}). At Arlington, seasonal variability was 3.5 times that of field variability (0.94 vs. 0.27 Mg ha^{-1}).

Corn/Soybean Yield Variability

This section deals with the yield response of corn and soybean grown in a corn/soybean rotation. At each location four plots were planted to corn one year and soybean the following year, while four other plots were planted to soybean one year and corn the following year. For both corn or soybean grown in a corn/soybean rotation there were eight plots: four plots the first year, four other plots the second year, the original four plots the third year, and so on. The individual plot data are not included in this text.

The averages, ranges and standard deviations obtained for both corn and soybean grown in a corn/soybean rotation were similar to those obtained for continuous corn and continuous soybean. Seasonal variability in corn yield was 3.7, 4.1, and 3.5 times that of field variability at Lamberton, Waseca, and Arlington, respectively. Seasonal variability in soybean yield was 2.0, 2.8, and 3.6 times that of field variability at Lamberton, Waseca, and Arlington, respectively (data not shown).

Corn yield in the corn/soybean rotation averaged 8.18, 8.92, and 9.71 Mg ha^{-1} at Lamberton, Waseca, and Arlington, respectively. The range between the highest and lowest yielding plots within a year averaged 1.21, 1.34, and 1.42 Mg ha^{-1} over the 10-year period at Lamberton, Waseca, and Arlington, respectively. The range between the highest and lowest yielding seasons over the 10-year period for each plot averaged 6.06, 7.03, and 5.79 Mg ha^{-1} at Lamberton, Waseca, and Arlington, respectively. Seasonal yield variability within a plot as measured by standard deviation averaged 1.98, 2.40, and 2.34 Mg ha^{-1} at Lamberton, Waseca, and Arlington, respectively. Field yield variability within a year as measured by standard deviation averaged 0.54, 0.58, and 0.66 Mg ha^{-1} over the 10-year period at Lamberton, Waseca, and Arlington, respectively (data not shown).

Soybean yield in the corn/soybean rotation averaged 2.74, 2.73, and 3.76 Mg ha^{-1} at Lamberton, Waseca, and Arlington, respectively. The range between the highest and lowest yielding plots within a year averaged 0.54, 0.42 and 0.61 Mg ha^{-1} over the 10-year period at Lamberton, Waseca, and Arlington, respectively. The range between the highest and lowest yielding seasons for each plot averaged 1.48, 1.66, and 3.28 Mg ha^{-1} at Lamberton, Waseca, and Arlington, respectively. Seasonal yield variability within a plot as measured by standard deviation averaged

Table 3. Continuous soybean yields over time at four plots in fields near Lamberton MN, Waseca MN, and Arlington WI.

Plot	1986	1987	1988	1989	1990	1991	1992	1993	1994	1995	10-yr avg.	10-yr range	10-yr std. dev	Avg. across years
							Mg ha^{-1}							
Lamberton														
1	2.4	3.0	1.8	2.0	2.8	3.0	1.7	2.2	2.2	2.6	2.37	1.30	0.5	
2	2.3	2.8	1.6	1.8	2.7	2.7	1.7	1.7	2.8	2.9	2.32	1.30	0.6	
3	2.6	2.6	1.9	1.8	2.6	3.2	1.4	1.9	2.9	2.8	2.36	1.78	0.6	
4	2.7	2.7	2.2	1.9	2.7	3.1	1.8	1.8	2.2	2.7	2.38	1.33	0.5	
Field avg.	2.5	2.8	1.9	1.9	2.7	3.0	1.6	1.9	2.5	2.8				2.36
Field range	0.4	0.4	0.5	0.3	0.2	0.4	0.5	0.5	0.7	0.3				0.40
Field std. dev.	0.2	0.2	0.2	0.1	0.1	0.2	0.2	0.2	0.4	0.1				0.19
								Avg. across plots:			2.36	1.46	0.52	
Waseca														
1	2.2	2.7	1.5	2.5	2.9	--	2.1	--	3.0	2.9	2.48	1.41	0.5	
2	2.2	2.8	1.8	2.5	3.1	--	1.9	--	3.1	2.8	2.52	1.38	0.5	
3	2.3	3.1	1.8	2.5	3.0	--	1.7	--	2.9	2.7	2.50	1.36	0.5	
4	2.5	2.9	2.3	2.4	2.4	--	1.6	--	2.4	2.6	2.36	1.32	0.4	
Field avg.	2.3	2.9	1.9	2.5	2.9		1.8		2.8	2.7				2.47
Field range	0.3	0.4	0.7	0.1	0.8		0.5		0.7	0.3				0.49
Field std. dev.	0.1	0.2	0.3	0.1	0.3		0.2		0.3	0.1				0.22
								Avg. across plots:			2.46	1.38	0.48	
Arlington														
1	3.3	4.0	0.8	4.4	4.5	4.4	3.0	2.6	3.0	4.0	3.39	3.70	1.1	
2	3.2	4.3	1.4	4.2	3.6	4.4	2.9	3.2	3.1	4.5	3.48	3.09	1.0	
3	3.6	4.4	2.0	4.0	4.2	4.4	3.1	3.0	2.9	4.3	3.60	2.42	0.8	
4	3.8	4.4	2.1	4.1	4.2	4.5	2.6	3.1	2.9	4.1	3.58	2.42	0.8	
Field avg.	3.5	4.3	1.6	4.2	4.1	4.5	2.9	3.0	3.0	4.2				3.51
Field range	0.6	0.5	1.3	0.4	0.9	0.1	0.5	0.7	0.2	0.5				0.56
Field std. dev.	0.3	0.2	0.6	0.2	0.4	0.0	0.2	0.3	0.1	0.2				0.27
								Avg. across plots:			3.51	2.91	0.94	

0.48, 0.53, and 0.99 Mg ha^{-1} over the 10-year period at Lamberton, Waseca, and Arlington, respectively. Field yield variability within a year as measured by standard deviation averaged 0.24, 0.19, and 0.27 Mg ha^{-1} over the 10-year period at Lamberton, Waseca, and Arlington, respectively (data not shown).

Range in Continuous Corn and Soybean Yields at Each Location

Over the 10-year period, the range in corn yield among the four plots (field range) expressed as a percentage of the field average each year averaged 23, 18, and 19% at Lamberton, Waseca, and Arlington, respectively (Fig. 1). In any given year, the range in corn yield expressed as a percentage of the field average was as low as 4% at Lamberton in 1989 and as high as 72% at Arlington in 1988. The range in corn yield expressed as a percentage of the field average was greater than 10% in 22 of the 29 growing seasons studied. The range in corn yield expressed as a percentage of the field average was greater than 15% and 25% in 15 and 7 of the 29 growing seasons studied, respectively.

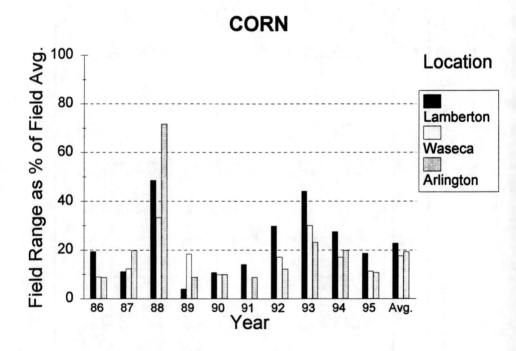

Fig. 1. Field range (maximum minus minimum yield of four plots each year) expressed as a percentage of the average corn yield at each of three locations from 1986 through 1995.

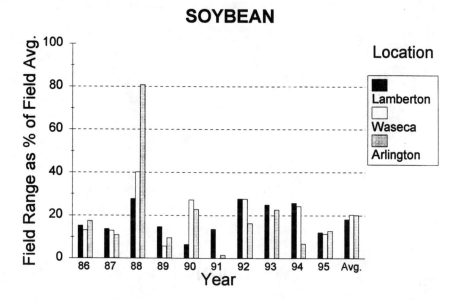

Fig. 2. Field range (maximum minus minimum yield of four plots each year) expressed as a percentage of the average soybean yield at each of three locations from 1986 through 1995.

Over the 10-year period, the range in soybean yield among the four plots (field range) expressed as a percentage of the field average each year averaged 18, 20, and 20% at Lamberton, Waseca, and Arlington, respectively (Fig. 2). In any given year, the range in soybean yield expressed as a percentage of the field average was as low as 2% at Arlington in 1991 and as high as 81% at Arlington in 1988. The range in soybean yield expressed as a percentage of the field average was greater than 10% in 23 of the 28 growing seasons studied. The range in corn yield expressed as a percentage of the field average was greater than 15% and 25% in 14 and 8 of the 28 growing seasons studied, respectively.

The field range was large compared to the field average in seasons with poor growing conditions. Field yield variability for both corn and soybean was greatest in 1988, when yields were depressed due to hot, dry growing conditions (Figs. 1 and 2). In 1993, a year with poor growing conditions for corn due to cool, wet conditions, field yield variability was also relatively large.

The range in corn yield for each of the four plots across the 10-year period (10-year range) expressed as a percentage of the season average was greater than 60% for all plots at all three locations studied (Fig. 3). Similarly, the range in soybean yield for each of the four plots across the 10-year period (10-year range) expressed as a percentage of the season average was greater than 50% for all plots at all three locations studied (Fig. 4).

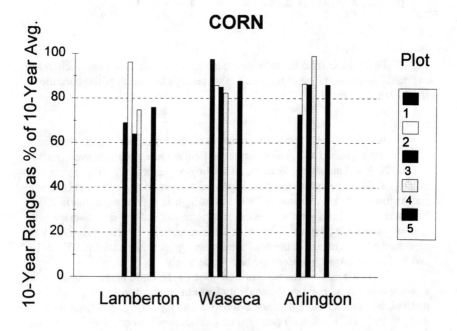

Fig. 3. Ten-year range (maximum minus minimum yield from 1986 through 1995 of each plot) expressed as a percentage of the average corn yield at each of three locations.

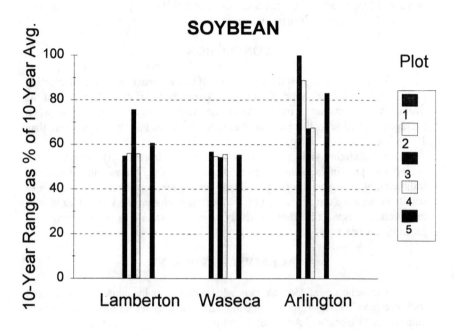

Fig. 4. Ten-year range (maximum minus minimum yield from 1986 through 1995 of each plot) expressed as a percentage of the average soybean yield at each of three locations.

Implications for Managing Spatial Variability

Yield variability was approximately three times greater for soybean and four times greater for corn from year-to-year than from plot-to-plot at the three locations studied. The 10-year time frame this study encompassed included two relative harsh growing seasons (1988 which was hot and dry, and 1993 which was cool and wet). To disregard these growing seasons as anomalies would be a mistake, as harsh climatic conditions resulting in poor crop production do occur regularly (U.S. Geological Survey, 1991).

During the 10-year period, soybean yields were not affected by site within the field (plot effect) at any of the three locations, and corn yields were affected by site within a field only at Waseca (Table 1). The range between the maximum and minimum corn yield of the four plots at Waseca was 8.2% when averaged over the 10 years. Yet, for both corn and soybean, the range in yield across the four plots exceeded 10% of the average yield in two-thirds of the growing seasons studied (22 of 29 for corn & 23 of 28 for soybean), 15% of the average yield in half of the growing seasons studied (15 of 29 for corn & 14 of 28 for soybean), and 25% of the average yield in one-fifth of the environments studied (7 of 29 for corn & 8 of 28 for soybean). Clearly, basing yield predictions on individual year data would result in quite different and perhaps erroneous conclusions than if yield predictions were based on longer-term (10 year) averages.

CONCLUSIONS

These results suggest the importance of taking great care when interpreting yield maps. In any one year, yield ranges (in one field) of more than 25% of the average yield occurred in one-fourth of the growing seasons studied, but when averaged over 10-years there was no reason to suspect that differences of more than 10% existed.

Emphasizing yield map variability observed in relatively uniform fields during poor growing seasons (when the field range is very large compared to the field average) is especially risky, and may lead to erroneous conclusions. These results underscore the necessity of in-season field observations to aid yield map interpretation, especially when relatively large yield variations occur during poor growing seasons.

ACKNOWLEDGMENTS

The authors would like to acknowledge and thank the numerous faculty and staff who have been instrumental in conducting the long-term rotation trials at the Lamberton, Waseca, and Arlington locations.

REFERENCES

Changon, S.A. 1996. The Great Flood of 1993: Causes, Impacts, and Responses. Westview Press. 319 p.

Crookston, R.K., J.E. Kurle, P.J. Copeland, J.H. Ford, and W.E. Lueschen. 1991. Rotational cropping sequences affects yield of corn and soybean. Agron. J. 83:108-113.

Meese, B.G., P.R. Carter, E.S. Oplinger, and J.W. Pendleton. 1991. Corn/soybean rotation effect as influenced by tillage, nitrogen, and hybrid/cultivar. J. Prod. Ag. 4:74-80.

Minnesota Department of Natural Resources, Division of Waters. 1989. Drought of 1988. 46pp.

SAS Institute, Inc. 1988. SAS/STAT guide for personal computers. Version 6.03 ed. SAS Institute, Inc., Cary, NC.

U.S. Geological Survey. 1991. National Water Summary 1988-1989. Hydrological Events and Floods and Droughts. Paper 2375, 591pp.

Dependence of Barley Growth on Soil Compaction and Methods of Fertilizer Application

U.V. Chigarev
L.A. Veremeychik

Belorussian AgroTechnical University
Minsk, Belarus

A.V. Skotnikov

Department of Soil, Water, and Climate
University of Minnesota
St. Paul, Minnesota

ABSTRACT

In this work we review the results of experience which measured growth of barley in its development periods as related to soil compaction and different methods of fertilizer application as a source of a yield variability. The analysis of storage by plants of dry matter during the whole vegetation period is also presented.

INTRODUCTION

Soil fertility is greatly determined by physical, mechanical and chemical properties, changes in which depend on weather conditions and mechanical exposure to agricultural machines and topographic relief. Soil is not a homogenous matter, therefore areas of the same field can vary in their characteristics and hence degree of fertility. These differences in fertility impose a challenge to improvement of biological activity by changing structure and through methods of applying mineral and organic fertilizers . In other words, each area is to be treated individually. This will allow profitable application of mineral fertilizers, cultivate environmentally clean products , maintain a steady level of soil fertility. The solution to this task depends on two factors. The first one is connected to the physical conditions of the soil (density, air permeability, moisture and other). The second is connected to the selection of a method of fertilizer application, as reactions of chemical elements greatly depend on the soil's physical and mechanical properties .

In this work we review the results of experience which measured growth of barley in its development periods as related to soil compaction and use of banded (banding), or broadcast fertilizer, with or without incorporation as a possible source of a yield variability.

METHODS AND EQUIPMENT

The research was conducted in a pot experiment. A sandy soil was used with agrochemical characteristics: pH - 6.74; organic matter content - 2.11%; P_2O_5 - 12.1 and K_2O - 16.6 mg/100g of soil.

Soil analyses were run according to the methods set by State Standards of Belarus. pH was defined from 1N KCl salt extract by electrometrical method (Arinushkina, 1970). Humus content analysis was run according to Turin's method which is based on humus carbon oxidization with dichromate. Mobile forms of potassium and phosphorus were determined using Kirsanov's method of HCl extract with further phosphor identification using a calorimeter and potassium identification using flame photometer.

The germination of the seeds was evaluated as 96% ; the weight of 1000 seeds was 40 grams. Before filling the pots the soil was mixed to produce homogeneous physical and chemical properties. 120 cylinder pot samples were filled . All possessed equal density, $P=1.07g/cm^3$; moisture content of 17.8%; and air permeability, with a value of B=853 size-free units . Air permeability was measured using a Glavlitmash special instrument brand 042.

The experiment was run using Honor variety of barley which was sown according to the scheme shown in Table 1.

All soil samples were divided into 5 groups of different soil compaction with 24 samples in each (Table 1). Each group was divided into 4 rows of 6 subsamples each so that the experiment could be repeated.

A row's number corresponds to a certain method of fertilizer application and a group number corresponds to a certain soil compaction. Each group's first row (30 samples) was a control - no fertilizer applied. In the remaining groups fertilizers were applied using three methods. Fertilizers were applied in the second row using a broadcast method with incorporation, i.e., fertilizers were homogeneously mixed with the whole soil in the pot to correspond to broadcasting of fertilizers on a field's surface and further incorporation into the soil using a plow or a cultivator.

Table 1. The design of the experiment.

ROWS	GROUP					Method of mineral
	I	II	III	IV	V	fertilizer application
	Seeds sprouting					
1	64	114	115	127	66	No application
2	118	138	119	107	77	Broadcasting with cultivation
3	103	125	114	101	92	Banding
4	104	110	105	95	78	Only broadcasting

In the third row of each group fertilizers were applied at a depth of 7.5 centimeters in each pot where they were placed as a continuous band (banded method).

In the fourth row fertilizers were homogeneously broadcast on the soil's surface in the pots (broadcast method).

Each treatment was accomplished in six replications.

Fertilizers were applied at the rate of $N_{90}P_{90}K_{120}$ kg of active matter to 1 ha. Nitrogen fertilizers were applied as ammonium nitrate; phosphorus, as double superphosphate; potassium as potassium chloride.

The second, third and fourth group samples were compacted by instrument to the following pressures : G_2=0.02 MPa, G_3=0.03MPa, G_4= 0.04 MPa. The fifth group samples were compacted using a standard hydraulic press with the pressure of G_5=0.2 MPa. After compaction density of the soil samples was as follows: P_2=1.18 g/cm^3 ; P_3=1.26g/cm^3; P_4=1.32 g/cm^3; P_5= 1.47g/cm^3.

Air permeability decrease with increase in compaction is caused by pore volume decrease. It was found that increase in density by 1.36 times (from 1.07 to 1.46 g/cm^3) leads to air permeability decrease by 3.87 times.

This influences, negatively, air conditions in which the root system development and chemical element reactions depend.

The first group's samples were not compacted, i.e., P=1.07 g/cm^3. Soil compaction presses had 25 pins 2.5 cm length which served simultaneously as markers for homogeneous seed sowing to the same depth.

Filling the pots , compacting the soil, and application of fertilizers were accomplished on April 21, 1995.

Observations of plants' growth dynamics, dry matter storage, and phenophases were made.

The uptake of nitrogen, potassium, calcium, phosphorus and magnesium in the barley was measured during the vegetation period. In our analytical work we used commonly recognized methods .

Common nitrogen, phosphorus, calcium, potassium and magnesium contents were determined in one set by wet ashing according to the Ginsburg-Scheglova method with further nitrogen determination by Kjeldahl; determination of phosphorus using Levitsky's colorimetric method; potassium- using the flame photometer; calcium and magnesium, by using trilonometrical method.

All pots were watered daily with equal amounts of water to provide equal soil moisture conditions.

Observations displayed phenophases of the crops in different variants taking place unevenly. Sampling was made once a month.

RESULTS AND DISCUSSION

In the first and the second group seed sprouting was observed on the fourth day; in the fifth group, on the ninth day. Thus significant soil compaction in group 5 influenced the seeds' germination period.

The results of seed sprouting in groups on May 10 are displayed in Table 1. In this case the average quantity of stocks in the row was taken into account, i.e., the general number of stocks was divided by six. The results in Table 1 show that soil compaction significantly influences seed sprouting.

Soil density in the first group was P= 1.07g/cm^3. This is less than the lower limit of optimal density, P=1.1 g/cm^3, for soils of this type. Undercompaction caused barley seed to spout more slowly in this group because of the weak seed contact with the soil.

Therefore, either overcompaction or undercompaction of soil leads to delay of barley seed emergence, with overcompaction influencing this process more negatively.

Seed sprouting results show their dependence on the fertilizers application method (Table 1). The best results showed broadcasting with incorporation method of fertilizers application. Lower results occur with the banding and broadcasting on the surface method of fertilizer application. It is evident that a high concentration of nutrient elements in the upper layer of soil negatively influenced barley seed germination. Fertilizer free treatment showed the lowest results.

The data in Table 2 show that, in the beginning of the vegetation period (on May 30,1995), the most height of crops occurs in the broadcasting with incorporation and banding method of fertilizer application. The broadcasting on the surface method showed significantly smaller height of crops. Vegetation processes were also weak in fertilizer free treatment, e.g. the height of plants was 21 cm in the fertilizer free condition with a soil density of 1.18 g/cm^3, while in broadcast with incorporation treatment it was 42.5 cm. Moreover, plants deprived of necessary nutrient elements had pale-green leaves and other signs of the shortage of the nutrient components that will influence general productivity. However with a soil density of 1.26 g/cm^3 we observed smaller height in the surface fertilized treatment - 38.5 cm while in the control it was 41 cm.

Therefore surface fertilization contributed to an increase in nutrient elements concentration during the initial period, that apparently suppressed growth and development of the plants. This is confirmed by plant height data recorded a month later.

The height of plants on June 10,1995 differ with banding and broadcast on the surface method of fertilizer application for different degrees of soil density (Table 2). With the density of 1.47 g/cm^3 the height in the broadcast method was 59.3 cm.

Considering dry matter storage (Table 2) we should note that fertilizers influenced the plants' weight favorably. In the broadcast on the surface method the weight of plants was much smaller according to the observations May 30,1995. However, a month later (June 30 1995) dry matter storage became equal in these treatments or even exceeded (with the soil density of 1.26 g/cm^3) the other treatments of fertilizer application. Maximum biomass storing in this period was observed with the soil density of 1.47 g/cm^3 in all variants.

Barley yield structure analysis was run and this data is displayed in Table. 3. The results show that maximum productivity of barley was much higher in the fertilized treatment. It is difficult to construct a regularity in this data depending on the method of fertilizer application.

Table 2. Growth and dry element storage depending on soil compaction and method of fertilizing.

Density g/cm^3	Method of fertilizing	Plant growth, cm		Dry elements storage, g/10 plants	
		05.30.95	06.30.95	05.30.95	06.30.95
1.07	none	25.5	40.1	9	32
	broadcast/cultiv	39	47.5	26	63
	banding	43.5	51.2	35	60
	broadcast only	38.5	53.7	24	60
1.18	none	21	37.5	5	22
	broadcast/cultiv	42.5	50.3	26	67
	banding	41.5	54.3	30	58
	broadcast only	40	54.7	25	53
1.26	none	41	35.3	25	18
	broadcast/cultiv	44.5	51.5	37	55
	banding	40	56.2	39	65
	broadcast only	38.5	53.4	25	68
1.32	none	24.5	38.2	8	23
	broadcast/cultiv	45.5	51.2	40	40
	banding	42.5	58.3	44	55
	broadcast only	39	54.1	20	65
1.47	none	23.5	40.3	7	25
	broadcast/cultiv	30	59.3	10	80
	banding	32.5	58.2	13	76
	broadcast only	30	58.8	10	75

Although the tendency for productivity to increase if banding or surface methods are used with densities of 1.07; 1.16; 1.26 g/cm^3, respectively, with soil density of 1.47 g/cm^3 maximum productivity was observed in treatments with banding (2.8) and broadcast with incorporation (2.7) methods.

It should be noted that extreme weather conditions influenced the growth of the crop and yield. Night frost in the period of emerging plants and severe heat in the final period of vegetation influenced crop development. The plants were short, grain ripening was slow. Especially slowed was the development of the control plants. The grain head was underdeveloped and the number of seeds was 7-8. Hence crop nutrient regulation may be one way of minimizing the external factors negative influence.

Den- sity g/cm^3	Methods of ferti- lizing	Bushiness of barley's head, cm		Plant height	Head length	Number of seeds/ head	Yield kg/m^2
		general	product	cm	cm		
1.07	1	1.5	1.3	38	10	13	0.04
	2	1.8	1.4	53	17	15	0.31
	3	2.4	1.7	56	21	18	0.26
	4	2.6	1.7	52	18	11	0.24
1.18	1	1.8	1.5	39	12	8	0.06
	2	2,4	1.8	58	20	16	0.38
	3	2.6	1.9	57	20	16	0.54
	4	2.7	2.4	56	16	14	0.34
1.26	1	2	1.7	41	12	7	0.06
	2	3.4	2	56	17	16	0.32
	3	3.6	2.2	53	17	14	0.52
	4	3.2	2.4	55	16	13	0.28
1.32	1	1.6	1.2	41	12	10	0.07
	2	4.3	2.7	58	17	12	0.32
	3	3.3	2.8	56	16	13	0.50
	4	3.3	2	54	17	12	0.24
1.47	1	1.8	1.5	41	14	8	0.05
	2	2.5	1.6	57	18	21	0.41
	3	2.4	1.6	52	18	17	0.32
	4	3.1	2	59	19	17	0.28

Table 3. Dependence of structure and quantity of barley yield on methods of fertilizing and soil compaction.

Methods of fertilizing - 1, 2, 3, 4 - are, respectively, none, broadcast with cultivation, banding, and broadcast only.

The data in Table 3 show that higher productivity was obtained in the banding fertilization method. It was 0.54 kg/m^2 with the soil density of 1.18 g/cm^3 and 0.52 kg/m^2 with a soil density of 1.26 g/cm^3. The second most productive treatment was the broadcast with incorporation method. However, with soil densities of 1.07 and 1.47 g/cm^3, this method was more effective than the banding method, with increases of 0.04 and 0.08 kg/m^2, respectively.

It is noteworthy that an increase in soil density caused a decrease of barley productivity in the banding treatment of fertilizer application. However, in the broadcast with incorporation method the highest yield was obtained in the treatment with the highest soil density, 0.40 kg/cm^2.

Table 4. Dynamics of nutrient element storage in barley plants depending on soil

CONTENT OF NUTRIENT ELEMENTS IN THE PLANT, mg/100g

Density g/cm3	Methods of fertilizing	05.30.1995					06.30.95				
		N	P_2O_5	K_2O	Ca	Mg	N	P_2O_5	K_2O	Ca	Mg
1.07	none	2.63	1.28	5.6	0.7	0.24	1.1	0.45	3.29	0.37	0.14
	broadcast with cultivat	4.48	1.01	9.41	0.86	0.23	1.63	0.23	3.54	0.45	0.14
	banding	4.49	1.01	9.84	1.00	0.24	1.00	0.48	1.9	0.12	0.11
	broadcast only	4.49	0.76	8.95	1.09	0.24	1.96	0.32	2.74	0.43	0.14
1.18	none	1.92	1.08	5.57	0.52	0.24	1.02	0.57	1.87	0.14	0.13
	broadcast with cultivat	4.48	0.96	9.47	0.86	0.22	2.03	0.42	3.47	0.34	0.13
	banding	4.44	0.98	8.69	0.86	0.21	2.38	0.45	3.74	0.37	0.15
	broadcast only	4.44	0.73	9.05	1.03	0.26	2.1	0.49	3.3	0.3	0.12
1.26	none	4.45	0.98	9.47	0.96	0.24	0.92	0.57	1.75	0.1	0.12
	broadcast with cultivat	4.51	1.08	9.7	0.89	0.2	1.92	0.48	2.97	0.34	0.13
	banding	4.67	1.11	9.25	1.02	0.23	1.9	0.5	3.15	0.35	0.13
	broadcast only	4.67	0.88	9.37	0.8	0.21	1.96	0.31	3.47	0.4	0.13
1.32	none	1.78	1.06	5.48	0.46	0.2	1.07	0.57	1.78	0.2	0.14
	broadcast with cultivat	4.16	0.88	9.56	0.8	0.19	1.87	0.44	3.19	0.4	0.12
	banding	4.65	1.27	9.36	0.96	0.19	1.87	0.48	3.82	0.28	0.12
	broadcast only	4.57	0.85	9.1	0.93	0.18	1.9	0.35	3.3	0.3	0.12
1.47	none	2.51	0.96	6.00	0.66	0.2	0.93	0.38	1.83	0.1	0.12
	broadcast with cultivat	4.55	0.93	9.37	0.96	0.21	1.58	0.3	2.9	0.3	0.08
	banding	4.78	0.94	9.29	0.86	0.23	1.44	0.3	2.52	0.28	0.11
	broadcast only	4.98	0.83	8.05	0.83	0.2	1.59	0.37	3.21	0.25	0.12

Control plants fell noticeably behind, their yield was almost ten times lower than in the fertilized treatment.

During the process of research observations were made on nutrient element absorption during the barley plants growth and development. The pace of nutrient element uptake by plants serves as a theoretical ground for the rational methods of the application of fertilizers. Barley, by the biological peculiarity of its nature, is very demanding of the conditions of cultivation. Dry matter storage depends on the uptake and storage of the nutrient elements. In each fertilization treatment with fast growth and maximum dry matter storage more intensive uptake of the nutrient elements by plants is observed (Table 4).

The content of nitrogen, phosphorus, potassium , calcium and magnesium decreased steadily with the transition from earlier to the later stages of plant development. Young plants contain larger quantities of nutrient elements that can be explained by the limited dry matter storage in them and relatively intensive uptake of nutrient elements.

The decrease of nutrient element content in barley plants with age takes place not because of stoppage of uptake but due to the more intensive storing of the dry matter during later vegetation periods.

According to the data on May 30,1995 nitrogen uptake by plants was higher in the fertilized treatment. It seems to be impossible to find any dependence on the method of the application of fertilizers. With the soil densities of 1.07 and 1.18 g/cm^3 the nitrogen content was almost equal in different treatment. With other degrees of soil density the content of nitrogen was a little higher in the banding and broadcast on the surface method comparing to the broadcast with incorporation method.

A regularity in the uptake of phosphorus in this period cannot be clearly observed. The content of calcium, however, is much higher in the all the fertilized treatments with the exception of the soil density of 1.26 g/cm^3.

The uptake of potassium was higher in the major number of treatments with the broadcast with incorporation method.

The same results were received on the uptake of calcium. A definite regularity of magnesium uptake can not be observed on this date.

The data for nutrient element content in barley on June 30.1995 shows that fertilizers influenced positively the storage of nitrogen by plants. Maximum nitrogen content with a soil density of 1.07 g/cm^3 was observed in the surface fertilizer treatment. With all other degrees of soil density the change in nitrogen content did not exceed the experimental error. During this period a regularity in phosphorus uptake can not be observed. Potassium content was higher in the major part of the fertilization treatment, excluding banding and broadcast on the surface application of fertilizer with a soil density of 1.07 g/cm^3. In this case the uptake of potassium was a bit lower than in the control. To determine the content of calcium and magnesium on this date was impossible.

Straw quality analysis results show that the application of fertilizers increased the content of nitrogen. potassium. calcium. This influence was not observed for phosphorus and magnesium content in the barley straw (Table 5).

Table 5. Yield quality dependence on soil compaction and methods of fertilizing.

Density g/cm³	Methods of fertilizing	Quality of barley straw					Quality of barley seeds					
		N	P₂O₅	K₂O	Ca	Mg	N	N	P₂O₅	K₂O	Ca	Mg
1.07	none	0.57	0.03	1.69	0.36	0.11	2.44	2.19	1.56	0.34	0.05	0.15
	broadcast with cultivat.	1.19	0.02	3.37	0.44	0.12	2.6	2.33	1.57	0.35	0.06	0.15
	banding	1.16	0.03	3.61	0.5	0.12	2.75	2.46	1.56	0.36	0.07	0.16
	broadcast only	0.74	0.02	3.4	0.39	0.10	2.65	2.36	1.54	0.33	0.06	0.15
1.18	none	0.66	0.04	1.32	0.32	0.14	2.00	1.78	1.37	0.34	0.04	0.14
	broadcast with cultivat.	0.67	0.02	2.66	0.4	0.09	2.42	2.17	1.42	0.33	0.05	0.14
	banding	0.84	0.03	4.08	0.34	0.09	2.62	2.36	1.39	0.25	0.05	0.14
	broadcast only	1.15	0.03	2.66	0.78	0.17	2.64	2.35	1.41	0.21	0.06	0.14
1.26	none	0.45	0.04	1.13	0.26	0.10	1.94	1.75	1.24	0.24	0.02	0.13
	broadcast with cultivat.	0.8	0.02	3.24	0.41	0.08	2.66	2.42	1.31	0.24	0.06	0.14
	banding	0.7	0.02	4.32	0.40	0.14	2.62	2.28	1.25	0.24	0.05	0.13
	broadcast only	1.44	0.03	3.21	0.43	0.11	2.62	2.38	1.31	0.2	0.04	0.14
1.32	none	0.44	0.04	1.22	0.26	0.09	2.13	1.93	1.28	0.24	0.03	0.14
	broadcast with cultivat.	0.98	0.04	3.44	0.57	0.10	2.55	2.31	1.33	0.21	0.05	0.12
	banding	0.84	0.02	3.76	0.66	0.11	2.63	2.38	1.32	0.22	0.05	0.14
	broadcast only	0.75	0.02	3.06	0.34	0.08	2.64	2.38	1.32	0.19	0.06	0.12
1.47	none	0.62	0.02	1.31	0.24	0.08	2.22	2.01	1.21	0.12	0.03	0.14
	broadcast with cultivat.	0.65	0.02	2.44	0.27	0.06	2.23	1.99	1.27	0.24	0.04	0.12
	banding	0.67	0.02	2.09	0.38	0.06	2.28	2.07	1.24	0.25	0.04	0.12
	broadcast only	0.62	0.02	2.96	0.36	0.07	2.24	2.02	1.26	0.28	0.03	0.12

Density g/cm^3	Method of fertilizing	pH	Organic matter	Content mg/100g	
			%	P$_2$O$_5$	K$_2$O
1.07	1	6.64	2.02	12.5	18.6
	2	6.7	2.3	20.4	24.9
	3	6.88	2.53	19.6	25.4
	4	6.9	2.44	22.6	31.4
1.18	1	6.82	2.3	12.7	13.4
	2	6.84	2.2	23.4	20.6
	3	7.11	2.45	32.8	25.8
	4	6.9	2.35	23.6	28.6
1.26	1	6.82	2.32	12.6	15.1
	2	6.83	2.3	18.9	24.3
	3	6.92	2.26	21.8	25.5
	4	6.82	2.35	24.2	37.2
1.32	1	6.52	2.4	12.2	14
	2	7.41	2.32	16.6	28.6
	3	7.01	2.18	24	26.8
	4	6.69	2.35	23.6	34.3
1.47	1	6.92	2.4	12.7	14.1
	2	6.81	2.4	19.8	23.8
	3	6.86	2.4	21.4	21.2
	4	6.56	2.44	22.1	29

Table 6. Agrochemical characteristics dependence on of soil compaction and method of fertilizing

Methods of fertilizing 1, 2, 3, 4 are respectively none, broadcast with cultivation, banding, and broadcast only.

In analyzing the quality of barley seed we should note that fertilizers significantly increased the content of general protein nitrogen. There was a tendency for phosphorus content to increase in the barley seed on the application of fertilizers. No regularity in increase is observed in the content of calcium. magnesium and potassium.

Maximum general protein nitrogen content was observed in the banding treatment of fertilizer application with a soil density of 1.27 g/cm^3 and was 2.75% and 2.46%, respectively. That was 0.15% higher than in the broadcast treatment and 0.10% higher than in the surface treatment. On soils with density of 1.18 ; 1.26; 1.32 g/cm^3 barley seeds contained more nitrogen in the banding and surface method of fertilization.

In all treatments with the soil density of 1.47 g/cm^3 the content of nitrogen was minimum.

Hence soil overcompaction influences negatively the seed quality.

After the experiment was finished sampling was made in two replications from each treatment. Before the experiment started the pH was 6.74. After the yield collection an alkaline tendency was noticed in the banding and surface treatment with the soil densities of 1.18 and 1.26 g/cm^3 (Table 6). With greater density surface fertilization caused acidification of the soil.

Humus content increased compared to the initial value of 2.11% in practically all samples taken after the yield collection. Maximum humus content was observed with the minimum soil density - 1.07 and 1.18 g/cm^3 in the banding treatment. With maximum density of 1.47 g/cm^3 humus content was almost the same in each treatment.

Phosphorus and potassium content remained practically unchanged during the vegetation period in the fertilizer free treatment compared to its initial value (P_2O_5 -12.1 and K_2O - 16.6 mg/100g soil). In all the fertilized treatments these numbers increased. Maximum phosphorus content was achieved in the banding treatment (soil density level was 1.18 g/cm^3) and was 32.8 mg/100g soil. A regularity of phosphorus content increase in the banding and surface treatment, regardless of the soil density, can be observed. The content of potassium in the soil is higher in all surface application treatments. Its content in the soil in some treatments was more than twice higher than in after treatments.

Preliminary data suggest that the banding and the broadcast on the surface method of fertilizer application promote the increase in potassium and phosphorus content in soil. The content of potassium in soil after the yield collection increased by more than two times in the surface treatment. Considering this we suggest that this will help decrease the fertilizer treatments using the banding and broadcast on the surface methods and attain the same yield as applying the whole treatment using broadcast with incorporation method.

CONCLUSIONS

The research results indicate preliminary conclusions :
1. Methods of the application of fertilizers influence emergence and initial growth of plants. The broadcast on the surface method. for example. created high concentration in the seed location area, which influenced, negatively, germination and development. The best germination was in the broadcast with incorporation method; lesser for the treatment with the banding method of fertilizer application.

2. Soil density influences the formation, amount and quality of yield of barley. The highest yield of barley was achieved in the banding method of the application of fertilizers and with soil densities of 1.18 and 1.26 g/cm^3. A further increase in density caused a decrease of yield.

3.   It was difficult to find any regularity in the uptake of the nutrient elements by the barley plants and any dependence on the level of soil density and methods of fertilizing. Fertilizers promoted a significant increase in general and protein nitrogen content in seed of barley as well as higher than in the fertilizer free treatment phosphorus content in barley. Maximum content of nitrogen was received in the banding treatment of fertilization with a soil density of 1.07 g/cm^3. The regularity of potassium. calcium and magnesium content could not be clearly seen.
4.   Agrochemical soil characteristics are dependent on soil density. The banding and the broadcast on the surface method promoted an increase in the content of potassium and phosphorus with potassium content increasing, by the end of the vegetation period, by more than two times in some treatments.

These conclusions need to be considered preliminary as one year of data is not conclusive. For the future it is recommended to enlarge the research so that its results can be more widely applied to agricultural production. It is expedient to study the energy aspects concerning different methods of fertilizer application for different crops and to elaborate standards of soil compaction by different agricultural machines and on this basis make recommendations on environmentally safe tillage of soils with different granulometric characteristics. This research may indicate how to reduce technological operations, decrease soil density, save fuel and reduce environmental problems. The choice of an optimal method of the application of fertilizers for various crops depending on the type, granulometric and agrochemical characteristics of the soil may save fertilizers, increase yield and improve quality of crops as well as promote soil fertility.
5.   The overcompaction of soil influences negatively soil air movement and the development of the root system. However, this problem needs to be studied more intensely as air permeability depends not only on density but also on humidity. Thus air permeability can serve as an indicator and its evaluation is necessary to determine the agrophysical state of soil.

ACKNOWLEDGMENT

This study was funded by Tyler Industries Inc. Many suggestions and comments from Don McGrath have been incorporated into research plan and authors acknowledge this contribution.

REFERENCES

Arinushkina, E.V. Chemical analysis of soils. Manual. 1970. Moscow University.

The Interaction Between the Spatial Variability of Velvetleaf Populations and Corn Grain Yield Potentials

S. A. Clay
K. Brix-Davis

South Dakota State University
Brookings, SD

ABSTRACT

Weed populations and yield potentials are spatially distributed in farmer's fields. The objective of this study was to determine if weed stress differentially influenced corn yields in areas of high, medium, and low grain yield potentials. High (20 to 24 plants/m row) and low (6 to 8 plants/m row) velvetleaf populations were chosen for the study. Corn grain yield was measured in the center, fringe, and outside of the weed infestation. As the yield potential decreased the impact of velvetleaf on grain yields increased. For example, the low velvetleaf infestation reduced yields 50% under low yielding conditions and 10% under high yielding conditions.

The Potential Contributions of Precision Farming to IPM

S. Daberkow
L. Christensen

USDA-Economic Research Service
Washington, DC

ABSTRACT

Based on a review of site specific and precision farming literature, a relatively small number of public and private R&D efforts are focusing on pest control. This made an informal survey of the leading precision farming researchers and agri-businesses tractable . Approximately 30 individuals were contacted. The individuals selected were from State Land-Grant Universities, Federal agricultural research agencies, agri-businesses providing precision farming technology to farmers, precision farming hardware and software vendors, and crop consultants. The respondents were asked about the potential for precision farming technologies to control pests and reduce environmental risks.

Precision farming can add a spatial element to conventional IPM programs, and will likely enhance IPM programs by helping: 1) to more precisely identify areas of a field where pests are present (based on predictive evidence such as soil tests, last year's infestation, previous crop, etc. or current year real-time observation) and assist crop scouts in identifying areas of crop stress; 2) quantify the economic significance of the pest and determine whether a chemical or non-chemical (e.g., biological) treatment is optimal; 3) the pesticide applicator locate and treat the economic pests in a timely manner when a rescue treatment is called for; and 4) identify the environmentally vulnerable parts of the field and adjust the pest treatment accordingly (i.e., karst areas, shallow aquifers, coarse textured soils, wetlands, nearby rivers or ponds, nearby crops sensitive to drift from pesticides and habitat of endangered species).

Weed Population Variability as Influenced by Different Sampling Approaches on a Field Wide Scale

G. Jason Lems
S. Clay
F. Forcella

South Dakota State University
Brookings, SD

ABSTRACT

Due to weed patchiness, it is difficult to estimate weed populations in large fields for precision herbicide application. This experiment compared several different scouting techniques for evaluating the spatial variability of 3 weed species (Setaria spp-, Cirsium arvense, and Ambrosia artemisifolia) populations in a no-till corn field. Weed populations were measured at 1300 sampling points in a 65 ha field on a 15- by 30-m grid. The population variability was compared with sampling schemes of using all grid points in the field, random sampling, and a W shaped pattern across the field. Preliminary results indicate that random sampling and the W pattern will dictate the entire field to receive a blanket treatment of herbicide according to conventional means. However, the grid sampling reveals the patchy nature of weeds and that approximately 50% of the field did not contain weeds and did not require herbicide treatment. This paper discusses how weed variability influences our attempt for precision herbicide applications.

Modelling the Patch Spraying Concept

M.E. R. Paice
W. Day

Silsoe Research Institute
Bedford, United Kingdom

ABSTRACT

A cellular weed demography simulation model has been developed which allows the economics and likely long term effects of Patch Spraying to be investigated. The model represents weed germination, spatially variable chemical control, seed production and seed mortality as a sequence of stochastic processes. Sub-models describe seed dispersal by natural effects, cultivation and combine harvesting.

Weed patchiness can be caused by spatial heterogeneity of conditions, by spread from an original colonist plant or by stochastic demographic processes which act to progressively increase population variance (in relation to the mean). Modeling shows that, in the latter case, seed dispersal tends to limit the level of patchiness which can develop. The mean spatial scale of patchiness increases with seed dispersal range.

The most substantial cost savings from Patch Spraying will be seen where it is possible to define large contiguous areas of the field which have not yet been invaded by the target weed species. In the rest of the field further savings might be obtained by treating the relatively small scale and low variance patchiness brought about by demographic and seed dispersal processes. This would require weed identification and treatment systems operating at relatively high resolution. The model suggests that in both cases the cost savings from Patch Spraying will be inversely related to seed dispersal range. For partially invaded fields, this largely reflects the importance of controlling weeds at the boundary of the colonized area, to prevent steady expansion into the weed free zone.

Patch work - a European Collaborative Project on Targeted Weed Control

J. V. Stafford

Silsoe Research Institute
Bedford, United Kingdom

ABSTRACT

A consortium of research groups from the United Kingdom, France, Denmark and Spain are applying precision agriculture concepts to weed control for two scenarios. The first topic is concerned with targeting herbicide to weed patches in cereal crops in Northern European conditions. The second concentrates on targeted non-chemical weed control in rowcrops in Southern Europe.

The system for cereals is based on generating weed maps. The maps, together with information on soil variability and crop agronomy, are used to build a treatment map to control a herbicide sprayer with a sectioned boom. Uncertainty or 'fuzziness' in the input data such as patch boundaries and infestation levels are taken into account by using a Bayesian probability approach in a decision support system to generate treatment maps. The treatment map is held in an on-board PC and is related to the sprayer position within the field by an enhanced GPS system. The PC generates control signals to drive the sprayer control system. A CAN data bus is used to pass data between PC, sensors, operator panel and control systems.

The rowcrop system is based on real-time detection and control of weeds, using a mobile robotic manipulator with 6 degrees of freedom which is tracked between the rowcrops. Weed detection is undertaken by image analysis using colour images captured by a CCD colour camera. End effector positioning is determined by a combination of inertial measurements and visual measurements (using a monochromatic CCD camera). The end effector is programmed to approach each group of weeds detected. Weed control is by means of an electrical discharge probe.

Making Site Specific Nitrogen Fertilizer Recommendations for Corn

T. M. Blackmer
J. S. Schepers
D. D. Francis

USDA-Agricultural Research Service
University of Nebraska
Lincoln, NE

ABSTRACT

Technologies to apply varying rates of N fertilizer have developed faster than methods of making site specific recommendations. Tools such as yield mapping are sometimes thought to be the answer to obtaining reliable N recommendations. However, yield maps alone do not provide sufficient information for accurate site specific recommendations. The objective of this study is to evaluate where in a field current N fertilizer recommendations worked and where they did not. The evaluation field was the Nebraska variable rate project located in Shelton, Nebraska. Grid soil samples, stalk nitrate samples, yield maps, remote sensing data and individual ground truthing information were collected in 1994 and 1995. Remote sensing data were used to identify portions of the field that were N deficient. Stalk nitrate samples were used to evaluate the accuracy of the N recommendations. A comparison of N recommendations based on yield was made to recommendations made by incorporating stalk nitrate data. Stalk nitrate data were used to identify portions of the field that were over-fertilized and where lower N recommendations might be appropriate for next year. Overall, stalk nitrate samples provided another tool that can improve N fertilizer recommendations for site-specific farming.

Similarities Between Organic Matter Content and Phosphorus Levels when Grid Sampling for Site-Specific Management

G. E. Varvel
T. M. Blackmer
D. D. Francis
J. S. Schepers

USDA-Agricultural Research Service
University of Nebraska
Lincoln, NE

ABSTRACT

Grid sampling may be considered a necessary evil in order to make variable rate phosphorus fertilizer recommendations. The sampling frequency and cost of obtaining a representative map of phosphorus availability may be prohibitive in many cases. This study was conducted to evaluate patterns in available phosphorus maps as influenced by sampling frequency. Results showed that a map of available phosphorus in surface soil followed a similar trend as for subsoil phosphorus. Phosphorus maps showed a strong resemblance to an organic matter map and an aerial photograph of soil color. Results suggest that if grid sampling is not practical, then an aerial photograph of soil color could be useful to identify typical areas within a field for soil sampling. The exception is where manures have been applied in the past, which may not have affected soil color.

Remotely Sensed Soil Organic Matter Mapping for Site-Specific N Management in Dryland Winter Wheat

C S. Walters
B. E. Frazier
B. C. Miller
W. L. Pan

J.R. Simplot
Colton, WA

ABSTRACT

The soils of eastern Washington have highly variable properties, crop productivity and input requirements, yet this crop land is managed uniformly. Site-specific farming may improve nitrogen (N) management if zones of similar N requirement can be established. Our objectives were to (i) define site-specific management zones based on remotely sensed soil organic matter (OM) levels, (ii) evaluate directed sampling within OM zones to establish zones of similar N requirement and (iii) assess the benefit of variable rate N management among N requirement zones for winter wheat production. Infrared bare soil photos of four sites were classified to define spectral classes representing distinct OM zones. Soil samples within each OM zone were analyzed for OM, N and preplant moisture and correlated to image band values. All image bands were significantly correlated to OM and classification defined zones with significantly different OM. The red image band values contain most of the information available indicating that standard color photos would be useful. Organic matter zones were regrouped into N fertilizer management zones called N-zones. Grain yield, protein and test weight were compared between site-specific N rates and standard N rates. Regrouped N-zones at each site contained one zone with significantly higher OM and soil N supply but not preplant moisture. Comparing variable to standard N management, yields were not significantly different in two of three years yet the total N applied in the variable application rates were reduced by up to 44%. Where high preplant residual N existed, the variable N rates did not reduce yields, but grain protein was lowered significantly to more desirable levels. Estimated excess N was correlated with grain protein and could be used for N sufficiency calculations post harvest. Preplant soil NO_3^- was highly correlated with organic matter percent among zones within sites. High preplant soil N variability among zones necessitates variable rate N management based on reasonable yield estimates and preplant soil N levels.

Computer Methods to Investigate Site-Specific Crop Management

M. D. Cahn
J. W. Hummel
D. G. Simpson

USDA-Agricultural Research Service
University of Illinois
Urbana, IL

ABSTRACT

Field evaluations are needed to determine sites where spatial application of inputs would be agronomically superior to uniform applications, but field experiments are expensive to implement, and many site-years are necessary for a fair evaluation. Experimental results from site-specific crop management research (SSCM) are often difficult to extrapolate to other sites where spatial patterns of soil properties may differ. Additionally, much of the equipment used for site-specific crop management, such as real-time sensors of soil nutrients, variable rate applicators, and navigation systems are still evolving. Conclusions about SSCM may change as equipment become more accurate for both measuring crop needs and for applying inputs. Computer simulations are an inexpensive alternative to field experiments, yet a powerful tool for investigating agronomic benefits of SSCM. Using Monte Carlo methods, computers can simulate field experiments on thousands of sites, or at a single site during many years. Simulations may provide a broad understanding of the factors important to SSCM. Such information could aid in developing meaningful hypotheses and experimental designs to evaluate SSCM practices in the field, as well as identify conditions and environments where SSCM would be superior to conventional, uniform management practices. Examples of computer simulated field experiments, comparing SSCM and conventional management practices will be presented.

Six Year Yield Variability Within a Central Iowa Field

T. S. Colvin
D. B. Jaynes
D. L. Karlen
D. A. Laird
J. R. Ambuel

USDA-NSTL-Agricultural Research Service
Ames, IA

ABSTRACT

Technologies to support precision farming (PF) began to emerge in 1989 when the global positioning system became available to a limited extent and was tested as a means for locating farm equipment within fields. Substantial PF technology is available, with costs rapidly decreasing, while capabilities are increasing. One major class of information that is missing is a method for determining how much material to apply or what action to take as a result of a condition at a position within a field. Yield variability, within fields, and its cause will need to be understood to provide that information. This study was designed to determine variability within one field. Crop yields showed a coefficient of variation ranging from near 12% in 1989 and 1992 to over 30% in 1990 and 1993. Relative rankings of yields between specific locations were not stable even after 6 years when rankings were recalculated after each new year. Many PF scenarios are based on the assumption of a stable yield pattern within a field. Only a few points in this field have exhibited a stable yield pattern during the study period. Perhaps stable patterns will eventually emerge, but the time frame for this to occur may be quite long. Overall, this study suggests that implementation of PF practices within the Clarion-Nicollet-Webster soil association area will reveal both difficulties and opportunities.

The Effect of Site Variability on the Design of Agricultural Field Trials

W. P. Dulaney
L. L. Lengnick
G. F. Hart

USDA-Agricultural Research Service
Kensington, MD

ABSTRACT

Experimental precision is increased by minimizing location effects which obscure or confound the detection of treatment differences. Location effects occur because the distribution of biophysical factors that control plant growth is rarely homogeneous--neither spatially nor temporally. However, if experimental blocks are sited on areas that are as homogeneous as possible, the power to detect experimental treatment differences is enhanced. This investigation attempts to evaluate whether Geographic Information Systems (GIS) and geostatistical techniques can be used in the design of agricultural field trials by providing improved models for evaluating the spatial and temporal variability of yield.

GIS data layers of corn grain uniformity trials and soil physical, chemical, and biological properties were used to determine and evaluate alternative design strategies. The principal design criterion was that at least six experimental blocks, each no smaller than 120m x 120m, needed to be located within the 16 ha field site. For each uniformity trial and soil parameter, a grid-cell GIS data layer was created using kriged block estimates (20m x 20m). A candidate design was determined for each parameter by identifying the location of six 120m x 120m experimental blocks with the lowest standard errors--i.e., the six most homogeneous blocks on the GIS data layer. The above experimental designs were superimposed, both singly and in combination, on 1993, 1994, and 1995 uniformity trial data as well as on an interannual yield variability data layer. By calculating the average yield standard error as blocked by these designs, the relative efficiencies of the designs was determined both between and across growing seasons.

Evaluation of Soil Sampling Methodologies - Does Grid Sampling Work?

P. Y. Gasser
H. Waddington

Ag-Knowledge
Ottawa, Canada

ABSTRACT

A field was divided into 240 foot grids and sampled using cell point and cell average methods. The 240 ft grid was then sub-divided into 120 ft grids and cell point sampled. A 360 ft grid cell point data set was generated from the 120 and 240 ft grid point data sets.

Grid maps were created using data from the cell average and cell point sampling at 240 foot spacing These maps were compared to grid maps generated from the 120 ft cell point data set. Grid maps created using the 360 ft cell point data were also compared to the 120 ft cell point grid maps.

The contouring and potential map functions in SPANS software were used to smooth the point data sets. Comparisons were made between "smoothed" maps and gridded maps. The effects of sample point spacing was also evaluated.

Index overlays were used for map comparisons.

Using DGPS and GIS for Knowledge-Based Precision Agriculture Applications

G.A. Johnson
M. D. Bickell

Southern Experiment Station
Waseca, MN

ABSTRACT

Precision agriculture goes beyond just variable placement of inputs. It is a dynamic, landscape-based crop management strategy based on information. This information allows one to gain knowledge and insight over time about a given system. Knowledge about the temporal and spatial variability of biotic and abiotic landscape attributes is critical and will help the producer design and manage cropping systems that are economically and environmentally sustainable. However, the lack of understanding of spatial and temporal interactions between landscape variables (soil biophysical properties, slope, aspect), pest populations, soil water relations, etc... has slowed the development and implementation of knowledge-based crop management systems. The newly acquired Agroecology Research Farm (AERF) at the University of Minnesota Southern Experiment Station will serve as a knowledge-based precision agriculture research site. The focus of this 160 acre field site is to investigate temporal and spatial relationships between biotic and abiotic landscape attributes. Field landscape data was collected using a DGPS and imported into a GIS for analysis. Data includes soil fertility, weed spatial distribution, disease and insect occurrence, soil water relations, topography and grain yield. This paper will demonstrate the use of DGPS and GIS for obtaining, managing, and analyzing information on a landscape level.

NESPAL's Precision Farming Initiative

D. Walters
C. Kvien

UGA-CPES-NESPAL
Tifton, GA

ABSTRACT

The National Environmentally Sound Production Agriculture Laboratory began a precision farming project in July 1994. The objectives are to identify, research, and implement precision farming technologies suitable to agriculture in the southeastern United States. In this overview of the project, results will be presented from various components which include soil sampling, yield mapping, variable rate technology, pest monitoring and mapping, and data management. Yields of corn, soybeans, canola, wheat, peanuts and cotton have been mapped and will be discussed, as will the potential of using new technologies for sampling and mapping pest populations. We will discuss the role of a geographic information system (GIS) for data management.

Essential Technology for Precision Agriculture

J.V. Stafford

Silsoe Research Institute
Wrest Park
Silsoe
Bedford   MK45 4HS
United Kingdom

ABSTRACT

The concept of precision agriculture - the targeting of inputs to field crops according to locally determined requirements - has developed rapidly because of the ease of access to a technically advanced positioning system (GPS). Technological advance in other areas such as sensing and control systems will show the way for precision agriculture systems to be implemented, though appropriate agronomy lags behind. The whole concept is dependent on acquisition and interpretation of data on spatial variability. From the understanding that follows, specifications for appropriate sensors will open up major opportunities for precise targeting of inputs.

INTRODUCTION

There can be few in the agricultural industry who have failed to notice the current high profile of precision agriculture as a concept. There are articles weekly on the subject in the farming press, numerous Internet pages are devoted to it, and scientific papers are appearing increasingly frequently. Yet the very *raison d'etre* for precision agriculture - that spatial variability in soil and crop factors is inherent in all agricultural fields - has been appreciated for decades, even centuries. The observation of Linsley and Bauer (1929) that the requirement for liming in agricultural fields varies across the field has been quoted in recent correspondence concerning patents in precision agriculture. It is only in recent years that appropriate technology has become available to enable the go-ahead farmer to begin to implement measures to take in-field spatial variability into account.

The introduction of the precision agriculture concept is just as revolutionary in agriculture as the introduction of "Just in Time" philosophy in manufacturing industry, as pointed out by Searcy (1995). Whereas in conventional arable agriculture the approach is "we will put effort into minimising the potential risks in crop production in order overall to maximise gross margins", precision agriculture says "we will determine the requirement of the crop and use technology to ensure that inputs work together to optimise the output produce". A brief definition of precision agriculture would therefore be "the targeting of inputs to arable crop production according to crop requirement on a localised basis".

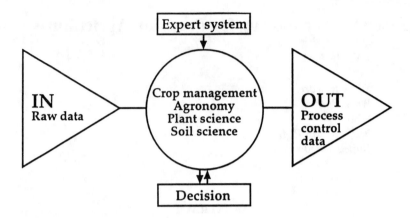

Fig. 1. Data flow for precision agriculture

The potential benefits of applying the concept of precision agriculture have been well rehearsed. Spatially variable application of inputs should lead to optimal use of those inputs and thus to improved gross margins and/or reduced inputs costs, environmental benefit in terms of reduced leaching of agrochemicals and reduced residues on crops, and reduced variation of produce quality.

After initial enthusiasm for the new technologies, there is increasing concern about and consequently research into the interpretation of spatially variable soil and crop data, crop management on a localised basis and the decision support systems needed to ensure rationally based spatially variable applications. The connection between technology functions and management functions may be illustrated by the "data flow" diagram shown in Fig. 1. Whilst crop management, decision support and the relevant sciences may be at the hub of precision agriculture, they are without foundation without real data on spatial variability. Similarly, they are a pointless exercise if the output process control data cannot be utilised in appropriate control, data transfer and precision application systems. Thus the essential technologies for precision agriculture, as illustrated in Fig. 2, are sensing and monitoring systems, and control, data transfer and precision application systems.

RESOLUTION, PRECISION AND ACCURACY

The concept of precision agriculture was spawned by technological advance in the precision with which soil and crop factors could be sensed, accuracy in positioning systems and equipment to apply inputs more precisely. The lead technology was the Global Positioning System giving promise of position resolution to a few metres.

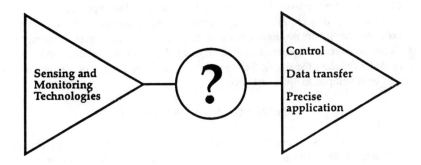

Fig. 2. Essential technologies for precision agriculture

The question has now however moved on to "what is the spatial resolution in spatially variable application that will optimise crop production in agronomic, economic and environmental terms?". In other words, we may be able to control weeds by targeting them individually within centimetres of crop plants because of technical advances (Brivot & Marchant, 1996), but would such operations be economically attractive in terms of improvement in harvested produce? Sensors may indicate that there is significant variability in mineral N over very short distances. But has the appropriate agronomy been developed to the extent that the information may be used to vary inputs over short distances? Much discussion of precision agriculture seems to regard it as axiomatic that spatial variability in yield and site variables necessarily implies that there is a significant spatial variability in optimum rates of inputs.

There are therefore important questions to be answered in terms of the spatial resolution that is appropriate for variable application of inputs. Stafford and Miller (1993) have shown that it is technically feasible to vary herbicide application at a spatial resolution of the order of 1 to 2 m. However, agronomic studies (eg. Miller et al, 1995) have shown that appropriate scales of application may be four or more metres. Equally, the appropriate spatial resolution (the appropriate management unit) needs to be determined for variable fertiliser or lime application.

Once the appropriate size of management unit has been determined in agronomic terms, there are still limitations on the accuracy and repeatability of spatial resolution obtainable by the relevant technology. There are significant problems in obtaining reliable position resolution with GPS (see next section) and other sensing systems, such as for crop or soil parameters, have limitations with regard to the spatial resolution obtainable. Stafford et al (1996) have analysed the spatial resolution obtainable in yield mapping by a conventional combine. They determined that yield variations on a scale of less than 15 to 20 m has little significance, and this has important implications for interpretation of variation. The economics of generating data to very high position resolution may also militate against the use of small management units in variable treatment. Manual soil sampling and analysis is notoriously expensive; geostatistical procedures may indicate a very fine grid for sampling for a particular factor (eg. mineral N) but the cost may be prohibitive. Using very high spatial position approaches to prevent spray overlap may require the use of survey quality kinematic GPS to obtain the necessary centimetre resolution, but cost may again prohibit implementation. It is likely, however, that where a sufficiently sound agronomic and economic case is made for reducing the size of the management unit then, in due course, technology will be developed to meet that need.

SENSING AND MONITORING

The successful implementation of precision agriculture requires significant data availability; crop factors, soil factors, field environment, cropping history, weather data. Much of this needs to be in the form of mapped data. Most of the factors pertinent to precision agriculture can currently be measured - but mostly only by manual methods, in-situ sampling, wet chemical analysis, etc. - approaches that are both costly and labour intensive. The prime requirement is for rapid low cost sensing techniques that can map the variation of factors across a field. The lack of suitable sensing techniques may be cited as the most critical factor in preventing the wider implementation of precision agriculture.

In the following sections, only a selection of the necessary sensing systems will be considered. The parameters that need to be sensed, accuracy and position resolution need to be defined by agronomists and plant scientists.

In-field Location and Positioning

In-field positioning is required in order to map the sensed soil and crop factors and for the control of application equipment (Stafford and Ambler, 1994). The position resolution required depends on the operation under consideration as indicated above, perhaps being:

variable fertiliser application:	30 m
yield mapping:	10 m
variable application of herbicide:	1 m
spray overlap avoidance:	10 cm
row crop planting:	10 cm
seed bed structure:	5 cm.

These figures are open to considerable debate but they illustrate the large range of position resolution that is required.

The Global Positioning System is generally considered to be the location system for precision agriculture but other positioning technologies such as radio systems (Palmer, 1995; Scorer, 1991), microwave systems (Monod, 1991) and laser (Gorham & MacLeod, 1991) have been advocated. Some of these systems depend on locally installed reference points. The attraction of the Global Positioning System is the establishment of global reference points (the constellation of satellites) with very accurately tracked positions.

Table 1 : GPS Error Sources - Meters

Source	GPS	DGPS
Satellite clock	15	0.1
Ephemeris	40	1.0
Orbit	5	0.1
Ionosphere	12	1.0
Troposphere	3	0.5
Multipath	2	2.8
Receiver noise	0.5	0.7
Total rms	44.8	3.3

GPS pseudo-range position computation with differential correction attains a position resolution of the order of 5 m. The significant error components contributing to this resolution are shown in Table 1. The problem however with such position computations is that they are subject to an error distribution with tails extending from sub-meter accuracy to ten or more metres. The five metre figure simply represents the two standard deviation parameter of the error distribution curve. These figures assume that the GPS receiver is locked into a constellation of satellites well positioned across the sky (giving low Dilution of Precision). In practice, satellite switch-over as satellites move below the elevation mask of the receiver, obscuration of satellites by trees and buildings (Lachapelle & Henriksen, 1995) and multipath reflections lead to significant degradation in position resolution. This lack of reliability in positioning resolution has serious implications for real time dynamic positioning of application equipment within the field. Serious positional error will lead to the treatment map that controls the applicator providing positionally misplaced information.

In order to improve the reliability of GPS, positioning studies have been undertaken to incorporate other information available on the field vehicle such as speed and heading, using techniques such as Kalman filtering and forward error estimation. However, for accurate and reliable positioning, kinematic GPS may have more potential, particularly when amalgamated with pseudo-range computation. Whilst kinematic GPS[+] has the potential for centimetre level positioning, it has a serious disadvantage for dynamic positioning of field vehicles. Signal loss leads to ambiguity in counting the carrier phase cycles. For most field operations, it would be unacceptable for the vehicle to stop in the middle of an

---

[+] In kinematic GPS, position is computed with reference to measurement of the phase shift in the carrier signals transmitted from the satellites.

operation whilst position was reinitialised after momentary loss of satellite signal. Techniques are being developed (commonly known as "on the fly" ambiguity resolution) to overcome the problem of cycle slip (e.g., Teunissen et al, 1995).

Yield Mapping

Much of the interest in precision agriculture in the farming community has been generated by the commercial availability of yield mapping on combine harvesters. In essence, this comprises a continuous grain flow measuring system and a position sensing system. It is illustrative of the mapping systems that are required for other soil and crop parameters, ie. two relatively low cost and reliable sensors forming a sensing system operating in a routine field operation with little or no intervention by the operator. Grain flow sensors may measure mass flow rate or volumetric flow rate; examples of both are commercially available in Europe (the Massey Ferguson gamma absorption mass flow sensing system, the RDS Ceres "Light beam interruption volumetric system" and the Claas paddle wheel/tipping bucket volumetric system). The advantage of mass flow sensing systems is that they are not dependent on variation in grain specific weight across the field and thus are potentially more accurate than volumetric systems.

Various techniques have been used for grain mass flow sensing including electrical capacitance (Stafford et al, 1996) and strain-gauged impact plate (Vansichen & DeBaerdemaeker, 1991). Rietz & Kutzbach (1996) have developed the light interception technique to improve accuracy in measuring volumetric flow and to compensate for errors when the combine is working on sloping ground.

Whilst most yield mapping systems appear to be accurate in terms of equating total grain flow to that measured over a weighbridge, the instantaneous accuracy of flow sensors has not been reported. A more important limitation on the spatial resolution achievable by combine-mounted yield mapping systems is the effect of the grain flow path through the combine on the instantaneous measured throughput. Lark et al (1996) analysed the convolution effects of the complex flow path by computing (*inter-alia*) the impulse response function of the system. This was determined from the flow sensor response to the combine entering the crop at an abrupt edge. They concluded that the flow sensor cannot be expected to resolve detailed yield variation over spatial intervals of less than around 15 m, although 20 to 25 m may be a more realistic scale of resolution.

Remote Sensing

The precision agriculture approach normally requires sensing at a large number of points in order to deduce spatial variability (represented for example by interpolation on to a map). Remote sensing techniques, where a relatively large area is sensed, can be more appropriate than spot sensing, in terms of cost and rapidity and ease of data collection. Here the term "remote sensing" encompasses both satellite imaging and near ground sensing - in this context it infers non-contact sensing.

The potential of satellite imaging to produce information on in-field crop condition has been recognised for some time. However, considerations of cost, timeliness and availability have inhibited its use for monitoring in-field spatial variability. Stevens (1993) has considered some of the logistical constraints on the use of such imaging techniques. More recently, efforts have been made to remove these constraints and, with improved satellite imaging in terms of spatial resolution, there are now moves to use satellite images at least in the decision making processes in precision agriculture. Indeed, a commercial service has recently been launched in the UK to provide farmers with enhanced maps derived from satellite imagery for individual fields as decision support. A research project at Nottingham University (UK) is studying the application of satellite imagery to assess the variability in crop condition during the growing season and link the information with that obtained from yield maps.

The basis of both satellite and near ground imaging is the variation in spectral reflectance of vegetative and non-vegetative material in the visible and near-infrared spectra. The most commonly used approach is to compute a vegetation index derived from reflection in the chlorophyll absorption band (around 680 nm) and the near infrared (750 to 900 nm). Such indices discriminate between vegetative and non-vegetative material and also indicate the condition of the vegetative material. Zwiggelaar (1996) amongst others has reviewed and made proposals for the most appropriate combination of spectral bands to use for various purposes for crop monitoring. He concluded that reflectance indices based on the chlorophyll and near infrared bands are optimal for within-field monitoring but need to be complemented with other data (such as crop spatial geometry). Such approaches are beginning to be applied to the sensing of in-field spatial variability in crops.

Christensen et al (1995) have used two spot radiometers mounted on a field vehicle to derive a vegetation index map by interpolation. To provide information on spatial variability of crop condition, Silsoe Research Institute studies are taking this approach further by using six radiometers on a tractor-mounted boom to rapidly scan a field and produce a "sample data" image. Such approaches could well be engineered to provide timely information on crop variability for generating treatment maps for crop inputs. Studies at Silsoe are being complemented by aerial imaging using a twin-video camera arrangement on a radio controlled model plane to provide larger area "snapshots" with sub-metre pixels.

Studies using imaging approaches to discriminate crop and weed species are showing promise for practical application to enable the targeting of herbicides to weeds and weed patches. Brivot and Marchant (1996) have successfully discriminated weeds in row crops from crop plants and soil, for real time operations. Bull& Zwiggelaar (1996) have shown that the use of image processing techniques has potential for discriminating weeds from crop and Molto et al (1995) have successfully segmented weeds from specific crop varieties. All these approaches have been based on capturing near-ground images (some in specific wavebands), with millimetre level resolution.

PRECISE CONTROL AND APPLICATION

Having generated suitable treatment maps, the targeted application of agrochemicals and fertiliser requires equipment that can control the rate of application automatically and also control the spatial placement. As indicated in an earlier section, the spatial resolution required may vary considerably. Although a number of systems are in place that can readily be adapted to variable application, there remains a requirement for development of equipment that can target application within a very close specification of rate, uniformity of distribution and definition of application area. An example is agrochemical sprayers where conventional nozzles apply material across an area typically 0.5 to 1 m wide. Targeted application at the plant level is therefore not possible. The control of application rate is another important area for development. Some consideration of the requirements for spatially selective herbicide sprayers has been given by Paice et al (1996).

Targeted application using treatment maps necessarily infers considerable data transfer and usage on the field machine. Rationalising of the data transfer process makes a data bus system almost essential. Efforts by ISO to develop a standard for a field vehicle databus, based on both American and German approaches[1], have continued for some years. The rapid commercialisation of precision agriculture makes urgent the need to complete such a data bus standard and associated data dictionary in order to handle logically and rationally the vast quantities of data "generated and consumed" in the implementation of precision agriculture.

CONCLUSION

The precision agriculture concept was conceived in the light of technological advance, particularly in positioning systems. As precision agriculture is implemented and the agronomy of localised application is developed, so the technical requirement for sensing and for application equipment can be more rationally specified in terms of crop requirement.

It is proposed in this paper that the critical path to implementation of precision agriculture is the sensing of spatial variability. Sensing systems with the necessary specification for generating spatially variable data have largely yet to be developed.

REFERENCES

Brivot, R, and J.A. Marchant. 1996. Segmentation of plants and weeds using infrared images. Acta Horticulturae, 406 (in press).

Bull, C R, and R. Zwiggelaar. 1996. The potential of discrimination between crops and weeds on the basis of their spectral reflectance characteristics. Crop Protection (submitted).

Christensen, S, K. Christensen, and A. Jensen. 1995. Spatial variability of spectral reflectance and crop production. Proceedings of the Seminar on Site Specific Farming, Koldkaergaard, Aarhus, 20-21 March 1995. Danish

---

[1]USA: Society of Automotive Engineers Recommended Practice J1939
Germany: DIN Standard 9684

Institute of Plant and Soil Science, SP Report No. 26, 56-66.

Gorham, B. and F. MacLeod. 1991. Lasers in farm management. Land and Mineral Surveying, 9 (1), 22-24.

Lachapelle, G, and J. Henriksen. 1995. GPS under cover: the effect of foliage on vehicular navigation. GPS World, Volume 6 (3), 26-35.

Linsley, C.M. and F.C. Bauer. 1929. Illinois Agricultural Experiment Station, Circular 346.

Miller, P.C.H., J.V. Stafford, M.E.R. Paice, and L.J. Rew. 1995. The patch spraying of herbicides in arable crops. Proceedings Brighton Crop Protection Conference, November 1995, British Crop Protection Council, 3, 1077-1086.

Molto, E., N. Aleixos, L.A. Ruiz, J. Váyquery, F. Fabado, and F. Juste. 1995. Determination of weeds and artichoke plants position in colour images for local herbicide action. 2nd International Symposium on Sensors in Horticulture.

Monod, N O. 1991. Localisation of agricultural machines. Paper presented to the International Symposium on Locating Systems for Agricultural Machines, Gödöllö, Hungary, June 1991, 6 pp.

Paice, M E R., P.C.H. Miller, and W. Day. 1996. Control requirements for spatially selective herbicide sprayers. Computers and Electronics in Agriculture, 14, 163-177.

Palmer, R J. 1995. Positioning aspects of site-specific applications. Proceedings of the 2nd International Conference on Site Specific Management for Agricultural Systems (Eds.) P.C. Robert, R. H. Rust and W.E. Larson, American Society of Agronomy, Madison, WI, USA, 613-618.

Rietz, P., and H.D. Kutzbach. 1996. Investigations on a particular yield mapping system for combine harvesters. Computers and Electronics in Agriculture, 14 (2/3), 137-150.

Scorer, A G. 1991. The development of Datatrack. Journal of Navigation, 44 (1), 37-47.

Searcy, S.W. 1995. Engineering systems for site-specific management: opportunities and limitations. Proceedings of the 2nd International Conference on Site Specific Management for Agricultural Systems, Editors: P. C. Robert, R. H. Rust and W.E. Larson, American Society of Agronomy, Madison, WI, USA, 603-612.

Stafford, J.V., B. Ambler, R.M. Lark, and J. Catt. 1996. Mapping and interpreting the yield variation in cereal crops. Computers and Electronics in Agriculture, 14 (2/3), 235-248.

Stafford, J.V., and P.C.H. Miller. 1993. Spatially selective application of herbicide, Computers and Electronics in Agriculture, 9 (3), 217-229.

Stevens, M.D. 1993. Satellite remote sensing for agricultural management: opportunities and logistic constraints. Journal of Photogrammetry and Remote Sensing, 48 (4), 29-34.

Teunissen, P.J.G., P.J. De Jonge, and C.C.J.M. Tiberious. 1995. A new way to fix carrier-phase ambiguities. GPS World, 6 (4), 58-61.

Vansichen, R., and J. De Baerdemaeker. 1991. Continuous wheat yield measurement on a combine. Proceedings of Symposium on Automated

Agriculture for the 21st Century. 16-17 December, Chicago, IL, USA, American Society of Agricultural Engineers, 346-355.

Zwiggelaar, R. 1996. A review of spectral properties of plants and their potential use for crop/weed discrimination in rowcrops. Crop Protection (submitted).

Soil Tillage Resistance as Tool to Map Soil Type Differences

J. van Bergeijk
D. Goense

Department of Agricultural Engineering and Physics
Wageningen Agricultural University
The Netherlands

ABSTRACT

Precision agriculture incorporates spatial knowledge of soil and crop conditions in the management decisions. In this paper, a method to improve determination of soil physical properties is proposed. Current practice is to analyze soil samples taken at several locations in a field. To obtain a sound coverage many soil samples have to be analyzed. To reduce these costs, information gathered automatically during the major soil tillage operation is used. Measurements of plow draft proved to be useful to define different regions in a field.

INTRODUCTION

The results of yield mapping on a Udifluvent soil type show high correlation between yield and soil type (Finke & Goense, 1993). While yield mapping, at least in cereals, can be done automatically by harvesting equipment, determination of soil physical properties is more time and resource consuming. In a precision agriculture context every bit of knowledge on the origin of crop variability can be useful for management. Not only differences noticed while harvesting a crop, but also differences in operating conditions during other field operations must be taken into account.

Especially the monitoring of soil tillage operations might give useful information to recognize different soil types within a field. Moldboard plowing in the autumn, at approximately 30 centimeters depth is general practice in Dutch agricultural, and is an operation that covers a whole field. With the electronic hitch control on modern tractors it offers the opportunity to log the signals on working depth and draft force to create a soil tillage resistance map. Previous work on this topic show the technical feasibility of the measuring method. Boundaries between adjacent soil types were detected although soil physical characteristics were not quantified. The resolution of the measurement method was high enough to show local distortions like tramlines in a field.

Literature in which plow draft is mentioned focuses mainly on the determination of factors influencing soil tillage draft. Research objectives were, for example, to find the optimal moldboard design for a certain soil type, to estimate needed tractor power for a given plow and soil condition, and to optimize the timing of tillage operations. The influence of velocity and depth on specific draft for four soil series has been measured in a study to compare different plow bottom designs (Reaves & Schafer, 1975). Eradat Oskoui et.al., report cone index, specific weight, moisture content, moldboard tail angle and plow speed as major factors influencing specific draft. The cone index, closely related to soil moisture content, contributed most to a specific draft estimate. In a study to investigate tillage draft and fuel consumption for different tillage implements for twelve soil series the specific draft force varied from 20 to 76 kN/m^2 (Bowers, 1989). The soil series ranged from loamy sands to clay loams with different soil moisture contents and tillage history. The influence of velocity and depth on tillage draft for four primary tillage tools on four Oklahoma soils has been investigated by Summers et.al. (Summers et.al., 1986). Their interpretation of the experiments led to regression type formulas that can only be compared with other regions if other factors like soil moisture content are in the same range. More generic relations between tillage operations and draft may be useful for machine management but show large ranges in estimated draft (Harrigan & Rotz, 1995). The above mentioned literature provides a reference to be able to estimate soil type from measured specific draft. The literature shows that estimation of physical soil properties based on plow resistance must take into account local soil moisture content, cone index and bulk density measurements.

OBJECTIVES

The ultimate goal is to use crop growth stimulation models as part of precision agriculture management packages. These models require spatial data on soil properties as bulk density, pore volume, structure specification and moisture status. While relations between specific plow draft and soil moisture status and specific plow draft and soil type interfere with each other, both specific draft and soil moisture content are measured. Main focus of this paper is on determination of specific plow draft. Soil moisture status has been measured and will be discussed briefly. A discussion on the contribution of measurement of plow resistance in a precision agriculture management system will end this paper.

METHOD

Data Acquisition System

The measurement system was built on a standard tractor and plow. The tractor, a Deutz 6.31, was equipped with a Bosch Hitchtronic system. The draft sensors in the left and right connection of the lower link arms and the position sensor on the lift axis from this electronically controlled hitch were connected to the data logger. The Rumptstad three furrow moldboard plow was fitted with RS480 bottoms. Additional sensors on the plow recorded plow width adjustment, actual plow width and plow depth. The working width of the first share on the plow can be adjusted by rotation of the plow frame. A sensor alongside the hydraulic cylinder recorded rotation angle of this width adjustment. Between rear tractor wheel and first plow share an ultrasonic sensor measured actual width of the first plow share at measuring the distance to the furrow wall. The plowing depth was measured with a sliding plate on the landside of the plow. This measurement method might give problems when plowing on bare soil but on a winter wheat stubble the plate remained clean. A complication of mounting sensors on a reversible plow is that either a double set of sensors is needed or the sensors have to be displaced to the working side. The sliding plate of the plow depth sensor rotated on an axis through the center of the plow frame and was in this way able to work on both sides of the plow. The ultrasonic sensor was mounted on a boom which displaced the ultrasonic sensor when reversing the plow. Figure 1 shows the locations of the sensors on the plow and in the hitch of the tractor.

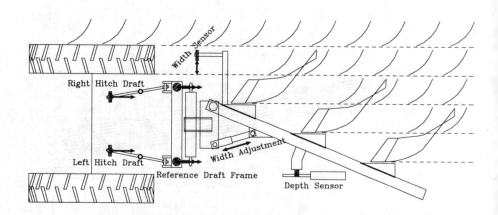

Fig. 1. Sensor mount for specific plow draft measurements.

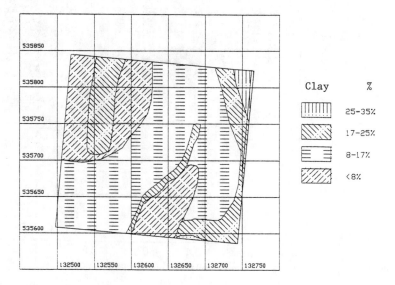

Fig. 2. Classical soil survey of experimental area.

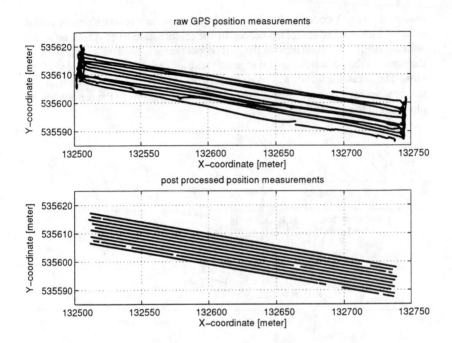

Fig. 3. Results of post processing position data of the first ten lanes.

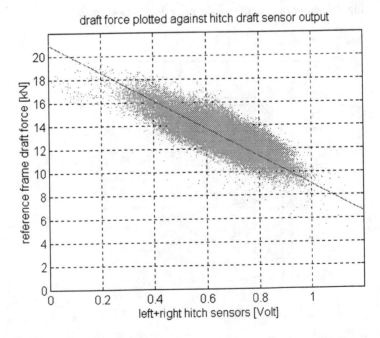

Fig. 4. Calibration of hitch draft sensors.

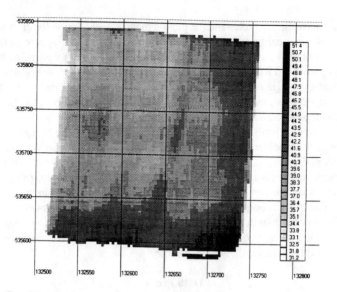

Fig. 5. Specific plow draft (kN/m²) measured with reference frame, X, Y-coordinates according to the RD grid projection (meter).

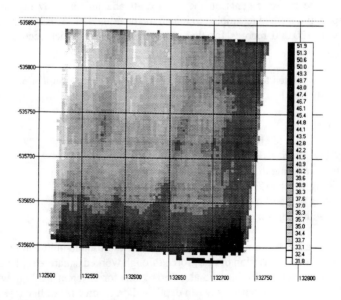

Fig. 6. Specific plow draft (kN/m²) measured with hitch draft sensors, X, Y-coordinates according to the RD grid projection (meter).

Between the tractor and the plow a measurement frame for draft force was mounted. This calibrated measurement frame provides the system with a hitch independent draft force measurement. Velocity measurements were taken from both driven wheel and a radar. A differential C/A code GPS receiver updated position information at 4 Hz rate. Finally, an electronic compass was added to augment GPS data. All analog signals were sampled at 40 Hz rate and stored on the hard disk of the data logger to enable post-processing.

The experimental field is in the north-west part of The Netherlands, in the Wieringermeer polder. The total examined area of the nearly square field was six ha (240 x 250 meters). The field belongs to a regional research station where a soil survey was carried out about twenty years ago. In Fig. 2, the results of this classical soil survey are presented. The scale is in meters and coordinates are from the Dutch national grid. The soil type varies between sandy loam and light clay with large small scale variations due to former sand banks and currents. During measurements with the plow samples of the top soil layer of the field were taken to determine soil moisture content and bulk weight volume.

Data Processing

Position, velocity, heading and wheel slip were reconstructed from the GPS, wheel and radar velocities and compass angle measurements. Additional information, like measured plow width and a straight start/stop line along the field boundaries was also used to shift positions until overlap was eliminated. The results from this post-processing were adjacent lanes with position, velocity, heading and slip information for every quarter second.

The measurements on the hitch draft and on the plow geometry were low-pass filtered and resampled to match the position data. This resulted in a data set with 120,000 records for the entire six ha. field. Each record contained information on:

• Position, related to square x,y grid of the Dutch 'RD' map projection.
• Velocity [m/s]
• Slip [%]
• Heading [rad]
• Left and right hitch draft sensor output [Volt]
• Hitch position sensor output [Volt]
• Reference frame draft [kN]
• Working width of first plowshare [m]
• Working depth of plow [m]
• Plow width adjustment sensor output [Volt]

The most common unit for plow specific draft found in literature is kN/m^2. It is calculated by division of draft force through worked square area, i.e., width x depth. Though this unit incorporates corrections for different working width and working depth the relation between depth and draft force is not linear [Reaves & Schafer, 1982]. The depth in our dataset varied between 23 and 33 centimeters. In this small range the specific plow draft is assumed to be independent of working depth. The working width of the plow is divided in two components. First, the width of the first share, which has been measured, and second, the width of the following two shares, which we assume constant at 0.8 meter. The equation to calculate specific plow draft for this plow is given by:

$$F_{s,v} = \frac{F_d}{(0.8 + \text{width}_{\text{firstshare}}) * \text{depth}}$$ [1]

where: Fs = specific plow draft [kN/m²]
F_d = plow draft [kN]
Width$_{\text{firstshare}}$ = working width of first plowshare [m]
depth = working depth of plow [m]

Another important factor influencing plow specific draft is velocity. While it is not expressed in the plow specific draft equation it is often specified in a separate notation valid for experiments carried out at the same velocity. For field scale experiments this approach doesn't work. The velocity in our dataset varied between 1.4 and 1.7 m/s. To be able to compare specific draft within one field a correction to a standardized velocity of 1.6 m/s was needed. According to ASAE draft is quadratic related to velocity [Harrigan & Rotz, 1995]. For velocities in the range 1.25 - 2.35 m/s the quadratic relation fitted data best [Summers et.al., 1986]. For different soil types both articles mention different parameters to relate specific draft to velocity. Because of the small velocity range present in our data we used a linear relation. The estimated correction factor is 5 kN draft increase at 1 m/s velocity increase based on figures from Reaves & Schafer and Summers et.al. Equation 2 displays the form of this velocity correction. A more exact determination of the specific draft - velocity relation should take local soil properties into account and has to be extracted from the data set itself. Possibilities to do so will be mentioned in the discussion at the end of this paper.

$$F_{s,\,1.6\,m/s} = F_{s,v} + 5.0*(1.6—V)$$ [2]

where: $F_{s,1.6\,m/s}$ = specific plow draft at 1.6 m/s [kN/m²]
$F_{s,v}$ = specific plow draft at velocity v [kN/m²]
V = actual velocity [m/s]

Visualization

The first method to analyze the data is to plot a colored indexed dot for every point in the dataset. This method reveals the smallest details measured with the system. Among these are small scale variations like the effect of tramlines. What is also visible are draft force shifts between adjacent lanes due to sensor drift and errors in the specific plow draft models (equations 1 and 2) While we assume these distortions don't represent different soil types a smooth operation is needed to filter these out.

Analysis of the raw data led to the implementation of an inverse distance weighing method with anisotrophy. Separate weighing distances were used for data points along the driving direction and data points across the driving direction. Typical weights were to apply an inverse distance weighing over 2.5 meter single side along driving direction and over 20.0 meter single side across the driving direction. With these parameters distortions from tramlines which were situated along the driving direction for this field were sufficiently suppressed.

RESULTS

Position improvements after post-processing for the first ten lanes are plotted in Fig. 3. The upper graph shows the positions according to the DGPS receiver including measurements taken while turning at the head sides of the field. The straight lines in the lower graph are the reconstructed positions of valid plow draft measurements. Without post-processing position errors up to five meters occur.

The noise on the plowdepth sensor remained within a range of two centimeters. The estimated accuracy after low-pass filtering is about one centimeter. The influence of the draft control of the hitch on the plowing depth is a point of concern. The plow depth varied four centimeters at the same hitch control settings in different areas of the field. As we will discuss later the hitch should maintain plowing depth constant but is affected by soil type. Exact determination of plowing depth is important in regard to equation 1. Small variations in depth have large impact on specific draft.

Measurements of the plowing width of the first plowshare with the ultrasonic distance sensor required severe filtering. Due to irregularities in the plow furrow many ultrasonic echos were lost. The filter algorithm first skipped all measurements which were logically out of range. Next step was to fit a smooth line through the measurements that remained. However the influence of errors in the plow width determination have low impact on specific draft. Visualization of the data under assumption of an equal plowing width of 1.2 meters over the entire field yielded to the same pattern. The width of the first share plowshare varied between 35 and 47 centimeters.

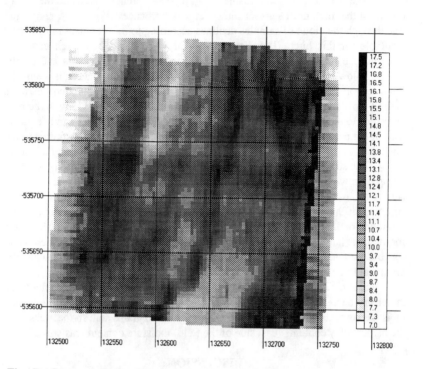

Fig. 7. Straw yield (ton/ha) August 1995. X,Y-coordinates in RD projection (meter).

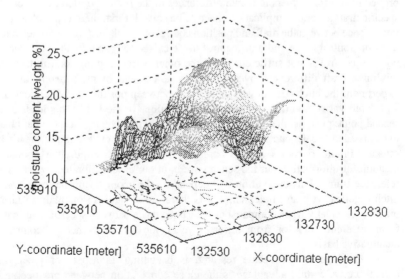

Fig. 8. Moisture contents during plowing (weight %).

The relation between measured draft by the reference frame and the sensor output of the hitch draft sensors shows large disturbances (Fig. 4). A regression line through the cloud of samples declines according to the technical information available for this hitch. The disturbances might be caused by hitch geometry affecting the hitch draft sensor output, by hysteresis in the sensors and by the relatively low measurement range. Because of the large disturbances it seems the hitch sensors are not suitable to monitor plow draft. However if we compare the map of plow specific draft according to the reference frame (Fig. 5) with the map according to the hitch sensors (Fig. 6) the same patterns appear. The sandy areas in the top-left part of the field are even more recognizable in the map according to the hitch draft sensors if we compare Fig. 6 to the conventional field survey in Fig. 2.

The straw yield map from August 1995 for this field (Fig. 7) shows a high correlation with the specific draft figures. Especially a band orientated east-west across the middle of the field and two spots in the lower left corner indicate structure differences which were not mapped in the conventional field survey. The resolution of the straw yield map is higher than the resolution of the grain yield map for the same harvest due to the measurement method and crop characteristics. Both specific draft and straw map show sharper borders between different areas while the grain yield map is smoothed.

Soil moisture contents at time of plowing are given in Fig. 8. The field was rather dry because of the dry season in 1995. The two sandy areas show lowest moisture contents down to 11 grams moisture per 100 grams soil. In the areas with the highest clay contents (east side of the field) moisture content was still 23%.

DISCUSSION

To monitor signals already present on the electronic hitch of the tractor proved useful for farmers to locate differences in the field For determination of specific draft several complications need to be solved. First, there is a problem of interference between the draft and position control of the hitch. A trade off between position control mode and draft control mode exists; for good quality plow work position control is not sufficient while the other extreme, draft control, tries to minimize draft differences over a field by adjustment of plow depth. In our experiment the hitch control was set at a mix between position and draft control; still depth varied significantly at different locations in the field. This leads to a second point: plow depth determination. An external plow depth measurement has to be installed to calculate specific draft. Third point is the calibration of the hitch sensors. Figure 6 shows a map based on hitch sensor output that contains meaningful information but the sensors were, in our case, calibrated by use of a reference frame. This calibration will depend on the geometry of the three point hitch together with an implement and is therefore difficult to maintain in farm practice, especially when comparisons have to be made between specific draft data from different fields or from different years, measurements need a common quantitative basis.

On-farm or farm-scale research is becoming an aspect of precision agriculture. A certain advantage is the close cooperation between practice and theory. A disadvantage is the difficulty to determine separate relations between different variables, as in this paper, the influence of velocity and depth on specific draft. Test plots with these variables varied over a wide range would give better correction factors but have the disadvantage whether they are portable to another field at another time. Another approach is to find algorithms that separate influences of single factors from the fieldscale experiment itself. For instance, from a comparison of adjacent plowed lanes that differ only in velocity a correction figure for velocity could be reconstructed. The found correction figures, valid primarily for that particular dataset, might even work out better compared to general figures made over larger, for instance, velocity ranges. The remaining question is whether enough different measurements were made to carry out this exercise. In addition to this, it is not farmers' practice to vary working conditions for experimental purposes. If velocity drops during plowing, it is likely to be correlated with a heavier soil structure where the tractor-plow combination has a difficult job to maintain velocity. In this study, the velocity range was small because the tractor was overdimensioned compared to the plow, a situation which is not optimal from a farm mechanization point of view.

CONCLUSIONS

With a three furrow plow the specific draft varied between 30 and 50 kN/m² on a six ha. field. Clay contents varied between 5 to 35%. The soil moisture contents range at time of plowing was 11 to 23%. Draft figures corresponded well with the clay contents of the soil survey but had a higher resolution.

To ensure a correct specific draft determination the standard electronic hitch has to be calibrated and an additional plow depth sensor is needed. Plow width determination had low impact on the specific draft and might be displaced by a once measured value. Velocity measurements were important to translate the specific draft data to a standard velocity. Post-processing of the measured positions was needed to detect borders of different areas and to detect small scale (< 15 meters) areas.

A map of plow resistance is suitable to locate different top soil types in a field. Consistency of the data over several years is yet unknown but the experiments will be repeated in the coming season. The value at moment is to use the information to reduce the amount of soil samples required to make a sound coverage of a field inventory.

ACKNOWLEDGMENTS

We wish to acknowledge Rumptstad BV for providing the plow and draft measurement frame. We are also grateful to the regional research station 'van Bemmelenhoeve' for adaptions made to the tractor and their flexibility in fieldwork. Additionally, we want to thank Michel Govers for his support concerning the electronic data acquisition system and the soil tillage laboratory for support concerning collection of soil physical properties.

REFERENCES

Bowers, C.G. 1989. Tillage draft and energy measurements for twelve southeastern soil series. ASAE vol 32 (5), p. 1492-1502.

Canarache, A. 1993. A preliminary model estimating soil specific resistance to plowing, Soil and Tillage Research, no. 27, p. 355-363.

Eradat Oskoui K., D.H. Rackham, and B.D. Witney. 1982. The determination of of plow draught - Part II. The measurement and prediction of plow draught for two mouldboard shapes in three soil series, Journal of Terramechanics, Vol. 19, No. 3, p. 153-164.

Finke, P.A., and D. Goense. 1993. Differences in barley grain yields as a result of soil variability. J.Agric. Sci., Cambridge, 120? p. 171-180.

Gebresenbet, G. 1991. Analysis of forces acting on mouldboard plows and seed drill coulters in relation to speed, depth and soil conditions: Static and dynamic studies. Uppsala, Sweden: Swedish University for Agricultural Sciences, 168 p.

Harrigan, T.M., and C.A. Rotz. 1995. Draft relationships for tillage and seeding equipment. Trans. ASAE 11(6):773-783.

Reaves, C.A., and R.L. Schafer. 1975. Force versus width of cut for moldboard bottoms. ASAE Paper No 74-1588.

Summers, J.D., A Khalilian, and D.G. Batchelder. 1986. Draft relations for primary tillage in Oklahoma soils. Trans.ASAE 29(1):37-39.

Development of a Texture/Soil Compaction Sensor

W. Lui
Shrini K. Upadhyaya
T. Kataoka
S. Shibusawa

Biological and Agricultural Engineering Department
University of California, Davis
Davis, CA 95616

ABSTRACT

Possible serious, widespread, long-term effects of soil compaction on environmental quality have become a serious issue in recent years. Soil texture and compaction level play an important role in precision farming since it is one of the key factors involved in deciding spatially modulated chemical application rate in agricultural production. This paper discusses the development of a soil texture/compaction sensor to map the variability of soil texture/compaction level in field.

BACKGROUND

Possible serious, widespread, and long-term effects of soil compaction on the quality of the environment have become a major concern in recent years (Gupta and Allamaras, 1987; van Owerkerk, 1995). Soil compaction can be caused by external loads applied by traffic or by the deterioration of soil structure caused by either a decrease in soil organic matter or application of certain tillage practices (Daniel et al., 1988; Voorhees, et al., 1988; Taylor and Gill, 1984; Gameda, et al., 1985; Negi, et al., 1981). It manifests itself as an increase in bulk density, decrease in porosity or void ratio, or change in pore size distribution (Soane and van Ouwerkerk, 1995). Soil compaction has been found to have a significant effect on four environmental quality issues - atmosphere, surface water, ground water, and soil resources. Atmospheric effects are related to increased emissions of greenhouse gases - CO_2, CH_4 , and N_2O (Horn et al., 1995). Increased CO_2 emission is due to increased tillage energy requirements in compacted soils (Hadas and Wolf, 1984; Wolf and Hadas, 1983, 1984). Increased emission of CH_4 and N_2O is caused by the anaerobic conditions present in moist and compact soils. Surface water contamination is related to decreased infiltration and increased runoff (LAWR - Cooperative Extension, 1984; Lipiec and Stepniewski, 1995; Fattah and Upadhyaya, 1996). Ground water contamination is caused by the increased fertilizer needs for crop production in compacted soils to maintain yield. This is because of reduced root growth and decreased ability of plants to absorb nutrients when grown in a compacted soil (Upadhyaya, 1992; Douglas and Crawford, 1993). Poincelot (1986) pointed out that the increased need for chemicals and water to maintain crop yield has implications on agricultural sustainability. Effects of soil compaction on soil resources are related to changes in soil physical, chemical and biological quality. Changes in soil physical quality are due to decreased saturated hydraulic conductivity of soil caused by reduced soil porosity. Effects on chemical quality of soil are due to increased need for fertilizer to maintain crop yield and decreased fertilizer use efficiency. Increased fertilizer need is caused by decreased root growth and proliferation (Russell and Gauss, 1974; Bowen, 1982). Adverse effects on soil biological aspects are due to a loss of habitat quality for macro-fauna and micro-fauna (LAWR- Cooperative Extension, 1984; Soane and van Ouwerkerk, 1995; Whalley et al., 1995). It should be noted that the adverse effects of soil compaction depend on soil type, crop type, and soil moisture (Soane, 1985). Since soil compaction is a potential serious concern it will be beneficial to develop a sensor to quantify soil compaction and map its variability. Soil texture/soil compaction level sensor can be useful in sustainable agriculture or alternate farming practices and in a rapidly evolving concept known as Site-Specific Crop management (SSCM) or precision farming. Both of these practices have gained in popularity due to increasing concerns in NO-3 contamination of ground water in recent years and possible health and environmental effects associated with the NO3 contamination.

Precision Farming or Site Specific Crop Management (SSCM) involves applying just the right amount of inputs (chemical, seed etc.) based on crop requirements taking into account the soil type, topography, soil fertility level, soil organic matter content, and soil moisture content (Robert et al., 1992). Development of sensors to determine soil properties in real-time is one major bottleneck in the application of SSCM(Gaultney, 1989). There is a need to develop a rapid sensor that can quantify soil physical conditions (texture, compaction level etc.). Such a device can also be helpful in mapping variability in soil texture/compaction within a field or a geographic area. These maps can be very helpful in alternate farming practices also since differences in water infiltration rate resulting from changes in soil physical conditions have been found to be important consequences of alternate farming practices.

Soil texture/compaction level sensing: Soil texture and compaction level influence the ability of soil to retain nutrients. In coarse textured or loose soils, nitrate may leach easily adding to soil water contamination. On the other hand, adding additional nitrogen may increase yield in a heavy textured soil without increasing ground water pollution, if the native nitrogen level is low. Thus, texture/compaction level information is critical in making decisions on spatially modulated chemical application rates. The SCS soil survey maps can be helpful in determining soil texture. However, spatial resolution of this approach is low and an in-situ texture sensor is desirable. Hellebrand (1993) suggested the use of a horizontally moving cone penetrometer to measure soil texture. Liu et al. (1993) used acoustic signals produced by cutting tools moving through the soil to determine its texture. The soil cone index values obtained using the ASAE (American Society of Agricultural Engineers) standard cone penetrometer can be useful in estimating the soil compaction level. However, it is a point measurement which tends to be highly variable. It is often misleading in the dry and cloddy soil conditions. Glancey et al. (1989) developed an instrumented chisel and used it to predict various tillage implement draft requirements in different soil types and conditions (Glancey et al., 1996). They found that draft requirements of this device depended on soil strength properties (dynamic failure stress values in shear and compression), or soil type and conditions (texture, compaction level or bulk density, and moisture content) when it was operated at a constant depth and forward speed. Individual draft data points contained high frequency variations which appeared like noise, but were actually related to the fracture phenomena in the soil as well as mechanical vibrations induced by tractor-implement combination and could not be used to predict soil texture and compaction level. Their results show that at least 40 consecutive draft data points should be averaged to obtain representative mean draft values (less than 5% error in the mean value at 95% confidence level). It should be noted that unlike soil cone index values which are point measurements (therefore, show high degree of variability), continuous measurement of draft requirements of a reference tillage tool provides a more reliable estimation of variation in soil condition. Moreover, it provides an integrated effect over a desirable depth. This device has the potential for use in soil texture/compaction level determination provided the draft values are properly adjusted for moisture content.

Soil moisture sensor: A soil moisture sensor is necessary to correct the reference tillage tool draft requirements to estimate soil texture/compaction level. A soil moisture sensor can also be very useful in irrigation management. Electrical conductivity methods, capacitance techniques, spectra-photometric techniques (optical, near-infrared, and microwave spectroscopy), nuclear magnetic reflectance (NMR), and nuclear scattering techniques have been used to rapidly measure moisture content of agricultural materials including soil with varying degrees of success (Young, 1975). Sudduth and Hummel(1993) and Upadhyaya et al.(1994) found that NIR absorbance of soil was highly correlated to soil moisture content. Time domain reflectometry also provides a rapid and accurate technique to measure soil moisture content as well as salinity (Dayton, 1992). This method requires perfect contact with electrodes and soil for accurate measurement of soil moisture content. Since water has a very high dielectric constant compared to soil, dielectric technique provides an inexpensive way of measuring soil water content. However, water also has a high loss factor (poor insulator) and its dielectric constant is influenced by temperature, pH, salinity, and bulk density. Using an inductor across the capacitor a resonant circuit can be developed. Moreover, there is no phase shift at resonance and a phase lock loop can be established (Retrokool Inc., Berkeley, CA[1]). Using a resonance and the phase lock technique, the effect of salinity and pH on dielectric measurement can be eliminated.

The objective of this study is to develop a soil texture/compaction sensor based on the draft requirements of an instrumented reference tillage tool to map the variability in soil texture/compaction level in fields.

---

[1] Mention of company names or trade names is not an approval of a specific product over similar products by the authors or the University California Davis.

Texture/Soil Compaction Sensor Development

The reference tillage tool developed by Glancey et al. (1996) will be used in this study. This reference tool has been upgraded with two load cells to measure draft and another one to measure vertical load, a radar gun to measure ground speed, a displacement sensor, and a GPS receiver to obtain spatial coordinates using a differential measurement technique [Fig. 1]. Moreover, a moisture sensor which measures changes in dielectric constant using a resonance frequency and phase lock technique (developed by Retrokool Inc.) is also incorporated into this device [Fig. 1]. These sensors have been interfaced to a PC using a 16 bit data acquisition board [Fig.2]. The basic working principle of this system is as follows:

The draft of a tillage tool depends on its geometry, soil properties, its operating depth, and speed. i.e.

$$D = f(w, Gi, BD, MC, TC, d, S)** \qquad (1)$$

where
D = draft of the tillage tool
w = operating width
Gi = all other geometric parameters such as length,
        curvature, cutting angle etc. (i = 1,2, ...)
BD = soil bulk density
MC = moisture content
TC = texture
d = operating depth
S = travel speed

Equation (1) is based on the assumption that the engineering properties of soil relating to soil cutting (cohesion, internal angle of friction, elastic and plastic parameters etc.) depend on soil physical conditions represented by its bulk density, moisture content, and its texture. If a reference tillage tool is selected with standard geometry (e.g. a 2-D version of a standard cone penetrometer, ASAE standard S313.2), and if the speed and depth of operation are held constant, then equation (1) becomes:

$$D = g(BD, MC, TC) \qquad (2)$$

If soil moisture is independently measured, then equation (2) can be used as an indicator of soil compaction (i.e. bulk density) and texture. Glancey et al. (1996) found that such a device could be used as an indicator of soil physical conditions and can be used in estimating draft requirements of various tillage tools. Using a similar approach one can show that the vertical load on the instrumented tine can also be described by an equation of the type shown in (2) above. In soil cutting problems it is not unusual to find the component equations due to moisture and due to texture and density multiply as shown below:

$$D = g1 (BD, TC) * g2 (MC) \qquad (3)$$

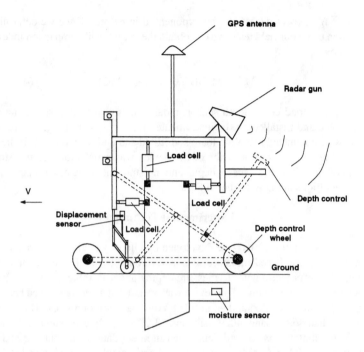

Figure 1. A schematic diagram of the texture/soil compaction sensor

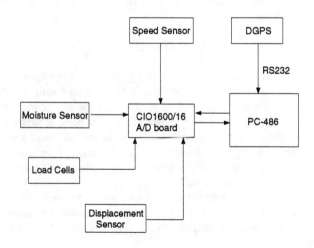

Figure 2. Various sensors and data acquisition system used on the texture/soil compaction sensor

We expect g2(MC) to be exponential in nature. Once we determine this function using our field tests, we can obtain the texture/soil compaction index (TCI) as follows:

$$TCI = g1(BD,TC)= D/ g2 (MC) \qquad (4)$$

The load cells, depth sensor, radar gun, and moisture sensor have been calibrated and installed on the instrumented tine as shown in Fig. 1. Preliminary tests were conducted in a Yolo loam soil (tilled and untilled) to check if the device is field ready. The performance of the device was satisfactory from a structural point of view. The draft, speed, depth, and moisture data were also obtained in these test conditions along with the GPS data.

Experimental Techniques

The instrumented tine developed by Glancey et al.(1996) was retrofitted with a dielectric based moisture sensor developed by the Retrokool Inc., a radar gun, a linear potentiometer to measure depth, and a Trimble navigation Inc. sub-meter accuracy GPS unit. The device was operated at a constant speed and 305mm (12 in.) depth in a tilled and an untilled Yolo loam soil on the UC Davis campus. The soil had 46.5% sand, 40.3% silt, and 13.2% clay. In each soil condition draft, moisture content, speed, and depth measurement data were obtained at 1.2 kHz along three, 30.5 m (100 ft) transects. Soil core samples were obtained at 125 mm (5in) below the ground surface to measure soil density and moisture content at 7.6 m (25 ft) intervals along each transect. The tests were conducted on May 29, June 1, 5, 10, and 17, 1996. Additional tests will be conducted in a Capay clay soil using a similar test plan.

Results and Discussion

The moisture and dry bulk density obtained from core samples were analyzed using analysis of variance (ANOVA) based on a split plot design. The dates of test was considered as blocks, the transects were considered as main treatments, and locations along the transects were considered as subtreatments. The analysis revealed that moisture content was significantly different among different dates whereas the dry bulk density was insignificant in both soil conditions. Transects were not significantly different in any of the cases. The location along the transect was significant for soil moisture content in the untilled soil conditions and for dry bulk density in the tilled soil conditions. It was insignificant in all other cases. The experiments were conducted on different dates to achieve different moisture contents while keeping dry bulk density constant. The design appears to have achieved this goal.

Figure 3a is a typical plot of transient draft signal over the transect. Figure 3b is a semivariogram of the same data. The semivariogram does not show a clear sill indicating cyclic nature of the data. This is further verified by the FFT of the draft data shown in Fig. 3c. Similar results were also obtained with the moisture signal (Figs. 4a through 4c). The speed and depth variation was found to be very

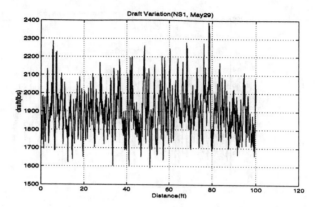

Figure 3a. Draft variation in an untilled Yolo loam soil on May 29, 1996

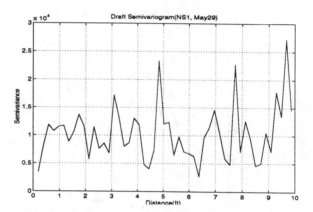

Figure 3b. Semivariogram of the data shown in Figure 3a

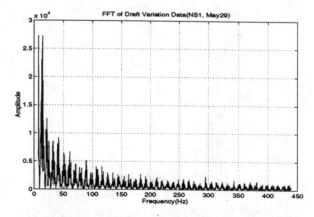

Figure 3c. Fourier transformation of the data shown in Figure 3a

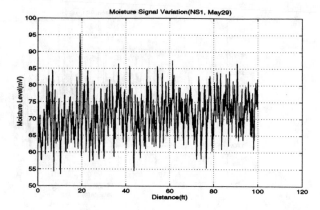

Figure 4a. Moisture Variation in an Untilled Yolo loam soil on May 29, 1996

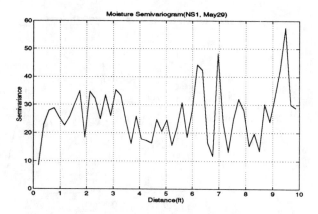

Figure 4b. Semivariogram of the data shown in Figure 4a

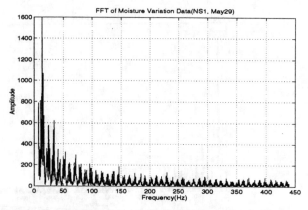

Figure 4c. Fourier transformation of the data shown in Figure 4a

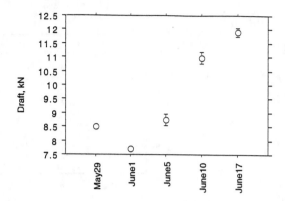

Figure 5a. Draft versus test dates in an untilled Yolo loam soil

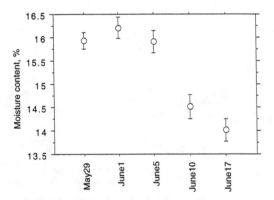

Figure 5b. Moisture content versus test dates in an untilled Yolo loam soil

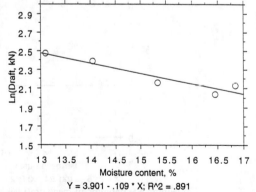

Y = 3.901 - .109 * X; R^2 = .891

Figure 5c. Variation of mean draft versus mean moisture content
in an untilled Yolo loam soil

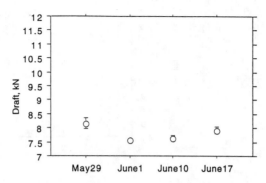

Figure 6a. Draft versus test dates in a tilled Yolo loam soil

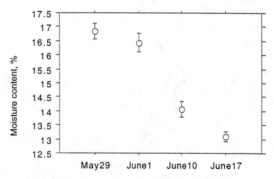

Figure 6b. Moisture content versus test dates in a tilled Yolo loam soil

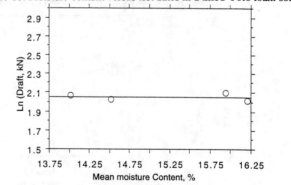

Y = 2.06 - 3.021E-4 * X; R^2 = 8.315E-5

Figure 6c. Variation of mean draft versus mean moisture content
in a tilled Yolo loam soil

Table 1. Experimental results in a Tilled and Untilled Yolo Loam soil on the UC Davis campus.

Soil Condition	Date of Test	Draft and Moisture data obtained from Sensors		Core sample data	
		Draft, kN*	Moisture content* signal, mV	Moisture content, %	Dry bulk density, kg/m^3
Tilled Yolo Loam	May 29	8.16± 0.72	81.26±3.31	15.94±0.68	1.54±0.06
	June 1	7.56±0.27	84.04±4.81	16.21±0.86	1.47±0.07
	June 5	------------	27.28±0.97	15.32±1.59	1.47±0.07
	June 10	7.63±0.37	38.20±1.09	14.52±1.01	1.50±0.14
	June 17	7.92±0.45	86.56±4.13	14.02±0.92	1.46±0.08
Untilled Yolo loam	May 29	8.50± 0.21	67.95±4.59	16.84±1.06	1.56±0.06
	June 1	7.72±0.27	65.87±19.0	16.44±1.32	1.54±0.10
	June 5	8.70±0.79	39.65±1.07	15.33±1.62	1.55±0.11
	June 10	10.97±0.74	51.48±2.48	14.04±1.04	1.52±0.07
	June 17	11.90±0.62	63.32±2.86	13.09±0.71	1.55±0.06

* Note : Mean ± Standard deviation

small in all tests. The coefficients of variations were determined for a few typical cases. The coefficients of variation were found to vary from 2 to 5% in the speed data and 1 to 2% in the depth data.

   Based on the cyclic nature of the data, we decided to average moisture and draft sensor data obtained over a 3 m distance (1.5 m on both sides of the location where manual data points were taken) and assign it to manually sampled location. The scatter plot of these data (i.e. draft vs. moisture content) looked very noisy. Since only moisture content significantly differed between date of tests, we decided to average the data by the date of test . These results are shown in Table 1. The results clearly show that the dielectric moisture sensor fluctuated from day to day. This is due to the temperature variation. A temperature sensor needs to be included to account for the temperature effect.

   Figure 5 a and b show the variation of draft and moisture content variation on test dates in a tilled soil. Figure 5c is the plot of natural logarithms of mean draft values versus the moisture content. In this tilled Yolo loam soil draft did not depend on the moisture content in the range of 12 to 17%. This is most likely due to the effect of tillage on breaking the soil particle to particle binding force (i.e. lack of adequate cohesion). Figures 6 a, b, and c are similar plots for the untilled Yolo loam soil. The mean draft values decreased as the moisture content increased (Fig. 6c). A linear regression between natural logarithm of mean draft values and mean moisture values showed that the regression was significant and the associated coefficient of determination (r2) was 0.89. The relationship was found to be

$$D = 49.4 \, e- 0.11\_\theta \qquad\qquad (5)$$

   Therefore from equation (4) the TCI value for the untilled soil is D/e- 0.11_θ =49.4. For the tilled soil the TCI in the moisture range of 12 to 17 % is 7.85.

CONCLUSIONS

Based on this study we reach the following conclusions:

1. An instrumented tine can be used to provide a soil texture/compaction index if it is operated at a constant speed and depth and if the draft readings are properly adjusted for the moisture content values.

2. In the untilled Yolo loam soil the relationship between draft and soil moisture was found to be an exponential decay function. In the tilled Yolo loam soil draft values did not depend on the moisture content values in the range of 12 to 17%.

3. The texture/soil compaction index (TCI) was found to be 49.4 kN for the untilled Yolo loam soil and 7.85 for the tilled Yolo loam soil in the range of moisture content values used in this study.

Further work is necessary to verify the results of this study in other soil types and conditions.

REFERENCES

ASAE Standards. 1995. Soil cone penetrometer. ASAE S313.2. ASAE St. Joseph, MI 49085.

Bowen, H. D. 1982. Alleviating mechanical impedance. *In* Arkin, G. F. and H. M. Taylor (Eds). Modifying the root environment to reduce crop stress. ASAE Monograph No. 34. ASAE, St. Joseph, MI.

Daniel, H., R. Jarvis, and L. A. G. Aylmore. 1988. Hardpan development in loamy sand and its effects on soil conditions and root growth. Proceedings of the 11th International ISTRO conf., Edinburgh, Scotland. 1:233-238.

Dayton, F. N. 1992. Development of time-domain reflectometry for measuring soil water content and bulk soil electrical conductivity. Ch 8. p 143-167. *In* Advances in Measurement of soil physical properties: Bringing theory into practice. Edited by G. C. Topp, W. D. Reynolds, and R. E. Green. SSSA Publication No. 30. Soil Science Society of America, Inc. Madison, WI.

Douglas, J. T. and C. E. Crawford. 1993. The responses of ryegrass sward to wheel traffic and applied nitrogen. Grass Forage Sci. 48:91-100.

Fattah, H. A. and S. K. Upadhyaya. 1996. Effect of soil crust and compaction on infiltration in a Yolo loam soil. Trans. ASAE. 39(1):79-84.

Gameda, S., G. S. V. Raghavan, E. McKyes, and R. Theriault. 1985. A review of subsoil compaction and crop response. Proceedings of the Int. Soil Dynamics Conf., Auburn, AL. 5:970-978.

Gaultney, L.D. 1989. Prescription farming based on soil property sensors. ASAE Paper No.89-1036. ASAE St. Joseph, MI.

Glancey, J. L., S. K. Upadhyaya, W. J. Chancellor and J. W. Rumsey. 1989. An instrumented chisel for the study of soil-tillage dynamics. Soil and Tillage Research 14:1-24.

Glancey, J.L., S.K. Upadhyaya, W.J. Chancellor, and J.W. Rumsey. 1996. Prediction of implement draft using an instrumented analog tillage tool. Soil and Tillage Res. (In Press).

Gupta, S. C., and R. R. Allmaras. 1987. Models to assess the susceptibility of soils to excessive soil compaction. pp 65-100. *In* Advances in Soil Sci. Vol. 6. Springler Verlag, NY.

Hadas, A., and D. Wolf. 1984. Soil aggregates and clod strength dependence on clod size, cultivation, and stress load rates. Soil Sci. Soc. of America. J. 43:1157-1164.

Hellebrand, H.J. 1993. Trends in sensors for spatially variable control of field machinery. ICPPAM Paper No. 93-1075. 5th International Conference on Physical Properties of Agricultural Materials (ICPPAM). Sept. 6-8, Bonn, Germany.

Horn,R., H. Domzal, A. Slowinska-Jurkiewich, and C. van Ouwerkerk. 1995. Soil compaction processes and their effects on the structure of arable soils. Soil and Tillage Res. 35: 23-36.

LAWR - Cooperative Extension. 1984. Water penetration in California soils. Technical Report. Joint Infiltration Committee. Dept. of LAWR, UC Davis. Davis, CA.

Lipiec, J., and W. Stepniewski. 1995. Effects of soil compaction and tillage systems on uptake and losses of nutrients. Soil and tillage Res. 35:37-52.

Liu, W., L. D. Gaultney, and M. T. Morgan. 1993. Soil texture detection using acoustic methods. ASAE Paper No. 93-1015. ASAE St. Joseph, MI.

Negi, S. C., E. McKyes, G. S. V. Raghavan, and F. Taylor. 1981. Relationships of field traffic and tillage to corn yields and field properties. J. Terramechanics 18(2):81-90.

Poincelot, R.P. 1986. Towards a more sustainable agriculture. AVI Publications Inc., Westport, Connecticut.

Robert, P.C., R.H. Rust, and W.E. Larson. 1992. Adapting Soil-specific crop management to today's farming operations. A Workshop on Research and Development Issues. April 14-16. Minneapolis, MN. 19p.

Russel, R.S., and M.J. Goss. 1974. Physical aspects of soil fertility. The response of roots to mechanical impedance. Neth. J. Agr. Sci. 22:305-318.

Soane, B.D. 1985. Traction and transport systems as related to cropping systems. Proceedings of the Int. Conf. on Soil Dynamics. Auburn, AL. pp 863-935.

Soane, B.D. and C. van Ouwerkerk. 1995. Implications of soil compaction in crop production for the quality of the environment. Soil and Tillage Res. 35:5-22.

Sudduth, K.A., and J.W. Hummel. 1993. Evaluation of reflectance methods for soil organic matter sensing. Transaction of ASAE. 34(4):1900-1909.

Taylor, J.H., and W.R. Gill. 1984. Soil compaction: state of the art report. J. Terramechanics. 21(2):195-213.

Upadhyaya, S.K. 1992. Deterring compaction of soil by heavy machines. California Agriculture 46(4):19-20.

Upadhyaya, S. K., S. Shafii, and D. Slaughter. 1994. Sensing soil nitrogen for site-specific crop management (SSCM). ASAE Paper 94-1055, ASAE, St. Joseph, MI.

Van Owerkerk, O. 1995. Soil compaction and the environment. Special issue of Int. Soil and Tillage Res. Org. 35: 1-113.

Voorhees, W.B., J.F. Johnson, and G.W. Randall. 1988. Maize growth and yield as effected by subsoil compaction and deep tillage. Proceedings of the 11th Int. ISTRO Conf, Edinburgh, Scotland. 1:335-340.

Whalley, W.R., E. Dumitru, and A.R. Dexter. 1995. Biological effects of soil compaction. Soil and Tillage Res. 35:53-68.

Wolf, D. and A. Hadas. 1983. Conventional versus controlled traffic and precision tillage for cotton. ASAE Paper No. 83-1040, ASAE, St. Joseph, MI.

Wolf, D. and A. Hadas. 1984. Soil compaction effects on cotton emergence. Trans. ASAE. 27:655-659.

Young, J.H. 1975.Moisture. Ch. 7. In Instrumentation and measurement for environmental sciences. Edited by Z. A. Henry. American society of Agricultural Engineers. St. Joseph, MI 49085. 10p.Table 1. Experimental results in a Tilled and Untilled Yolo Loam soil on the UC Davis campus.

New Mobile Soil Sampler Compared to Hand Probes and Augers for Fertility Evaluations

N. A. Wright

GEOPHYTA, Inc.
Vickery, Ohio, USA

ABSTRACT

A new mobile soil sampler (ARMS) was designed to reduce the time demands of intensive soil sampling. The objective of this study was to determine whether the use of seven different soil sampling devices would show any nutrient test changes on twelve different soil type-moisture locations. A randomized complete block design with four replications was used. The seven devices included a 1.9 cm and 3.2 cm diameter hollow tube probe, three diameters of a single flight auger (1.9 cm, 2.5 cm, and 3.8 cm), a 2.5 cm diameter double flight auger, and ARMS equipped with a 2.5 cm single flight auger. No significant soil sampler differences were detected for soil acidity, soil phosphorus, soil potassium, or organic matter. Significant differences between samplers were detected in calcium, magnesium, and soil recovery. A slight trend existed toward higher calcium and magnesium values for the two hollow tube probes compared to all augers. Higher soil recovery was detected for the hollow tube probes over all auger types.

BACKGROUND

Soil sample collection must become faster while increasing precision and accuracy. This requires a rethinking of the soil sample collection process to make it more efficient. This can be accomplished by studying the automation possibilities of soil sample collection to eliminate time consuming steps and operator fatigue associated with hand probing. Soil sample collection should utilize the same computerized technology and automation employed by field equipment applying agricultural chemicals.

Only in the recent literature has there been research performed to compare soil samplers of different designs. Most generally, the research only reported the determination of one chemical or physical property of the soil samples collected by each means. None compared the range of routine fertility soil tests.

Two citations were found to be of greatest importance to the research proposed here by Geophyta. Hassan et al. (1983) desired an understanding of the importance of soil sampling methods and salt contents of a soil profile. They collected soil samples with both a Veihmeyer tube with a 2.1 cm diameter and a bucket auger 7.9 cm in diameter. Chloride was determined on all samples. They concluded that measurements with larger samples were less variable than smaller samples. No follow-up studies were reported to test equal soil volume collection by the auger and probe. They stated that the minimum desired sampling volume is 50 cubic centimeters.

A more closely related study was performed by Shapiro and Kranz (1992). They compared a powered soil auger with a powered hollow tube probe for the collection of soil nitrate samples under field conditions. While the two samplers were of different sizes, equal volumes of soil were collected by varying the number of soil samples. Four locations were sampled under nitrogen rate studies to get a wide variation in soil nitrate. Shapiro and Kranz concluded that no differences existed in soil nitrate between the auger and probe methods of sampling. They did notice that soil moisture and texture affected the volume of soil removed by each method, but did not collect any quantitative data. They did not analyze the soil samples for any parameter other than nitrate. No follow-up studies were indicated although it was stated that further research is needed in this area.

RATIONALE

A new soil sampler design is required that can measure up to the intense demands of intensive soil sampling or grid sampling. Such a sampler must meet and exceed all five criteria for good sampling equipment outlined by Peck and Melsted (1973). First, it must take a small enough equal volume of soil from each site so that composites are of an appropriate size for processing. Second, it must be easy to clean. Third, it must be adaptable to all soil types. Fourth, it must be rust resistant and durable. Last, the soil sampler must be relatively easy to use and allow for rapid sampling.

New soil sampler designs must require only one operator for both the vehicle and sampler. Additionally, the operator should not have to dismount the vehicle to collect soil samples. Finally, new designs should allow for full automation of the soil sample collection process.

OBJECTIVE

Determine whether the use of seven different soil sampling devices (2 sizes of probe, 3 sizes of single flight auger, a double flight auger, and an ARMS soil accumulator) show any nutrient test changes on twelve different soil type/moisture level locations.

Soil volume and collection method differences from production type, hand operated samplers will be important in selecting auger size and design for refinements in a semi-automated sampler (ARMS).

RESEARCH METHODS

A randomized complete block design with four replications was employed to accomplish objective one and two. Locations were twelve different soil type - moisture content sites as presented in Table 1. The two higher moisture level sites, indicated by the code suffix "w" were created by application of water through a soaker hose assembly. This assembly permitted the establishment of a 0.5 by 0.5m wetted area to conduct the treatments. Four blocks or replications of seven different

Table 1. Experiment site descriptions with soil texture and moisture content.

Location	Code	Soil Name and Type	Moisture Content (%)
1	KbAd	Kibbie fine sandy loam	14.4
2	KbAw	Kibbie fine sandy loam	26.0
3	Le	Lenawee silty clay loam	17.2
4	Co	Colwood fine sandy loam	17.1
5	SoB	Spinks fine sand	7.1
6	Bt	Bono silty clay	36.6
7	FuAd	Fulton silty clay loam	20.0
8	FuAw	Fulton silty clay loam	27.7
9	To	Toledo silty clay	23.8
10	NpA	Nappanee silt loam	12.8
11	Mo	Mermill loam	24.6
12	Ht	Hoytville silty clay loam	17.2

soil sampling devices were used. Each replication block was oriented with crop row middles in fields with 0.76 m row widths. Therefore, each replication was separated by 0.76 m. A rubberized conveyor belt with punched holes served as the template from which to collect soil cores parallel to the crop rows and 10 cm apart. This allowed for the collection of all treatment cores for one block to be within 2.0 m. The seven devices included both a 1.9 cm and a 3.2 cm diameter hollow tube probe, three diameters of a single flight auger (1.9 cm, 2.5 cm, and 3.8 cm), a 2.5 cm diameter double flight auger, and finally, the ARMS prototype equipped with a 2.5 cm single flight auger. All soil samples were weighed after collecting a number of borings from each sampler to roughly equal the same soil volume. Table 2 presents this comparison.

Additionally, to provide a check of spatial variability within each soil type site, five 3.8 cm OD auger single boring samples were collected along side the treatment cores, spaced at twenty inches. This was done for each replication block to give a four by five, point sampled grid of each site.

Table 2. A comparison of theoretical soil volume collection for seven soil sampler designs.

Trt. #	Soil Sampler Type	Core Volume 20 cm deep	# Cores	Total Volume 20 cm deep
		cm3		cm3
1	1.9 cm ID hollow tube	57.8	5	289.2
2	3.2 cm ID hollow tube	160.9	2	321.8
3	1.9 cm OD single flight auger	57.8	5	289.2
4	2.5 cm OD single flight auger	102.9	3	308.7
5	3.8 cm OD single flight auger	231.7	1	231.7
6	2.5 cm OD double flight auger	102.9	3	308.7
7	2.5 cm OD single flt. auger-ARM	102.9	3	308.7

Nutrient analyses were performed on all samples for pHw, double buffer lime requirement, Bray P-1, 1 N ammonium acetate extractable potassium, calcium, and magnesium, as well as Loss On Ignition organic matter. Soil test procedures were performed according to North Central Regional Publication 221 (1988).

RESEARCH RESULTS AND DISCUSSION

Nutrient Stratification

To further aid interpretation of both the soil sampler and soil core composite geometry experiments, depth increment samples were collected for each soil type location. Twenty soil cores were collected with a 1.9 cm ID hollow tube probe and divided into two inch sections for nutrient analysis. Table 3 contains this data and shows a pronounced stratification for all soil tests at all locations. These locations served as an excellent resource to look at soil sampler comparisons. The range of test values will allow a more robust determination of equipment strengths and weaknesses.

Soil Sampler Comparisons

The results of the soil sampler comparisons provided some interesting conclusions which will assist in further refinements of the ARMS soil sampler. Table 4 contains the treatment means across all locations with associated Fisher least significant differences. These are the most important data for drawing the broadest possible conclusions since the data represent twelve different soil type/moisture combinations.

No significant soil sampler differences were detected for soil acidity, soil phosphorus, soil potassium, or organic matter. The 0.04 pH, 4 ug P/g, and 8 ug K/g variations among treatment means are the result of normal experiment variation. Therefore, when these nutrient tests are conducted, all soil samplers performed equally. This included the ARMS soil accumulator. These are probably the most common nutrient tests performed on soil samples for the purpose of fertilizer and pesticide applications. Also, this data indicates that if proper care is taken to control sampling depth and prevent soil spillage, then any of these samplers will perform adequately. Since no significant sampler effects were detected over all locations no further exploration of individual site differences will be considered here.

Significant differences between samplers were detected during the determination of calcium, magnesium, and soil recovery. While both calcium and magnesium may be of lesser importance compared to pH, P, and K, these two nutrients are being analyzed by an increasing number of laboratories.

Both values are used to calculate an estimated cation exchange capacity for use in fertilizer recommendations. A slight trend exists toward higher calcium and magnesium values for the two hollow tube probes compared to all augers. As the significance values indicate, differences of just 59 ug Ca/g and 13 ug Mg/g can be detected in this study. Also, higher soil recovery (g/cc) was detected for the hollow tube probes over all auger types. To obtain a better understanding of these

Table 3. Nutrient analysis of soil depth increments for ten soils.

Site #	Soil Code	Depth (cm)	pH	P	K	Ca	Mg	%OM
					---- ug/g ----			
1	KbAd	0-5	5.88	20	246	1058	175	4.34
		5-10	5.89	40	160	1026	162	3.65
		10-15	5.64	23	71	1039	154	3.03
		15-20	5.73	9	43	970	144	2.76
4	Co	0-5	6.47	12	95	2142	318	5.33
		5-10	6.43	9	68	2097	301	5.24
		10-15	6.44	9	52	2171	303	5.13
		15-20	6.66	7	47	2175	314	4.51
6	Bt	0-5	5.22	31	235	3349	418	10.70
		5-10	5.36	25	133	3556	413	9.92
		10-15	5.35	15	133	3976	425	9.46
		15-20	5.46	11	127	4232	413	7.91
9	To	0-5	6.73	58	381	2198	758	4.74
		5-10	6.85	36	247	2370	824	4.26
		10-15	7.08	37	200	2336	886	3.84
		15-20	7.00	28	206	2275	990	3.61
11	Mo	0-5	6.91	110	188	3627	346	8.99
		5-10	6.96	82	117	3558	337	10.48
		10-15	6.98	55	95	3855	362	9.68
		15-20	7.01	35	85	3980	342	8.59
3	Le	0-5	7.21	10	82	2155	433	4.10
		5-10	7.24	8	65	2249	450	3.79
		10-15	7.32	7	63	2342	475	3.80
		15-20	7.40	5	66	2565	530	3.82
5	SoB	0-5	5.69	230	323	347	58	2.18
		5-10	5.51	188	267	323	42	2.07
		10-15	5.39	184	221	237	27	1.66
		15-20	5.13	188	165	164	17	1.65
7	FuAd	0-5	6.04	23	270	1885	638	4.46
		5-10	6.27	9	161	2027	706	3.78
		10-15	6.42	7	154	2122	794	3.31
		15-20	6.32	6	156	1949	877	3.15
10	NpA	0-5	6.60	119	318	1534	230	3.84
		5-10	5.90	78	259	1322	243	3.53
		10-15	5.84	17	214	1342	301	2.96
		15-20	5.87	10	173	1433	362	2.80
12	Ht	0-5	6.28	96	171	2032	312	4.57
		5-10	6.64	91	104	1971	307	4.62
		10-15	6.76	52	96	2123	322	4.60
		15-20	6.85	46	88	2375	354	4.56

Table 4. A comparison of soil tests by sampling method, over all twelve
      locations.

Sampling Method	pH	P	K	Ca	Mg	OM	Rec.
			ug/g			%	g/cc
1.9 cm ID probe	6.36	44	160	2126	446	4.53	.32
3.2 cm ID probe	6.34	46	166	2102	444	4.54	1.30
1.9 cm OD single flt. auger	6.37	46	158	2038	424	4.62	0.82
2.5 cm OD single flt. auger	6.35	45	160	2071	431	4.60	0.98
3.8 cm OD single flt. auger	6.38	47	163	2035	423	4.69	1.10
2.5 cm OD dbl. flt. auger	6.37	47	165	2052	430	4.62	1.08
2.5 cm OD single flt. ARMS	6.35	48	164	2034	432	4.61	0.87
Flsd (0.05)	ns	ns	ns	59	13	ns	0.04

Rec. g/cc = Recovery of soil based on sampler volume

significant differences, treatment means within each location must be examined. Some loss of ability to detect treatment differences has occurred due to the loss of statistical degrees of freedom when considering each location. Table 5 contains soil calcium tests separated by location. Significant sampler differences only occurred for two soil types, a Spinks fine sand (SoB) and Hoytville silty clay loam (Ht). Opposite responses were detected for these two soil types. A very slight increase in calcium levels was detected for the augers compared to probes in the fine sand. The silty clay loam exhibited a decrease in calcium for the augers compared to the probes. When relating these responses to the depth increment calcium levels an explanation can be offered. For the Spinks fine sand, it is likely that the augers had difficulty retrieving the lower portions of the 20 cm depth due to soil looseness. In this case the additional friction of the probe side walls allowed for a more complete removal of soil to depth. This same explanation applies to the Hoytville silty clay loam since calcium increased with depth instead of decreasing with depth in the Spinks fine sand. Also, the Hoytville soil was very dry at 17.2% moisture content, preventing soil cohesion from allowing higher sample recovery at increasing depths.

This explanation for sampler differences concerning calcium is confirmed by significant differences in soil magnesium presented in Table 6 with the exception of the Kibbie fine sandy loam (KbAd) location. The Fulton silty clay loam (FuAd), Nappanee silt loam (NpA), and Hoytville silty clay loam all have an increase of magnesium with depth. The soil loss by the augers is still from the deeper increments. The elevated magnesium levels for the probes compared to the augers at the KbAd location cannot be explained since this nutrient decreases with depth for this site. While a significant sampler difference was detected at the KbAd site, it was a relatively small difference of 10 ug Mg/g and all augers did not perform different from the probes.

Table 5. Soil test calcium means by sampling method within each location.

Method	KbA-d	KbA-w	Le	Co	SoB	Bt	FuA-d	FuA-w	To	NpA	Mo	Ht
							ug Ca /g					
1.9 cm probe	1235	1255	2287	2254	231	3954	2032	2526	2408	1410	3592	2322
3.2 cm probe	1214	1248	2350	2210	246	3773	2048	2612	2330	1388	3459	2339
1.9 cm sf aug	1189	1228	2280	2175	242	3666	2056	2484	2302	1284	3458	2098
2.5 cm sf aug	1186	1217	2348	2250	268	3754	1959	2523	2395	1344	3466	2137
3.8 cm sf aug	1156	1216	2230	2173	261	3509	2027	2499	2325	1406	3534	2080
2.5 cm df aug	1199	1224	2260	2298	291	3620	2036	2485	2376	1359	3322	2156
2.5 cm ARM	1186	1255	2248	2203	246	3714	1986	2477	2300	1392	3222	2179
Flsd (0.05)	ns	ns	ns	ns	48	ns	ns	ns	ns	ns	ns	151

Table 6. Soil test magnesium means by sampling method within locations.

Method	KbAd	KbAw	Le	Co	SoB	Bt	FuAd	FuAw	To	NpA	Mo	Ht
							ug Mg/g					
1.9 cm probe	183	188	468	326	26	438	936	900	904	296	334	348
3.2 cm probe	181	191	474	319	31	423	941	919	883	284	324	354
1.9 cm sf aug.	174	187	460	312	31	425	887	850	868	262	316	318
2.5 cm sf aug.	175	187	476	330	33	434	846	887	892	267	320	329
3.8 cm sf auger	171	188	450	316	32	411	865	884	845	260	328	323
2.5 cm df auger	175	187	455	332	39	426	873	858	897	276	308	334
2.5 cm ARMS	178	190	455	321	31	424	912	888	888	260	304	335
Flsd (0.05)	10	ns	ns	ns	ns	ns	82	ns	ns	23	ns	27

Tables 4 and 7 provide the soil recovery values as determined by dividing the total air dry weight of soil collected by the theoretical volume of each sampler. The recovery values are an attempt to perform a fair comparison of soil samplers and to assist in interpretation of nutrient differences. There is no standard 'correct' value for soil recovery. These values were a useful tool. Ideally, all recoveries should be equal, but there are inherent geometric differences in all the samplers in this experiment. Higher recoveries were obtained with the probes compared to the augers as a whole. This is likely due to the side walls of the probes which allow the core to rest within the tube and be held by friction until extraction. With an auger, the "side walls" are the soil bore hole left by the inclined cutting edge. Consequently, the ability of the auger to effectively remove the soil is somewhat dependent on the firmness or density of the surrounding soil. Another observation is that soil recovery increases as the single flight auger size increases. This is due to the increased distance between auger flights, allowing the soil to move more freely up the incline. This suggests that a very smooth, highly polished flighting may improve soil recovery. A partial explanation for why the 2.5 cm ARMS auger had a lower recovery than the identical 2.5 cm single flight auger was that the soil accumulator had to be raised and moved to each boring location in the experiment, resulting in some loss of soil from the base of the accumulator. A comparison of samplers at the wet and dry sites can be seen in Table 7. Each is identified with a "w" or "d" at the Kibbie fine sandy loam (KbAd & KbAw) and Fulton silty clay loam (FuAd & FuAw) locations. Visually, the KbAw location was very near water saturation from irrigation. For this reason the augers could not transport soil easily without firm bore hole sides. This explains the lower auger recovery at KbAw compared to KbAd. The FuAw location appeared to only have the soil ped surfaces wetted from the same amount of irrigation. The increased recovery at FuAw compared to FuAd for all samplers was due to the better cohesion upon wetting and less cracks in the soil from a very dry summer.

Table 7. Soil recovery means by sampling method within each location.

Method	KbAd	KbAw	Le	Co	SoB	Bt	FuAd	FuAw	To	NpA	Mo	Ht
						g/cm3						
1.9 cm probe	1.27	1.26	1.33	1.27	1.37	1.07	1.31	1.55	1.39	1.39	1.20	1.43
3.2 cm probe	1.24	1.26	1.32	1.28	1.37	1.02	1.32	1.53	1.32	1.36	1.18	1.43
1.9 cm sf aug	0.96	0.74	0.91	0.87	1.00	0.61	0.77	0.86	0.73	0.94	0.70	0.81
2.5 cm sf aug	1.03	0.80	1.18	1.09	1.09	0.68	0.89	1.01	1.01	1.09	0.94	1.00
3.8 cm sf aug	1.17	1.12	1.14	1.05	1.16	0.90	1.11	1.22	1.16	1.08	0.99	1.07
2.5 cm df aug	1.08	0.85	1.16	1.16	1.33	0.89	0.95	1.03	1.01	1.28	1.05	1.15
2.5 cm ARM	0.90	0.80	0.84	0.70	0.86	0.73	1.02	0.96	0.99	0.85	0.78	0.94
Flsd (0.05)	0.23	0.19	0.19	0.13	0.14	0.12	0.22	0.14	0.17	0.10	0.17	0.21

While the text has concentrated on the significant sampler differences for soil calcium, magnesium and recovery, the fact remains that the majority of soil type locations exhibited no sampler differences. The thoughts discussed above will provide guidance for further understanding of how soil samplers perform. This discussion will also lead to improvements in soil sampling equipment. For example, the 2.5 cm ARMS accumulator may perform better with a 3.8 cm single flight auger and provide more uniform results. This is an area which needs to be explored in future research.

The need for replication blocking of soil based experiments was shown in the statistical analyses of both the sampler comparisons and the one boring point samples collected for spatial analysis. For the sampler study, replication blocks oriented parallel with previous crop rows removed a significant amount of variability not attributed to the sampling methods for pH, potassium, magnesium, and organic matter. No blocking effect was detected for phosphorus, calcium, or soil recovery. Soil type locations exhibited highly significant differences for all tests. The spatial sampling performed along with the sampler comparisons exhibited a slight significant difference in nutrient levels in the x-axis direction which is the direction perpendicular to the previous crop rows and the same as the replication block effect mentioned above. No y-axis, or parallel with the crop row effects were detected for any tests. These observations confirm that the experiment was set up correctly and gives more weight to the sampler comparisons.

Excerpted from: A DEVICE TO RAPIDLY AND PRECISELY SOIL SAMPLE FOR SITE SPECIFIC CROP MANAGEMENT - PHASE I - FINAL REPORT to the U. S. DEPARTMENT OF AGRICULTURE SMALL BUSINESS INNOVATION RESEARCH PROGRAM under Grant #94-33610-0093 by Nathan A. Wright, Research Director, Geophyta, Inc., Vickery, Ohio, USA

REFERENCES

Hassan, H.M., A.W. Warrick, and A. Amoozegar-Fard. 1983. Sampling volume effects on determining salt in a soil profile. Soil Sci. Soc. Am. J. 47:1265-1267.

North Central Regional 13 Committee. 1988. Recommended Chemical Soil Test Procedures. NCR Pub. 221(rev.). W.C. Dahnke (ed.). Fargo, ND.

Peck, T.R., and S.W. Melsted. 1973. Field Sampling for Soil Testing. *In* Soil Testing and Plant Analysis. L.M. Walsh and J.D. Beaton (eds.). American Society of Agronomy. Madison, WI.

Shapiro, C.A., and W.L. Kranz. 1992. Comparison of auger and core soil sampling methods to determine soil nitrate under field conditions. J. Prod. Agric. 5:358-362.

Methods for Characterization and Analysis of Spatial and Temporal Variability for Researching and Managing Integrated Farming Systems

J. R.Hess
R.L. Hoskinson

Idaho National Engineering Laboratory
Idaho Fall,. Idaho

ABSTRACT

A potato and wheat field comprising about 200 ha was characterized for spatial and temporal variations in yield, crop quality, microbial populations, soil mineral nutrient concentrations as well as physical properties, plant mineral nutrient concentrations and climate. Characterization methods employed a wide variety of georeferenced measurement technologies and sampling/analysis procedures. Site-specific yield measurements were compounded by factors affecting crop quality. Variations in soil nutrient concentrations were not always temporally predictable nor correlated with plant nutrient concentrations. Statistical and artificial intelligence analyses were able to begin establishing relationships between interacting cropping system factors.

INTRODUCTION

Agriculture is based on a system of harvesting solar energy through photosynthesis to produce food and fiber. Crop production ultimately depends on the size and efficiency of the photosynthetic system, and crop management practices are founded on these assumptions. Even with great differences in climate, agronomic practices, and crop cultivars, the size and duration of the leaf area, a measure of the primary photosynthetic system, could attribute for over half of the variation measured in grain yields (studies cited by Gardner et al., 1985). Management strategies, including precision agriculture, can greatly affect the size and efficiency of the crop photosynthetic system (Gardner et al., 1985; Birrell and Suddeth, 1995), as well as ameliorate climatic and other environmental effects impacting crop yield and quality (Rao et al., 1993). However, the next generation of gains in crop production will only be realized with an increased understanding of the dynamic relationship between the crop photosynthetic system and the environment.

Because of the complexity of agriculture cropping systems, it is unreasonable to expect that any single individual or institution could possess all capabilities required to solve the complex issues facing precision agriculture. To achieve the needed expertise, public and private sector partnerships have begun to form. Partnerships between universities, USDA, national laboratories, and private industry provide the knowledge base required for developing precision agriculture systems. One such precision agriculture systems study, conducted by partnerships including the Idaho National Engineering Laboratory, USDA-ARS, universities, and industry, is underwav at a site located near Ashton Idaho.

Methods for Characterizing Factors That Affect Crop Yield and Quality

Study Area

The study area included a Russet Burbank potato field and a soft-white winter wheat field that collectively comprised about 200 ha. The fields were irrigated by center pivot irrigation systems. Cropping zones not covered by the center pivots were, for the most part, irrigated by side-roll and/or hand-line irrigation systems. The climate in the study area is very cold and moist from September through March. In spring it is wet and somewhat warmer. Summers are dry in most years, and temperatures range from below freezing to more than 32°C. The windiest periods of the year are spring and fall. The average annual precipitation is about 40 to 56 cm, and the average annual temperature is about 5.5°C. Elevation is about 1,615 m. Soil characteristics are well drained fine silty loams to silty clay loams on top of unweathered bedrock. Frost free period is about 75 to 85 days, and cool temperatures restrict biological activity to only the summer months

Measurement of Site-specific Potato Yield and Quality

Site-specific potato yields were monitored in the potato field using a scale mounted into the sorting table of the potato harvester. The initial construction and first use of this potato yield monitor was described by Rawlins et al. (1994). Briefly, the scale replaced the chain carrier rollers in the sorting table of the harvester, and was coupled with chain speed and ground speed monitors. About every three seconds, integrated mass measurement data and the differential global positioning system (DGPS) location date were logged to an onboard computer system, as well as telemetered to a base location. Every truck load of harvested potatoes was also weighed prior to unloading at the potato storage facility, and selected loads were weighed on certified scales at a local elevator. Total mass recorded by the potato yield monitor averaged within 5% of the certified weights of selected truckloads.

The recorded data represented the gross mass of material collected by the harvester at each point location. Samples of harvested material (randomly collected at approximately two per hectare and georeferenced by global positioning system [GPS] time stamp) were analyzed to determine the tare of rocks and dirt. Variability in soil conditions and numbers of rocks resulted in spatial variations as great as 63.8% of the gross harvested mass representing "tare dirt" (Fig. 1). Schneider et al. (1996) reported similar discrepancies in potato yield measurements resulting from areas in the field with high numbers of rocks that were carried over the potato harvester. However, due to the sandy soils of the Prosser, WA area, they did not observe significant contributions to harvested gross mass by adhering field soil.

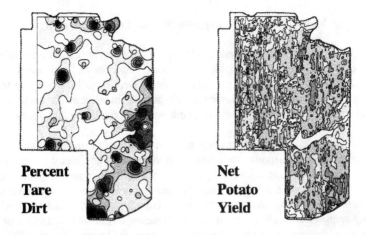

Fig. 1. Potato field "tare dirt" and yield maps. "Tare dirt" is a cumulative term **referring** to soil, rocks and other harvested material that is not potatoes. Spatial variability in percent tare dirt was a great as 63.8% of gross harvested mass; Spatial variability in net potato yield was as great as 52.3 Mg ha^{-1}. The map gray scale **ranges from** light (low values) to dark (high values). White map areas represent areas of no data.

Depending on soil types and harvest conditions, measurements of "tare dirt" may be necessary to accurately assess site-specific net potato yield (Fig. 1).

Samples collected from the potato harvester were also analyzed to determine potato sizes and grade, external and internal defects, injury/damage, rot, bruise, frost specific gravity and pathogenic diseases (USDA-AMS, 1983; USDA-AMS, 1991). Potato crop quality was highly variable across the field. Specific gravity (Fig. 2) is related to the storability and processability of the crop. Single percentage increases in potato solid content enhance potato crop storability, processability and ultimately economic return. Quality analyses also identified areas in the field that may produce "hot spots" in the potato storage facility. The term "hot spots" refers to areas in the potato pile of elevated temperature caused by increased biological activity. Quality factors such as rot, frost damage (Fig. 2), and/or other factors can contribute to the development of storage "hot spots", which can destroy the stored crop. To preserve the identity of site-specific potato quality in the storage facility, each truck carried a unique radio-frequency transponder identification tag that was logged with the potato harvest data points. Identity preservation of the crop increased the farmer's ability to manage the development of storage "hot spots" resulting from frost damage (Fig. 2), *etc.* and market the potato crop. Additionally, potatoes yielding particularly good or poor economic returns can be traced back to the position in the field where the potatoes were grown, allowing for review of the site-specific management practices that produced the crop.

Measurement of Site-specific Wheat Yield and Quality

Two grain combines were used to harvest the wheat field. Both combines were equipped with MicroTrak GrainTrak yield monitors and moisture sensors. Integrated yield data from the GrainTrak system were recorded every two to six seconds along with a DGPS location. Wheat yield measurements made by the GrainTrak yield monitors on the two combines were combined into a single data set.

Grain yield measurements only report a portion of the information required to determine crop response to nutrients, environment, and many other growth factors. To characterize assimilate partitioning within the grain and general crop quality, samples were collected from the clean grain elevators of both combines (while in motion) at approximately two per hectare from the east half of the wheat field. Samples were collected by inserting a sampling device into a trap door on the clean grain elevator, and each sample was georeferenced by recording the GPS time stamp. The collected wheat samples were delivered to the Idaho Grain Inspection Service in Pocatello, ID for analysis of protein (USDA-FGIS, Book V-A, 1995), moisture, shrunken and broken kernels, bulk density and dockage (USDA-FGIS, Book II, 1995). All quality parameters were highly spatially variable.

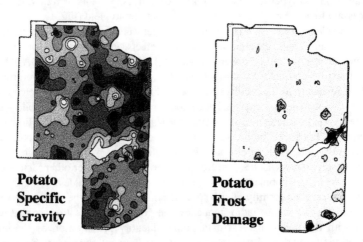

Fig. 2. Potato field quality maps of specific gravity and frost damage. Specific gravity has reference to total solids. A change of 0.0046 in specific gravity represents a 1.0% change in total solids. Spatial variability in specific gravity was as great as 0.027411 (or about 6.0% total solids). Spatial variability in percent of net harvest potatoes damage by frost was as great as 16.7%. The map gray scale ranges from light (low values) to dark (high values). White map areas represent areas of no data.

Grain bulk density varied spatially by as much as 25% based on a standardized bulk density of 77.2 Kg hectoliter^{-1} (60 lbs bu^{-1}) for wheat (Fig. 3), and percent moisture of total grain mass varied by 8.8%. In most cases, grain yield monitors are calibrated to read harvested grain mass, and site-specific grain volume can be calculated using the site-specific moisture correction. Yield monitor calibrations or data post-processing could correct for average grain bulk density deviations, but these methods would not account for site-specific bulk density deviations. Bulk density variations also affect volumetric grain yield monitors (Murphy et al., 1994). The importance of considering site-specific grain bulk density in yield mapping has yet to be determined, but as with moisture, if it could simply be done, making such corrections would be desirable.

Soft-white winter wheat is generally used in confectionery products, which require optimum grain protein levels between 9 and 10% (study cited in Mulla et al., 1992). Measured spatial variations in grain protein content varied well beyond the desired limits for soft-white wheat (Fig. 3), which significantly impacted the overall crop quality. Partitioning of photosynthate as starch or protein within the grain is primarily affected by climatic factors to the extent that genetic differences between soft white wheat cultivars are often concealed (Graybosch et al., 1996; Rao et al., 1993). The effect climatic factors have on protein levels in the grain can be ameliorated, however, by suitable agronomic and fertility management practices (Mulla et al., 1992; Rao et al., 1993). Site-specific monitoring of grain protein levels can lead to management strategies that will improve the overall quality of the grain crop.

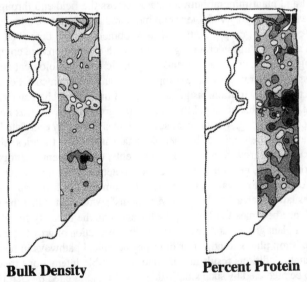

Bulk Density   **Percent Protein**

Fig. 3. Wheat field quality maps of grain bulk density and grain percent protein. Spatial variability in grain bulk density was as great as 20.0 Kg hectoliter~1 (15.5 lbs bu^{-1}); Spatial variability in percent protein of total grain mass, adjusted to 12% moisture, was as great as 5.4%. The map gray scale ranges from light (low values) to dark (high values). White map areas represent areas of no data.

Measurement of Spatial and Temporal Factors Affecting
Crop Mineral Nutrition

To establish general spatial soil mineral nutrient concentration baselines, both the wheat and potato fields were grid soil sampled. The wheat field was fertilized and sown with **winter white wheat** in October of 1994, and soil samples were collected in the spring of 1995. Soil samples were collected from the potato field in the spring of 1995 after the ground had been prepared for planting but prior to fertilization. The wheat field under the north pivot was sampled on a 0.4 ha grid and 1.6 ha grids were used for sampling the remainder of the wheat field and the entire potato field. A composite often 0.3 m soil cores collected from a 3 m radius at each GPS located grid intersect were collected. All soil samples were analyzed by a certified laboratory for pH, salts, sodium, cation-exchange capacity (CEC), excess lime, organic matter, organic nitrogen, nitrate, phosphorus, potassium, calcium, magnesium, sulfur, zinc, iron, manganese, copper, boron, and soil texture from CEC.

The soil nutrient status was also monitored throughout the growing season. Sixteen of the original potato field sample locations were relocated and sampled three additional times. In addition, spatial and temporal changes in plant nutrient status were characterized by potato petiole samples collected twice at the same sixteen soil sample locations during the growing season. Petiole samples were analyzed by a certified laboratory for nitrate, phosphorus, potassium, calcium, magnesium, sulfur, zinc, iron, manganese, copper and boron concentrations.

Soil and plant mineral nutrient status across the field and throughout the growing season was highly variable for all measured parameters. Under controlled laboratory conditions, crop growth and development can be described in detail (Gardner et al., 1985), but in cropping systems such descriptions are compounded by changing environmental factors and the genetic and physiological abilities of plants to respond differently to these complex cropping environments. For example, spatial and temporal variations as great as 3.5 pH units were characterized in the potato field, yet the change in pH throughout the growing season was not always spatially predictable (Fig. 4), nor were such changes obviously correlated to plant nutrient status as determined by petiole analysis. Many theories about the availability of plant growth factors (e.g., nutrients, environment, pathogenic and beneficial organisms, genetics, *etc.*) and their potential limiting effects on plant growth have been put forward. Inappropriately, some of these theories have been referred to as laws, such as the "Law of Minimums", which states that the factor in least abundance becomes the limiting factor and sets the capacity for yield. The complexity of plant growth and development, the interaction of growth factors, and the ability of crop plants to adjust or omit physiological pathways in response to environmental conditions result in a myriad of possible interactions that are too extensive to be predicted by such simplistic models (studies cited by Gardner et al., 1985). As mineral nutrition becomes spatially and temporally more refined, these often unpredictable variations confirm that the interactions of many factors supersede and/or confound the effect of an individual factor in crop production, which would expectedly have significant implications in site-specific crop nutrient management.

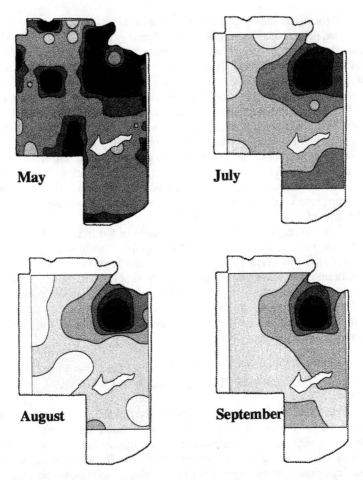

Fig. 4. Potato field map of spatial and temporal variations in soil pH, which variations were as great as 3.5 pH units. The map gray scale ranges from light (low values) to dark (high values). White map areas represent areas of no data.

In cropping systems, nutrient availability generally has a greater impact on determining plant nutrient status than absolute nutrient concentrations. The availability, or unavailability of nutrients can be dramatically affected by soil pH, CEC, microbial activity and other factors (Marscher, 1995). Variations in pH and soil organic matter change throughout the growing season, and such changes affect the availability of mineral nutrients (studies cited by Gardner et al., 1985 and Marscher, 1995). In temperate climates, a drop in pH will increase the availability of micronutrients such as manganese, iron, zinc and copper, and decrease the availability of nitrogen and phosphorus (studies cited by Marscher, 1995). In the current study, observed changes in levels of iron correlated well with changes in pH, but such a simple relationship can be misleading, as in the case of nitrogen, which lacked any apparent correlation to soil pH.

Nitrogen availability is not only determined by soil nitrogen concentration, moisture, pH and other factors, but is also altered by biological processes which are also affected by pH, moisture and even temperature and organic matter (Marscher, 1995). The total soil microbial population was characterized in the potato field. Soil samples were collected at the same 16 GPS located soil nutrient sample positions in the potato field. Microbial analyses revealed spatial variability as great as 1.1 X 10^{11} cfu g^{-1} soil for total heterotrophic bacteria and 2.0 X 10^{12} cfu g-l soil for total fungi. Microbes not only convert essential nutrients like nitrogen to a usable form, but can benefit or inhibit plant growth by immobilizing nutrients. Decay of crop residue by microbes can immobilize large amounts of nitrogen, while soil sterilization, which kills microorganism, can result in the release of toxic levels of manganese (Gardner et al., 1985). The introduction of microbial communities into the crop growth and development equation, substantially increases the complexity of defining factors affecting crop nutrition.

Advances in plant nutrition and commercial fertilizers are dominant factors attributing to the present day yields and crop quality of many cereal grains and other crops. Crop nutrition still remains one of the best hopes for the next generation of significant gains in crop production. However, simply using nutrient products more efficiently without achieving and employing an increased understanding of spatial and temporal crop nutrient requirements and the complex interactions affecting nutrient utilization will not yield the next generation of gains in crop productivity, quality or environmental surety.

METHODS OF ANALYSIS FOR SPATIAL AND TEMPORAL CHARACTERIZATION PARAMETERS

Agriculture is a system of complex interactions rather than one-to-one relationships. For new analysis tools to be of value, they must be capable of identifying relationships in the absence of exact matches and function iteratively as new data becomes available. Statistical analysis coupled with artificial intelligence tools comprise a promising tool set for meeting these requirements. Case-based reasoning is an artificial intelligence technique that uses "nearest neighbor matching and ranking" (Kolodner, 1993). This method allows a set of features to be entered (e.g., Why was the site-specific potato yield only 25.0 t ha^{-1}?; What is the best management practice to increase specific gravities?) and then matches those features against a corresponding set of stored features (e.g., water, nitrate, heat units [all the layers of data in the GIS maps], and/or information about features from many other data formats). As in nearly all instances, exact matches are seldom found that explain the response or suggest an action, but generally there are many conditions or situations that are quite similar. Case-based reasoning will identify comparable situations and conditions, and rank them according to their applicability, which is very similar to the way farmers draw upon their experience to make management decisions. Using an iterative process, statistics and case-based reasoning were used to analyze all the data collected from the potato and wheat fields. The analyses confirmed the incomplete nature of the data sets for quantitatively describing complex agronomic interactions, but the tools successfully developed relationships and correlations sufficiently to recommend wheat to be planted on the potato ground with little to no initial fertilizer required. These recommendations coincided with grower experience, since the residual nutrients from the previous year's potato crop are generally sufficient to supply the majority of the nutrient requirements for the wheat crop, thus one of the reasons for rotating small grains, and other relatively low input crops, behind potatoes. These correlations confirm that artificial intelligence tools, coupled with statistics, crop models and many other information technology tools, can become useful tools in understanding the complexity of cropping systems.

CONCLUSION

While researchers in the basic areas of science continue to emphasize molecular biology, physics, chemistry and other reductionist approaches to agriculture, precision agriculture is providing new data on cropping systems and crop production management possibilities. The technologies of precision agriculture do not change the agronomy and physiology of crop production, but comprise the tools that will allow the agronomist, soil scientist, crop physiologist and farmer to integrate fundamental and cropping system information to synthesize new levels of knowledge for maximizing agriculture productive efficiency while minimizing negative environmental impacts. The complexity of agriculture and diverse expertise needed for development of knowledge based precision agriculture systems, require that future development be done collaboratively in order to fully realize the expected benefits. Fortuitously, it is this very complexity that serves as the greatest catalysts for building the required partnerships, thus ensuring good probabilities of success for achieving the desired outcome.

ACKNOWLEDGEMENTS

The authors recognize the significant contributions of Idaho National Engineering Laboratory personnel Scott Bauer and Jerry Scott who were responsible for the hardware systems; Dave Hempstead, Kim Mousseau and Randy Lee for endless hours spent in data preparation and analysis. Though space limits the mentioning of names, significant contributions were made by the students and faculty of Ricks College; University of Idaho; Utah State University; Washington State University; USDA-ARS-Prosser, and USDA-ARS-System Research Lab. A special thanks is extended to the farmer and many industry cooperators for their endless tolerance, and financial and intellectual contributions to these research activities. Without those acknowledged, this project would not have been possible. A portion of this work was funded by the U.S. Department of Energy, Idaho Field Office Contract DE-AC07-94ID13223.

REFERENCES

Birrell, S.J., and K.A. Suddeth. 1995. Corn population sensor for precision farming. ASAE Paper #95-1334. ASAE. St. Joseph, MI.
Gardner, F.P., R.B. Pearce, and R.L. Mitchell. 1985. Physiology of crop plants. 2nd ed. Iowa State University Press, Ames, IA.

Graybosch, R.A., C.J. Peterson, D.R. Shelton, and P.S. Baenziger. 1996. Genotypic and environmental modification of wheat flour protein composition in relation to end-use quality. Crop Sci. 36:296-300.

Kolodner, J. 1993. Case Based Reasoning. Morgan Kaufmann Publishers, Inc., San Mateo, CA.

Marschner, H. 1995. Mineral nutrition of higher plants. 2nd ed. Academic Press, Inc., San Diego, CA.

Mulla, D.J., A.U. Bhatti, M.W. Hammond, and J.A. Benson. 1992. A comparison of winter wheat yield and quality under uniform versus spatially variable fertilizermanagement. Agric. Ecosyst. Eviron.38:301-311.

Murphy, D.P., E. Schnug, and S. Haneklaus. 1994. Yield mapping - A guide to improved techniques and strategies. pp. 33-48. *In* P.C. Robert et al., (eds.) 1994. Site-specific management for agriculture systems. Proc. 2nd international conference. ASAICSSA/SSSA Madison, WI.

Rao, A.C.S., J.L. Smith, V.K. Jandhyala, R.I. Papendick, and J.F. Parr. 1993. Cultivar and climatic effects on the protein content of soft white winter wheat. Agron. J. 85: 1023- 1028.

Rawlins, S.L., G.S. Campbell, R.H. Campbell, and J.R. Hess. 1994. Yield mapping of potato. pp. 59-68. *In* P.C. Robert et al., (eds.) 1994. Site-specific management for agriculture systems. Proc. 2nd international conference. ASAICSSAISSSA. Madison. WI.

Schneider, S.M., S. Han, R.H. Campbell, R.G. Evans, and S.L. Rawlins. 1996. Precision agriculture for potatoes in the Pacific Northwest. This proc. *In* P.C. Robert et al., (eds.) 1996. Proc. 3rd Intern. Conf. on Precision Agriculture. Minneapolis, MN. June 23-26, 1996. ASA/CSSA/SSSA, Madison, WI.

U.S. Department of Agriculture. 1995. Grain inspection handbook, book II, Grain grading procedures, Chapters 1 and 12. Federal Grain Inspection Service, U.S. Department of Agriculture, Washinaton, D.C.

U.S. Department of Agriculture. 1995. Grain inspection handbook, book V-A, NIRT wheat protein. Federal Grain Inspection Service, U. S. Department of Agriculture, Washington, D.C.

U.S. Department of Agriculture. 1983. United States standards for grades of potatoes for processing. Agricultural Marketing Service, U. S. Department of Agriculture. Washington, D.C.

U.S. Department of Agriculture. 1991. United States standards for grades of potatoes. Agricultural Marketing Service, U.S. Department of Agriculture, Washington, D.C.

A COMPARISON OF RAPID GPS TECHNIQUES FOR TOPOGRAPHIC MAPPING

R. L. Clark

Department of Biological and Agricultural Engineering
University of Georgia
Athens, GA

ABSTRACT

Dual frequency carrier phase GPS receivers were used to examine the feasibility of using GPS, particularly kinematic modes on moving vehicles with a one person crew, to develop topographic maps for precision farming. On Field 1, six field procedures using walking and driving techniques with post processed and real time differential corrections were used. It was found that the standard deviation of the elevation error between 10 true error points and the calculated topographic surface ranged from 3 to 9 cm for these modes, with an increased error for the kinematic modes. The kinematic mode with real time differential corrections was selected for three other fields, with a resulting standard deviation of elevation error between 4-7 cm for those fields. From these initial results it appears that kinematic GPS may be a viable alternative for topographic mapping for precision farming.

INTRODUCTION

An important issue in forming an on-farm geographic information system (GIS) for precision farming is the development and maintenance of base maps, and the accuracy with which they are generated. For example, the following base maps may be considered:
(1) topographic and physical features map,
(2) soil type,
(3) surface drainage, and
(4) sub-surface drainage

There will be a number of other maps which will become a part of the GIS for a farm, which may include information on planting density, variety planted, application rates for each fertilizer element, application rates for insecticides and herbicides, application rates for irrigation water, rainfall rates, and other data bases. Some of these maps will be static, that is, the data will be collected only once. Other maps will be very dynamic, with frequent updates, and it is expected that the GIS will look at a historical perspective of certain types of maps.
It is important that the processes used to develop these maps be economical in nature for precision farming to succeed. Along with economics, the precision of each map is an important issue. This author proposes that if at least one base map is constructed with high position and elevation precision, it can be used as a

standard against which all other maps can be compared and evaluated, and this base map may be used to adjust positional accuracy in subsequent maps. It is proposed that the topographic map, along with permanent physical features, be used as the standard, and that equipment and field procedures be used which will produce a topographic and physical features map which has high positional accuracy.

For large land areas, photogrammetry has been the technique which has been used for some time as the economical method to develop topographic maps. Conventional engineering surveying techniques for topographic mapping have included the use of transits, theodolites and total stations (McCormac(1991)), and these techniques can produce precise maps, but usually with a high cost. New technologies have recently become available, including robotic total stations, the Global Positioning System (GPS), and laser systems combined with GPS, which may offer viable economic alternatives for topographic mapping.

In the last year, GPS receivers have become commercially available with claims of high accuracy kinematic procedures. The objective of the work described herein was to evaluate the potential for using GPS to develop topographic maps for precision farming, with an emphasis on techniques which minimize the time while optimizing accuracy, and include the collection of data from vehicles which drive over the surface to collect the data needed to form the topographic map.

INSTRUMENTATION AND PROCEDURES

Instrumentation

The GPS receivers used in this study were as follows:

(a)   The base station consisted of a Leica Wild SR299E sensor with an AT202 antenna, a CR 344 controller, and a Wild GST05 tripod (Leica Inc(1995a and 1995b)). When real time differential corrections were used, a Pacific Crest Model RFM96W 35 watt transmitter was used to transmit the differential corrections (Pacific Crest(1995)).

(b)   The rover had a Leica Wild SR399E sensor with AT302 antenna, and a CR344 controller. For rapid static and stop-and-go procedures (described later), the rover antenna was mounted on a Wild Stop/Go kinematic pole. A Pacific Crest Model RFM96W 2 watt radio was used to receive differential corrections at the rover when real time differential corrections were used.

These GPS receivers are 9 channel L1 and L2 sensors capable of tracking C/A code, P1 code, and carrier phase on the L1 frequency, and P2 code and carrier phase on the L2 frequency. When the P code is encrypted, the sensor automatically switches to P-code-aided tracking. These receivers are capable of six measuring modes, depending upon the application, and are known as static, rapid static, reoccupation, stop and go, kinematic with static initialization, and kinematic with on-the-fly initialization. For this study, only the rapid static, stop and go, and kinematic techniques were considered. The field procedures for these measuring modes are outlined below. These measuring modes only work for maximum baselines of 15 km for rapid static mode, and about 10 km for the other modes. However, for topographic mapping purposes, this is not usually a limitation. The positional precision of differential GPS is usually a linear function of the distance between the base station and rover (baseline distance), therefore the user will normally need to have the base station set up relatively close to the field, or preferably, near the middle of the field to achieve maximum precision.

A Dell Latitude XP 4100CX 100 MHZ laptop with 8 meg memory and a 500 MB hard drive were used in the field to download and process the data.

Field Procedures

Four fields were selected for this study, identified as Field 1, 2, 3 and 4, and were located at the University of Georgia Plant Science Farm at the intersection of Highway 53 and Snows Mill Road west of Watkinsville, Georgia. All the fields had mowed grassed surfaces. One base station location was used for the study, and was located near the middle of the fields in a location with good satellite visibility.

On Field 1, several data collection modes were used:

(1)   The rapid static mode involved placing the rover at an unknown point for a period of time usually ranging from 3-10 minutes, and averaging the positional calculations over that time period. The Wild CR399 controller software contains what is called a stop and go indicator which indicates to the user when sufficient data has been collected to resolve the ambiguities on the carrier phase portion of the signal, which is required to obtain the precision advertised by Leica. For this study, with baselines less than 5 km, the time period required to obtain sufficient data was usually about 3 minutes. Two second epoch rates were used for data collection. Because of the time required for each point, the rapid static mode was not used in this study to obtain ground surface detail for the topographic maps, but it was used as described below to initialize other modes, and to collect true error points on each surface. Ten of these points were obtained in random locations on each surface, and the same ten points were used to check elevation accuracy for all of the techniques used to develop the topographic surface.

(2)   The stop and go mode consisted of two parts: (a) rapid static initialization on an unknown point, and (b) collection of data on a chain of unknown points by collecting two-two second epochs at each point. Both post processed and real time differential corrections were used for the stop and go mode.

(3)   Two kinematic modes were used with post processed differential corrections- (a) kinematic with rapid static initialization and (b) kinematic with on-the-fly resolution of ambiguities (AROF- ambiguity resolution on the fly). The kinematic with static initialization mode required that the vehicle remain at one position for a rapid static point.   The AROF

technique allowed the vehicle to begin moving when data collection began, and the software resolved the ambiguities after about 200 seconds of data had been collected. The AROF technique requires that the GDOP be less than 8 and there is a minimum of 5 visible satellites requirement. The software was capable of backing the ambiguity resolution to the beginning of data collection so that all points collected were of high accuracy.

(4)   The kinematic mode with static initialization was used with real time differential corrections. The kinematic with AROF mode only works if both receivers are CR399 sensors.

The real time differential correction mode required that both receivers maintain lock on a minimum of 5 satellites. This places a constraint on the time of day when a minimum of 5 satellites were available, and also required that none of the 5 available satellites be blocked by trees, hills, etc. A major advantage to real time differential corrections was that the operator knew at all times if the software was able to resolve ambiguities on the fly, instead of waiting for post processing. The ability to resolve the ambiguities on the carrier phase portion of the GPS carrier signal is the critical issue which determines if the positional results meet the stated precision. The Leica CR344 controller software constantly displays a term called "CQ", which stands for coordinate quality, which is in metric units. Though not clearly explained in the Leica literature, this coordinate quality parameter is evidently a statistical measure of horizontal baseline coordinate quality. Leica recommends that a maximum allowable CQ of 0.05 m be used, which indicates to the operator that the horizontal baseline positional accuracy is about 0.05 m. For these studies with real time differential corrections, it was a simple matter for the operator to observe that the CQ was equal to or less than 0.05 m for each position. However, for kinematic mode, the operator was driving the vehicle and therefore it was difficult to also observe the controller to guarantee that CQ was equal to or less than 0.05 m. However, casual observation of the CQ parameter on the move revealed that it was usually in the 0.01-0.03 m range.

The baseline precision advertised by Leica (1995) for each measuring mode is as follows:
   (a)   rapid static (5 mm + 1 ppm),
   (b)   stop and go with post processed differential corrections (1 to 2 cm + 1 ppm),
   (c)   kinematic (either static or AROF initialization) with post processed differential corrections (1 to 2 cm + 1 ppm), and
   (d)   stop and go or kinematic with real time differential corrections (1 cm + 2 ppm).

The term "baseline precision" refers to the accuracy of the GPS rover with respect to the base station. The term "ppm" refers to the distance between the rover and base station; 1 ppm means that the positional error will increase about 1 mm for every 1 km increase in distance between the rover and the base station. These figures refer to horizontal precision; Leica states that elevation error will increase by a factor of about 2.

For the rapid static and stop and go measuring modes, the antenna was mounted on a Leica stop/go kinematic pole with extension and with tripod legs. Use of this pole placed the antenna at a precise height vertically above a ground point, and allowed the user to position the antenna precisely over a ground point. The tripod legs held the antenna steady for rapid static points because the antenna could not move more than 10 mm during the time of initialization or ambiguities would not be resolved. Also, for stop and go points, the antenna could not move more than 10 mm while the two epochs of data were being collected, and the tripod enables this lack of motion. For the kinematic measuring modes, the antenna was attached to a magnetic base which was in turn placed on top of the canopy of a Kubota L245DT tractor. The 10 mm motion restriction does not apply to kinematic modes. It was considered that mounting the antenna on top of the tractor would allow for accurate control of the antenna height above the ground surface, due to tractor motion, and could lead to excessive elevation error.

An important issue in collecting data for a topographic map is the field procedure used to determine how many points to obtain, and where to obtain these points. A procedure frequently used is to layout a grid on a field, say a 30 m grid, and obtain data points at each grid intersection. However, this procedure may entirely miss some detail features which are important to the topographic map. This author considers that one of the most important applications of the topographic map is to enable the user to visualize surface water flow. From that perspective, if a field contains ditches, terraces, depressions, or hill tops, a sufficient number of data points should be collected to detail these surface water control features. A terrace will then require data points along the line of the top of the terrace, and along the line of each foot. A drainage ditch will be similar, requiring data points along the line of the bottom of the ditch, and along the line of the top of each bank. These lines are commonly called "break lines", and are important features to detail for a topographic map. Also, the location of depressions and hilltops and their elevation are important features which should be detailed.

For each measuring mode, the general field procedure was to first obtain data points around the field boundary. For the stop and go modes, these points were approximately 20 m apart, unless there was an obvious change in ground slope, in which case more data points were taken to detail the change in slope. Then data points were collected over the field surface. For the stop and go modes, an approximate visual 20 m grid was used by pacing, with additional points for break lines and other features as needed. For the kinematic modes, the controller software was set to collect a data point every time the vehicle moved 10 m in its trajectory. A maximum field speed of 7 kph was used, and sometimes the tractor was operated at a slower speed because of terrain. At 7 kph, with one point collected every 10 m, the field time required for each data point was about 5 sec. The tractor was driven in a spiral, starting with the field boundary and moving towards the field center. Two different approximate visual spiral spacings were used, 10 m and 20 m.

Following data collection for each mode, the data from both the rover and base station were downloaded to a computer if post processed differential corrections were used. If real time differential corrections were used, it was only necessary to download the rover data. The Leica Ski software was then used to change the horizontal position data from latitude and longitude to state plane coordinates, and the elevation data from ellipsoidal to orthometric elevations. The output of Ski was translated to an ASCII file, which was imported into another Leica software product called Wildsoft, where the data was translated to an AutoCad(1994) DXF file format for use in an AutoCad add-on product made by Leica called Wildsoft 2000(Leica Inc(1992)), which utilized another Leica product called CIP(Leica Inc(1991)). These programs allows the user to identify the points that are to be used in calculating a topographic surface, and those points that are to be used as true error points to check the final result, all from within AutoCad. This software was very flexible, allowing the user to set a number of parameters during the process of calculating the digital terrain model (DTM) which is used to construct the topographic lines for the topographic map. After calculating the DTM, the software found the elevation of the topographic surface at each of the true error point locations, and then calculated the standard deviation of the error between the true error point elevation and the surface elevation for all of the true error points. The concept followed in this study is that the rapid static technique used to find the true error points yields positional and elevation information which is somewhat more accurate than the other techniques and can therefore be used as the standard against which the surface is checked.

Based on observations of the advantages and disadvantages of the measuring modes on the first field (see discussion in results and discussion), the kinematic mode with real time differential corrections was the only mode used to obtain topographic points for the other three fields. Ten rapid static points were collected at random on each field to use to check the elevation accuracy of the topographic surface. The data for the true error points was collected after the kinematic data was collected so that these points could purposefully be collected in areas where the tractor did not drive to be sure that the true error points were on the calculated topographic surface, and not at or near the data points collected to generate the surface. Fields 2, 3 and 4 had terraces, which were detailed by placing the GPS antenna as close as possible over the left rear tractor tire as viewed from the tractor operator's seat. Then to detail features such as the top or bottom edges of a terrace, the tractor was driven so that the left rear tire followed the desired ground feature. The data was then processed as described in the previous section to obtain the topographic map and to check elevation accuracy.

RESULTS AND DISCUSSION

The area and elevation change for each field is presented in Table 1. Field 1 did not have any readily visible terraces, hilltops, or depressions. Field 1 did have some subtle hilltops and depressions which were difficult to locate visually. Fields 2 had one terrace, and fields 3 and 4 had three terraces each.

Figure 1 contains maps for each measuring mode used which show the location of the topographic data points used to generate the field, the location of the check points, and the topographic maps for each measuring mode. It can be seen that there are subtle differences between the maps, but the general shape is the same. Figures 2-4 present maps showing the data points and the topographic maps for Fields 2, 3, and 4.

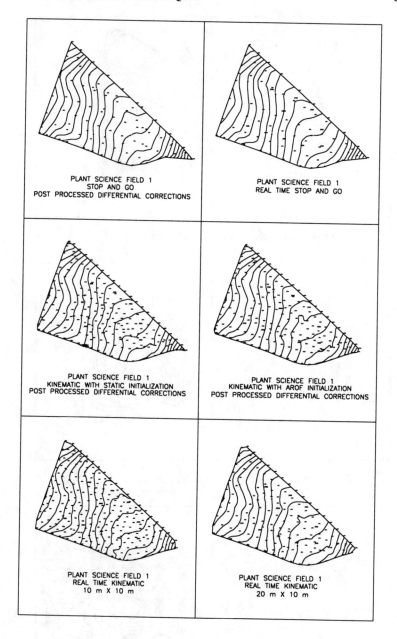

Figure 1. Topographic maps for six GPS measurement modes for Field 1 showing location of the data points and contours.

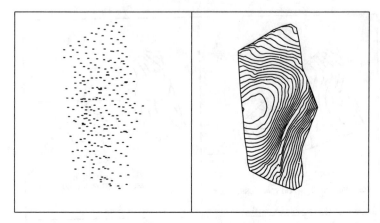

Figure 2. Data points (left) and contour map for Field 2.

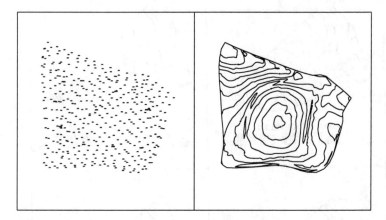

Figure 3. Data points and contour map for Field 3.

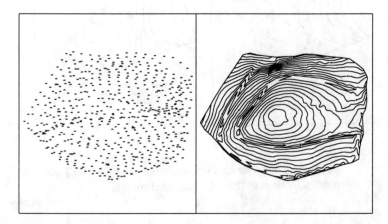

Figure 4. Data points and contour map for Field 4.

Table 1. General Field Characteristics.

	AREA		ELEVATION (m - orthometric)		DELTA ELEV. (m)
	HA	ACRES	MINIMUM	MAXIMUM	
FIELD 1	1.52	3.76	241.09	246.97	5.88
FIELD 2	1.63	4.03	240.48	247.05	6.57
FIELD 3	2.21	5.46	245.53	249.04	3.51
FIELD 4	2.90	7.17	244.17	250.39	6.22

Table 2 is a summary of the standard deviation of the difference in elevation between the calculated topographic surface and the true error points for each measuring mode. As hypothesized, the elevation error was lowest for the stop and go techniques, 3-4 cm, and 4- 8 cm for the kinematic techniques. It is possible that the increase in error with the kinematic techniques was due to vehicle motion, and the resulting lack of elevation precision as compared with the stop and go technique. It is expected that the acceptability of these error figures will depend significantly on the topographic relief in a field. For example, a large field in flat farming country which has a total relief less than 1 m would likely require at least the 4 cm accuracy to provide a topographic map which can sufficiently describe surface water flow. In fields with significant relief, it is expected that the 4-8 cm accuracy of the kinematic technique will provide an acceptable topographic surface. The large number of points acquired on Field 4 were needed to detail the three terraces on this surface, but the reduced error with an increased number of points implies that error can be minimized by acquiring a large number of points on a surface. The number of points required to minimize error needs further study.

Other factors given consideration in evaluating the measuring modes were: (1) time to collect the data, (2) time to process the data, (3) feedback to GPS receiver operator indicating that the techniques was working. The post processing techniques require that the operator download the data from both the rover and the reference station, enter the correct coordinates of the reference station, and process the data. Exact time records were not maintained for this study, but it was observed that the post processed differential correction process usually required almost as much time as the time required to record the data. For example, the time required to obtain the stop and go post processed data on Field 1 was about 1 hour; the time required to download and process the data was about 45 minutes. It should be noted that the downloading process could be faster if a PCMCIA memory card reader was attached to the computer, rather than downloading through the RS-232 port of the controller. The processing time could also be faster with a computer with a faster processor and hard drive. In addition, post processing generates a significant amount of data which in turn requires significant hard drive storage, particularly for the base station data.

Table 2. Standard Deviation of Difference in Elevation between Topographic
Surface and True Error Points. "P. P." means post processed differential
corrections. "R. T." means real time differential corrections.

Field	Measurement Mode	No. GPS Points	Error (cm)
Field 1	P. P. Stop and Go	74	3
	R. T. Stop and Go	65	4
	P. P. Kinematic/Static Init.	213	9
	P. P. Kinematic/AROF Init.	170	7
	R. T. Kinematic/Static Init. /10m x 10m	228	8
	R. T. Kinematic/Static Init. /20m x 10m	143	8
Field 2	R. T. Kinematic/Static Init.	252	6
Field 3	R. T. Kinematic/Static Init.	413	7
Field 4	R. T. Kinematic/Static Init.	590	4

Following the collection of data on Field 1, it was realized that there were
significant advantages to real time differential corrections: (a) it was not necessary
to record and download base station data and do post processed differential
corrections, which reduced the time to acquire and process the data approximately
in half, and (b) the operator knew on the fly if it was possible to resolve the
ambiguities to achieve the desired precision. The ability to immediately know that
the ambiguities were resolved was the most significant factor. The author has had
several experiences with post processed differential corrections in which it was
found that perhaps the rover was unable to see a sufficient number of satellites to
resolve ambiguities, but this was not known until post processing. It was then
necessary to return to the field to reacquire the data. The decision was made to only
use the kinematic with static initialization mode and real time differential
corrections for the other three fields.

The kinematic techniques can cut the time required to obtain sufficient data
significantly. Because a large number of data points can be collected rapidly with
the kinematic technique, there is a tendency for the operator to collect significantly
more points on a field surface, providing more surface detail. The main drawback
to the kinematic mode is that with the GPS antenna mounted on the vehicle, the
antenna position and elevation are affected by vehicle motion, such as bounce on
a rough surface. In addition, with the GPS antenna mounted on top of the vehicle,
it is difficult to accurately detail field features such as terraces and drainage ditches.
If the GPS antenna were mounted on its own wheel which could be guided over
selected field features, such as the top of a terrace or the bottom of a ditch, then

detailing could be more accurate. An alternative would be to use the stop and go technique with the stop/go pole to detail these types of features. It should be noted that the elevation error on Field 4 was 4 cm with the real time kinematic, comparable with the stop and go technique on Field 1, which may be attributable to the relatively large number of points collected on Field 4 as shown in Table 2. The results on Field 4 imply that if a sufficient number of real time kinematic points are acquired on a field, the elevation error will be comparable to the stop and go mode. Further studies of these results could clarify this point.

CONCLUSIONS

The main objective of this study was to determine if dual frequency high precision GPS receivers in kinematic mode can be used to collect ground surface 3D position data from which a topographic map can be developed with sufficient accuracy. The concept included mounting the GPS antenna on a vehicle driven over the surface to minimize the time required to collect the data, while hopefully maximizing the resulting data precision. For comparison purposes on one field, data was also collected with the antenna on a light weight tripod to obtain maximum accuracy of antenna height above the ground surface.

On Field 1, three modes with post processed differential corrections were used, and three modes with real time differential corrections were used. On three other fields, only the kinematic technique with real time differential corrections was used. It was concluded that there were significant advantages in the use of real time differential corrections with the GPS receiver mounted on a field vehicle. The real time differential corrections allowed the operator to instantaneously know that the ambiguities were resolved on the carrier phase part of the GPS signal to guarantee that the required precision was enabled. Mounting the GPS receiver on a moving vehicle minimized the time required to acquire data and enabled the collection of a larger number of points to detail the surface. The standard deviation of the elevation error between the true error points and the calculated topographic surface increased by 0-6 cm between the technique in which the tripod was used to place the antenna accurately above the surface, and the kinematic technique in which the antenna was on a moving vehicle. On the other three fields, only the kinematic mode with real time differential corrections was used, with the standard deviation for elevation error for ten true error points ranging from 4 to 6 cm.

For most agricultural applications, this increase in elevation error of 0-6 cm does not appear to negatively impact the kinematic technique and the time advantages of this technique. Though not thoroughly studied, the results implied that collection of a sufficient number of kinematic points minimized elevation error, making this error comparable to the results with the stop and go technique with the antenna mounted on a tripod. However the additional data points will require more field time to collect the data points.

REFERENCES

AutoCad. 1993. AutoCad, Release 12. Autodesk Inc., 2320 Marinship Way, Sausalito, CA 94965.

Leica Inc. 1991. Wildsoft/CIP Contour Interpolation Program. Leica, Inc., 3155 Medlock Bridge Road, Norcross, GA 30071.

Leica Inc. 1992. Wildsoft 2000. Leica, Inc., 3155 Medlock Bridge Road, Norcross, GA 30071.

Leica Inc. 1995a. GPS - SYSTEM 300, Dual Frequency GPS Surveying. Leica, Inc., 3155 Medlock Bridge Road, Norcross, GA 30071.

Leica Inc. 1995b. Real Time GPS Surveying. Leica, Inc., 3155 Medlock Bridge Road, Norcross, GA 30071.

McCormac, J. C. 1991. Surveying Fundamentals. Prentice Hall, Englewoods Cliffs, NJ 07632.

Pacific Crest. 1995. RFM96 Radio Modem Family. Pacific Crest Corporation, 2285 Martin Avenue, Suite A, Santa Clara, CA 95050-2715.

A Method for Direct Comparison of Differential Global Positioning Systems Suitable for Precision Farming

S. P. Saunders
G. Larscheid
B. S. Blackmore

Centre for Precision Farming
School for Agriculture Food and the Environment
Cranfield University
England

J. V. Stafford

Precision Agriculture Team
Silsoe Research Institute
Silsoe, Bedford
England

All precision farming operations require the use of a positioning system, typically GPS. The choice and use of the most suitable system is very important as any spatial positioning errors will result in inaccurate mapping and application processes. This paper details a test procedure for direct comparison between various makes of DGPS systems. Three systems have been compared to illustrate the procedure and a single data set is reported. Also a summary of problems experienced during tests are listed.

INTRODUCTION

A positioning system is essential to the implementation of precision farming. As DGPS receiver prices decrease and positioning resolution improves, the commercial viability of systems is increasing. Farmers on ever smaller holdings can now justify the price of a system. Many novel measuring and metering devices have already been developed but all processes rely on one factor, the positioning system.

The correct choice of positioning system is important. There is a vast array available on the market, offering many different specifications over a large price range and so the method of selection is very important.

The purpose of the study was to design and verify a series of repeatable tests pertinent to the requirements of precision farming. It should be feasible for anyone to undertake similar tests using these procedures. They should allow the monitoring of individual factors to compare GPS receiver systems and give the user the ability to choose the most suitable system.

A brief explanation of GPS requirements, systems tested, map datums, projections and GPS errors are given. The development of the methodology is described and the results from practical testing are reported. From these results, conclusions are drawn.

GPS REQUIREMENTS AND RECEIVER SPECIFICATIONS

Essentially, the GPS requirements for precision farming are, a position resolution in the range 0.2 metres up to 20 metres, reliably maintained in all areas of the field.

Three commercial GPS systems were tested. Their specifications included:

System 1 - $600 for receiver.
            8 channels (2 multiplexed).
            Differential correction by national radio carried signal ($1000 for
                receiver and $400 annual subscription).
            Commercially available for one year.

System 2 - $400 for receiver.
            10 channels (0 multiplexed).
            Differential correction by local base station.
            Commercially available for one year.

System 3 - $1500 for receiver.
            4 channels (2 multiplexed).
            Differential correction by local base station.
            Commercially available for two to three years.

A test track was used for certain practical tests. It included buildings one and both sides of the track, trees, direction changes (both marked and gradual) and a slope of approximately six degrees.

MAP PROJECTIONS, DATUMS AND GPS ERROR

A map projection is a systematic representation of all or part of the surface of a round body, using lines delineating meridians and parallels. On the Earth, a network of longitude and latitude (lat/lon) lines have been superimposed on the surface. From these rectangular grids have been developed, typically Universal Transverse Mercator (UTM) Eastings/Northings so that a point may be designated merely by its distance from two perpendicular axes. Euclidean distances are the distance in metres between two points with no reference to direction.

The map datum used for data in this report was Ordnance Survey of Great Britain (OSGB) 1936, which can be transformed from the WGS84 datum over the UK to an accuracy of 0.2 metres (OSGB 1995).

Selective Availability (SA) is a pseudo-random error superimposed on satellite signals for civilian use. The effect of SA can largely be removed using differential correction, but can be a source of GPS error. Other small errors are present mainly due to noise in the propagation medium (i.e. the ionosphere and troposphere), electronic noise in the receiver and the positioning solution algorithm the receiver uses.

DEVELOPMENT OF METHOLODOGY

The first stage was to identify features desired in the test. These were detailed as tests of accuracy, stability, repeatability, horizon changes and signal loss effects.

Next the variables present when using any DGPS system for precision farming were listed. For each variable, a method of stabilisation was decided upon, allowing a balanced and consistent comparison between systems. The variables were grouped into system, satellite, environmental, weather, time and operator categories.

Each of the desired features was then incorporated into a test with all relevant variable stabilising measures being applied. The tests were all conducted on the same day, (04/04/1996), sampling data at a frequency of one hertz and were as follows:

Test 1 - Static Accuracy

This test was used to show the outright accuracy of a system, allowing the 'wandering' of the signal from a set base point to be seen. Also the average of all data was plotted allowing any permanent offset to be seen.

A base point was established, accurately surveyed from a known Ordnance Survey (OS) benchmark. The systems were positioned around this point at two metre spacings, allowing sufficient distance to avoid interference between aerials.

All systems were switched on and given thirty minutes to settle, and to allow systems with old almanacs to update. Although Selective Availability is pseudo-random, it was believed that logging for thirty minutes would show maximum deviation from a base point. Comparative accuracys are quoted as ninety fifth percentile horizontal error (R95) about the mean of all data points.

Test 2 - Static Stability

Using the previously established base point, the systems were positioned as in test 1. Three tests were undertaken:

2a - Cold Settling Time

This test was used to show how long the operator must wait from switch on before field operations can be undertaken. The systems were logged from switch on (cold) and the cold settling time quoted as the time for the system to achieve ten consecutive points within a ten metre radius of the base point. It was ensured that all systems had updated almanacs before the test. This figure was chosen because it is the previously stated minimum requirement on positioning.

2b - Re-establish Settling Time - Differential Signal

This test showed how long a period of differential signal loss would affect the position recorded, or how soon reliable positional data can be regained after a differential signal loss.

Once again the systems were positioned around the base point at the standard 2 metre spacing. The systems were switched on and allowed thirty minutes to 'warm up'. Whilst being logged, the differential signals were disrupted for five minutes, and the re-establishing time quoted as the time for each system to position ten consecutive points within a ten metre radius of the base point.

To remove the differential signal, the aerials were removed and five minutes of this loss was deemed sufficient not to allow any system to remember previous differential signal.

2c - Re-establish Settling Time - Satellite Signal

This test showed how long a period of satellite signal loss, would affect the position recorded, or how soon reliable positional data can be regained after a satellite signal loss.

The same procedure as test 2b was applied but the satellite signal was removed. To achieve this each aerial was moved inside the test vehicle. Once again, settling time was quoted as the time to position ten consecutive points within a ten metre radius of the base point.

Test 3 - Dynamic Stability

These tests were used to simulate the loss of satellite and differential signals when working in a field situation and assess the severity of this loss on position.

All systems were attached to a test vehicle, shown in Fig. 1, switched on and allowed thirty minutes to 'warm up'. This vehicle was then driven at four kilometres per hour, from a known start point, around a surveyed test route, with all systems recording positional data, at a frequency of 1 Hertz. Two tests were undertaken :

Fig. 1 - Test Vehicle with Differential and Satellite Aerials Fitted

Test 3a - Loss of Differential Signal

For the first part of the test route, all systems were free to record as in normal field situations. At a known point (surveyed from the benchmark) the differential signal was removed as in test 2b, for two hundred metres (three minutes at four kilometres per hour). The signal was then restored for the rest of the circuit.

Test 3b - Loss of Satellite Signal

Using the same route as in test 3a, the satellite signal was removed for two hundred metres between the two known points. The signal was then restored for the rest of the circuit.

Data from both tests was plotted as a series of points on an outline of the roadway, digitised from an OS map. Error was quoted as maximum deviation (Euclidean distance) and a visual assessment can be made from the map.

Test 4 - Dynamic Repeatability

When working in the field, lines of data points from each pass of the machine should not cross each other as this will impair mapping effectiveness.

This test considers the deviation from a known route when all systems are logging effective signals and shows deviation from that route. Also, the test route used includes buildings, trees and horizon changes and their effect on signal can be assessed.

All systems were attached to the test vehicle, switched on and allowed thirty minutes to 'warm up'. This test vehicle was then driven, from a known start point, around the surveyed test route at four kilometres per hour.

Three laps of the test route were driven and plotted on a map. Error was quoted as maximum deviation (Euclidean distance) and is visible on the map.

Test 5 - Precision

This test considers the distance travelled before each system recognises movement in that direction, allowing the comparison of precision between systems.

All systems were attached to the test vehicle, switched on and allowed thirty minutes to 'warm up'. This vehicle was then driven from a surveyed point forwards for two metres with the relevant GPS time noted. It was stopped for one minute then driven a further four metres and stopped, then eight metres, sixteen metres and thirty two metres with stops of one minute at each.

The resulting maps were studied and the point at which the system first recorded movement and the distance recorded were noted.

The series of tests are abbreviated in Table 1 below :

Test	Category	Section	Type	Assessment
1	Accuracy	Static Accuracy	Static	R95 (metres)
2a	Stability	Cold Settling Time	Static	Time (seconds)
2b	Stability	Settling Time Differential	Static	Time (seconds)
2c	Stability	Settling Time Satellite	Static	Time (seconds)
3a	Stability	Loss of Differential	Dynamic	Euclidean dist.
3b	Stability	Loss of Satellite	Dynamic	Euclidean dist.
4	Repeatability	Laps of Test Circuit	Dynamic	Euclidean dist.
5 (metres)	Sensitivity	Forward Movement	Dynamic	D i s t a n c e

Table 1 - Summary of DGPS Tests

RESULTS AND DISCUSSION

To appraise the effectiveness of these tests, a practical assessment was undertaken on the three systems previously specified. Due to time constraints, only a single data set was taken but serves to illustrate their use.

Two laptop computers were used to record data, with any differences in time being corrected during post-processing. In all tests, systems were logged at one hertz and recorded data in lat/lon projection, which was converted to UTM Eastings/Northings allowing position representation in metres. In post-processing, Excel 5.0 was used for statistical evaluation and Surfer 6.01 used for all map creation.

Test 1 - Static Accuracy

All distance in metres System 3	System 1	System 2
R95 5.05 m	14.1 m	14.6 m
Euclidean Error of Mean 24.2 m	7.6 m	36.6 m

The 'R95' figure refers to the 95 percentile horizontal error and the 'Euclidean error of mean' is the distance between the established base point and the average of all data points recorded.

From Fig. 2, it can be seen that system 1 offers the most accurate mean, but system 3 the lowest R95 figure.

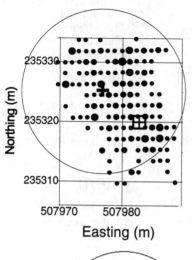

- · 1 Repetition
- : 2 Repetitions
- • 3 Repetitions
- ● 4 Repetitions
- ● 5 Repetitions

⊞ Base Point

+ Average of All Data Poin

Circle shows R95 About Offs
Mean

System 1

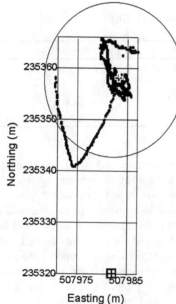

- · 1 Repetition
- • 2 Repetitions
- • 3 Repetitions
- ● 4 Repetitions
- ● 5 Repetitions

⊞ Base Point

✚ Average of All Data Points

Circle shows R95 About Offset
Mean

System 2

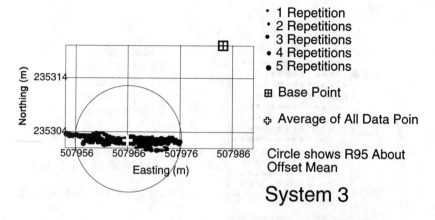

Fig. 2 - Plots From Static Accuracy Test Showing R95 Value As Circles

Tests 2a,b,c - Static Stability

All times in seconds	System 1	System 2	System 3
Cold Settling Time	49 s	81 s	143 s
Re-establish Time:			
- Loss of Differential	35 s	243 s	51 s
- Loss of Satellite	123 s	364 s	197 s

The longest cold settling time is two minutes and twenty three seconds, which is an acceptable time for an operator to wait before commencing work. The recovery after loss of differential signal is excessively high for system 2. If working at four kilometres an hour, this is equivalent to a distance of two hundred and seventy metres. The recovery after loss of satellite signal has the largest effect in all three systems, with system 1 travelling one hundred and thirty metres, system two travelling four hundred metres and system three travelling two hundred and twenty metres before reliable positional data can again be recorded. These figures are much larger than expected but highlight the possible problems of each system well. Figure 3 shows a typical 'cold settling' plot, illustrating how Euclidean distance and time relate.

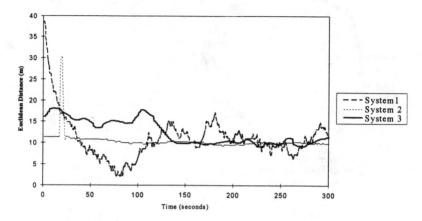

Fig. 3 - Euclidean Distance Against Time for 'Cold Settling'

Test 3a,b - Dynamic Stability

All distances in metres	System 1	System 2	System 3
Euclidean Error			
- Loss of Differential	9.0 m	23 m	Corrupt File
- Loss of Satellite	32 m	47 m	Corrupt File

The Euclidean error is the maximum distance, in metres, between the actual position and the recorded position at the worst point during the signal loss. The previous test (test 2) showed the recovery times when signals were lost and this test shows the worst case distance error during the signal loss. System three recorded corrupt data for no known reason during both tests. Due to time constraints it was not possible to repeat this test and investigate the reason for file data corruption but this is obviously unacceptable as any mapping or application function will be incorrect. Figure 4 shows a typical plot from dynamic stability testing with loss of differential signal.

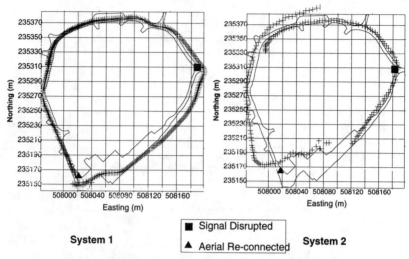

Fig. 4 - Dynamic Stability Test - Loss of Differential Correction Signal

Test 4 - Dynamic Repeatability

All distances in metres	System 1	System 2	System 3
Euclidean Error	8.7 m	12.1 m	Corrupt File

As with test 3, the Euclidean error refers to the maximum distance, in metres, between the actual position and the recorded position at the worst point during the test. Figure 5 shows a typical plot of three laps of the test route. System one and two exhibited a stable repeatable signal which was not distorted by trees. Slight horizon changes (slope of six degrees) have no effect as the system has the ability to recognise satellites other than the ones it is using. This allows it to easily switch to other satellites when some are lost due to the horizon.

Large horizon changes (driving between buildings) did affect both systems one and two, as this resulted in a much poorer visible satellite constellation geometry. System three again recorded a corrupt file which would result in unacceptable mapping or application processes.

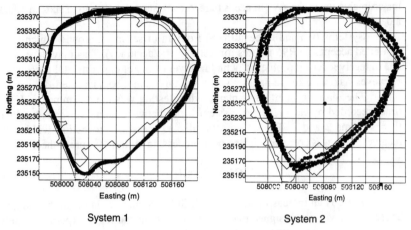

System 1   System 2

Fig. 5 - Dynamic Repeatability - Three Laps of the Test Route

Test 5 - Precision Test

All distances in metres	System 1	System 2	System 3
Actual distance for system response	2 m	8 m	16 m
Distance recorded	1.8 m	8.5 m	10 m

It can be seen that system 1 is very responsive to small changes in movement. From this, it uses a least squares positioning solution which offers motion detection almost immediately but with a less damped response. Systems 2 and 3 use a Kalman filter giving smoother movement but are less responsive to small movements. However, upon close inspection, it can be seen that system 2 offers a six percent error at the first detected movement whereas system 1 offers a ten percent error when first movement is detected.

SUMMARY OF RESULTS AND PROBLEMS ENCOUNTERED

For this single data set, system one using an alternative differential receiver appeared to give a better overall performance. It suffered from little static 'wander', was ready for operation from cold in the shortest time and was the least affected and recovered relatively quickly from differential or satellite signal loss. It gave the most repeatable stable signal and, in this case, would be the obvious choice for a precision farming system

The performance of system two was proportionately poorer than system one and if the choice was given, on the strength of this single data set, it would not be chosen for a precision farming application.

Likewise, system three did not perform as well as system one for the single data set taken. Its main failing was the corruption of signal which is wholly unacceptable when the system needs to record constant data for up to sixteen hours a day.

Systems two and three suffered from an offset mean around which both exhibited acceptable R95 figures for test 1 (static accuracy). This may have been due to receiver set-up problems, actual receiver errors or experimental procedure. However, as only a single data set was taken this cannot be quantified.

The main problem encountered when testing was the difference in UTC times recorded by each receiver. System 1 was between 4 and 12 seconds faster than systems 2 and 3. This is probably due to speed of processing within the receiver. This was corrected during post-processing within the spreadsheet.

CONCLUSION

There is an increasing amount of DGPS systems available on the market. These tests have been designed to give the user with no particular expertise in GPS the ability to choose a system most suitable for precision farming applications and offer a good overall representation of DGPS performance. They have been designed to allow users anywhere to test systems in a similar manner with the ability to include any circumstances peculiar to their region.

By stabilising the variables present, a 'back to back' comparison of systems can be made and from the results achieved in this paper, 'system 1' would be selected, offering the best overall performance for this single data set. It should be remembered that this methodology was developed with respect to the previously stated system requirements and systems may behave differently in other applications.

As an aid to simplified DGPS selection, these tests have been deemed successful.

REFERENCES

Ordanance Survey of Great Britain, 1995. National Grid / ETRF 89 Transformation Parameters. Geodetic Information Paper Number 2, Southampton, UK.

Centimeter Accuracy Differential GPS for Precision Agriculture Applications

Arthur F. Lange

Trimble Navigation, Ltd.
645 North Mary Avenue
Sunnyvale, CA

ABSTRACT

Depending on the particular equipment utilized, GPS is capable of a wide range of accuracy, from tens of meters to centimeters. This paper's goal is to give an overview of GPS and how high accuracy is obtained. Since GPS will be used in many precision agriculture applications including field preparation, variable rate planting, yield mapping, field scouting, pest management and custom chemical application, it is important to understand the limitations of GPS to derive full benefits from the technology.

Introduction

The explosion in interest in precision agriculture technology has been accompanied by a blossoming of uses for a number of enabling technologies, the two most important of which are the Global Positioning System (GPS) and Geographic Information Systems (GIS). While GIS technology offers tremendous capabilities for more informed Agriculture Management decision making, rendering competent decisions still depends on having reliable data. In order to realize the benefits of GPS, and not mis-apply the technology, it is important to understand is limitations. This paper describes how GPS works and how to obtain reliable data. Some precision agriculture applications can be performed with less accurate data which cost much less to acquire. Other applications, like chemical application or field preparation may require higher accuracy in order to prevent overlapping applications of chemicals.

Overview of GPS

The Navigation Satellite Timing And Ranging Global Positioning System, or NAVSTAR GPS, is a satellite based radio-navigation system that is capable of providing extremely accurate worldwide, 24 hour, 3-dimensional location data (latitude, longitude, and elevation). The system was designed and is maintained by the US Department of Defense (DoD) as an accurate, all weather, navigation system. Though designed as a military system, it is freely available with certain restrictions to civilians for positioning. The system has reached the full operational capability with a complete set of at least 24 satellites orbiting the earth in a carefully designed pattern.

The Fundamental Components of GPS

The NAVSTAR GPS has three basic segments: space, control, and user. The space segment consists of the orbiting satellites making up the constellation. This constellation is comprised of 24 satellites, each orbiting at an altitude of approximately 20,000 km, in one of six orbital planes inclined 55 degrees relative to the earth's equator. Each satellite broadcasts a unique coded signal, known as Pseudo Random Noise (PRN) code, that enables GPS receivers to identify the satellites from which the signals came, and makes positioning possible.

The control segment, under DoD's direction, oversees the building, launching, orbital positioning, monitoring, and provides two classes of GPS service. Monitoring and ground control stations, located around the globe near the equator, constantly monitor the performance of each satellite and the constellation as a whole. A master control station updates the information component of the GPS signal with satellite ephemeris data and other messages to the users. This information is then decoded by the receiver and used in the positioning process.

There are two classes of GPS service; the Precise Positioning Service (PPS) which is available only to government authorized users and the Standard Positioning Service (SPS) which is available for civilian use.

The user segment is comprised of all of the users making observations with GPS receivers. The civilian GPS user community has increased dramatically in recent years, due to the emergence of low cost portable GPS receivers and the ever expending areas of applications in which GPS was found to be very useful. Some of these applications are: surveying, mapping, agriculture, navigation and vehicle tracking. The civilian users of GPS greatly out number the military users.

The Limitations of GPS

Though GPS can provide worldwide, 3D positions, 24 hours a day, in any type of weather, the system does have some limitations. First, there must be a (relatively) clear "line of sight" between the receiver's antenna and several orbiting satellites. Anything shielding the antenna from a satellite can potentially weaken the satellite's signal to such a degree that it becomes too difficult to make reliable positioning. As a rule of thumb, an obstruction that can block sunlight can effectively block GPS signals. Buildings, trees, overpasses and other obstructions that block the line of sight between the satellite and the observer (GPS antenna), make it impossible to work with GPS. Urban areas are especially affected by these types of difficulties. Bouncing of the signal off nearby objects may present another problem called multi-path interference. Multi-path interference is caused by the inability of the receiver to distinguish between the signal coming directly from the satellite and the "echo" signal that reaches the receiver indirectly. In areas that posses these type of characteristics, inertial navigation techniques must be used to complement GPS positioning.

The receiver must receive signals from at least four satellites in order to be able to make reliable position measurements. In addition, these satellites must be in a favorable geometric arrangement. The four satellites used by the receiver for positioning must be fairly spread apart. In areas with a relatively open view of the sky, this will almost always be the case because of the ways these satellites were placed in orbit.

An additional limitation of GPS is the DoD policy of Selective Availability. This policy limits the full autonomous accuracy only to official government users. This policy and methods used to overcome it will be described later in this paper. If Selective Availability is eliminated, autonomous GPS without Selective Availability will not be accurate enough for many precision agriculture applications.

How a GPS Receiver Calculates Position

The position of a point is determined by measuring distances (pseudo-ranges) from the receiver to at least 4 satellites. The GPS receiver "knows" where each of the satellites is at the instant in which the distance was measured. These distances will intersect only at one point, the position of the GPS receiver (antenna). How does the receiver "know" the position of the satellites? Well, this information comes from the broadcast ephemeris that are received when the GPS receiver is turned on. The GPS receiver performs the necessary mathematical calculations, then displays and/or stores the position, along with any other descriptive information entered by the operator from the keyboard.

The way in which a GPS receiver determines distances (called pseudo-ranges) to the satellites depends on the type of GPS receiver. Basically, there are two broad classes: code based and carrier phase based.

Code-based receivers

Though less accurate than their carrier phase cousins, code-based receivers have gained widespread appeal for applications such precision farming. This popularity stems mainly from their relatively low cost, portability and ease of use.

Code-based receivers use the speed of light and the time interval that it takes for the signal to travel from the satellite to the receiver, to compute the distance to the satellites. The time interval is determined by comparing the time in which a specific part of the coded signal left the satellite with the time it arrived at the antenna. The time interval is translated to a range by multiplying the interval by the speed of light constant (c=298,000 km/second). Ranges from at least four satellites are needed in order for a receiver to produce a position fix. Position fixes are made by the receiver roughly every second, and the more advanced receivers enable the user to store the position fixes in a file that can be downloaded to a computer for post processing.

Under normal circumstances, autonomous standard position fixes (SPS) made by code-based receivers would be accurate to within 25 meters. The DoD however began imposing its selective availability (SA) policy in July of 1992, which limits position fix accuracy to within 100 meters. The purpose of SA is to deny potential hostile forces accurate positioning capabilities. Military P(Y)-code receivers are not affected by SA, but, as mentioned earlier are not available for the general public. In order to overcome the limited positioning accuracy, differential GPS (DGPS) techniques have been developed. DGPS enables the user to improve

SPS and also to remove the effects of SA and some other sources of error. These differential correction techniques can produce positions generally accurate to within a few meters.

There are also now code based receivers capable of sub-meter differential accuracy. Some sub-meter receivers require longer data collection times (up to ten minutes), and perform best under very favorable satellite geometry, and with an unobstructed view of the sky. Some newer receivers can provide a sub-meter accurate position each second. These receivers cost more at the outset, but provide a good return on the added investment by providing substantially higher productivity.

It is very important that users of code based receivers understand the position accuracy limitations of the receiver. Due to SA, each coordinate viewed on a non-differential GPS receiver's display is only accurate to within 100 meters. This accuracy can be improved on by taking an average of 200 or so repeated position observations of the same point. The resulting accuracy would still be below what many users would consider acceptable quality. In order to produce acceptable results, GPS data collected in the field <u>must</u> be differentially corrected either in real-time, or by post-processing the data.

Carrier phase receivers

The carrier phase receivers, used extensively in geodetic control and precise survey applications, are capable of sub-centimeter (cm) differential accuracy. These receivers calculate distances (called pseudo-ranges) to visible satellites by determining the number (N) of whole wavelengths and measuring the partial (phase) signal wavelength there are between the satellites and the receiver's antenna. Once the number of wavelengths is known, a pseudo-range may be calculated by multiplying 'N' by the wavelength of the carrier signal (L1 and/or L2, 19 cm and 24.4 cm respectively) plus the partial wavelength. It is then a straight forward (albeit complex) task to compute a baseline distance and azimuth between any pair of receivers operating simultaneously. With one receiver placed on a point with precisely known latitude, longitude, and elevation, <u>and</u> with the calculated baseline (distance between 2 points), the coordinate for the unknown point may be determined.

The relative cost of the carrier phase receivers is high, but technological advances have made the dual frequency (using both, L1 and L2) carrier phase receivers of today much more efficient than the single frequency (using L1 only) receivers that were state-of-the-art only a few years ago. With some of the newest dual frequency receivers very precise measurements ($\pm$ 1 cm) can be made in real-time. These receivers will be used in machine control agriculture applications requiring a high degree of accuracy.

Differential GPS

Differential GPS (DGPS) can be employed to eliminate the error introduced by SA and other systematic errors. Differential GPS requires the existence of a base station, which is simply a GPS receiver collecting measurements at a known latitude, longitude, and elevation. The base station's antenna location must be located precisely, using carrier phase GPS or other traditional surveying techniques. The base station may store measurements (for post processed DGPS), broadcast corrections over a radio frequency (for real-time DGPS), or both.

The assumption made with the base station concept is that errors affecting the measurements of a particular GPS receiver will equally affect other GPS receivers within a radius of 200-300 miles. If the differences between the base station's known location and the base station's locations as calculated by GPS can be determined, those differences can be applied to data collected simultaneously by receivers in the field. These differences can be applied in real-time (especially applicable for accurate navigation) if the GPS receiver is linked to a radio receiver designed to receive the broadcast corrections. In some GIS mapping applications, these differences are applied in a post-processing step after the collected field data has been downloaded to a computer running a GPS processing software package. GPS processing software is typically integrated with GPS hardware and thus is provided by the receiver manufacturer. As a rule, post-processed DGPS is considered slightly more accurate than real-time DGPS.

Base Stations - The Source For Reference Data For DGPS

There are many permanent GPS base stations currently up and running in the United States that can provide over an electronic bulletin board to the users of code based receivers the data necessary for post-processing differentially correcting positions.

In many parts of the US, the US Coast Guard's DGPS beacon system is operational. These stations are part of a large network of coastal and Mississippi river valley stations. These US Coast Guard beacons broadcast in the frequency range of 285 to 325 khz. The range of many of these stations is 200-300 miles. These stations can provide differential accuracy in the one meter range, depending on distance from the station.

Carrier phase base differential processing, for centimeter accuracy, does not have the range of code based differential GPS. To obtain centimeter accuracy, it is now necessary to have the carrier base station within approximately 50 km. Extending the range of carrier differential beyond 50 km is an active GPS research topic.

Precision Agriculture Considerations

Several key issues need to be explored when considering GPS as a tool for capturing coordinate data for a precision agriculture application. First and foremost, is to determine the position accuracy requirements. If the data will be used for site specific analysis that require position accuracy to be within a few feet, high-quality code based differential GPS receivers will be used. If better accuracy is required, on the order of 10 cm or better, as for a spray vehicle guidance system to replace foam markers, then carrier phase differential GPS techniques have to be employed.

Every GIS database must be referenced to a base map or base data layer, and the reference datum of the various data layers must be the same. Ideally, the database should be referenced to a large scale, very accurate base map. If instead the base map is smaller scale (quad scale or smaller) there could be problems when attempting to view the true spatial relationships between features digitized from a small scale map and features whose coordinates were captured with GPS. This can be a real problem if a grower decides to use a particular GIS data layer that was originally generated using small scale base maps as a base to which all new data generated is referenced. The best way to avoid such incompatibility one should consider developing an accurate base data layer, based on geodetic control and photogrammetric mapping.

Map Datum

Understanding the concept of a map datum is important to obtain useful results from any GPS mapping exercise. A datum is a mathematical model of the earth over some area. GPS, being a world wide system, has a datum applicable over the whole earth. It is called World Geodetic System, 1984 (WGS-84). There are many local datums like North America Datum, 1927 (NAD-27) and North America Datum, 1983 (NAD-83) that have been used to make local maps. A particular point on the surface of the earth will have different latitude and longitude coordinates depending on the reference datum. For precision agriculture applications, it is recommended that all data be in the NAD-83 or WGS-84 datum. Fortunately, NAD-83 and WGS-84 are very similar. If the GPS receiver is not set to the correct datum (NAD-83 or WGS-84) then data from a yield mapping exercise could be tens or hundred of meters different from the maps used to plot the field boundaries or the satellite images of fields. Of course, it is important to use the same datum for all data collections over several years if the data is going to be overlaid in a GIS.

Is GPS The Answer To All Our Mapping Needs?

GPS is a positioning system that can also be used as a real-world digitizer for mapping point and line features such as roads or wetland boundaries or the location of pest infestations. However, for large volume data collection which includes measuring many points simultaneously, one should consider photogrammetry or remote sensing satellite imaging as a more efficient data collection tool.

GPS is an important (future) tool for Precision Agriculture

Field portable GPS receivers are available today for rapidly mapping insect infestations and this data can be accurately communicated to the field manager who may employ a custom spray operator to apply the correct chemicals only where they are needed. In the future, the spray control operator or the farmer planting row crops will use centimeter accurate differential GPS to control, the vehicle's steering. Additionally, the GPS equipment will provide a permanent record back to the field manager with GPS data of where and when the treatment took place.

Multispectral Videography and Geographic Information Systems for Site-Specific Farm Management

G. L. Anderson
C. Yang

USDA-ARS, Remote Sensing Research Unit
Weslaco, Texas

ABSTRACT

Researchers are expending considerable effort to develop the technology and methodology needed to identify and map within field management zones for site-specific farming. Much of the research has focused on the use of either a high-density geographically referenced grid of soil samples or mechanical yield sensor measurements that record geographic positions and production levels. In either case, complex spatial models are generally used to extrapolate the various soil variables and production level information across the entire field. Both procedures produce a wealth of information, however, the analysis of soil samples tends to be quite expensive and the accuracy of mechanical yield measurements does vary. This study represents an ongoing effort designed to evaluate remote sensing as a tool for determining within field management zones. Color-infrared aerial photography and multispectral videography were used in concert to map and stratify two grain sorghum fields into regions or zones of homogeneous spectral response. A limited number of soil and plant samples were acquired to characterize the biotic and edaphic conditions within each zone. Results obtained during the first year of the study indicated that multispectral video can be used to develop within field management zones. Simple univariate analysis indicated that pH, Ca, and Fe were important variables affecting yield. Analysis of the yield data indicated that the economic returns from 17 % of the first field and 20 % of the second field were insufficient to recoup planting costs. Multispectral video also proved instrumental in modeling the spatial variability of yield. A significant negative correlation ($r^2>0.90$) was obtained between the red spectral band and crop yields for both fields. Stratification, in this case using image data, reduces the number of samples required to characterize a field by reducing the variance associated within each stratum. Image data also provided a comprehensive view of each field that maintained the spatial connectivity between sites, thus reducing the need for complex spatial modeling.

INTRODUCTION

Traditional farming practices use the field as the minimum area of management. However, soil type, fertility, soil moisture, and other characteristics can vary substantially within fields and affect crop production (Pfeiffer et al., 1993). The development of geographic information system (GIS), global positioning system (GPS), and remote observation spatial analysis tools enable management of farm operations on a site-specific basis. Site-specific farming aims to improve production efficiency by adjusting crop treatments to conditions existing at specific areas within fields (Vansichen and Baerdemaeker, 1991). Optimization of agricultural inputs can improve economic returns and reduce the introduction of undesirable residues into the environment (Schueller, 1991).

Information concerning the spatial distribution of soil physical properties and nutrients, plant production, and crop yield are needed to develop a site-specific farming plan (Yang, 1994 and Bullock et al., 1994). Ground samples are important contributors to our understanding of the dynamics of a field; however, soil and plant samples are time consuming to acquire and expensive to analyze. In addition, the accuracy of maps developed from point data depend on mathematical models to emulate the spatial variation of the variables of interest. Spatial modeling requires intensive sampling efforts that increase in complexity as within field variation increases. Therefore, there is a need to develop more reliable and efficient methods of mapping within field spatial variation.

Remote sensing technology has been used for many years to identify, map, and estimate a wide variety of agricultural variables. The AGRISTARS and LACIE programs produced volumes of information concerning the use of remote sensing technology in agriculture. Recently, satellite and videographic data have been used to map the yield and salinity patterns in salt-affected cotton and sugarcane fields (Wiegand et al., 1992, 1994[a], and 1994[b]) and to estimate forage production on a semi-arid grassland (Anderson et al., 1993). Remote spectral observations can rapidly provide information about the condition of plants during the growing season. Plants integrate growing conditions and express their responses through the canopies achieved. Thus, vegetation indices calculated from remote spectral observations which measure the amount of photosynthetic tissue in the canopies are good predictors of yield (Wiegand and Richardson, 1990, and Wiegand et al., 1991, 1992, and 1994[a]).

Image data also provides a comprehensive view of an area which maintains the spatial connectivity between sites, thus reducing the need for complex spatial modeling. Management zones, developed by grouping picture elements (pixels) into categories of similar spectral response, should reduce the number of ground samples required to identify the cause of within field variation. The advantage of combining data into discrete strata is the reduced variance associated within each stratum and the reduced number of samples required to characterize each zone. Therefore, spatial information technology can improve the efficiency of data collection and contribute to the development of integrated crop management systems that optimize economic and environmental benefits.

A major contribution of this research will be in understanding the advantages and limitations of remote sensing as an information source for site-specific farming. Successful development of meaningful plant production zones from remote observations will reduce the amount of time and resources spent on ground sampling and, thus, improve the economic viability of site-specific management systems. Improved cost efficiency for investigating or implementing site-specific management systems will benefit both researchers and land managers at the local, state, and federal levels.

This paper presents the preliminary results of a multi-year study designed to evaluate: 1) The advantages and limitations of using remote sensing to map and monitor within field crop growth and yield variations for grain sorghum; 2) Relate within field crop growth and yield variance to the spatial distribution of physical and chemical properties of the soil, key plant biochemical levels, and competing undesirable plant species; and 3) Determine the suitability, reliability, and cost efficiency of using remote sensing devices and various ground sampling methodologies for identifying and mapping specific factors contributing to within field variation.

METHODS

Acquisition of Aerial Videography and Photography

Two irrigated grain sorghum fields with obvious variability in growth were selected from a study area which included fields owned and operated by Rio Farms, Inc., Monte Alto, Texas. Color infrared (CIR) digital video imagery was acquired for these two fields on three different dates: May 5, May 16, and June 14, 1995, using a multispectral digital video imaging system (Everitt et al., 1995). The video system simultaneously obtains images in the green (0.555-0.565 μm), red (0.625-0.635 μm) and near-infrared (NIR) (0.845-0.857 μm) bands of the electromagnetic spectrum. Color-infrared photographs in 23 cm × 23 cm (9 in. × 9 in.) format were also acquired on May 8, 1995 for use as base maps for registering the digital video images.

Image Registration and Classification

Registering acquired images to a real-world coordinate system allowed the images acquired at different times and other georeferenced data to be superimposed and analyzed. Ground control points were established in the universal transverse Mercator (UTM) coordinate system with the NAD-83 datum for each of the two fields using a Trimble SE-4000 series real-time differential GPS, which provided sub-meter accuracy. The control points were dispersed throughout each of the two fields and could be located precisely and easily on the ground and on the CIR photographs. Although the control points could be used to directly register digital video images, the acquired video images covered an insufficient number of identifiable control points to obtain the desired registration accuracy. Therefore, photographs of the two fields were digitized (scanned) and registered to the control points using GRASS software (GRASS, 1993). After preprocessing the acquired digital video images for misalignment, they were registered to the georeferenced photographs by identifying reference points easily seen on both images. Fig. 1 shows the black-and-white prints of the CIR digital video images of the two fields acquired on May 16. After the digital video images were registered, they were classified into several categories using unsupervised classification procedures in GRASS. The classified images were then filtered to remove the small inclusions of other classes within the dominant class while maintaining much of the original spatial pattern. Fig. 2 shows the classified and filtered images for the two fields.

(a) Field 1   (b) Field 2
Fig.1. Digital video images showing the variability in plant growth
of the two grain sorghum fields.

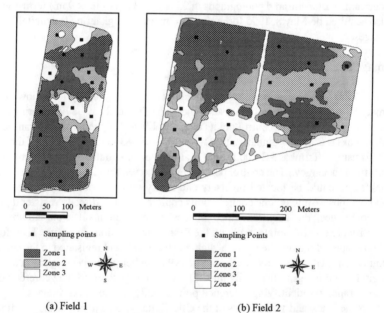

(a) Field 1   (b) Field 2
Fig.2. Classified grain sorghum spectral maps and sampling points
for the two fields.

The zone edges from each filtered image were extracted and then converted to vector format for inclusion in ARC/INFO (ESRI, Inc. 1995).

Determining Sampling Points

The location of sites for plant and soil samples was identified from the classified images of the two fields to determine plant growth differences among the zones and the sources of variation. Since only a limited number of different zones were mapped within the two fields, variability existed within each zone. Therefore, several sampling points were located within each zone as shown in Fig. 2. The total number of sampling points was 18 in field 1 and 19 in field 2, as determined using video images acquired on May 5 and May 16, 1995 for the respective fields.

Field Measurements

1) Plant sampling: Plant samples were collected from each of the two fields to determine plant height, leaf area index (LAI), and aboveground biomass. The GPS was used to pinpoint the exact location of the predetermined sampling points in the two fields. Quadrats measuring 1.93 m × 5.18 m and 2.03 m × 4.92 m were used to sample a 10-m^2 area at each sampling site for fields 1 and 2, respectively. The dimensions of quadrats differed for the two fields due to difference in planting row spacing. The number of plants within each quadrat was counted to calculate plant density. The aboveground portions of five representative plants within a quadrat were clipped for further analysis. Plant height measurements were made using a meter stick and plant leaf area was determined by measuring the total area of all the leaves of each individual plant using a leaf area meter. Plant dry weight was determined by drying the leaves and shredded stems for each plant at 68°C for four days and weighing them. LAI and biomass were then calculated from the measured plant dry weight, leaf area and density. Plant sampling was conducted from May 15 to 17, 1995 for field 1 and during June 8 and 9, 1995 for field 2. During the sampling periods, plants in field 1 were in the half-bloom stage, while those in field 2 were between half-bloom and soft-dough stages.

2) Leaf tissue sampling: Plant leaf tissues were collected from additional representative plants to determine plant nutrient levels. Five different locations within and around the 10 m^2 sampling area were selected and 5 to 7 flag leaves were clipped from each sublocation to obtain five leaf tissue samples at each sampling site. The leaf samples were analyzed to determine levels of nitrogen (N), phosphorus (P), potassium (K), sodium (Na), calcium (Ca), magnesium (Mg), zinc (Zn), iron (Fe), manganese (Mn), and copper (Cu).

3) Grain yield sampling: All the plant heads within a 2-m^2 area around each sampling point were clipped from the two fields shortly before harvest on June 22, 1995. Grain sorghum heads from each sampling area were threshed individually and the kernels from each head were bagged, dried at 68°C for four days, and then weighed to determine grain weight per head and calculate grain yield.

4) Soil sampling: Soil cores were taken on September 25, 1995, to a depth of 1.2 m (4 ft) from all the sampling points for the two fields using a hydraulic sampler. Each soil core was cut into four 30-cm segments and the top three segments were analyzed to determine soil texture, organic matter (OM), pH, electrical conductivity (EC), NO_3, P_2O_5, and water soluble and carbon dioxide extracts of K, Na, Ca and Mg. Levels of micronutrients Zn, Fe, Mn and Cu in the surface 30-cm segment were also determined.

Building GIS Database

Descriptive statistics, such as mean and standard deviation, for plant height, LAI, biomass, grain yield, leaf tissue nutrients, and soil nutrients were calculated for each sampling point of the two fields. The location coordinates for the sampling points and the associated mean and standard deviation values were properly formatted and converted to ARC/INFO point coverages. The GRASS video and classified images were also converted into ARC/INFO grid and polygon coverages. The sampling point coverage for each field was then overlaid on the respective management zone coverage for the field for analysis. ArcView was used to visualize and query these data, to generate statistics for each management zone, and to create charts and maps.

Extracting Spectral Data and Calculating NDVI

Digital count values for each 1-m^2 pixel in a 3 × 3 neighborhood surrounding the sampling points were extracted from each of the three digital video bands. The red and NIR digital video values were then transformed to NDVI for all pixels within each field:

$$NDVI = (NIR - Red)/(NIR + red) \tag{1}$$

Statistical Analysis

Multiple comparisons were made using the Fisher's least significant difference (LSD) procedure among the means of all plant growth, leaf nutrient and soil nutrient variables. Correlation matrices were calculated among plant growth variables, leaf nutrient variables, soil nutrient variables and spectral variables. Regression analysis was used to correlate grain yield to digital counts for each of the three bands and NDVI. All the statistical analyses were made using SAS software (SAS Institute Inc., 1988)

RESULTS

Comparisons of Plant Growth Variables among different Zones

Table 1 shows the means and standard deviations (SD) of plant height, LAI, biomass and yield among different management zones for fields 1 and 2, as well as the LSD statistical analysis results. All plant growth variables except LAI were

Table 1. Differences in Plant Growth Variables among Management Zones

Field	Zone	#	Area (ha)	Height Mean (cm)	SD	LAI Mean (m^2/m^2)	SD	Biomass Mean (kg/ha)	SD	Yield Mean (kg/ha)	SD
1	1	8	3.4	89[a]	14	3.5[a]	0.6	7414[a]	2120	4825[a]	777
	2	5	1.6	74[b]	9	3.2[a]	0.2	5195[b]	1486	2963[b]	1212
	3	5	1.0	48[c]	13	2.9[a]	0.4	2528[c]	611	40[c]	73
2	1	8	4.9	108[a]	4	2.6[a]	0.6	10030[a]	2028	4705[a]	400
	2	3	2.3	98[b]	9	1.6[b]	0.4	5301[b]	1778	2527[b]	557
	3	3	1.7	83[c]	4	1.7[b]	0.3	4574[b]	980	2539[b]	775
	4	5	2.2	65[d]	8	1.0[b]	0.4	1667[c]	424	619[c]	433

1. Values followed by the same letter within a column are not significantly different from one another at 0.05 probability level.
2. #=Number of samples.

significantly different among the three zones for field 1. The highest production level in the field was in zone 1 and lowest in zone 3. For field 2, plant height was significantly different among the four zones; LAI was significantly higher in zone 1 than in other zones; and biomass and yield were significantly different among all zones except between zones 2 and 3. Although an economic analysis of the data is not presented in this paper, it was determined that the return on the yield for zone 3 (field 1) and zone 4 (field 2) did not exceed the cost of planting. These two zones constituted 17% and 20% of the total area for fields 1 and 2, respectively. The low grain production in these two zones was mainly attributed to chlorotic conditions. In fact, most plants died as a result of chlorosis before reaching physiological maturity.

Comparisons of Plant Leaf Nutrient Levels among Zones

Table 2 shows the means of the ten plant leaf nutrients by zones for the two fields. For the first field, Ca was significantly different among the three zones with the highest level in zone 1; P, K and Na were significantly higher in zone 3 than in zones 1 and 2; Fe was lower in zone 3 than in zones 1 and 2, accounting for the chlorotic conditions of the plants in this zone; and N, Mg, Zn, Mn and Cu were not different among the three zones. For the second field, K level was significantly higher in zone 4 than in any other zone and Zn level was higher in zone 3 than the other zones; N, P, Mg and Fe were significantly different between at least two of the four zones; and Na, Ca, Mn and Cu were not significantly different among the four zones.

Comparisons of Soil Properties and Nutrients Among Zones

Table 3 shows the soil properties and nutrient levels in the upper 30-cm soil layer by zones for the two fields. For field 1, pH, as well as CO_2 extractable Ca and Mg were significantly higher in zone 3 than in zones 1 and 2; Fe, Mn and H_2O

soluble K were significantly lower in zone 3 than in zones 1 and 2; P, Cu, Zn, as well as CO_2 extractable K and H_2O soluble Ca were only significantly different between two of the three zones; and OM, EC and other nutrient levels were not significantly different among the three zones. For field 2, pH was significantly lower in zone 1 than in the other zones; Mn was significantly higher in zone 1 than in any other zone; OM, Zn, Fe, Cu, and H_2O soluble Ca and Mg were significantly different among some of the four zones; and EC and the remaining nutrient levels didn't differ significantly among the four zones.

Table 2. Plant Leaf Nutrient Levels by Management Zones

Field	Zone	N	P	K	Na	Ca	Mg	Zn	Fe	Mn	Cu
				(%)					(ppm)		
1	1	3.17[a]	0.35[b]	1.12[b]	0.046[b]	0.56[a]	0.28[a]	35[a]	124[a]	47[a]	22[a]
	2	3.18[a]	0.39[b]	1.25[b]	0.055[b]	0.37[b]	0.25[a]	40[a]	130[a]	53[a]	27[a]
	3	3.23[a]	0.52[a]	2.25[a]	0.084[a]	0.12[c]	0.25[a]	25[a]	86[b]	40[a]	24[a]
2	1	2.40[a]	0.23[a]	1.42[b]	0.062[a]	0.58[a]	0.27[b]	27[b]	111[ab]	59[a]	21[a]
	2	2.24[a]	0.16[b]	1.63[b]	0.077[a]	0.61[a]	0.32[ab]	26[b]	97[b]	57[a]	17[a]
	3	2.80[b]	0.21[ab]	1.43[b]	0.073[a]	0.68[a]	0.34[a]	53[a]	123[a]	97[a]	26[a]
	4	2.81[b]	0.27[a]	2.02[a]	0.050[a]	0.39[a]	0.26[b]	32[b]	98[b]	80[a]	13[a]

Values followed by the same letter within a column are not significantly different from one another at 0.05 probability level.

Table 3. Soil Properties And Nutrient Levels by Management Zones

Field	Zone	OM (%)	pH	EC (dS/m)	NO$_3$	P$_2$O$_5$	K H$_2$O	K CO$_2$	Na H$_2$O	Na CO$_2$	Ca H$_2$O	Ca CO$_2$	Mg H$_2$O	Mg CO$_2$	Zn	Fe	Mn	Cu
1	1	0.30[a]	7.04[b]	0.61[a]	8.7[a]	5.7[ab]	40[a]	54[ab]	84[a]	125[a]	26[b]	191[b]	12[a]	56[b]	0.23[b]	3.30[a]	3.78[a]	0.37[ab]
	2	0.45[a]	6.90[b]	0.79[a]	11.3[a]	13.6[a]	47[a]	83[a]	106[a]	156[a]	38[ab]	364[b]	14[a]	81[b]	0.39[a]	3.48[a]	4.19[a]	0.51[a]
	3	0.26[a]	8.06[a]	0.65[a]	8.6[a]	3.4[b]	16[b]	42[b]	80[a]	131[a]	49[a]	1976[a]	12[a]	145[a]	0.19[b]	1.93[b]	1.81[b]	0.33[b]
2	1	0.35[b]	7.35[b]	0.84[a]	11.7[a]	12.7[a]	35[a]	75[a]	106[a]	177[a]	41[b]	1034[a]	11[ab]	81[a]	0.26[a]	4.44[a]	5.46[a]	0.51[a]
	2	0.21[b]	8.00[a]	0.57[a]	9.3[a]	8.9[a]	26[a]	58[a]	103[a]	164[a]	37[b]	1956[a]	8[b]	96[a]	0.13[b]	2.11[a]	1.99[b]	0.29[b]
	3	0.31[ab]	7.77[a]	1.85[a]	14.3[a]	1.9[a]	23[a]	57[a]	283[a]	337[a]	48[ab]	1019[a]	12[ab]	90[a]	0.30[a]	5.69[a]	3.16[a]	0.41[ab]
	4	0.24[ab]	7.80[a]	1.67[a]	12.2[a]	2.4[a]	29[a]	58[a]	231[a]	271[a]	62[a]	1674[a]	15[a]	98[a]	0.25[a]	4.35[a]	3.43[a]	0.44[ab]

1. Values followed by the same letter within a column are not significantly different from one another at 0.05 probability level.
2. The unit for all the nutrient levels is ppm.

Table 4. Correlations Coefficients Between Plant Growth Variables and Spectral Variables

	Field 1				Field 2			
	Height	LAI	Biomass	Yield	Height	LAI	Biomass	Yield
NIR	0.37	0.34	0.36	0.34	0.60[*]	0.34	0.38	0.45
Red	-0.89[*]	-0.47	-0.79[*]	-0.93[*]	-0.87[*]	-0.83[*]	-0.92[*]	-0.95[*]
Green	-0.83[*]	-0.45	-.077[*]	-0.91[*]	-0.82[*]	-0.81[*]	-0.90[*]	-0.93[*]
NDVI	0.90[*]	0.53	0.83[*]	0.95[*]	0.93[*]	0.85[*]	0.92[*]	0.95[*]

Values followed by [*] are significant at 0.01 probability level.

Relationship between Plant Growth Variables and Spectral Variables

Table 4 summarizes the correlation coefficients between plant growth variables and the digital videographic spectral variables for the two fields. Among the four spectral variables, NDVI had the highest correlation coefficients with each of the four plant growth variables. Figure 3 shows the relations between yield and NDVI for the two fields. The two plots display obvious clusters, which separate the points among the zones except for zones 2 and 3 for field 2. The scatter plots between yield and the red band and those between yield and the green band displayed similar clusters, though the clusters on the plots between yield and the NIR band were not as distinct. NDVI was highly related to yield and could be also used to establish within-field management zones, but it was not ideal for estimating yield variations within zones due to the clusters. The scatter plots between yield and each of the spectral variables and the correlation matrix showed that the red band estimated yields better than any other variable. Figure 4 shows the relations between yield and the red band for both fields. A quadratic model and an exponential model were first fitted to the data for field 1, but both model forms provided erroneous estimates for yield in zone 3. Therefore, a segmented nonlinear model of the following form was used (SAS Institute Inc., 1988).

$$yield = \begin{cases} a + b \cdot red + c \cdot red^2, & if\ 0 < red \le red_0 \\ 0, & if\ red > red_0 \end{cases} \qquad (2)$$

To ensure the continuity and smoothness at the joint point, red_0, the two sections must meet at red_0 and the first derivatives with respect to red should be the same at red_0. These conditions imply that $red_0 = -b/(2c)$ and $a = b^2/(4c)$. Parameters b and c were then estimated by the nonlinear regression procedure and the model with its coefficients are given in Fig. 4 (a). A linear model was adequate to fit the data for field 2. However, when digital counts reached higher levels, the linear model

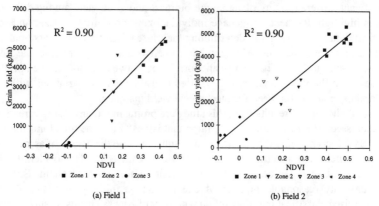

(a) Field 1   (b) Field 2

Fig. 3. Relations between grain yield and NDVI for the two fields.

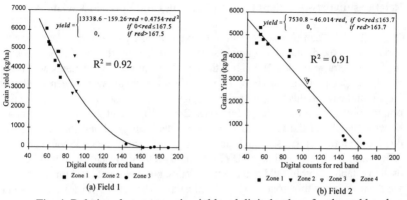

Fig. 4. Relations between grain yield and digital values for the red band
for the two fields.

produced negative yield estimates. Therefore, a segmented model with the coefficients shown in Fig. 4 (b) was used. The models of Fig. 4 can be used to generate grain yield maps for the two fields.

CONCLUSIONS

This study demonstrated that aerial digital videography is a useful tool for establishing within-field management zones for site-specific farming. The integration of aerial videography, GPS and GIS provides an effective way to collect, process and analyze spatial information. The delineation of management zones using unsupervised classification can reduce the number of ground samples required to determine spatial variability of nutrient levels across fields. The significant differences in plant height, biomass and grain yield among different zones highlight the need to manage individual zones differently. Leaf and soil nutrient data are important in determining the cause of the variability and making fertilizer recommendations. More work is needed to analyze and interpret the 1995 growing season data. The significant correlation between yield and the spectral values from the video imagery can be used to estimate yield variations within a field, thus reducing the need for mechanical yield mapping.

Although the results from this study are promising, many issues remain to be addressed. Below are three of the important issues. First, the time for acquiring imagery affects not only the classification results, but also the degree of association between yield and spectral observations. More work is needed to identify the optimum seasonal periods for image acquisition for grain sorghum. Second, although any reasonable number of classes can be initially requested in the unsupervised classifications, the optimum number of zones depends on such factors as the amount of crop variability and the physical size on the ground of practical management units. Third, additional research is needed to determine the stability of management zones based on remote imagery. The cost of data acquisition plus the frequency with which the procedures must be repeated will determine the appropriateness of using airborne imagery to establish within-field management zones.

ACKNOWLEDGMENTS

We thank David Escobar and Rene Davis for acquiring the videography and photography; Wayne Swanson and Americo Garza for ground data collection; Dale Murden, Mark Willis and Kenneth Berg for allowing us to use their fields; and Dr. Craig Wiegand, Steve Neck and Reginald Fletcher for assistance with field work. Thanks also go to James Everitt for his support for this study.

REFERENCES

Anderson, G.L., J.D. Hanson, and R.H. Haas. 1993. Evaluating landsat thematic mapper derived vegetation indices for estimating above ground biomass on semi-arid rangelands. Remote Sensing of Environment. 45:165-175.

Bullock, D.G., R.G. Hoeft, P. Dorman, T. Macy, and R. Olson. 1994. Nutrient management with intensive soil sampling and differential fertilizer spreading. Better Crops, Vol. 78, No. 4, p. 10-12.

ESRI, Inc. 1995. Understanding GIS, the ARC/INFO method. Environmental Systems Research Institute, Inc., Redlands, CA.

Everitt, J.H., D.E. Escobar, I. Cavazos, J.R. Noriega, and M.R. Davis. 1995. A three-camera multispectral digital video imaging system. In Proc. 15th Workshop Color Aerial Photography and Videography in the Plant Sciences, p. 244-252. Am. Soc. for Photogrammetry and Remote Sensing, Bethesda, MD.

GRASS. 1993. GRASS 4.1 User's reference manual. United States Army Corps of Engineers' Construction Engineering Research Laboratories, Champaign, IL.

Pfeiffer, D.W., J.W. Hummel, and N.R. Miller. 1993. Real-time corn yield sensor. ASAE, 1993 International Summer Meeting, Paper No. 931013, 25 p. St. Joseph, MI.

SAS Institue Inc. 1988. SAS/STAT User's Guide, Release 6.03 edition, p. 699-702. SAS Institute Inc., Cary, NC.

Schueller, J.K. 1991. In-field site-specific crop production. In: Automated Agriculture for the 21st Century, Proceedings of the 1991 Symposium, p. 291-292. ASAE Publication No. 1191, American Society of Agricultural Engineers, St. Joseph, MI.

Vansichen R., and J. De Baerdemaeker. 1991. Continuous wheat yield mesaurement on a combine. p. 346-355. In: Automated Agriculture for the 21st Century, Proceedings of the 1991 Symposium, ASAE Publication No. 1191, American Society of Agricultural Engineers, St. Joseph, MI.

Wiegand, C.L., and A. J. Richardson. 1990. Use of spectral vegetation indices to infer leaf area, evapotranspiration, and yield: I. Rational. Agron. J. 82:623-629.

Wiegand, C.L., J.H. Everitt, and A.J. Richardson. 1992. Comparison of multispectral video and SPOT-1 HRV observations for cotton affected by soil salinity. Int. J. Remote Sensing, 13:1511-1525.

Wiegand, C.L., J.D. Rhoades, D.E. Escobar, and J.H. Everitt. 1994[a]. Photographic and videographic observations for determining and mapping the response of cotton to soil salinity. Remote Sensing of Environment, 49:212-223.

Wiegand, C.L., D.E. Escobar, and S.E. Lingle. 1994[b]. Detecting growth variation and salt stress in sugarcane using videography. Proc. 14th Workshop on Aerial Photography and Videography in the Plant Sciences. Amer. Soc. Photogram. and Remote Sens.

Yang, C. 1994. Measurement and analysis of field spatial variability for site-specific crop management. Ph.D. Dissertation, University of Idaho, Moscow, ID.

Variability in Volume Metering Devices

L.L. Bashford
A. Bahri
K. Von Bargen
M.F. Kocher

Department of Biological Systems Engineering
University of Nebraska-Lincoln
Lincoln, NE

ABSTRACT

The inherent variability of seed and fertilizer application from volumetric metering devices is not readily recognized. The Canadian Prairie Agricultural Machinery Institute (PAMI) suggests a maximum coefficient of variation (CV) of 15% among outlets for seeding grain or applying fertilizer. PAMI does not report down-the-row variability of individual outlets. Parameters that influence variability of volumetric measuring external fluted wheels such as rotational speed of the metering wheel, product delivery rate, seed size, and cell collection lengths were examined.

   In the first study, external fluted wheel meters on four grain drills were tested for seed delivery variability for wheat and soybeans, both among the metering outlets and down-the-row for individual meters. Tests on two additional drills, one an air drill and the other with external fluted metering, used two sizes of soybean seeds and two travel speeds. For wheat, down-the-row CV ranged from 12.5 to 22.5% and the CV among metering units ranged from 12.5 to 21%. For soybeans, the CV ranged from 15.5 to 41.5% with the air drill having the lower CV. A faster travel speed gave a lower CV for both drills metering soybeans.

   In a second study, when metering wheat, the seeding rate variability due to cell size and seeding rate were evaluated. Each meter was evaluated with cells 0.48 or 0.96 m in length and seeding rates of 60, 80, 90, and 100 kg/ha. The down-the-row CV ranged from 10 to 28% with 0.48 m length cells, and from 4 to 22% with 0.96 m length cells. Some of these CVs may be too high for a metering mechanism such as the fluted wheel to be used in SSCM.

INTRODUCTION

   Site specific crop management (SSCM) is a technique used in precision agriculture when a product is applied to a field based on requirements of subunits of the field instead of an average over the whole field. The utilization of application technology to site specific farming requires knowledge about the application variations inherent in metering mechanisms.

   The Canadian Prairie Agricultural Machine Institute (PAMI) has published a number of reports on their evaluation of seeders and the coefficient of variation (CV) as related to the delivery rate of the drill. PAMI accepts 15% as the maximum acceptable CV in the seeding rate for their evaluation of seeders. The significance of the variation in metering can be stated very simply. As specified in ASAE S341.2 DEC92, a 15% CV means that at a setting of 100 kg/ha (89 lb/acre), the actual application rate would be expected to range between 85 and 115 kg/ha (76 to 102 lb/acre) on 68% of the area. This range of CV may not be acceptable when integrated into a SSCM program.

The delivery devices for precision planters have been modified over the years to eliminate the problems encountered with seed size variation. Air metering devices are independent of seed size and more adaptable to utilization of irregular shape and size seed.

The precision for drills will be more difficult to define as drills do not singulate seed but deliver a volume of seed. There have been some changes and developments in drill metering devices. However, the most common types of metering devices still use external flutes and opening gates to control the application rate.

The objective of this research effort was to evaluate the variation in the metering capabilities of external flute metering mechanisms on drills. Specifically, data were obtained to evaluate performance criteria as follows:
1. the ability of the drill to deliver the same volume of seed from each metering unit, and
2. the ability of each metering outlet to deliver a consistent number of seeds per unit row length (down-the-row).

METHODS AND PROCEDURES

There are no specific standards that define individual metering outlet evaluations on drills. In general, calibration of drills is based on the seeds delivered from all outlets. For precision planters, individual seeds are captured on an artificial adhesive surface. The seed spacing can then be easily measured. However, an adhesive surface is impractical for evaluation of volumetric meters.

Study 1

Bashford (1993) discussed a system consisting of a moving table with six rows of 15 catch boxes, or cells used to catch seed from individual metering units. The catch boxes were 102-mm by 102-mm (4-in by 4-in) and mounted on the moving table adjacent to each other in six rows, Figure 1. The seeds metered from each of six outlets were collected in the catch boxes. The catch boxes were attached to a chain driven by an electric motor through a variable speed gear box. This permitted the catch boxes to be pulled underneath the outlets to capture seed. A table speed of approximately 1.6 km/h (1 mile/h) was used for all tests. To eliminate any variation caused by the seed drop tubes, they were removed prior to the tests. Therefore, the moving table was set up immediately underneath the metering mechanisms.

The drill was driven by a treadmill. The operating procedure was to start the treadmill turning the drill wheels, and then start the moving table. As the boxes passed beneath the metering units, the seeds were caught in the catch boxes. After

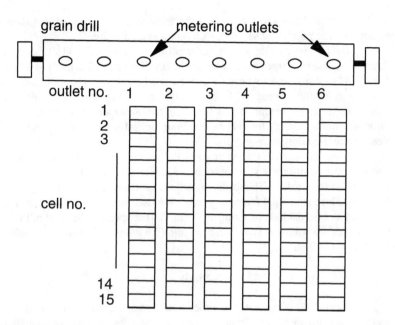

Figure 1. Stationary drill driven by a treadmill with the moving table of 15 rows of boxes (cells).

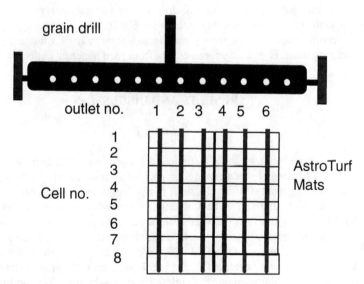

Figure 2. Drill pulled over two rows of eight mats (cells).

all boxes passed beneath the outlets, the table was first stopped and then the treadmill. The number of seeds in each box was then counted.

The first series of tests were accomplished on four drills, identified as 1, 2, 3, and 4. These tests were run at a simulated travel speed of 7.8 km/h (4.8 mile/h) for all drills. Wheat and soybean seeds were used for the tests at application rates of 67 kg/ha (60 lb/acre) and 117 kg/ha (105 lb/acre), respectively. The drill settings closest to the desired seeding rate as printed in the owner's manuals were used. No effort was made to verify the calibration. Four replications of each test were made. Seeds were used only once and discarded.

A second series of tests were run with two additional drills, identified as 5 and 6, and two different size soybeans. Jacques 201 was the larger bean and Jacques 245 the smaller. The seeding rate used for these tests was 112 kg/ha (100 lbs/acre). Three simulated forward speeds were used on an air metering drill, drill 5, and on an external flute metering drill, drill 6. The speed of the treadmill was set to simulate forward travel speeds of 5.8 km/h (3.6 mile/h), 7.0 km/h (4.3 mile/h), or 9.6 km/h (5.9 mile/h).

Study 2

Bahri (1995) presented a procedure developed to determine the seed metering performance for different rate settings and different collection length cell sizes. A single drill with external flutes was run outdoors over AstroTurf doormats. These mats, or cells, were 0.48 x 0.74 m, with 'hair' about 16 mm high. The mats were oriented with the 0.48 m side in the direction of travel. The mats were placed in line on a paved parking lot. The drill was then operated over the test course at 6.4 km/h (4 mile/h). Seeding rates were 60, 80, 90, and 100 kg/ha. Sixteen collection mats, or cells, were placed in two rows to collect the seeds from six metering units, Figure 2, for seed counts from each metering outlet.

RESULTS AND DISCUSSION

Study 1

Down-the-row variability for four drills with external flute meters was based upon the number of seeds captured in each of 15 boxes (cells) per row, Table 1. Each cell represents 0.49 m (19.2 in.) of row length at the simulated travel speed of 7.8 km/h (4.8 mile/h). The most uniform down-the-row delivery was achieved by drill 3 with an average CV of 13.73% and a range from 12.55% to 14.88%. Among outlet variability for six outlets based upon the seed delivered to each box varied from an average CV of 14.96% for drill 3 to 19.06% for drill 4, Table 2. Drill 3 had the lowest average CV of 14.96% while drill 4 had the lowest CV range of 3.75%. Combining among the outlets and down-the-row observations resulted in overall CVs of 18.76, 19.06, 15.20, and 19.05% for drills 1, 2, 3, and 4, respectively.

Illustrated in Tables 3 and 4 are the results for metering soybeans. Drill 1 had the lowest average down-the-row CV, 17.20%, and the smallest CV range,

Table 1. Coefficients of variation for down-the-row variability at each of six outlets on four different drills metering **wheat** seed at 67 kg/ha.

Drill	Outlet Number						Average
	1	2	3	4	5	6	
1	19.47	19.79	18.47	18.07	16.56	18.67	18.51
2	17.25	19.05	18.21	22.50	18.53	18.05	19.93
3	13.18	13.21	14.59	12.55	14.88	13.99	13.73
4	17.95	21.56	17.85	19.99	16.99	19.76	19.02

Table 2. Fifteen observations of the coefficient of variability among the outlets of each of four different drills metering **wheat** seed at 67 kg/ha.

Cell Number	Drill			
	1	2	3	4
1	21.11	19.65	14.12	18.49
2	19.72	16.64	12.51	19.17
3	17.93	21.69	16.45	20.15
4	18.69	20.41	17.42	19.51
5	20.31	19.86	16.00	17.76
6	17.76	18.86	15.16	21.26
7	16.94	15.66	14.37	18.26
8	18.81	20.38	14.42	17.93
9	15.33	16.50	15.41	17.02
10	19.50	18.67	14.81	19.45
11	20.15	18.05	14.71	20.06
12	18.26	14.02	13.44	17.51
13	19.04	19.51	15.51	18.78
14	18.15	19.34	15.20	19.53
15	16.07	20.15	14.89	21.04
Average	18.52	18.63	14.96	19.06

Table 3. Coefficients of variation for down-the-row variability at each of six outlets on four different drills metering **soybean** seed at 117 kg/ha.

Drill	Outlet Number						Average
	1	2	3	4	5	6	
1	17.48	18.34	18.29	17.10	16.51	15.53	17.20
2	26.51	22.71	22.05	25.19	22.38	22.06	23.48
3	29.73	24.01	20.39	23.28	25.45	22.43	24.22
4	41.59	37.78	40.22	40.21	36.16	37.88	38.97

Table 4. Fifteen observations of the coefficient of variability among the outlets of each of four different drills metering **soybean** seed at 117 kg/ha.

Cell Number	Drill			
	1	2	3	4
1	18.18	21.45	22.36	37.95
2	13.98	20.11	30.34	43.92
3	18.83	23.49	18.46	30.26
4	18.41	26.82	27.10	39.61
5	19.20	24.24	27.07	35.87
6	17.89	25.37	22.77	36.16
7	20.07	22.69	24.72	37.51
8	18.11	22.84	26.65	41.42
9	15.56	23.07	29.53	36.73
10	16.96	23.49	31.03	39.84
11	14.51	22.96	18.07	43.20
12	17.61	16.43	21.22	47.77
13	16.91	27.02	20.98	38.93
14	17.46	23.81	24.18	32.46
15	15.49	26.25	28.00	37.33
Average	17.28	23.34	24.83	38.60

2.81%, Table 3. Drill 3 had the largest down-the-row CV range of 9.34%. Drill 4 had the largest average down-the-row CV of 38.97%. Drill 1 maintained approximately the same average down-the-row CV for wheat, 18.51%, as for soybeans, 17.20%. The average down-the-row CV change from metering wheat to metering soybeans for drill 2 was 19.93 to 23.48%, for drill 3 was 13.73 to 24.22%, and for drill 4 was 19.02 to 38.97%, respectively. Therefore, seed characteristics do have a profound effect on metering capability of some drills. Combining among outlets and down-the-row observations resulted in overall CVs of 17.60, 23.58, 25.24, and 38.96% for drills 1, 2, 3, and 4, respectively. Drill 1 had a lower overall CV for soybeans than for wheat. The average down-the-row CVs, Table 4, illustrate the same response as Table 3 for each of the four drills, with drill 1 having the lower CV and drill 4 the higher CV.

When metering two different sizes of soybean seeds, the down-the-row CV for drill 6 was 27.16% for the larger seed and 27.70% for the smaller seed, Table 5. Because the simulated travel speed was similar to that used for the drills in Table 3, the results are comparable to the drills in Table 3. Drill 5, the air drill, had much lower CVs of 11.21% and 12.23%, than drills 1 through 4 and 6. The CV ranges for drills 5 and 6 were narrower than for drills 1 to 4 as tested with soybeans. The CVs across all bean sizes and observations were 15.93 and 28.62% for drills 5 and 6, respectively.

Illustrated in Table 6 are the CVs for the two drills, each operating at two forward speeds and metering two sizes of soybean seed. Simulated down-the-row cell lengths were approximately 0.37 m (14.4 in.), 0.44 m (17.2 in.), and 0.60 m (23.6 in.) for simulated forward travel speeds of 5.8 km/h (3.6 mile/h), 7.0 km/h (4.3 mile/h), and 9.6 km/h (5.9 mile/h), respectfully. The CVs are based on observations across all outlets and cells. Bean size seemed to have little significance on the variability in metering of both drills. The CV for the two drills had a tendency to decrease as the simulated travel speed increased.

Study 2

Among outlet variability for six outlets on a drill observed over eight cells 0.48 m in length and four different seeding rates is illustrated in Table 7. Illustrated in Table 8 are the CVs for down-the-row variability for each of six outlets on a drill as influenced by four different seeding rates using eight cells 0.48 m in length. Illustrated in Tables 9 and 10 are results similar to Tables 7 and 8, respectively, except the cell lengths were 0.96 m in length. It can be observed that the longer cell lengths result in lower among outlet and down-the-row variability. The overall CV was 27.82% for the 0.48 m cell length, 23.76% for the 0.96 m cell length.

It is obvious that the CV for seed delivery was less than PAMI's recommended maximum of 15% in very few instances. In the context of SSCM and the desire to accurately meter seeds, this type of metering mechanism may have too much inherent variability to obtain a CV less than 15% for small cells, while on a larger scale, CVs may be less. The CV for volume metering units may be in excess of the range of product application rate required. If the desired application rate is 100 units/area using a meter with a CV of 15%, the application rate will range from 85 to 115 units/acre over 68% of the area and will exceed this range

Table 5. Coefficients of variation for down-the-row variability at each of six outlets for two different drills metering two different sizes of **soybean** seed at 7.0 km/h.

Bean Size	Drill	Outlet Number						Average
		1	2	3	4	5	6	
Larger	5	9.20	10.78	13.26	11.22	13.09	9.69	11.21
	6	26.42	31.94	24.52	25.10	28.25	26.71	27.16
Smaller	5	10.47	11.99	13.62	12.96	13.31	11.01	12.23
	6	26.73	28.53	33.29	29.34	25.20	23.12	27.70

Table 6. Coefficients of variation for two drills, each operating at two different forward speeds and metering two different sizes of **soybean** seed. The CV for each drill represents the combination of the observations among outlets and down-the-row variations.

Bean Size	Drill	Simulated Forward Speed, km/h		
		5.8	7.0	9.6
Larger	5	13.20	11.30	
	6		27.30	23.40
Smaller	5	13.20	12.40	
	6		28.20	21.70

Table 7. Coefficients of variation among six outlets of a single drill with eight 0.48 m cell lengths and four different seeding rates metering **wheat**.

Mat Number	Seeding Rate, kg/ha			
	60	80	90	100
1	22.80	15.75	35.24	14.63
2	24.82	28.62	15.08	13.17
3	15.13	26.88	12.95	24.01
4	20.52	18.11	19.50	11.04
5	20.12	20.76	8.18	11.23
6	14.63	7.71	27.06	21.71
7	29.68	26.02	21.98	17.31
8	21.45	12.60	10.49	22.18
Average	21.14	19.55	18.81	16.91

Table 8. Down-the-row coefficients of variation for each of six outlets on a single drill as influenced by four different seeding rates using eight 0.48 m cell lengths metering **wheat**.

Outlet Number	Seeding Rate, kg/ha			
	60	80	90	100
1	23.77	10.57	21.70	15.72
2	17.54	18.62	12.41	22.47
3	22.72	22.96	14.74	17.08
4	24.97	22.05	12.75	8.11
5	23.42	15.04	18.09	13.71
6	30.93	28.99	21.00	21.77
Average	23.89	19.70	16.78	16.48

Table 9. Coefficients of variation among six outlets of a single drill with four 0.96 m cell lengths and four different seeding rates metering **wheat**.

Mat Number	Seeding Rate, kg/ha			
	60	80	90	100
1+ 2	14.21	17.72	22.44	10.89
3+4	10.86	19.63	9.13	14.46
5+6	13.10	16.74	13.93	15.14
7+8	23.06	15.30	13.15	15.31
Average	15.30	17.35	14.66	13.95

Table 10. Down-the-row coefficients of variation for each of six outlets on a single drill as influenced by four different seeding rates using four 0.96 m cell lengths metering **wheat**.

Outlet Number	Seeding Rate, kg/ha			
	60	80	90	100
1	21.34	8.80	5.93	10.82
2	7.32	12.39	3.69	18.45
3	14.49	22.63	13.33	7.47
4	19.53	15.53	5.46	4.55
5	17.95	14.36	14.06	13.41
6	8.00	13.59	6.46	15.92
Average	14.77	14.55	8.15	11.77

over 32% of the area. If the desired application rate is changed to 90 units/area, the application rate will range from 76 to 103 units/area over 68% of the area Therefore, it is likely that the original and new application rates will not be different.

A drill with a 15% CV will not provide reasonable precision in product application for development of SSCM recommendations. Application rate recommendations for SSCM obtained from studies involving equipment with these magnitudes of CV would likely be erroneous. There is a real need to establish an acceptable CV and develop metering mechanisms that can meter with acceptable uniformity. Obviously, the CV cannot be zero, but the influence of variability in any product delivery system must be identified and factored into the development of SSCM recommendations and decision support systems.

CONCLUSIONS

Not all outlets on a drill with volumetric external fluted metering units will deliver a consistent volume of seed among outlets or down-the-row. Delivery differences were observed when metering wheat and soybeans. The outlet CV's for wheat ranged from 13.18% to 22.50% and for soybeans from 15.53% to 41.59% and suggests that external fluted metering devices are not adequate for SSCM. How variability in seed or fertilizer application affects yield variability or other evaluation criteria are not known. This type of variability in a SSCM protocol cannot continue to be ignored.

REFERENCES

ASAE S341.2 DEC92. Procedure for measuring distribution uniformity and calibrating granular broadcast spreaders. ASAE Standards, 1993. St. Joseph, MI 49085.

Bahri, A. 1995. Modulating wheat seeding rate for site specific crop management. Unpublished Ph.D. thesis. University of Nebraska-Lincoln, 157p.

Bashford, L.L. 1993. External flute seed metering evaluation related to site specific farming. ASAE Paper No. 93-8517. ASAE, St. Joseph, MI.

Mass Flow Measurement with a Curved Plate at the Exit of an Elevator

G.J. Strubbe
B. Missotten
J. De Baerdemaeker

Agricultural Engineering Department
Katholieke Universiteit
Leuven, Belgium.

ABSTRACT

The grain flow at the exit of an elevator pulsates and disperses over a considerable wide angle. A smoothed and concentrated flow was achieved by an appropriate guidance fixture. The guidance by a curved chute allows parallel streamlines if the flow at the entrance is tangential to it. It causes also a centrifugal force that restricts inclination influences. Guidance, however, causes friction losses.

The force executed by a grain flow on a curved chute can be used as a measure for the mass flow rate. It is possible to make a measurement independent of influence of friction by the right choice of the direction in which the force is measured. A combination of acceptable guidance and friction compensation permitted a mass flow measurement independent of grain properties.

INTRODUCTION

Harvesting circumstances are characterised by continuous changing crop properties and changing field conditions that influence the behavior of the threshing and separating process in a combine. Typical influences are changes in kernel/straw ratio, weed infestation, windrowed crop and variations in maturity and moisture content. Most yield variations go together with these changing crop conditions. This results mostly in both a fluctuation of the grain flow rate in the combine and a fluctuation of the grain properties and the quality of the grain sample to the graintank.

Currently, most mass flow measuring devices on combines are installed at the exit of the clean grain elevator (Vansichen & De Baerdemaeker, 1991, 1992; Yang,1992; Stafford et al.,1993; J Deere,1995). Although this might be the best place to measure mass flow, the local granular flow pattern does show problems. Slope variations and composition and properties of the grain can easily influence the flow characteristics and thus the mass flow reading.

Reliable yield mapping starts with a reliable grain flow monitor. Many manufacturers of grain flow meters claim excellent accuracy (1 to 2%) if the total surface is large. The accuracy of the yield determination, however, drops considerably when looking to the field at small grids (Auernhammer et al.,1993).

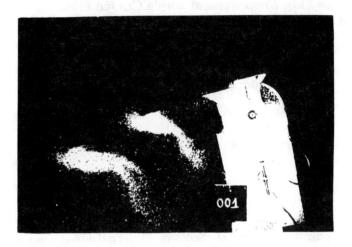

Fig. 1. Flow stream at the exit of an elevator.

grids are considerably variable in crop condition, field condition, in grain yield and in straw yield. A first step towards differentiation between these influences is to measure the grain flow rate on a combine independently of fluctuations in grain properties and independently of changes in the grain sample.

Flow characteristics at the elevator exit

Grain elevators on combines transport packets of grain on the paddles. Rotation at the upper elevator shaft move these packets of grain completely over to the paddle outside tip, where they get the momentum for the throwing action at the elevator exit. A typical free flow pattern at the elevator exit is shown on Fig. 1. Characteristic for this flow pattern is its pulsation that is synchronised with the paddle frequency and also the dispersion of the kernels. This dispersion is dependent on several construction parameters, granular characteristics, and also dependent on the elevator slope.

At the end of the rotational movement, most kernels reach a stable position on the paddle before they leave the elevator.They are concentrated in a dense triangle that is pushed by the paddle and that leaves the elevator tangentially to the exit. This dense triangle deforms slightly during its flight. It shows a very small amount of kernels at the top that go higher than the tangent to the exit. These kernels still have a radial velocity at the front face of the grain packet before they leave the paddle.

At the back side of the triangle is a considerable number of kernels that form a tail. The flow pattern is directly related to the position of the kernels on the paddle before they leave. There is an influence of the volume of kernels and there is the limitation of free movement on a paddle. The movement of kernels on a rotating paddle also depends on the shape and dimensions of the paddle, the

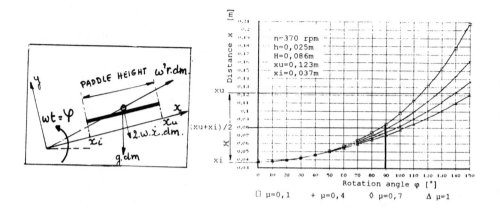

Fig. 2a: Forces on a free mass point.
on a rotating elevator paddle.

Fig. 2b: Displacement of a mass point
as a function of the angle of rotation

elevator inclination, the rotation angle and the friction coefficient. An equation of motion was set up for a straight paddle with design parameters defined in Fig. 2a. The relation between the displacement, x, of a free mass on the paddle and the angle of rotation, ω, is shown in Fig. 2b. The curves are given for different values of the friction coefficient. A kernel that starts at the inner paddle tip x_i, reaches the outer paddle tip x_u after a rotation angle of 110° to 150°. Most elevators have a rotation angle ωm of 90°. A kernel that starts the rotation at x_i reaches only half of the paddle before it is at the exit. Kernels that start at x=50mm reach the outer paddle tip after a rotation angle of about 90°. If the kernels are evenly divided on the paddle before the rotation starts and if they can move freely, then at least 15% to 20% of the kernels take part in the tail. These kernels follow a path at the elevator exit that has a considerably downward component.

A curved plate for mass flow detection, assembled at the elevator exit, will not guide all the kernels. An uncontrolled amount can pass under the detector without any contact with the sensor plate. This would not be a problem if the distribution in the flow would remain constant. This distribution, however, is dependent on the friction, as shown in Fig. 2b. It is also dependent on the inclination and on the flow rate. Additional guidance is required to have a better controlled and concentrated flow pattern at the elevator exit if the force on the chute is to be a good measure for the flow rate.

Flow concentration

Poor concentration of the flow at the entrance of the curved chute results in a lower bulk velocity in the chute due to kernels that makes contact only at a distance further into the chute. This concentration is therefore of mayor importance if any reliable measurement is required. Better concentration of the flow can be reached with several adaptions of the construction, but not all adaptions are effective. The installation of a higher rpm only marginally helps, because the relation between displacement x and the rotation angle ω is independent of the angular velocity. A greater rotation angle can be considered, but the exit of the chute should remain sufficiently inclined to the horizontal. The position of the elevator to the vertical limits therefore the maximal allowable rotation angle ωm. Also the free space at the chute exit should remain great enough. For a chute of 90°, the maximal rotation angle in the elevator is about 110°.

The following design changes were effective:

1.   Curved elevator paddles.
The replacement of straight elevator paddles with curved paddles, as shown in Fig. 3, avoids that kernels could be located at the under side of the straight paddle. Also, they conduct the packets of grain better to the outside of the paddle. These curved paddles give less return flow in the elevator.

2.   Use of deflector plates.
Straight or curved deflector plates could be installed at the upper and at the lower side of the elevator exit to concentrate the flow. The upper deflector plate was effective and slightly deviated the flow downwards. A deflector at the under side also did concentrate the flow. This deflector, however, is very sensitive to build-up of material and mud, since its position is too near to horizontal.

3.   Non-tangential position of the deflector.
The upper deflector plate was assembled with an offset to the tangential to allow part of the grain to leave the paddle before contact with the deflector starts. This allowed the flow to be smooth and become less pulsating at the exit of the deflector. The same function with less smoothing can be achieved with a gap between the paddles and the upper elevator housing plate. The gap gradually increases with the rotation angle.

4.   An impeller rotor.
A rotor of 150mm diameter was installed at the lower side of the elevator exit. The tip velocity of the rotor is slightly higher than the paddle tip velocity. The drive was taken from the upper elevator shaft. The rotor avoided the tail to move downwards and considerably improved the concentration of the flow at the exit of the elevator.
        A combination of mentioned changes considerably improved the concentration of the flow, as shown in Fig. 4.

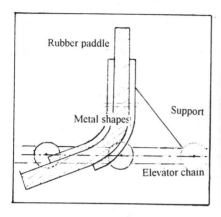

Fig. 3: Curved elevator paddles

Fig. 4: Bulk flow at the elevator exit after design changes.

Friction effects

Guidance of the flow results in higher friction losses. This can be seen from the glumy look of the inner surface of the chute.

The velocity in a circular chute was calculated to estimate the friction losses.

Consider a mass point that moves with a velocity v in the chute, as shown in Fig. 5. This mass executes a force N normal to the chute bottom:

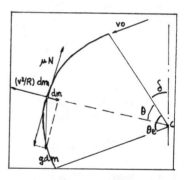

Fig. 5: Forces on the mass flow in a circular chute.

$$N = [(v^2/R) - g.\cos(\delta+\theta)] \qquad (1)$$

The change in velocity dv/dt is the difference between the tangential component of the gravity force and the friction force $\mu.N$. Since $v.dt = R.d\theta$, the change in velocity becomes:

$$d(v^2) = -2\mu v^2 + 2.g.R.[\sin(\delta+\theta) + \mu.\cos(\delta+\theta)].d\theta \qquad (2)$$

Three installation characteristics are important: $Uo = vo^2/(g.R)$, the inclination angle δ and the total chute angle θe. With $U = v^2/(gR)$ can (2) be written as a simple differential equation:

$$dU = -2.\mu.U + 2.[\sin(\delta+\theta)] + \mu.\cos(\delta+\theta)] \qquad (2A)$$

The velocity can be obtained as a function of the chute angle θ after integration. In most installations conditions the entrance velocity vo is high so that the second part in g can be neglected. The exit velocity then becomes then an exponential function of the friction coefficient:

$$ve/vo = \exp(-\mu.\theta e) \tag{3}$$

The exit velocity strongly depends on the friction, even if the entrance velocity is very high. Also the bulk velocity in a chute changes considerably with a rather small change in friction coefficient as shown on Fig. 6.

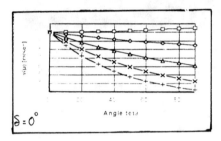
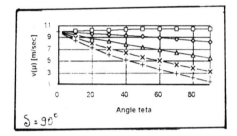

μ=0   ◇ μ=0,15   Δ μ=0,4   × μ=0,7   + μ=1

Fig. 6: Behavior of the granular velocity v(μ) in a circular chute.
vo=10m/s; Chute radius R=0,5m.

The force on a chute

The momentum balance on the chute can be derived from Fig. 7. The force measured in a direction α is:

$$F = Q.vo.\sin\alpha + Q.ve.\sin(\theta e-\alpha) - G.\cos(\delta+\alpha) \tag{4}$$

G is the weight of grain in the chute and ve is the exit velocity. Both values are strongly dependent on friction. The influence of friction can be eliminated if the design is so that:

$$\theta e = \alpha \tag{5}$$
and
$$\delta+\alpha = 90° \tag{6}$$

The direction to measure the force for elimination of friction is the perpendicular to the tangent to the exit. This installation condition is shown in Fig. 8. Condition (6) requires that the force should be measured horizontally. This method for elimination of friction is valid for any continuously curved shape of a chute. The force becomes then:

$$F = Q.vo.\cos\delta \tag{7}$$

If the force is measured in the horizontal direction, then the influence of gravity is eliminated. This avoids any shift in zero reading due to dirt that sticks to the chute. The horizontal force, however, is dependent on the inclination angle. Also, (5) and (6) considerably restrict the choice of θe and δ. The installation conditions (5) and (6) are therefore only advised for stationary installations where δ can be fixed, and on mobile installations where the effect of the inclination angle is compensated by an inclination sensor and corresponding algorithm for correction of $\cos\delta$.

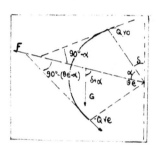

Fig. 7: Forces of Impulse momentum.   Fig. 8: Exact elimination of influence
                                                              of friction.

The direction to measure the force with optimal compensation of friction under other conditions can be derived from numerical experiments with the force as function of the friction coeffcient.

Quantification of influence of friction

As important as the optimal direction to measure the force is the estimation of how much friction influence is still left when using this optimal direction. To quantify the influence of friction $\mu = 0.4$ used as a reference value. This value is frequently reported as friction coefficient of dry grain on steel. F(0.4) is considered a reference value for F. Common values of μ are considered in the range between 0.15 and 0.7. A measure for the influence of friction is defined as:

$$\Delta F(0,15) = \left| \frac{F(0.15) - F(0.4)}{F(0.4)} \cdot \frac{10}{0.15 - 0.4} \right| \% \tag{9}$$

$$\Delta F(0,7) = \left| \frac{F(0.7) - F(0.4)}{F(0.4)} \cdot \frac{10}{0.7 - 0.4} \right| \% \tag{10}$$

These dimensionless measures indicate the percentage of change of F per change of .1 in the value of μ. The measures are given in the left columns of Table 1 for optimal directions α_{opt} to measure the force and for a freely chosen direction $\alpha = 45°$. Results show that the influence of friction on F is reduced to an acceptable level simply by the right choice of the direction in which this force is measured.

Table 1: The effect of influence of friction on the force for practical values of the installation conditions.

Uo $(=v_o^2/gR)$	δ (°)	θe (°)	αopt (°)	EFFECTS for αopt		EFFECTS for $\alpha=45°$	
				$\Delta F(0,15)$ (%)	$\Delta F(0,7)$ (%)	$\Delta F(0,15)$ (%)	$\Delta F(0,7)$ (%)
0.20	90	55	37	0.37	0.20	0.92	1.59
1,02	90	57	43	0.28	0.37	0.04	6.67
	0	90	90	0.00	0.00	8.72	6.96
5,1	90	75	64	0.34	0.48	3.32	1.98
	0	90	90	0.00	0.00	7.28	5.07
12,74	90	81	74	0.31	0.28	4.75	3.08
	0	90	90	0.00	0.00	6.95	4.68
20.39	90	83	79	0.08	0.32	6.18	3.95
	0	90	90	0.00	0.00	6.86	4.57

Measurement results

A test rig was built for continuous metering of a granular material up to 25 T/H. A combination of auger and elevator on an 8055 New-Holland combine was installed and equiped with the adaptions to improve the flow at the entrance to the sensor. The force was measured in the direction to optimally compensate for influence of friction.

Four types of kernels were used, of which the properties are summarised in Table 2.

Table 2: Kernel types and properties.

Kernel type	dimensions (mm)	Specific weight (kg/m³)	Friction coefficient
Polyester	5*3*6	710	0,26
Wheat, 11% moisture	6*3*4	760	0,46
Wheat, 21% moisture	6*3*4	670	0,62
Corn	11*6*5	710	0,53

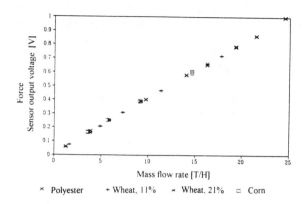

Fig. 9. Measurement results on a grain elevator.

The measurement results given in Fig. 9 confirm that the influence of grain properties was eliminated. Friction influences can be effectively eliminated to an acceptable level in all installation circumstances where the velocity in the chute remains high enough.

CONCLUSION

A mass flow measurement at the exit of a grain elevator was obtained independently of influences of bulk properties. This result was achieved by smoothing and guiding the grain flow at the elevator exit. A curved chute was installed at the elevator exit. The force executed by the flow on the chute was considered a measure for the mass flow rate. This force was measured in a direction that optimally compensates for influence of friction. The influence of friction was reduced to an acceptable level. The optimal direction to measure the force depends on several installation conditions. It can be made practically independent of the mass flow rate and is minimally influenced by the combine slope. The mechanics of friction compensation apply to stationary as well as mobile installations.

Influence of friction on the force on a curved chute can be reduced to an acceptable level in all installation conditions if the entrance velocity is taken high enough. This is the case on most combines.

REFERENCES

Auernhammer H., M. Demmel, K. Muhr, J. Rottmeier, and K. Wild. 1993. Yield measuring on combine harvesters. ASAE paper 93-1506.
Deere J. Company,1995. Leaflet of the Greenstar.
Stafford, J.V., B. Ambler B. and.A. Sudduth. 1993. Yield determination using an instrumented Claas combine. ASAE paper 93-1507.

Vansichen R., and J. De Baerdemaeker J., 1991. Continuous wheat yield measurement on a combine. Automated agriculture for the 21st century, Chicago, December 1991. ASAE-symposium.

Vansichen, R., and J. De Baerdemaeker. 1992. An impact type grain flow sensor for combines. Agricultural engineering international conference, June 1992, Uppsala-Sweden.

Yang W. 1992. Kapazitive Korndurchsatzmessung auf Mahdreschern. VDI-tangung, October 1992.

Accuracy of grain and straw yield mapping

B. Missotten
G. Strubbe
J. De Baerdemaeker

Department of Agro-Engineering and Economics
Katholieke Universiteit Leuven
Leuven, Belgium

ABSTRACT

The minimum grid size for yield mapping is limited to the variability of some field characteristics and the accuracy performance of the sensors used. The overall accuracy of the yield data gathered with a combine instrumented with a curved plate grain yield sensor, a cutting width sensor and a speed sensor were checked on different grid sizes in a field. In addition a straw yield sensor has been developed and evaluated.

INTRODUCTION

A yield map plays a double role in site specific farming. First, before starting any site specific management action the yield differences within the fields have to be mapped. Second, validation of any site specific action can be evaluated by another yield map of the crop which has been locally treated. Financial validation calculations are only possible when the yield can be measured with a defined accuracy. In most cases yield differences are being detected with harvesting machines like combines instrumented with sensors to map the local yields. The accuracy of these maps depends on the chosen grid size and the accuracy of the yield measurement. The grid size of the yield map should be chosen according to the size of the local management tools like spraying widths or spreader widths but also according to the accuracy of the yield measurement. It is not appropriate to have the yield on 100 m² field with an accuracy of +/-20 %. This would make management decisions and validations unreliable. In the following paragraphs the causes of errors of yield maps are summarized. Afterwards field measurements and accuracy estimates of the yield mapping system of the K.U.Leuven (Katholieke Universitieit Leuven) are given.

Errors of grain flow measurement are due to uncontrolled influences on the sensor behavior and therefore depend on the type of grain yield sensor used. Auernhammer et al. (1994) compared three commercial grain flow sensors in a study based on accuracy of the measured grain tank loads in the field compared to the weights at the scale. All three sensors had an accuracy of 1% on a total area of minimum 40 ha. However errors of yield measurement on surfaces corresponding to the weight of one full grain tank increased up to 11 % for all of the sensors. Lamb et al. (1994) compared the measurements of a continuous yield meter with the results of a plot combine which determined grain yield in 15.2 m segments. On this small grid the errors of the continuous yield meter went up to 25 %.

Errors in continuous flow meter readings are caused by changing crop properties like kernel size, moisture content differences, dust and mud, weed infestations, machine vibrations, etc... For each change in these harvesting conditions the yield sensor will react strongly or weakly depending on the interaction between the type of disturbance and the measurement principle.

The accuracy of the grain flow measurement and the yield map will depend also on the farmers willing to recalibrate the instrument. Therefore it would be very useful to have an instrument for which the sensitivity to changes in crop conditions is minimized.

Accuracy of continuous yield measurement in ton/ha is determined by the grain flow sensor but also by the errors of the speed measurement, the estimation or measurement of the cutting width and the localization errors. In row crops like corn, cutting width can easily be estimated by counting the harvested rows. In other crops like wheat a simple estimation, with a button indicating 1/4 to 4/4 of the cutting width, can lead to large errors. Take for instance a header of 6 m with a reel divided in 4 parts. If not using 0.75 cm (1/8) of the header, the error on the cutting width and also on the yield measurement is 12.5 %, not taken into account the errors of the other sensors like the grain flow sensor. Therefore it is advisable to measure the cutting width continuously and automatically (Vansichen & De Baerdemaeker, 1992).

At last, grain yield can only be detected after it has been separated from the material other than grain. This causes not only a time lag but also smoothing through the different crop flows through a combine (Vansichen & De Baerdemaeker, 1992, Birrell et al.,1994). Noise on the grain flow signal prevents inverting a first or higher order model of the grain flow for reconstructing the true yield variations in the crop. This can cause large errors in yield at crop entrance and exit or at rapid speed changes.

The objective of this research was to investigate the overall accuracies of yield maps generated with a curved plate grain yield sensor, a cutting width and a speed sensor. A GPS-system has been used for positioning. In addition a straw yield sensor has been developed. The straw yield map was compared to the local grain yield of fields. These fields were monitored (treated and sampled) by partners in a European project (ITCF, WAU, Cemagref, SRI and K.U.Leuven). Both yields were later used as input for nutrient balances of the fields by partners in the same project, but this is beyond the scope of this paper.

MATERIALS AND METHODS

A New Holland TX-64 combine was instrumented with the sensors to gather spatial data of grain and straw yield (Fig.1). The grain and a straw flow sensor measure the mass flow in kg/s. Ultrasonic distance sensors measure the cutting width and a radar sensor measures speed. Inclination sensors were present but the slope data were not used for compensation of inclination influences on the measured yield data because the fields were flat.

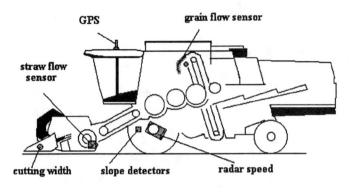

Fig. 1. Instrumented combine for grain and straw yield mapping

A detailed description of the sensors used and the way they were calibrated is given in the next paragraphs.

Grain yield sensor

The grain yield sensor is a mechanical type of sensor consisting of a 90° curved plate, placed at the exit of the clean grain elevator. The force on the plate induced by the grain stream is a measure for the mass flow from the elevator into the grain tank.

At the exit of the elevator and before the sensor a rotor (Strubbe et al., 1996) was used to condition the grain stream. A special support was designed to exclude influences from translational vibrations on the sensor reading and to reduce the influence of driving uphill or downhill on the zero reading of the sensor.

The force measurement in the flow sensor is described by Strubbe et al. (1996). It minimizes the sensitivity of the sensor to changes in kernel characteristics such as : friction coefficients, kernel size, shape,...

The calibration of the grain yield sensor was done in the field by harvesting stretches at different speeds and cutting widths in order to generate different grain flow rates. The integrated sensor signal was compared to the weights of the collected grain. An estimate of the accuracy of the grain yield sensor was obtained.

Straw yield sensor

Straw flow measurements for speed control of the combine have been described before. Good results of the straw flow measurement using the torque detection of the auger were obtained (Huisman, 1983). The same type of auger torque measurement of the 6.1 m header was chosen to measure straw yield. The sensor consists of a sprocket mounted on two springs (Fig. 2). The sprocket is loaded by the force at the pull side of the chain depending on the torque needed to drive the auger in the header of the combine. The small bending (4 mm) of the

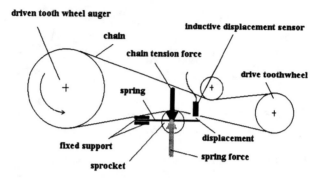

Fig. 2. Straw flow sensor : torque measurement on the auger drive of the header.

springs is a measure for the amount of straw harvested. The displacement of the springs are measured by means of a linear inductive distance sensor. Springs with different spring constants were available to make the sensor more or less sensitive depending on the magnitude of the flow in the different crops. The straw sensor has been tested in barley, wheat and peas. Cutting height was kept constant while harvesting.

The sensor was calibrated by gathering the straw of a stretch of 25 m on a cloth. This was repeated several times per field. The 25 meters were driven at different speeds in order to obtain different magnitudes of the straw flows. The integrated sensor signal during a period of harvesting the stretch was compared with the weight of straw on the cloth to construct a calibration curve for the particular crop and spring constant of the sensor.

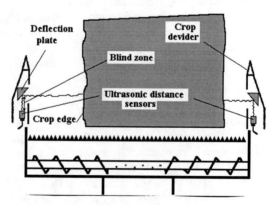

Fig. 3. Top view on the header with cutting width measurement by means of two ultrasonic distance sensors (range 6 m). These sensors are integrated in the crop dividers with the deflection plates.

Cutting width and speed detection

The cutting width was measured by two ultrasonic distance transducers. The sensors measure the width of the cutterbar not utilized (Fig. 3).

In order to cope with the blind zone (0.45 m) of the sensor, where no detection is possible, deflection of the ultrasonic signal on a plate was used. In that way the sensors could easily be incorporated in the crop dividers.

A Dickey John radar speed sensor was used for speed detection. The sensor was faced backwards to the driving direction to prevent influences of rising stubbles shortly after the cutter bar.

The position signal came from the GPS-receiver, Navstar XR5-M. A portable computer inside the combine cab was used for logging the data. Data were logged every second.

RESULTS

As mentioned in the introduction any practical use of yield maps for further interpretation depends on the tolerances that can be attributed to yield data. Those tolerances were frequently considered at each field since several sensors were combined in the yield measurement. These accuracies should in a certain way determine the grid size for converting the data into maps.

Grain flow measurement

Data for the calibration were gathered for peas, wheat, barley and corn crops in France, the Netherlands and Belgium. In the corn crop large differences of grain humidity between fields were measured. The calibration data of the curved plate yield sensor on all these fields are shown in Fig. 4.

Although the harvest conditions varied severely (different crops, different moisture content...) no significantly different calibration curves could be established. This means that recalibration of this sensor, due to differences in crop, variety, moisture content.... is not necessary. This shows the reliability and accuracy in practical use.

Accuracy of the grain yield sensor was checked on different sizes of harvested areas, ranging from 120 m² up to 2000 m². The error percentage increases with decreasing harvested area. This error is due to inaccuracies in the procedure and errors due to the sensor behavior. For an harvested area of 400 m² , which matches the grid size of 20 m x 20 m for soil sampling on the experimental sites, the maximum error was 5 %. For a surface of 2000 m² the maximum error decreased to 3 %. The error on estimating the yield on a entire 6 ha field in the Netherlands was 1.7 %.

Integrated sensor signal (volt)

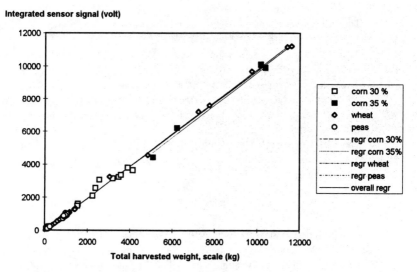

Fig. 4. Calibration curves from the curved plate yield sensor in different crops and harvest conditions.

Harvest area measurement

To determine the harvested area, the speed and the cutting width were logged every second. The maximum error of the speed measurement in the different crops was 2.5 % on tracks ranging from 20 m to 100 m length.

For the cutting width, measured by an ultrasonic distance sensor, the accuracies were determined by comparing readings from the sensor with measurements with a ruler. Errors less than 5 % were found in a standing wheat crop (Fig. 5).

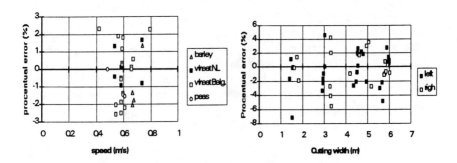

Fig. 5. Accuracy of speed and cutting width measurement. The cutting width was tested in a straight wheat crop.

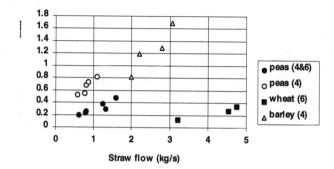

Fig.6. Results of straw flow calibration measurements by means of a torque detection on the auger (6.1 m header) in pea, barley and wheat crops. The figures between the brackets indicate the thickness of springs used (e.g., 4 & 6 indicate a 4 and 6 mm spring, 4 means two springs of 4 mm).

Recalling the 5% error on 400 m² for the grain flow measurement in combination with 5% and 2.5 % for cutting width and speed measurement respectively results in an overall yield measurement (ton/ha) accuracy of 7.5 % on a 400 m² (20 x20 m) grid. On the total harvested area of the wheat field in the Netherlands an error for area measurement of 1.5 % was obtained. The total measured yield in ton/ha deviated 0.25 % from the value determined with the weight of the trucks divided by the area calculated from a ground plan.

Straw flow measurement

The springs in the straw sensor were changed for different crops to adapt the sensor sensitivity to the straw mass flows in the different crops (Fig. 6).

For a given crop type, like wheat, the mass flow was different in Belgium and France compared to the much higher flows (yields) in the Netherlands. In general it was possible to use the weak springs for barley and peas. The stiff springs were used for the wheat fields.

Yield maps

The spatial yields were measured on selected fields in France, the Netherlands and Belgium.

On the first field in Bertem (Belgium) a barley crop was grown. The yield maps show correlation with the topographic data of the field. In the region of the steep slope the best soil layers have been washed away over years resulting in lower grain and straw yields in these areas.

The pea crop in France was lodged quite close to the ground On some spots it was difficult to measure the cutting width by means of ultrasonic sensors. Therefore it was decided to use the full width of the header while harvesting and to leave a strip of a meter of unharvested crop. Comparing the yield maps with the soil structure show that the grain and straw yield drop considerably in the area where the soil contains a lot of stones.

A wheat crop in Wieringermeer (Netherlands) was also mapped. In this field on two spots (in the north and the south of the field) the crop density was visibly less. As shown in Fig. 7 the grain and straw yield dropped considerably in these areas. It was noticed that the straw yield decreases much more in those areas than the kernel yield. According to the soil maps these yield differences are most likely caused by a sandy soil texture compared to the clay on the rest of the field.

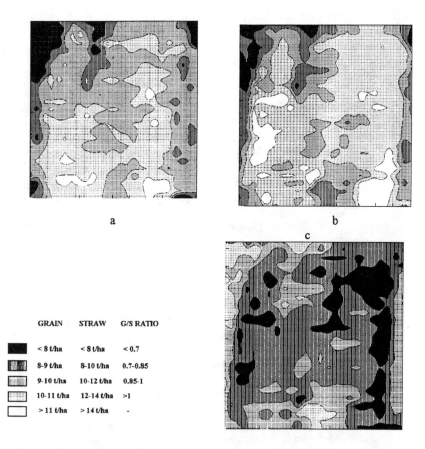

Fig. 7. Grain (a), straw (b) and ratio (c) yield map of the wheat crop on the 6 ha field in Wieringermeer (Netherlands)

Local grain and straw yield are the result of the interaction of soil conditions (texture, fertility, structure), infestations (weed, diseases) and weather conditions. Probably a combination of several reasons in a dry year (1995) can explain a lower grain and straw yield and a higher grain/straw yield ratio in the dry soil regions on that field (sandy regions). Since fertilizer application effect on yield strongly depends on the humidity of the soil, grain/straw yield mapping is an interesting tool for finding the regions where differences in soil moisture exist.

As shown above, grain/straw ratio maps can give more information on the underlying cause of yield differences. Not only soil moisture relation differences could be detected but also weed infestations. In a weed spot, grain yield will decrease. The straw yield, however, will increase because a torque measurement cannot make a difference between crop and weed. On top of this the mostly green weed has a much higher friction which causes a more than normal increase in torque expected for the increase in mass (crop and weed stems).

A second advantage of this straw yield measurement is that the detection takes place directly after the crop has been cut. This makes it possible to delineate lower or higher yielding areas easier and more accurate.

DISCUSSION AND CONCLUSION

1. Accuracy of the grain flow sensors depends on the unit area considered because of averaging on a larger number of different measurements. With the curved plate sensor an accuracy of 5 % on 400 m² was reached. The error for a field of 6 ha decreased to 1.7 % .

2. When choosing the grid size of a yield map, total accuracy of the whole system as a combination of sensors and their errors should be taken into account. The system tested with a experimental curved plate sensor, a cutting width sensor and a speed sensor gave an overall accuracy of 7.5 % on 400 m² which makes reliable financial calculations on this grid size possible.

3. Yield maps showed good correlation with differences in soil conditions like chemical composition, texture and especially soil water relations. As a consequence the yield maps were useful in locating the regions where other management can be considered. Even better contouring of these areas was made possible with the straw sensor.

4. A new dimension can be reached for the yield map when a straw sensor is added. Grain/straw yield ratio can give information on the cause of the yield differences. Interpretation of yield differences becomes easier and better decisions can be made.

ACKNOWLEDGMENTS

This research project was carried out with support from funds by the European Union under contract AIR3-CT94-1204 : Reduced Fertilizer Input by an integrated, Location Specific Monitoring and Application system. The New Holland (Belgium) company provided equipment support. Thanks also to the partners in the EU-project (ITCF, Cemagref, WAU, SRI, K.U.Leuven).

REFERENCES

Auernhammer, H., M. Demmel, T. Muhr, J. Rottmeier, and K. Wild. 1994. Site specific yield measurement in combines and forage harvesting machines. Ag Eng Milano '94 Report no. 94-D-139.

Birrell S.J., S.C. Borgelt, and K.A. Sudduth. 1994. Crop yield mapping : comparison of yield monitors and mapping techniques. Proceedings of Site-Specific Management for Agricultural Systems. Second International Conference. March 27-30, 1994 Minneapolis.

Huisman, W. 1983. Optimum cereal combine harvester operation by means of automatic machine and treshing speed control. Department of Agricultural Engineering. Agricultural University, Wageningen, the Netherlands Ph.D., 1983.

Lamb, J.A., J.L. Anderson, G.L. Malzer, J.A. Vetch, R.H. Dowdy, D.S. Onken and K.I. Ault. 1994. Perils of monitoring grain yield on-the-go. Proceedings of Site-Specific Management for Agricultural Systems. Second International Conference. March 27-30, 1994 Minneapolis.

Strubbe, G., B. Missotten, and J. De Baerdemaeker. 1996. Mass flow measurement with a curved plate at the exit of an elevator. This conference proceeding.

Vansichen, R., and J. De Baerdemaeker. 1992. Measuring the actual cutting width of a combine by means of an ultrasonic distance sensor. Prague. September 15-18, 1992, Proceedings Trends in Agricultural Engineering.

Vansichen, R. and J. De Baerdemaeker. 1992. Signal processing and system dynamics for continuous yield measurement on a combine. AgEng Uppsala June 1-4, 1992, International conference on agricultural engineering, paper no. 9206-01.

Corn Population and Plant Spacing Variability: The Next Mapping Layer

D. Easton

Easton Goers, Inc.
2699 Hwy. 141
Bagley, Iowa

ABSTRACT

While it has long been known that the errors of planting can haunt you all season long, we have never been able to readily assess the damages by collecting population and plant spacing data in a practical manner.

Now, we have invented (Patent Pending) a handheld device that conveniently measures plant population and calculates plant spacing variability as you walk down the row. This provides a site specific data set which allows a farmer or consultant to assess the effect of plant stand on yield potential and use the information in crop management decisions throughout the season. This information can be used to study planter settings, attachments, planter speeds, variety differences, tillage systems, pest effects, calculate economic threshold, etc. The device features an onboard computer to perform statistical analysis and provide real time readout, as well as an output port for linking to computers, dataloggers, or GPS equipment.

Perhaps more importantly, the application of this new technology to row equipment such as cultivators and combines, combines with GPS positioning capabilities, will allow a farmer to create field mapping layers that can be overlaid onto yield maps to study crop stand effects on yield. As yield mapping becomes increasingly available to farmers, there will be a demand for more layers of data in order to understand the cause and effect relationship of multiple management/ environment/ crop interactions. Real time plant populations and plant spacing data could be the next layers in GPS computer mapping technology.

BACKGROUND

The ability to plant corn kernels at a precise spacing has been desirable for many decades. Years ago, accurate seed placement was requisite to complete weed control by cross cultivation. As hybrids responded to higher populations and improved means of weed control, the mechanical seed placement devices gave way to metering mechanisms which attempted to give uniform distribution within the row. But even a perfect planter will not give a perfect stand, due to seed, soil, and pest variables. Though we assume a uniform stand is the ideal, the ability to quantify the effects of non-uniformity has been impractical for all but the most intensely investigated fields. Recently, seventy seven Indiana corn fields were intensely investigated by Purdue University for plant space variation.

Spacing Results from 77 Fields (Purdue)

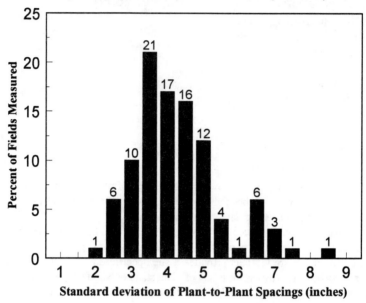

This research has shown that, in the conditions studied, increased variability in plant spacing causes a decrease in yield. This investigation was at the expense of many hours of painstaking measurements by graduate students with backaches. Their measurements of successive plant spacings were manually gathered and logged for analysis. The summation of their research is that when the standard deviation of plant spaces reaches 10% of the average plant spacing, the yield loss due to variability will be about 100kg/ha (2.5 bu/acre for each inch of standard deviation). Increases in the deviation will give proportional reduction in the yield.

Precision agriculture demands that a variable so fundamental in maximizing economical crop production as population be measured more that a few places per field, more than a few times per year. It is now evident that spacing variability is also an influential quantity, and to really know what is going on in the field, both quantities must be measured thoroughly. Historical means of collecting this information are clearly not up to the task.

A NEW MEASURING DEVICE

In order to quickly gather more complete population information, we began to design a hand operated electronic device for counting plants as it measured distance down the row. During the design phase, it seemed reasonable to also gather spacing information on the plants as they were being counted for the population calculation. With this additional goal, there were now two main challenges in making a functioning prototype - correctly identifying the target plants, and getting sufficient distance resolution. A description of the production Space Cadet will show how the challenges were resolved.

A ground engaging wheel of one meter circumference directly drives a small 60 toothed disc. An optointerrupter reads the teeth and sends an electrical signal to a microcontroller, each pulse incrementing the distance counter by nearly 17 mm. Plants are sensed by a small pivoting arm. When the traveling arm passes a stationary plant, the arm pivots. A light beam is allowed to pass within a second optointerrupter. When the plant slips off the end of the arm, the beam is again interrupted, and the plant is sensed by the controller. We use this method rather than direct sensing with a light beam because it is less sensitive to counting leaves and weeds, more compatible with the operating conditions, and easier to adjust for species and stage of growth.

Distance resolution much finer than 17 mm is preferred, and the microcontroller has the means readily available to perform an accurate interpolation. The LCD module for data presentation requires a negative voltage for contrast, so the controller generates a 4 kHz square wave to drive a charge pump inverter. A counter within the controller sums the quantity of these 250 microsecond timing periods between each successive distance pulse from the wheel. Each time a plant is sensed, the controller calculates what portion of the 17 mm distance must be added to the number of whole pulse distances since the last plant. It would be possible to use a larger vane with more teeth, a gear driven vane, or a more expensive measuring means, but this system is easily executed by the microcontroller, and requires nothing but software. Resolution is now well under 0.5 mm.

An audio annunciator provides a beep each time a plant is sensed or when a control switch is actuated. Plants which are very close together may not all be detected by the arm. When the operator, aided by the annunciator, identifies that a plant has not been sensed, a press of the INSERT key allows the number of plant spaces to be corrected. The data will be more representative if the "inserted" plant is in close proximity to another plant.

The device can accumulate information for up to 99 plant spaces per sample, giving a live readout of population per acre, standard deviation of plant spacing, and the number of plant distances measured. Each time a sample is terminated, these quantities are stored in RAM for later recall. Up to 99 samples can be stored. An RS-232 connection can provide a batch download, or the data can be downloaded as it is generated. When using this last method, greater amounts of raw data are recorded, such as the actual space measured between each pair of plants. A GPS receiver can also provide the location stamp for each sample as the data is delivered to the datalogger.

USES FOR THE DATA

With the ease of collecting masses of plant stand data via the Space Cadet, it is reasonable for a producer to measure the differences in stand between different units of a planter, different areas or soil types in a field, or different levels of pest competition. By mapping the stand data, areas of a field with consistent stand problems might be identified as soil or pest management opportunities. Equipment manufacturers can save time in measuring the efficacy of new designs. Farmers can quickly assess the effects of changes such as planter adjustments or planting speeds, the addition of planter attachments such as row cleaners or seed firmers, changes in varieties or planting dates, etc. Seed producers, particularly when using the live download capability, can have an instant plant-by-plant record of test plots for stand management or yield interpretation.

IMPLEMENT MOUNTED DEVICE

Yield and moisture data at harvest are important inputs for precision agriculture. As variability of these data become evident with GPS-linked monitoring systems, logic indicates that it would be very beneficial to also correlate these data with the quality and quantity of plant stand. Overlaying maps of yield and plant stand data may reveal important information about a field's landscape personality. For example, the soil type with higher clay content may be yielding less because of plant stand qualities rather than pest competition or a lack of fertility.

As cropping inputs are increasingly varied across the field, data collection regarding the varied inputs also becomes an increasingly valuable management tool. The most obvious variable to benefit from established stand monitoring would be variable seeding rates. It would be highly desirable to critically evaluate the effects of varied planting rates on grain yield, moisture content, fertility levels, net income, etc. Our technology will provide the data layer to make such evaluations.

There are some significant differences between gathering data by a hand operated device and one mounted to a combine or other implement. First, the data collected on a vehicle such as a combine could not be readily corrected for observed omissions as the hand-held unit can be. While the hand-held Space Cadet offers a high degree of accuracy desirable for research purposes, work conducted at the University of Missouri would indicate that accuracy within 5% is certainly possible with a vehicle-mounted device, and this level would be adequate for crop management decisions.

Another difference is that plant sensing and distance sensing mechanisms are likely to be physically separated though still united electronically. While most modern tractors and combines already have a speedometer, a more accurate distance sensing device may be required for this application. The plant sensors could be mounted near the snouts of the cornhead, and the data from these two components is united at the microcontroller.

Mature plants generally have a much more rigid posture of both stalks and leaves than those sensed early in the season with the hand-held unit but are subject to weakening by stalk diseases. It would be desirable to be able to apply sufficient sensing pressure to deter the inclusion of corn leaf and weed tissue, yet make accommodation for the possible occurrence of corn stalks weakened by diseases. The inability to adjust for weak stalks may hamper the flow of stalks into the cornhead and slow the harvesting operation, though our current design is both unobtrusive and placed where the gathering chains can assist in stalk flow.

Another difference is the speed at which the machine operates. Most field operations are performed much faster than the walking pace used with the hand-held Space Cadet. Though this isn't a big problem for the electronics, the sensing mechanism must be able to overcome inertial limitations in order to identify each plant.

Since it is preferred to gather data on all rows of a large corn head or cultivator, the number of plants sensed can become a computational burden for a low end microcontroller. Sensing on numerous rows also invites the desire to have the sensitivity of the row units adjustable from the cab.

Our second generation sensing mechanism is intended to incorporate all these features, and to provide a viable opportunity to map plant quantities and plant spacing across entire fields.

PATENT COVERAGE

Work began on the Space Cadet well before the recent mushrooming of interest in precision farming. As the design developed, so did the expectation of its usefulness, not only for farmers and their consultants, but also to researchers and seed and equipment producers. Once the first prototype was running, application for patent was made. At the time, there was no apparent competition or previous art in the measurement of either population or spacing variability. Since then, several other designs have been shown, but only for population measurement with larger plants.

In late March, the US Patent Office allowed all claims in our application, protecting the ideas first embodied in the Space Cadet, namely the automated collection of data on average plant population and individual spacing variability.

Digital Control of Flow Rate and Spray Droplet Size from Agricultural Nozzles for Precision Chemical Application

D. Ken Giles

Biological & Agricultural Engineering Department
University of California, Davis, CA

Graeme W. Henderson

Capstan Ag Systems, Inc.
Pasadena, CA

Kent Funk

RHS Spraying & Fertilizing, Inc.
Hiawatha, KS

ABSTRACT

Pulse-width modulation of solenoid valves coupled to agricultural spray nozzles was used to control spray liquid flow rate independently of liquid supply pressure. Droplet size was correspondingly controlled by adjustment of the liquid supply pressure. Flow rate, and subsequently application rate, could be controlled over a continuous 10:1 range at a fixed pressure and over a 30:1 range with variable pressure. At fixed pressures, response time for flow control changes can be greater than 5 Hz. The technique was implemented onto conventional spray rate controllers and interfaced to a GPS guidance system to allow geographically-based setpoints for spray rate and droplet size.

BACKGROUND

Variable rate technology inherently requires that flow or discharge rates (mass or volume per unit time) of agrochemicals be adjusted to achieve desired application rates (mass or volume per unit land area). When active ingredients are mixed into carrier liquids prior to discharge, the mixing rate of active ingredient concentrate can be controlled as in "injection" type systems or the emission rate of the batch-prepared "tank mix" can be controlled. While injection systems offer the advantage of reducing disposal, contamination and cleaning requirements, their response time may be inadequate for high-resolution spatial variation in application rates. More commonly, rate control is achieved through adjustment of the flow rate of tank mix through a sprayer system.

Conventional Flow Control

The classic technique for controlling the flow rate of liquid from a spray nozzle is to vary the pressure of the liquid supply. Such an approach introduces three undesirable effects. First, the system response can be inadequately slow (<1 Hz) since motorized servo valves are used to adjust pressure and the plumbing and piping systems can further damp the dynamic response. Secondly, wide ranges in pressure are required for useful flow control ranges since flow is proportional to the square root of pressure Finally, supply pressure affects nozzle performance; specifically, the droplet size spectrum and the distribution pattern of the spray cloud are altered. Standard design criteria (ASAE, 1985) suggest that "Operating pressure may also be changed to adjust volume while operating within the recommended pressure range if the volume change is under 25 percent. A greater range affects drop size and pattern excessively." While improved nozzle designs have reduced the sensitivity of droplet size and pattern on liquid pressure, useful flow control ranges are still limited.

Pulse-Width Modulation Flow Control

Variable-duration, pulsed spray emission was developed for flow rate control with conventional agricultural spray nozzles (Giles and Comino, 1989, 1990). Nozzle flow was modulated by intermittent operation of an electrically-actuated solenoid valve coupled to the inlet of the spray nozzle. Controlling the relative proportion of time during the which the valve was open and the valve was closed controlled the volumetric flow rate from the nozzle. This general approach, commonly called pulse width modulation (PWM), is typically used in industrial control systems using electrical, hydraulic or pneumatic actuators.

A linear relationship between duty cycle (the relative proportion of time during which the valve is open) and flow could be achieved. Through proper physical design of the device, compatibility with standard agricultural nozzles could be achieved since the flow control actuator was not physically part of the nozzle; rather, it was immediately upstream of the spray orifice.

The undesirable effects on pattern and droplet size from pressure control of flow are caused by reducing the instantaneous rate of liquid flow through the nozzle orifice. The premise of the PWM approach is that the nozzle operates at the full design pressure and flow during periods while the valve was open. This maintains the full flow spray characteristics of the nozzle. To reduce the temporally-averaged flow rate through the nozzle, the valve is closed for increasingly longer periods of time, rather than reducing the continuous flow rate of the liquid through the nozzle.

Independent Flow and Droplet Size Control

The independence between PWM-controlled nozzle flow rate and resulting spray droplet size has been experimentally confirmed (Giles and Ben-Salem, 1992; Giles et al., 1995). Experimental data (Giles, 1996) for a standard 8004 spray nozzle (80° spray angle, 1.5 l/min @ 280 kPa) is shown in Fig. 1. PWM at 10 Hz

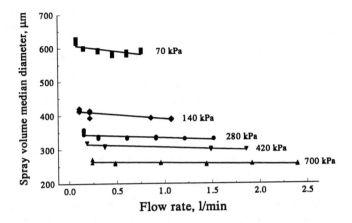

Fig. 1. Spray volume median diameter for pressures from 70 to 700 kPa and flow rates achieved by 10% - 100% modulation of control valve with an 8004 nozzle (Giles, 1996).

and duty cycles from 10% to 100% were used to modulate flow while liquid was supplied to the valve at 70, 140, 280, 420, and 700 kPa. The horizontal lines indicate the relative independence between spray droplet size, as indicated by the volume median diameter of the emitted spray cloud, and the nozzle flow over a 10:1 range at fixed liquid supply pressures.

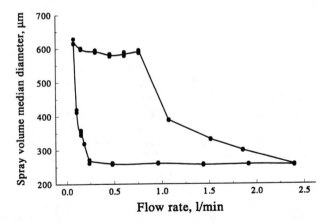

Fig. 2. Observed droplet size - flow rate control envelope for pressures from 70 to 700 kPa and 10 - 100% modulation of control valve with 8004 nozzle (Giles, 1996).

From the observed data in Fig. 1, a flow rate and droplet size control envelope can be generated using the boundary points of the response lines as shown in Fig. 2. Within the envelope, any combination of flow and droplet size can be achieved by exciting the valve/nozzle device with a corresponding PWM duty cycle and liquid supply pressure. For example, with flowrates below 1 l/min, the spray vmd can be adjusted between 250 and 600 µm by selection of appropriate PWM signals and liquid pressures. Alternatively, in applications where droplet size may be of no concern, the combination of a 10:1 flow control range from the PWM signal and the 3.2:1 flow control range from a ten-fold pressure range results in a 32:1 total range of flow rate control.

SYSTEM DEVELOPMENT

The feasibility of using PWM actuation for flow and droplet size control on a complete application control system was established through design and field testing of two prototype systems. The initial tests examined the control response rate, stability and uniformity of spray deposit. Design and testing of the commercial prototype examined retrofit capability with existing technology and integration into GPS-based, variable-rate application systems.

Feasibility Study

A model rate control system was installed on a truck-mounted, commercial sprayer. The test system was designed to maintain a specified application rate by adjusting PWM duty cycle in response to speed changes. Vehicle speed was determined from a pulse stream from a proximity switch on a ground wheel. Twelve flow control valves were installed on a 6 m spray boom at 50 cm spacings. In operation, the controller sampled the ground speed at 4 Hz, determined the corresponding duty cycle from a calibration curve of flow vs. duty cycle and output the duty cycle to the control valves. Actuation frequency was 10 Hz. A 4:1 flow control range was nominally established in the calibration curve.

Dynamic response time was determined by inputting step changes in ground speed into the controller. The resulting boom flowrate was measured with a digital flowmeter interfaced to a laptop computer sampling at 10 Hz. Additionally, gradually increasing and decreasing (ramp) speed changes were input to the system. Typical results are shown in Fig. 3 and indicate stability during the ramp changes and rapid (>3 Hz) response over the 4:1 range.

Uniformity of deposition was a concern since the PWM technique resulted in intermittent flow from the nozzles. While aerodynamic mixing from boom travel could provide some damping from non-uniform deposition, the control system was designed such that adjacent valves received PWM signals approximately 180° out of phase. The offset, with proper overlap designed into nozzle spacing, would reduce potential gaps in deposition. Uniformity of deposition was investigated by placing 4 m strips of water sensitive paper perpendicular and parallel to the direction of vehicle travel. The paper changed color in response to spray deposition. Deposition was determined through

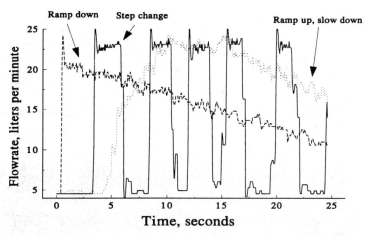

Fig. 3. PWM application rate control system response to speed change inputswith twelve 8008 nozzles at constant liquid supply pressure.

measurement of the color change.

Deposition from the PWM rate control system was compared to that from a commercial, pressure-based, rate controller (MT-5000 Trak Net, Micro-Trak, Mankato, MN). The PWM system used 8008 nozzles operating at 280 kPa. The commercial system used 8004 nozzles operating at variable pressures in response to ground speed (nominally 280 kPa for a 12.9 km/hr ground speed). Both control systems were mounted on the sprayer and calibrated to maintain an application rate of 150 l/ha. Spray passes were made with each system operating at ground speeds of 6.4, 12.9 and 19.3 km/hr. Three replications were made. Uniformity of deposition was determined through calculating the coefficient of variation of deposit on the cards at 2.5 cm intervals. Results (Table 1) showed variation to be significantly greater from the PWM technique in some cases.

Table 1. Deposition variability from sprayer control systems. Entries are means of coefficients of variations with standard deviations shown in parentheses.

| | Coefficient of variation , per cent | | | | | |
| | 6.4 km/hr | | 12.9 km/hr | | 19.3 km/hr | |
Controller	Along	Across	Along	Across	Along	Across
Pressure-based	19.4 (10.0)	15.1 (2.7)	10.0 (2.4)	10.1 (2.6)	10.0 (3.6)	12.3 (3.2)
Pulse-width modulation	19.3 (3.7)	17.7 (3.3)	19.3 (2.9)	17.2 (2.6)	13.6 (0.9)	17.0 (2.6)
Test for difference between controller systems ($t_{df=10}$)	0.01 (p< 0.98)	2.15 (p<0.18)	37.3 (p<0.01)	23.44 (p< 0.01)	5.83 (p<0.04)	5.17 (p<0.05)

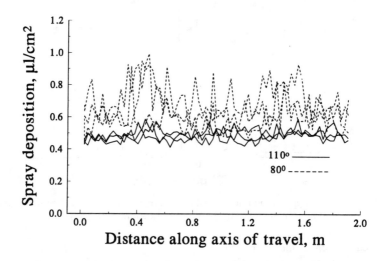

Fig. 4. Spray deposition uniformity along the axis of travel from 110° and 80° fan
nozzles at 12.9 km/hr with valve actuation of 50% duty at 10 Hz. Three
sample lines, spaced 25 cm apart, were used. Coefficients of variation: 80°
nozzles = 19.2%, 18.3%, 13.9%; 110° nozzles = 6.0%, 7.1%, 6.8%.

While the initial test compared nozzles with equivalent divergence angles (80°),
further testing investigated the use of wider angle (110°) nozzles to produce more
overlap between adjacent nozzles. The resulting deposition was significantly more
uniform as shown in Fig. 4.

The initial feasibility testing established that the PWM technique could
function for flow control actuation for spray rate control. The dynamic response
and range of flow control were demonstrated. The non-uniformity of deposition,
while higher than that from continuously operating nozzles, appeared reasonably
acceptable.

Integration into Commercial Spray Rate Controllers

The PWM technique and the associated hardware essentially comprise an
actuator which can be used in existing, commercial spray rate controllers. Such
an approach maintains the closed-loop control design and the flow control
hardware such as flow meters and pressure regulating valves which would be
present on an existing installation. The PWM technique has been commercialized
as a retrofit kit as described here (Synchro System, Capstan Ag Systems, Inc.,
Topeka, KS).

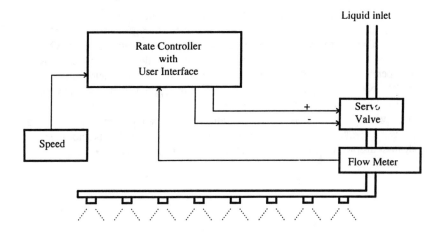

Fig. 5. Typical configuration for a pressure-variation based, spray rate controller.

A typical, commercial spray rate controller is conceptually shown in Fig. 5. As typically used for speed compensation, the rate controller receives current ground speed data and calculates a corresponding liquid flow rate to achieve the desired application rate. The flow rate setpoint is then maintained by a closed loop with a pressure control servo valve as the actuator and a flow meter as the feedback device. While an in-line servo valve is shown, an alternate configuration is to have the valve in a pump by-pass line. As discussed earlier, the flow rate in such systems is controlled through regulation of the liquid supply pressure to the nozzles.

The retro-fit implementation of a PWM actuator system into the conventional rate controller in shown in Fig. 6. The closed flow control loop within the rate controller is retained along with the speed measurement system and the user interface. The rate controller servo output lines are removed from the pressure regulating valve and redirected to an electronic interface unit which interprets the servo commands and adjusts the duty cycle of the PWM signal in accordance with the demand for increased or decreased flow. The interface unit also generates PWM signals which are 180° out-of-phase for adjacent valves. The time constant of the PWM adjustment routine can be tuned for compatibility with that of the rate controller.

Droplet size control is achieved through a closed loop pressure controller upstream of the liquid supply to the spray boom. A pressure sensor (for feedback) is added to the liquid line and the existing servo pressure control valve (actuator) is coupled to the controller. The desired droplet size defines a corresponding liquid pressure setpoint based on the nozzle characteristics. The pressure setpoint is then maintained by the controller. The setpoint can be provided from a potentiometer knob or an analog input.

Integration into GPS-based Controllers

The control system based on the retrofit implementation of the PWM flow technique, as represented in Fig. 6, was further developed to demonstrate independent rate and droplet size control based on setpoints determined from a pre-defined field map and GPS positioning of the spray vehicle. While variable-rate technology has been extensively investigated, control of droplet size has not been demonstrated in real time applications. Such capability may reduce off-target movement, e.g., "spray drift" in sensitive areas by allowing the operator to specify desired droplet sizes to be used in specified geographic areas.

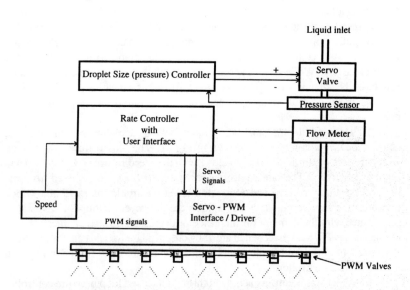

Fig. 6. Integration of PWM flow and droplet size control technique into a typical pressure-variation based, spray rate controller.

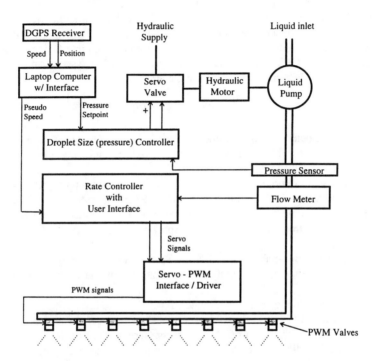

Fig. 7. Integration of PWM flow and droplet size control technique into a position-sensitive application controller using DGPS navigation.

The system was developed as illustrated in Fig. 7. A differential GPS receiver supplied ground speed and lat/long data to a laptop computer. The computer generated position-specific flow rate and pressure setpoints based on user-defined application rate and droplet size maps. The pressure control loop was modified to use an electrically-actuated hydraulic valve for controlling a centrifugal spray pump. The pressure setpoint was maintained by the closed-loop controller which received the setpoint from a laptop computer. The commercial rate controller was retained for closed loop flow control. The rate setpoint was automatically adjusted by altering a pseudo speed input to the controller. The system was installed on a commercial trailer sprayer. Rate changes of up to 4:1 and spray pressure changes of 70 to 700 kPa were demonstrated with 1-2 Hz response.

CONCLUSION

Pulse-width modulation of nozzle flow allowed dynamic ranges up to 10:1 at fixed liquid pressure. The resulting independence between pressure and flow allowed independent control of spray droplet size. The technique was retro-fit into existing commercial controllers designed pressure-variation flow control. Position-sensitive setpoints for chemical application rate and spray droplet size were used to achieve precision spray application through interfacing the control system into a GPS-based vehicle spray system. Such a system can be used to reduce spray drift in sensitive geographic areas.

REFERENCES

American Society of Agricultural Engineers. 1985. Guide for preparing field sprayer calibration procedures. ASAE Engineering Practice ASAE EP367.1. ASAE, St. Joseph, MI.

Giles, D.K.; Comino, J.A. 1989. Variable flow control for pressure atomization nozzles. J. of Commercial Vehicles, SAE Trans. 98 (2): 237-249.

Giles, D.K.; Comino, J.A. 1990. Droplet size and spray pattern characteristics of an electronic flow controller for spray nozzles. J. of Agric. Eng. Res. 47: 249-267.

Giles, D.K.; Ben-Salem, E. 1992. Spray droplet velocity and energy in intermittent flow from hydraulic nozzles. J. of Agric. Eng. Res. 51: 101-112.

Giles, D.K.; Young, B.W.; Alexander, P.R.; French, H.M. 1995. Intermittent control of liquid flow from fan nozzles in concurrent air streams: wind tunnel studies of droplet size effects. J. of Agric. Eng. Res. 62: 77-84.

Giles, D.K. 1996. Independent control of liquid flowrate and spray droplet size from hydraulic nozzles. 1996. Paper Presented at ILASS-96, 9th Annual Conference on Liquid Atomization and Spray Systems, San Francisco, CA.

Use of Injection for Site-specific Chemical Application

S. J. Nuspl

Field Control Systems, Inc.
White Bear Lake, MN

W. W. Rudolph

Midwest Technologies, Inc.
Springfield, IL

R. Guthland

Spray-Air USA, Inc.
Spray-Air Canada, Ltd.

INTRODUCTION

In an injection system the chemical to be applied is mixed with the carrier during a spraying operation. Compared with a tank mix system, injection offers advantages to the application operator (Rudolph & Searcy, 1994) . However, if injection is to be used for site-specific applications, there are some complicating factors which must be taken into account. This paper presents experiences with these factors.

These results are based on work performed in a USDA SBIR[1] project titled "Remote Digital Mapping for Site-Specific Weed Management". We refer to the system being developed as Map 'N Zap (Hanson et al., 1995).

Project Overview

The Map 'N Zap system is based on the Hanson Repeatable Pattern Field Spraying Control patent (Hanson., Lowell D. & John R. Schafer, 1991). The system starts with getting an aerial photograph of the field to be sprayed and entering it into a special MapInfo program. Then a weed specialist can designate weed areas based on the image and on other information he may have about the field. A program generates a traversal pattern representing the planned path through the field and a controller file. The controller is copied to a PC card and put into a field control computer on the sprayer. The Map 'N Zap controller tells the

---

[1]SBIR is an acronym for Small Business Innovative Research. All Federal agencies with a research program are required to budget a percentage of their research funds to SBIR projects.

injection system the type and amount of chemicals to be injected into the carrier, based on the sprayer's location in the field.

During the 1995 field season a number of tests were conducted using the Mid-Tech TASC-6300 (Midwestern Technology, Inc., Springfield, IL) injection controller on a Willmar Air-Trak (Willmar Mfg., Willmar, MN.) sprayer. Two corn fields in Lyon and Yellow Medicine counties, Minnesota, were sprayed with Banvel and Stinger to control broadleaf annual weeds and Canada thistle (*Cirsium arvense* L.) on June 3, 1995. Banvel was broadcast and Stinger was applied on patchy infestations. Stinger was applied on 31 and 29 percent of the two fields, respectively, but the thistle infestations occupied less than these areas.. The TASC-6300 two pump injection system accurately delivered the Banvel at 0.6 litres/ha (8 oz/ac) and Stinger at 0.4 litres/Ha (5.3 oz/ac). This paper deals with factors and results related to the use of injection as the means of applying chemicals in a site-specific manner.

INJECTION SYSTEM ISSUES

The use of injection is essential for practical, site-specific applications. Without injection, the sprayer needs a full set of distribution lines, nozzles and a control system for each chemical to be applied. Injection also has well-known benefits: ease and safety of chemical handling, less rinsate to dispose of and increased efficacy of chemicals, especially when performing a multi-product application.

An injection system has some unavoidable issues which must be considered. With an ideal system, the chemical would be injected into the carrier at the nozzle with uniform mixing; the application on the crop would be precisely in the field position where the chemical is injected. Such a system is not even theoretically feasible, let alone economical.

In most injection systems, chemicals are injected into a single carrier pipe that is then forked into smaller pipes up to the pipe which has the spray nozzles attached. The injection point must be upstream far enough from the first fork point to ensure that the chemical mixes properly with the carrier for uniform spraying.

Following are some of the key factors involved in using injection:

Uniform mixing across boom
Product delivery at a uniform rate across the boom is the most important consideration. The design of the plumbing system in the sprayer determines the uniformity.

Pipeline effect
If a drop of full strength chemical is injected into the carrier at the point of injection and if we assume the drop stays together, it will take a spray-rig specific given time for this drop to be emitted at a spray nozzle. The distance covered by the sprayer while the chemical is in the plumbing is referred to as the "pipeline effect" (Fig. 1). If one were injecting a dye into the carrier, the difference in the sprayer downrange field position at which the dye is injected and the position at which the dye starts to appear in the

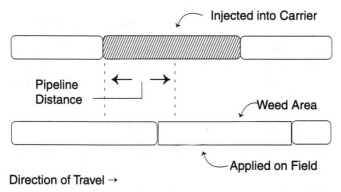

Fig. 1. Pipeline Effect.

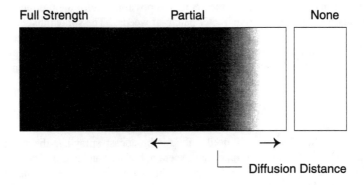

Direction of Travel →

Fig. 2. Diffusion Effect.

nozzles can be considered as the "pipeline distance". The "pipeline volume" is the product of the flow rate at the nozzles and the time associated with pipeline distance. A site-specific injection sprayer can be designed to minimize the pipeline effect, but never remove it entirely.

Diffusion along traversal

The same factors that result in uniform mixing across the boom also cause diffusion along the length of the plumbing and consequently along the downrange distance of the sprayer. When a unit of chemical is injected and begins to mix with the carrier, some particles will move downstream faster and some slower. The net effect is that the rate at which the product is emitted at the nozzle will first be a low concentration and gradually increase (Fig. 2). Using the dye example, the dye would at first appear very weak and gradually build up to full strength concentration. The same gradual change in concentration also applies when the chemical is turned off.

There is a second diffusion factor related to the sprayer plumbing. The boom section feeder lines into the boom are closer to some nozzles than to others. The closest nozzles will emit chemical before those that are farthest from the feeder line. This results in a "W" pattern along the width of the boom. For our purposes, this effect can be considered part of diffusion.

The downrange distance between the start of the appearance of the dye (assume a concentration of 10%) to the time that it is full strength (assume 90%) is the "diffusion distance". The "diffusion volume" is the product of the flow rate at the nozzles and the time required to cover the diffusion distance. Again, a site-specific injection sprayer can be designed to minimize the diffusion factor, but never remove it entirely.

The diffusion distance limits the precision of the sprayer. For example, if the diffusion distance is 20 meters it is impossible to turn the product on and off to apply chemical to a five meter patch of weeds. The spray must be turned on early enough to be at full strength when the weed patch is encountered. The diffusion distance reduces the "precision" at which the application is made. It is most important that we not have over-expectations of how precisely chemicals can be applied.

Additional Spray Control Implications

When injection is used only for broadcast spraying, the pipeline and diffusion factors are not important except at the start and end of field operations. But, if site-specific applications are to be made, the designer of the sprayer is challenged with producing a good balance between uniform product delivery and minimized pipelining and diffusion effects.

The operator of the sprayer must know the sprayer characteristics, particularly the pipeline distances for preferred application rates. If he is controlling spraying manually, then he must look ahead the pipeline distance for upcoming weed patches. If an automatic controller is used, the controller must automatically factor the "pipeline distance" into the controller/injector and sprayer operation.

If injection is to be used in site-specific spraying there are a number of further considerations.

Operator load

Driving a sprayer requires a lot of attention. The operator is constantly occupied with many tasks: following his pre-planned path, watching the booms to ensure they are not too high or low as the sprayer moves over uneven ground , watching speed and monitoring equipment instruments to ensure all is normal. With all that going on, the operator may not have time to monitor and control even one chemical. The only practical option is to automatically control the site-specific applications based on the position in his field traversal pattern.

Traversal path

With the availability of accurate GPS, one could try to use the position in the field and a spray map for chemical application. When injection is used such an approach will not be satisfactory unless the pipeline effect is factored in. The pipeline effect forces use of a planned path through the field. Within long passes a "positional offset" could be used. But, when moving from one pass to the next, the chemical for the new pass must already be injected while the sprayer is applying chemical from a current pass. Field Control Systems (FCS) has developed a program that allows the definition of traversal paths. These paths are then combined with areas to be sprayed to produce a controller file for activating the injection.

Inject at pipeline distance

When a weed patch is coming up, the controller must start injecting the chemical into the carrier at the sum of the pipeline and diffusion distances in order to have full strength at the start of the patch. When the end of the patch is approaching, injection is stopped at the pipeline distance away from the end. If the patch is shorter than the pipeline distance, the injection may actually have stopped before the sprayer arrives at the patch.

TEST RESULTS

During the summer of 1995 two sets of test sprayings were conducted: one set for calibrating the characteristics of the sprayer and one set for application of Stinger for thistle control in corn. For both cases a red dye was used as an indicator of proper operation of the injection.

The following steps were involved in going from an aerial image to the actual spraying of the chemicals:

Map 'n Zap operational steps

Aerial photography: An aerial photographer was contracted to take pictures of the fields using both color and infrared film. The film was processed within a few days turn-around.

Scanning and geo-referencing: The images were scanned into the computer and were geo-referenced using an FCS developed MapInfo® (MapInfo Corp., Troy, NY) based program.

Weed patch identification: The weed patches were then identified and entered as polygon overlays on top of the field.

Traversal pattern: The planned traversal through the field during the application was entered as another overlay. The traversals represented a sequence of "cells" in this case 18.3 meters by 18.3 meters (60 ft. by 60 ft). Each cell was either sprayed as a whole or not.

Spray pattern and controller file generation: The traversal pattern and the weed patch polygons were then used to form a spray map which in turn resulted in a compact controller file to be transferred to the spray rig.

Willmar Air-Trak sprayer with Mid-Tech TASC-6300: The application of the chemicals and dye was done with a Willmar Air-Trak sprayer. The controller was a small-screen control computer with a key pad. It was interfaced to the TASC-6300 for monitoring operations and for control of the injection.

Calibration test results

The pipeline and diffusion factors were calibrated by using a bright red dye as a chemical and a linear path marked with flags for distance and with paper towels for checking dye intensity. At an application rate of 56.8 litres per hectare (15 gallons per acre), the pipeline effect was about 73.1 m (240 ft.) and the diffusion distance was about 18.3 m (60 ft).

As part of the test runs, a small field was laid out with a specific set of weed cells marked by flags. This map was entered into the system and a controller map was generated. When used in the field, the sprayed areas were observed as pink clouds of spray and corresponded very well with the marked weed patch areas.

CONCLUSIONS

Using a properly designed injection sprayer for site-specific applications can be very effective. Compared to broadcasting, 30 to 80% reduction in amounts of chemicals used can be anticipated. On the corn fields we observed a savings of $33.30 / ha ($13.50 an acre) in reduced cost of the Stinger chemical.

It is not realistic to expect an injection system to accurately carry out spot spraying of herbicide on small, scattered weed patches. With the equipment we used, weed patches on a specific pass should be 60 to 100 meters (200 to 328 ft) in length and some herbicide will be applied beyond target weed infestations. However there is a 30 to 80 percent savings potential in chemical costs and land and water chemical exposure with site-specific injection systems.

To accurately spray the desired patches, the pipeline effect must be taken into account. Otherwise strips of weed patches may not be covered and chemical will be applied where it is not needed. Diffusion needs to be known so that the chemical will be at the full recommended rate uniformly across the boom when the area to be sprayed is entered.

REFERENCES

Hanson, L.D., P.C. Robert, and M. Bauer. 1995. Mapping wild oats infestations using digital imagery for Site-Specific Management. Proc. of Site-Specific Management for Agricultural Systems. March 27-30, 1994. p. 495-503. Minneapolis, MN. ASA, CSSA, SSSA, Madison, WI.

Hanson, L. D., and J. R. Schafer. 1991. Repeatable Pattern Field Spraying Control. U.S. Patent 5,050,771. Date issued 24 September.

Rudolph, W.W., and S.W. Searcy. 1994. Strategies for Prescription Application Using the Chemical Injection Control System with Computer Commanded Rate Changes. ASAE Paper No. 94.1585. American Society of Agricultural Engineers, St. Joseph, MI USA.

Design of a Centrifugal Spreader for Site-Specific Fertilizer Application

Robert Olieslagers
Herman Ramon
Josse De Baerdemaeker

Department of Agro-Engineering and Economics
K.U.Leuven, Kardinaal Mercierlaan 92
B-3001 Heverlee, Belgium

ABSTRACT

The success of centrifugal spreaders (80 % market share in Europe) can be explained by their low price, easy maintenance and a working width exceeding up to 15 times the machine width. The main disadvantage of this machine is the high sensitivity of the spread pattern to flow rate variations. To reduce the effort of prototype spreader design and tests ('trial-and-error' procedure), a simulation model has been developed. The model allows a goal-oriented design which is essential when making a spreader for site-specific purposes.

INTRODUCTION

Site-specific fertilizer application is a small but important part of the extended and complex field of site-specific farming. Therefore, it is important to adapt the techniques for site-specific fertilizing. Despite all kinds of available new distribution methods, broadcast spreading of dry fertilizer remains a very important way of fertilizing. Broadcast spreading can be realized by using a pendulum, a pneumatic or a centrifugal spreader. Pendulum spreaders have a working width up to 14m, limited by the pendulum length and the pendulum angular velocity. Pneumatic spreaders can go up to a higher working width of 24 m. As wind influences on the fertilizer distribution are reduced, the only important disturbances here can be caused by boom movements or irregular internal flow distributions. Therefore, to assure an accurate fertilizer distribution, the boom length is limited. Centrifugal or spinning disc spreaders can realize a working width up to 36 m. To avoid fertilizer damage and herewith irregular particle size distributions, disc rotation speed has to be limited. Overall, centrifugal spreaders are very popular because of their high working width, low price, small size and easy maintenance. At the moment, centrifugal spreaders are not adapted to site-specific application since their distribution pattern is significantly influenced by the flow rate. Adaptation of the mass flow induces a change of the spreading width and the shape of the transverse distribution pattern. This causes a bad overlap of the adjacent patterns. Design of centrifugal spreaders for site-specific (constant working width for varying flow rate) as well as for homogeneous (reduction of 'trial-and-error' design tests) applications can be based on modeling of the distribution patterns.

OBJECTIVES

The objective of this paper is to evaluate the development of an efficient centrifugal spreader for site-specific fertilizing. This spreader allows a continuous adaptation of the fertilizer flow rate without its distribution pattern being influenced. The determination of the fixed and continuously adaptable spreader parameters for given particle characteristics, a given working width and a given time dependent mass flow rate, is realized by means of a validated simulation model. The model construction as well as its validation will be discussed briefly.

REVIEW OF LITERATURE

Calculation of particle trajectories on a spinning disc as well as through the air has often been examined. Patterson and Reece (1962) and Inns and Reece (1962) described particle trajectories on the disc for single particle motions. Mennel and Reece (1963) described single particle trajectories through the air. These equations formed the basis for further development of the simulation model. Much later, several investigators analyzed the distribution of fertilizer as a function of particle characteristics (Hofstee, 1993), or created semi-empirical simulation models (Adjroudi, 1993), (Heppler, 1993) for calculations of distribution patterns. These simulation models are accurate, but require a lot of measurements and they apply to only one type of fertilizer spreader with a limited number of spreader adjustments. For design purposes however, it is interesting to have a deterministic general model of a spreader.

SIMULATION MODEL

The construction of the general simulation model, used for the spreader design, is discussed by Olieslagers et. al (1994, 1996). The particle trajectories on the disc are calculated by a differential equation in which several spreader adjustments and particle characteristics are variable or design parameters. These are given in Fig. 1. Important spreader parameters are the disc radius r_d, cone angle of the disc α, distance between two discs for a twin disc spreader A_d, disc angular velocity ω, height of the disc above the ground h, vane pitch r_p, vane length r_v, vane shape, position ϕ_{0o} and dimensions r_{bo}, r_{eo} and ϕ_{do} of the orifice above the disc. Particle characteristics of importance are the friction coefficient between particle and vanes f_v or disc f_d.

The calculation of the particle trajectory on the disc uses the initial particle position on the disc ϕ_i as an input parameter and gives the outlet velocity v_u, the angle with which the particle leaves the disc, referred to the radial direction β_u and the position referred to the driving direction ϕ_u at which the particle leaves the disc. With this information, the particle trajectory through the air was calculated by means of 2 differential equations, one for the horizontal and one for the vertical component of the particle trajectory. In these differential equations, the height of the disc h as well as the particle density ρ_p, the drag coefficient of the particle in the air C_d, the particle size distribution D and the air density ρ_a are variable parameters

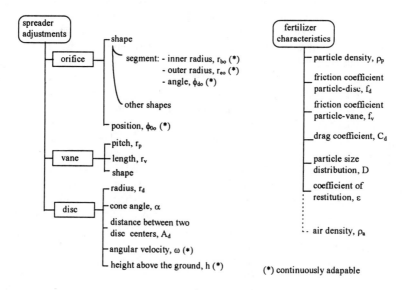

Figure 1: Model parameters

(Fig. 1). The position of the particle in the field is calculated. The particle trajectory calculations are repeated for all different initial particle positions on the disc as determined by the orifice position and dimensions. This results in a fertilizer distribution pattern.

MODEL CALIBRATION AND VALIDATION

A major goal of the validation was to compare simulated and measured distribution patterns for a whole range of spreader and fertilizer characteristics. The use of a commercial spreader considerably limits the number of possible spreader adjustments. To be able to perform spreading tests for gradual changes of individual spreader adjustments, a prototype spreader was build. This prototype spreader allows a change of disc velocity, disc radius, disc cone angle, disc height, vane shape, vane position, vane length, orifice shape and orifice position. So, every spreader adjustment included in the model, could be realized in practice.

Model calibration

- Particle distribution along the edge of the disc ('tangential distribution pattern')

To be able to examine the particle flow on the disc separately from the flow in the air, a circular tray was constructed which surrounded the disc. The tray allowed a quick registration of the 'tangential' distribution pattern, with respect to the registration of the total or 'spatial' distribution pattern. The tray is divided into 22 compartments (10° per compartment). The calculated and measured tangential distribution pattern were analyzed for several spreader adjustments and particle characteristics where the effect of one parameter at the time on the particle distribution was systematically examined. Differences between measurements and simulations demonstrated equivalent tendencies for all spreader adjustments. Fig. 2 shows this difference for a measurement with an orifice opening ϕ_{do}=60°, *spreader adjustment A* (r_b=0.05m, r_e=0.1m, ϕ_{oo}=-45°, r_p=0m, r_d=0.3m, α=0°, ω=516 rev/min, h=0.962m, symbols explained in Fig. 1) and Kemira CAN 27%N fertilizer (ρ_p=1800 kg/m³, ϕ_d=0.3, f_v=0.3, C_d=0.44, known size distribution D, ρ_a=1.2 kg/m³). For measurements and simulations, discussed in the text, adjustment A is used. The measured pattern is wider than the simulated one. This important difference is caused by particle interactions, which are not included into the model. These interactions affect the initial conditions of the differential equation.

By means of an optimization routine, an adaptation of the initial conditions of the simulation model was performed. In the initial simulation model, the area in which all particles started their trajectory on the disc, determined by ϕ_{di}, ϕ_{0i}, r_{ei} and r_{bi} (Fig. 3), were set equal to the characteristics ω_{do}, r_{bo}, r_{eo} and ω_{0o} of the segment orifice. Based upon measurements, the initial conditions ω_{di}, r_{bi}, r_{ei} and ω_{0i} were estimated for 4 mass flow rates, ω_{do}=15°; 30°, 45° and 60°, adjustment A and Kemira CAN 27%N fertilizer. The inner radius r_{bi} remained the same as r_{bo}. The estimated angle ω_{0i} differed 55° from ω_{0o} for all 4 mass flows. The angle ω_{di} as well as the outer radius r_{ei} became much larger than ω_{do} and r_{eo}. The correspondence between measured and simulated tangential distribution pattern after estimating the dimensions of the initial particle area on the disc for ω_{do}=60°, is shown in Fig. 2.

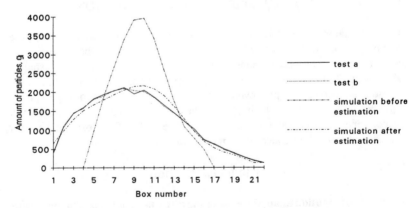

Figure 2: Comparison between measured and simulated tangential distribution patterns for ϕ_{do}=60°, adjustment A and Kemira CAN 27%N

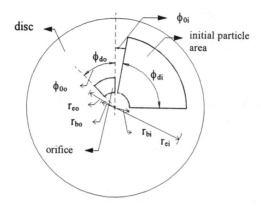

Figure 3: Relation between orifice dimensions and initial particle field on the disc.

This estimation procedure was executed for all spreader adjustments and 4 fertilizer brands (2 spherical types, two irregular types) as well as 2 types of plastic particles with known characteristics. For each of the spreader adjustments, the same tendency as a function of the mass flow of a gradually higher initial particle area dimensions ω_{di} and r_{ei} referred to the orifice dimensions ω_{do} and r_{eo} and a constant difference between positions ω_{oi} and ω_{oo} is found. This is mainly caused by particle interactions, which was confirmed by the visual analysis discussed next.

- Visual analysis of the particle flow on the disc

A visual analysis of the particle flow on the disc and against the vanes was performed by means of photographs. This is a cheap method to analyze the spread of the particles against the vanes and on the disc. The visual analysis showed that there was a sudden spread out of the falling particle flow as it touched the disc, as shown schematically in Fig. 4a and 4b. This explains in part the difference between ω_{do} and ω_{di}, as well as the difference between ω_{oi} and ω_{oo}. The impact against the vanes results in a radial spread of the particle flow (Fig. 4c) as well as a tangential spread of the falling particle flow in the turning direction (Fig. 4d). The former causes the difference between r_{ei} and r_{eo}, while the latter explains part of the difference between ω_{do} and ω_{di}. The required adaptation to the model of the dimensions and position of the initial particle area could be explained qualitatively on the basis of visual analysis.

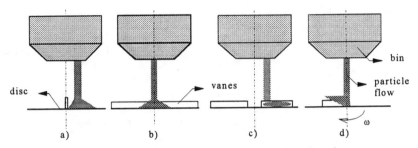

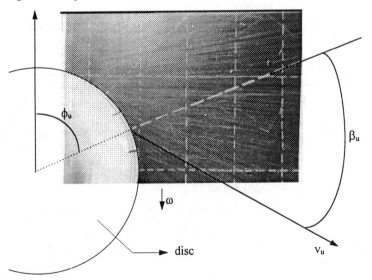

Figure 4: Spread out of the fertilizer flow on the disc and against the vanes

- Particle velocity measurements at the edge of the disc

By means of long exposure photographs (diaphragm open for 1/15s) perpendicular to the disc surface, the horizontal particle velocity at the edge of the disc was measured. The long exposure creates lines on the images describing the particle trajectory through the air (Fig. 5). The outlet angle β_u, with respect to the radial direction was measured. For radial vanes ($r_p=0$), this angle is directly related to the particle velocity v_u by equation (1).

$$ \tag{1} $$

Comparison between simulated and measured velocities at different angles ϕ_u (Fig. 5) showed a good correspondance.

Figure 5:Image of particle outlet angles at the edge of the disc

Model validation: Spatial particle distribution

For the validation of the calculation of fertilizer particle trajectories through the air, static spatial distribution patterns were measured for the whole range of spreader adjustments mentioned before. These patterns were measured in a large, standardized, humidity and temperature controlled test hall for fertilizer spreaders at the Danish Institute of Animal Science in Horsens, Denmark. The prototype spreader was placed at different positions (step=1m) on the centerline of the hall, perpendicular to a double row of collection trays with a total length of 60m. The amount of particles in each tray was weighed automatically. During 15 seconds, fertilizer was spread at each position on the centerline. The sum of all transverse distribution patterns results in a static spatial distribution pattern, represented in Fig. 6a (ϕ_{do}=60°, adjustment A, Kemira CAN 27%N). For the same spreader adjustments and fertilizer type, a spatial pattern was calculated (Fig. 6b) and compared with the measured pattern. Figure 6c shows the difference between measured and simulated fertilizer distribution. After estimating the initial conditions on the disc, based on the measured tangential distribution patterns, the simulated spatial distribution pattern approximates the measured one quite well.

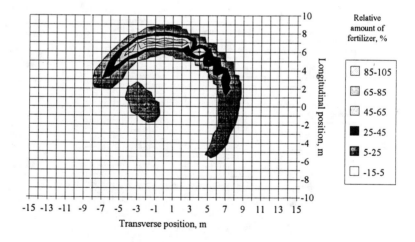

Figure 6a: Top view of the measured static spreader pattern for ϕ_{do}=60°, adjustment A and Kemira CAN 27% N

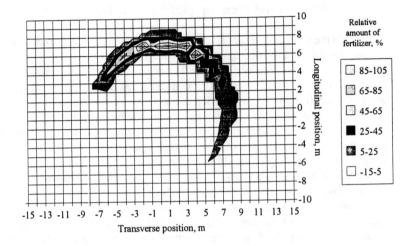

Figure 6b: Top view of the simulated static spreader pattern for $\phi_{do}=45°$, adjustment A and Kemira CAN 27% N

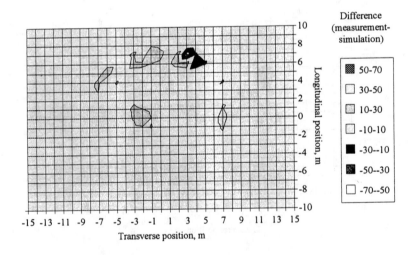

Figure 6c: Difference between measured and simulated pattern

After estimation of the initial parameters for different mass flows (ω_{do}=15°,30°,45° and 60°), these parameters did not have to be changed for a varying disc height h, angular velocity ω, disc radius r_d, orifice position ϕ_{oo} and spherical particles. The fact that no further adaptation of the model is required indicates that the calculation of particle trajectories through the air can be approximated very well by the model. This was expected as almost no particle interactions occur during fertilizer movement in the air. Non-radial placed vanes ($r_p \neq 0$), other shapes of orifices and non-spherical fertilizer particles require a re-estimation of the dimensions of the initial particle area on the disc. These re-estimations will be executed in later research. The need for re-estimation here can be explained by the different levels of fertilizer spread against the vanes for other vane positions and orifice shapes and by an other air resistance for non spherical particles.

These results show that the simulation model is a valid tool for spreader design with which influences of a variable disc height h, disc angular velocity α, disc radius r_d, orifice position ω_{oo} and particle size distribution on the fertilizer distribution can be examined for different mass flows. Model based design of a spreader can be performed by relating continuous changes of disc height, disc angular velocity and orifice position to compensate for the continuously varying spreading width and shape of the distribution pattern, perpendicular to the driving direction (transverse distribution pattern) caused by flow rate changes.

SPREADER DESIGN FOR SITE-SPECIFIC PURPOSES

A specific design for site-specific fertilizing was based on the simulation model. The coefficient of variation (CV) for adjacent transverse distribution patterns, which is a standard criterion for spreader evaluation, was taken as the cost function. The program permits changing the cost function by introducing other criteria such as minimum or maximum value of the transverse pattern. By means of an optimization routine, represented in Fig. 7, spreader adjustments were determined for which the cost function was minimized. For a given spreader configuration (r_d=0.3m, r_p=0m, α=0°, r_{bo}=0.05m, r_{eo}=0.1m, segment shape of the orifice, and Kemira CAN 27%N fertilizer (spherical, known particle distribution, Cw=0.44, ρ_p= 1800 kg/m^3), the optimum disc height h, disc angular velocity ω and orifice position ω_{oo} for a given orifice opening of ω_{do} =30° and a given working width of 15m were determined. These three spreader adjustments are continuously adaptable during fertilizer spreading. An increase of the disc height has an equivalent effect on the distribution pattern as an increase of the disc angular velocity. Therefore, only the adaptation of the disc angular velocity and the orifice position for a fixed disc height is considered here. The choice of a parameter for control purposes depends on the simplicity and precision with which this parameter can be controlled. An adaptation of the disc angular velocity, disc height as well as orifice position are easy to manipulate in practice.

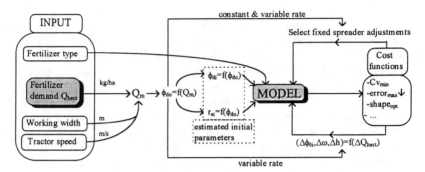

Figure 7:   Overview of the design strategy for centrifugal spreaders with a
            segment shaped orifice

        Table I demonstrates the effect on the CV of manipulating zero, one or two
spreader adjustments continuously for a given working width and for 4 different
flow rates, corresponding to 4 orifice openings ϕ_{do}. For a constant working width
of 15m, a travel speed of 12 km/h and Kemira CAN 27%N, an orifice opening of
ϕ_{do}=15°, 30°, 45° and 60° corresponds to 90, 275, 500, and 750 kg/ha fertilizer,
which means a variation from 1 to 8.2. The chosen standard adjustment for a
constant ω and/or ϕ_{0i} is ϕ_{0o}=45°. Table I shows that only a continuous adaptation of
ω and ϕ_{0i}, where ϕ_{0i}=ϕ_{0o}+55° for this adjustment as mentioned before, gives a CV
below 10% for all 4 flow rates. The average CV values for the standard setting of
ϕ_{do}=45° are given in the last column. Olieslagers et al. (1994) showed by means of
theoretical simulations that site-specific fertilizing with a centrifugal spreader
where only the mass flow rate is changed results in unacceptable deviations of the
applied amount of fertilizer from the required amount. The results here confirm that
keeping ϕ_{0o} and ω constant results in an average CV > 10% as shown in Table I.

Table I: Coefficient of variation CV for different adjustments of ω and ϕ_{0o} with a
standard adjustment of ϕ_{do}=45°

	ϕ_{do}, deg	ϕ_{0i}, deg	ω, rev/min	CV, %	average CV, %
$\Delta\phi_{0o}$	15	-10	500	8.13	
$\Delta\omega$	30	-5	500	6.66	
	45	**-10**	**450**	5.58	
	60	20	550	4.80	**6.29**
$\Delta\phi_{0o}$	15	-15	450	15.82	
ω=const.	30	-15	450	14.16	
	45	-10	450	5.58	
	60	-10	450	10.18	**11.44**
ϕ_{0o}=const.	15	-10	500	8.13	
$\Delta\omega$	30	-10	475	9.28	
	45	-10	450	5.58	
	60	-10	450	10.18	**8.29**
ϕ_{0o}=const.	15	-10	450	23.04	
ω=const.	30	-10	450	14.39	
	45	-10	450	5.58	
	60	-10	450	10.18	**13.30**

Up to now, the CV is used as a standard criterion for spreader evaluation. When evaluating the difference between the variable effective and required fertilizer amount, the standard deviation might be a better criterion as this gives the total and not the relative amount of fertilizer overdose or underdose. The adaptation of this cost function can easily be realized in the design procedure.

CONCLUSIONS

The simulation model for the calculation of fertilizer distribution patterns coming from a spinning disc is an important tool which permits to determine spreader configurations in a way that the required amount of fertilizer is approximated well in the field. The determination of fixed and continuously variable spreader adjustments can be easily performed for variations of fertilizer flow rate, a given swath width and given fertilizer characteristics. The time consuming measurements can be reduced as only a limited number of validations is required. Moreover, the validations can be based on the measured tangential distribution patterns. Registration of these patterns is very fast compared to the registration of the transverse distribution pattern and requires only a limited space. For a number of spreader adjustments, further validation is required. Spreader design for site-specific application resulted in a continuous adaptation of several spreader adjustments to compensate for undesired influences of flow rate variations on the spreader pattern. Future work will be focused on a quantification of particle interactions, further validation for a number of spreader adjustments and spreader design, based on an extended cost function, including criteria for flow rate variations in the transverse direction.

ACKNOWLEDGMENTS

The authors thank K. Persson, M.H. Jorgensen and H. Skovsgaard of the Danish Institute of Animal Science. Their contribution allowed a registration of the static spreader patterns which were essential for the model validation.

REFERENCES

Adjroudi, R. 1993. Comportement d'un flux de particules solides heterogènes sous l'action d'un lanceur rotatif. (Behavior of a flow of heterogenic dry fertilizer, spread by a spinning disc spreader.) Ph.D. Thesis, Institut Nationale Agronomique, Paris-Grignon, France.
Heppler, K. 1993.Parameterstudien zur Granulatausbringung mit Schleuderscheiben. (Parameter study of the particle flow of spinning disc spreaders.) Ph.D. Thesis, Facultät für Maschinenebau der Universität Karlsruhe (TH), Germany.
Hofstee, J.W. 1993. Physical properties of fertilizer in relation to handling and spreading. Ph.D. Thesis, Agricultural University, Wageningen, The Netherlands.
Inns, F.M., and A.R. Reece. 1962. The theory of the centrifugal distributor II: Motion on the disc, off-centre feed. J. Agricultural Engineering Research,

7(4):345-353

Mennel, R.M., and A.R. Reece. 1963. The theory of the centrifugal distributor III: Particle trajectories. J. Agricultural Engineering Research, 8(1):78-84

Olieslagers, R., H. Ramon, H. Delcourt, and L. Bashford. 1994. The accuracy of site-specific fertilizer application by means of a spinning disc fertilizer spreader. Reviewed by A.S.A., Proc. Site-Specific Management for Agricultural Systems, ASA, CSSA, SSSA, Madison, WI, USA, 709-721.

Olieslagers, R. H. Ramon, and J. De Baerdemaeker. 1996. Calculation of Fertilizer Distribution Patterns from a Spinning Disc Spreader by means of a Simulation Model. Journal of Agricultural Engineering Research , 63(2), 137-153

Patterson, D.E., Reece, A.R., 1962. The theory of the centrifugal distributor I: Motion on the disc, near-centre feed. J. Agricultural Engineering Research, 7(3):232-240

Center-Pivot Irrigation System Control and Data Communications Network for Real-Time Variable Water Application

R. W. Wall

Department of Electrical Engineering
University of Idaho
Boise, Idaho

B.A. King

Biological and Agricultural Engineering Department
University of Idaho
Aberdeen, Idaho

I.R. McCann

Department of Agricultural Mechanisation
Sultan Qaboos University
Sultanate of Oman

ABSTRACT

A multi-microprocessor based distributed control system manages spatially variable water and chemical application. The microprocessors are networked together in a master slave configuration communicating over the 480VAC power cable. The control system offers high reliability, ease of use and effective management of water, chemicals and energy.

BACKGROUND

The recent development and commercial success of conventional ground-based variable rate chemical application equipment has encouraged the development of center pivot and linear move irrigation systems capable of spatially variable water and chemical application. The application of chemicals through an irrigation system known generically as chemigation, is commonly practiced as an economical and effective means of applying appropriately labeled chemicals. The ability to spatially vary water and chemicals to address spatial differences such as soil texture, topography, and crop health and development throughout the growing season has the potential to increase water and chemical use efficiency, improve yield and quality, and reduce water quality degradation.

King et al. (1995) modified a 3-span linear move irrigation system to provide step-wise variable water application along the system lateral, thus effectively providing two-dimensional control of water and chemical application. Variable water application is achieved by replacing the single sprinkler package with a dual sprinkler package sized to provide 1/3 and 2/3 of the original sprinkler flow rate. Each sprinkler has on-off control in multiple sprinkler zones along the system length. This arrangement provides for step-wise variable application depths of 0, 1/3, 2/3, and full along the system lateral. Spatially variable chemical application is achieved in the same step-wise proportions as water application by adjusting the chemical injection rate proportional to total system flow rate thus maintaining constant chemical concentration in the applied water. Fraisse et al. (1995b) evaluated the use of on-off pulsing as a means of obtaining a wide range of water application rates from a linear move irrigation system. Camp and Sadler (1994) reported development progress on a variable rate center pivot. They employed three manifolds along the system lateral in 10 meter lengths to provide application rates from zero to 100% in eight step-wise increments.

The objective of this research was to enhance the capabilities of the variable rate irrigation system of King et al. (1995) and test the control system on a full scale commercial center pivot irrigation system. The control scheme and its features and capabilities are presented.

CONTROL SYSTEM DESIGN CONSIDERATIONS

In spatially distributed systems, distributed control is more efficient than central control for both reliability and cost. Distributed control allows computer intelligence needed for control and instrumentation to be placed as physically close to the point of actuation or measurement as feasibly possible. Reliable communications is paramount for the distributed control technology to link the numerous points of control with points of instrumentation. Converting measurements to digital signals before transmission reduces sensitivity to noise and transmission losses. Placing control processors in close proximity to the actuators achieves similar noise immunity and signal quality. Bi-directional network communications increases reliability by incorporating automatic error detection and correction to all data exchanges as well as the ability to provide real-time feedback that the desired control was actually implemented.

Complexity, response, and resolution are defining characteristics for all automated systems. The requirements in these areas dictate the necessary computing performance. For precision agriculture applications, and in particular, spatially variable irrigation, computational complexity is not nearly as critical as data storage capacity because formula type algorithms are replaced with look-up tables. Irrigation system speed and map resolution, both angular and radial, determine the rate of change of the system dynamics. Satisfying the Nyquist sampling theorem (Defetta et al., 1988), a fundamental law in control theory, requires that the control system response be at least twice as fast as the inputs can potentially change. As a rule, all elements of the entire control system must reach steady state before the input changes. Hence, all valve operations must be completed and the variable speed drives must reach steady state before the next set of valve on-off settings are initiated by a pivot angle change. If this cannot be guaranteed, then one or more of three possible recourses need to be pursued: a faster response control system, slower rotational speeds, and-or reduced angular and lateral resolution.

SYSTEM CAPABILITY

A demonstration system has been implemented near Aberdeen, Idaho on a commercial 392m center pivot that irrigates agricultural land with sufficient spatial variations in soil texture, soil depth, topography, and yield potential to warrant a spatially variable irrigation and chemigation. The primary control objectives are: 1. safe-fail operation (reliability), 2. ease of use, 3. water conservation, 4. chemical conservation, and 5. energy conservation. This ranking gives highest priority to the revenue producing activities of growing crops and the lowest priority to conservation activities. Ranking the irrigation objectives defines the constraints for both the formulation process and the control implementation of process automation.

The block diagram in Fig. 1 depicts the hardware organization for the control system currently implemented at Aberdeen. System instrumentation consists of two, 0-100 psig, pressure transducers and one, 0-1000 gpm, flow meter. The system controls the flow rate of the water supply pump and chemical injection pump with variable speed electronic drives. This distributed control system uses an Echelon CSMA-CA (Carrier Sense - Multiple Access with Collision Avoidance) bi-directional network communicating directly over the 480VAC power line.

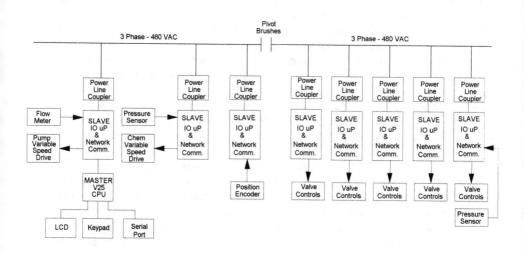

Fig. 1. Irrigation control system block Diagram

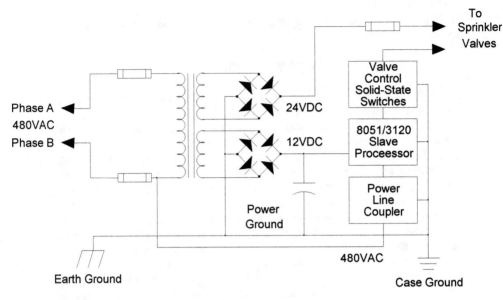

Fig. 2. Power supply for a valve control node

Figure 2 shows that a 480VAC transformer steps down the voltage to 24VAC and 12VAC. The 12VAC is rectified and filtered to power the microprocessor boards described below. The 24VAC is full-wave rectified but not filtered for operation of DC solenoids on pilot operated diaphragm valves. The 480VAC connects to a communications coupler board which separates the 60Hz power from the high frequency communications signals. Power and network interfaces for other nodes are very similar to Fig. 2.

Central to the control system is a master computer which contains the control algorithms and water and chemigation application maps. This master CPU communicates over the bi-directional network to numerous slave microprocessors distributed at points or "nodes" needed for control and instrumentation. The user interfaces with the irrigation control system with a keypad and LCD display located with the master CPU. Water and chemical application maps and logged operational data can be transferred between the master CPU and a laptop computer through the serial communications port using any one of many commercially available modem software packages.

A microprocessor supervisor IC monitors the activities of the 8051 µP and initiates a processor reset if the 12VDC supply voltage dips too low or the 8051 µP ceases normal operations. The 8051 µP communications with the master CPU through a 3120 µP dedicated to managing the network messages. The 8051 has an RS232 serial port for diagnostics and future interfaces to other telemetering devices.

The valve control nodes are distributed along the 392m center pivot, each controlling sprinkler valves for two spans using two valve control boards. Each valve control board with seven high current outputs for switching solenoid operated valves interfaces to a microprocessor board over the 14 digital IO lines. The control provided by the five valve control nodes results with the center pivot being divided

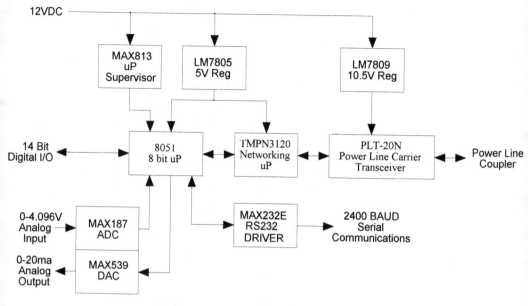

Fig. 3. Block diagram of the slave microprocessor units.

into 32 control zones along its length. Each of the zones controls multiple valves to provide a step-wise variable application rate of 0, 1/3, 2/3, and full along the center pivot lateral.

A pulsed method of flow control was investigated based upon work by Fraissie et al. (1995a). The decision to use step-wise flow control is influenced largely by reliability issues with solenoid operated valves. Pulsing operates valves significantly more often than does step-wise control which reduces reliability. When the valve technology produces an economical and reliable valve that can be continually operated, the electronics used in the step-wise approach is fully capable of supporting the pulsed approach with only software modifications.

For the step-wise flow control implemented at Aberdeen, the five valve control nodes control 62 groups of valves along the pivot with up to six valves per group. System software limits the number of groups that can be switched on and off in a set time interval. This is done to minimize pressure fluctuations and allow the water pump variable speed drive to maintain pressure with the changes in flow and the chemical injection pump variable speed drive to maintain a constant concentration in the applied water..

An instrumentation node is placed at the pivot point for sensing pivot angle using an eleven bit optical encoder. This angular position encoder interfaces to the microprocessor board shown in Fig. 3 using the 14 digital IO lines. Another instrumentation and control node is placed at the variable speed chemigation pump drive. Variable speed drive control is provided by the 4-20 ma analog output signal. If the chemical pump is physically located at the pivot point, the two nodes just described can be combined. A pressure sensor or other type of analog sensor can be placed at these nodes if desired.

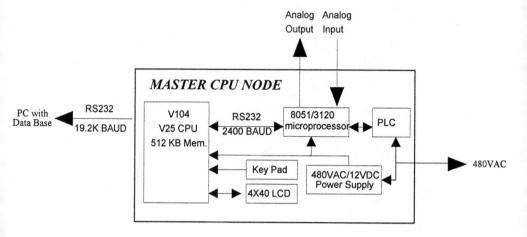

Fig. 4. Block diagram of electronics for master CPU node

Another control node is placed at the water pump variable speed drive and also uses the 4-20 ma analog output for control. A pressure transducer is installed here as well. Since all IO and network processor boards are identical, theoretically any node can interface to the master CPU which contains the application map information and control algorithms as shown in Fig. 4. However, this interface is usually assigned to the most convenient node.

MASTER CPU CONTROL

While the master CPU node does not directly control any valves or pumps, it does manage and supervise all control operations. This supervision is accomplished by exchanging messages with each slave processor node in the control system. The master CPU communicates with the network through the 8051 IO processor then through the 3120 communications processor as shown in Fig. 4.

The NEC - V25 (similar to an i286) based master CPU is extremely more powerful than either the network communications processor or the IO processor used at each node. The master CPU has sufficient speed and memory to maintain multiple maps for water and chemical applications and log actual operational conditions.

The three main operational tasks for the master CPU are: determine which valves need to be turned on or off based upon position and water application map information, adjust the chemigation rate based upon either measured or calculated flow rate, and adjust the water pump speed to regulate the water pressure for a minimum set-point value at one or more locations. The distributed processing software organization structure allocates IO operations to the 8051 processor, the network communications to the 3021 processor, and the water and chemical management to the master V25 CPU.

A four-line LCD, 16 key keypad, and a serial port constitute the user interface managed by the master CPU. The keypad has numbers '0' through '9', an 'ENT' key, 'CLR" key and keys labeled 'A', 'B', and 'C'. The LCD has three main display formats shown in Figs. 5 through 7; the AUTO mode, MANUAL mode and the MAINTENANCE mode displays. Pushing the 'A' key causes the system to immediately begin operating in the AUTO mode, the 'B' key is for operating in the MANUAL mode and the 'C' for performing system maintenance. For the AUTO and MANUAL modes, the LCD displays the all information from the systems instrumentation and control settings. From the MAINTENANCE menu, the raw analog input and output data as well as accumulated communications 'no-responses' are displayed for system diagnostics.

The AUTO and MANUAL mode displays are very similar, the top line of the LCD shows the node currently exchanging information with the master CPU, the mode and the center pivot angle. For the AUTO mode, the top line also displays which map is being used to control water application. The master CPU can store four variable application rate maps and one fixed default base application rate map. Although the fixed base map applies water and chemicals uniformly on the entire field, constant pressure and chemical concentration is maintained. The first line of the LCD also displays the day of the month, hour, and minute. The date is maintained by a real-time clock on the V25 processor board.

The second line of the LCD for the AUTO mode shown in Fig. 5 describes the data displayed on the third row. The "GPM" variable is the calculated flow based upon which valves are open and the pressure at the end of the pivot. The "FLOW" variable is the measured flow rate at the water pump. GPM and FLOW should correlate very closely. "PSI1" is the measured pressure at the water pump and "PSI2" is the pressure at the end of the pivot. "CHEM" is the percentage of predetermined full flow that the chemigation pump must operate to maintain constant chemical concentration.

The last row of the LCD shows the valve status for each of the 32 zones on the pivot. The data is grouped to represent which node has control of which zones. A "0" for a particular zone indicates that the sprinklers are off. A "1" indicates that the 1/3 flow sprinkler is on and a "2" indicates that the 2/3 flow sprinkler is on. A "3" indicates the both the 1/3 and the 2/3 flow sprinklers are on. For the current demonstration system, zones 0 and 31 (first span and end gun) are either on or off indicated by a "1" or a "0".

```
N: A   [AUTO MAP 2]   Ang: 312.24   28:15:32
  GPM | FLOW | PSI1 | PSI2 | CHEM
  924   933     43     41     92%
1330330 333333 333333 333333 333301
```

Fig. 5. Example AUTO mode display format

```
N: F  [   MANUAL   ]   Ang: 312.24   28:15:32
Enter a zone number (0-31):_
  924    933      43      41       92%
1330330 333333 333333 333333 333301
```

Fig. 6. Example MANUAL mode display format

While operating in the MANUAL mode (Fig. 6), the first line shows the node communicating with the master, pivot angle, date and time just as in the AUTO mode. The third and fourth line of the LCD are identical to AUTO mode. The second line alternately prompts the user to enter a zone number from 0 to 31 and then a valve setting from 0 to 3. When the program switches from AUTO mode to MANUAL mode, the sprinkler setting last determined from the active map and the pivot angle are fixed until the user either specifies manual changes or switches back to the AUTO mode.

When the LCD is in the maintenance mode, the system continues to operate in the AUTO or MANUAL mode, which ever was selected last. The user is presented with five command options which prompts for additional information to complete the intended operation. Using commands in the maintenance mode allows the user to transfer data to and from the system master CPU and a laptop computer, activate one of the five stored maps, and clear the data log memory. The diagnostics displays allow the designers and maintenance personnel to observe information transactions over the network as well as monitor specific control variables.

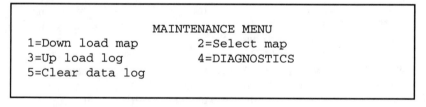

```
                    MAINTENANCE MENU
  1=Down load map        2=Select map
  3=Up load log          4=DIAGNOSTICS
  5=Clear data log
```

Fig. 7. Maintenance mode LCD format

SYSTEM OPERATIONS

Much of the system operations can be deduced from the functional descriptions discussed above. However the functional descriptions describe the operations from a static perspective. While an eleven bit angular encoder is used, the resolution used for the application map is one degree. The pivot being divided into 32 valve control zones results in approximately 12.6m radial increments. Considering that the maximum angular velocity of the center pivot is one revolution in approximately 18 hours, means that potentially, the system is moving one degree every three minutes. Using the extreme scenario where all valves go from off to on or vise versa, 64 valve group or zone operations must occur in three minutes or one valve group changes every 2.8 seconds. This means that the variable speed drives for both the water and chemigation must be able to go from fully off to fully on in three minutes as well.

Theoretically, commands to turn all valves off or on can be accomplished in less than one second as there is a broad cast message that allows all nodes to receive the data and execute simultaneously. However, the master CPU program currently limits the system to only one zone operation per span every five seconds to limit dynamic pressure fluctuations and allow the variable speed drives to respond. Since there are 64 zone operations required to switch all valves off or on, the algorithm requires 10.7 minutes. Hence the maximum rotational speed for this scenario is one degree every 10.7 minutes or one revolution every 64.2 hours. Recall, that this scenario assumes all valves are switching on or off. If only half the valves are expected to change between angular increments, then the rotational speed can be doubled and so on. The preceding discussion demonstrates the relationship of angular and application map resolution to process speed.

The water application map stored in the data base can be used to create a software ramp to effectively reduce the flow step size when such radical control is needed. The more gradual the spatial variations, the faster the pivot can travel and still maintain variable rate control. The master CPU program computes new flow rates after a message is sent to each valve control node. If the computed flow rate changes new messages are sent that adjusts the controls for the chemigation and water pump variable speed drives before a message is sent to the next valve control node.

Each time the pivot rotates one degree, 49 variables are stored in non-volatile memory for later transfer to a laptop computer and post processing. The 512KB of memory on the master CPU allows data for seven rotations to be stored before new data begins to over-write older data. The information currently logged include pivot angle, date and time, calculated and measured water flow, percent chemigation, pressure from two pressure transducers, and computed flow at each of the 32 zones along the pivot. When the data is retrieved by a laptop computer using standard modem software, it is saved in comma delimited format suitable for importing directly into most spread sheet applications software.

CONCLUSIONS

The demonstration center pivot control system implements spatially variable water and chemical application. The control scheme uses a distributed multi-processor based system communicating on a power line carrier network. The communications provides bi-directional data transfers for closed-loop pressure control.. The pivot is divided into 32 control zones which can be controlled to apply water in step-wise variable amounts. The implemented angular resolution of the application maps is one degree resulting in a minimum spatial resolution of approximately 6.8m by 13.7m at the end of the pivot. Data logged will be collected over the 1996 growing season to evaluate the effectiveness of this precision irrigation process and the water application control system described in this paper.

REFERENCES

Camp, C.R., and E.J. Saddler. 1994. Center pivot irrigation systems for site specific water and nutrient management. ASAE Winter meeting, Atlanta, GA, paper no. 94-1586.

Defetta, D.J., J.G. Lucus, and W. Hodgkiss, 1998. Digital signal processing - system design approach. John Wiley, pp. 13,60,105, 332.

Fraisse, C.W., H.R. Duke, and D.F. Heermann. 1995a. Laboratory evaluation of variable water application with pulse irrigation. Trans.ASAE, Vol.38(5), pp. 1363-1369.

Fraisse, C.W., H.R. Duke, and D.F. Heermann. 1995b. Simulation of variable application with linear-move irrigation systems. Trans. ASAE, Vol.38(5), pp. 1371-1376.

King, B.A., R.A. Brady, I.R. McCann, and J.C. Stark. 1995. Variable rate water application through sprinkler irrigation. Site-specific management for agricultural systems. ASA. Madison, WI. pp. 485-493.

Automatic Steering of Farm Vehicles Using GPS

M. O'Connor
T. Bell
G. Elkaim
B. Parkinson

*Stanford University Stanford
California*

ABSTRACT

Operating agricultural equipment accurately can be difficult, tedious, or even hazardous. Automatic control offers many potential advantages over human control; however, previous efforts to automate agricultural vehicles have been unsuccessful due to sensor limitations. With the recent development of Carrier Phase Differential GPS (CDGPS) technology, a single inexpensive GPS receiver can measure a vehicle's position to within a few centimeters and heading to within 0.1°. The ability to provide accurate real-time information about multiple vehicle states makes CDGPS ideal for automatic control of vehicles.

In this work, a CDGPS-based steering control system was designed, simulated, and tested on a large farm tractor. A highly simplified vehicle model proved sufficient for accurate controller design. After various calibration tests, closed-loop heading control was demonstrated to a one-σ accuracy of better than 1°, and closed-loop line tracking to a standard deviation of better than 2.5 cm. Future plans for research include the use of a pseudo-satellite to eliminate any position bias and extending the current control system to control a towed implement.

INTRODUCTION

Autonomous guidance of agricultural vehicles is not a new idea, however, previous attempts to control agricultural vehicles have been largely unsuccessful due to sensor limitations. Some control systems require cumbersome auxiliary guidance mechanisms in or around the field (Young et al., 1983; Palmer, 1989) while others rely on a camera system requiring clear daytime weather and field markers that can be deciphered by visual pattern recognition (Brown et al., 1990; Brandon & Searcy, 1992). With the advent of affordable GPS receivers, engineers now have a low-cost sensor suitable for vehicle navigation and control. GPS-based systems are already being used in a number of land vehicle applications including agriculture. Meter-level code-differential techniques have been used for geographic information systems (LaChapelle et al, 1994; Pointer & Babu, 1994; Lawton, 1995) driver-assisted control (Vetter, 1995), and automatic ground vehicle control(Crow & Manning, 1992).

Using precise differential carrier phase measurements of satellite signals,

CDGPS-based systems have demonstrated centimeter-level accuracy in vehicle position determination (Cohen, 1995) and 0.1° accuracy in attitude determination (Cohen et al., 1994).

EXPERIMENTAL SETUP

System integrity becomes impeccable with the addition of pseudo-satellite Integrity Beacons (Pervan et al, 1994). The ability to accurately and reliably measure multiple states makes CDGPS ideal for system identification, state estimation, and automatic control. CDGPS-based control systems have been utilized in a number of applications, including a model airplane (Montgomery & Parkinson, 1996), a Boeing 737 aircraft (Cohen, 1995), and an electric golf cart (O'Connor et al., 1995).

This paper focuses on the automatic control of a farm tractor using CDGPS as the only sensor of vehicle position and attitude. An automatic control system was developed, simulated in software using a simple kinematic vehicle model, and tested on a large farm tractor.

The primary goal of this work was to experimentally demonstrate precision closed-loop control of a farm tractor using CDGPS as the only sensor of vehicle position and attitude. This section describes the hardware used to do this.

Fig. 1 - Experimental Farm Tractor

Vehicle Hardware

The test platform used for vehicle control testing was a John Deere Model 7800 tractor (Fig. 1). Four single-frequency GPS antennas were mounted on the top of the cab, and an equipment rack was installed inside the cab. Front-wheel angle was sensed and actuated using a modified Orthman electro-hydraulic steering unit. A Motorola MC68HC 11 microprocessor board was the communications interface between the computer and the steering unit (Fig. 2). The microprocessor converted computer serial commands into a pulse width modulated signal which was then sent through power circuitry to the steering motor; the microprocessor also sampled the output of a feedback potentiometer, the only non-GPS sensor on the vehicle, attached to the right front wheel. The 8-bit wheel angle potentiometer measurements were sent to the computer at 20 Hz. through the serial link.

GPS Hardware

The CDGPS-based system used for vehicle position and attitude determination was identical to the one used by the Integrity Beacon Landing System (IBLS) (Montgomery & Parkinson, 1996) (Fig. 2). A four-antenna, six-channel Trimble Vector receiver produced attitude measurements at 10 Hz. A single-antenna nine-channel Trimble TANS receiver produced carrier- and code-phase measurements at 4 Hz. which were then used to determine vehicle position. An Industrial Computer Source Pentium-based PC running the LYNX-OS operating system performed data collection, position determination, and control signal computations using software written at Stanford.

The ground reference station (Fig. 2) consisted of a Dolch computer, a single-antenna nine-channel Trimble TANS receiver generating carrier phase measurements, and a Trimble 4000ST receiver generating RTCM code differential corrections. These data were transmitted at 4800 bits/sec through Pacific Crest radio modems from the ground reference station, which was approximately 800 m from the test site to the tractor.

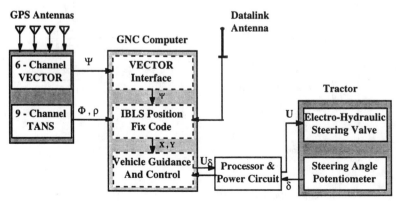

Fig. 2 - Vehicle Hardware Architecture

VEHICLE MODELING

Performing a valid tractor simulation required a good model of dynamics and disturbances. Ground vehicle models in the literature range from simple to complex, and no single model is widely accepted (Owen, 1982). The most sophisticated models are not always appropriate to use (El-Gindy & Wong, 1987), especially since controller and estimator design require a simple, typically linearized, model of plant dynamics.

Kinematic Model

The simplest useful model for a land vehicle is a kinematic model, which is based on geometry rather than inertia properties and forces. Assuming no lateral wheel slip, constant forward velocity, actuation through a single front wheel, and a small front wheel angle, the latter two equations of motion shown in Figure 3 can easily be derived. The first equation of Fig. 3 can be applied if we also assume small heading deviations from a desired path. Although some of these assumptions are violated for a tractor making large turns on loose soil, the control system based on this model was able to compensate for the small modeling errors.

The kinematic equations were derived in state-space form for ease of controller and estimator design. The state vector is composed of the lateral position deviation from a nominal path (y), heading error (ψ), and effective front wheel angle (δ).

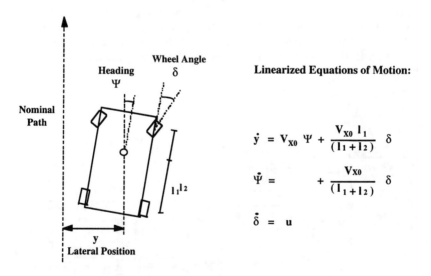

Linearized Equations of Motion:

$$\dot{y} = V_{X0} \, \Psi + \frac{V_{X0} \, l_1}{(l_1 + l_2)} \, \delta$$

$$\dot{\Psi} = \qquad\quad + \frac{V_{X0}}{(l_1 + l_2)} \, \delta$$

$$\dot{\delta} = u$$

Fig. 3 - Vehicle Kinematic Model

Steering Calibration

Initially, calibration tests were used to create two software-based "look-up" tables, one which linearized the output of the steering potentiometer versus the effective front wheel angle and the other which linearized the computer-commanded wheel-angle rate to the actual wheel-angle rate (Fig. 4).

To calibrate the potentiometer readings of effective front wheel angle, steady turn tests were performed to find the heading rate (dT/dt) of the tractor at various potentiometer readings. For each test, the tractor was driven in a circular path with a constant front wheel angle and constant forward velocity while GPS heading data was taken and stored. By compiling all these tests, a function was generated that related steady-state heading rate to potentiometer reading. The kinematic model described above suggests that the effective wheel angle is directly proportional to the steady-state heading rate, so the data collected during the tests was also used to generate the "Table 1" shown in Figure 4.

Calibration of the commanded wheel angle rate was simpler. Constant steering slews were commanded by the computer at varying levels of actuator authority (u) while wheel angle data was taken and stored. The time rate of change of the effective wheel angle was later estimated for each steering slew and stored in "Table 2" (Fig. 4).

CLOSED-LOOP HEADING RESULTS

The first controller designed, simulated, and tested on the tractor performed closed-loop heading. The computer code was written so a user could command a desired heading using a keyboard input. The computer would then send the appropriate commands to the electro-hydraulic actuator to track the desired heading. The first tests were closed-loop heading tests designed to verify the kinematic vehicle model. These initial tests also yielded a better feel for tractor disturbances.

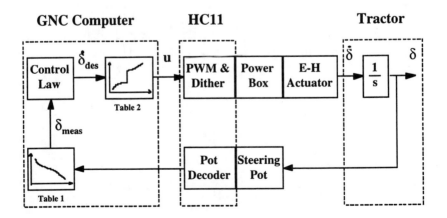

Fig. 4. Steering inner loop with "look-up" tables.

Heading Controller Design

A hybrid controller was designed to provide a fast response to large desired-heading step commands. A non-linear "bang-bang" control law generated actuator commands when there were large errors or changes in the vehicle heading or effective wheel angle states. Typically, these large changes occurred in response to a large heading step command. When the vehicle states were close to zero, a controller based on standard Linear Quadratic Regulator (LQR) design (Bryson & Ho, 1975) was used.

"Bang-bang" control is a standard non-linear control design tool based on phase-plane technique (Graham & McRuer, 1961) . Unlike linear feedback controllers, bang-bang controllers use the maximum actuator authority to zero out vehicle state errors in minimum time just as a human driver would. For example, in response to a commanded heading step increase of 90°, a bang-bang controller commands the steering wheel to hard right, holds this position, and then straightens the wheels in time to match the desired heading. In contrast, a linear controller would respond to the step command by turning the wheels to hard right, then slowly bringing them back to straight, asymptotically approaching the desired heading.

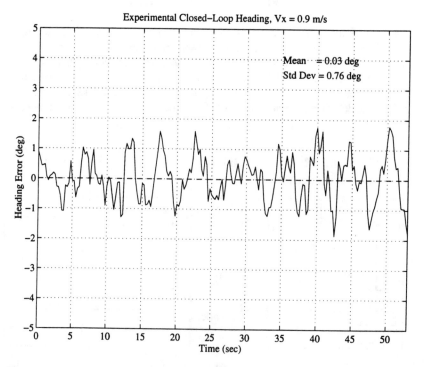

Fig. 5 - Closed-Loop Heading, Regulator Performance

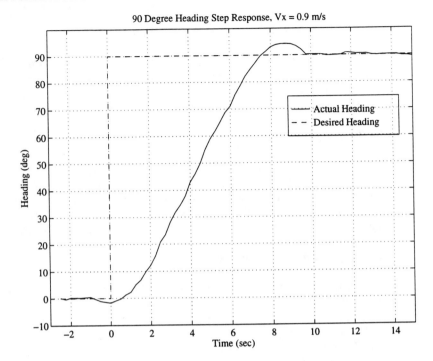

Fig. 6. Closed loop heading, step response.

The drawback to bang-bang control is that when state errors are close to zero, the controller tends to "chatter" between hard left and hard right steering commands. For this reason, a linear controller was used for small deviations about the nominal conditions.

Experimental Heading Results

During the heading tests, the tractor was driven over a bumpy field at a nearly constant velocity of 0.9 m/s. The driver commanded an initial desired heading and a number of desired heading step commands through the computer.

The tractor tracked the commanded headings very accurately, even in the presence of ground disturbances. Figure 5 shows a plot of CDGPS heading measurements during the longest closed-loop heading trial recorded. Over about one minute, the mean heading error was 0.03° and the standard deviation was 0.76°. From separate tests, the expected sensor noise was zero mean with approximately 0. 1° standard deviation, so the true system heading error standard deviation was almost certainly less than 1°.

A plot of the time response for a 90° step in commanded heading is shown in Fig. 6. The rise time of the controller for this particular command was approximately 7 seconds, and the settling time was less than 10 seconds. An small overshoot of about 4° occurred at the end of the heading step response.

CLOSED-LOOP LINE TRACKING RESULTS

After performing closed-loop heading, the next step toward farm vehicle automation was straight-line tracking. These series of tests were designed to simulate tracking a row. To track a straight line, vehicle position was fed back to the control system along with heading and effective wheel angle.

Line Tracking Controller Design

As in the closed-loop heading case, the line tracking controller was implemented as a hybrid controller with various modes. To get the vehicle close to the beginning of the "field" and locked on to each line or "row", a coarse control mode was used based on the closed loop heading controller described above. Once a line was acquired, a precise linear controller based on LQR techniques took over.

Experimental Line Tracking Results

Two line-tracking tests were performed on the same field as the closed-loop heading experiments. The vehicle forward velocity was manually set to first gear (0.33 m/s), and the tractor was commanded to follow four parallel rows, each 50 meters long, separated by 3 meters (Fig. 7). Throughout these tests, the steering control for line acquisition, line tracking, and U-turns was performed entirely by the

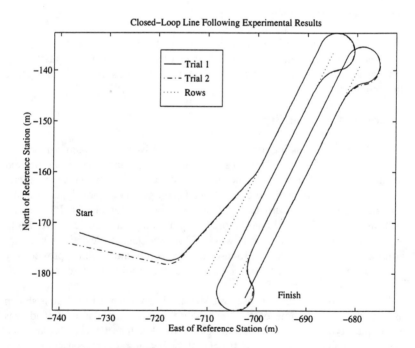

Fig. 7. Trajectory for closed-loop line following.

control system. CDGPS integer cycle ambiguities were initialized by driving the tractor as closely as possible to a surveyed location and manually setting the position estimate.

It is important to note that Fig. 7 shows the CDGPS measurements taken during the two tests, not the "true" vehicle position. In fact, there was a small. steady position bias (about 10 cm) between the two trials due to the unsophisticated method that was used for GPS carrier phase integer cycle ambiguity resolution. A more sophisticated method involving pseudolites or dual frequency receivers would have eliminated this bias and is a topic of future research.

Line tracking measurements for both trials are shown in Fig. 8 and summarized in Table 1. Since the plots show CDGPS measurements and not "truth", they represent the error associated with the control system and physical vehicle disturbances. The tractor controller was able to track each straight line with a standard deviation of better than 2.5 cm., the vehicle lateral position error never deviated by more than 10 cm, and the mean error was less than 1 cm for every trial.

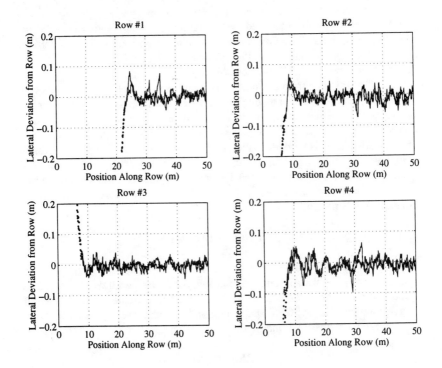

Fig. 8. Line following experimental results.

Table 1 - Summary of Line Following Control System Errors

Measured Lateral Position (cm)	Row #1	Row #2	Row #3	Row #4
Trial 1, mean	0.10	-0.85	-0.32	-0.66
Trial 2, mean	0.27	-0.58	-0.35	-0.68
Trial 1,1-σ	1.98	2.01	1.33	2.45
Trial 2,1-σ	1.86	2.13	1.39	1.93

CONCLUSION

This research is significant because it is the first step towards a safe, low-cost system for highly accurate control of a ground vehicle. The experimental results presented in this paper are promising for several reasons. First, a farm tractor control system was demonstrated using GPS as the only sensor for position and heading. Only one additional sensor—the steering potentiometer—was used by the controller. Second, a constant gain controller based on a very simple vehicle model successfully stabilized and guided the tractor along a straight, predetermined path. Finally, it was found that a GPS controller could guide a tractor along straight rows very accurately. The lateral position standard deviation was less than 2.5 cm. in each of the 8 line tracking tests performed

Transitioning from automatic control of a lone farm tractor to automatically controlling the same tractor towing an implement is a large step since the combined system will have more complex dynamics and larger physical disturbances acting on it. Guiding a vehicle along curved paths will also present a challenge that has not been addressed. This work describes a control methodology that was successfully employed to control a real farm tractor to high accuracy. This same methodology, combined with a more sophisticated dynamic model may be sufficient to control the more complicated tractor-implement system. Further research is currently underway to explore this possibility.

ACKNOWLEDGMENTS

The authors would like to thank several groups and individuals who made this research possible. At Stanford, Ben Jun, Andy Barrows, Dave Lawrence, Stu Cobb, Boris Pervan, and Clark Cohen were all extremely helpful. Trimble Navigation provided the GPS equipment used to conduct the experiments. Funding was provided by the FAA and by Deere and Company.

REFERENCES

Brandon, J. R., and Searcy, S. W. 1992. Vision assisted tractor guidance for aagricultural vehicles. International Off-Highway and Powerplant Congress and Exposition, Milwaukee, WI. Sept. 1992. PA, pp. 1-17.

Brown, N. H., Wood, H. C., and Wilson, J. N. 1990. Image Analysis for vision-based agricultural vehicle guidance. Optics in Agriculture, Vol. 1379, pp. 54-68.

Bryson, A.E., and Ho, Y.C. 1975. Applied optimal estimation. Hemisphere Publishing Corp.

Cohen, C. E., et al. 1995. Autolanding a 737 using GPS Integrity Beacons. Navigation, Vol. 42, No. 3, Fall 1995, pp 467-486.

Cohen, C. E., Parkinson, B. W., and McNally, B. D. 1994. Flight tests of attitude determination using GPS compared against an Inertial Navigation Unit. Navigation, Vol. 41, No. 1, Spring 1994, pp 83-97.

Crow, S. C., and Manning, F. L. 1992. Differential GPS control of Starcar 2. Navigation, Vol. 39, No. 4, Winter 1992-93, pp. 383-405.

El-Gindy, M., and Wong, J. Y. 1987. A comparison of various computer simulation models for predicting the directional responses of articulated vehicles. Vehicle SystemDynamics, Vol. 16, 1987, pp. 249-268.

Graham, D., and McRuer, D. 1961. Analysis of nonlinear control systems. John Wiley & Sons, Inc.

Lachapelle, G., Cannon, M. E., Gehue, H., Goddard, T. W., and Penney, D. C. 1994. GPS systems integration and field approaches in precision farming. Navigation, Vol. 41, No. 3, Fall 1994, pp. 323-335.

Lawton, Kurt. 1995. GPS system in a box. Farm Industry News, Vol. 28, No. 8, July/August 1995, p. 10.

Montgomery P.Y., and Parkinson, B. W. 1996. Carrier differential GPS for takeoff and landing of an autonomous aircraft. Proceedngs of ION National Technical Meeting, Santa Monica, CA, Jan. 1996.

O'Connor, M.L., Elkaim, G.H., and Parkinson, B. W. 1995. Kinematic GPS for closed-loop control of farm and construction vehicles. Proceedings of ION GPS-95, Palm Springs, CA, Sept. 1995, pp 1261-1268.

Owen, G. M. 1982. A tractor handling study. Vehicle System Dynamics, Vol. 11, pp. 215-240.

Palmer, R. J. 1989. Test results of a precise, short range, RF Navigational/ Positional System. First Vehicle Navigation and Information Systems Conference - VNIS '89, Toronto, Ont., Canada, Sept.1989, pp 151-155

Pervan, B. S., Cohen, C. E., and Parkinson, B.W. 1994. Integrity monitoring for precision approach using kinematic GPS and a ground-based Pseudolite. Navigation, Vol. 41, No. 2, Summer 1994, pp 159-174.

Pointon, J., and Babu, K. 1994. LANDNAV: A highly accurate land navigation system for agricultural applications. Proceedings of ION GPS-94, Salt Lake City, UT, Sept. 1994, pp. 1077-1080.

Vetter, A. A. 1995. Quantitative evaluation of DGPS guidance for ground-based agricultural applications. Trans. ASAE, Vol. 11, No. 3, 1995, pp. 459-464.

Young, S.C., Johnson, C. E., and Schafer, R. L. 1983. A vehicle guidance controller. Trans.ASAE. Vol. 26, No. 5, 1983, pp. 1340- 1345.

Applications of Image Understanding Technology in Precision Agriculture: Weed Classification, and Crop Row Guidance

F. Sadjadi

Lockheed Martin Corporation
3333 Pilot Knob Road
Eagan, Minnesota

ABSTRACT

In this paper two applications of image understanding technology, weed recognition, and crop row guidance in precision agriculture are presented. Weed recognition is an important application of image analysis whereby in an automated sensing and classification fashion the weed types are classified for on line weed and location specific chemical applications. Row guidance is the automatic navigation of the off-the-road vehicles such as cultivators in such a way as to position the vehicle precisely on the crop row edges.

INTRODUCTION

Computer image understanding refers to the description of an image-an array of pixels- as well as the scene that the image represents. This description is in terms of the objects that are in the scene and the relationships that may exist among them. With the proliferation of computers, the availability of the low cost imaging sensors, and signal processing hardware, image understanding technology is playing an increasingly important role in the precision agriculture.

IMAGING AND MACHINE CLASSIFICATION OF WEED

The cost of overuse of herbicide has health, economical, and even legal ramifications. The ill effects of the chemical sprays on human and animals are becoming more understood. The economical cost of the overuse, even though is not high for small farms, for large acreage it is high enough that has prompted the farmers and cooperatives to explore means for reducing their usage. The legal issues are also on the horizon. The voters in California were recently asked in one of their propositions to decide on the overall ban for the application of herbicides. Even though the proposition was defeated, few people believe that its proponents will give up and go away.

The above forces, health, economy and law has led to the development of a number of systems for reducing the amount of the chemical usage and at the same time effectively suppressing weeds, thus keeping up the crop yield to make the investment in this technology feasible.

Computer-based image understanding technology offers a useful solution to this problem in terms of weed imaging and classification. One of major problems in computer image understanding is the problem of segmenting an image into its meaningful parts. This problem is even more challenging for the case of the agricultural field imagery due to their inherent complexities. However, making use of the spectral properties of the crops, background, and weeds can make the segmentation more manageable.

Several studies [3,4] has shown that the ratios of the near infrared reflectance to that of the reflectance in the red region of the electromagnetic waves has a high sensitivity to the presence or lack of green vegetation. From these studies it can be seen that the ratio of NIR to R reflectance is high for the living plant (due to the steep jump in the curve at longer wavelengths), whereas for soil and crop residue the ratio is relatively small (due to the nearly flat reflectance curves). It is interesting to note that one can see a bump in the curve at about 550 nm wavelength (in the green of the visible spectrum). A noticeable jump in the curve can be seen from Red to Near infrared. This larger effect translates into a much more reliable living plant detection sensor. This observation has led to design and development of vegetation-sensitive sensors.

In the case of crop fields the vegetation-sensitive camera can be useful for scene sensing. However, for the golf course/turf applications, both target and background are living plants, it is not clear whether or not the use of the NIR/R ratio is necessary or even desirable over a standard black-and-white camera equipped with NIR filter.

The following figure (Fig. 1) illustrates the steps involved in the weed imaging and classification application: a texture segmentor will separate the weeds from the grassy or bare background, A statistical classifier (minimum probability of total error classifier) is then used to classify the weed segments into various desired classes based on the stored weed class statistics that are stored in the *texture model base*.

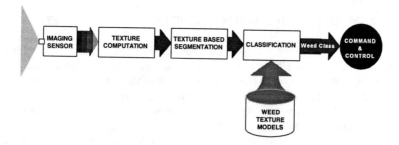

Figure 1. The approach for Weed Classification

Figure 2. Near IR image of a weed scene and its 2D wavelet Transform

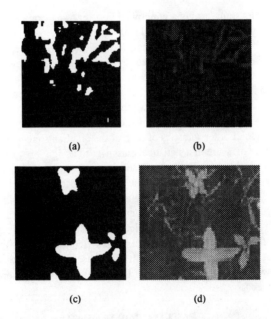

Figure 3. The images of a weed field (b) broad leaf and (d) grassy weed and their wavelet-
based texture difference segmented output (c) and (a).

A texture (Duda & Hart, 1973) measure that conveys both statistical and structural information is used in this approach. This texture measure is based on the wavelet (Daubechies, 1992) transformation.

Wavelet transforms are linear and square integrable transforms whose basis functions are called wavelets. In general given a function s(t) its wavelet transform is :

$$W_s(a,b) = \int s(t)\, g_{ab}(t)\, dt$$

$$g_{ab}(t) = g[(t-b)/a] / \sqrt{a}$$

where g(t) is the mother wavelet. For example for the case of Morlet wavelet one has :

$$g(\;) = \exp[-(t/t_0)^2/2]\exp(j2\pi f_0 t)$$

Then the wavelet transform is a windowed Fourier transform except the gaussian window is adaptive by dilation, a, and shift operation, b.

At each pixel in the sensed image a window is positioned. The size of the window is a parameter to be determined. A two-dimensional wavelet transform is performed on the pixels inside this window. Fig. 2 shows the typical wavelet transform of a two-dimensional image. The top left section contains the energy of the original image. The top right shows the vertical edges, bottom left shows the horizontal edges and the bottom right shows the diagonal edges of the original image. The second order statistics of each of the quadrants are then computed and used as a texture measure in a statistical classifier to assign label to the considered pixel in the original image. The texture measure is also used to segment the original image by systematically computing the texture differences between windows corresponding to neighboring pixels. Figure 3 shows typical weed images of grassy (b) and broad leaf (d) weeds. The wavelet-based texture difference segmented images are shown in Fig. 3 (a) and Fig. 3 (c) respectively. As can be seen the wavelet-based texture difference segmentor separates the weed from the background as well separating weeds from each other and the background. The size of the window used in the estimation of the texture statistics is dependent on the types of the weed.

VEGETATION DIRECTED CROP ROW GUIDANCE SYSTEM

Systems that detect vegetation are well known devices, light reflection devices, acoustic devices and infrared imaging sensors.

Automatic guidance systems for farm vehicles employ detection system to generate signals which define rows in a field, including rows of matter, and employ this information to generate guidance signals which would direct the wheels or tracks of the farm vehicle along rows or on a row (Reid & Searcy, 1988). Cultivated fields are never smooth and farm vehicles move in a path or line best defined as being jagged. After vegetation is planted in a row in a field, the crop becomes a wide column which could overlap into adjacent rows. On occasion some of the vegetation crop may grow outside of a row or even between rows. During the early growth season the undesired vegetation or weeds can exceed the presence of the desired vegetation emerging in the rows. The crop row guidance is accomplished through the use of an imaging sensor positioned on a tractor.

Figure 4 displays the steps that are involved in the vegetation-directed row guidance. The scene in front of the tractor is sensed via imaging sensor(s). The imaging sensors could be NIR/R sensors or a simple CCDs equipped with NIR filters. A typical image is shown in Fig. 5 (a).

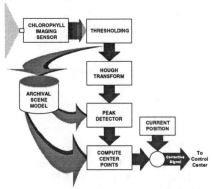

Fig. 4. The Approach for Crop Row Guidance

Fig. 5. The images of (a) original row crop in near IR, (b) the thresholded images, (c). The Hough transform of the thresholded image, and (d) the extracted lines.

The output of this sensor is then thresholded (Fig. 5 (b)) and then a Hough transform is performed. Hough transform (Hough, 1962) s a voting mechanism whereby all the pixels in the imagery are passed through a set of different lines each with different slope and intersects. Then a histogram of the all of the slope - intersects are formed (Fig. 5 (c)). The slope-intersect pairs that are associated with the largest number of pixels (the highest peaks in the Hough space) are then detected by a peak detection operation. The slope-intersect pairs associated with the largest peaks (Fig. 5 (d)) are then used to predict crop row lines. Then, by comparing the current headings with those of the predicted lines an error signal is generated. This error signal is then used in a feedback mechanism to correct the vehicle's direction of motion.

SUMMARY

In this paper two applications of image understanding technology in precision agriculture, weed imaging and classification, and crop-row guidance were discussed.

Weed imaging and classification approach uses both the spectral and textural properties of the vegetation for segmentation and classification. Spectral information is obtained through the use of the NIR/R and NIR imaging sensors. Textural properties, however is exploited by the multi-dimensional wavelet decomposition technique.

Crop-directed row guidance approach also uses the spectral properties of the vegetation in separating crops from the background. The output of which is then used in a robust line segmentation technique, Hough transform, to capture the major crop rows whose direction are used for guiding the vehicle in the filed.

For both applications the results of applying the techniques on real imagery were provided.

REFERENCES

Duda, R.O., and P.E. Hart. 1973. Pattern classification and scene analysis. Wiley & Sons Publication, New York.

Daubechies, I. 1992. Ten lectures on wavelets. SIAM Publ., Philadelphia, PA.

Hough, P.V. C.. 1962. Methods and means for recognizing complex patterns. U. S. Number 3069654 (December 1962).

Reid, J.F., and S.W. Searcy. 1988. An algorithm for separating guidance information from row crop images. Transactions of ASAE 31(6).

Corn Plant Population Sensor for Precision Agriculture

C. E. Plattner

Agricultural Engineering Department, University of Illinois
1304 W. Pennsylvania Avenue
Urbana, Illinois

J. W. Hummel

USDA Agricultural Research Service
1304 W. Pennsylvania Avenue
Urbana, Illinois

ABSTRACT

A photoelectric emitter and receiver pair produced the signal used to measure the in-row distance between plants to provide information on plant spacing, skips, and doubles. Data were collected with the sensor mounted on a corn-head on a combine, and in corn in the early growth stages, 4 - 8 leaves per plant. Plant spacing and stalk diameter were used in filtering to remove erroneous plant counts due to weeds and plant leaves.

INTRODUCTION

Plant population is known to have a considerable effect on corn yield (Nielsen 1995). Accurate spacial plant population data could be an important factor in site-specific crop management. The data could help quantify the plant population-yield relationship, improve the interpretation of yield maps, be useful in identifying problem areas in a field, and aid in planter performance analysis. A real-time corn population sensor could provide this data including information on average plant spacing, variation of plant spacing, number of doubles, and number of skips. Without the sensor, getting a true measure of plant population and spacing uniformity manually is tedious, time consuming, and expensive (Fee 1994).

Several attempts to develop a corn population counter have been reported in the literature. Easton Goers, Inc. (1995) developed a hand-held corn population analyzer. The device included a measuring wheel which rolled along the corn row and sensed plants with a spring-loaded wire arm, providing population data on up to 99 plants. Birrell and Sudduth (1995) developed a mechanical stalk sensor mounted in front of the gathering chains on one side of a combine's row divider. Using filtering techniques, sensor population results were within 5% of actual populations. The major source of error was multiple counts in areas of heavy weed infestation. Deere and Company (Gore 1996) engineers built a very similar stalk counter and obtained similar results. An inherent problem with a mechanical-arm design population sensor was the inability to detect two stalks closer than 40 mm, in addition to its sensitivity to weeds.

SENSOR DESIGN

The photoelectric sensor used for detecting plant stems incorporated a Mini-beam® Model SM31L emitter and receiver pair (Banner Engineering, Minneapolis, MN). These were small, rugged, self-contained, modulated infrared (880 nm) devices with a fast response time (1 ms). The thin, vertically aligned light beam was projected by the emitter across a row of corn, activating the receiver, spaced 220 mm away, on the opposite side. As the emitter-receiver pair was traversed down the row, a plant stem would block the beam of light, detecting the location of the plant. For harvest stage data collection, a pair of spring-loaded fingers were mounted on the corn head to deflect leaves out of the beam of light.

Plant Diameter and Spacing Measurement

To obtain a distance measurement of a plant diameter or spacing from the light/dark photoelectric sensor output, both low and high resolution distance inputs were investigated. The distance input was a signal with a frequency proportional to the travel velocity, with a resolution in pulses per unit distance.

When a high resolution input was used, i.e., 10 or more pulses per stalk diameter of the smallest plant measured, the length of the dark/light period was obtained by counting the number of pulses from the distance input over the duration the sensor was activated/deactivated. The number of pulses counted divided by the resolution (pulse/unit distance) of distance input resulted in the length of the period.

If a low resolution distance input was used, the distance measurement of a plant diameter or spacing was calculated by multiplying the period of time the sensor was dark/light by the travel velocity. The interval of time was measured by counting the number of pulses at a known clock frequency which occurred over the duration the sensor was dark/light. The travel velocity was calculated from the frequency and the resolution of the distance input.

With the latter approach to measuring plant diameters and spacings, the frequency of the distance input was calculated by measuring the amount of time it took for a specified number of pulses to occur. By timing a specified number of pulses, a more precise frequency measurement was made versus the alternate method of counting the number of pulses over a specified time. Because the velocity varied, the frequency measurement was updated during every light period of at least 5 distance input pulse in length.

Signal Processing and Hardware

A signal processing circuit received the outputs of the distance sensor and the photoelectric sensor and converted the signals into a usable form to interface to the data acquisition board. One part of the circuit converted the sinking output of the photoelectric sensor into a logic signal and produced an interrupt signal using two non-retriggerable one-shot monostable multivibrators to indicate a dark/light status change.

The second part of the circuit was used to measure the distance input frequency when the low resolution input was used. Two D-type flip-flops controlled two counters; one counter incremented down five pulses while the other simultaneously measured the time it took the first counter to complete the countdown.

A DaqBook 100 data acquisition board (IOtech, Inc., Cleveland, OH) was used to measure the light and dark intervals and control the signal conditioning circuit in measuring the distance input frequency. A Toshiba T4400C 80486 laptop computer with a parallel connection to the DaqBook was used for all testing of the population sensor.

Software

A compiled Microsoft C program was written with the DaqBook software drivers to arm, read, and reset the five available counters in the DaqBook, and then record individual plant diameters and spacings to a file . The heart of the program was an acquisition loop which repeatedly measured each plant diameter and its following spacing interval, filtering out plant diameters which were below a minimum allowable value. When the low resolution distance input was used, the travel velocity was estimated during the loop. The acquisition was terminated by either specifying a number of plants to acquire, or by an input from the keyboard.

EQUIPMENT AND PROCEDURE

Laboratory Tests

To perform laboratory tests, a test stand was built that had 10 wooden dowel rods of various diameters mounted around the perimeter of a 0.556-m diameter disc. The nominal diameters of the wooden dowels ranged from 6 mm to 25 mm and the tangential speed of the rods could be varied from 3 kph to 22 kph.

Two distance inputs were tested. One distance input was from a permanent magnet and coil adjacent to a 60-tooth gear mounted on the drive shaft of the disc providing an output resolution of approximately 0.034 pulses per mm. The alternate distance input was from a rotary optical shaft encoder (Litton Encoder, Chatsworth, CA) with a soft rubber, 60-mm diameter wheel mounted on its shaft, attaining a linear distance resolution of approximately 10 pulses per mm.

The laboratory tests investigated how plant population accuracy was affected by the type of distance input used (low, high resolution), acquisition speed (3.0 kph, 5.5 kph, 7.0 kph), quality of beam alignment (good, poor - 10° misalignment), aperture size (0.5 mm, 1.0 mm), lateral position of the plant relative to the emitter and receiver (20 mm from emitter, middle, 20 mm from receiver), and multiple filtering events (20 additional "weeds" present, absent). The standard test, to which all others would be compared, consisted of the high resolution distance input, 5.5 kph acquisition speed, good beam alignment, 1.0 mm aperture, the plant passing midway between the emitter and receiver, and no multiple filtering events. Six complete experiments were conducted. In each experiment two factors were varied, testing all possible combinations of the two factors while all other factors remained fixed at their default levels.

Ideally, a randomized complete block design would have been used to investigate the effects of the test factors on the measurement of plant diameter and spacing interval. However, restrictions on randomization of some of the factors precluded the use of this experimental design. To test the factors that existed with the restrictions on randomization that existed, it was assumed that time had no effect on the tests. To check this assumption, the standard test was performed at the beginning, middle, and end of the test sequence.

Early Growth Stage Tests

A two-wheel push cart was built to test the sensor on early growth stage corn. It transported the signal processing circuit, the data acquisition hardware, laptop computer, distance sensor, and the power supply down a 76 cm row of corn.

The sensor was mounted on skids which followed the soil surface and kept the light beam at approximately 55 mm above the soil surface, clearing soil clods in the row, but still lower than most leaves. Horizontal 1.0 mm x 6.4 mm apertures were used on both sensor components.

For a distance input, the optical shaft encoder used in the laboratory tests with the soft rubber wheel was driven off the tire of the cart. Having a high resolution distance input, approximately 10 pulses per mm, was very important when using the push cart since the cart's low mass, in combination of manual movement over uneven soil, resulted in an nonuniform forward velocity.

In a greenhouse, a bed of early growth stage corn, averaging 7 visible leaves was used for testing the sensor. Four 7.3-m rows of corn containing approximately 24 plants each at 30 mm average spacing in rows 76 cm apart were tested. The location of each plant in the four rows was manually measured with a fiberglass tape, recording the beginning edge of each plant stem in the direction that the data would be acquired relative to the first plant in the row. The diameter of each plant in two rows was recorded with a digital caliper.

Harvest Stage Tests

For field testing, the population sensor was installed on a CaseIH 1620 Axial-Flow combine equipped with a 1040 CaseIH four-row corn head. The sensor was located slightly ahead of the gathering chains on the combine head on one outside row to scan the plant before it was disturbed, because an accurate velocity of the combine relative to the plant was crucial.

To protect the beam from leaves and weeds during plant spacing measurements, two pairs of break away fingers were built, one set mounted on each side of the row. The sensor light beam was located between and behind the upper and lower fingers. When a stalk was encountered, the spring-loaded fingers rotated rearward, allowing the stalk to pass.

For a distance input, a Radar II ground speed sensor (DICKEY-john, Auburn, IL) provided a low output resolution of approximately 0.1 pulse/mm.

Three different test speeds were used: 3.0 kph, 5.5 kph, and 8.0 kph. Each replication consisted of three tests, each at a different speed, with the run order randomly chosen within the replication. Three replications were conducted with small (1.0 mm x 13 mm) apertures in both the emitter and receiver. Another three replications were conducted with no apertures used. Finally, one replication with no apertures and with the fingers removed was conducted. In each replication, a small test of approximately 10 plants with their leaves stripped off was conducted, to provide a clean control test without any false counts.

For testing the effect of aperture and acquisition speed, the statistical design was a split-plot in time with replications nested within the aperture effect. Because of the split-plot in time, there was a restriction on randomization on the aperture effect.

During each test at least 4 m of untouched plant stand was harvested and the plant diameter and spacing interval recorded. A typical test would include approximately 24 plants. The location of each plant was measured with respect to the beginning side of the first plant using a fiberglass tape. Diameters of each plant were recorded for one replication with an aperture and for one replication without an aperture.

RESULTS AND DISCUSSION

Laboratory Tests

The diameter estimation was the most sensitive to the various factors tested. Estimated diameter changed as much as 20% when two factors (beam alignment and travel speed) were changed. The estimated spacing, on the other hand, changed less than 0.5 % across all of the factors tested.

Speed had a very significant effect on the average diameter measurement. The average diameter increased more than 9% for the standard test configuration as the speed increased from 3 kph to 8 kph. The speed effect occurred because the photoelectric sensor's response time to turn on was greater than the response time to turn off (a result of the demodulation scheme). The difference caused the elapsed time between when the leading edge of the object was sensed (transition) and when the sensor activated to be shorter than the elapsed time between when the trailing edge of the object was sensed (transition) and when the output deactivated.

An analysis of variance (ANOVA) was used to test for significant differences in estimated stalk diameter. The Waller-Duncan multiple range test was used to identify differences among the estimated diameters due to the variables being tested (Table 1). None of the factors tested resulted in a significant effect on plant spacing, defined as a plant diameter and the next spacing interval, since a complete cycle (one dark and one light interval) was made that cancelled out the effects of response time.

Poor beam alignment caused the dowel rod diameters to appear 8.2 % larger than with good beam alignment. With poor beam alignment, which results in low excess gain, partial blockage of the light beam by an object reduced the light impinging on the receiver to a level below the threshold. Changes in excess gain level, due to beam alignment, produced different transition points relative to the leading and trailing edges of objects passing through the light beam.

Quality of beam alignment had a significant effect on estimated diameter, as indicated by comparing the results of the poor beam alignment test with the first standard test (Table 1). With misalignment of 10°, the LED alignment indicator on the receiver showed a very low excess gain. Upon completion of the poor beam alignment test, the sensor was realigned, but the new setting was slightly different from the setting used for the first standard test. Differences between the estimated diameters for the second and third standard tests (Tests 6 and 9, Table 1) were not statistically significant, but the estimated diameters for both tests were significantly different from those obtained in Test 1. This comparison suggests that even though the beam alignment was good and the LED alignment indicator had shown a high excess gain for all three standard tests, optimum alignment was difficult to discern with the alignment indicator.

The effect of resolution of the distance input on estimation of stalk diameter can be seen by comparing Tests 6 and 9 (Table 1), where the high-resolution distance input was used, with Test 7, where the low-resolution distance input was used. Differences in estimated diameter were not significant, according to the Waller-Duncan multiple range test. All tests were performed at constant travel speeds with no induced fluctuations.

The lateral position of the plants passing through the emitter and receiver pair had a statistically significant effect on estimation of stalk diameter. When the plant's path was centered between the photoelectric pair in the second and third standard tests (Tests 6 and 9), the estimated stalk diameter was significantly different than when the plants passed near the receiver (Test 3) and near the emitter (Test 4), according to the Waller-Duncan multiple range test (Table 1). But, there was no significant difference in diameter estimation between the dowel rods passing close to the receiver or close to the emitter (Test 3 and Test 4, Table 1). The greatest change was when the plants passed very close to the emitter, causing the average diameter estimation to increase 3.7% as compared to the estimated diameter at the middle position. With the plants passing close to the receiver, the average estimated diameter was 2.5% higher than for the standard tests in the

Table 1: Laboratory comparison of variable effects on plant stalk diameter estimation

Test No.	Condition	Mean, normalized diameters[1]	
1	Standard	1.007	c[2]
2	Poor alignment	1.090	a
3	Objects near emitter	0.994	c d
4	Objects near receiver	0.982	d
5	Small aperture	1.050	b
6	Standard	0.954	e
7	Low-resolution distance input	0.951	e
8	Multiple filtering events	1.043	b
9	Standard	0.961	e

[1] Plants were simulated by using dowel rods whose nominal diameters ranged from 6.35 mm to 25.4 mm.
[2] Estimated diameters with the same letters do not differ significantly at the 5% level, based on the Waller-Duncan multiple range test.

middle position.

The effect of reduced aperture size (0.5 mm) was quite similar to a poorly aligned beam for the same reason, i.e., both changes reduced the excess gain. The average estimated diameter with the small aperture (Test 5, Table 3) increased 9.3% from the standard tests (Test 6 and 9) with the 1 mm aperture which was statistically significant in the Waller-Duncan multiple range tests.

The small aperture (Test 5) was used for comparison with the multiple filtering events test (Test 8). A small aperture was necessary to preclude flooding of light around the small objects used in this test comparison. The results of the Waller-Duncan multiple means comparison (Table 1) showed no significant difference in estimated stalk diameter with twenty smaller objects to be filtered out (Test 8) as compared to Test 5 without any filtering events. From the close similarity of the results of the two tests, it is evident that the processing speed of the computer and acquisition hardware was sufficient to filter out real-time high levels of small, consecutive, close-together objects.

Early Growth Stage Tests

Field conditions in a green house, obviously, only approximate those encountered in typical fields. Uniform terrain, and the absence of weeds and clods were conditions that existed in the greenhouse that will reduce the applicability of these results to actual field conditions.

The first filter used in post-processing was a minimum diameter filter, which removed any plants whose stem diameters were too small to likely be an actual corn plant. The minimum diameter acceptance level was set at 4.6 mm, which was approximately an aperture width larger than two standard deviations below the actual mean plant diameter.

The second filter used in post-processing was a distance filter which removed any plant within 50 mm of a previous plant. This filter eliminated most false counts due to leaves that occurred just before or just after a plant stem.

Of the eight tests conducted, the sensor had only one erroneous count in each of the two tests, and missed counting only one plant in two other tests. The plant spacing estimation results showed that the standard deviation of the mean spacing error was ±3.1% (Plattner 1996).

Although the distance filter eliminated several false counts, there were two situations where filtering errors could still occur. First, a leaf of significant apparent size, occurring before the actual plant stem would be recorded as the plant while the actual stem would be filtered out. Second, errors could also result due to the elimination of multiple plants if they were spaced closer together than the distance filter range. In the greenhouse bed, the latter error possibility did not occur since there were no close-spaced plants.

Analysis of the plant diameter data which was collected on two test in each of two rows, shows that the sensor's ability to estimate a plant's diameter was not nearly as good as its ability to estimate plant spacing. Even after minimum diameter and distance filters were applied, the average plant diameter error was ±7.0%. A maximum plant diameter filter was applied to remove plant diameter measurements that were more than two standard deviations greater than the mean

(15.9 mm) of the manually measured data. Errors occurred when the width of the leaf was measured instead of the stem, when measurement at leaf attachment points increased the apparent plant diameter, or as mentioned before, when the actual plant was eliminated by the distance filter, and the leaf hanging over on the leading side of the plant was recorded as a plant diameter.

Harvest Stage Tests

The most extensive and realistic field test was performed at harvest. Many factors were present at harvest that were not experienced in the laboratory, e.g., weeds, leaves, actual stand variations, outdoor conditions, and machine effects. The effect of travel speed, aperture size, and the mechanical fingers were tested.

During testing, a real time minimum plant diameter software filter was set at 7.6 mm and 12.7 mm when the standard (1.0 mm) aperture and no aperture were used, respectively. This preliminary filter was kept small to retain all real plants, but still reduced the initial data file size by deleting any objects that were obviously too small to be a plant. This filter proved to be quite effective, usually removing a greater number of plants (false plants and weeds) from the data set than there were actual plants in the row.

As in the early growth stage field tests, a minimum diameter filter and a distance filter were applied in post-processing. Both operated as described previously.

The limit levels of the filters were optimized for each set of test conditions, according to the size of aperture and the presence or absence of the mechanical fingers. The limit levels were determined by comparing the sensor data with the actual manually measured data within each testing condition. The optimum filter limit levels minimized the total errors due to false counts included in, and actual plants eliminated from, the data set.

The profiled data of a test run (Fig. 1, Test No. 6) illustrates typical data sets. Several filtering errors are evident: Extra plants were not able to be filtered out (2.5 m, 3.7 m) and a double was eliminated by the distance filter (3.0 m).

The average plant spacing was measured most accurately when the sensor was equipped with the standard aperture and mechanical fingers resulting in a 6.2% error (one standard deviation). With no aperture the error was 8.4%, and with no aperture and the fingers removed, the error increased to 9.0%. The data from the test with the standard aperture and mechanical fingers are in Table 2.

The "control" runs, with the leaves stripped off the plants, had the purpose to remove the effect of leaves, the primary source of false counts. As desired, there normally were not any false counts. While no formal statistical test was performed on the effect of leaves, it was very evident from the control test that nearly all false counts can be attributed to plant leaves blocking the sensor's light beam.

Table 2: Comparison of manually measured and sensor estimated plant spacing at harvest stage.

Speed, kph	Test No.	No. of Plants		Manually Measured Spacing, mm		Sensor[1] Estimated Spacing, mm		Sensor Spacing Error, %
		Manual	Sensor	Mean	Std. Dev.	Mean	Std. Dev.	
3	1	22	25	215.6	113.3	192.0	105.1	-10.9
	6	23	25	198.1	103.8	184.2	109.5	7.0
	11	20	23	230.1	123.2	200.2	111.9	-13.0
	12[2]	12	12	207.3	144.7	216.4	158.8	4.4
5.5	2	22	23	204.5	101.1	192.2	67.3	-6.0
	4	23	23	195.1	76.6	192.3	68.6	-1.4
	9	23	23	195.2	100.0	203.1	105.6	4.1
	7[2]	10	10	248.4	91.3	248.1	82.7	-0.1
8	3	21	23	204.2	58.0	210.4	75.3	3.0
	5	22	22	220.3	134.1	221.2	141.5	0.4
	10	22	22	176.6	61.1	179.2	78.2	1.4
	8[2]	8	10	193.5	53.3	193.2	55.4	-0.2

[1] Sensor was fitted with 1 mm apertures and spring-loaded leaf deflectors.
[2] All lower leaves removed from corn plants prior to test.

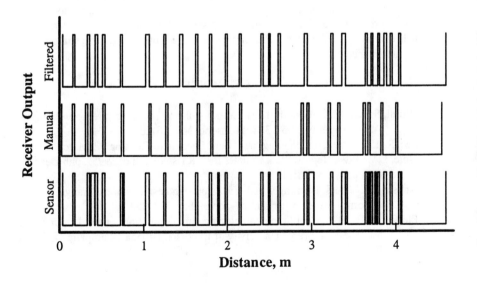

Figure 1: The raw sensor data, manually measured data, and the filtered data profile from a typical harvest stage test (Test No. 6).

When the remaining plant's diameters (after the diameter and distance filters eliminated most of the false counts) were compared to the manually measured plant diameters, there were large differences. There were many estimated plant diameters that were a few times larger than their actual size. Leaves attached at the base of the plant and hanging straight down along the plant can make the estimated diameter quite bigger than the actual stalk size. An even larger effect, causing an increased estimated diameter, resulted from the spring-loaded mechanical fingers bending the plants slightly until they exerted a horizontal force strong enough to overcome the spring. While this was happening the stalk would remain in the light beam longer than normal.

A maximum diameter filter was applied so that only plant diameters below a determined limit were analyzed. The maximum diameter limit was set at approximately two standard deviations above the mean, resulting in an 8.3% error in estimating stalk diameter. Often the filter removed more than a fourth of the plants in the test, resulting in significant reductions in the error in estimating stalk diameter.

SUMMARY

A corn population sensor was developed and tested. Laboratory testing revealed that travel speed was a significant factor in the estimation of plant diameter as a result of the photoelectric's response time (correctable in software). Laboratory testing showed that plant population accuracy was not effected by the type of distance input, multiple filtering events, aperture size, quality of beam alignment, or lateral plant position. The sensor was able to estimate average plant spacing at early growth stage with a ±3.1% error. At harvest stage, there was a ±6.2% error when estimating average plant spacing. In both field test, filtering was able to remove small stalks such as weeds, but large corn leaves were still a source of interference

REFERENCES

Birrell, S., and K.A. Sudduth. 1995. Corn population sensor for precision farming. ASAE Paper No. 95-1334. St Joseph, MI: ASAE.

Fee, R. 1994. Precision planting time trials. Successful Farming 92(March):34-37.

Gore, L.M. 1996. Report: Stalk Counter for VRT study fall of 1995. John Deere & Company. Moline, IL

Nielsen, R.L. 1995. Planting speed effects on stand establishment and grain yield of corn. J. Prod. Agric. 8(July-September):391-393.

Plattner, C.E. 1996. Real time photoelectric corn population sensor. Unpublished MS Thesis. Library. University of Illinois at Urbana-Champaign, Urbana, IL.

Nozzle Selection and Replacement Based on Nozzle Wear Analysis

K. Ballal
P. Krishnan
J. Kemble
A. Issler

Agricultural Engineering Department
University of Delaware
Newark, DE

ABSTRACT

Overapplication of pesticides results in increased production costs and may also cause environmental damage due to excess chemicals contaminating both ground and surface water. The specific objectives of this research were to compare the wear rates of brass, nickel-coated brass, plastic and stainless steel 8004 fan nozzles at 137, 275 and 551 kPa (20, 40 and 80 psi) and to analyze the wear data and select the appropriate types of nozzle using cost-benefit analysis. The nozzles were placed randomly, using the randomized completely blocked design on a test stand which consisted mainly of a 208L (55 gal) tank with a pumping system to recirculate the mixture of water and abrasive(Georgia Kaolin Clay). After the experiment was run for a predetermined interval of time, the equipment was stopped. The nozzles were removed and the flow rates were measured using another test stand with an electromagnetic flowmeter. Brass had the highest wear rate percentage at 137 kPa(27% wear after 120 hours) and 275 kPa nozzle pressures(33% wear after 100 hours). At 551 kPa nozzle pressure, after 36 hours, the wear rate for brass nozzle was 22%. At 137 kPa(20 psi) nozzle pressure, after 120 hours , the wear rate for nickel-coated brass nozzles was 15.82% as compared to 16% wear for stainless steel nozzles. At 275 kPa(40 psi) nozzle pressure, after 58 hours, the wear rate for nickel-coated brass nozzles was 15% as compared to 13% wear rate for stainless steel nozzles. Since the cost of nickel-coated brass nozzles is one- third less than that of stainless steel nozzles, it is economical to use nickel-coated brass nozzles. At 551 kPa (80 psi) nozzle pressure, after 36 hours, stainless steel nozzles had the least wear percentage(15%) as compared to nickel-coated brass (20%). Hence at higher pressures it is profitable to use stainless steel nozzles. The wear rates for plastic nozzles were very high at each of the selected pressures.(20% at 137 kPa, 27% at 275 kPa and 31% at 551 kPa)

INTRODUCTION

The Environmental Protection Agency (1990) estimated that US users of pesticides spent about $7.4 billion for pesticides in 1988. Due to increasing concern about crop production costs and environmental pollution, it is essential to spray pesticides with precision and care. A very large portion of pesticides are applied with sprayers through hydraulic pressure nozzles. These nozzles are of many different types(flat spray tips, wide angle full cone spray tips, etc.), capacities and can be made up of different materials like brass, stainless steel, ceramic, plastic and nylon. Since nozzle orifices greatly influence application rate, it is important to know the wear rates of nozzle orifices. Factors which influence nozzle wear include spraying pressure, duration of test, type and concentration of material used in the spray mixture, time of use of abrasive before it is changed during the test, type and size of nozzle, shape and material of the orifice. Much of the difference in nozzle wear rates is due to the different operating conditions used when testing nozzles.

There is sufficient evidence to prove that spray tips may be the most neglected component in today's farming; yet they are among the most critical items influencing the proper application of valuable agricultural chemicals. Sprayer calibration clinics conducted in Ohio revealed that more than one-third of the sprayers surveyed were overapplying chemicals. The major reason for overapplication was worn nozzles. For example, a 10 percent over application of chemical on a twice-sprayed 1000 acre farm could represent a loss of $2,000-$10,000 based on today's chemical investments of $10.00-$50.00 per acre. This does not take into account potential crop damage. The best way to determine if a spray tip is excessively worn is to compare the flow rate from the used tip with that from a new tip of the same size and type. Spray tips are considered excessively worn and should be replaced when their flow exceeds the flow of a new tip by more than 15%.It would be of interest to the farmers and to chemical spray applicators to know when to replace the worn out nozzles with new nozzles. Replacement of nozzles would save them time and money. Manufacturers also want their nozzles to have long life and high quality to confront the market competition.

The specific objectives of this research were:

1.   To compare the wear rates of brass, nickel-coated brass, plastic and stainless steel nozzles.
2.   To analyze the wear data and select the appropriate types of nozzle using cost-benefit analysis.

LITERATURE REVIEW

There has been considerable interest in nozzle wear rates and there are several publications regarding the wear rates of various nozzles for different operating conditions. Wilson (1943) measured percentage increase in flow rates through brass discs after spraying various carriers with fungicides at 2760 kPa (400 psi) pressure. He used 120 g of carrier/L (1.0 lbs/gal) of water and found that the flow rate increases within a range of 1.8% to 48% of the initial flow rate after spraying for 30 minutes. Novak and Cavaletto (1988) also conducted wear tests using a herbicide (atrazine) and fan tips. Their tests with 0.76 L/min(0.2 gpm) tips operated at 276 kPa (40 psi) indicated usage times of about 100, 200, and 400 hours before 10% increase in flow rate of brass, nylon, and stainless steel tips, respectively.

Reed (1984) reviewed the results of nozzle wear tests conducted by several different researchers. The results were represented by a wear number which was defined as the relative wear life of a spray tip material using brass as the reference. Brass had a wear number =1. A comparison of the relative wear rates between brass and stainless steel fan tips by five different researchers revealed a range of wear numbers indicating from no difference in wear life to stainless steel lasting 19 times longer than brass. Reed (1984) also reported that the wear rate of an 8001 brass fan spray tip (Spraying Systems Co.) was greatly influenced by pressure. For example, the relative wear life of brass tips was five times longer at 138 kPa (137 kPa) than at 414 kPa (60 psi). A comparison of relative wear between 8001 stainless steel and brass fan spray tips showed that stainless steel had 9.5 times longer life when operating pressure was 138 kPa (137 kPa) but only four times longer when operated at 414 kPa (60 psi). Obviously, operating conditions can greatly influence the results while measuring nozzle wear rates.The literature on nozzle wear rates indicates considerable differences among reported results. Much of the difference in nozzle wear rates is due to the different operating conditions used when testing nozzles. There is considerable need for reliable information on wear rates of nozzles.

Nozzle wear affects the flow rate, droplet size and the spray pattern. The relative wear rates of nozzle tip materials vary at different times of usage. It is found that the percentage increase in flow rates varies directly with approximately the square root of the time of use. When nozzles wear, they no longer produce the spray pattern essential for uniform coverage (Ozkan et al., 1993). It is found that for the nozzles and the amount of wear tested, there is little difference in the width of spray deposit patterns of new and worn nozzles. However, there are greater differences between new and worn nozzles in volumes of liquid collected in the centers of the patterns than at the edge of the patterns (Ozkan et al., 1991). Similarly, nozzle wear may affect droplet sizes which may be an important factor in achieving satisfactory pest control. Deposition efficiency of droplets on targets is affected by droplet size. (Ozkan & Reichard, 1993).

EQUIPMENT AND PROCEDURE

Nozzle Wear Test Stand

A nozzle wear test stand used to wear out the nozzles was procured from USDA-ARS laboratory, Wooster, Ohio.The test stand consists mainly of a 208 L (55 gal) tank with pumping system to recirculate the mixture of water and abrasive (Georgia Kaolin Clay flat D^R) through the nozzles (Reichard et al., 1991). Twenty pounds of Kaolin clay was mixed with forty gallons of water to form a mixture. Up to six nozzles can be mounted on each of the three pipes at the top of the tank. Quick Teejet nozzle assemblies were used so that the spray tips could be rapidly removed to measure the flow rates with water. The spray mixture was delivered from a Tee Valve to the center of each pipe containing nozzles. The valve allows the operator to direct the liquid flow to any of the desired pipes containing the nozzles. A diaphragm type Model PA40 Bertolini pump was used to supply the liquid to the nozzles. It was driven with a belt drive from a 2.2 kW (3 hp) motor.

A bypass flow type, pressure regulator was used to maintain the desired pressure at the nozzle and return the excess flow to the tank.

It is essential to maintain uniform mixture of water and the abrasive during nozzle wear tests. A mechanical agitation system was used. Four paddles, each 5.1 cm (2 in.) wide and 9.5 cm (3.75 in.) long, are mounted on a shaft that is driven at 250 r/min through a chain drive from a 0.75 kW (1 hp) motor. The paddles are twisted so that their tips are at about a 20° angle to the axis of the shaft. During operation, the tank was covered with a clear, rigid, plastic cover. The cover is fabricated so that it could be easily removed or replaced. It also contains a soft gasket to seal the cover and flange around the top of the tank. The test stand was mounted on wheels so that it could be easily moved around. All electric controls were mounted on the frame of the test stand. Either of the motors used to drive the agitator and pump could be controlled independently or from a timer. The agitator was started for about 10 minutes before the beginning of the test to ensure a uniform mixture of water and abrasive. An electro mechanical timer was used to stop the motors at the end of a test(Reichard et al., 1991).

Nozzle Flow Rate Measurement Test Stand

An experimental nozzle flow rate measurement test stand with an electronic flowmeter was built in-house. It was used to measure the flow rates of water delivered by the nozzles. The nozzle-flow-rate test stand is basically a pumping system that recirculates clean water through a nozzle mounted over a 57 L (15 gal.) tank. A cast iron roller pump capable of delivering 24.7 L at 0 kPa (6.5 gpm at 0 psig), 17.8 L at 345 kPa (4.7 gpm at 50 psig), 13.3 L at 689.5 kPa (3.5 gpm at 100 psig) running at 1800 rpm was used. The pump is driven with a belt drive from a 2.2 kW (3 hp) motor. The system has a pulsation dampener and high and low pressure regulating system. A Brooks instrument electromagnetic flowmeter with 0.38 cm (0.15 in.) diameter flow tube and a pressure gage were mounted in line with the nozzle. The meter is set up for a range of 0.6-6.8 L/min (0.18-1.8 gpm) but could be set up for other ranges. The electrical unit displays flow rate in mL/min but it could be programmed to display other engineering units. The flow-rate-test stand was also mounted on wheels, so that it could easily be moved around.

Test Procedure

The four nozzles used in the study were 8004 fan nozzles made of brass, nickel-coated brass, plastic, and stainless steel. The tests were performed at three different nozzle pressures of 137, 275, and 551 kPa (20, 40, and 80 psi). Three nozzles each of brass, nickel-coated brass, plastic, and stainless steel nozzles were used in the study for each pressure condition. A total of twelve nozzles were used at each pressure condition. Probably the major reason for the large differences reported in nozzle wear rates is due to the great variation in abrasive materials and lack of consistency in carrying out the tests. Also of importance is the time of use and the number of recirculations of the mixture before it is used. Since many manufacturers use various kaolin carriers for their wettable powder formulations of pesticides, kaolin was selected to mix with water for the nozzle wear tests. Up to six nozzles can be mounted on each of the three pipes. Since there were four types of nozzles and eighteen nozzle positions, two alternatives were considered:

1. To use all eighteen nozzle positions by having different number of nozzles on each pipe.
2. To use lesser nozzle positions and block the remaining nozzle positions with stoppers.

The first approach would lead to an unbalanced design because the number of nozzles used under each type would be different. The second approach was used in the formulation of the problem. Four nozzles were mounted on each of the three different pipes on the nozzle wear test stand. The selection of the nozzles was random. The two central positions in each pipe were blocked with stoppers so that there is no flow of the abrasive mixture through the orifices. A randomized completely blocked design is used when variability arising from known nuisance sources can be systematically controlled. In this case there are four nozzles and three pipes. If the pipes differ, then it will contribute to the variability observed in the flow rate. As a result, the experimental error would reflect both random error and the variability between the pipes. Since the objective of the design was to make the experimental error as small as possible, the nozzles were placed in such a manner that there was a nozzle of each type in each of the three pipes. In such a design any variability in the pipe affects all the nozzles equally. The agitator was started for about 10 minutes before the beginning of the test to ensure a uniform mixture of water and abrasive. The abrasive mixture was allowed to pass through the nozzles for a certain duration of time after which the system was automatically shut off by a preset timer. Then the nozzles were removed and cleaned with a plastic wire brush and water. At this juncture the flow rate of each of the nozzles was determined with the help of an electromagnetic flowmeter. Three readings were taken for each nozzle so that the error in the readings was minimized. The order and the identity of the nozzles were maintained throughout the testing period for each nozzle pressure. The flow through each nozzle was measured before the experiment was started at each pressure level. At a nozzle pressure of 551 kPa, the experiment had to be stopped after 60 hours of operation. This was due to the fact that the equipment could not withstand the high pressure. Also, the abrasive mixture was replaced at each pressure level.

The wear rate was calculated as follows:

Wear rate = ((Measured flow - Initial flow)/ Initial flow)*100

Prior research conducted has shown that much of the wear takes place initially, i.e., less increase in flow rate with time (Ozkan & Reichard, 1993). Hence, the flow rates of the nozzles were measured at very short intervals in the initial stage of the study. The wear rate data was analyzed using SAS (Statistical Application Software). This was done for two reasons:

1. To find out whether there was a location effect. For example, it was necessary to know whether the wear rate for any nozzle tip is significantly different from the wear rate of a similar nozzle tip.

2.   Whether the wear rates between any two nozzles were significantly different for any given time period.The above two effects were observed for every ten hours of usage for each pressure level.

A cost-benefit analysis was carried out for the selection of nozzles after analyzing the data.This would give the chemical spray applicators a better idea of the optimal pressure to be used in their operation.

RESULTS AND DISCUSSION

Figures 1, 2, and 3 show the wear rate (flow rate increase with time) for brass , nickel-coated brass, plastic and stainless steel nozzles at 137, 275 and 551 kPa(20, 40 and 80 psi) respectively. For all pressures, the flow rates of brass tips increased more rapidly with time of use than with any of the other materials. Also, for all the pressures tested, the flow rates of stainless steel tips increased less with time than with any of the other tip materials. The curves for wear rate for plastic tips were always between brass and stainless steel tips at pressures 137 and 275 kPa. At a nozzle pressure of 551 kPa, plastic tips had the highest wear rate. For all the nozzle tips the wear rate increased at a high rate in the beginning and then increased at a lower rate. Towards the end of the tests, the wear rate for nickel - coated nozzle was greater than the wear rate for stainless steel tips. At pressures 137 and 275 kPa, the slopes for the curves for plastic, nickel-coated brass and stainless steel tips were generally about the same. A comparison of nickel-coated brass and stainless steel tips indicated that nickel-coated brass tips had longer usage times at 137 kPa, whereas stainless steel tips had a higher usage time at pressures 275 and 551 kPa. Another important point noted was that for pressures 275 and 551 kPa, during the initial period, the wear rate for nickel-coated brass nozzles was less than any other nozzle tip.

Wear Rates at 137 kPa (20 psi) Nozzle Pressure

At a nozzle pressure of 137 kPa, brass had the highest wear rate percentage (27% wear after 120 hours). For the same time period, the wear rate for nickel-coated brass nozzles was 15.82% as compared to 16% wear for stainless steel nozzles. Since the cost of nickel-coated brass nozzles is one- third less than that of stainless steel nozzles, it is economical to use nickel-coated brass nozzles. Plastic nozzle tips had a wear rate of 20%.

Wear Rates at 275 kPa (40 psi) Nozzle Pressure

At a nozzle pressure of 275 kPa, brass had the highest wear percentage of 33% after 100 hours. After 58 hours, the wear rate for nickel-coated brass nozzles was 15% as compared to 13% wear rate for stainless steel nozzles. At the end of 100 hours, the wear rate for nickel-coated brass nozzles was 19% whereas for stainless steel nozzles, the wear rate was 15% for the same time period. Plastic nozzle tips had a wear rate of 27%.

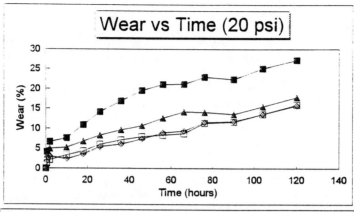

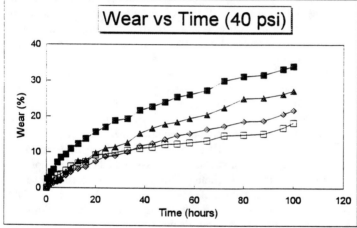

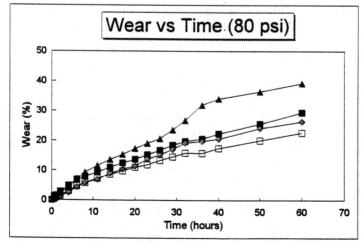

□ - steel;   ◇ - nickel coated brass;   ■ - brass nozzle;   △ - plastic

Figures 1,2 &3: Nozzle wear vs Time for Pressures 20, 40 and 80 psi

Wear Rates at 551 kPa (80 psi) Nozzle Pressure

After 36 hours, brass had a wear rate of 22% at a nozzle pressure of 551 kPa. Stainless steel nozzles had the least wear percentage(15%) as compared to nickel-coated brass (20%). Hence at higher pressures it is profitable to use stainless steel nozzles. Plastic nozzles had the highest wear rate of 31%. As the nozzle pressure increased, the wear rate for plastic tips increase had a higher rate than for any other nozzle tip.

SAS Output

The SAS results indicate that there does not exist any location effect. The p values for the location effect are greater than 0.05 which suggests that the location of the nozzles did not make any significant difference in the wear rate measurement. (Aside: Location effect will be significant only if the p value is less than 0.05). The wear rate for the nozzles was not significant during the initial time period. The p values were greater than 0.0825 (0.05/6 because there are six possible combinations to be tested). As the time increased beyond 30 hours the wear rate between brass and other nozzles becomes significantly different whereas the wear rate between the other nozzles was not significantly different. The wear rate between plastic and nickel -coated nozzle and plastic and stainless steel nozzle was significantly different at a higher p value of 0.1. Hence we can conclude that at 137 kPa pressure, the wear rate of nickel-coated brass nozzle and stainless steel nozzle was not significantly different. Hence both these nozzles behave similarly. At 275 kPa nozzle pressure, the wear rate between brass and other nozzle tips was significantly different from the beginning. After 40 hours, plastic behaved differently from nickel- coated and stainless steel nozzles. After a period of 80 hours, the wear rates for nickel- coated brass nozzle tips and stainless steel nozzle tips become significantly different. At a nozzle pressure of 551 kPa, the wear rate between any two nozzles was significantly different from the beginning of the experiment. In other words each nozzle behaved differently from the other.

Cost- Benefit Analysis

A cost-benefit analysis was carried out for the selection of nozzles at different pressures for every twenty hours. It takes twenty hours for one application for a 500 acre (220 hectares) farm using a thirty feet length boom at a speed of 6 mph(9.6 km/hour). The cost of two chemicals were taken into account. The cost of the nozzle, though negligible as compared to the cost of the chemicals was added to the final cost. Eighteen nozzles are used during an application.

Following is the cost per nozzle:
Brass - $2.5, Nickel-coated brass - $4.5, Plastic - $1, Stainless steel - $6

Sample calculation at Nozzle Pressure 137 kPa (See Fig. 4)

Brass nozzles: The first column represents the flow rate in liters/min at time t=0. The second column represents the flow rate in liters/min at time t=20 hours. The third column indicates the excess chemicals in gallons used after 20 hours. It is calculated as follows:

$$(1.2367-1.1039)/3.785*60*20 = 42.10 \text{ gallons}$$

20 psi

Time	Material	Initial flow lpm	Final flow lpm	Excess gal	Excess $	Excess $	Nozzle to be used
Time=20 hours	brass	1.039	1.2367	42.10	1615	1163	
	nickel	1.195	1.437	13.41	581	437	Nickel-coated
	plastic	1.082	1.597	23.68	901	647	
	steel	1.092	1.633	17.15	747	564	
Time=40 hours	brass	1.039	1.2997	61.13	2324	1668	
	nickel	1.195	1.22	24.19	983	723	Nickel-coated
	plastic	1.082	1.19	34.24	1294	927	
	steel	1.092	1.933	26.66	1102	816	
Time=60 hours	brass	1.039	1.3433	75.90	2875	2061	
	nickel	1.195	1.243	31.48	1255	917	Nickel-coated
	plastic	1.082	1.223	44.70	1685	1205	
	steel	1.092	1.2057	30.91	1260	929	
Time=80 hours	brass	1.039	1.3633	82.24	3111	2229	
	nickel	1.195	1.2833	44.26	1731	1257	Nickel-coated
	plastic	1.082	1.2067	49.05	1846	1321	
	steel	1.092	1.246	43.37	1725	1260	
Time=100 hours	brass	1.039	1.375	85.95	3249	2328	
	nickel	1.195	1.294	47.65	1857	1347	Stainless-steel
	plastic	1.082	1.244	51.36	1933	1382	
	steel	1.092	1.252	45.27	1796	1310	
Time=120 hours	brass	1.039	1.404	95.14	3592	2572	
	nickel	1.195	1.323	56.85	2200	1591	Nickel-coated
	plastic	1.082	1.2746	61.06	2294	1640	
	steel	1.092	1.298	56.05	2198	1597	

40 psi

Time	Material	Initial flow lpm	Final flow lpm	Excess gal	Excess $	Excess C	Nozzle to be used
Time=20 hours	brass	1.591	1.8397	78.85	2984	2139	
	nickel	1.628	1.7492	38.43	1513	1102	Nickel-coated
	plastic	1.591	1.745	49.14	1850	1323	
	steel	1.578	1.724	46.29	1834	1337	
Time=40 hours	brass	1.591	1.9433	111.69	4209	3012	
	nickel	1.628	1.812	58.34	2256	1630	Nickel-coated
	plastic	1.591	1.8345	77.20	2896	2068	
	steel	1.578	1.759	57.38	2247	1632	
Time=60 hours	brass	1.591	2.008	132.21	4974	3556	
	nickel	1.628	1.89	79.89	3059	2203	Stainless-steel
	plastic	1.591	1.91	101.14	3788	2704	
	steel	1.578	1.782	64.68	2519	1826	
Time=80 hours	brass	1.591	2.087	157.25	5907	4222	
	nickel	1.628	1.93	95.75	3650	2624	Stainless-steel
	plastic	1.591	1.989	126.18	4722	3369	
	steel	1.578	1.814	74.82	2897	2085	
Time=100 hours	brass	1.591	2.1317	171.42	6436	4598	
	nickel	1.628	1.9812	111.98	4256	3055	Stainless-steel
	plastic	1.591	2.023	136.96	5124	3656	
	steel	1.578	1.865	90.99	3500	2525	

80 psi

Time	Material	Initial flow lpm	Final flow lpm	Excess g	Excess $	Excess $	Nozzle to be use
Time=20 hours	brass	2.273	2.595	96.92	3733	2672	
	nickel	2.307	2.5713	83.79	3205	2307	Stainless-steel
	plastic	2.27	2.6628	124.53	4661	3326	
	steel	2.2956	2.515	79.07	3056	2208	
Time=40 hours	brass	2.273	2.776	159.47	5990	4281	
	nickel	2.307	2.78	149.96	5672	4064	Stainless-steel
	plastic	2.27	3.04	244.12	9119	6502	
	steel	2.2956	2.66	125.04	4770	3429	
Time=60 hours	brass	2.273	2.939	211.15	7917	5653	
	nickel	2.307	2.916	193.08	7279	5209	Stainless-steel
	plastic	2.27	3.16	282.17	10537	7512	
	steel	2.2956	2.779	162.77	6176	4431	

Figure 4: Cost-Benefit Analysis for Nozzle Pressures 20, 40, and 80 psi

The next two columns give the cost of the excess chemicals used $37.28/gallon and $26.56/gallon respectively.

For example , the total excess cost = 42.10 *37.28 + 2.5*18 = $1615

The last column represents the recommended nozzle based on the cost-benefit analysis. This procedure was carried out for all the three pressures at a time interval of 20 hours.

CONCLUSIONS AND RECOMMENDATIONS

From the results discussed, we conclude the following:

1. The relative wear rates of nozzle tip materials can vary with the percentage flow rate increase selected for the comparison
2. At nozzle pressures 137 and 275 kPa brass nozzles had the highest wear rate.
3. At 551 kPa nozzle pressure, plastic nozzles had the highest wear rate.
4. At a nozzle pressure of 137 kPa and during the initial period of 275 kPa, nickel-coated brass nozzles had the lowest wear rate.
5. Towards the end of 275 kPa test and at 551 kPa nozzle pressure, stainless steel nozzles had the lowest wear rate.6. At pressures 137 and 275 kPa, the wear rates for plastic nozzles was always between nickel-coated brass and stainless steel nozzles

RECOMMENDATIONS

1. At a nozzle pressure of 137 kPa, nickel-coated brass nozzles are recommended because of the lower cost associated with it.
2. At a nozzle pressure of 551 kPa, stainless steel nozzles are preferred.
3. At 275 kPa nozzle pressure, upto 40 hours, it is preferable to use nickel-coated brass nozzles. For the remaining period, stainless steel nozzles should be used.
4. High phosphor nickel-coated brass nozzles should be used because of the better resistance to wear.

REFERENCES

Ozkan, M.E., D.L. Reichard, and K.O. Ackerman. 1991. Effect of nozzle wear of spray pattern. Trans. ASAE 35(4):1091-1096.

Ozkan, H.E., and D.L. Reichard. 1993. Effect of nozzle wear on flow rate, spray pattern and droplet size distribution of fan-pattern nozzles. Trans.ASAE 38(4):1071-1080.

Reichard, D.L., M.E. Ozkan, and R.D. Fox. 1991. Nozzle wear rates and test procedure. ASAE 34(6) : 2309-2316.

Reed, T. 1984. Wear life of agricultural nozzles. ASAE Paper No. AA84-001. ASAE, St. Joseph, MI .

Dynamics of Peanut Flow Through a Peanut Combine

B. Boydell
G. Vellidis
C. Perry
D.L. Thomas
R. Hill

NESPAL & Biological and Agricultural Engineering
The University of Georgia
Coastal Plains Experiment Station
Tifton, GA

R.W. Vervoort

Crop and Soil Science Department
University of Georgia
Athens, GA

The accuracy of yield monitors and any resulting yield maps relies not just on the ability of the monitor to record incoming grain flow but also on the systems ability to relocate these yield measurements back to the space from which they originated (Eliason et al. 1995). The time lag between when a crop is collected at the header and when it is measured at the yield monitor has previously been described by Borgelt and Sudduth (1992), Murphy et al. (1995) and Searcy et al. (1989) and is a critical parameter in the monitoring system for the accurate mapping of yield. Commercially available yield monitoring systems for grain currently use a combine specific time lag measurement along with a record of combine speed to reposition the measured yield a calculated distance back along the combine travel path. The measurement error of a transfer model of this kind is a function of all location and yield sensing errors. Additionally, if there is mixing within the combine of crop from different locations in the field, the resulting memory of yield may mask the boundaries of rapid yield changes and further degrade accuracy. Searcy et al. (1989), Vansichen and De Baerdemaeker (1991) and Birrell et al. (1995) investigated the use of a first order transfer function to account for convolution within grain combines. On each occasion there was an accuracy increase for grain combines from deconvolution with the first order model. Birrell et al. (1995) however, concluded that this increased accuracy resulting from deconvolution may be insignificant for grain combines when weighed against that of simple time subtraction, taking into account the relative complexity of the deconvolution and the fact that there are additional sources of yield measurement error. As part of the development of a peanut yield monitor, experiments were conducted on a two-row peanut combine to determine the duration of time lag between pickup and yield measurement, and to characterize the convolution of peanut flow within the combine.

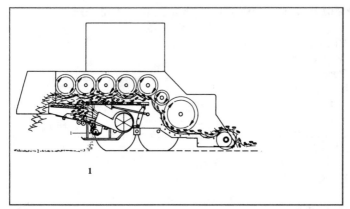

Figure 1 Internal schematics of a peanut combine. Floor auger and common location of yield monitoring devices (1). (diagram reproduced courtesy of KMC inc. Tifton GA.)

Mechanical peanut combine harvesting

To prepare for harvest peanut plants are dug, the nuts shaken free of soil and the whole plant inverted before being laid back on the soil surface to dry down to a moisture content suitable for harvest. With the dried peanut plants arranged in rows, a combine will operate in much the same way as a grain combine fitted with a pickup reel for harvesting windrows. This pickup reel feeds the vine-like bushes onto a throat elevator where they are drawn into a series of rotating drums and over sieves where the pods are separated from the vegetative section of the plant. Through these sieves, the nuts fall into a collecting hopper where a lateral floor auger (see Fig. 1) moves them across the bottom of the machine and into an air-duct which then delivers them up and into a collecting basket on the top of the combine. Experimental yield monitoring devices for the peanut combine that are being developed by the University of Georgia and Auburn University have generally been located in and around this lateral floor auger.

The "Blue peanut experiment" : Investigation into the time delay of flow through a peanut combine

Measurement of the time lag between peanut pickup and transport to the point of yield measurement (lateral floor auger) was determined in the field in 1995 by University of Georgia researchers. Inverted peanuts were sprayed with dye[1] such that there were two colored bands of nuts in each harvested bed (two rows, Fig. 2.). The lengths of the bands and band separation times were uniform. Initial determination of these distances was based on normal harvest speed and calculated

---

[1]Hi-Light™ Blue indicator for spray applications. Becker Underwood. 801 Dayton avenue, P.O. Box 667, Ames. IA 50010

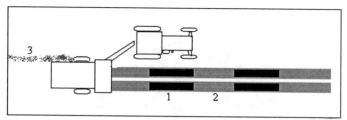

Figure 2 Field arrangement of one treatment. Two rows of peanuts in the field where 1 denotes colored peanuts, 2 demotes non-colored peanuts and 3 identifies the peanuts after combining and deposition on the soil surface by the lateral move floor auger 4.

such that each band and separation would be 20 seconds long. Peanuts were allowed to fall immediately to the soil surface after exiting the floor auger by disconnecting the fan and ducting that would normally deliver peanuts into the overhead basket. The resulting linear pile of pods were collected and bagged in 1.14m sections which corresponded to 1 second of travel time. The total number of pods and the weight of both colored and non-colored peanuts in each bag was determined. This experiment was repeated five times resulting in the measurement of ten separate colored bands.

RESULTS AND DISCUSSION

The five replicates within the "blue peanut" experiment were conducted on peanut yields that averaged, within row, from 3886 to 2382 kg/ha (3460 to 2121 lbs/ac) with the overall experimental average being 3045 kg/ha (2711 lbs/ac). The state averages for Georgia in the 1995 growing season was 3053 kg/ha (2718 lbs/ac) with 95% confidence that the average would fall between 3290 and 2190 kg/ha (2930 and 1950 lbs/ac). These state averages indicate that the yields under which the experiment was conducted are representative of common field conditions.

The actual mass (grams) of peanuts delivered was recorded as well as the number of pods. Due to the fact that farmers payments depend more often on mass than individual pod count, the mass was used to calculate blue peanuts as a percentage of total delivery. (Figs. 3a & 3b)

Throughout the experiment, breakthrough curves of the outflow were of consistently similar shape. Variation between curves was not significantly linked to the experimental feed rates at the 0.05 level. (Fig. 4.)

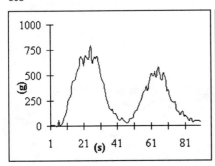

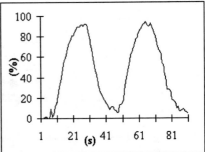

Figure 3a. The actual mass of blue peanuts (grams) delivered per second in row 32 over time.

Figure 3b. Blue peanuts as a percentage of the total mass of peanuts delivered over time.

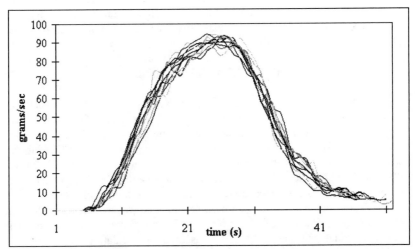

4. Percentage of blue peanuts in total outflow for all 10 pulses against time after pulse delivery. There is a 95% confidence that the predicted outflow curve (average) has a correlation that is between 99.65% and 99.43% with all curves and the mean residual error at any point is from 0.44% to 0.11%. There is an R^2 of only 0.08 and a Prob > F of 0.76 at the 95% level for a simple relationship between feed rate and residual error. The experiment was conducted over individual pulse feed rates that ranged from 3886 kg/ha to 2382 kg/ha (3460 lbs/acre to 2121 lbs/acre) against time.

Time lag

The time lag duration for this combine under normal operating conditions was calculated as being the difference between mean peanut delivery time "μd" and mean outflow time "μo". From these experiments it was calculated as being 14 seconds (Fig. 5).

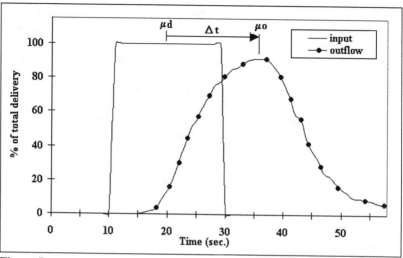

Figure 5. Actual blue peanut delivery (input) vs measured blue peanut outflow from the combine. Mean delivery is μd, mean outflow is μo and Δt is the time lag.(μo - μd = Δt)

Convolution of flow within a peanut combine

While the extent of time lag was suitably determined in the "blue peanut" experiment, the high degree of smoothing that occurred between the input curves and output curves indicates convolution of flow within the combine. Observations of flow convolution are common in runoff hydrology (Singh, 1988) and are explained by the multiple flow paths, each having different travel speeds between the point of rainfall and the point of runoff measurement. Similarly, a peanut traveling through a peanut combine has alternative flow routes between the pickup reel and the floor auger. The route taken is not a purely random process and is determined by the strength of connection between the pod and the vegetative sections of the peanut plant. If the connection is weak, a pod is likely to separate from the plant early in the threshing operation leaving it free to be walked directly to the floor auger. If, however, it has strong connections to the plant, the pod may not be separated until late in the combine process where it passes over a series of saws designed to cut the pods free. The net result is that pods from the same plant may enter the floor auger at significantly different times and mix with pods collected at different locations in the field.

When monitoring yield, the magnitude of error in a site specific yield measurement is dependent on both the degree of short term yield variation and the magnitude of within combine flow convolution. If the field has no variation in yield, or if the combine performs no convolution on the harvested product, a direct time correction will accurately reposition the yield values back in space. If, however, there is significant convolution within the combine, a deconvolution transformation will have to be applied to the outflow to accurately represent the yield in the field.

To test the impact of this convolution on yield measurements for peanuts, a time deduction was applied to the yield measurements (Fig. 6) and the average residual error between the monitored and known yield was calculated (Table 1). From this experiment it appears that on the borders of sharp yield changes, there is approximately a 10 second period during which the true nature of yield is masked by residual flow. Areas where there are rapid yield changes, such as at the ends of rows and in those portions of the field that have been subjected to catastrophic insect pressure or heavy weed competition, will not be represented accurately on yield maps. The often discrete boundaries of these areas will be smoothed into local yield resulting in the appearance of yield depressions instead of discrete yield holes. While field yield values will remain accurate to the capacity of the sensor, no confidence can be placed in yield estimates for periods of time less than 10 seconds which in the case of a two-row combine relates to 11.5 meters of row (38').

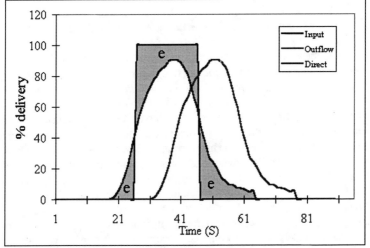

Figure 6. Known flow into the combine is denoted "input" and is square shaped. The bell shaped curves are the measured outflow of peanuts after passing through the combine. The curve that has been relocated in time by simple subtraction of a time lag component is denoted "Direct". Shaded regions denoted with the identifier "e" are areas of residual error between known input and the time corrected measured output.

Deconvolution of peanut yield measurements in the Fourier domain

Using the data obtained previously in the time delay experiments, it may be possible to partially reconstruct the input values of peanut yield as it was before convolution within the machine. There was no significant difference between the shapes of the outflow curves and that of the "average" curve, and there was no significant correlation between the curve shape and the feed rate (yield) within the tested ranges (see Fig. 4). This indicates that there is a fairly rugged transfer function within the combine that is relatively unaffected by feed rates under normal operating conditions. Consequently, it may be possible to model the transfer function within the combine, and in doing this, derive the true input yield from the monitored (convoluted) output flow.

The inverse Fourier transform of the product of two Fourier transforms is equal to the convolution of the two functions (Haberman, 1987). The measured outflow, OUT, may be deconvoluted to give true yield, YIELD, by transforming the known inflow, INFLOW, and outflow data into the Fourier domain and dividing output by input.

$$F_f = \frac{F_{OUT}}{F_{INFLOW}} \tag{1}$$

In which F_{OUT} and F_{INFLOW} are the Fourier transforms of OUT and INFLOW respectively. The result of this division is the Fourier transform of the transfer function, F_f. Thus by transforming the measured OUT and dividing by F and performing an inverse transformation one can reconstruct the YIELD.

$$YIELD = [\frac{F_{OUT}}{F_f}]^{-F} \tag{2}$$

In which $[..]^{-F}$ is the inverse Fourier transform operation. The degree of accuracy of this deconvoluted yield measurement, YIELD, should be independent of the rate of change of yield in the field since there is no longer a mixing component. Results of this transformation and the site specific accuracy are presented in Fig. 7 and Table 1.

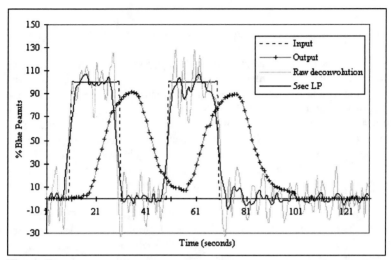

Figure 7 Graph presentation of the square input of blue peanuts, bell shaped convoluted output, raw deconvolution with high frequency noise and the 5 second LP yield estimates after deconvolution and smoothing with a 5 second low pass filter.

Table 1. Residual errors as a % of total flow after subtraction of simple time delay and deconvoluted yield estimations from the known square input pulses. Deconvolution columns 0 LP..9 LP represent duration in seconds of the low pass filter used to smooth over out the effects of high frequency noise.

	Simple time delay	Deconvolution (Duration of low pass filter)					
	0 LP	0 LP	2 LP	3 LP	5 LP	7 LP	9 LP
Average	24.80	14.50	10.15	8.51	6.69	7.14	8.35
s.d.	14.80	13.18	11.39	11.69	10.90	12.08	13.46
CI. 95%	7.60	2.27	1.96	2.01	1.87	2.08	2.31
Upper	32.40	16.77	12.11	10.52	8.56	9.21	10.66
Lower	17.20	12.24	8.20	6.50	4.82	5.06	6.03

Extent of flow convolution within a peanut combine

The convolution that occurred within this peanut combine between pickup and the yield monitoring site was significantly large to reduce confidence in the accuracy of site specific yield measurements. After a simple time delay correction of 14 seconds, the average error is 24.8% with 95% confidence that it will fall in the range 32.4~17.2%. A maximum average error of 15% (95% confidence) is not recovered for more than 10 seconds after the yield change, and maintained only while the yield remains constant. Deconvolution of the output flow produced a curve with high frequency noise, however, it approximated the sharp boundaries of the true yield very successfully indicating a capacity to provide site specific yield readings. After smoothing with a 5 second low pass filter, average errors were reduced from 14.5% to 6.69% with 95% confidence that the error will remain within the range 8.56~4.82%. The five second filter reduces site specific resolution to 5.7 m (19').

SUMMARY

It is evident from this research that the two-row peanut combine used in the blue peanut experiment subjects harvested product to significant convolution between pickup and transport to the site of likely yield monitoring. A simple time lag correction will not recover the site specific or short term accuracy of yield measurements. The distance and time period required to achieve an error less than 15% (95% confidence) is greater than 10 seconds for simple time lag correction while it is 5 seconds for deconvoluted data. The net result is that smaller regions of yield variability may be recognized with greater confidence using the deconvolution method than with the simple time delay method.

An obvious alternative method for acquiring non-convoluted yield measurements is to take them before the yield is subjected to convolution. Relocation of the yield monitoring site from the clean grain area into the throat elevator is one method to minimize convolution. While it would be difficult to discriminate between vegetative and reproductive product at the front end of a combining process, it may be the most accurate solution if indeed there is a demand for more site specifically accurate yield monitors.

ACKNOWLEDGMENTS

We would like to thank the staff of the Biological and Agricultural Engineering department and NESPAL on the CPES-Tifton for many tedious hours of peanut counting and for assistance with the work used as the basis for this paper. This work would not have been accomplished without the cooperation of the landholder, Mr Tim Ross, the manufacturer of our two row peanut combine KMC Inc., Tifton, GA. Also John Deere of East Moline, IL who supplied the tractor used throughout experimentation.

The mention of commercially available products is for information only and does not imply endorsement.

REFERENCES

Birrell, S.J., S.C. Borgelt, and K.A. Sudduth, 1995. Crop Yield Mapping: Comparison of Yield monitors and Mapping techniques. P. 15-31. *In* P.C Robert et al.. (ed.) Proc. of second International conference on Site-Specific management for Agricultural Systems. ASA-CSSA-SSSA, Madison, WI.

Borgelt, S.C., and K.A. Sudduth. 1992. Grain flow monitoring for in-field yield mapping. International Summer Meeting of The American Society of Agricultural Engineers, Charlotte, NC. 21-24 June. ASAE Paper 921022.

Eliason, M., D. Heaney, T. Goddard, M. Green, C. McKenzie, D. Penny, H. Gehue, G. Lachapelle, and M.E. Cannon, 1995. Yield Measurement and Field Mapping With an Integrated GPS System. P. 49-58. *In* P.C Robert et al.. (ed.) Proc. of Second International conference on Site-Specific management for Agricultural Systems. ASA-CSSA-SSSA, Madison, WI.

814 BOYDELL ET AL.

Haberman, R., 1987 Elementary applied partial differential equations, 2nd ed. Prentice Hall, N.J. 07632.

Murphy, D.P., E. Schnug, and S. Haneklaus, 1995. Yield mapping - A guide to Improved Techniques and Strategies. P. 33-47. *In* P.C Robert et al. (ed.) Proc. of second International conference on Site-Specific management for Agricultural Systems. ASA-CSSA-SSSA, Madison, WI.

Searcy, S.W., J.K. Schuller, Y.H. Bae, S.C. Borgelt, and B.A. Stout. 1989. Mapping of spatially variable yield during grain combining. Trans. ASAE 32(3):826-829.

Singh, V.P. 1988. Hydrologic systems, Volume 1, Rainfall-runoff Modeling. Prentice Hall N.J. 07632.

Vansichen, R., and J. De Baerdemaeker. 1991. Continuous wheat yield measurement on a combine. P. 346-355. *In* Automated Agriculture for the 21st Century, Proceedings of the 1991 symposium. ASAE Publ.. 11-91, ASAE, St. Joseph, MI.

Precision GPS Flow Control for Aerial Spray Applications

I. W. Kirk
Agricultural Research Service, US Department of Agriculture
Southern Crops Research Laboratory
Areawide Pest Management Research
2771 F&B Road
College Station, TX

H. H. Tom
Agricultural Research Service, US Department of Agriculture
Southern Crops Research Laboratory
Areawide Pest Management Research
2771 F&B Road
College Station, TX

INTRODUCTION

GPS parallel swath guidance systems for spray aircraft have been rapidly adopted by the agricultural aviation industry. These computerized systems facilitate automation of various monitoring and control functions associated with aerial spray applications. Automatic control of spray flow rate, adjusted for ground speed, introduces a level of precision to aerial spray application that was heretofore unavailable.

Automatic control of flow rate will permit more uniform chemical application which could result in reduced chemical usage while still achieving the desired biological effect. Conventional practice has been to set spray flow rate based on nominal airspeed and swath width to give the desired spray application rate. With this practice, errors in application rate are introduced by variation in ground speed due to airspeed differences in 1) upwind and downwind spray passes and 2) entry and exit to and from the spray pass over obstructions such as trees or powerlines on field perimeters.

OBJECTIVE

The objective of this study was to determine the uniformity of spray application rate in various simulated aerial spray application situations with a GPS-based flow controller, compared to conventional aerial spray applications without a flow controller.

EXPERIMENTAL METHODS

A prototype SATLOC Flow Control/Monitor (SFC/M), an optional sub-system of the SATLOC AIRSTAR GPS Guidance System for agricultural aircraft, was installed in a Cessna AgHusky aircraft. A one-mile-long flight path was established on a concrete runway with 50-foot-high markers 50 feet beyond each end of the flight path. Spray passes were made with the flow controller operating at target application rates of three and five gallons per acre, both upwind and downwind, under two ambient wind conditions and two flight path conditions. Flight and flow data were recorded on 1-second intervals with the SATLOC system for at least three replications of each flight condition. Wind speed and direction, monitored on the flight line with a Young anemometer, were averaged for 30 seconds and recorded with a Campbell data logger.

RESULTS

Spray rates were maintained by the flow controller to the target rate, within an average of less than 0.05 gpa, on four of the eight spray conditions summarized in Table 1. Spray rates on the other four spray conditions ranged within an average of less than 4 percent of the target rate. Spray rates computed for flight conditions in the study, without flow control, show errors in application rates, just due to the wind conditions, of up to 13 percent. The effects of the flow controller on maintaining target spray rate over barriers at the ends of the flight line could not be accurately assessed from the data collected.

Table 1. Spray rate control with SATLOC Flow Control/Monitor with airspeed controlled at 120 mph.

GPS Ground Speed, mph (s)	Wind Velocity on Flightline, mph	Flight Direction	Controlled Spray Rate, gpa (s)*	Spray Rate Without Flow Control, gpa
		Target Spray Rate = 3.0 gpa		
112 (2.6)	5.5	Upwind	3.0 (0.25)	3.2
132 (2.0)	6.5	Downwind	3.0 (0.08)	2.7
110 (3.5)	11.0	Upwind	3.1 (0.37)	3.3
139 (2.8)	11.7	Downwind	3.1 (0.18)	2.6
		Target Spray Rate = 5.0 gpa		
112 (1.4)	5.6	Upwind	5.2 (0.29)	5.4
131 (1.0)	5.6	Downwind	5.0 (0.12)	4.6
111 (2.0)	10.5	Upwind	5.1 (0.28)	5.4
136 (2.7)	10.9	Downwind	5.0 (0.16)	4.4

* s = sample standard deviation

ACKNOWLEDGMENT

Appreciation is expressed to SATLOC Inc. for supplying equipment used in these studies. Trade names are mentioned solely for the purpose of providing specific information. Mention of a trade name does not constitute a guarantee or warranty of the product by the U. S. Department of Agriculture and does not imply endorsement of the product over other products not mentioned.

Effectiveness of AgLeader® Yield Monitor for Evaluation of Varietal Strip Data

Thomas L. Krill

Department of Extension
Ohio State University
Columbus, Ohio

OSU Extension-Van Wert
Van Wert, Ohio

ABSTRACT

For the past two years, Ohio State University Extension-Van Wert in cooperation with Farm Focus, Inc. has harvested the varietal strip plots of the Van Wert/Paulding Counties Replicated Varietal Testing Program at the Farm Focus site using a yield monitor. In both 1994 and 1995, 70 strips, measuring approximately 1,000 ft. by 15 ft., of each corn and soybeans were harvested using a John Deere 6620 combine equipped with an AgLeader yield monitor. Each strip was harvested as an individual load in the yield monitor and therefore remains as a unique entity. After harvesting, each strip was individually weighed and sampled for moisture in the field using a calibrated electronic weigh wagon and a portable moisture tester. The yield monitor load data were then compared to the actual weight measured by the weigh wagon and moisture reading of the sample drawn. The result for corn is 140 comparisons of actual vs. yield monitor results over 2 years and 100 different hybrids. In soybeans, because 1994 presented some technical difficulties for the yield monitor and the data was determined invalid, only 70 comparisons of actual vs. yield monitor results over 56 different varieties exist. Though not conclusive, this data provide insight into the considerations and limitations an individual should consider when using a yield monitor for data collection.

INTRODUCTION

The Ohio State University Extension office in Van Wert County has been involved in varietal testing for the past 26 years. In 1994, after an advisory committee recommendation, this educational organization expanded its educational efforts to also encompass the emerging agricultural practices known as site specific management (SSM). The 1994 growing season also marked the first season in which one complete replication of all varieties in the varietal testing program were located at the Farm Focus site. With the tradition of varietal testing, the developing SSM program, and the Farm Focus location for one complete replication, it seemed only natural to attempt to collect data using the newly developed yield monitoring technology.

The Van Wert/Paulding Counties Replicated Varietal Testing Program is a side by side strip yield test with a common variety or hybrid "tester" regularly inserted. All varieties or hybrids are planted in groups of ten with the "tester" strategically located in both the 3 and 8 positions. All yields collected are then indexed to the nearby "tester" for analysis. Each group of ten varieties or hybrids is then replicated over 6 sites across the two county area. In 1994 and 1995, one complete replication of all 7 groups was planted side by side at the Farm Focus site for corn and soybeans. This accounted for the testing of 56 different varieties or hybrids with 14 strips of the "tester" variety or hybrid located within. The 1994 soybean results were determined to be unusable due to harvest difficulties. In 1995, 12 additional strips were located adjacent to the Van Wert/Paulding Counties Replicated Varietal Testing Program containing 6 different imidazalinone resistant or tolerant hybrids (IMI corn). Also 21 strips of popcorn were located on the Farm Focus site.

OBJECTIVES

- To work into the existing Van Wert/Paulding Counties Replicated Varietal Testing Program
- To compare the estimated weight calculated by the yield monitor with the actual weight as measured by the weigh wagon for individual strips
- To identify if variation in accuracy is varietal or hybrid related

PROCEDURES

Strips of both corn, popcorn, and soybeans were planted side by side using the procedures identified by the Van Wert/Paulding Counties Replicated Varietal Testing Program. These procedures basically requires consistent treatment of the entire site using best management practices for all strips over their entire length and assigns the varieties or hybrids into specific strips. All varieties or hybrids were selected for evaluation in the Van Wert/Paulding Counties Replicated Varietal Testing Program by the company producing the seed and each company was limited to two entries. At the Farm Focus site, all strips were exposed to a minimum tillage situation with herbicidal weed control. Corn, including IMI and popcorn, was planted using a 6 row unit planter on 0.76 meter (30 in.) adjusted to 72,000 seeds/hectare (29,000 seeds/acre). Soybeans were solid seeded using a 5 meter (15 ft) drill on 0.19 meter (7.5 in.) spacing adjusted to 445,000 seeds/hectare (180,000 seeds/acre). 1995 corn strips, including IMI and popcorn, measured 335 meters (1,100 ft.) in length. 1994 corn was 312 meters (1,030 ft) with 1995 soybeans only 303 meters (1,000 ft.) in length.

All strips were harvested under normal harvest conditions using a John Deere 6620 combine equipped with an AgLeader 2000 yield monitor. The yield monitor was field installed in 1994 following the installation instructions provided. It's initial operation was during the 1994 winter wheat harvest and it was determined that all systems were operating properly. The system was equipped with a Differential Corrected Global Positioning System (DGPS) receiver and programmed to record data on one second intervals. Corn, including popcorn, was harvested using a six row John Deere corn head and soybeans with a 5 meter (15 ft.) grain platform. Operating ground speed was adjusted according to crop conditions and machine capacity following normal operating procedures.

Each strip was harvested, weighed, and recorded individually. Strips were each harvested with the combine traveling in the same direction. Completing the harvest of each strip, the grain was weighed and sampled for moisture using a calibrated electronic weigh wagon and portable moisture tester. Prior to each harvest season, the weigh wagon was calibrated at 5 pound increments to weights exceeding the anticipated strip weights following the manufacturer's calibration procedures. The portable moisture tester also went through a similar procedure following manufacturer's instructions. Within the yield monitor, each strip was treated as an individual load and therefore remained a unique entity. The yield monitor was annually calibrated, post harvest, using the manufacturer's procedures and all known weights of the harvest season including the strip data. The yield monitor was used for approximately 80 hectares (200 acres) of corn and 80 hectares (200 acres) of soybeans in both 1994 and 95 with the majority of those measurements referenced to a known, scaled, weight.

Analysis was completed following the harvest of all data. Estimated weight was defined as the weight of grain as "estimated" by the yield monitor Actual weight was defined as the "actual" weight as reported by the scale on the electronic weigh wagon. These numbers were available individually for each strip harvested. Percent error was used as the measurement of accuracy and was calculated by the absolute value of the difference between the estimate and actual weights divided by the actual weight. Percent error was expressed as a percentage.

RESULTS

(see charts in appendix)

Conclusions and Limitations

The data collected indicated that the AgLeader 2000 yield monitor did an acceptable job of estimating the quantity of harvested grain. These data were collected successfully during the Van Wert/Paulding Counties Replicated Varietal Testing Program at Farm Focus. The percent error calculated in this study indicated a significantly less error for corn when compared to soybeans. The monitor also appeared to be more accurate on uniform varietal or hybrid strips that strips of many varieties or hybrids. The AgLeader 2000 yield monitor can be an appropriate tool for yield calculation when installed properly and used within the capabilities of the yield monitoring system.

For accurate interpretations of grain yield, it is essential to complete the calibration procedure. The calibration procedure described in the AgLeader yield monitor instructions provides a satisfactory methodology for yield monitor calibration. This calibration procedure will necessitate the use of a scale to determine the actual weight of several loads. In this study, calibrations were the result of a minimum of 25 known loads with actual weights. Calibration is essential for precise measurement of grain and requires the use of a calibrated scale.

      Finally, this study only evaluated the AgLeader 2000 yield monitor's ability to calculate the yield over a complete strip. No attempt was made to determine the accuracy of the yield monitor within the spacial segmentation of the strip. Additional study must be completed before any statement regarding the spacial accuracy of the yield monitor within a given strip can be made. The AgLeader 2000 yield monitor is a beneficial tool for production agriculture when properly installed and used within its capabilities.

Appendix

1995 Popcorn Harvest

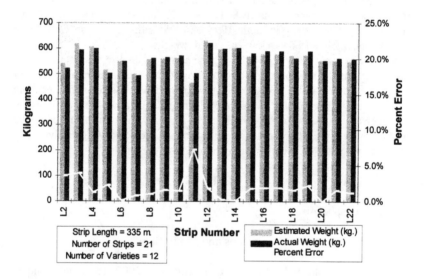

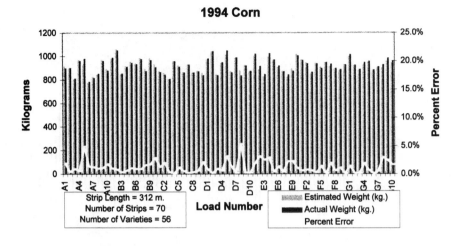

1994 Corn

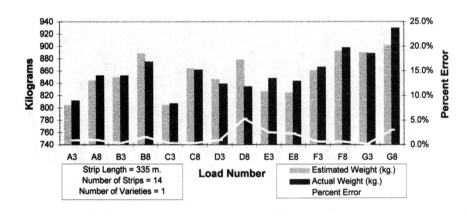

1994 Corn (Tester Only)

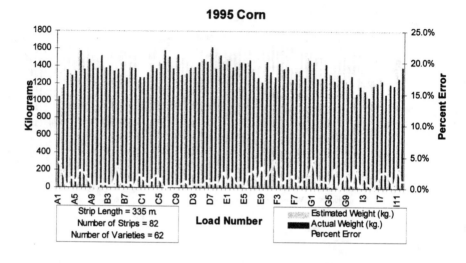

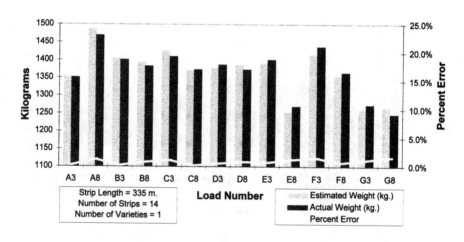

1995 Soybeans

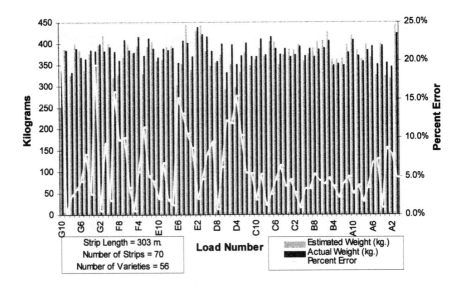

Strip Length = 303 m.
Number of Strips = 70
Number of Varieties = 56

Load Number

Estimated Weight (kg.)
Actual Weight (kg.)
Percent Error

1995 Soybeans (Tester Only)

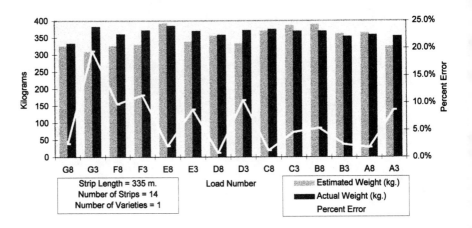

Strip Length = 335 m.
Number of Strips = 14
Number of Varieties = 1

Load Number

Estimated Weight (kg.)
Actual Weight (kg.)
Percent Error

A Site-Specific Center Pivot Irrigation System for Highly-Variable Coastal Plain Soils

E.J. Sadler
C.R. Camp
D.E. Evans
L.J. Usrey

USDA-ARS
Coastal Plains Soil, Water, and Plant Research Center
2611 West Lucas St.
Florence, SC

ABSTRACT

The Coastal Plains Soil, Water, and Plant Research Center has been monitoring spatial yield in a test field since 1985, using a conventional corn-wheat-soybean rotation most of that time. Observations of variation in soil and crop response that correlate with yield variation suggest that crop water relations may be the key feature that causes spatial variability in yield for the Southeastern Coastal Plain. Experience with mechanistic modeling indicates that for normal weather years, the final yield is particularly sensitive to variations in soil water, presumably because the surface soil is sandy and rooting volume is limited. These conclusions, plus difficulties encountered in scheduling irrigation under a center pivot on typically variable soils, led the USDA-ARS to design and build a site-specific center pivot capable of differentially irrigating 100-m^2 areas. A 3-tower commercial center pivot was modified by adding 39 9.2-m manifolds in 13 sections, 3 to a section. The manifolds and nozzles were sized 1x, 2x, and 4x, so that octal combinations would provide up to 7x the minimum application depth for a given outer tower speed. At 50% speed, the application depths are 0 to 12.5 mm in 1.8-mm increments. A programmable controller was attached near the pivot end of the boom, so that it was proximal to but avoided the pivot control panel when the system rotated. The individual manifolds were controlled by a program residing in the programmable controller, which obtained pivot position and other information via radio modem link with the pivot control panel. Water and nitrogen application has been accomplished using this system on a replicated field experiment. Experience gained during this phase will guide modification of a similar pivot for site-specific water, nutrient, and pesticide management on a typically variable Coastal Plain field.

INTRODUCTION

The southeastern US Coastal Plain is comprised of nearly level, sandy surface soils overlying a sandy clay subsoil (Pitts, 1974; USDA-SCS, 1986). The terrain is marked by numerous Carolina Bays, which are shallow (<3 m) depressions of varying size and indeterminate origin. Surface texture within the depressions is generally finer than that outside, with the deeper depressions tending toward clays, and the shallower ones, loams. The bulk of the soils outside the bays is sandy loam or loamy sand, with extensive inclusions of sands. Much of the Coastal Plain also has an eluviated E horizon of similar texture to the A, but with essentially no organic matter, and increased density (1.7 to 1.8 g cm^{-3} being common). Thickness of the A horizon and both existence and thickness of the E horizon are important distinguishing characteristics of the soils for a given area. The sandy soils and root-restricting eluviated horizons combine to make nonirrigated crop production a challenge in the area. Management practices to increase rooting depth include subsoiling to a depth of about 0.4 m beneath the crop row.

Climate in the area is warm, humid, and frequently cloudy. Average annual rain for Florence, SC, is 1100 mm/yr. Most summertime rain occurs in thunderstorms, such that June, July, and August are the months with highest monthly total rain, from 100 to 150 mm/month. However, each month during the growing season has 110-yr record low totals in the 20-mm range, and record high totals in the 250-mm range. Such variability in rainfall, coupled with the poor water relations described above, means that yield-reducing drought stress is common in an area one would otherwise assume had plentiful rain. Sheridan et al. (1979) reported that 22-d droughts during the growing season occur, at 50% probability, every other year. Such drought dramatically reduces crop growth and yield.

Observations of spatial patterns in crop growth, particularly during periods of drought, suggest water management may be the key for managing soil variability in the Coastal Plain (Karlen et al. 1990; Sadler et al., 1995a; 1995b). Similarity of yield patterns during drought and non-drought years (though with differing means) supports this assumption. Further support comes indirectly from the apparent non-correlation between yield and fertility patterns. Local experience in scheduling irrigation for a center pivot sited on variable soils had set the stage for the problems one would encounter when attempting to spatially manage soil water (Camp et al., 1988).

Consequently, in 1991, an irrigation system design team drew up specifications for a computer-controlled, variable-rate center pivot (see Camp & Sadler, 1994). Two commercial machines were acquired (description below), and modifications were made to achieve this objective. The machine was first used during the 1995 season, and demonstrated under the controlled conditions of a replicated experiment on a reasonably uniform field. The second machine, which will be modified during 1996 based on experiences with the first, will represent the final stage of the process -- that of variable-rate management of water, fertility, and pesticides on a highly variable Coastal Plain soil.

Concurrent with the above process, three other research groups have been working toward similar goals. Lyle (personal communication, 1992) described a multiple-orifice emitter design that could be individually switched to provide a series of step-wise incremental flow rates. This was part of the Low-Energy Precision Application (LEPA) system. Duke et al. (1992) and Fraisse et al (1992) described an alternative time-slice approach in which variable rates were achieved by switching sprinklers on and off for varying proportions of a base time period, usually 1 min. The advantages of this design are that a continuous range of application rates can be obtained using a single nozzle, where the others' systems require additional nozzles, manifolds, and switches to achieve additional increments of rate. The disadvantage is that the on/off sprinkler may be either in phase or out of phase with the start-stop motion of the irrigation tower, impressing additional variability in application depth. This disadvantage is minimized when the wetted radius is larger, the alignment of the irrigation machine is controlled very closely, and the base time period of the sprinkler is small relative to the duration of tower stoppage. Stark et al. (1993) used a similar concept with a patented (McCann & Stark, 1993) control system for a variable-rate linear-move system, in which individual conventional sprinklers were controlled by computer. Three sprinkler sizes (1/4, 1/4, ½ of full flow) provided 1/4, ½, 3/4, and full irrigation. This system was installed on a field-scale center pivot, and uniformity of application was reported. Further developments on a linear move system were reported by King et al. (1995).

The objective of this presentation is to describe the variable-rate center pivot machine developed at Florence, and to illustrate its capabilities to the precision agriculture audience.

MATERIALS AND METHODS

The design specifications of the center pivot irrigation machine were to achieve practical variable-rate control on control elements of approximately 100 m² area. The unit needed to be able, at normal operating speeds (usually 50% duty cycle of the outer tower), to apply sufficient water to replace an average daily potential ET. For practical reasons, 8 discrete increments of the potential ET were considered as a minimum working approximation of true variable-rate irrigation. Control of the application rates was to have been achieved using computerized maps, so that algorithms, yet to be developed, could be employed to choose application depths based on historical yields, soil maps, or on-board sensors. Further, the variable-rate modifications were to be implemented on commercially-available systems. An overview of the concept is described in Camp and Sadler (1994).

The hardware for this system is described in Camp et al. (1996), and will be briefly summarized here. Two small, 3-tower, 137-m commercial center pivots were purchased in 1993 (Valmont Irrigation, Inc, Valley, NE). In anticipation of increased load, a heavier truss design was requested. Otherwise, the unit was conventional in both design and control. A set of overhead sprinklers and a set of LEPA quad sprinkler heads on drop tubes were installed on both machines, to provide immediate ability to irrigate, albeit uniformly.

The design and modification of the manifolds and sprinklers for the first commercial pivot were done in cooperation with The University of Georgia Coastal Plain Experiment Station, Tifton, GA. More details of the hydraulic design can be found in Omary et al. (1996). In brief, the length of the truss was logically segmented into 13 sections 9.1 m (30 ft) long (see Fig. 1). Each of these sections had 3 parallel, 9.1-m manifolds, each with 6 industrial spray nozzles on 1.5-m spacing. Water was supplied to each set of 3 manifolds directly from the boom via 5-cm (2 in) ports, drop pipes, a tee, and hoses. Each individual manifold had a solenoid valve, pressure regulator, low-pressure drain, and air entry port. The 3 manifolds and their nozzles were sized to provide 1x, 2x, and 4x of a base depth at the position of the section, which meant that all actual flow rates were larger at the outer end to account for the greater area subtended per unit angle traveled. Octal combinations of the 3 manifolds provided 0x, 1x, 2x,...7x the base depth. The 7x depth was set to 12.5 mm (0.5 in) at 50% duty cycle on the outer tower. The small size of the unit, 120 m, meant that a full circle could be irrigated in less than 4 hr at 100% duty cycle, so that a 17% setting could still complete the circle in less than 24 hr.

All solenoids were controlled using a programmable logic controller (PLC: GE-Fanuc model 90-30, Charlottesville, VA[1]) mounted on the mobile unit, about 5 m from the pivot point, far enough to clear the tripod and control box. The PLC had an on-board 80386 PC with hard drive, floppy drive, serial ports, and peripheral connectors. Software was written in Visual Basic (Microsoft Corp., Redmond, WA) to convert a map of control values to on-off settings in the directly-addressable solenoid control registers of the PLC. In order to determine location from the C:A:M:S (Valmont Irrigation, Inc.) controller, a communication link between the mobile PC and the stationary C:A:M:S had to be established, which

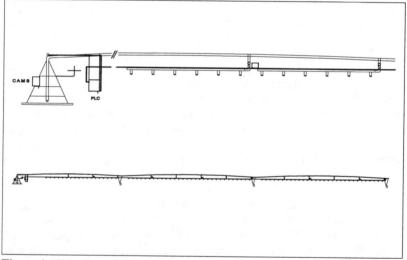

Figure 1. Side view of site-specific center pivot, and close up of tripod and example section.

---

[1]   Mention of trade names is for information purposes only. No endorsement implied by USDA-ARS or any cooperator of preference over other equipment that may be suitable for the application.

was done with short-range radio-frequency modems (900 MHZ, broad-band modems; Comrad Corp., Indianapolis, IN). The on-board PC repeatedly interrogates the C:A:M:S unit to determine, primarily, the angle of the pivot, but also other parameters to provide assurance the system is functioning properly. From the angle and the segment position on the truss, the position in polar coordinates was fixed. (The angle reported was found to be systematically in error, so a correction was determined with surveying techniques, and built into the software.) When the location has been determined, the program checks whether a boundary has been crossed. If not, the interrogation cycle repeats until something needs to change. When a boundary is crossed, the expected application map is checked, the appropriate table lookup is performed, and the solenoid registers set accordingly.

Injection of nutrients into the irrigation water was accomplished using a 4-head, 24V DC variable-rate pump (Ozawa Precision Metering Pump, model 40320), check valve, and nurse tank connected to the stationary vertical riser at the pivot point. Because the flow rate of water could vary depending on the spatial application schedule, the amount of fertilizer injected into the boom needed to vary proportionately in order to hold the concentration constant. This was done by having the PC on board the PLC calculate the aggregate flow rate, calculate the required injection rate, compute the 0-5V DC voltage setting required to provide that, and report that to the operator. For this season, the operation was monitored and controlled manually using operator inputs to a CR7X (Campbell Scientific Inc., Logan, UT). For later operations, the 0-5 V DC value will be connected directly to the pump. Spatially-variable application of nutrients was done using a minimal, spatially-variable irrigation, but with uniform concentration. Concurrent spatial control of water and nutrients would require distributed control of multiple injection points. Although sophisticated, controlling such a system would not be particularly difficult. The requirement for multiple pumps made this option cost-prohibitive at this time.

Demonstration in a Replicated Field Experiment

The pivot described above was sited on a relatively uniform soil area (USDA-SCS, 1986), chosen specifically for proving the technology under conditions somewhat more controlled than the highly variable long-term spatial yield field. The primary experimental objectives were to test rotation and irrigation effects on a corn-soybean rotation vs continuous corn under conservation tillage. A secondary objective was to test subsoiling against not doing so, in the possible trade-off of irrigation to manage water rather than subsoiling to increase the rooting depth.

There were 144 treatment plots in total: 4 replications x 3 rotations (corn-corn, corn-soybean, soybean-corn) x 2 tillage (subsoiled, non-subsoiled) x 3 water managements (rainfed, tensiometer, crop stress) x 2 nitrogen (single sidedress, incremental appl.). In 1995, the latter two water management treatments were both operated based on tensiometers. The individual plots were laid out in a regular 7.5° by 9.1-m (30-ft) pattern, which made the minimum plot length 10 m in section 7, and 15 m in section 13. As seen in Fig. 2, the four replicates were sited in the outer

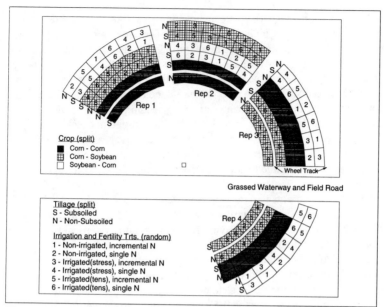

Figure 2. Plot plan for replicated field experiment used to test site-specific center pivot under controlled conditions.

annulus, on the most uniform soil areas. The outer rings were used so that the planting and other operations could be done without sharp turns. All operations were done on the circle rather than with straight rows. This, often done in commercial operations, greatly simplified matters in this experiment.

Table 1 lists fertilization operations during this season. All nitrogen fertilization was accomplished through injection of urea-ammonium-nitrate (UAN 24S) through the system. Included are time, amount of N applied, and amount of irrigation to achieve the fertilization. To prevent spray drift, 1.5" layflat hose was placed around the 2x nozzles and extended to the ground. The 2x nozzles provided 3.6 mm of irrigation at 50% duty cycle and 1.8 mm at 100%.

Table 1. Fertilization via injection through site-specific center pivot.

Date	N Application	Speed, #passes	Irrigation amount	Treatments fertilized
6/1/95	22.5 kg/ha	50%, 1	3.6 mm	All
6/7/95	112.3 kg/ha	100%, 2	3.6 mm	Single
6/8/95	22.5 kg/ha	50%, 1	3.6 mm	Incremental
6/12/95	22.5 kg/ha	50%, 1	3.6 mm	Incremental
6/14/95	22.5 kg/ha	50%, 1	3.6 mm	Incremental
6/19/95	22.5 kg/ha	50%, 1	3.6 mm	Incremental
6/21/95	22.5 kg/ha	50%, 1	3.6 mm	Incremental

Observations from Season's Use

The site-specific center pivot evolved from the basic commercial machine in March, 1995, to a functioning, proven technology by summer's end. Control software was primitive and fragile initially, but similarly evolved through modification and experience such that operation was possible via the remote C:A:M:S unit by the end of the summer. Prior measurements of system uniformity had demonstrated acceptable distribution within control elements as well as expected border effects between elements with contrasting application depths (Omary et al., 1996). Results from the season presented no evidence that uniformity or border width had changed. Surface redistribution had been a concern during design, because of the small wetted radius of the sprinkler, but even the collection into layflat hose for fertilization did not cause excessive local ponding and runoff.

Changes for the Future

The next steps will include outfitting the second pivot with variable rate irrigation and fertilization based on experiences gained with the first pivot, outfitting both with low-volume pesticide variable-rate application equipment, adding sensors to the machines to detect stress, and modifying software to accommodate irregular soil unit boundaries.

REFERENCES

Camp, C.R., G.D. Christenbury, and C.W. Doty. 1988. Chapter 5. Florence, SC. p 61-78. *In* C.R. Camp and R.B. Campbell (ed.) Scheduling irrigation for corn in the Southeast. ARS-65, USDA-ARS. 184 pp.

Camp, C.R., and E.J. Sadler. 1994. Center pivot irrigation system for site-specific water and nutrient management. ASAE Paper No. 94-1586. ASAE, St. Joseph, MI.

Camp, C.R., E.J. Sadler, D.E. Evans, L.J. Usrey, and M. Omary. 1996. Modified center pivot irrigation system for precision management of water and nutrients. ASAE Paper No. 962077, ASAE Annual Meeting, Phoenix, AZ, July 14-18, 1996.

Duke, H.R., D.F. Heermann, and C.W. Fraisse. 1992. Linear move irrigation system for fertilizer management research. Proc. International Exposition and Technical Conference, The Irrigation Association. p 72-81.

Fraisse, C.W., D.F. Heerman, and H.R. Duke. 1992. Modified linear move system for experimental water application. Advances in planning, design, and management of irrigation systems as related to sustainable land use. Leuven, Belgium. Vol 1, p 367-376.

Karlen, D.L., E.J. Sadler, and W.J. Busscher. 1990. Crop yield variation associated with Coastal Plain soil map units. Soil Sci. Soc. Am. J. 54:859-865.

King, B.A., R.A. Brady, I.R. McCann, and J.C. Stark. 1995. Variable rate water application through sprinkler irrigation. P 485-493. *In* Site-specific management for agricultural systems. 2nd International Conf.,

Bloomington/Minneapolis,   MN,   27-30   Mar.   1994.
ASA/CSSA/SSSA/ASAE, Madison, WI.

McCann, I.R., and J.C. Stark. 1993. Method and apparatus for variable application of irrigation water and chemicals. U.S. Patent No. 5,246,164, September 21, 1993.

Omary, M., C.R. Camp, and E.J. Sadler. 1996. Center pivot irrigation system modification to provide variable water application depth. ASAE Paper No. 962075, ASAE Annual Meeting, Phoenix, AZ, July 14-18.

Pitts, J.J. 1974. Soil survey of Florence and Sumter Counties, South Carolina. USDA-SCS. U.S. Gov. Print. Office, Washington, DC.

Sadler, E.J., P.J. Bauer, and W.J. Busscher. 1995a. Spatial corn yield during drought in the SE Coastal Plain. p. 365-382. *In* Site-specific management for agricultural systems. 2nd International Conf., Bloomington/Minneapolis, MN, 27-30 Mar. 1994. ASA/CSSA/SSSA/ASAE, Madison, WI.

Sadler, E.J., W.J. Busscher, and D.L. Karlen. 1995b. Site-specific yield on a SE Coastal Plain field. p. 153-166. *In* Site-specific management for agricultural systems. 2nd International Conf., Bloomington/Minneapolis, MN, 27-30 Mar. 1994. ASA/CSSA/SSSA/ASAE, Madison, WI.

Sheridan, J.M., W.G. Knisel, T.K. Woody, and L.E. Asmussen. 1979. Seasonal variation in rainfall and rainfall-deficit periods in the Southern Coastal plain and Flatwoods Regions of Georgia. Georgia Agric. Exp. Sta. Res. Bull. 243, 73 ppg.

Stark, J.C., I.R. McCann, B.A. King, and D.T. Westermann. 1993. A two-dimensional irrigation control system for site-specific application of water and chemicals. Agronomy Abstracts 85:329.

USDA-SCS. 1986. Classification and correlation of the soils of Coastal Plains Research Center, ARS, Florence, South Carolina. USDA-SCS, South National Technical Center, Ft. Worth, TX.

Site-Specific Sugarbeet Yield Monitoring

J. D. Walter
V. L. Hofman
L. F. Backer

Agricultural Engineering Department
North Dakota State University
Fargo, ND

ABSTRACT

A site-specific sugarbeet yield monitoring system was developed for the 1995 harvest campaign. A handheld computer, GPS receiver, and a slide bar weight sensing configuration were used during the sugarbeet harvest to collect site-specific yield information. An average accuracy within 2.3% of actual weight was achieved and site-specific sugarbeet yield maps were developed to illustrate yield variability.

INTRODUCTION

Site-specific or precision farming technology provides a crop producer with information and application tools to manage spatial variability within fields. Information tools such as grid soil sampling, yield mapping, field scouting and satellite images are presently used to identify field variability. Once identified, variable rate applications of fertilizer, seed, and pesticides allow producers to tailor crop inputs to maximize profits and minimize environmental damage.

Site-specific yield mapping is one of the basic building blocks of a precision farming program. It can help identify and quantify crop responses to soil types, nutrient levels, plant populations, chemical rates, crop diseases and many other applied or existing crop factors. Yield mapping also enables the producer to evaluate the economic return of site-specific applications. Identifying high yielding areas enables the producer to assess the qualities in these areas for future crops. If excess nutrients remain, lower fertilization rates or higher plant populations can be used to maximize return. Low yield areas can indicate weed pressure, poor drainage, soil compaction, or insect/disease infestations.

In the past several years, site-specific grain yield monitoring equipment has been readily available for installation on combines. This equipment provides respectable accuracy for measuring yield on a variety of grains. Only limited work has been accomplished with bulky crops such as sugarbeets or potatoes. The high value of these crops, coupled with their sensitivity to management practices, makes them ideal candidates for site-specific yield monitoring.

Yield monitoring research has been conducted on both potatoes and sugarbeets in recent years. Potato yield monitoring conducted in Idaho and Washington (Campbell, et al., 1994) used idler wheels attached to load cells to measure the potato weight on a moving conveyor. Accuracies of 1% of a normally loaded conveyor were achieved with a harvester. But, harvester movement produced errors as high as 10%. Preliminary sugarbeet yield monitoring was conducted near Crookston, Minnesota during the 1994 harvest campaign by the Agricultural Engineering Department at North Dakota State University (Hofman, et al., 1995). An average system accuracy within 3.8% of actual weight was achieved.

Recently, increased concern over maintaining legal road weights has prompted sugarbeet growers to desire the actual weight of trucks leaving the field. A field system to accumulate weight and provide total truck weight will allow growers to maximize individual truck weights without exceeding legal road limits.

OBJECTIVES

The objectives of this study were to:
1. develop/modify a sugarbeet yield sensing system compatible with current harvester designs
2. provide a truck weight totalizing feature for maintaining legal road weights
3. collect site-specific sugarbeet yield information
4. develop sugarbeet field yield maps.

EQUIPMENT AND METHODS

Initially, the prospect of modifying an existing grain yield monitor to determine sugarbeet yield seemed attractive. But, the proprietary systems would require additional interfacing and dependence on detailed product support. Considering the recent progress in potato yield mapping, potato yield monitoring hardware would allow simple adaptation and provide more flexibility towards sugarbeet yield mapping. HarvestMaster Inc. supported the 1994 potato yield mapping efforts and planned to release a limited number of potato yield mapping systems for continued product testing in 1995.

The HarvestMaster yield mapping hardware system was purchased with the intent of modifying it for use in site-specific sugarbeet yield mapping. The following hardware components were used to support the 1995 sugarbeet yield monitoring trials:

- HarvestMaster HM-500 yield mapping hardware including:
  - signal conditioning and conversion unit (SCCU)
  - magnetic speed sensor
  - two 500 pound Weighbar™ load cells
- HarvestMaster Pro2000 handheld computer
- Concord, Inc. BR6-183 Differential Global Positioning System (DGPS) receiver
- Used sugarbeet outlet conveyor section for laboratory testing
- Six row WIC sugarbeet lifter.

Two weight sensing configurations were developed and evaluated on the test conveyor system. Both configurations used load cells under the link chain conveyor to convert weight information into an electrical signal for processing. The first configuration consisted of a 152 mm (6 in) idler wheel directly attached to the load cell (Fig. 1). One instrumented idler replaced one original idler on each side of the conveyor. It was hypothesized that the idler wheel would exhibit considerable impact loading from individual chain links. A slide type weight sensing assembly could reduce the impact loading and yield more accurate results. Therefore, the second configuration consisted of a 50.8 mm (2 in) wide by 609.6 mm (24 in) long slide bar covered with 9.53 mm (0.375 in) thick ultra high

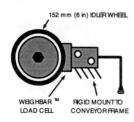

Figure 1. Idler wheel weight sensing assembly.

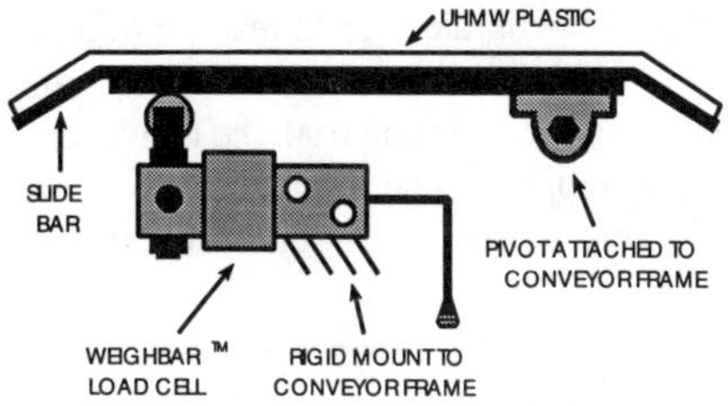

Figure 2. Slide bar weight sensing assembly.

molecular weight (UHMW) plastic (Fig. 2). One end of the slide bar was fixed to the conveyor frame in a pivot assembly while the other end was vertically constrained by a load cell. The slide bar assembly replaced two existing idlers on each side of the conveyor.

A custom mounting rail provided easy installation to existing conveyor frame mounting holes and the ability to quickly interchange the weight sensing configurations. Proper installation was critical to reduce tension effects on the load cells due to the conveyor chain. A horizontal tangent was maintained across the top of the weight sensing assemblies and the top of adjacent conveyor idler wheels. In addition, a minimum of one idler wheel between the sensing assembly and either end of the conveyor was maintained.

The equipment setup and connection diagram is shown in Figure 3. The SCCU sampled the sugarbeet weight at a rate of 100 samples per second on each load cell. To minimize vibration effects, a 400 point moving average of the load cell signal was processed by the SCCU. A magnetic speed sensor was installed on the drive shaft of the conveyor to detect conveyor speed. The BR6-183 Differential Global Positioning System was connected to the SCCU via an RS-232 serial cable. The six channel DGPS receiver provided latitude, longitude, and velocity information to the SCCU every second. Finally, the SCCU was connected to the handheld computer with an RS-232 serial cable. The computer requested, processed, and recorded the weight, conveyor speed, harvester speed, latitude, and longitude information once per second to derive the site-specific sugarbeet yield information.

Laboratory Testing

Calibration, testing, and evaluation of the weight sensing configurations were accomplished on a laboratory test conveyor. Conveyor speed calibration was achieved by spraying one chain link with marking paint and manually measuring the time for one revolution with a constant conveyor speed. Actual conveyor speed was derived from the measured time period and the length of conveyor chain. This actual conveyor speed was compared to the speed generated by the magnetic pickup. Adjustments to the user programmed speed calibration factor properly scaled the generated speed reading to equal the actual speed.

Weight calibration was also required for proper system operation. Before each weight calibration trial, the conveyor was operated at a constant speed with no weight applied. The load cell weight on the computer was "retared" or zeroed under this no load condition. Weight calibration was achieved by loading the

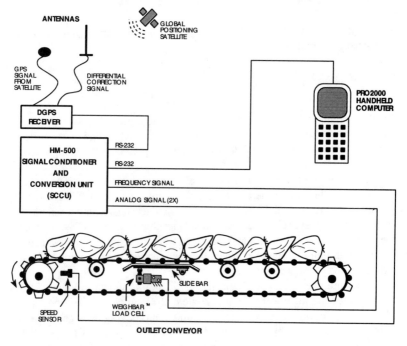

Figure 3. Sugarbeet yield monitoring equipment setup and connection diagram.

moving conveyor belt with known weights and allowing the computer to accumulate the load cell weight. The user programmed weight calibration factor was adjusted to scale the accumulated weight to equal the actual weight that passed over the weight sensing system.

Numerous test trials were conducted with both weight sensing configurations. Conveyor speed was held near 0.613 m/s (1.8 ft/s) to simulate typical sugarbeet harvester conveyor speed. Weights were also placed on the moving conveyor with varied intensities to simulate actual harvester conditions. Empty conveyor weight was observed and retared occasionally throughout the testing.

Field Testing

Field testing was performed exclusively with the slide bar weight sensing configuration. The weight sensing assembly was installed in the truck outlet boom conveyor of a WIC sugarbeet harvester owned and operated by A.W.G. Farms of Crookston, Minnesota. The outlet conveyor mounting position provided the best placement for accurate calibration and true representation of delivered tare and sugarbeet weight. Installation of the weight sensing configuration was performed before preharvest and the physical slide system was in use through the duration of the 1995 harvest campaign. The slide system supported both forward and reverse operation of the conveyor to facilitate use of the onboard storage tank.

Although the SCCU was weatherproof, it was mounted near the top of the lifter to minimize damage from dirt and debris. The GPS and FM correction antennas were mounted near the SCCU at the top center of the lifter. The DGPS unit itself was placed inside the SCCU enclosure. An RS-232 serial cable and 12V system power cable were routed to the tractor cab for connection to the handheld computer and a 12V power supply, respectively.

Conveyor speed calibration was identical to the laboratory speed calibration. Weight calibration was also similar except it was performed on individual truck weights. The total sugarbeet and tare weight was required from each truckload in order to obtain accurate weights for calibration. Regular truck weighing procedures were first completed at the piling station. Next, each truck used for calibration was allowed special provisions by American Crystal to dump tare dirt at the piling station and obtain empty truck weight. The difference between full truck weight and empty truck weight provided total delivered sugarbeets and tare dirt as measured by the yield monitoring system.

Normal harvest procedure was modified slightly for calibration purposes and for obtaining accurate and complete site-specific yield information with the monitoring system. The harvester was completely emptied into the truck to prevent erroneous weight accumulations whenever the harvester stopped moving in the field. Also, the operation of the onboard storage tank on this harvester would cause inaccurate site-specific yield information and was generally not operated during data collection. Numerous trucks during preharvest and full harvest were used for calibration and validation of system accuracy. Once an accurate system calibration factor was determined, site-specific sugarbeet yield data was collected.

RESULTS AND DISCUSSION

Approximately 17 weight trials were conducted with both the idler wheel and slide bar on the laboratory test conveyor. Total weight passed over the scale during each trial ranged from 127 to 136 kg (280 to 300 lbs). The laboratory testing resulted in lower weight errors achieved by the slide bar configuration compared to the idler wheel configuration (Table 1).

The conveyor was operated unloaded to obtain tare readings with both the sensing configurations. Table 1 shows the resulting maximum bounds of the tare readings. The slide bar did exhibit lower variation under the no load situations, but not to the degree anticipated. Upon further investigation of the raw unloaded conveyor weight signal, cyclical patterns were noticed in the tare weight signal. The period of these patterns coincided with one revolution of the conveyor. This cyclical pattern was attributed to the partially used conveyor chain and drive sprockets employed to construct the laboratory conveyor. On sugarbeet harvesters, the sprockets and conveyor chains are typically replaced as a set and would wear equally throughout their usable lifetime. Therefore, lower tare readings were expected on the harvester.

Table 1. Laboratory test results.

Configuration	Absolute Error %	Standard Deviation %	Tare Reading kg/s (lb/s)
Idler wheel	3.49	2.83	± 0.58 (1.28)
Slide Bar	2.45	1.94	± 0.53 (1.17)

Table 2. Field test results.

Configuration	Absolute Error %	Standard Deviation %	Tare Reading[1] kg/s (lb/s)
Slide Bar	2.28	1.98	± 0.36 (0.79)

[1] Conveyor operation only

The laboratory results strongly favored the slide bar assembly and it was used exclusively for field testing due to time constraints during harvest. Sixteen truckloads were used to determine the system accuracy. Measured truck weights were accumulated by the weighing system while actual truck weight was determined by the piling station scale. Truck loads contained approximately 13608 kg (30,000 lbs) of sugarbeet and tare weight. The measured truck weights were within 6.4 kg (14 lbs) to 938.9 kg (2,070 lbs) of actual truck weight. This resulted in errors ranging from 0.05% to 6.94% with an average error of 2.28% as shown in Table 2. Typical sugarbeet flow across the conveyor ranged from 18.1 kg/s (40 lb/s) to 27.2 kg/s (60 lb/s).

Similar to the laboratory testing, the conveyor was operated empty to obtain tare readings. As shown in Table 2, the tare reading with the conveyor operating alone was ± 0.36 kg/s (0.79 lb/s). The range increased to ± 0.53 kg/s (1.17 lb/s) with the harvester operating at rated RPM and the conveyor operating. Finally, a tare reading range of ± 0.56 kg/s (1.24 lb/s) was recorded with the conveyor and harvester operating as well as moving through the field at nominal field speed. As expected, the harvester's conveyor operating alone exhibited lower tare readings than laboratory results, but machine vibration had a significant effect on the weighing system. The large mass of the sugarbeet harvester combined with the smooth soil terrain of the sugarbeet fields helped to dampen the effects of field movement on the weighing system.

After successful system calibration and testing in field conditions, site-specific yield information was collected. Communication difficulties with the DGPS receiver resulted in numerous computer system stalls, but partial site-specific sugarbeet yield information on two fields were obtained (Figures 4 and 5). Figure 4 represents a total of six passes through the field. A small field ditch traversed the upper portion of the field and is illustrated by the lower sugarbeet yield in a small cross section through the field. Figure 5 accounts for approximately 16 passes or 4.05 ha (10 acres) of a sugarbeet field. Sugarbeet yield in the upper half of Figure 5 was generally higher than the lower half. The erratic position data on the middle right portion of Figure 5 occurred when battery power was lost in the handheld computer.

Individual sugarbeet yield readings reached as high as 78.5 t/ha (35 tons/acre) during the yield monitoring trials. The 78.5 t/ha (35 tons/acre) yield may not fairly represent the actual sugarbeet yield. First, the yield measured by the monitoring system contains both the sugarbeet weight and the tare dirt weight. Wet field conditions will increase the tare dirt weight on individual sugarbeets and result in higher than actual yields. Second, the inherent flow of the sugarbeets through the harvester was not steady. Visual observation of the sugarbeets moving on the conveyor and the real time yield readout illustrated this condition. High yield sites on the computer would almost always be followed by lower than average yield sites which indicated uneven sugarbeet flow. A longer recording interval, such as three or five seconds, would reduce the effect of the uneven sugarbeet flow on the yield sites. Finally, individual yield points also may have been affected by natural weight variations in the conveyor.

The extremely moist soil conditions during the later portion of the harvest contributed to an increased amount of dirt clinging to the conveyor chain links. Even though the amount of dirt seemed trivial, the dirt was weighed each conveyor revolution and therefore exhibited an accumulated effect on the measured truck weight. The yield monitoring system was retared more frequently under these conditions to ensure accurate weight readings.

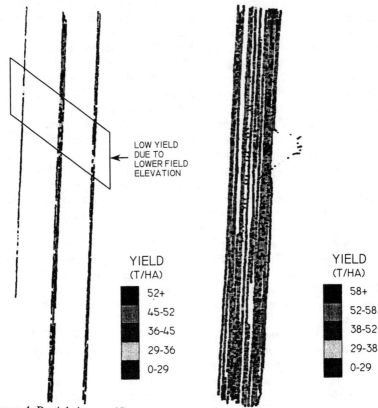

Figure 4. Partial site-specific
sugarbeet yield map of field 1.

Figure 5. Partial site-specific
sugarbeet yield map of field 2.

During site-specific yield monitoring, a time lag exists in the data between when the sugarbeet is actually removed from the soil to when the sugarbeet weight is measured. Measured time lag for this particular harvester was approximately 18 seconds. Time lag was not corrected during field data collection due to the DGPS communication difficulties. Data post-processing could be used to coordinate the yield information with the actual site-specific field position. But, the small portions of field yield sites recorded in 1995 are not suitable for site-specific management decisions and therefore the time lag correction was not post-processed. Upon close inspection of Figures 4 and 5, one end of each pass contains extremely low yield data which represents the time lag effect. Also, several instances of extremely low and high sugarbeet yield sites can be attributed to the use of the onboard storage tank to harvest through extremely wet portions of the field.

The UHMW plastic on both slide bars had to be replaced once during the harvest season due to excessive wear. The 9.53 mm (0.375 in) thick plastic harvested approximately 141.6 ha (350 acres) of sugarbeets before the chain link entry point on both plastic slides was completely worn through. The slide bar system caused no other operational difficulties during the harvest.

CONCLUSION

Two sugarbeet weight sensing configurations were developed, tested, and evaluated on a laboratory test conveyor. The idler wheel configuration produced an accuracy within 3.5% of actual weight while a 2.5% error was achieved by the slide bar system. The slide bar weight sensing configuration was installed in the outlet conveyor of a sugarbeet harvester and tested during the 1995 sugarbeet harvest campaign. Sugarbeet truck weights were accumulated and recorded on sixteen trucks. An average error of 2.3% of actual weight was determined. Site-specific sugarbeet yield data was collected on several partial fields and yield maps were developed. Sugarbeet yield maps were able to illustrate the variation of sugarbeet yield throughout the field.

Several areas of the sugarbeet yield monitoring system require additional attention.

1. A trouble free DGPS communication link must be established.
2. Continued research into weight sensing configurations is necessary. Slide bars employing UHMW plastic exhibit significant wear through the harvest season. Idler wheels provide more durability and simplicity, but weight errors are greater. A weight sensing configuration using multiple idlers on one load cell may provide more accurate yield information.
3. Electronic filtering of the load cell signal can reduce unwanted vibrations and improve system accuracy.
4. Changing the data collection interval from once per second up to once every three to five seconds should be considered. Using the one second recording interval, the ten acre sugarbeet field map required approximately 700Kb of storage space. The smaller harvester width already produces approximately three times the data recorded by yield monitors on some large combines.
5. Alter the weight sensing location or include software and hardware capabilities to allow the harvest operators to use the onboard storage tank.
6. Integrate the time lag correction into the field data system.

ACKNOWLEDGEMENTS

This project was made possible by partial funding from the Sugarbeet Research and Education Board of Minnesota and North Dakota, American Crystal Sugar Company - Hillsboro, and the North Dakota Power Use Council. Cooperation from the American Crystal piling station at Crookston, Minnesota and the use of North Dakota Agricultural Experiment Station equipment and research facilities was greatly appreciated. Also, special thanks to A.W.G. Farms of Crookston for providing the sugarbeet lifter and assisting in field installation and testing.

REFERENCES

Campbell, R.H., S.L. Rawlins, and S. Han. 1994. Monitoring Methods for Potato Yield Mapping. ASAE Paper No. 94-1584. American Society of Agricultural Engineers, St. Joseph, MI.

Hofman, V.L., S. Panigrahi, B.L. Gregor, and J.D. Walter. 1995. Weighing Sugarbeets on the Lifter. 1994 Sugarbeet Research and Extension Reports. Vol 25, p231-5.

Multispectral Remote Sensing and Site-Specific Agriculture: Examples of Current Technology and Future Possibilities

E. M. Barnes
M.S. Moran
P.J. Pinter, Jr.
T. R. Clarke
USDA-ARS
US Water Conservation Laboratory
Phoenix, Arizona

ABSTRACT

Multispectral data can meet many of the information requirements of site-specific farming. Examples from the literature are presented where multispectral data has been applied to agricultural management problems. Some of the examples are illustrated using remotely sensed estimates of green leaf area index for a cotton field during the 1994 growing season.

INTRODUCTION

Remote sensing has shown potential for use in agricultural management for a number of years; however, the availability of fine spatial resolution, near real-time data has limited its application in the past (Jackson, 1984). New companies that provide aircraft-based imagery to meet the resolution and temporal requirements for agricultural management are now emerging. The promise of commercially available, high-resolution satellite imagery will also provide additional sources of remotely sensed data (Fritz, 1996).

Advances in precision farming technology (GIS, global positioning systems, and variable rate equipment) provide the tools needed to apply information from multispectral images to management problems. There is still considerable work to be done before the full benefits of remotely sensed data can be realized, but there are applications that can benefit from this data at the present time. The purpose of this paper is to provide an overview of how remotely sensed data can be utilized in site-specific agricultural management.

Vegetation Spectral Response

Digital imagery is obtained in distinct areas of the electromagnetic spectrum. Sensors used in vegetation monitoring are typically in the green, red, and near infrared portions of the spectrum. The importance of these spectral areas is illustrated by the high-resolution spectral response for a cotton canopy at different stages of development in Fig. 1. As the canopy develops, there is a definite increase in reflectance in the near-infrared (~725 to 900 nm), as the internal leaf structure of the plant reflects more of the energy in this portion of the spectrum compared to a bare soil. There is also a development of a green peak (~ 550 nm) and decrease in red reflectance (~650 to 790 nm) due to chlorophyll reflectance and absorption respectively. Thermal imagery (8,000 to 12,000 nm) has also been proven useful in monitoring vegetation, as this imagery can be used to determine surface temperature. Any stress which lessens a crop's transpiration ability will result in a relative increase in the surface temperature of the leaves. Additional factors which impact the spectral response of crops to stress are presented by Jackson et al. (1986).

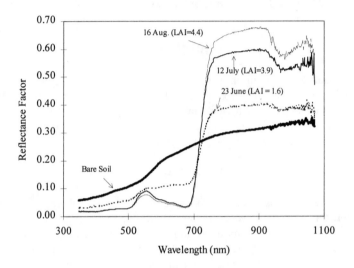

Fig. 1. High resolution reflectance spectra for both a bare soil and a cotton canopy on different dates. Measurements of green leaf area index (LAI) are shown for the dates the spectra were acquired.

The spectral response of vegetation has been used to formulate several vegetation indices, such as the Soil Adjusted Vegetation Index (SAVI, Huete, 1988) expressed as

$$\text{SAVI} = \frac{\text{NIR} - \text{RED}}{\text{NIR} + \text{RED} + \text{L}}(1 + \text{L}), \tag{1}$$

where L is a unitless constant, and NIR and RED the near-infrared and red reflectances, respectively. A good introduction to the interpretation of vegetation indices is provided by Jackson and Huete (1991), and the functional relationship between different indices is reviewed by Perry and Lautenschlager (1984). Vegetation indices are often well correlated with measures of plant density; as a crop's canopy develops, less bare soil is apparent, and thus a decrease occurs in red reflectance with an increase in NIR reflectance.

Previous Applications of Remote Sensing for Farm Management with Implications to Site-Specific Agricultural Management

Several methods have been developed to assess both soil and crop conditions using multispectral imagery. Some studies using remote sensing for soil properties, pest detection, and water stress are presented in the following sections.

Soil Properties

Soil physical properties such as organic matter have been correlated to specific spectral responses (Dalal and Henry, 1986; Shonk et al., 1991). Therefore, multispectral images have shown potential for the automated classification of soil mapping units (Leone et al., 1995). Such direct applications of remote sensing for soil mapping are limited because several other variables can impact soil reflectance such as tillage practices and moisture content. However, bare soil reflectance could have an indirect application in interpolating the results of gridded soil samples. For example, Fig. 2 shows a gray-scale image in the red portion of the spectrum. Percent sand and clay in the top 30 cm of the soil horizon is displayed over the approximate location of point samples taken by Post et al. (1988). Note that the brighter portions of the image correspond to areas of high sand content.

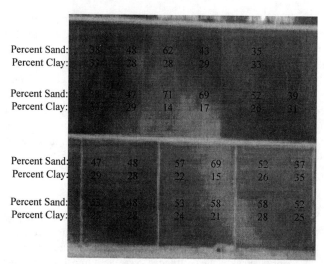

Fig. 2. Gray-scale image of a fallow field in the red portion of the spectrum with point measurements of percent sand and clay shown over the approximate sampling locations.

Vegetation spectral response has also been used to infer other soil conditions. Wiegand et al. (1994) showed a vegetation index was useful in mapping soil salinity over a sugar cane field. The nitrogen status of crops has also been estimated using remotely sensed data (Blackmer et al., 1995; Filella et al., 1995). Yang and Anderson (1996) describe methods to utilize multispectral images of vegetated fields for the determination of within-field management zones for application to site-specific farming.

Pest Detection

Sprayer mounted sensors have been found useful for the control of herbicide applications (such as Shearer and Jones, 1991). Brown and Steckler (1995) developed a method to use digitized color-infrared photographs to classify weeds in a no-till corn field. The classified data were placed in a GIS and a decision support system was then used to determine the appropriate herbicide and amount to apply.   Penuelas et al. (1995) used reflectance measurements to assess mite effects on apple trees. Powdery mildew has also shown to be detectable with reflectance measurements in the visible portion of the spectrum (Lorenzen and Jensen, 1989). The ability to detect and map insect damage with remotely sensed imagery implies that methods can be developed to focus pesticide applications in the areas of fields most infected, thus decreasing the damage to beneficial insects.

Water Stress

The difference between remotely sensed surface temperature and ground-based measurement of air temperature has been established as a method to detect water stress in plants (Jackson et al., 1981). More recently, methods to integrate spectral vegetation indices with temperature have been used to improve remotely-sensed estimates of evapotranspiration (Carlson et al., 1995; Moran et al., 1994).   Moran et al. (1994) defined a Water Deficit Index which uses the response of a vegetation index to account for partial canopy conditions, so that false indications of water stress due to high soil background temperatures were minimized. Spectral indices have also been used to determine "real-time" crop coefficients to improve irrigation scheduling (Bausch, 1995).

EXAMPLE DATA SET

Some of the capabilities of multispectral images are illustrated using a subset of data from the Multispectral Airborne Demonstration at Maricopa Agricultural Center (MADMAC) conducted during the summer growing season in Arizona (Moran et al, 1996). Images were acquired in four spectral bands (green, red, near-infrared and thermal) at a spatial resolution of 2 m, from April to October, 1994. The data presented here correspond to a field planted with two varieties of an upland cotton (Gossypium hirsutum L.) on April 5, 1994. The field was used in studies of irrigation efficiency by Watson (J.W. Watson, 1994, personal communication), and was divided into 12 irrigation borders. The irrigation levels and border layout are pictured in Fig. 3.

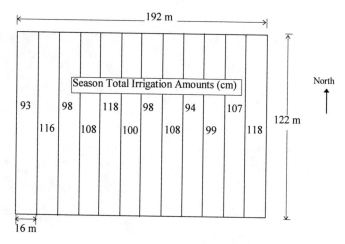

Fig. 3. Field layout and irrigation levels of a cotton field at Maricopa, AZ, in 1994.

The near-infrared and red images were calibrated to units of reflectance and then the SAVI was calculated. The SAVI was further modified to units of green leaf area index (LAI) using an empirically derived relationship (Moran, et al., 1996), where

$$LAI = -3.45 \ln(1 - SAVI) - 0.58. \tag{2}$$

Gray-scale representations of the LAI over the cotton field for different dates in 1994 are pictured in Fig. 4. Darker shades of gray represent higher values of LAI. The maximum LAI was 4.9 and occurred in the August 2 image. The average field conditions for the dates shown in Fig. 4 are listed in Table 1.

Table 1. Average field conditions corresponding to the dates of each image in Fig. 4.

Date	Percent Cover	Average Height (cm)	Growth Stage
7 June	25	25	15-Leaf
14 June	30	30	Flowering
6 July	50	30	
12 July	80	50	
21 July	80	70	
2 August	95	85	
16 August	100	100	Mature Bolls
23 August	100	100	
31 August	100	100	
8 September	100	100	

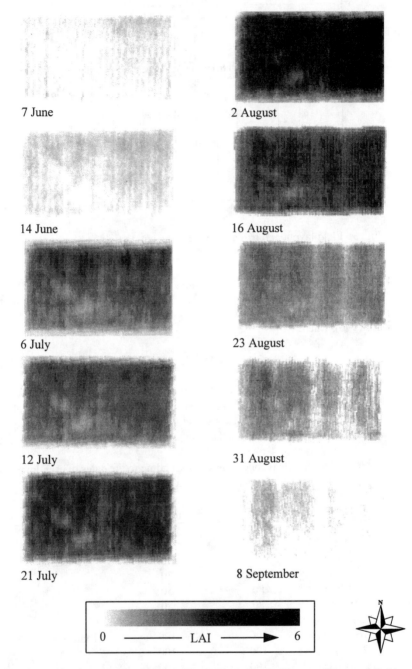

7 June

2 August

14 June

16 August

6 July

23 August

12 July

31 August

21 July

8 September

0 ——— LAI ——➤ 6

Fig. 4. Remotely sensed estimates of leaf area index (LAI) over the cotton field
with the irrigation treatments shown in Fig. 3 for ten days during 1994.

The first point of interest in Fig. 4 is the center and lower right hand portion of the field that had a consistently lower LAI throughout the season. Similar patterns were visible in images from previous years. It is likely that this response was due to a higher sand content than that of the surrounding portions of the field, indicating precision applications of herbicide and fertilizers may be advantageous. The tendency of the southern portion of the field to have a lower LAI, especially early in the season, may be an indication of non-uniform irrigation applications or variation in soil type. The impact of the different irrigation levels on LAI was evident by August 2 (irrigation levels were essentially the same until July 9).

The images indicated LAI reached its maximum value for most of the field on August 2. Beginning August 16, LAI decreased, which is also the time white fly and leaf perforator damage was first observed in the field. The damage appeared to begin on the eastern side of the field and spread to the west as indicated by the more rapid decrease in LAI in the east for later dates (note that defoliant was not applied until September 9).

From this example, it can be seen that multispectral images of red and NIR reflectance are useful for monitoring changes in vegetation patterns and development. It should be noted that the data presented in Fig. 4 were based on images calibrated to units of reflectance (that is, sensor characteristics and solar illumination conditions have been accounted for). Without this calibration, temporally-consistent estimates of LAI would not have been possible.

FUTURE POSSIBILITIES

Several applications have been developed to use remotely sensed data to infer both plant and soil characteristics. Three approaches of development appear to be emerging in the application of remote sensing and site-specific agriculture. In one approach, multispectral images are used for anomaly detection; however, anomaly detection does not provide quantitative recommendations that can be directly applied to precision farming. A second approach involves correlating variation in spectral response to specific variables such as soil properties or nitrogen deficiency. For example, in the case of nitrogen deficiency, once site-specific relationships have been developed, multispectral images can then be translated directly to maps of fertilizer application rates.

The third approach is converting multispectral data to quantitative units with physical meaning (such as LAI or temperature) and integrating this information into physically based growth models. For example, Moran et al. (1995) utilized remotely sensed estimates of LAI and evapotranspiration as inputs to a simple alfalfa growth model. The remotely sensed estimates were used to adjust the model's parameters throughout the season and resulted in improved predictions. Other applications of growth models with remotely sensed data are under development (Mougin et al., 1995; Carbone et al., 1996). Using remotely sensed inputs to growth models also provides a means to obtain predictions over large areas, which will increase the application of these models to site-specific agricultural management.

The latter two approaches have potential for incorporating remote sensing into decision support systems in a geographical information system environment (for example, Brown and Steckler, 1995). Further development will ultimately allow farm managers to make informed decisions about site-specific applications of farm materials.

ACKNOWLEDGMENTS

We would like to thank Jack Watson and Mike Sheedy of the University of Arizona's Cooperative Extension Service for supplying the information on their irrigation experiment; and Tom Mitchell and Jiaguo Qi for their image processing work. Thanks are also due to Christopher Neale and his staff in the Department of Biological and Irrigation Engineering at Utah State University; Roy Rauschkolk, Bob Roth, Pat Murphree and MacD Hartman of the University of Arizona Maricopa Agricultural Center; many hard-working technicians from the USDA-ARS U.S. Water Conservation Lab.

REFERENCES

Bausch, W.C. 1995. Remote sensing of crop coefficients for improving the irrigation scheduling of corn. Agric. Water Management 27:55-68.

Blackmer, T.M., J.S. Schepers, and G.E. Meyer. 1995. Remote sensing to detect nitrogen deficiency in corn. p. 505-512. In P.C. Robert, R.H. Rust and W.E. Larson (ed.) Proc. of Site-Specific Management for Agric. Systems, Minneapolis, Minn, 27-30 March 1994. ASA-CSSA-SSSA, Madison, WI.

Brown, R.B. and J.-P. G.A. Steckler. 1995. Prescription maps for spatially variable herbicide application in no-till corn. Trans. ASAE 38:1659-1666.

Carlson, T.N., W.J. Capehart, and R.R. Gillies. 1995. A new look at the simplified method for remote sensing of daily evapotranspiration. Remote Sens. Environ. 54:161-167.

Carbone, G.J., S. Narumalani, and M. King. 1996. Application of remote sensing and GIS technologies with physiological crop models. Photogram. Eng. Remote Sens. 62:171-179.

Dalal, R.C. and R.J. Henry. 1986. Simultaneous determination of moisture, organic carbon and total nitrogen by near infrared reflectance spectrophotometry. Soil Sci. Soc. Am. J. 50:120-123.

Filella, I., L. Serrano, J. Serra, and J. Penuelas. 1995. Evaluating wheat nitrogen status with canopy reflectance indices and discriminant analysis. Crop Sci. 35:1400-1405.

Fritz, L.W. 1996. The era of commercial earth observation satellites. Photogram. Eng. Remote Sens. 62:39-45.

Huete, A.R. 1988. A soil-adjusted vegetation index (SAVI). Remote Sens.

Environ. 25:89-105.

Jackson, R.D., S.B. Idso, R.J. Reginato, and P.J. Pinter, Jr. 1981. Crop temperature as a crop water stress indicator. Water Resour. Res. 17:1133-1138.

Jackson, R.D. 1984. Remote sensing of vegetation characteristics for farm management. SPIE 475:81-96.

Jackson, R.D., P.J. Pinter Jr., R.J. Reginato and S.B. Idso. 1986. Detection and evaluation of plant stresses for crop management decisions. IEEE Transactions on Geoscience and Remote Sensing GE-24: 99-106.

Jackson, R.D. and A.R. Huete. 1991. Interpreting vegetation indices. Preventive Veterinary Medicine 11:185-200.

Leone, A.P., G.G. Wright and C. Corves. 1995. The application of satellite remote sensing for soil studies in upland areas of Southern Italy. Int. J. Remote Sens. 16:1087-1105.

Lorenzen, B. and A. Jensen. 1989. Changes in leaf spectral properties induced in barley by cereal powdery mildew. Remote Sens. Environ. 27:201-209.

Moran, S.M., T.R. Clarke, Y. Inoue and A. Vidal. 1994. Estimating crop water deficit using the relationship between surface-air temperature and spectral vegetation index. Remote Sens. Environ. 49:246-263.

Moran, S.M., S.J. Maas, and P.J. Pinter, Jr. 1995. Combining remote sensing and modeling for estimating surface evaporation and biomass production. Remote Sensing Reviews 12:335-353.

Moran, S.M., T.R. Clarke, J. Qi, and P.J. Pinter Jr. 1996. MADMAC: A test of multispectral airborne imagery as a farm management tool. p. 612-617. *In* Proc. of the 26th Symposium on Remote Sens. Environ., March 25-29, 1996, Vancouver, BC.

Mougin, E., D.Lo Seen, S. Rambal, A. Gaston, and P. Hiernaux. 1995. A regional Sahelian grassland model to be coupled with multispectral satellite data. II: Toward the control of its simulations by remotely sensed indices. Remote Sens. Environ. 52:194-206.

Penuelas, J., I. Filella, P. Lloret, F. Munoz and M. Vilajeliu. 1995. Reflectance assessment of mite effects on apple trees. Int. J. Remote Sens. 16:2727-2733.

Perry, C.R. and L.F. Lautenschlager. 1984. Functional equivalence of spectral vegetation indices. Remote Sens. Environ. 14:169-182.

Post, D.F., C. Mack, P.D. Camp and A.S. Suliman. 1988. Mapping and characterization of the soils on the University of Arizona Maricopa Agricultural Center. Proc. Hydrology and Water Resources in Arizona and the Southwest, Arizona-Nevada Academy of Science 18:49-60.

Shearer, S.A. and P.T. Jones. 1991. Selective application of post-emergence herbicides using photoelectrics. Trans. ASAE 34:1661-1666.

Shonk, J.L., L.D. Gaultney, D.G. Schulze and G.E. Van Scoyoc. 1991. Spectroscopic sensing of soil organic matter content. Trans. ASAE 34:1978-1984.

Wiegand, C.L., D.E. Escobar, and S.E. Lingle. 1994. Detecting growth variation and salt stress in sugarcane using videography. p. 185-199. *In* Proc. 14th Biennial Workshop on Color Aerial Photography and Videography for Resource Monitoring. American Society for Photogrammetry and Remote Sensing.

Yang, C. and G.L. Anderson. 1996. Determining within-field management zones for grain sorghum using aerial videography. *In* Proc. of the 26th Symposium on Remote Sens. Environ., March 25-29, 1996, Vancouver, BC.

Variable Rate Application Technology: An Overview

R.L.Clark
R.L.McGuckin

Department of Biological and Agricultural Engineering
University of Georgia
Athens, GA

ABSTRACT

This paper provides a brief overview of precision farming, followed by an overview of the main components of variable rate equipment, then concludes with a pointer to a WWW site which contains a summary of the current commercially available equipment for variable rate application of seeds and chemicals. The equipment reviewed includes: computer/controllers, liquid sprayers, granular fertilizer applicators, air sprayers and spreaders, and drills and planters.

INTRODUCTION

Precision farming is a farming system concept which involves the development and adoption of knowledge-based technical management systems with the main goal of optimizing profit. This management system will enable micro-management concepts, that is, the ability to appropriately manage every field operation at each location in the field, if it is technically and economically advantageous to manage at that level. The system will likely include the ability to vary or tailor the rate of application of all inputs such as tillage, seeds, weed, insect and disease control, cultivation and irrigation.

It will be possible to implement precision farming at many different levels. In its most extensive form, it will include precise micro-management of every step of the farming process. It is expected that the advisability of micro-management will be dependent upon many factors, such as soil type, crop, seasonal weather, and other factors. For example, in a dry year, it may be possible to control insects by spraying only small areas where the insects are known to exist; in a wet year, it may be advisable to uniformly spray the whole field.

Technically, one important aspect of the development of precision farming concepts is the development of the hardware and software necessary to vary the rate of the application of agricultural inputs. A number of research projects have been conducted in this area, and several companies have been developing variable rate application equipment in recent years. The objective of this paper is to provide a brief overview of precision farming systems, then outline the main components which are usually found in variable rate application equipment, followed by a review of the commercially available equipment on the market today.

Precision Farming System Overview

An overview of the precision farming system of the future is depicted in Fig. 1. The brain of the system is a geographic information system (GIS), which will form the knowledge base and decision making parts of the precision farming system. The technical and economic decisions related to the farming operation will be governed by this knowledge based GIS. A GIS will be made up of layers of related information, and the GIS will allow a quantitative study of the relationships

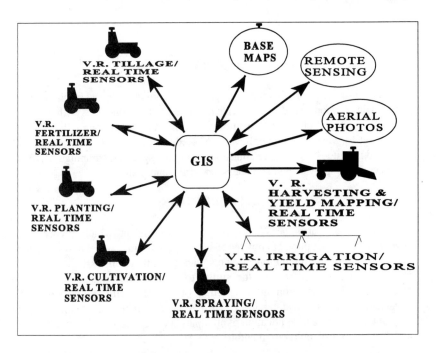

Fig. 1. Overview of the Precision Farming System
    Note 1: Each line represents a two way flow of information
    Note 2: V.R. means "variable rate"
    Note 3: Each field operation will include real time sensors,
            such as sensors for position, moisture, nitrogen, flow rate, etc.

between the layers. For example, the GIS may contain the following layers: (1) field topography, (2) soil types, (3) surface drainage, (4) sub-surface drainage, (5) soil testing results, (6) rainfall, (7) irrigation, (8) actual chemical application rates, and (9) yield. Some of these layers will be entered once; some will be entered annually or even more frequently. The GIS will then allow a study of the relationship between these layers of information to determine cause and effect and to base decisions upon this knowledge.

As indicated in Fig. 1, each field operation may include variable rate technology. Tillage depth may be varied according to field location; for example, subsoiling depth may be dependent on field location. Seeding rates may vary according to field location, which may depend on factors such as topography and soil type. Fertilizer application rates may vary in relationship to factors such as soil type and the results from either real time or pre-application testing. Application of insecticides may be dependent on insect location from either scouting reports or from aerial imaging. In like manner, the application of all inputs to the crop production process may vary with field location.

Overview of The Components of Variable Rate Application Equipment

The main components which make up a variable rate application system are shown in Fig. 2. Not all systems will necessarily contain all of the components shown. As variable rate technology develops, other system components may be included.

The central component of variable rate application equipment is the computer/controller. This device receives information from several sources which will in turn be used to control the application equipment. The controller may receive information from the application equipment and other sensors to maintain a database on the actual application rate as a function of field position.

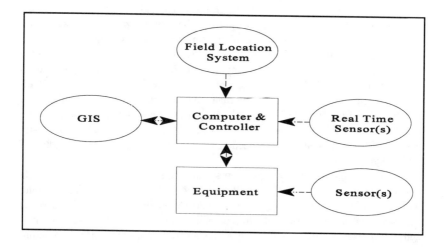

Fig. 2. General Overview of Variable Rate Equipment

A key component for all precision farming operations is the technology to determine the instantaneous position of equipment as it operates in the field, and to provide this information in a computer compatible format. The technology which has rapidly gained acceptance as the optimum system is the Global Positioning System (GPS). A stand-alone GPS receiver can have instantaneous errors as high as 100 m, which is unacceptable for precision farming. Fortunately, several systems to calculate what is known as "differential corrections" have been designed, which can allow the GPS system on a farm vehicle to achieve position accuracies in three basic accuracy ranges: (1) 2-5 meters, (2) sub-meter, or (3) in the sub-decimeter range, depending on the technologies used. These maximum error figures relate to horizontal position, and vertical position (elevation) error is usually 1.5-5 times the horizontal position error. Most precision farming operations do not require vertical position information; the main application requiring vertical as well as horizontal information is to develop topographic maps. Most precision farming operations will require real-time differential corrections so that vehicle position information will be accurate when the vehicle is operating in the field. For precision farming applications, the GPS positioning technology should be thought of as RT-DGPS, that is, the farmer should always be using real-time differential corrections to minimize position error.

Information contained in the geographic information system related to a specific field operation is downloaded to the system computer before field operations commence. The computer/controller will continuously control variable application rates based upon knowledge gained both from the geographic information system, from a knowledge of field location as provided by RT-DGPS, and perhaps from real time sensors. For example, assume that the desired fertilizer application rate is known to be a function of results from soil analysis tests, field location, and crop. The soil analysis test results as a function of field location would be entered into the GIS and downloaded to the computer/controller of the fertilizer applicator. If one crop is being grown in the field being fertilized, then the operator may simply enter the crop from the computer/controller keyboard. However, if two crops are grown in alternating strips, this information would be entered into the GIS as a function of field position, then also downloaded to the variable rate application (VRA) computer/controller. When the equipment is operating in the field, the VRA computer/controller will be receiving RT-DGPS receiver position information and will match required application rate and crop as a function of field location to control the applicator equipment. It may also be possible to have a real time soil sensor which will provide information on-the-fly about fertilizer application rate needed, rather than using pre-application soil sampling/analysis techniques.

The application equipment may also have sensors which provide quantitative information on the actual application rates. This information, along with RT-DGPS position, can be recorded to maintain a historical record of application rates. This historical information may allow the farmer to analyze cause and effect in the precision farming system, and perhaps can influence future decision making processes implemented in the computer/controller. For example, assuming that sufficient information has been gathered over several years, the farmer may have historical records on the effect of all of the inputs to his system for a specific field, including the crop yield. The GIS would then allow an analysis of cause and effect, based upon many factors, and allow fine-tuning of chemical application rates in subsequent seasons.

Eventually the RT-DGPS system may also be used for vehicle guidance. Most farm vehicle guidance systems today are visual prompting systems for the vehicle operator which can establish accurate vehicle position for application swaths. In the future, the guidance system may automatically guide the application vehicle.

A VR Sprayer Scenario

To provide a specific illustration, consider the diagram of a relatively simple liquid sprayer VRA system as depicted in Fig. 3. The following discussion is provided as one scenario for each component, but there may be alternative sensors and methods of control. A radar based ground speed sensor would be used to provide true ground speed to the computer/controller since application rate is a function of speed. This system depicts the use of a direct injection sprayer, which is the direction in which sprayer technology is proceeding. With this type of sprayer, the operator does not mix the chemical(s) in the main tank, rather, the chemical(s) remains in a container, where it may be pumped as needed into an injector where the chemical(s) is automatically mixed with water on-the-fly. There are many advantages to this system as compared with tank-mixing, such as safety, managing mixed chemicals, and automation. The injector pump may be designed to provide precise control of the injection rate of the chemical concentrate to the injector.

The water tank may have a level sensor which will allow the computer/controller to determine the amount of water remaining in the tank in gallons. The total flow rate of the fluid going to the boom(s) will be controlled by the flow control valve, which in turn is controlled by the computer/controller. The actual total fluid flow rate will be monitored by the fluid flow rate sensor, and this information will be used by the computer/controller for fine adjustments in the flow control valve. The fluid flow rate and the vehicle position will be continuously recorded in the computer as the vehicle sprays to provide a historical record for the GIS about where and how much chemical was dispensed. The boom valve will be used to turn the boom on or off to provide fast accurate control of the application area.

To further illustrate this system, assume that you may be in the middle of the cotton season, and the cotton is being scouted on a normal cycle for insects. When the scout goes to the field, he/she may carry a portable GPS unit. When an insect infestation is identified, the scout could walk around each infested area with the GPS unit, thereby recording the location of the areas of infestation. Assume that the scout finds two such infested areas. The scout would inform the farmer that infested areas were located, and the farmer would download the map which shows the infested areas. The map would include not only the insect found, but the estimated insect density.

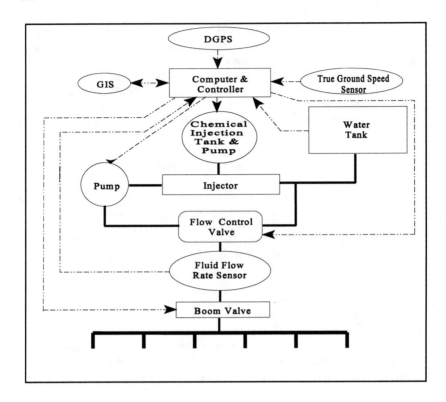

Fig. 3. General Components of a Variable Rate Sprayer

        The farmer would then enter this infestation map into his existing GIS for
that field. The GIS software would examine the data as related to appropriate
information such as current and forecast weather conditions, crop age, and the
history of this crop, including other chemical applications. The GIS software would
be designed to model the growth of the crop and the expected effect of this insect
on crop yield. The objective would be to determine the cost effectiveness of
spraying with several possible scenarios: (1) uniform spraying of the entire field,
(2) spraying of the infected areas only, or (3) no spraying. Assume that this
intelligent system indicates that the farmer should just spray the infected areas. The
farmer would then download several important maps to the computer/controller on
the spray vehicle. The GIS information would likely consist of several maps: (1)
a map giving the coordinates of the field boundaries, which may exclude areas
within the outer boundary (waterways, roads, etc.), (2) a map giving the coordinates
of the crop boundaries, (3) a map giving the location of each crop row, and (4) a
map giving the location of the infected areas, and the name of the insect. It will be
assumed that the insecticide application rate may be varied within each infected
area. Information on the total amount of water and chemical concentrate required
for the spot spraying would also be downloaded to the computer/controller.

When the vehicle operator starts, the software in the computer/controller will examine the data downloaded from the GIS. The computer display will provide instructions about which chemical concentrate to load onto the vehicle, and how much concentrate and water are needed. The operator will then place the chemical concentrate tank onto the vehicle and hook it to the computer/controller. The computer will read information from a microchip on the concentrate tank and will check to be sure that this is the correct chemical for this crop and insect, along with determining the appropriate application rate. Also, the computer will check a concentrate tank sensor to be sure that the concentrate tank has sufficient chemical for the operation. To load the water tank, the operator will attach a water hose to the tank, which will have a valve on the inlet line controlled by the computer/controller. If the tank has insufficient water, the inlet valve will be opened. When the computer senses that the tank has sufficient water for the operation, the inlet valve will be closed.

If the vehicle operator is spraying many fields over a wide area, the system can incorporate a road map of the area which can be displayed in the cab. The RT-DGPS system will be used to display the actual vehicle location on the map, and the mapping system will be used to determine the optimum route to the field. When the vehicle arrives at the field, the display system will automatically change scale to show a map of the field, including the location of the infested areas. The display will provide directional information to the operator indicating which rows the operator should drive down to spray. The spray booms will be automatically extended when the RT-DPGS information says that the vehicle is within the field. The operator will proceed down the first row to which he has been directed. As the vehicle approaches the boundary of the infested area, the main pump will be automatically started, the boom valve will be opened, and the sprayer will begin to dispense water. The injector pump will begin at the appropriate time, depending on the lag time for the concentrate to enter the injector and arrive at the nozzle. As the vehicle approaches the boundary where the spray application will stop, the injector pump will stop to allow the appropriate time for the chemical to clear out of the boom. The boom valve will be closed when the vehicle reaches the other boundary of the infested area.

This process will continue until the operator has sprayed the infested areas. Note that it will be necessary to only drive selected rows to cover the infested areas, not the entire area.

When the field is finished, the operator will download the information on the actual rates applied as a function of field location. This data will be entered into the GIS for use in further operations as needed. This data may also be useful for further studies of the effectiveness of this chemical, and may be used in subsequent years to modify decisions.

Variable Rate Application Equipment

Because of space limitations, this paper does not contain a listing of the companies who currently market variable rate application equipment. This information is available on the WWW site listed in the summary below. This site contains a tabular summary of currently available variable rate application equipment, and information on how to contact these companies. Company names reported herein were found through magazine ads, scientific publications and word of mouth. The authors do not wish this report to be viewed as a complete review of all the possible companies working in the precision farming arena, as the concentration of this paper is only on the VRA aspects. Because the VRA technology is a rapidly developing field, it was found that some companies were unwilling to divulge engineering details related to their equipment because of patent rights.

For each company only the equipment or systems that are directly related to VRA are listed. We apologize to any companies producing VRA equipment who were inadvertently excluded from this review. We invite any companies not included herein to forward technical information on their VRA equipment for inclusion in future review publications of this nature. The mention of brand names is for information only and does not imply endorsement by the University of Georgia.

SUMMARY

In this paper the precision farming system of the future was briefly outlined. The brain of the system is a geographic information system which will enable knowledge-based farming decisions to optimize net profit. An important aspect of the technology is the ability to vary the rate of application of all inputs, that is, to tailor or prescribe the application to various sites throughout each field, including tillage, fertilizer and lime application, planting, cultivation, and spraying. The components usually found in variable rate application equipment were outlined and discussed in some detail. The paper appendix contains two summary tables which provide information on most companies involved in producing variable rate application equipment.

Most of the commercial ventures to date have focused on the variable rate equipment for application of liquid and granular materials. There remain many unanswered questions about how to implement this technology. It was pointed out that the GIS is the brain of the system, but this aspect of the technology is still in the infancy stage. A critical aspect of the electronic technologies is standardization, ranging from physical connections which can withstand the farming environment, to standardization of data format. It will be critical to develop the technologies to make them simple to use and user friendly, as well as economical. Much technical development work remains before the precision farming system of the future can be implemented. In the final analysis, it must be shown that precision farming pays- particularly economically, environmentally, and from the viewpoint of the conservation of our natural resources.

For more information on precision farming, contact the University of Georgia precision farming web site at:
  http://www.bae.uga.edu/dept/research/precision/index.html.
Links to other similar web sites, and information on precision farming are provided at this site.

Development of a Precision Application System for Liquid Animal Manures

D. R. Ess

Department of Agricultural and Biological Engineering
Purdue University
West Lafayette, IN

B.C. Joern

Department of Agronomy
Purdue University
West Lafayette, IN

S.E. Hawkins

Purdue Agricultural Centers
Purdue University
West Lafayette, IN

ABSTRACT

An applicator was modified to facilitate the precision management of liquid animal manures. A set of load cells mounted under the unit's slurry tank provided a means of measuring the load within the tank and application rates in the field. In addition, the applicator was equipped with an electrohydraulic control system that will permit automatic, map-based control of manure output.

INTRODUCTION

Most research effort in precision farming has been focused on the management of manufactured inputs such as chemical fertilizers and pesticides. Precision management of animal manures is not common practice because, due to production, treatment, and storage factors, animal manures are highly variable and present special challenges. Precision application of a variable product to variable soils is a daunting task. The technology for accurately monitoring and metering manure is available (Krause and Peters, 1986), but the combination of a positioning system, a computerized interface, a satisfactory control algorithm, and monitoring and metering technologies has not been achieved.

Animal manures are a complex combination of compounds that are managed as a unit. It is not feasible to separately manage the N, P, K and micronutrient components. Each nutrient component must, however, be accounted for in a comprehensive manure management plan (Purdue University Cooperative

Extension Service, 1994). Under current manure management planning approaches, application rate recommendations are made on a whole-field basis. Appropriate nutrient input levels are based on soil type, soil slope, soil test results, crop type, and expected crop yield. Even if appropriate input levels are established, manure application rates are often determined from nutrient concentrations that are estimated rather than measured. Mahamane (1993) found that former feedlots and manure disposal areas close to the farmstead were an important source of variability for non-mobile nutrients such as phosphorus.

Initial economic analyses of precision farming have focused on management of one or two nutrients and have shown mixed results (Mahamane, 1993; Wollenhaupt & Buchholz, 1993; Reetz & Fixen, 1995). Lowenberg-DeBoer et al. (1994) indicate that precision management of one or two inputs will seldom pay the cost of data collection and analysis. The potential for precision application of manure to be cost effective is impacted by factors well beyond crop response to nutrient inputs, however. Large, concentrated animal feeding operations that are required to obtain National Pollution Discharge Elimination System (NPDES) permits are subject to inspection and civil and criminal penalties for violations of the Clean Water Act (Wyant, 1995). Civil fines can range up to $25,000 per day for a willful or negligent violation and criminal fines range from $5,000 to $50,000 per day, plus up to three years' imprisonment.

In the view of the U.S. EPA, land application of animal manures at "agronomic rates" should not be subject to permit requirements (Weitman, 1995). It is, however, incumbent upon the land applicator to be able to justify and verify application rates. A system is needed to permit easy, accurate documentation of land application rates and sites.

Identification of suitable and unsuitable manure application sites is a challenge to livestock producers. Although site features such as waterways and road ditches can be easily identified from the tractor seat, separation distances cannot. Separation distances can be measured and marked prior to manure applications, but procedures such as taping and flagging are tedious and the markings are typically temporary. Failure to measure or mark sites can lead to setbacks that are too small and manure application in unsuitable areas or to setbacks that are unreasonably large and overapplication on "safe" sites. The development of a relatively easy, permanent method of marking and adhering to application sites could reduce the potential for environmental damage resulting from misapplication of animal manure. Global positioning system hardware and appropriate mapping software could facilitate site mapping and application map creation.

Research in the state of Indiana has shown that managing manure as a fertilizer resource for crop production can increase the return to the producer, minimize the pollution potential of manure, and enhance overall production efficiency of an animal-crop farming system (Huber et al., 1993). Precision management of manure has the potential to further improve farming system production efficiency and reduce farmer exposure to potentially devastating legal action resulting from unintended, but inappropriate manure applications.

EQUIPMENT AND INSTRUMENTATION

In October 1993, Stephen E. Hawkins and Brad C. Joern, developed the concept for a pull-type liquid manure applicator that would serve as a fundamental component of a precision manure management system. The project that evolved from the concept had the following objectives:

1   develop a manure applicator with the ability to weigh output or measure flow rate per unit area covered;

2   create a data collection system that would permit the documentation of application rates and locations; and

3   develop a system to permit the automatic control of application location within a mapped site.

The applicator was intended to serve as a testbed for precision manure application concepts. The unit, designed to permit multiple loading and unloading modes while maximizing material homogeneity, included the following components:

•   a PTO-driven slurry pump,
•   an auxiliary-engine-driven vacuum pump,
•   a liquid recirculation system to enhance material uniformity, and
•   a set of load cells for weighing the contents of the slurry tank.

The applicator, a significantly modified version of a commercially-available Balzer Model 2250 spreader (Balzer Mfg. Corp., Mountain Lake, MN), was completed in November of 1994. The applicator has a tank capacity of 8700 L and weighs 4545 kg empty. The vacuum pump is powered by a 14.9-kW auxiliary engine. The unit, depicted in Fig. 1, is equipped to permit surface application or injection of liquid manure. The tank can be unloaded by the slurry pump, the vacuum pump, or by gravity.

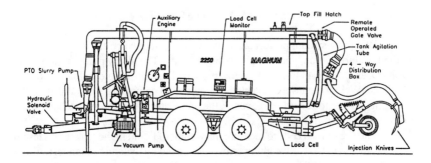

Fig. 1. Modified Balzer Model 2250 liquid manure applicator.

Weighing System

The applicator's weighing system consists of a set of four load cells, a digital scale indicator, and a portable computer. The load cells, J-STAR® DB Series weighbeams (J-STAR Electronics, Inc., Fort Atkinson, WI), are mounted between the unit's tank and independent frame. Measured loading is displayed by a J-STAR® EZ 210 Electronic Scale Indicator. The indicator/monitor, powered by an auxiliary 12-volt battery, is mounted on the applicator within a weatherproof enclosure.

Data from the scale indicator are transmitted via a J904 port (RS-232) in asynchronous ASCII format at a rate of 4 to 8 times per second. Weight data are captured and recorded 3 times per second by a remotely-mounted TelePad SL portable computer (TelePad Corporation, Reston, VA). A program written in Microsoft QuickBASIC is used to create data files containing weight readings and the associated recording times for use in spreadsheet analyses or map creation. The digital scale indicator produces weight readings in 4.5-kg increments.

Electrohydraulic Control System

The Balzer Model 2250 liquid manure spreader is equipped with a 15-cm brass gate valve mounted on the unit's main discharge tube and controlled by a hydro-pneumatic actuator. The actuator consists of a gas spring that holds the gate in the closed position until being acted upon by fluid provided by a tractor's hydraulic system through a directional control valve (DCV). A poppet-type, solenoid-operated auxiliary directional control valve (Fasse Valves, Kearney, NE) was installed on the applicator to work in conjunction with a tractor's hydraulic system. The valve has a flow rating of 83 L/min and was designed for use with closed-center hydraulic systems.

Flow from a tractor's remote hydraulic outlet is directed to the auxiliary DCV. The direction of flow through auxiliary valve is controlled by a pair of 12-volt solenoids actuated by a tractor-cab-mounted toggle switch. An isolator circuit (Fig. 2), installed to permit computer control of the auxiliary DCV, facilitates automatic control of gate valve position without any direct electrical connection between the computer and the toggle switch.

RESULTS AND DISCUSSION

Weighing system performance was tested using water discharged by three methods - slurry pump, vacuum pump, and gravity. The slurry pump, powered by a John Deere Model 4440 diesel tractor, produced the most rapid unloading rate -- 84.1 kg/s. Weight data were collected and plotted to demonstrate the ability of the weighing system to record data even at the applicator's maximum discharge rate. Results are shown in Fig. 3.

The use of load cells to measure the weight of liquid in the tanker was tested to assess the potential for weighing errors introduced by dynamic loading resulting from the movement of the unit over land. Dynamic loading effects were measured by operating the applicator with its tank filled with water to three different levels

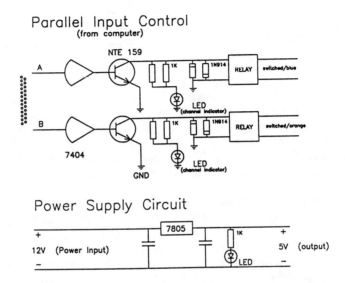

Fig. 2. Isolator circuit used for computer control of a solenoid-operated directional control valve.

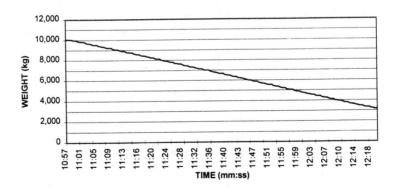

Fig. 3. Tank weights recorded during unloading of water using the slurry pump.

Table 1. Percent error in measured liquid weights resulting from dynamic
loading effects.

	Fill Level %		
Travel Speed (m/s)	12	50	85
1.34	5.6	2.8	3.1
2.68	7.3	2.9	3.9

(12%, 50%, and 85% of full capacity). The unit was pulled over a hard-surfaced course of varying slope and roughness at speeds of 1.34 m/s and 2.68 m/s. The amount of error was defined as the difference between minimum and maximum weights recorded by the weighing system during one travel circuit with a fixed amount of liquid in the tank. The amount of error was divided by the measured static weight of water for each tank fill level to produce the percent error values provided in Table 1.

Weighing error increased at each loading level as travel speed increased. The absolute magnitude of weighing errors decreased as tank fill levels were reduced, but error as a percent of the weight of liquid increased.

The electrohydraulic control system was tested to ensure its effectiveness. The system was found to be capable of controlling flow from the applicator's slurry pump at the highest system discharge rate. Flow was diverted from the unit's injectors to the bypass/agitation system when the gate valve was closed.

CONCLUSION

A commercially-available liquid manure spreader was modified to permit the measurement of liquid application rates and the automatic control of material discharge. Testing demonstrated the effectiveness of the application rate measurement system at water discharge rates up to 84.1 kg/s. The unit's weighing system was affected by errors resulting from travel-induced dynamic loading, but the magnitude of such errors was less than eight percent of measured liquid weight. Discharge from the applicator could be controlled by a microcomputer interfaced with an electrohydraulic control assembly.

FUTURE DEVELOPMENT

The liquid manure applicator described herein will undergo further development as it is integrated into a precision manure application system. The envisioned system will enhance environmental stewardship by permitting essentially automatic documentation of manure application rates and locations and control of application locations. A portable computer interfaced with a global positioning system receiver will be used prior to manure application at a site to produce a map of the area. Mapping software will be used to record boundaries and features and to note any site and/or environmental factors that would prevent safe application of liquid manure (e.g., surface tile inlets, water wells, road ditches, sinkholes, intermittent streams, waterways, etc.). Appropriate manure-spreading separation distances marked on the map will define areas that are to receive no manure application. Actual application rates will be recorded and incorporated into a geographic information system.

The capability of measuring and controlling discharge will be enhanced by the metering of liquid manure based on measured nutrient contents. Computer programs such as AMANURE (Sutton et al., 1993) will provide the basis for prescribing a manure application rate for each sub-field-sized management unit within an application site. The prescribed rates can then be assigned site-specifically to permit control of manure application rates.

Improved nutrient management and reduced environmental impact related to land application of liquid animal manures should result from the work. The effort is a complement to site-specific farming system implementation that is currently underway at Purdue University.

ACKNOWLEDGMENTS

The authors would like to express their gratitude to Wayne Werne, Purdue University Department of Agronomy, Michael Roberts, Alabama A&M University Department of Plant and Soil Science, and Steve Shillington, Purdue University Department of Agricultural and Biological Engineering, for their assistance in developing and evaluating the performance of the liquid manure applicator.

Mention of trade names and specific products is done for the benefit of the reader and does not imply the endorsement of the products by Purdue University.

REFERENCES

Huber, D.M., A.L. Sutton, D.D. Jones, and B.C. Joern. 1993. Nutrient management of manure to enhance crop production and protect the environment. p. 39-45. *In* J.K. Mitchell (Ed.) Integrated Resource Management and Landscape Modification for Environmental Protection. ASAE, St. Joseph, MI.

Krause, R., and H. Peters. 1986. Sensors and actuators for speed controlled application by manure tankers. Grundlagen der Landtechnik 36(4):97-104.

Lowenberg-DeBoer, J., R. Nielsen, and S. Hawkins. 1994. Management of intra-field variability in large scale agriculture: a farming systems perspective. pp. 551-555. *In* Proceedings of the Systems-Oriented Research in Agriculture and Rural Development Symposium, Montpelier, France.

Mahamane, I. 1993. An evaluation of soil chemical properties variation in northern and southern Indiana. Ph.D. thesis, Department of Agronomy, Purdue University, West Lafayette, IN.

Purdue University Cooperative Extension Service. 1994. Swine Manure Management Planning. Purdue University, West Lafayette, IN.

Reetz, H.F., Jr., and P.E. Fixen. 1995. Economic analysis of site-specific nutrient management systems. pp. 743-752. *In* P.C. Robert, R.H. Rust, and W.E. Larson (Eds.) Site-Specific Management for Agricultural Systems. ASA-CSSA-SSSA, Madison, WI.

Sutton, A.L., D.D. Jones, D.M. Huber, and B.C. Joern. 1993. Integrated swine manure nutrient management. p. 29-38. *In* J.K. Mitchell (Ed.) Integrated Resource Management and Landscape Modification for Environmental Protection. ASAE, St. Joseph, MI.

Weitman, D. 1995. Federal water quality policy and animal agriculture. p. 1-5. *In* C.C. Ross (Ed.) Proceedings of the Seventh International Symposium on Agricultural and Food Processing Wastes. ASAE, St. Joseph, MI.

Wollenhaupt, N.C., and D.D. Buchholz. 1993. Profitability of farming by soils. p. 199-211. *In* P.C. Robert, R.H. Rust, and W.E. Larson (eds.) Soil Specific Crop Management. Department of Soil Science and Minnesota Extension Service, University of Minnesota.

Wyant, S. 1995. Do point-source rules apply to you? Soybean Digest 55(1):A20.

Estimation of Quality of Fertilizer Distribution

J. Kaplan
J. Chaplin

Department of Biosystems and Agricultural Engineering
University of Minnesota
St. Paul, MN

ABSTRACT

Differential and integral methods are presented to quantify the distribution unevenness of fertilizer during field operation of chemical applicators. The coefficient of variation obtained using these methods was compared to results from field tests. This comparison showed that both proposed methods can be used to estimate the quality of material distribution in the field.

A new spread pattern test, which works for collecting tray of any size has been developed.

INTRODUCTION

The coefficient of variation of the fertilizer distribution in the field can be used as a parameter of quality of fertilizer distribution (Jensen & Pesek, 1962). Commonly used methods of measuring distribution uniformity for broadcast spreaders do not provide an estimation of the unevenness in the area of the field. ASAE Standard S341.2 "Procedure for measuring distribution uniformity and calibrating granular broadcast spreaders"(1992) presents a method to determine the effective swath width, the application rate and the uniformity of distribution across the swath (spread pattern uniformity). The International Standard ISO 5690/1-1982(E) "Equipment for distributing fertilizer. Test methods"(1983) recommends the same tests be conducted and in addition, the unevenness along the swath be determined. However these standards do not provide methods for estimating unevenness over an entire area of the field.

As a result of interest in site specific farming and application of agricultural chemicals according to crop/soil requirements, a revised procedure is needed to determine the unevenness of a fertilizer distribution that accounts for variability of factors that have influence on the distribution process.

LITERATURE REVIEW

ASAE Standard S341.2 (ASAE 1995) allows a choice of collecting tray size. Spread pattern evaluation and collection methods have been investigated by Parish (1986). The results indicated that the method of collection can influence the observed application rate, the effective swath width, amount of skewing, and the coefficient of variation across overlapped swath.

Ostanin and Yanishevsky (1973) showed that as the size of the sample plot is reduced, the fertilizer distribution unevenness had a reduced effect on the yield.

Both ASAE and ISO standards recommend testing spreaders with the same level of material in the hopper. Coates (1992) investigates the effect of swath width, feed gate opening, and hopper level on the spread pattern of a pendulum spreader (oscillating spout). Coefficients of variation increased as the hopper emptied and decreased as swath width decreased.

According to the recommendations in ASAE S341.2, spreaders should be tested on smooth surfaces. Parish (1991) showed that operating a rotary spreader on a rough surface had a detrimental effect on the spread pattern uniformity. The coefficient of variation, as determined by ASAE Standard S341.2 increased from 10% for a spreader operating on a smooth surface to 30% for the same machine on a bumpy surface.

ASAE Standard S341.2 recommends a constant travel speed. Parish and Chaney (1986) and Parish (1987) showed that under normal conditions the working speed of the spreader tends to vary, and the speed affects the performance of spreaders significantly.

According to the recommendations of ASAE S341.2, the coefficient of variation should be determined using the effective swath width. Marchenko and Chernicov (1977) showed that the swath width of the spreader changes considerably during normal use. Dorr et. al. (1992) presented the results of measuring the spreading performance of boom spreaders. They concluded that 10% of the field had either overlapping or underlapping patterns.

Roth et al. (1985) showed that spread patterns are frequently asymmetrical and rough.

Based on a review of the literature, the following sources of variation have a direct effect on material distribution or the ability of measurement:

1. Driving precision, 2. Field surface and test conditions, 3. Metering efficiency and variation with hopper level, 4. Systematic errors associated with machine calibration, and 5. Size of the collecting tray.

OBJECTIVES

The following objectives were established:

- to develop a statistically sound differential model that explains the spatial variability of material applied with a field applicator,
- to develop a method of integral estimation,
- to make comparison among results obtained by the differential model, the integral method and the field estimation.
- to estimate influence of the collecting tray size on the quality parameters
- to develop a refined spread pattern test procedure.

DIFFERENTIAL MODEL

The application rate of the fertilizer on each sample plot of a field can be represented by

$$X = X_s \pm \Delta X_w \pm \Delta X_d \pm \qquad (1)$$

where X_s is desired application rate,

ΔX_w is the deviation of the application rate as a result of a uneven distribution in the area of adjusting,

ΔX_d is the deviation of the application rate as a result of the feed drift along the path of spreader unloading,

ΔX_r is the deviation of the application rate as a result of inaccurate adjustment,

ε is an error.

The adjustment area is defined as a small part of the field, large enough for testing and adjusting the machine according to manufacturer's recommendations (International Standard ISO 5690/1-1982).

Setting the spreader can be conducted on a small plot of the field, when the level of fertilizer in a spreader hopper is a constant. Therefore, all assumed deviations are independent. In this case, the variance of the application rate is the sum

$$D_x = D_w + D_d + L \qquad (2)$$

(subscripts correspond to those of deviations in Eq. (1)).

A previous investigation J. Chaplin et al., (1994) showed that the coefficient of variation of the application rate due to uneven distribution in the adjustment area can be determined by the following equations:

$$CV_w^2 = CV_a^2 + CV_t^2 \qquad (3)$$

$$CV_a = \sqrt{\left(\frac{B_w}{bq^2} \sum_{i=1}^{B/b} \sigma^2_i \right)} \qquad (4)$$

$$CV_t = \sqrt{\left(\frac{B_w}{bq^2} \sum_{i=1}^{B_w/b} \Delta^2_i \right)} \qquad (5)$$

$$q = \sum_{i=1}^{B/b} \overline{X}_i \qquad (6)$$

where CV_a is the coefficient of variation of the application rate along the swath (longitudinal unevenness),

CV_t is the coefficient of variation of the application rate across the swath(lateral unevenness),

B_w is the effective swath width,   B   is the overall swath width,

b   is the collecting tray width, σ_i is the standard deviation of the weight in the I-th longitudinal row of trays

$\overline{X}_i$ is the average weight in the tray of the I-th longitudinal row,

Δ_i is the deviation of the I-th observation from the overall mean.

Equations (4) and (5) do not account for the variations in swath width that do not influence the longitudinal unevenness. The same can not be said about the lateral unevenness, CV_t. Each value of the swath width corresponds to the defined value of CV_t.

Investigators Marchenko et al., (1977) and Kaplan et al.(1994) showed that variations of the swath width are normally distributed. Since the normal distribution is symmetrical, for any positive deviation of the swath width in any place in the field, an equivalent negative deviation could be found elsewhere. Distances between successive swaths can be grouped in pairs for each measured swath width. Since the lateral unevenness is being considered, the average distribution pattern is used. Therefore, after regrouping, the magnitude of the lateral unevenness remains constant over the whole field. This allows the distance between the axes of the first and the third swaths to be fixed and lateral position of the middle swath to be changed in increments of i to investigate the lateral unevenness as a function of driving precision.

The mean lateral unevenness for the entire field can be determined using

$$CV_t^2 = \sum_{i=1}^{n} P_i CV_{ti}^2$$
(7)

where P_i is the probability of finding the unevenness CV_{ti} (as defined from the parameters of a normal distribution), I is the number of a middle pattern shift, that is equal the collecting tray width b, and n is the total number of these shifts. Lateral unevenness caused by movement of the central spread pattern was conducted for several spreaders at different application rates (Kaplan and Rumyantcev, 1986). When the magnitude of lateral movement of the middle swath was smaller than the overlap, the value of the lateral unevenness can be approximated by

$$CV_t = 0.85V_{t0} + X / \overline{B}_w$$
(8)

where CV_{t0} is the coefficient of variation with zero displacement of the middle swath.

Assuming normal distribution for the population of lateral displacements of the middle pattern, and using Eq. (8), the average coefficient of variation can be determined as follows:

$$\overline{V}_t^2 = 2\int_0^\infty (0.85V_{t0} + X / \overline{B}_w)^2 \frac{1}{\sigma_B\sqrt{2\pi}} \exp(-X^2 / 2\sigma_B^2)dx =$$

$$= 0.72V_{t0}^2 + 1.36V_{t0}\omega + \omega^2 \qquad (9)$$

where

$\omega = \sigma_B / \overline{B}_\backslash$, and σ_B is standard deviation of B_w.

If $0.2V_{t0} \le \omega \le 0.4$ Eq. (9)is approximated by

$$\overline{V}_t = 0.82V_{t0} + 0.95\omega \qquad (10)$$

Therefore, substituting Eq. (10) into Eq. (3), the coefficient of variation CV_w, can be determined by

$$V_w = \sqrt{(0.85V_{t0} + 0.95\omega)^2 + V_a^2} \qquad (11)$$

For $\omega \le 0.2V_{t0}$, $CV_t = CV_{t0}$, and

$$V_w = \sqrt{V_{t0}^2 + V_a^2} \qquad (12)$$

These equations can be used to determine the coefficient of variation of the application rate in an adjustment area while taking into account the variability of the swath width.

Let us estimate the effect of two other sources of variations, feed drift and systematic errors in adjustment. Since the coefficient of variation CV_w is independent from ΔX_d and ΔX_r, it can be seen from Eq. (2) that

$$\frac{D_x}{X_q^2} = \frac{V_w^2 X_q^2 + \sigma_d^2 + \sigma_r^2}{X_q^2}$$

or

$$V^2 = V_w^2 + V_d^2 + V_r^2 \qquad (13)$$

where

$$V_d = \frac{\sigma_d}{X_q},$$ is the coefficient of variation of application rate due to a feed drift,

$$V_r = \frac{\sigma_r}{X_q},$$ is the coefficient of variation of application rate due to systematic error,

X_q is a desired application rate.

Metering efficiency at various hopper levels determines the drift in application rate. If the field has K plots, each equal to the adjustment area, and the mean application rate in the each plot is $\overline{X}_i$, then the standard deviation of application rate due to a feed drift, σ_d, can be defined

$$\sigma_d^2 = \frac{1}{K-1} \sum_{i=1}^{K} \Delta X_i^{'} \tag{14}$$

where $\Delta X_{id} = \overline{X} - \overline{X}_i$ is the deviation from the average application rate for i-th plot.

The systematic error is derived from the deviation of the mean swath width from the set width and errors in the application rate (feeder adjustment). The effective swath width is known as B_w. After spreading, the mean swath width is $\overline{B}_w = B_e + \Delta B$, where ΔB is a average error of B. The actual application rate is computed as:

$$\overline{X} = X_q \frac{B_e}{B_e + \Delta B}$$

and, the coefficient of variation due to the deviation of the swath width, CV_{rB},

$$C V_{rB}^2 = \left(\frac{\Delta X}{X_q} \right)^2 = \left(\frac{\Delta B}{B_e + \Delta B} \right)^2 = \frac{\Delta B^2}{\overline{B}_w^2} \tag{15}$$

The application rate is adjusted for the effective swath width. The feeder adjustment error in the application rate $\Delta X_r = \overline{X} - X$, does not depend on width variations caused by irregular driving. Therefore, the fraction of CV_r^2 due the systematic error, CV_{rf}^2, can be expressed as

$$C V_{rf}^2 = \frac{\Delta X_{rf}^2}{X_r^2} \tag{16}$$

and

$$CV_r^2 = CV_{rB}^2 + CV_{rf}^2 \tag{17}$$

Equations (3) - (16) can be used to determine the coefficient of variation for the field while taking into account all mentioned factors. This estimation can help to find the means and methods to improve the design of spreaders.

INTEGRAL ESTIMATION

Often an overview is necessary where all sources of variation are lumped together. An integral estimation provides such an overview. Let us consider the following procedure of integral estimation of the fertilizer distribution quality .

1.   After setting the spreader application rate, a test should be conducted under field conditions for an entire spreading cycle. During the cycle, the spreader pattern test must be conducted several times as the machine empties its hopper.

2.   After the spreading is completed, swath spacing should be measured and the parameters of swath spacing ($\overline{B}_w, \sigma_b$) should be determined.

3.   The fertilizer distribution is represented by several spread pattern replicates, which are arranged on an axis in the order in which they were obtained. The distance between adjacent patterns is equal to the mean swath spacing. Each pattern should be plotted according to the method of spreading, i. e., either progressive or one directional.

4.   The driving precision is taken into account using the displacement of the even patterns on the equivalent shift. Let us find the order of the pattern displacement.

Following any swath after spreading, we can observe four situations: 1)the left adjacent swath comes closer to us and a right one moves away, 2) the right adjacent swath comes closer to us and the left one moves away, 3) both adjacent swaths come closer, and 4) both adjacent swaths move away. All of these situations can be taken into account by the following order of the pattern displacement: right, left, left, right.

Consequently, for this situation at least eight patterns are needed. For a double or triple repetitions 16 or 24 spread patterns are needed. The displacement order of patterns is shown on Fig. 1. The magnitude of an equivalent pattern shift, x_e, can be determined using normal distribution and function $CV_t = f(X)$. For a first approximation, the value of equivalent shift x_e can be defined from Eq.(8) and Eq.(10)

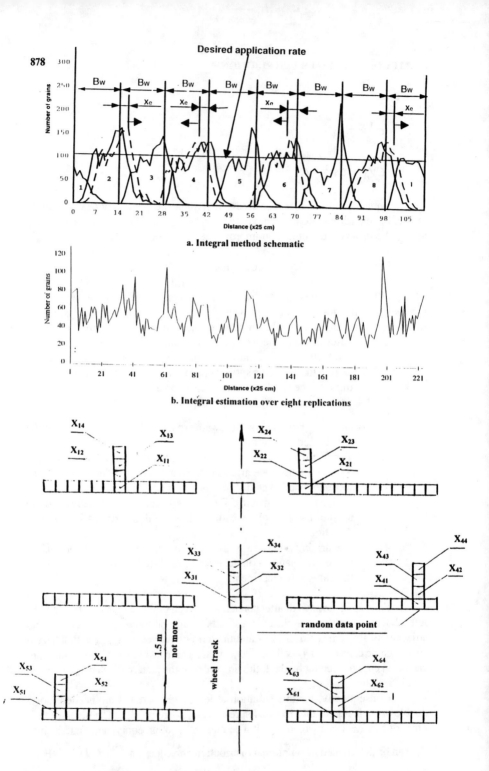

a. Integral method schematic

b. Integral estimation over eight replications

Figure 1
NEW SPREAD PATTERN TEST SCHEMATIC

$$V_t = 0.85V_{t0} + \frac{x_e}{\bar{B}_w} \approx 0.82V_{t0} + 0.95\omega$$

or

$$x_e \approx 0.95\omega\bar{B}_w - 0.03V_{t0}\bar{B}_w \tag{17a}$$

5.   The coefficient of variation of the fertilizer distribution in the field is estimated by the use of deviations of the cumulative distribution of the entire set from the set application rate along the line of patterns placement.

THE COLLECTING TRAY SIZE AND THE VALUE OF THE C.V.

The calculated value of the coefficient of variation depends on the collecting tray size. If the size of the collecting tray is increased, the high frequencies of the fertilizer distribution are filtered, reducing the variance and, therefore, the coefficient of variation. Thus, for white noise

$$V_1 / V_2 = \sqrt{b_2/b_1}, \tag{18}$$

where V_i and b_i are C.V. and the width of the i-th collecting tray respectively.

Sometimes, equation (18) is used to show the difference between values of the CV and to determine an appropriate value of the coefficient of variation knowing V_1, b_1 and b_2 White noise, however, does not represent observed distributions.

Another factor, which influences the coefficient of variation is a border effect around sample plots. Here changes in appreciate rate are "smoothed" across the border. Ostanin et. al. (1973) showed that if the size of the sample plot is reduced, the depressing influence of the fertilizer distribution unevenness on the yield is also reduced. Therefore, the estimation of the coefficient of variation should be conducted with a variance weight function. That part of the variance resulting from higher frequencies of fertilizer distribution should be given a smaller weighting.

On investigation of the data presented by Ostanin et. al. (1973) a linear weighting function $\mu(b_t)$ was shown to be valid.

$$\mu \cong kb_t \quad \text{at} \quad 0 \leq b_t \leq b_{tm}$$

$$\mu \cong 1 \quad \text{at} \quad b_t \geq b_{tm} \tag{19}$$

$k = 1/b_{tm}$, $b_{tm} = 1.2 - 1.5$ m.
where b_t is the length of the side of a square collecting tray.

Then, the effective variance value which can be used for correction of the application rate and determination of the yield and fertilizer loss is

$$D_e = (D_1 - D_2)kb_t(\frac{1+2}{2}) + (D_2 + D_3)kb_t(\frac{2+3}{2})+...$$

$$...+(D_{n-1}-D_n)kb_t(\frac{2n-1}{2})+D_n = kb_t[\sum_{i=1}^{n} D_i + 0.5(D_1-D_n)]$$

(20)

where D_i is the variance , which was obtained using trays of side ib_t.

From (20) the effective coefficient of variation

$$V_e = \sqrt{kb_t[\sum_{i=1}^{n} V_i^2 + 0.5(V_1^2 - V_n^2)]}$$

(21)

In order to obtain variances of fertilizer weight from all collection areas a modified spread pattern test is conducted. Here the test uses an array of trays (rather than a line). The array is placed across the path of the spreader. The total depth of this band must be b_{tm}. After spreading the weights of fertilizer in adjacent cells are added. We also could use an alternative method. Here two or three longitudinal rows of collecting trays of length b_{tm} are placed at random along the row. The values of selective variances are determined by gradually adding weights from the adjacent cells over all repetitions.

$$D_1^L = \frac{1}{n}\sum(\bar{x}_1 - x_{i1})^2; \quad D_{12}^L = \frac{1}{n}\sum(\frac{\bar{x}_1 + \bar{x}_2}{2} - \frac{x_{i1} + x_{i2}}{2})^2;$$

$$D_{123}^L = \frac{1}{n}\sum(\frac{\bar{x}_1 + \bar{x}_2 + \bar{x}_3}{3} - \frac{x_{i1} + x_{i2} + x_{i3}}{3})^2 \ etc.$$

where $\bar{x}_j$ is the mean application rate in the j-th collecting tray of longitudinal row,

x_{ij} the application rate in the j-th collecting tray of the i-th repetition.

Values of D^L can be determined by proportion. As an example

$$\frac{D_{123}^L}{D_1^L} = \frac{D_{e3}}{D_3}$$

(22)

where D_3 is the variance of the fertilizer weight in three adjacent cells, and D_{e3} , can be found using equations (20) and (21).

TEST PROCEDURE

Experimental validation of these methods was conducting using laboratory and field tests. Experimental data from the field test (J. Chaplin et al., 1994) was used to verify integral and differential estimations of the fertilizer distribution quality.

Laboratory tests were conducted to validate the selective method presented in Equation (22). A manually-operated rotary spreader was used for this test. The effective swath width with this spreader ranged from 2.0 to 3.5 m. Collecting trays were made from thin (1 mm) dense cardboard . The size of the collecting trays were 100x100x50 mm. They were placed across the swath of the spreader in an array 3.0m x 0.5m. The direction of the spreader was controlled using a track. Material from the track was also collected and weighted. Twenty four repetitions were conducted.

RESULTS AND DISCUSSION OF FIELD TEST

The effective swath width was determined by overlapping individual distribution patterns and by using an overall average distribution. The range of variation of the effective swath with over all replications was from 6.27 m to 7.30 m. This range of effective swath width is a result of using a single distribution pattern as required by the ASAE Standard.   If the effective swath width is determined using the overall average distribution, this value is 6.77 $\pm$ 0.09 m.

The definition of equivalent pattern shift was conducted using a normal distribution of the swath width and the average distribution pattern. The values of x_{Bi}/σ_{ei} was close to 0.9 for all values σ_b . This allow the equivalent pattern shift to be approximated as 0.9 times the standard deviation of swath width. This is close to the value calculated using Eq.(17a).

The variability determined by either the differential or the integral method presented in this paper are approximately 50% greater than the variability determined using the ASAE Standard. This is due to the added variability explained by the new methods. The test shows that a set of eight independent observations of the spread pattern is sufficient to determine field or sub-plot variability.

The coefficient of variation obtained by the integral method was 0.34. The field estimation of the coefficient of variation resulted in 0.33. No significant difference (α =0.05) in variability was found between this and the field test.

This work showed that field variations could be quantified equally well by either of the proposed methods. The error attributable to driving did not significantly influence the lateral distribution in this test. The integral method is not as computationally intensive as the differential method.

THE COLLECTING TRAY SIZE
AND QUALITY PARAMETERS

In a laboratory verification of the integral estimation using a hand powered spreader a series of distribution pattern was collected in 16 replicates. For the method verification, the equivalent swath width was 2 m, the equivalent pattern shift was 0.2 m, and the lateral unevenness was 31.1%. Material distribution was determined for five sample plots ranging in size from 0.1x0.1 m to 0.5x0.5 m in 0.1 increments. The following parameters were used to describe the distribution: average M,number of grains / 0.1 m^2 , St. Dev, CV, %. Two points in each row were selected at random, and longitudinal data was defined from this column of collecting trays. This is shown in Figure 2. The coefficient of variation for each size of sample plot was determined using Eq (22). The results are in Table 1.

Table 1. Parameters of the application rate distribution in the different area of the field.

The size of the sample plot, cm	Parameters of the application rate distribution			Approx. coefficient of variation (Eq. 22)		
	M,	St. Dev	CV,%	1 rep.	2 rep.	3 rep
10x10	13.4	8.27	63.5	63.5	63.5	63.5
20x20	12.84	6.31	49.2	52.6	52.7	51.4
30x30	13.15	5.86	44.6	44.1	45.5	44.9
40x40	13.43	5.51	41.1	39.4	41.0	41.6
50x50	13.75	5.23	37.9	36.9	36.9	36.7

The average difference between the CV, as determined using the full data set, and those values found using Eq. (22) was less than 2.6%. This demonstrated that the new spread pattern test is valid.

CONCLUSION

Two methods of analysis (integral and differential) , were developed to determine the field variability of field spreaders. The results showed that both methods can be used to estimate the quality of material distribution. They also form the basis for a modified spread pattern test that takes into account lateral and longitudinal variability, driving precision systematic errors and feed rate drift.

REFERENCES

ASAE Standards 42th Ed.,1995. S341.2. Dec 1992. Procedure for measuring distribution uniformity and calibrating granular broadcast spreaders. St. Joseph, MI. Pp. 177-179.

Chaplin, J., E. Roytburg, and J. Kaplan. 1994. Measuring the spatial performance of chemical applicators. *In* Proceedings of the 2nd International Conference on Site Specific Farm Management. Bloomington MN, March 27-30.

Coates W. 1992. Performance evaluation of a pendulum spreader. Applied Engineering in Agriculture. ASAE 8(3):285-288.

Dorr G.J., and D.J. Pannel. 1992. Economics of improved spatial distribution of herbicide for weed control in crops. Crop protection (11): 385 - 391.

Hemingway, R.G. Soil Sampling errors and Advisory analyses. Agriculture Science. June 1955, Vol. 46, Part 1, pp. 1-7.

International Standard ISO 5690/1-1982 (E). 1983. Equipment for distributing fertilizer. Test methods. Agricultural machinery. ISO Standard Handbook 13, 1983. Pp. 373-386.

Jensen, D., and J. Pesek. 1962. Inefficiency of fertilizer use resulting form non-uniform distribution. Soil Sci., 26(2)

Kaplan, I.G., and I.V. Rumyantcev. 1986. Distribution quality estimation for fertilizer spreader. Mekh. I Electriph. S. H. (7): 6-8.

Marchenko, N.M., and B.P. Chernikov. 1977. Moscow. The ways for spreader development. Mekh. I Elektriph. S. H. No 7. p 3-6.

Ostanin, A.U. and F.V. Yanishevsky. 1973. The influence of an uneven fertilizer distribution on cereals yuield. Agrochimia. No. 1, pp. 11-19.

Parish, R.L. 1986. Comparison of spreader pattern evaluation methods. Applied Engineering in Agriculture. ASAE 2(2):89-93.

Parish, R.L., and P.P. Chaney 1986. Speed effect on drop-type spreader application rate. Applied Engineering in Agriculture.ASAE 2(2) :94-96.

Parish, R.L. 1987. The effect of speed on performance of a rotary spreader for turf. Applied Engineering in Agriculture. ASAE 3(1): 17-19.

Parish, R.L. 1991. Effect of rough operating surface on rotary spreader distribution pattern. Applied Engineering in Agriculture.ASAE 7(1):61-63.

Roth, L.O., R.W. Whitney, and D.K. Kuhlman. 1985. Application uniformity and some non-symmetrical distribution patterns of agricultural chemicals. Transactions of ASAE 28(1): 47-50.

Building a Yield Map from Geo-referenced Harvest Measurements

S.C. Nolan
G.W. Haverland
T.W. Goddard
M. Green
D.C. Penney

Alberta Agriculture, Food and Rural Development
Edmonton, Alberta, Canada

J.A. Henriksen
G. Lachapelle

Dept. of Geomatics Engineering
University of Calgary
Calgary, Alberta, Canada

ABSTRACT

Yield mapping at field scales is an essential component of our evaluation of the potential for varying fertilizer rates within farm fields in Alberta, Canada. We monitored yield using standard harvesters outfitted with a high-precision Differential Global Positioning System (DGPS). However, the raw yield-position point data were unacceptable for the creation of a yield map. We used a series of corrections and data eliminations (36%) to "clean" the data. These included: eliminating DGPS errors and overtravel areas, offsetting a grain flow transportation delay between crop cutting and the actual yield monitor reading (throughput lag), and error reduction in yield. This reduced the coefficient of variation by 32% and increased the yield average by 10%. Inaccurate lag correction combined with adjacent harvest tracks having opposite directions of travel created streaking in the yield map. Attempts to eliminate the remaining streaks required smoothing and de-densifying the data to the point where the map became unrepresentative.

INTRODUCTION

An important step toward managing fertility at the sub-field scale is producing field scale yield maps. Three years of yield data ('93-'95) were monitored and located using a high precision Differential Global Positioning System (DGPS) to create a field yield map in southern Alberta, Canada. Initial yield maps produced from the "raw" yield-position data (e.g. Fig. 1) were judged to be inaccurate. For instance: low yields are apparent at left and right field edges, a yield spike occurs in the upper right portion of the field, and the yield map has a streaky or striped appearance, features we know to be wrong. This paper

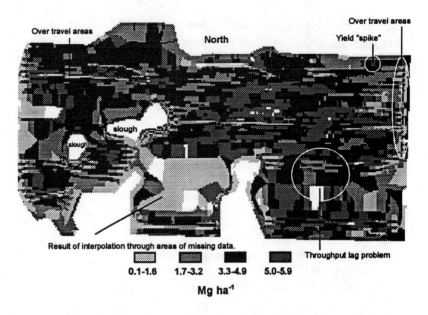

Fig. 1. Initial yield map interpolated from the raw yield-position data using an inverse distance weighting function with two nearest neighbors. Examples of irregularities and problems areas are noted.

documents our progress in rectifying irregular measurements of yield- position point data to produce an accurate yield map.

BACKGROUND

Our 28 ha site is near Stettler, Alberta, Canada (51° N, 115 W) on hummocky terrain. In 1995, a barley crop was swathed and then harvested with a Case IH 1680[1] harvester outfitted with an Ag-Leader Technology Yield Monitor 2000™. Three-dimensional Differential GPS (DGPS) positioning was achieved with two 12-channel C/A-code narrow-correlator-spacing NovAtel RT-20™ sensors and Model 501 antennas (Fenton et al. 1991, Van Dierendonck et al. 1992). Real-time position precision was calculated within 5 cm horizontally (x and y coordinates) and within 7 cm vertically (z coordinate) on level terrain (Lachapelle et al., 1994) using a receiver on the harvester interfaced with the yield monitor. Another receiver acted as a base station with known coordinates located on a survey pin. Yield and position were simultaneously measured at a one second interval. At each interval, the yield monitor measured yield (bu ac[-1]) passing a pressure transducer in the clean grain elevator.

---

[1]Reference to brand names does not imply endorsement.

RESULTS AND DISCUSSION

DGPS Errors

DGPS errors occur when an irregular and large magnitude variation occurs in the DGPS data due to factors such as weak satellite geometry or a discontinuous radio-modem link. Resulting yield data can be mis-positioned on the order of 25 m. Errors can be identified from outliers of elevation, slope, speed, or turn rates.

Corrections

We eliminated data where slopes were greater than 50% between adjacent points, since the maximum field slope was 22% (assuming that if z coordinates were wrong, then the x and y coordinates must be also). We could have also identified incorrect x and y coordinates using turn rates outside the range of normally occurring harvester rates. However, we used these criteria to identify turns at field edges (see below). Figure 2 shows that some anomalous turn rates occurred in the center of the field and not just at the edges and could have been included as DGPS errors. We eliminated elevations that were less than 812 m or greater than 831 m (outliers on histogram) and speeds that were greater than 4 m sec $^{-1}$. Since the harvester was stationary for an unusually long time due to equipment problems, we first eliminated that portion of the total data where speeds were zero (13.7%). Of the remaining data 1.4% were then eliminated for DGPS errors.

Yield Errors

a) Grain flow transportation delays (throughput lags)

The delay between crop cutting and monitor sensing (throughput lag) resulted in misrepresentation of the time when yields are recorded compared with the time (or position) of yield in the field. This lag varies with make and model of harvest equipment, speed, incline attitude, load and whether the harvester is moving into or out of a crop. We measured the lag when moving into and out of the unharvested field edge; defining "lag" as the time when the harvester begins harvest until a relatively stable yield was recorded. We measured the lag as long as 35 seconds (Fig.3), which resulted in yield displacements of up to 49 m at an average travel speed of 1.4 m sec^{-1}. Since adjacent harvest tracks have opposing directions of travel, yield areas were "stretched" up to 98 m. Figure 1 illustrates the effect of the yield displacement due to variations in throughput lag .

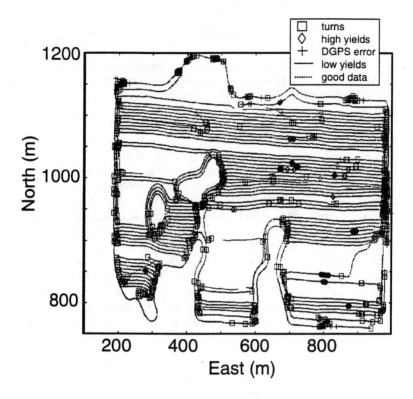

Fig. 2. Yield data located by DGPS with errors represented by symbols. Turns represent rates greater than 24 deg sec^{-1}.

b) Monitor errors

The yield monitor average was 22% higher than the field average. Anomalously high yields resulted from harvester plugging, then unplugging; caused by speed changes, or by windrows of swathed grain concentrated by corners.

c) Over-travel areas

Large areas of falsely low yields were recorded when the harvester passed over previously harvested areas: e.g. harvester idling, headland turns, repeated trajectories around wetlands or other obstacles and swath overlaps. This error is distinct from correctly low yield measurements around weedy or lodged crop patches, for example.

d) Noise

Inaccuracies in the yield monitor were caused by such factors as vibration and uneven grain flow.

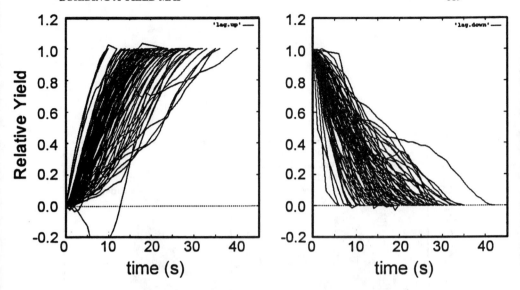

Fig. 3. The variation in time due to grain flow transportation delay (throughput
lag) between initial yield monitor recording and relative stability of
measurement (left graph) and between stability and time to measure a
zero yield (right graph). Graphed data represent 20% of total raw yield
data.

Corrections

Yield positions were offset by the average throughput lag of 15 seconds (21
m at average speed of 1.4 m sec[-1]). Yield values were scaled by 1.32 to account for
monitor over-prediction, assuming a linear relationship between monitor and weigh
wagon yield. We also assumed an average 5% overlap in cutting width to calculate
yields. A cut off threshold for unusually high yields was determined using a
histogram (Fig. 4a), and eliminated 0.8% of the remaining data. Over-travel areas
were identified and eliminated using: low speeds of stationary harvester (13.7% of
data), zero yields from a falsely reading header down sensor (10.8% of remaining
data), threshold turning rates of + or - 24 deg sec[-1] (2.0% of remaining data), and
low yields from histogram (7.1% of remaining data, Fig.4a). We chose to maintain
sharp changes in yield due to weeds or soil changes and reduced noise only
minimally using a two nearest neighbor interpolation.

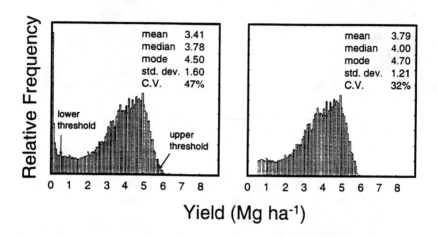

Fig. 4.   A histogram of raw yield data (left graph) illustrating the wide
             standard deviation, skewed distribution toward low yields, and
             excessively high yield values, and a histogram of "cleaned" yield
             data (right graph).

INTERPOLATION

Initial interpolation of the "cleaned" data was unsatisfactory because of
streaking. To correct this we tried two strategies. First, we de-densified the data
by ignoring points within six m of each other (distance between harvester tracks)
and then smoothed the result using the thin plate spline technique of GRASS (U.S.
Army Corps of Engineers, 1993). This reduced the streaking but did not eliminate
it (Fig. 5). As well, this technique arbitrarily ignored 80% of the cleaned data.
Further smoothing increased the differences between our interpolated and actual
extreme yield values from ±0.02 Mg ha^{-1} to ±0.11 Mg ha^{-1}. To also illustrate the
difficulties arising from missing data (compare Fig. 2 with Fig. 5), we purposely
interpolated across areas of missing data. We then unsuccessfully attempted to re-
grid the data to six m from a surface constructed using a two nearest neighbor
interpolation. This used 50% measured and 50% interpolated data, and ignored
only 40% of the cleaned data. The result was a visually acceptable map with
smooth contours and few irregularities (not shown), however it did not resemble
the raw yield map.

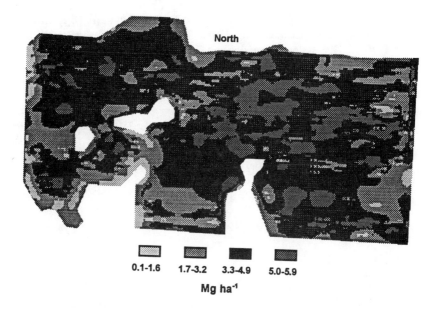

Fig. 5.   A yield map interpolated from corrected yield-position point data
                after ignoring data closer than six m apart and smoothing using a
                thin plate spline function.

CONCLUSION

Our raw yield data irregularities were unacceptable for the creation of a
yield map. A series of corrections and data eliminations (35.8%) were used to
"clean" the data. Figure 4 illustrates the raw and corrected yield histograms,
showing the elimination of low and high yields. The coefficient of variation was
thus reduced by 32%, and the yield average increased by 10%. We believe that this
is a more accurate distribution of yield.

Interpolation of the cleaned yield-position data to create a map was not as
straightforward. To eliminate streaking, we smoothed and de-densified the data to
the point where the map became unrepresentative. We believe that the biggest
problem in our ability to produce a yield map was the variable throughput lag which
resulted in the translation of yields along the harvest tracks. Correction of the lag
using the average lag time was insufficient due to the wide range of the variability
and the fact that adjacent tracks have opposing directions of travel, resulting in
distortion of the yield pattern. An accurate correction of the lag variability will ease
many of the difficulties that we faced.

REFERENCES

Fenton, P., Falkenberg, W., Ford, T., Ng. K., and Van Dierendonck, A.J. 1991. NovAtel's GPS sensor, the high performance OEM sensor of the future. Proceedings of GPS-91. Sept. 1991. The Institute of Navigation, pp 49-58.

Lachapelle, G., Cannon, M.E., Gehue, H., Goddard, T.W., and Penney, D.C. 1994. GPS system integration and field approaches in precision farming. J. of the Institute of Navigation. 41:323-335.

U.S. Army Corps of Engineers. 1993. Geographic Resource Analysis Support Software (GRASS 4.1) User's Manual. Construction Engineering Research Laboratories, Champaign, Illinois.

Van Dierendonck, A.J., Fenton, P., and Ford, T. 1992. Theory and performance of narrow correlator spacing in a GPS receiver. J. of the Institute of Navigation. 39:265-283.

Yield And Residue Monitoring System

A.V. Skotnikov

Department of Soil, Water, and Climate
University of Minnesota
Saint Paul, Minnesota

D.E. McGrath

Tyler Industries Inc.
Benson, Minnesota

ABSTRACT

The proposed yield and residue monitoring system includes a local automated system for stabilizing the combine's cutting height, harvesting width, and the harvester load. This permits stabilizing a transport delay and makes the yield and residue monitor more precise due to counting of grain losses and straw.

INTRODUCTION

Yield monitors are designed to obtain data for developing yield maps of different agriculture crops. These maps are necessary to calculate fertilizer and chemical applications, seeding rates; make decisions about soil sampling, and irrigation.

Today practically every combine manufacturer provides their own yield monitor (YieldMonitor, 1993). There are several commercial yield monitors - Grain-Track, Ag-Leader, (MF, 1993), etc.

These devices measure and indicate yield, moisture content, combine speed, grain flow, harvested area per hour, and also calculate, and record average yield, average moisture content, moist and dry grain volume, harvesting date, time of each unloading, and also performances for a whole field.

A common disadvantage some of these devices have is that they do not count grain losses during harvesting. These losses may vary from 5 to 16%, depending on harvesting conditions and the field relief. Due to these variations, accuracy of information on actual yield is lowered. Besides, the transport delay (time from crop cutting to getting grain into the bin) of threshed grain of the above mentioned devices is considered to be constant. This ultimately leads to a considerably distortedyield map.

The most comprehensive is the yield monitoring system (Reitz & Kutzbach, 1994). It comprises a combine speed sensor, grain flow sensor, grain moisture sensor, grain loss sensor, a sensor of a turn angle of rear wheels, a GPS aerial and receiver. Outputs of these sensors are inputs to the I/0 board installed in an on-board computer. Information from these sensors is supplied to the on-board computer through the I/0 board and is displayed on a monitor screen and is inputted in memory of the on-board computer as well. Coordinates of yield determination points are gathered through the GPS aerial and receiver and are recorded as well. Then using the gathered yield information with its respective coordinates, entered into the computer along with harvesting width, and transport delay, a yield map is compiled.

But as noted transport delay of threshed grain is considered to be constant. Transport delay depends on moisture, relief, ratio of straw and grain, and harvester load (kg/sec). The straw-grain ratio is dependent on the cutting height. Moreover, the combine does not operate at a full reaper width due to driving inaccuracy. This varying harvesting width adds an additional error to yield determination. Therefore the mass feeding into the combine varies as the transport delay and losses, respectively. This system assumes a constant transport delay due to lack of tools to evaluate performance parameters such as harvesting width, cutting height, torque on the threshing cylinder or harvester load. An error in estimation of yield for particular spot may be as much as 35 %.

A more precise yield map may be developed if the transport delay is stabilized. Other important information for improving site-specific crop management is residue measurement of lost grain, and threshed straw. Presence of this information permits us to calculate residue decomposition, snow accumulation on the field, sowing, fertilizer and chemical application rates.

The proposed system (Skotnikov & McGrath, 1996) is intended to obtain a more complete picture of yield and residues by counting grain losses, distributed straw and remaining stubble with respective transport delay and to provide stabilization of transport delay, during harvesting.

DISCUSSION

The defined problem is solved by installation on the combine (Fig. 1, 2) a portable on-board computer (1) with a keyboard (2) and a monitor (3). In the computer (1) an I/0 board (4) is installed and it (computer) is connected to a relay interface (5). Additionally, the combine is equipped with the following sensors: odometer (radar) (6); grain flow (7); grain moisture (8); grain losses (9); cutting height (10); tension of conveyor chain (11); turn angle of rear wheels (12); harvesting width (13); torque on the shaft of the combine's threshing cylinder (14); a GPS aerial (15) and receiver (16). Outputs of these sensors are inputs to the I/0 board. In turn the relay interface (5) is connected to corresponding solenoids (17-22) already installed on the combine by manufacturer. Solenoids (17-22) are moving hydraulic valves (24,26,28) controlling hydraulic cylinders (25,27,29) (Fig. 1).

All these sensors and others mentioned above are organized into three local automated systems: stabilizing of cutting height, stabilizing of harvesting width, and stabilizing of torque on the shaft of the combine's threshing cylinder (harvester load).

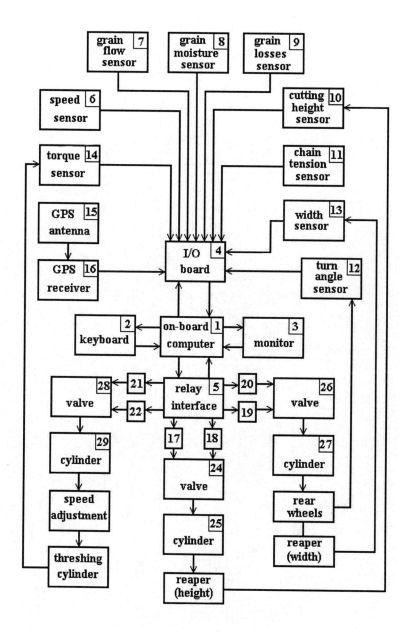

Fig. 1. Block-diagram of a yield and residue monitoring system.

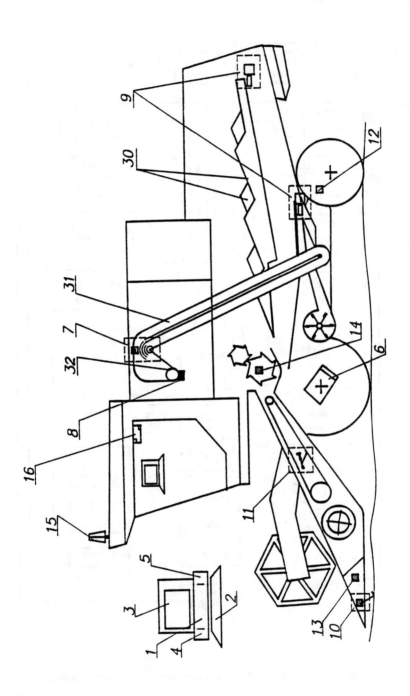

Fig. 2   Automated harvester.

The common work of all automated systems permits simplifying of the operator's work, stabilizes transport delay and monitors more precise yield and residue.

Consider the work of each local automated system and the system of yield and residue monitoring in total:

The automated system of stabilizing of cutting height (Fig. 1) comprises the sensor of cutting height (10), the speed sensor (6), the outputs of which are inputs to the I/0 board (4), computer (1), relay interface (5), solenoids (17) and (18) of the hydraulic valve (24), which control the hydraulic cylinder (25) moving up and down the combine header.

This system works as follows. Signal from the cutting height sensor (10) is supplied to the I/0 board (4), transformed to a digital signal, and compared to the data entered using keyboard (2) by the computer (1). The discrepancy signal through the relay interface (5) is supplied to the solenoid (17) or (18) of the hydraulic valve (24) controlling the hydraulic cylinder (25) which moves the cutting head of the combine along the vertical direction. The actual height of cutting depends on combine speed and cutting bar design. Therefore a signal of speed sensor (6) serves to correct the cutting height. The cutting height sensor (10) can be implemented as an ultrasonic sensor or an electro-mechanical device (a spring-loaded lever, upper end of which is connected to a potentiometer).

Additionally, the signal obtained from the sensor (10) is used to record condition of the residues (the stubble height and distribution of thrown straw), and to determine the grain/straw ratio in order to calculate the transport delay. This system also helps to stabilize the transport delay and simplify the operator's work. Another feature of this system is that a farmer in order to prevent wind erosion or manage snow accumulation can create a desired residue map and enter it in computer (1). In this case the system will determine a combine location by means of DGPS and change a stubble cutting height in accordance with entered program.

The automated system of stabilizing the harvesting width or autoguidance (Fig. 1) contains the sensor (12) of turn angle of rear wheels, and the reaper's harvesting width sensor (13). Outputs of these sensors are inputs to the I/0 board installed in computer (1). Outputs of the relay interface (5) are inputs for solenoids (19) and (20) of the hydraulic valve (26) controlling the hydraulic cylinder (27) turning through trapeze for the rear wheels.

Sensor (12) is the electromechanical device moving proportionally to the rear wheel's turn angle.

The reaper's harvesting width sensor (13) may be implemented mechanically, optically (recognition of video image), piezo-electrically or by acoustic signal (bar with piezoelements or microphones) or an ultrasonic sensor. In the suggested system the signal of this sensor is used not only for calculation of harvested area, but as input signal for an autoguidance system.

This system works as follows: The signal from the sensor (13) is supplied to the I/0 board (4), transformed to a digital signal, and compared to the data entered by operator using keyboard (2) by the computer (1). The discrepancy signal through the relay interface (5) is supplied to solenoid (19) or (20) of the hydraulic valve (26) controlling the hydraulic cylinder (27), which through trapeze, turns the rear wheels. Sensor (12) in this system serves to return wheels to neutral position after disappearance of discrepancy signal from sensor (13).

More stable harvesting width provides a more constant mass flow and stable transport delay. That eases operation of combine, allows the operator to control the process of threshing and make the necessary adjustments "on the go", e.g., adjust a fan rotation, vary gap in the jalousie.

The automated system of combine load stabilization (Fig. 1) contains a sensor of chain tension of the feeding conveyor (11), and a torque sensor (14) on the threshing cylinder shaft, outputs of which are inputs to the I/0 board and computer (1). Outputs of the relay interface (5) are inputs to the solenoids (21) and (22) of the hydraulic valve (28), controlling hydraulic cylinder (29) and inclining the plate of the hydraulic transmission pump and changing the combine speed.

The chain tension sensor (11) can be implemented as a two-arm spring-loaded lever. One arm of the lever is moved along the lower branch of the feeding conveyor, and another end of the arm is loaded by spring from one side and is connected to the shaft of the potentiometer from another side.

The torque sensor (14) is installed on the threshing cylinder shaft (Fig. 2) between the driving pulley and the threshing drum.

The automated system of stabilizing harvester load works as follows: Before harvesting the operator enters in the computer a desired harvester load, 90 %, for instance. Percentage of load will depend on final losses of grain. If losses are high it suggests that cleaning facilities are overloaded and it may be necessary to reduce percentage of harvester load. Then during harvesting a signal from sensors (11) and (14) through I/0 board (4), is transformed to a digital signal, and compared to the data entered using keyboard (2) of the computer (1). The discrepancy signal through the relay interface (5) is supplied to solenoids (21) or (22) of the hydraulic valve (28) controlling the hydraulic cylinder (29) inclining the plate of hydraulic transmission pump and changing the combine speed. A signal from sensor (11) is mostly used to calculate the harvester load, amount of spread straw after deduction of grain from this volume, and, consequently, transport delay. Plus, a signal from sensor (11) is also used to register empty spots in the field. But, at the same time, signals from sensors (11) and (14) are compared in the computer to make a decision about adjusting harvester's speed. For example, a signal of sensor (14) is reached at its critical high value. Decision to continue work in this regime or reduce speed is taken in accordance with a signal of sensor (11). If it is below the critical value the harvester continues to work in previous regime; if the signal of sensor (11) is a high critical value, the system will reduce the harvester's speed.

This system stabilizes the harvesting mass flowing to threshing and cleaning and, consequently, prevents plugging up of the threshing cylinder and feeding conveyor.

A radar installed on most harvesters may be used as the odometer and speedometer sensor (6). The grain feed sensor (7) can be installed on the elevator (31) (Fig. 2).

The grain moisture sensor (8) is installed in the grain flow in an auger (32) in the combine's bin (Fig. 2). The sensor determines the grain moisture proportionally to capacity of the capacitor. More precise data can be received by means of a microwave device.

The grain loss sensor (9) is a set of piezoelectric plates connected to a conversion unit and installed along the straw walker (30) (Fig. 2) along every key and along the clearing width at the end of the extension on screens of the slipping board (depending on the combine design). The sensor functions as follows: When grain strikes the piezoelectric plates, voltage spikes are formed therein which are transformed by unit into the level respective number of strikes. This level is entered into the I/0 board (4), computer (1), and is displayed in the monitor (3) as well.

The yield and residue monitoring system (Fig. 2) functions as follows: Parameters of the harvested crops, the combine parameters, dependence of the transport delay on the grain/straw ratio, on moisture of the harvested crop and parameters of harvesting (technological adjustments for particular crop) are entered into the computer (1) (Fig. 1). Before starting, the combine operator enters the required level of the combine loading (torque on the cylinder's shaft), the cutting height, the coverage width, the crop name, the distance for obtaining and entering data on yield and residue. Then he switches ON the DGPS aerial (15) and receiver (16), threshing units of the combine, drives of all necessary operating mechanisms and begins harvesting.

The harvested area data are obtained by multiplying with a respective time sequence the data obtained from the width sensor (13) and odometer (6). The yield harvested from the area per time unit is determined using grain flow sensor (7), grain moisture sensor (8) and loss sensor (9) along with the respective transport delay. The transport delay is computed in accordance with entered data on type of crop, straw/grain ratio, moisture, technological adjustments. Thus one obtains the harvested (physical and dry) and grown (plus losses) yield from the area unit. Signals of cutting height sensor (10) and width sensor (13) with respective transport delay are used to determine stubble residue. In case of spreading straw on a field one can use data from sensor (11). All data, by DGPS, is connected to geographical coordinates.

The necessary adjustments can be obtained empirically. We give one possible method. First of all, it is necessary to spray by metal- containing paint a crop strip (reaper width × 2m) in contrasting areas of the field. Then it is necessary to install a metal detector on the exit of an auger in the bin and connect it to I/0 board. We assume that all described above sensors will be installed as well. During harvesting the operator may manually mark in computer the starting point of the painted strip harvesting. The metal detector will register a transport delay for different harvesting conditions. After accumulating sufficient data it will possible to develop the required adjustments.

The whole system obtains data on the grain harvested from the determined area, on the averaged and current moisture, on the overall and current grain loss, on the actual yield (+ loss), and on the residue. This obtains a more stabilized transport delay and also stable parameters affecting thereon, and also to make the combine operator's work easier due to automated stabilization of the cut height, auto driving and controlling the combine motion speed as well as to prevent the threshing machine from being plugged up.

The above mentioned advantages of the described system allow the combine operator to pay more attention to quality of the harvesting process and to tune the combine for minimum losses in the course of harvesting and to maintain maximum productivity. In the future, the proposed system, after developing of appropriate software will permit automation of the tuning mechanism while minimizing the grain loss.

CONCLUSIONS

1.   The suggested system permits a practically completely automated harvester, and will result in more precise yield and residue maps.
2.   The use of a portable computer with internal I/0 board and relay interface instead of existing processors will result in more flexible and easily updatable automated and monitoring systems; also to make the systems compatible with each other and with other farmer's equipment.
3.   The proposed method of obtaining transport delay adjustments is more precise (in comparison to recommendations in yield monitor manuals) and can be used by manufacturers of yield monitors.

REFERENCES

MF Yield Mapping System. Massey Ferguson Group Limited, 1993.
Reitz, P., and H.D. Kutzbach. 1994. Data acquisition for yield mapping with combine harvesters. pp. 42 - 47. Proceedings 5th International Conference on Computers in Agriculture, Orlando, Florida.
Skotnikov, A.V., and D.E. McGrath. 1996. On-board device for yield and residue monitoring. Patent of Belarus. (2553-01, 22.11.94)
Yield Monitor, Ag-Leader Technology, 1201 Airport Road, Ames, IA 50010, 1993.

Comparison of AgLeader Yield monitor and Plot Combine yields

K. Brix-Davis
J.A. Schumacher
D. E. Clay

South Dakota State University
Brookings, SD

ABSTRACT

As yield monitors become widely accepted and used, more farmers will implement on-farm research. Yield monitor data accuracy and interpretation will be essential in understanding the field information. The objective of this study was to evaluate the accuracy of the yield monitor data recorded on the-go to yield data collected from a particular area. The field was harvested with a 2-tow plot combine (Massey Ferguson) and an 8-row 1660 International axial flow combine that used an Agleader yield monitor. The plot combine harvested 200 areas that had dimensions of 25 ft by 44 inches. In adjacent areas yields were measured with the AgLeader yield monitor. The paper will discuss the error associated with yields measured by the AgLeader.

Weed Detection in Cereal Fields Using Image Processing Techniques

J.V. Benlloch
A. Sanchez
M. Agusti
P. Albertos

D.I.S.C.A.
Valencia, Spain

ABSTRACT

Recent studies have demonstrated a patchy distribution of weed species in cereal crops, the degree of patchiness varying between species and fields. Agrochemical usage can be reduced by identifying the location and features of weed patches in the field followed by a selective herbicide application.

Bearing in mind that real-time patch spraying is still a difficult process, the objective of this study were to develop a procedure which used digital image processing techniques for detecting broadleaf weeds in cereal crops, under actual field conditions This research is part of the E.E.C. Project "Patchwork", whose main goal is to develop systems and techniques for carrying out spatially variable treatments.

The proposed method included the following steps: first, a normalized difference index image, using green and red information, was defined in order to improve the contrast between vegetation and soil. Then, a global thresholding followed by a labeling process was applied to discriminate plants from the soil background. Finally, plants located between the crop rows were considered as weeds provided that their shape features were distinct from those of cereal leaves. For accomplishing this last step, a shape analysis has been carried out for each object.

Two test sets of data were used to evaluate the procedure. The first was composed by 35 mm. slides, whereas the second was formed by color images acquired with an industrial video camera. In both cases, images represented areas of about 0.25m2 and were taken in a Danish field of spring-barley, under natural lighting conditions.

Initial experiments showed this method to be satisfactory and that information of weed infestation in each sample (number of seedlings, area, coverage...) may be useful to define a weed map. Thus, taking into account other sources of information such as visual mapping or knowledge bases including historical data, a control map may be designed.

ADAR Digital Aerial Photography Applications In Precision Farming

B. Burger

Positive Systems, Inc.
Whitefish, MT

ABSTRACT

Geographic Information System (GIS) technology has contributed significantly to managing the complex spatial data generated through new precision farming techniques. As efficiencies in computer processing continue to provide faster, less expensive easier to use hardware and software alternatives. demand has increased for digitally captured aerial photographs that can be acquired quickly, repeatedly, and at a reasonable cost. Combining these attributes with capabilities for streamlined integration with leading image processing software makes digital aerial imagery a natural component of numerous precision farming solutions.

Today's Digital Aerial Photography technology combines the detail offered through low altitude flying with color, color infrared. and four-band multispectral image capture at resolutions similar to film-based aerial photography. With no film or film processing required, digital serial photography can provide extraordinarily rapid data turnaround and allow for immediate data integration into GIS or image processing systems through industry compatible file formats. Image data can be repetitively collected on a weekly or monthly basis as required, to meet critical needs for both short term decision making and long term planning.

Positive Systems, Inc. (Whitefish, Montana) has developed the ADAR System family of Digital Aerial Photography Systems to address the needs of applications demanding fast, efficient cost effective acquisition of aerial imagery captured in fully digital formats. Since 1991, ADAR Systems have been utilized within a variety of agricultural applications that continue to expand directly into the realm of precision farming.

California-based Datron Transco Inc. currently utilizes the ADAR System 5500 in weekly flights over high value cash crops to identify problem areas associated with vegetative stress due to disease malfunctioning irrigation equipment, and misapplication of pesticides and herbicides. Providing growers of these crops with high resolution digital images on a weekly basis was previously impossible utilizing film-based equipment due to the inherent delays associated with film extraction, film processing and digitizing of hardcopy photographs. Detecting these anomalies from the air at the earliest possible stage help meet the overall goal of rapid, effective response in the field.

New agribusiness applications benefiting from digital aerial photography continue to be identified. These include the enhancement of GPS-based yield monitoring and soil sampling data, as well as verification of crop damage from hail, flood or other natural disasters for crop Insurance purposes. Digital aerial photography has proven cost effective not only in meeting today's evolving requirements in site specific agriculture applications, but also in helping to define the applications of tomorrow as well.

Estimating Plant N Status from Leaf and Canopy Reflectance Data

C. S. T. Daughtry
C. L. Walthall
J. E. McMurtrey III

USDA-Agricultural Research Service
Beltsville, MD

ABSTRACT

Numerous reports have shown that total biomass and grain yields of corn and wheat are correlated to concentrations of leaf N and leaf chlorophyll. Because leaf reflectance is related to chlorophyll concentration, it follows that leaf reflectance is also correlated to biomass and grain yield. Chlorophyll meters use the relationship between leaf reflectance and leaf chlorophyll to assess relative leaf chlorophyll concentration and infer plant N status. However, there is a need for rapid, non-contact methods of assessing crop N status across fields.

The reflectance of plant canopies is more complex than the reflectance of a collection of leaves. It is affected by 1) the reflectance and transmittance of the scene elements (leaves, stems, soil), 2) the amount and distribution of the scene elements, 3) the view and illuminations angles, and 4) the atmospheric path. We used canopy reflectance model (SAIL) to simulate canopy reflectance for a wide range of leaf chlorophyll concentrations and view and illuminations conditions. Model results were compared to measured data. We will discuss the feasibility of remotely monitoring N status of crop canopies.

The Center for Precision Farming, School of Agriculture Food and Environment

S. Blackmore

Silsoe College
Bedfvord, United Kingdom

ABSTRACT

The Center for Precision Farming has been set up within the School of Agriculture Food and Environment to coordinate the Precision Farming activities. The School is a faculty of Cranfield University and encompasses Silsoe College, Shuttleworth College and The Soil Survey and Land Research Center.

Cranfield University is a leading international center for the generation and application of knowledge in engineering, science, manufacturing and management, for government agencies and blue-chip industry world-wide. Cranfield earns more contract research income from UK industry than any other British university: half as much again as Imperial College, and as much as Cambridge and Oxford Oxford.

Controlling Variable Rate Applications on Self-propelled Irrigation Systems

G.W. Buchleiter
H.R. Duke
D.F. Heermann

USDA-Agricultural Research Service
Ft. Collins, CO

ABSTRACT

Self-propelled irrigation systems are a versatile platform for data collection, analysis and implementation of precision farming under irrigated conditions. The development of both hardware and software for variably applying water and chemicals such as fertilizer or herbicides, with self-propelled sprinklers is demonstrated. A linear move machine was modified to apply water and chemicals independently and in differing amounts along the mainline by dividing each span into 2 separate manifolds that supply water or chemicals to individual sprinkler heads. Different application amounts along the mainline were achieved with a programmable auxiliary controller that operated electric valves at the manifold inlets for different amounts of time in a 60 second cycle The appropriate auxiliary controller program was selected by operating relays in the main sprinkler control panel that are connected to the input ports of the auxiliary controller. The main sprinkler control panel was linked via radio with a central microcomputer which allowed the user to graphically create, store, and run programs that change speed and pulsing patterns to variably water and chemicals throughout a field.

Linear Move Irrigation System Position as Determined with Nondifferential Economic GPS Unit

D.F. Heermann
G.W. Buchleiter

USDA-Agricultural Research Service
Colorado State University
Ft. Collins, CO

ABSTRACT

A four span linear irrigation system is being equipped with eight segments that can be controlled individually for variable application of water. The control is accomplished by pulsing the individual segments to obtain the desired treatment. Equally important to the control along the lateral is the control in the direction of travel. A position sensor on the system is required for controlling the application appropriately in the direction of travel. The use of economical GPS units is investigated. The use of typical GPS differential technology would provide acceptable accuracy but would be quite expensive for commercial application of the variable rate sprinkler system. Experimental data collected with the economical GPS unit has shown that normal operation would result with a range in error of 30 meters. This is unacceptable for controlling variable rate application. The data was analyzed by calculating 30 minute running averages and it was found that the running average had a range of error equal to 3 meters. The running time of each tower can be obtained easily from the linear move system. This running time when multiplied by a speed of the tower provides an estimate of the speed. The major problem with this estimate is the change in speed due to a differential slip and thus a change in speed. An algorithm has been determined that uses the 30 minute average of the GPS data to adjust the speed during operation. With this algorithm the expected error is reduced to 1 meter which is assumed to be adequate for controlling the variable rate application in the direction of travel. The changes from one rate to another are not instantaneous since the overlap and spray distance of individual sprinklers are about 3 meters.

GPS for Precision Farming: A Dense Network of Differential Reference Stations

J.S. Speir

Oklahoma State University
Stillwater, OK

ABSTRACT

This research developed a low-cost, Wide Area Differential GPS (WADGPS) capability utilizing a dense network of multiple reference receivers (RR). The research focused on determination of possible resolution with the dense network of reference receivers as applied to precision farming applications. "Resolution" is defined as repeatability vs. accuracy at a location once a Differential GPS (DGPS) spot measurement has been made with the dense network of RR. The improved resolution offers potential innovative solutions to farmers faced with the need of increasing accuracy as a way of reducing labor, chemical and fertilizer costs and at the same time, providing documentation for new regulatory requirements.

GPS location measurement to 30 meters is possible in native mode and two meter accuracy is achievable over limited distance with DGPS, using a single RR. GPS signals are affected by atmospheric signal propagation effects, satellite orbital errors, receiver noise, clock synchronization, etc. This research determined resolution improvement with nullification of these errors, using correction data provided by a "dense network of multiple RR." The attainable precision will be mapped against agriculture requirements for continuous yield sensors, remote sensing, variable rate treatment VRT and GIS.

Use of High-Resolution Global Positioning Systems in a Site-Specific Crop Management Project in Ontario

I.P. O'Halloran
R.G. Kachanoski
D. Aspinall
P. Von Bertoldi

University of Guelph
Guelph, Ontario, Canada

Development of site-specific cropping systems requires a knowledge of the spatial variation in both soil properties (such as soil fertility) and plant responses to those management factors which can be variably applied. Considerable attention has been placed on the most appropriate sampling schemes to determine the spatial pattern of a given soil property, and in particular the increases or decreases in information obtained with sampling grids of different dimensions. Such studies are limited in their usefulness in that the appropriate sampling scheme will vary with site, and may not be the same for different soil properties. In other words, sampling schemes may be both site and parameter specific. In Ontario there is considerable evidence that spatial variations in soil properties (e.g., soil fertility, pH, organic carbon, etc.), crop growth and crop response to fertilizer inputs are correlated to spatial variations in elevation. The objective of this project is to determine the usefulness of a high-resolution global positioning system (HR-GPS) for obtaining topographical information of a field for site-specific management of N fertilizer application. Twenty-four farms sites from across Ontario were selected, and fields sampled in 30 x 30 m grids for soil test P and K, pH, organic C and texture. Nitrate-N soil test samples were taken along a transect (10 m sampling interval) in each of three 0 kg/ha N fertilizer check strips. Check strips were used to determine crop response to fertilizer N across the field. Each sampling point is geo-referenced to allow the study of the relationships between topographic position, soil properties and crop yields. Yield maps of corn at each site were developed using on-the-go yield monitors and GPS units. The HR-GPS is a fast, very precise (within centimeters) method of obtaining elevation data. Results have confirmed that topography influences crop yield and crop response to N fertilizers, even in areas with relatively small (less than 1 m) changes in elevation.

Impact of Dry Fertilizer Multiple Product Blending on Variable Rate Application Precision

W.H. Thompson
D. McGrath

EDAPHOS Ltd./Tyler Industries
St. Paul, MN

ABSTRACT

Most variable rate granular fertilizer application systems in use provide single bin fertilizer application capability. That is, only one fertilizer product or multiple product blend is applied at variable rates following a single predetermined variable rate map.

Results are presented for application precision achieved with single bin systems that incorporate two- and three-product blends (N-P205-K20) relative to the original predetermined variable rate nutrient requirement layers from 42 fields (three nutrient maps per field). Alternative management strategies intended to maximize agronomic precision and economy of the single bin system including a 2-bin double blend approach and different methods of creating variable rate maps are presented.

The single blend approach induces high levels of imprecision relative to the original variable rate nutrient requirement layers and requires multiple pass applications to achieve high precision. The two bin approach (two complementary three-product blends) provides high precision for two nutrients and greatly improves application precision for the third nutrient of a three nutrient recommendation. Blend application precision (relative to predetermined nutrient requirement maps) is directly related to spatial correlation across paired variable rate nutrient requirement layers, where each layer represents natural and management induced variability across a farm landscape.

REVOLUTION, EVOLUTION OR DEAD-END: ECONOMIC PERSPECTIVES ON PRECISION AGRICULTURE

J. Lowenberg-DeBoer
M. Boehlje

Department of Agricultural Economics
Purdue University
W. Lafayette, IN

Precision farming has been variously labeled a revolution, an evolutionary step and a technological dead-end. Some economists have seen profit potential, others have focused on the possibility of major changes in farm structure while a third group views precision farming as a small part of an overall trend toward larger farms and more vertical integration in agriculture. The objective of this paper is to outline economic perspectives on three themes:

1) the search for profitable farm level precision technology, 2) the economics of information about precision farming, and 3) the potential role of precision technology in the industrialization of North American crop production. This paper draws on the growing body of theory and data on the economics of precision farming.

Precision farming is information technology applied to agriculture. Similar concepts are discussed under various labels, including variable rate technology (VRT), site specific management (SSM), prescription farming and farming-by-the-foot. Both VRT and SSM refer primarily to use of precision technology for spatial management. For this paper, precision farming is broadly defined as electronic monitoring and control applied to agriculture, including site specific application of inputs, timing of operations and monitoring of crops and employees. This paper focuses on agronomic crops, but many of the same arguments could be applied to horticultural crops or livestock.

The discussion will introduce some important concepts of information that will be used in later sections, and then move to a review of farm level studies of profitability of precision farming. It will then turn to a brief discussion of the economics of information on the techniques and procedures of precision farming. The focus will then turn to the structural implications of precision farming, and finally to some of the broader impacts or economic benefits of this technology to suggest hypotheses for further study.

Concepts of Information and Knowledge

The essence of precision agriculture is obtaining more data on production processes and converting that data into information that can be used to manage and control those processes. Concepts of information are useful in our understanding of the potential impact and opportunity of precision farming in two dimensions: 1)

What is the information content, costs and benefits of various tools and techniques of precision farming? 2) Where and how do farmers obtain information about the tools and technologies of precision farming?

The concepts of information and knowledge mean different things to different people. One common model of the role of information is summarized in Fig. 1. The premise of this model is that collecting, sorting and analyzing data creates information, and accumulation of information eventually results in knowledge and possibly even wisdom. In this model knowledge and wisdom are used to guide decisionmaking. This model has been labeled "empirical".

Most of the scientific advances of the Twentieth Century have come from a different view of how knowledge is created. The classic "logical positivist" model gave theory a primordial role. Theory was used to specify hypotheses and to guide the collection of data to be used to test those hypotheses. A theory which withstood empirical testing became part of knowledge. In this view, data collection without a theoretical framework is at best unproductive and at worst the source of erroneous conclusions. It is unproductive because the analysis tends to get lost in the mountains of data generated by unfocused collection efforts. It may be dangerous if relationships in the data are misinterpreted because no theory is used to guide the analysis.

Alternative model - An alternative model of information and knowledge which may be more consistent with the scientific method of using concepts and theory to confront data is summarized in Fig. 2. In this model knowledge is the broad based concepts, theories, principles and models that are necessary to understand a particular phenomena. Knowledge can be applied broadly across many sets of facts and circumstances or contexts. It is not data specific or unique, but helps one sort through the vast quantities of data available to determine what is relevant.

In this alternative approach data are more specific than knowledge; they are individual numbers or observations. Data can be quantitative or qualitative in nature. At the extremes data is distinguishable from knowledge in that data is specific while knowledge is general. Clearly this clean distinction becomes fuzzy at times.

Information is different from data or knowledge in two important dimensions: first it is context specific, and second it is decision focused. In essence, if knowledge and data are combined and applied to a specific context (for example, a specific crop and parcel of land) and a specific decision (the proper level of fertilizer to apply to obtain a particular yield of a particular crop), they are transformed into information. Information becomes more valuable as it results in improved decision making and better physical and financial performance.

Learning is the dynamic process by which through observation, experience or inductive and/or deductive reasoning the data and knowledge set at a particular point in time are augmented. This learning process then has the potential to generate new or more valuable information.

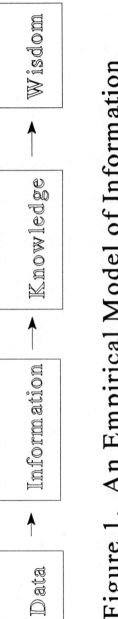

Figure 1. An Empirical Model of Information

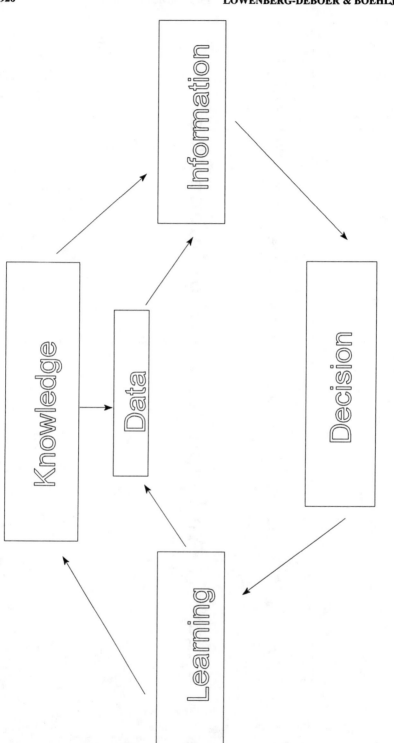

Figure 2. An Alternative Model of Information

Information Attributes - Information has many attributes that determine its value. It must be <u>timely</u> -- appropriate to the decision context and not out-of-date. It must be technically <u>accurate</u> and scientifically sound. It must be <u>objective</u> and unbiased, and/or value judgements must be explicitly identified. It must be <u>complete</u> (as opposed to partial) so as to be useful in a decision, or its partial or incomplete nature must be clearly specified. It must be <u>understandable</u> -- communicated in such a way that the user can comprehend it. And finally it must be <u>convenient</u> -- available when and where and at what time the user needs or wants it. These attributes will determine the value of information.

With the increased context specificity and decision focused nature of information in recent years, it has become more valuable. And as information becomes more valuable, the incentive for the private sector to provide that information and capture some of that value increases. Consequently, growth in the private sector data gathering and information service firms is not surprising given the growing value of information.

Information and Management - Information has always been important to agriculture, but its role has not always been as obvious. Today, information can be a significant source of strategic competitive advantage. Those firms that can obtain superior information can act on that information and improve their performance compared to those firms with inadequate access to the latest and best information and technology. Thus superior (better in terms of context specificity and decision focused) information is a source of competitive advantage to the supplier of that information -- allowing them to extract value or income from the user of that information by charging fees or maintaining or improving related product sales. And it also provides a competitive advantage for the user of that information -- in this case the producer -- in terms of better performance and higher profits compared to other producers.

Not only is the relative role and importance of information and knowledge changing, the sources of that information for farmers is also going through a transformation process. Farmers have access to more information from the private sector (or from internal sources on the part of large scale or integrated producers) and less from the public sector. In many cases, providers of key farm inputs such as pharmaceuticals and chemicals have become critical suppliers of information along with those inputs, leaving the traditional Extension Service and Land Grant/USDA complex at a significant disadvantage in terms of providing the latest technology and information. And in some geographic regions larger and more educated producers, who are becoming a larger proportion of U.S. producers, rate traditional public sector information sources, such as county extension agents and even University specialists, significantly lower than many other sources of information for production, marketing or financial decisions (Ortmann, et. al., 1993). These dramatic changes -- both in the importance of information and the preferred provider of that information raise a number of questions about the changing role of the public sector in the knowledge, data and information industries.

Studies of Farm Level Profitability

The framework for economic decision under conditions of spatial variability is relatively well developed, but until recently practical applications were limited because the cost of collecting spatial data and variable rate application. In 1962 Jensen and Pesek examined the yield and profits lost from non-uniform distribution of fertilizer on a uniform soil. Feinerman et al. (1983, 1989) and Letey et al. (1984) developed an approach for optimizing irrigation of soils that showed spatial variability in water infiltration. They assumed a uniform application of water over the whole field. Extending their model to allow for variable rate application the objective function might be expressed for the one variable case with discrete management areas as:

$$1) \qquad Profit = \sum_{i}^{n} \sum_{j}^{m} [\, p \cdot f_{ij}(x_{ij}) - r \cdot x_{ij} - g - v\,] - F$$

Where:
p = output price
f_{ij} = continuous, differentiable, crop response function that depends on the characteristics of the management area i,j
xij = input on management area i,j
r = input cost
i,j = location coordinates of the management area
n = number of rows in the management area grid
m = number of columns in the management area grid
g = quasi-fixed costs for intensive data collection and analysis in each management area
v = quasi-fixed costs for variable rate input application
F = fixed cost for the field

Quasi-fixed costs are constant if a given technique or input is used, but zero if that technique or input is not used. A variant of the model could treat the intensity of data collection as a decision variable, instead of as a quasi-fixed cost. The assumption here is that all management areas are of equal size.

This formulation is very flexible. Soil test levels and other site specific characteristics can be included as response function parameters. Multiple inputs and outputs can be incorporated. One of those outputs could be the environmental consequences of production. Using integrals the problem could be expressed in terms of continuous variability. Quality differences in the output could be modeled by making price a function of the site specific output characteristics. The size of the management area can be optimized by making N a decision variable. For ease of exposition, the discussion below uses the one input case with discrete management areas and single, homogeneous output.

Assuming concave crop response functions and variable rate input application, the interior solution would require that marginal value product equals input price in every management area:

$$2) \qquad p \cdot df_{ij}(x_{i,j})/dx_{ij} = r, \text{ for all j,i}$$

This applies the usual profit maximizing rule of equating marginal value product to input price to each individual management area. The usual whole field profit equation would be:

3) $N \cdot (p \cdot f(x) - rx) - F$

where: $f(x)$ = the average crop response function per unit of area and
       x = the uniform input rate used per unit of area
       N = total number of units of area in the field

The whole field profit maximizing condition would be:

4) $p \cdot df(x)/dx = r$

The whole field profit maximizing condition differs from the site specific condition mainly in the response function specification. The whole field approach relies on a single, "average" crop response function and assumes it represents every area of the field. The marginal product of this average response function is used in estimating the average marginal value product. The site specific function in profit equation (2) allows response to vary between sites and requires that the site specific marginal value product equals the input cost.

The input use found in solving the marginal value equation (2) will be profit maximizing if it provides a positive return, that is, if profit equation (1) is positive when evaluated at the levels indicated by solving equation (2) for the x_{ij}. The choice between whole field technology and site specific management can be made by comparing profit equations (1) and (3) evaluated at the input level determined by equations (2) and (4), respectively, and choosing the approach which provides the greater profit:

5) $N \cdot (p \cdot f(x^*) - rx^*) - F \overset{>}{\underset{<}{=}} \sum \sum [\, p \cdot f_{ij}(x_{ij}^*) - r \cdot x_{ij}^* - g - v\,] - F$

where: x^*, x_{ij}^* = profit maximizing input levels for the whole field and variable rate application approaches, respectively.

In the technology choice, the question boils down to whether the yield increase $(\sum \sum p \cdot f_{ij}(x_{ij}^*) - N \cdot p \cdot f(x^*))$, plus input savings $(N \cdot r \cdot x^* - \sum \sum r x_{ij}^*)$ is greater than the quasi-fixed costs of intensive data collection and analysis (g) and variable rate application (v).

Precision farming technology is not an all or nothing choice. It can be adopted piecemeal. One particularly interesting partial adoption case is the uniform application with site specific information explored by Feinerman and colleagues. In that case the profit equation is:

6)   Profit $= \sum_{i}^{n} \sum_{j}^{m} [\, p \cdot f_{ij}(x)] - r \cdot x - g\,] - F$

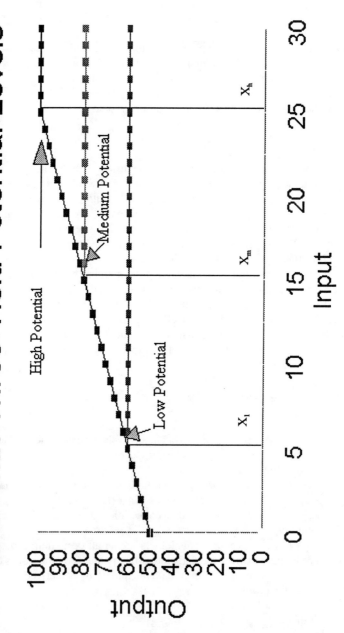

Figure 3

In this case the variable rate application fee (v) drops out of the profit equation, but the costs of gathering site specific information is still incurred. Assuming a rectangular field the profit maximization condition is:

7) $\quad [\sum \sum p \cdot df_{ij}(x)/dx_{ij}]/(m \cdot n) = r$

That is to say, the average marginal value product equals the input cost. It differs from whole field management (4) in that the site specific information approach averages over N individual marginal products, while the whole field profit maximization condition uses the marginal product of an average response function.

The choice between full adoption with both intensive data collection and variable rate application boils down to whether the value of the yield increase with full adoption ($\sum p \cdot fij(xij) - \sum \sum pfij(x)$), plus the input cost savings (m·n ·r·x- $\sum \sum r \cdot xij$) is greater than the cost of variable rate application (v). Intensive data collection and analysis costs (g) is incurred in both the full and partial adoption scenarios and thus is not a factor in the technology choice between these two options.

As in the case of Feinerman's irrigation water problem, the management choice depends crucially on the shape of the response function, for example, if the crop response is roughly approximated by a linear response and plateau (LRP) function (Fig. 3), where site specific differences are mainly in the yield plateau level. A linear response is used here for simplicity; a response curve with diminishing marginal returns would lead to almost the same conclusions if it culminated in a yield plateau. Jomini (1990) provides a review of LRP crop response function research. Babcock and Blackmer (1994) provide a more recent example of use of this type of functional form. Figure 3 illustrates that case in which three levels of yield potential are present. The response to the input is similar across all areas, but in some zones other constraints limit yields. For instance, in the low yield potential area production might limited by droughty soils. Unless the input cost is very high or output price very low, the profit maximizing solution with an LRP crop response is to set input use to correspond to the beginning of the yield plateau. In the precision farming case, input use would be set to X_l for the low yield potential areas, X_m for the medium yield potential areas, and X_u for the high potential areas.

In the LRP case, the full site specific yield increase could also be achieved by setting the uniform input level at X_m and the technology choice reduces to whether or not the additional input cost incurred by over applying the input on some low yield potential sites (m·n·r·x*- $\sum \sum$ r·x$_{ij}$*) is greater than the cost of variable rate input application (v). It is conceivable that if the number of low potential sites is small and the cost of variable rate application is substantial, that it would be more profitable to over apply in a few management area rather than pay variable rate application fees.

A key assumption in the LRP example is that over application of the input does not cause yield reductions. This may be a reasonable assumption for some crop inputs, such as phosphorous (P) and potash (K). Without explicit consideration of environmental effects in these cases, the profit maximization may lead to over application and greater environmental problems than a whole field approach.

Early Economic Evaluation - Because of data problems and other limitations, early economic studies of precision farming technology have not made full use of the site specific crop response model specified in equation (1). Instead early studies concentrated on partial budgeting studies comparing whole field and various SSM approaches using observed or simulated yields with various discrete input combinations.

$$
8) \qquad N \cdot (p \cdot Y_o - r \cdot x^*) - F \overset{>}{\underset{<}{=}} \sum \sum [\, p \cdot Y_{ijo} - r \cdot x_{ij} - g - v \,] - F
$$

where: Y_o, Y_{ijo} =   yields observed or simulated for whole field and site specific management, respectively

The optimization using site specific response information implied by profit maximization conditions (2) and (4) is only used in some more recent work (Schnitkey et al, 1996; Hertz, 1994). This section will review the partial budgeting studies and examine the implications of some studies which have attempted to broaden the information set to include site specific crop response.

The results from eleven recent field crop partial budgeting studies are summarized in Table 1. Overall, five studies found SSM not profitable, four had mixed or inconclusive results, and two showed potential profitability. Because results hinge so crucially on how the costs of sampling and variable rate application are treated, these are specified for each study in addition to the crop, inputs managed, duration, and grid cell area used. They are divided between real and simulated crop yields, but there were no clear differences in profitability results between these two approaches. The Reetz study (1994), which found SSM profitable on the example "Larson farm", was omitted because it merged two farms under different managers and attributed all profitability differences to SSM.

What factors accounted for profitability of SSM in the partial budgeting studies? The two studies showing potential profits from SSM (Fiez et al., 1994; Hayes et al., 1994) had two elements in common: both omitted costs of SSM and both focused on nitrogen management. In lieu of explicit SSM cost accounting, both studies observed that increased revenues could be documented and that these might cover costs. The Hayes et al. (1994) study assumed that target yields could be reached with adequate nitrogen, whereas the Fiez et al. (1994) article documented higher average yields where nitrogen was not limiting, but also noted great variability in nitrogen response.

The common thread among those studies that found SSM not to be profitable was that SSM costs outstripped its benefits. The information investment and management costs of SSM were distributed across no more than two nutrients in any of these studies. In some cases, high initial soil fertility levels meant that yield gains were negligible, so that the main benefits derived from saving on low-cost fertilizers (Lowenberg-DeBoer et al., 1994a; Mahaman, 1993). In the case of Wibawa et al. (1993), information quasi-fixed costs were overestimated both due to the very small grid cells and because all fixed costs were charged in one year.

Table 1. Profitability Conclusions from 11 Site Specific Management Partial Budgeting Studies

Study	Crop	Inputs Managed	Treatment of Sampling & VRT Cost ($)	SSM Profitable?
Empirical Yields				
Carr et al.	Wheat, barley	N,P,K	Not included	Mixed.
Fiez et al.	Wheat	N	Not included	Yes, potentially
Hammond	Potato	P,K	Variable & fixed	Inconclusive (costs only)
Lowenberg-DeBoer et al.	Corn	P,K	Variable & fixed custom rates	No, but might for low-soil test fields
Wibawa et al.	Wheat	N,P	Variable & fixed w/ 1 yr. amort.	No (but over-ests. annual fixed costs)
Wollenhaupt & Buchholz	Corn	P,K	Variable & fixed w/ 4-yr. amort.	Mixed; deps. on yield gain
Wollenhaupt & Wokowski	Corn	P,K	Variable & fixed w/ 4-yr. amort	Mixed; deps. on sampling density & amort. period
Simulated Yields				
Beuerlein & Schmidt	Corn, soy	P,K	Variable & sample; no equip.	No, but more efficient fertilizer use
Hayes et al.	Corn	N	Not included	Higher revenue has potential to cover costs
Hertz & Hibbard	Corn	P,K	Variable & fixed custom rates	No, but close to uniform in profitability
Mahaman	Corn	P,K	Variable & fixed custom rates	No if 1-yr sample amort.; yes if 4-year sample amort.

SOURCE: J. Lowenberg-DeBoer and S.M. Swinton, "Economics of Site Specific Management in Agronomic Crops," Staff Paper 95-14, Department of Agricultural Economics, Purdue University, West Lafayette, IN. 1995.

The studies with mixed profitable and unprofitable results are perhaps the most revealing. Profitability turns on the degree of yield gain attainable (Wollenhaupt & Buchholz, 1993), the sampling density (Hertz & Hibbard, 1993; Wollenhaupt & Wolkowski, 1994), and the amortization period over which the soil test results and maps are valid (Mahaman, 1993; Wollenhaupt & Wolkowski, 1994).

More Information - Two studies which attempt to make better use of site specific crop response information in the economic analysis of precision farming are Hertz (1994) and Schnitkey et al. (1996). Hertz shows that fields with more soil variability and a higher average level of soil fertility benefit most from precision fertilizer application. He also found that when whole field fertilizer recommendations were applied to specific sites less than half of the field studies showed a gross benefit greater than the assumed $7.41/ha ($3/a) cost of variable rate application. When site specific fertilizer recommendations were used based on a crop response function the number of fields showing returns over VRT costs doubled. It should be noted that several sources indicate that the typical cost soil sampling, mapping, developing recommendations and VRT application is substantially more than the $7.41/ha ($3/a) assumed by Hertz (e.g., Lowenberg-DeBoer & Swinton, 1995; Giacchetti, 1996), thus Hertz may overestimate the proportion of fields with profitable SSM for fertilizer.

The main contribution of the Schnitkey et al (1996) study is to divide the return from site specific management into two parts: benefits from gathering site specific information and the profits from variable rate input application. The information gathering benefits are calculated assuming that fields are grid soil sampled and that a uniform fertilizer application rate is chosen based on the grid data. For a sample of 18 corn-soybean rotation fields in Ohio their simulation results indicate that the benefits of information gathering exceed those from variable rate application.

Overall available studies show that farm level profitability of precision farming remains elusive for producers of bulk commodities. Profitability has been easier to achieve with higher value products (sugar beets, potatoes, seed crops). Partial budgeting studies show that stand alone precision farming technology for fertilizer may not cover additional costs of soil sampling, mapping and variable rate application for many farmers. The studies emphasize the importance of the precision farming system, in which data collection and analysis costs are spread over many inputs, not just fertilizer. The Hertz and Schnitkey et al studies show the economic importance of making better use of site specific crop response information in developing variable rate input recommendations. In terms of the information attributes outline above, the information used in the precision farming applications evaluated with partial budgeting was incomplete. It does not develop and make use of site specific crop response information.

Economics of Information about Precision Farming

Precision farming uses crop information as an input to fine tune production strategies, but like all new technologies potential users also need information about the technology to make adoption and adaptation decisions. This information can come from many sources -- conferences, field days, university specialists, consultants, etc. Each source has characteristics that determine its economic value. The economics of information on precision farming is important to those firms and entrepreneurs who see business opportunities and profit potential in providing this information, as well as those firms who are merchandising precision farming technology and want to develop effective marketing strategies to inform producers of their product characteristics and features and benefits.

A recent survey at Purdue University attempted to measure some of those characteristics and their economic value for a group of farmers and agribusiness people who participated in the Precision Decisions Conference held in Champaign, IL, in Nov., 1995. Some 394 surveys were mailed to farm and retail agribusiness participants in the conference. The questionnaire asked respondents about their sources of information about precision farming, the expenditures on that information and the evaluation of those sources on the characteristics of timeliness, accuracy, objectivity, completeness, clarity and convenience. The evaluation of characteristics was on a scale of 1 to 5 with 5 being the most highly rated. As of April 25, 1996, 140 questionnaires had been returned (36%). Most of the respondents are from Illinois and Indiana, with a sprinkling of responses from Ohio, Michigan, Kentucky, Missouri, Iowa and Wisconsin. Of those 140 respondents, 96 are farm operators. The average corn, soybean and wheat acreage is 723 ha (1787 acres). They expect this acreage to grow to an average of 1000 ha (2472) acres by the year 2000.

Descriptive statistics for this survey are presented below. The proposed framework for in-depth analysis is a modified hedonic pricing model. Hedonic pricing is a statistical technique used to decompose product prices into their component values. For example, hedonic pricing has been used with land prices to determine the value of soil fertility, location and other factors in the overall price (Palmquist & Danielson, 1989; Miranowski & Hammes, 1984; Gardner & Barrows, 1985, Herriges et al., 1992). Usually, the dependant variable is the price of the item. Because of the difficulty in defining the unit of information, the dependant variable in this case will be the share of precision farming information budget going to a given information source. The independent variables are the ratings given by respondents to the attributes of the various information sources.

Preliminary observations based on the descriptive statistics for farm operators (Table 2) suggest that many respondents had clear opinions on the attributes of precision farming information from different sources. A few respondents were only able to differentiate between information sources; for them all attributes of the same source had the same rating, but rating differed between sources. Farm magazines were used by 100% of the respondents for precision farming information and this accounted for 23% of the information budget. Field days and conferences were used by almost 94% of the respondents and they accounted for 27% of the budget. The other most commonly used information sources for precision farming were newsletters (82%) and other farmers (79%). Their own records, university specialists, sales personnel, computerized information services (e.g., DTN.) and consultants were used by about 50% of the farm operators responding.

Table 2. Summary of Results of Precision Farming Information Sources Survey as of 25 April, 1996.

	Percent Using Source	Percent of Budget	Average Attribute Score						
			Time	Accuracy	Objectivity	Completeness	Clarity	Convenience	Overall
Newsletters	82.29%	5.06%	3.75	3.50	3.47	3.08	3.38	3.90	3.47
Radio and TV	26.04%	0.66%	2.86	2.41	2.48	2.08	2.39	3.00	2.32
Own Records	59.38%	11.61%	2.15	2.53	2.87	2.46	2.47	2.22	2.36
County Ext.	22.92%	0.48%	3.03	3.49	3.44	3.29	3.15	2.82	3.11
University Specialists	56.25%	2.93%	3.91	3.89	3.57	3.75	3.71	3.16	3.70
Field days/Conf.	93.75%	26.66%	3.49	2.93	2.19	2.90	3.10	3.13	2.85
Salesmen	52.08%	1.96%	3.52	3.18	3.17	2.94	3.24	3.50	3.26
Other Farmers	79.17%	3.35%	3.52	3.18	3.17	2.94	3.24	3.50	3.26
Own Employees	21.88%	1.33%	2.87	2.69	2.92	2.65	2.75	3.27	2.81
DTN, etc.	50.00%	9.45%	3.70	3.35	3.13	3.05	3.17	3.75	3.29
INTERNET	26.04%	4.59%	3.55	3.07	2.93	2.95	2.89	3.30	3.07
Consultants	46.88%	7.06%	3.80	3.67	3.42	3.66	3.56	3.42	3.58
Lenders	9.38%	0.20%	2.09	1.98	2.26	2.13	2.17	2.15	2.00
Farm Magazines	100.00%	22.58%	3.64	3.82	3.55	3.45	3.73	4.00	3.91

SOURCE: Sources and Characteristics of Precision Farming Information, survey conducted by J. Lowenberg-DeBoer and M. Boehlje, Purdue University, 1996.

1)   Scored on a five point scale, 1=low, 5=high.
2)   Budget for information about precision farming. Does not add to 100% because of expenditures on miscellaneous other information sources amounting to less than 3% of budgets.

The highest overall information attribute score was for farm magazines. The second highest score was for university specialists. Other categories with high overall scores were newsletters and consultants. The respondents assessment by individual attribute can be summarized as follows:

Timeliness - University specialists received the highest overall score for timeliness of precision farming information. Other sources with high timeliness scores are farm magazines, newsletters, and consultants.

Accuracy - University specialists have the highest average attribute score for accuracy. Farm magazines have the second highest accuracy score. Other sources with high accuracy scores are newsletters, county extension and consultants.

Objectivity - University specialists have the highest average rating for objectivity. Other sources with high objectivity scores are farm magazines, newsletters, county extension and consultants.

Completeness - University specialists have the highest average score for completeness. Consultants have the second highest completeness score and farm magazines are rated third.

Clarity- Farm magazines have the highest average clarity rating, followed by university specialists and consultants.

Convenience - Farm magazines rate highest in the convenience category, followed by newsletters, computerized information services, other farmers and consultants.

Among the limitations of the survey are definition problems, the non-random nature of the sample and lack of questions on information dynamics. Judging from notes on the questionnaires some respondents may not have clearly distinguished between farm magazines and newsletters. Others mixed ag chemical dealers and consultants. Some information sources overlapped. For example, some respondents noted that they were in contact with university specialists only at conferences and field days.

A random sample of all farmers and agribusiness personnel would be a very inefficient approach to collecting data on sources of precision farming information. Only a small percentage of farmers and agribusiness people are actively seeking and using this information. Thus is was decided to focus on a group that was known to be seeking this information. Among the farmer respondents, they are probably operators of larger than average farms and have more education than the typical farmer. There may be some self selection in the responses because the questionnaire is clearly send out by university personnel, so those that respond are likely to think favorably of university efforts.

The relatively low ranking of conferences and fields is puzzling. Most respondents use this source of information and expend a significant portion of the information budget on it, but they do not rank it highly. One explanation is that conferences and field days are particularly valued for networking that may contribute to future information. The questionnaire did not cover the dynamic aspects of information, including the potential future contacts through networking.

Information Technology and Industrialization

Precision farming has the potential to facilitate and possibly even generate significant structural change in agriculture -- structural changes that are often described as the industrialization of agriculture. Changes are occurring in the structure of agriculture -- not only in size and ownership of farm firms but also in the linkages/coordination of farm production activities with input suppliers and product purchasers. More and more of these linkages are occurring through personal negotiated/contractual/ownership arrangements rather than impersonal open markets. Although numerous forces and drivers are contributing to these structural changes, information and knowledge play a significant role. As in other industries characterized by negotiated/personal linkages, those individuals with unique and accurate information and knowledge have increasing power and control in the food production system. And with power and control is the capacity to garner profits from and transfer risk to others with less power.

The increasing role that knowledge and information play in obtaining control, increasing profits and reducing risk is occurring for two fundamental reasons. First, the food business has become an increasingly sophisticated and complex business in contrast to producing commodities as in the past. This increased complexity means that those with more knowledge and information about the detailed processes as well as how to combine those processes in a total system (i.e. a food chain approach) will have a comparative advantage. The second development is the dramatic growth in knowledge of the chemical, biological and physical processes involved in agricultural production. This vast expansion in knowledge and understanding means that those who can sort through that knowledge and put it to work in a practical context have a further comparative advantage. Thus the role of knowledge and information in success in the agricultural industry is more important today than ever before.

The logical question for individuals in the food manufacturing chain is how to obtain access to this knowledge and information. Historically, particularly for the independent producers in the farm sector, this knowledge and information has been obtained from public sources as well as from external sources such as genetics companies, feed companies, building and equipment manufacturers, packers and processors, etc. In general, independent producers have obtained knowledge and information from external sources in much the same fashion as they have sourced physical and financial resources and inputs. In contrast, ownership/contract coordinated production/processing/distribution systems have sourced their knowledge and information from a combination of internal and external sources. Many of these firms or alliances of firms have internal research and development staffs to enhance their knowledge and information base. And the knowledge they obtain is obviously proprietary and not shared outside the firm or alliance; it is a source of strategic competitive advantage.

Furthermore, the research and development activities in coordinated systems are more focused on total system efficiency and effectiveness rather than on only individual components of that system; it is focused on integrating the nutrition, genetics, building and equipment design, health program, marketing strategy, etc. rather than on these areas or topics separately. And in addition to more effective research and development, such alliances or integrated firms have the capacity to implement technological break-throughs more rapidly over a larger volume of output to obtain a larger volume of innovator's profits. In the case of a defective new technology, ownership/contract coordinated systems generally have more monitoring and control procedures in place and can consequently detect deteriorating performance earlier and make adjustments more quickly compared to a system with impersonal market coordination.

As knowledge and information becomes a more important source of strategic competitive advantage, those who have access to it will be more successful than those that do not have access. Given the declining public sector funding for research and development and knowledge and information dissemination which has been the major source of information for independent producers, the expanded capacity of integrated systems to generate proprietary knowledge and technology and adapt it rapidly enables the participants in that system to more regularly capture and create innovator's profits while simultaneously increasing control and reducing risk. This provides a formidable advantage to the ownership/contract coordinated production system compared to the system of independent stages and decision making.

Potential Impacts of Precision Farming

The potential economic impacts of precision farming are profound with respect to changing structure and the industrialization of agriculture. But the firm level studies suggest limited cost savings of precision farming, implying a relatively slow adoption rate unless other benefits or impacts are identified. What might be the broader set of economic benefits or impacts of precision farming? Or stated different what are some of the economic impacts of precision farming that merit assessment if we are to more completely understand its potential and rate of adoption.

1)   **Cost Reduction/Efficiency Increases** - The improved measurement of soil characteristics and weather patterns that is part precision farming has the most direct and obvious payoff in terms of cost reductions and efficiency increases from more accurate use of inputs such as fertilizer, seed, chemicals and other inputs and the systematic measurement of the impacts of these inputs on yield and profitability. In essence, precision farming is one step closer to the manufacturing mentality of production agriculture. Precision farming combined with creative ways to schedule and sequence machinery use including 24 hr/day operations, moving equipment among sites and deployment based on weather patterns also has the potential to increase machinery utilization and lower per acre machinery and equipment costs as well.

2)   **Span of Control** - A key concern in crop operations is the perceived and, in many cases, real limit on size of operation because of the difficulty of monitoring progress and performance on large geographically dispersed acreages. The fundamental argument is that if plant growth processes can only be monitored by people with unique skills and those resources are costly or expensive to train, the monitoring process limits the span of control to what one individual (or at least a few) can oversee personally. If electronic monitoring systems can be developed that monitoring the processes of plant growth (whether it be machinery operations or the growth process of the crop or the level of infestation of insects or weeds), fewer human resources are needed for this task and generally larger scale is possible. An analogy is the transformation from the labor intensive corn processing or feed milling plants of the past to the electronically controlled and monitored plants and mills of today with computer based monitoring and control systems and few employees producing significantly more output. Crop production can and will move more and more in that direction with improved electronic monitoring and control systems which expands the span of control.

3)   **Differentiated Products** - As has been suggested earlier, a segment of agriculture will move from commodity to differentiated product production. One dimension of that differentiation may be the production process itself -- for example the use of chemicals during only certain stages of the plant growth process. And with more specificity required in the raw material to meet qualified supplier requirements, increased measurement and monitoring of both the growth process and the end production will be important for quality control and compliance. In fact precision farming in its broader context of measuring, monitoring and controlling the plant growth process is expected to have more payoff in differentiated production rather than commodity product production because it has the potential to not only lower cost but to simultaneously enhance revenue by producing a higher valued product.

4)   **Food Safety** - One of the most difficult risks for a food processor to manage is the potential of contamination in raw materials. And for a branded product food company, a food safety scare can be disastrous. The improved measurement and monitoring of the soil preparation, growth, harvesting, storage and handling and processing processes that have the potential to be part of precision farming in the future will enable trace-back from end-user through the production/distribution chain which is the only secure method of guaranteeing food safety. If food safety concerns continue to increase and consumers demand more documentation that food products are in fact safe, precision farming has the potential to become one of the most effective ways of providing that documentation and reducing the risk of food contamination.

5)   **Environmental Benefits** - Much has been asserted about the benefits of precision farming in terms of more accurate and precise application of chemicals and fertilizer to better match plant needs and thus reduce leaching and runoff into ground and surface water. Undoubtedly this potential exists, but caution should be exercised in this assertions. Without improved measurement and monitoring of chemical and fertilizer uptake by the plant and movement in the soil, we are not sure of the environmental impact. What if the precision farming recommendations are for the highest application of chemicals or fertilizer on the soils closest to a stream or with a shallow water table and heavy rains occur after application? No doubt site specific farming and precision agriculture have the potential to reduce environmental degradation, but we need to measure and monitor this phenomena to be sure we are obtaining the expected results.

CONCLUSIONS AND IMPLICATIONS

In this communication, information has been defined as a combination of data and knowledge brought to bear on a specific decision situation. Information is context specific and decision focused. The information model presented differs from the empirical model often seen in precision farming circles in the role it accords to the theory which guides data collection, hypothesis testing and interpretation. The empirical model holds that the process starts with data collection, data is sorted and analyzed to yield information, information is accumulated to become knowledge. The history of science suggests that the empirical model is an attractive, but unproductive paradigm. At best is the empirical model tends to get lost in the mountain of data generated by unfocused collection efforts. At worst, it results in misinterpretation because no theoretical framework is used to guide analysis.

Information can be characterized by various attributes, including: timeliness, accuracy, objectivity, completeness, clarity, and convenience. These attributes can be applied both to the crop information used by precision farming and information about precision farming techniques.

A review of partial budgeting studies of precision farming suggests that profitability remains elusive for producers of bulk commodities. Part of the problem seems to be related to the limited scope of the precision technologies evaluated. They were all variable rate fertilizer studies that spread the cost of intensive data collection over only a few relatively inexpensive inputs. Profitability will be improved in a precision farming system which spreads data collection and other costs over all inputs. The few available studies which go beyond partial budgeting suggest that making better use of site specific crop response information will be crucial. In terms of the information attributes, the crop response information used in the partial budgeting studies was incomplete and thus less valuable than it might be.

In the area of information about precision farming, a recent survey suggests that farmers are quite willing and able to evaluate the attributes of information sources. A preliminary assessment of their evaluations indicates that in addition to the static attributes of information outline here, it will be necessary to consider dynamic attributes related to developing networks for on-going information. This may be particularly important for private sector information providers, since the dynamic attributes seem to be particularly important for the conferences organized by these groups.

The debate about the role of information technology in the industrialization of agriculture does not limit the economic impact of precision farming technology to fine tuning crop inputs. It includes use of this technology to increase the span of control, by allowing managers to monitor larger acreages and more employees electronically, by permitting the development of differentiated products by more closely controlling crop growth, by enhancing food safety with improved monitoring, and through environmental benefits.

Agriculture has always relied on information. Farmers watched their crops. They sought out information on new practices and techniques. They always need price information. In this sense, precision farming is not a revolution. Precision farming is an evolution toward use of more information, because electronic monitoring has made it less expensive. The available economic analysis can not disspell the possibility that precision farming is a technological deadend. Precision farming profits are hard to measure. They suggest that precision farming profitability can be improved by using more of the available data in more complete systems.

REFERENCES

Babcock, B.A., and A.M. Blackmer. 1994. The ex post relationship between growing conditions and optimal fertilizer levels. Review of Agricultural Economics, 16 p. 353-362.

Beuerlein, J., and W. Schmidt. 1993. Grid soil sampling and fertilization. Ohio State University Extension, Agronomy Technical Report 9302.

Carr, P.M., G.R. Carlson, J.S. Jacobsen, G.A. Nielsen, and E.O. Skogley. Farming by soils, not fields: A strategy for increasing fertilizer profitability. J.Prod.Agric. (January-March 1991)4: 57-61.

Feinerman, E., E. Bresler, and G. Dagan. 1989. Optimization of inputs in a spatially variable natural resource: Unconditional vs. conditional analysis. J. Environ. Econ. and Manage. 17, p. 140-154.

Feinerman, E., J. Letey, and H.J. Vaux. 1983. The economics of irrigation with nonuniform infiltration. Water Resources Research, 19 , p. 1410-1414.

Fiez, T.E., B.C. Miller, and W.L. Pan. 1994. Assessment of spatially variable nitrogen fertilizer management in winter wheat. J. Prod. Agric. 7. 17-18, 86-93.

Gardener, K., and R. Barrows. 1985. The impact of soil conservation investments on land prices. Am. J. Agric. Econ. 67 , p. 943-947.

Giacchetti, N.. 1996. The pricing paradox. Farm Chemicals, Mid-March, p. 38.

Hammond, M.W. 1993. Cost analysis of variable fertility management of phosphorus and potassium for potato production in central Washington. P. 213-228. In P. C. Robert et al (Eds.) Proceedings of Soil Specific Crop Management: A Workshop on Research and Development Issues.

Workshop held April 14-16, 1992, Minneapolis, MN. Madison, WI. ASA, CSSA, and SSSA.

Hayes, J.C., A. Overton, and J.W. Price. 1994. Feasibility of site-specific nutrient and pesticide applications. P. 62-68. *In* K.L. Campbell, W.D. Graham, and A.B. Bottcher, eds., Environmentally Sound Agriculture. Proceedings of the Second Conference, April 20-22, 1994, Orlando, FL. St. Joseph, MI: American Society of Agricultural Engineers.

Herriges, J.A., N.E. Barickman, and J.F. Shogren. 1992. The implicit value of corn base acreage. Am. J. Agric. Econ. 74, p. 50-58.

Hertz, C. 1994. An economic evaluation of variable rate phosphorous and potassium fertilizer application in continuous corn. M.S. thesis, Univ. of Illinois, Champaign-Urbana, Dept. of Ag. Economics.

Hertz, C.A., and J.D. Hibbard. 1993. A preliminary assessment of the economics of variable rate technology for applying phosphorus and potasium in corn production. Farm Economics 93-14, Dep. of Agric. Econ., Univ.of Illinois, Champaign-Urbana.

Jensen, D., and J. Pesek. 1962. Inefficiency of fertilizer use resulting from nonuniform spatial distribution: I. Theory and II. Yield Loss Under Selected Distribution Patterns. SSSA Proceedings, Vol. 26, p. 170-178.

Jomini, P. 1990. The economic viability of phosphorus fertilization in Southwestern Niger: A dynamic approach incorporating agronomic principles. Ph.D. Dissertation, Dep. of Agric. Econ., Purdue Univ., West Lafayette, IN.

Letey, J., J. Vaux, and E. Feinerman. 1984. Optimum crop water application as affected by uniformity of water infiltration. Agron. J., 76 , p. 435-441.

Lowenberg-DeBoer, J., R. Nielsen and S. Hawkins. 1994a. Management of intra-field variability in large scale agriculture: A farming systems perspective. Proceedings of the International Symposium on Systems Research in Agriculture and Rural Development, Montpelier, France, Nov., p. 551-555.

Lowenberg-DeBoer, J., and S. Swinton. 1995. Economics of site-specific management of agronomic crops. 1995. Purdue Univ., Dep. of Agric. Econ., Staff Paper 95-14.

Mahaman, M.I. 1993. An evaluation of soil chemical properties variation in northern and southern Indiana. Ph.D. thesis, Dep. of Agron., Purdue Univ., W. Lafayette, IN.

Miranowski, J., and B.Hammes. 1984. Implicit prices of soil characteristics for farmland in Iowa. Am. J. of Agric. Econ. 66, p. 745-749.

Ortmann, G.F., G.F. Patrick, W.N. Musser, and D.H. Doster. 1993. Use of private consultants and other sources of information by large cornbelt farmers. Agribusiness (4) July , p. 391-402.

Palmquist, R.B., and L.E. Danielson. 1989. A Hedonic study of the effects of erosion control and drainage on farmland values. Am. J. of Agric. Econ. 71. p. 55-62.

Reetz, H. F., Jr., and P.E. Fixen. 1995. Economic analysis of site-specific nutrient management systems. *In* P.C. Robert et al. (Eds.) Site Specific Management for Agricultural Systems, Madison, WI, ASAA-CSSA-SSA,

p. 743-752.

Schnitkey, G., J.Hopkins, and L. Tweeten. 1996. An economic evaluation of precision fertilizer applications on corn-soybean fields. Selected Paper for the American Agricultural Economics Association meetings, San Antonio, TX.

Wibawa, W.D., D.L. Dludlu, L.J. Swenson, D.G. Hopkins, and W.C. Dahnke. 1993. Variable fertilizer application based on yield goal, soil fertility, and soil map unit. J. of Prod. Agric. 6: 255-261.

Wollenhaupt, N.C., and D.D. Buchholz. 1993. Profitability of farming by soils. P. 199-211. *In* P. C. Robert et al. (Eds.) Proceedings of Soil Specific Crop Management: A Workshop on Research and Development Issues. Workshop took place April 14-16, 1992, Minneapolis, MN. Madison, WI: American Society of Agronomy, Crop Science Society of America, and Soil Science Society of America.

Wollenhaupt, N.C., and R.P. Wolkowski. 1994. Grid soil sampling for precision and profit. Unpublished manuscript.   Department of Soil Science, University of Wisconsin, Madison, WI. Modified from a paper prepared for 24th North Central Extension-Industry Soil Fertility Workshop, St. Louis, MO, October 26-27, 1994.

Calculating Profitability of Grid Soil Sampling and Variable Rate Fertilizing for Sugar Beets

Doug Lenz

CENTROL Crop Consulting of Twin Valley
Buxton, ND

ABSTRACT

A mathematical formula was used to analyze nearly 900 commercial fields that were grid soil sampled in the fall of 1995 to estimate expected profitability of variable rate fertilizer application for sugar beets. Profitability was analyzed on the basis of previous crop, nitrogen level and grid size for each field. The trends discovered may help farmers, dealers and consultants select the fields that are more likely to profit from intensive fertility management.

BACKGROUND

Intensive soil sampling and variable rate fertilizer application in the Red River Valley of Minnesota and North Dakota is now done on approximately 30% of sugar beet acres. Research conducted at the University of Minnesota at Crookston and at North Dakota State University has shown an increase of $48-$72 per acre in net income using grid sampling and variable rate spreading as compared to conventional practices. This increase in profit came from an average of 1.12 ton/A increase in yield and 0.37% increase in recoverable sugar per acre. The cost of grid testing is between $10 and $15 per acre and the cost of variable rate spreading is $8 to $12 per acre.

Due to the extra cost involved and skepticism on the amount of variability thought to occur within fields, there are still questions whether grid sampling is cost effective. For those producers who had fields grid sampled, occasionally the soil test results were only marginally variable throughout the field. The determination whether or not to variable rate apply fertilizer or to conventionally broadcast the field at a single rate and blend was hard to quantify. In response to the need to quantify the variability of grid test results, a mathematical formula was developed to calculate the expected profitability of variable rate fertilizer application.

METHODS

The calculations are based on the nitrogen response of sugar beets and the amount of nitrogen variability found within the field. The nitrogen levels from all the grid points within the field were averaged. This average was used as the basis for the whole field and assumed the nitrogen recommendation for the conventional application would have been based on that average. Calculations were made of the number of acres that would have been either over-or-under-fertilized comparing the nitrogen level of each grid against the field average. The nitrogen response for sugar beets is 6 pounds of nitrogen for each ton of yield. A 20 ton yield goal would call for a total of 120 #/A of nitrogen. A yield loss was calculated for grids that would have been under fertilized by at least 20#/A.

Grids that would have been over fertilized by at least 40 lbs/A of N had a factor of .3% lower sugar content as compared to grids at the target N level. A conservative value of $45 per ton of sugar beets was used in the calculations. Nitrogen response was the only nutrient used in the calculations even though the fields were tested for other nutrients. The actual cost for grid testing and a $10/A cost for variable rate spreading were used.

RESULTS

A total of 897 field were analyzed using the profitability formula. The average grid size was 4.1 acres. According to the formula 625 or 69.7% of the fields paid to grid sample and variable rate apply fertilizer. Average net return per acre for all fields was $40.78. The average net return per acre for fields with a grid size of 4 acres or smaller was approximately $50/A. The fields with a grid size of greater than 4 acres had an averaged net return of $35/A. Table 1 is a summary of results sorted by previous crop and the field average nitrogen level.

Table 1. Profitability of grid soil sampling and variable rate fertilizer application in sugar beets, fall 1995, based on a calculated formula.

					Net Profit Per Acre By			
Previous Crop	Ave. Net	Ave N Level	% Paid	Total Fields	Average Residual Nitrogen			
					<30	30-90	90-150	150-210
Fallow	$47.22	89	91	33	($19.51)	$33.32	$ 71.26	$ 62.01
Potato	$58.45	79	89	70	($ 3.99)	$42.65	$ 90.95	$ 69.33
Beans	$36.96	54	73	41	($14.03)	$45.47	$106.61	
Wheat	$43.52	52	71	596	($ 3.92)	$62.01	$106.83	$119.87
Barley	$21.63	23	51	152	($ 5.37)	$75.15	$131.98	
Average	$40.78	51	70	892	($ 4.73)	$59.61	$100.28	$111.28
			Percent paid		31%	95%	99%	100%

Fallow ground had the highest residual nitrogen and a corresponding high incidence of profitability. Conversely barley ground had the lowest nitrogen level and the lowest incidence and degree of profit. Fields that tested within 30 pounds above or below the target nitrogen level of 120 lbs/A had a net profit of $100.238 per acre. Fields that tested below 30 pounds had a negative return on investment 69% of the time. Generally, as residual nitrogen increased so did the percent of fields that paid as did the level of profit for those fields.

CONCLUSIONS

Profitability of grid soil sampling and variable rate fertilizer application is dependent on the amount of nitrogen variability. The amount of variability within a field has a strong correlation to the level of residual nitrogen. Fields with relatively high nitrogen levels tend to have a higher degree of variability as compared to fields with low nitrogen carryover. The previous crop has an influence over the amount of residual nitrogen and therefore profitability. Fields tested using a grid size of 4 acres or less tend to be more cost effective than larger grid sizes.

P And K Grid Sampling : What Does It Yield Us?

G. W. Rehm
J. A. Lamb
J. G. Davis
G. L. Malzer

Department of Soil, Water, and Climate
University of Minnesota
St. Paul, Minnesota

ABSTRACT

At its inception, the concept of grid sampling appeared to be a very reasonable and practical approach to eliminating reported variability in soil tests over time. This concept also appeared to have potential for improving farm profitability while having positive environmental implications. Research conducted in fields of cooperating farmers using four rates of applied phosphate across a landscape has focused on providing answers to several questions that have evolved from the practice of grid sampling in Minnesota. If taken from the same locations, results of analysis for P and K were not substantially different when samples were collected in June and October. Collection of samples from a grid spacing of 18m x 18m showed that variability of soil test P within a grid cell can be as great as variability of soil test P in any given field. Therefore, collection of soil samples from multiple sites rather than one site within a grid cell is suggested for obtaining a more accurate nutrient status of a grid cell. Considering grid cell sizes, the limited data suggest that smaller grid cells may increase the accuracy of prediction. The increased accuracy in prediction, however, must be balanced against the added costs of sampling and laboratory analysis.

BACKGROUND

Using production inputs in the most efficient way has always been a key management strategy for producing the most profitable yields. The development of technology that provides for changing fertilizer application on-the-go has strong implications for increasing profitability in crop production. Yet, there are many unanswered questions associated with this emerging science.

There's general agreement that a more detailed sampling of fields will result in more precise fertilizer recommendations. The past practice of collecting cores at random across a field or a landscape has resulted in highly variable results when these samples have been analyzed for the immobile nutrients over a period of years. This inconsistency with time has caused many crop producers to lose confidence in soil testing as a management tool.

Utilizing modern technology, grid sampling has rapidly become an accepted practice for achieving a more accurate measure of the variability of nutrient levels across a landscape. Regardless of grid size used, a widely accepted pattern or system for sampling grid cells has not emerged. Wollenhaupt, et al (1994) has suggested one scheme. This information base for sampling schemes which might be used in the grid sampling approach is very limited.

Grower acceptance of grid sampling and the associated variable rate fertilizer application will be strongly related to economics. The size of the grid cell chosen by those who collect soil samples will have a major impact on the cost of using this technology. Yet, a wide variety of sizes are sampled when the grid cell sampling technique is used. A review of the literature shows that very little research has been conducted to identify an appropriate yet economical size of grid cell that can be used in a variety of field situations. Franzen and Peck (1995) showed that fertilizer was applied more accurately if smaller grid cells were used for making fertilizer recommendations.

Recognizing the needs described above, this study was conducted to: 1) measure changes in soil test values during the growing season, 2) determine an accepted pattern that could be used in sampling grid cells, and 3) describe an optimum size of grid cells that might be used in sampling for variable rate fertilizer application.

MATERIALS AND METHODS

This study was conducted in fields of cooperating farmers in Sibley and Renville Counties in central Minnesota. In Sibley County, the experimental area was approximately 4.2 ha in size. Approximately 3.6 ha were used for the experimental area in Renville county. Soils in the experimental areas at each site were variable and typical of the soils in central Minnesota.

At each site 0-20-0 (triple superphosphate) was broadcast in strips (4.6m in width) to supply 15, 30, and 45 kg P ha[-1]. An appropriate control (no P applied) was also included. Each rate of phosphate was applied six times in the experimental area. Both sites received adequate N and K to provide for optimum corn yield. All fertilizer was incorporated by the use of the chisel plow.

Soil samples (0-15 cm) were collected from the control strips at various times and analyzed for pH, P (both Olsen and Bray procedures), K, and organic matter content. The distance between sample locations was 18m at both experimental sites.

Initial samples were collected from the Sibley County site in July of 1993. This was followed by sample collection from the same locations, in each strip, in October of 1993, June of 1994, and October of 1994. For the Renville County site, initial samples were collected in June of 1994 followed by samples collected in October 1994, June 1995, and October 1995..

Corn yields were measured with a two-row combine equipped with a weigh cell. The harvest length at each site was 18 meters.

Grain yields were related to soil test values for P as measured by either the Bray or Olsen procedure via regression analysis. The data collected allowed for statistical analysis of cells of various sizes.

RESULTS AND DISCUSSION

Changes of soil test values with time have been a major concern for agricultural professionals who have become involved in the collection of soil samples from a number of various sizes of grid cells. One purpose of this study was to monitor changes in soil test properties over time.

From a fertilizer use standpoint, growers are most interested in soil tests for P (Bray or Olsen procedures) and K. Results from both the Sibley and Renville County sites show high correlation among three sampling times. The "r" values for the Sibley County site are summarized in Table 1. The average soil test values for each time of sampling are listed in Table 2.

If samples were collected from the same site, soil test values for P and K were relatively constant in the control strip over time. Values measured in June were nearly the same as values measured in October and again in the following June. The results from these two experimental sites show that consultants and others responsible for collection of soil samples can collect the soil samples at one or more times during the growing season. The absence of variability of soil test values over time should help to restore farmer confidence in soil testing as a management tool to predict fertilizer needs.

Table 1. Correlation coefficients showing the relationship for soil tests for P (Bray, Olsen) and K over time at the Sibley County site.

| Measurement | Date | Sampling Date | |
		Oct. (1993)	June (1994)
		- - - - - - - - - - - "r" - - - - - - - - - - -	
Olsen P	June (1993)	.98	.96
"	Oct. (1993)	-	.97
Bray P	June (1993)	.97	.96
"	Oct. (1993)	-	.95
Soil Test K	June (1993)	.97	.95
	Oct. (1993)	-	.94

Table 2. Average soil test values for various sampling dates at two locations.

Soil Property	6/93	10/93	6/94	10/94	6/94	10/94	6/95
Sibley County:							
pH	7.3	7.3	7.1	6.9	-	-	-
P (Bray), ppm	28	27	25	27	-	-	-
P (Olsen), ppm	21	21	21	24	-	-	-
K(1M NH$_4$C$_2$H$_3$O$_2$), ppm	255	262	215	269	-	-	-
Renville County:							
pH	-	-	-	-	7.7	7.4	7.5
P (Bray), ppm	-	-	-	-	15	13	13
P (Olsen), ppm	-	-	-	-	14	11	14
K (1M NH$_4$C$_2$H$_3$O$_2$), ppm	-	-	-	-	152	207	237

A wide variety of patterns can be used for collection of samples from individual grid cells. These patterns are independent of the size of the grid cell. To evaluate the effect of sampling pattern on soil test results for P, the experimental site was divided into cells which were 91m x 91m. Three sampling patterns were selected. These were: 1) center point, 2) multi-point, and 3) all points. The data from the 18m x 18m intensive sampling pattern were used for this portion of the study. Cores were taken from the center point of the cell for the center point sampling system. Various organized sampling patterns were used for the multi-point sampling system. Using the various patterns, values from 5 to 9 locations within a cell were averaged. Twenty-five sample points were averaged to obtain a value for the all point sampling system.

The impact of soil sampling pattern on the soil test values for P (Bray procedure) is summarized in Table 3. At both sites, soil test values were similar when the multi-point and all point patterns were used. For six of the seven 91m x 91m cells, P values were substantially different when the mid-point sampling pattern was used.

The differences in soil test values associated with the three sampling patterns translate into substantial differences in phosphate fertilizer recommendations (Table 4). The recommendations are averages for the four cells at the Sibley County site and three cells at the Renville County site. The cell size is 91m x 91m.

Table 3. Soil test values for P (Bray procedure) as affected by sampling pattern used at two locations. Grid size was 91m x 91m.

Site	Sampling Pattern	Cell Number			
		1	2	3	4
		ppm P			
Sibley Co.	multi-point	7.5	41.5	45.2	27.2
	all points	5.7	41.1	40.9	26.7
	mid-point	0	22.0	49.0	13.0
Renville Co.	multi-point	18.8	14.6	16.5	-
	all points	20.1	13.4	16.9	-
	mid-point	26.0	2.0	1.0	-

Table 4. Effect of soil sampling pattern on average soil test values for P (Bray procedure) and phosphate recommendations for corn at two sites.

	Site			
Sampling Pattern	Renville County		Sibley County	
	P test (ave of all cells)	P_2O_5 Rec.* (ave of all cells)	P test (ave of all cells)	P_2O_5 Rec.* (ave of all cells)
	ppm	kg ha^{-1}	ppm	kg ha^{-1}
multi-point	16.7	22	30.4	17
all points	16.7	21	28.8	17
mid-point	9.7	64	21.0	34

* Yield Goal = 9.4 Mg ha^{-1}

The data collected from the two sites show that collection of soil from more than one location in a grid cell provides a more accurate picture of the nutrient status of the soil in that cell. This conclusion is consistent with past recommendations for sampling fields. When soil samples were collected at random from fields, accuracy was improved as the number of cores collected from a field increased.

Response to applied phosphate was consistent with past measured responses of phosphate applied to corn. There were substantial increases in yield when soil test P levels were low or very low. Smaller, but significant, increases were measured when the soil test P level was in the medium range. These data suggest that established response curves for the response of corn to phosphate can be used to compute costs and returns to grid sampling and variable rate fertilizer application.

The response of corn to phosphate fertilization was also used to evaluate the effect of sampling pattern and grid cell size on the ability to predict a response to applied phosphate. Two sampling patterns (all point, mid-point) were compared. Two sizes of grid cells (55m x 55m, 91m x 91m) were also evaluated.

Because of the limited number of data points available from each site, it is difficult to arrive at firm conclusion regarding the effect of sampling pattern and grid cell size on the ability to accurately predict the response of corn to phosphate fertilization. In general, there was no response to applied phosphate when soil test levels for P were in the high and very high range (Table 5). There was one exception. A response was measured in Renville County when the mid-point pattern showed a very high soil test for P.

In Sibley County, there was also a measured response when the average of all samples showed a high soil test for P. In this particular cell, there was a dramatic change from very low to high values in the same cell (91m x 91m). The response in the low testing area was large enough so that the overall response for the grid cell was positive.

There was no response to applied phosphate when the soil test P level in Sibley County was in the medium range. There was a measured response at this range at the Renville County site. The explanation for this difference is not obvious from the data collected but might be explained by differences in environmental conditions.

The effect of grid cell size on predicting the response to phosphate is summarized in Table 6. For the Sibley County site, there was good accuracy in prediction if the 55m x 55m cell size is compared to the 91m x 91m size. The use of the smaller grid size was not completely accurate at the Renville County site. Considering the smaller cell size, four cells showed a high soil test for P and two of these showed a response to phosphate use. Likewise, there was no response for one cell which had a low test for P. This is no explanation for these unexpected observations at this time.

In general, the small grid cell size can be expected to be more accurate in measuring variability in soil test levels. The use of the smaller grid cells, however, adds considerably to the cost of sampling. Therefore, cost considerations should have a major impact on the choice of the size of the grid cell.

Table 5. The effect of sampling pattern on prediction of the response to phosphate fertilization when 91m x 91m grid cells were used.

| Soil Test Category | County | | | | | |
| | Sibley | | | Renville | | |
	# cells	response	no response	# cells	response	no response
Midpoint:						
v. high	1	0	1	1	1	0
high	1	0	1	-	-	-
medium	1	0	1	1	1	0
low	1	1	0	1	1	0
v. low	-	-	-	-	-	-
All Point:						
v. high	2	0	2	-	-	-
high	2	1	1	-	1	1
medium	-	-	-	2	2	0
low	-	-	-	-	-	-
v. low	-	-	-	-	-	-

Table 6. The effect of grid cell size on prediction of the response to phosphate fertilization when the all point sampling pattern was used.

| Soil Test Category | County | | | | | |
| | Sibley | | | Renville | | |
	# cells	response	no response	# cells	response	no response
55m x 55m:						
v. high	12	0	12	2	0	2
high	-	-	-	4	2	2
medium	-	-	-	3	3	0
low	2	1	0	3	2	1
v. low	-	-	-	-	-	-
91m x 91m:						
v. high	2	0	2	-	-	-
high	2	1	1	1	1	0
medium	-	-	-	2	2	0
low	-	-	-	-	-	-
v. low	-	-	-	-	-	-

SUMMARY

There were three major questions associated with grid sampling used with variable rate fertilizer application when this study was started. The results reported have provided some answers .

If soil samples are collected from the same location, soil test values for pH, P, and K do not change substantially with time. Therefore, consultants who provide the sampling service can collect samples throughout the growing season, if needed.

A variety of patterns can be used to collect samples from grid cells. However, collection of soil from several locations in a cell is important.

Neither of the grid cell sizes used in this study was able to predict the response to phosphate fertilization with complete accuracy. The number of data points needed to reach a conclusion regarding grid cell size is limited. A definite answer to this concern is not yet available.

REFERENCES

Franzen, D.W., and T.R. Peck. 1995. Field soil sampling density for variable rate fertilizer application. J. Prod. Agric. 8:568-574.

Wollenhaupt, N.C., R.P. Wolkowski, and M.K. Clayton. 1994. Mapping soil test phosphorus and potassium for variable rate fertilizer application. l J. Prod. Agric. 7:441-448.

Agronomic Benefits of Varying Corn Seed Populations: A Central Kentucky Study[1]

R.I. Barnhisel
M.J. Bitzer
J.H. Grove

Agronomy Department
University of Kentucky
Lexington, KY

S.A. Shearer

Biosystems Agricultural Engineering Department
University of Kentucky
Lexington, KY

INTRODUCTION

Over the past three years, a precision agriculture has been conducted in central Kentucky in which corn seeding rates have been adjusted according to soil depth. Although there are a few similar studies reported in the literature (e.g., Denholm, et al., 1993; Sadler et al., 1995), none of those reported varied seed population according to a soil property such as topsoil thickness. Reichenberger (1996a, 1996 b) described studies in Indiana where yields increased 0.125 to 0.878 Mg/ha as a result of varying the seeding rate.

In 1993, a grower in Hardin Co., KY, Kevin Clark, was interested in our evaluating the potential of a variable-rate planter drive for a corn planter. His perception was that the thin soils were not producing as much when compared with deeper soils. The objectives of this project were as follows: 1) to evaluate the cost effectiveness of varying corn population according to soil depth; 2) to determine whether total yields are higher where populations are varied across the field in contrast to a constant population.

METHODS AND MATERIALS

Four fields were chosen as sites for the variable seeding rate experiment; three of which were in Hardin Co., KY on a farm operated by Kevin Clark. The remaining field was located on the Woodford Research Farm, operated by the University of Kentucky. The Hardin Co. fields are similar and consisted of three soil series: Crider, fine-silty, mixed, thermic Rhodic Paleudalfs; Pembroke, fine-silty, mixed, mesic Mollic Paleudalfs; and Huntington, fine-silty mixed, mesic Fluventic Hapludolls. The Woodford Co. site was mapped as Maury, fine, mixed

[1] Funded in part by KY Corn Growers Association.

mesic, Typic Paleudalfs.

The major difference between Crider and Pembroke is that the latter has redder Bt horizons. Both may occur on karst uplands with slopes ranging from 2 to 12%. Huntington soils occur in depressions or on flood plains. In these fields, the depressions or sinkholes consist of Huntington whereas either eroded Crider or Pembroke occur on the slopes and the ridge tops consist of more typical versions of either soil. In many cases in Hardin Co., as was true in all three fields of this study, the slopes were so short and the depressions so small that the field was mapped as large areas of either Crider or Pembroke due to constraints in delineating areas larger than about one hectare (or two acres). Had selection of the fields been based solely on the soil survey maps, only one seeding rate would have been predicted.

Figure 1 illustrates a small portion of the Hardin Co. field from which data were collected in 1995. This is not drawn to scale, as the width of each variety of each seeding rate treatment is only 3 meters (four 30-inch rows). The length of this strip was 396 m (or 1300 ft). Two hybrids, DeKalb 646 and Pioneer 3140, were planted each year with an 8-row planter, 4 rows of each variety. Two constant seeding rates of 49.4 K/ha and 64.2 K/ha were seeded with one 8-row variable strip between these two strips. The variable seeding rates were 44.5 K/ha, 56.8 K/ha, and 69.2 K/ha, planted on the shallow sideslopes, ridgetop positions, and depressional sites and rates specified by the Rawson controller. Soil samples were collected from each of the regions indicated in this figure. These samples were analyzed for organic matter, using the "LECO" method by Nelson and Sommers (1982) and texture by the hydrometer method, Gee and Bauder (1986).

The Woodford Co. site had three strips seeded across the field with planting rates of 44.5 K/ha, 56.8 K/ha, and 69.2 K/ha, respectively. This portion of the 20 ha field was gridded into 30 x 30 m (100 x 100 ft) cells to facilitate soil sampling. At this site, Pioneer 3279 was seeded.

At the Hardin Co. location, known lengths of 15.24 m (50 ft) from the first two rows were established within each of the sections where the seeding rate was varied according to soil depth. Grain yields were measured with a 2-row MF 8 plot combine equipped with an Ag Leader 2000 yield monitor and Larson Systems GPS data logger. The remaining two rows were harvested continuously using the data logger and post-processing software to generate yield maps of these narrow strips across the field. At the Woodford Co. site, grain yield was measured from the center two rows with the same combine, but only using the GPS data logger system, however, grain yield was calculated at 15.24 m intervals (50 ft) from these data.

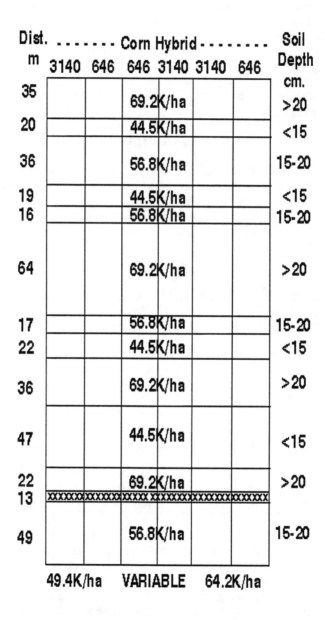

Fig. 1. Plot diagram illustrating - Hardin Co. site with variable seeding rates as a function of soil depth.

The topsoil depths at the Clark Farm in Hardin Co. were grouped into three ranges: < 15 cm, 15 to 20 cm, and > 20 cm (<6 in., 6 to 8 in. and > 8 in.) - depths from which the 15.24 m yield measurements were taken. The entire strip was placed into these three topsoil ranges, shown in Fig. 1, as shallow, medium, and deep. In one case, shown as "x", the area was variable and yields were not measured except by GPS techniques. At each border between soil depths, a transition zone occurred as well. These transition zones were the regions where the seeding rate was changed and were excluded from the yield data for finite lengths. Once the data was collected, yields were computed for each strip, variety, soil depth. Economics of the results were based on $1 per 1000 seeds and $118 per Mg of grain ($3.00 per bu).

RESULTS AND DISCUSSION

As indicated in the Method and Materials section, corn seeding rates for the two varieties were varied dependent on the soil depth and were placed in ranges of either < 15; 15-20; or > 20 cm. Data for organic matter and clay content from these probed areas are found in Table 1. In general, the sites with thinner topsoil had lower organic and higher clay contents. Many of these differences, although small, were significant at the 5% level. In spite of past erosion, all textures were silt loam.

Table 1. Soil characteristics of three fields at Clark Farms, Hardin Co., Kentucky.

Year	Topsoil Thickness	Organic Matter (%)	Clay Content (%)
1993	< 15 cm	1.0 c	18.3 a
	15-20	1.1 b	17.8 ab
	> 20	1.2 a	16.1 b
1994	< 15 cm	0.9 c	20.0 a
	15-20	1.0 b	17.5 b
	> 20	1.1 a	12.5 c
1995	< 15 cm	1.0 b	21.7 a
	15-20	1.0 b	17.6 b
	> 20	1.1 a	14.6 c

The seeded populations versus the plant population at harvest are presented in Table 2 for both hybrids. In general, the final stand, as a percentage of the seeding rate, decreased with increasing population. Stand counts were not made at planting. It was assumed that the planter delivered the rates specified by the Rawson controller. The germination listed on the seed tag for both varieties was 95%; hence at harvest all but one population were lower than the assumption of 95% germination. The recommended seeding rate for these soils in Kentucky is 64.2 K seeds/ha. The harvest population was 12 to 14% less than the seeding rate depending on the variety.

Table 2.　　Comparison of seeding rate versus harvest population.

Seeds/ Hectare (1000)	DeKalb 646		Pioneer 3149	
	Ave[†]	%	Ave	%
44.5 T [††]	40.3	90.6	42.5	95.5
56.8 M	50.6	89.1	53.4	94.0
69.2 D	57.9	83.7	62.4	90.2
49.4	43.8	88.7	45.5	92.1
64.2	55.5	86.4	57.6	88.3

[†]Average population for 1993-95 for each hybrid at harvest
[††]Seeded rate for T = thin, M = medium, and D = deep soils, respectively

　　　In Table 3 are given yield averages across all seeding rates and cultivars for the thin, moderate, and deep soils. Each soil depth exhibited a significantly different yield. Based on these data, one would also expect corn yields to vary according to soil depth when seeded at constant seeding rates as can be seen in Table 4.

Table 3. Effect of soil depth on corn yield[†], 1993-1995.

Soil Depth cm	Yield Mg/ha	Yield bu/a
< 15	5.39c	86c
15-20	6.90b	110b
> 20	8.59a	137a

[†] These are averages for the strip in which the seeding rate was varied.

　　　Table 4 gives the detailed yields for all years, both hybrids, and all seeding rates at the three soil depths. In general, low seeding rates out-yielded high seeding rates on thin topsoiled areas, while the opposite was true on deep topsoiled areas. Close inspection of these data indicates that this trend held for all combinations of depth and seeded population, for either hybrid and in all years, with the exception of the DeKalb 646 hybrid planted at 49.4 seeds/hectare in 1995. In this case, the yield was not related to soil depth. In general, yields were highest in 1994 and lowest in 1995, with a few exceptions for individual hybrids and seeding rate combinations. The low yields in 1995 were related to a later planting date due to a wet spring and a short period of high temperatures and reduced water availability in July.

Table 4. Effect of variable and constant seeding rates on corn yield (Mg/ha), according to soil depth.

Seeds Hectare (1000)	- - - DeKalb 646 - - - -			- - - - Pioneer 3140 - - - -			Grand AVE.
	1993	1994	1995	1993	1994	1995	
44.5 T†	6.52	5.33	3.45	6.65	6.02	4.45	5.39
56.8 M	7.08	7.90	5.83	6.90	8.03	5.77	6.90
69.2 D	9.28	9.09	7.46	9.16	9.16	7.46	8.59
Average	7.65	7.46	5.58	7.59	7.97	5.90	7.02
49.4 T	5.39	5.52	4.95	6.15	5.77	4.26	5.33
49.4 M	6.84	7.40	4.89	7.59	6.96	5.52	6.52
49.4 D	8.34	8.34	4.95	9.41	8.53	6.77	7.71
Average	6.84	7.09	4.93	7.71	7.09	5.52	6.52
64.2 T	5.27	6.02	3.26	4.95	6.40	3.14	4.83
64.2 M	7.46	7.53	3.76	7.21	7.90	4.26	6.33
64.2 D	8.72	8.78	7.21	8.53	8.40	7.02	8.09
Average	7.15	7.46	4.77	6.90	7.59	4.83	6.40
Grand Averages	7.21	7.34	5.08	7.40	7.53	5.39	6.65

† The depth of topsoil; T = thin or < 15 cm; M = medium or 15-20 cm; and D = deep or > 20 cm.

Table 5. Economics of varying seeding rates for corn over a three-year period.

Seeding Rate	Yield Mg/ha	Return[1] $/ha	Seed[2] $/ha	Return[3] $/ha	Net Diff.
Variable[4]	7.02	$828.40	$56.80	$771.60	--
49.4 K/ha	6.52	$769.40	$49.40	$720.00	$51.60
64.2 K/ha	6.40	$755.20	$64.20	$691.00	$80.60

[1]Return from corn grain $118/Mg or $3/bu
[2]Cost of seed per hectare $1.00/1000 seeds
[3]All other fix and variable rates are assumed to be equal, hence return is from grain sales minus seed costs.
[4]Variable seeding rates were 44.5 K/ha, 56.8 K/ha, and 69.2 K/ha or 18 K/a for thin, medium, and deep soil, respectively. The area of the fields at each level of topsoil averaged 40, 26, and 34% for the thin, medium, and deep soil depths, respectively.

The bottom line in this study is: Does varying the seeding rate result in an increased profit? These analyses are given in Table 5. Returns were calculated with several assumptions as given in the footnotes to Table 5. Corn was valued at $118/Mg ($3.00/bu) for comparison purposes. This was a higher value for grain than this grower received, but well within 1996 contract prices. The return was greater for variable seeding rate when compared with constant seeding rates of 49.4K/ha and 64.2 K/ha. In the case of the 64.2 K/ha seeding rate, the increased return was due to both a higher yield and a lower seed cost. One can adjust these differences in return according to current grain prices and seed costs. In this case, if the grower planted over 405 hectares (1000 acres), his return over the recommended 64.2 K/ha (26 K/a) seeding rate would be $32,650 which would pay for all the added equipment needed to implement variable seeding rates.

We quickly add caution to the above economics; in Kentucky, we are reasonably sure this applies to rolling upland sites. However, we do not know if yield differences would be as large (maybe greater) than measured on the fields in this area. As a result of this uncertainty, a major expansion of this project was initiated in 1996 and includes 9 new locations across the major corn producing regions of the state. Sites were selected on level as well as undulating topographies.

One test site, the Woodford Co. Research Farm, was actually seeded in 1995, with a slight variation, that being only one corn hybrid was planted and three seeding populations were used across a field about 500 m long. Three "constant" seeding rates of 44.5 K/ha, 56.8 K/ha, and 69.2 K/ha were used. The yields were plotted as a function of distance in Figure 2. These data were collected by the same plot combine used in Hardin Co. The first two data points on the left hand side are low due to the time needed to "load" the combine. For the next 150 meters, the yield from the 44.5 K/ha rate were always equal or greater than those of other two populations, especially the last few meters of the 150 m section. The yields then dropped sharply over the next 30 m distance in which the highest seeded rate (69.2 K/ha) was the highest yielding strip. With greater distance, the yield from 44.5 K/ha again exceeded that from the 69.2 K/ha rate. These lines crossed once more near the 365 m and merged into one line at 425 m.

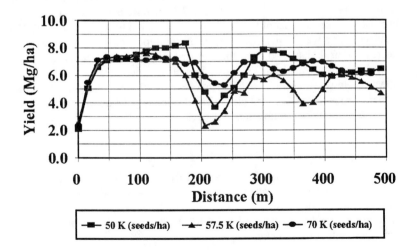

Fig. 2. Effect of seeding rates on corn yield over a 500 meter distance at the Woodford Co. research farm.

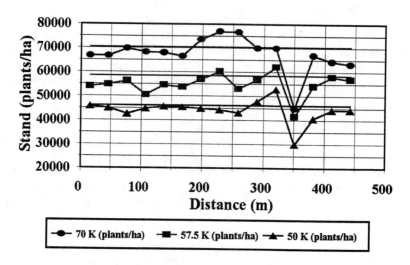

Fig. 3. Comparison of seeded versus harvest populations over a 500 m distance at the Woodford Co. research farm.

It is interesting to note that at no time did the yields from the 56.8 K/ha seeding rate exceed both of the other rates, even though this is the recommended seeding rate on this soil.

Figure 3 presents the seeded versus stand counts at harvest. It is obvious that there were fluctuations at all three populations. Unfortunately, stand counts were not performed. Between 180 and 200 meters, the stand for the highest population exceeded that of the seeding rate. It should be noted that at the 350 m distance, there was a sharp decline for all three populations. This occurs at a depression or drainage corridor that perhaps should have been considered for a grass waterway. Although excessive erosion was not occurring, the stand was affected. However, only for the 56.8 K/ha seeding rate was it likely that this decline in stand produced a decreased yield. The decline in yield occurred over a greater distance than did the decline in population.

SUMMARY

In Kentucky, especially on rolling topography, this initial study indicates major economic benefits may be gained from a precision agriculture approach for corn production. Increased returns exceeded $80/ha. Expansion of this study is underway to validate if similar benefits may be realized throughout the Commonwealth of Kentucky.

REFERENCES

Denholm, K.A., J.L.B. Culley, J.D. Aspinall, and E.A. Wilson. 1993. Soil landscape relations and their influence on yield variability in Kent Co. Ontario. p. 355-356. *In* Proceedings of First Workshop. Soil Specific Crop Management. P.C. Robert, R.H. Rust, and W. E. Larson (eds). ASA/CSSA/SSSA.

Nelson, D.W., and L.E. Sommers. 1992. Total carbon, organic carbon and organic matter. pp. 539-579. *In.* Page. A.L., R.H. Miller, and D.R. Keeney (eds) Methods of Soil Analysis Part 2-- Chemical and Microbiological Properties. Agronomy Monograph No. 9 (2 Ed) ASA/SSSA. Madison, WI.

Gee, G.W., and J.W. Bauder. 1986. Particle-Size Analysis. pp. 383-411. *In* Klute, A. (ed) Methods of Soil Analysis Part 2-- Physical and Mineralogical Methods. Agronomy Monograph No. 9 (2 Ed) ASA/SSSA.Madison, WI.

Ruchenberger, L. 1996 a. Precision farming turns a profit. Farm Journal Mid-February issue, (Data accessed via internet: www.farmjournal.com).

Ruchenberger, L. 1996 b. A vote for variable seeding. Farm Journal Mid-February issue 1996, (Data accessed via internet: www.farmjournal.com).

Sadler, E.J., P.J. Bauer, and W.J. Busscher. 1995. Spatial corn yield during drought in the SE Coastal Plain. pp. 365-381. *In* Site Specific Management for Agricultural Systems. P.C. Robert, R.H. Rust, and W.E. Larson (eds).

Spatial Variability of Profitability in Site-Specific N Management

G. L. Malzer
P. J. Copeland
J. G. Davis
J. A. Lamb
P. C. Robert

Department of Soil, Water, and Climate
University of Minnesota
St. Paul, Minnesota

T. W. Bruulsema

Potash and Phosphate Institute of Canada
Guelph, Ontario

ABSTRACT

The potential profitability of site-specific N management depends on predicting the spatial variability of profitability associated with N use. Four experiments conducted over two years in Minnesota showed that yield response and profitability due to N use vary widely across landscapes. The potential profitability of site-specific N rate management ranged from 11 to 72 $ ha^{-1} above conventional uniform rate recommendations. The ability to attain maximum profitability will depend upon accurate N rate predictions that are spatially targeted to responsive and profitable areas of the field.

INTRODUCTION

Site-specific N rate management (variable rate application) has been proposed as a method that can improve fertilizer use efficiency, thereby increasing the profitability to the producer and minimizing potential environmental concern. Although this concept appears to have merit, relatively limited research information is available to conclusively defend or refute either of these potential benefits. Site-specific management could increase the net return to the producer either by reducing fertilizer costs in areas of a field that are currently being over-fertilized, or by increasing yields in areas of the field that would otherwise be under-fertilized.

Field experiments were established in 1994 and 1995 to evaluate the potential economic benefit associated with site-specific N rate management. The results presented should be regarded as the potential net return that could be expected with ideal site-specific management conditions. The actual return to a producer will depend on the cost of the technology, and the ability to interpret field variability in such a way that will allow the most profitable rate of N application without over-fertilization.

MATERIALS AND METHODS

Field Procedures

Experiments were conducted for one year on four commercial dryland corn (*Zea mays* L.) fields in southwestern and south-central Minnesota during 1994 and 1995. The sites, approximately 5 ha in size, were chosen to reflect typical topographical features of the area. Soils at these sites were predominantly clay loams and loams. All field operations except application of N and yield determinations were performed by the producer cooperators. The previous crop was soybeans at three locations and sugarbeets at the fourth.

The experimental design was a split-block in space with six replications of six N treatments (0, 67, 101, 135, 168, and 202 kg ha^{-1}). Each treatment was 5.6 to 6.1 m wide and extended 274 to 305 m. Treatments were randomized within each block and were fall applied as anhydrous ammonia. A toolbar equipped with a heat exchanger (Hiniker cold flow) and computerized flow sensing and variable metering system were used to ensure a uniform rate of application within each strip.

Grain harvest was accomplished utilizing a plot combine equipped with a ground distance monitor and computerized Harvest Master load cell. Harvest yields were obtained from the center 1.7 m of each treatment strip. Yield estimates from 7.6 m of each end of the treatment strip were discarded to avoid border effects. Grain yield determinations were recorded every 15.2 m within each treatment strip and all yields were adjusted to 15.5% moisture.

Statistical Analysis

Semivariograms of treatment grain yields indicated strong spatial correlations in all fields and spatial trends in some fields. Jack-knife R^2 values, however, were considered to be inadequate to interpolate and estimate grain yields for areas where a particular treatment was not applied in the field. Interpolation errors were eliminated by using actual grain yield observations. Prior to economic evaluation of crop response data, whole- field spatial trends (if present) were removed utilizing a nearest neighbor analysis described by Mulla et al. (1990). Crop response to applied N was first evaluated on a whole-field basis, followed by crop response and economic evaluation on the smallest measured unit in which all treatments were represented (subblock). Subblocks were one replication wide and 15.2 m in length.

A quadratic polynomial model ($Y = ß_0 + ß_1X + ß_2X^2$) was used to describe the crop response to applied fertilizer N in each level of analysis. The intercept and linear coefficients of this model were used to group similar subblocks together into subregions which could potentially be used as management zones. The model intercepts (check plot yields) were divided into three equal-sized classes, and the linear coefficients (responsiveness to applied N) were grouped into four equal-sized classes. The quadratic polynomial was again used to fit the yield response data of all subblock data that had been grouped into the 12 potential subregions. These 12 regression functions were used to calculate a weighted whole field crop response and economic potential.

The economic optimum N rate (EONR) was defined as that N rate where the marginal cost was equal to the marginal return. Fertilizer N prices at the time of application ($0.264 kg^{-1} in 1994 and $0.374 kg^{-1} in 1995) and an average sale price for corn after harvest ($78.6 Mg^{-1} in 1994 and $118 Mg^{-1} in 1995) were used in the analysis. These functions were also used to determine the grain yield at the EONR and the marginal increase in value from applying N fertilizer. If the regression function predicted the EONR to be higher than 202 kg ha^{-1} the EONR was placed at 202 kg ha^{-1}. The marginal increase in value from applying N fertilizer was calculated as the gross value of crop fertilized at EONR minus the value of the crop at 0 N, minus the cost of fertilizer at EONR.

RESULTS AND DISCUSSIONS

Growing conditions for corn production in southern Minnesota were very good in 1994 and 1995. Grain yields were much higher in 1994 vs 1995. Average check plot yields were 9.7 Mg ha^{-1} in 1994 and 6.2 Mg ha^{l} in 1995 while grain yields at the EONR were 12.4 Mg ha^{-1} and 10.0 Mg ha^{-1}, respectively. The higher check plot yields in 1994 could be a reflection of the poor growing season, and limited mineralization of soil N during the 1993 season. The polynomial regression equations for the whole field at the four locations are presented in Table 1.

Similar regression equations were calculated for each of the subblocks that make up the whole-field data set. A wide range of values were observed for all coefficients of the quadratic model. A strong inverse correlation ($r = -.94$ to $-.97$) was found between the linear coefficient and the quadratic coefficient. Because of that correlation it was concluded that the grain yield at 0 N (intercept) and the slope of the linear coefficient explained most of the response variability. The variability of the intercept and the linear coefficient (Table 2) were divided into three and four classes respectively to generate 12 potential subregions in the field that exhibited similar crop response to applied N.

Table 1. Whole-field regression equations for yield response

n	Equations	R^2	C.V.	RMSE
			%	Mg ha^{-1}
	Hanska 1994			
684	Yield=10.88+0.0239(N rate)-0.0000851(N rate)^2	0.36	6.4	.77
	Hector 1994			
612	Yield=8.62+0.0334(N rate)-0.0000936(N rate)^2	0.47	10.3	1.11
	Hanska 1995			
612	Yield=6.17+0.0330(N rate)-0.0000914(N rate)^2	0.69	8.3	.69
	Morgan 1995			
612	Yield=6.22+0.0332(N rate)-0.0000600(N rate)^2	0.72	10.0	.89

Table 2. Variability of the intercept and linear coefficients of subblocks.

Location	Year	n	Intercept			Linear coefficient		
			Min.	Max.	Mean	Min.	Max.	Mean
			Mg ha^{-1}			Mg ha^{-1} kg^{-1}		
Hanska	1994	114	8.1	12.6	10.9 (0.94)†	-0.012	0.068	0.024
(0.018)								
Hector	1994	102	6.3	11.6	8.6 (1.12)	-0.05	0.084	0.032 (0.027)
Hanska	1995	102	4.4	7.6	6.2 (0.65)	-0.008	0.081	0.033 (0.016)
Morgan	1995	102	4.1	9.7	6.2 (0.96)	-0.031	0.096	0.033 (0.022)

† Values in parenthesis are the standard deviation of the Mean

Although the subblocks provide a fine scale of spatial resolution, they contained minimal data sets, and the interpretations of the regression analysis were sometimes influenced by only one or a few data points. Grouping of subblocks into larger areas that represented similar crop response areas reduced the standard errors associated with measured and derived parameters, and also assisted in delineating larger areas of the field that should be managed in a similar manner. As the scale of interpretation changes, the resolution and sensitivity of interpretation will also change. An example of the different types of response equations generated by grouping subblocks together is presented in Fig. 1. The wide variety of different crop response functions within a given field suggested that portions of each field were still responding to N at the highest rate of N application while other portions of the field needed little or no additional N.

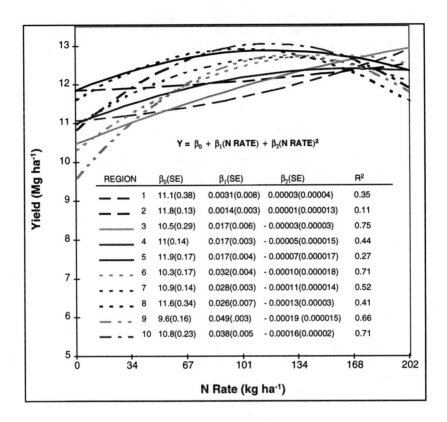

Figure 1. Yield response curves at Hanska 1994.

An economic evaluation using marginal cost analysis for N costs and value of the corn was conducted on each regression equation to evaluate how the economic optimum N rate (EONR) and the potential increased value (profitability) might change in different areas of the field (Table 3). Summaries of both the subblocks and the subregions are included to provide information related to changes in the scale of interpretation.

Table 3. Marginal cost analysis of subblocks and subregions for defining economic optimum N rate (EONR) and increased value (profitability) at EONR due to fertilizer N

Locations		Subblocks		Subregions	
		EONR ($kg\ ha^{-1}$)	Inc. Value ($\$\ ha^{-1}$)	EONR ($kg\ ha^{-1}$)	Inc.value ($\$\ ha^{-1}$)
Hanska 94	Min	0	0	0	0
	Max	202	322	202	217
	Mean	123	112	100	104
	SE†	53	78	4.5	8.4
Hector 94	Min	0	0	107	67
	Max	202	568	202	509
	Mean	152	224	162	217
	SE	47	124	4.0	16.2
Hanska 95	Min	96	104	123	174
	Max	202	706	202	706
	Mean	159	324	163	317
	SE	34	123	2.1	16.3
Morgan 95	Min	0	0	144	276
	Max	202	867	202	682
	Mean	185	447	188	438
	SE	30	158	2.6	15.0

† SE=Standard error of the mean for subblocks, and estimated standard error of the weighted means for subregions.

There was a wide range of economic optimum N rates and of potential increased value to the producer at each location. Grouping subblocks into subregions increased the value of the minimums and decreased the value of the maximums, but had relatively little influence on the mean.

The grouping of response data into response regions will be useful only if it gives an accurate estimate of the observed data collected in the subblocks, and if it gives a reliable estimate of the error within a subregion. This was tested by comparing the degree of areal association, and by comparing the spatial patterns of the difference between two maps.

In general, the areal association, as indicated by the correlation coefficient r, (Table 4) was high for EONR, grain yield at EONR, and increased value due to N (IVN). These whole-map correlations between subblocks and subregion values, however do not explicitly take into account the spatial arrangements or patterns of values in the field. Two maps may be highly correlated, but the spatial arrangement of differences may be different.

Spatial correspondence between subblock maps and subregion maps were tested by the map difference approach of Cliff (1970). The basic rationale is that patterns of differences between two maps will be random if the maps are the same and non-random if they are different. Acceptance of the null hypothesis of randomness would declare that the maps are the same. The null hypothesis was accepted in 10 of 12 comparisons, suggesting that the grouping done by the response area method was generally reliable (Table 4).

Table 4. Similarity of subregion maps and subblock maps for EONR, predicted yield at EONR, and increase in crop value due to application of N fertilizer.

Location	Correlation coefficient, $r^{\dagger}$			Result of z tests(α=0.01) for null hypothesis that maps are the same[‡]		
	EONR	Yield@ EONR	IVN[§]	EONR	Yield@ EONR	IVN
Hanska94	0.34	0.32	0.88	Accept	Accept	Accept
Hector94	0.45	0.67	0.89	Accept	Accept	Reject
Hanska95	0.72	0.81	0.89	Accept	Accept	Accept
Morgan95	0.50	0.70	0.81	Reject	Accept	Accept

[†] r   calculated as covariance of response area values and subblock values divided by product of standard deviations of subregion values and subblock values.

[‡] z   test based on comparison of number of observed differences and similarities and expected number of expected differences and similarities in a random pattern. Null hypothesis accepted based on evidence of random spatial patterns for over-estimations and under-estimations of response area maps.

[§] INV=Increase in value from N at EONR

The response region method as an estimate of field potential is useful only when compared to other N management scenarios. The approach utilized in these experiments allows for the evaluation of a virtually unlimited number of management combinations. For the purpose of this paper, one comparison will be made between the "ideal" crop response region and the current existing uniform rate recommendation that would be made by the University of Minnesota (Table 5). Table 5 compares the yield and economic potential of site-specific N management at the subregion level with the results that would be expected with the current University of Minnesota uniform rate of application.

Table 5. Weighted whole field means for estimated yield and economic returns using variable N rates at the economic optimum vs uniform recommended rates of application.

Location	Variable-subregions			Uniform-whole field			
	N rate	Yield	IVN	N rate	Yield	IVN	Potential Benefit
	kg ha^{-1}	Mg ha^{-1}	$ ha^{-1}	kg ha^{-1}	Mg ha^{-1}	$ ha^{-1}	$ ha^{-1}
Hanska94	100	12.5	104	146	12.6	93	11
Hector94	162	11.9	217	168	11.6	190	27
Hanska95	163	9.4	317	146	9.1	284	33
Morgan95	188	10.5	438	146	9.8	366	72

IVN=Increased value due to N fertilization

Current N recommendations over-fertilized one field by ~45% and under-fertilized a second by ~30% with the weighted average for the other two locations being similar. The expected yields using each technique were very similar except for the field that was under-fertilized. Potential added value associated with site-specific N rate management appears to be in the order of 10-20% of the current N profitability estimates. Care, however, should be used in this interpretation because this is the potential increase in profitability. Existing methodologies for developing site-specific N rate recommendation maps have seldom provided more than 50% of the potential profitability.

Acknowledgements

Funding for this project was provided by the Department of Energy via the Legislative Commission on Minnesota Resources.

REFERENCES

Cliff, A.D. 1970. Computing the spatial correspondence between geographical patterns. Transactions Inst. Brit. Geogr. 50:143-154.

Mulla, D.J., A.U. Bhatti, and R. Kunkel. 1990. Methods for removing spatial variability from field research trials. Adv. Soil Sci. 13:201-213.

An Economic Evaluation of Precision Fertilizer Applications on Corn-Soybean Fields

G. D. Schnitkey
J. W. Hopkins
L. G. Tweeten

Agricultural Economics Department
The Ohio State University
Columbus, Ohio

Precision farming is the practice of varying input application rates across a field. Compared with traditional single rate applications, precision farming is economically attractive when a field's soil characteristics vary so that varying input rates across a field based on soil characteristics results in more efficient input use. Soil characteristics have long been known to be heterogenous across a field (McBratney & Whelan, 1995) and the concept of precision farming is not new (Sawyer1994). Recent interest in precision farming likely results from advances in computer technologies that allow for easier capture and analysis of spatial variability in fields and from advances in application technologies that allow for variable application rates across a field.

While current developments conceivably allow all inputs to vary across a field, much of the interest has focused on fertilizer applications. The relative popularity of fertilization-related precision technologies occurs because of the robust knowledge behind fertilizer-soil nutrient-yield relationships and the relative importance of fertilizer among total crop production expenses. Also, since the 1950s tests of a soil's phosphorus and potassium levels have been used to adapt fertilization rates to individual fields (Fixen & Reetz, 1995). Recent research shows that soil phosphorus and potassium levels vary within a field. Therefore, optimal fertilization rates could change across a field. Moreover, improving nutrient use efficiency is seen as a way to improve the quality of the nation's waters (National Academy of Sciences). Therefore, precision farming takes on land stewardship dimensions.

In this paper, we evaluate the returns from precision fertilizer applications on fields where corn and soybeans are grown in rotation. Our major contribution is to divide these benefits into two parts: information gathering and precision application. Information gathering through practices such as grid soil sampling must occur before fertilizer can be applied using a precision technology. However, information gathering on its own may have value even if the application technology is not implemented. "Information" value occurs if the single application rate considering the collected information differs from the application rate that does not use the information. These conceptual issues are developed in the next section.

We also contribute to the methodologies for evaluating fertilizer decisions. Fertilizer applications have long been known to have dynamic considerations because soil nutrient levels in future years depend on cultural practices in previous years.   Optimal fertilizer use has been studied using dynamic programming considering one nutrient in a monocultural setting (e.g., Kennedy, 1986; Taylor, 1983). We extend these methods by considering the two major nutrients having strong carryover relationships (phosphorus and potassium) in a corn-soybean rotation. These advances allow optimal decisions to be examined in a setting very close to actual practice.

Most previous studies of precision fertilizer applications have not used a dynamic framework, instead relying on a partial budgeting approach (Griffith, 1995; Reetz & Fixen, 1995; Hibbard et al, 1993; Wollenhaupt & Buchholz, 1993). Therefore, results from these studies are suspect because they do not consider the dynamic nature of fertilizer response.   We further extend previous studies by considering variability in soil nutrient levels. Conceptually, returns from precision fertilizer applications should increase as the variability of soil nutrient levels across the fields increase. We develop these linkages by showing the impact of variability on returns.   Moreover, we estimate returns from 18 grid sampled fields in Northwest Ohio. These estimates illustrate the variability in returns resulting from applying precision technologies.

Methods for Evaluating Elements of Precision Technologies

We develop return measures for three fertilization strategies. The first is an "average" strategy in which the application rates of phosphorus and potassium are the same across the field. These rates are determined based on the average phosphorus and potassium nutrient levels in the field. This strategy mimics old recommendations suggesting that farmers should collect samples from different areas in the field, mix the samples, and base fertilizer recommendations on the mixed sample. The second strategy is an "information" strategy in which the application rates of phosphorus and potassium are the same across the field. The information rate is determined considering differences in soil nutrient levels across the field. The third strategy is a "precision" application strategy in which the fertilization rates vary across the field.

The difference in returns from the information and average strategies allows analysis of the value of collecting additional information concerning soil nutrient levels. Collecting information is economically beneficial if the costs of information collection are less than the difference in returns from the information and average strategies.   The difference between the precision and information strategies determines the economic feasibility of precision application technologies. Adopting precision application technologies is warranted if the cost of the application technologies is less than the difference between returns from the precision and information strategies.

We capture differences in soil nutrient levels by dividing a field into n plots. Each plot has two production functions:

$$(1) \quad y_{i,j} = f_j (x_{i,p}, x_{i,k}, s_{i,p,t}, s_{i,k,t}) \qquad\qquad j = c, b$$

where $y_{i,j}$ is yield from plot i for crop j, j is a crop index equaling c for corn and b for soybeans, $f_j(\cdot)$ is a production function for crop j, $x_{i,p}$ and $x_{i,k}$, respectively, are applications of phosphorus and potassium to plot I, and $s_{i,p,t}$ and $s_{i,k,t}$, respectively, are soil nutrient levels of phosphorus and potassium of plot I at the beginning of the growing season t. Each plot also has four soil nutrient carryover relationships capturing changes in soil nutrient levels from one growing season to the next:

(2) $s_{i,m,t+1} = g_{m,j}(x_{i,m}, s_{i,m,t}, y_{i,j})$ $m = p, k$

where m is a nutrient index equaling p for phosphorus and k for potassium, $g_{m,j}(\cdot)$ is the carryover relationship for nutrient m given that crop j is grown. We presume that corn and soybeans are grown in rotation. Mathematically, we state this as:

(3) $j = \begin{cases} c \text{ when t is even} \\ b \text{ otherwise} \end{cases}$

    Decisions for each strategy are derived using an objective that maximizes the discounted present value of returns. We use expected prices of p_j, input prices of w_p and w_k, and a discount rate of d.

Precision strategy. Under the precision strategy, a plot's fertilization rates depend only on its soil nutrient levels. Since a plot's fertilization rates do not depend on other plot's nutrient levels, each plot's optimal rates are found separately from one another. These optimal rates are obtained using dynamic programming by solving the following Bellman equation:

(4) $V_i(s_{i,p,t}, s_{i,k,t}) = \max_{x_{i,p}, x_{i,k}} E[p_j f_j(\cdot) - w_p x_{i,p,t} - w_k x_{i,k,t}]$
$+ B\, E V_i(s_{i,p,t+1}, s_{i,k,t+1})$

where $V_i(\cdot)$ is the equilibrium recursive objective function giving the maximum discounted returns from plot i, E is an expectations operator, and B is a discount rate (i.e, $B = 1/(1+d)$). This objective function is subject to the state transition equations in (2) and (3).

    Solving this model yields four decision rules: phosphorus applications in corn and soybeans and potassium application in corn and soybeans Returns with precision applications for a field equal the sum of discounted returns from all the plots:

(5) $V_p(S_t) = \sum_{i=1}^{n} \rho_i V_i(s_{i,p,t}, s_{i,k,t})$

where $V_p(S_t)$ give the present value of precision applications for the field, S_t is a soil nutrient vector containing each plot's soil nutrient levels (i.e., $S_t \in \{s_{1,p,t}, s_{1,k,t}, s_{2,p,t}, s_{2,k,t}, \ldots, s_{n,p,t}, s_{n,k,t}\}$) and ρ_i is the proportion of plot i in the entire field.

Information strategy. Under the information strategy, a single rate is determined using all values in the soil nutrient vector. Solutions to this problem are obtained by solving the following Bellman equation:

(6) $V_I(S_t) = \max_{x_p, x_k} \sum_{i=1}^{n} E\rho_i p_j f_j(\cdot) - w_p x_p - w_p x_p + B E V_I(S_{t+1})$

where $V_I(S_t)$ gives the present value of discount returns for the information case, the transition equations for each element of S_{t+1} are given in equation (2), and the value of j is given in (3).

A key difference between the information and precision strategies is the dependence of the decision rules. Optimal decisions depend on all soil nutrient levels in the field under the information strategy while optimal decisions depend only on the plot's nutrient levels under the precision strategy.

Average strategy. Under the average strategy, fertilization decisions are based on the average soil nutrient levels for the field. The average soil phosphorus, $s_{a,p,t}$, and potassium, $s_{a,k,t}$, levels are found by averaging all plots' soil nutrient levels. Given these levels, decisions are taken from the precision strategy's decision rules. Implicit in this approach is the assumption that the decision-maker views the field as being homogeneous. Given this approach, the discounted returns from applying the average information strategy $V_a(S_t)$ can then be stated as:

$$(7) \quad V_a(S_t) = \sum_{t=0}^{\infty} \sum_{i=1}^{n} B^t [\, \rho_i (\, p_j \, f_j(\, x^{p,m,*}(s_{a,p,t}), x^{k,m,*}(s_{a,k,t}), s_{i,p,t}, s_i$$
$$- w_p \, x^{p,m,*}(s_{a,p,t}) - w_k \, x^{k,m,*}(s_{a,k,t})\,]$$

where $x^{p,m,*}(\cdot)$ and $x^{k,m,*}(\cdot)$ are optimal decision rules from the precision strategy. Transition equations are given in (2) and (3).

We state returns from gathering information as the difference between discounted returns from the information and average strategies stated on a yearly equivalent basis (i.e., $dV_I(S_t) - dV_a(S_t)$). Returns from precision applications are stated as the difference in returns from the precision and information strategies (i.e., $dV_p(S_t) - dV_I(S_t)$). We do not include costs of either information gathering or precision applications in the calculation of returns. Therefore the differences given above serve as a benchmark for comparing the costs to the returns. If costs are less than the returns, the respective activity is economical.

Empirical Setting and Numerical Approach

We estimate returns by first solving the dynamic programming problems in (4) and (6), yielding decision rules for the precision and information strategies. Solving these problems requires numerical specifications of the production function in (1) and state transition equations in (2), as well as numerical specifications of prices and the discount rate. When specifying the production function, we use specifications for an average soil in Northwest Ohio from a computer program used to make fertilizer recommendations for the Ohio Cooperative Extension Service. In the specifications, phosphorus applications are stated in pounds of P_2O_5 per acre, potassium applications are stated as pounds of K_2O per acre, phosphorus soil nutrient level are stated as pounds of Bray P per acre, and potassium soil nutrient levels are stated as pounds of K per acre. The production function has the following form:

$$(8) \quad y_{i,j} = \alpha_j (1 - 10^{(-\beta_{1,j} x_{i,p} - \beta_{2,j} s_{i,p,t})})(1 - 10^{(-\beta_{3,j} x_{i,k} -}$$

where α_j and $\beta_{i,j}$ are parameters. Parameters for corn are $\alpha_j = 164$, $\beta_{i,j} = .0091$, $\beta_{2,j} = .043$, $\beta_{3,j} = .008$, $\beta_{4,j} = .0064$. Parameters for soybeans are $\alpha_j = 45$, $\beta_{1,j} = .0071$, $\beta_{2,j} = .054$, $\beta_{3,j} = .0157$, $\beta_{4,j} = .0038$. Nutrient carryover follows:

$$(9) \quad s_{i,m,t+1} = s_{i,m,t} + \delta_{m,j,1} \, x_{i,m} - \delta_{m,j,2} \, y_{i,j} - \delta_{m,j,3}$$

where the δ are parameters. The set $\{\delta_{m,j,1}, \delta_{m,j,2}, \delta_{m,j,3}\}$ equals $\{.1, .037, 0\}$ for phosphorus carryover in corn, $\{.10, .08, 0\}$ for phosphorus carryover in soybeans, $\{.25, .0675, 5\}$ for potassium carryover in corn, and $\{.25, .35, 5\}$ for potassium carryover in soybeans. Prices used in deriving decision rules are a \$2.20 per bushel corn price, a \$6.00 per bushel soybean price, a \$.20 per pound P_2O_5 price, and \$.13 a per pound K_2O price. The discount rate is 5 percent.

Given the decision rules, estimates of returns are obtained using simulation. In addition to the numeric specification given above, simulating returns requires an initial distribution of soil nutrient levels. We use two different initial distributions. In the first results section, we use hypothetical soil nutrient levels for two plots to illustrate steady-state nutrient values and impacts of alternative soil nutrient values on decisions and returns. Using only two plots simplifies presentation of results. In the second results section we use soil nutrient levels obtained from eighteen grid sampled fields in Northwest Ohio. These fields were soil sampled every 2.5 acres in 1992. Results for these eighteen fields provide realistic estimates of returns.

Results for Two Hypothetical Plots

For the precision strategy, the steady state soil phosphorus value is 42 pounds per acre when corn begins the year and 46 pounds when soybeans are grown. The nutrient values cycle between these two levels depending on the crop grown during the year. Phosphorus applications are 93 pounds in the corn year and 3 pounds in the soybean year. The steady state soil potassium value is 279 pounds when corn begins the year and 278 pounds when soybeans begins the year. Potassium applications are 58 pounds in the year corn is grown and 89 pounds when soybeans are grown. Steady state nutrient levels match recom-mendations from agronomists for "critical" soil test levels, levels causing no nutrient deficiencies in growing crops. Steady state application rates for the rotation on average match agronomic recommendations.

Optimal decisions under the precision strategy always lead to these steady state soil nutrient levels. The bold lines in Figure 2 show phosphorus soil levels over time under the precision strategy for two plots: one having an initial soil phosphorus level of 20 and the other a level of 60. For the plot with a 20 pound level, phosphorus applications are above the steady-state application rates, causing the plot's phosphorus soil level to increase over time, reaching the steady-state levels by the sixth year. For the plot with the 60 pound initial level, phosphorus applications are below the steady state application rates, again leading to the steady state levels. These "build-up" and "draw-down" results closely match agronomic recommendations.

The process of reaching the potassium steady state values are more protracted, as illustrated by the bold lines in Figure 2. The build-up and draw-down processes require 30 years before the steady state levels are reached. A more rapid convergence for phosphorus suggests that reaching steady state levels is more critical for phosphorus than for potassium.

Figures 1 and 2 compare soil nutrient levels over time for the precision, information, and average strategies. The single fertilizer rate average and information strategies do not reach the steady state values within 50 years. The information strategy results in higher soil nutrient levels than does the average strategy. Higher soil nutrient levels result from higher application rates, particularly in early years of the simulation.

In general, average soil nutrient levels are lower under the precision strategy than under the information strategy. Under the information strategy, there are economic incentives to raise the "low" plots nutrient levels in order to gain higher returns from the low plots. This occurs even though little to no yield increase occurs on the "high" nutrient plot.

Table 1 shows the effect of fertilization strategies on net returns. Panel A, B, and C of Table 1 demonstrate the effect on net returns to precision application when one plot's soil phosphorus level is fixed and the other plot's soil phosphorus level varies. Overall, net returns from precision applications and from information gathering increase as the difference in phosphorus soil levels increase. Increasing returns for increasing differences in soil test levels is consistent with intuition. If fields are close to being homogeneous, treating them as heterogeneous results in little gain.

Often, there are significant increases in returns from information over average application. Gains are particularly pronounced when one plot's soil nutrient level is above the steady state and one is below, seen best in Panel A and C. Higher gains in net revenue are the result of increased fertilizer applications, which bring up the lower soil test level and increases crop yield.

Returns For Alternative Fields

Differences in returns for the three strategies are shown in Table 2 for the eighteen grid sampled fields. Table 2 also shows the average, standard deviation, and minimum levels for phosphorus and potassium for the plots. The average phosphorus soil level for all eighteen fields is 111 pounds which is high compared to steady state values in the 40 pound range. Standard deviations for the plots are also quite high. This high standard deviation, along with the fact that 13 of the 18 plots have minimum phosphorus tests below the steady state value, suggests that significant portions of these fields may have phosphorus nutrient levels below steady state values. A similar phenomenon exists for potassium levels. Given this situation, there is an incentive to raise nutrient levels in order to avoid high yield penalties associated with low soil nutrient values.

Differences in returns from the precision and information strategies average $3.28 per acre on an annual equivalent basis. There is significant variability in these returns, ranging from a low of $.74 per acre to a high of $5.36 per acre. Higher differences are associated with fields that have high phosphorus soil levels relative to the minimum phosphorus level. The average difference between the information and average scenario is $5.74. Variations across the fields are much more

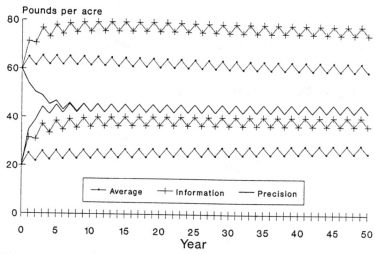

Figure 1. Soil Phosphorus Levels
Over Time.

Initial P=20, K=200 for plot 1
Initial P=60, K=300 for plot 2

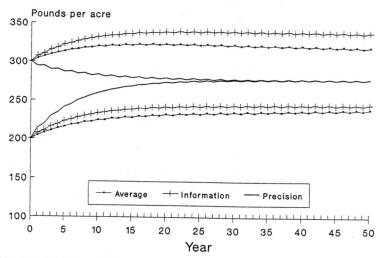

Figure 2. Soil Potassium Levels
Over Time.

Initial P=20, K=200 for plot 1
Initial P=60, K=300 for plot 2

Table 1. Expected Yearly Difference in Returns from Precision,
Information, and Average Strategies.

Plot 1's Phosphorus Level	Information Less Average		Precision Less Information		Precision Less Average
Panel A. Plot 2's Soil Phosphorus Level = 20.					
	-------------------------- $ per acre --------------------------				
10	$ .02	+	$ .42	=	$ .44
20	.03		.31		.33
30	.01		.48		.49
40	.09		.86		.95
50	.46		1.37		1.83
60	1.35		1.99		3.34
70	3.03		2.69		5.72
80	5.86		.3.39		9.26
Panel B. Plot 2's Soil Phosphorus Level = 45.					
	-------------------------- $ per acre --------------------------				
10	.87		1.37		2.24
20	.25		.93		1.17
30	.04		.56		.59
40	.02		.33		.35
50	.02		.36		.38
60	.02		.65		.66
70	.21		1.05		1.26
80	.69		1.58		2.27
Panel C. Plot 2's Soil Phosphorus Level = 60.					
	-------------------------- $ per acre --------------------------				
10	3.26		2.07		5.33
20	1.41		1.59		3.00
30	.47		1.13		1.60
40	.09		.72		.81
50	.02		.42		.43
60	.03		.31		.33
70	.01		.46		.47
80	.07		.77		.84

[1]Plot 1 and Plot 2's potassium levels are 200 and 300 pounds per acre, respectively.

Table 2. Nutrient Levels and Estimated Returns for 18 Grid Sampled Fields in Northwest Ohio

Field	Phosphorus			Potassium			Estimated Returns	
	Average	Standard Deviation	Minimum	Average	Standard Deviation	Minimum	Information Less Average	Precision Less Infor.
		pounds per acre			pounds per acre		$ per acre	
1	165.	45.	74.	299.	77.	166.	3.02	3.64
2	99.	78.	38.	310.	66.	151.	15.02	3.47
3	142.	75.	34.	313.	69.	166.	10.46	6.96
4	112.	57.	40.	295.	128.	92.	10.44	5.36
5	75.	44.	16.	292.	62.	185.	5.45	3.41
6	73.	36.	18.	386.	92.	262.	4.23	3.55
7	81.	60.	26.	395.	115.	303.	6.68	2.71
8	63.	38.	22.	298.	72.	192.	4.47	2.75
9	102.	30.	58.	516.	57.	388.	2.60	1.96
10	89.	68.	20.	385.	99.	246.	12.60	4.83
11	70.	26.	40.	457.	43.	385.	.94	1.23
12	84.	24.	32.	405.	90.	266.	1.71	2.88
13	240.	64.	140.	528.	160.	309.	3.52	2.54
14	70.	50.	26.	410.	97.	178.	6.16	3.19
15	42.	46.	12.	243.	66.	158.	2.27	1.68
16	257.	134.	84.	539.	110.	353.	10.23	4.91
17	201.	56.	96.	472.	89.	346.	3.36	3.40
18	36.	11.	16.	437.	50.	370.	.14	.74
Average	111.	52.	44.	388.	86.	251.	5.74	3.28

pronounced than in the precision less information strategies. Higher information returns are associated with higher standard deviations of a field's soil nutrient levels.

SUMMARY AND CONCLUSIONS

Our results suggest that gathering information will increase returns by an average of $5.74 per acre per year on eighteen grid sampled fields. Precision applications further increase returns by $3.28 per acre per year. A complication of costs from previous studies suggest that information gathering using grid sampling will cost about $1.00 per acre and precision applications will cost about $3.00 per acre (Lowenberg-Deboer & Swinton, 1995). Given that returns exceed costs, we conclude that both aspects of the technology have economic merit. Most of the benefits can be gained by collecting information and modifying fertilizer applications as a result of the information. Low returns from precision technologies exists in some fields. Therefore, precision fertilizer applications will not have positive benefits in all fields.

Our results also suggest that precision fertilizer applications will result in lower average soil nutrient values. To the extent that lower nutrient levels result in less effluents from land in production agriculture, precision fertilizer applications will have environmental benefits.

REFERENCES

Fixen, P.E., and H.F. Reetz, Jr. 1995. Site-specific soil test interpretation incorporating soil and farmer characteristics. *In* P.C. Robert et al (Eds) Site Specific Management for Agricultural Systems. ASA, Madison, WI.

Griffith, D. 1995. Incorporating economic analysis into on-farm GIS. *In* P.C. Robert et al (eds.) Site specific management for agricultural systems. ASA. Madison, WI.

Hibbard, J.D., D.D. White, D.A. Hertz, R.H. Hornbaker, B.J. Sherrick. 1993. Preliminary economic assessment of variable rate technology for applying P and K in corn production. Presented at the Western Agricultural Economics Association meeting, Edmonton, Alberta.

Kennedy, J. O. S. 1986. Dynamic programming: Applications to agriculture and natural resources. London: Elsevier.

Lowenberg-DeBoer, J., and S.M. Swinton. 1995. Economics of site-specific management in agronomic crops. East Lansing: Dep. Agric. Econ. Michigan State University.

McBratney, A.B., and B.M. Whelan. 1995. Continuous models of soil variation for continuous soil management. *In* P. C. Robert et al. (eds.) Site specific management for agricultural systems. ASA, Madison, WI.

National Research Council. 1993. Soil and water quality: An agenda for agriculture. Washington, D.C.: National Academy Press.

Reetz, H.F., and P.E. Fixen. 1995. Economic analysis of site-specific nutrient management systems. P.C. Robert et al. (eds.) *In* Site specific management for agricultural systems. ASA, Madison, WI.

Sawyer, J.E. 1994. Concepts of variable rate technology with considerations for fertilizer application. J. Prod. Agric. 7:195-201.

Taylor, C.R. 1983. Certainty equivalence for determination of optimal fertilizer application rates with carry-over. Western J. Agric. Econ. 3:64-67.

Wollenhaupt, N.C., and D.D. Buchholz. 1993. Profitability of farming by soils. *In* P. C. Robert et al (eds.) Proceedings of Soil Specific Crop Management: A Workshop on Research and Development Issues. ASA, CSSA, SSA. Madison, WI:

An Economic Analysis of Variable Rate Nitrogen Management

C. Snyder
T. Schroeder

Department of Agricultural Economics
Kansas State University
Manhattan, Kansas

J. Havlin
G. Kluitenberg

Department of Agronomy
Kansas State University
Manhattan, Kansas

INTRODUCTION

Precision farming is an emerging technology that prescribes inputs based on site-specific soil and crop characteristics. This technology enables one to spatially measure, monitor, and manage factors that influence crop yield, and to place inputs where they are most needed. Given that farmers face increasing input costs and a highly competitive market structure, it is important they maximize efficiency. A site-specific farming system provides the manager with increased information which may improve efficiency.

Analysis of crop production, regardless of the technology, requires determining the responsiveness of yield to inputs. The interaction between soil, seed, applied inputs, and the environment must be managed to optimize yield. One of the greatest challenges in precision farming is linking spatial yield data to agronomic practices to explain and manage variability (Lowenberg-DeBoer et al., 1994). The success of site-specific management is contingent on the reliability of quantifying the spatial distribution in crop yield as well as field and soil characteristics. Since input application decisions are based on yield goal, and field and soil characteristics, it is vital to understand the relationships between these parameters and yield.

The economic analysis of production primarily evaluates enterprise selection and input (or resource) allocation. A production function describes the relationship that transforms inputs (resources) into outputs (commodities) (Debertin, 1986). Two important issues that surface when estimating a production function are 1) what variables to include in the model, and 2) the functional form. The variables to include in a site-specific production function are those measurable factors expected to spatially affect yield. These include field characteristics, nutrient status and inputs applied (i.e. seed, fertilizer). The functional form of the model depends upon the expected nature of the relation between yield determinants and yield.

Benefits from precision agriculture are expected to be derived from savings in input costs and/or increased yields. Using site-specific agronomic data, the objective of this analysis is to illustrate an economic framework that will aid in the development and adoption of precision farming technologies. A site-specific production function was estimated to evaluate the impact of field conditions on yield variability. Results are used to conduct an economic comparison of variable and uniform nitrogen (N) management.

MATERIALS AND METHODS

A three-year study was conducted to compare variable and uniform N management on continuous corn under center pivot irrigation in central Kansas. Site 1 is 49.8 hectares and site 2 is 65.2 hectares. Additional agronomic information on this study can be found in Redulla et al. (1996, these proceedings). In the spring of 1993, organic matter, pH, and soil texture were determined from soil samples (0-15cm) collected on a 55 by 55 m grid(Table 1). Elevation was also determined in each grid cell at each site.

Spatially distributed yield goal and soil nitrate (NO_3) were used to prescribe spatially variable N fertilizer rates. In the fall of 1993, grain yield was measured with a combine equipped with a yield monitor and global positioning system (GPS). Yield goals were based on the actual yield maps obtained with minor adjustments for previous yield history. In the spring of 1994, grid-based soil samples (0-1.22m) taken as a single core within each cell, were collected to measure NO_3, and depth to clay (Table 1).

The Kansas State University Soil Testing Laboratory N recommendation model was used to determine the N application rate for the VRT cells and is described by:

$$\text{Nrec} = (24 \times \text{YG} \times 1.1) - (0.938 \times NO_3) \qquad (1)$$

where Nrec = N fertilizer recommendation (kg N/ha)
         YG = yield goal (Mg/ha)
         1.1 = textural adjustment factor for sandy soil
         NO_3 = profile soil nitrate content (kg N/ha, 0-0.6 m soil depth)

A new Nrec was calculated each spring based on yield goal and spring profile NO_3. Because of differences in yield goal and profile NO_3, differences in the VRT Nrec existed between 1994 and 1995. Compared to 1994, 29% of the VRT cells had a higher Nrec in 1995, 35% had a lower Nrec, and 36% remained the same at site 1. At site 2, 48% of the VRT cells had a higher Nrec in 1995 than 1994, 17% had a lower Nrec, and 35% remained the same. The 1995 Nrecs were higher than 1994 Nrecs because yields were higher in 1994 than in 1993 which impacted yield goal (Table 1).

Environmental conditions had a significant influence on the differences in yield between years at both sites. Average annual rainfall at both sites was 94, 43, and 76 cm in 1993, 1994 and 1995, respectively (Kansas State University Weather Data Library, 1995). In 1993, the heaviest rains were received in July while the heaviest rains in 1995 came in May. At each site, yields averaged 2 to 3 Mg/ha less in 1995 than 1994 (Table 1). Grain yields were reduced in some areas at each site

Table 1. Definition and summary statistics of variables used in the production function.

Variable	Definition	Site 1				Site 2			
		Mean	St. Dev.	Min.	Max.	Mean	St. Dev.	Min.	Max.
Y93	1993 corn yield (Mg/ha)	10.3	1.3	5.7	12.4	7.6	1.7	3.8	11.1
Y94	1994 corn yield (Mg/ha)	11.6	1.3	8.2	15.3	10.8	1.6	5.6	14.4
Y95	1995 corn yield (Mg/ha)	8.4	1.9	0.3	11.2	8.1	1.9	0.2	12.1
NAVAIL94[1]	1994 total N available (kg N/ha)	294.4	26.4	191.0	358.6	277.7	35.9	172.7	349.2
NAVAIL95[1]	1995 total N available (kg N/ha)	266.1	21.5	191.2	325.4	264.1	23.8	181.5	332.9
NAVAIL24	1994 square of NAVAIL94	78E+3	13E+3	33E+3	11E+4	70E+3	17E+3	27E+3	11E+4
NAVAIL25	1994 square of NAVAIL95	64E+3	98E+3	33E+3	95E+3	63E+3	11E+3	29E+3	99E+3
NOVER94[2]	1994 N over-applied (kg N/ha)	11.8	24.4	0.0	158.9	25.9	35.6	0.0	150.2
NOVER95[2]	1995 N over-applied (kg N/ha)	15.1	23.0	0.0	112.1	20.1	26.7	0.0	134.0
NUNDER94[3]	1994 N under-applied (kg N/ha)	-9.7	14.1	-58.5	0.0	-8.1	17.2	-83.4	0.0
NUNDER95[3]	1994 N under-applied (kg N/ha)	-3.4	6.1	-25.6	0.0	-7.2	12.3	-72.7	0.0
ELEV	Elevation (meters above sea level)	-1.3	2.7	-6.1	5.6	-4.1	2.8	-10.9	3.0
CHGELEV	Relative change in elevation (meters)	0.1	1.3	-3.2	3.0	-0.04	1.4	-3.1	5.3
OMA	Soil organic matter (%) in 0-15 cm	0.8	0.3	0.2	2.3	0.7	0.4	0.2	2.2
PHA	Soil pH index in 0-15 cm	6.5	0.5	5.0	7.2	6.1	0.6	4.9	7.4
SANDA	Soil texture, % sand in 0-15 cm	85.4	6.6	58.5	96.2	92.7	3.9	74.5	98.6
CLAYA	Soil texture, % clay in 0-15 cm	7.2	3.2	0.5	16.9	4.5	2.1	0.0	12.3
DPTHCLAY	Depth to clay layer (cm)	19.4	27.5	0.0	121.9	19.8	34.5	0.0	121.9
CLAY	Dummy variable =1 for clay layer, 0 otherwise	0.4	0.5	0.0	1.0	0.4	0.5	0.0	1.0
NO3S94	Spring 1994 soil nitrate (ppm)	6.1	1.5	4.1	9.4	6.5	1.3	3.9	9.8
NO3S95	Spring 1995 soil nitrate (ppm)	6.6	1.4	3.7	9.1	5.5	3.6	1.0	23.6

[1] N available is calculated as the sum of N applied plus NO_3 present in the soil
[2] N over-applied is positive difference between N applied and the N recommendation
[3] N under-applied is negative difference between N applied and the N recommendation

from excessive spring rainfall or weed pressure in 1995, and to a lesser extent in 1993. In terms of growing conditions, 1994 was the best year for corn.

Although Nrecs for the VRT treatments represent a continuous range of N rates, VRT Nrecs were grouped into six N rates to facilitate N application. The VRT N rates were 146, 179, 213, 236, 269, and 314 kg N/ha and were allocated to VRT cells based on the Nrec for that cell. The uniform N rate in 1994 and 1995 was a field average recommendation of 258 kg N/ha. A completely randomized block design with 12 replications was used to divide the fields into uniform and variable N treatments. Each experimental plot was composed of six 55 by 55 meter (0.30 hectare) contiguous cells. Treatment effects on grain yield were measured in the fall of 1994 with a yield monitoring combine equipped with GPS. The same process was repeated in 1995 providing two years of data.

To evaluate the influence of N and other soil and field characteristics on grain yield, a production function was developed. Given the narrow range of applied N, relatively simple functional forms were selected for the production function. This was done because N, which is known to be nonlinearly related to yield, may not be visibly nonlinear within the narrow range of recommended N application rates. A linear and quadratic function were estimated using data from both uniform and variable treatments. The quadratic model best fit the data and was selected for use in further analysis (Eq.2). A separate production function was estimated for each site and each year. The quadratic function has commonly been used in quantifying corn yield response to N (Bundy and Andraski (1995); Cerrato and Blackmer (1990); Featherstone et al. (1991); Schlegel and Havlin (1995)).

The quadratic model used was:

$$
\begin{aligned}
Y = \; & \beta_0 + \beta_1 NAVAIL + \beta_2 NAVAIL2 + \beta_3 NOVER + \beta_4 NUNDER \\
& + \beta_5 ELEV + \beta_6 CHGELEV + \beta_7 OMA + \beta_8 PHA + \beta_9 PHA2 + \beta_{10} SANDA \qquad (2) \\
& + \beta_{11} CLAYA + \beta_{12} CLAYA2 + \beta_{13} DPTHCLAY + \beta_{14} CLAY
\end{aligned}
$$

Definitions and summary statistics of the variables in equation 2 are presented in Table 1.

The relationship of most interest in the evaluation of VRT N management is the interaction of N and yield. Total N available (NAVAIL) was calculated as the sum of N applied and NO_3 present in the soil. The difference between N applied and the Nrec provides an absolute amount of over or under application of N for each cell. The magnitude of misapplication on the VRT cells is very slight (17 kg N/ha maximum) since the exact Nrec was not applied when grouping the Nrec into six variable N treatments. Variation in the magnitude of misapplication on the uniform cells is greater due to differences in the field average recommendation and the actual Nrec for each uniform treatment cell. A differentiation was made between over and under application of N because using less N results in a savings in fertilizer expense but may also imply production loss and a reduction in revenue. In turn, over-application of N may imply an input expense coupled with reduced yield and a loss in revenue. Over-application of N may also result in an increased potential for N contamination of groundwater.

Therefore, misapplication of N was split between over and under-application to obtain a sensitivity for how yield responds with respect to misapplication of N.

RESULTS AND DISCUSSION

At site 1, NAVAIL was positively correlated with yield and statistically significant at the 5% level in 1994 (Table 2). However, in 1995, it was negatively correlated and not significant. Similarly, NAVAIL was positively correlated with yield at site 2 in 1994 and was statistically significant at the 1% level, but was not significant in 1995. The NAVAIL2 term was statistically significant in 1994 but not in 1995 at both sites. The N over-applied (NOVER) term was highly significant and negatively correlated to yield at each site and for each year. The magnitude of the relationship between over-application and yield was similar across years and sites. Using site 1 results from 1994, the model indicates that for every kilogram of N over-applied, yield fell by 0.016 Mg/ha (Table 2). Nitrogen under-applied was negatively related to yield with the exception of site 2 in 1995. Site 1 results indicate that for every kilogram of N under-applied, yield decreased by 0.00457 Mg/ha in 1994 and 0.044 Mg/ha in 1995.

At site 1, elevation and yield were negatively correlated, while at site 2, elevation and yield were positively correlated (Table 2). A change in elevation variable (CHGELEV) was formulated as the relative change in elevation with respect to the neighboring cells. A positive value of CHGELEV indicates that elevation in a cell is higher relative to the surrounding cells. A positive coefficient on CHGELEV indicates a positive correlation of yield to relatively higher areas in the field. At site 1, yield was greater on the relatively higher cells in each year while at site 2, yield was lower on the higher cells in 1994 yet higher on these same cells in 1995. This variable may be capturing the interrelationship between environment and slope.

This study also examined the net returns to VRT over uniform N management. Net return is defined as the additional return from VRT less additional VRT costs. Additional return is the change in VRT income derived from a change in yield and changes in fertilizer cost as compared to uniform N application. Additional returns are calculated for each cell and totaled for the field. The analysis of net return was based on variable or uniform N applied over the entire field. To estimate the effects of VRT and uniform N application on yield for the entire field, the prescribed N application rates along with the parameter estimates from the production function (Table 2) were used to determine expected yields for each cell and each site (1994-95). From the expected total N application and predicted yield response, additional returns to VRT N management can be determined. The difference between expected returns from VRT management and additional costs of VRT management indicate the profitability of VRT N management.

Table 2. Corn yield production function coefficient estimates, by site and year.

Variable	Site 1		Site 2	
	1994	1995	1994	1995
Intercept	8.478	-13.282	4.069	11.62
NAVAIL	0.078**	-0.039	0.073***	0.045
NAVAIL2	-0.151E-03**	0.117E-03	-0.135E-03**	-0.378E-04
NOVER	-0.016***	-0.016***	-0.024***	-0.015***
NUNDER	-0.457E-02	-0.044**	-0.985E-02*	0.032***
ELEV	-0.122***	-0.025	0.249***	0.074
CHGELEV	0.061	0.145	-0.483***	0.918E-02
OMA	-0.676E-02	-1.828***	-0.546**	0.166
PHA	-1.779	8.378	3.310	0.569
PHA2	0.125	-0.653	-0.288	-0.049
SANDA	-0.019	0.451E-03	-0.117***	-0.155***
CLAYA	0.010	-0.089	-0.140	-0.060
CLAYA2	-0.380E-02	0.197E-02	0.256E-02	0.403E-02
DPTHCLAY	-0.272E-02	0.538E-02	0.176E-02	0.653E-02
CLAY	0.280	0.426	0.102	0.104
R^2	0.16	0.24	0.49	0.34
Obs.	161	161	162	162

*** =significant at 0.01 level, ** =significant at 0.05 level

If the entire field is farmed using VRT N management, N applied is determined from the Nrec model (Eq.1) and added to the sampled NO_3 content (Table 1) in the soil at planting to determine a predicted total N available. These new N variables, NREC94 and NREC95 (Table 3), take the place of NAVAIL94 and NAVAIL95 in the production function matrix (Eq.2) to predict yield because NAVAIL for each treatment was based on one half of the field. Over or under-application variables are zero with VRT N management since N is applied at the precise Nrec levels in each cell.   If the entire field is farmed using uniform N management, N is applied at a uniform rate across the field based on a field average yield goal and soil NO_3 content. Predicted N available under uniform N management is the uniform Nrec plus the sampled NO_3 content in the soil at planting. These variables are defined in the model as NUNI94 and NUNI95 (Table 3). The amount of N applied remains constant across the field so any variation in NUNI94 and NUNI95 reflects changes in soil profile NO_3. Additionally, over or under application of N occurs in many cells with uniform N management. The difference between the uniform rate applied and the Nrec for each cell represents the amount of over or under-application of N for each cell. NOVRUNI4 and NOVRUNI5 (Table 3) summarize the predicted over-application of N in 1994 and 1995, respectively. Predicted over-application at site 1 averaged 18.7 and 21.9 kg N/ha in 1994 and 1995. Predicted over-application at site 2 averaged 40.1 and 26.1 kg N/ha in 1994 and 1995. Maximum predictions of over-application were 112 kg

N/ha or more at each site. NUNDUNI4 and NUNDUNI5 (Table 3) summarize the predicted under-application of N in 1994 and 1995, respectively. Predicted under-application averaged 12.9 kg N/ha in 1994 and 4.1 kg N/ha in 1995 at site 1. Likewise, predicted under-application at site 2 averaged 11.0 kg N/ha in 1994 and 7.4 kg N/ha in 1995. These new uniform N variables, NUNI94, NOVRUNI4, NUNDUNI4, NUNI95, NOVRUNI5, and NUNDUNI5 take the place of NAVAIL94, NOVER94, NUNDER94, NAVAIL95, NOVER95, and NUNDER95 in the production function matrix (Eq.2) to predict yield.

Comparing the two N management treatments, the predicted amount of total N applied was always less under VRT N management than uniform N management (Table 4). In 1994, 3% less N was applied at site 1 while 13% less N was applied at site 2. In 1995, 8% less N was applied at both sites. As a result, a savings in N expense occurred at both sites when VRT N management was applied.

Net returns per hectare were calculated as the difference between VRT and uniform N management (Table 5). Changes in returns were based on predicted differences in corn yield and N use under each N management treatment. A range of corn and N prices were used. Results from site 1 indicate positive net returns to VRT N management over uniform N management. Given $108/Mg corn price and $0.64/kg N, average returns at site 1 were $27.15/ha in 1994 and $64.55/ha in 1995. At site 2, opposite results were obtained between 1994 and 1995. At the same price levels, site 2 returns averaged $96.32/ha in 1994, but ($42.42/ha) in 1995. These results may reflect the difference in growing conditions between years.

The additional costs of VRT consist of additional N application charges, soil sampling, lab analysis, labor, and data management costs (Table 6). These reported costs are based on surveys and estimates from precision farming users. Given an estimated additional cost of $42.76/ha, VRT N management was not an economically viable option at site 1 in 1994, but it was able to recover costs in 1995. At site 2 in 1994, the estimated net return from VRT N management more than covered the additional cost of VRT, but a negative net return occurred in 1995. The differences in results across years and across sites, emphasize the importance of evaluating economic consequences of VRT N management on a field by field basis. Using only two years of data, VRT N management was a viable option on each field during one of the years analyzed. Coupled with future reductions in technology costs associated with precision agriculture, VRT N management may be a profitable investment over traditional N management but not necessarily at all times or in all locations.

CONCLUSION

The ability of precision farming technology to monitor and manage factors that cause yield variability provides farmers with a higher level of management information and control. Variable N management is expected to reduce the number or areas in the field that are either over or under applied.

Table 3. Definition and summary statistics of predicted N management variables (kg N/ha).

Variable	Definition	Site 1				Site 2			
		Mean	St. Dev.	Min.	Max.	Mean	St. Dev.	Min.	Max.
Site 1:									
NREC94	1994 N available under VRT management	292.3	39.9	143.7	354.3	260.0	56.7	131.3	376.9
NREC95	1995 N available under VRT management	254.4	29.5	160.0	327.1	251.2	36.3	156.0	341.9
NUNI94	1994 N available under uniform management	298.2	3.3	293.6	305.6	289.1	3.2	282.8	294.7
NUNI95	1995 N available under uniform management	272.1	2.9	266.3	279.6	269.8	8.1	259.9	310.5
NOVRUNI4	1994 N over-applied on uniform cells	18.7	29.8	0.0	158.9	40.1	41.7	0.0	154.6
NOVRUNI5	1995 N over-applied on uniform cells	21.9	25.9	0.0	112.1	26.1	27.9	0.0	111.6
NUNDUNI4	1994 N under-applied on uniform cells	-12.9	16.8	-58.5	0.0	-11.0	20.1	-83.4	0.0
NUNDUNI5	1995 N under-applied on uniform cells	-4.1	8.9	-58.0	0.0	-7.4	14.5	-72.7	0.0

Table 4. Total N applied (kg) under different N management scenarios.

Location	Method of N management		Change in total VRT N applied over uniform application	
	VRT	Uniform		
Site 1:				
1994:	16,496	16,916	(420)	-3%
1995:	15,720	16,916	(1,196)	-8%
Site 2:				
1994:	15,086	17,021	(1,935)	-13%
1995:	15,777	17,021	(1,244)	-8%

Results from this study indicate that less total N fertilizer was used with VRT than uniform N management. The net return to VRT N management was positive at least once for each field during the two years studied. The results indicated potential for profitable use of precision N management. The economic benefit of precision farming depends on identifying places in a field where additional input use will increase yield greater than the added costs, and in identifying places where reduced input use will reduce costs while maintaining yield potential.

Table 5. Net returns ($/ha) from VRT N management over uniform N management.

Cost of N ($/kg)		Price of Corn ($/Mg)		
		$96.00	$108.00	$120.00
Site 1:				
0.55	1994:	$24.01	$26.46	$28.93
	1995:	$56.97	$62.48	$67.98
0.64	1994:	$24.68	$27.15	$29.62
	1995:	$59.04	$64.55	$70.06
0.73	1994:	$25.37	$27.84	$30.29
	1995:	$61.12	$66.63	$72.13
Site 2:				
0.55	1994:	$77.67	$85.62	$93.58
	1995:	($38.12)	($44.59)	($51.06)
0.64	1994:	$88.36	$96.32	$104.27
	1995:	($35.94)	($42.42)	($48.89)
0.73	1994:	$91.77	$99.73	$107.68
	1995:	($33.77)	($40.22)	($46.69)

Table 6. Additional per hectare costs associated with VRT management.

Additional Spreader Charge	$7.41
Soil Sampling Costs	$24.70
Labor ($12/hr @ 6 samples/hr)	$2.00
Data Management($6.18/ha maps + $2.47/ha record keeping)	$8.65
Total Additional Cost of VRT Practices (per hectare):	$42.76

REFERENCES

Bundy, L.G. and T.W. Andraski. 1995. Soil yield potential effects on Performance of soil nitrate tests. J. Produc. Agric. 8:561-567.

Cerrato, M.E., and A.M. Blackmer. 1990. Comparison of models for describing corn yield response to nitrogen fertilizer. Agronomy Journal, 82:138-143.

Debertin, D. L. 1986. Agricultural production economics. Macmillian Publishing Co., New York, New York.

Featherstone, A.M., J.J. Fletcher, R.F. Dale, and H.R. Sinclair. 1991. Comparison of net returns under alternative tillage systems Considering spatial weather variability. J. Produc. Agric. 4:166-173.

Kansas State University Soil Testing Lab. 1994. Unpublished research. 2308 Throckmorton Hall, Manhattan, Kansas 66506.

Kansas State University Weather Data Library. 1995. Personal communication February 6, 1995 with Mary Knapp. 212 Umberger Hall, Manhattan, Kansas 66506.

Lowenberg-DeBoer, J., S. Hawkins, and R. Nielsen. 1994. Economics of precision farming. Unpublished paper, Purdue University.

Schlegel, A.J., and J.L. Havlin. 1995. Corn response to long-term nitrogen and phosphorus fertilization. J. Produc. Agric., 8:181-185.

Weed Managing Model for Patch Spraying in Cereal

T. Heisel
S. Christensen
A.M. Walter

Danish Institute of Plant and Soil Science
Dept. of Weed Control and Pesticide Ecology
Flakkebjerg, DK-4200 Slagelse

ABSTRACT

A weed managing model for patch spraying is described. The model includes competition between crop and weed mixtures, herbicide performance and economic optimization of herbicide dose. Weeds were mapped in an 8 ha winter barley field. The results showed savings from 66 to 75% with patch spraying compared to whole-field spraying using manufacturers dose recommendation.

BACKGROUND

Patch spraying is an option for sustainable farming with positive effects on environment and economy. With a large within-field variation in weed occurrence and density patch spraying may lead to herbicide savings and diminishing herbicidal loadings to the environment. Several field studies have shown that most weed species are aggregated and often occur in patches within arable fields (Marshall, 1988; Wilson & Brain, 1990; Mortensen et al., 1992). Spatial dispersion of weed species influences the economic results of patch spraying, as aggregation reduces the impact of weed competition (Brain & Cousens, 1990). Herbicide efficacy varies among weed species, which must be taken into account with a patch spraying strategy.

Two concepts of weed patch spraying have been suggested. Weed monitoring and spraying are either carried out simultaneously (the real-time concept) (Felton et al., 1991) or weed monitoring precedes the spraying (the mapping concept) (Nordbo et al., 1995). With the real-time concept no mapping or positioning is required and only one field run is needed. The real-time concept, however, is not yet developed to manage weeds in cereals. Weed management and decision making on herbicide choice may require knowledge about the whole-field weed population, to insure a satisfactory level of control. This knowledge does not exist in real-time. With the mapping concept information from previous years (e.g. crop yield, weed distribution or soil properties) can be utilized in a weed managing model in addition to competition and economic parameters. Further, weed maps and treatment maps may be useful documents for prescription farming.

Using the mapping concept, a decision support system is necessary for optimizing weed control. Wiles et al. (1992) and Audsley (1993) used a threshold concept to simulate economic benefit of patch spraying with a single species population. Johnson et al. (1995) used a mean threshold value for varying mixtures of broad leaved and grass weed species to simulate the spatial variation in the need for weed control. Weed populations rarely consist of single weed species or a uniform mixture of species with a constant threshold.

The ultimate goal of patch spraying is to select an efficient herbicide and an economically optimum dose for each part of the field, by including the yield variations in the calculations. This goal requires a position related decision support system and advanced sprayers; sprayers that operate with several herbicides and a single nozzle dose system (Paice et al., 1995). So far these sprayers are not available commercially. However, conventional sprayers could be used to vary the dose of one herbicide (mixture) suitable for most of the weeds present.

In Denmark a plan of action from the Ministry of Environment has set a goal by 1997 (Thonke, 1988). The target is a 50% reduction of the amount of pesticide use and a 50% reduction in a so-called 'treatment index' (TI), compared to the average level used from 1981 to 1985. A TI of 1 is defined as one treatment of a field with a 'normal dose' of pesticide. A normal dose equals one field run with the manufacturers recommended dose. This means that treatment of a field with 70% of a normal dose implies a TI of 0.7. The mean dose of a patch spraying strategy would equal the TI.

This paper describes a desk study, based on a field survey. The study uses a decision model to optimise the herbicide dose for a mixture of broad leaved species with different competitive abilities and dose-response parameters. The model takes economic consequences of market prices for herbicide and crop, and position specific crop yield into account, and gives a continuous herbicide dose as output. Classification of the dose into four or five is analysed together with the sensitiveness of the model to herbicide price.

MATERIALS AND METHODS

The object model in Fig. 1 describes the concept and modules of studying and describing a weed population in an 8 ha winter barley (*Hordeum vulgare* L.) field at Risø National Laboratory, Roskilde in 1994.

The 'Field survey' object provides the input data to the decision model. 'Field survey' consists of a weed survey at 231 equidistant points in a 20 m x 20 m grid. No herbicide application was performed before weed surveying. At each location species composition and species density were measured within a 0.25 m² circular frame. In 1993 crop yield was spatially measured in winter barley in the same field using GPS and a Massey Ferguson Yield-metre.

The 'Point model' object consists of three sub-objects. The first sub-object 'Competition model' is based on the model derived by Berti & Zanin (1994). It converts species competitiveness and weed density into density equivalents showing the economic impact of the actual species. The actual crop yield level for each sampling point is used. The second sub-object 'Herbicide choice' finds a suitable herbicide for the mean species density by running the Danish advisory system PC-Plant Protection (Rydahl & Thonke, 1993). Herbicide dose is optimized economically in the sub-object 'Herbicide dose'. Dose-response parameter values from PC-Plant Protection are used together with the density equivalents of the species as weights. The differential net return of increasing herbicide doses is maximised in order to find the dose with the maximum net return. The 'Point model'

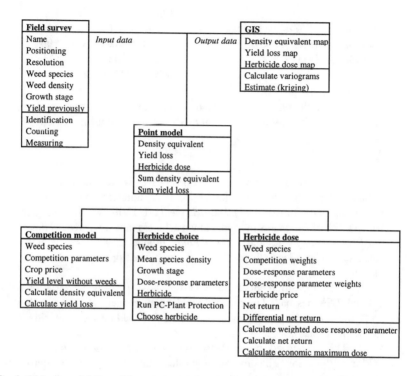

Fig. 1. Object model describing classes, attributes and operations (Rumbaugh et al., 1991) used for simulations and mapping weeds.

object calculates the total density equivalent, the total potential yield loss and the economic optimal herbicide dose in each point, considering weed density, weed species competitiveness, crop yield, crop and herbicide price and dose-response parameters.

The 'GIS' object converts the calculated point data of potential yield loss and herbicide dose into a new set of data with estimated yield loss and herbicide dose in all regions (cells) using the geostatistical procedure block kriging (Hawkins & Cressie, 1984). The field was divided into 2448 cells with a 6 m x 6 m cell size to fit a conventional 12/18/24 m boom sprayer with 2/3/4 boom sections and an on/off system with a dual independent nozzle system (Paice et al., 1995). The calculated dose (z) is a continuous variable. Using a common sprayer with two tanks and two sets of nozzles (one tank with 1/3 of the maximum calculated dose (Max) and the other with 2/3 Max) each block was divided into four dose categories. To investigate the gain of a denser classification, five dose categories were investigated as well. The doses were classified by:

Four dose classes: *Five dose classes:*

0 Max: $z = 0$	0 Max: $z = 0$
1/3 Max: $0 < z < 1/3$Max	1/4 Max: $0 < z < 1/4$Max
2/3 Max: $1/3$Max $< z < 2/3$Max	½ Max: $1/4$Max $< z < 1/2$Max
1/1 Max: $2/3$Max $< z$	3/4 Max: $1/2$Max $< z < 3/4$Max
	1/1 Max: $3/4$Max $< z$

where Max is the maximum calculated dose. A sensitivity analysis of herbicide price was done by running the 'Point model' with a 4 times increased price.

RESULTS

Mean and max number of occurring weeds are shown in Table 1. The number of points with densities above 0 is indicated together with Lloyds' patchiness index (Lloyd, 1967). *Stellaria media* (L.) Vill. was the dominant species with a mean of 244 plants m^{-2} and a low patchiness index, whereas *Galium aparine* L., *Elymus repens* L. and *Chenopodium album* L. were sparse with a high patchiness index. Generally the field had a high weed pressure with a mean of 328 plants m^{-2}.

Yield measurements from 1993 (not shown) revealed a variation between 2.4 and 7.5 tones ha^{-1}. The reduced yields were concentrated to a hollow at the north-west end whereas the major part of the field had only small yield variations (5 - 7.5 tones ha^{-1}).

The mean weed species densities (Table 1) were used as input for running PC-Plant Protection. The program suggested a tank mixture of tribenuron-methyl (as Express, 500 g a.i. kg^{-1}, Du Pont) and fluroxypyr (as Starane 180, 180 g a.i. l^{-1}, DowElanco) as the most suitable herbicide. Normal dose (N) of the mixture is 15 g Express and 0.6 l Starane 180 per hectare. The parameters of this herbicide mixture were used in the algorithm for optimizing herbicide dose.

Output visualized through the 'GIS'-object is shown in Figs. 2-5. Predicted yield loss (Fig. 2) shows that there is a strong need for control in a hollow in the north-west end and in the larger south-east area (dark areas). A high *Tripleurospermum inidorum* Scultz Bip. infestation cause the potential yield loss in the hollow whereas *S. media* is the primary cause in the south-east area. Predicted yield loss varies from 100 to 1600 kg ha^{-1}. The calculated dose (Fig. 3) reflects the predicted yield loss. Doses from 0.1 to 0.44 of the normal dose (N) were required, with the highest doses in the areas with high predicted yield loss (compare Figs. 2 and 3). Dose partitioning in four classes, after rescaling from 0 to 1/1 Max (equals 0.44 N), is mapped in Fig. 4. The acreage of the four doses 0, 0.15, 0.29 and 0.44 N is 0, 0.04, 5.89 and 2.07 ha, respectively.

Table 1. Statistical summary of weed surveys in a field with winter barley at Risø. The total number of sampling points was 231. Patchiness index according to Lloyd (1967).

	Points with density> 0	Plants m⁻² Max	Mean	Patchiness index
Viola arvensis Murray	31	40	1.2	10
Elymus repens L.	3	100	0.6	126
Galium aparine L.	5	8	0.1	23
Poa annua L.	81	800	18.9	15
Stellaria media (L.) Vill	224	1680	243.8	3
Chenopodium album L.	7	720	3.4	198
Capsella bursa-pastoris (L.) Medicus	35	48	1.9	10
Tripleurospermum inidorum Schultz Bip.	57	360	8.1	26
Papaver rhoeas L.	70	360	4.6	28
Myosotis arvensis (L.) Hill	21	36	0.9	16
Silene noctiflora L.	31	60	1.1	16
Brassica napus L.	118	76	5.1	3
Lamium purpureum L.	162	284	30.8	4
Veronica spp.	90	116	7.6	6
All weed species	231	1700	328.2	2

Table 2 summarizes the treatment index (TI) and the cost of the different treatment strategies for the field. PC-Plant Protection reduces TI by 27%. Patch spraying with four doses reduces TI by 66%, whereas five doses give a 68% reduction. Using a continous dose on the field reduces TI by 75%. Weed control becomes ineconomic in a larger part of the field when the price of the herbicide is increased. With a four time higher herbicide price TI on the field was reduced by 88%, i.e. the price increase resulted in a 50% reduction of TI (Table 2).

Table 2. Treatment index (TI) and cost (US$/ha) for five different treatment strategies. Price sensitivity analysis under dotted line. Patch treatment on the basis of output from object model 'GIS' (Fig. 1). 1/1Max equals 0.44 normal dose (N) and TI. 1 US$ = 7 DKR.

Treatment strategy	TI (field)	Expense (US$/ha)
Normal (1/1 N)	1.000	33.3
PC-Plant Protection	0.735	244
Patch, classified (0, 1/3, 2/3, 1/1Max)	0.338	11.3
Patch, classified (0, 1/4, 1/2, 3/4, 1/1Max)	0.320	10.7
Patch, continous dose	0.263	8.7
Patch, continous dose, 4 × herbicide price increase	0.128	17.0

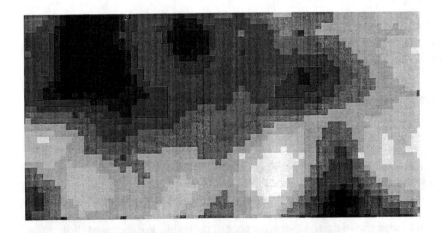

0   213   427   640   853   1067   1280   1493 (kg/ha)   N↓

Fig. 2. Predicted yield loss (kg grain ha^{-1}) caused by broad leaved weeds. Block size is 6 m × 6 m.

0   0.059 0.118 0.178 0.237 0.296 0.355 0.414 (TI)   N↓

Fig. 3. Treatment index (TI) map of tribenuron-methyl + fluroxypyr.

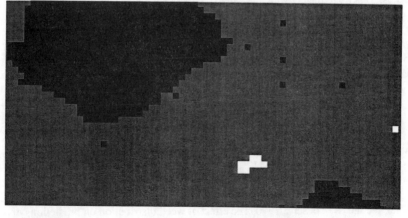

0.15 0.30 0.44 (TI)   93.60352   N↓

Fig. 4. Treatment index (TI) map derived from Fig. 3 through classification in 0, 1/3, 2/3 and 1/1 of the maximum treatment index.

DISCUSSION

The decision model in this paper integrates the effect of all weed species in all sample points. This procedure may reduce errors from identification and counting, and it also seems to reduce the relative impact of each species. Many attempts have been made to map weed species in arable fields (e.g. Heisel *et al.*, 1996) with the best estimator kriging. If the particular species is frequent, a usable map can be obtained, whereas rare species are difficult to map. Thus, adding kriged weed species maps to build a treatment map can lead to problems with rare competitive species. This problem is overcome with introduction of this point model decision system.

The model used for creating the treatment map is parameter sensitive. Competitiveness and dose-response parameters for each weed species are based on experience build up through years of dose-response surveys (Rydahl & Thonke, 1993). Attention should be given to determination of these parameters.

When introducing a patch spraying model and applying herbicide in four dose classes, the treatment index (TI) could be reduced by 66% compared to manufacturers recommendation (normal dose). Little was gained by using five classes, whereas a reduction of TI by 75% was obtained when using an infinitely variable system. This leads to the conclusion, that the technical innovations should concentrate on developing a simple 4 dose class spraying system to begin with. When such a system is operative, innovations should continue in developing an infinitely variable system. There is a potential economic and environmental gain in patch spraying which so far has not been utilized.

Pesticide price taxes could be a means to reduce pesticide pressure on the environment. By running the model again with a four times higher herbicide price, generally all doses were halved, which is reflected in the treatment index (Table 2), too. All other parameters were unchanged. Obviously, the results are field specific, i.e., dependent on present weed types, crop and weed competition, etc. However, it shows that herbicide price affects the benefit of patch spraying significantly.

Patch spraying reduces the treatment index (TI) from 66 to 75% compared to normal dose recommendation. Theoretically this implies, that nationwide use of patch spraying satisfies the Danish Plan of action from the Ministry of Environment mentioned earlier.

REFERENCES

Audsley, E. 1993. Operational Research Analysis of Patch Spraying. Crop Protection 12, pp 111-119.
Berti, A., and G. Zanin. 1994. Density equivalent: A method for forcasting yield loss caused by mixed weed populations. Weed Research 34, pp 326-333.
Brain, P., and R. Cousens. 1990. The effect of weed distribution on predictions of yield loss. Journal of Applied Ecology 27, pp 735-742.
Felton, W.L., A.F. Doss, P.G. Nash, and K.R. McCloy. 1991. A microprocessor controlled technology to selectively spot spray weeds. Procedings ASEA Symposium 1991: Automated agriculture for the 21st century, pp 427-432.

Hawkins, D.M., and N. Cressie. 1984. Robust kriging; a proposal. Mathematical Biology 16, pp 3-18.

Heisel, T, C. Andreasen, and A.K. Ersbøll. 1996. Annual weed densities can be mapped with kriging. Weed Research 36(4):325-337.

Johnson, G.A., D.A. Mortensen, and A.R. Martin. 1995. A simulation of herbicide use based on weed spatial distribution. Weed Research 35, pp 197-205.

Lloyd, M. 1967. Mean crowding. Journal of Animal Ecology 36, pp 1-30.

Marshall, E.J.P. 1988. Distribution patterns of plants associated with arable field edges. Journal of Applied Ecology 26, pp 247-257.

Mortensen, D.A., G.A. Johnson, and L.J. Young. 1992. Weed distribution in agricultural fields. P. 113-213. *In* Proceedings of Soil Specific Crop Management, ASA, CSSA, SSA, Madison, WI.

Nordbo, E, S. Christensen, and K. Kristensen K. 1995. Weed patch management. Zeitschrift für Pflanzenkrankheiten und Pflanzenschutz 102, pp 75-85.

Paice, M.E.R, P.C.H. Miller, and J.D. Bodle. 1995. An experimental sprayer for the spatially selective application of herbicides. J. Agric. Eng. Res. 60, pp 107-116.

Rumbaugh, J, M. Blaha, W. Premerlani, F. Eddy, and W. Lorensen. 1991. Object-oriented modelling and design. Prentice-Hall International, Inc., New Jersey, 500 pp.

Rydahl, P, and K.E. Thonke. 1993. PC-Plant Protection: optimizing chemical weed control. Bulletin OEPP/EPPO Bulletin 23, pp 589-594.

Thonke, K.E. 1988. Research on pesticide use in Denmark to meet political needs. Aspects of Applied Biology 18, 327-329.

Wiles, L.J., G.G. Wilkerson, and H.J. Gold. 1992. Value of Information About Weed Distribution for Improving Postemergence Control Decisions. Crop Protection 11, pp 547-554.

Wilson, B.J., and P. Brain. 1990. Long-term stability of distribution of *Alopecurus myosuroides* Huds. within cereal fields. Weed Research 31, 367-377.

Returns to Farmer Investments in Precision Agriculture Equipment and Services

Scott M. Swinton
Mubariq Ahmad

Department of Agricultural Economics
Michigan State University
East Lansing, MI

ABSTRACT

Michigan farmers interviewed report many unexpected costs and benefits from site-specific crop management (SSCM). Existing profitability analyses based on agronomic data are shown to be very incomplete. Guidelines are offered for how to draw inferences from incomplete data and how future economic research can take fuller account of costs and benefits of SSCM.

INTRODUCTION

To taste the promise of site-specific crop management (SSCM), farmers are called upon to make costly investments in equipment and services. These investments may include 1) data gathering using yield monitors, soil sensors, positioning systems, and soil or pest sampling equipment, 2) data processing and output using computers and printers, and 3) computer-driven controllers applying variable rate technology (VRT) fertilizer applicators or seed planters.

Evaluation of these investments calls for thorough cost and benefit accounting. This may be done at the level either of private profitability or of public welfare. Taking a primarily private perspective in this paper, our specific objectives are:

1. To identify what farmers have found affects profitability of SSCM, based on focus group interviews conducted in Michigan,
2. To link those findings to principles of investment analysis,
3. To evaluate what can and cannot be inferred about returns to investments in SSCM from experimental agronomic data,
4. To highlight areas for further economic research into SSCM net returns.

WHAT FARMERS WANT TO KNOW

A reasonable place to begin examining returns to SSCM is by asking farmers and agribusiness representatives both what they have encountered and what they expect to encounter. Six focus group meetings were held across southern Michigan between January and March, 1996 (for details, see Swinton et al., 1996). Three were held with farmers having some experience with SSCM, two with farmers interested but inexperienced with SSCM, and one with agribusiness representatives who provided or expected soon to provide SSCM services. The meetings ranged in size from four to seven participants. Participants were recruited by field crops extension agents; they operated 380 to 3,000 acres, and virtually all were cash crop farmers.

The focus group discussions were loosely organized around a set of discussion points which were handed out to participants at the beginning of each meeting. The points relevant to assessing profitability of SSCM included:

◆   What unexpected costs have you encountered in implementing SSCM?
◆   What problems have you encountered in implementing SSCM?
◆   What actual benefits have you realized from this technology?
◆   What information would you have desired if you were again faced with the decision to utilize or not utilize this technology?

Key Results from Focus Group Interviews

Focus group participants experienced with SSCM were chiefly experienced with yield monitors. The only one using VRT had a year of experience with spatial nitrogen sensing and application.

Costs and Problems with SSCM

When asked about cash costs, farmer participants pointed to the high cost of grid soil sampling ($4.75 to $10.00/acre) above all else. Their other major concern was the unforeseen cost of additional equipment (e.g., more powerful computers, monitors, printers) and software upgrades needed to make the new components run. For example, growers who had invested $3,000 to $9,000 in yield monitoring systems found that if they wished to analyze their own data, it was not enough to buy a $1,000 mapping program; they typically found it necessary also to purchase a computer with more random access memory, a high-quality color monitor, and a color printer. The cost of learning to use the new technology was mentioned by both producers and agribusiness persons.

Some growers and most agribusinesses were concerned with the cost of VRT controller technology. Representatives of two agribusinesses noted that their companies could not justify providing VRT fertilizer spreading with purpose-manufactured applicators like AgChem's Soilection™ spreader until they could be assured of spreading over some minimum area, like 10,000 acres.

Apart from cash costs, the non-cash costs associated with delays and repair problems elicited more discussion than any other topic. Incompatibility of SSCM system components was the lead complaint. Equipment obsolescence was a related problem, since obsolescence often results from incompatibility with new equipment or software. Equipment unreliability was also encountered almost universally; several farmers told anecdotes about early yield monitors that did not function properly or required excessive calibration to work right. As a result, some farmers perceived SSCM to be a risky technology in the sense that breakdowns were likely. Even when equipment operated properly, there were problems with loss of the GPS signal and poor calibration of yield monitors at some combine operating speeds.

Linked to the reliability problem were farmers' own mistakes. Several cited the significant learning time required to make effective use of SSCM equipment and software--and these were nearly all cash crop farmers who had committed or planned to commit a portion of their off-season to learning about SSCM. Farmers complained that reliability and learning problems were aggravated by lack of local vendor support. The weakness of vendor support was indirectly acknowledged by agribusiness representatives who noted the difficulty in becoming proficient at using the new SSCM technologies in the face of rapid technological change and system incompatibilities due to many different manufacturers.

Major Benefits Encountered or Expected
The benefits of SSCM were expected more often than realized. Participants chiefly expected increased yield benefits from a) seed population and variety control, b) better lime management to avert herbicide injury, notably for soybeans), and c) better control of nitrogen and phosphorus fertilizers. Several also cited the potential for cost savings on inputs not needed. However, since only one farmer was currently using VRT, few of these input control benefits had been realized. Several farmers felt yield monitor data would pay off in negotiations with landlords by giving the grower a clearer sense of how much he could afford to bid to rent a piece of land. Two others planned to use yield monitors to evaluate experiments with crop varieties and lime applications. Nonetheless, many farmers felt that a major benefit was simply having better data for decision making which would pay off in unexpected ways.

Potential benefits that have been cited elsewhere that were not mentioned by farmers included yield risk management and improved water quality from reduced leaching and runoff of excess nutrients. (By contrast, the agribusiness representatives were acutely aware of these environmental benefits.)

Most of the farmers who owned yield monitors were willing to wait for three to five years before using their yield information to manage inputs. While many hoped the yield monitor might pay off rapidly from a one-time discovery such as the need to add drainage tile to a field, a few were skeptical that yield monitors could contribute any direct value beyond an improved awareness of productivity patterns.

Role of Expected Profitability in Affecting Adoption of SSCM

The focus group interviews revealed a fairly widespread willingness to experiment with SSCM so long as the cost is modest. Several interested farmers were willing to buy a yield monitor for $3,000 - $5,000, but were reluctant to invest in grid soil sampling or VRT input application until they could see evidence of clearer benefits. It appeared that compared with the early-adopter group, a larger group of farmers were still waiting for evidence of SSCM profitability before they would do more than low-cost experimentation with it.

The interviews highlighted several key factors driving the profitability of SSCM practices adopted so far. First, benefits are ill-defined and typically not realized during the first year or two. This makes cost containment especially important. Second, costs are high. Even yield monitoring, which is viewed as relatively low-cost, typically triggers unexpected costs including added equipment and software needs, hardware and software incompatibilities, delays and learning time, and lack of repair support. Nutrient mapping and VRT are much more costly.

These comments suggest that SSCM profitability analyses should focus on 1) valuing benefits of SSCM, 2) identifying optimum grid sampling density, and 3) quantifying hidden costs (due to learning time, equipment calibration, equipment and software incompatibilities, and delays resulting from poor access to qualified technical support). Benefits valuation should be tied to specific investments so that SSCM benefits can be disaggregated into those resulting from yield mapping, nutrient mapping, or VRT, or some combination of these.

Given these farmer experiences with SSCM costs and benefits, and given the growing body of agronomic research data, what conclusions can we draw about profitability of investments in SSCM?

ANALYZING INVESTMENTS IN SITE-SPECIFIC FARMING

Investment analyses are appropriate for studying profitability when benefits and costs are spread over time. Typical examples from farming include the purchase of durable equipment that will yield services over many years. In instances where time is not an important factor, partial budgets or enterprise budgets can be adequate. For example, a partial budget would be appropriate to study profitability of nitrogen management using a pre-sidedress nitrate test, since the benefits are expected to occur chiefly in the same season.

Standard investment analysis using capital budgeting is done by projecting future cash flows from changes in costs and revenues due to the investment of interest. Net present value (NPV) is the most common measure used in investment analysis.[1]

Unexpected Costs and Benefits

The challenge is to account fully for all costs and benefits. In reviewing published studies on profitability of SSCM, Lowenberg-DeBoer and Swinton (1996) found that none accounted fully for the hidden costs identified by the Michigan farmer focus groups, namely learning, incompatibilities, early

---

[1]The equation below illustrates how to compute the NPV of an investment by summing up net cash flows that vary from the time of the investment (year 0) until the end of the investor's planning horizon (in year T).

$$NPV_T = (R_0 - C_0) + (1+r)^{-1}(R_1 - C_1) + \ldots + (1+r)^{-T}(R_T - C_T)$$

R_t represents revenue cash flows that vary due to the investment, while C represents cost cash flows that vary due to the investment. The term $(1 + r)^{-t}$ converts a future cash flow into present dollars, based on the discount rate r. The planning horizon for an investment in depreciable equipment is typically the expected useful lifetime until it either wears out or becomes obsolete.

obsolescence, and repair delays. Apart from full cost accounting, an important issue is the acreage over which the fixed cost of an equipment investment can be spread. More acres means lower average fixed cost. Omitting relevant costs or overestimating acreage can lead to overestimating the profitability of SSCM.

By contrast, benefits are typically underestimated, because they can take many forms. Benefits of SSCM can be divided into those that affect profitability, business risk, and environmental quality. We can list these as:

A. Profitability
1)   Within-season benefits (increased yield or reduced cost),
2)   Carry-over benefits to subsequent years (e.g., improved yields from ameliorating low fertility patches),
3)   Effects of managing multiple inputs with the same SSCM investment,
4)   Effects on multiple outputs (e.g., crops in rotation on the same field),
5)   One-time benefits not related to VRT (e.g., yield maps resulting in a decision to add drainage tile to improve drainage),
6)   Off-field benefits from field data, such as for land rent negotiation or sale of data to seed or herbicide companies (Nowak, 1996).

B. Reduction of income risk (when stable yields stabilize net revenues over time)

C. Environmental benefits (notably from improved water quality due to less nutrient and pesticide leaching and runoff),

Among published studies, the only benefits typically measured are single-year yield and cost changes for one crop using one to two inputs. An important exception is Schnitkey et al.'s (1996) examination of carryover fertilizer effects in a corn-soybean rotation. In general, overlooking benefits leads to underestimating returns to investments in SSCM.

The "bottom line" question becomes, "Are the omitted costs offset by the omitted benefits?" While we cannot measure what we cannot see, we can identify different types of SSCM investments and identify which costs and benefits are relevant to which investment type. From this, we can begin to infer whether omitted information bias in studies so far has been large or small.

Types of SSCM Technology for Benefit-Cost Analysis

From the standpoint of farmer benefits and adoption, SSCM technologies divide into two types. Input control technologies such as VRT (whether based on sensors or input management maps) fall in one category. These technologies can be evaluated based on the value of gains from inputs controlled. By contrast, yield monitoring and mapping technologies monitor performance, they do not control inputs *per se*. The information they provide, however, can prompt one-time field investments (e.g., adding drainage tile). The fact that yield monitors are the SSCM technology most widely purchased by farmers makes it especially important to evaluate them properly. The differences in kinds of benefits to evaluate is illustrated indicatively in Table 1.

All the profitability research of which we are aware is based on mapping and VRT technologies. Since most farmers using these technologies purchase services on a custom-hire basis, partial budgets are a suitable "first cut" at profitability analysis. However, partial budgets require that current year costs and benefits be known. In a research setting *where custom hire costs are not known,* a suitable alternative is break-even benefit analysis, identifying the maximum custom rate a farmer could pay while still breaking even (so costs do not exceed the yield gain and input savings from VRT). From a farmer perspective *where benefits are not known,* the analogue is break-even cost analysis. This identifies the minimum revenue gain the farmer must earn in order to offset the added costs incurred. With custom-hired services, farm size is irrelevant, as are many of the learning and other noncash costs borne by the service provider.

Whether this "first cut" captures most of the relevant information depends on whether there are important carryover or rotation effects, as well as nonmonetary factors such as risk reduction and environmental benefits. As Nowak (1996) has pointed out, soil nutrient maps may also have value as a record of and justification for chemical application practices, thereby demonstrating compliance with environmental regulations and insulating the farmer against environmental liability risk. Such benefits can be hard to quantify.

No investment analysis has been done for a yield monitor, so far as we know. One important reason is that benefits are hard to predict. Yield map information sometimes triggers new investments in the land (recall the poor drainage example), but these benefits must be ascribed both to the new investment (e.g., the new drainage tile) as well as to the information system which brought to light the need for it. Good data on the whole farm is required

Table 1: Indicative benefits from three kinds of SSCM practices.

SSCM practice	Single season	Rotation & carry-over	One-time changes	Off-field data value	Risk reduc-tion	Water quality
Yield monitor		✔	✔✔	✔		
Nutrient maps & VRT	✔✔	✔*		✔	✔	✔✔
Nitrate sensor	✔✔			✔	✔	✔✔

* Depends on which inputs are being controlled.

to attribute information system costs properly to new investments. Another imputed benefit is the sale of data to interested third parties, such as seed companies. However, markets for yield monitor information are not yet developed, so it is hard to estimate how much a farmer might earn from the sale of yield data. The value of yield maps for bidding on land rental contracts is equally hard to estimate, although some farmers consider it important. In general, yield mapping benefits must be measured empirically on individual farms.

So What Can Be Concluded about Profitability?

In spite of all the unknowns, we can make some clear inferences about profitability and adoptability of SSCM practices. Recall that most of the unexpected costs of SSCM are associated with equipment ownership. When services are custom hired, virtually all costs are monetary. So where financial benefits exceed custom hire costs, we can conclude that a practice is profitable. So far, there appear to be no significant environmental costs associated with SSCM, so we can also assume that there are no major environmental negatives.

Custom-hire nutrient mapping and VRT fertilizer applications are one SSCM technology pair which meets the data needs for a profitability analysis. VRT nitrogen application to sugar beet in the Red River Valley is a case in point. Partial budget accounting of VRT nitrogen spreading versus uniform rates shows single-year revenue gains of some $74 acre^{-1} due to gains in both yield quantity and quality (in the form of reduced sugar loss to molasses). These gains are offset by increased costs of $26 acre^{-1} for fertilizer, grid sampling and VRT spreading, leaving a net margin of $48 acre^{-1} (Anonymous, 1996). The research, by Larry Smith at the University of Minnesota's Crookston experiment station, showed consistent net margins in 1994 and 1995; it is matched by similar results by Al Cattanach at North Dakota State University.[2] These results suggest that grid soil testing and VRT spreading of a single input --nitrogen-- can be profitable for fields going into sugar beet, a high-value field crop. It remains an open question whether the same would be true of other crops in rotation with sugarbeet, such as wheat or dry beans.

Other nutrient mapping and VRT application research applied to one or two inputs has been inconclusive or failed to find profitability (Hertz and Hibbard, 1993; Wollenhaupt and Buchholz, 1993; Wollenhaupt et al., 1994; Lowenberg-deBoer et al., 1994; Mahaman, 1993). These were all applied to relatively low-value cereal crops with one or two low-cost inputs such as phosphorus or potassium. Given that their profitability analyses omitted other potential benefits such as carry-over and rotation effects, off-field data value, risk reduction, and improved water quality, the published results have a fairly narrow interpretation: VRT for the specified input(s) in the specified crop was not profitable in a single-season analysis. More data on omitted benefits and

---

[2]Berglund, Dennis. Personal communication, March 8, 1996. Centrol Consulting, Twin Valley, MN.

costs are needed before a general verdict can be reached about SSCM on these farms.

The same is true of the benefits from farmer investments in yield monitoring. Subsequent evaluation of the net benefits from these investments will have to be based on survey data that go beyond current anecdotes.

Two key parameters could change the profitability of SSCM investments. Technological innovation could reduce costs, particularly if reliable sensor technologies can replace costly manual data gathering (e.g., for soil nutrients and weeds). Public policy changes could also affect costs. If SSCM practices such as grid soil sampling are deemed beneficial for water quality, the U.S.D.A. Consolidated Farm Services Agency may share costs with farmers. Foreseeable changes would appear to lower farm-level costs, making SSCM more attractive.

CONCLUSION & SUMMARY

To evaluate fully the returns on investments in site-specific crop management requires more information than the yield and cash cost data that are typically available from agronomic experiments. Farmers report a variety of unexpected, often nonmonetary costs due to learning, incompatibilities among software and equipment, and delays in obtaining repairs and spare parts. But the farmers and other observers have also identified a variety of additional potential benefits, including carry-over effects, off-field data sales, yield risk reduction, and water quality improvements.

Although comprehensive cost and benefit accounting is difficult, estimating returns to investments in SSCM centers on several key factors. The first determinant of profitability is crop value. High-value crops like sugar beet and potato have the greatest potential to benefit from SSCM (especially when farmers are rewarded for both quantity and quality). The number of inputs controlled is also important, for it allows the fixed cost of data gathering to be spread across more possible ways to boost yield.

Future economic research into SSCM needs to go beyond partial budget analyses to delve into the many data gaps highlighted here. These include the effects of SSCM systems on nutrient carryover in crop rotations, dynamic yield risk, and water quality. Better cost measurement is essential for investment appraisal. It should include the cost of learning how to use SSCM hardware and software, the risk of premature obsolescence, the incidence of incompatible components, and proper allocation of centralized information costs across farm enterprises. Obsolescence and incompatibilities will likely diminish in importance as SSCM technologies mature, but they matter for early adopters. Technological change will offer interesting new research opportunities in areas such as multiple input control and sensor-based technologies (e.g, for nitrogen and weeds).

Filling these data gaps will demand closer collaboration among economists, agronomists and engineers. Experiments designed to measure and/or manage soil nutrient variability will fail to meet the economic data needs listed above unless the experiments are designed collaboratively.

But more experimental data is only half the economic research agenda. New methods must also be brought to bear. Surveys will be needed to inventory the kinds of benefits (and related costs) that farmers are obtaining from yield monitor information. Despite potential environmental benefits from SSCM, there has been no attempt to assess their value. A wide array of economic tools exists for estimating nonmarket values. Up to now, there existed scarcely any data on environmental effects of SSCM. This is rapidly changing. As data become available, nonmarket valuation methods should be brought to bear to estimate environmental benefits from SSCM.

REFERENCES

Anonymous. 1996. Grids' value for beets. Sugarbeet Grower (Feb.): 14-15.

Hertz, C.A., and J.D. Hibbard. 1993. A preliminary assessment of the economics of variable rate technology for applying phosphorus and potassium in corn production. Farm Economics: Facts and Opinions. P 93-14. Department of Agricultural Economics, Cooperative Extension Service, University of Illinois, Urbana-Champaign, IL.

Lowenberg-deBoer, J., R.L. Nielsen, and S.E. Hawkins. 1994. Management of intra-field variability in large scale agriculture: a farming systems perspective. Paper presented at the International Symposium on Farming Systems Research in Agriculture and Rural Development, Montpellier, France, November 21-25. 1994.

Lowenberg-DeBoer, J., and S. Swinton. 1996 (forthcoming). Economics of site-specific management in agronomic crops. In F. J. Pierce, P. C. Robert, and J. D. Sadler (eds.) The State of Site-Specific Management in Agriculture, Madison, WI: ASA-CSSA-SSSA.

Mahaman, M.I. 1993. An evaluation of soil chemical properties variation in northern and southern Indiana. Ph.D. Thesis, Department of Agronomy, Purdue University, W. Lafayette, IN.

Nowak, P. 1996 (forthcoming). A sociological analysis of site-specific management. In F. J. Pierce, P. C. Robert, and J. D. Sadler (eds.) The State of Site-Specific Management in Agriculture. Madison, WI: ASA-CSSA-SSSA.

Schnitkey, G., J.W. Hopkins, and L.G. Tweeten. 1996. An economic evaluation of precision fertilizer application in corn-soybean fields. Paper presented at the Third International Conference on Precision Agriculture, Minneapolis, MN, June 23-26, 1996.

Swinton, S.M., S.B. Harsh, and M. Ahmad. 1996. Whether and how to invest in site-specific crop management: results of focus group interviews in Michigan, 1996. Staff Paper 96-37. Department of Agricultural Economics, Michigan State University, E. Lansing, MI.

Wollenhaupt, N.C., and D.D. Buchholz. 1993. Profitability of farming by soils. P. 199-211. In P. C. Robert, R. H. Rust, and W. E. Larson (eds.) Proceedings of Soil Specific Crop Management: A Workshop on Research and Development Issues. Madison, WI: ASA-CSSA-SSSA.

Wollenhaupt, N.C., R.P. Wolkowski, and M.K. Clayton. 1994. Mapping soil test phosphorus and potassium for variable-rate fertilizer application. J. Prod. Agric. 7: 395-396, 441-448.

Cost Analysis of Variable Rate Application of Nitrogen and Phosphorus for Wheat Production in Northern Montana

D.S. Long
G.R. Carlson

Northern Agric. Res. Center
Montana State Univ.
Havre, Montana

G.A. Nielsen

Plant, Soil, and Environmental Sciences
Montana State Univ.
Bozeman, Montana

Most farm fields are fertilized with a single rate despite the fact that they commonly have smaller parts that differ in crop requirements and productivity. Consequently, conventional uniform fertilization may rob growers of profitability by overtreating the low yielding parts of a field and undertreating the high yielding parts (Larson and Robert 1991). Variable rate application has been proposed which revolves around maximizing return in each part of a field by adjusting fertilizer rates to the productivity levels.

Few studies have been reported that document the profitability of variable rate fertilizer application in dryland wheat production systems in the northern Great Plains. Farmers and fertilizer dealers need this information to decide whether to invest in the time and technology required for variable rate fertilizer application. Carr et al. (1991) in Montana did not find significant dryland wheat yield improvement or increased net returns with variable rate fertilizer application by soil map unit. However, significant returns were produced from optimum fertilizer rates other than the ones recommended for soil map units by published fertilizer guidelines. Wibawa et al. (1993) in North Dakota found that any cereal grain yield increase from variable rate fertilizer application was not enough to offset the added cost of soil sampling and testing that is associated with this management.

The objective of the present study was to determine whether variable rate fertilizer application, based on yield goals and composite soil sampling of smaller field areas differing in productivity, is more profitable than conventional uniform rate application. A field experimental design was used that accommodates whole production fields and commercial-scale farm equipment for planting and harvesting.

MATERIALS AND METHODS

We conducted three fertilizer trials in dryland wheat production fields in northern Montana. The first trial involved N fertilization that was conducted in 1994 and comprised a 41 ha field in a summer fallow rotation. The second trial involved N fertilization that was conducted in 1995 and comprised a 21 ha field in a chemical fallow rotation. The third trial involved P fertilization that was conducted in 1995 and comprised three 15 ha fields in a chemical fallow rotation. The fallow period extends from harvest of spring wheat in late-summer to seeding of the next crop in spring of the third year.

Fertilizer Treatment Rates

To establish the range in treatment rates for the N trials, soil N was determined for the top 60- and 120-cm of the profile at numerous grid locations in the fields. Likewise, treatment rates were established in the P trial by determining soil P (by Olsen method) for the top 15-cm of the profile at grid locations. A randomized block design was used with three replications of five to six treatments. Fertilizer treatments included 0-, 22-, 45-, 67-, 90-, and 112-kg of N/ha for the 1994 N trial, 0-, 22-, 45-, 67-, and 90-kg of N/ha for the 1995 N trial, and 0-, 12-, 24-, 36-, 48-, and 60-kg of P_2O_5/ha for the 1995 P trial. N treatments were broadcast as urea fertilizer in 18 m wide strips lengthwise across a field using a truck mounted, fertilizer applicator as illustrated in Fig. 1 for the 1994 N trial. The fields were then planted to spring wheat. Similarly, P treatments were banded as triple super phosphate in 15 m wide strips across the field using a Concord air seeder which also planted the field in the same operation.

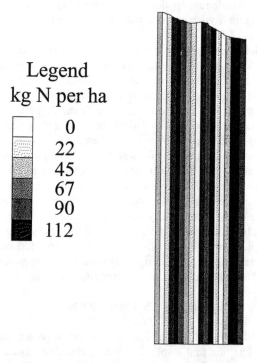

Fig. 1. Treatment N rates of urea fertilizer that were applied in strips across the field for the 1994 N trial.

Crop Yield Data Collection

We cut each 18-m wide N treatment strip and each 15-m wide P treatment strip into two parallel swaths using a combine harvester equipped with a high precision GPS receiver and yield monitor. Grain yield observations of each swath in a strip were randomly allocated to one of two independent groups of data: termed either uniform fertilization or variable fertilization. Yield was sampled at a rate of one observation per second and hence, several thousand observations were available for the analyses. Grain was manually sampled from the combine's elevator at numerous locations in a swath and analyzed for protein content using laboratory NIR methods.

Recommended Field Rates

We computed the N recommendations for the fields with the N trials using the equation:

$$N_R = (0.042 \times YG) - N_S$$

where N_R is the N recommendation in kg/ha, YG is the yield goal in kg/ha, and N_S is the test value for soil N in kg/ha. Yield goals were based on the historical yields that had been obtained in a field. Two recommendations were computed using soil test N from each of the top 60- or 120-cm of soil. All samples were averaged in the field to arrive at a N recommendation for uniform fertilization as illustrated in Fig. 2A for the 1994 N trial. For variable fertilization, a field was first divided into management units (Fig. 2B) based on appearance of crop growth differences in air photos taken 2-3 weeks before harvest, and soil N test levels. An average value was then computed from the test results of the available samples in each management unit to arrive at multiple N recommendations for variable fertilization.

The field for the P trial was divided into three management units based on expected yield differences between depressions, slopes, and knolls indicated by land contours in a USGS 1:24,000 scale topographic map. We evaluated the economic potential for variable rate P fertilization by examining the crop yield response to added phosphate. The criterion used was the difference between the yield of each phosphate treatment and the control yield treatment. It was our opinion that a wheat yield increase from 134- to 336-kg/ha is a marginal P response and over 336 kg/ha is a P response.

Economic Analysis

We analyzed only the yield observations of the treatment rates that closely matched the recommended field rates. The method of comparing uniform fertilization vs. variable fertilization is illustrated in Fig. 2C with results for the 1994 N trial. The fertilizer rates for uniform fertilization involved the yield observations from three continuous swaths within an entire field. The fertilizer rates representing variable fertilization involved the yield observations from two or three swaths within each management unit in a field. Swaths between neighboring

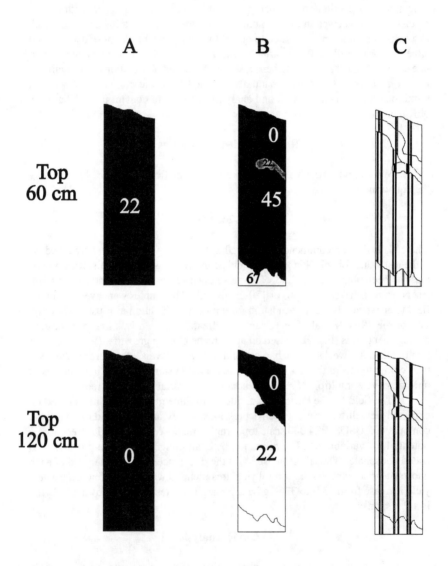

Fig. 2. Nitrogen recommendations for uniform fertilization (A) and variable fertilization (B) based on soil test N and treatment N rates that matched the recommended rates (C).

areas differing in a N or P recommendation were not continuous in one strip, because treatments were applied at constant rates across a field. This procedure of comparing uniform vs. variable fertilization is similar to the one proposed by Mulla et al. (1992) whereby fertilizer treatments are applied in two parallel strips: one fertilized uniformly along the entire length and another fertilized variably according to soil fertility status.

Gross returns were computed by multiplying the wheat yield times the Portland market quotes for dark northern spring wheat (Table 1). The value of wheat for the 1994 N trial was based on average 1994 quotes and for the 1995 N and P trials it was based on average 1994-1995 quotes. The average grain protein in a strip was used to establish wheat prices yielded from uniform or variable fertilization. Net return was computed by subtracting the cost of fertilizer, soil testing, and fertilizer application from gross return. Returns were averaged for uniform- and variable fertilization using the PROC MEANS routine of SAS (SAS Institute Inc., 1988). Yield observations in a swath assigned to uniform or variable fertilization were considered to be repeated measures of one independent random sample. Hence, average net returns of uniform vs. variable fertilization were not statistically contrasted.

Table 1. Average Portland market quotes for dark northern spring wheat in years 1994 (for 1994 N trial) and years 1994-1995 (for 1995 N and P trials).

Protein	Quote
---%---	-$/kg-
Average 1994	
<12.0	0.141
12.0	0.164
12.5	0.170
13.0	0.175
13.5	0.185
14.0	0.194
14.5	0.199
15.0	0.204
Average 1994-1995	
<12.0	0.138
12.0	0.161

Table 2. Yield goals for the field and its management units (MU) used for the 1994 trial, and corresponding soil test N and N recommendations based on samples from top 60- or 120-cm of soil.

	Whole Field	Management Unit				
		Depressions		Slopes		Knolls
		MU 1	MU2	MU4		MU5
Yield Goal (kg/ha)	2352	3024	3024	2352	2352	1680
Top 60-cm of Soil Profile						
Soil Test N (kg/ha)	68	63	118	55	119	53
N Recommendation (kg/ha)	31	64	9	44	0	18
Top 120-cm of Soil Profile						
Soil Test N (kg/ha)	129	105	249	78	340	79
N Recommendation (kg/ha)	0	22	0	21	0	0

RESULTS AND DISCUSSION

N Trial in 1994

About 10 cm of rain fell during the 1994 growing season resulting in an average grain yield of 2267 kg/ha and average protein content of 14%. The average 1994 Portland market price for grain at this protein content was $0.159/kg. Five management units were delineated in the 1994 trial based on crop appearance in a color air photo taken before harvest, and soil profile N levels. Lighter areas tended to be higher drier sites, darker areas lower and wetter sites. Management Unit 1 is located in concave depressions where profile N is low and Unit 2 in concave depressions where profile N is high. Unit 3 is found in slope positions where profile N is low and Unit 4 in slope positions where profile N is high. Unit 5 is

The field had little history of N fertilization and exhibited wide variation in soil N levels (Table 2). Soil test N averaged 68 kg/ha in the top 60-cm of the profile and 129 kg/ha in the top 120 cm. Based on these average levels and a 2352 kg/ha yield goal, the N recommendations were 31 kg/ha relative to the 60 cm depth and 0 kg/ha relative to the 120 cm depth. Soil test N averaged 63-, 118-, 55-, 119-, and 53-kg N/ha in the top 60 cm, and 105-, 249-, 78-, 340-, and 79-kg N/ha in the top 120 cm of soil in order of Units 1, 2, 3, 4, and 5. Based on yield goals of 3024-, 3024-, 2352-, 2352-, and 1680-kg/ha, in order of Units 1, 2, 3, 4, and 5, the N recommendations for variable fertilization were 62-, 0, 0-, 42-, and 16-kg N/ha relative to the top 60 cm of soil, and 22-, 0-, 0-, 21-, and 0-kg N/ha relative to the top 120 cm. The treatment N rates closely matching these recommended N rates were 0-, 22-, 45-, and 67-kg of N/ha.

About 4,000 observations were analyzed for comparing uniform and variable fertilization (Table 3). Uniform fertilization returned about $372/ha at a cost of about $17/ha whereas variable fertilization returned about $443/ha at a cost of about $33/ha when N recommendations were based on tests from the top 60 cm of soil. Despite about $16/ha greater treatment cost, variable fertilization netted about $71/ha more than uniform fertilization. Alternatively, when N recommendations were based on tests from the top 120 cm, uniform fertilization returned about $327/ha at a cost of about $1/ha whereas variable fertilization returned about $344/ha at a cost of about $25/ha. In this case, variable fertilization netted about $17/ha more than uniform fertilization despite about $24/ha greater treatment cost.

Table 3. Average returns (with standard deviation) and costs in the 1994 N trial for uniform- and variable-rate fertilization ($/ha) with respect to treatment N rates matching N recommendations derived from soil test N from the top 60- and 120-cm of soil.

	Gross Return	Cost*	Net Return
	Top 60-cm of Soil Profile		
Uniform (22 kg N/ha)	388.86±66	17.00	371.86
Variable (0 to 67 kg N/ha)	476.42±74	33.46	442.96
	Top 120-cm of Soil Profile		
Uniform (0 kg N/ha)	327.85±112	1.33	326.52
Variable (0 to 22 kg N/ha)	369.70±67	25.28	344.42

*Nitrogen $0.407/kg, soil sampling 41 ha field @ $55 per composite sample (uniform) and $48 per composite sample (variable), and fertilizer application charge of $6.54/ha (uniform) and $12.35/ha (variable).

In this case study, variable fertilization gave greater returns than uniform fertilization because fertilizer could be distributed according to spatial patterns in soil test N levels and yield goals. For instance, the areal frequency distribution of soil test N within the field was dominated by high values representing small areas of the field comprised by Units 2, 3 and 4. This skewed the average level of soil N for the whole field upwards thus causing the amount of N recommended for the crop to be underestimated in the large portion of the field dominated by Unit 5 that was deficient in soil N. In contrast, variable fertilization was able to account for the areal frequency distribution of soil test N within the field by basing the N recommendations on the available samples within smaller areas, or management units. This neutralized the influence of the extremes in the distribution of soil test N and hence, reduced the likelihood of underfertilization.

N Trial in 1995

Rainfall in the 1995 growing season amounted to about 30 cm which elevated grain yields and depressed protein contents. Consequently, the field trial averaged 3333 kg/ha yield and 11.5% protein content. Portland market quotes for dark northern spring wheat at 12% protein content, available only for 1994 and 1995, averaged $0.161/kg. Three management units were delineated in the 1995 trial based on appearance of crop ripening differences in a color air photo taken before harvest. Lighter areas tended to be higher drier sites, darker areas lower and wetter sites. Management Unit 1 is mainly found in concave depressions, Unit 2 is located in slope positions, and Unit 3 is found on convex knolls.

The field has a long history of regular N fertilization and was relatively uniform in soil test N. The soil N test values are listed in Table 4 for the field and management units. Soil test N averaged 51 kg/ha in the top 60 cm and 80 kg/ha in the top 120 cm of soil. Using a 2333 kg/ha yield goal for the whole field, the N recommendations for uniform fertilization were 47 kg of N/ha based on soil test N in the top 60 cm and 18 kg N/ha based on soil test N in the top 120 cm of soil. Soil test N averaged 53-, 50-, and 48-kg N/ha in the top 60 cm and 80-, 86-, and 72-kg N/ha in the top 120 cm of soil in order of Units 1, 2, and 3. Using yield goals of 3000 kg/ha for Unit 1, 2333 kg/ha for Unit 2, and 1667 kg/ha for Unit 3, the N recommendations for variable fertilization were 73-, 48-, and 22-kg N/ha relative to the top 60 cm, and 46-, 12-, and 0-kg N/ha relative to the top 120 cm of soil. The treatment N rates closely matching these recommended N rates were 0-, 22-, 45-, and 69-kg of N/ha.

Table 4. Yield goals for the field and its management units (MU) used for the 1995 N trial, and corresponding soil test N and N recommendations based on samples from top 60 and 120-cm of soil.

		Management Unit		
	Whole Field	Depressions	Slopes	Knolls
		MU1	MU2	MU3
Yield Goal (kg/ha)	2333	3000	2333	1667
		Top 60-cm of Soil Profile		
Soil Test N (kg/ha)	51	53	50	48
N Recommendation (kg/ha)	47	73	48	22
		Top 120-cm of Soil Profile		
Soil Test N (kg/ha)	80	80	86	72
N Recommendation (kg/ha)	18	46	12	0

About 3,000 total observations were analyzed for uniform and variable fertilization. Costs and returns are listed in Table 5. Uniform fertilization returned about $479/ha at a cost of about $32/ha whereas variable fertilization returned about $551/ha at a cost of about $47/ha when N recommendations were based on tests from the top 60 cm of soil. Returns exceeded treatment costs such that variable fertilization netted about $62/ha more than uniform fertilization. Meanwhile, uniform fertilization returned about $462/ha at a cost of about $22/ha whereas variable fertilization returned about $472/ha at a cost of about $32/ha when N recommendations were based on tests from the top 120 cm. Returns from variable fertilization were not enough to offset treatment costs which resulted in a negative net return.

In this case study, variable fertilization gave greater returns than uniform fertilization even though soil test N was relatively uniform among the management units within the field. This was because more fertilizer could be placed in certain parts of the field where yield goals were assumed to be higher than the single average yield goal assumed for the whole field.

Table 5. Average costs and returns in the 1995 N trial for uniform- and variable-rate fertilization ($/ha) with respect to treatment N rates matching N recommendations derived from soil test N from the top 60- and 120-cm of soil.

	Gross Return	Cost*	Net Return
Top 60 cm of Soil Profile			
Uniform (45 kg N/ha)	479.24±77	32.27	446.97
Variable (22 to 67 kg N/ha)	551.46±122	45.75	505.71
Top 120 cm of Soil Profile			
Uniform (22 kg N/ha)	462.89±74	21.01	441.88
Variable (0 to 45 kg N/ha)	470.54±94	32.89	437.65

*Nitrogen $0.575/kg, soil sampling 21 ha field @ $55 per composite sample (uniform) and $48 per composite sample (variable), and fertilizer application charge of $7.41/ha (uniform) and $12.35/ha (variable).

P Trial in 1995

At seeding time, soil N averaged 105±60 kg of N/ha and 27 kg of fertilizer N/ha was applied uniformly across the field. This amount of N was adequate for obtaining a yield goal of 2352 kg/ha (35 bu/ac) and 14% grain protein content in normally dry years. However, heavy rainfall was received in the 1995 growing season that resulted in abnormally high average yield of 3333 kg/ha (51 bu/ac) and low average protein content of 11.5%. Portland market quotes for dark northern spring wheat below 12% protein content, available only for 1994 and 1995, averaged $0.131/kg.

Three management units were delineated based on land contours: MU1 in concave depressions, MU2 in slope positions, and MU3 on convex knolls. Color air photos were not used for this purpose because, likely due to high rainfall, appearance of crop ripening differences were lacking before harvest. Plant available P in the top 15-cm of soil averaged 8.5 ppm in the entire field, 9 ppm in MU1, 7 ppm in MU2, and 8 ppm in MU3. According to current MSU recommendations, the field and its management units were low in plant available P and 39 kg of banded P_2O_5/ha were required to supply enough P for the crop. We did not attempt to compare returns from uniform- and variable fertilization using soil test P values and current MSU recommendation methods because of this relatively uniform distribution of soil test P, and the fact that these methods are

Table 6. Average yield and yield response (YR), and gross return and cost associated with YR by P rate and management unit.

P_2O_5	Yield	YR	Gross Return	Cost
kg/ha	-------------kg/ha------------		-------------$/ha------------	
		Depressions		
0	3716			
13	3783	70	10	23
27	3689	27	3	31
40	3716	00	0	38
54	3810	94	13	46
67	3944	288	40	53
		Slopes		
0	3386			
13	3749	363	50	23
27	3575	188	26	31
40	3608	222	31	38
54	3695	309	43	46
67	3742	356	49	53
		Knolls		
0	3292			
13	3433	141	19	23
27	3447	155	21	31
40	3575	282	39	38
54	3729	437	60	46
67	3857	564	78	53

based on soil test levels alone and do not consider differences in yield potentials within fields.

Table 6 lists the mean yield difference between each P treatment rate and the control yield treatment with respect to management unit. In general, a yield response to added P fertilizer is evident within all management units which increases in order of depressions, slopes, and knolls. Mean yield for the 13 kg/ha P treatment deviated from the overall trend within MU1 and MU2 because of extra soil moisture from snow that accumulated in stubble along the east edge of the field-strips where this treatment was located. The decreased yield response in depressions may have resulted from cold, wet conditions in spring and early summer that limited root growth and slowed P movement from the soil to the plant. The net effect of low temperature and low soil P availability was to limit adventitious root formation and tillering, which in turn limited grain yields.

The highest yield response in each management unit corresponded with the highest treatment rate of 67 kg/ha where the difference was 228 kg/ha in depressions, 356 kg/ha in slopes, and 564 kg/ha in knolls. However, the yield increase derived from added fertilizer is profitable only in MU3 where gross return offset treatment cost by $25/ha. Any contribution of MU3 to profitability likely will be small because this unit comprises less than 8% of the field. This suggests that profitability of variable rate P fertilization will depend upon the degree of yield response to added phosphate among management units testing low in plant available P and amount of field area offered by the P responsive management units.

This case study did not investigate the interactive role P plays with N in increasing grain yield. P influences uptake of N through its stimulating effect on initiation, number, and length of wheat-roots and resultant increase in soil volume explored. Therefore, N fertilization coupled with adequate P would likely result in additional grain yield and protein content above the response to P fertilization alone. Initially no response to available N would be obtained until the P needs for the crop have first been satisfied. Furthermore, due to limited mobility in the soil profile, there is higher potential for an investment in P to favorably impact the net return of subsequent crops. P uptake may well have been impeded by cold soil conditions under the climatic conditions experienced in this case study. Thus single-year accounting does not always assess the full economic impact of a P application regardless of whether it is uniform or variable rate in nature. Difference in return between this variable- and uniform-fertilization is likely too small to justify variable rate P application for a year such as 1995.

CONCLUSIONS

Opportunities for increased profit appear to exist in Montana dryland wheat production systems using variable rate N fertilization based on yield goals and composite soil sampling of smaller field areas differing in productivity.

Maximizing returns with variable rate N fertilization depends on recommendations that closely meet actual crop requirements.

Profitability of variable rate P fertilization will depend upon the degree of yield response to added phosphate among management units testing low in plant available P.

More years are needed to test feasibility of variable rate P and N fertilization before statistically-based conclusions can be made.

ACKNOWLEDGEMENTS

This research was funded by the Montana Fertilizer Tax Fund Committee and the Montana Agricultural Experiment Station. Production fields for the fertilizer trials were provided by the Mattson Farm of Chester, Montana, and the Vaughn Farm and Peterson Farm of Havre, Montana.

REFERENCES

Carr, P.M., G.R. Carlson, J.S. Jacobsen, G.A. Nielsen, and E.O. Scogley. 1991. Farming soils, not fields: A strategy for increasing fertilizer profitability. J. Prod. Agric. 4:57-61.

Larson, W.E. and P.C. Robert. 1991. Farming by soil. In R. Lal and F.J. Pierce (ed.) Soil Management for Sustainability. Soil and Water Conservation Society of America, Ankeny, IA. pp. 103-112.

Mulla, D.J., A.U. Bhatti, M.W. Hammond, and J.A. Benson. 1992. A comparison of winter wheat yield and quality under uniform versus spatially variable fertilizer management. Agric. Ecosystems, and Env., 38:301-311.

SAS Institute Inc. 1988. SAS procedures guide, Release 6.08 edition. Cary, NC. 441 pp.

Wibawa, W.D., D.L. Dludlu, L.J. Swenson, D.G. Hopkins, and W.C. Dahnke. 1993. Variable fertilizer application based on yield goal and soil map unit. J. Prod. Agric. 6:255-261.

Grid Soil Testing and Variable Rate Fertilizer Application Effects on Sugarbeet Yield and Quality

Allan Cattanach

North Dakota State University
University of Minnesota
Fargo, North Dakota

Dave Franzen

North Dakota State University
Fargo, North Dakota

Larry Smith

Northwest Experiment Station
University of Minnesota
Crookston, Minnesota

The adoption of the practices of grid soil sampling and variable rate fertilizer application by Minnesota and North Dakota sugarbeet producers has increased dramatically since 1993. A limited number of growers used these practices on a few hundred hectares of sugarbeet in 1994 and about 8,000 hectare in 1995. A 1995 sugarbeet production practice survey indicates more than 62,500 hectares were grid sampled in the fall of 1995 for 1996 production. Nearly 40% of the acreage in some counties was grid sampled. Grower acceptance of these practices has progressed more rapidly than the research data base required to guide use of the technology.

Three field scale trials to investigate this technology were initiated in 1993 and 1994. The objectives of the studies were to (1) determine field variability of nitrate-nitrogen (NO_3-N), phosphorus, and potassium and (2) compare variable rate nitrogen (N) fertilizer application based on grid soil testing, versus conventional fertilizer application based on whole field sampling, for effects on sugarbeet yield and quality.

Two of the test sites were located at the Northwest Experiment Station, University of Minnesota, Crookston, and the third site on the Don Bradow farm, Breckenridge, Minnesota. All three field sites were in commercial sugarbeet rotations and had been planted to small grains the year prior to sugarbeet. Field sizes ranged from 67-90 hectare. Nitrate-nitrogen content was determined at the 0-15, 16-60, and 61-105 cm depths in mid October the year preceding sugarbeet. Six to eight soil cores were taken from each grid. Grid size ranged from 1.25-1.60 hectares. The conventional sampling consisted of 30-40 soil cores taken in a random pattern across the field and then composited. Soil analysis was done by AgVise, Inc., Northwood, North Dakota. Mapping was done by Centrol, Inc. of Morris and Twin Valley, Minnesota. At each site the fields were divided into strips receiving either a single rate nitrogen application based on the conventional soil test or a variable rate nitrogen application based on grid soil testing. As nitrogen was the only variable being tested, a blanket application of phosphate was added to ensure its adequacy over the test area. Fertilizer was applied the last week in October at all sites. Population at harvest at all sites was approximately 90,000 plants ha^{-1}.

Each field was harvested on or about October 1. Two root quality samples were taken from each truck load of beets at the Northwest Experiment Station and from each harvest pass across each grid at Breckenridge. Beet root quality was determined at the American Crystal Sugar Company Tare Lab. From 80 to 240 composite root samples were analyzed for sucrose content, impurities, and dirt tare for each location. All loads of beets were delivered to the same piler to eliminate differences in tare determination.

Data for phosphorus and potassium variability will not be discussed in the paper.

Results and Discussion

Grid soil sampling gave a far more accurate estimate of total soil NO$_3$-N in the 0-105 cm soil profile than did the conventional random soil test (Tables 1-3). This was particularly true for NO$_3$-N at 61-105 cm depths, which is a major problem in the production of high quality sugarbeet. For example, the conventional soil test in the fall of 1993, at the Northwest Experiment Station site, showed 150 kg ha^{-1} of total NO$_3$-N in the 0-105 cm soil profile. Grid soil sampling showed a range from 38 kg ha^{-1} to 267 kg ha^{-1}. Nitrate-nitrogen in the 61-105 cm depth ranged from 7 kg ha^{-1} to 192 kg ha^{-1} whereas the conventional test showed 83 kg ha^{-1}.

Table 1. Fall 1993 Total Soil NO$_3$-N by Grid (kg ha^{-1})
(Northwest Experiment Station)

0-60 cm 35 / 61-105 cm 61 / 0-105 cm 96	41 / 45 / 86	52 / 54 / 105	75 / 134 / 209	24 / 16 / 39
69 / 78 / 147	67 / 65 / 132	75 / 192 / 267	85 / 123 / 208	31 / 29 / 60
69 / 141 / 211	62 / 157 / 218	68 / 57 / 125	95 / 146 / 241	50 / 34 / 84
47 / 27 / 74	47 / 13 / 60	32 / 29 / 61	63 / 27 / 90	31 / 7 / 38

Conventional soil test 150 kg ha^{-1} (0-60 cm: 67, 61-105 cm: 83)

Table 2. Fall 1994 Total Soil NO$_3$-N by Grid (kg ha^{-1}) (Breckenridge)

0- 60 cm 71 61-105 cm 80 0-105 cm 151	46 48 94	35 55 90	95 119 214	57 63 120	Slough
41 41 82	88 104 192	36 36 72	72 77 149	53 53 105	78 102 180
41 41 82	88 104 192	36 36 72	72 77 149	53 53 105	78 102 180
35 44 79	82 84 166	49 49 98	38 38 76	41 41 82	72 74 146

Conventional soil test of 83 kg ha^{-1} (0-60 cm: 52, 61-105 cm: 45)

Table 3. Fall 1994 Total Soil NO$_3$-N by Grid (kg ha^{-1}) (Northwest Experiment Station)

0-60 cm 50 61-105 cm 13 0-105 cm 63	44 7 51	65 13 78	53 20 73	48 13 61	55 10 65
57 3 60	68 13 81	58 3 61	66 13 79	77 10 87	73 64 137
57 7 64	64 34 98	69 24 93	75 64 139	84 10 94	54 13 67
100 37 137	99 30 129	81 34 115	60 101 161	43 84 127	53 10 63

Conventional soil test 129 kg ha^{-1} (0-60 cm:76, 61-105 cm:53)

The current N recommendation for a 45 Mg ha^{-1} high quality sugarbeet crop in Minnesota and North Dakota is 134 kg ha^{-1} of soil N plus added fertilizer N in the 0-105 cm soil profile. These guidelines assume a 34 kg ha^{-1} level of residual N in the 61-105 cm depth that is subtracted from the soil test value, and that 80 percent of the remaining N is available for plant growth. Conversely, in the 1995 trials, an adjustment was made for deep soil N below 34 kg ha^{-1} in the 61-105 cm depths. If the soil NO_3-N was below 34 kg ha^{-1}, the amount below 34 kg ha^{-1} was calculated and 80 percent of that value was added to the surface application. Tables 4,5 and 6 show the recommended rates of nitrogen by grid compared to the conventional soil test recommendation for each of the three sites.

Using these N recommendations, the conventional random soil test would have resulted in under-fertilization of 65 and 79 percent of the field in 1994 and 1995, respectively, at the Northwest Experiment Station and 50 percent of the 1995 Bradow site. The remainder of each site received adequate or excess nitrogen fertilization.

Table 4. Nitrogen Recommendation by Grid (kg ha^{-1}),
Northwest Experiment Station, 1994.

Conven.	78	84	67	0	111
Variable	29	43	0	0	103
Conven.	0	0	47	0	84
Variable	87	87	102	72	103

Conventional Soil Test Recommendation was 28 kg ha^{-1} (amount
spread on conventional areas versus the amount variable applied).

Table 5. Nitrogen Recommendation by Grid, (kg ha^{-1}),
Breckenridge, 1995.

Variable	27	77	83	0	54	Slough
Variable	86	0	96	28	66	0
Conven.	86	0	96	28	66	0
Variable	92	12	78	93	86	30

Conventional soil test recommendation was 54 kg ha^{-1} (amount
spread on conventional area versus the amount variable applied).

Table 6. Nitrogen Recommendation by Grid (kg ha^{-1}),
Northwest Experiment Station, 1995.

Variable	100	112	97	93	102	99
Conven.	102	92	101	75	76	37
Variable	99	71	84	35	69	96
Conven.	32	38	54	29	52	101

Conventional soil test recommendation was 41 kg ha^{-1} (amount
spread on conventional areas versus the amount variable applied).

At all three locations grid soil sampling and variable rate fertilization increased yield and quality of the sugarbeet crop over conventional soil testing and single rate nitrogen application. Sugarbeet yield increased between 2.1-2.73 Mg ha^{-1}, with recoverable sucrose content increasing between 0.44-0.47 percent. Recoverable sucrose ha^{-1} increased by an average of 553 kg^{-1}ha (Table 7). Adjusting for the increased cost of grid soil sampling and analysis, fertilizer and variable rate fertilizer application, an average net return of $143 ha^{-1} was realized (Table 8).

Table 7. Effect of Soil Testing and Fertilization Method on Recoverable Sucrose Per Hectare.

Location	Grid & Variable	Conventional	Difference
	----------Recoverable Sucrose kg ha^{-1}------------		
Crookston, Northwest Experiment Station,1994	8462	7820	+642
Crookston, Northwest Experiment Station,1995	7813	7327	+486
Breckenridge, MN, 1995	6270	5740	+530
MEAN	7515	6962	+553

Table 8. Effect of Soil Testing and Fertilization Method on Return per Hectare.

Location	Grid & Variable	Conven.	Gross Difference	Net Difference*
	-----------------------$/ha----------------------			
Crookston,Northwest Experiment Station,1994	2485	2245	+240	+180
Crookston,Northwest Experiment Station,1995	2302	2120	+183	+120
Breckenridge, MN 1995	1733	1548	+185	+128
MEAN	2173	1970	+203	+143

*Adjusted for extra costs of soil sampling, sample analysis, and variable rate fertilization.

SUMMARY

Relative to these particular studies, the following conclusions can be drawn:

1)   Grid soil testing more accurately reflected the NO$_3$-N needs for a 45-Mg ha^{-1} sugarbeet crop than did the conventional method used by most sugarbeet producers at all three locations.

2)   Grid sampling and variable rate fertilization produced an average increase in net return of $143 ha^{-1} because of the variability in soil NO_3-N in these fields. It appeared that conventional random soil testing overestimated the amount of available NO_3-N in these fields.

A special thanks to Harold Stanislowski, Otter Tail County Extension Educator, for assistance and to Larry Sax and Dave Genereux of Centrol, Inc. for mapping and soil sampling fields. Thanks to John Lee, AgVise, Inc. for providing the soil sample analysis and Charles Hotvedt, American Crystal Sugar Co. for root analysis.

Partial funding support for this project was provided by the Sugarbeet Research and Education Board of Minnesota and North Dakota.

Mathematical Model and Algorithm of Optimization of Yield Production

A.V. Skotnikov

Department of Soil, Water, and Climate
University of Minnesota
St. Paul, MN

D.E. McGrath

Tyler Industries Inc.
Benson, Minnesota

ABSTRACT

It is assumed that data for soils properties (21 parameters) and yield are available in a small field for several previous years at many locations (X, Y, Z coordinates). The goal of the research is to develop a mathematical model that will find an optimal set of fertilizer recommendations to be applied to each area of the field for maximum yield or maximum profit.

INTRODUCTION

There are a lot of different programs, software-based expert-support or decision-support systems for agriculture and natural resource management. All these systems address a specific problem or specific parameter, yield mapping for example, rather than the whole system for SSCM. We are suggesting a different and more comprehensive solution. It is combination of a geostatistic, statistic, and economic analysis permitting us to optimize a strategy of yield production.

A user may work with this program according to the following scenario. It is supposed that at the moment the user has a set of soil sampling data from a certain field. It is necessary to identify the field from which soil samples were taken and the date of sampling entered in the computer. Then the user chooses:

- Crop cultivated in this field
- Seed variety sown in this field as well as their parameters : price, the price of sowing, crop selling price, etc.
- Technologies he uses including their cost and equipment including its use and maintenance.
- Fertilizers applied, pesticides and herbicides including the cost of applying.

Then the user chooses optimizing parameters :
- fertilizer optimizing parameters;
- additional optimizing parameters;
- range of change of fertilizer parameters;
- range of change of additional parameters.

The program will create a statistical model describing dependence of the yield on the fertilizers applied. Statistical model parameters are entered into the data base.

Further the user can calculate maximum yield. As this takes place, optimal fertilizer recommendations and maximum yield are entered into the database. Then one can calculate profit.

Information about the received values of fertilizers applied can be printed. Information about several variants of expenses close to maximum profit can be displayed. In this case one can choose the most reasonable variant of management to achieve expected profit. Then the expected profit variants received can be printed .

In this paper we consider only the mathematical model and algorithm for fertilizer application optimization.

DISCUSSION

Developing statistical models requires a substantial data base. In our case it is assumed that results of soil sampling analysis and yield for several years are available. These data may have the following structure. For each field's point there are 21 (or less) soil composition parameters (clay, %; silt, %; sand, %; humus soluble, humus common, carbonates (CO_2), sesquioxides, acidity pH, soil compaction, air permeability, moisture, ammonium NH_4, nitrate NO_3, phosphorus P-Bray or Olsen, potassium K, calcium Ca, microelements Zn, Mn, residue, 6 parameters of any fertilizer or chemicals applied to this point, yield value in this point and the point's X, Y, Z coordinates. Each point we identify with the small field area surrounding this point.

Let us suppose that at the moment a user has a set of data for the soil composition and the decision is made on the crop which will be cultivated.

The goal of the research is to choose for each point an optimal, in certain sense, set of fertilizers that are to be applied to the area, related to this point. Optimal conditions are defined as "maximum yield", "maximum profit" or "minimum quantity of fertilizers for a defined profit rate".

To find optimal variants one is to solve several tasks, namely:

1.   Extract so-called anomalous data which differ from others. This extraction, as a rule, originates from some mistakes of measurements.

2.   To find for each field area considered, similar areas where the selected crop was cultivated during previous years, i.e., to make a cluster analysis.

3.   To develop a statistical model for the yield dependence of fertilizers applied in each area considered using similar areas .

4.   To find an optimal combination of fertilizers providing "maximum yield "and calculate this maximum yield for each considered area.

5.   To calculate income, expenses, and profit for "maximum yield" variant.

6.   To find a combination of fertilizers providing the maximum profit (if possible) and calculate this profit for each designated area.

Let us consider these tasks.

1. The task of defining abnormal data is solved by Grabbs criteria. The decisive rule of this criteria for exclusion of extremal member for each data set $(x_1, x_2,...,x_n)$ is based on the statistic $T = |x_{extr} - m|/s$, where

x_{extr} - extremal value of data set;

m - mean of data set;

s - mean deviation.

Critical value of the T statistic is computed by the formula

$$v = u * \sqrt{\dfrac{2(n-1)}{2n+u^2+\dfrac{3+u^2+2u^4}{12n-30}}}$$

where

u - quintile of level $(1-\epsilon)/2$ for the standard normal distribution law, n - number of data points.

Let us consider, that $\epsilon = 0.05$, i.e., abnormal data will be defined with the preciseness

$(1 - \epsilon) \times 100\% = 95\%$, given that u = 1.65.

If $T > v$, the influential observation is removed, otherwise it is retained. If data set has several abnormal values then Grabbs criteria is applied to the maximum (minimum). If it is abnormal the criteria is applied to the second in magnitude until there are no outliers left. In this case the data for the soil composition are read from the floppy disk and for each soil characteristic, an abnormal value is defined. If abnormal values exists, each value is replaced by the sample mean for the parameter considered.

2. For each area it is necessary to find similar soil composition from the previous years or other fields.

The task of choosing similar areas is achieved using cluster analysis. From the formal mathematical point of view this task is formulated and solved in the following way.

There are sets of stochastic data taken from L groups W_1, W_2, ...W_L, with dimensionality of each set equal to p. A priori probabilities of groups P_1, P_2, ...P_L are unknown. Data from group W_i is a stochastic vector X_i of dimensionality p with the unknown probability density $g_i(x)$. Index "i" refer to the group number, i.e., the whole field has L different groups of similar in soil composition. Dimentionality of the set is equal to a number of parameters, characterizing the soil, in our case p = 21. Probabilities $P_i = 1/L$, i.e., all soil composition sets are equally probable.

For each i, $1 < i < L$ define aggregate :

$W_i(d_i) = \{x: \ P_i/g_i(x) \geq d_i\}$,　　　　　　　　　　　　　(1)

where numbers d_i are chosen so that the probability of X_i to fit into $W_i(d_i)$ is strong enough:

$$P\{\ x_i \in W_i(d_i)\ \} \ge 1-e, \qquad (2)$$

where e >0 - defined relatively small positive number .

Let us consider that e = 0.05, i.e., we refer the point to a certain set of soil composition with probability $(1-e)\times100\% = 95\%$. The unknown probabilities densities $g_i(x)$ we evaluate by means of histograms; $_id$ values are equal to the quintile of standard Gauss distribution, i.e., 1.65.

We can observe here non-classified stochastic samples A of volume n
$$A = \{\ x_j,\ 1 \le j \le n\ \},$$
where x_j - data from a certain group $W_j(d_j)$, $d_j \in \{\ 1,...,L\ \}$. Index "j" refers to the point number.

Actually the sample A is a combination of areas of the considered field and each area from A is to be transferred to the group with similar soil composition.

Group $d_1, d_2,... d_n$ numbers , from which data $x_1 , x_2,... x_n$ are registered , are unknown. We have to find these numbers.

The task is to :

1)   classify sample A with minimum error probability, i.e., to evaluate numbers $d_1, d_2, ...d_n$;

2)   to constitute decisive rules for classification of new received points.

Mathematical solution of this task is the following:

Points from W_j are close to the mathematical expectation m_j (group W_j mean) and are far from the points from aggregates $W_q, j \ne q$. Therefore the decision rule is used

$$d = d(x,A) = \arg \min |\ x - m_j^*\ |, \qquad (3)$$
$$j \in \{1, 2,...L\},$$

where means $\{m_j^*\}$ depend on A and are defined by means of the following iteration procedure:

Let us define $N_L = \{1,2,...,L\}$, t - iteration number (t=0, 1, 2,...), m_j^t-evaluation of m_j^* on the t-th iteration

$Dt = (d_1^t, d_2^t,... d_n^t)$ - vector of solutions on the t-th iteration,

where $d_j^t \in N_L$ - solution about group number for the point x_j on t-th iteration.

With t=0 we define the vector of a priori solutions D^0. On the t-th iteration let vector D^t be computed. The iteration number (t+1) is contained in the following. Let us compute centers' evaluations .

$$m_j^{t+1} = \sum_{k=1} \delta(k-d_j t) \times x_k / n , \qquad 1 \le j \le L \qquad (4)$$

where $\delta(z) = \begin{cases} 1, z = 0 \\ 0, z \ne 0 \end{cases}$ - Heavyside's function.

Then we define the vector of solutions :

$$d_j^{t+1} = \arg\max |x_k - m_j^t + 1|, \quad 1 \le k \le n. \tag{5}$$

If $D^{t+1} = D^t$, the procedure is ended: $D = D^t$ is accepted as optimal classification of A (in view of minimum error probability) and substitute this value into decision rule (3)

$$\{ m_j^* = m_j^t + 1, \quad 1 \le j \le L \}.$$

This is the end of the formal explanation of the cluster analysis algorithm.

From the practical point of view this algorithm is accomplished as follows.

We suppose that the whole field has L different sets of soil composition and all these compositions are equiprobable, i.e. $P_i = 1/L$, $1 < i < L$. The program automatically selects L different soil compositions and then defines for them class centers m_j^0, and then new points from L groups by formulas (4)-(5) are classified.

Thus we have constituted the rule according to which a lot of samples of soil composition and it's fertility parameters for previous years are divided into a certain number of groups with minimum error probability. Besides, the rule has been constituted which allow us to refer this point to one of the groups with minimum error probability .

3. Now we have data from the group with similar soil composition. In this circumstance yield will depend only on the amount of fertilizers, i.e. from six parameters. Now we can develop a model of yield dependence of the set of fertilizers. This dependence is called regression. We have no reason to suppose that only linear regression takes place:

$$y_j = f_1(x) = \sum_{k=1}^{6} a_k x_{kj} + a_0 , \tag{6}$$

where y_j - yield value in j - th point,

$\quad\quad x_{kj}$ - the value of k - th fertilizer in j - th point,

$\quad\quad a_k$ - some parameters , usually called coefficients.

It is supposed that yield dependence y of fertilizers' parameters x_k can be either linear (6), or non-linear of the following types:

$$y_j = f_2(x) = a_1 e^{a_2 x_{1j}} + a_3 e^{a_4 x_{2j}} + ... + a_{11} e^{a_{12} x_{6j_1}}$$

$$\tag{7}$$

$$y_j = f_3(x) = (a_1 e^{a_2 x_{1j}} + a_3 e^{a_4 x_{2j}} + ... + a_{11} e^{a_{12} x_{6j}})^{-1}$$

$$\tag{8}$$

$$y_j = f_4(x) = \frac{a_1 e^{a_2 x_{1j}} + a_3 e^{a_4 x_{2j}} + ... + a_{11} e^{a_{12} x_{6j}}}{b_1 e^{b_2 x_{1j}} + b_3 e^{b_4 x_{2j}} + ... + b_{11} e^{b_{12} x_{6j}}}$$

(9)

$$y_j = f_5(x) = a_1 x_{1j}^{a_2} e^{a_3 x_{1j}} + a_4 x_{2j}^{a_5} e^{a_6 x_{2j}} + ... + a_{16} x_{6j}^{a_{17}} e^{a_{18} x_{6j}}$$

(10)

At first, using stochastic model identification methods we choose one of the models (6) - (10), which describes available data with a maximum probability. This selection is accomplished using least squares method. The gist of it is as follows

We must choose the necessary yield dependence y_j on fertilizers x_j by solving the task

$$\sum_j [y_j - f_s(x)]^2 \to \min_s,$$

(11)

It is necessary to find the number s, $1 < s < 5$ for which the left part of ratio (11) is minimum. Summation in (11) is accomplished according to the number of j points, which was chosen on the step of the cluster analysis for the point considered for the new field .

The selection of the necessary model number "i" is accomplished by straight sorting. It is noteworthy that when computing the left part (10) it is necessary to evaluate unknown parameters a_k and b_k. The evaluation of these parameters is accomplished also by using the minimum squares method .

4. Now we have a model of yield dependence y of fertilizers values
$x = (x_1, x_2, x_3, x_4, x_5, x_6)$ in a certain type $y = f(a,x)$, in which "a" are found assessments of unknown parameters of this dependence. Then it is necessary to find an optimal composition which will maximize the chosen criteria.

Let us consider the search of the optimal values of fertilizers applied to maximizing yield, i.e., maximizing $y = f(a,x)$.

For that we use the method of probabilistic branches and limits. The gist of the method is as follows.

From all points in previous years close to the considered point of soil composition (such points are already determined by means of cluster analysis) we take the point with maximum yield. Thus we have "good enough" set of fertilizers.

In the vicinity of this set using imitation modeling methods we generate new sets of fertilizer values in which the components are equidistributed in congruent segments. These segments constitute 1/3 from the common range of values, i.e., the segment's limits with the center in z point within the common interval (a,b) will be as follows: max [a,z-0.33×(b-a)], min [b,z+0.33×(b-a)]. In each set we calculate possible yield values and then narrow ranges of possible values in the set on 1/3 of the previous set's ranges. This is done until, for the given quantity of points, we find no one greater yield value (the given quantity of numbers is a parameter of the algorithm and is determined in the program).

5. Let us consider now the task of determination of the set of fertilizers maximizing profit received from the examined field.

First we consider an algorithm of calculation the profit received from the field. For this field it is known a number of points and the area related to the point. Profit value Pr is calculated in the following way:

$Pr = I-E$

where Pr - profit from the field ($),

I - income received from selling a crop with a certain yield ($),

E - the sum of expenses for producing a certain yield ($).

$I = Y_{yi} \times V_g$

V_g - gross yield gathered from the field (kg).

$V_g = \Sigma\, V_i$,

where V_i the yield collected from the area related to the i- th point (i-th area) (kg)

$V_i = V_{pi} \times F_i$,

in which V_{pi} - yield collected from the i- th area (kg/m^2) ,

F_i - i-th area (m^2),

$E = E_s + E_p + E_f + E_t + E_g + E_w$

in which

E_s - expenses of crop sowing, ($)

E_p - expenses of soil processing, ($)

E_f - expenses of applying fertilizers, ($)

E_t - expenses of applying pesticides, ($)

E_g - expenses of yield gathering, transportation, drying, refinement , storage, etc., ($)

E_w - possible expenses of irrigation,($).

$$E_s = E_{sa} \times \underset{i}{\Sigma F_i} + \underset{i}{\Sigma Y_i} \times H_{bi} \times F_i ,$$

in which

E_{sa} - sowing expenses per area unit ($/m^2),

Y_i - the price of the selected type of seeds for the i-th plot ($/kg)

H_{bi} - seeds sowing norm for the i-th plot (kg/m^2)

$E_p = \Sigma\, E_{pi} \times F_i$,

in which i - number of points (areas).

E_{pi} - expenses per area unit for each soil processing ($/m^2).

$E_f = \Sigma\, V_{gi} \times F_i + E_{ap} \times \Sigma F_i$,

in which V_{gi} expenses for one type of fertilizers for i-th area ($/m^2).

E_{ap} - expenses per area unit of fertilizers being applied ($/m^2).

$E_t = \Sigma\, Y_k \times H_{ti} + E_{at} \times \Sigma\, Fi$

in which

E_t - expenses of applying chemicals ($/m^2),

Y_k - the price of a certain chemical ($/kg)

H_{ti} - dose of the chemical applied to i-th area ,(kg/m^2).

$E_g = \Sigma\, V_i \times Y_g \times F_i$

in which Yg - the price of yield harvesting ($/m^2).

6.   The optimizing of crop management for obtaining maximum profit is accomplished as follows. We make several iterations of the following kind; we reduce the volume of the most expensive fertilizer ($/acre) by 5 percent and calculate profit. If the profit decreases the procedure of the obtaining of maximum profit is finished. If the profit increases, then we reduce the volume of the second most expensive fertilizer by 5 percent. We do so until the profit growth stops.

   To accomplish this algorithm a program for Windows was developed.

CONCLUSIONS

Develop a mathematical model and algorithm which will
- extract so called anomalous data,
- make a cluster analysis for each field's area,
- develop an individual statistical model for each area of the yield dependence on fertilizer or chemical applied,
- find an optimal combination of fertilizers providing "maximum   yield " and calculate this maximum yield for each considered area,
- calculate income, expenses, and profit for "maximum yield" variant,
- find a combination of fertilizers providing the maximum profit (if possible) and calculate this profit for each designated area.

Economic Opportunity in Yield Variability

T.S. Colvin
D.L. Karlen

ISDA-ARS
National Soil Tilth Lab
Ames, IA

The important reason for adopting precision farming techniques and technologies will be the economic return that can be generated. If we truly have precision, then there should also be environmental benefits which may also represent an economic opportunity, but that is not considered in the current project. This study was designed to explore the magnitude of potential net return that may exist and might be captured by the adoption of precision farming on one field. Partial budgets were generated for seven years of historical yields on 224 locations within the field. At least 30% of the locations were found to have negative returns in every year. Unfortunately the locations with negative returns were not always in the same place in different years. Five to ten percent of the locations were found to have negative returns that were so large that they should not have been farmed. The net returns at those locations did not even cover the fixed costs of land and machinery. The application of precision farming concepts is new enough that it will be difficult to generalize the results of the project to other fields at this time. More analyses will be needed to determine underlying management principles.

SESSION V

ENVIRONMENT

Potential Environmental Benefits of Adopting Precision Agriculture

J. L. Baker
T.S. Colvin
D.B. Jaynes

ABSTRACT

Agricultural chemical transport is affected by three sets of factors: chemical properties, hydrologic conditions, and management practices. Precision agriculture in terms of timing, placement and rate of inputs (including tillage) has the potential to reduce off-site transport of pollutants with surface runoff and subsurface drainage. Losses of chemicals are strongly influenced by the rate and method of application. Experimental and modeling data are presented to show that losses of chemicals not naturally present in the soil are roughly proportional to the amount applied. Furthermore, for naturally occurring chemicals, losses are roughly proportional to the amount present in the soil. Placement of chemicals relative to "zones" of higher interaction with runoff and leaching water can affect their susceptibility to be lost. Needs for more information on the interaction between practices and chemical transport, and more options for chemical application are expressed.

Application of Site-Specific Farming in a Sustainable Agriculture Project at Chesapeake Farms

D. Raymond Forney

Chesapeake Farms
7321 Remington Drive
Chestertown, MD

Larry D. Gaultney
DuPont Ag Products
Stine-Haskell Research Center
Newark, DE

ABSTRACT

The Chesapeake Farms Project occupies four adjacent two- to nine- hectare watersheds and four hectares of replicated plots on a 1336 hectare production farm and wildlife preserve in the coastal plain of Maryland. The project was established to promote farming practices that are productive, economically viable, environmentally sound, and socially acceptable. The project compares four different crop rotation systems chosen by an advisory board composed of industry leaders, government scientists, environmental activists, and farmers. The systems are designed to represent realistic production choices for farmers in the mid-Atlantic region. Precision farming technology including GIS, DGPS, soil nutrient and pH mapping, and yield mapping, has been applied to fine-tune the best management practices for each of the production systems on the farm. In addition to monitoring all cropping inputs and yields, and studying the effects of farming methods on water quality, the project also serves as a testing and demonstration site for some of the new technologies.

BACKGROUND

The means to productivity and profitability of farming systems are of utmost concern in U.S. agriculture and have been the subject of on-going debate for several years (CAST, 1990; Harwood, 1990; Keeney, 1990). One major challenge is the integration of economic and ecological impacts of alternative farming systems (Costanza, 1992), and approaches to the quantitation of sustainability are still evolving (House of Representatives, 1990). The importance of long term evaluation of entire crop rotations, from both agronomic and economic perspectives, has been well established (Madden & Dobbs, 1990; Lohr et al, 1992; Bushnell et al, 1991). However, previous studies of whole-farm productivity and profitability have obtained varying results, particularly those related to sustainable farming methods (Chase and Duffy, 1991; Hansen et al, 1990). Since findings

---

among locations and over time are inconsistent, it is necessary to conduct continuous comparative studies in different agro-climatic zones. Farming practices must change if environmental quality is to be improved (Mussman, 1991), and changes must be accomplished with economic viability if we are to preserve a rural infrastructure considered invaluable to society (House of Representatives, 1990).

During the past three years, representatives of the EPA, the USDA, University of Maryland, University of Delaware, Cornell University, the Rodale Institute, DuPont Agricultural Products, and local farmers have established a major effort to foster the goals of sustainable agriculture. A Farmer Advisory Panel of prominent area farmers with diverse enterprises meets regularly to review and plan project activities. The project mission is to promote a more sustainable agriculture by means of on-farm research and demonstration of farming systems that are productive, economically viable, environmentally sound, and socially acceptable. The project objectives are: 1) establish a private/public partnership in a jointly planned research, demonstration and education program to promote sustainable agriculture, 2) evaluate and demonstrate the profitability, ecological impact and sustainability of selected cropping systems, and 3) promote the recognition of agriculture as a crucial and responsible resource use, and enhance public confidence in the farmer as an agricultural producer and steward of the land, through promotion and adoption of project findings.

Four cash grain farming systems are evaluated and demonstrated on field-scale watersheds and a replicated small-plot experiment. The systems represent a continuum of increasing reliance on rotation diversity, in-season management and labor, and decreasing reliance on purchased inputs.

After establishing the site and rotations in 1993, 1994 was our first full year of production under planned management practices (we intend to run the project a minimum of six years).

For two years now, an additional group of collaborators have enabled a modest evaluation of certain aspects of precision agriculture in conjunction with A Sustainable Agriculture Project at Chesapeake Farms. These include ESRI, who have supplied ARC/VIEW software and support; CASE IH, who have supplied the use of a combine equipped with an AgLeader 2000 yield monitor and its software; Starlink, Inc., who have supplied the use of a DNAV-212 DGPS receiver system and its software; Topsoil Testing, Willard's Agri-Service, and Skybit, who furnished detailed soil sampling, mapping and analysis; and AGRIS, who have supplied the use of yield mapping software. All have provided tremendous support for our efforts. In addition to our sustainable agriculture project watersheds, we have applied the technologies to Chesapeake Farms production acreage. This report summarizes our 1995 corn and soybean harvest results, within a glimpse of how the various technologies are being employed.

SITE CHARACTERIZATION

ESRI's ARC/VIEW has been used to create georeferenced maps of our site (Figure 1). The field boundaries were originally entered using digitizing technology and aerial photographs, and watershed boundaries were drawn based on visual examination of 30-cm aerially-derived contour maps. The boundaries were then

Figure 1.

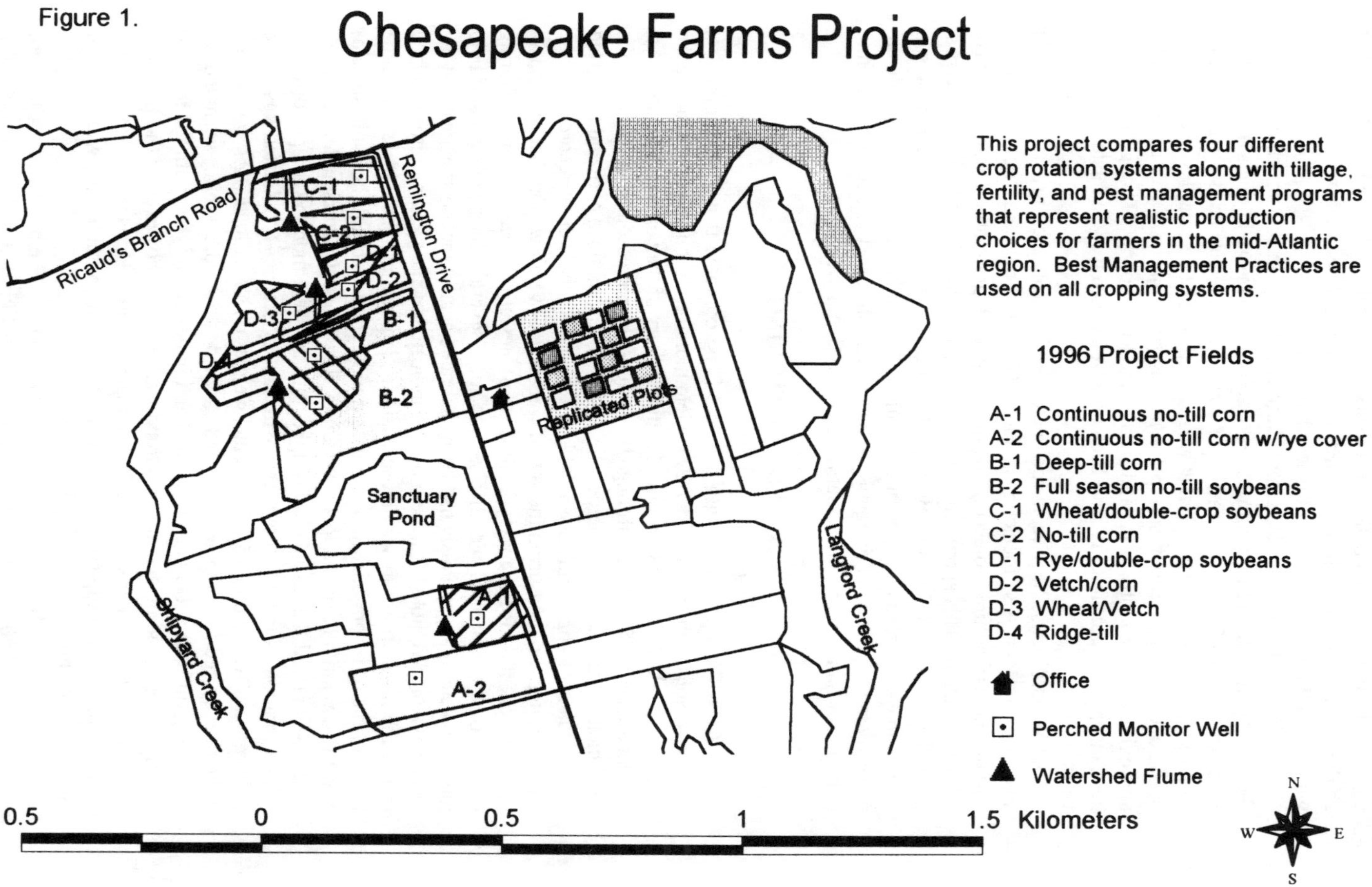

Chesapeake Farms Project
This project compares four different crop rotation systems along with tillage, fertility, and pest management programs that represent realistic production choices for farmers in the mid-Atlantic region. Best Management Practices are used on all cropping systems.
1996 Project Fields
A-1 Continuous no-till corn
A-2 Continuous no-till corn w/rye cover
B-1 Deep-till corn
B-2 Full season no-till soybeans
C-1 Wheat/double-crop soybeans
C-2 No-till corn
D-1 Rye/double-crop soybeans
D-2 Vetch/corn
D-3 Wheat/Vetch
D-4 Ridge-till
Office
Perched Monitor Well
Watershed Flume
Ricaud's Branch Road
Remington Drive
Replicated Plots
Sanctuary Pond
Langford Creek
Shipyard Creek
C-1
C-2
D-2
D-3
B-1
B-2
A-1
A-2
0.5
0
0.5
1
1.5 Kilometers
N
W
E
S

corrected by ground-based collection of survey points using Starlink's DNAV-212 DGPS. This technology has provided a far-more precise characterization of our watersheds than was possible with aerial imagery.

Topsoil Testing Service sampled each of our sustainable agriculture project fields and prepared grid-map summaries of soil pH and macronutrients (not shown). Since some of the fields are as small as 1.2 hectares, standard grids were too large, and we sampled on the principle that we required at least 16 samples in each field. In field C-1, a pattern of soil pH was observed that allowed us to amend approximately one-third of the field with lime (not shown). Though not shown here, other patterns of fertility have been observed which we will relate to in future years.

Willard's Agri-Service, in conjunction with Skybit, Inc., also performed soil sampling and analysis, focusing on Field 37 (not shown, and not part of the sustainable agriculture project). This field is in cash grain production, in a corn-soybean rotation. This approximately 21-hectare field has five different soil types, according to USDA soil survey maps (Figure 2). Visual examination of soil fertility maps revealed no apparent correlations with soil type (not shown). We are presently in the process of bringing georeferenced soil survey and fertility data into our GIS for more thorough evaluation, along with our yield maps.

YIELD MONITORING

Case IH supplied us with a model 2166 combine that was equipped at Hoober & Sons with an AgLeader 2000 yield monitor. At our site, we attached the Starlink DNAV-212 DGPS system. Taking delivery of the combine the latter part of September, 1995, we went through the various setup procedures described in the yield monitor manual. We entered a field of relatively uniform corn and harvested six bin loads of approximately 3200 kg each, noting the weight recorded by the yield monitor and measuring the actual weight in a weigh wagon that previously had been checked for accuracy against the scales at the local elevator. During harvest of the first load we checked moisture several times with a hand-held moisture meter that previously had been calibrated against the moisture tester at the local elevator. We noted that the moisture indicated by the yield monitor matched that of the moisture tester very closely and made no adjustments. We varied the speed of harvest from 6.8 km/h (the maximum our operator felt he could sustain) down to 1.6 km/h to obtain loads at various flow rates. Upon completing the six loads, we entered the actual load weights into the yield monitor and performed a full calibration. Figure 3 shows that with factory settings ("Indicated kgs"), the yield monitor was within 5% of the actual weights on all six loads. As we reduced flow rate by reducing speed from 6.8 km/h down to 3.2 km/h, we noted an increasing difference between the actual and indicated weights, with the monitor slightly overestimating weight. This trend was reversed, however, at the 1.6 km/h speed, when the monitor slightly underestimated weight. Upon performing the full calibration, all of the adjusted weights indicated by the monitor were within 1% of the actual weights.

Soybean calibration was performed October 11. We harvested six loads of approximately 1360 kg each, varying speed from 5.4 km/h down to 1.3 km/h. We again noted that the monitor moisture sensor was reading very close (within 1/2 moisture point) and made no adjustment to it. Figure 4 shows that with factory settings, the yield monitor was within 5% of the actual weights on all six loads, and seemed slightly more accurate at the lower speeds. Upon performing the full calibration, the adjusted weights indicated by the monitor were greatly improved or essentially the same in five of the six loads, while a slight loss of accuracy apparently occurred for the slowest flow rate.

During corn harvest, we recorded every combine bin load as a load in the yield monitor, and kept track of which truck received each load. At season's end, we reconciled the yield monitor data against our elevator receipts by summing each group of weights for bin loads making up a truck load, and averaging moisture for the bin loads making up each truck load. Wherever apparent variances of greater than 10% appeared for a truck load, we checked our records for bookkeeping errors, and in some cases found possible errors that imposed uncertainty. Each of these cases was discarded from the data set, leaving 61 truckloads of corn for which we are quite confident that our elevator receipts can be properly matched against our yield monitor records. Figure 5 shows that across the season, the monitor was quite accurate in measuring the corn yields, being within 5% of the elevator scales on all but seven of the truckloads. Among those seven, the yield monitor underestimated yield 5 times and overestimated yield twice. The matches between elevator records and the yield monitor for truckloads seven and nine (Figure 5) were disturbingly different, and we have been unable to determine any record keeping errors that can account for the differences. The moisture sensor in the yield monitor was within 5% of that at the elevator on all but eight loads. Of these eight, the yield monitor sensor underestimated moisture only twice, and overestimated moisture six times. Moisture variance of 5% or more occurred on only two of the seven loads where yield varied by 5% or more. The yield monitor estimated the total corn harvest within less than 1% of the elevator weigh tickets (not shown).

Soybean yield monitoring accuracy varied across the harvest season somewhat more than that for corn (Figure 6). We experienced the anticipated moisture sensor problems when operating in weedy soybean fields, especially where Eastern black nightshade was present as in the third truckload (Figure 6). Among the 14 truckloads of soybeans, the yield monitor underestimated yield relative to the elevator by greater than 10% on four loads, and overestimated yield only once and only by 1.5%. The yield monitor moisture sensor varied from the elevator by more than 10% on seven of the 12 loads; on six of these moisture was overestimated by the yield monitor. Moisture variance exceeded 10% on all four loads where yield variance exceeded 10%. The yield monitor underestimated the total soybean harvest by 9.2% (not shown).

During harvest of truckload 9, the operator noted the yield monitor was indicating an erroneously high moisture level, probably due to sticky weed residue. After checking the soybean moisture manually, he set the moisture in the yield monitor manually to the appropriate level (14.7%) for the affected yield monitor loads. This caused moisture data in memory for all the previous soybean loads to change: some loads changed to 14.7%, others were reduced by 14.7%. To correct these changes, we manually reentered all the soybean summary data prior to that date, and the presently reported data reflect those changes.

Figure 7 is an excellent example of a yield map produced with AgLINK software from yield data obtained with the CASE combine, the AgLeader 2000 yield monitor, and the Starlink DNAV-212 DGPS receiver. While not a specific part of our sustainable agriculture project, field 37 is under our farm management and represents a good place to explore precision agriculture applications. According to soil-type map (Figure 2) the field includes five different soil types. These range in corn production potential from approximately 6200 to 8500 kg/ha, according to University of Maryland soil productivity classification.

Production practices for the 1995 corn crop on field 37 (following a 1994 crop of no-till soybeans) included deep ripping with a DMI Tiger ripper (35 cm), disking, and seedbed preparation with a Landsman combination tillage tool. UAPX 15538, a TOPCROSS Blend of high-oil corn, was sown at 60,515 per ha, with liquid starter fertilizer on April 27. Weed control was accomplished with a postemergence spray of Accent/Banvel/atrazine tankmix, followed by a cultivation. A sidedress application of a UAN solution supplying 151 kg/ha of N was dribbled prior to cultivation.

The early growing season was excellent in both soil moisture and temperature. As the corn approached silk and tassel, however, several weeks of extreme heat and drought stressed the crop. This stress was relieved by rainfall toward the end of the pollination period, and the remainder of the growing season was favorable. The corn was apparently sensitive to the mid-season stress, and yield on field 37 varied from 3150 kg/ha or less on up to 9450 kg/ha or more. The contour region map (Figure 7) shows especially well that the regions of different yield seem to correspond to the soil type regions on the soils map. We have intensively soil sampled this field and mapped regions of pH, P, K, C.E.C., and O.M., and are pursuing the utilization of ESRI GIS to statistically verify potential correlation among yield, soil type, and these other variables.

SUMMARY

With two years' experience now in yield monitoring and mapping (and two-years' yield histories on several fields), we have made a good initial exploration of the technology. We aim for our efforts to contribute to the development, refinement, adoption, and optimum utilization of the technology, toward enhancing the sustainability of grain farming. Our next phases include the analysis of soil type, fertility, and yield data. Further measurements of these parameters over the coming years will add to the database and to the likelihood we can identify management factors that may be optimized for improved productivity, profitability, and environmental compatibility. We also are engaged in an evaluation of remote sensing. Combined with the georeferenced ground-based scouting possible with equipment such as the new Field Pack II version of DNAV-212 from Starlink, georeferenced multispectral images obtained via aircraft or satellite may serve well in guiding IPM scouts to particular problem areas of fields.

Figure 2 Field 37 Soil Types and Productivity Potentials

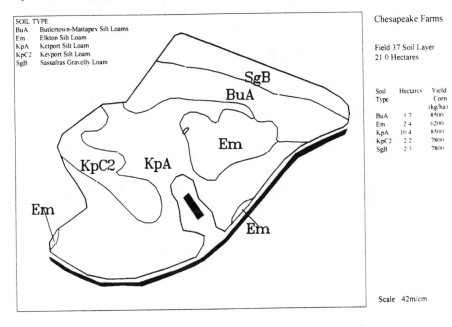

Figure 7 Field 37 Yield Regions

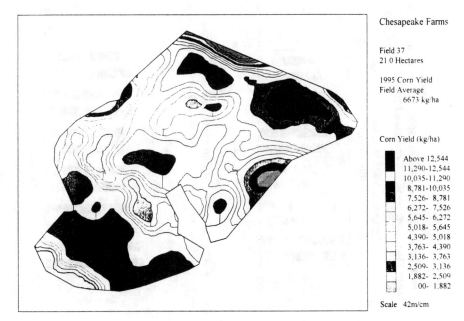

Figure 3. Ag Leader Yield Monitor 2000 Calibration at Chesapeake Farms, 1995, Corn. Case IH 2166 Combine with 6-row corn head.

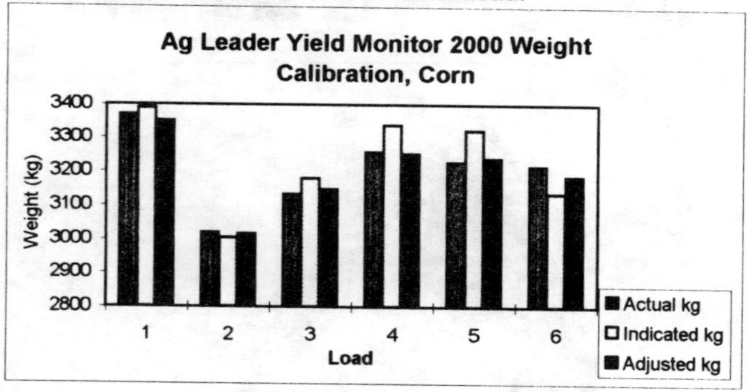

Field 52 Load	km/h	Actual kg	Indicated kg	Ind/Act % Diff	Adjusted kg	Adj/Act % Diff
1	6.4	3370	3390	0.58	3352	-0.52
2	6.8	3021	3005	-0.53	3018	-0.11
3	5.6	3134	3181	1.52	3149	0.49
4	4.8	3259	3341	2.52	3254	-0.14
5	3.2	3229	3326	2.99	3241	0.38
6	1.6	3218	3137	-2.51	3188	-0.93

Ag Leader Yield Monitor 2000 Weight Calibration, Corn

Figure 4. Ag Leader Yield Monitor 2000 Calibration at Chesapeake Farms, 1995, Soybeans. Case IH 2166 Combine with 20' model 1020 grain head.

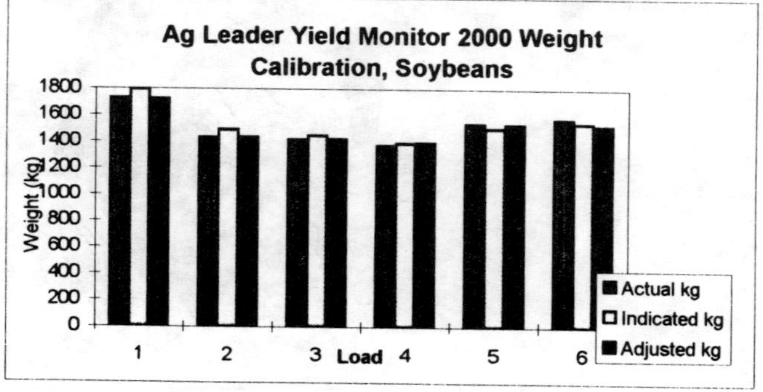

Field 60 Load	km/h	Actual kg	Indicated kg	Ind/Act % Diff	Adjusted kg	Adj/Act % Diff
1	5.5	1725	1793	3.94	1724	-0.05
2	4.8	1429	1492	4.38	1435	0.38
3	4.2	1416	1449	2.37	1420	0.32
4	2.7	1372	1391	1.39	1394	1.62
5	2.1	1538	1505	-2.18	1535	-0.21
6	1.3	1576	1545	-1.93	1527	-3.08

Ag Leader Yield Monitor 2000 Weight Calibration, Soybeans

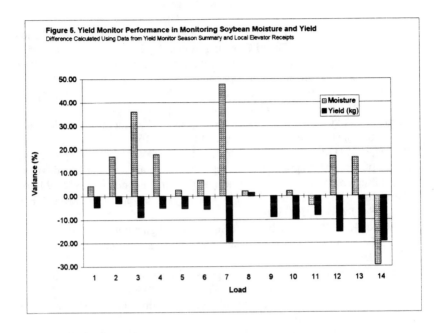

Figure 5. Yield Monitor Performance in Monitoring Soybean Moisture and Yield
Difference Calculated Using Data from Yield Monitor Season Summary and Local Elevator Receipts

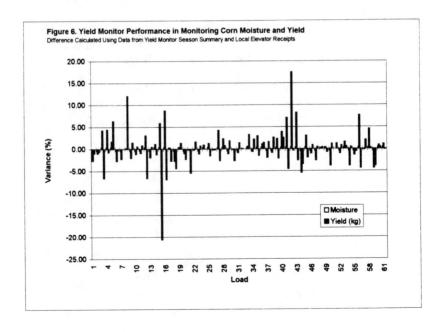

Figure 6. Yield Monitor Performance in Monitoring Corn Moisture and Yield
Difference Calculated Using Data from Yield Monitor Season Summary and Local Elevator Receipts

REFERENCES

Bushnell, J., C. Francis, and J. King. 1991. Design of Resource Efficient, Environmentally Sound Cropping Systems. J. Sust. Agric. 1(4): 49-65.

Chase, C., and M. Duffy. 1991. An economic comparison of conventional and reduced-chemical farming systems in Iowa. Am. J. Alt. Ag. 6:168-173.

Costanza, R., ed. 1992. Ecological Economics: The Science and Management of Sustainability. Columbia University Press, New York.

Council for Agricultural Science and Technology. 1990. Alternative Agriculture -- Scientists' Review. CAST Special Pub. No. 16.

Hanson, J.C., D.M. Johnson, S.E. Peters, and R.R. Janke. 1990. The profitability of sustainable agriculture on a representative grain farm in the mid-Atlantic Region, 1981-89. NE J. Agric. and Resource Economics. 19(2): 90-98.

Harrington, L.W. 1992. Measuring Sustainability - Issues and Alternatives. J. Farm Sys Res. - Ext. 3(1): 1-19.

Harwood, R.R. 1990. A History of Sustainable Agriculture. p. 3-19. *In* Clive Edwards et al. (eds.) Sustainable Ag Systems. Soil and Water Cons. Society, Ankeny, IA.

House of Representatives. 101st Congress 2d Session. Oct. 22, 1990. Food, Agriculture, Conservation, and Trade Act of 1990. Report 101-916.

Keeney, D. 1990. Sustainable Agriculture: Definitions and Concepts. J. Prod. Agriculture 3: 281-285.

Lohr, L., O. Hesterman, J. Kells, D. Landis, and D. Mutch. 1992. Methodology for Designing and Evaluating Comparative Cropping Systems. J. Farm Sys Res. - Ext. 3(1): 105-129.

Madden, J. P., and T.L. Dobbs. 1990. The Role of Economics in Achieving Low-Input Farming Systems. P. 459-477. *In* Clive Edwards et al (eds.) Sustainable Agriculture Systems. Soil and Water Conservation Society, Ankeny, IA.

Mussman, H.C. 1991. The President's Water Quality Initiative. P. 67-77. *In* D.T. Smith (ed.) Agriculture and the Environment: The 1991 Yearbook of Agriculture. U.S. Govt. Printing Office, Washington, D,C.

Mechanisation For Sustainable Arable Farming Systems: A Precision Farming Perspective

Fabio R. Leiva
Facultad de Agronomia
Universidad Nacional de Colombia
Bogota, Colombia

Joe Morris
Silsoe College
Cranfield University
Silsoe, Bedford, England

Mechanisation can contribute directly and indirectly to sustainable agriculture. This paper shows links between mechanisation and sustainable arable farming systems, and the findings of a study undertaken to assess sustainability in a British arable farm. The concepts of farming systems and site specific crop management proved to be useful to understand farming circumstances and to assess their sustainability.

INTRODUCTION

The concept of sustainable development has emerged in the last fifteen years in an attempt to deliver desirable economic, social and environmental conditions which endure over time (Turner, 1995; IUCN, 1991). For its part sustainable agriculture (SA) involves producing quality food and non-food products now and into the future in an environmentally benign, socially acceptable and economically efficient way. The preservation and enhancement of environmental quality and the improvement of the social welfare and particularly rural livelihoods are central to SA.

LINKS BETWEEN MECHANISATION AND SUSTAINABLE AGRICULTURE

Mechanisation is an important element of modern farming and one which has diverse consequences to sustainability particularly in terms of impacts on incomes, employment, energy use, soils and other environmental issues, especially through the application of chemical agents. Table 1 illustrates links between mechanised operations and sustainable criteria. Soil degradation and increased use of fossil fuels on farming are the two major direct environmental effects of using farm machinery. Indirect effects such as water pollution are associated with the mechanised application of pesticides and fertilisers. Deleterious effects have potential to affect not only the farming system itself but also the off-farm environment (Fig. 1).

TABLE 1. Links Between Mechanisation And Sustainable Criteria.

COMPONENT	LINKS WITH MECHANISATION	REFERENCES	Links With Other Components	REFERENCES
SOIL				
PHYSICAL STRUCTURAL DAMAGE AND COMPACTION	Tillage intensity	Spoor (1982)	Erosion	Spoor and Tatham (1994), Voorhees, Young and Lyles (1979).
	Machinery traffic	Watts and Dexter (1994) Soane and van Ouwerkerk (1994), Hakansson, Voorhees and Riley (1988), Soane, Dickson and Campbell (1982), Soane et. al. (1980a, 1980b),		Brussard and Van Faasen (1994)
			Soil Biological Activity	Chamen et. al. (1992), Eradat-Oskoui and Voorhees (1990)
			Energy Use	
			Air Pollution (N_2O, CO_2)	Arah et. al. (1991), Eradat-Oskoui and Voorhees (1990), Bakken, Borresen and Njos (1987).
			Soil Fertility,	Torbert and Wood (1992), Petelkau and Dannowski (1990)
			Premature Wear Of Farm Machinery	Soane, Dickson and Campbell (1982)
EROSION	Intensity of tillage: soil disturbance and cover management	Cannel and Howes (1994), Chisci (1994), Allmaras et. al. (1991),	Soil Productivity Decrease	Pierce (1991), Boardman and Evans (1994).
			Pollutant Transport to Water	Logan (1990), Sibessen et. al. (1994), Harrod, Carter and Hollis (1990)
			Siltation, Nuisance	Boardman and Evans (1994).
CHEMICAL NUTRIENT DECLINE	Tillage intensity, ploughing	Elliot et. al. (1994), Jenkinson (1988).	Structural Stability, Erosion and Compaction	Payne (1988) Soane (1990)
			Soil Water Availability	Gregory (1988)
			Soil Nutrients	Wild and Jones (1988)
			Soil Carrying Capacity	Conway and Pretty (1991)
			Air Pollution	Watson et. al. (1993)
			Water Pollution	Addiscot and Dexter (1994), Goss et. al. (1993)
			Biological Activity	Hendrix et. al. (1990)
CHANGE IN pH	Tillage effect on soil drainage	Lal (1994)	Soil Nutrients, Biological Activity Soil Pollution (heavy metals)	Brady (1986) MAFF (1991)

WATER

AVAILABILITY	Tillage practices	Lal, 1994	Soil biota	Harris (1988)
POLLUTION				
SURFACE, GROUNDWATER				
N, P	Fertilising practices	Moller and Svenson (1991), Logan (1990)	Human Health	Dudley (1990)
	Tillage effect on run-off and leaching	Addiscot and Dexter (1994), Sharpley and Smith (1994), Goss et. al. (1993), Logan (1990).	Flora And Fauna	Conway and Pretty (1991)
PESTICIDES	Application equipment	Matthews and Hislop (1993), Matthews (1992)	Human Health	Roberts (1991), Conway and Pretty (1991)
	Tillage effect on run-off and leaching	Addiscot and Dexter (1994), Hall and Mumma (1994), Czapar, Kanwar and Logan (1994)	Flora And Fauna	Edwards, Pimentel and Lehman (1993)

PESTICIDES ON NON-TARGET ORGANISMS

PESTICIDES	Application equipment (exo- and endodrift)	Miller (1993).	Human Health	Conway and Pretty (1991), Berry (1988)
			Flora And Fauna	MAFF (1992), Edwards, Pimentel and Lehman (1993)

ENERGY USE AND AIR POLLUTION

DEPLETION OF FOSSIL FUELS	Engine based machinery use and manufacture	CEC (1988), Stout (1990)	Air Pollution	Taylor, O'Callaghan and Probert (1993)
GREENHOUSE GASES (CO_2)	Emissions from engine based machinery	Taylor, O'Callaghan and Probert (1993)	Global Warming, Energy Use, Depletion Of Fossil Fuels	Watson et. al. (1993), WRI (1995)

SOCIAL

FOOD QUALITY/QUANTITY	Timeliness in mechanised operations	Witney (1988)		
HEALTH	Accidents with farm machinery, application of agrochemicals	Monk et. al. (1984), Chester (1993)		
EMPLOYMENT	Displacement of labour by farm machinery	Murphy (1995), Witney (1988), Bizwanger (1985)		
LAND TENURE (size)	Machinery requirements	Murphy (1995)		

ECONOMIC SUSTAINABILITY

ON FARM	Incomes, costs, profitability in using farm machinery	Nix (1995), Murphy (1995)		
OFF FARM	Farm machinery industry.	CEC (1992)		

Soil Compaction

Soil degradation by compaction is associated with machinery traffic and is a widespread problem affecting an estimated total of 68.3 million ha., 33 million ha. (48%) of it in Europe (Oldeman, Hakkeling & Sombroek, 1991). Soane and Ouwerkerk (1995) presented evidence of possible long term detrimental damage to the environment as a result of soil compaction.

The extent of man induced soil compaction depends upon soil specific characteristics and conditions, such as state of looseness, moisture content (Spoor, 1975) and levels of soil organic matter (Soane, 1990), and upon the type, distribution and intensity of the machinery traffic (Hakansson, Voorhees & Riley, 1988; Soane, Dickson, Campbell, 1982; Soane et. al., 1980a). Mechanisation, however, can help to alleviate soil compaction by using tillage implements (Bowen,

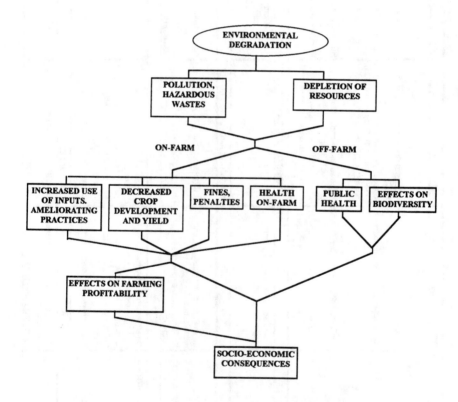

Fig. 1. Likely consequences of environmental degradation

1981), although there is evidence of compaction persistence even after loosening (Hakansson & Reeder, 1994).

Three primary ways of reducing the risk of soil compaction of field by agricultural vehicles are reduction in the number of passes of machinery; reduction of the vehicle mass and the contact pressure of wheel system; and confinement of traffic to permanent or temporary wheel tracks (controlled traffic) (Soane et. al. 1982). Reduction of number of passes of machinery may be achieved with careful planning of the timing and intensity of tillage operations and combination of operations. Average contact pressure can be reduced by reducing loads, using bigger tires and decreasing inflation pressure of rubber tires (Vermeulen & Perdok, 1994; Chamen et. al., 1992), or using tracklayers (Erbach, 1994).

Heavy machines, especially harvesters, pose risk of subsoil compaction, which may become permanent and amelioration by mechanical loosening may be impossible or expensive (Dickson, 1994; Hakansson & Reeder 1994). Some countries have developed recommendations regarding maximum limits on the axle or wheel load (Hakansson & Petelkau, 1994).

'Controlled' or 'zero-traffic' systems possibly using gantries or adapted tractors can avoid soil compaction, improve soil structure and afford significant savings in energy and potential increases in crop yield (Chamen et. al., 1992; Watts & Dexter, 1994). The system may also reduce significantly annual labour and machinery costs per ha (Chamen, Audsley & Holt, 1994).

Soil Erosion

Soil is a non-renewable resource and erosion involves the degradation of the soil capital. Furthermore, off-farm effects of erosion may be worse than those on-farm (Lal & Pierce, 1991; Boardman & Evans, 1994). Worldwide, Oldeman et. al. (1991) estimated that 1094 million ha. are degraded by water erosion and 548 million ha. are degraded by wind erosion, jointly accounting for the 84% of world's degraded land. Erosion in Europe is believed to be increasing (Chisci, 1994; Boardman & Evans, 1994).

Erosion risk depends on a complex of factors (Evans, 1980; Wilson & Cooke, 1980). Erosion in western Europe affects mainly the sandy, loessial and chalky soils devoted to continuous arable production and may be associated with the increased area of winter cereal crops, more mechanisation, consolidation of land into larger fields (Morgan & Rickson, 1990) and tractor wheel ruts and 'tramlines' (Spoor & Tatham, 1994).

Mechanisation can help to alleviate soil erosion. Appropriate tillage and especially conservation tillage (CT) can improve water and soil conservation and reduce the risk of soil erosion (Cannel & Howes, 1994; Chisci, 1994; Allmaras et. al., 1991; Lal et. al., 1990). The efficiency of soil water use can be increased by decreasing soil evaporation and improving soil structure by CT and mulch farming techniques (Lal, 1994). CT can contribute to increased SOM (Ball, 1994; Cannell and Howes, 1994). Soil organisms and particularly earthworms tend to increase under CT (Ball, 1994; Patterson et. al., 1980). CT may save labour and energy like machinery itself and fuels and lubricants (Patterson et. al., 1980). Yet, as a whole CT systems do not usually generate savings of energy because of increases in the use of pesticides and sometimes fertilisers (O'Callaghan, 1994; Lal et. al., 1990).

Conversion from conventional tillage to CT may decrease soil losses in surface runoff and in turn the risk of surface water pollution (Lal, 1994; Sharpley and Smith, 1994). However, CT may increase the leaching potential of nitrate (Sharpley & Smith, 1994) and pesticides (Czapar, Kanwar & Fawcett, 1994; Hall & Mumma, 1994). Increases of pesticide and nitrate leaching under CT have been attributed mainly to preferential macropore flow and high water content (Lal, 1994; Sharpley & Smith, 1994). Chemical transport, however, is also influenced by soil characteristics cultural practices and rainfall intensity, application rates, time and methods (Sharpley & Smith, 1994; Logan, 1990) and the nature of the substance (Czapar et. al., 1994; Hall & Mumma, 1994).

In Britain, direct drilling (DD) has shown decreased risk of nitrate leaching because of lower mineralisation of organic nitrogen (Goss et. al., 1993; Ball, 1994), however, in the long term weed infestation, soil compaction, straw disposal and surface acidity are probably major constraints for adopting DD (Ball, 1994; Christian, 1994; Patterson et. al., 1980). There is renewed interest in the applicability of CT in western Europe (Tebrugge, 1994), especially within crop rotations.

Soil Organic Matter (SOM)

SOM is perhaps the most important single indicator of soil quality and productivity and closely related to other soil physical, chemical and biological properties (Cannel & Howes, 1994). Oxidation of SOM releases nitrates (Goss et. al., 1993) which may leach and pollute water sources, and in temperate agriculture emits limited amounts of CO_2 (Watson et. al., 1993).

Soil tillage and particularly ploughing encourages mineralisation of SOM (Addiscot & Dexter, 1994; Goss et. al., 1993). Not all the changes, however, can be attributed to the direct action of tillage. Crop harvest also removes soil organic residues. After an initial decline under arable farming the SOM tends to a steady state which depends on the soil's properties, the climate and the management (Jenkinson, 1988).

SOM can be conserved by avoiding excessive tillage and using cultivation practices which improve the management of residues (Cannel & Howes, 1994; Jenkinson, 1988).

Change In Soil pH

Changes in soil pH may affect availability of nutrients and soil environment for higher plants and micro-organisms. In the UK change of soil pH is not a major concern in arable farming. Soil acidity, however, may be a constraint in peaty soils and locally in sandy soils, while salination is associated with thin soils over chalk or limestone (McGrath & Loveland, 1992). Mechanisation practices may help to avoid undesirable changes in soil pH. Deep cultivation and incorporation of gypsum and organic amendments may be useful for reclaiming salt-affected or sodic soils (Lal, 1994).

Water Availability and Water Quality

Efficient use of water on farms and the risk of water pollution by soil nutrients (phosphates and nitrates) (Logan, 1990) and pesticides (Brown, Carter & Hollis, 1995) are major concerns for the sustainability of arable farming. Modifications of soil structure and crop residues as a result of tillage operations are likely to affect movement and availability of water for crops (Lal, 1994). Adequate timing and placement of fertilisers and pesticides are likely to be relevant factors in ensuring increased efficiency and in turn less losses of such inputs (Goss et. al., 1993; Logan, 1990).

Risks of Pesticides Use

Pesticides can be lost from the 'target' through evaporation and atmospheric drift or can reach soil and move to water sources via run-off or leaching (Brown, Carter & Hollis, 1995). Evaporation and drift are likely to occur mainly with sprayed pesticides (Miller, 1993; Matthews, 1992). The fate of pesticides depends on characteristics of the chemicals, application methods and site conditions (Brown, Carter & Hollis, 1995; Matthews, 1992). The importance of the equipment on the quality of pesticide application is well recognised (Matthews, 1992; Matthews & Hislop, 1993). Dose of the pesticide, timing and placement of the pesticide are key factors to achieve high efficiency in crop protection and to reduce losses and risk of pest resistance. Optimal timing is dependant on the work rate of the sprayer (Robinson, 1993). Droplet sizing and distribution influence the performance of spraying pesticides and are determined by the type and pressure of the spraying device (Lefebvre, 1993).

Energy Use and Air Quality

The use of energy is a key element of SA. Modern agriculture is heavily dependant on non-renewable energy sources such farm machinery, fertilisers and pesticides. Farm machinery may contribute to increased energy efficiency both directly and indirectly. Direct savings of energy can be achieved by appropriate selection and operation of the mechanised system with respect to soil cultivation and traffic systems, matching tractor and implement, and proper maintenance, adjustments and use (CEC, 1988). Shallow cultivation is regarded to be a relevant practice to save energy and costs (Patterson et. al., 1980). Using adequate weights and proper tires are also key factors to improve the pull ability of the tractor (Gee-Clough, 1980) and, in turn, energy efficiency. Mechanisation influences energy use indirectly through efficient use of inputs such as fertilisers, particularly N-fertiliser which is regarded to be the major consumer of energy on farming (Taylor, O'Callaghan & Probert, 1993) and pesticides.

In the UK, the contribution of agriculture to global atmosphere impacts such as global warming in terms of CO_2 is about 1 %, while in terms of nitrous oxide (N_2O) is about 50% (Dept. Env. Prot., 1993). Considering that denitrification in soil contributes about 47% of the total emission of N_2O (Dept. Env. Prot. 1993), farm machinery is not considered a main polluter in terms of N_2O or CO_2.

Social

Farm mechanisation may impact on social issues such as food quality and quantity, land structure, incomes, employment and welfare. Mechanisation contributes to increased crop production and timeliness of operation (Witney, 1988), and has encouraged in recent years the increase in field and farm size and specialisation of production system (Murphy, 1995). Mechanisation can improve labour productivity, release people from tedious tasks and minimise drudgery. However, this often means less workers on farms and the threat of rural unemployment, especially as farmers respond to increased pressure on profitability (Murphy, 1995).

Introduction of new technologies, requires skilled people and training needs which could lead to improve farmers and operators knowledge and better understanding of farming systems. Shortage of useful life of machines, soil degradation (Spoor, 1991) and misuse of pesticides (Hislop, 1993) are consequences of lack of well trained operators. Use of toxic substances, noise and machinery may pose risk for human health directly to people engaged in farming and indirectly through environmental pollution (Table 1). Training serves to improve safety, minimise health risk (Hislop, 1993; Pfalzer, 1993) and accidents (Hank et. al., 1984) and avoid excessive costs of operation.

Economics

From an economic viewpoint four main aspects of mechanisation must be considered; the costs of the mechanised operation itself, the influence on crop development and yields, the effect on the use of other inputs, and the economic consequences of direct and indirect environmental impacts.

Labour and machinery account for over one third of the total costs in arable farming in the UK (Witney, 1988; Murphy, 1995). Thus, profitability of arable farming is highly dependent on the appropriate selection of the mechanised system. Timeliness of operations is a key criterion for farm machinery selection; bigger machines with high capacity decrease the risk of untimeliness but have higher depreciation costs than smaller machines (Witney, 1988) and may increase the risk of soil compaction.

Tillage operation costs alone account for about 30% of mechanisation costs in arable farms. There is much scope for direct savings, particularly through less intensive cultivation practices, decreased working depths (Patterson et. al., 1980) and reduced traffic systems (Chamen et. al., 1994).

SUSTAINABILITY OF CURRENT ARABLE PRACTICES

The previous discussion shows complex processes and interactions between physical, environmental, social, and economic factors which determine sustainable agricultural mechanisation. This highlights the importance of a holistic and integrated approach as the only way to attempt to identify sustainable practices and to propose improvements where practices are non sustainable.

In this context an English heavy land farm was selected as a case study to examine the sustainability of mechanised practices. The cropping pattern included winter wheat (WW), oil seed rape (OSR) and beans. Five sustainability criteria were examined, namely energy, nitrate leaching, pesticide leaching, soil organic matter (SOM) and compaction. Spreadsheet and LP models were developed as tools for the assessment. Soil erosion was not a major issue.

Table 2 shows a summary of indicators of sustainability and the levels found for the study farm. At the present, many of the critical sustainable levels are unknown. One of the outputs of the current study will be to define them. The main conclusions arising from the case study were:

- The most energy demanding inputs in the farm are fertiliser (50-70% of the total energy) and diesel fuel for farm machinery (20% of the total energy). Though direct energy consumption in soil tillage operations was modest, there is scope for direct energy saving in such operations. Contribution of tillage practices to energy efficiency should be also seen in terms of proper crop development, improved use of other inputs and soil and water conservation.
- Excluding solar energy, the balances of energy for all the crop assessed were positive. Emissions of pollutant gases into the atmosphere were not a major issue in the case study farm.
- The farmer's profits were not sensitive to current relative price of energy inputs.
- The economic assessment of practices aimed at increasing energy efficiency, and particularly improved use of N-fertiliser and alternative tillage systems was recommended. Precision farming techniques and less intensive cultivation could contribute to this goal.
- The risk of groundwater pollution by nitrate leaching was low, but there was risk of surface water pollution mainly through the drainage system. Improving application of N-fertiliser and encouraging SOM pools are likely to decrease the risk of nitrate leaching. Precision farming techniques and reduced tillage systems could help.
- Some of the pesticides used in the farm, such as IPU, mecocrop, prochloraz and metalaxil pose risk of water pollution above the EC Directive (0.1 µg per litre). Use of alternative pesticides or mixtures, and improved applications including precision farming techniques are likely to contribute to decrease the risk of pesticide leaching.
- Use of the plough is likely to be one of the main causes of the relatively low level of SOM. The use of legumes, increased organic inputs and reduced/zero tillage could contribute to increased SOM level.
- There were not many differences in terms of risk of compaction among the different crops. Ploughing and harvesting were the major contributors to soil compaction. Use of tramlines is an advantage on the studied farm, but the traffic intensity is high. The economic appraisal of the economics of reduced and zero traffic system was recommended.
- Profitability on the case study farm is relatively high for the region, but the farm failed to cope with some of the main indicators of sustainability, although critical levels are ill-defined.

TABLE 2. Sustainability components and indicators

COMPONENT	INDICATOR	SUSTAINABLE VALUE	LEVEL FOR THE STUDY FARM
Energy	Total E. input (GJ/ha) E. intensity (GJ/t crop) Total E. output (GJ/ha) E. balance (GJ) E. ratio (GJ in/GJ out)	? ? ? ? ?	9 (Beans) - 20 (OSR) 2.4 (Beans) - 5.9 (OSR) 61.9 (Beans) - 115.3 (WW) 52.8 (Beans) - 97.4 (WW) 4.0 (OSR) - 6.8 (Beans)
Nitrate leaching	Kg NO_3-N/Ha mg NO_3-N/L	? 11.3	11-54 in WW, higher in OSR 20-50
Pesticide leaching	Vulnerability assessment µg/L	'Low', 'Moderate' 0.1	'High' for some products 1.4 -36
Soil Organic Matter	%w/w	?	2.3
Compaction	rut length (Km/ha) % cover by wheel/tracks wheel intensity (tKm/ha) field load intensity(th/ha)	? ? ? ?	32.9 (WW) - 36.9 (OSR) 127 (WW)- 145 (2nd.WW) 98 (WW) - 112 (2nd.WW) 21.0 (WW) - 22.2 (OSR)
Profitability	GM (£/ha)	674.7 - 964.3 [1]	730

CONCLUDING REMARKS

Mechanisation has potential to contribute to the sustainability of modern arable farming systems, in social, environmental and economic terms. This contribution is greatest where the mechanisation system suits site specific circumstances. Results of the present study suggest that the main opportunities for sustainability of mechanised practices are reduced tillage and direct drilling, zero and reduced traffic, and precision farming techniques regarding tillage and application of fertilisers and pesticides.

REFERENCES

Addiscot, T.M., and A.R. Dexter. 1994. Tillage and crop residue management effects on losses of chemical from soils. Soil and Tillage Res. 30:125-168.

Allmaras, R.R., G.W. Langdale, P.W. Unger, R.H. Dowdy, and D.M. van Doren. 1991. Adoption of conservation tillage and associated tillage systems. pp.53-83. In Soil management for sustainability. R. Lal and F.J. Pierce (Eds.). Soil and Water Conservation Soc.

Arah, J.R.M., K.A. Smith, and I.J. Crighton. 1991. Nitrous oxide production and denitrification in Scottish arable soils. J. Soil Sci. 42:351-367.

Bakken, L.R., T. Borresen, and A. Njos. 1987. Effect of soil compaction by tractor traffic on soil structure, denit6rification and yield of wheat (Triticum aestivum L.) . J. Soil Sci. , 28:541-552.

---

[1] All arable and Top ten farms for the Region, respectively (Murphy, 1995).

Ball, B.C. 1994. Experience with minimum and zero tillage in Scotland. pp. 15-24. *In* Experience with the applicability of no-tillage crop production in the western-European countries. Proc. of EC-Workshop. Giessen.

Berry, C.L. 1988. Pesticides. Human toxicology. 7:5, 433-436.

Bizwanger, H.P. 1985. Agricultural mechanisation. A comparative historical perspective. World Bank Staff Working Papers. No. 673. 80 pp.

Boardman, J., and R. Evans. 1994. Soil erosion in Britain: a review. pp. 3-12. *In* Conserving soil resources. European perspectives. R.J. Rickson (Ed.) CAB International.

Bowen, H.D. 1981. Alleviating mechanical impedance. pp.21-57. *In* Modifying the root environment to reduce crop stress. G.F. Arkin and H.M. Taylor (Eds.).

Brady, N.C. 1986. The nature and properties of soils. 9th Ed. Macmillan Publ. Co. 750 pp.

Brown, C.D., A.D. Carter, and J.M. Hollis. 1995. Soils and pesticide mobility. pp. 132-184. *In* Environmental behaviour of agrochemicals. Wiley.

Brussaard, L., and H.G. van Faassen. 1994. Effects of soil compaction on soil biota and biological processes. pp. 215-235. *In* Soil compaction in crop production. B.D. Soane and C. van Ouwerkerk (Eds.). Developments in Agr. Eng. 11. Elsevier.

Cannel, R.Q., and J.D. Howes. 1994. Trends in tillage practices in relation to sustainable crop production with especial reference to temperate climates. Soil and Tillage Res. 30, 2-4:243-282.

Commission of the European Communities. 1988. Energy savings in agriculture and mechanisation. CEC. G Pellizzi, AG Cavalchini and M Lazzari (Eds.). 143 pp.

Commission of the European Communities. 1992. Possibilites offered by new mechanisation systems to reduce agricultural production costs. G. Pellizzi et. al. (Eds.) CEC. 180 pp.

Chamen, W.C.T, E. Audsley, and J.B. Holt. 1994. Economics of gantry and tractor-based zero-traffic systems. pp. 569-595. *In* Soil compaction in crop production. B.D. Soane and C. van Ouwerkerk (Eds.). Developments in Agr. Eng. 11. Elsevier.

Chamen, W.C.T., G.D. Vermeulen, D.J. Campbell, and C. Sommer. 1992. Reduction of traffic-induced soil compaction: A synthesis. Soil and Tillage Res., 24:303-318.

Chester, G. 1993. Operator exposure to pesticides. *In* Application technology for crop protection. G.A. Matthews and E.C. Hislop (Eds.) CAB International pp. 123-144.

Chisci, G. 1994 Perspectives on soil protection measure in Europe. pp. 339-353. *In* Conserving soil resources. European perspectives. R.J. Rickson (Ed.) CAB International.

Christian, D.G. 1994. Experience with direct drilling cereals and reduced cultivation in England. pp. 25-32. *In* Experience with the applicability of no-tillage crop production in the western-European countries. Proceedings of EC-Workshop. Giessen.

Conway, G.R., and J.N. Pretty. 1991. Unwelcome harvest. Earthscan Publications Ltd. 645 pp.

Czapar, G.F., R.S. Kanwar, and R.S. Fawcett. 1994. Herbicide and tracer movement to field drainage tiles under simulated rainfall conditions. Soil and Tillage Res. 30,1:19-32.

Dickson, J.W. 1994. Compaction by a combine harvester operating on moist, loose soil. Soil Tillage and Res. 29:145-150.

Department of the Environmental Protection.1993. Digest of environmental protection and water statistics.DOE. London: HMSO. 149 pp.

Dudley, N. 1990. Nitrates and the threat to food and water. Green Print London. 118 pp.

Edwards, C.A., D. Pimentel, and M. Lehman. 1993. The impact of pesticides in the environment. In The pesticide question, environment, economics and ethics. 13-46.

Elliot, E.T. et al. 1994. Terrestrial carbon pools in grasslands and agricultural soils: Preliminary data from the Cornbelt and Great Plain regions. pp. 179-191. In Defining soil quality for a sustainable environment. J.W. Doran, D.C. Coleman, D.F. Bezdicek, and B.A. Stewart (Eds.). SSSA Special Publ. 25.

Eradat-Oskoui, K.E., and W.B. Voorhees. 1990. Economic consequences of soil compaction. Paper No. 901089. ASAE.

Erbach, D.C. 1994 . Benefits of tracked vehicles in crop production. pp. 501-520. In Soil compaction in crop production. B.D. Soane and C. van Ouwerkerk (Eds.). Developments in Agr. Eng. 11. Elsevier.

Evans, R. 1980. Mechanics of water erosion and their spatial and temporal controls: an empirical viewpoint. pp. 109-128. In Soil erosion. M.J. Kirby and R.P.C. Morgan (Eds.) Wiley.

Gee-Clough. 1980. selection of tyre sizes for agricultural vehicles. J. Agric. Eng. Res. 25:261-278.

Goss, M.J., K.R. Howse, P.W. Lane, D.G. Christian, and G.L. Harris. 1993. Losses of nitrate-nitrogen in water draining from under autumn-sown crops established by direct drilling or mouldboard ploughing. J.Soil Science. 44: 1, 35-48.

Gregory, P.J. 1988. Water and crop growth. pp. 338-377. In Russell's soil conditions and plant growth. A Wild (Ed.)

Hakansson, I., and H, Petelkau. 1994. Benefits of limited axle load. pp. 479-499. In Soil compaction in crop production. B.D. Soane and C. van Ouwerkerk (Eds.). Developments in Agric. Eng. 11. Elsevier.

Hakansson, I. and R.C. Reeder. 1994. Subsoil compaction by vehicles with high axle load: extent, persistence and crop response. Soil-and-Tillage-Res. 29: 2-3, 277-304

Hakansson, I., W.B. Voorhees, and H. Riley. 1988. Vehicle and wheel factors influencing soil compaction and crop response in different traffic regimes. Soil and Tillage Res. 11:239-282.

Hall, J.K., and R.D. Mumma. 1994. Dicamba mobility in conventionally tilled and non-tilled soil. Soil and Tillage Res. 30,1:3-17.

Harrod, T.R., T.D. Carter, and J.M. Hollis. 1990. The role of soil organic matter in pesticide movement via runoff, soil erosion and leaching. Advances in soil organic matter research. Proceedings of A Symposium. Colchester. WS Wilson (Ed.) 129-138.

Hendrix, P.F., D.A. Crossley, J.M. Blair, and D.C. Coleman. 1990. Soil biota as components of sustainable agrosystems. pp. 637-654. *In* Sustainable agricultural systems. C.A. Edwards, R. Lal, P. Madden, R.H. Miller and G. House (Eds.).

Hislop, E. C. 1993. Application technology for crop protection: An introduction. pp. 3-11. *In* Application technology for crop protection. G.A. Matthews and E.C. Hislop (Eds.). CAB International.

IUCN 1991. Caring for the Earth: a strategy for sustainable living. 228 pp.

Jenkinson, D.S. 1988. Soil organic matter and its dynamics. pp. 564-607. *In* Russell's soil conditions and plant growth. A Wild (Ed.).

Lal, R., and F.J. Pierce. 1991. The vanishing resource. pp. 1-5. *In* Soil management for sustainability. R. Lal and F.J. Pierce (Eds.). Soil and Water Conservation Soc.

Lal, R., D.J. Eckert, N.R. Fausey, and W.M. Edwards. 1990. Conservation tillage in sustainable agricultural systems. pp. 203-225 *In* Sustainable agricultural systems. C.A. Edwards et. al. (Eds.).

Lal, R. 1994. Water management in various crop production systems related to soil tillage. Soil and Tillage Res. 30:169-185.

Lefebvre, A.H. 1993. Droplet production. pp. 35-54. *In* Application technology for crop protection. G.A. Matthews and E.C. Hislop (Eds.). CAB International.

Logan, T.J. 1990. Sustainable agriculture and water quality. pp.582-613. *In* Sustainable agricultural systems. C.A. Edwards, R. Lal, P. Madden, R.H. Miller and G. House (Eds.).

MAFF. 1991. Code of good agricultural practice for the protection of soil. Ministry of Agriculture Fisheries and Food. Welsh office. 55 pp.

MAFF. 1992. Pesticides, cereal farming and the environment. The Boxworth Project. P. Greig-Smith, G. Frampton and J. Hardy (Eds.). HMSO London. 288 pp.

Matthews, G.A., and E.C. Hislop. 1993. Application technology for crop protection CAB International. 359 pp.

Matthews, G.A. 1992. Pesticide application methods. 2nd Edition. Longman. 405 pp.

Mc Grath, P.S., and P.S. Loveland. 1992. The soil geochemical atlas of England and Wales. Blackie Academic and Professional.

Miller, P.C.H. 1993. Spray drift and its measurement. pp. 101-122. *In* Application technology for crop protection. G.A. Matthews and E.C. Hislop (Eds.). CAB International.

Moller, N., and E.T. Svenson. 1991. Modern techniques in application of granular fertilisers. Proceedings Fertiliser Society. 1991. No. 311, 27 pp.

Monk, A.S., D.D.V. Morgan, J. Morris, R.M. Radley. 1984. The cost of farm machinery accidents. Occas. Paper. No. 13. Silsoe College.

Morgan, R.P.C., and R.J. Rickson. 1990. Issues on soil erosion in Europe: The need for a soil conservation policy. pp. 591-603. *In* J. Boardman , I.D.I. Foster and J.A. Dearing (Eds.), Soil Erosion on Agricultural Land. Wiley.

Murphy, M.C. 1995. Report on farming in the Easter countries. Cambridge University Press. 277 pp.

Nix, J. 1995. Farm management pocketbook. 26th Ed. Wye College.

O'Callaghan, J.R. 1994. Resource utilisation and economy of soil tillage in crop production systems. Soil and Tillage Res.. 30:327-343.

Oldeman, L.R., R.T.A. Hakkeling, and W.G. Sombroek. 1991. World map of the status of human-induced soil degradation: An explanatory note. ISRIC, Wageningen, Netherlands/UNEP, Nairobi, Kenya, 34 pp.

Patterson, D.E., W.C.T. Chamen, and C.D. Richardson. 1980. Long term experiments with tillage systems to improve the economy of cultivation of cereals. J. Agric. Eng. Res. 25, 1-35.

Payne, D. 1988. Soil structure, tilth and mechanical behaviour. pp. 378-341. *In* Russell's soil conditions and plant growth. A Wild (Ed.).

Petelkau, H., and M. Donnowski. 1990. Effect of repeated vehicle traffic in traffic lanes on soil physical properties, nutrient uptake and yield of oats. Soil and Tillage Res. 15:217-225.

Pfalzer, H. 1993. Safety aspects and legislation trends. pp. 13-32. *In* Application technology for crop protection. G.A. Matthews and E.C. Hislop (Eds.). CAB International.

Pierce, F.J. 1991. Erosion productivity impact prediction. pp. 35-52. *In* Soil management for sustainability. R. Lal and F.J. Pierce (Eds.). Soil and Water Conservation Soc.

Roberts, T.R. 1991. Pesticides in water: human health, agriculture and environmental aspects. pp. 429-443. *In* Chemistry, Agriculture and the Environment. M.L. Richards (Ed.). Cambridge Royal Soc. of Chemistry.

Robinson, T.H. 1993. Large-scale ground-base application techniques. pp. 163-186. *In* Application technology for crop protection. G.A. Matthews and E.C. Hislop (Eds.). CAB International.

Sharpley, A.N., and S.J. Smith. 1994. Wheat tillage and water quality in the Southern Plains. Soil and Tillage Res. 30(1):33-48.

Sibessen, E., A.C. Hansen, J.D. Nielsen, and T. Heidman. 1994. Runoff, erosion and phosphorus loss from various cropping systems in Denmark. pp. 87-93. *In* Conserving soil resources. European perspectives. R.J. Rickson (Ed.) CAB International.

Soane, B.D., and C. van Ouwerkerk. 1995. Implications of soil compaction in crop production for the quality of the environment. Soil and Tillage Res. 35:5-22.

Soane, B.D., and C. van Ouwerkerk. 1994. Soil compaction in crop production.

Soane, B.D., P.S. Backwell, J.W. Dickson and D.J. Painter. 1980a. Compaction by agricultural vehicles: A review. I. Soil and wheel characteristics. Soil and Tillage Res. 1:207-237.

Soane, B.D., P.S. Backwell, J.W. Dickson and D.J. Painter. 1980b. Compaction by agricultural vehicles: A review. II. Compaction under tyres and other running gear. Soil and Tillage Res. 1:373-400.

Soane, B.D., J.W. Dickson, and D.J. Campbell. 1982. Compaction by agricultural vehicles: A review. III. Incidence and control of compaction in crop production. Soil and Tillage Res. 2:3-36.

Soane, B.D. 1990. The role of organic matter in soil compactibility: A review of some practical aspects. Soil Tillage and Res. 16:179-201.

Spoor, G., and S.E.B. Tatham. 1994. Minimising soil erosion and runoff in tractor wheeling. pp. 365-370. *In* Proceedings of 13th International Conference ISTRO. Vol. 1.

Spoor, G. 1975. Physical conditions and crop production. MAFF. Technical Bull. 29, 128 pp.

Spoor, G. 1982. The causes and nature of soil damage. The Agric. Eng. 37(1):4-7.

Spoor, G. 1991. Ploughing and non-ploughing techniques. Soil and Tillage Res. 21:177-183.

Stout, B.A. 1990. Handbook of energy for world agriculture. Elsevier. 504 pp.

Taylor, A.E.B., P.W. O'Callaghan, and S.D. Probert. 1993. Energy audit of an English farm. Applied energy 44: 315-335.

Tebrugge, F. 1994. Experience with the applicability of no-tillage crop production in the western-European countries. Proceedings of EC-Workshop. Giessen. 162 pp.

Torbert, H.A., and C.W. Wood. 1992. Effects of soil compaction and water-filled pore space on soil microbial activity and N losses. Communication. Soil Sci. Plant Anal. 23:1321-1331.

Turner, R.K. 1995. Sustainability principles and practice. pp. 3-36. *In* Sustainable environmental economics and management: principles and practice. R.K. Turner (Ed.).Wiley.

Vermeulen, G.D., and U.D. Perdock. 1994. Benefits of low ground pressure tyre equipment. pp. 447-478 *In* Soil compaction in crop production. B.D. Soane and C. van Ouwerkerk (Eds.). Developments in Agr. Eng. 11. Elsevier.

Voorhees, W.B., R.A. Young, and L. Lyles. 1979. Wheel traffic consideration in erosion research. Transactions of ASAE 32:786-790.

Watson, R.T., H. Rode, H. Oeschger, and U. Siegenthaler. 1993. Greenhouse gases and aerosols. pp. 1-40. *In* Climate change. The IPCC Scientific Assessment. J.T. Houghton, G.J. Jenkins and J.J. Ephramus (Eds.). Cambridge Univ. Press.

Watts, C.A., and A.R. Dexter. 1994. Traffic and seasonal influences on the energy required for cultivation and on the subsequent tilth. Soil and Tillage Res. 31:303-322.

Wild, A., and L.H.P. Jones. 1988. Mineral nutrition of crop plants. pp. 69-112. *In* Russell's soil conditions and plant growth. A Wild (Ed.)

Wilson, S.J., and R.U. Cooke. 1980. Wind erosion. pp. 217-251. *In* Soil erosion. M.J. Kirby and R.P.C. Morgan (Eds.). Wiley.

Witney, B.D. 1988. Choosing and using farm machines. Longman.

World Resources Institute. 1995. World resources 1992-93. Oxford Univ. Press.

Soil-Specific Production Strategies and Agricultural Contamination Levels in Northeast Kansas

S. Koo
J.R. Williams

Department of Agricultural Economics
Kansas State University
Manhattan, Kansas

INTRODUCTION

Agricultural activities are widely considered to be among most significant sources of nitrate (NO_3) contamination (Fletcher, 1991). Environmental degradation, particularly of water, can occur from excessive use or improper handling or application of nutrients (USDA, 1995). A major concern is the increase in fertilizer applications; the U.S. per-acre application rate doubled between 1965 and 1984 (Nielsen & Lee, 1987). However, the relative use of nitrogen (N) increased much more rapidly. Nitrogen use in 1960 was about 37 percent of total commercial nutrient use, but in 1993 was 11.4 million tons, or 55 percent of total commercial nutrient use (USDA, 1995). Stochastic characteristics of nature itself also contribute significantly to contamination from agriculture. The recent study by Wu et al. (1994) shows that natural factors cause variation in the amounts of nitrogen runoff and percolation to the water resource from many cropping systems used in the High Plains. Factors that affect water quality vary with climate, topography, soils, and other region- and site-specific parameters (Crowder & Young, 1988). Properties of the site, such as soil and geologic formations, are important in determining potential for nutrient and pesticide movement to surface or groundwater.

The U.S. Environmental Protection Agency estimates that between 50 and 70 percent of assessed surface waters are impacted adversely by agricultural nonpoint source pollution. The potential for such pollution is especially high in the Midwest because of the extensive use of N fertilizers and pesticides (Prato et al., 1995). In the Corn Belt, Great Plains, the lower Mississippi Valley, and the Pacific Northwest, application of N fertilizer is concentrated on three major crops: corn, sorghum, and wheat. The potential for NO_3 leaching below the zone of rooting is greatest where these major crops are grown and most commercial fertilizers are applied (Knox & Moody, 1991).

The relationship between nitrogen and nitrate contamination is somewhat different than the relationship between other chemical use and contamination. Other inorganic chemicals such as herbicides and insecticides have a partially proportional relationship between the amount of input and the level of contamination. In addition, N produces complex outcomes that involve more conflicts in both economic and environmental decision-making processes because of its multiple sources, such as inorganic purchased fertilizer, organic animal manure, cover crops, and crop residues. The adverse effects of N contamination include not only the degradation of surface and groundwater quality by its runoff and leachate, but also the negative economic impact of lowered soil productivity.

This study focuses on NO_3 contamination of water resources from typical agricultural production systems in the northeast Kansas. The general objectives were: (1) to estimate the economic and environmental outcomes of conventional and alternative cropping systems and (2) to develop a mathematical risk model to select the optimum cropping system under different attitudes to risk and multiple economic and/or environmental objectives. Soil-specific factors are some of the most important components in producing various levels of crop yields and contaminants by soil types. The soil-specific factors, including soil property and hydrologic characteristics, have major influences on making individual decisions or relevant policies for the specific region. Two different soil types typical of the northeast Kansas were considered in this study.

Northeastern Kansas lies in the transition zone between the subhumid Corn Belt and the semiarid Great Plains. A variety of crops are produced in this region. The primary reason for choosing this area is that it is located close to towns with large populations such as Kansas City and Lawrence which rely on surface water for drinking. Another reason is that the typical and conventional production system in northeast Kansas includes many row crops, such as corn and sorghum. These crops have relatively high requirements of inorganic fertilizer use and, therefore, high potential for producing nitrate contamination.

DATA AND METHODOLOGY

Simulation of Crop Yields and Contaminant Loadings

Diebel et al. (1993) conducted on-farm interviews in 1993 with 15 northeast Kansas farmers using alternative cropping practices. They collected information about crop rotations, operation schedules, yields, and equipment needs from each participant. Average characteristics of the 332 farms in the Kansas Farm Management Association in the 14-county study area for the period 1986-1990 were used to determine the representative farm. The average dryland crop area is 260 hectares, with 40 percent owned and 60 percent rented. The survey data and ongoing research at the Corn Belt Experiment Station provided information to construct field operation schedules that were used to estimate costs in the study.

Five kinds of farming systems were identified, including one conventional and four alternative farms. The conventional farm is representative of commercial farms in northeast Kansas. The conventional and alternative cropping systems are illustrated in Table 1. The 260 hectares of the conventional farm are distributed among four major crops: wheat, grain sorghum, soybeans, and corn. These crops are components in the five crop rotations common in northeast Kansas. Alternative system 1 has 87 hectares each planted to wheat, sorghum, and soybeans. The total acreage in Alternative system 2 is divided equally between sorghum and wheat, 130 hectares each. In Alternative system 3, alfalfa accounts for 155 hectares, with

Table 1. Fertilizer application rates and tillage practices for the cropping systems.

Cropping System†	Fertilizer	Rate per Hectare	Number of Tillage Operations (Type)
Conventional			
C-Sb	Anhydrous (C)	100.87 kg	7 (disc and chisel)
Sg-Sb	Anhydrous (Sg)	100.87 kg	7 (disc and chisel)
W-Sg	28-0-0 (W)	240.98 kg	6 (disc and chisel)
	Anhydrous (Sg)	135.30 kg	
W-Sb	Potassium (W)	67.25 kg	5 (disc and chisel)
	Ammonium (W)	128.89 kg	
C-C	Anhydrous (C)	201.75 kg	4 (disc and chisel)
	18-46-0 (C)	78.46 kg	
Alternative 1			
W-Sg-Sb	18-46-0 (Sg)	33.62 kg	8 (disc)
	Urea (Sg)	112.08 kg	
	18-46-0 (W)	33.62 kg	
	Urea (W)	78.46 kg	
Alternative 2			
W-Sg	34-0-0 (W)	168.12 kg	8 (disc)
	18-46-0 (W)	112.08 kg	
Alternative 3			
A-A-A-W-Sb	18-46-0 (W)	33.62 kg	13 (disc and plow)
Alternative 4	None		13 (disc and plow)
C-Sb-C-Sb-A-A-A			

† A = alfalfa, C = corn, Sb = soybeans, Sg = grain sorghum, and W = wheat.

the rest used for wheat and soybeans. Alfalfa was harvested three times in the second year and once in the third year. Alternative system 4 consists of corn, soybeans, and alfalfa with a 7-year rotation. Harvesting of alfalfa occurs three times in years 2 and 3.

The Conventional System uses the most fertilizer of all the systems. Alternative systems 1, 2, and 3 use several chemicals; however, the amounts applied are smaller than those in the Conventional System.

The chemical transport/crop growth simulation model, Erosion Productivity Impact Calculator (EPIC), was used to simulate potential crop yields and contaminant loadings under each system. EPIC is a comprehensive model developed to determine the relationship between soil erosion and soil productivity throughout the United States. The EPIC physical components include hydrology, weather simulation, erosion-sedimentation, nutrient cycling, plant growth, tillage, and soil temperature (Williams et al., 1990). Three major components in the simulation for this study are crop yields, NO_3 runoff, and NO_3 leachate.

The model requires information regarding weather, soil, and tillage operations and chemical application schedules for each cropping system. To obtain the historical distributions of crop yields and contaminant loadings, the weather simulation component of EPIC was run for 20 years. Two representative soil types,

Pawnee and Grundy, were selected to assess soil-specific factors affecting crop yields and contamination.

Calculations of Net Return and Contaminant Loadings

EPIC simulated each cropping system once for every year in the rotation sequence. The starting dates were lagged each year, so that each crop was simulated for each of the 20 years in the study period. The annual contaminant levels and crop yields were calculated by weighting by the contribution of each crop to total hectares grown. To calculate annual net returns from each system, price and cost information from enterprise budgets was combined with the crop yields simulated by EPIC. Only the variation of crop yields is considered in the calculation of net returns, because costs and prices are held constant over the 20 years. Net return in this study is net return to management. Variable costs include the costs of labor, seed, herbicide, insecticide, fertilizer, fuel, oil, equipment repair, custom hire, and interest on variable cost. The fixed costs include real estate taxes, interest on land mortgages, share rent, depreciation and interest on machinery loans, and insurance and housing. Data regarding these costs and prices were collected from Kansas Farm Management Association, Kansas Agricultural Statistics, northeast Kansas cooperatives, and the previously mentioned producer survey.

Table 2 contains the 20-year average net returns and average contaminant loadings for Conventional and Alternative Systems for each soil type. The minimum and maximum values, standard deviations, and coefficients of variation (CV) also are included. CV is the standard deviation divided by the mean and represents a measure of relative risk. According to the expected utility theory, given the probability distributions of economic goods, decision makers who are risk averse would maximize expected value and minimize variance. In this study, a manager who is generally risk averse would prefer the system with the combination of the highest mean for net return and the lowest standard deviation.

The results indicate there is no cropping system which meets this criteria. Alternative System 3 has the highest net return. It also has the highest NO_3 leachate but lowest NO_3 runoff for Pawnee soil. Alternative System 1 would be preferred by extreme risk averse manager because this system has the smallest minimum observation of net return. However, this system has relatively high levels of NO_3 runoff. Alternative System 2 has the lowest levels of NO_3 leachate for the Pawnee soil but it also has the lowest net return. Alternative System 4 has the lowest NO_3 leachate of system on the Grundy soil. These results indicate that there is not a single system that minimizes contaminants and maximize net returns while giving a low level of risk.

Table 2. Twenty-year average net returns and average contaminant loadings for conventional and alternative cropping systems for each soil type.

Variable†	Pawnee Soil					Grundy Soil				
	Cropping Systems					Cropping Systems				
	C	A1	A2	A3	A4	C	A1	A2	A3	A4
NR	92.7	89.2	21.4	218.8	107.0	149.3	157.9	33.6	249.1	148.2
SD	77.1	64.5	51.3	95.1	72.4	67.2	88.4	45.3	51.2	75.3
CV	83.2	72.3	240.0	43.5	68.0	45.0	56.0	135.0	20.6	50.8
Max	220.9	189.3	80.4	304.9	200.7	249.5	450.9	95.1	311.9	225.9
Min	-71.6	-40.2	-96.5	-115	-54.4	-48.5	-17.9	-51.9	98.3	-41.7
$NO_3 R$	4.2	4.3	3.9	3.4	3.7	6.1	6.3	4.7	4.6	4.9
SD	1.6	1.8	2.0	1.2	1.3	2.5	2.4	2.3	1.7	1.7
CV	37.0	40.3	51.0	36.1	35.8	40.4	38.6	48.7	37.0	35.1
Max	6.8	9.0	10.5	5.8	5.6	10.2	11.0	9.5	8.2	7.7
Min	1.0	1.3	1.0	1.0	0.7	0.8	1.0	0.5	1.0	1.0
$NO_3 L$	25.8	20.6	15.5	35.5	23.8	6.2	3.9	3.3	5.5	1.9
SD	19.6	16.6	16.2	17.4	13.1	8.5	6.1	5.4	5.7	3.4
CV	75.9	80.3	104.6	49.2	55.0	137.0	156.2	162.6	103.9	182.2
Max	71.9	56.0	58.0	70.5	59.2	36.5	27.4	20.5	19.4	11.7
Min	0.9	0.7	0.5	3.6	0.3	0.0	0.0	0.0	0.0	0.0

† NR = Net Return ($/ha), $NO_3 R$ = NO_3 Runoff (kg/ha), $NO_3 L$ = NO_3 Leachate (kg/ha), SD = Standard Deviation, CV = Coefficient of Variation, Max = Maximum, and Min = Minimum.

Mathematical Risk Model

When the decision maker faces an environment of multiple goals that cannot be determined by mean-variance criteria, one approach to be considered is multiobjective programming. Underlying the assumption and theory regarding individuals' risk averting behavior, goal programming (GP) is formulated to find the optimal production system when considering multiple objectives. In the case of GP, goals and constraints have the same mathematical structure and appear exactly the same because both of them are equalities. The difference between them lies in the meaning attached to the right-hand side of the equalities (Romero & Rehman, 1989). In this study, weighted goal programming (WGP) is used to minimize the weighted sum of deviations between the individual objective functions and their corresponding goals.

The WGP model that minimizes the sum of the percentage deviations from target levels of net returns, contaminant loadings, and the associated risk for this study is as follows:

$$Minimize \ W_{RE}^-(\frac{y_{RE}^-}{TRE})(\frac{100}{1}) \ + \ W_{RECV}^+(\frac{y_{RECV}^+}{TCV_{RE}})(\frac{100}{1})$$

$$+ \ W_{iCL}^+(\frac{y_{CL}^+}{TCL_i})(\frac{100}{1}) \ + \ W_{iCV}^+(\frac{y_{CV}^+}{TCV_i})(\frac{100}{1}) \tag{1}$$

subject to

$$\sum_{j=1}^{n} RE_j X_j \ - \ (y_{RE}^+ - y_{RE}^-) \ = \ TRE \tag{2}$$

$$\sum_{j=1}^{n} CV_{REj} X_j \ - \ (y_{RECV}^+ - y_{RECV}^-) \ = \ TCV_{RE} \tag{3}$$

$$\sum_{j=1}^{n} CL_{ij} X_j \ - \ (y_{iCL}^+ - y_{iCL}^-) \ = \ TCL_i \ \ for \ i=NR \ and \ NL. \tag{4}$$

$$\sum_{j=1}^{n} CV_{ij} X_j \ - \ (y_{iCV}^+ - y_{iCV}^-) \ = \ TCV_i \ \ for \ i=NR \ and \ NL. \tag{5}$$

$$\sum_{j=1}^{n} l_j X_j \ = \ 1 \tag{6}$$

and

$$x_j \geq 0; \ \ y^{+\prime}s \geq 0; \ \ y^{-\prime}s \geq 0. \tag{7}$$

where: W_{RE}^- = penalty weight for underachieving the target level of net return,
$\quad\quad\ W_{CL}^+$ = penalty weight for exceeding target loading level of the contaminant i,
$\quad\quad\ W_{RECV}^+$ = penalty weight for exceeding target CV (Coefficient of Variation) level of net return,
$\quad\quad\ W_{CV}^+$ = penalty weight for exceeding target CV level of the contaminant i,
$\quad\quad\ y^+$ = auxiliary variable for exceeding the targets,
$\quad\quad\ y^-$ = auxiliary variable for under achieving the targets,
$\quad\quad\ RE_j$ = expected net return of the jth cropping system,
$\quad\quad\ CL_{ij}$ = expected contaminant i's loading of the jth cropping system,
$\quad\quad\ CV_{ij}$ = contaminant i's CV of the jth cropping system,
$\quad\quad\ TRE$ = target level of net return,
$\quad\quad\ TCL_i$ = target level of contaminant i's loading,
$\quad\quad\ TCV_{RE}$ = target level of net return's CV,
$\quad\quad\ TCV_i$ = target level of contaminant i's CV, and
$\quad\quad\ i's$ = contaminant loadings on NO$_3$ runoff (NR) and NO$_3$ leachate (NL).

Inclusion of the coefficients of variation to account for risk factor is from Fiske et al. (1994). The penalty weights are determined arbitrarily by the decision makers. The magnitudes of the penalty weights are in proportion to the relative importance or priority among the components in the model; e.g., net return, contaminant loadings, and CV's in this study. Various scenarios targeting multiple goals can be examined by modification of right-hand side values of the constraints or by combining penalty weight scenarios. Given the penalty weights and target levels of net return, NO_3 loadings, and the associated risks, the optimum cropping system chosen must minimize the underachieved net returns, and overachieved NO_3 loadings and risks from the targets. A mixed integer programming technique allows the model to choose only one optimum system.

Three goal scenarios are represented in Table 3. A, B, and C represent a base, a 30 percent increase from base, and a 30 percent decrease from base, respectively, for each selected soil type. Base scenarios consist of the median values of the average and CV values of net returns, and potential NO_3 loadings over the cropping systems. Each contaminant with its CV (risk factor) is combined with net return and its CV for the individual scenarios. All the contaminants can be included at the same time, but this was avoided because of the model restrictiveness in this study. Penalty weight scenarios represent the importance attached by the decision maker to each attribute in the objective function (Romero & Rehman, 1989).

Each penalty weight scenario is combined with goal scenarios: for example, AP1 in Table 4 implies that the base scenario (A) is combined with the first penalty scenario (1) under Pawnee soil (P). The following numbers were used to construct penalty weight scenarios: 1, 3, 6, and 9. Scenarios 1 to 8 focus on the situation in which the decision maker places higher weights, that is higher priorities, on target levels of net return and contaminant loadings than risk factors (CV's). Risk factors receive more emphasis in scenarios 9 to 16.

RESULTS AND CONCLUSION

Tables 4 and 5 represent the results from the WGP, including the optimal cropping systems and those weighted deviations under the alternative scenarios constructed in Table 3. Weighted deviation is the objective function value, which is the minimized sum of the deviations from the targets. Under the AP scenarios (base scenario for Pawnee soil), Alternatives 3 and 4 are dominant. Alternative 3 is the optimum system with the maximized expected net return, when the target contaminant is NO_3 runoff. Alternative 4 is the optimum when NO_3 leachate is the target contaminant while maximizing the expected net returns. When the targets are increased by 30 percent (scenario BP), Alternative 3 is more dominant than Alternative 4. Under the CP scenarios, three alternative systems compete. Alternative 3 is dominant in the cases for which NO_3 runoff is included in the model. Alternatives 1 and 4 compete when NO_3 leachate is included in the scenarios. However, Alternative 4 is more frequent when risk factors are emphasized more than target amount.

For Grundy soil (Table 5), Alternatives 1, 3, and 4 are optimal under the base scenario (AG1 to AG16). Alternative 4 is optimal when NO_3 runoff is

Table 3. Composite goal and penalty weight scenarios.

Scenarios	TRE† TCV_NL ($/ha)	TCV_RE	TCL_NR (kg/ha)	TCV_NR	TCL_NL (kg/ha)	
Goal‡						
AP	105.8	103.9	3.9	48.1	24.3	74.9
BP	137.5	135.1	5.1	53.4	31.5	97.3
CP	74.1	72.8	2.7	28.8	17.0	52.4
AG	147.6	63.1	5.3	41.1	4.2	152.3
BG	191.9	82.0	6.9	53.3	5.4	197.9
CG	103.3	41.2	3.7	28.7	2.9	106.5
Penalty Weight						
1	9	3	6	1		
2	9	1	6	3		
3	6	3	9	1		
4	6	1	9	3		
5	9	3			6	1
6	9	1			6	3
7	6	3			9	1
8	6	1			9	3
9	3	9	1	6		
10	1	9	3	6		
11	3	6	1	9		
12	1	6	3	9		
13	3	9			1	6
14	1	9			3	6
15	3	6			1	9
16	1	6			3	9

†TRE = target level of net return, TCV_{RE} = target level of net return's CV, TCL_{NR} = target level of NO_3 runoff, TCV_{NR} = target level of NO_3 runoff's CV, TCL_{NL} = target level of NO_3 leachate, and TCV_{NL} = target level of NO_3 leachate's CV.

‡ A, B, and C represent the base scenario, 30 percent increase from the base, and 30 percent decrease from the base, respectively. P and G represent Pawnee and Grundy soil, respectively. These goal scenarios are combined with the alternative combinations of the penalty weight scenarios.

included in the model. Alternative 1 is dominant in other cases. Alternative 3, however, is optimal for all the cases of Scenario B (BG's). Under Scenario C, Alternative 3 is again the optimal except for cases CG5 to CG8, which target levels of net return and NO_3 leachate have the higher weights.

The overall results from the WGP model indicate that alternatives 1, 3, and 4 may be environmentally more beneficial than the conventional system, while providing economic incentives. Alternative 2 is not chosen at all under the alternative combination of scenarios of this study. The chemical- and tillage-intensive conventional system is not selected under the alternative scenarios. Although some of the optimal systems are found to be similar for a specific soil (Pawnee) and for certain target scenarios (Scenario B), different solutions exist for each set of scenarios across the soil types, implying the site- and soil- specific importance of environmental policies for various agricultural production systems.

Table 4. Optimal cropping systems under the alternative scenarios† for Pawnee soil.

Scenarios	Optimal Weighted System	Deviation	Scenarios	Optimal Weighted System	Deviation	Scenarios	Optimal Weighted System	Deviation
AP1	A3	0.00	BP1	A3	0.00	CP1	A3	168.25
AP2	A3	0.00	BP2	A3	0.00	CP2	A3	226.13
AP3	A3	0.00	BP3	A3	0.00	CP3	A3	237.90
AP4	A3	0.00	BP4	A3	0.00	CP4	A3	295.78
AP5	A4	0.00	BP5	A3	74.98	CP5	A1	192.43
AP6	A4	0.00	BP6	A3	74.98	CP6	A4	263.07
AP7	A4	0.00	BP7	A3	112.47	CP7	A1	257.18
AP8	A4	0.00	BP8	A3	112.47	CP8	A1	367.65
AP9	A3	0.00	BP9	A3	0.00	CP9	A3	196.88
AP10	A3	0.00	BP10	A3	0.00	CP10	A3	243.32
AP11	A3	0.00	BP11	A3	0.00	CP11	A4	282.62
AP12	A3	0.00	BP12	A3	0.00	CP12	A3	330.14
AP13	A4	0.00	BP13	A3	12.50	CP13	A4	85.81
AP14	A4	0.00	BP14	A4	22.23	CP14	A4	165.87
AP15	A4	0.00	BP15	A3	12.50	CP15	A4	108.70
AP16	A4	0.00	BP16	A4	22.23	CP16	A4	188.76

† For example, AP1 stands for the base scenario (A) with Pawnee soil (P) and the first penalty weight scenario (1).

Table 5. Optimal cropping systems under the alternative scenarios† for Grundy soil.

Scenarios	Optimal Weighted System	Deviation	Scenarios	Optimal Weighted System	Deviation	Scenarios	Optimal Weighted System	Deviation
AG1	A4	0.00	BG1	A3	0.00	CG1	A3	173.74
AG2	A4	0.00	BG2	A3	0.00	CG2	A3	238.57
AG3	A4	0.00	BG3	A3	0.00	CG3	A3	244.40
AG4	A4	0.00	BG4	A3	0.00	CG4	A3	309.23
AG5	A1	5.33	BG5	A3	10.55	CG5	A4	129.67
AG6	A1	15.99	BG6	A3	10.55	CG6	A4	244.39
AG7	A1	5.33	BG7	A3	15.82	CG7	A4	129.67
AG8	A1	15.99	BG8	A3	15.82	CG8	A4	244.39
AG9	A4	0.00	BG9	A3	0.00	CG9	A3	218.06
AG10	A4	0.00	BG10	A3	0.00	CG10	A3	265.17
AG11	A4	0.00	BG11	A3	0.00	CG11	A3	315.32
AG12	A4	0.00	BG12	A3	0.00	CG12	A3	362.42
AG13	A1	31.97	BG13	A3	1.76	CG13	A3	88.98
AG14	A1	31.97	BG14	A3	5.27	CG14	A3	266.94
AG15	A3	32.29	BG15	A3	1.76	CG15	A3	88.98
AG16	A1	47.96	BG16	A3	5.27	CG16	A3	266.94

† For example, AG1 stands for the base scenario (A) with Grundy soil (G) and the first penalty weight scenario (1).

One of the greatest limitations of this study revolves around data availability. Prices and costs are held constant in this study. Crop yield is a factor only in the distributions of net returns. However, this assumption can cause some bias to support certain cropping systems, because prices and costs also have random characteristics and can be functions of crop yields over time. Another data limitation arises because this study assumes a representative farm case, the average of 14 counties in the study area; therefore, the simulation model does not use specific field observations.

REFERENCES

Crowder, B.M., and E.C. Young. 1988. Managing farm nutrients: Tradeoffs for surface- and ground-water quality. Agricultural Economic Report No. 583, U.S. Dep. Agric., Economic Research Service, Washington, DC.

Diebel, P. L., R.V. Llewelyn, and J.R. Williams. 1993. An economic analysis of conventional and alternative cropping systems for northeast Kansas. Report of Progress 687, Agric. Exp. Stn., Kansas State Univ., Manhattan, KS.

Fiske, W.A., G.E. D'Souza, J.J. Fletcher, T.T. Phipps, W.B. Bryan, and E.C. Prigge. 1994. An economic and environmental assessment of alternative forage-resource production systems: A goal-programming approach. Agricultural Systems 45.

Fletcher, D.A. 1991. A national perspective. Chapter 2 in Nitrogen Managing for Groundwater Quality and Farm Profitability. Soil Science Society of America, Madison, WI.

Knox, E., and D.W. Moody. 1991. Influence of hydrology, soil properties, and agricultural land use on nitrogen in groundwater. Chapter 3 in Nitrogen Managing for Groundwater Quality and Farm Profitability. Soil Science Society of America, Madison, WI.

Nielson, E.G., and L.K. Lee. 1987. The magnitude and costs of groundwater contamination from agricultural chemicals -- A National Perspective. Agricultural Economic Report No. 576, U.S. Dep. Agric., Economic Research Service, Washington, DC.

Prato, T., F. Xu, S.X. Wu, and J.C. Ma. 1995. Costs and returns of alternative farming systems in Missouri MSEA Projects. Proceeding of the Clean Water-Clean Environment-21st Century, Volume III: Practices, Systems & Adoption. ASAE. Kansas City, MO.

Romero, C., and T. Rehman. 1989. Multiple criteria analysis for agricultural decisions. Elsevier, New York.

U.S. Department of Agriculture. 1995. Agricultural resources and environmental indicators. Economic Research Service, Natural Resources and Environment Division, Agricultural Handbook 705.

Williams, J.R., C.A. Jones, and P.T. Dyke. 1990. EPIC-Erosion/Productivity Impact Calculator 1. Model Documentation. Tech. Bull.No. 1768. Agricultural Research Service. U.S. Department of Agriculture, Washington, DC.

Wu, J., D.J. Bernado, H.P. Mapp, S.Geleta, M.L. Teague, K. B. Watkins, G.J. Sabbage, R.L. Elliot, and J.F. Stone. 1994. An evaluation of nitrogen runoff and leaching potential in the high plains. Selected paper presented at the 1994 AAEA Meetings.

Missouri Precision Agriculture Research and Education

N. R. Kitchen

Soil and Atmospheric Sciences
University of Missouri-Columbia
Columbia, Missouri

K. A. Sudduth

USDA-Agricultural Research Service
Cropping Systems and Water Quality Research Unit
Columbia, Missouri

S. J. Birrell
S. C. Borgelt

Biological and Agricultural Engineering
University of Missouri-Columbia
Columbia, Missouri

ABSTRACT

A number of projects related to different facets of precision agriculture are being carried out by an interdisciplinary team of scientists and engineers with the University of Missouri and the USDA Agricultural Research Service. This paper provides an overview of those projects.

BACKGROUND

Precision agriculture research at the University of Missouri began in the 1980's with the development of concepts for grid soil sampling (Buchholz, 1991; Wollenhaupt and Buchholz, 1992). The current precision agriculture research program developed as a part of the Missouri Management Systems Evaluation Area (MSEA) project, and has focused on developing and evaluating precision farming technology on the claypan soils of northern Missouri. Today, the Missouri precision agriculture program exists as an interdisciplinary, multi-organization group of scientists and engineers investigating the methods, technologies, and impacts of precision farming, with emphasis on the application of these systems to several climate-soil regions of Missouri.

"Precision agriculture" really defines itself. It means farming with *preciseness*. It has been given many different titles, including "precision farming", "prescription farming", "farming by soil", and "variable-rate management" to name a few. Unfortunately, many have tried to define precision agriculture by one of its technologies or activities. For example, some may say precision farming is "Global

Positioning Systems" (GPS); others have claimed grid soil sampling for nutrient mapping to be the definition of precision agriculture. Any information gathering, management planning, or field operation that improves the understanding and management of soil and landscape resources so that cropping inputs or management practices (e.g., seed, fertilizers, herbicides, tillage, etc.) are utilized more efficiently than with conventional "one-fits-all" strategies could be called "precision farming".

A complete precision agriculture system must include four basic functions: (1) spatial and temporal data acquisition; (2) data mapping and interpretation for management planning; (3) control of field operations; and (4) evaluation of the agronomic, environmental, and economic effects of precision agriculture. Each cropping year is an iteration through the cycle which includes these four phases of the system.

While the main focus of our research group's activities has been the acquisition and interpretation of spatial within-field data, research and demonstration activities have been conducted in each of these four phases. This paper explains the precision agriculture research and education activities in Missouri within the context of this four-phase system.

PHASE 1: DATA ACQUISITION

Grain Yields

We developed and tested yield data collection systems for both a research combine (initiated in 1992) and the MSEA farm cooperator's combine (initiated in 1993) (Birrell et al., 1996). These systems include continuous grain flow and grain moisture sensors, along with Global Positioning System (GPS) receivers to track combine location during harvest. Yields of corn, wheat, soybeans, and grain sorghum have been mapped for a number of fields, totaling over 300 ha. Most of these areas have been mapped for at least two years, and some for as many as four years. Without exception, yields in the most productive areas of each field were at least double those of the lowest yielding areas (Birrell et al., 1995). Yield trends from successive crops on a field can be quite different due to climatic differences. A dry period early in the 1992 season stressed the crop and the highest yields generally occurred in the drainage areas of the study fields, while for the excessively wet 1993 season the highest yields were obtained in the higher parts of the fields. Dry growing conditions in 1994 again created yield patterns similar to those in 1992. Although moisture appeared to be the dominant factor, soil nutrients, particularly phosphorus, may have affected yield in parts of the fields.

Soils

Four fields at the MSEA site have been intensively sampled to characterize the variation of soil nutrients and the distance over which soil samples are spatially related. We have found all of the soil nutrients to be highly variable and to have different spatial trends, except for calcium, magnesium and CEC which have tended to have the same trends. In many cases the levels of important soil nutrients ranged from very low, which would require fertilization, to very high where no benefit from fertilization would occur. While some nutrients were spatially correlated across the whole field, for others the range of spatial influence was generally 100 m or less. This means that claypan soil fields should be soil sampled at an intensity greater than every 100 m to obtain reasonable data for creating soil fertility maps (Birrell et al., 1995; Birrell et al.,1996).

We have been involved in the development of sensor technology for the measurement of soil moisture, soil organic matter and soil nitrate (Hummel et al., 1996). Soil organic matter and moisture have been successfully estimated using a portable near infrared (NIR) sensor in the laboratory. Work is ongoing to adapt this sensor for on-the-go field mapping. Laboratory development of a soil nitrate sensor based on ion-selective field effect transistors (ISFETs) is also continuing.

Mapping of the topsoil thickness above a claypan layer provides important information on the water retention capacity of the claypan soil profile. Topsoil depth is correlated with yield variations in water-limited growing seasons, which are not uncommon in Missouri. We have found that soil conductivity as measured by electromagnetic induction (EM) sensing can be used to estimate topsoil depth (Sudduth and Kitchen, 1993; Doolittle et al., 1994). To exploit this relationship for precision management, we have developed an automated sensing system for on-the-go whole-field mapping of topsoil depth. While yield by EM model r^2 values have been fairly low, EM sensing has helped to explain crop production variability for most crop years on several different soil types (Kitchen et al., 1995a).

We have proposed a theoretical relationship between EM and grain production (Fig. 1). When EM readings are low, the soil is sandier and has high hydraulic conductivity but low water holding capacity. When EM readings are high, the soil has greater clay content and lower hydraulic conductivity but high water holding capacity. Somewhere in the middle both hydraulic conductivity and

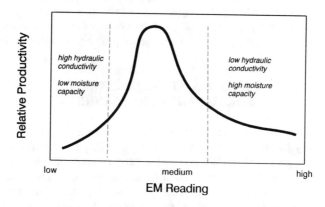

Fig. 1. Theoretical relationship of EM reading to productivity.

water holding capacity are such that total soil water available is optimized and crop production is greatest. At both extremes, total water for plant growth is less and there is a relative decrease in crop productivity.

Remote Sensing

Satellite-based remote sensing has been applied to agriculture for well over 20 years. However, the spatial resolution needed for precision agriculture is only now becoming available. In the last several years there has been developing interest in remote sensing systems designed specifically for precision farming. A number of organizations have tested airplane-based systems which give producers a picture of within-field crop stress variations through the growing season. Plans are for the images to be returned to the grower within two or three days of acquisition. These images then can be used by the grower to scout the field and determine the cause of the crop stress.

We have found that remotely sensed images of crop canopy variations can be related to measured grain yield variations (Sudduth et al., 1996), and may help to locate areas of heavy weed infestation. Historical aerial photographs have also been used to discover the past management history of fields, such as cases where large fields have previously been farmed as a number of separate, smaller fields. These patterns of past management can often be related to within-field patterns of soil nutrient levels.

Other Factors

Yields may also be affected by within-field variations in such factors as crop plant population and weed density. We are collecting data on both corn population (Birrell and Sudduth, 1995) and weeds from the combine at harvest. In some instances, weed density has shown correlation with yield differences. There may also be a relationship between weed density and soil pH variations, due to the effects of pH on the efficacy and degradation of herbicides. Correlations have also been found between population and yield, particularly in those areas where plant numbers were reduced as a result of emergence problems or seedling death from insects.

PHASE 2: DATA MAPPING AND INTERPRETATION FOR MANAGEMENT PLANNING

Data mapping and interpreting the relationships between various mapped data layers is key for understanding within-field variability and for developing site-specific management plans. We have investigated a number of analysis techniques, ranging from simple linear correlation to nonparametric regression and neural network analysis, in an attempt to relate yield variability to variability in soil and topographic properties (Drummond et al., 1995; Sudduth et al., 1996).

It was difficult to interpret the results from linear correlation and regression analysis in a way that yielded meaningful information. We concluded that these problems were likely due to a complex and nonlinear functional relationship between yield and soil properties. Also, the form of the function was likely different from region to region within the field, due to different factors controlling the expression of yield. To deal with these problems, we investigated the use of projection pursuit regression, a nonparametric regression method, and neural network analysis. Predicted yield maps obtained with these methods agreed reasonably well with measured yields, and response curves developed with these methods agreed in general with our observations of yield-limiting behavior on the study fields (Sudduth et al., 1996).

The results of these studies, coupled with our observation that different critical factors often appear to be limiting yields in different parts of the same field, led us to develop a new model for management planning relative to variable-rate fertilizer application.

Management Planning: A Case for "Precise" Variable-Rate Fertilizer Application

The Current Model

The current approach for site-specific fertilizer recommendations in crop production is to map fields by soil type or by grid-soil sampling. A map of nutrient needs for the crop is then generated based on a generalized response relationship between the soil test, fertilizer additions, and yield increase developed for a state or region (see Fig. 2).

Step 1. Measure and map grid-soil sample or soil type and possibly yield information.

↓

Step 2. Site-specific fertilizer plan developed using grid-soil sampled or soil type information. In some cases yield goal might be modified as shown from yield mapping.

↓

Step 3. Whole-field nutrient plan developed and fertilizers variably applied.

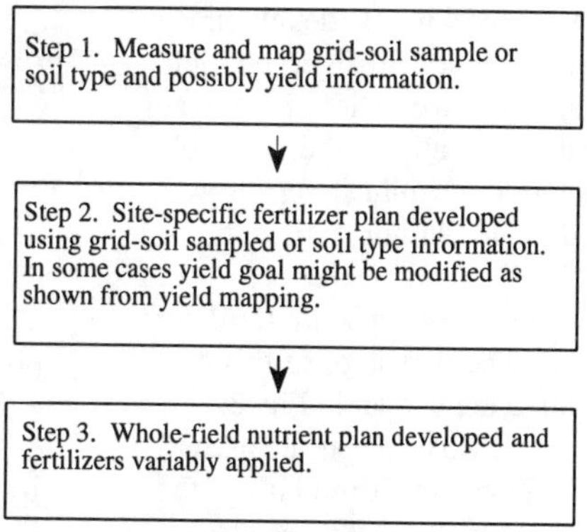

Fig. 2. Current method for site-specific fertilizer recommendations.

This approach is simplistic when compared to the diversity of variability that exists within fields. Our research has demonstrated that nutrient availability is only one of many factors that controls yield. For claypan soils in Missouri, limited plant available water during the growing season will often result in yield variability within fields. Topsoil depth above the claypan is a good indicator of plant available water. Other factors that will influence yields include, but are not limited to, soil pH, soil organic matter, soil compaction, nutrient toxicity, weed and insect pests, crop population, tree hedges along field edges, and poor surface drainage.

When considering these non-nutrient factors, the usual approach as shown in Fig. 2 is really only a beginning. Such use of grid-soil sample maps as the basis for site-specific fertilizer plans will not likely improve utilization of fertilizer nutrients in areas of a field where factors other than nutrients are controlling grain yield. Continued reliance on this strategy may actually hinder the adoption of site-specific management when positive returns are not realized by farmers.

The Proposed New Method

With multiple factors controlling crop growth and productivity within fields, spatial data needs to be sorted and interpreted in a sequence of decisions that will allow for isolation, evaluation, and fertilizer prediction by those areas within a field that are alike in the factor(s) controlling yield. While this requires more processing and careful examination of mapped information, such an approach is necessary in order to recognize and manage the complex nature of field variability in a true site-specific management manner. Presently we are working on developing a method that will interpret field mapped information in a way that will allow the multiple factors controlling crop growth and yield within a field to be isolated and analyzed for developing site-specific fertilizer plans. The five steps of this new method are shown in Fig. 3.

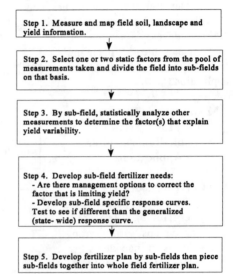

Two concepts are used as guidelines in formulating the new method: (1) the need to use a specific soil, yield, or landscape measurement for variability anal-ysis and the development of vari-able nutrient application plans is based upon whether or not it relates to a crop-growth or yield-controlling factor(s) that can be corrected with management, and (2) mapped information needs to be sorted and analyzed in such a way as to give field-specific sensitivity for predicting fertilizer input needs. With this new method, good quality yield maps from combine yield monitoring are necessary.

Fig. 3. Improved site-specific fertilizer recommendation method under development.

The goal of this new method is to develop and validate a procedure for site-specific fertilizer recommendations that is agronomically, economically, and environmentally precise. The method we are working on provides a framework for systematic analysis of soil and landscape variations and their impact on crop production. A site-specific fertilizer plan derived from using this method will refine fertilizer recommendations and result in improved crop nutrient utilization.

PHASE 3: CONTROL OF FIELD OPERATIONS

Variable-rate fertilizer and herbicide application equipment is available from a number of manufacturers. Equipment for commercial applicators has been on the market for several years, and other equipment targeted at the individual farmer is now becoming available. Variable-rate planters and grain drills are now also available, and allow producers to change seeding rates on-the-go in response to changes in field conditions.

For our research and demonstration work, we have integrated equipment and control systems that allow simultaneous on-the-go changes of four production inputs at planting. Fertilizer (N, P, and K) and seeding rates are controlled by a computer installed in the tractor cab, based on GPS location data and pre-defined maps of desired application rates.

PHASE 4: EVALUATION OF AGRONOMIC, ENVIRONMENTAL, AND ECONOMIC EFFECTS

A necessary step in evaluating the agronomic effects of precision agriculture is understanding the spatial relationships of yields and the various critical factors which may affect yields on a whole-field or a sub-field basis. As such, much of the work described earlier under "Phase 2" also has application for this section. In addition, we have conducted studies to compare the effect of yield-map derived, variable-rate nitrogen (VRN) application to whole-field, single-rate application, both on claypan and alluvial soil sites in Missouri (Kitchen et al., 1995b). Except for one site-year, yields were neither helped nor hurt with VRN as compared to a "best years/best areas of field" single rate. However, VRN did reduce the residual soil nitrate following corn harvest during some years in the less productive portions of the landscape.

This cropping season (1996) we have initiated a whole-field study comparing variable nutrient management (nitrogen, phosphorus, potassium, and lime) to conventional, single-rate management. Like many researchers, we realize that the evaluation of precision management systems as compared to conventional management systems is of great importance for the continued development and acceptance of precision agriculture.

EDUCATION OPPORTUNITIES

As is the case in many areas of the country, producers in Missouri have shown an increased interest in precision agriculture and its technologies over the last couple of years. Most of the research activities listed above have been incorporated into traditional extension education programs such as field days, workshops, producer conferences, and state extension publications. As an outgrowth of our research program, funding has been sought and plans are in place for a more formalized extension program in precision agriculture to begin in late 1996. This program will be led by a full-time Precision Farming Extension Specialist. Initially the extension program will focus on precision technology workshops and statewide demonstrations of variable rate fertilizer applications and the benefits to water quality.

On campus, students have also shown an increased interest in precision agriculture. Beginning in January 1997, a topics course on precision agriculture is planned. We expect that the curriculum in precision agriculture will grow quickly as many graduating students seeking agricultural jobs will need to demonstrate skills and knowledge in this area.

SUMMARY

Precision agriculture presents an exciting opportunity to crop producers in Missouri and around the country. There is currently much activity in both the public and private sectors to provide technologies, services, and management systems for precision agriculture. In Missouri, we have attempted to develop an integrated precision agriculture research program which can make a national impact while at the same time addressing issues unique to Missouri producers.

REFERENCES

Birrell, S.J., and K.A. Sudduth. 1995. Corn population sensor for precision farming. Paper No. 951334, Am. Soc. of Agricultural Engineers, St. Joseph, MI.

Birrell, S.J., K.A. Sudduth, and N.R. Kitchen. 1995. Within-field variations in yields and soils: implications for improved management. p. 158-163. *In* Proc. 5th Annual Missouri Water Quality Conference, Columbia, MO, Feb. 2, 1995. Univ. of MO, Columbia, MO.

Birrell, S.J., K.A. Sudduth, and S.C. Borgelt. 1996. Comparison of sensors and techniques for crop yield mapping. Computers and Electronics in Agriculture 14:215-233.

Birrell, S.J., K.A. Sudduth, and N.R. Kitchen. 1996. Nutrient mapping implications of short-range soil variability. *In:* Proc. 3rd International Conf. on Precision Agriculture, Minneapolis, MN, June 23-26, 1996 (this volume).

Buchholz, D.D. 1991. Missouri grid soil sampling project. p. 6-12. *In* Proc. 21st North Central Extension-Industry Soil Fertility Workshop, St. Louis, MO. Oct. 28-29, 1991. Postash and Phosphate Inst., Manhattan, KS.

Doolittle, J.A., K.A. Sudduth, N.R. Kitchen, and S.J. Indorante. 1994. Estimating depths to claypans using electromagnetic induction methods. Journal of Soil and Water Conservation 49:572-575.

Drummond, S.T., K.A. Sudduth, and S.J. Birrell. 1995. Analysis and correlation methods for spatial data. Paper No. 951135, Am. Soc. of Agricultural Engineers, St. Joseph, MI.

Hummel, J.W., L.D. Gaultney, and K.A. Sudduth. 1996. Soil property sensing for site-specific crop management. Computers and Electronics in Agriculture 14:121-136.

Kitchen, N.R., K.A. Sudduth, S.J. Birrell, and S.T. Drummond. 1995a. Spatial prediction of crop productivity using electromagnetic induction. p. 299. *In* Agronomy Abstracts, 1995 Annual Meetings, Oct. 29-Nov. 3, 1995, St. Louis, MO. ASA, Madison, WI.

Kitchen, N.R., D.F. Hughes, K.A. Sudduth, and S.J. Birrell. 1995b. Comparison of variable rate to single rate nitrogen fertilizer application: corn production and residual soil NO_3-N. p. 427-441. *In:* P.C. Robert, R.H. Rust, and W.E. Larson (ed.) Site-specific management for agricultural systems. Proc. 2nd Intl. Conf., Minneapolis, MN. 27-30 March 1994. ASA, CSSA, and SSSA,

Madison, WI.

Sudduth, K.A., and N.R. Kitchen. 1993. Electromagnetic induction sensing of claypan depth. Paper 931550. Am. Soc. of Agricultural Engineers. St. Joseph, MI.

Sudduth, K.A., S.J. Birrell, and S.T. Drummond. 1996. Acquisition and interpretation of spatial data for precision farming. p. 600-603. *In:* Proc. 26th International Symposium on Remote Sensing of Environment, Vancouver, BC, Canada, March 25-29, 1996.

Sudduth, K.A., S.T. Drummond, S.J. Birrell, and N.R. Kitchen. 1996. Analysis of spatial factors influencing crop yield. *In:* Proc. 3rd International Conf. on Precision Agriculture, Minneapolis, MN, June 23-26, 1996 (this volume).

Wollenhaupt, N.C., and D.D. Buchholz. 1992. Profitability of farming by soils. p. 199-211. *In* P.C. Robert et al. (ed.) Soil specific crop management. ASA, CSSA, and SSSA, Madison. WI.

Variable Nitrogen Management for Improving Groundwater Quality

C. A. Redulla
J. L. Havlin
G. J. Kluitenberg
Dept of Agronomy
2004 Throckmorton Plant Sciences Center
Kansas State University
Manhattan, KS

N. Zhang
M. D. Schrock
Dept of Biological and Agricultural Engineering
Seaton Hall
Kansas State University
Manhattan, KS

INTRODUCTION

The predominant management practice is to apply crop production inputs as uniformly as possible and adjust rates between fields (Sawyer, 1994). However, emergence of precision farming technologies has caused farmers and the crop input industry to consider non-uniform factors affecting crop yield.

One of the more important farm inputs is nitrogen (N) fertilizer. Uniform fertilization may reduce profitability by underfertilizing high yielding areas and overfertilizing low yielding parts of a field (Carr, et al., 1991; Larson and Robert, 1991). Underapplication of N fertilizer results in reduced yield potential, while overapplication may result in enhanced nitrate leaching potential. Nitrogen fertilizer contributes significantly to soil nitrate-nitrogen (NO_3) contamination of groundwater, especially under irrigated cropping systems (Madison and Brunett, 1985).

Low-cost computer power, availability of GIS software, and development of global positioning system (GPS) and remote sensing technologies have advanced precision farming or variable rate technology where the application rate varies with the distribution of the production potential and other factors (Wibawa et al., 1993). In addition to reducing N leaching potential, this technique may also improve N fertilizer use efficiency.

A USDA-funded project initiated in 1993 is underway in central Kansas to evaluate variable N management under center-pivot irrigated corn. The spatial distribution in yield and soil profile N were used to develop spatially variable N fertilizer rates. The objectives of the project were: 1) to quantify the influence of N management on crop yield, fertilizer N recovery, and potential for NO_3 contamination of groundwater and 2) to characterize the spatial distribution of soil properties that influence the heterogeneity in crop yield.

MATERIALS AND METHODS

Two center-pivot irrigated continuous corn (*Zea mays* L.) fields located in south central Kansas were used as project sites. The soil at Site 1 is an undulating sandy, mixed, thermic Psammentic Haplustalf while the soil at Site 2 is a mixture of the previous soil and a sandy, mixed, thermic Typic Ustipsamment (Table 1).

Table 1. Description of experimental sites. The soil physical and chemical characteristics are spatial means for the 0-0.6 m soil layer.

	Site	
	1	2
Area (ha)	50	65
Soil taxonomy	Psammentic Haplustalf	Typic Ustipsamment
Soil texture		
Percent sand	83.6	90.9
Percent clay	9.3	5.6
pH	6.0	5.9
Bray 1-P (mg/kg)	34	26
NH_4OAc-K (mg/kg)	291	208
Organic matter (g/kg)	8	7

Corn yields in 1993 through 1995 were measured using a combine equipped with a grain flow sensor and a GPS. Yield maps (Fig. 1) were developed using 55 m by 55 m cells. Yield goal maps were computed by adjusting the yield maps such that average yield for each field matched historical yield records. This computation was done by multiplying the yield from each cell by a single correction factor. Yield goal maps for 1994 were made by adjusting the 1993 yield maps to match historical yield records. Yield goal maps for 1995 were made by adjusting the average of 1993 and 1994 yield maps to match historical records.

In 1994 and 1995, soil NO_3^- was measured in soil samples (0-1.2 m) collected prior to planting and after harvest on a 55 m by 55 m grid. Soil NO_3^- values were block kriged using 55 m by 55 m blocks to create soil NO_3^- maps (Fig. 2).

Yield goal maps and soil nitrate maps were used to develop variable fertilizer N recommendation maps. The Kansas State University Soil Testing Laboratory N recommendation model is:

Nrec   = (24 x YG x 1.1) - (0.94 x NO_3^-)

where

Nrec   = N fertilizer rate recommendation (kg N/ha)
YG   = yield goal (Mg/ha)
1.1   = textural factor for sandy soil
NO_3^- = soil NO_3^- content (kg/ha; 0-0.6 m soil layer)

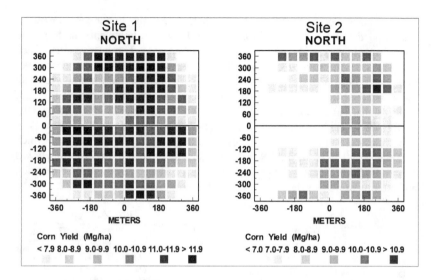

Fig. 1. Corn yield maps in 1993.

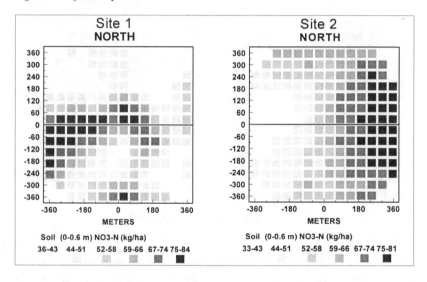

Fig. 2. Soil NO$_3^-$ maps from samples collected prior to corn planting in 1994.

This model is typically used on a whole field basis. In our case, the model was applied to each 55 m by 55 m cell. In 1994, two treatments, uniform and variable application of N fertilizer (Fig. 3), were imposed on the two sites. Each plot was composed of 6 contiguous cells in the west-east direction. The design was completely randomized with 12 replications. The cells in the variable treatment plots used the N rate as recommended by the N recommendation map. The rest of the cells, those in the uniform treatment plots and the peripheral cells, received the uniform N rate. The same plot plan was used in 1995 but with slightly lower fertilizer rates. The treatments only involved preplant broadcast urea which comprised < 48% of the total N fertilizer applied. The rest of the N fertilizer was applied uniformly, namely, at preplanting with a herbicide, at planting as a starter, and after planting by sidedressing and by fertigation.

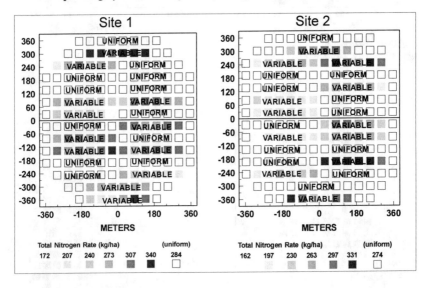

Fig. 3. Plot plan for 1994.

RESULTS AND DISCUSSION

Yield maps for Site 1 and Site 2 from 1993 through 1995 show both temporal and spatial variability (Fig. 4). The temporal variability at both sites was predominantly the result of year-to-year variations in weather. Excellent weather conditions in 1994 led to record high grain production at both sites. Lower yields at both sites in 1993 and 1995 were the result of excessive precipitation and unusually low temperatures during portions of the growing season. Both sites contain poorly drained areas that were waterlogged for extended periods of time. The sandy surface soil at both sites is underlain by intermittent clay layers that result in localized drainage problems. Waterlogged conditions slowed crop development, increased weed pressure, and resulted in complete stand loss in some areas. At both sites, comparison of 1993 and 1995 yields with 1994 yields (Fig. 4)

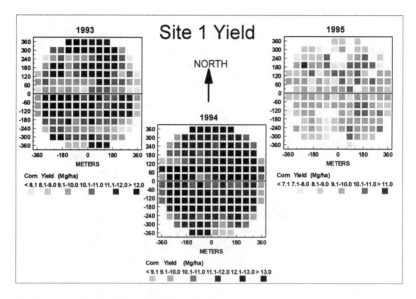

Fig. 4a. Corn Yield at Site 1 for 1993 to 1995.

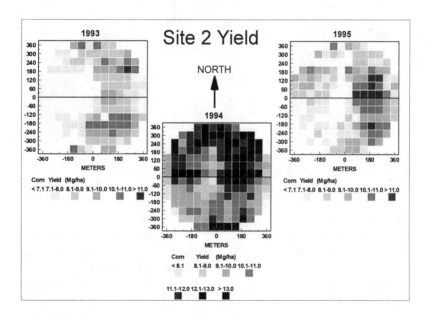

Fig. 4b. Corn yield at Site 2 for 1993 to 1995.

reveals that many of the poorly drained areas (extreme low yields) are areas with excellent yield potential. Corn yields at Site 1 in 1995 were lower than in 1993 because stand loss was so severe that the entire field was replanted, resulting in a shortened growing season.

Distinct spatial patterns in yields are evident at both sites. The persistent east-west trend in yield at Site 2 may be the result of extreme land leveling done on the west half of the field when the irrigation system was installed. The yield patterns at Site 2 correlate well with texture of the 0 - 0.6 m soil layer with lower yields associated with sandier soil.

The three-year sequence of yield maps demonstrates the challenge of variable-rate N management at these sites. How do we decide on a spatial N application pattern when the yield patterns appear to vary so much from year to year? Temporal yield instability is particularly troublesome in this case because the N recommendation model used in this study causes spatial patterns in N fertilizer recommendation to be closely coupled with spatial yield patterns. Comparison of Fig. 1 and Fig. 3 reveals this close coupling. We are encouraged, however, that, despite encountering some of the most extreme yield conditions on record in this region, there is still temporal stability in the spatial yield patterns. This is particularly true for Site 2. These results clearly indicate that multiple years of yield information may be needed to establish the time-persistent component of the spatial pattern in yield goal or yield potential for a particular field.

Grain yields were statistically similar for uniform and variable N treatments for both sites and both years (Table 2). The variable N treatment resulted in higher N efficiency (kg yield/kg N) than the uniform N treatment although only at Site 2 in 1994 were statistically significant differences observed. The slightly higher N efficiency with variable N treatment was related to both slightly higher yields and slightly lower fertilizer N applied than in the uniform N treatment.

Table 2. Some crop yield results and fall residual soil NO_3^- in 1994 and 1995.

Parameter	Uni-form	Varia-ble	CV (%)	Pr>F
--------------Site 1 (1994)--------------				
Mean yield (Mg/ha)	11.9	12.0	8.5	0.79ns
N efficiency (kg yield/kg N)	41.6	43.3	10.6	0.46ns
Soil Nitrate (kg NO_3^--N/ha; 0-1.2 m)	126	135	23.0	0.45ns
--------------Site 1 (1995)--------------				
Mean yield (Mg/ha)	9.0	9.2	8.6	0.47ns
N efficiency (kg yield/kg N)	34.9	37.7	11.5	0.15ns
Soil Nitrate (kg NO_3^--N/ha; 0-1.2 m)	207	185	24.0	0.28ns
--------------Site 2 (1994)--------------				
Mean yield (Mg/ha)	11.0	11.3	10.9	0.46ns
N efficiency (kg yield/kg N)	39.2	45.3	13.6	0.02*
Soil Nitrate (kg NO_3^--N/ha; 0-1.2 m)	149	135	40.4	0.60ns
--------------Site 2 (1995)--------------				
Mean yield (Mg/ha)	8.4	8.5	15.8	0.93ns
N efficiency (kg yield/kg N)	34.9	37.7	14.6	0.30ns
Soil Nitrate (kg NO_3^--N/ha; 0-1.2 m)	120	120	48.6	0.99ns

The residual soil NO_3^- measured after corn harvest was similar for the two treatments at both sites and both years (Table 2). At Site 1, the soil NO_3^- measured after harvest was significantly higher in 1995 than in 1994, probably due to lower crop N uptake in 1995. At Site 2, however, even though the average 1995 corn yield was lower than the 1994 yield, the residual soil NO_3^- in 1995 was still slightly lower than in 1994. Leaching losses may have been greater at Site 2 than at Site 1 because of the coarser soil at Site 2. Enhanced leaching could have prevented the buildup of NO_3^- evidenced at Site 1.

The general distribution of after-harvest soil NO_3^- concentrations was similar to the preplant soil NO_3^- concentrations (Fig. 5) especially at Site 2 (Fig. 5b). The decrease in soil NO_3^- concentration between the two sampling dates may be due to overwinter denitrification and leaching. However, on some portions at Site 1 (e.g. southwest portion in Fig. 5a), the spring soil NO_3^- concentration was greater than that of the preceeding fall. These especially occured on cells where fall soil NO_3^- concentrations were low. Net N mineralization for these cells may be the reason for the higher NO_3^- concentration in the spring soil sampling compared to that of the previous fall.

In 1994, measured grain yield (Y) was greater than the yield goal (YG), (Y-YG)>0, for cells in which the N fertilizer rate was <290 kg/ha (Site 1) or < 260 kg/ha (Site 2). These results are presented in Fig. 6 (solid lines). For cells receiving N fertilizer at higher rates, (Y-YG) was negative; that is, yield goal was not realized for these cells. Deviations from the horizontal line (Y-YG) = 0 indicate lack of agreement between the yield map and the yield goal map. The spatial distribution of yield goal incorrectly predicted the spatial distribution of yield. The lack of agreement between Y and YG at both sites in 1994 is consistent with our earlier observations of how weather conditions affected yield in 1993 and 1994. Recall that the yield goal map for 1994 was based entirely on 1993 yield data. Many low-yielding cells in 1993 that resulted in low yield goals for 1994 produced yields in excess of the yield goal in 1994 and were probably underfertilized. Likewise, high-yielding cells in 1993 that resulted in high yield goals for 1994 produced yields below the 1994 yield goal and were probably overfertilized. This exercise shows how error in predicting the spatial distribution of yield causes error in predicting the spatial distribution of yield goal which then causes error in the N recommendation map.

In 1995, poor weather conditions caused corn yields to fall below yield goals in all cells. This is evident from the fact that the line (Y-YG) is below the horizontal line (Y-YG) = 0 across all nitrogen rates. Despite the overall lower yields, the slope of the (Y-YG) lines for Sites 1 and 2 are considerably smaller than the slopes obtained in 1994. This indicates that, in 1995, at both sites, the spatial distributions of yield matched the spatial yield goal distributions more closely than they did in 1994. Additional data will need to be added to the time sequence of yield maps (Fig. 4) before it will be possible to discern if prediction of yield patterns can be refined to the point that we approach the condition of (Y-YG) = 0. Both sites were probably overfertilized in 1995 because the yields consistently fell short of yield goals across all N fertilization rates. Increased fall residual soil NO_3^- levels (Table 2) at Site 1 in 1995 support this conclusion; however, increased residual soil NO_3^- levels were not observed at Site 2 in 1995. Despite overfertili-

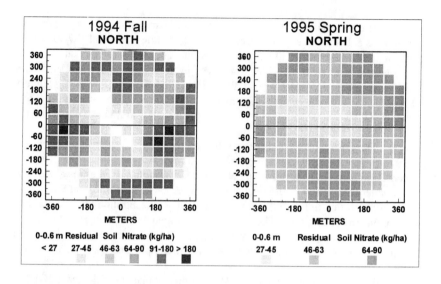

Fig 5a. Residual soil nitrate at two sampling dates in Site 1.

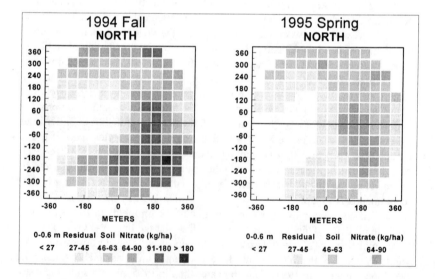

Fig. 5b. Residual soil nitrate at two sampling dates in Site 2.

zation across both fields, the improved spatial agreement between Y and YG (reduced slopes in Fig. 6) probably resulted in N recommendation maps with better spatial N distributions than the 1994 N recommendation maps.

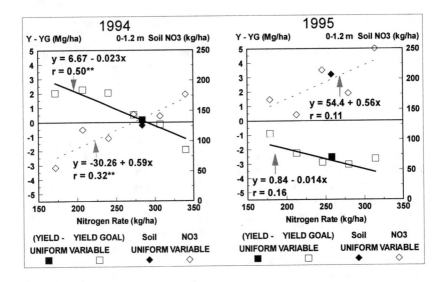

Fig. 6a. Relationship between the difference of yield and yield goal and the residual soil nitrate to total applied N fertilizer at Site 1.

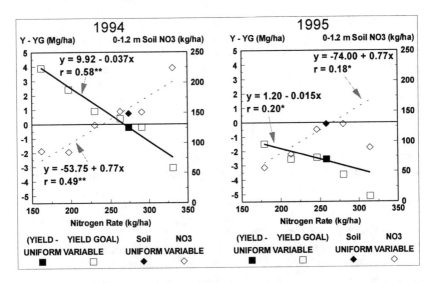

Fig. 6b. Relationship between the difference of yield and yield goal and the residual soil nitrate to total applied N fertilizer at Site 2.

As might be expected, residual soil NO_3^- present after harvest in 1994 (Fig. 6, dashed lines) consistently increases as (Y-YG) decreases. This trend supports the earlier observation that overfertilization occured in cells where Y<YG and underfertilization occured in cells where Y>YG. We anticipate that the dashed lines would become more horizontal as the amount of N fertilizer applied more closely matches crop N uptake. Unfortunately, there appears to be little change in the slopes of these regression lines from 1994 to 1995, even though we hypothesized earlier that the 1995 N recommendation maps provided better spatial N distributions. Additional analysis with soil profile NO_3^- data from the spring of 1995 is needed to determine how much of the trend in post-harvest residual soil NO_3^- level is "carry over" from 1994.

SUMMARY

The long term goal of the project is integrating yield mapping into variable N application technology to increase fertilizer N use efficiency and reduce NO_3^- leaching. Since 1993, yields and other crop and soil parameters were monitored in two center-pivot irrigated corn fields. In 1994 and 1995, two treatments, uniform and variable application of N fertilizer, were imposed. The yield within a site varied from year to year (temporal variability) which depended greatly on the weather during the growing period. Spatial variability was also observed on both sites where yield ranged from 5 to 15 Mg/ha. Temporal stability in spatial yield patterns was observed, especially at Site 2. However, additional years of yield mapping at both sites appears to be needed to establish the time-persistent component of the spatial yield patterns that can be used as a basis for spatially-variable N fertilizer recommendation. The yield and the residual soil NO_3^- measured after harvest were statistically similar for both treatments at both sites and years. The N use efficiency was slightly higher with variable N treatment than with uniform N treatment although the difference was generally statistically nonsignificant. There was an indication that in 1995, at both sites, the spatial distributions of yield matched the spatial YG distributions more closely than they did in 1994. This was probably due to the fact that the 1995 YG was based on two years of yield data while the 1994 YG was based only on one year of yield data.

REFERENCES

Carr, P.M., G.R. Carlson, J.S. Jacobsen, G.A. Nielsen, and E.O. Scogley. 1991. Farming soils, not fields: A strategy for increasing fertilizer profitability. J. Prod. Agric. 4:57-61.

Larson, W.E. and P.C. Robert. 1991. Farming by soil. In R. Lal and F.J. Pierce (ed.) Soil Management for Sustainability. Soil and Water Conservation Society of America, Ankeny, IA. pp. 103-112.

Madison, R.J. and J.O. Brunett. 1985. Overview of the occurences of nitrates in groundwater of the US. US Geol. Survey Water Supply Paper 2275. pp. 93-105.

Sawyer, J.E. 1994. Concepts of variable rate technology with considerations for fertilizer application. J. Prod. Agric. 7:195-201.

Wibawa, W.D., D.L. Dludlu, L.J. Swenson, D.G. Hopkins, and W.C. Dahnke. 1993. Variable fertilizer application based on yield goal, soil fertility, and soil map unit. J. Prod. Agric. 6:255-261.

Assessment of Variable Fertilizer-N Form Management in Indian Soils for Groundwater Quality

A.K. Bhattacharyya

Jowa Harlal Nehru University
New Delhi, India

ABSTRACT

All India consumption of fertilizer-N kg/ha steadily increased from 0.44 in 1951-52 to 52.05 in 1994-95. To meet the target production of 240 mt of food-grains by 2000 AD, consumption of more than 20 mt of NPK nutrients will be required through fertilizers against the current consumption of 13.5 mt. Different states of India consume N-fertilizers in different amounts and within the state, districts vary in the use of the same item. Upland crops use more efficiently N-fertilizers in comparison to lowland crops which suffer from various pathways of N-losses from the soils viz.volatilisation, clay-mineral fixed NH_4-N, denitrification, microbial immobilisation, leaching and run-off. Methods for increasing N-use efficiency include split application, proper placement & use of nitrification inhibitors, slowly releasing N-fertilizers and urea briquettes or super granules. Ammonifiction and nitrification processes influence the leaching loss of N from soils to contaminate groundwater quality. Nitrate released into aqua bodies poses several environmental problems ranging from eutrophication to causation of diseases. Ingestion of excessive nitrate causes methemoglobinema. Infants are more susceptible to it than adults. Another manifestation of toxicity is formation of N-nitrosamines reported to be carcinogenic, mutagenic and teratogenic. Results from field experiments have shown that high content of NO_3-N in ground water does not necessarily come from the application of high doses of the N-fertilizers as in the case of Mahindergarh of Haryana State while a lower contamination of NO_3- N has been reported from the groundwater of Ludhiana District of Punjab State which received higher doses of N-fertilizers. The details are discussed in this paper.

Precision Fertilizer Application-Fertilizer Regulatory Considerations

D. L. Terry

University of Kentucky
Lexington, KY

ABSTRACT

Background information on fertilizer regulatory laws in the United States and specifically the Uniform State Fertilizer Bill of the Association of American Plant Food Control Officials (AAPFCO) provided a springboard for our consideration of the fertilizer regulatory aspects of precision fertilizer application (PFA) within precision agriculture practices. The difficulties of applying fertilizer regulatory laws to PFA were discussed. The difficulties cited were labeling the fertilizer applied to each management site, supporting claims made for fertilizers used in PFA programs and the sampling of the fertilizers applied in a PFA program. The results from a survey of the US and Canada indicated that there is a lack of consensus on how to regulate this practice. Regulatory programs must be tailored to meet the needs of PFA technology without sacrificing consumer or industry protection.

SESSION VI

TECHNOLOGY TRANSFER

Using Decision Cases to Enhance Technology Transfer in Precision Agriculture

R. Kent Crookston

Department of Agronomy & Plant Genetics
University of Minnesota
St. Paul, MN

ABSTRACT

Decision cases were developed over 70 years ago at Harvard University and are used to teach management and decision making throughout business schools in the U.S. Decision cases are one of the best ways to research and teach complex systems that cannot be reduced to limited variables. The researcher of a decision case will identify:

- a decision maker (this person's identity and role is key).
- a dilemma or issue confronting the decision maker.
- essential information for analysis and appraisal of the dilemma.
- the objectives of the decision maker, and options available to her/him.
- an interpretive note (insights gathered while researching the case; shared with case teachers, but not with students).

At the University of Minnesota, we have developed a large number of agricultural decision cases. Most recently we have approached the challenge of yield map interpretation and analysis with decision cases. We have used a yield-map case in professional workshops and in the classroom. We have observed that case users become participants in their own learning, they cannot be passive spectators. We have seen that cases draw out the collective wisdom of the group, that they are an ideal means of developing synergy from a background of differences. Trainees working with decision cases develop skills in problem definition and informed decision making.

Background

The effective transfer of technology represents a challenge to all branches of agriculture in all parts of the world. The development of precision agriculture appears to have intensified the challenge. Massive amounts of data, combined with a wide array of collection and delivery options represents a difficulty for researchers, educators, consultants, dealers, and farmers alike. We're seeing a situation wherein the application of a technology precedes the researching of that technology unlike anything we have experienced before. The result is a mix of enthusiasm, entrepreneurship, skepticism, confusion, and questions that often focus on a single issue: how to make sound decisions?

At the University of Minnesota, we have recently been developing a method of dealing with the dilemma of decision making in agriculture. We are researching and teaching complex systems and issues by means of decision cases.

Validating Experience

Scientists know that the experiences of farmers, the results of their trial-by-error efforts, have been extremely important in the development of agriculture as we know it today. I have spent several years working with agricultural researchers in developing countries. It is my observation that the family farm has been a valuable research enterprise ever since agriculture began. Each generation has studied its alternatives, and made its decisions. There were no research grants or publications, there were no statistical evaluations of results. But if it were not for rural people's experiential decision making, most of the modern agriculture that we take for granted today would be unknown.

Yet, the experiences of today's farmers are given minimal attention by scientists and their publications. Why? The standard explanation is that an individual farmer's experiences and conclusions are unique to a specific site and situation. These experiences cannot be tested or verified as to repeatability. In other words, observations made by farmers on their own farm are usually considered too subjective.

By contrast, scientists makes every effort to eliminate bias from the design and management of their research. Randomized replicated plots help to overcome unplanned variability, and whatever variability persists can be measured or estimated. Limited-variable studies allow scientists to assign significance to some variables and to omit others from further consideration.

Farmers make no structured effort to eliminate subjectivity from their observations, and find that cold objectivity often does not fit with family or community relationships and obligations. This results in a dilemma. Every year, thousands of farmers have highly valuable experiences which receive limited exposure off the farm. Agricultural researchers have not yet found an effective way to capture those valuable on-farm experiences without the subjectivity and bias problem.

Decision cases represent a solution to this dilemma. A properly developed decision case can take a farmer's many years of work and experience (which a scientist cannot duplicate) and put that experience into a format and context that can be evaluated and used professionally. Decision cases are one of the best ways to research complex systems that cannot be reduced to single variables.

What is a decision case

Decision cases were pioneered more than 75 years ago by the Harvard Graduate School of Business Administration. Today, decision cases are used in most leading business schools throughout the world. In the late 1980's the University of Minnesota began using decision cases for research and education in agriculture. The approach has been highly successful and is becoming the subject of considerable interest by agricultural scientists.

It should be noted that case-type exercises are not new to agriculture; simulations and field-based problems have been a part of agricultural education for a long time. However, agricultural cases have typically been descriptive in nature and have often been based on fabricated or hypothetical situations.

The term "case study" or "case" has a variety of meanings. Depending on the profession, a case study can refer to a <u>legal case</u>, a <u>clinical case</u>, an <u>appraisal case</u>, or a <u>descriptive case</u>. A <u>decision case</u> is similar to, yet different from, each of these.

A decision case is a documentation of reality, the written product of investigation into an actual situation. This is one reason I believe decision cases qualify as legitimate instruments of research. A valid discovery cannot be fabricated or manufactured. If scientific data have integrity, they will stand up under scrutiny. Similarly, a good decision case will be based on documentable reality and observation, not on supposition or conjecture.

A decision case is based on a dilemma. This must be a genuine dilemma for which there is no obvious, rational, or logical solution. While working to resolve an engaging dilemma, case users identify relevant facts, analyze them, and draw conclusions about the cause of the problem as well as actions that might be taken. Sharon McDade (McDade, 1988) notes that "the most interesting and powerful cases are those that allow for several equally plausible and compelling conclusions, each with different implications for action. 'Real life' is ambiguous, and cases reflect that reality. A 'right' answer or 'correct solution' is rarely apparent."

A decision case focuses on a specific decision maker. Case users need to be able to relate to this decision maker. As they consider the decision maker's objectives and options, they realize that their own biases are irrelevant. If significant differences of opinion exist within a group that is working to solve the decision maker's dilemma, the result is often synergy. Synergy results in creativity and new insights. This often leads to new hypotheses for deductive research.

A good decision case is publishable, based on anonymous peer review (Simmons et al., 1992). Reviewers are asked to determine whether the case deals with issues that are current and of interest to a wide audience, is well written, is based on sound objectives, contains sufficient information and documentation to meet the stated objectives, and has been interpreted competently.

A Tradeoff

Thomas Bonoma (Bonoma, 1985) describes two divergent paths of scientific investigation. The more popular path involves "controlling situational events in order to observe the validity of empirical deductions." The other, which he describes as less popular but equally valid consists of "reasoning from individual and naturally occurring but largely uncontrollable observations toward generalizable inductive principles." Bonoma suggests a major tradeoff between "precision in measurement and data integrity" versus "currency, contextual richness or external validity."

Note (Fig. 1) that Bonoma places case research just above the line which separates science from non-science. Note also, however, that much of the

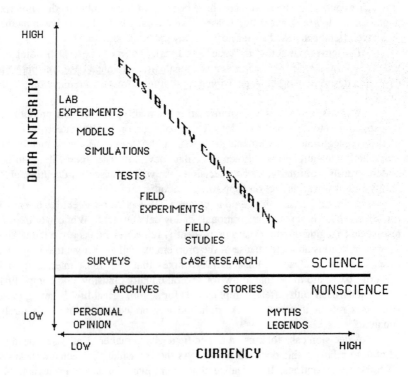

Fig. 1. A knowledge-accrual triangle (from Bonoma, 1985).

non-science has very high currency or contextual relevance across settings and time. Bonoma suggests that it is not possible to do "good" research that has both strengths. It is my opinion that Bonoma's suggested tradeoff represents reality, but that this should not inhibit the use of case studies any more than the use of controlled experiments. The fact that a decision case is based on an event that cannot be replicated nor repeated should not be considered a weakness. It is, in fact, this feature that helps make decision cases uniquely valuable. There is much to be gained from life's rare and singular experiences, many of which cannot be understood if removed from their social context.

A new paradigm

Decision cases require agricultural scientists (researchers and educators) to project themselves into the shoes of nonscientific decision makers (farmers, agricultural agents, consultants, etc.), and to evaluate specific decisions or dilemmas facing these people. When scientists do this, they experience a paradigm shift. The new paradigm reveals the validity of experience, the power of social values, and the subtle importance of ethics. The new paradigm may also reveal the futility of fixed replications over years, or limited variables, or even statistics.

This new paradigm could help us incorporate relevance into agricultural research. With this new paradigm we could begin to question a professional approach based almost entirely on statistically-significant, limited-variable, hypothesis-driven, deductive work; work which does not accommodate holism, nor take into consideration the populist perspective.

A Proposal

I propose that agriculture learn from the business world and incorporate decision cases into its research and education efforts. I am confident that quality decision cases will be an excellent complement to agriculture's data-based research programs, and that they will lend holistic practicality to the rush of data being made available with precision agriculture.

Minnesota faculty have built cases around farmers, scientists, business people and politicians (Crookston & Stanford 1989; Crookston & Stanford 1992; Crookston et al., 1993; Taack et al., 1994). Some of these people have been invited to participate with groups of students or professionals assembled to work their cases. Invitees have benefitted from debate and discussion of their dilemmas, and from the synergy that occurred when diverse viewpoints were focused on recommending a solution.

But the real benefit of decision cases is realized by their users (students). Decision cases are based on the principle of participative learning. Cases are a highly effective means of providing students with skills in analysis of problems, synthesis of action plans, and development of maturity, judgment, and wisdom (Dooley and Skinner 1977; Gragg 1954; Hammond 1976). These are skills that are acutely needed to direct the research efforts of scientists who otherwise gravitate toward theoretical academic pursuits, and approval (via technical publications) of intellectual colleagues.

I am confident that if decision cases were included in our precision agriculture education programs, better research, better information, better technology transfer, and better decisions would be the outcome.

REFERENCES

Bonoma, T.V. 1985. Case research in marketing: opportunities, problems, and a process. J. of Marketing Research. 22:199-208.

Crookston, K., and M. Stanford. 1989. AgriServe Crop Insurance. College of Agriculture decision cases #2. College of Agriculture, University of Minnesota, Saint Paul, MN. 55108.

Crookston, R.K., and M.J. Stanford. 1992. Dick and Sharon Thompson's "problem child": A decision case in sustainable agriculture. J. Nat. Resour. Life Sci. Educ. 21:15-19.

Crookston, R.K., M.J. Stanford, and S.R. Simmons. 1993. The worth of a sparrow. J. Nat. Resour. Life Sci. Educ. 22 (2)134-138.

Dooley, A., and W. Skinner. 1977. Case casemethod methods. Academy of Management Review. April, 1977.

Gragg, C.I. 1954. Because wisdom can't be told. Harvard business school publications on case development and use (9-451-005). Publishing Division, Harvard Business School, Boston, MA. 02163.

Hammond, J.S. 1976. Learning by the case method. Harvard business school publications on case development and use (9-367-241). Publishing Division, Harvard Business School, Boston, MA. 02163.

McDade, S. 1988. An introduction to the case study method: preparation, analysis, and participation. Notes on the case method. Institute for Educational Management, Harvard College, Boston, MA. 02163.

Simmons, S.R., R.K. Crookston, and M.J. Stanford. 1992. A case for case study. J. Nat. Resour. Life Sci. Educ. 21:2-3.

Taack, D.L., H. Murray, and S.R. Simmons. 1994. Minto-Brown Island Park: A case study of farming the urban-agricultural interface. J. Nat. Resour. Life Sci. Educ. 23:98-103.

Precision agriculture and risk analysis: An Australian example

S.E. Cook
R.J. Corner

CSIRO Division of Soils
Private Bag,
PO Wembley,
Western Australia 6014

G. Riethmuller
G. Mussel

Agriculture Western Australia
Dryland Research Station
Merredin, Western Australia 6415.

M.D. Maitland

Box 63, Wyalkatchem,
Western Australia 6485

ABSTRACT

A potential benefit of precision agriculture is the capability it gives us to analyse the risk of change. We illustrate how GIS-based modeling of expected outcomes from alternative fertilizer plans can be used to identify the comparative risks associated with each. The major source of uncertainty in our example appeared to be climatic variation, which increased the risk of variable rate application relative to its benefits.

INTRODUCTION

A common question asked about precision agriculture is how profitable it is. Early studies on the economics of precision agriculture by Wollenhaupt and Buchholz (1992) indicated that the financial benefits could be elusive, and that they depended largely upon seasonal variation.

The question may be misdirected because precision agriculture is not a technique, but a management process through which farmers can acquire benefit. Risk is an important factor in this process because in deciding whether to change, the farmer needs to know not only the scale of potential benefit but also the likelihood of achieving it. This being so, a major advantage of precision agriculture could be that it provides farmers with an analytical capability to assess the likelihood of benefit and hence improve decision-making.

In this paper we examine the issue of risk in an Australian context and illustrate how variable rate management can be used to evaluate the risks associated with alternative fertilizer strategies.

Precision Farming and Risk Management

Risk management describes the process of evaluating uncertainties within a system and selecting management options which generate preferred expected outcomes (Royal Society, 1992). Precision farming offers farmers two practical aids to improved risk management: The first is the increase in information about field performance from which risk can be evaluated more precisely. The second is the additional control of inputs which can be used to modify expected outcomes.

Financial and environmental adversities are increasing pressure on Australian farmers to initiate changes in order to increase profitability (Chisholm, 1992). However, change implies additional uncertainty, or risk. An essential part of a risk management strategy is the representation of expected outcomes from which the decision maker can select a preferred option. Using yield mapping farmers can now examine risks within their fields by monitoring field performance and using variable rate technology (VRT) to manipulate inputs according to expected outcomes. The objective here is to illustrate how this could be done.

METHOD

Following a brief description of our experimental field site, we illustrate the application of precision farming to risk management by the following steps:
- First, we develop four alternative management options for applying P fertilizer and writing these in a form which can run in GIS.
- Next, for each option, we describe the expected outcome as a map of expected profit.
- Finally, we summarise the risk associated with each outcome by comparing the distributions of expected with actual profit, as measured over a series of trial strips within the field.

Experimental Site

The field we used for the trial is located at Newdegate Research Station (119°E; 33°S). This was cropped for narrow leaved lupin (*L. Angustifolius*, c.v. Gungurru) in 1995, the year of variable application. We applied superphosphate (approximately 9% soluble P) to this field at rates of 0, 7, 10, 14 and 23 kgP/ha, varying rates between randomly ordered strips across the field. Average rainfall over the growing season (May to October) is 219 mm. Yield and grain moisture content were measured using an AgLeader AL2000 yield monitor, calibrated on site. Differential GPS was provided by FUGRO Omnistar using a wide area solution from stations based in Perth and Kalgoorlie. Yield data was downloaded using the AL2000 software and manipulated subsequently in ARCInfo GIS.

Alternative Management Options

We examined the expected outcomes from four simple management options. These options were developed in consultation with the farmer and local agronomists. They are not intended to define the full range of behavior of the system but to illustrate the relative risks associated with each. The options we chose were:

Uniform application at a rate of approximately 10 kgP/ha.
1. Varying rate according to information of prior yield.
2. Varying rate according to information of variable fertility.
3. Varying rate according to information of both prior yield and fertility.

These options are described in detail below:

Option 1 is conventional practice and assumes that the farmer has no yield map and poor information about spatial variation within the field. Optimum levels of applied P are determined by an empirical model based on the standard Mitscherlich response curve (Bowden & Bennett, 1976), calibrated locally by an extensive series of crop trials (Bowden et al., 1993). The response curve is described by the equation:

$$Y = A - B * e^{-cP} \qquad (1)$$

The optimum level of P to apply is determined as the position of maximum economic return, taking account of potential yield (A), the efficiency of applied P (B) and the shape of the response curve (c). In practice, A is estimated by the farmer on the basis of prior experience and the expectation of rainfall during the impending growing season. B is derived from an assay of $NaHCO_3$ (0.5M) extractable P in the surface soil, and calibrated from crop trials. The c coefficient describes the curvature of the line, and represents a broad range of effects, including the reaction of P fertilizer and the soil, the uptake of soil P by the plant. Curvature is estimated from either a Phosphate Retention Index (Allen & Jeffrey, 1990), or an estimate of reactive Fe (Bolland et al., 1996) and is required because of the widespread problem of P sorption and diffusion on aluminium and iron oxides. The curve is scaled for economic effects by using the ratio of values for fertilizer and product.

Option 2 represents the situation where the farmer has a map of prior yield but poor information about the spatial variation of P response within the field (Fig. 1a). This option relies on an intuitive interpretation of prior performance. A reasonable response might be to consider areas that yielded poorly last year to be of low potential and worthy of low rates of P, and to consider areas which yielded above average in 1994 to be being the optimum on the response curve. The intention is to divert investment to areas which yielded moderately well, in the expectation of increased overall profit.

The third option illustrates the situation where the farmer wants to vary P application within the field, but wants to use existing guidance on P applications, such as the 'Decide' model described for option 1. We therefore estimated optimum rate according to information derived from the soil map and gridded soil samples. The soil map identifies three broad zones of high, medium and low potential (assigned 2.5, 1.5 and 0.75 t/ha); soil bicarbonate P (mg/kg), and the c coefficient, which in this case was estimated from reactive iron, which varied between 100 and

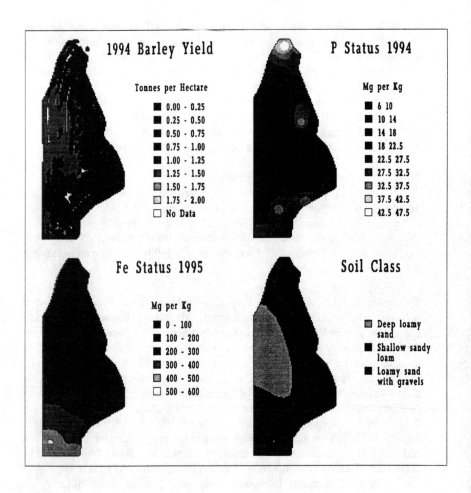

Fig. 1. Maps of data used in analysis.

600 mg/kg (Fig. 1). These data were used to evaluate an optimum level of P fertilizer as it varied over the field.

The fourth option uses spatially variable information of both prior yield and soil fertility over the field in a single estimate of the likely response to P. This raises a difficulty of combining information of uncertain meaning within a single numerical estimate. We tackled this problem using a simple 'expert system' approach to represent the way an expert might evaluate information about prior yield and soil characteristics. Further, we reduced the complexity of the problem to a single question: Do we believe that a given level of fertilizer (about 10 kg/ha) is appropriate, given evidence of prior yield, soil P status, soil Fe status or soil type?

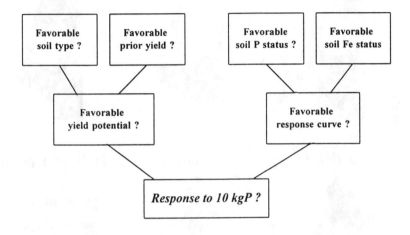

Fig. 2   Simple rule-based system to estimate the likely response to P

Information is organised in a series of IF..THEN... type rules as shown in Fig. 2. Evidence of soil type and prior yield was used to infer yield potential. Evidence of soil P and Fe status was used to infer whether the response curve is favorable. Since we did not expect clearly positive or negative answers to these questions, we expressed the answers as a conditional probability that a site would respond to P, given evidence of soil type, prior yield, soil P status and soil Fe status. Details of the Bayesian algorithms on which this method is based are described in Cook *et al.*, (1996).

Values of prior crop yield, soil type, soil P and reactive iron were input for each grid cell and evaluated by the rules to produce an estimate of the probability of a response to 10 kgP/ha. The output is a probability map to which, for the purpose of this exercise, we assigned fertilizer levels from the range 0, 7, 10, 14 and 23 kgP/ha.

Comparison of expected outcomes for each option

A map of expected yield was generated for each option using the Mitscherlich function (Eq. 1) and estimates of A, soil P, fertilizer P and c, where fertilizer P varies for each of the management options. Maps of expected yield were converted to maps of expected profit (Fig. 3) using current prices for lupin grain ($112/t) and cost of P ($1.50/kgP/ha). Other farming costs were taken as a constant across all options.

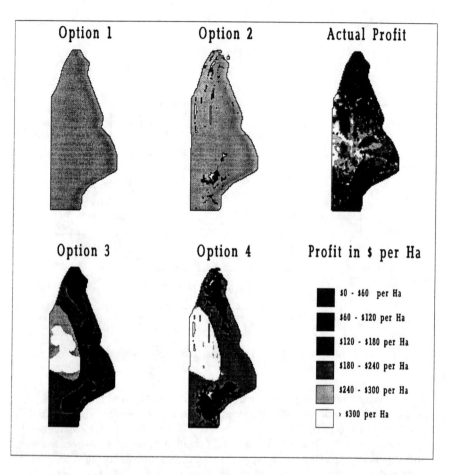

Fig. 3. Maps of expected and actual profits

We compared the expected with actual profit for each option to give a visual assessment of their relative performance. This was possible because we had previously varied rates of P application over the field. Comparison was made by retrieving yield data from pixels which had received approximately the desired rate of P for each option. We represented risk as a frequency:magnitude curve, which plots the decreasing probability of achieving a given net return as its magnitude increases.

RESULTS

The expected outcome from each option is shown as maps of expected and actual profit (Fig. 3). The relative performance of each option, expressed as the discrepancy between expected and actual outcome, is shown in Fig. 4. The probabilities of achieving a given profit (or loss) for each option is shown in Fig.5.

Expected Outcomes

The expected outcome from option 1 is a uniform profit, representing the average response from an average input over soils of average character. Deviations from the average are unspecified.

With the exception of small areas which yielded poorly in 1994, the expected outcome from option 2 is also uniform. This results from the attempt to allocate resources to the medium range of prior yield.

We expected a wide variation in yield from option 3. Most of the variation is caused by variation in expected potential, which divides the field into 3 zones. Within these, our expectation is conditioned by the status of soil P and Fe.

Expected variations in option 4 is also driven strongly by perceived variations of potential associated with the 3 soil map units, but in this case it is also conditioned by evidence of prior yield and soil P status.

Comparison of Expected and Actual Profit

Figure 3 shows the expected and actual profits in 1995. This contrasts markedly with the 1994 pattern. The shallow gravelly soils at each end of the field yielded poorly as in the previous year, but the areas of poor yield is substantially reduced. The yield pattern over the central area contrasts markedly with the 1994 pattern, the area which yielded most then producing a very low yield in 1995 as a result of waterlogging. Areas which performed moderately well in 1994 produced the highest yields in 1995.

Estimates of the average profit for each option suggest that, in this year, variable rate applications would produce a profit of about $83/ha, which is lower than would be expected after uniform application, which was about $88/ha. The estimates of profit from the variable rates all approximated to within $1/ha of one another.

Risk Analysis

More detailed comparison of the outcomes from each option explain more about the disappointing performance of variable rate application. The probability plot of absolute discrepancy between expected and actual yields (Fig. 4), shows that with the exception of a small area which received the additional P it required, option 2 shows a large consistent negative discrepancy, suggesting that the additional investment did not produce the desired response over the mid-range of prior yields. This reflects the pre-existing level of P status (Fig. 1).

Options 3 and 4 performed reasonably well over the mid-range area, but performed very badly over about 10% of the area, which was the area affected by waterlogging. This area returned very little of the investment it received in expectation of strong performance. The direct and indirect losses imposed a penalty on overall profitability.

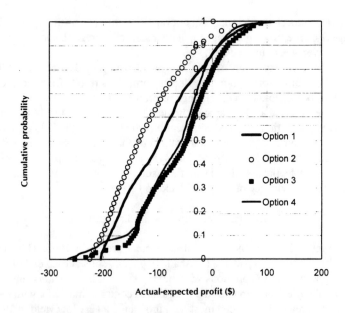

Fig. 4. Comparison of expected and actual profit.

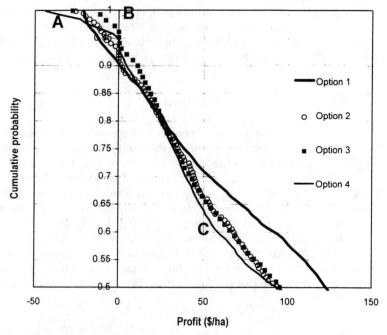

Fig. 5. Cumulative probability of achieving a given profit.

Comparisons of the frequency curves of profit and loss distributions provide more details. On the extreme left hand side of the curve (marked A on Fig. 5), a small number of sites suffer major losses for options 3 and 4, while those of option 1 and 2 are moderated by the fact that investment is constrained. Option 4 recovers some of these losses by reducing the sites in the $-20 to 0 range (marked B).

The major divergence in profit distribution is seen in the $0-100 profit range (marked C), where the uniform rate increases more rapidly than the variable rate options. The poor returns from variable rate application could be caused by two connected effects. The first is a poor response from additional investment at the upper end of the response curve; the second is the withdrawal of investment from areas which were of better than expected potential.

DISCUSSION AND CONCLUSIONS

The distributions of profit and discrepancy with expected profit indicates three reasons for the disappointing performance of variable rate applications:

- Areas we expected to be of high potential yielded very poorly.
- Areas we expected to be of low potential yielded moderately well
- The overall response to additional P was poor, which reduced the advantage of variable rates relative to the risk of mis-predicting potential yield.

Uniform treatment offers a degree of conservatism which, in this instance, capitalised on risk aversion.

The failure of variable rate management to increase overall profitability does not imply that precision agriculture is not of benefit. In this example it illustrates the conservatism of the uniform rate. It also identifies the benefits of accurate prediction and the costs of inaccurate prediction. Were the overall level of P status lower, the relative benefits of precise placement of fetilizer would be enhanced, and so increase the advantage of VRT over uniform application.

The major source of uncertainty in this example arose from variation in potential associated with soil water conditions. We would expect similar problems over much of the Western Australian wheatbelt because of the apparent dominance of water use effects on yield (Hamblin & Kyneur, 1993; French and Schultz, 1984; and Hamblin and Kyneur, 1993) and incidence of waterlogging (McFarlane & Wheaton, 1990). The farmer might benefit from a more accurate assessment of these uncertainties before modifying investment patterns on the basis of expected yield potential. This could be approximated more precisely using simulation models based on CERES (Jones & Kiniry, 1986) or SWIM (Ross, 1991).

The ability to represent explicitly the outcomes from a range of options enables the farmer to evaluate the performance of alternative strategies, which can be used to guide changes in management practice. The analysis of outcomes will provide the farmer with a much clearer idea of the risks associated with each. We expect this to provide a significant benefit to the long-term development of farmers' risk management.

ACKNOWLEDGMENTS

The authors acknowledge the financial assistance of the Grains and Research Development Corporation of Australia. We also wish to thank FUGRO Australia for supplying differential GPS, and colleagues Drs. Mike Wong in CSIRO and Bill Bowden in Agriculture Western Australia for assistance with interpreting the Decide model.

REFERENCES:

Allen, D.G., and R.C. Jeffery. 1990. Methods of analysis of phosphorus in Western Australian soils. Report of Investigation No. 37. Chemistry Centre of Western Australia, Perth.

Bolland, M.D.A., R.J. Gilkes, R.F. Brennan, and D.G. Allen. 1996. Comparison of seven phosphorus sorption indices. Aust. J. Soil Res. 34: 81-89.

Bowden, J.W., and D. Bennet. 1976. The 'Decide' model for predicting superphosphate requirements. In: Proceedings of Phosphorus in Agriculture Symposium. Australian Inst. Agricultural Science, Parkville, Victoria, Australia.

Bowden, J.W., C. Shedley, and S.J. Burgess 1993. Soil test and phosphorus rate. Agriculture Technote No. 5/93. Agriculture Western Australia, Perth.

Chisholm, A. 1992. Australian agriculture: A sustainability story. Aust. J. Agric. Economics 36: 1-29.

Cook, S.E., R.J. Corner, G. Grealish, P.E. Gessler, and C.J.Chartres. 1996. A rule-based system to map soil properties. Soil Science Society of America Journal. In press.

French, R.J., and J.E. Shultz. 1984. Water use efficiency of wheat in a mediterranean-type environment. (i) The relationship between yield, water use and climate. Austr. J. Agric. Res. 35: 743-764.

Hamblin, A., and G. Kyneur. 1993. Trends in wheat yields and soil fertility in Australia. Bureau of Resource Sciences Report. Aust. Government Publishing Service, Canberra.

Jones, C.A., and J.R. Kiniry. 1986. CERES-Maize. A simulation model for maize growth and development. Texas A& M Univ. Press.

McFarlane, D., and G.A. Wheaton. 1990. The extent and cost of waterlogging. Western Australian J. Agriculture 31:44-47.

Ross, P.J. 1991. SWIM- Soil Water Infiltration Model. CSIRO Division of Soils, Adelaide.

Royal Society. 1992. Risk: Analysis, Perception and Management. Report of a Royal Society Study Group. Royal Society, London.

Wollenhaupt, N.C., and D.D.Buchholz. 1992. Profitability of farming by soils. In (Eds.) P.C. Robert, R.H. Rust and W.E. Larson. Proceedings of Soil Specific Crop Management. Amer. Soc. Agronomy, Madison.

GOSSYM/COMAX/WHIMS: A Site-Specific Farming Management Tool

Mariquita Y.L. Boone

Dep. of Plant and Soil Sciences
Mississippi State University
P.O. Box 9555, Mississippi State, MS, USA

D.C. Akins
J.M. McKinion

USDA-ARS Crop Simulation Research Unit
P.O. Box 5367, Mississippi State, MS, USA.

M. Kikusawa

Center for Information Science
Fukui Prefectural University
4-1-1 Kenjojima Matsuoka-cho, Fukui 910-11
Japan

ABSTRACT

Farmers need information that minimizes the risks of making wrong decisions. GOSSYM/COMAX is currently used in commercial cotton production to simulate the basic biological and physical processes involved in the growth, development and yield over a wide range of soils and climates. A primary goal of GOSSYM is to effectively manage production inputs--irrigation, nitrogen fertilizers and plant growth regulators. In conjunction with COMAX, it can provide recommendations on optimal applications of irrigation and fertilizers, timing of plant growth regulator applications and crop termination for maximal economic growth production and minimal nitrogen residuals in the soil profile at the end of the growing season. WHIMS, scheduled for release in the Midsouth area in 1996 as part of the GOSSYM/COMAX system, is a decision aid for the application of pesticides for insect control. Since most models cannot predict the disturbances occurring in the field, periodic plant mapping data, scouting reports and visual field observations are used to adjust the simulation results. The ability of GOSSYM/COMAX/WHIMS to anticipate the impact of current proposed management tactics or future weather influences is essential in precision agriculture. In addition, the biologically important and dynamic information they provide when used in combination with other technologies such as geographic information systems (GIS), variable rate machineries and global positioning systems (GPS) could further optimize site-specific farming practices to address some of the social, ethical and environmental issues that face the declining number of farmers.

INTRODUCTION

US agriculture is one of the most productive in the world. This high level of productivity, achieved through heavy use of chemicals for fertilizers and pest control, irrigation water and modern varieties, has also caused the loss of rural communities, degradation of the resource base, increased reliance on purchased inputs and increased regulations (Bezdicek and DePhelps, 1994). Urban migration of rural people has become common because of a lack of rural economic opportunity. "Biotechnology and infotechnology may cause even greater changes than those produced by the combined mechanical and chemical innovations of the twentieth century" (Stauber, 1994). With reduced rural population and increased concerns about food safety and the environment, a new concept for agriculture is emerging. Crop simulation models, geographic information systems (GIS), global positioning systems (GPS), intelligent implements (II) and site specific management (SSM) farming techniques are the technologies behind precision agriculture.

Precision, prescription, site-specific crop management--all refer to a management system of production agriculture using diverse technologies to increase field productivity and protect the environment. These technologies should improve input efficiency, maximize farm profitability and provide computerized field histories. GPS can guide: a) in grid sampling to produce an accurate soil data base, weed and insect pest population data; and b) variable rate farm machineries that increase chemical application efficiency as a result of adjustments due to soil properties, fertility or pest population density. However, specific data on crop growth and development status are not provided by these technologies. Crop models running on real time can furnish data on irrigation and nitrogen stress status, and other crop data essential for achieving greater accuracy in the field and increased efficiency in the use of management practices. Results obtained from these models can be used to manage row spacing; population density; nitrogen fertilizer, irrigation water, and plant growth regulators applications; and harvest timing, to name a few.

CROP MODEL: GOSSYM/COMAX/WHIMS

Crop models have been developed to serve as educational, research and management tools. Each crop model varies in each degree of sophistication and reflects the current understanding of crop growth and development. There are crop models for cotton, wheat, soybean, rice, peanut, corn, potato, etc. Some are appropriate more for strategic planning while others like GOSSYM and GLYCIM, cotton and soybean crop models respectively, are suited for tactical purposes.

GOSSYM/COMAX, the cotton simulation model and *CO*tton *MA*nagement eXpert system, is a result of continuing research efforts since the early 1970's by a multi-disciplinary team at Mississippi State University, Clemson University and the USDA-ARS Crop Simulation Research Unit (CSRU). Because GOSSYM is a process-level, material-balance computer model, it requires constants and rate coefficients that are obtained under closely controlled environmental conditions, called the Soil Plant Atmosphere Research (SPAR) units (Phene, et. al., 1978, McKinion, 1986). GOSSYM simulates the basic biological and physical processes involved in the growth, development and yield of cotton over a wide range of soils and climates. It uses the farmer's cultural practices (i.e., planting density, row spacing, variety, preplant fertilizer applications, date of crop emergence, in-season fertilizer, irrigation and plant growth regulators applications), soil physical characteristics (i.e., bulk density, hydraulic properties and initial fertility and moisture status), and daily weather data (i.e., maximum and minimum temperature, total solar radiation, rainfall and wind) obtained from a weather station accessed by the user's computer via a modem. It provides a status report on simulated plant height; number of fruiting nodes and vegetative nodes; number of squares, green bolls, open bolls and abscised fruits; and carbohydrate, nitrogen and water stresses as often as daily. A summary table at the end of a full-season run presents the date of maturity, plant height, LAI, yield and the number of nodes, squares, green bolls and open bolls at each designated developmental event.

GOSSYM can be subdivided into two subsystems: the above-ground (primarily plant physiological) and the below-ground (soil) processes. Figure 1 presents the structural framework of GOSSYM. The major subroutines are: CLYMAT--reads all the weather information and calls the subroutines that keep track of the time (DATES) and that calculate the soil temperature (TMPSOL); SOIL--the controlling program for all the soil subroutines; CHEM--the controlling program for all the plant growth regulator subroutines; PNET--calculates the photosynthesis rate/dry matter production; GROWTH--calculates potential and actual daily growth rates of each of the organs on the plant including roots (RUTGRO); and PLANTMAP--simulates plant morphogenesis and abscission of leaves and fruits (ABCISE) (Baker, et. al., 1983, Baker and Landivar, 1991).

The SOIL subroutine is basically derived from the 1973 work of Lambert and Baker known as RHIZOS (Boone, et. al., 1995). It consists of the following subprograms: FRTLIZ--initializes nitrogen and organic matter content of the profile at planting and distributes applied fertilizers in the profile; GRAFLO--moves the water into the profile after a rain or an irrigation event by gravitational flow and moves nitrogen in solution by mass flow; RRUNOFF--estimates the amount of runoff from a rain or irrigation for the day; ET--estimates evapotranspiration; UPTAKE--calculates the amount of soil water taken up from the soil region where the roots are present; CAPFLO--rewets the soil by capillary flow in response to soil moisture gradients and moves nitrogen by mass flow; NITRIF-calculates the mineralization of organic matter and urea and the nitrification of ammonia; RUTGRO--calculates the potential and actual growth of roots and average soil water potential, and calls RIMPED, which estimates the effects of bulk density on the capability of the root to elongate; and TMPSOL--calculates the soil temperature by layer. A spatial description of roots, water, nitrogen and temperature is achieved through the use of two-dimensional geometry.

COMAX, added in 1984, is designed to improve the efficiency in using the GOSSYM simulation and to automate the use of GOSSYM model for making management recommendations for the cotton crop. It consists of an inference engine, a knowledge base, a file management system and a mouse-driven graphical user interface (GUI). Knowledge of the plant and soil processes reside in

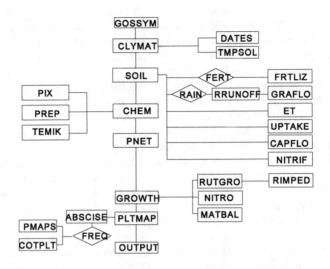

Fig.1. Flow chart of GOSSYM cotton crop simulation
        model showing the organization of the model and
        the program flow.

GOSSYM and is accessed by the COMAX knowledge base. To make management
recommendations, COMAX uses a set of rules based on the expert knowledge of
the GOSSYM model builders on the use and interpretation of model results. To
increase the flexibility of the system over the wide variety of cultural practices
encountered across the Cotton Belt, COMAX uses local expertise supplied by users
to adapt the rules to different cultural conditions.

     COMAX provides three types of analysis of field profiles: advisors,
scenarios and risk analysis. The advisors use a set of if-then rules to formulate
different strategies for fertilizing and irrigating. It then uses GOSSYM simulation
runs to evaluate the impacts of certain cotton management decisions. After a series
of optimization runs, specific recommendations on irrigation or fertilization are
generated. Strategies for nitrogen analysis include stress relief and risk analysis
while for irrigation, three are included--long-term, short-term and water
conservation.

     Since the PIX, PREP and defoliation advisors are under development,   the
scenario options can be used as GOSSYM information managers. The PIX and
Crop Termination Scenario options run the GOSSYM simulator for several weather
scenarios with a set of user-specified inputs and provide a summary output
comparing the selected scenarios (AGBIT, 1995).

     rbWHIMS is a rule-based expert system designed to assist farm managers,
consultants, or extension personnel in making decisions on cotton arthropod pests
in the Midsouth. First developed in 1991, the system has seen numerous changes
from its original form using the ART-IM™ expert system shell, character based
interface, and rule base of nine pest species. The current version, released in 1995,

is written following the object-oriented programming paradigm in the C++ programming language. rbWHIMS has a Windows™ based graphical user interface and makes management recommendations on 13 arthropod pest species.

rbWHIMS models dynamic pest associations through the season by partitioning the season into nine distinct plant growth stages. These stages are determined by crop phenological events that have significant effects on pest population dynamics and management. Pest management is very data intensive and demands high quality information acquired with minimal use of resources. Therefore, the system has been designed to utilize the latest methods in sampling and statistical research, including line-intercept and quadrat sampling methods.

The system is designed to compliment the decision-making process by expanding the expertise of the user. For example, the greatest possibility to squander or conserve resources directed to pest management exists when field conditions do not lead to definitive actions on specific pests, and it is at these times that a "second opinion" is most valuable. During these times, it is possible for farm managers to take action when none is needed, not take action when it is needed, or take the wrong action in a given situation. These actions may be ineffective, wasting time and money, or they may make matters worse by producing the opposite effect desired or by adversely affecting other parts of the ecosystem.

Today, GOSSYM is the only cotton crop model used routinely in commercial crop production. It is used as a decision aid for fertilizer application, irrigation, crop growth regulator applications, crop growth terminator applications, planting and harvesting. The GOSSYM model has been validated against numerous comprehensive data sets. The validation tests consist of checking model prediction against actual phenological events such as time of first square, time of first bloom, and time of first open boll. More detailed tests compare model predictions of plant height, number of main stem nodes, leaf area index, stem weight, leaf weight and fruit weight over time to weekly field plot measurements of the same parameters. GOSSYM has been validated across multiple varieties, regions, climates and soils. The GOSSYM/COMAX system now has detailed soil physical property information on over 350 cotton soils. Figure 2a, 2b, 2c present a comparison of the performance of GOSSYM to actual field data. Yield data comparison for several variety classes across the US is shown in Table 1.

Despite the widespread use of GOSSYM-COMAX, it is still far from perfect. Factors unaccounted for by the model can influence the growth of the crop. These include herbicide injury, nematodes, plant pathogens, insect pests, deficits of mineral nutrients other than nitrogen, and weather events, (e.g., high winds which cause lodging and hail damage). Consequently, periodic plant mapping data and visual field observations can be used to adjust the simulation results and still use the model effectively when non modeled issues corrupt the simulation.

The hardware requirements of GOSSYM-COMAX are: 1) 66-MHz or higher, P5 or 80486 processor, 2) 8 megabyte RAM, 3) 170 megabyte hard disk, 4) SVGA graphics board and SVGA color monitor, 5) high density floppy disk drives (1.2 or 1.44 megabyte), 6) serial and parallel ports for communications and printer, 7) 9600 baud Hayes-compatible modem, 8) 80 column Epson or IBM compatible printer, 9) MS-DOS Version 5.0 or later, 10) Windows 3.1 or higher, and 11) mouse, Microsoft compatible (AGBIT, 1995).

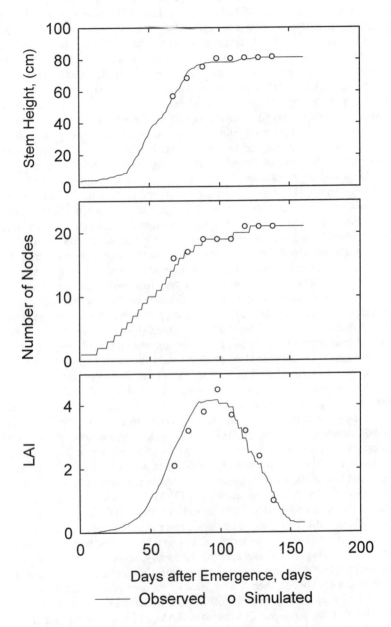

Fig. 2a. Plots of the stem height, number of nodes and LAI on a per plant basis.

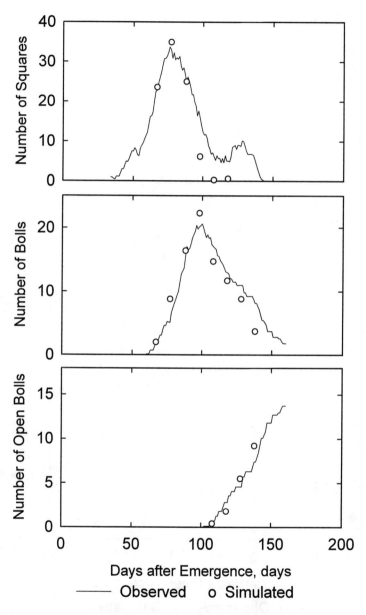

Fig. 2b. Plots of the number of squares, bolls
and open bolls on a per plant basis.

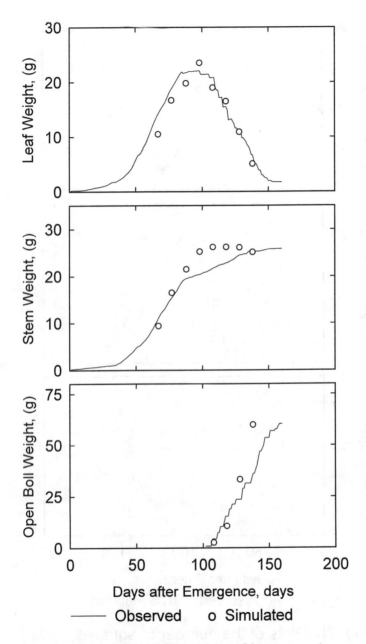

Fig. 2c. Plots of the weights of leaves, stems and open bolls on a per plant basis.

Table 1. Validation of GOSSYM in terms of yield prediction expressed in terms of average absolute ratio of actual over simulated yield.

Cultivar class	n	Yield ratio*
GC 510	8	0.86
DES 119	16	0.83
EARLY	15	0.82
FULL	8	0.80
MID	36	0.81
ST1	3	0.92
ST2	3	0.87
OVERALL	89	0.83

*ratio of maximum (actual or predicted) over mimimum (actual or predicted) yield.

CONCLUSION

Current agricultural practices estimate the amount of critical agronomic inputs by averaging the needs for a field (or a farm). Based from on-site soil sampling and scouting reports, fertility and pest infestation status are determined and recommended amounts of a product (water, herbicide, fertilizer, insecticide) are applied homogeneously throughout relatively large, and often heterogeneous areas. The generalized prescriptive nature of management recommendations of agronomic input (fertilizers, herbicides, insecticides, plant growth regulators and harvest aid chemicals) limit their effectiveness and result in less than optimal use of resources.

GOSSYM/COMAX/WHIMS simulates cotton growth, development and yield over a wide range of soils and climate. Mechanistic and process-oriented crop models using: a) real-time daily climatic data obtained from a weather station accessed by the user's computer via a modem; b) soil information such as textural classification, bulk density, soil-water retention and initial soil fertility nitrogen status; and c) cultural inputs like timing, amounts and methods of application of water, fertilizer and plant growth regulators can provide relevant information for the day-to-day or weekly management of the crop. Its use in combination with other technologies such as geographic information systems (GIS), variable rate machineries, global positioning systems (GPS) and scouting reports could further optimize site specific farming practices to address some of the social, ethical and environmental issues that face the declining number of farmers.

Other Decision Support System Development at CSRU

Farmers and consultants have expressed their interest in seeing GOSSYM/COMAX/WHIMS integrated with GPS and GIS. They have also suggested that better plant mapping methods be explored to facilitate their scouting.

Scouting is the observation of production fields during the course of the growing season to check for harmful insects, weeds, nutrients levels and anything else that would cause a decrease in production so as to make recommendations to improve yields (T.W. Oswald, 1996, personal communication). Normally, plant mapping is employed in cotton fields and is quite labor intensive and time consuming (Boone, et.al., 1995)

To improve the efficiency and speed of plant mapping, COTTON TALK, a speech interface for cotton plant mapping, has been developed using the IBM™ Continuous Speech Series (ICSS) speech recognition toolkit. It allows the user to speak directly to the computer while examining and mapping the plant by speech input. It also provides appropriate audio responses to help confirm the user's previous input. This system is intended to be an improvement over the current plant mapping method by eliminating the paper data sheets and subsequent data entry (Liang, et.al., 1996).

CropView is a GIS-based, multi-crop, decision-support system currently in prototype form (R. Olson, 1996, personal communication). It integrates simulation and decision-support technology with a spatially-registered database and a user-friendly graphical user interface. The simulation models will allow short-term (i.e., a week in advance) predictions of agronomic needs at within-field scales. Other modules will take this information and provide control programs for variable-rate field implements, such as fertilizer and pesticide applicators. The system is object-oriented, modular and is implemented in the PC-based ArcView GIS programming system (Earth Sciences Resources Inc., Redfield, CA).

"AgBook™, developed by AGBIT and marketed as part of the GOSSYM/COMAX software, is a computerized record keeping program designed to assist in managing the farming business. It runs under Microsoft Windows and has user friendly screens and menus to help maintain records on every cropping operation including pesticide and fertilizer use, purchases and quantity on hand. Pesticide information includes date, rate and cost of application, chemical name and all data required by State and Federal guidelines concerning Restricted Use Pesticide. It includes Profit Analysis data and graphics that enable users to determine production costs as well as break-even yields and current profit margin (AGBIT, 1995)."

REFERENCES

AGBIT, Inc.. 1995. GOSSYM-COMAX User's Manual. Starkville, MS

Baker, D.N., J.R. Lambert, and J.M. McKinion. 1983. GOSSYM: A simulator of cotton crop growth and yield. S.C. Agric. Exp. Stn. Bull. 1089.

Baker, D.N., and J.A. Landivar. 1991. The simulation of plant development in GOSSYM. p. 153-170. *In* T. Hodges (ed.) Predicting Crop Phenology. CRC Press, Boston, MA.

Bezdicek, D.F., and C. DePhelps. 1994. Innovative approaches for integrated research and educational programs. Am. J.of Alternative Agriculture. 9:3-8.

Boone, M.Y.L, D.O. Porter, and J.M. McKinion. 1995. RHIZOS 1991: A simulator of row crop rhizosphere. National Technical Information Service, US Department of Commerce, Springfield, VA.

Liang, C., J.L. Willers, S. Bridges, and J.M. McKinion. 1996. COTTON TALK: A speech interface for cotton plant mapping. *In* Beltwide Cotton Conf., Nashville, TN. 8-12 Jan. 1996. National Cotton Council of America, Memphis, TN.

McKinion, J.M. 1986. SPARNET: A data acquisition/analysis computer network. Computers and Electronics in Agriculture. 1:163-172.

Phene, C.J., D.N. Baker, J.R. Lambert, J.E. Parsons, and J.M. McKinion. 1978. SPAR--A soil-plant-atmosphere-research system. Trans. of ASAE. 21:924-930.

Stauber, K.N. 1994. The future of Agriculture. Amer. J. of Alternative Agriculture. 9:9-15.

Site-Specific Crop Management - A System Approach

A. Skotnikov
P. C. Robert

Department of Soil, Water, and Climate
University of Minnesota
St. Paul, MN

ABSTRACT

Today site-specific crop management (SSCM) is a reality. But most of components or equipment for SSCM work as separate entities, not joined in a common system. We propose a system approach outlining theoretical and practical aspects for integrating these components. We give an example of structure of an integrated system and indicate needed equipment.

INTRODUCTION

Site-specific crop management (SSCM) refers to a rapidly developing system that promotes variable agricultural management practices within a field according to site conditions. This is a new interdisciplinary concept based on a systems approach to problem solving.

One of the proposed definitions of SSCM is the following: site-specific crop management is an information and technology based agricultural management system to identify, analyze and manage site-soil spatial and temporal variability within fields for optimum profitability, sustainability, and protection of the environment [1].

Scientists from many Universities, research institutions, private and public sectors are working in different aspects of SSCM: yield monitoring and mapping; soil resource variability; managing variability; engineering technology; profitability; environment; technology transfer.

Practically the full set of equipment necessary for SSCM has been developed, including variable rate seed drills, sprayers, fertilizer applicators, irrigation systems; differential global positioning systems (DGPS); yield monitors and mapping software packages. Some of the equipment necessary for SSCM is in various stages of development: automated soil samplers [2] and work station for soil analysis [3], applicators and tenders with variable size of compartments [4], residue monitors [5], cultivators.

Most of this equipment is uncoordinated and has not been combined into a comprehensive system. Every piece of equipment has its own "smart box" processor or board computer with a software package not usually compatible with any other. Elaboration of programs for seed planting, applying chemical and fertilizer, and tillage are still in some stage of development and usually accomplished by different services. Presently the scientific base for development of such programs is not elaborated and no common procedures for collecting initial information and its processing exist. There are a number of software-based expert support or decision support systems for agriculture and natural resource management. All these systems address a specific problem or specific parameter, yield mapping for example, rather than the whole system for SSCM. Therefore farmers find them expensive and difficult to use.

Our goal is using a system approach to demonstrate methods of software and hardware integration, necessity of development new types of equipment and comprehensive management for this technology, making them more relevant, useful and affordable.

DISCUSSION

The analysis of the existing situation in SSCM permit us to make the following preliminary conclusions:
- there is the need in specialized agricultural computer with external I/0 board and set of sensors and relay interface;
- it is necessary to integrate decision support system and include in it modules for data base creation, mapping, geostatistical, statistical and economical analyses, soil sampling, seed planting, fertilizer and chemical application, and yield map interpretation;
- it is necessary to develop a special service capable to consult in all aspects of SSCM;
- initiate education and training of specialists for SSCM.

The possible configuration or structure of an integrated system needed for SSCM equipment is presented in Fig. 1.

Let's briefly discus the necessity of new equipment included in suggested structure.

First of all - it is necessary to have compatible set of automated soil sampler and work station for soil samples analysis. Such equipment will significantly reduce the cost of soil sampling and analysis, permit us accumulate more sufficient initial data base, develop a special strategy for soil sampling based on our previous experience [3], and using differential global positioning system (DGPS).

Existing equipment for variable rare of fertilizer application (VRA) has a significant disadvantage - constant size of compartments or bins for applying components. This reduce the productivity significantly, because when one of the compartments is empty, the whole unit has to stop for refill. In addition to supplying by fertilizer VRA, the farmer or dealer usually keeps a fleet of tenders near the field. This in turn, increases the cost of operation. To improve the performance and productivity of equipment for applying fertilizers it is necessary to have changeable size of compartments on VRA and supplying tender. A possible technical decision are presented in [4]. The same design will be suitable for seed drills.

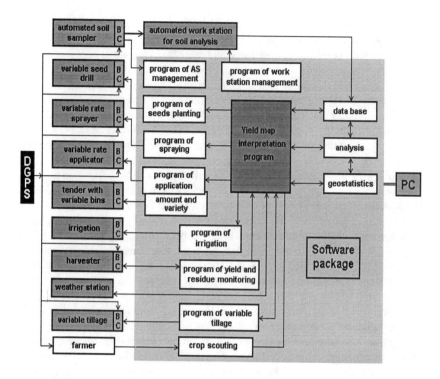

Fig. 1. The integrated structure of needed for SSCM equipment and software.

Existing yield monitor systems does not compute residue and transport delay of threshed grain is considered to be constant. But transport delay depends on moisture, ratio of straw and grain, slope, and harvester load (kg/sec). The straw-grain ratio is dependent of the cutting height. Besides, the combine does not operate for the full reaper width due to driving inaccuracy. This varying harvesting width add an additional error in yield determination as well. Therefore the mass feeding into the combine varies as the transport delay and losses respectively. These systems does not provide constant transport delay due to lack of tools to stabilize the performance parameters such as the harvesting width, the cutting height, torque on the threshing cylinder or harvester load. All this gives an error in estimation of yield for particular area up to 35 %. Residue monitoring will aid in developing better model of soil moisture and nutrients cycling. Some of technical decisions for improving yield and residue monitoring are suggested in [5].

Every piece of equipment from Fig. 1 has to have its own automated system. But any automated system consists of sensors, data acquisition board, processor, relay interface and actuators. If the farmer will use a portable rugged computer with I/O board, set of sensors and relay interface (BC on Fig. 1), it will be possible to eliminate different "smart boxes". The farmer will able to use the same computer with every piece of his equipment. In case of DGPS it will be possible to use only GPS antenna, receiver and software package. In other cases it will be even possible to economize on specialized processors, sensors and relays already installed on the equipment. For example, to activate the automated system on the seed drill, the farmer will need to transfer the computer from a sprayer or harvester, connect the I/0 board with corresponding sensors, and relay interface with necessary electric valves, insert a diskette with programs of sensors recognition, management and seed planting and then he will able to plant in an automated regime. The use of the computer will significantly reduce the common cost of equipment involved in SSCM, simplify the upgrading of automated systems and improve their compatibility, help to collect, store and process data for further needs.

Another real necessity of SSCM is an integrated decision support system. It must be a reliable, self teaching, easy-to-use system, combining expert assessments with statistical models. It is practically impossible to develop optimum management based on only one or several operations or parameters. It is necessary to consider the whole system of crop production and factors influencing it.

Our hypothesis is that by knowing initial spatial conditions of the field, management, quality of equipment work, weather during the growing season and final crop yield, it will possible to find out why reductions in yield occurred and how to avoid them in the future. A statistical models of yield crop dependence on different parameters and programs for input applications can be developed.

Crop yield must be the main criterion for comparison of results , different managerial decisions, crop growth technologies and development of programs for input application. Potential environmental impact may be a limiting factor in the development of programs for fertilizer and chemical applications. Maximum profit will be the main economical criteria.

The base unit will be a set of software developed for the farm-based computer that will be linked electronically to a WEB site with common data base for SSCM decisions support system. The software package should include modules for data base creation, mapping, geostatistical, statistical and economic analyses, soil sampling, seed planting, fertilizer and chemical applications, irrigation, yield map interpretation.

We will consider, mainly, an algorithm for developing a yield map interpretation module.

We recommend starting with a data base connected to available GIS data and choose parameters which have influence on yield. The more parameters available for the user and involved in explanation of yield results, the more accurate and precise the expert system output will be. All parameters must be based on or connected to a field map.

The initial set of parameters is field characteristics (Fig. 2). All characteristics involved in further analysis may consist of two groups: mandatory (white) and optional (gray). Some of them are quantitatively assessed (white triangle), while other are qualitative (gray triangle). The qualitative assessment may have several grades for a particular parameter. For example, for two grades (irrigation) it can be yes or no; for three (carbonates) - low, medium, deep; for four (relief) - shoulder, backslope, footslope, toeslope. For some crops (alfalfa for seed production, for example) presence of bees is important. In this case it is necessary to include this information in the data base as well.

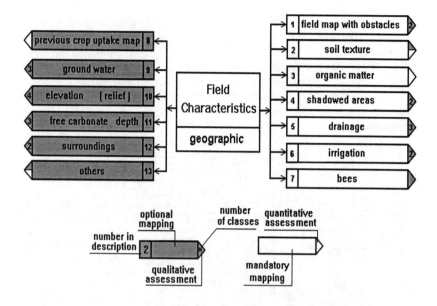

Fig. 2. Field characteristics

The second set of parameters is soil analysis (Fig. 3). This set of data has mandatory and optional geographic parameters.

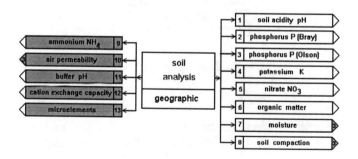

Fig. 3. Soil analysis

The next set of parameters relates to a field management (Fig. 4). It contains information about all operations to produce a crop and results of crop scouting.

The next set of parameters is related to field management (Fig. 4). It is necessary to provide certain information about every operation involved in crop growing. The results of a crop scouting may require additional actions. These additional actions and results of crop scouting must be mapped as well.

First - it is necessary to specify operation and equipment.

Mark one of three categories "early", "in time", and "late". If each of these categories applies it is necessary to give a date for the record and further comparison.

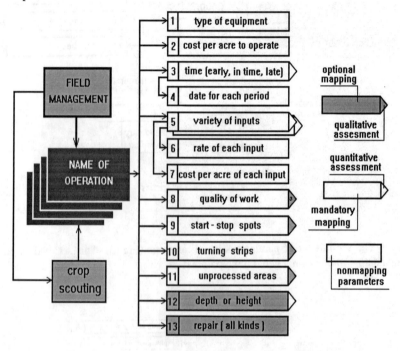

Fig. 4. Field management.

Define the input for every operation. For planting it is crop variety or varieties; for chemical application type of chemical, for tillage, type of tillage and depth.

It is necessary to eliminate start - stop events and turning strips from further consideration. The quality of all applications in these areas is usually lower due to overdose or underdose of materials. In order to do this it will be necessary to identify them on the field map.

Obstacles need to be indicated in mapping because these areas are poorly processed or unprocessed and will be excluded from yield map analysis. They may be analyzed as potential for yield increase.

It is also necessary to mark all locations of equipment repair or malfunction where user would have quality reduction.

During the entire growing season, it is necessary to accumulate weather data (Fig. 5) such as daily temperature and growing degree days, precipitation, evapotranspiration, moisture, etc. This data may provide an explanation of yield results and will accumulate important information about the influence of temperature and precipitation in different growing periods, help to develop soil moisture model. In the data base of an expert system it will create additional layers of information.

The program will provide a blank field map with all obstacles created during first step. The user will manually create new overlays of wash outs, standing water spots, chemical spills, quality of tillage, etc. All parameters will be connected to geographical coordinates.

Interactions Between Farm Managers and Information Systems with Respect to Yield Mapping

G. Larscheid
B. S. Blackmore

Centre for Precision Farming
School of Agriculture, Food and Environment
Cranfield University
Silsoe, Bedford. England

Developments in the area of Precision Farming have led to many of the hardware requirements being available. The target of further activities is to develop an interactive information system which supports the use of this hardware. ISO 11783 Part 9 (1994) is a data dictionary which specifies mobile data communications in agriculture. In addition, Hansen (1994) describes an interface for the communication between management information systems and mobile process control systems.

Alongside this kind of interactive information flow there appears to be a second kind of interactive information flow which takes into account the data transfer between information systems on farm computers and farm managers. Since an information system has to support farm managers in their decision making process the information has to be customized. Decision analysis must be carried out as part of the development of an information system. Farm managers must be able to use this information system for: (I) Data entry, organization and storage; (ii) Data analysis and interpretation; (iii) Data integration and implementation.

For the data entry a data dictionary was developed, which defines a structure for naming files according to their content. The files are organized in a database and stored in a normal Directory / File structure. Due to the fact that spatial data relies on visual analysis, guidelines for map presentation were established, as they provide a straight forward visual comparison. In order to perform automated map interpretation, AI tools have to be employed. For the integration and implementation of spatial and temporal data, methodologies are rare and need further development. In this study four different yield maps from one field were combined. The authors conclude that further developments especially in the software area have to consider farm managers decision making processes.

INTRODUCTION

To date, farm managers have always tended to treat their fields as a whole unit although they have been fully aware that variability (spatial and temporal) exists within their fields. Nowadays, technologies such as DGPS and spot sensors offer the possibility to quantify some of these variabilities.

Precision Farming is mainly concerned with managing variability by using information technology tools to acquire and handle information.

In particular the description of field variability is highly complex, because it takes place in a natural environment and includes many unknown variables (weather, pest and diseases, etc.). One major problem appearing nowadays is the transfer of this highly complex information between the farm, the farm manager and the computer.

The aim of an information system is to gain decision support which leads to an increase of productivity in farm business. But before decisions can be supported, they have to be analyzed and identified. Due to the fact that decisions come from an individual farm manager, an interactive information flow between the farm manager and the information system must be made possible. As information is personal, information systems must be customized to be able to make full use of the information they hold.

This paper sets out to study the farm manager's role in the Precision Farming environment and how a farm manager interacts with software tools in order to transfer information.

INTERACTIVE INFORMATION FLOW

The literal meaning of interactive can be seen as a two-way flow of information. In the area of arable management, it can be considered as an information flow between the farm manager and the farming environment. Farm managers base their decisions on gathered information from the farming environment in the broader sense as in Figure 1 described.

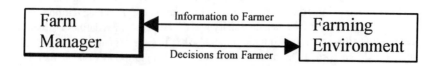

Figure 1: Interactive Information Flow

Within farming business a farm manager has to make many decisions that are based on values and beliefs. Jule et. al. (1995) describe in their paper three levels of decisions that can be taken on the farm: Strategic, Practice and Operational. *Strategic* decision making will only occur from year to year and will affect the whole purpose of the farm enterprise and is likely to be personal to an individual farm manager. *Practices* are the management options that apply to a particular sector. *Operational* elements describe the particular field operations.

Any decision requires a certain amount of information acquisition in order to support itself. Usually farm managers take these decisions based on their education and experience. In addition, they tend to establish their decisions beforehand by searching for advice among their farming community.

Blackmore et. al. (1994) consider three possible levels of information technology that can be adopted in farm business. (I) Technology Adoption Level 1: Traditional practice with no information technology (IT), (ii) Technology Adoption Level 2: Management information Systems, "What If" models and limited instrumentation and (iii) Technology Adoption Level 3: Full spatial understanding and treatment. According to these levels the interactive information flow between farm managers and their environment is going to be analyzed in the following chapters.

Information Flow: Technology Adoption Level I

Up to the mid-eighties, commercial farm management decisions were entirely based on experience, which were supported by advisors, suppliers, neighbors, etc.. Interactive communication takes place mainly on oral exchange of information with advisors, suppliers, etc.. Figure 2 describes the situation in the decision making process.

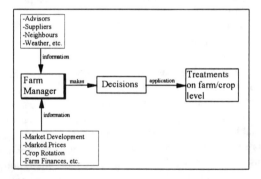

Figure 2: Information Flow - Technology Adoption I

Information Flow: Technology Adoption Level II

Technology Adoption Level II describes the level at which most of the European commercial farmers currently are practicing. Arable farming software tools on spreadsheet basis are accepted among the farmer community. Figure 3 describes the implementation.

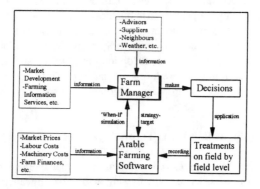

Figure 3: Information Flow - Technology Adoption Level II

Modular structured programs provide the possibility to incorporate strategy options for different needs in form of "What If" Models. This narrows the risk of the decision making process by reducing the unknown variables through a target driven simulation of the process beforehand and a record keeping of the treatments after their application.

Technology Adoption Level III

Blackmore et. al. (1994) describe technology level 3 as the ultimate goal in Precision Farming which would never happen as there are too many unknowns in the entire process. Although it is known that this target cannot be completely reached, researchers are trying to approach it with the speed provided by the technology available. As far as the interactive information flow between the farm manager and his environment is concerned, technology adoption level three has to be considered as well. The possibility to produce yield maps and the acceptance among farm managers to adopt this technology is evidence enough to encourage further activities. Gedes (1994) conducted a survey to question farm managers in England, who have been using yield monitors with DGPS, in order to reveal their opinion about this new technology. In farm manager's decision making process, technology adoption level III can be implemented such as described in Figure 4.

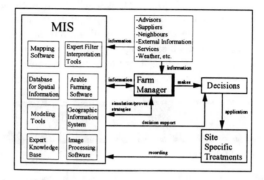

Figure 4: Information Flow - Technology Adoption Level III

The major feature of this level is the Management Information System (MIS). The MIS incorporates all IT-tools involved in the entire decision making process. Due to the fact that most of the tools nowadays are too complicated to be driven by farm managers, an MIS in this form is unlikely to be installed on farm computers. The MIS in this project consists of the following software tools: Mapping Software; Expert Filter Interpretation Tools; Database for Spatial Information; Arable Farming Software; Modeling Tools; Geographic Information System; Expert Knowledge Base; Image Processing Software.

The MIS is part of the interactive information flow according to technology adoption level 3 and, therefore, information must flow in either direction (see Figure 4). The overall performance of the interactive information flow of the MIS can be divided into three major tasks.

MAJOR TASKS OF THE MIS

- Data Entry, Organization and Storage
- Analysis and Interpretation
- Integration and Implementation

In the following paragraphs these three major tasks of the MIS are being investigated and ways, to approach these by putting them into practice, are being proposed. In order to reveal important factors several farm managers are involved in the project.

Data Entry, Organization and Storage

Raw data can be acquired from a number of sources as mentioned earlier. Data is recorded in different formats and requires manipulation to change data into information depending on its use. One problem which occurs during this process is the generation of data files which are probably not worth keeping on a long term basis. The issue addressed is the storage of data files which are useless or can easily be retrieved from the original dataset.

The first step in order to organize map data is to name all datasets according to a data dictionary which was developed especially for this purpose. Figure 5 shows the structure and an example. The application of an unique format which considers only file names, enables recognition of files exactly only by seeing the actual file name with its extension. Despite new operation systems which allow more than eight characters for filenames, the overall limit was set to eight characters.

Field	Map type	Map specification	No. of Derivation from original data
3 characters	1 character	2 characters + I for index	up to 1 integer

Example:

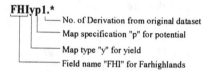

Figure 5: Structure and example of the data dictionary

The following step is to create a database containing all spatial datasets. This was performed by developing a database in Excel with the following information: Farm Name; Field Name; Year; File Name; File Extension; File Size; Map Type; File Type; Source File and Data source as table fields. An Excel database allows sorting and querying particular data. The extraction of for instance yield maps from one particular field over a certain period of time can be achieved.

The first three fields in the table are representing simultaneously the directory where the actual file is held. An ordinary Directory / File Structure is employed to store the files on hard disk. Figure 6 gives the structure and an example.

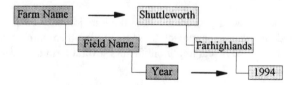

Figure 6: Directory / File - Structure

The database and the files are installed on farm managers' computers so that they can keep records updated and have access to their datasets.

Presentation and Analysis

Guidelines for map presentation are necessary to provide a straight forward comparison of maps. To compare, for example, yield maps of one field from different years, they have to be produced applying the same processing techniques and classifications.

The guidelines established for the presentation of yield maps, in agreement with the farmers, are as follows:

Map Projection: Universal Transverse Mercator (UTM) - British Grid,
                     true co-ordinates in meters rather than Lat./Lon.
Color-Scale: Deep blue for low yielding areas continuously to light
                     blue, light red and deep red for high yielding areas
Yield-Scale: Unique for similar crops related to a particular farm
Map-Info-Box: Date of harvest, crop, average yield

Figure 7 shows an example of a yield map which is produced according to the previous mentioned guidelines. Despite the map in Figure 7 appears in Black and White rather than the original colors Blue and Red, the detection of yield trends is still possible. This is advantageous as not every farm manager has access to a color printer.

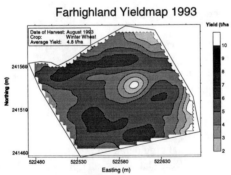

Figure 7: Yield map 1993 Farhighlands

Analysis of spatial data is nowadays mainly based on a visual investigation. Automated analysis tools are not available off the shelf and most farm managers are not able to do statistical analysis, for instance, in a spreadsheet program, because they are neither computer experts nor statisticians. In order to do visual analysis it is essential to keep a certain map presentation at least on farm level.

Figure 8 shows a set of yield maps from one field over four different years where the map presentations conform to the guidelines established.

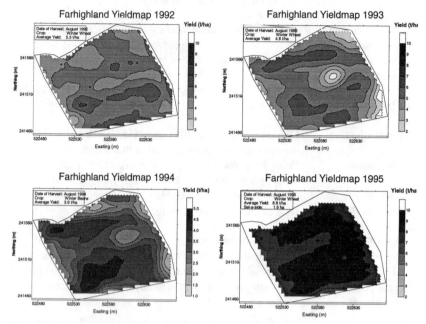

Figure 8: Farhighland yield maps from 1992 to 1995

Due to the same map presentation, visual analysis is feasible without misleading effects.

In the above displayed set of yield maps, it can be seen that the yield scale for same crops is unique (2 - 10 t/ha for Winter Wheat; 1 - 5 t/ha for Winter Beans). These ranges were set according to the overall yield potential for this particular crop during the actual time period when the data was gathered. Obviously, the range will differ from farm to farm.

The procedure of changing data into information can be divided into three different stages which are: Raw Data; Presentation of Data; Interpretation of Data.

Raw data is untreated and obtained directly from the source, i.e. sensor. But systematic errors can distort the result. If we consider raw data from a yield monitor and a DGPS, we know that inherent errors decrease the accuracy.

A straight forward display of untreated yield data is very difficult to interpret and thus of little use for farm managers.

Presentation of data is the step where certain data treatments are mostly applied for better visualization. Detection and removal of systematic errors as well as interpolation, especially for spatial data, makes visual analysis much easier. If visual analysis is employed to establish high or low yielding areas based on a number of yield maps, yield from different crops can be represented in normalized form, such as difference from the mean as displayed in Figure 9.

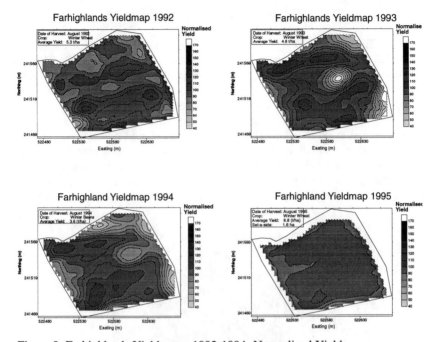

Figure 9: Farhighlands Yield maps 1992-1994 -Normalized Yield-

In Figure 9, the yield normalization was achieved in Excel by setting the average yield for each individual dataset to one hundred percent and changing all yield readings accordingly. In order to obtain equal presentation, the outcoming values were classified in a range between 40 and 170 percent with 10 percent spacing. This procedure provides two main possibilities. Firstly, spatial variation in each individual yield map can still be detected. Secondly, visual comparison of yield maps over more than one year adds the second possible degree of variability caused by temporal changes.

Basically, it has to be noted that data preparation for processed analysis differs from data preparation for visual analysis. If data is used for processed analysis, it has to remain as accurate as possible.

For visual analysis the level of map accuracy has to be considered before producing a map. Factors such as farm manager's understanding of spatial variability or present technology level (machinery) on the farm determine the appropriate level of map accuracy. Neither high resolution nor over simplified presentation can be considered advantageous.

Interpretation of Data is the implementation of expert knowledge into data presentation. This requires information about the further use or intention of the presentation. In the case of a yield map, values can be classified into three yield classes (i.e. low, medium, high), which provide farm managers with the information that only they might require.

Integration and Implementation

The major advantage comes into account when farm managers are able to establish their decisions by combining different data layers within a GIS. Then, the results are based on a number of processed values describing the relationships rather than expectations.

The map represented in Figure 10 is an example of a yield trend map created by combining yield maps of four different years within a GIS. To achieve this, all datasets had to be interpolated exactly at the same grid points. Then, the four datasets were reduced to two by combining two sets at a time whereas the normalized yield value for each interpolated grid point was averaged. The outcoming two datasets were combined in exactly the same way which has led to the final result displayed in Figure 10. A further classification into five trend-classes was applied.

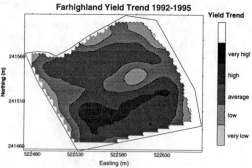

Figure 10: Farhighland Yield Trend (1992-95)

Once the yield trend zones are established further investigations have to show whether the occurring variability at a particular location is caused either by spatial or temporal influences. Therefore, statistical methods can be adopted. The aim is to reveal limiting factors for crop growth.

Nowadays, methodologies for combining different spatial data layers are rare. In order to develop these methodologies, relationships between different factors in arable farming processes have to be investigated and established.

CONCLUSION

The implementation of an MIS in farm manager's decision making process will not change the main structure of the procedure. Farm managers want always to be able to create their own decisions supported by MIS rather than adopting decisions proposed by the MIS.

Due to the fact that information is personal, information from the MIS has to be customized. Both farm manager's IT capability and understanding of spatial and temporal variability determine the success of the entire Precision Farming enterprise on an individual farm.

The structured development of a database for spatial data is essential for an assessment of variability (spatial and temporal) levels within fields. The variability level in accordance with the farm strategy will justify site specific field treatments.

Map presentation and interpretation has to be customized according to factors such as farm manager's understanding of variability, farm manager's IT capability and farm machinery. These items have to be considered beforehand in terms of map accuracy and resolution.

ACKNOWLEDGMENTS

Thanks to Massey Ferguson and Shuttleworth Farms for the use of their data.

REFERENCES

Blackmore, B.S., P.N. Wheeler, R.M. Morris, J. Morris, R.J.A. Jones. 1994. The role of precision farming in sustainable agriculture: A European Perspective. Scottish Natural Heritage; TIBRE Project.

Gedes, A. 1994. Precision farming in the office: Investigating the development of a GIS-based field information system. Unpublished.

Hansen, M. 1994 Interface Description communication between Management Information Systems and Mobile Process Control Systems. DIN 9684/5.1 Draft proposal.

ISO/ CD 11783 - 9. 1994 Data Dictionary for Mobile data communications in Agriculture.

Jule, I.J., G. Jahns, and S. Blackmore. 1995. Information framework for precision farming. Published at the International Conference "Agricultural and Biological Engineering - New Horizons, New Challenges". University of Newcastle upon Tyne, UK 20-23 September 1995. Session 5: Trends in Agrotechnology.

Site Specific Management: Educating through Cooperating Farmers

Thomas L. Krill

Department of Extension
Ohio State University
Columbus, Ohio

OSU Extension-Van Wert
Van Wert, Ohio

ABSTRACT

In response to local interest in site specific management, Ohio State University Extension-Van Wert initiated an educational demonstration project in 1994. The objectives of this project were to: 1)demonstrate soil nutrient variability in Van Wert County, 2)demonstrate yield variability in Van Wert County, 3)introduce area farmers to the equipment of site specific management, and 4)share the data collected and experiences with area farmers. During the spring of 1994, four farmers were selected to participate in the project each agreed to provide a 40 acre field for three years and pay a small fee to help offset the project costs. It was agreed that each 40 acre field provided be treated uniformly by each cooperating farmer using their best field management practices. OSU Extension agreed to harvest the crop using a combine equipped with a yield monitor using the global positioning system with differential correction(DGPS). OSU Extension would also soil sample the field annually using 360 ft. by 360 ft. grids and input the data into a geographic information system (GIS). Information collected from this project, with two of its three years now complete, is available for public educational and research efforts and has brought an important and local perspective to agricultural decision makers as they determine how site specific management will be incorporated into the future of production agriculture.

INTRODUCTION

In the Fall of 1993, a local agricultural producer spoke up during an advisory committee meeting. "I heard about a farmer taking a computer and putting it into his tractor and combine." The advisory committee fell silent and a new educational program was born. The members of the Ohio State University Extension Advisory Committee in Van Wert County, Ohio had just begun what would become their Site Specific Management educational effort that would grow to involve the entire state of Ohio and beyond. Though not the first documented educational effort by any means in Site Specific Management, this grass roots effort has lead to a successful and continuing program in Site Specific Management education and applied research.

Technology Transfer

Following the proud extension tradition of technology transfer, the Ohio State University Extension-Van Wert began the educational process. The extension service in the United States was founded in 1914 with the passage of the Smith-Lever Act. A cooperative effort of federal, state, and county government, the extension service is charged with the responsibility of linking the research and educational resources of the land grant universities with the people. The mission of Ohio State University Extension is to help people improve their lives through an educational process using scientific knowledge focused on identified issues and needs. Technology transfer is an important part of this mission.

Technology transfer is the process through which a new innovative idea is passed from its conception into accepted practice by the general public. The diffusion and adoption process has been well documented (Rogers and Burdge, 1972; Rogers, 1983, and Kaimowitz 1990). For this process to occur and the technology to be successfully transferred, individuals must first have a technology to consider. If validated, considered adaptable by the producer, the technology will then be scrutinized as the producer looks for information relevant to the new technology. The educational systems of both private and public organizations can often act as this change agent. As change agents, our challenge is often to impact the rate of this information dissemination or diffusion process through education. After collecting information, the producer will go through the persuasion stage as they decide to adopt the technology. If never persuaded, the adoption process falters and the innovation dies. Following initial adoption called persuasion, the producer must still integrate the technology with other systems and reinforce the technology by repetitive use of it for successful adoption.

Individuals have been classified as: 1) innovators, 2) early adopters, 3) early majority, 4) late majority, or 5) laggards when it comes to adopting new technology. As change agents, the "early adopters" are of particular interest in impacting the rate of change. These individuals are characterized as being progressive, but are also well integrated members of the local society. Early adopters are the embodiment of successful and discrete use of new technology. As such, they are respected and often asked for advice and information by other members of the community. (Rogers and Burdge, 1972) Early adopters are often the intended audience of educational programs relative to new technologies.

Educational programming directed at the early adopter should attempt to incorporate the principles of learning as identified in educational theory. The principles of practice, effect, and association are of particular interest in agricultural education settings and their traditional problem solving approach. (Crunkilton and Krebs 1982). First, individuals will have better retention of the information presented when they can participate in the actual practice of the task. Self activity is essential for learning. Secondly, if the individuals are satisfied with the learning experience, they are more likely to retain the information presented and seek additional information on their own. Individuals need to see the results of what their learning experience could bring to them personally. Finally, items presented together will tend to be learned and remain together. As educators, it is important to remember that teaching is not just transferring of knowledge but an actual change in behavior of the student.

The County

Van Wert County is located in northwestern Ohio and a productive part of the Eastern U.S. Cornbelt. The county is composed of 1060 sq.km. (409 sq. mi.) and is the home for just over 30,000 people and 890 farms. Approximately 96% of the land area in Van Wert County is used for production agriculture. The county is centered around the City of Van Wert and its 11,000 citizens. The number one industry in the county is agriculture accounting for an estimated 53 million dollars of annual cash receipts. Van Wert County is dominated by cash grain operations consisting predominately of corn, 32,400 hectares (80,000 acres), soybeans, 46,500 hectares (115,000 acres), and soft red winter wheat, 16,200 hectares (40,000 acres). (Van Wert Area Chamber of Commerce 1995)

Originally settled by the American Indians, the first European visitors to the area came as a military expedition in 1794 from Cincinnati. The area was almost entirely covered with a dense, swampy, deciduous forest. The county was first settled in the southern portion and settlers moved north as the great swamp in the north was drained. The land was known for its vast amounts of hardwood. Much of this hardwood was removed and shipped for industrial purposes, while other less valuable timber was removed and burned to expose the productive soils for agricultural production. Van Wert County was founded in April of 1820.

The soils of Van Wert County are mostly deep, dark colored, nearly level, fertile soils well suited for cash grain crops. The southern portion of the county is formed predominately from glacial till deposited by the Wisconsin age glacier that once covered the entire county. The northern areas were in turn formed by lake water covering this portion of the county following the glacial retreat and is a part of the Lake Plain area of northwestern Ohio. These soils are therefore mostly formed through water worked till, lacustrine sediments, and old beach ridge deposits. All land elevations within the county would fall within a 75 meter (240 ft.) range with the lowest elevation in the northeastern corner and highest elevation on the Fort Wayne terminal moraine in the southwestern corner. In general Van Wert County, is commonly described as flat, poorly drained, and consisting of productive soils. (USDA 1972)

The two predominant soil associations in Van Wert County are the Hoytville association and the Pewamo-Blount association. The Hoytville association is common in the northern section of the county and the Lake Plain area. It comprises 32% of the county and is described as "Dark-colored, very poorly drained soils that formed in water worked glacial till on the lake plain (page 3, USDA 1972)". In the glacial till region of the southern portion of the county, Pewamo-Blount association comprises the major portion. This association is the most prevalent association compromising approximately 45% of the county. The Pewamo-Blount association is described as "very poorly drained and somewhat poorly drained, nearly level to gently sloping soils on glacial till uplands (page 4, USDA 1972)".

Site Specific Management

Site Specific Management (SSM) was defined for the project as:
A management concept which recognizes variability within the soil and crop environment and maximizes economic production while minimizing environmental impact for a specific location.
(Krill 1994)

The key points of this definition are:
1) SSM is management.
2) SSM must have variability.
3) SSM must have specific location.
4) SSM will maximize net economic return.
5) SSM will minimize environmental impact.

The Project

In the winter of 1994, a group of individuals representing local producers, local extension professionals, and state extension professionals met to discuss and plan an educational program for Van Wert County, Ohio to educate local producers on the emerging agricultural practice known as Site Specific Management. This meeting was the result of a previous OSU Extension-Van Wert agricultural advisory committee meeting which recommended to examine the educational possibilities of site specific management in Van Wert County. In the meeting a series of objectives were drafted for this special project.

Objectives

Site Specific Management
Special Project
OSU Extension-Van Wert

1) Demonstrate soil nutrient variability in Van Wert County
2) Demonstrate yield variability in Van Wert County
3) Introduce area farmers to the equipment of site specific management
4) Share the data collected and experiences with area farmers.

To meet the challenge presented by these objectives, it was determined that this project was going to require a cooperative effort involving local producers, extension professionals, and private industry. Four local producers were sought across the county each providing approximately a 16 hectare (40 acre) field for three years involved in a corn soybeans rotation. They would be responsible for all crop production activities and costs excluding harvest, expected to treat the field uniformly during the three years, anonymously share all data relative to the field, and provide financial assistance to offset the operational costs of soil sampling and combine operation. OSU Extension professionals would provide technical assistance, be responsible for soil grid sampling of each field annually, coordinate all harvest activities including the operation of the combine, and maintain, analyze, and disseminate all information collected. Private industry, public organizations, and foundations would be sought to provide the equipment necessary for the project.

In the spring of 1994, four cooperating farmers were identified and accepted into the project. Cooperating farmers were approached by project committee members after being identified as leaders within the agricultural community. Each potential cooperating farmer accepted the invitation with different perceptions on the technology. In selecting cooperating farmers the county was segmented into quarters with one farmer selected from each section, northeast, southeast, southwest, northwest. This also insured that both the lake plain area of northern Van Wert County with its Hoytville association soils and the glacial till area of southern Van Wert County with its Pewamo-Blount association soils were represented. For the 1994 crop season, two fields were planted to corn and two fields to soybeans. One corn and one soybean field were located in the northern section and one each in the southern section.

During the summer of 1994, equipment was procured for the project. A John Deere 6620® combine was acquired and equipped with an AgLeader 2000® yield monitor and UniLink® differentially corrected global positioning system receiver (DGPS). Software for the project was selected and both DataVision® and CropSight® were selected. Mapsight® was selected as the navigation software for soil sampling and this process also used the UniLink® DGPS receiver.

In the fall of 1994, the crops were harvested and soil samples pulled. The grain was harvested using the combine with yield monitor. The yield monitor data was collected on 1 second intervals using a 5 meter (15 ft.) grain platform or 6 row corn head. Following harvest, soil samples were collected from each site. Samples were taken using a center point grid sampling technique on 110 meter (360 ft.) grids. Six soil cores were collected to a depth of 15 cm. (6 in.) around a 2 meter (6 ft.)radius and represented the 1.2 hectare (3 acre) grid. The six cores were combined into a single sample and sent to the Ohio Agricultural Research and Development Center, Research Extension Analytical Laboratory for analysis. All data were compiled at the Ohio State University Extension office in Van Wert County.

In 1995 similar procedures were carried out on the cooperating farmer sites. The two soybean plots were planted to corn. One of the corn plots was planted to soybeans while the other corn plot was planted to soft red winter wheat. This was a deviation from the original project but necessitated by the cooperator farmer's situation and approved by the committee. Harvest was completed in 1995 using the same combine and yield monitor. Following harvest, grid soil samples were again collected using the same center points and procedures as 1994. Soil sampling of the wheat field was delayed until the fall to maintain consistency between all plots.

Initial plans by the committee for 1996 are to continue the original procedures. The fields are intended to be planted to corn or soybeans with a corn and soybean field each in the northern and southern section. Soil sampling will be completed following the fall harvest using the established procedure.

The Educational Results

Soil variability was discovered in Van Wert County. Initial soil samples in 1994 and 1995 soil sample results indicated spacial variability. The samples were evaluated for the elements phosphorus(P), potassium(K), calcium(Ca), and magnesium(Mg). Testing also included the base saturation for potassium, calcium, and magnesium, organic matter (OM), lime test index (LTI), and pH. Variability did exist in the elemental areas of P, K, Ca, and Mg; however, the indicated levels generally exceeded the response range for these nutrients as recommended. pH variability was also discovered and was determined to be within a treatable region. (Vitosh, M. L., Johnson, J. W., Mengel, D. B. 1995) Detailed examination of the data collected from the soil samples will be carried out following completion of the project and collection of the third year soil samples.

The 1994 results of these findings were presented to Van Wert County agricultural producers. Preceding public presentation, the data were first presented to the cooperating farmers. Response to the data was received with much interest and surprise. The variability exceeded the expectations of each cooperating farmer. To confirm the laboratory results, one cooperating farmer requested that the field be resampled and sent to another private laboratory for analysis at his expense. Results of this analysis were similar to first. During the annual countywide Agronomy Day the data were also anonymously revealed to the county. Interest was intense and was again met with surprise. Popular opinion was concerned with the variability present in Van Wert County soils where little visible variation is present.

A similar procedure was used in 1995 to release the data. 1995 data demonstrated similar variability across each of the fields. Producer interest in the results continued to remain high. In 1995, a local crop consultant established a grid soil sampling service. To date, this service has sampled over 5,000 hectares (12,000 acres) of crop land. In 1996, a local fertilizer dealership invested in the equipment necessary to perform grid soil sampling. Another local fertilizer dealership has invested in retrofitting an existing dry fertilizer applicator to variable rate application technology. Variability was discovered in the perceived consistent, flat, agronomic soils of Van Wert County and the local producers have advanced along the adoption process.

Fall harvest in 1994 also revealed that yield variability existed in Van Wert County. The cooperating farmer fields were each harvested by the extension agent of Van Wert County using the combine described above. During harvest the extension agent was often accompanied by the cooperating farmer, his staff, or local neighbors. Interest was high but skeptical in the yield monitoring combine. Variation reported by the yield monitor was a surprise to all individuals who participated in harvest. All fields were harvested and machine yield monitor calibrations checked and adjusted prior to the release of any data. Data was presented in the form of field yield maps to first the cooperating farmers and then anonymously to the public during the Agronomy Day program. Unlike the soil sample information presented, audience attention immediately turned to attempting to identify the causes of the yield variability. Initial producer attention turned immediately to soil type but was somewhat subdued when it was revealed that one

40 acre field was composed of just one soil type. Complete and detailed analysis of the yield data will be initiated after the completion of the project during the 1996 harvest season. Correlation analysis will also be performed between the fertility and yield data at that time.

Prior to the 1995 harvest, 4 yield monitors were purchased for combines operating in Van Wert County. One of these monitors was purchased by a cooperating farmer. The 1995 harvest again indicated variability in yield throughout the cooperating farmer fields. Producer interest continued with more emphasis not on the actual yield variability but on what caused the yield variability. Through an informal survey of local agricultural equipment establishments it is estimated that an additional 12 yield monitors have been purchased in Van Wert for installation prior to the 1996 harvest season. Yield variability has been found in Van Wert County.

Van Wert County and other agricultural producers have been introduced to both the equipment of site specific management and the local results of site specific data collection. The equipment obtained for this project and data collected during this project have been displayed at local events including the annual Farm Focus Show hosted by the county and attended by 20,000 people from across the region. Educational efforts have been conducted to inform local agricultural producers of the new technology known as site specific management. These efforts have included traditional methods but have been highlighted with local experiences and data and the opportunity to see the actual equipment.

In summary, the objectives of the original project are well on their way of being met even before the project completion. The project has also seen the development of several related spin-off projects in Van Wert County and statewide projects for Ohio. General opinion among the cooperating farmers was that they expected to see some variability within their fields but the magnitude of the variability was surprising. The results have caused each cooperating farmer to rethink some of their agronomic practices and consider additional site specific activities. Countywide, the interest in site specific management is growing with producers and industry responding by the investment in site specific management resources.

Summary

When trying to introduce a new technology into an agricultural audience it is important to remember the principles of learning. This paper has not been about the effects of site specific management but rather the educational process used in the technology transfer process. By definition, site specific management is more attractive to locations with variability; yet, in Van Wert County where variability is often hard to visualize, a successful educational program is in progress. The program is successful because it practiced the principles of learning. Local producers were involved in the project whether as actual cooperating farmers, committee members, or just local producers. Information has been shared and opportunities given for hands-on experiences with both the equipment and data. The students were able to get involved in the educational process and thus the principle of practice. The local data has made it easier for the agricultural public to make the association between the cooperating farmers and their own agricultural operations. By making this association, students could relate to how this practice could effect their own personal situation and thus the principle of effect. Attempts were made to show site specific management as more than just one activity but a series of events using different methodologies, techniques, and equipment. Site specific management was shown as a process and the student realized that several components were needed to accomplish the process and thus the principle of association. Well planned educational program can have an impact on the technology transfer process and this program appears to have had an impact on the adoption diffusion process.

REFERENCES

Crunkilton, J.R., and A.H. Krebs. 1982. Teaching agriculture through problem solving 3rd ed. Interstate Printers and Publishers, Inc. Illinois.

Kaimowitz, D. 1990. Making the link: Agricultural research and technology transfer in developing countries. Westview Press: Colorado.

Krill, T.L. 1994. An introduction to site specific management. Agricultural Finance Seminar. The Ohio State University Extension. Ohio.

Rogers, E.M. 1983. Diffusion of Innovation.The Free Press. New York.

Rogers, E.M., R.J. Burdge. 1972. Social Change in Rural Societies 2nd ed. Prentice Hall. New Jersey.

United States Department of Agriculture (USDA). 1972. Soil Survey of Van Wert County, Ohio. USDA-SCS.

Van Wert Area Chamber of Commerce, Agribusiness Committee. 1995. Van Wert County Agriculture. Van Wert Area Chamber and Development Center.

Vitosh, M.L., J.W. Johnson, D.B. Mengel, 1995. Tri - State Fertilizer Recommendations for Corn, Soybeans, Wheat and Alfalfa, Extension Bulletin E-2567. Michigan. Michigan State University.

On-Farm Research Opportunities Through Site-Specific Management

H. F. Reetz

Midwest Director
Potash & Phosphate Institute
1497 N 1050 East Road
Monticello, Illinois, USA

ABSTRACT

The new management tools associated with site-specific management systems offer some new opportunities for research to be conducted under real farm situations. Such research will likely be more readily accepted and the results more easily adapted to other farms. GIS data bases, GPS positioning, and geo-statistical analytical procedures along with computer-aided interpretation of the results should help make this type of research acceptable in the scientific community as well.

BACKGROUND

New Tools for Research

The adoption of new technology and management systems is often delayed by the need for transfer of experience and information from small-plot research systems to field scale implementation. Often equipment constraints further delay the process. The use of global positioning satellites (GPS) and geographic information systems (GIS) can help eliminate much of the delay by allowing research to be conducted on field-scale plots with field-scale equipment commonly used by farmers. Small research plots and statistical design associated with such plots are largely designed to remove within-field variability as a factor in determining response to treatments.

With GPS and GIS, within-field variability can be managed and often utilized as a part of the experimental design. Intensive management systems that are developed based on the GPS and GIS information can take advantage of the measured variability within a field and develop a management system to best utilize the resources available. Note that the goal of site-specific management is not to eliminate within-field variability, but rather to manage that variability for optimum return on resources invested.

Management systems must be studied in detail before reliable recommendations can be made for implementation. Site-specific management systems are no exception. But traditional small plot research, or field-scale on-farm research does not adequately meet the information needs for evaluation. Site-specific management studies require new approaches to data collection and new methods for analysis. Fortunately, the new tools of site-specific management help meet these requirements.

New Opportunities

The ability to accurately document specific locations within a field where data are being collected is not only a benefit for farmers' record keeping, but also a key to the opportunity to move accurate research to a field-scale basis. Using GPS to document precise locations and GIS to record and relate the data collected, farmers and researchers can capitalize upon the variability in the field rather than try to eliminate it. Treatments can be applied with field-scale equipment in large plots that can be harvested with field scale harvesters equipped with yield monitors. Such capabilities open new opportunities for farmers to do more reliable research on their own farms, and for researchers to move their studies to the farm.

University and industry researchers may find that their role is changed in field-scale studies. By moving to farmers' fields, they can often arrange for the farmer to plant, manage, and harvest the research plots. The researcher can focus his resources and time to planning the studies and interpreting the data, and to detailed observations. There may be less need for small plot equipment, reducing the overhead costs of conducting field research. The results from field-scale studies on cooperator farmers' fields may be more readily acceptable to other farmers and thus more quickly implemented.

Small plot studies are still important for comparisons of treatments and detailed evaluation under controlled conditions. The large plot studies with field-scale equipment provide an additional perspective from which to evaluate treatments and management systems. Usually a combination of both will be needed to fully evaluate treatment effects and potential responses to implementing the management system.

SITE-SPECIFIC FIELD RESEARCH

USB Project

Beginning in 1995, a project was initiated to compare site-specific precision management with field-average management in a soybean/corn rotation system. Sponsored by the United Soybean Board in a grant to the Foundation for Agronomic Research, the project is the nucleus of a multi-state cooperative project that is supplemented by various industry, government agency, producer check-off, and experiment station funding sources.

Project activities and funding are being coordinated by the staff of the Potash & Phosphate Institute. Data management, mapping, and information coordination are being handled by the University of Illinois Department of Crop Sciences. As the project expands beyond the initial 10 sites in Illinois and Indiana, researchers from other universities will become involved.

Cooperation with Other Projects

Several ancillary projects will be conducted in cooperation with this master project, studying specific factors in the management system. Fertilizer rate studies, plant population studies, and weather monitoring have been incorporated into some of the locations. Data collected for the project are being used to test crop growth models. A National Oceanic and Atmospheric Administration (NOAA) project to monitor weather, crop development, and carbon dioxide and water exchange in the crop canopy has been linked to this project.

Project Procedures

Farm Cooperators

The field studies are being hosted by selected farmer cooperators. Each farmer has agreed to provide an average of approximately 30 hectares of soybeans and 30 hectares of corn in a soybean and corn rotation system. The farmers are supplying the equipment and labor for the crop production system. Each has already installed a yield monitor and has collected at least 2 years of yield maps for the fields being used. Data from previous years, including soil tests, crop practices, yields, etc., are being incorporated into the data base so that there is a substantial historical record to begin the project.

NRCS

The Natural Resources Conservation Service (NRCS) state office in Illinois has coordinated the development of digitized Order 1 soil surveys for the fields used in the project. This information is providing more details about the soils than is available in the standard soil survey reports. Their objective is to determine the value of this additional information in management decisions and to estimate the costs associated with the development of the Order 1 survey.

Other Projects and Funding Sources

Several other projects have been identified that will be linked to this project to provide a broader base of information and more efficient use of the money invested in site-specific research. Projects in South Dakota, Florida, Arkansas, Virginia, and Ohio will be linked for 1997. Additional opportunities are being explored. The data requirements can generally be satisfied with minimal adjustment to the protocol for a planned field study on site-specific management. The goal is to gain GPS/GIS data bases of soil survey, soil test information, production inputs and practices, and yield for a wide range of farms across the soybean/corn production belt. These data will then be used to compare site-specific management with field-average management on the basis of the best agronomic plan, economics, and potential environmental impact.

Future Direction

This project is planned for a 5-year period to provide a range of weather scenarios. During that time frame, procedures for studying site specific management systems will be evaluated. This information will be used to help formulate more detailed studies on specific practices. Some of the components being studied are the size of sampling grid most appropriate for site-specific management, procedures for mapping to gain most useful information, statistical procedures for accurate assessment of the information, methods for economic analysis and methods to determine potential environmental benefits of site-specific management. Procedures for handling large data sets form multiple sources are also being developed, along with procedures for disseminating the results to various cooperators.

Additional sources of funding and in-kind services are being sought and additional projects that can be linked to the data base are being explored.

Internet Access

An Internet website for the USB project is being developed and will be linked to the research pages for the StratSoy site when available. The World Wide Web address is:

http://www.ag.uiuc.edu/stratsoy.html

The North Carolina Precision Farming Project: Managing Crop Production with Precision Technologies Using On-Farm Tests

R. W. Heiniger

Crop Science Department
North Carolina State University
Raleigh, North Carolina

ABSTRACT

The North Carolina Precision Farming Project was started in 1995 to demonstrate the use of precision farming technologies and techniques to grain producers and to conduct research into methods and techniques for improving management using these new technologies. Two on-farm sites were selected based on opportunities to examine identified sources of variability in soil types, nutrients, or other management factors. Farm cooperators secured the necessary equipment and software for yield monitoring and recording farm management information. Procedures for sampling soil, soil moisture, insect, disease, weed, and crop parameters were developed to help monitor crop growth and environment at the sites.

Yield and soil data collected in the initial year of this project show a strong link between yield and soil pH on the soils common in eastern North Carolina. Management factors such as tillage practices (no-till or conventional tillage), variety selection, and plant populations were also found to be related to yield results. Opportunities for increasing crop yield and profit were found at both sites. The farmer cooperators found the information generated using precision farming techniques to be valuable and were eager to adopt the techniques in their farming operation.

Plans for future years include developing insect, disease, and weed monitoring techniques which will allow for grid based scouting and mapping of these variables. Nitrogen management in wheat using grid based information (tiller counts or reflectance measurements) also looks promising. Satellite photos will also be used to assess water and pest stresses on the crop. All of these techniques will be tested and evaluated. Promising processes will be incorporated into the project with the goal of developing a complete precision farming system.

INTRODUCTION

From the moment the first seed was dropped into the soil, farmers have recognized the special relationship that exits between plant, soil, and environment. The goal of farm managers and agronomists has been to enhance that relationship by improving plant management and soil fertility. Unfortunately, dramatic changes in soil and environment within farm fields often makes it difficult to fit the plant to its environment. Agronomists have long recognized that variability exits in almost every aspect of the crop environment (Pierce et al., 1994; McGraw and Hemb, 1994; Mortensen et al., 1994). Within a farm field, changes in soil types, nutrient levels, terrain, cropping histories, pest infestations, and microclimate all impact crop yield and productivity. In the past, farmers have compensated for some of these variables by differential application of cropping practices. However, as farm size and farm mechanization has increased, the ability of the farmer to apply site-specific practices to small field areas has declined. Today, as farm input costs increase in respect to crop prices and concerns mount over the damage to water quality and environment from non-point source pollution, farmers must improve the efficiency of crop production to remain competitive.

A leap forward in technologies such as global positioning, graphical information systems, yield monitoring, crop modeling, and computerized data acquisition and decision support has made it possible for producers to recognize multiple crop production variables within small areas of a field and to customize their management to those variables (Robert et al., 1992). A seamless system of information gathering and mapping for each field location combined with variable rate application methods to allow on-the-go application of crop inputs has the potential to improve input efficiency, field profitability, and environmental stewardship. Precision farming technologies and methods must lead to increased profitability of fields or better environmental stewardship for them to be adopted by farmers. Unfortunately, previous research has been inconclusive in demonstrating the benefits of precision farming. For instance, while several researchers have found large variation in soil nutrient levels and soil physical properties (Pierce et al., 1994; McGraw and Hemb, 1994), little work has been done on defining the amount of variability needed to properly utilize the new site-specific technologies or how precise the application of inputs must be to get the full benefit from using precision farming. Wolkowski and Wollenhaupt (1994) did an analysis that found there was a potential for increasing input costs by using site-specific management.

One of the problems in quantifying the benefits of precision farming is determining what management and fertility practices to use for each of the sites based on the variability present. Almost without exception, there has been little success in matching the soil variability with yield levels monitored at each site (Sadler et al., 1994). This is primarily due to the overwhelming influence of weather variability. Interpretation of site-specific data and the extension of that information to making fertility and management recommendations depends upon the successful quantitative description of the causes of yield variation within soil types under the climate present. Finally, while the technology needed for site-specific management has improved, more needs to be done. Precision farming requires extensive use of new technologies. Computers are necessary to manage large datasets containing soil and crop information, sensors are needed to measure yields on-the-go, geostatistical software is required to analyze the acquired data, and controllers must vary application rates by location within a field. Crop status monitoring must be improved by incorporating remote sensing techniques into site-specific measurements. Weed and pest control techniques in the form of integrated pest management (IPM) systems must be integrated into the precision farming concept by developing new hardware and software interfaces. Some parts of this system are well established, some continue to mature in both technology and cost, while other technologies are yet to be developed (Searcy, 1994).

OBJECTIVES

The primary goal of this project is to evaluate the profitability and other potential benefits from using precision farming technologies and methods. In particular we wanted to: (i) evaluate the use of Global Positioning Systems (GPS) for use in precision farming systems, (ii) develop methods of seamlessly collecting, analyzing, and imputing site-specific information into graphical information systems (GIS) databases given the equipment and technology available, (iii) collect site-specific yield and soil information and develop methods for using this information to modify management decisions and the resulting inputs needed for crop productivity, and (iv) analyze the economic and environmental impacts of precision farming systems.

METHODS

Two farms in the tidewater area of North Carolina, Coastal Carolina Farms and Open Ground Farms, were selected as experimental and demonstration sites. These farms were picked based on their size, the variability in soil types, willingness to cooperate on experimental trials, and knowledge of computer applications. Coastal Carolina Farms is a 1 215 ha farm owned by Spurgeon Foster located south of Columbia, NC in an area of organic soils. Soil types range from a Hyde silt loam, organic matter > 20%, to a Weeksville sandy loam, organic matter < 5%. Fields on Coastal Carolina Farms vary in size each being subdivided into cuts 100 m wide.

Open Grounds Farm is a 16 200 ha corporate farm found on the Carteret peninsula in eastern North Carolina. It lies between the Neuse River and Pamlico Sound to the west and north and the Cape Lookout National Seashore to the east. Protected wetlands surround the farm, with water from the farm flowing into the South River which empties directly into the Pamlico Sound. A wide range of soil types are found on Open Ground Farms ranging from a Belhaven silty clay loam to a Ponzer fine sandy loam. The farm is laid off in 260 ha blocks. Each block is further subdivided into rectangular fields 100 m wide by 1 605 m long. On both Open Ground and Coastal Carolina Farms the fields or cuts are nearly level. Each field or cut is separated from the other by drainage ditches which dictate the pattern of mechanical tillage and fertility operations.

To meet the objectives of the project, several technologies were used. For soil sampling, a 4-wheel-drive all terrain vehicle was equipped with a portable computer system based on a 486 DX75 laptop computer. This was linked to a global positioning system with integrated Coast Guard Beacon receiver. The use of the Coast Guard Beacon system for differential corrections was facilitated by the location of a beacon at Fort Macon, NC which is less than 16 km from Open Grounds Farm and approximately 72 km from Coastal Carolina Farms. FARMGPS software (Farmer's Software Association and Red Hen Systems, Fort Collins, Co.)

was used in conjunction with MAPINFO (MapInfo Corporation, Troy, NC), a desktop GIS system, to link the GPS coordinates to sample data taken at selected sites in the field. For field scouting, a handheld DGPS data gathering system was developed around a 486 minicomputer and a lightweight integrated GPS and beacon receiver carried in a backpack. For yield mapping, AGLEADER 2000 yield monitors (AgLeader Corporation, Des Moines, IA) were mounted on three of the 15 combines owned by Open Grounds Farm and on the combine used by Coastal Carolina Farms. Again, differential GPS units provided location data for the yield monitors. MapInfo was used as the base GIS system for layering soil, yield, insect, weed, and other maps and for developing recommendations for applying fertilizer, seed, and chemical inputs.

The precision technologies described above were utilized to gather site-specific information. Initially, this information included soil fertility, soil type, and grain yield. Soil properties were measured on grids of 0.4 ha or less using point sampling which were done on a systematic unaligned grid (Wollenhaupt and Wolkowski, 1994). Selected field areas were measured on 10 m x 10 m grids to help determine the accuracy of sampling techniques and patterns. Grain yield was measured every second as the combine harvested the crop. On Open Ground Farms over 162 ha were sampled for soil properties and nutrients, and over 2 025 ha of yield data were collected. On Coastal Carolina Farms, 101 ha were analyzed for soil texture, physical features, and sampled for soil nutrients with yield data collected on approximately 405 ha.

Maps of soil properties, soil nutrients, and grain yield were analyzed to determine how the crop responded to soil and environment. Geostatistical techniques were used on the 10 m x 10 m sample data to determine the spatial pattern of soil nutrient variability. Semivariograms for pH, P, K, and selected micronutrients were plotted using GS+ version 2.3 (Gamma Design Software, Plainwell, MI). These semivariograms were analyzed for anisotrophy and used to determine sampling accuracy and maximum grid size. Punctual kriging (Isaaks and Srivastava, 1989) was used to interpolate soil nutrient values for areas between samples. Kriged maps were then visually compared with the raw yield data to determine if similar patterns existed. When the yield and nutrient patterns within a field were found to be visually similar, yield data within a 15 m radius of the point from which the soil sample was taken was regressed against the soil nutrient data to determine if they were correlated. Other management factors such as changes in tillage patterns and crop varieties were also examined to determine if they had any influence on yield patterns.

RESULTS AND DISCUSSION

At both locations the differential GPS systems chosen performed very well. Tests using surveyed benckmarks showed that the reported position was within 1 to 2 m of the true position 98.2% of the time. This resulted in position data for the soil and yield samples which was adequate for the type of testing performed.

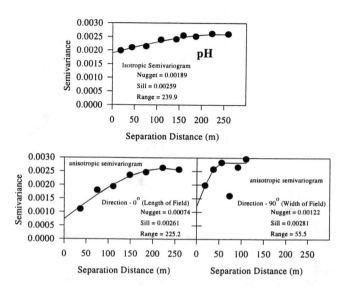

Fig.1. Isotrophic and ansiotrophic semivariograms for pH. Soil test data were taken from four fields (cuts) located on Open Ground Farms.

Based on the 10 m x 10 m sampling data, semivariograms of pH, P, and K from both Coastal Carolina Farms and Open Ground Farms showed strong anisotrophy with the direction of maximum variability across the 100 m width of the field or cut and the direction of minimum variability along the length of the field or cut (Fig. 1). While the nugget and sill parameters were similar, the primary difference between the semivariograms was in the range, 55.5 m across the cut compared to 225.2 m in the direction of the cut. Because traffic patterns are restricted to always following the length of the cut, these semivariograms indicate that the pattern of variability is associated with mechanical operations performed in farming the land. As a result, grid sizes are restricted by the variability found across the field or cut. Potential exists for using retangular rather than square grids.

Using 0.4 ha grids, soil nutrient analysis showed variability in soil pH in several blocks or fields in both locations. Soil pH ranged from 4.2 to 6.4 at Coastal Carolina Farms and from 4.4 to 6.9 at Open Ground Farms (Fig. 2).
Other nutrients such as phosphorus and potassium also showed a large amount of variability, but most of the test results were in the sufficient to excessive range. This means that little, if any, advantage would be gained by using variable rate technology for making phosphorus or potassium applications. Comparisons with yield results showed little correlation between yield results and either phosphorus or potassium. However, in several fields or cuts there was a visual similarity between the pH pattern and the yield pattern (Fig. 3).

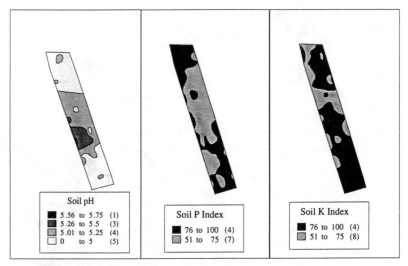

Fig. 2. Spatial patterns for pH, phosphorus, and potassium found in a 17 ha field on Open Ground Farms.

Fig. 3. 1995 wheat yield pattern found on a 17 ha field on Open Ground Farms.

When yields from a 15 m radius of a given soil test site were regressed on the soil pH, a significant correlation was observed (Fig. 4). These results indicate that, depending on the field, a slight 0.2 to 0.3 increase in pH could increase yields 0.67 to 1.34 Mg ha^{-1}. Many soils in the southeastern US have acidic subsoils with only an 20 to 25 cm layer of topsoil. The low native pH of the soil material and non-uniform applications of lime have resulted in a large amount of variability in pH within fields.

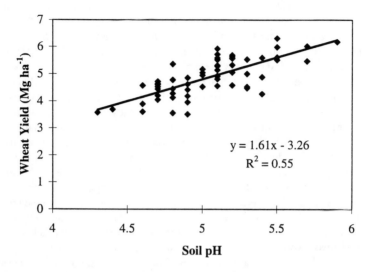

Fig. 4. Linear relationship between wheat yields and pH

Tillage also had an impact on the yield patterns observed. A 260 ha block was mapped which had two different tillage practices applied side-by-side. One half of the fields within the block were tilled before planting, while the remaining fields were planted without tillage using a no-till John Deere planter. Grain yield maps of the block showed a seven to nine bushel increase in corn yields in the area where the no-till planter was used (Fig. 5). Open Grounds Farm has been moving toward no-till over the last several years as a method of reducing erosion and preventing excessive runoff of rainfall. However, there have been doubts about whether grain yields on the no-till fields were sufficient to justify the expense. This data has helped show that no-till is profitable. Yield maps also showed patterns that were influenced by the varieties planted. Each cut in one block was planted to a different variety. Different varieties were found to have different yield patterns. This information was valuable in determining which variety performed best.

CONCLUSIONS

The future for these new technologies in agriculture lie in their ability to provide useful information to farm managers at Open Grounds Farm and to Spurgeon Foster at Coastal Carolina Farms. In a profitable farming operation, new technologies must be able to pay their way. This means that these systems must provide either a savings in the amount and, therefore, the cost of inputs applied or an increase in grain yield as a result of improving farm management. Several such potential benefits were noted in this project.

Probably the most important finding was the correlation between grain yield and soil pH. This was noted at several locations and has since been confirmed by results from growers fields across the region. This correlation is probably due to the acidic subsoils of this region and the shallow layer of pH neutral topsoil. This relationship indicates that variable rate technology based on grid soil sampling could lead to lowered costs for liming, improved recommendations for optimum soil pH levels, and, most importantly, to increases in grain yield. Estimates show that a 0.67 to 1.34 Mg ha^{-1} increase could be achieved. Continuing studies in 1996 have been initiated to quantify these potential benefits.

Other benefits to precision technologies were found in the ability to assess farm management through site-specific yield information. The economic benefits of no-till systems were quantified in the study at Open Ground Farms. Furthermore, the performance of corn hybrids was determined over a large number of acres. This information helped determine the environmental conditions which best suited individual varieties. There were several other ways that precision technologies lead to decreased costs or increased yields. For instance, both farmers found that they could reduce fertilizer use because they had more confidence in their knowledge of soil fertility within a field. They were able to use the grid based soil test records and yield maps to help design fertility programs for each field. Normally, they would have used their best judgment about how much fertilizer to apply between soil tests. Now they have specific yield records that can tell us how much phosphorus and potash was removed by the crop and how much they should apply. Information on grain moisture within the field helped determine what preparations needed to be made for post harvest grain handling. The ability to work directly with farmers and farm managers was instrumental in determining just how precision technologies and methods would be useful.

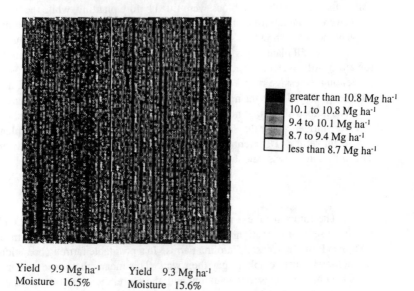

greater than 10.8 Mg ha^{-1}
10.1 to 10.8 Mg ha^{-1}
9.4 to 10.1 Mg ha^{-1}
8.7 to 9.4 Mg ha^{-1}
less than 8.7 Mg ha^{-1}

Yield   9.9 Mg ha^{-1} Yield   9.3 Mg ha^{-1}
Moisture   16.5%   Moisture   15.6%

Fig. 5. 1995 corn yields on a 260 ha block on Open Ground Farms. The left half of the block has been in continuous no-till while the right half is conventionally tilled using standard practices for the area.

Despite having been in operation for only one year, this project achieved a number of secondary goals. GPS systems were evaluated which resulted in recommendations for cost effective systems for North Carolina. Soil sampling procedures were tested which helped identify optimum grid sizes and sampling methods. GIS software was tested along with techniques for handling large databases of yield and soil information. Working with the North Carolina Department of Agriculture, soil testing information was digitized to allow for easy input into computer databases.

Probably the most important measure of the success of the project was the response from the participating producers. Both farms not only indicated a willingness to continue the project, but were eager to expand the process to included variable rate technologies and grid-based weed and insect scouting. The enthusiasm of these growers over the information they gathered helped encourage others to purchase yield monitoring systems and to contract for grid soil sampling services. This indicates the value, both real and intrinsic, of precision farming systems.

REFERENCES

Isaaks, E.H., and R.M. Srivastava. 1989. An introduction to applied geostatistics. Oxford University Press. New York, NY.

McGraw, T., and R. Hemb. 1994. The amount of variability of fertility in the Minnesota river valley watershed in 1993 as determined from grid testing results on 52,000+ acres in commercial fields. *In* Site-specific management for agricultural systems. P.C. Robert, R.H. Rust, and W.E. Larson (eds.). 2nd Intern. Conf. on Site-Specific Management for Agricultural Systems. 27-30 March. Minneapolis, MN.

Mortensen, D.A., G.A. Johnson, D.Y. Wyse, and A.R. Martin. 1994. Managing spatially variable weed populations. *In* Site-specific management for agricultural systems. P.C. Robert, R.H. Rust, and W.E. Larson (eds.). 2nd Intern. Conf. on Site-Specific Management for Agricultural Systems. 27-30 March. Minneapolis, MN.

Pierce, F.J., D.D. Warncke, and M.W. Everett. 1994. Yield and nutrient variability in glacial soils of Michigan. *In* Site-specific management for agricultural systems. P.C. Robert, R.H. Rust, and W.E. Larson (eds.). 2nd Intern. Conf. on Site-Specific Management for Agricultural Systems. 27-30 March. Minneapolis, MN.

Robert, P.C., R.H. Rust, and W.E. Larson. 1992. Adapting soil-specific crop management to today's farming operations. 14-16 April. Minneapolis, MN.

Sadler, E.J., W.J. Busscher, and D.L. Karlen. 1994. Site-specific yield histories on a SE coastal plain field. *In* Site-specific management for agricultural systems. P.C. Robert, R.H. Rust, and W.E. Larson (eds.). 2nd Intern. Conf. on Site-Specific Management for Agricultural Systems. 27-30 March. Minneapolis, MN.

Searcy, S.W. 1994. Engineering systems for site-specific management: opportunities and limitations. *In* Site-specific management for agricultural systems. P.C. Robert, R.H. Rust, and W.E. Larson (eds.). 2nd Intern. Conf. on Site-Specific Management for Agricultural Systems. 27-30 March. Minneapolis, MN.

Wolkowski, R.P., and N.C. Wollenhaupt. 1994. Use of row-placed fertilizer with corn (*Zea mays* L.) for the management of the effects of field P and K variability. *In* Site-specific management for agricultural systems. P.C. Robert, R.H. Rust, and W.E. Larson (eds.). 2nd Intern. Conf. on Site-Specific Management for Agricultural Systems. 27-30 March. Minneapolis, MN.

Wollenhaupt, N.C., and R.P. Wolkowski. 1994. Grid soil sampling. p. 6-9. *In* D.L. Armstrong (ed.) Better Crops with Plant Food. Vol. 78. No. 4. Potash and Phosphate Institue. Norcross, GA.

Building Total Crop Production Management Solutions Using OLE Automation for Third-Party Interoperability

T.S. Macy
A.J. Dondero

Agris Corporation
Roswell, GA

Agris has implemented a crop recordkeeping system that is built entirely upon OLE Automation servers. One of the specific reasons for selecting this architecture was to reduce the complexity of working with third-party vendors for such things as Task Controller Interfaces and Agronomic Modelling. The OLE Automation interface that Agris has implemented has been offered to the Ag Electronics Association Software Council for inclusion as part of the standard for interoperability.

This paper is a presentation of the Agris OLE Automation interface, including the properties and methods for the proposed AEA server object, as well as the protocol through which a third-party client may query a recordkeeping server, or a recordkeeping client may utilize third-party servers for things such as fertilizer recommendations, irrigation scheduling, etc.

Agris anticipates that universities will begin to develop third-party modelling extensions to commercial recordkeeping solutions, rather than their own foundation solutions.

Precision Agricultural Management: Practical Consideration and Technological Issues for Small Farmers and Producers

T.U. Sunday

Iowa State University
Ames, IA

ABSTRACT

Since the beginning of the Green Revolution, efficient production of food and fiber has depended on the transfer of indigenous knowledge within farm families and the use of uncomplicated tractors equipped with accessories that small farmers could easily manipulate and maintain. By contrast today's agriculture is driven by policies and institutions, and there is a trend towards loading farm tractors with cab-mounted precision controls (e.g., GPS), digital computers, and custom-made GIS databases and maps. The recent advancements in computer-based technologies have ushered in a new era of precision agriculture under terms such as site-specific farming, soil-specific crop management, precision farming, etc. The shift towards precision agriculture is driven in part by the complex challenges of environmental quality, and by the demands for more globally competitive and profitable methods of producing food and fiber. The use of precision controls and farm-level spatio-temporal data in agrochemical management can increase production efficiency and reduce runoff and leaching losses to surface and ground water. The potential savings to be derived from precision agricultural management are considerable on the farm-scale and huge on the regional or global scale, let alone the improvements in environmental quality. However, for small farmers the adoption of precision agriculture particularly the use of GPS and GIS raises many technological, social and management issues. The success of precision agriculture for small farmers depends heavily on a good understanding of the operational and technological requirements, as well as on the appropriate interpretation and proper use of information and data for management decision-making. This paper examines the many practical issues and implications of precision agriculture for small farmers and producers. It also offers suggestions that can reduce the number of technological problems, management decisions, and frustrations associated with precision

Maximizing the Utility of Precision Agricultural Technologies While Improving Producer Safety

J.M. Shutske
J. Chaplin
W. Wilcke
R. Ruan

Biosystems and Agricultural Eng. Department
University of Minnesota
St. Paul, MN

ABSTRACT

There continues to be concern and controversy about the cost/benefit relationship of precision farming technologies such as GPS and mapping software for small and medium sized crop producers One potential solution that could help alleviate some of these concerns is to use devices employed m precision farming to accomplish other functions. This paper will examine the use of GPS, electronic sensors, and other precision farming hardware and software to improve the safety of people operating crop production equipment Farming continues to be one of the most deadly industries in the U.S. The problems associated with farm injuries and fatalities cost producers at least $5 billion annually. Specifically, this paper will present a summary of research in the University of Minnesota's Department on Biosystems and Agricultural Engineering examining the use of GPS, mapping software, cellular communications, and electronic sensors to enhance operator safety on tractors, combines, and other field equipment. This hardware and software can be used to provide a more timely response by emergency medical personnel in the event of an injury in the field. A GPS-based early alert system could be triggered by the machine operator or by software that monitors patterns in machine activity and movement through the field to determine when an accident has occurred. *New sensors* and electronic control systems could also be used to detect the presence of people in unsafe areas or near unsafe machine components. This paper also outlines the need to integrate new electronic technologies to take advantage of their potential safety-related benefits

SUMMARY OF WORKGROUPS: PRECISION
AGRICULTURE RESEARCH & DEVELOPMENT
NEEDS

Compiled by: P.C. Robert, Conference Chair
Workgroup moderators: members of the Agricultural Experiment Station
                          Research
Committee NCR-180 Site-Specific Management

Fifteen workgroups met on Tuesday morning from 10:45 to 12:00. The workgroup membership was assigned at registration based on affiliation to form groups that would provide a mix of participants. Each workgroup task was to 1) list and 2) rank research and development needs in precision agriculture.

Since only five workgroups ranked the R&D needs, results are presented in two tables. The first table ranks needs compiled from the five workgroups that had prioritized the needs, using the top three identified issues. The second table ranks the R&D needs using the most frequently named topics. The number of votes is given in parenthesis. Table three gives a list - without ranking - of factors that had few occurrences.

Table 1. R&D needs ranking from workgroup lists with ranking of issues.

Rank	Needs
1	• Decision support systems, including agroecosystem modeling, risk assessment, confidence level, and management prescription. [3][1] • Selective soil sampling strategies adapted to field conditions. [3]
3	• Profitability: cost-benefit analyses. [2] • Technology accuracy and reliability assessment: data quality and measurement errors of yield monitors, planters, and applicators. [1]
5	• Understanding of interactions between natural resource factors and crops. [1] • New generation of site-specific soil and crop management recommendations for tillage, planting, nutrient, and pest management. [1] • On-farm experimental methods: design and data analysis. [1] • Soil and plant sensors for yield monitoring of major crops, soil nutrients, soil water, pests, and plant population. [1]

(1) Frequency of occurrence. For example, [3] = cited by 3 workgroups

Table 2.   R&D needs ranking based on the first twelve most frequently named topics

Rank	Needs
1.	• Decision support systems, including agroecosystem modeling, risk assessment, confidence level, and management prescription. [12](1)
2.	• Selective soil sampling strategies adapted to field conditions. [9] • Understanding of interactions between natural resource factors and crops [9] • Soil and plant sensors for yield monitoring of major crops, soil nutrients, soil water, pests, and plant population. [9] • Technology accuracy and reliability assessment: data quality and measurement errors of yield monitors, planters, and applicators. [9]
6.	• Profitability: cost-benefit analyses. [8]
7.	• Education and outreach: college students, agribusiness representatives, and farmers. [7] • Remote sensing: airborne and spaceborne, field scouting, correlation to real-time soil and crop management, image interpretation and accuracy, and economics. [7]
9.	• Understanding of interactions between natural resource factors and crops. [6]. • New generation of site-specific soil and crop management recommendations for tillage, planting, nutrient, and pest management. [6]
11.	• On-farm experimental methods: design and data analysis. [5] • Development of precision management for other crops and practices (e.g., pasture, pests, and manure). [15]

(1) Frequency of occurrence. For example, [12] = cited by 12 workgroups.

Table 3. Other R&D needs. No ranking

• Development of standards for machinery, sensors, software, and data. [3]
• Use of soil maps for precision agriculture [2]
• Study of long term effects of precision agriculture practices
• Development of GIS tools
• Study of spatial variability of crop quality
• Sociological impact of precision agriculture

APPENDIX II - PARTICIPANT LIST

Aal, Ervin
Tyler Industries
E. Hwy 12, Box 249
Benson, MN 56215
(320) 843-3333 f: 843-2467

Acock, Basil
USDA-/ARS/-BARC
Sensing & Model
Bldg 007, Rm 008 BARC-W
10300 Baltimore Ave.
Beltsville, MD 20705
(301)504-5827: f 504-5823
bacock@asrr.arsusda.gov

Ahlrichs, John
Cenex Land O'Lakes
PO Box 64089
St. Paul, MN 55164
(612) 451-5533; f 451-4561
jahlr@cnxlol.com

Albaugh, Mike
John Deere Co.
2001 W. 94th St.
Minneapolis, MN 55431
(612) 887-6268; f 887-6385
ma00525@deere.com

Allmaras, Ray
USDA-ARS
439 Borlaug Hall
University of Minnesota
St. Paul, MN 55108
(612) 625-1742;f 625-2208
allmaras@soils.umn.edu

Alms, Maggie
Blue Earth Agronomics, Inc.
Box 230
Lake Crystal, MN 56055
(507) 947-3362; f 947-3404
bluesoil@ic.mankato.mn.us

Alton, Vanessa
Soil Stewardship Group
2136 Jetstream Rd.
London, Ontario
Canada N5V 3P5
(519) 457-2575

Amerman, Dick
USDA-/ARS/-NPS
Rm 233 Bldg 005 BARC-W
Beltsville, MD 20705
(301) 504-6441; f 504-5467
ramerman@asrr.arsusda.gov

Anderson, Gerry
USDA-ARS
2413 E. Hwy. 83
Weslaco, TX 78596
(210) 969-4834; f 969-4893
gl-anderson@tamu.edu

Anderson, Lynda
Terra Industries
PO Box 70
Dewey, IL 61840
(800) 252-7690;
f(217) 897-6410

Anderson, Mike
Univ. of California-Davis
Dept. of Vihiculture & Enology
Davis, CA 95616
(916) 752-0358; f 752-0382
mmanderson@ucdavis.edu

Anderson, Noel
Concord Inc.
3000 7th Ave., N.
Fargo, ND 58102
(701) 280-1260; f 280-0706
75032.2040@compuserve.com

Arslan, Selcuk
Iowa State
Buchanan Hall 2771
Ames, IA 50013
(515) 296-4409
sarslan@ia.state.edu

Asbell, Rob
DeKalb Agra
Box 127
Waterloo, IN 46793
(219) 837-8611; f 837-7220

Ascheman, Robert
Ascheman Assoc.
Consulting
2921 Beverly Dr.
Des Moines, IA 50322
(515) 276-7371; f 276-8708
rascheman@delphi.com

Aspinall, Doug
Ontario Min. of Agriculture
Box 1030 52 Royal Rd.
Guelph, Ontario
Canada N1H 1G3
(519) 767-3575; f 767-3567

Baker, James
Iowa State University
Davidson Hall
Ames, IA 50011
(515) 294-4025; f 294-2552
jlbaker@iastate.edu

Barber, Daniel
DowElanco
9330 Zionsville Rd.
308/2E
Indianapolis, IN 46268
(317) 337-4578; f 337-4567

Barnes, Edward
USDA-ARS-U.S. Water Cons.
4331 E. Broadway Rd.
Phoenix, AZ 85040
(602) 379-4356; f 379-4355
ebarnes@uswcl.ars.ag.gov

Bashford, Leonard
University of Nebraska
203 LW Chase Hall
Lincoln, NE 68583
(402) 472-1627

Bauer, Marvin
University of Minnesota
1530 N. Cleveland Ave.
St. Paul, MN 55108
(612) 624-3703; f 625-5212
mbauer@forestry.umn.edu

Bauer, Scott
Lockheed Martin
Idaho Tech.
Box 1625
Idaho Falls, ID 83415-3790
(208) 526-9714; f 526-2818
sgb@inel.gov

Bausch, Walter
USDA-ADS-NPA
Water Research AERC-CSU
Fort Collins, CO 80523-1325
(970) 491-8511; f 491-8247

Baxter, Paul
Baxter & Associates
99 Reservoir Rd.
Norris, TN 37828
(423) 574-5968; f 576-6196

Beck, Robert
Cenex Land O'Lakes
Box 64089
St. Paul, MN 55164
(612) 451-5383; f 451-4561
bbeck@cnxlol.com

Beeman, Tracy
Box 105
Broadview, MT 59015
(406) 667-2292

Beeson, Michael
Great Plains Agronomics
7977 Cty Rd. 15
Wahpeton, ND 58075
(701) 274-8818

Bell, David
Byron Equipment Co.
N901 Hwy. 26 N.
Watertown, WI 53098
(414) 262-8620; f 262-8630

Bell, Thomas
Stanford University
Gravity Probe
c/o Hansen Labs
Stanford, CA 94305
(415) 725-6378; f 725-8312
tombell@leland.stanford.edu

Bellows, Bob
Cargill. Inc.
1st and Main
Fonda, IA 50540
(712) 288-4453; f 288-6131

Benlloch, Jose'
D.I.S.C.A.
Box 22012
Valencia, Spain 46071
(346) 387-7211; f 387-7219
jbenlloc@disca.upv.es

Bennett, John
Noetix Research Inc.
978 Merivale Rd.
Ottawa, Ontario
Canada   K1B 4L3
(613) 729-4899; f 722-3798
noetix@magi.com

Bennett, Jon
Delta Farm Press
Box 1420
Clarksdale, MS 38614
(501) 338-2784; f 627-1977
elurdan@aol.com

Bergeijk, Jaap Van
Wageningen Agri Univ.
Pomenweg 4
The Netherlands   6703 MO
+31 317 482889; 317
484819
jaap.vanbergeijk@aenf.wau.nl

Berry, Joseph
1701 Lindenwood Dr.
Fort Collins, CO 80524
(970) 490-2155; f 490-2300
joeb@cnr.colostate.edu

Beverly, Reuben
University of Georgia
1109 Experiment St.
Griffin, GA 30223
(770) 412-4765; f 412-4764
rbeuerl@gaes.peachnet.uga.edu

Bhattacharyya, A.K.
Jawaharlal Nehru University
School of Environmental Sc.
New Delhi-67 India
H110067
91-011--6475453; --
6865886
asimb@jnu.niv.ernet.in

Bickell, Matt
Southern Exp. Station
35838 120th St.
Waseca, MN 56093
(770) 412-4765; f 412-4764

Bickerton, Thomas
WAUB-USDA
12th & Independence Ave.
Room 5133
Washington, DC 20250
(202) 720-5913; f 690-1805

Birrell, Stuart
University of Missouri
241 Agric. Eng. Bldg.
Columbia, MO 65211
(573) 882-1135; f 882-1115

Blackmer, Alfred
Iowa State University
2218 Agronomy hall
Ames, IA 50010
(515) 294-7284; f 294-3163
ablackmr@iastate.edu

Blackmer, Tracy
USDA-ARS-NPA
University of Nebraska
119 Keim Hall, E. Campus
Lincoln, NE 68583-0915
(402) 472-8494; f 472-0516
tblacke@unlinfo.unl.edu

Blackmore, Simon
Silsoe College-Univ.
England, Silsoe, Bedford
England MK45 4DT

Bodie, Cameron
Flexi-Coil Ltd.
PO Box 1928
Saskatoon, Sask
Canada S7K 3S5
(306) 664-7600; f 664-7626

Bodinnar, Nigel
Pivot Ltd.
Box 400
Strathfieldsaye, Victoria
Australia 3551
+054 393210; f +054
393209

Boisgontier, Denis
Inst. Technique des Cereales
Station Experimentale
Boigneville
France 91720
+33 164992211;+33
164993330

Bolte, Charles
AgSource Soil & Forage Lab.
106 N. Cecil St.
Bonduel, WI 54107
(715) 758-2178; f 758-2620

Booker, Jill
Texas A&M
6500 Amarillo Blvd
Amarillo, TX 79106
(806) 354-5808; f 354-5829
j-booker@tamu.edu

Booltink, H.W.G.
Wageningen Agriculture Univ.
P.O.B. 37
Wajeninje
Netherlands 6703 MD
+31 317482422;
f +31 317482419

Boone, Mariquita
Mississippi State University
Box 5367
MS State, MS 39762
(601) 324-4342; f 324-4371

Borgelt, Steve
University of Missouri
Agric. Eng. Dept
Columbia, MO 65211
(573) 882-7549; f 884-5650

Barnhisel, Richard
University of Kentucky
Agronomy Dept.
Lexington, KY 40546
(606) 257-8627; f 257-2185
rbarnhis@aa.uky.edu

Bower, Matthew
University of Maryland
Bldg. 142, Rm 0501
College Park, MD 20742
(301) 405-6109; f 314-9023
mhbow@glue.umd.edu

Boydell, Broughton
University of Georgia
UGA-NESPAL
Tifton, GA 31793
(706) 542-0911; f 542-0914
bboydell@uga.cl.uga.edu

Braaten, David
ProGold
101 North 10th Street
Fargo, ND 58102
(701) 298-4044; f 298-4059

Bremer, Chuck
Pioneer Hi-Bred Int'l, Inc.
Box 1536
O'Fallon, IL 62269
(618) 624-8222; f 624-8283
bremerc@phibred.com

Brennan, Mark
TASC
55 Walkers Brook Dr.
Reading, MA 01867
(617) 942-2000; f 942-7100
mubrennan@tasc.com

Bridle, Tracy
Communication Systems Int.
6009 1 A St. SW
Calgary, Alberta
Canada T2H 0G5
(403) 259-3311; f 259-8866
info@csi-dgps.com

Brix-Davis, Kalyn
SDSU Plant Sciences
219 Ag Hall Box 2207A
Brookings, SD 57007
(605) 688-5122; f 688-4602

Brumback, Tom
Pioneer Hi-Bred Intern'l. Inc.
7300 N. W. 62nd Ave.
Box 1004
Johnston, IA 50131-1004
(515) 270-4220; f 270-4312
brumackt@phibred.com

Bruulsema, Tom
Potash & Phosphate Inst.
18 Maplewood Dr
Guelph, Ontario
Canada NIG 1L8
(519) 821-5519; f 821-6302
bruulsem@sentex.net

Bryan, Jeanne
Farm Journal
222 S Jefferson
Mexico, MO 65265
(573) 581-9641; f 581-9646
jkatbryan@aol.com

Buchleiter, Gerald
USDA-ARS-NPA
AERC-CSU
Fort Collins, CO 80523-1325
(970) 491-8511; f 491-8247

Budding, Jeanette
Farm Industry News
7900 International Dr.
Ste.300
Minneapolis, MN 55425
(612) 851-4682; f 851-4601

Burkhart, Roger
Deere & Co.
John Deere Rd.
Moline, IL 61265
(309) 765-4365; f 765-5168
roger@90.deere.com

Burt, Doug
Washington State University
24106 N. Bunn Rd.
Prosser, WA 99350
(509) 786-9338
dburt@mail.wsu.edu

Busch, Randy
Dairyland Laboratories
217 E. Main
Arcadia, WI 54612
(608) 323-2123; f 323-2184

Buss, David
Datawise
N1926 County Rd II
Waterloo, WI 53594
(414) 478-2091; f 478-3523

Byrne, Tom
McKensey & Comp.
2600 DeMers Ave., Ste. 105
Grand Forks, ND 58201
(701) 746-8580; f 746-8681

Cambouris, Athyna
Agriculture & Agri Food
2560 Boul.
Hechelaga, Quebec
Canada
(418) 657-7980; f 648-2402
cambourisa@em.agr.ca

Campbell, Jack
Canadian Agra Corporation
6212 Rime Village Dr. #102
Huntsville, AL 35806
(205) 971-5432; f 971-5350
cconsult@traveller.com

Campbell, Ron
Harvest Masters
1740 N. Research Parkway
N. Logan, UT 84341-1941
(801) 753-1881; f 753-1896

Carbonell, Javier
Cenicana
AA 9138 Cali-Colombia-SA
Cali, Valle Colombia
+ 6648025; f 6641936

Carlson, Alan
PO Box 246
Brandon, SD 57005
(605) 582-3647

Carlson, Ken
AGSCO, Inc.
Box 13458
Grand Forks, ND 58108
(701) 775-5325

Carroll, Bill
Midwest Technologies, Inc.
2733 E. Ash SE
Springfield, IL 62703

Carter, Lyle
USDA-ARS
17053 Shafter Ave.
Shafter, CA 93263
(805) 746-8004; f 746-1619
lcarter@lightspeed.net

Casady, William
University of Missouri
205 Agric. Engineering
Columbia, MO 65211
(573) 882-2731; 884-5650

Catt, Corey
Brown Seed Farm
N1279 530th St.
Bay City, WI 54723
(715) 594-3003; f 594-3758

Cattanach, Allan
North Dakota State Univ.
227 Walster Hall Box 5758
Fargo, ND 58105
(701) 231-8596; f 231-7861

Chaplin, Jonathan
University of Minnesota
1390 Eckles Ave.
St. Paul, MN 55108
(612) 625-8146; f 624-3005
jchaplin@gaia.bae.umn.edu

Cheng, H. H.
University of Minnesota
Dept.of Soil, Water & Climate
St. Paul, MN 55108
(612) 625-9734; f 625-2208
hcheng@soils.umn.edu

Choriki, Michael
B & C Ag Consultants
PO Box 1184
Billings, MT 59103
(406) 259-5779; f 259-1038

Christensen, Lee
USDA-ERS
Room 424
1301 New York Ave., N.W.
Washington, DC 20005-4788
(202) 219-0474; f 219-0418
leec@econ@ag.gov

Clark, Rex
University of Georgia
Driftmeier Eng. Center
Athens, GA 30602
(706) 542-0864; f 542-8806
rclark@bae.uga.edu

Clarke, James
ADAS Boxworth
Boxworth, Cambridge
United Kingdom CB1 4YH
James_clarke@adas.co.uk

Clausen, Heidi
"The Country Today"
Newspaper
521 Clayton Ave. W. #4
Clayton, WI 54004
(715) 948-4187; f 833-7439
countrytdy@aol.com

Clay, Sharon
SDSU Plant Science
219 Ag Hall
Box 2207A
Brookings, SD 57007
(605) 688-5122; f 688-4602

Colvin, Tom
USDA-ARS-NSTL
2150 Pammel Dr.
Ames, IA 50011
(515) 294-5724; f 294-8125
colvin@nstl.gov

Cook, Simon
CSIRO (Division of Soil)
Private Bag (P O Wembley)
Wembley, W. Australia
Australia 6014
+6193870138
simonc@per.dms.csiro.au

Cooker, Dennis
Poole Chemical Co., Inc.
Box 10
Texline, TX 79087
(806) 362-4261

Copeland, Philip
University of Minnesota
439 Borlaug Hall
St Paul, MN 55108
(612) 625-1767; f 625-2208
pcope@soils.umn.edu

Copeland, Steve
University of Minnesota
439 Borlaug Hall
St. Paul, MN 55108
(612) 625-2747; f 625-2208
copeland@soils.umn.edu

Cora, Jose
Michigan State University
A567 PSSB Crop & Soil Dept.
East Lansing, MI 48824
(517) 355-9289; f 355-0270
corajose@pilot.msu.edu

Court, Bruce
Soil Stewardship Group
2136 Jetstream Rd.
London, Ontario
Canada N5V 3P5
(519) 457-2575

Cox, David
Davco Farming
Box 972
Ayr, Queensland
Australia 4807
+6177827575; f
6177827576

Cox, Graeme
University of S. Queensland
Toowoomba, Queensland
Australia 4350
+6176311713
coxg@foes.usq.edu.au

Cresswell, Mark
Bourgault Industries Ltd.
Box 39
St. Brieux, Saskatchewan
Canada S0K 3V0
(306) 275-2300; f 275-2307

Croegaert, Mike
DMI, Inc.
PO Box 65
Goodfield, IL 61742
(309) 965-2233; f 965-2684

Crookston, Kent
University of Minnesota
1991 Upper Buford Circle
411 Borlaug Hall
St. Paul, MN 55108
(612) 625-8761; f 625-1268

Cugnasca, Carlos
Universidade De Sao Paulo
Av. Prof. Luciano Gualberto
Trav. 3, NS 158, San Paulo
Brazil 05508-900
+ 55118185366;
f +55118185718

Cutting, Paul
SW Wisc. Technical College
1800 Bronson Blvd.
Fennimore, WI 53809

Daberkow, Stan
USDA-Econ. Research Serv
1301 New York Ave., N.W.
Washington, DC 20005
(202) 219-0461; f 219-0418
daberkow@econ.ag.gov

Daniels, Richard
Ag-Tech
W5607 Aebly Rd.
Monroe, WI 53566
(608) 328-1441; f 328-1579

Dau, Gary
Glenn Brothers LLC
RR 1 Box 102
Stanford, IL 61774
(309) 379-6581; f 379-6581
glennbro@ice.net

Daughtry, Craig
USDA-ARS
10300 Baltimore Ave.
Beltsville, MD 21043
(301) 504-5015; f 504-5031

Davis, J. Glen
University of Minnesota
1991 Upper Buford Circle
St Paul, MN 55108
(612) 625-1767; f 625-2208
gdavis@soils.umn.edu

DeBoe, Joyce
University of Minnesota
405 Coffey Hall
St Paul, MN 55410
(612) 625-8198; f 625-2207

Deboer, J. Lowenberg
Purdue University
West Lafayette, IN 47907
(317) 494-4230

DeBower, Mark
TEAM Technologies
602 Ansborough Ave.
Waterloo, IA 50701
(319) 235-6507; f 235-0603
markd@teamnet.net

Demetriades, Tanvir
LI-COR, Inc.
4421 Superior St.
Lincoln, NE 68504
(402) 467-3576; f 467-2819
tdshah@env.licor.com

Denholm, Ken
Agriculture Canada
70 Fountain St., E.
Guelph, Ontario
Canada N1C 1B1
(519) 826-2080; f 826-2090
denholmk@em.agr.ca

Despain, Al
Agri Northwest
Box 2308
Tri-Cities, WA 99302
(509) 735-6461; f 735-6471

Dickson, Stephanie
Dickson Precision Farming
2312 Boysenberry Ln #3
Springfield, IL 62707
(217) 793-4020; f 793-4020
sdickson@fgi.net

Dikici, Huseyin
University of Minnesota
Dept of Soil Water & Climate
St Paul, MN 55108
(612) 625-1767; f 625-2208
hdikici@soils.umn.edu

Dlugosz, Steve
Countrymark Coop
950 N. Meridan
Indianapolis, IN 46204
(317) 972-3210; f 972-5042

Dobberstein, Dennis
9679 Sandpit Road
Larsen, WI 54947
(414) 876-3381; f 779-6979

Dohnalik, Martin
Badger Northland, Inc.
1215 Hyland Ave.
Kaukauna, WI 54130
(414) 766-4603; f 766-5011

Dondero, Albert
AGRIS Corporation
300 Grimes Bridge Rd.
Roswell, GA 30075
(770) 518-5185; f 643-2239

Donovan, William
S. Florida Water Mgmt. Dist.
Box 24680
W Palm Beach, FL 33416
(407) 687-6761; f 687-6896

Dorman, Paul
AGRIS Corporation
300 Grimes Bridge Rd
Roswell, GA 30075
(770) 518-5185; f 643-2239

Dowbenko, Ray
Viridian, Inc.
3500 Manulife Pl. 10180 101st
Edmonton, Alberta
Canada T5J 3S4
(403) 493-8636; f 493-8792

Dowdy, Robert
University of Minnesota
1991 Upper Buford Circle
St. Paul, MN 55108
(612) 625-7058; f 625-2208
bdowdy@soils.umn.edu

Downs, George
SATLOC, Inc.
4670 S. Ash Ave.
Tempe, AZ 85282
(602) 752-7439; f 752-7457

Drake, Jim
Cargill, Inc.
15407 McGinty Rd. W.
Wayzata, MN 55391
f (612) 742-4519

Drollinger, Darrin
Equipment Manfacturers Inst.
10 S. Riverside Plaza, #1220
Chicago, IL 60606-3710
(312) 321-1470; f 321-1480

Drummond, Scott
Univ. of Missouri-Columbia
269 Ag. Eng. Bldg.
Columbia, MO 65211
(573) 882-1146; f 882-1115

Duerrstein, Georg
Satcon System GmGH
Bundessti 7
97537 Obertheres
Germany
+49 95277072; f 95277350

Dulaney, Wayne
USDA-ARS-Remote Sensing
10411 Fawcett St., #202
Kensington, MD 20895
(301) 504-6076; f 504-5038
wdulaney@rsrlgis.arsusda.gov

Durand, Larry
University of Manitoba
Room 362 Ellis Bldg
Winnipeg, Manitoba
Canada R3T 2N2
(204) 474-6035; f 275-8099

Durgan, Michael
Cenex Land O'Lakes
980 Hoyt Ave., W.
St. Paul, MN 55117
(612) 448-1727

Easton, Dan
Easton Agri-Consulting
2699 Hwy 141
Bagley, IA 50026
(515) 427-5268; f 427-5269

Eghball, Bahman
USDA-ARS
108 Keim, East Campus
Lincoln, NE 68583
(402) 472-0741; f 472-0516
beghball@unlinfo.unl.edu

Elkins, Ron
Racal-LandStar S.E. Region
3624 Westchase
Houston, TX 77042
(713) 784-4482; f 784-8162

Ellingson, Ron
American Crystal Sugar Co.
101 N. 3rd St.
Moorhead, MN 56560
(218) 236-4405; f 236-4485
rellings@acs.usa.com

Elliott, Lloyd
USDA-ARS
3450 S.W. Campus Way
Corvallis, OR 99331
(541) 750-8722; f 750-8750

Ellis, Charles
University of Missouri
880 W. College
Troy, MO 63377
(314) 528-4613; f 528-4613
ellisc@ext.missouri.edu

Ellsbury, Michael
USDA-ARS-NPA
Rt. 3
Brookings, SD 57006
(605) 693-5212; f 693-5240

Elms, Michael
Texas Tech University
Box 42122
Lubbock, TX 79409
(806) 744-6247; f 742-0775

Engel, Richard
Montana State University
Plant, Soil & Env. Science
Bozeman, MT 59717-0312
(406) 994-5295; f 994-3933

Engelhart, Dennis
IMC/Kalium
Box 486
Ankeny, IA 50921
(515) 964-0222; f 964-5293

Engelstad, Myles
Northrup King Co.
7500 Olson Memorial Hwy.
Minneapolis, MN 55427
(612) 755-0276; f 755-4795

Eraud, Oliver
Isagri
BP 333
Beauvais, Oise
France 60026
+33 44064044;
f +33 44453150

Ernst, Steve
Lockheed Martin
Ms U1L29, Box 64525
St. Paul, MN 55164
(612) 945-2392; f 939-4418

Evans, Mike
Ag-Chem Equipment Co., Inc.
202 Industrial Park
Jackson, MN 56143
(507) 847-2690; f 847-4940

Evans, Robert
Washington State University
24106 N. Bunn Rd.
Prosser, WA 99350
(509) 786-9281; f 786-9370

Evott, Steven
USDA-ARS
Drawer 10
Bushland, TX 79012
(806) 356-5775; f 356-5750
srevett@ag.gov

Ewaschuk, Carolyn
Contact Pedology Services Ltd
609 26th Ave., N.W.
Calgary Alberta
Canada, T2M2E6
(403) 284-2463; f 284-2463

Fairchild, Dean
Cargill, Inc.
PO Box 9300 MS19
Minneapolis, MN 55440
(612) 742-2061; f 742-7313

Falen, Christi
University of Idaho
Box 1827
Twin Falls, ID 83303-1827
(208) 736-3600; f 736-0843
cfalen@uidaho.edu

Fenton, Thomas
Iowa State University
2407 Agronomy
Ames, IA 50011
(515) 294-2414; f 294-3517
tefenton@iastate.edu

Ferguson, Richard
University of Nebraska
PO Box 66
Clay Center, NE 68933
(402) 762-4431; f 762-4422
scrc007@unlvm.unl.edu

Fiala, Michael
AgriData, Inc.
2600 DeMers Ave. Ste 105
Grand Forks, ND 58201
(701) 746-8580; f 746-8681

Filip, S. P.
Mississippi State University
105 Bellwood Dr.
Starkville, MS 39754
(601) 325-3282; f 325-3853
fto@abe.msstate.edu

Fixen, Paul
Potash & Phosphate Inst.
Box 682, 305 5th St.
Brookings, SD 57006
(605) 692-6280; f 692-6280
pfixen@itctel.com

Fleming, Kim
University of Nebraska
5830 Queens Dr.
Lincoln, NE 68516
(402) 472-3674; f 472-3858
strc017@unlum.unl.edu

Fly, Carl
Del Norte Tech
38411 Rocky Hock Rd.
Wakefield, VA 23888
(800) 477-3524

Fogel, Bob
DeKalb Agra
PO Box 127
Waterloo, IN 46793
(219) 837-8611; f 837-7220

Fohner, George
Resource Seeds, Inc.
Box 1319
Gilroy, CA 95021
(408) 847-1051; f 847-0604

Follman, Randy
AgriData, Inc.
2600 DeMers Ave. Ste. 105
Grand Forks, ND 58201
(701) 746-8580; f 746-8681

Foord, Karl
Minnesota Extension
500 E. Depue Ave.
Olivia, MN 56277
(320) 523-2522; f 523-5244
kfoord@mes.umn.edu

Forney, Raymond
DuPont Ag Products
Chesapeake Farms
7321 Remington Dr.
Chestertown, MD 21620
(410) 778-0141; f 778-6741

Francis, Dennis
USSA/ARS/NPA
119 Keim Hall
University of Nebraska
Lincoln, NE 68583
(402) 472-8494; f 472-0516
dfrancis@unlinfo.unl.edu

Fransen, John
Ag-Chem Equipment Co.
5720 Smetana Dr.
Minnetonka, MN 55343
(507) 847-2690
jfrasen@agchem.com

Franzen, Dave
North Dakota State University
Box 5758
Fargo, ND 58105-5758
(701) 231-8884; f 231-7861
dfrazen@ndsuext.nodak.edu

Freeman, Paul
Rt 3 Box 114
Starbuck, MN 56381
(320) 239-4456

Friis, Ege
Danish Inst. of Plant & Soil
Research Centre Foulum
Bx. 23
DK-8830 Tjele
Denmark
+45 89991811
f +45 89991619

Fuchs, Dennis
Stearns County SWCD
110 2nd St., S. Suite 128
Waite Park, MN 56387
(320) 251-6718; f 291-9171

Fuller, Les
University of Manitoba
Rm 362 Ellis Bldg
Winnipeg, Manitoba
Canada R3T 2N2
(204) 474-9319; f 275-8099

Fuxa, Max
Cenex Land O'Lakes
1731 31st St. S.
Moorhead, MN 56560
(218) 233-7218; f 233-8321
mfuxa@cnxlol.com

Gaebel, Jim
The Toro Company
300 W. 82nd St.
Bloomington, MN 55420
(612) 887-8897; f 887-8695
jim.gaebel@toro.com

Gandrud, Dale
Gandy Company
Box 528
Owatonna, MN 55060
(507) 451-5430; f 451-2857

Gardisser, Dennis
University of Arkansas
2301 S. University
Box 391
Little Rock, AR 72203
(501) 671-2241; f 671-2303
dgardisser@uaex.edu

Gasser, Pierre-Yves
Ag-Knowlege
1-117 Charlotte St.
Ottawa, Ontario
Canada K1N 8K4
(613) 789-9603; f 241-3115
pyg@agriprecision.com

Gaudet, Rick
Monsanto
800 N. Lindbergh Blvd.
St. Louis, MO 63167
(314) 694-3048; f 694-2306

Gaultney, Larry
DuPont Agric. Products
Box 30
Newark, DE 19714
(302) 366-6587; f 366-5467

Ge, Jianlin
Badger Northland, Inc.
1215 Hyland Ave.
Kaukauna, WI 54130
(414) 766-4603; f 766-5011

Gerhards, Roland
University of Nebraska
Dept of Agronomy
362C Plant Science
Lincoln, NE 68583
(402) 472-9563; f 472-7904
roland@mortsun.unl.edu

Gibbons, Glen
GPS World Magazine
859 Willamette St.
Eugene, OR 97401
(541) 984-5286; f 344-3514
mggpsworld@aol.com

Giebink, Bruce
University of Minnesota
452 Borlaug Hall
St. Paul, MN 55108
(612) 625-4749; f 625-2208
bgiebink@soils.umn.edu

Giles, Ken
University of California-Davis
Bio. & Ag Eng. Dept.
Davis, CA 95616
(916) 752-0687; f 752-2640
dkgiles@ucdavis.edu

Girgin, Burhan
Washington State University
Dept. of Crop & Soils
Pullman, WA 99164-6420
(509) 335-2381
girgin@wsuvm1.wsu.edu

Glenn, Brad
Glenn Brothers/ LLC
RR1 Box 102
Stanford, IL 61774
(309) 379-6581; f 379-6581
glennbro@ice.net

Golly, Todd
Ag Prophets
33229 215th St.
Winnebago, MN 56098
(507) 893-3151; f 893-3152
tgolly@winnebago.polariste
l.net

Gotway, Carol
Univ. of Nebraska-Lincoln
103 Miller Hall Box 830712
Lincoln, NE 68583
(402) 472-2903; f 472-5179
cgotway@unl.edu

Grau, Scott
Cargill, Inc.
14300 34th Ave. N. # 130
Plymouth, MN 55447
(612) 472-4514; f 472-7909
scott.grau@cargill.com

Grenier, Gilbert
Enita de Bordeaux
B.P. 201
Gradignan
France 33175
+33 57350774
f +33 57350789
grenier@enitab.fr

Grey, Travis
CSI
6009 1A St., S.W.
Calgary, AB, Alberta
Canada T2H OG5
(403) 259-3311; f 259-8866
info@csi-dgps.com

Griebenow, Bret
Lockheed Martin Idaho
Box 1625
Idaho Falls, ID 83402
(208) 526-0389; f 526-4366
bret@pmafire.inel.gov

Griggs, Ray
Texas Agr. Exp. Station
808 E. Blackland Rd.
Temple, TX 76502
(817) 770-6631; f 770-6561
griggs@brcsuno.tamu.edu

Gritzner, Janet
South Dakota State University
Brookings, SD 57006
(605) 688-4184; f 688-5880

Groen, Brant
Willmar Technical College
Box 1097
Willmar, MN 56201
(320) 231-7647; f 231-2965
bgroen@hut.tec.mn.us

Gunderson, Larry
BNC Clean Water
Partnership
301 S. Washington
St. Peter, MN 56082
(507) 931-4140; f 931-2654
bnccwp@ic.mankato.mn.us

Guthne, Scott
Country Mark Coop Inc.
10525 N. Allen Rd.
Breckenridge, MI 48615
(517) 842-3104

Haberstroh, Brian
Midwest Ag Service
600 20th St., S.W.
Jamestown, ND 58401
(701) 252-0580; f 252-8178

Han, Shufeng
Washington State University
24106 N. Bunn Rd.
Prosser, WA 99350
(509) 786-9236; f 786-9370
shan@beta.tricity.wsu.edu

Hansen, Mark
NuWay Cooperative
PO Box 399
Sherburn, MN 56171
(507) 764-6541; f 764-4276

Hanson, David
Monsanto Co.
8668 Flamingo Dr.
Chanhassen, MN 55317
(612) 368-2846; f 368-3529
dghans@ccmail.monsanto.co

Hanson, Glenn
Midwest Ag Services
600 20th St., S.W.
Jamestown, ND 58401
(701) 252-0580; f 252-8178

Hanson, Lowell
Field Control Systems
Box 10851
White Bear Lake, MN 55110
(612) 426-8764
lhanson@soils.umn.edu

Hanson, Richard
Case Corporation
7 S. 600 County Line Rd.
Burr Ridge, IL 60521
(708) 887-3971; f 887-2101
rhanson@casecorp.com

Hanuschak, George
USDA-National Agric. Statistic
3251 Old Lee Hwy Room 305
Fairfax, VA 22030-1504
(703) 235-5218; f 235-3386
ghanuschak@nass.ag.gov

Hardt, Ivan
GPS Crop Systems
PO Box 5167
Cedar Rapids, IA 52406
(319) 363-6074; f 363-6438
cropgps@aol.com

Harford, Doug
Harford Farms, Inc.
650 S. Baker Rd.
Mazon, IL 60444
(815) 448-2137; f 448-2139
harford@interserv.com

Harland, Michelle
Argriculture Canada-PFRA
200- 101 Rt 100
Morden, Manitoba
Canada R6M 1Y5

Harman, Wyatte
Texas Agriculture Exp. Sta.
808 E. Blackland Rd.
Temple, TX 76502
(817) 770-6656; f 770-6678
harman@brcsuno.tamu.edu

Harms, Charles
Vigore Agribusiness
2428 Glick St.
Lafayette, IN 47905
(317) 477-1542; f 447-1629

Harrington, Pat
University of Illinios
1102 S. Goodwin Ave
Turner Hall
Urbana, IL 61801
(217) 333-4424; f 333-9817

Harroun, Joe
Cargill, Inc.
2301 Crosby Rd.
Wayzata, MN 55391
(612) 742-6476; f 742-7909
joe_harroun@cargill.com

Hart, Galen
USDA-ARS-BARC
Bldg. 007, Rm 008 BARC-W
Beltsville, MD 20705
(301) 504-5058; f 504-5031
ghart@asrr.arsusda.gov

Hart, William
The University of Tennessee
Box 1071
Knoxville, TN 37901-1071
(423) 974-7266; f 974-4514
whart@utk.edu

Hartsock, Nathaniel
Hartsock Ag
21290 St. Rt. 104
Chillicothe, OH 45601
(614) 775-1383; f 775-1383

Hartson, Craig
Computer Application Systems
Box 251
Signa Mountain, TN 37377
(423) 752-1787; f 752-1788
casi@chattanooga.net

Hatley, Elwood
Penn State University
116 ASI Building
State College, PA 16802
(814) 863-1013; f 863-7043

Hauwiller, Joe
Cargill, Inc.
15407 McGinty Rd., W.
Wayzata, MN 55391
(612) 742-2673; f 742-4519

Hayes, Adam
Ontario Min. of Ag & Food
Main St. E.
Ridgetown, Ontario
Canada N0P 2C0
(519) 674-1621; f 674-1600

Heard, John
Manitoba Agriculture
Box 1149
Carman, Manitoba
Canada R0G 0J0
(204) 745-2324; f 745-2299

Heermann, Dale
USDA-ARS
Colorado State University
Fort Collins, CO 80523
(970) 491-8229; f 491-8247

Heiniger, Ronnie
N. Carolina State University
207 Research Station Rd.
Plymouth, NC 27962
(919) 793-4428; f 793-5142

Heisel, Torben
Danish Inst. of Plant & Soil
Forsoegsvej 1
Slagelse Denmark 4200
+45 53586300;
f +45 53586371

Helmer, Tom
Ciba Crop Protection
410 Swing Road
Greensboro, NC 27407
(910) 632-2421; f 632-2421
thelmer134@aol.com

Henderson, Graeme
Capstan Ag. Systems
1777 La Cresta Dr.
Pasadena, CA 91103

Hendrickson, Larry
Case Corporation
7 S. 600 County Line Rd.
Burr Ridge, IL 60521
(708) 887-3924; f 887-2223

Hergert, Gary
University of Nebraska
Rt 4 Box 46A
North Platte, NE 69101
(308) 532-3611; f 532-3823
wcrc002@unlvm.unl.edu

Herring, Matthew
University Extension
Box 71
Union, MO 63084
(314) 583-5141; f 583-5145
herringm@ext.missouri.edu

Hess, J. Richard
Lockheed Martin
 Idaho Tech.
PO Box 1625
Idaho Falls, ID 83405-3710
(208) 526-1211; f 526-4325
jrh@inel.gov

Hill, Andy
DeKalb Agra, Inc.
PO Box 127
Waterloo, IN 46793
(219) 837-8611; f 837-7220

Hirose, Tom
Noetix Research, Inc.
978 Merivale Rd.
Ottawa, Ontario
Canada K1B 4L3
(613) 729-4899; f 722-3798
hiroset@magi.com

Hnatowich, Garry
Saskatchewan Wheat Pool
103-111 Research Drive
Saskatoon Saskatchewan
Canada 57N3R2
(306) 668-6643; f 668-5564

Hockel, Mark
United AgTech
Box 331
Trimont, MN 56160
(507) 539-6441; f 639-2210

Hodupp, Rich
Michigan State University
1575 Suncrest Drive
Lapeer, MI 48446
(812) 667-0344; f 667-0355
hodupp@msue.nsu.edu

Hoff, David
University of Minnesota
211 Hill Blvd.
Crookston, MN 56716
(218) 281-6702; f 281-5080

Hofman, Vernon
North Dakota State University
1221 Albrecht Dr.
Box 5626
Fargo, ND 58105
(701) 231-7240; f 231-1008
vhofman@ndsuext.nodak.edu

Hollands, Kevin
Centrol Inc.
102 East Main
Twin Valley, MN 56584
(701) 739-1637

Hood, Fred
Pioneer Hi-Bred Int'l. Inc.
7300 N. W. 62nd Ave.
Box 1004
Johnston, IA 50131-1004
(515) 270-5912; f 270-4312
hoodcf@phibred.com

Hoormann, Rich
Farmland Industries
55 Westport Plaza
St. Louis, MO 63146
(314) 275-7130; f 275-7211

Horstmeier, Greg
Farm Journal
Box 958
Mexico, MO 65265
(573) 581-9643; f 581-9646

Horton, Maurice
USDA/CSREES
329-J Aerospace Center
Washington, DC 20250
(202) 401-4504; f 401-1706
morton@reeusda.gov

Hoskinson, Reed
Lockheed Martin
 Idaho Tech.
PO Box 1625
Idaho Falls, ID 83405
(208) 526-1211; f 526-0603
hos@inel.gov

Huberty, Brian
USDA-NRCS-
Midwest Region
2820 Walton Commons Dr.
#123
Madison, WI 53704
(608) 224-3014; f 224-3010
brian.huberty@mw.nrcs.
usda.gov

Huffaker, Miles
Rutgers Cooperative Ext.
51 Cheney Rd. Suite 1
Woodstown, NJ 08098
(609) 769-0090; f 769-1439.

Huffman, Jerry
Dow Elanco
2009 Fox Dr.Suite D
Champaign, IL 61821
(217)352-3865; f 352-4844

Humburg, Daniel
South Dakota State Univ.
Box 2120
Brookings, SD 57007
(605) 688-5658; f 688-4917
humburgd@mg.sdstate.edu

Hummel, John
USDA-ARS
1304 W. Penn Ave.
Urbana, IL 61801
(217) 333-0808; f 244-0323
jhummel@u.uc.edu

Hunst, Michael
USDA-NASS-Minnesota
Box 7068
St. Paul, MN 55107
(612) 296-3896; f 296-3192
mhunst@mass.usda.gov

Huotari, John
Crop Growers Software
524 E. Mendens Hall, Ste F
Bozeman, MT 59715
(406) 582-1246; f 582-8522
jhuotari@cgro.com

Illich, John
Agri Imagis Inc.
701 33rd Ave., N. #119
Fargo, ND 58102
(701) 298-3566
illich@badlands.nodak.edu

Irwin, E. Dale
Cargill Limited
4096 Meadowbrook Dr.
Unit 127
London, Ontario
Canada N6L1G4
(519)652-8009; f 652-1202

Jacquin, Gerard
Cemagref
Domaine des Palaquine,
Montold
03150 Varennes-sur-Allier
France 98160
+33 70450312; f 70451946
montoldre@danube.cemagref.fr

Jaynes, Dan
USDA-ARS
2150 Pammel Dr.
Ames, IA 50011
(515) 294-8243, f 294-8125
jaynes@nstl.gov

Jarvis, Jeff
Terra Industries
3525 Terra Court
Sun Prairie, WI 53590
(608) 249-8500

Jenny, Richard
AGSCO, Inc.
Box 13458
Grand Forks, ND 58208
(800) 859-3047;
f 701-775-9587

Johnson, Dale
John Deere Ag Serv. Group
501 River Dr
Moline, IL 61265
(309) 765-7278; f 765-7083

Johnson, Gregg
Southern Exp. Station
35838 120th St.
Waseca, MN 56093
(507) 835-3620; f 835-3622

Johnson, Richard
John Deere
501 River Drive
Moline, IL 61265
(309) 765-7273; f 765-7083

Johnson, Richard
USDA-ARS-SRRC
Box 19687
New Orleans, LA 70179
(504) 286-4515; f 286-4217
rjohnson@nola.srrc.usda.gov

Johnston, Adrian
Agriculture Canada
Box 1240
Melfort, Saskachewan
Canada S0E 1A0
(306) 752-2776; f 752-4911
johnstona@em.agr.ca

Jordan, Ed
Texas Agri. Exp. Station
808 E. Blackland Rd.
Temple, TX 76502
(817) 770-6654, 770-6561

Juerschik, Peter
Inst. of Ag. Eng.
Bornim Atb
Max-Eyth-Allee 100
D-14469Potsdam-Bornim
Brandenburg
Germany D-14469
+49 3315699420

Kachanoski, R. Gary
University of Guelph
Dept. of Land Resource Sci.
Guelph, Ontario
Canada N1G 2W1
(519) 824-4120; f 824-5730

Kain, James
TASC
55 Walkers Brook
Reading, MA 01810
jekain@tasc.com

Kantz, Brian
Farm Chemicals Magazine
37733 Euclid Ave.
Willoughby, OH 44094
(216) 942-2000; f 942-0662
agcedit@meisterpubl.com

Kaplan, Joseph
University of Minnesota
1390 Eckles Ave
St Paul, MN, 55108
(612) 625-1708; f 624-3005

Karlen, Douglas
USDA-ARS,
National Soil Lab
2150 Pammel Dr.
Ames, IA, 50011
(515) 294-3336; f 294-8215
dkarlen@nstl.gov

Kataoka, Takashi
Univ. of California-Davis
Davis, CA, 95616
(916) 752-8400; f 752-2640

Keiser, Jeff
Terra Industries
430 W. Carmel Dr.
Carmel, IN, 46032
(317) 844-8221; f 571-0258
jeffk@in.net

Kerr, Greg
629 Sunset Ln.
RiverFalls, WI, 54022
(715) 672-8304

Kerr, Mark
AGSCO, Inc.
Box 13458
Grand Forks, ND 58208
(701) 775-5325; f 775-9587

Khakural, Bhairav
University of Minnesota
439 Borlaug Hall
1991 Upper Buford Circle
St. Paul, MN 55108
(612) 625-4721, f 625-2208
khakural@soils.umn.edu

Kiernan, Rob
Rinex Technology
173 Petra St.
E. Fremantle,
West Australia
Australia 6158
(619) 319-3434; f 339-4781
rkiernan@rinex.com.au

Kimball, Brad
Northrup King Co.
Box 415
Hampton, IA 50441
(515) 456-2592; f 456-2024

King, Bradley
University of Idaho
1693 S. 2700 W
PO Box AA
Aberdeen, ID 83210
(208) 397-4181; f 397-4311
bradk@uidaho.edu

Kirick, Daniel
Kirick Agronomy Services
3105 2nd. St., S.E.
St. Cloud, MN 56304
(320) 259-0668; f 259-0668

Kirk, Ivan (Buddy)
USDA-ARS
2771 F&B Road
College Station, TX 77845
(409) 260-9584; f 260-9386

Kitchen, Newell
Univ. of Missouri-Columbia
240 Agricultural
Engineering
Columbia, MO 65211
(573) 882-1138; f 882-1115

Klingberg, Kevan
Chippewa Co. Land
Conservation
711 North Bridge #011
Chippewa Falls, WI 54729
(715) 726-7921
chippewa1@aol.com

Kluempke, Donald
Alexandria Tech School
32936 420th St.
Melrose, MN 56352
(800) 253-9884

Knobbe, Curtis
University of Nebraska-Lincoln
325 Belgrade Ave. #4
Mankato, MN 56003
(507) 345-4963; f 345-4027

Koehler, Dwight
Koehler Search &
  Placement
5324 Nichols St.
Omaha, NE 68132
(402) 553-6947; f 553-6947

Kohls, Cheryl
Cenex Land O'Lakes
Box 64089
St. Paul, MN 55164
(612) 451-4306; f 451-4344
ckohl@cnxlol.com

Kohut, Connie
Agrium, Inc.
8825 88th Ave.
Edmonton, Alberta
Canada T6C 1L5
(403) 468-1765; f 463-8552
ckohut@agrium.com

Koo, Seungmo
Kansas State University
Dept. of Agriculture
400 Waters
Manhatten, KS 66506
(913) 776-8635; f 532-6925

Kosokowsky, Murray
Bourgault Industries Ltd.
Box 39
St. Brieux, Saskatchewan
Canada S0K 3V0
(306)275-2300; f 275-2307

Kosuge, Nobumasa
Chisso Corporation
7-3, Marunouchi 2-Chome
Chiyoda-Ku, Tokyo 100
Japan

Krause, Mark
North Dakota State Univ.
Dept of Ag Economics
Fargo, ND 58105
(701) 231-8935; f 231-7400
mkrause@ndsuext.nodak.edu

Krumpelman, Michael
119 Redwood Rd.
Columbia, MO 65203
(573) 882-1146

Kreun, Stewart
Telxon Corp.
5929 Baker
Minnetonka, MN 55345
(612) 933-1450; f 933-3177
skreun@telxon.com

Krill, Tom
Ohio State University
1055 S. Washington St.
Van West, OH 45891
(419) 238-1214; f 238-3276
krill.8@osu.edu

Krishnan, P.
University of Delaware
Ag. Engineering Dept.
Newark, DE 19717-1303
(302) 831-1502; f 831-3651
baba@brahms.udel.edu

Krohn, Mark
COASH, Inc.
Box 110
Albion, NE 68620
(402) 395-5051; f 395-2242

Krueger, David
North Carolina State Univ.
Box 7620 Williams Hall
Raleigh, NC 27695
(919) 515-5817; f 515-5855

Krug, Kyle
TEAM Technologies
602 Ansborough
Waterloo, IA 50701
(800) 728-8326

Kuehl, Brian
Terra Industries
15325 25th R St., S.E.
Amenia, ND 58004
(701) 347-4279; f 347-4279

Kutz, Lawrence
Auburn University
Agricultural Eng. Dept
Auburn, AL 36849-5417
(334) 844-4180

Kvien, Craig
UGA-CPES-NESPAL
PO Box 748
Tifton, GA 31793
(912) 386-7204; f 386-7005

Kwolek, Tom
Cargill, Inc.
1542 Castle Dr.
North Mankato, MN 56003
(507) 386-7482; f 386-7483

Lamb, John
University of Minnesota
1991 Upper Buford Circle
St. Paul, MN 55108
(612) 625-1772; f 625-2208
jlamb@soils.umn.edu

Lamker, Don
Cargill, Inc.
Box 190
Blue Earth, MN 56013
(507) 526-2290; f 526-5778

Lange, David
Mississippi State University
Box 9555
Mississippi St., MS 39762
(601) 325-2311; f 325-8742
dlang@dorman.msstate.edu

Larscheid, Georg
Silsoe College
Cranfield Univ
Silsoe, Bedford
England K45 4DT

Larson, Jim
Haug Implement Co.
Box 1055
Willmar, MN 56201
(320) 235-8115

Larson, William
University of Minnesota
1991 Upper Buford Circle
St. Paul, MN 55108
(612) 624-8714; f 625-2208

Lauer, Joe
University of Wisconsin
1575 Linden Dr.
Madison, WI 53706
(608) 263-7438
jglauer@facstaff.wisc.edu

Lee, David
Rutgers Cooperative Ext.
51 Cheney Rd. Suite 1
Woodstown, NJ 08098
(609) 769-0090; f 769-1439

Legg, Leslie
Precision Agriculture Center
439 Borlaug Hall
1991 Upper Buford Circle
St. Paul, MN 55108
(612) 624-4224; f 624-4223
llegg@soils.umn.edu

Lems, Jason
South Dakota State Univ.
Plant Science Dept.
219 Ag Hall
Box 2207A
Brookings, SD 57007
(605) 688-5122; f 688-4602

Lenertz, John
Willmar Manufacturing
6269 Chaska Rd.
Excelsior, MN 55331
(612) 470-6032; f 474-8840

Lewis, Scott
Northstar Technologies
30 Sudubry Rd.
Acton, MA 01720
(508) 897-6600; f 897-7241
slewis@northstarcmc.com

Lindsay, Dennis
Lindsay Farms
3389 208th St.
Masonville, IA 50654
(319) 927-3931; f 927-3931
lindfarms@aol.com

Liston, Devon
USDA-ARS
119 Keim Hall East Campus
Lincoln, NE 68506
(402) 472-1594; f 472-1506
dliston@unlgradl.unl.edu

Little, Mark
Crop Growers Software
524 E. Menden Hall Suite F
Bozeman, MT 59715
(406) 582-1246; f 582-8566
mlittle@agro.com

Long, Dan
Montana State University
HC 36, Box 43
Havre, MT 59501
(406) 265-6115; f 265-8288

Lorenc, Allan
Rawson Control Systems, Inc.
116 2nd St., S.E.
Oelwein, IA 50662
(319) 283-2225; f 283-1360
rawson@trxinc.com

Lowrie, Don
Cargill Ltd.
Unit 127-4096
Meadowbrook Dr.
London, Ontario
Canada N6L- 1G4
(519) 842-0315; f 688-2153

Lu, Ningping
Los Alamos Natl Laboratory
CST-7, Mail Stop J514.
LANL
Los Alamos, NM 87545
(505) 667-1476

Lu, Yao-Chi
Agricultural Res.Service
Rm 12, Bldg. 007, BARC-W
Beltsville, MD 20705
(301) 504-5821; f 504-5823
ylu@asrr.uso1

Lucas, John
Charles Sturt University
PO Box 588
Wagga Wagga
NSW, Australia 2678
(616) 933-2177; f 933-2924
jlucas@csu.edu.au

Lunde, Russ
Simplot Canada Limited
Box 10
Tuxford, Saskatchewan
Canada S0H 4C0
(306) 693-6355; f 693-8022

Luttickeu, Ruth
Fa Holzl
Rauhod 2
Schouslett
Germany 83137
+08 07518820

Mahurin, Robert
USDA-ARS
Univ. Missouri-Columbia
269 Agric. Eng.
Columbia, MO 65211
(573) 882-1148; f 882-1115

Mailander, Mike
Louisiana State University
167 E.B. Doran
Baton Rouge, LA 70803
(504) 388-1058; f 388-3492

Mallarino, Antonio
Iowa State University
Dept. of Agronomy
Ames, IA 50011
(515) 294-6200; f 294-3163
apmallar@iastate.edu

Malzer, Gary
University of Minnesota
Dept of Soil Water & Climate
St Paul, MN 55108
(612) 625-6728; f 625-2208
gmalzer@soils.umn.edu

Manchur, Lawrence
Manitoba Agriculture
27 2nd Ave., S.W.
Dauphin, Manitoba
Canada R7N 3E5
(204) 622-2009; f 638-2854

Marcynuk, Don
Vansco Electronics
1305 Clarence Ave.
Winnipeg, Manitoba
Canada R3T 1T4
(204) 453-3339; f 452-7156
marcynuk@mbnet.mb.ca

Marrah, Tom
Satloc, Inc.
12079 Kingston Place
Algonquin, IL 60102
(847) 658-2740; f 658-2745

Marsh, Brian
Kansas State University
Box 151
Powhattan, KS 66527
(913) 474-3469; f 474-3313
bmarsh@oznet.ksu.edu

Mattson, Marv
Univ. of Minnesota-
Crookston
101 Owen Hall
Crookston, MN 56716
(218) 281-8134

McBratney, Alex
University of Sydney
AOS, McMillan Bldg.
NSW Australia 2006
+61 23513214

McCann, Blair
University of Saskatchewan
51 Campus Dr.
Saskatoon, Saskatchewan
Canada S7N 5A8
(306) 966-4291; f 966-6881
mccann@sask.usask.ca

McCarty, Charlie
Cargill Central Research
2301 Crosby Rd.
Wayzata, MN 55391
(612) 740-6430; f 742-7909
charlie-mccarty@cargill.ca

McClellan, Phillip
Virginia Tech-
Bio Systems Eng.
306 Seitz Hall
Blacksburg, VA 24061
(540) 231-7602; f 231-3199
philbert@vtvm1.cc.ut.edu

McCoy, Tom
Cargill, Inc.
15407 McGinty Rd. West
Wayzata, MN 55391
(612) 742-5839; f 742-7313

McFadden, Bryan
Space Imaging
9351 Grant St. Suite 500
Thornton, CO 80229
(303) 254-2062; f 254-2211
bmcfadden@spaceimage.com

McGrath, Don
Tyler Industries
E. Hwy 12
Benson, MN 56215
(320) 843-3333, f: 843-2467

McGraw, Tom
Midwest Independent Soil
RR2 Box 156
Buffalo Lake, MN 55314
(320) 833-2200; f 833-2289

McKenzie, R. Colin
Alberta Ag Food &
Rural Devel.
SS #4 CDC-South
Brooks, Alberta
Canada T1R 1E6
(403) 362-3391; f 362-2554
mckenzic@agri.gov.ab.ca

McLachlin, Ian
Soil Stewardship Group
2136 Jetstream Rd.
London, Ontario
Canada N5V 3P5
(519) 457-2575

McMaster, Greg
USDA-ARS,
Great Plains Sys.
PO Box E 301 S. Howes
Fort Collins, CO 80522
(970) 490-8340; f 490-8310
greg@gpsr.colostate.edu

McNamara, Ed
Goodhue Co. SWCO
Box 335
Goodhue, MN 55027
(612) 923-4777; f 923-5288

McNelly, Paul
Minnesota Dept. of Agric.
90 West Plato Blvd.
St. Paul, MN 55316
(612) 297-7285; f 297-2271

McQuinn, A.E.
Ag-Chem Equipment Co., Inc.
5720 Smetana Dr., #100
Minnetonka, MN 55343
(612) 945-5803; f 933-7432

Melander, Marlin
Hovick Manufacturing
4350 48th Ave., N.
Fargo, ND 58104
(800) 373-4084; f 368-5321
marlin@horvick.com

Merk, Larry
Agri Northwest
Box 2308
Tri-Cities, WA 99302
(509) 735-6471; f 735-6471

Meyer, Brad
Concord Inc.
3000 7th Ave., N.
Fargo, ND 58102
(701) 280-1260; f 280-0706

Meysenburg, Dan
COASH, Inc.
Box 110
Albion, NE 68620
(402) 385-5051; f 395-2242

Mieth, Norman
Terra International, Inc.
709 Centennial Rd.
Box 385
Wayne, NE 68787
(402) 375-3170; f 375-5763

Milby, Ron
GROWMARK, Inc.
1701 Towanda Ave.
PO Box 2500
Bloomington, IL 61702
(309) 557-6251; f 557-6860

Miller, Mark
GIS Consultant
229 Victoria Dr.
Alexandria, MN 56308
(320) 763-6984

Miller, Neil
Agri-Business Consultants
258 N. East St.
Brighton, MI 48116
(810) 220-1571;
f (517)4826944

Miller, William
University of Nebraska
304B Filley Hall
Lincoln, NE 68583
(402) 472-0661; f 472-3460
agec098@unlvm.unl.edu

Milleville, Andrew
Case Corp.
7 S. 600 County Line Rd.
Burr Ridge, IL 60521
(708) 887-5429; f 887-2223
amilleville@case.corp.com

Millex, Robert
Univ. of California-Davis
812 Cambridge
Fort Collins, CO 80525
(970) 491-6516
robm846@aol.com

Minnichsoffer, Tony
Ag Retailer Magazine
32410 Nottingham Court
Lindstrom, MN 55045
(612) 259-8213; f 259-8213
tonyjminn@aol.com

Missotten, Bart
K.U. Leuven
Kardinaal Mercierlaan 92
Bloke, B-3001 Heverlee
Belgium

Mohamed, S. B.
University of Newcastle
Dept. of Agriculture
Newcastle Upon Tyne
England NE1 7RU
s.b.mohamed@newcastle.ac.uk

Molin, Jose
Univ. of Nebraska-Lincoln
L.W. Chase Hall-East Campus
Lincoln, NE 68583
(402) 472-6045

Monson, Robert
Ag-Chem Equipment Co., Inc.
5720 Smetana Dr., Ste 100
Minnetonka, MN 55343
(612) 945-5826; f 945-2385
rmonson@stthomas.edu

Montgomery, Bruce
Minnesota Dept. of Agriculture
90 West Plato Blvd.
St Paul, MN 55107
(612) 297-7178; f 297-2271

Moore, David
Cargill, Inc.
Box 9300
Minneapolis, MN 55440
(612) 742-2146; f 742-4519
david_moore@cargill.com

Morey, R. Vance
University of Minnesota
213 Bio Ag Eng.
1390 Eckles Ave.
St. Paul, MN 55108
(612) 625-8775; f 624-3005
rvmorey@gaia.bac.umn.edu

Morris, D. Keith
Purdue University
1146 Ag & Bio Eng. Bldg
W. Lafayette, IN 47907
(317) 494-1187; f 496-1115
morrisk@ecn.purdue.edu

Morse, Richard
Cargill Hybrid Seeds
3864 Pre Emption Rd.
Geneva, NY 14456
(315) 789-8626; f 789-8626

Mortensen, David
University of Nebraska
Dept. of Agronomy
Lincoln, NE 68583
(402) 472-1543; f 472-7904
dave@mortsun.unl.edu

Mueller, Tom
Michigan State University
Dept of CSS, PSSB A567
East Lansing, MI 48824
(517) 355-9289; f 355-0270
muelle26@pilot.msu.edu

Mulla, David
University of Minnesota
Dept. of Soil, Water & Climate
St. Paul, MN 55108
(612) 625-6721
dmulla@soils.umn.edu

Munn, Peter
Vansco Electronics
1305 Clarence Ave.
Winnipeg, MB
Canada R3T 1T4
(204) 452-6776; f 452-7156

Nafziger, Emerson
University of Illinois
1102 S. Goodwin
Urbana, IL 61801
(217) 333-4424; f 333-5299

Nelson, Jim
Agri Business Group
3905 Vincennes Rd. 402
Indianapolis, IN 46268
(317) 875-0139; f 875-0507

Newcomb, Tom
American Crystal Sugar Co.
RR 2
Hillsboro, ND 58045
(701) 430-0158

Nichols, Robert
Cotton Incorporated
4504 Creedmoor Rd.
Raleigh, NC 27612
(919) 510-6113; f 510-6124

Nielsen, Gerald
Montana State University
Plant, Soil & Environ. Science
Bozeman, MT 59717
(406) 994-5075; f 994-3933

Nielsen, Norman
Trimble Navigation
Inman, NE 68742
(402) 394-5405; f 394-5013
nielsencom@aol.com

Nipper, Steven
USDA-DNR/
Conservation Service
1605 Arizona St.
Monroe, LA 71202-3697
(318) 387-8683; f 388-4275.

Nolan, Sheilah
Alberta Agric, Food &
Rural
#206 7000- 113 St.
Alberta, Canada TGH 5T6
(403) 427-3719; f 422-0474

Nolin, Michel
Agric. & Agri-Food Canada
350 Franquet St. Entree 20
Ste-Foy, Quebec
Canada G1P 4P3
(418) 648-7730; f 648-5489

Nusbaum, Scott
Farm Works Software
6795 S. State Rd. #1
Hamilton, IN 46742
(219) 488-3388; f 488-3737
farmwork@farmworks.com

Nuspl, Steve
Field Control Systems
Box 10851
White Bear Lake, MN 55110
(612) 631-8239
nuspl-sj@ix.net.com

Nutting, Michelle
Viridian, Inc.
Manulife Pl. 10180 101st
Edmonton, Alberta
Canada T5J 3S4
(403) 493-8738; f 493-8792

Nykaza, Scott
Asgrow Seed Company
2605 E. Kilgore
Kalamazoo, MI 49002
(616) 384-5569

O'Connor, Michael
Stanford University
Gravity Probe B
c/o Hansen Labs
Stanford, CA 94305-4085
(415) 723-9350; f 725-8312
mocconor@leland.stanford.edu

O'Halloran, Ivan
University of Guelph
Guelph, Ontario
Canada N1G 2W1
(519) 824-4120; f 824-5730
iohallor@lrs.uoguelph.ca

O'Neal, Monte
Purdue University
213 West University Apts.
West Lafayette, IN 47406
(812) 857-3020

Oldham, Larry
University of Minnesota
439 Borlaug Hall
1991 U. Buford Circle
St. Paul, MN 55108
(612) 625-3717; f 625-2208
loldham@soils.umn.edu

Olechowski, Henry
Ontario Min. of Ag, Food
Guelph Ag Center Box 1030
Guelph, Ontario
Canada N1H 6N1
(519) 767-3257; f 837-3049

Olieslagers, Robert
K.U. Leuven
Kardiaal mercierlaan 92
Bloke, B-3001 Heverlee
Belgium

Oolman, John
Agri-Growth, Inc.
RR1 Box 33
Hollandale, MN 56045
(507) 889-4371; f 889-4381
agrigrowth@deskmedia.com

Ortega, Rodrigo
Colorado State University
Dept. of Soil & Crop Sciences
Fort Collins, CO 80521
(970) 491-6517; f 491-0564

Otter-Nacke, Susanne
Claas KGaA
Am Turmchen 9
Gutersloh Germany, 33332

Oyarzabal, Emilio
Monsanto
4123 77th St.
Urbandale, IA 50322
(515) 278-2101

Ozkan, Erdal
Ohio State University
590 Woody Hayes Dr.
Columbus, OH 43210
(614) 292-3006; f 292-9448
ozkan.2@osu.edu

Pachepsky, Yakov
Duke University
Bldg 007 Rm 008 BARC-W
Beltsville, MD 20705
(301) 504-7468; f 504-5823

Paris, Tom
Paris & Sons
Box 95
Masonville, IA 52402
(319) 927-2364; f 927-6935

Parker, Duane
Countrymark Cooperative, Inc.
950 N. Meridian St.
Indianapolis, IN 46040
(317) 972-5334; f 972-5042

Parkinson, Bradford
Stanford University
HEPL/GP-B MC #4085
Stanford, CA 94035
(415) 725-4107; f 725-8312

Parks, Sid
GROWMARK, Inc.
1701 Towanda Ave
Po Box 2500
Bloomington, IL 61702
(309) 557-6253; f 557-6860

Paulinski, Dennis
Lockheed Martin
MS U1L29, Box 64525
St. Paul, MN 55164
(612) 945-2388; f 939-4418

Pearse, Bruce
Farmwork Canada
Route 2
Sunderland, Ontario
Canada L0C 1HO

Perger, Mark
Precison Farming Australia
5A Walton St.
Corrigin, Western Australia
Australia 6375
+61 90632636

Perry, Calvin
University of Georgia
Bio & Ag Engineering Dept.
Box 748
Tifton, GA 31793-0758
(912) 386-3377; f 386-3958

Petersen, Dugan
FEI, Inc.
46304 Jeffery St.
Hartford, SD 57033
(605) 528-7117
dptersen@ldeasign.com

Petersen, Gary
Penn State University
116 ASI Building
University Park, PA 16802
(814) 865-1540

Petersen, Perry
Terra Industries
600 4th St.
PO Box 6000
Sioux City, IA 51102-6000
(712) 277-5432; f 277-7383

Peterson, Todd
Pioneer Hi-Bred Int'l. Inc.
4445 Corporate Drive
W. Des Moines, IA 50266
(515) 267-6707; f 226-2939
petersonta@phibred.com

Pevear, Bill
TASC
55 Walkers Brook Dr.
Reading, MA 01867-3297
(617) 942-2000; f 944-1313
wlpevear@tasc.com

Pfost, Donald
University of Missouri
205 Agricultural
Engineering
Columbia, MO 65211
(573) 882-2731; f 884-5650

Pierce, Francis
Michigan State University
564 PSSB Crop & Soil Dept.
East Lansing, MI 48824
(517) 355-6892; f 355-0270

Plattner, Chad
University of Illinois-
Urbana
1304 W. Pennsylvania
Urbana, IL 61801
(217) 244-8196; f 244-0323
cplattne@uiuc.edu

Pocknee, Stuart
NESPAL-UGA
Moore Hwy
Tifton, GA 31794
(912) 386-7057; f 386-7005
spocknee@uga.cc.uga.edu

Pointon, John
Omnistar, Inc.
8200 West Glen
Houston, TX 77063
(713) 785-5850; f 785-5164
dgps@omnistar.com

Porter, Paul
University of Minnesota
Box 428
Lamberton, MN 56152
(507) 752-7372; f 752-7374
pporter@maroon.tc.umn.edu

Posselius, John
New Holland N. America, Inc.
500 Diller Ave.
New Holland, PA 17557
(717)355-3663; f 355-1939

Pottinger, Don
Ag-Chem Equip. Co., Inc.
5720 Smetana Dr., #100
Minnetonka, MN, 55343
(612) 945-5803; f 933-7432

Quesnel, Gilles
Ontario Ministry of Ag, Food
Main St., East
Ridgetown, Ontario
Canada N0P 2C0
(519) 674-1616; f 674-1600

Rains, Douglas
American Crystal Sugar Co.
111 Mill St.
Crookston, MN 56716
(218) 281-0110

Rajasekaraw, B.
CEISIN
2250 Pieree Rd.
Saginaw, MI 48710
(517) 797-2749
raja@ciesin.org

Rambo, Greg
Hydro Agri North America Inc.
100 N. Tampa St., Ste 3200
Tampa, FL 33602
(813) 222-5700; f 875-5735

Rawlins, Stephen
Appropriate Systems
2638 Eastwood Ave.
Richland, WA 99352
(509) 627-4943; f 627-1841
srawlins@new.net

Redulla, Cristoti
Kansas State University
Throck Morton Plant Center
Manhattan, KS 66506
(913) 532-6366; f 532-6094
redulla@ksu.edu

Reetz, Harold
Potash Phosphate Institute
1497 N. 1050 East Rd.
Monticello, IL 61856
(217) 762-2074; f 762-8655
hreetz@uiuc.edu

Regan, Tanya
Case Corp.-Adv. Farming Sys.
7 S. 600 County Line Rd.
Burr Ridges, IL 60521-6995
(708) 789-7138; f 789-7174
tregan@casecorp.com

Rehm, George
University of Minnesota
Dept. of Soil, Water & Climate
439 Borlaug Hall
St. Paul, MN 55108
(612) 625-6210; f 625-2208
grehm@mes.umn.edu

Reichenberger, Larry
Successful Farming
PO Box 117
Andale, KS 67001
(316) 445-2589; f 444-2536

Reifsteck, John
892 Co. Rd. 900, E.
Champaign, IL 61821
(217) 485-5358; f 485-3101
johnr@nesa.uiuc.edu

Reyes, Carlos
Monsanto
700 Chesterfield
St. Louis, MO 63195
(314) 537-7592; f 537-6950

Riley, Brad
Poole Chemical Co., Inc.
Box 10
Texline, TX 79087
(806) 362-4261; f 362-4366

Rioux, Romain
Government du Canada
1642 ree de la Ferme
LaPocatere, Quebec
Canada G0R 1Z0
(418) 856-3141; f 856-5374

Roach, Matt
Monsanto
7617 Winston Dr., N.E.
Cedar Rapids, IA 52402
(319) 377-7607; f 373-8833
mnroac@monsanto.com

Robert, Pierre
University of Minnesota
Dept. of Soil, Water & Climate
St. Paul, MN 55108
(612) 625-3125; f 624-4223
probert@soils.umn.edu

Roberts, Roland
University of Tennessee
308 Morgan Hall
Knoxville, TN 37996-4500
(423) 974-7480; f 974-7448.

Roberts, Terry
Postash & Phosphate
Institute
Suite 704 CN Tower
Midtown Plaza
Saskatchewan
Canada S7K 1J5
(306) 652-3535; f 664-8941
ppic@sasknet.sk.ca

Romenelli, A. J.
GIS Solutions, Inc.
2387 W. Monroe, Ste 137
Springfield, IL 62704
(217) 546-3652; f 546-3839

Rooney, Tom
INSAT, L.C.
PO Box 2172
Sioux City, IA 51104
(712) 239-4775

Rosen, Carl
University of Minnesota
Dept. Soil Water & Climate
St Paul, MN 55108
(612) 625-8114; f 625-2208
crosen@soils.umn.edu

Rota, Lou
Northstar Technologies
30 Sudbury Rd.
Acton, MA 01720
(508) 897-0770; f 897-7241

Rust, Richard
University of Minnesota
Dept. of Soil, Water & Climate
St. Paul, MN 55108
(612) 625-8101; f 625-2208
rrust@soils.umn.edu

Sabbe, Wayne
University of Arkansas
276 Altheimer Dr.
Fayetteville, AR 72704
(501) 575-3910; f 575-3975
ws25038@uafsysb.uark.edu

Sadjadi, Firooz
Lockheed Martin
3333 Pilot Knob Rd.
Eagen, MN 55171
(612) 456-7526

Sadler, E. John
USDA-ARS
2611 W. Lucas St.
Florence, SC 29501
(803) 669-5203; f 669-6970

Sakae, Shibusawa
Univ. of California-Davis
Dept. Bio. Agr. Engr.
Davis, CA 95616
(916) 752-8400; f 752-2640

Saraiva, Antonio
Univesidade DeSao Paulo
Av. Prof Luciano Gualberto
Trav. 3, NS 158, San Paulo
Brazil 05508-900

Sarrazin, Philippe
ENESAD
735 Woodland Ave.
E. Palo Alto, CA 94303
(415) 604-3225; f 604-1088

Saunders, Stuart
Cranfield University
Silsoe, Bedford
United Kingdom 01525
+01 525863081:f
525863000

Sawyer, Brenda
Contract Pedology Ser. Ltd.
Box 6097
Innisfail, Alberta
Canada T4G 1S7
(403) 227-1526; f 227-1526

Sawyer, John
GROWMARK, Inc.
1701 Towanda Ave.
PO Box 2500
Bloomington, IL 61702
(309) 557-6250; f 557-6860

Schaat, Bradley
Monsanto Co.
538 N. 4210, E.
Rigby, ID 83442
(208) 745-8927; f 745-0613

Schaefer, Don
1408 N. E. Applewood Crt.
Lee's Summit, MO 64086
(816) 246-9056; f 246-9056
Stouffer@qni.com

Schepers, James
USDA-ARS-NPA
119 Keimn Hall
Univ. of Nebraska-
East Campus
Lincoln, NE 68583-0915
(402) 472-1513; f 472-0516

Schiebe, Frank
SST Development Group, Inc.
824 N. Country Club Rd.
Stillwater, OK 74075-0918
(405) 924-6868

Schmidt, Berlie
USDA/CSREES
329 Aerospace Center
Washington, DC 20250
(202) 401-6417; f 401-1706
bschmidt@reeusda.gov

Schmiess, Sheldon
CENTROL
PO Box 198
Cottonwood, MN 56229
schmiess@juno.com

Schneider, Sally
USDA-ARS
24106 N. Burr Rd.
Prosser, WA 99350
(509) 786-9242; f 786-9277
sschneid@asrr.arsusda.gov

Schnitkey, Gary
Ohio State University
324 Ag Adm. Bldg.
2120 Fyffe Rd.
Columbus, OH 48105
(614) 292-6409; f 292-0078

Scholl, Kevin
Mowers Soil Testing PLUS
117 E. Main
Toulon, IL 61483
(309) 286-2761; f 286-6251

Schoper, Robert
Farmland Industries, Inc.
Rt 1 Box 83
Jeffers, MN 56145
(507) 628-4418; f 628-4418

Schroeder, Dirk
Inst. of Plant Nut. & Soil Sc.
Bundesallee 50
D-38116 Braunschweig
Germany
+49/531-596-243;f:596-377

Schueller, John
University of Florida
PO Box 116300
Gainesville, FL 32611
(352) 392-0828; f 392-1071

Schuler, Ronald
Bio-Systems Engineering
460 Henry Mall Rm 115
Madison, WI 53706
(608) 262-0612

Schumacher, Joe
S.D.S.U. Plant Science
219 Ag Hall Box 2207A
Brookings, SD 57007
(605) 688-5122; f 688-4602

Schwartzbeck, Richard
National Crop Insurance
7201 W. 129th St. #200
Overland Park, KS 66213
(913) 685-2767

Schwarz, Greg
224 S. 5th St.
LeSeuer, MN 56058
(507) 665-2195

Scott, Jerry
Lockheed Martin
Idaho Tech.
PO Box 1625
Idaho Falls, ID 83415-3815
(208) 526-2836; f 526-3459
jbs1@inel.gov

Searcy, Stephen
Texas A & M University
Agric. Eng. Dept.
College Station, TX 77843
(409) 845-3668; f 847-8627
s-searcy@tamu.edu

Seymour, Anthony
Nuffield Farming Scholars
North Goodlands
Kalann.E
Australia 6468

Seymour, Stephanie
Nuffield Farming Scholars
North Goodlands
Kalann.E
Australia 6468

Shanahan, John
Colorado State University
CO2 Plant Science Building
Soil & Crop Sciences
Ft. Collins, CO 80523
(970) 491-6201; f 491-0564

Shoji, Sadao
Chisso Corporation
5-13-27 Nishitage, Taihaku-Ku
Sendai Miyagi
Japan 982
+81 222450120
f 222450120

Shribbs, John
Zeneca
108 Wyndham Way
Petaluma, CA 94954
(510) 231-5006; f 231-1255

Shutske, John
University of Minnesota
1398 Eckles Ave.
St. Paul, MN 55108
(612) 1250; f 624-3005

Sims, Albert
University of Minnesota
Northwest Experiment Station
Crookston, MN 56716
(218) 281-8619; f 281-8603
asims@mail.crk.umn.edu

Sinclair, Arnie
Ag-Chem Equipment Co., Inc.
5720 Smetana Dr., #100
Minnetonka, MN 55343
(612) 945-5803; f 933-7432

Skotnikov, Andrey
University of Minnesota
1991 Upper Buford Circle
St. Paul, MN 55108
(612) 625-8101; f 624-4223

Slater, Glen
University of Nebraska
PO Box 66
Clay Center, MN 68933
(402) 762-4445; f 762-4422
scrco24@unlvm.unl.edu

Smith, David
Massey Ferguson
Banner Lane
Coventry
England CV4 9GF
+01 203851137;f
203852659

Smith, Gene
Kalium Chemicals
Box 56
Prairie DuChien, WI 53821
(608) 326-5301

Smith, Jodie
Space Imaging
9351 Grant St. Suite 500
Thornton, CO 80229
(303) 254-2059; f 254-2215
jsmith@spaceimage.com

Smith, Kory
Mycogen Seeds
Box 209
Willmar, MN 56277
(612) 523-1331; f 523-5486

Smith, Lyall
Simplot Canada Limited
Brandon, Manitoba
Canada R7A 7C4
(204) 729-2815; f 729-8144

Snyder, Carol
Farmers Software Ass.
920 N Shields
Ft. Collins, CO 80521
(970) 407-0873

Sobolik, Chris
Hy Way Farms
3501 Jersey Ridge Rd. #707
Davenport, IA 52807
(319) 344-0040; f 765-7083

Solohub, Michael
University of Saskatchewan
51 Campus Dr.
Saskatoon, Saskatchewan
Canada S7H 0S2
(306) 966-6834
solohub@sask.usask.ca

Spencer, Joseph
SVS Environmental Systems
723 The Parkway, Ste 200
Richland, WA 99352
(509) 943-0560; f 943-5528

Srinivasan, Ancha
Regional Science Institute
4-13 Klta Nishi 2, Kita-ku
Sapporo, Hokkaido
Japan 001
+81 117176660
f: 117573610

Srinivasan, R.
Blacklund Research Center
808 E. Blackland Rd.
Temple, TX 76502
(817) 770-6670; f 770-6561
srin@brcsuno.tamu.edu

Stafford, John
Silsoe Research Institute
Silsoe, Bedford
U K MK45 4HS
john.stafford@bbsrc.ac.uk

Stiles, Steve
Brookside Laboratories, Inc.
308 S. Main St.
New Knoxville, OH 45871
(419) 753-2448; f 753-2949
sstiles@blinc.com

Stouffer, Bert
1408 N. E. Applewood Crt.
Liberty, MO 64086
(816) 246-9056; f 246-9056

Stouffer, Rob
1408 N. E. Applewood Crt.
Lee's Summit, MO 64086
(816) 246-9056; f 246-9056
stouffer@qni.com

Strand, Glenn
Tyler, Industries
E. Hwy 12
Box 249
Benson, MN 56215
(320) 843-3333; f: 843-2467

Strubbe, Gilbert
Katholieke University
Leuven
Kardinaal Mercierlaan, 92
B-3000-Heverlee-Leuven
Belgium

Sudduth, Kenneth
USDA-ARS
269 Ag. Engr. Bldg.
University of Missouri
Columbia, MO 65211
(573) 882-4090; f 882-1115

Suhr, Friedrich
Kincardine Agra Farming
Box 460
Kincardine, Ontario
Canada N2Z 2Y9
(519) 396-5279; f 396-9777
farming@canadian.agra.com

Sutton, Craig
AGRIS Corporation
300 Grimes Bridge Rd
Roswell, GA 30075
(770) 518-5185; f 643-2239

Swinton, Scott
Michigan State University
Dept. of Agric. Economics
East Lansing, MI 48824
(517) 353-7218; f 432-1800
swinton@pilot.msu.edu

Tardif, Francois
University of Guelph
Dept. of Crop Science
Guelph, Ontario
Canada N1G 2W1
(418) 824-4120; f 763-8933
ftardif@crop.uoguelph.ca

Taylor, Randy
Kansas State University
Extension Ag Engineering
237 Seaton Hall
Manhattan, KS 66506
(913) 532-5813; f 532-6944

Taysom, Dave
Dairyland Laboratories
217 E. Main
Arcadia, WI 54612
(608) 323-2123; f 323-2184

Teegardin, Norm
Farm Works Software
6795 S. State Rd. #1
Hamilton, IN 46742
(219) 488-3388; f 488-3737
farmwork@farmworks.com

Telck, Alan
Holly Sugar Corporation
2 N. Cascade Ave.
Colorado Spring, CO 80901
(719) 471-0123; f 630-3252
atelck@ix.netcom.com

Teoli, William
Ciba-Geigy Corporation
Box 18300
Greensboro, NC 27419
(910) 632-7706; f 632-2012

Terry, David
University of Kentucky
3521 Coltneck Ln
Lexington, KY 40502
(606) 257-2668; f 257-7351
dterry@ca.ulky.edu

Thompson, Wayne
EDAPHOS Ltd./Tyler Ind.
1200 Hawthorne Ave., E.
St. Paul, MN 55106-2131
(612) 772-3527; f 776-5349
wayne.thompson@soils.umn
.edu

Thylen, Lars
Swedish Inst. of Ag. Eng.
Box 7033
Uppsala, Sweden S-75007
+46 18303375; f 18300956
lars.thylen@jti.slu.se

Tillotson, Patricia
Advanced Geographics
135 Blemont Dr.
Wilmington, DE 19808

Tim, U. Sunday
Iowa State University
215 Davidson Hall
Dept of Ag & Biosystems
Ames, IA 50011
(515) 294-0466; f 294-2552
tim@iastate.edu

Tranel, Dean
Iowa State University
2104 Agronomy Hall
Ames, IA 50011
(515) 294-6834; f 294-9985
dmtranel@iastate.edu

Tricka, Paul
Hunting Elevator Company
2120 17th St., N.E.
Rochester, MN 55906
(507) 282-6651

Tupper, Gordon
Mississippi State University
Box 197
Stoneville, MS 38776
(601) 686-3270

Turner, Simon
ADAS
Woodthorne
Wolverhampton, Staffs
UK WV6 8TQ
+01 902693126; f
902693166
simon_turner@adas.co.uk

Tweeton, Theresa
Cenex Land O'Lakes
PO Box 64089
St. Paul, MN 55164
(612) 451-4922
ttwee@cnx.lol.com

Tyler, David
Ashtech
90 W. Central Ave.
Belgrade, MT 59714
(406) 388-1956; f 388-1883
david@ashtech.com

Upadhyaya, Shrini
University of California-Davis
Bio and Agr Eng Dept
Davis, CA 95616
(916) 752-8770; f 752-2640
Skupadhyaya@ucdavis.edu

Urbanowicz, William
Vigoro Industries
PO Box 58
Edison, OH 43320
(419) 947-5380; f 947-9671

Valco, Tommy
Cotton Incorporated
4505 Creedmoor Rd.
Raleigh, NC 27612
(919) 510-6123; f 510-612ℓ

VandenHeuvel, Richard
Cenex/LOL
Box 64089
St. Paul, MN 55164
(612) 451-4358; f 451-4561

VanderKant, Chris
Soil Stewardship Group
2136 Jetstream Rd.
London, Ontario
Canada N5V 3P5
(519) 457-2575

VanKessel, Chris
University of Saskatchewan
Dept. of Soil Science
Saskatoon, SK
Canada S7N 5A8
(306) 966-6854; f 966-6881

Vankoughnet, Brent
Agri Skills, Inc.
200 5711 Portage Ave.
Headingley, Manitoba
Canada R4H 1E7
(204) 888-2474; f 837-1342
Buank@magic.mb.ca

VanMaanen, Jason
Kent County Fertilizers Ltd.
Box 760
Ridgetown, Ontario
Canada N0P 2C0
(519) 674-5491; f 674-0322

Varvel, Gary
USDA/ARS/NPA
119 Kiem Hall Univ of NE
Lincoln, NE 68583
(402) 472-5169; f 472-0516
gvarvel@unlinfo.unl.edu

Verhagen, Jan
Agric. Univ. Wageningen
67ØØA Wageningen
Netherlands

Verhallen, Anne
Ontario Min of Agriculture
Main St., E.
Ridgetown, Ontario
Canada N0P-2C0
(519) 674-1621; f 674-1600

Von Bargen, Kenneth
University of Nebraska
6110 Elkcrest Circle
Lincoln, NE 68516
(402) 489-3552; f 472-6338

von Bertoldi, Peter
University of Guelph
131 Richards Bldg.
Guelph, Ontario
Canada N1E 6R1
(519) 824-4120; f 824-5730
pvonbert@irs.uoguelph.ca

Voorhees, Ward
USDA-ARS-MWA
NC Soil Conserv.
803 Iowa Ave.
Morris, MN 56267
(320) 589-3411; f 589-3787

Wagar, Tim
Minnesota Ext. Service
SE District Office
Rochester, MN 55901
(507) 280-2866

Waits, David
SST Development Group, Inc.
824 N. Country Club Rd.
Stillwater, OK 74074
(405) 377-5334; f 377-5746

Walker, Bruce
Agri-Food Canada
10130-103 Street, Ste 1295
Edmonton, Alberta
Canada T4X 1J4
(403) 495-6122; f 495-5344
walkerbd@em.agr.ca

Wall, Richard
University of Idaho
1910 University Dr. ET201
Boise, ID 83725
(208) 385-4073; f 385-1178
rwall@ce.uidaho.edu

Walley, Fran
University of Saskatchewan
Dept. of Soil Science
Saskatoon, Saskatchewan
Canada S7N 5A8
(306) 966-6854; f 966-6881
walley@sask.usask.ca

Walters, Craig
J. R. Simplot
Box 145
Colton, WA 99113
(509) 539-5778; f 229-3731

Wang, Wenwel
Univ. of Missouri-Columbia
159 Agric. Eng. Bldg.
Columbia, MO 65201
(573) 882-1116; f 882-1115

Wanzel, Robert
Clear Window Multi Media
15444 Clayton Rd., Ste 314
St. Louis, MO 63011
(314) 527-4001; f 527-4120
bobw@tetranet.net

Warncke, Darryl
Michigan State University
582 Plant & Soil Sc. Bldg
East Lansing, MI 48824
(517) 355-0210; f 355-0270
warncke@msue.msu.edu

Waterer, John
Cargill Ltd.
300 - 240 Graham Ave.
Winnipeg, Manitoba
Canada R3C 4C5
(204) 947-6385; f 947-6495

Waters, Debbie
University of Georgia
NESPAL/ Box 748
Tifton, GA 31794
(912) 386-7372; f 386-7005

Watkins, Hal
GROWMARK, Inc.
1701 Towanda Ave.
PO Box 2500
Bloomington, IL 61702
(309) 557-6244; f 557-6860

Weisberg, Paul
Flexi-Coil Ltd.
2414 Koyl Ave.
Saskatoon, SK.
Canada S7L 7L5
(306) 664-7686; f 664-6688

Weiss, Michael
USDA-ERS
1301 New York Ave., N.W.
Room 532
Washington, DC 20005-4788
(202) 219-0462; f 219-0477
mdweiss@econ.ag.gov

Welacky, Tom
Agriculture & Agri Food Can.
Greenouse & Processing Crops
Harrow, Ontario
Canada N0R 1G0
(519) 738-2251; f 738-2929
welackyt@em.agr.ca

Weller, Monte
New Holland N. America, Inc.
500 Diller Ave.
Box 1895
New Holland, PA 17557
(717) 355-1876; f 355-1939

Wells, Natasha
NESPAL-UGA
Moore Hwy
Tifton, GA 31794
(912) 386-7057; f 386-7005

Wheeler, Jim
INSAT, L.C.
PO Box 2172
Sioux City, IA 51104
(712) 239-4775

Whelan, Brett
University of Sydney
AOS, McMillan Bldg.
New South Wales
Australia 2006

White, Mark
Countrymark Coop
1217 S. Swegles
St. Johns, MI 48879
(517) 331-2174; f 224-2819

Wilkerson, John
University of Tennessee
Agric. Eng. Dept
PO Box 1071
Knoxville, TN 37901
(423) 974-7266; f 974-4514

Williams, Breadan
U of Melbourne-
Longerenong
RMR 3000
Horshiam, Victoria
Australia 3401
bwilliam@vcah.edu.au

Williams, Robert
USDA-ARS
Box 1430
Durant, OK 74702
(405) 924-5066; f 924-5307

Williford, Ray
USDA-ARS
Box 36
Stoneville, MS 38776
(601) 686-5352; f 686-5372
rwillifo@ag.gov

Willis, David
Agassiz Crop Management Inc.
RR 5 Box 126A
Thief R. Falls, MN 56701
(218) 681-6970; f 681-6173
agman@iom.net

Wilson, Robert
University of Nevada
PO Box 613
Eureka, NV 89316
(702) 289-4459; f 289-1462
rwilson@fs.scs.unr.edu

Wolkowski, Richard
University of Wisconsin
Dept. of Soil Science
1525 Observation Dr.
Madison, WI 53706
(608) 263-3913; f 265-2795
rpwolkow@facstaff.wisc.
edu

Wollenhaupt, Nyle
Ag-Chem Equipment Co., Inc.
5720 Smetana Dr., #100
Minnetonka, MN 55343
(612) 945-5803; f 933-7432

Woods, Samuel
OSU-ATI
1328 Dover Rd.
Wooster, OH 44691
(330) 264-3911; f 262-7634

Wright, Jerry
University of Minnesota
PO Box 471- WCES
Morris, MN 56267
(320) 589-1711; f 589-4870

Wrona, Anne
National Cotton Council
1918 N. Parkway
Memphis, TN 38112
(901) 274-9030; f 725-0510
awrona@cotton.org

Yang, Chenghai
USDA-ARS
2413 E. Hwy. 83
Weslaco, TX 78596
(210) 969-4824
yang@pop.tamu.edu

Zenk, David
Dunwoody Institute
818 Dunwoody Blvd.
Minneapolis, MN 55403
(612) 374-5800; f 374-4128

Zhang, Mingchu
University of Alberta
Dept. of Renewable Resources
Edmonton, Alberta
Canada T6G 2T3
(403) 492-2409; f 492-1767
mzhang@rr.ualberta.ca

Zwaenepoel, Philippe
Cemagref
6 Palaquins
03150 Montoldre
France 03950
+33 70450312; f 70451946
montoldre@danube.cemagref.fr

Adcon Telemetry, Inc.
1001 Yamato Rd.
Boca Raton, FL 33428

Mascoe, John
(407) 989-5309; f 989-5310

Ag/INNOVATOR Online
774 S. River Rd. W.
Linn Grove, IA 51033

Gepner, Dan
(712) 296-3615; f 296-3615

Mangold, Grant
(712) 296-3615; f 296-3615

Ag-Chem Equip. Co. Inc.
5720 Smetana Dr. Ste 100
Minnetonka, MN 55343

Mann, John
(612) 945-5803

Agribusiness
Commercializations
Development Center
901 N. Colorado
Kennewick, WA 99336

Bariault, Dan
(509) 735-1000; f 735-6609

Eakin, Dave
(509) 735-1000; f 735-6609

Patton, Tom
(509) 735-1000; f 735-6609

Agri-Logic, Inc.
1195 W. Private Rd. 660 N
Brazil, IN 47834

Bell, Denny
(812) 448-8590; f 442-8214

AGRIS Corporation
300 Grimes Bridge Rd.
Roswell, GA 30075

Jackson, Jerry
(770) 518-5185; f 643-2239

Macy, Ted
(770) 518-5185; f 643-2239

Pommier, Chuck
(770) 518-5185; f 643-2239

Agronomy Service Bureau
Route 2 Box 166A
Oran, MO 63771

Holmes, Bill
(573) 262-3474; f 262-2290
asb@asbl.com

Zakaluk, Rob
(573) 262-3474; f 262-2290
asb@asbl.com

Ag Leader Technology
Box 2348
Ames, IA 50010

Gunzenhauser, Bob
(515) 232-5363; f 232-3595

Spencer, Alan
(515) 232-5363; f 232-3595

Zielke, Roger
(515) 232-5363; f 232-3595

AGVISE Laboratories
Box 510
Northwood, ND 58267

Nordgaard, John
(701) 587-6010; f 587-6013

Lee, John
(701) 587-6010; f 587-6013

Pauly, Gary*
(701) 587-6010; f 587-6013

Ashtech Agricultural Div.
90 W. Central Ave.
Belgrade, MT 59714

Colliver, Corey
(406) 388-1993; f 388-1883

Communication Systems Int'l.
6009 1A St., S.W.
Calgary, Alberta
Canada T2H 0G5

DenHollander, David
(403) 259-3311; f 259-8866
info@csi-dgps.com

Hrynyk, Randy
(403) 259-3311; f 259-8866
info@csi-dgps.com

Concord Environmental
Equipment
Route 1 Box 78
Hawley, MN 56549

Wamre, Dennis
(218) 937-5100; f 937-5101

Correct Net
709 10th St.
Boone, IA 50036

Mathias, John
(515) 432-8602; f 432-9031

Control Technology, Inc.
Box 471
Waseca, MN 56093

Brace, Howard
(507) 835-4477; f 835-7207

Ristau, Bruce
(507) 835-7404

ERDAS, Inc.
2801 Buford Hwy Ne.,
Ste 300
Atlanta, GA 30329-2137

Epler, Bruce
(404) 248-9000; 248-9909
Hayes@erdas.com

Johnson, Amy
(404) 248-9000; f 248-9909

ESRI
380 New York St.
Redlands, CA 92373

Crandall, Max
909-793-2853; f 307-3039
mcrandall@esri.com

Vaaler, Dan
(909) 793-2853; f 307-3039

Farmers Software Assoc.
PO Box 660
Fort Collins, CO 80522

Burgess, Ken
Dillon, Mike
Havermale, Neil
Herrington, Rich
Williams, Lori

Farm Works Software
6795 S. State Rd. #1
Hamilton, IN 46742

Psurny, Mike
(219) 488-3388; f 488-3737
farmwork@farmworks.com

Parker, Ron
(219) 488-3388; f 488-3737
farmwork@farmworks.com

FMS Marketing, Inc.
1200 E. Haven Ave.
New Lenox, IL 60451

Groninger, Shaun
(815) 485-4955; f 485-4011
fmsharvest@aol.com

Geophyta, Inc.
2685 CR 254
Vickery, OH 43464

Wright, Nathan
(419) 547-8538; f 547-8538

HarvestMaster, Inc.
1740 N. Research Park Way
Logan, UT 84341

Hollist, Ray
(801) 753-1881; f 753-1896
si0sd@cc.usu.edu

Campbell, Ron
(801) 753-1881; f 753-1896
si0sd@cc.usu.edu

Highway Equipment Co
616 D. Ave. N.W.
Cedar Rapids, IA 52405

Stamats, Bill
(319) 363-8281; f 363-8284

Richards, Marty
(319) 363-8281; f 363-8284

INSAT, L. C.
Box 2172
Sioux City, IA 51104

Cobb, Kevin
(712) 239-3727; f 239-4775

K & L Agri-Sales, Inc.
4547 Calvert St.
Lincoln, NE 68506

Brinkhoff, Matt
(402) 489-6460; f 489-1292

Kalium Chemicals
901 Meadow Rd.
Litchfield, MN 55355

Olberding, Paul
(320) 693-2969; f 693-8885

Kluwer Academic Publishers
101 Philip Dr.
Norwell, MA 02061

Esser, Teresa
(617) 871-6600; f 871-6528
tesser@wkap.com

Larson Systems, Inc.
129 Welch Ave. #101
Ames, IA 50014

Larson, Donald
(800) 525-3995; f 233-5818

Leica, Inc.
3155 Medlock Bridge Rd.
Norcross, GA 30071

Kirkland, Debbie
(770) 447-6361; f 447-0710

MicroImages, Inc.
201 N. 8th St.
Lincoln, NE 68508

Reyal, Kevin
(402) 477-9554; f 477-9559
microimages.com

Micro Trak Systems, Inc.
Box 99
Eagle Lake, MN 56024

Hoehn, Rob
(507) 257-3600; f 257-3001

Midwest Technologies Inc.
2733 E. Ash St.
Springfield, IL 62703

Porter, Scott
(217) 753-8424; f 753-8426

Northstar Technologies
30 Sudbury Rd.
Acton, MA 01720

Murray, Amiee
(508) 897-0770; f 897-7241

NovAtel
6732 8th Street N.E.
Calgary, AB
Canada TZE 8M4

Burnell, Kirk
(403) 295-4595; f 295-4901
kburnell@novatel.ca

Reinhardt, Greg
(800) 280-2242;
f (403) 295-4901

Omni Advertising, Inc.
1358 Courthouse Blvd.
Inv. Gr. Hgts., MN 55077

Schoenecker, Bill
(612) 455-2426; f 455-0246
omni@spacestar.com

Potash & Phosphate Inst.
655 Engineering Dr. Ste 110
Norcross, GA 30092-2843
(770) 825-8074; f 448-0439

Precision Ag Center
439 Borlaug Hall
1991 Upper Buford Circle
St. Paul, MN 55108

Legg, Leslie
(612) 624-4224; f 624-4223

Racal Survey USA, Inc.
3624 W. Chase Dr.
Houston, TX 77042

Sugden, Jim

Raven Industries
205 E 6th St Box 5107
Sioux Falls, SD 57117

Choudek, Dale
(605) 331-0335; f 331-0426

McAvoy, Dale
(605) 331-0335; f 331-0426

Rawson Control Sys., Inc.
116 2nd St. S.E.
Oelwein, IA 50662

Gauquie, Sharon
(319) 283-2225; f 283-1360
rawson@trxinc.com

RDI Technologies, Inc.
Suite One Green Lake Mall
Spicer, MN 56288

Allen, Ryan
(320) 796-0019; f 796-0048
rditech@msn.com

Soderholm, John
(320) 796-0019; f 796-0048
rditech@msn.com

Rockwell International
267 Red Pine Trail
Hudson, WI 54016

Downing, Greg
(319) 359-1179

Elliott, Craig
(319) 395-1179

Norton, Todd
(319) 259-1179

Satcon System GmbH
Bundesstr 7
97537 Obertheres

Welz, Ralph
Germany

Satloc, Inc.
4670 S. Ash Ave.
Tempe, AZ 85282

Steinbrecher, Bill
(602) 831-5100; f 752-0457

Stewart, Collin
(602) 831-5100; f 752-0457
larry-buchanan@trimble.com

Servi-Tech. Laboratories
Box 169
Hastings, NE 68902

Hopkins, Bryan
(402) 463-3522; f 463-8132
edu.1617@tcgcs.com

McNickle, Keith
(402) 463-3522; f 463-8132
edu.1617@tcgcs.com

Space Imaging
9351 Grant St #500
Thornton, CO 80229

Liedtke, Jeff
(303) 254-2000; f 254-2215

Stratman, Timothy
(303) 254-2066; f 254-2211
tstratman@spaceimage.com

Spectrum Technologies
23839 W. Andrew Rd.
Plainfield, IL 60544

Thurow, K. Michael
(800) 248-8873;
f (815) 436-4460

Spot Image Corporation
1897 Preston White Dr.
Reston, VA 22091

Caraginnis, Dan
(703) 715-3100; f 648-1813

McKeon, John
(703) 715-3100; f 648-1813
mckeon@spot.com

Lees, Rob
(703) 715-3100; f 648-1813

Trimble Navigation
645 N. Mary Ave.
Sunnyvale, CA 94086

Huber, George
(408) 481-2994; f 481-6074

Siefken, Sid
(408) 481-2994; f 481-6074

Buchanan, Larry
1202 Lake View Dr.
Montgomery, TX 77356
(713) 363-4700; f 292-8876

Tyler Industries
(EDAPHOS)
1200 Hawthorne Ave. E.
St. Paul, MN 55106

Ellingson, Jon
(612) 772-3527; f 776-5349

Lent, Kevin
(612) 772-3527; f 776-5349

Master, Paul
(612) 772-3527; f 776-5349

Moore, Lee
(612) 772-3527; f 776-5349

Wingra Data Systems
2433 University Ave.
Madison, WI 53705

Amir-Fazli, Andrew
(608) 238-2880; f 238-2087

Zycom
Hangar 172Y
Hanscom Field
Medford, MA 01730

Gvili, Mike
(617) 274-1222; f 274-9866
bgroen@hut.tec.mn.us

Madden, Kimbly
(617) 274-1222; f 274-9866